H.-Peter Berlien · Gerhard J. Müller (Eds.)

**Applied Laser Medicine**

Springer-Verlag Berlin Heidelberg GmbH

H.-Peter Berlien · Gerhard J. Müller (Eds.)

# Applied Laser Medicine

With 633 Figures, 176 in Color
and 175 Tables

 Springer

Editors:

Professor Dr. H.-Peter Berlien
Klinikum Neukölln
Klinik für Lasermedizin
Rudower Straße 48
12351 Berlin
Germany

Professor Dr.-Ing. Gerhard J. Müller
Laser- und Medizin-Technologie GmbH
Fabeckstraße 60–62
14195 Berlin
Germany

Co-Editors:

Dr. Hans Breuer
18 Constantia Avenue, Stellenbosch
7600, Rep. of South Africa

Tetsuya Okunaka, MD, PhD
Assistant Professor
Department of Surgery
Tokyo Medical University
6-7-1, Nishishinjuku, Shinjuku-ku
Tokyo 160-0023 Japan

Neville Krasner, MD, FRCP(G), FRCP
Consultant Gastroenterologist
Aintree Centre for Gastroenterology
and Liver Disease
Fazakerley Hospitel, Lower Lane
GB-Liverpool L9 7 AL

David Sliney, PhD
Manger, Laser/
Optical Radiation Programm
US Army Center for Health Promotion
and Preventive Medicine
ATTN. MCBH-TS-OLO (Bldg. E-1950)
Aberdeen Proving Ground
MD 21010-5422, USA

Parts of this book is a translation from the German book: "Angewandte Lasermedizin" published by ecomed-Verlag, Landesberg

Translated by Dr. Hans Breuer

ISBN 978-3-540-67005-6      ISBN 978-3-642-18979-1 (eBook)
DOI 10.1007/978-3-642-18979-1

Cataloging-in-Publication Data applied for
Bibliographic information published by Die Deutsche Bibliothek
Die Deutsche Bibliothek lists this publication in the Deutsche Nationalbibliografie;
detailed bibliographic data is available in the Internet at <http://dnb.ddb.de>.

http://www.springer.de

© Springer-Verlag Berlin Heidelberg 2003
Originally published by Springer-Verlag Berlin Heidelberg New York in 2003

Production: ProEdit GmbH, 69126 Heidelberg, Germany
Cover: Erich Kirchner, Heidelberg, Germany
Typesetting and Repro: AM-productions GmbH, 69168 Wiesloch, Germany

Printed on acid-free paper      21/3150/ML – 5 4 3 2 1 0

# Preface

The history of laser applications in medicine starts almost with the invention of the laser itself.

It was only a few months after Maiman's invention when this new high-intensity light source was used for medical applications. Light as a therapeutic tool had long been used in medicine, especially in ophthalmology and dermatology. Therefore, these disciplines were the first to take advantage of this new tool. Although the early results were not as promising as expected, a new field for medical diagnosis and treatment had been defined. Most of the difficulties in the 1960s and 1970s were due to the fact that this new light source was developed primarily for research purposes in physics, and hence medical applications depended on the availability of appropriate laser systems. Indeed, during these years the laser was a case of "solution found, problem sought".

Forty years after the invention, laser applications in medicine have been accepted by physicians and are widespread throughout the world. In the 1980s, professional R&D centers for even better investigations of the potential of laser applications in medicine had been established. The first center of this kind was the Laser Medicine Center (LMZ), established in Berlin in spring 1985, which contributed a great deal to the understanding of laser interaction in biologic matter.

In close cooperation with the German Society for Lasers in Medicine and Surgery (DGLM) and other national and international societies, the LMTB (the former LMZ) launched a continuous medical education program for laser applications in medicine. For this purpose, a loose-leaf book entitled "Angewandte Lasermedizin – Handbuch für Praxis und Klinik" was published in German. This book has been used in the training of more than 3,000 physicians within the continuous education program since 1986.

In the early 1990s a concerted action program sponsored by the European Commission was carried out in Europe for the dissemination of the safe use of lasers in medicine, and in 1996 a similar program was developed with Russia. It was this German--Russian cooperation that led to a Russian version of the German original loose-leaf handbook. The editors realized an ever-growing demand for an English version of this handbook, too. Therefore, the editors are very grateful to the Springer publishing house for supporting the English edition of the German original. To be precise, this English version, like the Russian one, is not just a simple translation but a completely revised handbook, and especially with respect to the clinical chapters it provides up-to-date information and experience in the most important areas of applications.

As today's medical lasers represent rather "designed solutions to identified problems" various complex tissue effects are used in laser medicine. This book supplies the clinician and the scientist with technical and medical background, the basics of biomedical photonics and provides practical guidelines for and clear descriptions of established treatment methods. The principal editors would like to express their sincere gratitude to the corresponding editors H. Breuer (South Africa), N. Krasner (UK), T. Okunata (Japan) and D. Sliney (USA), to the authors and to all scientific and industrial partners, especially the companies Aesculap AG, Dornier Medizintechnik, Carl Zeiss, Siemens, Biolitec, W.O.M. World of Medicine, Hüttinger Medizintechnik and Dr. Hielscher for their continuous support for the LMTB, Berlin.

Berlin, January 2003

On behalf of the editors
GERHARD J. MÜLLER
H.-PETER BERLIEN

# Contents

# Part IV
# Laser Safety in Medicine

# List of Contributors

Algermissen, B., Dr. med. Dipl.-Biochem.
Klinikum Neukölln
Klinik für Lasermedizin
Rudower Str. 48
12351 Berlin, Germany

Beier, Ch.
Zentrum der Dermatologie und Venerologie
Klinikum der Johann Wolfgang Goethe-Universität
Theodor-Stern-Kai 7
60590 Frankfurt am Main, Germany

Berlien, H.-Peter, Prof. Dr.
Klinikum Neukölln
Klinik für Lasermedizin
Rudower Str. 48
12351 Berlin, Germany

Binding, U., Dr. rer.nat.
Laser- und Medizin-Technologie GmbH, Berlin
Fabeckstr. 60-62
14195 Berlin, Germany

Biamino, G., Prof. Dr.
Praxis Prof. Dr. Matzei, Prof. Dr. Schofer
Othmarscher Kirchenweg 168
22763 Hamburg, Germany

Brodzinski, T., Dipl.-Phys.
Schott Glaswerke
Karl-Bosch-Str. 10
65203 Wiesbaden, Germany

Dörschel, K., Dr. rer. nat.
Laser- und Medizin-Technologie GmbH, Berlin
Fachbeckstr. 650-62
14195 Berlin, Germany

Eichler, K.
Institut für Diagnostische
und Interventionelle Radiologie
Klinikum der Johann Wolfgang Goethe-Universität
Theodor-Stern-Kai 7
60590 Frankfurt am Main, Germany

Engelmann K.
Institut für Diagnostische
und Interventionelle Radiologie
Klinikum der Johann Wolfgang Goethe Universität
Theodor-Stern-Kai 7
60590 Frankfurt am Main, Germany

Fuchs, B.
Klinikum Neukölln
Klinik für Lasermedizin
Rudower Str. 48
12351 Berlin, Germany

Frank, F., Dr.-Ing.
Sarreiterweg 13
85560 Ebersberg, Germany

Fritsch, C., MD
Department of Dermatology
Heinrich Heine University
Moorenstr. 5
40225 Düsseldorf, Germany

Glotz, M., Dr.
Hüttinger Medizintechnik GmbH & Co.KG
Am Gansacker 1b
79224 Umkirch, Germany

Gottschalk, W., Dr.
Laser 2000 GmbH
Argelsrieder Feld
82234 Wessling, Germany

Greve, P., Dr. rer. nat.
Carl Zeiss
Abteilung QM
Carl-Zeiss-Str. 1
73447 Oberkochen, Germany

Hauptmann, G., Dipl.-Ing. (FH)
Schwedensteinstr. 13
81827 München, Germany

Helfmann, J., Dr. rer. nat.
Laser- und Medizin-Technologie GmbH, Berlin
Fabeckstr. 60-62
14195 Berlin, Germany

Herrig, M.
SLG
Prüf- und Zertifizierungs GmbH
Burgstödter Str. 20
09232 Hartmannsdorf, Germany

Hessel, S. Dr.-Ing.
MBB-Medizintechnik GmbH
Postfach 801168
81611 München, Germany

Heß, S.
Institut für Diagnostische
und Interventionelle Radiologie
Klinikum der Johann Wolfgang Goethe-Universität
Theodor-Stern-Kai 7
60590 Frankfurt am Main, Germany

Hopf, J.U.G, MD
Department of Otorhinolaryngology
Head and Neck Surgery
University Medical Center Benjamin Franklin
Free University of Berlin
Hindenburgdamm 30
12200 Berlin, Germany

Hopf, M., MD
Department of Otorhinolaryngology
Head and Neck Surgery
University Medical Center Benjamin Franklin
Free University of Berlin
Hindenburgdamm 30
12200 Berlin, Germany

Ismail, M.S., MD. PhD
Klinikum Neukölln
Klinik für Lasermedizin
Rudower Str. 48
12351 Berlin, Germany

Jovanovic, S., PD Dr.
Universitätsklinikum Benjamin Franklin
Klinik für Hals-Nasen-Ohrenheilkunde
Hindenburgdamm 30
12200 Berlin, Germany

Kato, H., MD, PhD. FCCP
Department of Surgery
Tokyo Medical University
6-7-1, Nishishinjuku, Shijuku-ku
Tokyo 160-0023, Japan

Kaufmann, R., Prof. Dr. med.
Zentrum der Dermatologie und Venerologie
Klinikum der Johann Wolfgang Goethe-Universität
Theodor-Stern-Kai 7
60950 Frankfurt am Main, Germany

Knappe, V., Dipl. Ing.
Laser- und Medizin-Technologie GmbH, Berlin
Fabeckstr. 60-62
14195 Berlin, Germany

Krabatsch, T., Dr.
Deutsches Herzzentrum Berlin
Augustenburger Platz 1
13353 Berlin, Germany

Krampe, C., Dipl. Mineralogin
MBB Medizintechnik GmbH
Postfach 801168
81611 München, Germany

Krasner, N., MD, FRCP(G), FRCP
Aintree Centre for Gastroenterology and Liver Disease
Fazakerley Hospital
Lower Lane
Liverpool L9 7AL
UK

Lang, K.
Department of Dermatology
Heinrich Heine University
Moorenstr. 5
40225 Düsseldorf, Germany

Lehmann, P.
Department of Dermatology
Heinrich Heine University
Moorenstr. 5
40225 Düsseldorf, Germany

Mack, M.G.
Institut für Diagnostische
und Interventionelle Radiologie
Klinikum der Johann Wolfgang Goethe-Universität
Theodor-Stern-Kai 7
60590 Frankfurt am Main, Germany

Mrochen, M., PH. D.
Aufenklinik Universität Zürich
Frauenklinik Str. 24
8091 Zürich

Müller, G., Prof. Dr.-Ing. Prof. h.c. Dr. h.c. Dr. h.c.
Laser-und Medizin-Technologie GmbH, Berlin
Fabeckstr. 60-62
14195 Berlin, Germany

Müller, U., Dr.
Klinikum Neukölln
Klinik für Lasermedizin
Rudower Str. 48
12351 Berlin, Germany

Müller-Stolzenburg, N., Dr. med. Dr. rer. nat.
Jägerstr. 11
12209 Berlin, Germany

Neuse, W.H.G.
Department of Dermatology
Heinrich Heine University
Moorenstr. 5
40225 Düsseldorf, Germany

Okunaka, T., MD, PhD, FCCP
Department of Surgery
Tokyo Medical University
6-7-1, Nishishinjuku, Shinjuku-ku
Tokyo 160-0023, Japan

Osterhaus, A. Dr. Dr.
Möserstr. 46
49074 Osnabrück, Germany

Philipp, C.M., Dr. med.
Klinikum Neukölln
Klinik für Lasermedizin
Rudower Str. 48
12351 Berlin, Germany

Poetke, M., Dr. med.
Klinikum Neukölln
Klinik für Lasermedizin
Rudower Str. 48
12351 Berlin, Germany

Roggan, A., Dr. rer. nat.
CELON AG
medical instruments
Rheinstr. 8
14513 Teltow, Germany

Ruzicka, T.
Department of Dermatology
Heinrich Heine University
Moorenstr. 5
40225 Düsseldorf, Germany

Schaldach, B., Dr. rer. nat.
Laser- und Medizin-Technologie GmbH, Berlin
Fabeckstr. 60-62
14195 Berlin, Germany

Scharschmidt, D.
Klinikum Neukölln
Klinik für Lasermedizin
Rudower Str. 48
12351 Berlin, Germany

Scheinert, D., MD
Herzzentrum Leipzig
Klinische und Interventionelle Angiologie
Strümpellstr. 39
04289 Leipzig, Germany

Scherer, H., Prof. Dr.
Department of Otorhinolaryngology
Head and Neck Surgery
University Medical Center Benjamin Franklin
Free University of Berlin
Hindeburgdamm 30
12200 Berlin, Germany

Schmitz, S.D.
Klinikum Neukölln
Klinik für Lasermedizin
Rudower Str. 48
12351 Berlin, Germany

Schulte, K.W.
Department of Dermatology
Heinrich Heine University
Moorenstr. 5
40225 Düsseldorf, Germany

Schönborn, K.-H., Dr.
W.O.M. World of Medicine GmbH
Kaiserin-Augusta-Allee 113
10553 Berlin, Germany

Scholz, C., Dr.-Ing.
W.O.M. World of Medicine GmbH
Kaiserin-Augusta-Allee 113
10553 Berlin, Germany

Schwarzmaier, H.-J., Dr.
Klinikum Krefeld
Lutherplatz 40
47805 Krefeld, Germany

Sedlmaier, B.
Department of Otorhinolaryngology
Head and Neck Surgery
University Medical Center Benjamin Franklin
Free University of Berlin
Hindenburgdamm 30
12200 Berlin, Germany

Seiler, T., MD, Ph.D.
Universität Zürich
Department of Ophthalmology
Zürich, Switzerland

Selman, St. A., MD FACS
Departments of Urology and Surgery
Medical College of Ohio, Toledo Ohio, USA

Senz, R., Prof. Dr. rer. nat., Dipl-Ing.,
Technische Fachhochschule Berlin
Luxemburger Str. 10
13353 Berlin, Germany

Shumaker, B., Dr.
2694 Cosa Mesa Road
Waterford, MI 48329
USA

Siebert W., PD Dr.
Orthopädische Kliniken Kassel
Wilhelmshöher Allee 345
34131 Kassel, Germany

Sliney, D.H., Ph.D
Laser/Optical Radiation Program
US Army Center for Health Promotion
and Preventive Medicine
Aberdeen Proving Ground, MD 21010-5422
USA

Steiner, R., Prof. Dr.
Institut für Lasertechnologien
und Messtechnik in der Medizin
Universität Ulm
Helmholtzstr. 12
89081 Ulm, Germany

Steiner, W., Prof. Dr.
Klinik für HNO
Georg-August-Universität
Göttingen, Germany

Straub, R.
Institut für Diagnostische
und Interventionelle Radiologie
Klinikum der Johann Wolfgang Goethe-Universität
Theodor-Stern-Kai 7
60590 Frankfurt am Main, Germany

Tempelhoff, W. von, Dr. med.
Klinikum Krefeld
Lutherplatz 40
47805 Krefeld, Germany

Ulrich, F., Prof. Dr.
Klinikum Krefeld
Lutherplatz 40
47805 Krefeld, Germany

Urban, P.,
Klinikum Neukölln
Klinik für Lasermedizin
Rudower Str. 48
12351 Berlin, Germany

Vogl, T.J., Prof. Dr.
Institut für Diagnostische
und Interventionelle Radiologie
Klinikum der Johann Wolfgang Goethe Universität
Theodor-Stern-Kai 7
60590 Frankfurt am Main, Germany

Waldschmidt, J. Prof. Dr. med.
St. Joseph Krankenhaus
Abteilung Kinderchirurgie
Bäumerplan 24
12101 Berlin, Germany

Wilder-Smith, P., PD, Dr. med. dent.
Universitiy of California,
1002 Helath Sciences Road East,
Irvine, CA 92612-1475, USA

Werner, J.A., Prof. Dr.
Klinikum und Poliklinik für HNO
Deutschhausstr. 3
35037 Marburg, Germany

Wondrazek, F., Prof. Dr. rer. nat.
Riegelstraße 27
85276 Pfaffenhofen, Germany

Zangos, St.
Institut für Diagnostische
und Interventionelle Radiologie
Klinikum der Johann Wolfgang Goethe-Universität
Theodor-Stern-Kai 7
60590 Frankfurt am Main, Germany

Zgoda, F.
Laser- und Medizin-Technologie GmbH, Berlin
Fabeckstr. 60-62
14195 Berlin, Germany

# Part I
# Basics

# I-1
# Basic Physics

G. Müller

## Contents

## Structure of Matter

Everyday experience suggests that it is possible to subdivide matter indefinitely, e.g. to divide a piece of iron or sugar by cutting, grinding, grating, hammering or similar mechanical activities into smaller and smaller pieces.

Neglecting the practical difficulties of doing this, we now definitely know that we would encounter a well-defined limit at about 1 nm. These smallest building blocks are called molecules. They can be subdivided further only by non-mechanical means, i.e. chemical or physical processes. Atoms are the smallest complete building blocks of all the matter surrounding us.

The concept of the atom is very old, but it became physical reality only about 100 years ago. Although atoms are the smallest complete building blocks of matter, they are composed of even smaller particles. To subdivide atoms into these particles, very elaborate and non-mechanical methods are required.

Initially, atoms were visualized as tiny droplets; their composition remained unknown. Rutherford's scattering experiments demonstrated the composition of the atoms. He directed α-particles toward a thin gold foil – about 2000 atomic layers thick – and from the resulting scattering pattern derived the mass distribution inside the atom. Finally, it was shown that the atom is composed of negatively charged electrons, positively charged protons and electrically neutral neutrons.

Protons and neutrons are concentrated in the centre of the atom; they make up the atomic nucleus. The electrons orbit the nucleus at a – relatively speaking – very great distance. The space between electrons and nucleus is empty (Fig. 1).

The numbers of electrons, protons and neutrons are related to each other. This relation is the ordering principle of the Periodic Table of the elements. The number of protons determines the actual element. For example, hydrogen has only 1 proton and iron has 26 protons. Using this simple model, atoms can be visualized (see Fig. 2 for representations of hydrogen and uranium atoms).

At present, 116 elements have been found; they are arranged in the Periodic Table of the elements.

Presentation of the atom

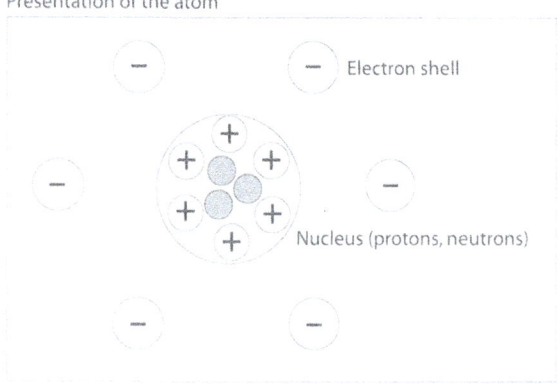

TV tower Stuttgart, height 211m

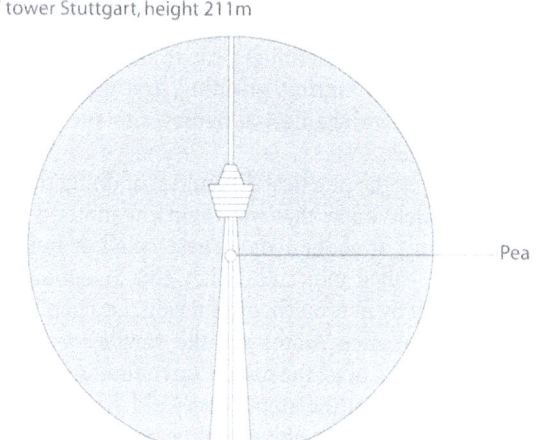

**Fig. 1.** The atomic nucleus will have the size of a pea if the atom's diameter is enlarged to 211 m

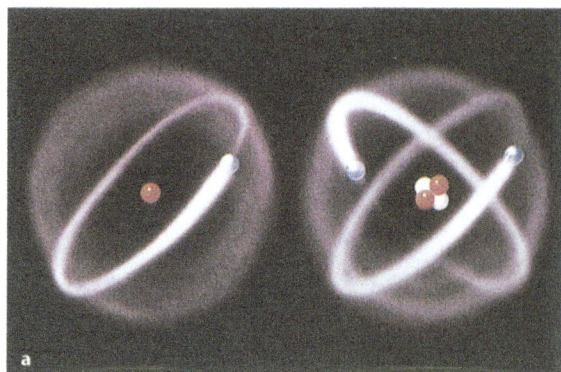

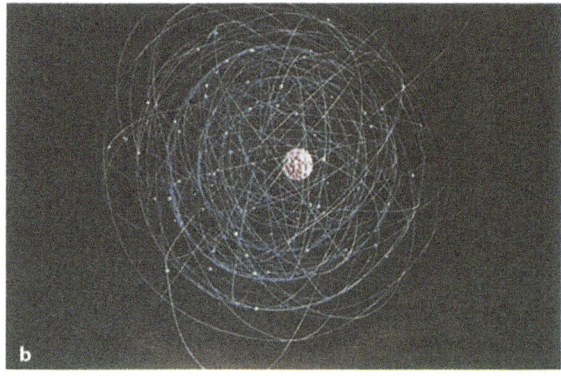

**Fig. 2. a.** A hydrogen atom (**left**) consists of a nucleus containing a single proton (**red**) and an electron (**blue**) which orbits the nucleus. **b.** A uranium atom is made up of a nucleus of 92 protons and about 146 neutrons and 92 electrons which orbit the nucleus (not to scale)

## Electromagnetic Waves and Photons

The electromagnetic field of light is composed of a periodically changing electric field and an orthogonal magnetic field (Fig. 3).

The electromagnetic wave does not require a medium for its propagation as does a sound wave, which needs air. Light moves fastest through a vacuum. The maximum speed of all electromagnetic waves is the speed of light:

$$c = 299\ 792\ 458\ \text{m s}^{-1}$$

Besides speed, wavelength $\lambda$ and frequency $\nu$ characterize an electromagnetic wave. All these quantities are related:

$$c = \lambda \cdot \nu$$

Figure 4 presents the spectrum of the electromagnetic waves.

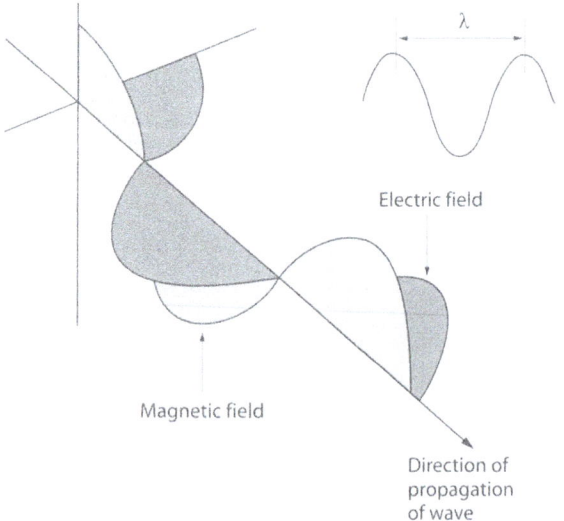

**Fig. 3.** Schematic presentation of light as an electromagnetic wave

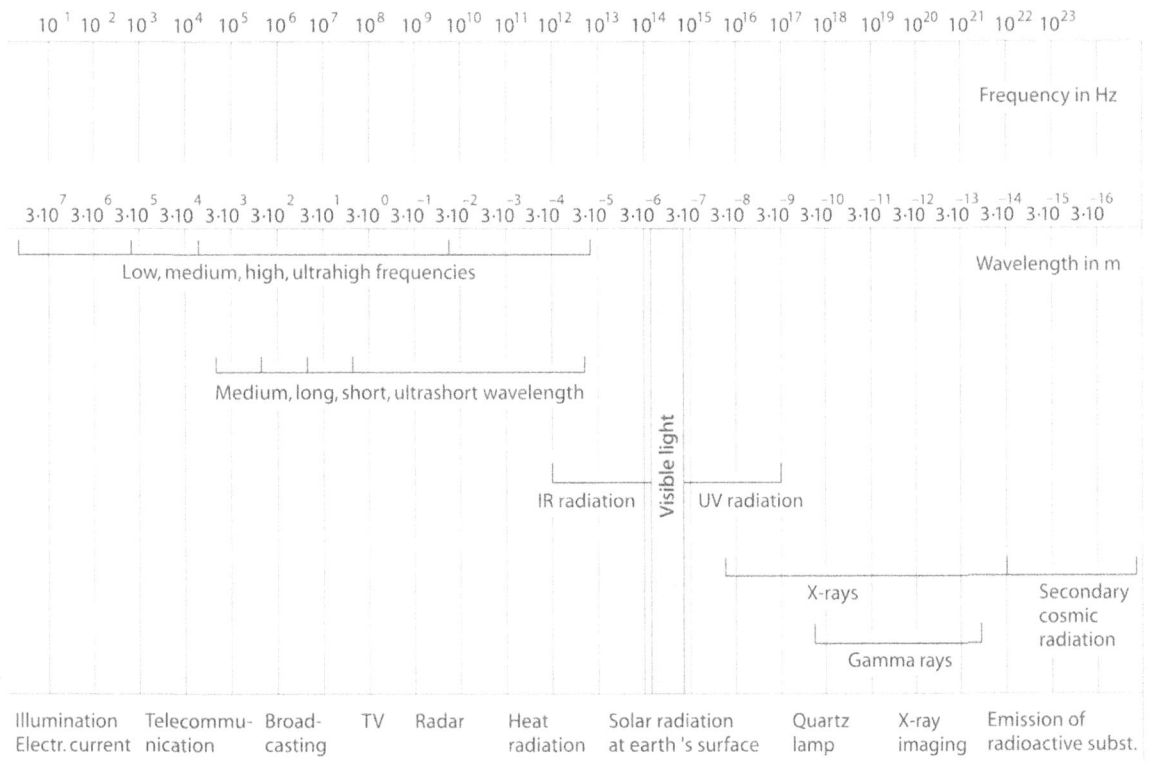

**Fig. 4.** Electromagnetic spectrum

The region of the visible light is extremely narrow. In physics its adjacent regions are also called light. Infrared radiation (IR) borders the long-wavelength limit of the visible spectrum, and ultraviolet radiation (UV) its short-wavelength limit.

Electromagnetic waves show all the properties of a wave (interference, diffraction, refraction). Nevertheless, any wave of the electromagnetic spectrum may also be considered a particle, called a photon. On the other hand, it is possible to assign wave properties to any particle (de Broglie wave). This is named the wave-particle duality. To describe laser processes, either the particle or the wave image is used, depending on the point of view.

## Origin of Light

Following Rutherford, Bohr formulated the following postulates concerning the orbits of the electrons around the nucleus:

1. The electrons occupy certain fixed orbits (principal quantum number).
2. The electrons can only jump from orbit to orbit. Doing this, they will either emit energy (emission of radiation) or absorb energy (absorption of radiation), e.g. in the form of light.

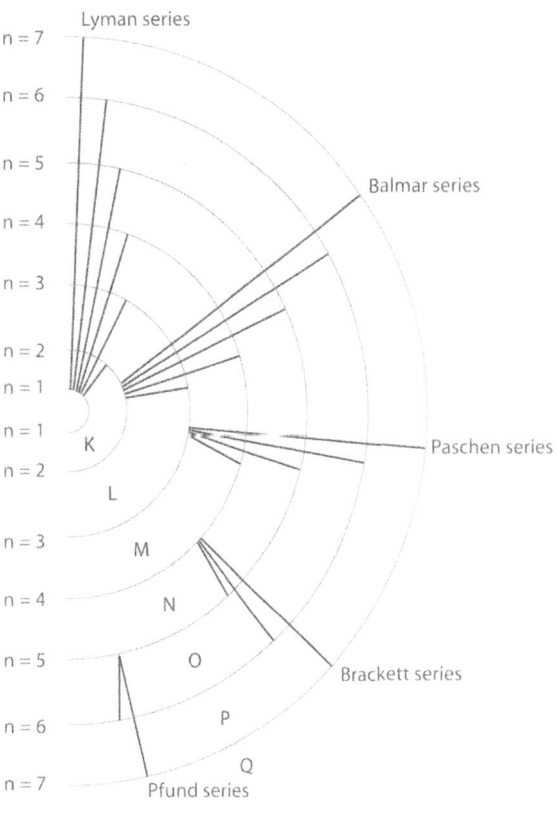

**Fig. 5.** Schematic presentation of the origin of spectral lines of hydrogen

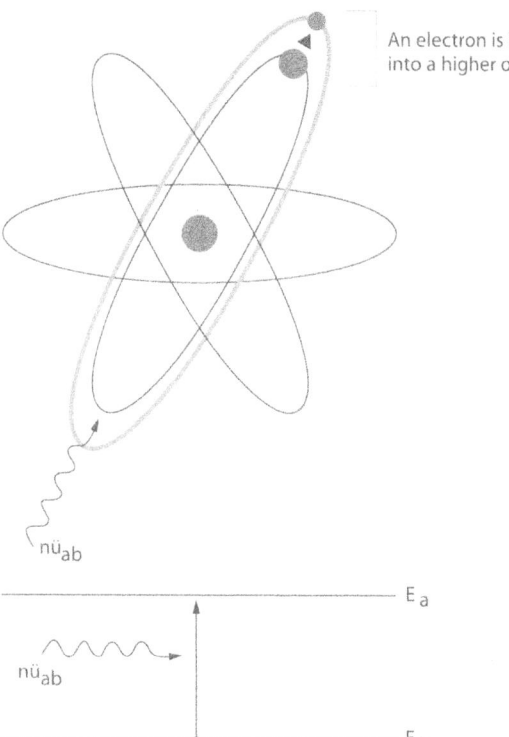

Fig. 6. Representation of an excited atom

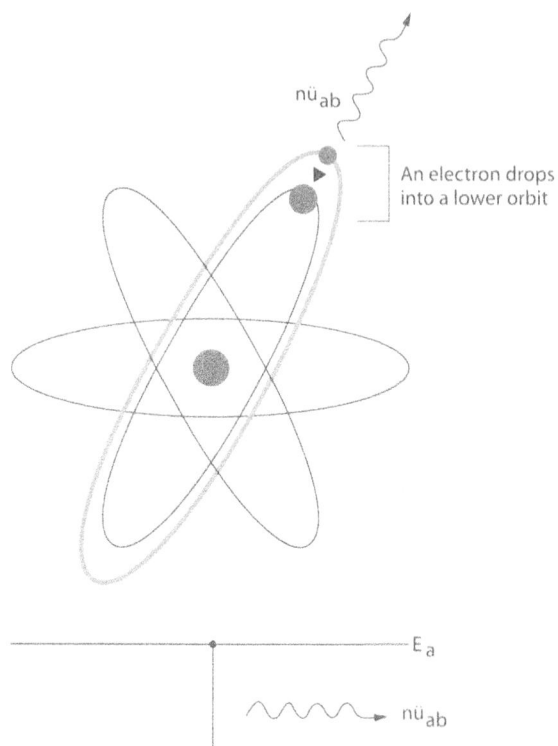

Fig. 7. Representation of spontaneous emission

If an electron moves into an orbit of higher energy, it has energy $E_a$. It has the energy $E_b$ if it occupies an energetically lower orbit.

In order to describe clearly the processes of absorption and emission, physics employs a simple one-dimensional representation. The energy of the electrons is plotted on a linear ordinate, while the energy assigned to individual allowed orbits is plotted as a horizontal bar, called energy level (Fig. 6).

In this representation absorption is the transition from a lower to a higher energy level. The absorbed radiation particle – the photon – needs to have just this energy difference to make absorption possible.

The difference between both energies yields:

$$E_a - E_b = h\nu$$

where $E_a$ is the energy of the electron in the excited state, $E_b$ the energy in the lower state, h is Planck's constant $6.26617 \times 10^{-34}$ Js, and $\nu$ is the frequency in Hz.

Electrons strive to decay spontaneously after a short holding time (about $10^{-8}$ s) in the excited state $E_a$, into a state of lowest energy which is called the ground state. In the ground state the atom is stable.

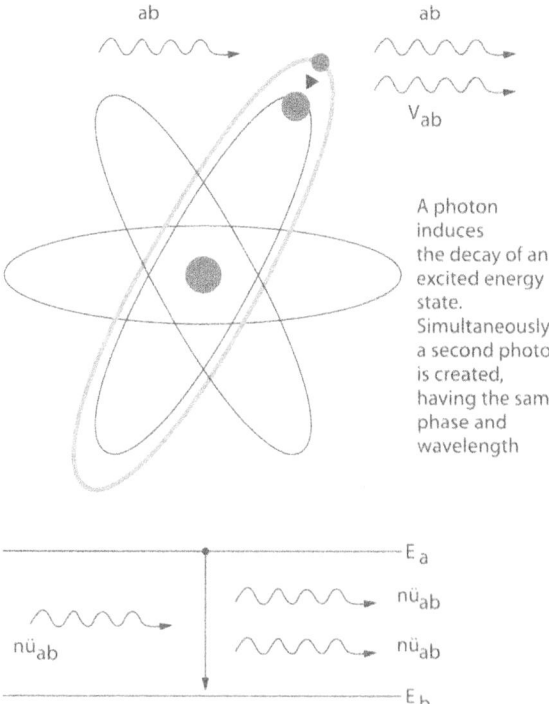

A photon induces the decay of an excited energy state. Simultaneously a second photon is created, having the same phase and wavelength

Fig. 8. Representation of stimulated emission

If an excited atom is struck by a photon having exactly the energy $E_{ab}$ then decay into the state $E_b$ will be accelerated.

In this case, the original photon will remain; during the decay of the atom into the state $E_b$ an identical photon will be released (Fig. 8). This process is called induced emission. Both photons have the same energy; they are emitted into the same direction. They have some place in space and time.

All these phenomena are conditions for the existence of the laser.

# I-2
# Basic Physics of Lasers

## I-2.1
## Amplifier with Feedback

K. Dörschel and G. Müller

Stimulated emission is an amplification process for photons, e.g. light (see Chap. I-1). Depending on the output wavelength, its technical manifestation is a laser (light amplification by stimulated emission of radiation) or a maser (microwave amplification by stimulated emission of radiation) or a graser (gamma ray amplification by stimulated emission of radiation).

At the beginning of the 20th century, long before the invention of laser (1960) and maser (1954), it was already possible to create and amplify long-wave electromagnetic radiation (radio-frequency, broadcasting). An important breakthrough was achieved by Meissner developing the feedback amplifier. Feeding back a small percentage of the signal's output into the input of the amplifier yielded a large amplification by a relatively low power amplifier.

Such an arrangement of amplifier and feedback is a self-exciting (or free-running) system, an oscillating circuit (Fig. 1).

If the feedback energy fed into the input of the feedback amplifier exceeds the losses occurring inside, the signal within the amplifier, and consequently its output signal will increase indefinitely at least theoretically. In practice, saturation effects will limit the output signal to very large values. At the same time the output signal becomes independent of the input signal. The feedback amplifier, itself having only limited amplification, has turned into a free-running oscillator. This state of self-excitation can be achieved

only by a feedback amplifier, due to the ever-present internal noise of the amplifier.

The laser may be understood in analogy with the features of radio-frequency techniques described above. This means that the laser is only a radio-frequency feedback amplifier with different building blocks – due to the extremely high frequencies (millionfold higher than in broadcasting). In most cases, the low-power amplifier must be boosted via feedback to such a large overall amplification that the oscillator (through amplification and feedback) transforms into a free-running state. This is called induced emission (Fig. 2). The laser can also reach this state spontaneously via internal noise – spontaneous emission. This is the beginning of laser action.

To build a laser it is necessary to find materials, named laser media, where the effect of induced emission yields a sufficiently high amplification. These substances constitute the amplifiers. In addition, an efficient feedback system must be established. To realize this, an arrangement of mirrors reflects the light leaving the laser medium back into it. Such an arrangement of mirrors is called a resonator (Chap. I-2.2). By utilizing additional building blocks placed into the feedback loop, the output signal can be influenced in the same way as in the radio transmitters described above. (Examples: narrow-band or tunable filters for setting the output frequency; switches used for short-term changes of the quality factor in order

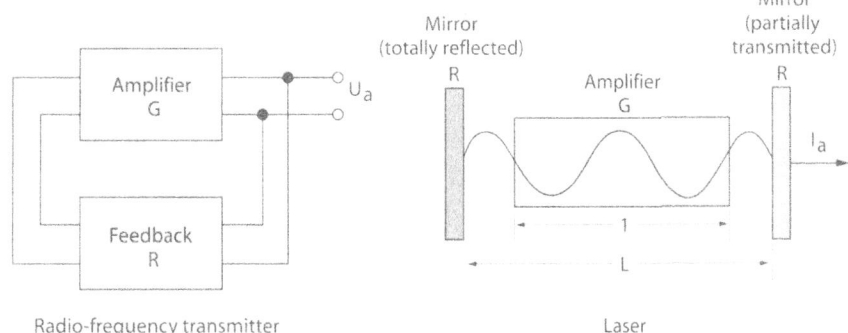

**Fig. 1.** A laser consists of an amplifier (induced emission) and a feedback arrangement (mirror) in analogy to a high-frequency transmitter (broadcasting, TV)

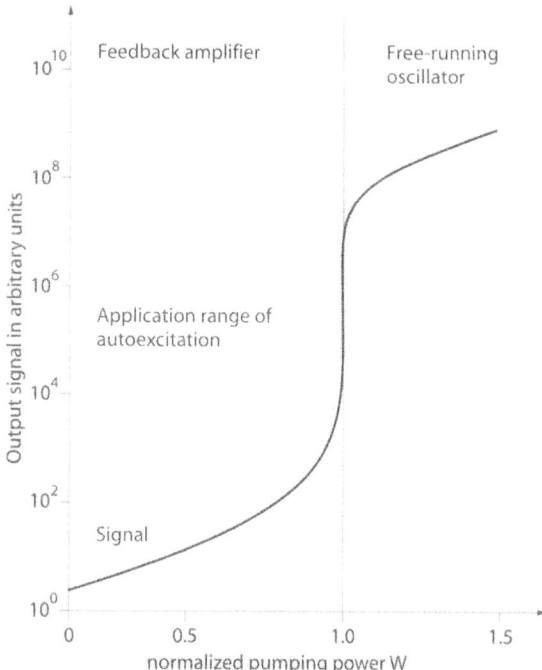

**Fig. 2.** Output intensity of a feedback amplifier as a function of normalized pumping power W. Starting at W = 1 the power supplied by the amplifier exceeds its losses and the system will operate as a free-running oscillator

to release within an extremely short period of time all the energy stored inside the oscillator.)

The various components and features of the special, high-frequency emitter "laser" will be described in the following chapters.

# I-2.2
# Laser Design

K. Dörschel

## Contents

## The Laser Process

The process of stimulated emission is the basis of laser amplification. To utilize this process it is necessary first to raise an electron of an atom (ion, molecule, solid) from a lower energy level into a higher one (Chap. I). To attain laser amplification this condition must be achieved not just for a single atom but for an entire large assembly of atoms. Consequently, the number of atoms with populated upper levels must always be larger than the occupation number of the lower laser levels. This is called "population inversion".

## How is it possible to achieve population inversion?

Thermal heating is not feasible because, according to Planck's radiation law (Fig. 1), higher energy levels are never more populated than lower ones. Irradiation with light (optical pumping) of a system having

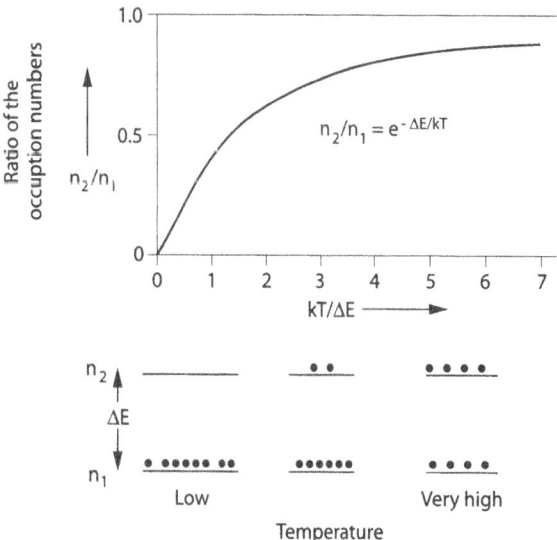

**Fig. 1.** According to Planck's radiation law the ratio $n_2/n_1$ of the occupation numbers shifts toward 1 during thermal heating. It is impossible to reach population inversion

only two energy levels will achieve merely an equal population of both levels. This feature is independent of the pumping intensity. The reason is that a high pumping intensity will not only populate the upper levels but, at the same time, induce an equal amount of emission, thus depopulating the upper levels. Consequently, optical pumping cannot achieve population inversion of a two-level system. However, three-level and multilevel systems will behave differently.

## Three-Level Systems

Pumping from the lowest level 1 to the uppermost level 3 of a three-level system (Fig. 2), a spontaneous emission from level 3 may fill the intermediate level 2. If the lifetime of this level is relatively long, its population number will increase with time. Utilizing a very high pumping intensity, it is possible to raise the population number of this second level above that of the lowest laser level (ground state).

The population inversion will decay rapidly as soon as the laser action begins. In general, the pumping intensity is insufficient to maintain the population inversion permanently. Three-level lasers are mainly pulsed lasers.

## Four-Level Lasers

Extending a three-level system by an additional level 2' situated between levels 1 and 2 (Fig. 3) can bypass the problems resulting from the short-term population inversion of the three-level system. If the additional level 2' has a very short lifetime and if the laser transition takes place from 2 to 2', then level 2' will empty steadily into the ground state as soon as the laser action begins. For this configuration even a very low pumping intensity will assure a steady population inversion between levels 2 and 2'. Thus, four-level lasers may be run in continuous-wave mode (cw).

All excitation modes must be set up in such a way that the population changes of the various levels are a circular process. Only then will the changes terminate at the ground state and the upper levels again be ready for a new pumping cycle. In many instances, these pumping cycles will terminate – at least partially – at so-called "metastable triplet levels" (Fig. 4). In general, these levels will not decay to the ground state, and with time all atoms will be pumped into this metastable state. Then these levels are no longer available to the laser pumping cycle and the laser action ceases. This problem may be, at least partially, bypassed by continuously exchanging the laser medium via pumping over. Another possibility is the addition of a so-called buffer gas. In this case, the lifetime of the metastable levels will be shortened by collisions with the buffer gas.

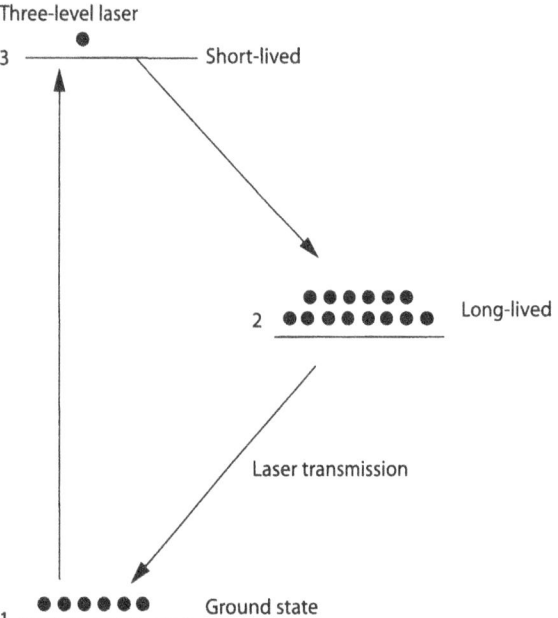

**Fig. 2.** For a three-level laser system it is possible to achieve a larger population of level 2, as compared to the ground state, by very intense pumping from level 1 to level 3

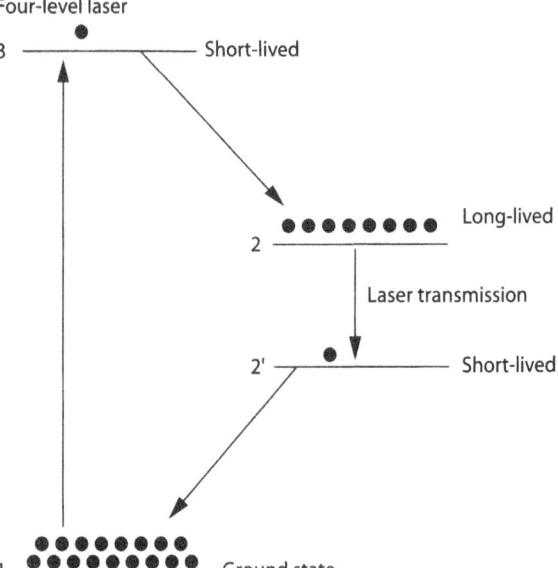

**Fig. 3.** For a four-level laser system it is possible to achieve, even by weak pumping into the long-lived level 2, a population inversion as compared with the short-lived level 2'; due to its short life level 2 empties immediately

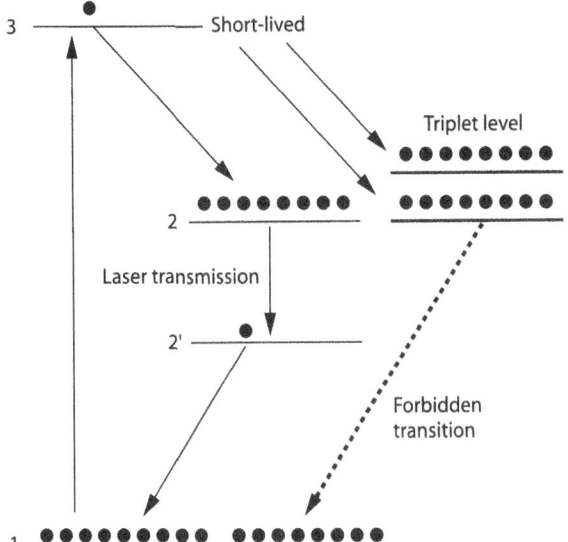

**Fig. 4.** If the lower laser level empties partly into a metastable triplet level, the laser process ceases after a short time; with increasing time the entire population will accumulate in the triplet level

## Laser Media

Substances can be used as laser media only if a population inversion can be generated in them. Many of the following substances are suitable:

1. Free atoms, ions, molecules and molecular ions in gases or vapours.
2. Dye molecules dissolved in liquids.
3. Atoms and ions incorporated within a solid.
4. Doped semiconductors.
5. Free electrons.

The number of media suitable for laser action and the number of laser transitions is at present so large that they fill an entire book [1]. For the element neon alone, about 200 different laser transitions have been observed.

Depending on the kind of medium, we distinguish between the gaseous and liquid state, and between semiconductor and solid-state lasers. As a curiosity, it may be noted that human breath; consisting of carbon dioxide, nitrogen and water vapour, is suitable as laser medium, forming a weak $CO_2$ laser. Even some varieties of dry gin were utilized as laser media because they contain a sufficient amount of quinine fluorescing in the blue region of the visible spectrum.

By now, laser lines are known from the ultraviolet spectral range (100 nm) to the far infrared. There, the laser transforms seamlessly into the maser. An intensive search is underway to find lasers in the X-ray region. However, only about two to three dozen laser types have achieved real practical significance.

Medical applications are presently restricted to $CO_2$, Ar and Kr lasers, cw and pulsed Nd:YAG lasers, cw and pulsed dye lasers, HeNe and GaAs lasers. Laser types like the eximer laser, frequency-doubled Nd:YAG laser, Er:YAG laser and metal vapour laser are just entering the medical arsenal.

It is also possible to distinguish between lasers emitting discrete lines, i.e. within only a very narrow wavelength band and those emitting within a much wider wavelength range.

Free atoms and ions, having well-defined energy levels, emit discrete laser lines. The well-known solid-state lasers also emit discrete laser lines (ruby laser, Nd:YAG laser).

In addition, other solid-state lasers have been developed (colour centre lasers, alexandrite lasers, diamond lasers) which allow a continuous change of emission wavelength within a fairly wide spectral range. This refers especially to dye lasers, where this feature is developed furthest. Semiconductor lasers do not produce narrow and well-defined laser lines due to the band structure of the energy levels of semiconductors.

**Fig. 5.** The most popular medical laser types

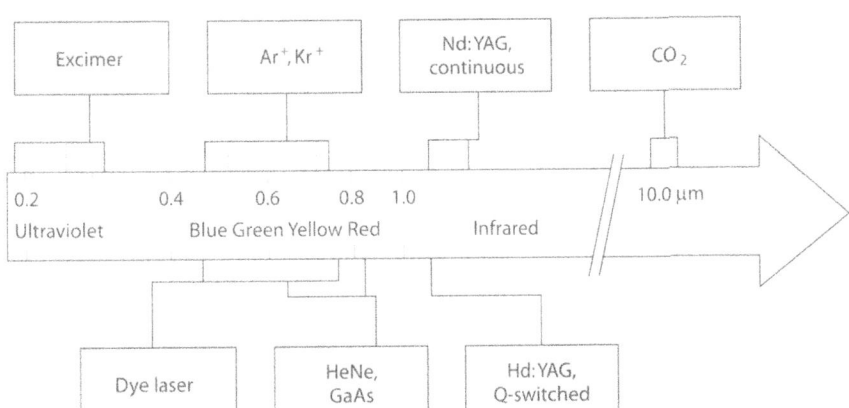

## Excitation Mechanisms

As mentioned above, laser action will take place only if the population densities of two energy levels are inverted. To achieve this population inversion, energy in a suitable form must be supplied. This can be achieved in very different ways, each one in principle independent of the specific laser process. Nevertheless, the appropriate excitation process must be carefully chosen and optimised with respect to the laser type. The essential processes to accomplish this are excitation by very intense light, called "optical pumping" and excitation inside an electrical gas discharge. Excitation of a semiconductor is achieved directly by an electric current. Chemical reactions are also suitable for excitation.

## Optical Pumping

By irradiating the laser medium with intense light it is possible to populate higher energy levels. This procedure is called optical pumping. Suitable sources of light are very intense flash tubes, continuously emitting high-pressure lamps and even other lasers. Because lamps emit radiation within a wide spectral range, optical pumping is especially suited for laser media having an abundance of excitation levels, even excitation bands. This is so because only those wavelengths will contribute toward pumping which exactly represent the energy difference between two levels. Consequently, suitable pumping lasers (argon lasers) will emit excitation bands. Since the laser transition utilizes only a fraction of the excitation energy, the resulting laser wavelength is always longer than the pumping wavelength.

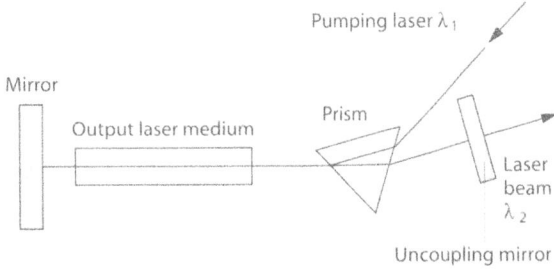

**Fig. 6.** Example for the collinear pumping of one laser by another pumping laser. The laser wavelength always exceeds the pumping wavelength, consequently the pumping beam and laser beam can be separated by the prism

## Gas Discharge

To achieve population inversion in gaseous or vaporous laser media, gas discharges may be employed. During gas discharge the neutral gas partially breaks down into ions and electrons. The field strength of the discharge will accelerate the electrons they will collide with atoms and ions. Much of the kinetic energy of the electrons is transferred to the collision partner. This energy may be directly or indirectly used to populate the upper laser levels. The current densities of the gas discharge employed can reach very high values (Ar-ion laser > 100 A/cm$^2$). Therefore, expensive cooling of the discharge tube becomes necessary. In order to enclose the discharges in very narrow channels, large magnetic fields are needed. The magnetic coils must also be cooled.

Because it is difficult to shift the maximum of the electron energy distribution into the region of the laser levels to be excited (about 20 eV for inert gases), a trick is employed to achieve the necessary population inversion. A pumping gas having a metastable level is mixed with the active medium. Collisions of the second kind will then excite the upper laser level. This collisional excitation will be efficient only if the metastable and the upper laser level have about the same energy. The population of this metastable level is increased by radiation transitions from other levels excited by the gas discharge; it quasi stores the excitation of many levels.

If a metastable pumping atom collides with a laser atom in its ground state, the excitation energy is transferred to the laser atom (Fig. 7).

The efficiency of the different laser types varies greatly. As an example, consider the ion laser: initially the ionisation energy must be supplied, followed by the excitation energy in the ionised state. But only a fraction of the pumping energy involved is usable for the laser transition. Much more favourable is the situation, for example, for the $CO_2$ laser. The upper laser level is reached by supplying much less energy. Figure 8 presents some examples of the ratio between pumping energy and laser energy.

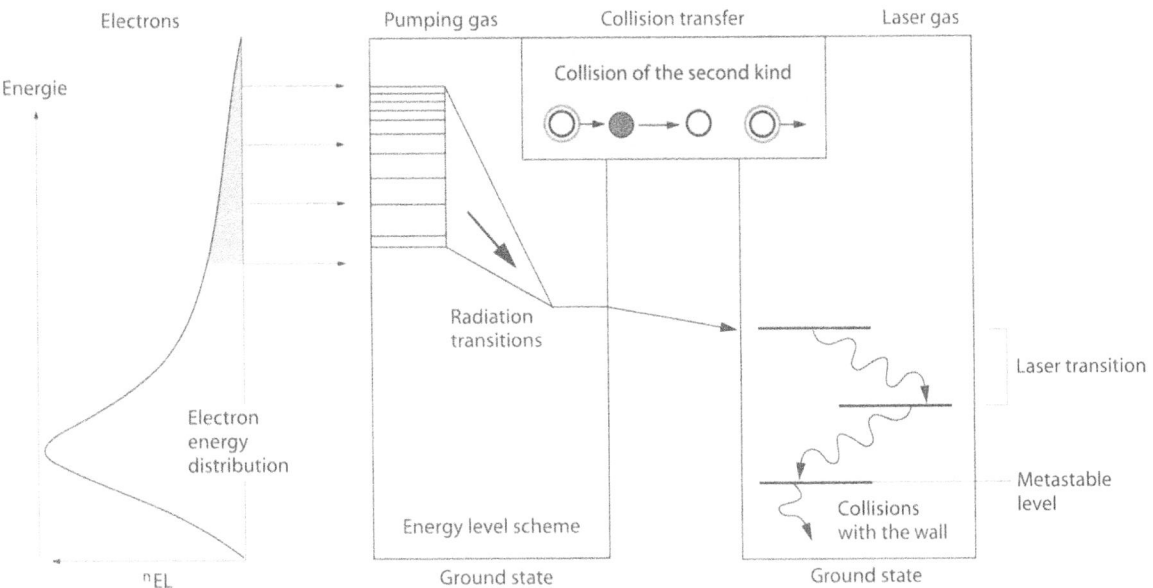

**Fig. 7.** Schematic presentation of the excitation process utilizing a pumping gas. The small fraction of the high energy electrons of the gaseous discharge excites the higher levels of the pumping gas. The levels decay into a metastable level and quasi-accumulate there. Collisions transfer the excitation energy to the upper laser level

**Fig. 8.** Some examples of the ratio between pumping energy and laser energy

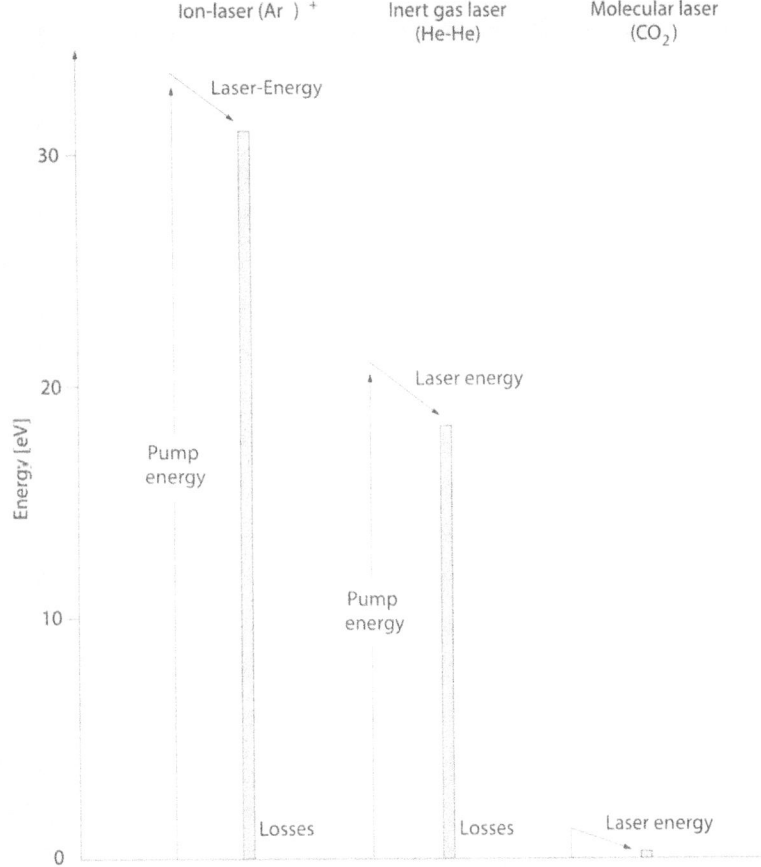

## Optical Resonators

As with all resonators, the optical resonator is expected to respond to a certain excitation frequency with a maximum oscillation amplitude. Since, for practical reasons, the geometric dimensions of an optical resonator cannot be of the same order of magnitude as the wavelength, an optical resonator has to be excited by very high harmonics. Another important function of the laser resonator is the feedback of photons into the laser medium; the probability of a stimulated emission is proportional to the residence time inside the laser medium; this represents the gain of the laser process.

In general, the laser resonator consists of two mirrors arranged in parallel and facing each other. They may be plane or curved. The various resonator types differ in their radius of curvature and separation of the mirrors (Fig. 9). Most popular is the confocal resonator.

The confocal resonator has the least diffraction losses. This resonator employs two concave mirrors having the same radius of curvature $b$. The length of the resonator $L$ corresponds to the radius of curvature ($L = b$). Since the focal length f of a spherical concave mirror is half its radius of curvature ($f = b/2$) the focal points of both resonator mirrors coincide.

To describe the radiation inside the resonator, the following information is needed:
1. The intensity as a function of the wavelength.
2. The geometric intensity distribution inside the resonator.

Both factors lead to the concept of "mode". The laser modes are the eigenfrequencies of the laser resonator.

In the first case, they are called longitudinal modes, in the second, transverse modes.

## Longitudinal Mode

As with any other resonator, the only excitable eigenfrequencies of the optical resonator correspond exactly to multiples of half the wavelength, determined by the geometric dimensions of the resonator (Fig. 10).

The laser resonator must adhere to the following relation:

$$n \cdot \lambda/2 = L,$$

where n = 1, 2, 3, ..., $\lambda$ = wavelength and $L$ = resonator length.

For laser resonators the ordinal number n is very large, such that the frequency difference $\Delta v$ between neighbouring longitudinal modes is

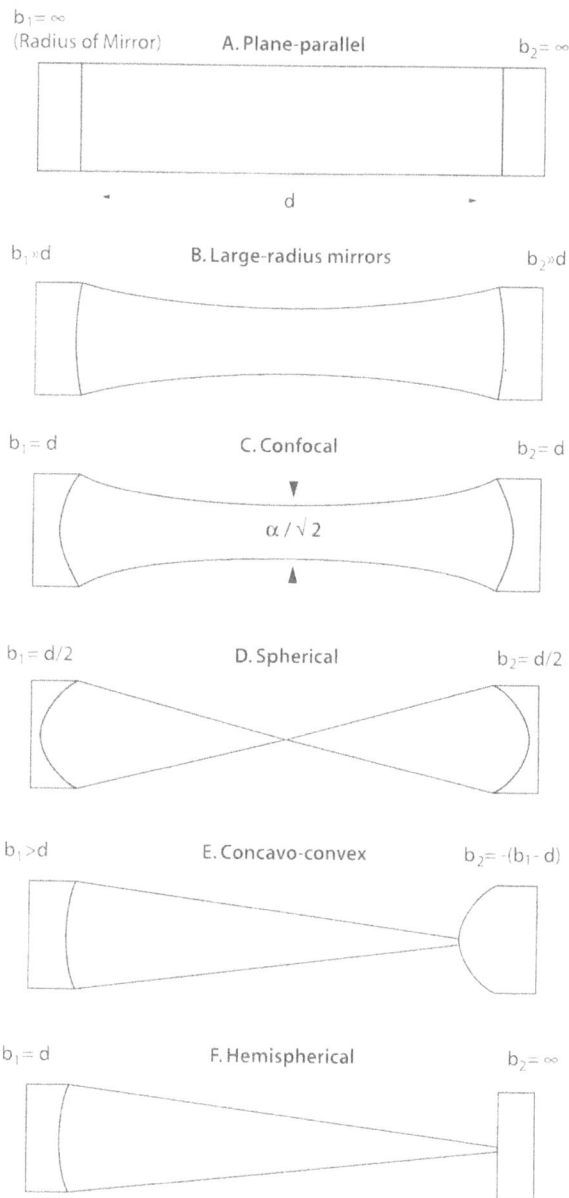

**Fig. 9.** Various types of laser resonators

$$\frac{\Delta v}{2L}$$

For a resonator length of 0.5 m neighbouring modes are separated by $\Delta v$ = 300 MHz.

Out of the large number of possible eigenfrequencies for the optical resonator, only some are excited with an appreciable intensity. They are those situated in the area which is determined by the Doppler maximum of the laser medium and the loss characteristic of all resonance losses (Fig. 11). Only for the frequencies $\Delta v_D$ does the gain overcompensate for the losses and laser action is achieved.

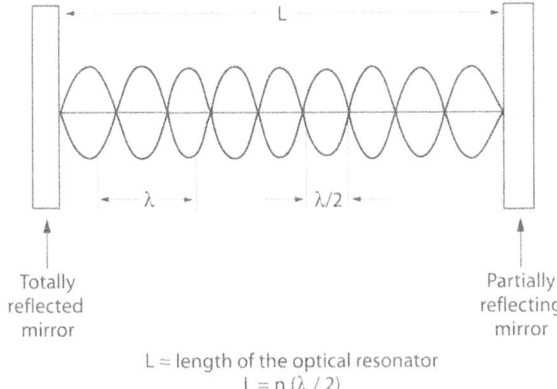

**Fig. 10.** An optical laser resonator can only amplify laser frequencies having an amplitude node at the mirrors (standing waves). This condition is met only if the separation is an integer multiple of half the wavelength

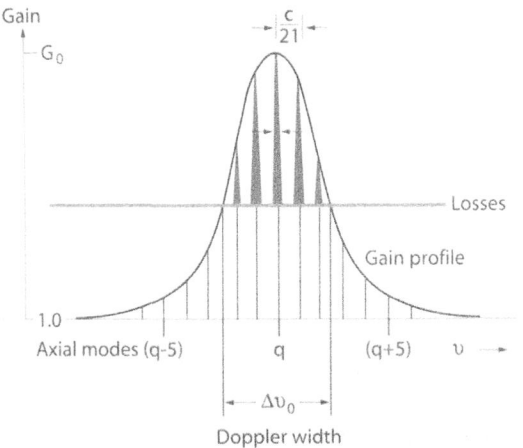

**Fig. 11.** Origin of the axial (longitudinal) modal spectrum. Laser activity is observed only where the gains within the Doppler width exceed the losses

In general, the number of oscillating axial eigenfrequencies is determined by the ratio of the Doppler width $\Delta v_D$ to the separation of the modes $c/2L$. To assure that only a single frequency is self-exciting, the condition $L < c/2\,\Delta v_D$ must hold. Assuming a typical Doppler width of $\Delta v_D = 1.6 \times 10.9$ Hz for the active medium of a gas laser, the resonator length must be chosen to be less than $L = 15$ cm.

## Transverse Electromagnetic Mode

Besides the longitudinal modes, there are transverse electromagnetic modes (TEM): these modes describe the spatial intensity distribution of the radiation inside the resonator. The principle mode $TEM_{00}$ is the

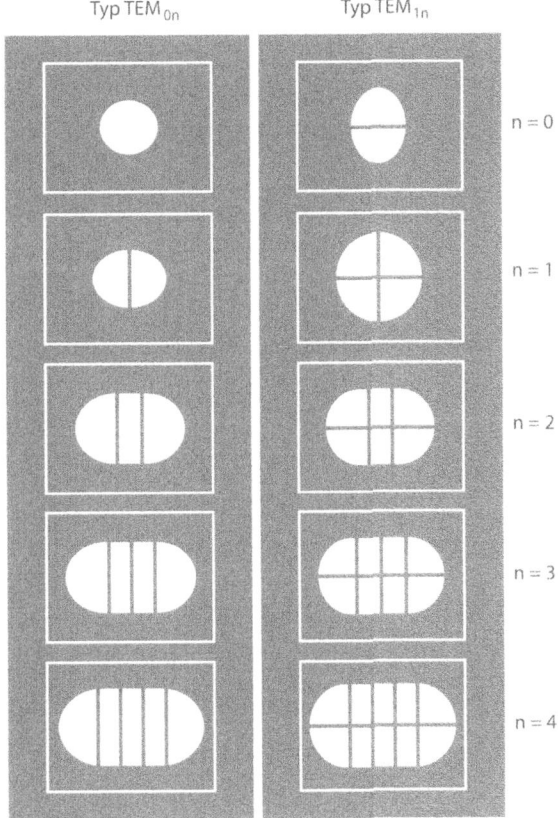

**Fig. 12.** Intensity distribution within a laser beam having various modal indices

lowest order of all modes. For higher modes at least one value for 1 or **n** is different from zero.

For higher TEM modes the intensity distribution of the laser radiation will split into $1 + 1$ or $n + 1$ partial beams, respectively (Fig. 12).

The optimal approach for discussing the ray trajectories inside confocal resonators is to investigate the principal mode $TEM_{00}$ because its field distribution can be described by a simple Gaussian. Figure 13 presents a typical ray trajectory of the principal mode and its pertinent characteristic data. At the position z = 0 is a characteristic constriction, the so-called "beam waist". For the principal mode its diameter $d_0$ has a simple and obvious meaning: it is the radial distance where the intensity has dropped to $1/e^{-2}$ of its maximum. This number can be considered an approximate value for the modal diameter.

The diameter of the mode increases with increasing distance from the beam waist according to the relation

$$W(z) = W_0 (1 + z'^2)^{1/2}; \quad z' = 2\,z/b.$$

At the position of the mirror the modal diameter has already increased by a factor of $\sqrt{2}$. The intensity

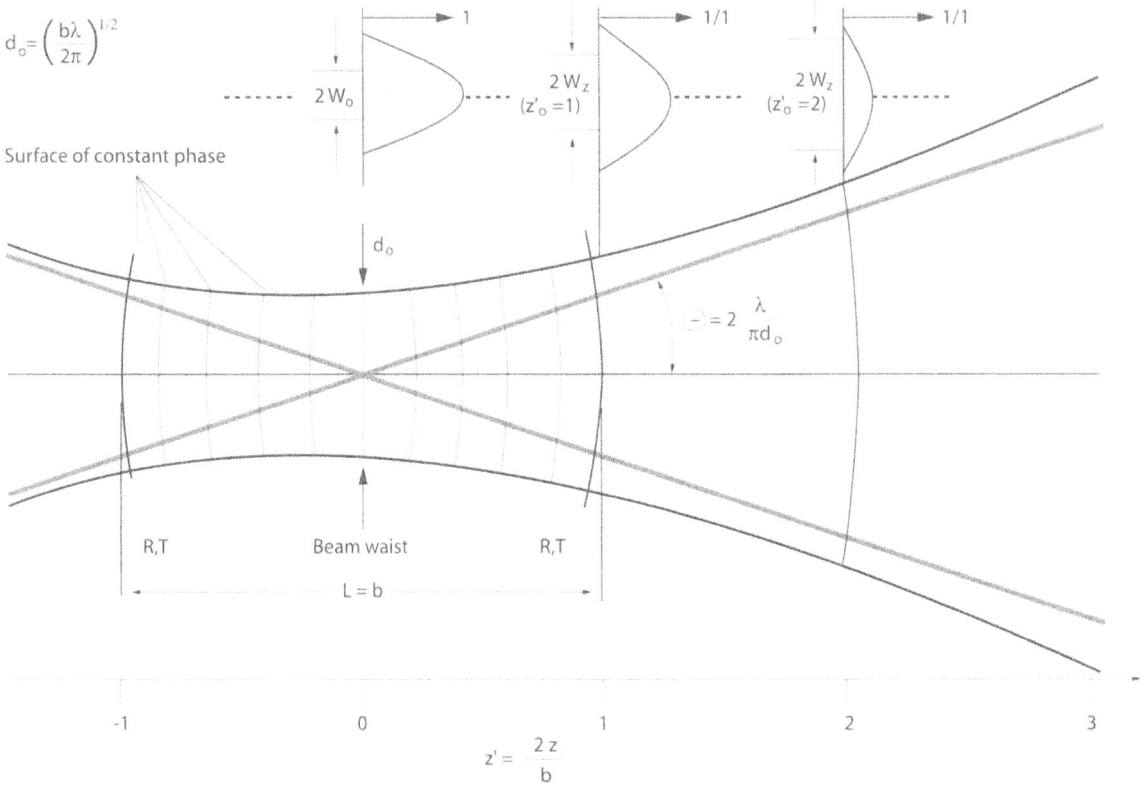

**Fig. 13.** Ray trajectories of the principal mode inside a confocal resonator. This resonator is characterized by its diameter $W(z)$ (the $1/e^{-2}$ value of the intensity) and its radius of curvature for the areas of constant phase. **Top:** field distributions for three different planes $z_0$.

of the mode remains constant at each position z, except for a W(z)-dependent change of scale.

The aperture angle θ can be interpreted graphically as the diffraction angle at the beam waist corresponding to an aperture having the diameter $d_0$. The angle θ is calculated according to the following equation:

$$\theta = 2 \cdot \lambda / \pi\, d_0$$

The relations are similar for nonconfocal resonators; however, their mathematical descriptions become more complex.

## References

1   Beck R, Englisch W, Gürs K (1976) Tables of laser lines in gases and vapours. Springer series in optical science, vol. 2. Springer, Berlin Heidelberg New York

# I-2.3
# Laser Radiation

K. Dörschel

Laser radiation, emitted by a laser resonator, displays three important characteristics (Fig. 1):

1. The radiation is coherent, i.e. all the wave trains are exactly in phase, in time as well as in space.

2. The radiation is well collimated, i.e. the radiation beam is almost parallel. The diameter of a laser beam increases very slowly with distance.

3. The laser radiation is monochromatic, i.e. all wave trains have the same wavelength, frequency and energy.

In addition, a laser beam can achieve very high radiant power.

Each one of these characteristics may be achieved by other light sources, but the laser is the only source of light combining all the above characteristics simultaneously.

In describing a laser beam, it is important to present the local intensity distribution inside the beam besides its coherence, collimation and monochromaticity. In addition, it is essential to know the influence of the optical elements – lenses and/or mirrors – on the beam shape.

Since the laser beam is an image of the laser radiation inside the resonator, the intensity distribution of the resonator's transverse electromagnetic mode (TEM) can be observed in the laser beam (Chap. I-2.2).

Consequently, a resonator working in the fundamental mode ($TEM_{00}$) will emit a laser beam having a fundamental mode $TEM_{00}$, etc. In this case, the inten-

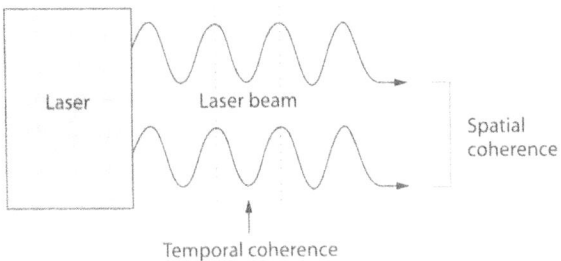

**a.** Coherence

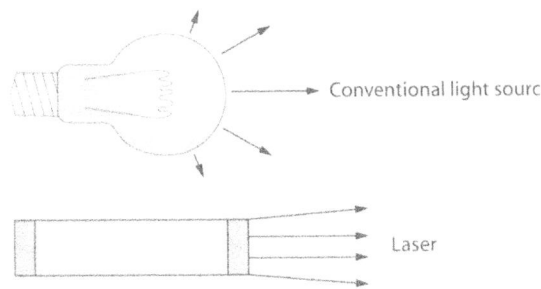

**b.** Collimation

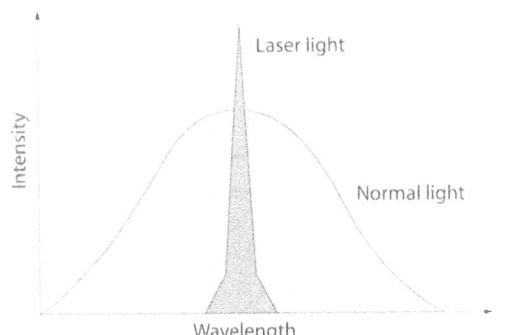

**c.** Monochromatism

**Fig. 1a, b, c.** Representation of coherence, collimation and monochromatism

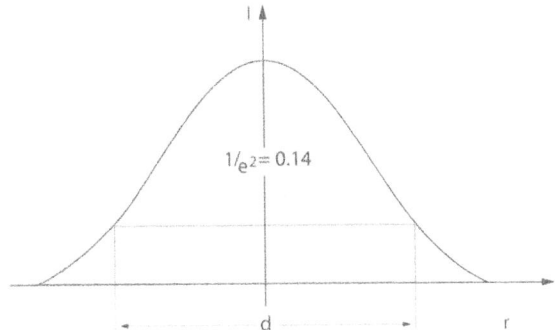

**Fig. 2.** The diameter **d** of a laser beam in its fundamental mode is that value where the radiation flux density $J$ is $1/e^2$, about 0.14 of its maximum value

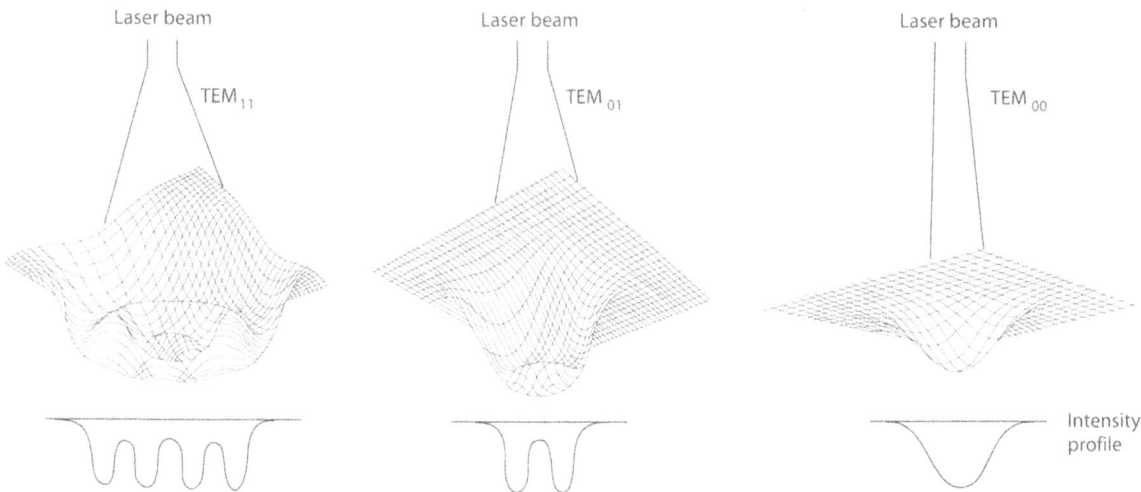

**Fig. 3.** Schematic presentation of the radiation flux in transverse modes and with no preferential direction in the resonator

sity profile of the laser beam at any position will be a Gaussian. In order to apply the term beam diameter to such an intensity distribution, it is necessary to define beam diameter as that value where the intensity of the laser radiation has decreased to $1/e^2$, about 0.14 of the maximum (Fig. 2).

For higher modes, $TEM_{mn}$, a clear definition of the diameter becomes impossible; nevertheless, it can be taken for granted that the diameter of a higher $TEM_{mn}$ mode is larger by a factor of $\sqrt{m}$ or $\sqrt{n}$, respectively, if produced by the same resonator (Fig. 3).

If a lense is placed into a laser beam, the laws of geometric optics will yield only approximate or even significantly wrong descriptions for the ray trajectories beyond the lens. In this case, one must refer to the theory of imaging Gaussian beams or, for multiple mode laser beams, to more complicated theories (Iffländer and Weber 1986).

There is an important difference between imaging a Gaussian bundle of rays and rays according to geometric optics: the image of a radiation waist $d_0$ situated at a distance $f$ in front of a lens appears at a distance $f$ behind the lens. Its diameter is $d = 4\lambda f/\pi d_0$. In contrast to this, geometric optics yields a symmetric figure from $2f$ to $2f$ and the image size remains unchanged (Fig. 4).

Often, position and diameter of the radiation waist to be imaged are unknown. In this case, the diameter of the radiation waist can be estimated more easily by determining the angular aperture $\theta$ (Fig. 5).

It is: $d = 2\theta f$, where $f$ is the focal length of the lens.

The following equation allows calculation of $\theta$ for a Gaussian-shaped beam:

$$0 = 2\lambda/\pi\alpha_D$$

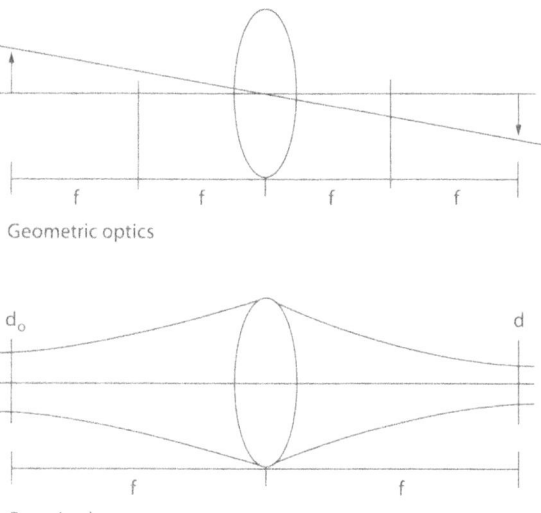

Geometric optics

Gaussian beam

**Fig. 4.** In geometric optics an image with magnification 1 is located at an image distance 2f, when the distance of the object is 2f

The radiation energy of medical lasers is mostly between 0.1 and 100 W. Concentrating the laser's radiation flux at the focal point of a lens, an enormous radiation flux density (measured in $Wm^{-2}$) or radiation intensity (measured in W/sr) can be achieved. The Table presents radiation flux densities at the focal point of a lens for some lasers and other light sources.

Laser action on tissue is dominated by the laser beam parameters radiation flux density and exposure time. The radiation flux density is thus the most important parameter for the irradiation of tissue by a laser beam of a given wavelength. The radiation flux density $M$ is:

**Fig. 5.** The diameter of the beam waist **d**, occurring while focusing with a lens having a focal length **f**, can be estimated from the focal length and the divergence 2 θ in front of the lens

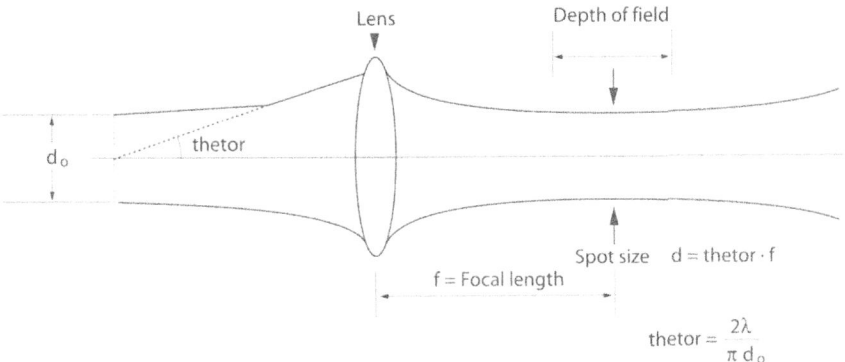

**Fig. 6.** Relation between radiation flux density and focal size

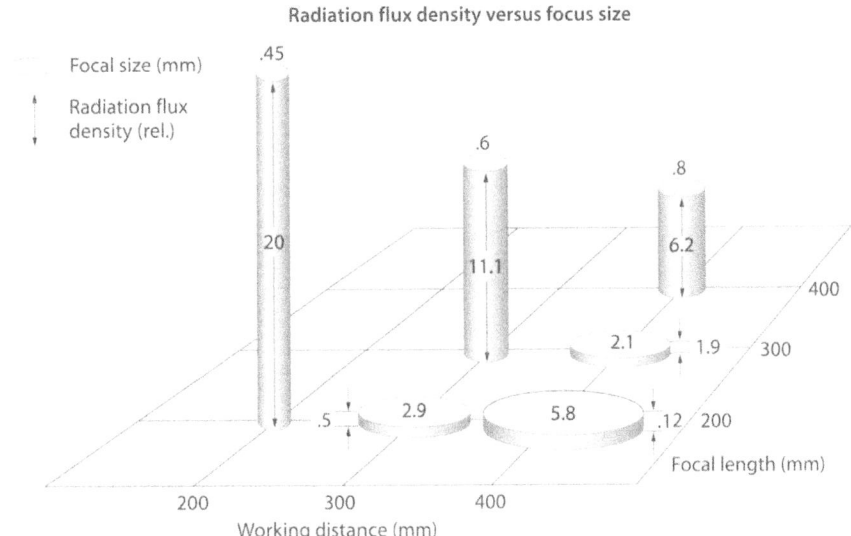

**Table 1.** Radiation flux densities at the focal point for lasers and other light sources

| Light source | Radiation flux (W) | Radiation flux density (Wcm⁻²) |
| --- | --- | --- |
| Sun | $10^{26}$ | $5 \times 10^2$ |
| Hg extrahigh-pressure lamp | 5 | $10^3$ |
| Incandescent lamp 100 W | 3 | $10^{-2}$ |
| HeNe laser | $10^{-3}$ | $4 \times 10^4$ |
| Ar⁺ laser | 10 | $4 \times 10^8$ |
| Nd:YAG laser with focusing hand-piece | 100 | $4 \times 10^5$ |
| $CO_2$ laser for metal work | $10^4$ | $10^8$ |
| Pulsed laser | $10^9$ | $10^{14}$ |

$$M = \frac{\text{beam cross-section}}{\text{radiation flux}}$$

In order to avoid using lasers incorrectly, the user must clearly understand the terms and relations between laser radiation flux, radiation flux density, focal area, focal diameter and focal length. Figure 6 demonstrates again the relation between radiation flux density inside a laser beam and the distance behind a lens.

## References

1. Kogelnik H, Li T (1966) Laser beams and resonators. Proc IEEE, vol 54 No 10
2. Iffländer I, Weber H (1986) Focussing of multimedia laser beams with variable beam parameters. Optica Acta 8: 1083-1090

# I-2.4
# Laser Types

# I-2.4.1
# Nd:YAG Lasers

B. Schaldach

The Nd:YAG laser is at present the most important solid-state laser. It covers a wide range of applications, e.g. in metal working, medicine and engineering.

The essential advantages of the Nd:YAG laser are its simple and compact construction and its high average power output. As opposed to the argon-ion laser (see Chap. I-2.4.2), the excitation of the Nd:YAG laser is more efficient, about 2 – 15%, depending on the excitation process. Figure 1 presents a survey of the loss mechanisms for Nd:YAG lasers incorporating gas discharge tubes.

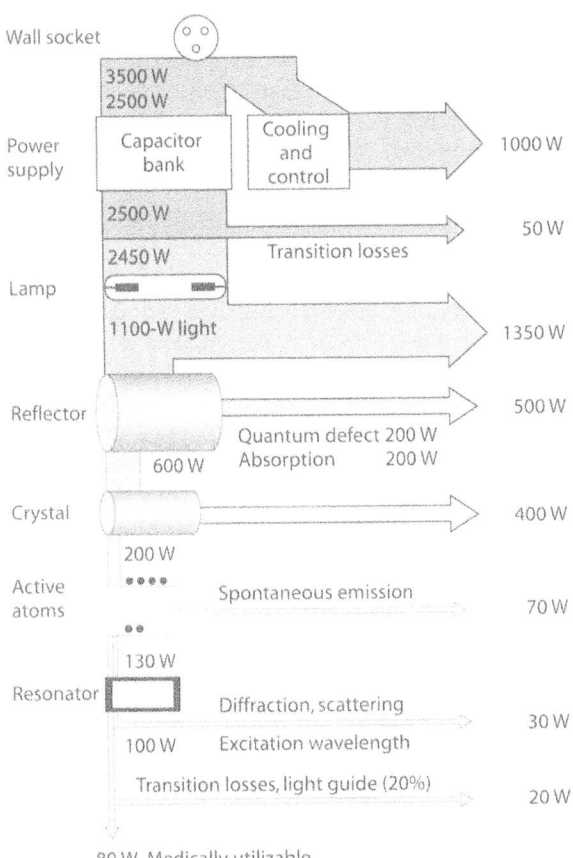

**Fig. 1.** Schematic survey of the losses for a Nd:YAG laser

The neodymium ($Nd^+$) ion, implanted into various host crystals, is the source of the laser radiation. Among the many crystals which have been investigated, one material stands out: $Y_3Al_5O_{12}$, yttrium-aluminium-garnet, a crystal with a garnet structure. It has a relatively high thermal conductivity – larger than other $Nd^{3+}$ host crystals –, high mechanical stability, good optical qualities, and can be grown to convenient crystal sizes. Currently, crystals are grown up to a length of 150 mm. For lasers the crystal rods have typical diameters of 3 – 7 mm and lengths of 90 – 150 mm. Inside the YAG crystals the $Nd^{3+}$ ions have a concentration of about 1.5 vol%. Higher dopant concentrations are unusual due to the different volumes of the $Nd^{3+}$ ions and the $Y^{3+}$ ions.

Although there are other host crystals with properties superior to those of the YAG, no single one presents such a combination of so many advantageous features as the YAG laser. The Nd:YAG laser is a four-level laser (see Chap. I-2.2). Figure 2 indicates the energy levels and the laser transition.

Initially, the $Nd^{3+}$ ions are in the ground state 1. Absorbing light from the pumping lamps they are lifted into the excitation bands 2. By rapid, non radiative transition, the upper metastable (long-life) laser level 3 becomes occupied. So, an occupation inversion develops between transitions 3 and 4, which enables a laser emission at 1.064 µm, in the near infrared, to take place. Some commercially available lasers employ a different transition, 3 – 4, emitting radiation having a wavelength of 1.32 µm. Figure 3 presents relative intensity versus wavelength. Obviously the two spectral lines at 1.064 and 1.0645 µm fuse due to their close proximity. A much weaker emission line is situated at the wavelength 1.320 µm.

The laser-active transition can be pumped with arc discharge lamps filled with noble gases (e.g. Kr or Xe), because the appropriate emission bands at 0.5 – 0.9 µm agree relatively well with the pumping levels. The pumping arrangement may be a double elliptic reflector made of a high-reflectivity, gold-coated material. A cylindrical YAG rod is placed at the common focal point. The two arc discharge lamps are centred at the other two focal points.

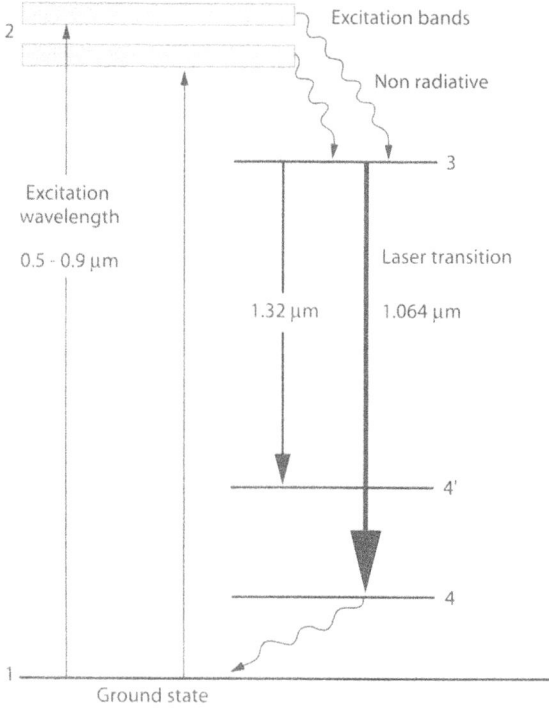

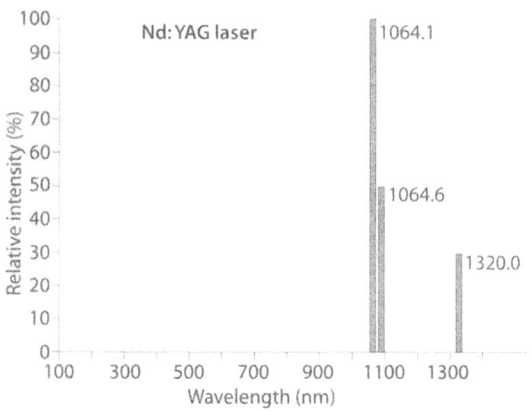

**Fig. 3.** Relative intensities of the most important wavelengths of the Nd:YAG laser

**Fig. 2.** Position of the energy levels and the laser transition

This arrangement ensures a high pumping efficiency. The system is cooled by passing purified water around the rod and lamps (see Fig. 4).

An alternative to this arrangement is pumping by using another laser (e.g. diode laser, λ is about 805 nm). This leads to an efficiency higher than 15% in the power range 0.1 – 10 W. The very high price of laser diodes prevents their use in commercial solid-state lasers.

There are various ways to construct the resonator: in most cases the active medium serves also as the optical resonator. The resonator mirrors are vacuum-coated directly onto the terminal faces of the crystal rod. One of the coated sides is highly reflective; the main part of the laser light leaves through the low-reflectivity side. This has the disadvantage that it is impossible to place any additional optically active elements into the resonator.

The second method utilizes an open (outside) resonator. In this arrangement it is possible to influence the radiation field inside the resonator by apertures, modulators or optical switches (see Fig. 4 in Chap. I-2.4.5).

Nd:YAG lasers are operated in either pulsed or in continuous mode; they deliver average output powers of up to 1000 W.

Pulsed lasers mostly utilize xenon-discharge lamps. By using Q-switching, typical pulse energies of up to a few Joule can be reached. Modern flash lamps achieve lifetimes of more than $10^7$ discharges.

Continuously pumped lasers can also employ Q-switching. In this case, much higher repetition rates – in the MHz range – are possible. However, for pulsed Nd:YAG lasers, the achievable repetition rates are almost 3 orders of magnitude lower. Due to its reduced efficiency, cw pumping achieves a considerably lower pulse power and pulse energy. Approximately constant pulse power can be delivered for a a repetition frequency which is associated with the lifetime of the upper laser level (for Nd:YAG this is about 250 μs). Above this frequency the power decreases: between single pulses the population inversion is insufficient. Q-switches for the pulsed operation mode are often electro-optical switches; the cw operation mode employs acusto-electrical switches which are operated at very high temperatures. It remains to be mentioned that trains of very short pulses, far below 1 ns, can be generated by mode-locking.

Placing wave-selecting elements (e.g. filters) into the resonator will lead to Nd: YAG lasers emitting also other wavelengths. In addition, nonlinear optical crystals will help to generate higher harmonics, i.e.:

1st harmonic – 1064 nm
2nd harmonic – 532 nm
3rd harmonic – 355 nm
4th harmonic – 266 nm

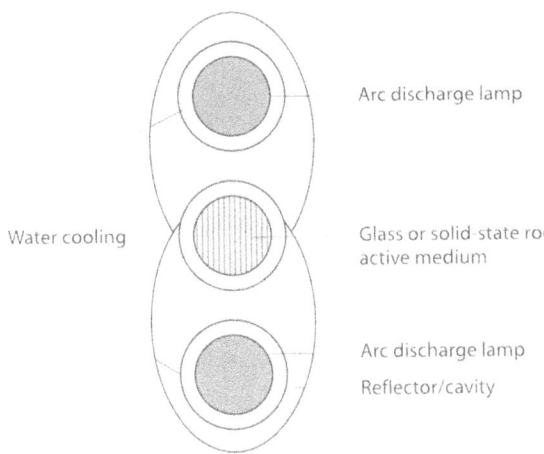

Water cooling

Arc discharge lamp

Glass or solid-state rod/
active medium

Arc discharge lamp

Reflector/cavity

**Fig. 4.**  Pumping arrangement with elliptical reflector

## Further Reading

Abraham NB et al. (eds) (1985) Physics of laser sources. NATO ASI Series (B): Physics, vol 132. Plenum Press, New York

Andrews DL (1986) Lasers in chemistry. Springer, Berlin Heidelberg New York

Arrechi FT, Schulz-DuBois EO (eds) (1972) Laser handbook. North Holland, Amsterdam

Bergmann-Schaefer (1978) Lehrbuch der Experimentalphysik, vol III. De Gruyter, Berlin

Dürr U (1983) Vibronisierte Festkörperlaser: Der Übergangsmetallionen-Laser. Laser Optoelektronik 1: 31-39

Hermann W (1984) Laser für ultrakurze Lichtimpulse. Physik-Verlag, Weinheim

Kulina P (1986) Festkörperlaser und –materialien. Laser Mag 4: 20-22

Luthey W, Stadler M, Weber HP (1987) Erbiumlaser für Mikrochirurgie. Laser Optoelektronik 2: 158-159

Shen H, Zhou Y et al. (1986) High power 1.3423 µm Nd:YAG cw laser. Optic Laser Technol 8: 193-197

Weber J, Herziger (1978) Laser Grundlagen und Anwendungen. Physik-Verlag, Weinheim

Wilson J, Hawkes JFB (1987) Lasers. Principles and applications. Prentice-Hall

Winburn CD (1987) What every engineer should know about lasers. Marcel Dekker, New York Basel

# I-2.4.2
# Argon and Krypton Ion Lasers

N. Müller-Stolzenburg

The argon laser is the best-known example of an ion laser having a noble gas as its active medium. The gas is contained at a pressure of about 0.5 mbar inside a plasma tube with an inner diameter of about 3 mm. Figure 1 indicates the general layout of a gaseous ion laser.

The laser tubes must be characterized by good thermal conductivity and robustness against impact by ions. Besides quartz, materials like graphite and berylliumoxide are utilized. By arranging magnetic coils around the discharge tube (see Fig. 2) the plasma is concentrated at the centre of the tube. This reduces the load on the tube caused by collisions with the ions, and it simultaneously improves the efficiency of the laser process. Because the discharge in the narrow discharge tube works like an ion pump, the Ar ions accumulate in front of the anode and consequently are no longer available for further excitation. To assure continuous laser action, a return channel is needed to allow the gas to return towards the cathode. To select a special wavelength an etalon or dispersion prism can be placed into the resonator.

Figure 3 indicates the energy level scheme of an Ar ion laser. The argon atoms are ionized by an electric discharge; the generated Ar ions are excited from the ground state into higher energy levels by an additional collision with an electron.

The cross-section for this second excitation is very small; high electron current densities are needed to sustain the process. For this reason, the cathode region has a larger diameter than the remainder of the discharge tube (see Fig. 1).Thus various excited states of the Ar ion will be occupied ($4p^4D^0$, $4p^2D^0$, $4p^2P^0$, $4p^2S^0 1/2$). Transitions from these levels into the $4s^2p1/2$ or the $4s^2p3/2$ level result in laser action. A number of discrete wavelengths between 250 and 530 nm are available (see Fig. 4). The two most powerful lines are emitted at 488 nm in the blue and at 514.5 nm in the green (in each case up to 15 W output power). Double ionized Ar atoms can emit laser lines in the near-UV at 351 or 363 nm.

The energy is supplied by a stationary electric discharge. Since the active laser medium, the Ar ions, must be first produced from Ar atoms, the efficiency of the argon laser is low (ca. 0.1%); plenty of waste heat is generated and must be removed by water cooling.

The krypton ion laser works on the same principle. The discharge tube is filled with the gas krypton. The Kr ion laser also emits several wavelengths. Laser lines in the spectral range from 350 to 800 nm are available. The most intense lines are at 530.9 nm in the green, 568.2 nm in the yellow-green and 676.4 nm in the red. The accessibe power for a $Kr^+$ ion laser is 5

**Fig. 1.** Basic design of a gas ion laser

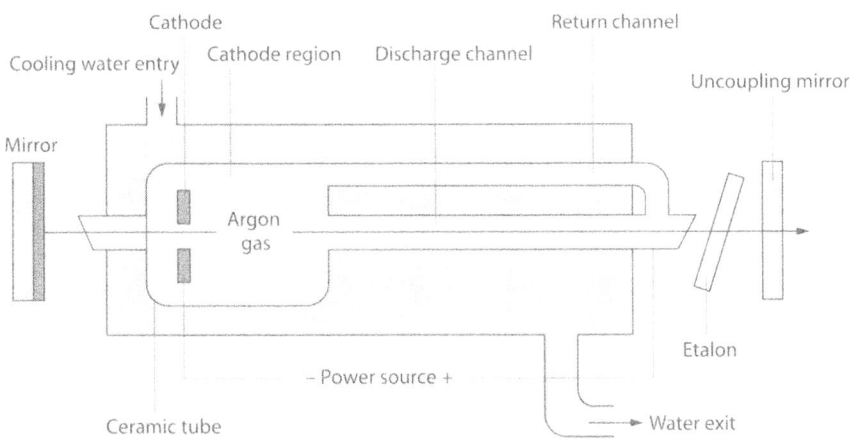

Brewster window                                          Brewster window

Resonator mirror        Cooling water        Discharge path    Gas-discharge tube

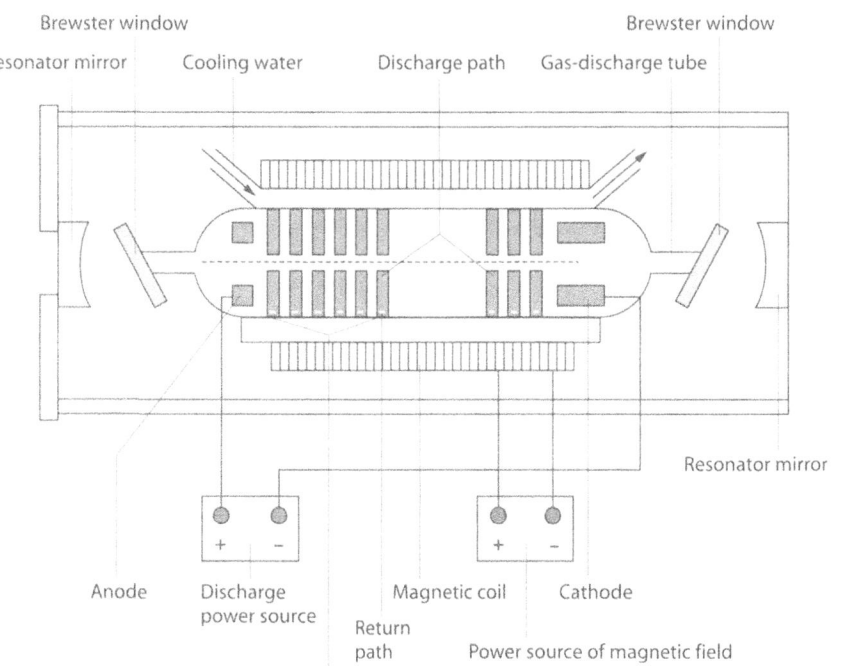

**Fig. 2.** Design of a gas ion laser in greater detail

Resonator mirror

Anode        Discharge        Magnetic coil        Cathode
             power source
                        Return
                        path          Power source of magnetic field

Alternate layers of graphite and BeO discs of the segmented laser

to 10 W. In Figures 4 and 5 the relative intensities of the lines for the Ar and the Kr ion lasers are indicated.

Gas ion lasers are a continuous lasers employing acusto-optical modulators however, light pulses with duration in the ps region can be achieved. Gas ion lasers are relatively expensive, comparatively vibration-sensitive, and the serviceable life of the laser tubes is between 1000 and 10000 operating hours.

**Fig. 3.** Energy level scheme of an Ar ion laser. The first collision with an electron (e-impact) ionizes the Ar atom, the second electron impact raises the Ar ion from its ground state into higher, excited states. Transitions from these excited states of the Ar ion into the $4s^2p$ levels lead to the emission of laser radiation

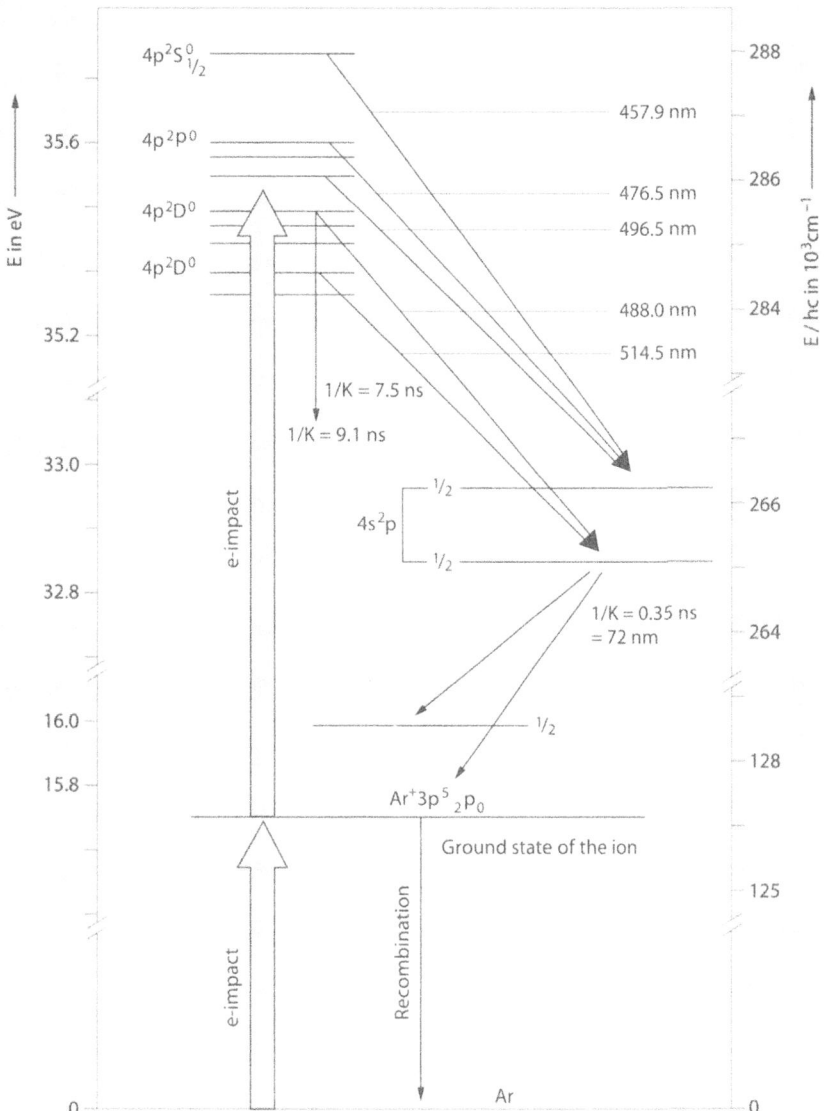

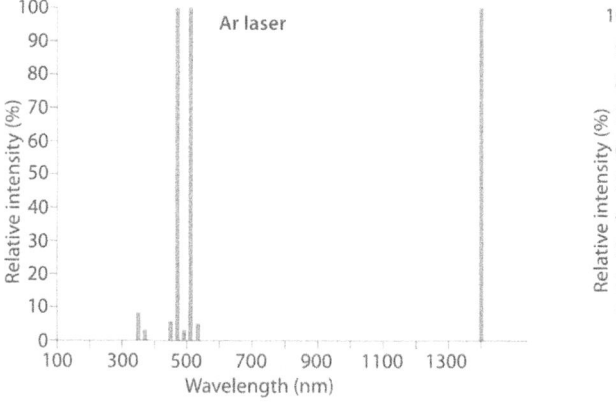

**Fig. 4.** Relative intensities of the lines for the Ar ion laser

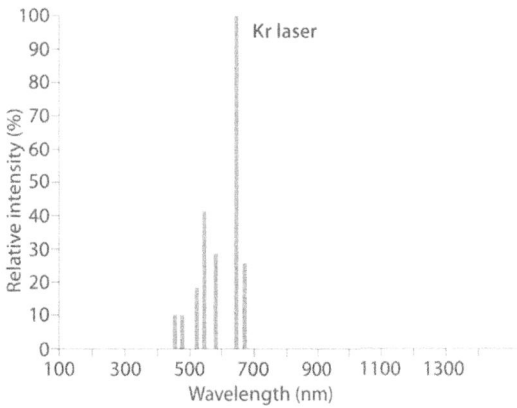

**Fig. 5.** Relative intensities of the lines for the Kr ion laser

# I-2.4.3
# HeNe Lasers

N. Müller-Stolzenburg

The HeNe laser was the first continuous-laser built and available commercially. Figure 1 indicates the basic design of an HeNe laser. The active medium is a mixture of helium and neon gases in proportion 10:1, contained inside a glass tube at low pressure (typically 1 mbar). The actual laser transition occurs in the neon.

The resonator mirrors can be placed directly onto the discharge tube. However, it is possible that the impact of the atoms of the laser gas damages the mirrors. To prevent this, the resonator mirrors are mounted separately from the discharge tube. If the laser tube is closed by a so-called Brewster window, as

indicated in Fig. 2, one direction of polarization of the laser radiation will be amplified loss-free while all other directions are suppressed.

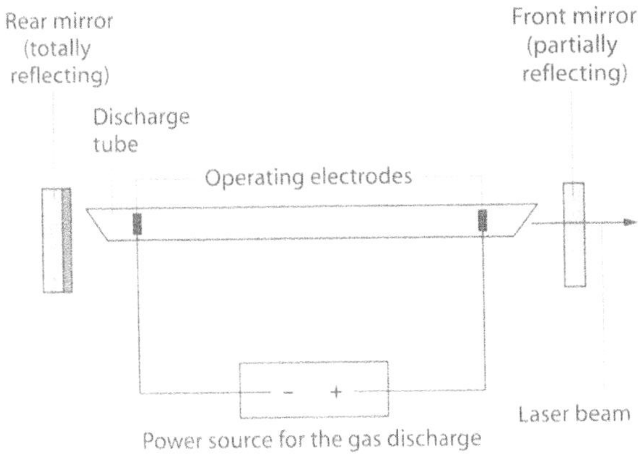

Due to the weak amplification of the HeNe system, Brewster windows are employed in many instances. This will cause the polarization in one direction of the laser radiation.

The energy level scheme of HeNe lasers is indicated in Fig. 3. The energy is supplied by an electric discharge: the metastable levels $2^1$s and $2^3$s of the He atoms are excited by electron impact; the energy is then transferred by collision with Ne atoms to the energetically neighbouring 3s and 2s levels of the neon. Because these energy levels of He and Ne are almost identical, the energy transfer is highly efficient. This type of energy transfer, also named collision of the second kind, is schematically indicated in Fig. 3 of Chapter I-2.4.2.

The 3s and 2s energy levels of Ne which are populated in this way are situated above lower and unpopulated levels of Ne (e.g. 2p and 3p). The result is a population inversion relative to these levels, making possible the emission of laser radiation.

The HeNe system offers about 200 potential laser transitions; only a few are exploited in commercially available lasers. Besides the well-known red laser line at 632.8 nm, there are two lines in the infrared at 1152 and 3391 nm. By now there are commercially available lasers emitting lines in the green at 543 nm, in the yellow at 594 nm and in the orange at 604 and

**Fig. 1.** Basic design of an HeNe laser

**Fig. 2.** Typical design of a discharge tube with a Brewster window

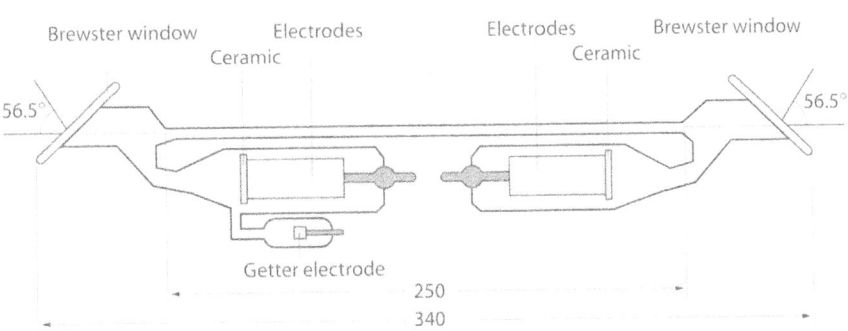

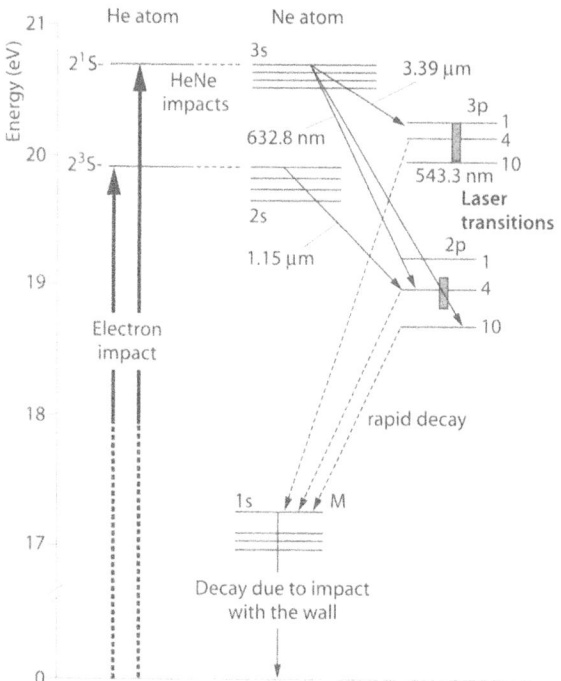

**Fig. 3.** Energy level scheme of a HeNe laser

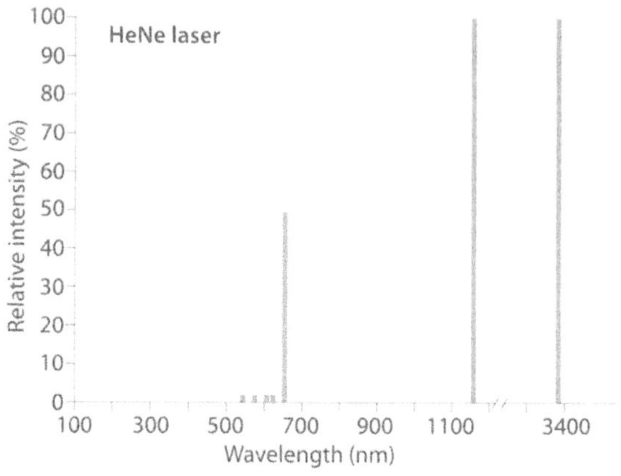

**Fig. 4.** Relative intensity of the lines of the HeNe laser

612 nm. As indicated in Fig. 3, the 3.391-µm line competes with lines in the visible spectrum, which are of main interest. Consequently, one tries to suppress the 3.391-µm line.

The relative intensities of the lines of the HeNe laser are indicated in Fig. 4. Typical output power of HeNe lasers is 1 mW for the green, the yellow and the orange line. The red line is available at significantly higher power (up to 100 mW).

HeNe lasers emit a continuous beam. Their low output power is offset by reliability, practicality and relatively low price. HeNe lasers display an extremely narrow line width and an excellent beam quality. Due to these advantages, they have found a very wide range of applications.

# I-2.4.4
# CO$_2$ Lasers

M. Glotz

**Contents**

## Introduction

The carbon dioxide laser or the CO$_2$ laser is one of the most important lasers used in the industrial processing of materials (laser machining); it is also of great importance in medicine. Owing to its high overall efficiency, it can achieve extremely high power outputs, making it ideally suitable for industrial applications and the construction of very compact medical devices.

## Active Medium

The name CO$_2$ laser designates a molecular gas laser in which a gas made up of CO$_2$ molecules is stimulated to emit useful coherent light with a wavelength range of 9 to 11μm. This type of light is localized within the infrared spectrum and is thus invisible to the human eye. The laser wavelength most often used is 10.6 μm because it is the most intensive. The laser gas is a mixture with a helium content of 60-80%; the rest of N$_2$ and CO$_2$ in a ratio of about 5:1. The added gases have different functions: they serve to increase pressure, stabilize the electrical discharge process, cool the laser gas, store energy and transfer such energy to the laser gas. Collisions produced by the laser gas also serve the purpose of reducing the population density in the lower laser level.

## Excitation

The electrons released by the gas discharge and the collisions of the N$_2$ molecules with the CO$_2$ molecules excite the latter. The nitrogen molecules are strongly and directly excited by the gas discharge, enabling them to store their energy for a long time before transferring it to the CO$_2$ molecules. It greatly benefits the laser that the energy levels of these two molecules are almost identical, which means that very little energy is lost during the energy transfer. Upper and lower laser levels of the CO$_2$ molecule are characterized by vibrational levels, which the three-atom molecule can occupy. An additional feature of an effective

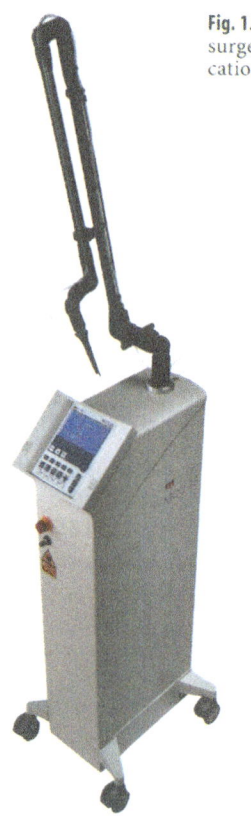

**Fig. 1.** Advanced $CO_2$ laser unit for laser surgery including aesthetic laser applications

## Designs

Essentially, the design of the laser depends on the power output desired. Consequently, the gas flows and the types of electric excitation differ. For the most powerful industrial-type lasers (used for processing materials), the laser gas is circulated at high speed. The gas flow passes through a gas cooler before it is fed once again into the laser light-generation process. Regeneration is achieved by adding fresh laser gas at a rate of about 20l/h. In medical technology applications, laser power outputs in the range of approx. 10 to 100 W are used; mainly within the power range of up to 25 W. The advanced, compact medical lasers employ a laser tube that is sealed after being filled with gas, and then used for many years. Regeneration is carried out through a catalyst incorporated into the laser tube. In this way, a gas life span of some 1000 hours can be achieved. These lasers are appropriately called sealed-off lasers.

There are two different electric excitation options: direct-current and high-frequency gas discharge. Since both methods have their advantages and disadvantages, both are used in practice.

## DC Excitation

Electrodes directly fused into the laser tube create a discharge in the laser gas by a high-voltage current. The resulting glow discharge stimulates (excites) the $CO_2$ molecules in the discharge area, located between the electrodes, through electron collisions produced by the $N_2$ molecules. The laser-generation process is continuous during DC excitation, so that the generated laser beam becomes available as a true continuous-wave (cw) beam. Additionally, the DC power supply units can be pulsed to produce short laser spikes. The pulse duration lasts from ms to s. During the buildup phase of this pulsing process a laser power peak is generated. If, therefore, a very short pulse width on the supply unit is selected in such a way that it corresponds to the laser peak generated during the building-up process (approx. 1 ms), a so-called super pulse with a higher peak pulse power (up to ten times more) can thus be generated. In the medical field, this type of pulse is particularly useful for tissue cutting and removal at high pulse repetition rates. The main advantage of this type of pulse is that the thermal effect of the laser is minimal but nonetheless available. It should be stressed that the superpulse technique can be used only with DC supply units, whose lower cost, compared with high-frequency systems, is an added advantage.

laser is its ability to generate a low population density on the lower laser levels. This is achieved by collisions with helium atoms which "empty" the lower levels. In addition, the strong cooling effect of the helium also reduces the thermal population density of the lower laser states (or energy levels). The high population density achieved at the upper level and the low population of the lower level are essential prerequisites for an effective laser.

Since several rotational levels are superimposed on the vibrational levels, this results in numerous densely packed energy states leading to a band spectrum of the laser line. In practice, however, the laser line with the highest occupation number (i.e. population density) gains the upper hand. For scientific (e.g. spectroscopic) applications, it is nonetheless possible to force other laser lines off a band. In the medical field, however, only the transitional state of one further vibrational level is sometimes used, corresponding to a wavelength of 9.6 µm. This wavelength is particularly useful for dental applications, because of higher absorption by the bony (hard) tissue.

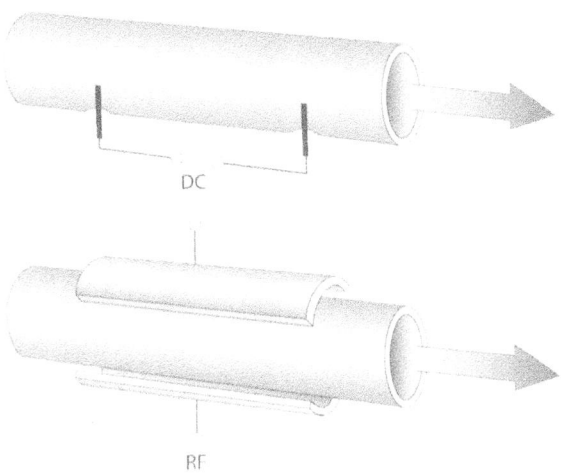

**Fig. 2.** Principle of laser-tube construction – DC vs. HF excitation

## HF Excitation

High-frequency (RF) excitation of the laser gas is a process that takes place through capacitor electrodes located outside the laser tube. They are usually semispherical and come attached to the exterior of the gas-filled laser tube. The electrical alternating (pulsating) field being generated between them lies within the MHz range and stimulates the laser gas. The user can regulate the laser output power by adjusting either the RF amplitude or the pulse duty factor accordingly. In medical applications, $CO_2$ lasers are usually regulated by output power modulation. The great advantage of RF excitation, as opposed to DC excitation, is that no metal electrodes are needed inside the tube, so electrode burnoff is avoided in the first place and no gas impurities are produced by residual burnoff particles. This technique brought about the decisive breakthrough needed for building high-performance $CO_2$ lasers. The efficiency of RF excitation, however, is somewhat lower than that of DC excitation.

## Laser Light

The coupling out of the laser light is done through a partially reflecting mirror of the laser resonator. These mirrors are the optical and physical windows that the gas-filled laser tube has to the outside world. In addition to their optical function, these mirrors also serve the mechanical purpose of sealing off the laser tube, thus making it gas-tight. The mirrors are attached to the laser tube with a gas-tight adhesive or sealing material. When using high-performance

lasers, however, there is the danger that residual amounts of energy absorbed in the mirror substrate or in the dielectric layers may lead to a build-up of heat and ultimately destroy the mirrors. Consequently, cooled high-performance laser mirrors are used that guarantee minimal residual absorption and prevent thermal damage or destruction. When very powerful high-performance lasers are used – with an output power of 20 kW or more – aerodynamic windows generated by a supersonic gas flow are employed. Here, the gas flow forms a barrier between the laser gas and the outside atmosphere. Aerodynamic windows are totally free from any laser beam-related heat problems and are formed by the gas nozzle in such a way that the laser beam profile is in no way adversely affected. In spite of being difficult and costly to build, they are ideal from an optical point of view. Moreover, this technique also allows the use of highly reflecting metal mirrors inside the laser resonator. Such mirrors ensure a very high degree of reflection of the $CO_2$ laser light and can also be easily cooled, thanks to their good thermal conductivity. In this type of laser resonator design, the laser light is coupled-out through an opening in the deflecting mirror placed. This opening is in front of the aerodynamic window, thereby reflecting most of the laser light back into the resonator so that the cycle can start anew. Only a small portion of the light is allowed to exit the laser through the opening and the aerodynamic window. As soon as it exits from the window, the laser beam is ready for any application.

## Beam guidance

The flexible transmission of the laser light takes place mostly through articulated mirror arms; for special applications, waveguides are employed. The articulated arms, used in the medical field, usually incorporate seven swivel joints, each with a laser mirror placed inside the joint at a 45° angle. Having an overall length of 1.30 m, the articulated arms must satisfy high standards, such as torsional rigidity, mechanical ease of movement and flexibility. This is why carbonfibre composite materials are used nowadays, as these materials are lightweight yet very rigid. The high quality of the laser beam allows transmission of the light to the user's hand piece without any significant losses. High-quality laser light signifies a small beam parameter product (i.e. beam divergence multiplied by beam diameter). At a first approximation, this product is conserved in the optical image, meaning that a laser beam with a low product generates a smaller focal point than a higher-product beam. Medical laser resonators emit light in the fundamental mode, i.e. light having the smallest possible beam parameter product. For the user, this means a laser beam with a very small focus diameter, allowing high-precision applications even in microsurgery. If, in addition, a micromanipulator is used for optimum beam guidance, focuses as low as 100 μm are possible.

In the dental field, flexible waveguides connecting the laser system to the user's hand piece are currently preferred. These waveguides serve as a substitute for the optical fibres that are still being developed. They are doing an excellent job as flexible light transmitters up to a laser output power of 10 W.

## Effects

The effects produced by the laser light on the material or the tissue depend on the wavelength of the light, its duration (time characteristics) and the transmission and absorption characteristics of the materials used. In material processing, the polarization direction is also important. The characteristics of the materials used are described in terms of reflection, transmission and absorption. In laser medicine, in particular, the $CO_2$ laser is subject to a very high degree of absorption at a wavelength of 10.6 μm, due to the high water content of the tissue. This means that the laser does not penetrate well and the absorbed laser energy vaporizes and carbonizes the upper tissue layer. For impact times above 1 ms, an additional thermal effect can be observed in the tissue lying underneath. This effect is used for coagulation purposes. At impact times below 1 ms, the laser produces practically no noticeable thermal effects at all. Consequently, this "athermal" quality is often used in cosmetic surgery for skin resurfacing (removing of folds and wrinkles).

## Application tools

In addition to the laser light source and the light transmission system, the applicator – as the application tool – plays a crucial role in the effects produced on tissue. In the industrial and medical fields, lenses as well as mirrors are used for focusing the beam. In high-performance lasers, mirror systems have a distinct advantage over lenses: absorption, and therefore the thermal stress produced, is quite low. This is of no great consequence in medical applications, however. In microsurgery, for example, mirror systems are incorporated into micromanipulators that are used in connection with surgical microscopes. The micromanipulator itself is built into the observation beam path of the microscope, thus allowing a highly precise guidance of the tiny laser beam focus with the help of a joystick. The mirrors serve to focus the $CO_2$ laser

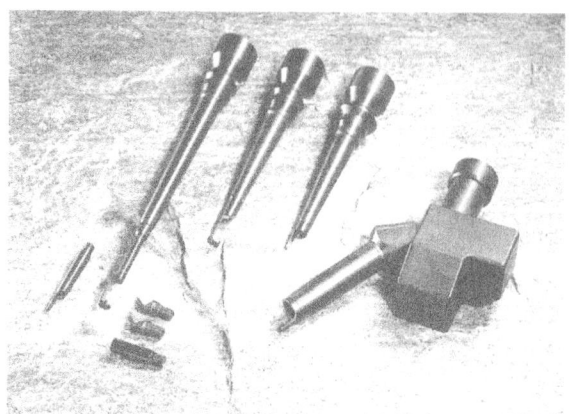

**Fig. 3.** Focusing hand pieces and scanners for medical $CO_2$ lasers

beam and the visible pilot laser beam together in one point.

For focusing hand pieces, ZnSe lenses are used. Mirror systems would be too costly and are not really needed. In the medical field, the range of application tools includes focusing handpieces with and without backstop or corner mirro, cannulas with straight or angular radiation outlet, and scanners.

## Further Reading

Eichler J, Eichler HJ (1991), Laser. Springer, Berlin Heidelberg New York
Hügel W (1992), Strahlwerkzeug Laser. Teubner Studienbücher
Scholz C (1992), Neue Verfahren der Bearbeitung von Hartgewebe in der Medizin mit dem Laser. ecomed Fachverlag
Witteman WJ (1987), The $CO_2$ laser. Springer Series in Optical Sciences. Springer, Berlin Heidelberg New York

# I-2.4.5
# Dye Lasers

K. Dörschel and J. Helfmann

## Contents

## Introduction

Dye lasers, in contrast to most other lasers, offer the possibility of shifting the output wavelength. The wavelength range for one dye is about 50 to 100 nm. Employing presently available dyes, it is possible to cover the entire wavelength range from 400 to 900 nm.

## Basic Design

Two types of dye laser design can be distinguished: firstly, the pulsed system with a dye-filled cuvette as presented in Fig. 1.

Figure 1 indicates an example of a dye laser. The dye molecules are excited by a flash lamp arranged parallel to the cuvette. An elliptical mirror surrounds the flash lamp and optic cell, focusing all the light of the flash lamp onto the cuvette. To increase the service life of the dye, the contents of the optical cell is pumped over and cooled.

Secondly, continuous-wave dye lasers employ a so-called jet, where the dye molecules are pumped

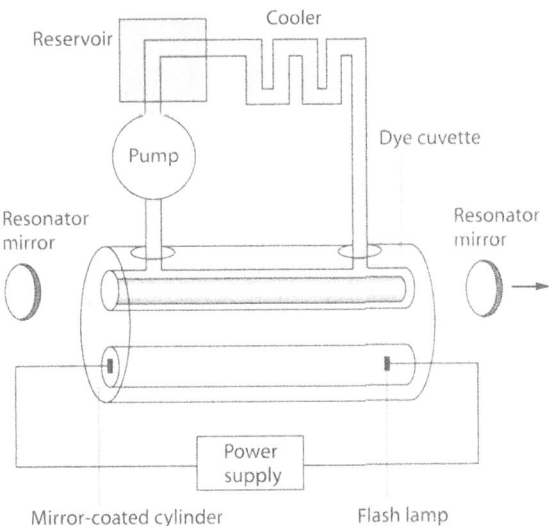

**Fig. 1.** Design of a flash lamp-pumped dye laser

through a small flat nozzle and are excited as a free beam by the pumping laser.

## Laser Media

Various dyes are employed as laser media. A typical representative of these aromatic molecules with conjugated double bonds, is rhodamine 6G (R 6G); its structural formula is indicated in Fig. 2.

Since these dyes can be applied only for a limited wavelength range, various dyes are employed to cover the complete visible spectral range.

Figure 3 is an overview of the dyes for the visible spectral range and the neighbouring IR. A comprehensive compilation of all presently applicable dyes can be found in [1].

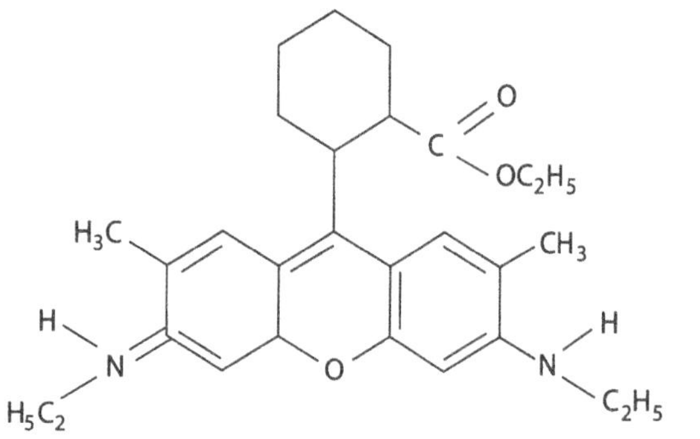

## Laser Process

The dye laser works as a four-energy-level laser, as described in Chapter I-2.2 and indicated in Fig. 4.

Absorbing the pumping wavelength, the dye molecules are lifted from the ground state $S_0$, into the higher vibrational states $S_1$. From there, they rapidly reach the lowest level of the singlet state $S_1$, via radiationless transitions. For all the atoms studied, most electrons occupy the state $S_1$. From this state, $S_1$ laser transition occurs into the various rotation-vibration levels of $S_0$, i.e. the various transition possibilities from $S_1$ to $S_0$ constitute the laser transition.

Since the energy width of the individual rotation-vibration levels are larger than their energy separation, the individual energy states overlap, resulting in a quasicontinuous laser spectrum.

The difference between absorption and fluorescent spectra is indirectly indicated in Fig. 4. The excitation energy needed to pump from the ground state $S_0$ into the excited state $S_1$ is larger than the energy of the laser transition, i.e. the pumping wavelength is smaller than the desired dye wavelength.

The absorption spectrum can be extracted from Fig. 5, e.g. the argon wavelengths at 488 and 514 nm are well suited for excitation.

**Fig. 2.** Structural formula of the dye rhodamine 6G

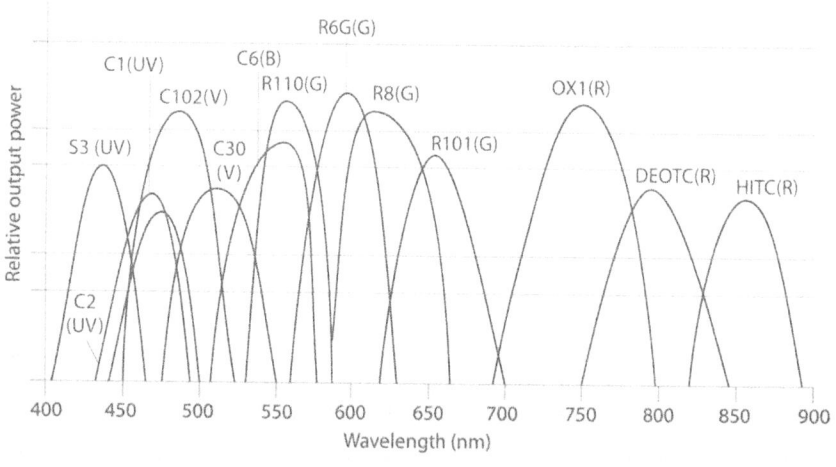

**Fig. 3.** Laser media for the visible and the neighbouring infrared spectral region

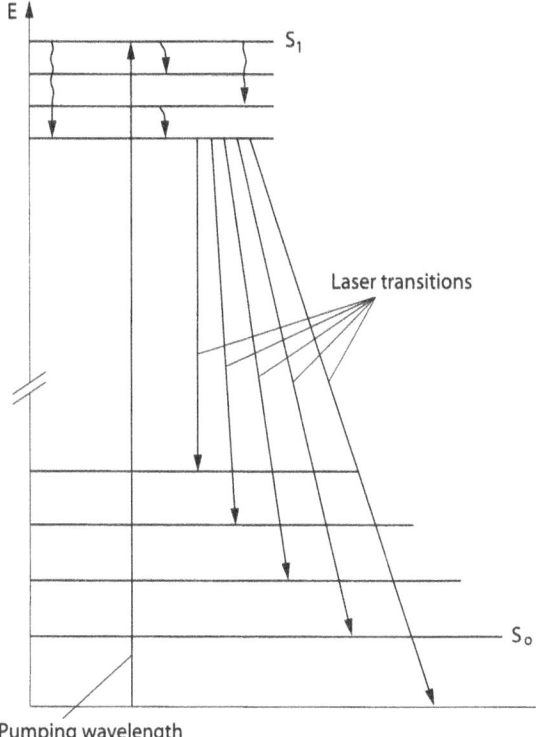

**Fig. 4.** Laser action of a dye laser

## Excitation Mechanisms

There are several possibilities to initiate the above described laser process. In principle, there is a difference between pumping with a pulsed or with a continuous source.

### Excitation by Using a Continuous Light Source

To operate a dye laser in a continuous-wave (cw) mode, the employment of a cw laser is a basic necessity. In this case, argon ion lasers are used.

An important disadventage is the high cost of procuring a noble gas ion pumping laser, see also Chapter I-2.4.3 (HeNe laser).

### Excitation by Using a Pulsed Light Source

To operate a pulsed dye laser system, various pulsed lasers and also flash lamps with sufficient intensity can be utilized. For commercial applications excimer lasers and Nd:YAG lasers dominate.

Table 1 summarizes the data of the most important dye laser systems.

**Table 1.** Data for some dye lasers

| Pumping laser | Pulse duration | Pumping power or pumping energy | Dye power dye energy |
|---|---|---|---|
| Ar$^+$ | cw | 10 W | 3 W (R 6G) |
| Flash lamp | 0.2 - 1\µs | 30 J | 1 J |
| Excimer | 20 s | 400 mJ | 100 mJ |

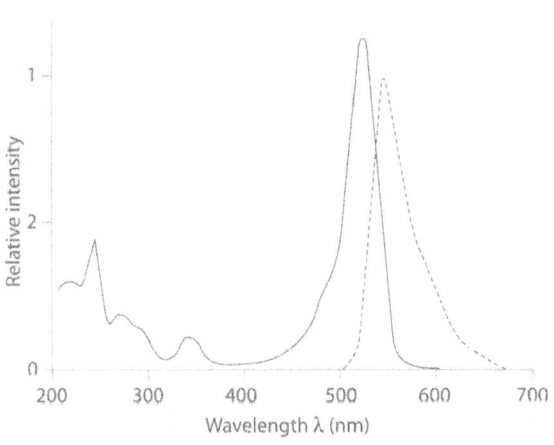

**Fig. 5.** Absorption and fluorescent spectrum of R 6G

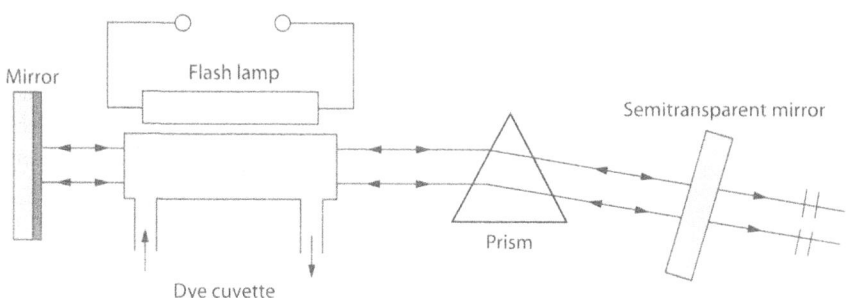

**Fig. 6.** Resonator of a pulsed dye laser

## Resonators

The following describes various designs of resonators for pulsed and continuous-wave dye lasers.

### Resonators for Pulsed Dye Laser Systems

Figure 6 presents a resonator for a flash lamp-pumped dye laser. Besides the resonator contains a prism as a wavelength-selecting component, i.e. by turning the prism the output wavelength can in rough steps be varied.

### Resonators for cw Dye Laser Systems

Figure 7 presents the typical design of a cw dye laser resonator.

The pumping laser, in this case an argon ion laser, is focused by a mirror into the dye beam. With the help of a second mirror, a parallel beam is created which subsequently passes through several wavelength-selecting components. The actual laser resonator consists of the two mirrors.

## References

1. Arecchi FT, Schulz-Dubois EO (1972) Liquid lasers. In: Laser Handbook, vol. 1, Chap. B 3. North-Holland, Amsterdam
2. Demtröder W (1977) Grundlagen und Techniken der Laserspektroskopie. Springer, Berlin Heidelberg New York
3. Weber H, Herziger G (1978) Laser: Grundlagen und Anwendungen. Physik-Verlag, Weinheim

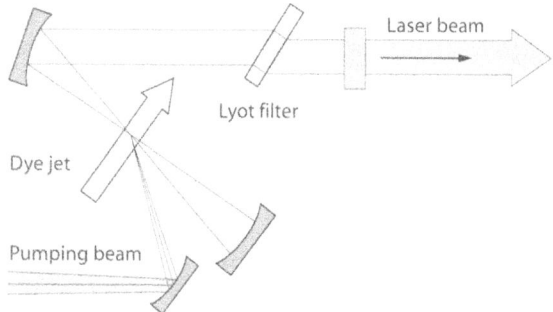

**Fig. 7.** Resonator of a cw dye laser

# I-2.4.6
# Excimer Lasers

H. Kar

## Contents

## Introduction

The excimer laser is a gas laser emitting in the UV-wavelength range from 157 to 351 nm. The name excimer laser is derived from **excited dimer** (halogen combinations with noble gases are called dimers).

Figure 1 shows the schematic design of an excimer laser; Figure 2 presents a medical excimer laser itself.

## Laser Media

The excimer laser utilizes as its active medium a mixture of a noble gas (argon, krypton or xenon), a halogen (chlorine or fluorine) and a buffer gas (helium or neon). The various wavelengths emitted depend on the combination of the noble gases and the halogen (see Table 1).

Figure 3 indicates the most important laser lines and their relative intensities.

A typical excimer laser operates at a total pressure of the laser medium of 2 – 3 bar and at a repetition rate of up to 1000 Hz.

Typically, an excimer laser system achieves a pulse duration in the region from 10 to several 100 ns; presently the average power output is up to 200 W.

The laser medium must be exchanged after an appropriate operating time of the laser; it depends essentially on the noble gas – halogen combination used. A gas filling of xenon chloride and operating at 308 nm may be used for about $10^7$ pulses. This corre-

**Table 1.** Relation between gas mixture, wavelength and typical energy per pulse

| Wavelength (nm) | Typical energy per pulse (mJ) | Gas mixture |
|---|---|---|
| 157 | 5 | $F_2$ |
| 193 | 200 | ArF |
| 222 | 35 | KrCl |
| 248 | 250 | KrF |
| 308 | 150 | XeCl |
| 351 | 80 | XeF |

Schematic design of an excimer laser

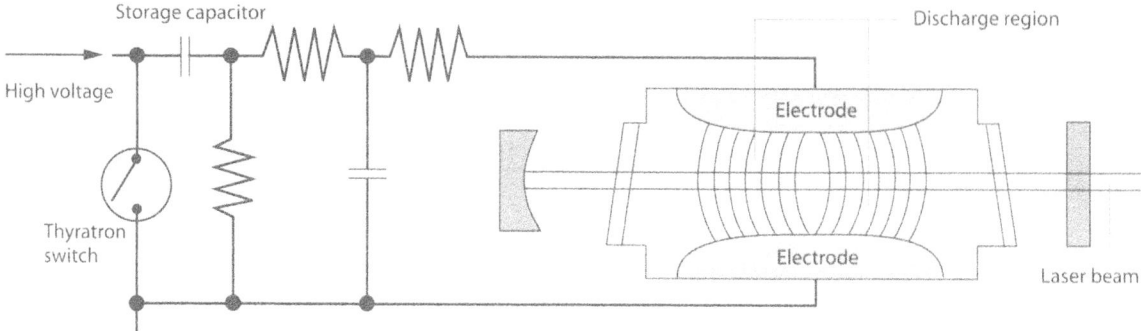

**Fig. 1.** Schematic design of an excimer laser

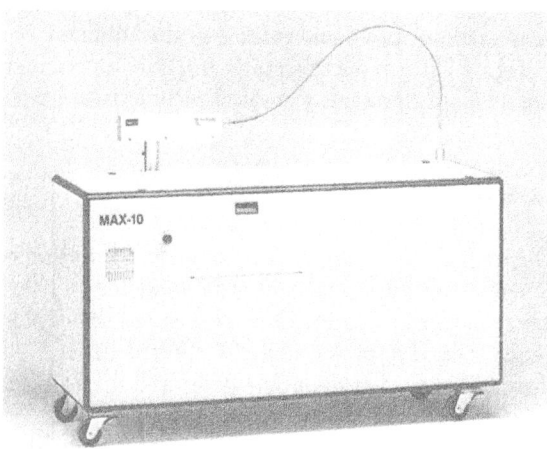

**Fig. 2.** A medical excimer laser

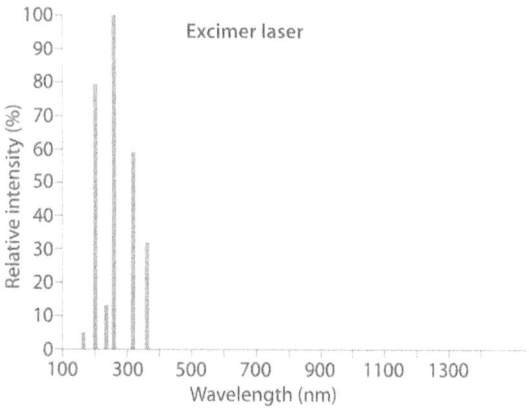

**Fig. 3.** The most prominent laser lines of the excimer laser and their relative intensities

sponds to an operating time of about 10 – 20 days, assuming that the laser is running 4 h per day at 30 Hz. An argon fluoride filling working at 193 nm will barely last 1 day under the same conditions.

Changing the wavelength of the excimer laser is very restricted in practice. An excimer laser operates without a problem only if operated with mixtures containing the same halogen. An exchange, e.g. between the mixtures containing argon, krypton and xenon chloride, will cause no complications. Problems are caused by the general handling of chlorine and fluorine. During exchange of the laser gas mixture, the halogens chlorine and fluorine are pumped through and finally discharged by a filter system. Note that these filters must be completely replaced after a certain number of pumping cycles (about 30 – 100). Safety considerations require the use of a gas mask during change of filters or gas bottles.

## Laser Process

Figure 4 shows a typical potential curve for a halogen rare gas molecule RGH in its excited and in its ground state. The electronically excited RGH forms the upper laser level with a lifetime of several ns. The laser radiation is emitted during transfer from the upper laser level, corresponding to the noble gas halogenide, to the lower laser level. The latter corresponds to the non-excited noble RGH.

At the lower laser level the noble gas halogenide is no longer stable, having a lifetime of only several ps, i.e. it decays rapidly – after the laser process – into individual atoms (noble gas and halogen).

The lifetime of the excimer molecule during its excited state is 1000 times longer as in its non-excited state. The separation between the potentials of upper and lower laser level corresponds to the emitted laser wavelength.

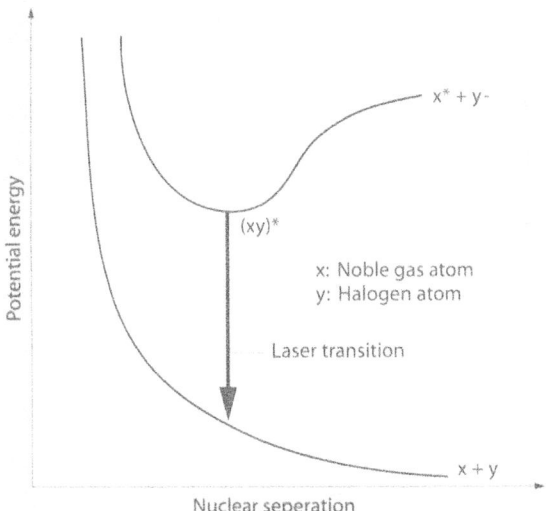

**Fig. 4.** Typical potential curves of a noble gas-halogenide (* indicates excited; - indicates negatively charged)

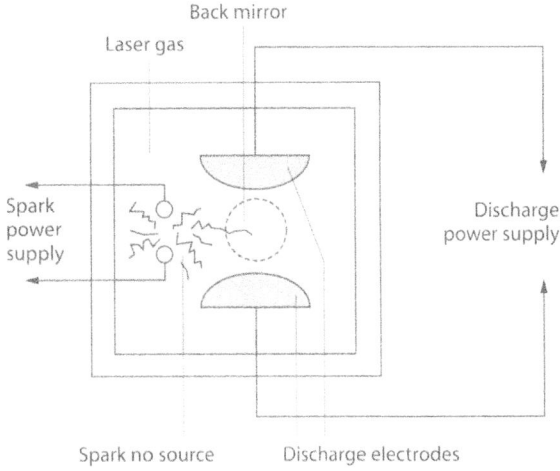

**Fig. 5.** Cross-section through an excimer laser excited by a transverse discharge. The main discharge between the discharge electrodes is pumping the laser. The predischarge – spark gap – generates free charge carriers inside the discharge volume

Different gas mixtures (see Table 1) lead to different potential separations, which cause the respective wavelengths of the excimer laser.

## Excitation Mechanism

Presenting a line of sight parallel to the laser beam, Fig. 5 indicates an excimer laser pumped by a transverse discharge and having a UV preionization. The excimer laser is a pulsed high-pressure gas discharge laser. An electric discharge excites the noble gas and halogen atoms between the electrodes. A homogeneous discharge over the entire electrode length of 1 m is typically achieved by igniting a predischarge and a main discharge. The predischarge generates free charge carriers – electrons – inside the discharge volume, causing a uniform ignition of the main discharge. In order for the excited excimer molecules to be generated most effectively, a buffer gas is added to the noble gas – halogen mixture. For instance, helium or neon are added, although these atoms do not participate directly in the laser process.

With the help of this excitation mechanism, an efficiency of 2 – 3% is achieved at present. This means that 97 – 98% of the supplied energy does not participate in the laser process, but produces heat only. Consequently, the excimer laser needs a cooling system; at present, air- and water-cooled systems are commercially available.

After a certain number of emitted radiation pulses (about $10^8$) excimer lasers must be returned to the manufacturer for a general overhaul. This number of pulses corresponds to an operation time of about 1 to 2 years if the laser is working daily for 4 h at 30 Hz.

## Resonator

An excimer laser may employ either a stable or an unstable resonator. In general, highly reflective aluminium mirrors or dielectrically coated mirrors are used. Aluminium mirrors are suitable for all wavelengths, while dielectrically coated mirrors can be employed for a single wavelength only. The excitation process of the excimer laser generates residues in the resonator. The residues are formed during the predischarge and consist of abrasion products from the electrode material; This electrode material will soil the resonator mirrors; they must be cleaned at regular intervals.

## Laser Beam Features

The excimer laser emits a mixture of modes; consequently, it is a multimode laser with a uniform beam profile. The local intensity is the same at each position within the laser beam. Employing a stable resonator, the divergence of an excimer laser is typically several mrad, resulting in poor focus control. Using an unstable resonator leads to a very low beam divergence, thus allowing very good focus control. The wavelength of the generated laser radiation is below 400 nm and thus, invisible, and unable to penetrate standard glass.

# I-2.4.7
# Erbium: YAG Laser

F. Frank, F. Wondrazek

## Contents

## Introduction

Since the end of the 1980s, high-power erbium lasers have been available and under investigation for a variety of medical applications.

The advantage of this laser system with wavelengths in the near-infrared around 3 µm is the very high absorption by water of biological tissue as well as hard tissue such as, for example, tooth and bone substances. The high absorption coefficient results in extremely small optical penetration depths and so in tissue ablation with minimal thermal damage. Another positive aspect for the medical application of near-infrared lasers is the fact that no mutagenic effects can occur, in contrast to the application of ultraviolet lasers.

The relatively simple technical setup and the reliability of solid-state lasers in connection with the possibility of laser light delivery with flexible waveguides have led to a rapid development of modern commercial erbium laser systems for medical applications in the past years. They offer a wide range of applications in dermatology, cosmetic surgery, dentistry and orthopaedics. A variety of beam transmission systems and applicators are available. By using simple treatment techniques rapid and precise ablation of tissue is possible with minimal thermal damage and controlled minimal coagulation zones.

## Laser Process

Responsible for the laser process in erbium lasers are erbium ions ($Er^{3+}$) in the solid-state materials yttrium aluminium garnet ($YAG \triangleq Y_3Al_5O_{12}$) or yttrium scandium gallium garnet (YSGG). For laser emission in the medically interesting 3-µm region, an erbium concentration of 30-50 % ($4\text{-}7 \times 10^{21} cm^{-3}$) is suitable.

In most applications, erbium lasers are optically pumped with high-power xenon or krypton lamps. For a more efficient coupling of lamp light and laser-active ions, sometimes a co-doping with chromium ($Cr^{3+}$) and thulium ions ($Tm^{3+}$) is carried out.

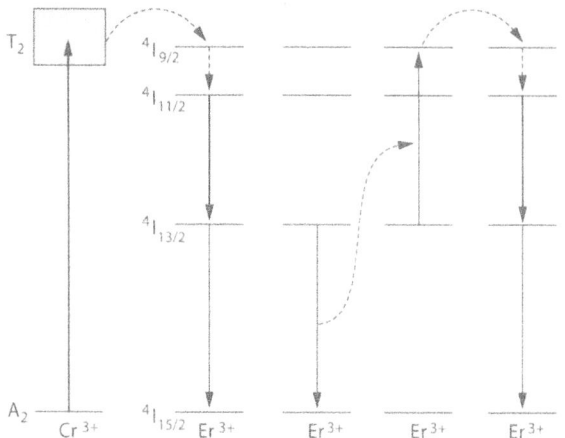

**Fig. 1.** Energy levels of erbium lasers

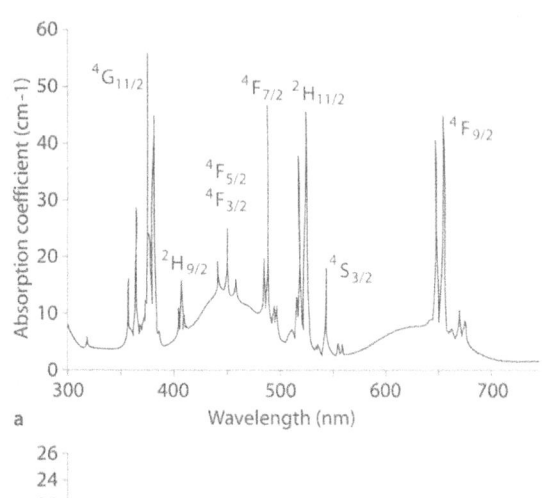

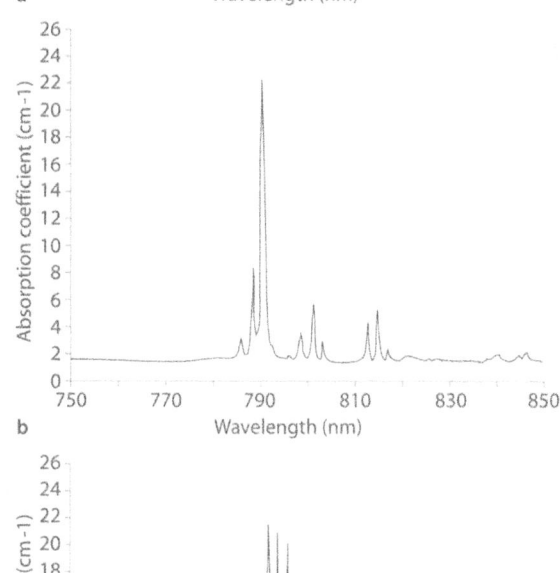

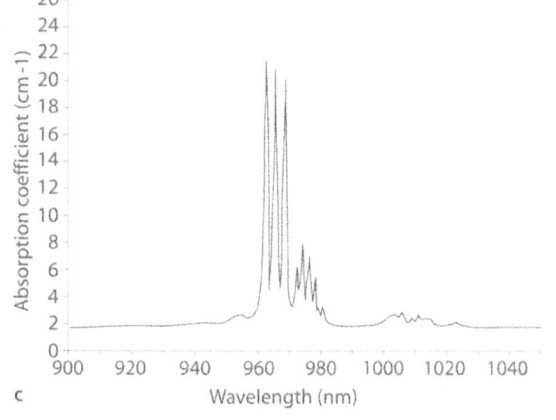

**Fig. 2. a-c** Spectral absorption of erbium laser material

The lasing process in a $Cr^{3+}$ co-doped erbium laser is schematically depicted in Fig. 1. The pump light is absorbed either by the broad absorption bands A2 → T2 of the $Cr^{3+}$ ions at wavelengths of 450 and 640 nm or directly by energy levels of the $Er^{3+}$ ions hown in Fig. 2. From all these excitation levels there occur rapid radiationless relaxation processes to the upper laser level $^4I_{11/2}$ of the $Er^{3+}$ ions. For most medical applications the laser transition to the lower laser level $^4I_{13/2}$ of the $Er^{3+}$ ions is stimulated, which results in laser wavelengths around 3 μm. The exact laser wavelengths depend on the detailed composition of the laser crystal material of the various erbium lasers:

| Laser material | Wavelength |
| --- | --- |
| Er:YAG | 2.94 μm |
| Er: YSGG | 2.78 μm |
| Cr:Er:YSGG | 2.80 μm |
| Ct:Tm:Er:YAG | 2.64 μm |

The depopulation of the lower laser level $^4I_{13/2}$ occurs directly by radiationless relaxation to the ground level $^4I_{15/2}$ or – supported by the high erbium ion concentration – by a resonant energy transition of the energy difference between $^4I_{13/2}$ and $^4I_{15/2}$ to a neighbouring $Er^{3+}$ ion. This so-called upconversion process leads to an excitation of a second $Er^{3+}$ ion into the upper laser level. It is responsible for the relatively high efficiency of erbium lasers.

Due to the long lifetime of the lower laser level $^4I_{13/2}$, of about 2 ms, a high excitation is necessary in the four-level laser system to achieve an inversion population. The resulting thermal load, caused by lamp pumping of erbium laser systems, allows laser operation only in the pulsed mode.

## Technical Aspects

The erbium laser materials are solid-state crystals. The excitation is realized by optical pumping with high-pressure xenon or krypton flash lamps. A typical setup of a solid-state laser is presented in Fig. 3. The light of the electrically excited flash lamp is collected by an elliptical or closed-coupled reflector to the laser rod which has a typical diameter of 3 – 7 mm and

**Fig. 3.** Principle setup of a solid-state laser

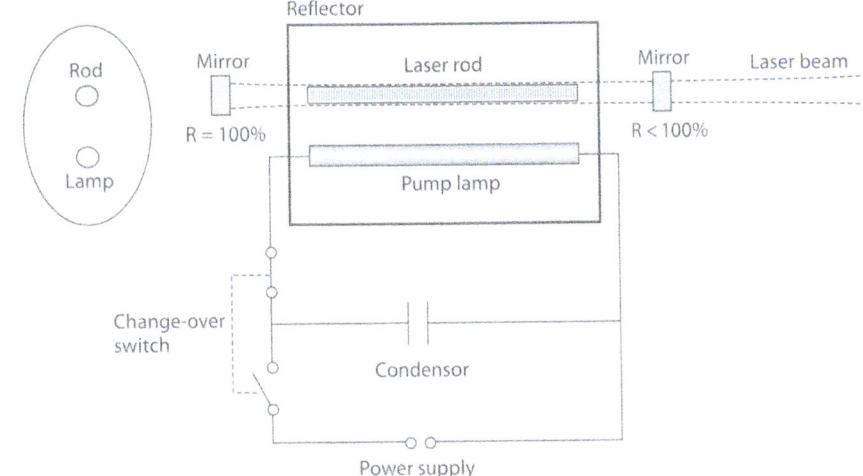

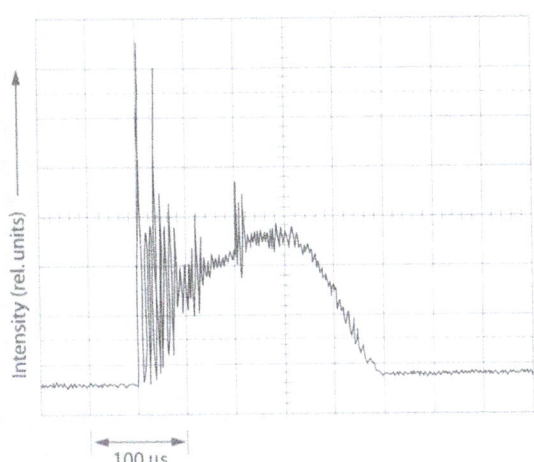

100 μs

**Fig. 4.** Time characteristics of a typical erbium: YAG laser pulse

**Fig. 5.** Commercial erbium: YAG laser

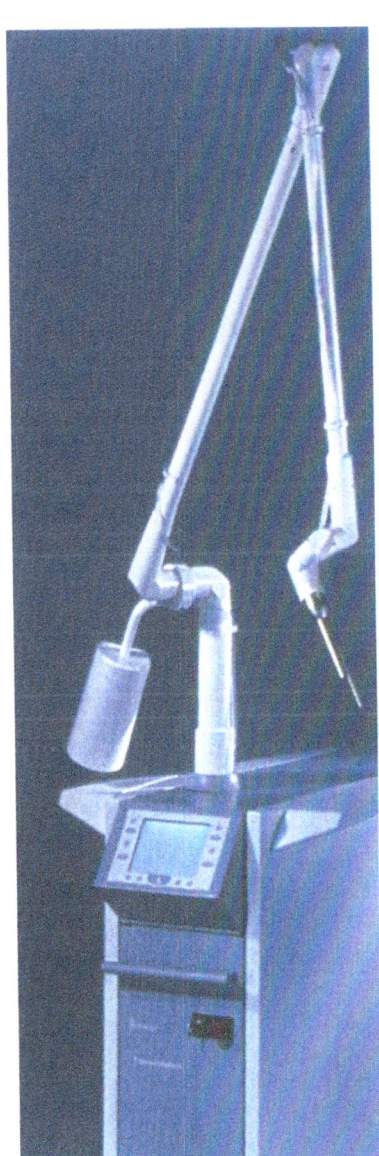

lengths of 40 – 100 mm. The typical optical resonator length is about 30 cm. The exact positions, radii of curvature and transmissions of the laser mirrors, the so-called resonator design, are determined by resonator stability conditions under the aim of a parallel laser beam with small diameter. This means a sufficiently high beam quality for the required application, compromising the variable thermal lens of the laser rod due to the thermal loading by absorption of lamp light.

The relatively long lifetime of the lower laser level of $Er^{3+}$ ions of about 2 ms allows only pulsed operation of flash lamp-pumped erbium lasers. Typical pulse durations of free-running systems are between 100 μs and 1 ms. The laser output is characterized by short spikes (ca. 1 μs), especially at the beginning of the pulse (Fig. 4). These relaxation oscillations at the rising part of a laser pulse are a consequence of light

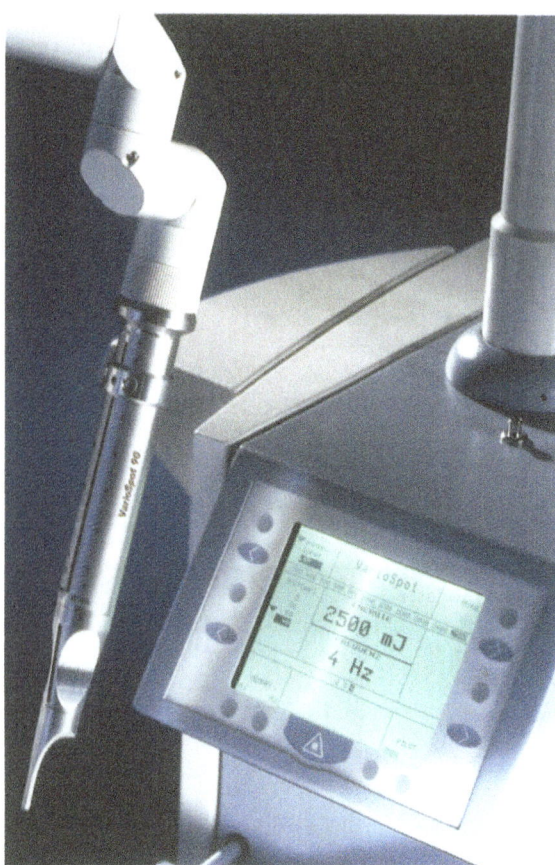

**Fig. 6.** Commercial applicator for erbium lasers

| Laser Handpiece VarioSpot | | | | |
|---|---|---|---|---|
| Spot ⌀ mm | | | | |
| | 2.5 | 3.6 | 5.0 | 7.2 |
| | 5.0 | 10.0 | 20.0 | 40.0 |

| Energy density (J/cm²) | | | | | Ablation depth in skin (µm)* |
|---|---|---|---|---|---|
| 2.0 | 100 | 200 | 400 | 800 | 2 |
| 3.0 | | 300 | 600 | 1.200 | 6 |
| 4.0 | 200 | 400 | 800 | 1.600 | 10 |
| 5.0 | | 500 | 1.000 | 2.000 | 14 |
| 6.0 | 300 | 600 | 1.200 | 2.400 | 18 |
| 8.0 | 400 | 800 | 1.600 | | 26 |
| 12.0 | 600 | 1.200 | 2.400 | | 42 |
| 16.0 | 800 | 1.600 | | | 58 |
| 20.0 | 1.000 | 2.000 | | | 74 |
| 24.0 | 1.200 | 2.400 | | | 90 |

Setting of laser energy (mJ)

*(data of Hohenleutner et al)

**Fig. 7.** Energy density and ablation depth in skin using variable hand piece spot diameters

generation by the stimulated emission process in lasers. They are especially noticeable for high gain laser materials and at multimode operation.

The spiking of pulsed solid-state lasers is the reason for desirable as well as undesirable effects of erbium laser radiation. Thus, the spikes limit the maximum pulse energy which can be produced in optical resonators and transmitted via optical fibres without destruction of optical resonator components or the fibre material. On the other hand, the ablation of biological material with erbium laser pulses is supported by intense short spikes, whose energy density reaches the ablation threshold.

The pulse energies of modern flash lamp-pumped medical erbium lasers vary between 10 and 3000 mJ. The repetition rates can be set from a single shot to about 50 Hz. The maximum average power is limited to about 30 W. The efficiency reaches 1 – 2%. An example of a clinical erbium:YAG laser system offered by Dornier MedizinLaser GmbH is shown in Fig. 5.

The efficiency of erbium lasers with low average power can be increased by diode or laser pumping the laser material. Laser pumping also reduces the thermal load on the laser crystal and so a cw operation is possible. First laser-pumped erbium laser systems for medical application are already under investigation.

The laser light is administered to the patient either by an articulated arm or by a zirconium fluoride or crystalline sapphire fibre often with a rigid waterproof applicator made of waterless silica. A variety of applicators cover the requirements of the surgeon (Fig. 6). A hand piece with variable spot diameter enables the user to vary the energy density and thus the ablation depth, e.g. in skin (Fig. 7). For large-area applications, as well as to avoid thermal damage of the residual tissue, various scanning systems are available.

## Summary

The solid-state erbium lasers, especially the 2.94 µm Er:YAG laser, emit light at wavelengths which have a very high absorption coefficient and thus a very small penetration depth in water (Fig. 8). This, in connection with the proper technical know-how, enables the construction of laser systems for rapid and precise ablation of biological tissues with minimal thermal injury and controlled minimal coagulation zones. A wide range of applications in the fields of dermatology, cosmetic surgery and dentistry have been established or are under investigation, as in orthopaedics.

Modern high-power erbium laser systems for medical applications are compact and reliable units with accessories suitable for all practicable situations. Average output powers of up to 30 W are achieved with

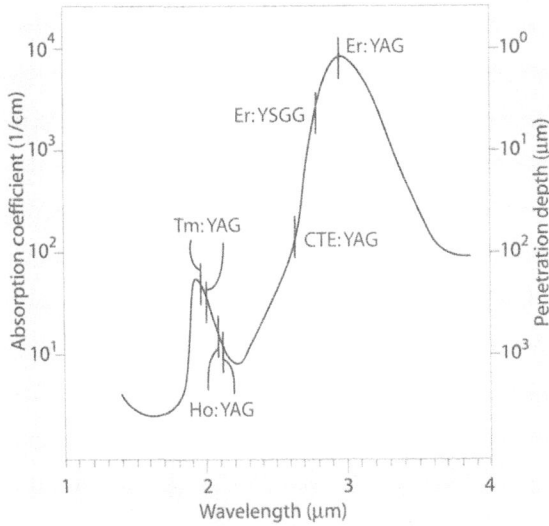

**Fig. 8.** Absorption coefficient and penetration depth of water at erbium laser wavelengths

efficiencies of 1 – 2%. The energy transport from the laser system to the patient is done by an articulated arm or by a flexible waveguide made of zirconium fluoride or crystalline sapphire. A variety of new applicators, especially arthroscopic tools, are continually extending the possible medical applications.

## Further Reading

Dinerman BJ, Moulton PF (1994) 3-μm cw laser operation in erbium-doped YSGG, GGG and YAG. Opt Lett 19: 1143

Eichler J, Eichler H-J (1991) Laser. Springer, Berlin Heidelberg New York

Eichler J, Seiler T (1991) Lasertechnik in der Medizin. Springer, Berlin Heidelberg New York

Esterowitz L (1990) Diode-pumped holmium, thulium, and erbium lasers between 2 and 3 μm operating cw at room temperature. Opt Engin 29, 6: 676

Frenz M, Romano V, Zweig AD, Weber HP (1989), Instabilities in laser cutting of soft media. J Appl Phys 66, 9: 4496

Hibst R (1997) Technik, Wirkungsweise und medizinische Anwendungen von Holmium- und Erbium-Lasern. In: Müller GJ, Berlien H-P (eds) Fortschritte der Lasermedizin. ecomed, Landsberg

Hohenleitner et al. (1997) Fast and effective skin resurfacing with an Er:YAG laser: determination of ablation rates and thermal damage zones. Lasers Surg Med 20: 242

Huber G, Duczynski EW, Petermann K (1998) Laser pumping of Ho-, Tm-, Er-doped garnet lasers at room temperature. IEEE J Quant Electr 24: 920

Jensen T, Diening A, Huber G, Chai BHT (1996) Investigation of diode-pumped 2.8-μm Er:LiYF$_4$ lasers with various doping levels. Op Lett 21, 8: 585

Kintz GJ, Allen R, Esterowitz L (1987) CW and pulsed 2.8-μm laser emission from diode-pumped Er$^{3+}$: LiYF$_4$ at room temperature. Appl Phys Lett 50, 22: 1553

Koechner W (1996) Solid-state laser engineering. Springer, Berlin Heidelberg New York

Moulton PF, Manni JG, Rines GA (1988) Spectroscopic and laser characteristics of Er, Cr:YSGG. IEEE J Quant Electr 24, 6: 960

Peuser P, Schmitt NP (1995) Diodengepumpte Festkörperlaser. Springer, Berlin Heidelberg New York

Schründer S (1989) Gepulste Festkörperlaser im nahen IR mit Thulium, Holmium und Erbium dotierten Laserkristallen. In: Müller GJ, Berlien H-P (eds) Angewandte Lasermedizin. ecomed, Landsberg

Steiner R, Pohl H-J, Mironov IA (1998) 3-μm-laser converter for medical applications. Proc SPIE 3262: 161

Stoneman RC, Esterowitz L (1992) Efficient resonantly pumped 2.8-μm Er$^{3+}$:GSGG laser. Opt Lett 17: 816

Vogler K, Reindl M (1996) Improved erbium laser parameters for new medical applications. Biophoton Int 40

Walsh JT, Flotte TJ, Deutsch TF (1989) Er:YAG laser ablation of tissue, lasers in surgery and medicine. 9: 314

Zweig AD, Frenz M, Romano V, Weber HP (1988) A comparative study of laser tissue interaction at 2.94 μm and 10.6 μm. Appl Phys (B) 47: 259

# I-2.4.8
# Holmium:YAG Laser

F. Frank, F. Wondrazek

### Contents

## Introduction

Since the early 1970s, holmium lasers with wavelengths around 2 μm and high output powers up to 50 W have been known. The disadvantage of these experimental setups was that they were working only at very low temperatures, for example at 77 K, the temperature of liquid nitrogen.

The development of compact, mobile holmium laser systems for medical applications in the past years was made possible by selecting suitable laser materials with high efficiencies and adequate thermal properties. They can be operated around ambient temperatures in the pulsed mode if pumped by flash lamps, or in the cw mode if diode-pumped.

The relatively simple technical setup and the reliability of solid-state lasers in connection with the possibility of laser light delivery by flexible, standard silica waveguides have supported the spread of modern holmium lasers in the medical field.

The strong absorption and low scattering of holmium laser light at wavelengths around 2 μm in biological tissue ensure a good cutting effect by photoablation even in hard tissue, with minimal thermal damage to the surrounding tissue. By using the appropriate laser parameters and suitable applicators, a superficial coagulation is also possible. A further positive aspect for the application of near-infrared lasers in the medical field is the minimal risk of generating mutagenic effects, in contrast to ultraviolet laser applications.

In the past years, the development of a variety of applicators has opened a wide range of uses, especially in the fields of neurosurgery, ENT, urology and orthopaedics, for example, laser arthroscopy and laser discectomy.

## Laser Process

In the course of the development of an efficient holmium laser with wavelength in the near-infrared region, a variety of laser materials have been tested since the early 1970s. Host materials such as, for example, yttrium aluminium fluoride (YFL), yttrium scandium gallium garnet (YSGG), yttrium gallium garnet (YGG), yttrium scandium aluminium garnet (YSAG) and yttrium aluminium garnet (YAG), have been doped with the laser-active holmium ions ($Ho^{3+}$) and cooped with various concentrations of, for example, chromium ($Cr^{3+}$), thullium ($Tm^{3+}$) and erbium ($Er^{3+}$) ions.

The most efficient laser material for high-power xenon or krypton lamp-pumped lasers proved to be Cr, Tm, Ho: YAG, i.e. laser-active holmium ions ($Ho^{3+}$) with a concentration of about 0.5 % AU (Atomic Units) in the host material yttrium aluminium garnet (YAG) codoped with thullium ions ($Tm^{3+}$) with a concentration of about 6 % AU and for lamp pumping with chromium ions ($Cr^{3+}$) with a concentration of about 1 % AU.

The laser process for such a lamp-pumped holmium laser system is schematically shown in Fig. 1. The pump light is absorbed by the broad absorption bands $A_2 \rightarrow T_2$ of the $Cr^{3+}$ ions at wavelengths of 420 and 600 nm or directly by energy levels of the $Tm^{3+}$ ions. Example: Transition from the low-energy level $^3H_6$ of $Tm^{3+}$ to the excited level $^3F_4$ at a wavelength of about 780 nm. This is often used for diode-pumping holmium lasers (Fig. 2). From all the excited levels rapid radiationless processes occur to the energy level $^3F_4$ of the $Tm^{3+}$ ions. The highly excited $Tm^{3+}$ ion is deactivated from the $^3F_4$ to the $^3H_4$ level. The energy difference of the levels $^3F_4$ and $^3H_4$ very well suits the energy difference between the low-energy level $^3H_6$ and the $^3H_4$ level of a $Tm^{3+}$ ion. So the energy of the deactivated $Tm^{3+}$ ion is, with a high probability, resonantly transferred to a neighbouring $Tm^{3+}$ ion which is excited from the low energy level $^3H_6$ to the $^3H_4$ level. This resonant energy transfer process is called cross-relaxation. The quantum efficiency is raised to 2 by the cross-relaxation process, i.e. a single pump photon excites two $Tm^{3+}$ ions to the $^3H_4$ level. The $^3H_4$ excitation energy migrates between neighbouring $Tm^{3+}$ ions via energy resonant processes. If a holmium ion is reached, the energy can be transferred without radiation to the quasi-resonant upper laser level $^5I_7$ of the $Ho^{3+}$ ion. The laser radiation terminates at the $^5I_8$ level of the $Ho^{3+}$ ion. The 2.1 μm wavelength of the laser transition is situated at the upper end of the fluorescence spectrum of Cr, Tm, Ho: YAG (Fig. 3).

The lower laser level of the $Ho^{3+}$ ion has only a small energy difference to the ground state and is consequently highly populated at ambient temperatures.

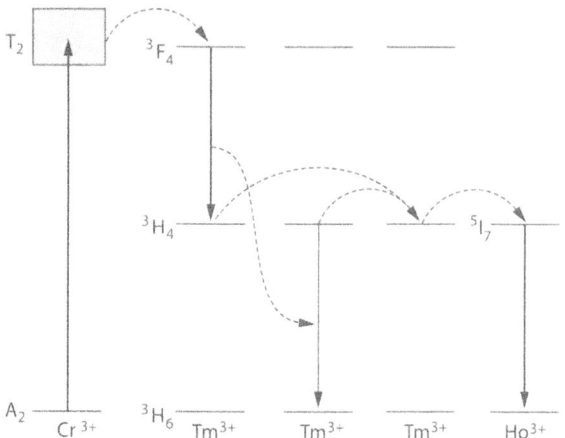

**Fig. 1.** Energy levels of holmium lasers

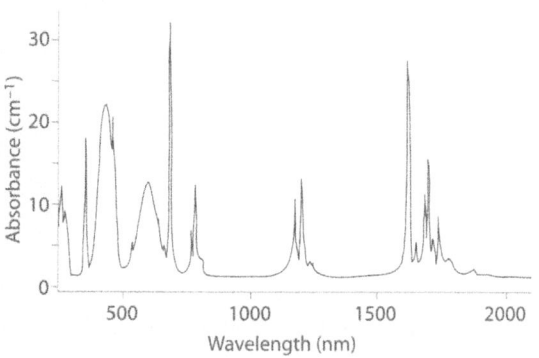

**Fig. 2.** Spectral absorption of holmium laser material

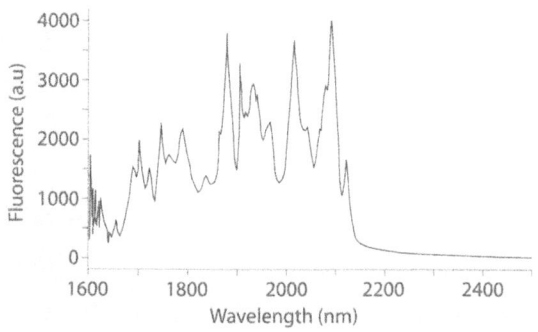

**Fig. 3.** Spectral fluorescence of holmium laser material

Holmium laser systems operating at about room temperatures are quasi-three-level laser systems; a very high excitation is thus necessary to reach a population inversion. The resulting thermal load to the laser material in lamp-pumped systems allows laser operation only in the pulsed modes. The output of quasi-three-level lasers is very temperature-dependent due to variation in the thermal population of the lower laser level. The output power of holmium lasers is reduced by about 3% per degree of rising temperature.

## Technical Aspects

Holmium lasers are solid-state lasers. The excitation of the Cr, Tm, Ho:YAG laser crystal in medical high-power laser systems is realized by optical pumping with high-pressure xenon or krypton flash lamps. A schematic setup of a solid-state laser is shown in Fig. 4. The light of the electrically excited flash lamp is collected by an elliptical or closed-coupled reflector to the laser rod which has a typical diameter of 3 – 7 mm and a length of 40 – 120 mm. The typical length of a single optical resonator is about 30 cm. For some very high-power applications, medical holmium laser systems with multiple optical resonators are available. The exact positions, radii of curvature and transmissions of the laser mirrors – the so-called resonator design – are determined by resonator stability conditions aiming for a parallel laser beam with small diameter, i.e. a sufficiently high beam quality for the required application, compromising the variable thermal lens of the laser rod due to the thermal loading by absorption of lamp light.

The quasi-three-level system of holmium lasers allows pulsed operation of flash lamp-pumped high power systems only at ambient temperatures. Typical pulse durations of free-running systems are between 100 µs and 1 ms. The laser output is characterized by short spikes (ca. 1 µs), especially at the beginning of the pulse. These relaxation oscillations at the rising end of a laser pulse are a consequence of light generation by the stimulated emission process in lasers. They are conspicuous especially for high gain laser materials and at multimode operation. The spiking of

pulsed solid-state lasers is the reason for desirable as well as undesirable effects of holmium laser radiation. On the one hand, the spikes limit the maximum pulse energy which can be produced in optical resonators and transmitted via optical fibres without destruction of optical resonator components or fibre materials. On the other hand, the ablation of biological material with holmium laser pulses is supported by intense short spikes, the energy density of which reaches the ablation threshold.

The pulse energies of modern flash lamp-pumped medical holmium lasers vary between 0.2 and 3 J. The maximum repetition rate depends on the residual temperature of the laser rod, which is determined by an effective cooling system and especially by the diameter of the rod. Holmium laser systems with a laser rod of small diameter have a shorter thermal relaxation time and can be operated at higher repetition rates. In modern medical holmium lasers the repetition rates can be set from single shot to about 30 Hz. This results in average powers of up to 45 W. The efficiency of modern holmium lasers, for example the clinical holmium: YAG laser system offered by Dornier MedizinLaser GmbH, reaches 2%.

The efficiency of holmium lasers can be raised by diode pumping of the laser material. Diode pumping also reduces the thermal load on the laser crystal; even cw operation at room temperature is possible. Diode-pumped holmium laser systems for medical applications with output powers up to about 1 W are already under investigation.

The administration of the 2.1 µm wavelength radiation of holmium lasers to the patient is realized by standard waterless silica fibres with core diameters between 200 and 600 µm. A variety of applicators cover the requirements of the surgeon in the fields of orthopaedics, neurosurgery, general surgery, ENT and

**Fig. 4.** Principle setup of a solid-state laser

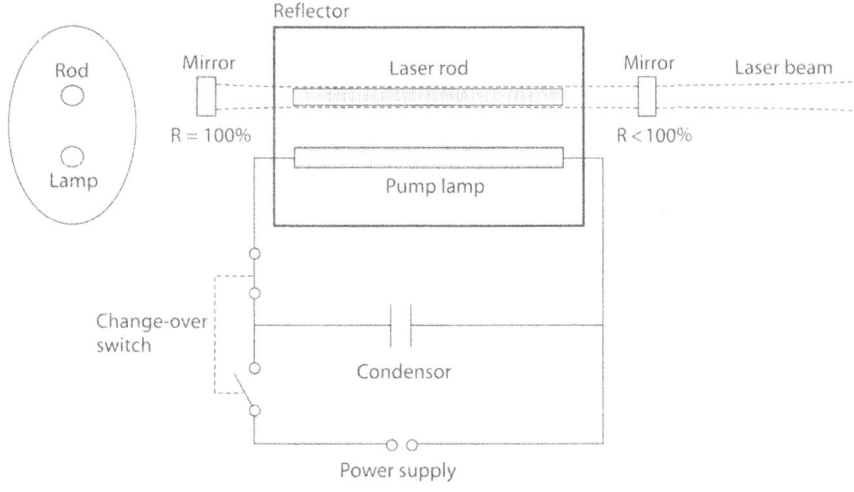

urology. With these applicators, tissue effects such as photoablation (vaporization, cutting) in the contact mode, or superficial coagulation in the non-contact mode, can be achieved. Some examples of applicators for discectomy and arthroscopy offered by Dornier MedizinLaser GmbH are shown in Fig. 5. Application parameters of holmium lasers used in urology are presented in Fig. 6.

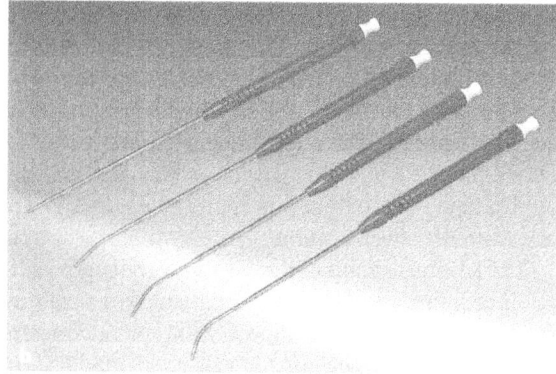

**Fig. 5.a,b** Commercial applicators for holmium lasers

### Application Parameters for Holmium Lasers in Urology

| Tissue | Energy (Joule) | Pulse Rate (Hz) |
| --- | --- | --- |
| **Calculus** | | |
| Ureter | 0.2 – 0.8 | 3 –16 |
| Renal pelvis | 0.4 – 1.4 | 6 –16 |
| | | |
| **Neoplasm** | | |
| Ureter | 0.6 – 1.2 | 10 –16 |
| Renal pelvis | 0.6 – 1.0 | 10 –12 |
| Bladder tumor | 1.0 | 10 |
| | | |
| Incision | 1.2 – 1.4 | 14 –16 |

References: D.Bagley, E.Erhard: Techniques in Urology, Vol.1, No.1 (1995)
D. Johnson: Lasers in Surgery and Medicine, 14: 213 – 218 (1994)

**Fig. 6.** Application parameters of holmium lasers in urology

## Summary

The radiation of holmium:YAG lasers having a wavelength of 2.1 μm is strongly absorbed by water (Fig. 7). The penetration depth in biological tissue of about 1 mm, is small enough for a proper photoablation. Moderate energy densities are attainable with a flash lamp-pumped pulsed holmium laser with an average power of 45 W. The strong absorption and the low scattering of holmium laser radiation thus ensure a good cutting effect even in hard tissue, with minimal thermal damage to the surrounding tissue. By reducing the energy density below the ablation threshold by – changing the laser parameters or by choosing adequate applicators – a superficial coagulation is also possible.

The multiple tissue effects in connection with the compactness and reliability of a solid-state laser with efficiency around 2%; the convenience of delivery by standard flexible silica fibres; and a variety of suitable instruments have opened a wide range of medical applications. For example in neurosurgery, ENT, urology and orthopaedics, for instance, laser arthroscopy and laser discectomy. A further extension of holmium lasers in the medical field can be predicted by the development of new applicators as well as efficient diode-pumped holmium lasers.

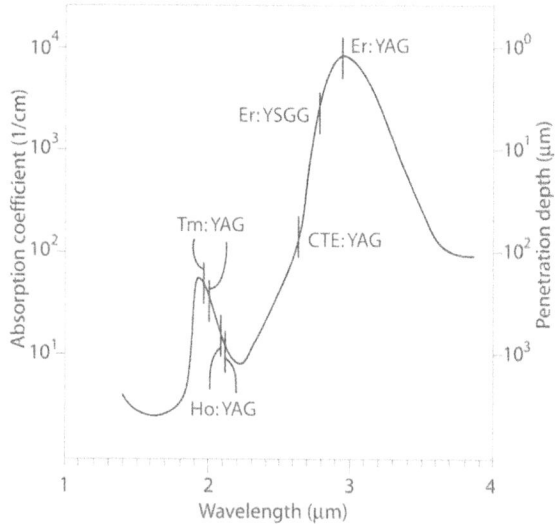

**Fig. 7.** Absorption coefficient and penetration depth in water at the holmium laser wavelength

## Further Reading

Bagley D, Erhard M (1995) Use of the holmium laser in the upper urinary tract. Techni Urol 1, 1: 25

Barnes NP, Gettemy DJ (1981) IEEE J Quant. Electr. 17, 7: 1303

Barnes NP, Gettemy DJ (1990) Performance of Ho:YAG as a function of temperature. Appl Op 29, 3: 404

Becker T, Huber G, Heide H-J v.d., Mitzscherlich P, Struve B, Duczynski EW (1990) 30-Hz operation of 2 µm Ho- and Tm-lasers. Opt Comm 80, 1: 47

Bowman SR, Lynn JG, Searles SK, Feldman BJ, McMahon J, Whitney W, Epp D, Quarles DJ, Riley KJ (1972) High-average-power operation of a Q-switched diode-pumped holmium laser. Opt Lett 18: 1724

Devor DP, Soffer BH (1972) 2.1-µm laser of 20-W output power and 4% efficiency from $Ho^{3+}$ in sensitized YAG. IEEE J Quant Electr 8, 2: 231

Duczynski EW, Huber G, Ostroumov VG, Shcherbakov IA (1986) cw double cross pumping of the $^5I_7 - ^5I_8$ laser transition in $Ho^{3+}$ doped garnets. Appl Phys Lett 48, 23: 1562

Esterowitz L (1990) Diode-pumped holmium, thulium, and erbium lasers between 2 and 3 µm operating cw at room temperature. Opt Engineering 29: 676

Gürs K, Beck R (1975) Ho-laser with 50-W output and 6.5% slope efficiency. J Appl Phys 46, 12: 5224

Hibst R (1998) Technik, Wirkungsweise und medizinische Anwendungen von Holmium- und Erbium-Lasern. In: Müller GJ, Berlien HP (eds) Fortschritte der Lasermedizin. ecomed, Landsberg

Huber G, Duczynski EW, Petermann K (1988) Laser pumping of Ho-, Tm-, Er-doped garnet lasers at room temperature. IEEE J Quant Electr 24, 6: 920

Johnson D (1998) Use of the holmium:YAG laser for treatment of superficial bladder carcinoma. Lasers Surg Med 14: 213

Koechner W (1996) Solid-state laser engineering. Springer, Berlin, Heidelberg New York

Lilge L, Radtke W, Nishioka NS (1989) Pulsed holmium laser ablation of cardiac valves. Lasers in surgery and medicine 9: 458

Nabors CD, Fun TY, Chei HK, Turner GW, Eglash SJ (1993) Holmium laser pumped by 1.9 µm diode laser. CLEO 93: Techn. Digest 16

Peuser P, Schmitt, NP (1995) Diodengepumpte Festkörperlaser. Springer, Berlin Heidelberg New York

Quarles GJ, Rosenbaum A, Marquart CL, Esterowitz L (1989) High-efficiency 2.09 µm flashlamp-pumped laser. Appl Phys Lett 55, 11: 1062

Schründer S (1989) Gepulste Festkörperlaser im nahen IR mit Thulium, Holmium und Erbium dotierten Laserkristallen. In: Müller GJ, Berlien H-P (eds) Angewandte Lasermedizin. ecomed, Landsberg

Shannon DC, Vecht DL, Re S, Kane TJ, Wallace RW (1992) High average power diode-pumped lasers near 2 µm. LEOS 92, Conf. Proc. 302

# I-2.4.9
# Ruby Laser

F. Frank, F. Wondrazek

## Contents

## Introduction

The historically first laser, invented in 1961 by Th. Maiman, was a solid-state ruby laser. Since this time, great efforts have been made to understand the laser properties of ruby lasers and to build the first commercial ruby laser systems for various applications, also in the medical field. This ruby laser system had low pulse repetition rates of 1 – 2 Hz as well as very low efficiencies, far below 1 %, which resulted in voluminous systems. Spatial intensity spikes prevented the coupling of the laser light into flexible optical fibres by destruction of the fibre material. All these negative properties restrained the commercial spread of ruby lasers.

In the past few years, new technologies have opened the way for the development of modern ruby laser systems which have overcome most of these drawbacks. Modern flash lamp-pumped ruby laser systems for medical application are compact and reliable solid-state lasers operating in a pulsed mode, with a flexible silica fibre light delivery equipment.

The red light of ruby lasers at a wavelength of 694 nm is only weakly absorbed by nonpigmented biological tissue, but selectively absorbed by structures with high melanin concentrations. This property has promoted the at present most important medical application of ruby laser light: rapid and gentle hair removal in the field of cosmetic surgery by thermal destruction of deep-lying hair follicles without damage to the surrounding skin. A successful and long-acting epilation is possible by the right choice of laser parameters and the use of suitable applicators. Further medical applications of ruby lasers are under investigation.

## Laser Process

The laser process in ruby lasers was investigated in detail in the early 1960s and is understood very well. The host material of the solid-state crystals is sapphire ($Al_2O_3$) which is doped with approximately 0.05 wt% $Cr_2O_3$. The laser-active chromium ions ($Cr^{3+}$) substitute some aluminium ions ($Al^{3+}$) in the crystal

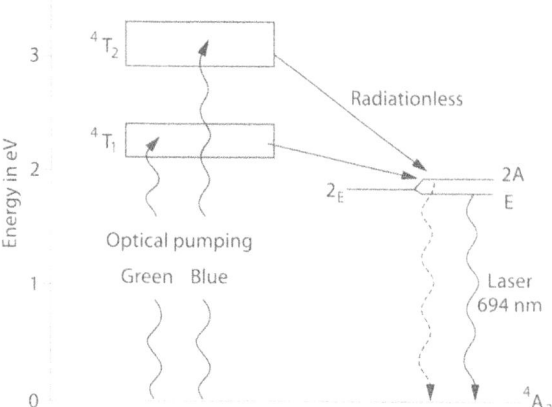

**Fig. 1.** Energy levels of ruby lasers

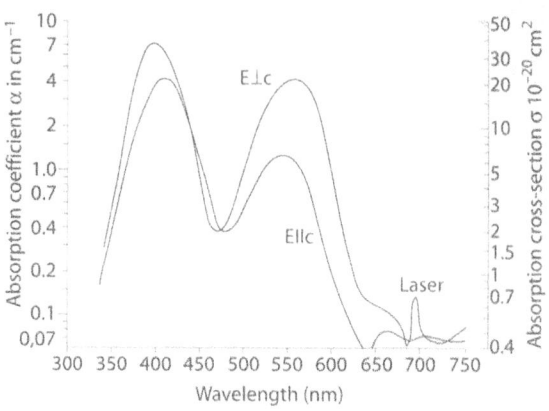

**Fig. 2.** Spectral absorption of ruby crystals

lattice and are responsible for the red colour of the ruby crystals.

The simplified energy level diagram of a ruby laser ($Cr^{3+} : Al_2O_3$) is shown in Fig. 1. By optical pumping with high-pressure flash lamps, electrons of the inner shells of $Cr^{3+}$ ions are raised from the ground level $^4A_2$ to the broad excitation levels, $^4T_2$ and $^4T_1$. The corresponding strong absorption bands of ruby crystals in the blue range of the spectrum around 400 nm or the green range around 550 nm can be seen in Fig. 2, where the absorption spectra for polarized light; the electrical field parallel or orthogonal to the c-axis of the crystal are depicted. From the excitation levels $^4T_2$ and $^4T_1$, the $Cr^{3+}$ ions proceed by rapid radiationless relaxation processes to the metastable state $^2E$, which is split into the two close-neighbouring sublevels, $^2A$ or E. The two sublevels are connected by very rapid quasi-resonant energy exchange processes by which a thermal equilibrium population between the $^2A$ and E levels is attained. At room temperature, about 15 % more $Cr^{3+}$ ions are in the lower energy level $^2A$, so under normal operating conditions, the inversion population between the upper laser level E and the lower laser level, the ground state $^4A_2$, reaches the laser threshold first; laser light with a wavelength of 694 nm at the red end of the visible spectrum is generated by ruby lasers.

Since the terminating lower laser level is the ground state of the $Cr^{3+}$ ions, a ruby laser acts as a three-level laser. In three-level lasers, a minimum of 50 % of the laser-active ions must be excited to reach population inversion. This can be done in optical pump processes using lamps only in the pulsed modes with high-power flash lamps. Advantageous for the peak load of the pump lamps is the relatively long lifetime of the upper laser level $\bar{E}$ of approximately 3 ms, which allows the accumulation of laser-active $Cr^{3+}$ ions in the upper laser level. A disadvantage of a three-level laser is the so-called self-absorp-

tion by which the laser light is absorbed by $Cr^{3+}$ ions in the ground state, $^4A_2$. This absorption line at the wavelength 694 nm between the levels $^4A_2$ and $\bar{E}$ can be seen in Fig. 2.

## Technical Aspects

Ruby lasers are solid-state lasers. The excitation of the $Cr^{3+} : Al_2O_3$ laser crystal in medical high-power laser systems is realized by optical pumping with high-pressure flash lamps. The first commercially available ruby lasers used a helical flash lamp surrounding close-coupled the laser rod. Modern ruby laser setups use straight flash lamps arranged parallel to the laser rod, as in nearly all modern solid-state lasers (Fig. 3). The light of the electrically excited flash lamp is concentrated by a closed-coupled diffuse reflector to the laser rod with typical diameters up to 10 mm and lengths up to 120 mm. As a consequence of the diffuse light reflection, the laser material is excited homogeneously and a smooth beam profile, almost without spatial hot spots, is produced. This allows the generation of laser beams with high energy densities without the destruction of optical components of the resonator or the beam delivery system. A typical length for the optical resonator is about 30 cm. The exact positions, radii of curvature and transmissions of the laser mirrors, the so-called resonator design, are determined by resonator stability conditions. The aim is a parallel laser beam with small diameter, i.e. a sufficiently high beam quality for the required application, compromising the variable thermal lens of the laser rod due to the thermal loading by absorption of lamp light.

Three-level laser systems, as the ruby laser, can only be operated in the pulsed mode if pumped with flash lamps due to the very high excitation necessary to reach population inversion. Typical pulse durations

**Fig. 3.** Principle setup of a
solid-state laser

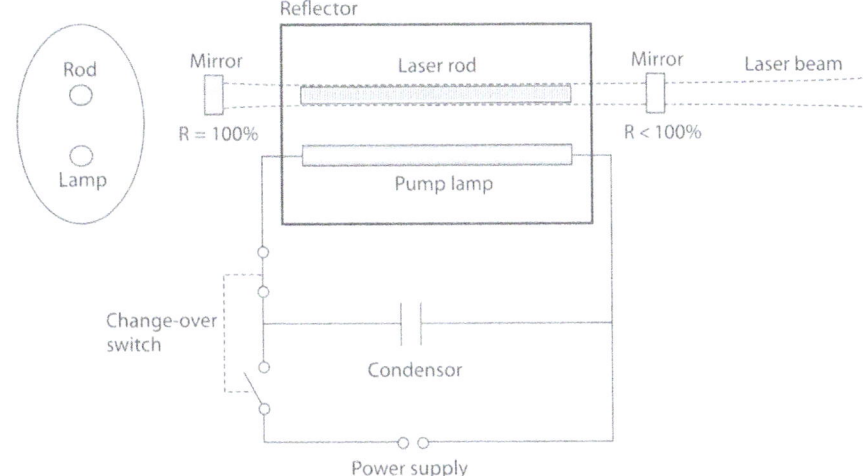

**Fig. 4.**
Commercial ruby
laser

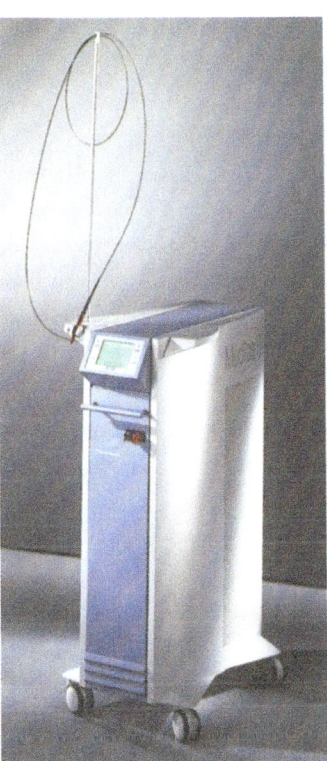

high gain laser materials as, for example, ruby and at multimode operation.

The spikes of pulsed solid-state lasers limit the maximum pulse energy which can be produced and transmitted via optical fibres without destruction of optical resonator components or fibre material. The spiking of a laser pulse must also be taken into account considering laser – tissue interactions. Thus, for example, depending on the absorption properties of the biological material, the short intense spikes can reach the ablation threshold of the material, resulting in tissue vaporization.

The pulse energies of modern flash lamp-pumped medical ruby lasers vary between ca. 1 and 15 J. The maximum pulse repetition rate is limited by thermal effects in the laser rod to about 10 Hz. Average powers of up to 20 W are realized in modern medical ruby lasers with efficiencies around 1%. An example of a clinical ruby laser system offered by Dornier Medizin-Laser GmbH is shown in Fig. 4.

The delivery of the laser light of ruby lasers at a wavelength of 694 nm is realized by standard silica fibres usually with a core diameter of 600 µm. Special applicators, for instance collimating hand pieces, cover the requirements of the surgeon, especially in the field of cosmetic surgery.

of free-running ruby lasers are between 500 µs and 10 ms. The laser output is characterized by short spikes with durations between 200 and 500 ns. The spiking of a ruby laser pulse with an energy just above laser threshold is shown in Fig. 4. This is in contrast to the smooth discharge of a flash lamp with a duration of about 1 ms. The relaxation oscillations, especially at the rising end of a laser pulse, are a consequence of light generation by the stimulated emission process in lasers and especially noticeable for

## Summary

Medical ruby laser systems are at present preferentially applied in the field of cosmetic surgery, especially for rapid and gentle hair removal from large body areas. The treatment technique is based on the selective heating and damaging of the hair follicle by selective absorption of pulsed laser radiation. This is called selective photothermolysis. The red light of the

Laser Parameters for Hair Removal

| Wavelength: | 630 – 1000 nm:<br>– little absorption by hemoglobin<br>– effective absorption by melanin |
| Pulse duration: | 1 – 20 ms:<br>– selective photo-thermolysis<br>(hair follicle is heated more<br>than epidermis) |
| Fluence | 10 – 40 J/cm$^2$:<br>– dependent on skin type |
| Spot diameter | 5 – 10 mm:<br>– large spot size provides deeper<br>penetration by scattering |

**Fig. 5.** Ruby lasers parameters for hair removal

ruby laser with a wavelength of 694 nm penetrates deep into the weakly pigmented skin, but is selectively absorbed by the melanin in the hair follicle. A long laser pulse duration of several ms is especially used to achieve optimal laser energy absorption by the follicle but to minimize absorption by surrounding structures. Biophysical considerations and clinical experience have shown that special laser parameters are necessary to obtain successful and long-acting epilation. The desirable laser application parameters in Fig. 5 are met without difficulties by modern medical ruby laser systems with adequate applicators.

The compactness and reliability of a solid-state laser together with the convenience of light delivery by standard flexible silica fibres will, in connection with the development of suitable new applicators, provide further possibilities for ruby laser systems in the medical field.

**Further Reading**

Cronemeyer DC (1966) Optical absorption characteristics of pink ruby. J Opt Soc Am 56, 12: 1703
Eichler HJ, Salle J (1989) Laseroptik und –elektronik. In: Reuber C (ed) Handbuch der Informationstechnik und Elektronik, vol 8. Hüthig, Heidelberg
Eichler J, Eichler H-J. (1991) Laser. Springer, Berlin Heidelberg New York
Evtuhov V, Neeland TK (1966) Pulsed ruby lasers. In: Levine AK (ed) Lasers, vol. 1. Marcel Dekker, New York
Hillenkamp F, Pratesi R, Sacchi CA (1980) Lasers in biology and medicine. Plenum Press, New York
Högele A, Reindl M (1998) Ruby laser – the first laser is coming back to medicine. Laser Med 30, 1: 64
Kneubühl FK, Sigrist MW (1998) Laser. Teubner, Stuttgart
Koechner W (1991) Solid-state laser engineering. Springer, Berlin Heidelberg New York
Maiman TH, Hoskins RH, D'Haenens IT, Asawa CK, Evtuhov V (1961) Spectroscopy and stimulated emission in ruby. Phys Rev 123: 1151
Schawlow AL (1961) Advances in quantum electronics. Columbia Univ. Press, New York
Shapshay SM (1987) Endoscopic laser surgery handbook. Marcel Dekker, New York

# I-2.4.10
# Diode Lasers

V. Knappe

## Contents

## Introduction

Diode lasers are finding more and more applications in medicine, as they display several advantages compared with other lasers. They are among the most efficient converters of electrical energy into coherent radiation. Besides an efficiency of about 40 %, they offer wavelength tunability (temperature-dependent, 0.3 nm $K^{-1}$), compactness and a long lifetime.

Diode lasers are frequently employed as diagnostic or therapeutic instruments, or as positioning tools for medical devices (MRT, X-rays, lasers). Diode lasers for use in diagnostic work at their lower output power range of up to 1 W and are mainly utilized as devices to illuminate structures in biological tissues and/or to determine, as a Doppler probe, the speed of moving particles (e.g. erythrocytes). Following the recent development in high-power diode lasers, many new laser configurations with outstanding features have become available. The possibility of transmission via light guides has increased the range of therapeutic applications in medicine.

Due to their small size, power output (range up to about 20 W) and stability, diode lasers offer a good alternative to the customary Nd:YAG laser systems with low output power. The penetration depths of diode lasers and Nd:YAG lasers in tissue vary with the optical properties of the tissues. Using diode lasers, various tissue reactions, like hyperthermia, coagulation and vaporization, can be induced. In addition, treatments in ophthamology and photodynamic therapy are possible.

## Laser Process

Semiconductors are compounds with a specific electric resistivity in the range from $10^{-3}$ to $10^{9}$ $\Omega$ cm; consequently they are placed between conducting metals and insulators. This, in contrast with other solid-state laser crystals, which have to be optically excited, allows an excitation by electric charge carriers. Appropriate semiconductor compounds are the elements of group II to group VI of the Periodic Table. Mixed (isomorphic) crystals of the groups III – V are of special significance. Among others, the following compounds are of importance:

- Gallium arsenide                                        GaAs
- Indium gallium arsenide                          InGaAs
- Gallium aluminium arsenide              GaAlAs
- Indium gallium aluminium arsenide    InGaAlAs
- Indium gallium arsenide phosphide    InGaAsP

Diode lasers employ semiconductor crystals as active media which, after excitation, will emit coherent radiation in the VIS or IR region. Excitation can be induced by injecting charge carriers at the so-called p – n junction after applying an external voltage.

Semiconductor crystals are mainly composed of heterostructures and the active semiconductor materials are built into passive layers. These elements can either emit electrons (donors) or bind electrons (acceptors). The energy band model offers a simplified scheme to present the electron energy states of the solid. Here, the electrons occupy permitted energy bands which are separated by forbidden energy bands. Doping produces regions of charge carrier excess (conducting band of the n type) and regions partly depleted of electrons which are bound to the acceptors (valence band of the p type). These defects behave like positively charged electrons. The junction between an n- and a p-semiconductor type is called the p – n junction; its energy level is determined by the potential barrier caused by the different charge carrier distributions (Fig. 1).

Applying a voltage $V_{pn}$ – about the size of the separation of the energy band of the p – n juction – to the p – n junction reduces the potential barrier. Consequently, electrons and defect electrons are injected into the p – n junction, generating an inversion within a narrow region. In suitable semiconductor materials, e.g. GaAs, a so-called direct-bandgap semiconductor, the probability of radiative transition recombination is high. Interaction between the recombination radiation and the electrons of the conductivity band causes stimulated emission of further photons having the same frequency. In this case, the plane-parallel end windows of the semiconductor form a Fabry-Perot laser resonator. In general, these end windows do not carry an additional reflecting layer because the large refractive index of the semiconductor ($n = 3.5$ for GaAs) already leads to a Fresnel reflectivity of about 32 % at the semiconductor-air junction [6]. If the concentration of the injected charge carriers is sufficiently large, an optical amplification with feedback will occur and, finally, laser radiation will be emitted through the junction of the active materials.

To reach the charge carrier density needed for the emission of laser radiation, a threshold current density $I_{thr}$ must be exceeded. $I_{thr}$ depends on temperature and material. For $I < I_{thr}$ spontaneous emission occurs having a large spectral width. Exceeding $I_{thr}$, the optical output power rises rapidly and linearly with increasing current, whereby the emitted laser light is better focused and of low spectral width (Fig. 2; [7]).

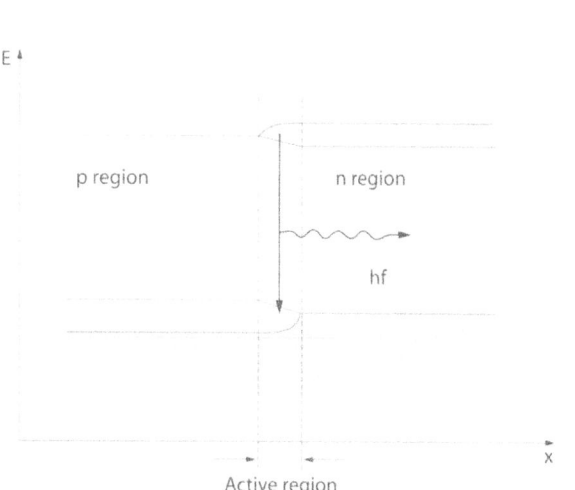

**Fig. 1.** Energy-band model of the p – n junction operating in forward direction. Hatched regions are occupied by electrons. (After [4])

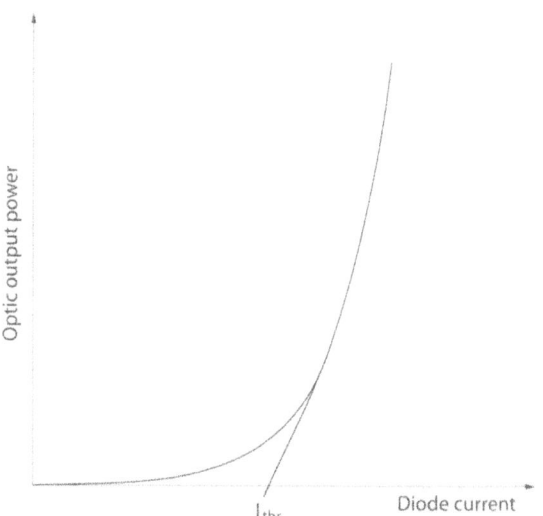

**Fig. 2.** Current-power characteristics. $I_{thr}$ depends on material and temperature

## Structure of Diode Lasers

The power dissipated inside the semiconductor must be kept at a low level, especially while operating continuously, because with increasing temperature the threshold current density will rise. This can be achieved either by cooling or by employing a special crystal structure.

If the injected charge carriers inside the semiconductor crystal are not confined, then the spread of the laser radiation perpendicular to the p–n junction plane will be larger than the thickness of the active layer. In this case, only low values for the charge carrier density inside the active layer can be achieved for a given current density.

For so-called double hetero-structure lasers, continuous operation at room temperature is possible. There, the charge carriers on both sides of the p – n junction will be enclosed by barrier layers of GaAlAs. The active region is then typically a layer of GaAs or GaAlAs, about 200 nm thick, having a lower share of aluminium than in the neighbouring enclosing layers. The lattice constants of the single epitactic layers must be adjusted to one another in order to prevent nonradiative recombination at the heterostructure junction. This embedding of the recombination region into a semiconductor material having a larger band gap achieves a suitable formation of potential barriers, and consequently confines the injected charge carriers to a very limited region. This waveguide allows efficient energy extraction from the active amplifier region without penetration of the highly doped and strongly leaking regions above and below the confinement layers. This causes a reduction in the threshold current density, making continuous operation at room temperature possible [7].

Equally as important as the vertical confinement of the charge carriers is the lateral limitation of the active region. Otherwise, the light would separate into single filaments, resulting in an unstable, incoherent radiation. This kind of radiation may cause local damage to the facets of the mirror and destroy the diode. In this case, so-called strip geometry lasers (Fig. 3) offer a solution by forming only one beam filament. These lasers are subdivided into gain-guided and index-guided diode lasers.

In the case of the gain-guided laser, diode lateral limitation is achieved by an appropriately shaped electric contact strip placed on the surface of the diode in the direction of the resonator axis. The laterally isolated regions, which are not pumped, cause internal losses due to the lateral temperature and charge carrier density profiles created, which will provide guidance of the optical field.

In the case of the index-guided laser diode, charge carriers and photons are enclosed by an appropriate

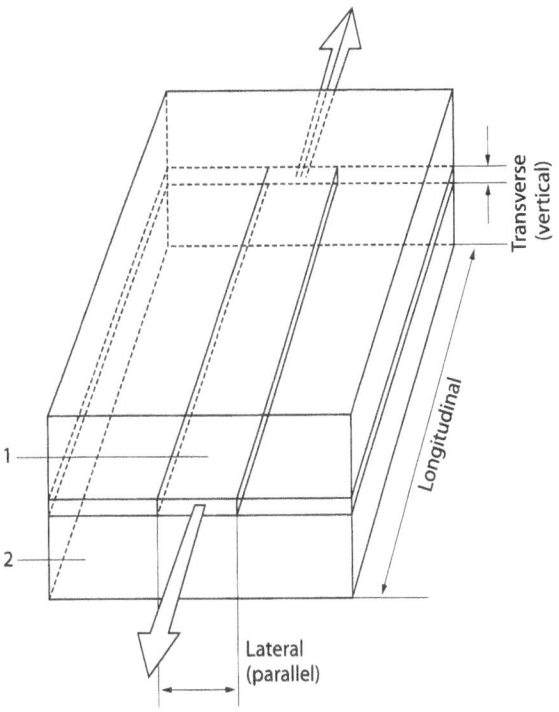

**Fig. 3.** Diagram of the spatial boundaries of the active region of a strip-structure laser diode. It is limited longitudinally by resonator mirrors, vertically by the double heterostructure and laterally by the additional strip structure. 1 Active region; 2 terminal areas with resonator mirrors. (After [1])

two-dimensional structure. The profile of the index of refraction will achieve the guidance. This structural design offers some advantage compared with the gain-guided laser diodes, with regard to a lower threshold current, field distribution and spectrum.

## Structural Improvements to Increase Power

To achieve high-power output for the laser diodes, two problems must be overcome: firstly, the high photon density and secondly, the high electric current density [7].

High optic flux is related with the problem of maximal mirror load. For a typical facet area of 0.1 μm × 5 μm, even a relatively low diode power output of 10 mW will generate an optic power density of 2 MW cm$^{-2}$. The following solutions are possible:

- Reducing the surface recombination rate.
- Extending the mirror area in the lateral and vertical directions.

It is possible to reduce the surface recombination rate significantly by introducing a dielectric coat, e.g. made of $Al_2O_3$, $SiN_4$ or $SiO_2$. This coating will also improve the heat dissipation of the mirror surface, increase the protection against oxidation and optimize the beam uncoupling.

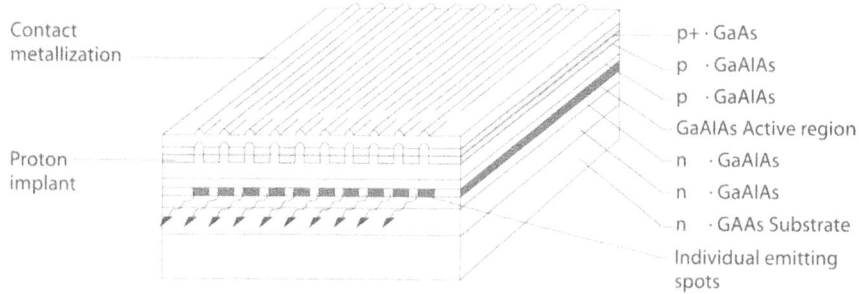

**Fig. 4.** Multiple-strip gain-guided diode laser array. (After [2])

To extend the mirror in the lateral direction, multistrip arrays have been developed. These are periodically arranged and parallel strip lasers on a common substrate, a typical array having about 10 to 20 strips (Fig. 4). Since the laser mode is not very well delineated, neighbouring strip lasers show phase coupling if their separation is sufficiently small (typically 10 µm). The long-range field shows an angular-dependent intensity distribution corresponding to the phase relations.

Broad-area diode lasers offer continuous emission area having a lateral aperture of up to about 500 µm, due to the highly developed epitaxial techniques. Compared with strip lasers, the intensity distribution in the long-range field is uniform and the laser operation stable.

The widening of the active mirror area in the vertical direction leads to so-called quantum well (QW) lasers. These are laser diodes having an extremely narrow active zone, of the order of magnitude of the de Broglie wavelength of the charge carriers (5 to 10 nm). The charge carriers move parallel to the active zone, the energy states in the conductivity and in the valence band may be considered to be discrete. The QW zone is embedded in waveguide layers containing varying concentrations of aluminium. Thus, a certain index of refraction profile will develop in the vertical direction. The guidance of the optical waves is not limited to the active region, but predominates in the nonabsorbing layers of the waveguide. This causes an enlargement of the mirror surfaces in the vertical direction. The threshold current of these laser diodes is low, because the population inversion between the single energy levels is generated only in a small active volume. Consequently, high efficiency is achieved which, in turn, leads to reduced heat generation [7].

Intense development has yielded further improved semiconductor structures promising high output power for single diodes. Traditional semiconductor lasers are restricted by the layer materials all having about the same lattice constants, otherwise lattice mismatch would result, accelerating degradation of the laser diode. Using thin layers (less than some 10 nm) tension inside the crystals caused by lattice mismatch can be compensated. These strained-layer techniques allow new combinations of layers, resulting in an improved quantum well (QW) structure. These structures are made up of very thin layers and have a pronounced waveguide-like buildup. Quantum-well laser diodes having a strained-layer construction are at present the most popular commercially available laser diodes.

The high electric current density causes a problem because significant ohmic losses will occur inside the contact layers. Heat will develop and the temperature will increase in the active as well as in the surrounding structural layers. The maximum optic output power depends greatly on the temperature of the crystal, emphasizing the importance of efficient heat conduction and cooling. Optimizing heat dissipation and cooling is not only necessary for the performance of a laser diode but very important with respect to its ageing process.

## High-Power Diode Lasers

GaAlAs diode lasers have achieved significance in the field of high-power lasers. The gain-guided structure type, as well as the index-guided structure type including QW layer design, are available. High-power diode lasers can be subdivided into various power classes. The already-mentioned multistrip arrays, as well as the broad-array diode lasers, belong to the lower high-power class, generating a maximum power output typically between 200 mW and 2 W. Bundling a few of such diodes results in a diode laser with a wide emission area of, e.g. 500 µm and a maximum output power of a few W. A simple Peltier element is sufficient to cool and control the temperature of this laser type [7].

To achieve higher power output (e.g. about 20 W), a larger number (20 or more) of such multistrip arrays or broad-area laser diodes have been arranged side by side over a width of typically 10 mm (Fig. 5).

Heat flow considerations determine the separation of the individual elements; often the fill factor is about 20%: Employing an appropriate and efficient cooling system, it is possible to increase this fill factor for higher power classes.

To increase the power output of diode lasers even further, the diodes can be arranged in stacked linear array bars. The packing density, and thus the output power density, are determined essentially by the thickness of the linear array bar, the spacer for the electric contacts and the thickness of the heat sink.

A design similar to the continuously emitting array bars is applied to the quasi-cw array bars. These diodes are utilized to pump pulsed solid-state lasers with a pulse length of ca. 200 – 400 μs and high peak pulse power. For such a pulse length a rapid temperature increase at the p – n junction can be observed. Consequently, the power limits to this long-pulse operating mode are similar to those for continuous

emission. This explains the term 'quasi-cw' Commercially available linear quasi-cw array bars easily exceed 100 W quasi-cw pulse power, corresponding to a pulse energy of more than 20 mJ at 200 pulse length. The mark-to-space ratio can reach up to 50 %, e.g. 2.5 kHz at 200 μs [7].

## Features of Laser Radiation

To apply the diode laser in medicine, it is necessary in general to collect the laser radiation and to feed it into a glass fibre. In this way, it can easily be guided to the treatment region. In contrast to pumping a solid-state laser, it is not sufficient to simply collimate the strongly diverging radiation before feeding it into glass fibres. It is important to achieve an approximately Gaussian intensity profile before focusing it onto the terminal face of the fibre. To achieve this, the asymmetric radiation beam profile emerging from the emission gap of the diode has to be improved by an optical system. This applies also to diode laser radiation, which is not transmitted by glass fibres but applied directly – focused or defocused – to the tissue.

The beam properties depend on the crystalline structure. The continuous or pulsed radiation emitted by gain-guided multistrip arrays displays a multi-longitudinal mode spectrum. The emission line width of an array is approximately 2 nm (about 1000 GHz), displaying 10 to 15 longitudinal modes inside this bandwidth. The line widths of these modes are between 10 MHz and 5 GHz, depending on the power they contain.

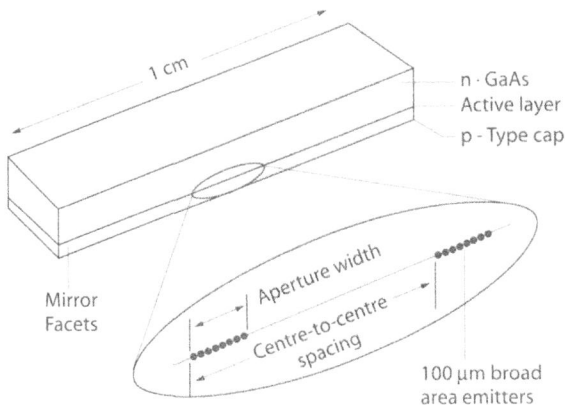

**Fig. 5.** Schematic view of a monolytic linear array bar. (After [3])

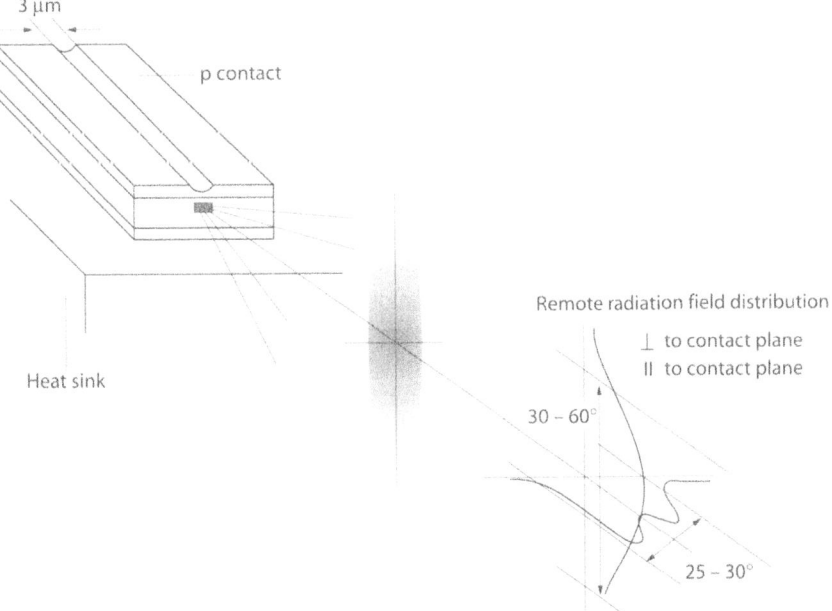

**Fig. 6.** Beam profile of an oxide strip-structure laser diode, gain-guided. (After [1])

The gain-guided laser diodes emit in the vertical direction an almost Gaussian intensity profile; the lateral planes display in general a complex intensity distribution. They are astigmatic and display a coherence length of only 0.5 mm. Figure 6 indicates the beam profile of a gain-guided oxide strip-structure laser diode.

The index-guided laser diodes generate in general a singe longitudinal mode. For a single-strip laser diode both radiation planes display an almost Gaussian intensity profile. However, the two radii differ and an elliptic beam profile emerges. Based on their better beam quality, they display a coherence length of up to 10 m and low astigmatism [7].

Laser diodes emit a broad band and incoherent radiation below, as well as above, the threshold current. The band width of this spontaneous emission is ca. 25 nm (FWHM), it contains 1 – 2 % of the maximum optic power output.

The centre of the emission spectrum of the laser diode depends on the temperature of the crystal. The range of change is about 0.25 to 0.3 nm/K for GaAlAs laser diodes. The spectrum shifts toward longer wavelengths with increasing temperature. This shift is a consequence of the temperature dependence of the band separation and the index of refraction; it may be controlled by changing the temperature of the heat sink. In this way it is possible to fine-tune the emission spectrum onto a certain central wavelength. In the same way, a wavelength shift can be induced by changing the operating current, which will, in turn, change the temperature. The pulsed quasi-cw laser diode also displays a wavelength shift, if the duty cycle is changed while holding the operating current constant [7].

## Ageing

An important quality indicator for a laser diode is its lifetime, especially since a laser diode that has once failed in general cannot be repaired. The lifetime of a cw high-power laser diode can reach $10^4$ h and more; quasi-cw laser diodes may achieve more than $10^{10}$ single pulses. Operational reliability can be described as the probability that under normal operating conditions the laser diode will perform flawlessly for a given time span. In general, failure to emit laser light is not sufficient to call a laser diode malfunctioning. The criterion for malfunction of a laser diode could rather be: the operating current necessary to achieve a given constant optic output power $P_{opt}$ exceeds the original normal value $I_B$ by a factor of 1.2 to 1.5.

The reason for and the frequency (probability) of the occurrence of certain types of degradation depend on the kind of laser diode, its internal crystal structure, the layout of the total chip onto the heat sink and on the kind of material used. GaAlAs lasers will degrade in a different way from, e.g. InGaAsP laser diodes.

Open laser diodes are frequently on offer the emitting surfaces of which are exposed to the environment and must be protected against dust and aerosols. Even the smallest deposit at the facets of the mirror may cause absorption of the emitted radiation and, consequently, lead to local heating or back reflection. Both effects will rapidly melt the facets.

Laser diodes may also be destroyed by overvoltage or overcurrent and especially by static charging. Laser diodes are much more sensitive towards static charging as compared with C-MOS components. This must be taken into account when designing power sources to operate the laser diodes [5].

## Cooling

Peltier elements are well suited to cool diode lasers with a power output of up to about 5 W. In this case, the Peltier current can be used to control and stabilize the temperature of the diode laser and, consequently, its emission wavelength. To apply this cooling method (thermoelectric cooling) the laser diode is soldered into a carrier (submount) which acts as a heat sink. The carrier, in turn, is connected by a layer of solder or adhesive to the cold electrode of the Peltier element. The hot side of the Peltier element dissipates the heat into its surroundings. If several laser diodes, each having its own cooling, are integrated to achieve higher output power, the heat dissipation can be improved by employing a ventilator.

For higher power outputs, entirely different cooling systems can be employed. Due to their small dimensions, laser diodes are well suited for the integration of many elements onto very small areas. Although the efficiency of the electro-optical conversion may be very high, appropriately large power densities for the waste heat will occur. This high thermal load, together with the need to operate the laser

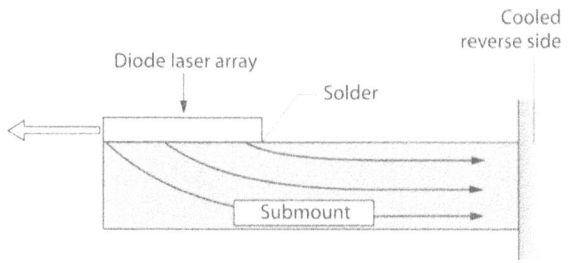

**Fig. 7.** Construction of a linear array onto a liquid cooled heat sink. (After [7])

diodes at a stable temperature, limits the density of integration for high-power lasers.

Simple cooling by liquids is achieved by soldering the diode laser onto a submount of solid copper (which has a high thermal conductivity coefficient), which is then bathed in a coolant (Fig 7).

An optimal but more elaborate cooling arrangement employs the microchannel technique, which was initially developed to cool ICs. It can achieve an extraordinarily low thermal resistance. The microchannel cooler is made of a material with good thermal conductivity (like copper or silicon) into which channels are cut. Typical values for the channels: width and separation are some 10 to 100 µm, depth is a few 100 µm. The heat produced by the laser diodes is conducted by the bars separating the channels into the coolant which fills the channels (Fig. 8).

This method allows heat losses corresponding to average power densities of more than 1 kW cm$^{-2}$ to be removed without exceeding the actual temperature of the crystal [8].

Employing optimized microchannel cooling, it is possible to achieve values for the thermal resistance of below 0.1 K/(W cm$^{-2}$). The thermal resistance depends on the pressure (about 1.5 to 3 bar) driving the coolant through the channels, besides the nature of the coolant material itself. Increasing the pressure leads to a lower thermal resistance.

Microchannel cooling of diode lasers calls for exceptional cleanliness of coolant (often a water-glycol mixture) and cooling system in order to prevent blocking of the fine channel structures by microorganisms or algae. It is necessary to continuously filter out the abrasion particles arising from the cooling circuit. From time to time, the direction of circulation is reversed to dislodge any blockages [7].

# Summary

Progress in semiconductor techniques and the further development of high-output power diodes has awakened interest in applying diode lasers in medicine. Due to their compactness and high efficiency, laser diodes offer many advantages as compared with conventional laser systems. They can be applied in the consulting room as well as in the hospital. Diode lasers are efficient converters of electrical energy into coherent radiation. Excitation is induced by injection of charge carriers at the p – n junction after application of an external voltage. The optical power output depends on the induced current density. A certain threshold current density must be exceeded in order to emit laser light of narrow spectral width and high power. Special crystal structures allow optimization of the optical output power with respect to the injected current. It is essential to keep the heating of the semiconductor, caused by ohmic losses, low because threshold current density, emission wavelength and lifetime of the diode are temperature-dependent. The narrow emission gap of the diode lasers causes an asymmetric (elliptic) and strongly divergent beam profile, frequently characterized by multi-longitudinal mode spectra. To obtain a better beam quality, optic systems collimate the light and achieve an almost Gaussian intensity profile.

Cooling and/or control of the stable operating temperature are essential to guarantee acceptable lifetime and power stability of a diode. Various cooling systems are in use depending on power and on the integration density of several laser diodes.

## References

1. Brunner W, Junge K (1989) Wissensspeicher Lasertechnik. VEB Fachbuchverlag, Leipzig
2. Evans D (1993) SDL Inc, San Jose, USA
3. Geels RS, Welch DF, Scifres DR, Bour DP, Treat DW, Brigans RD (1993) 20-W cw monolithic visible diode array. CLEO 93 Tech Digest CThQ3: 478
4. Hering E, Rolf M, Stohrer M (1989) Physik für Ingenieure. VDI-Verlag, Düsseldorf
5. Hodgson DJ (1994) How power-supply selection can improve laser-diode performance. Laser Focus World Jan: 129
6. Kneubühl K, Sigrist MW (1991) Laser. Teubner Studienbücher: Physik, 3rd edn.
7. Peuser P, Schmitt N (1994) Diodengepumpte Festkörperlaser.
8. Tuckermann D (1984) Heat-transfer microstructures for integrated circuits. Thesis, Stanford University, California

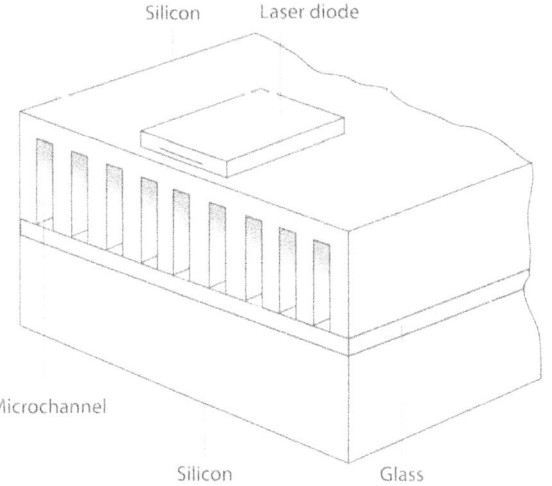

**Fig. 8.** Principle design of a diode laser microchannel cooler. (After [7])

# I-2.4.11
# Metal-Vapour Lasers

W. Gottschalk

## Contents

## Introduction

Metal-vapour lasers have a high output power of about 1 to 20 W in the visible and infrared spectral ranges. Due to its high efficiency (1%) in the visible spectral range, as compared with other gas lasers, and the available wavelengths, this laser is a potential light source for various therapeutic applications.

A few years ago such lasers became commercially available; however, they were characterized by unreliability, complex handling and extensive service requirements. In the meantime these systems have overcome their teething problems.

In particular, the copper-vapour lasers present a number of advantages in the fields of dermatology and head, neck and throat diseases, and have become increasingly important. The new modular techniques, which will be presented below, have further increased their therapeutic range of applications.

Originally, the gold-vapour laser was predestined for photodynamic therapy using Photofrin. The rapid development of diode laser systems, however, will displace it in the near future. In addition, the new and extended development of photosensitizers will shift the therapeutically required wavelength more and more into the direction of the infrared spectrum.

## Design of a Metal-Vapour Laser

Figure 1 displays the schematic design of a metal-vapour laser. The laser-active material (gold, copper, barium, etc.) is contained inside a plasma tube which is arranged and isolated within a glass tube and closed by a window. Inside this arrangement and at low pressure a discharge will be ignited with the assistance of an auxiliary gas. After some time (30 – 90 min.) the temperature inside the plasma tube rises (maximum 1600 °C) and vaporizes the metal, originally a solid. The discharge excites the metal-vapour and laser radiation will be emitted. A mirror arrangement collimates the radiation, because metal-vapour lasers emit so-called superradiance. This means that its internal amplification is so high that laser radiation arises even without a resonator.

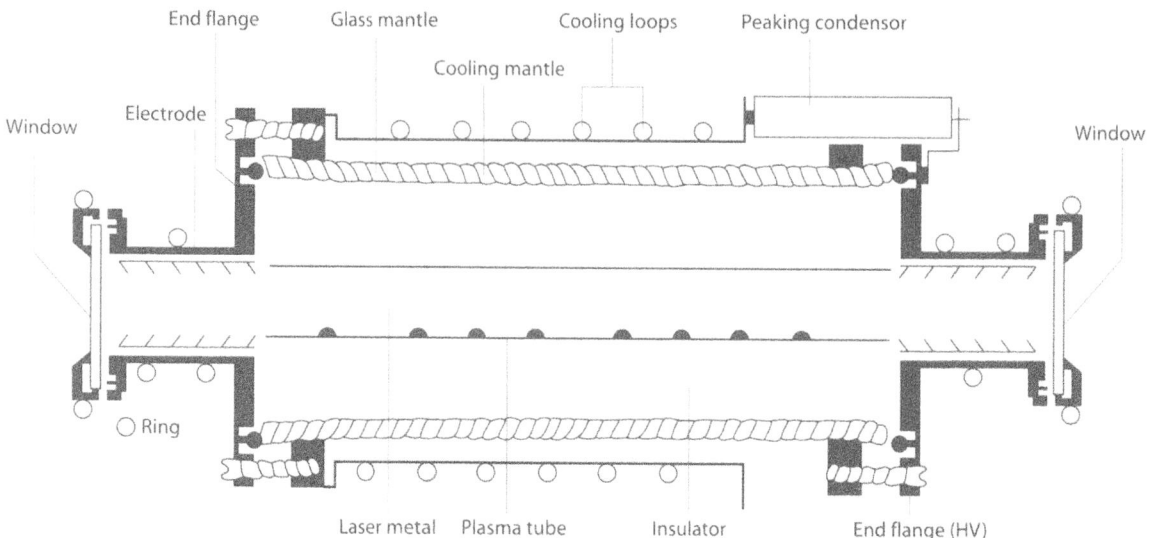

**Fig. 1.** Design of a metal-vapour laser

## Specifications for Metal-Vapour Lasers

Metal-vapour lasers are pulsed systems having a pulse-repetition frequency between 5 and 20 kHz because the lower laser level is metastable.

As already indicated, various metals are suitable as active media for this laser type. Table 1 gives an overview.

Of all these laser types, mainly copper-vapour lasers and gold- and barium-vapour lasers have achieved importance for clinical applications.

**Table 1.** Properties of laser-active materials for metal-vapour lasers

| Metal | Wavelength (nm) | Output power (W) | Plasma temperature (°C) |
|---|---|---|---|
| Copper | 510.578 | Up to 100 | 1420 |
| Copper-bromide | 510.578 | Up to 10 | 600 |
| Gold | 312.628 | 1 or 4 | 1570 |
| Manganese | 524.1290 | 2 or 5 | 1060 |
| Lead | 723 | 4 | 830 |
| Barium | 1130.1500 | 1.2 | 710 |
| Strontium | 645 | 1 | 630 |
| Iron[a] | 501.868 | - | 1650 |
| Cobalt | 408.735 | - | 1690 |

[a] Presented in the literature, not available commercially.

## Metal-Vapour Lasers as Pumping Light Source for Other Lasers

The copper-vapour laser can act as an excellent pumping light source for other lasers, as can the Ar-ion laser. The US company CJ-Laser (distributed by LASER 2000, Wessling) has developed and marketed systems with a previously unknown versatility. It is possible to integrate into a laser up to four additional lasers. The cooperation of all modules is displayed in Fig. 2.

This all-inclusive system provide the therapist with numerous wavelengths. Table 2 displays a listing.

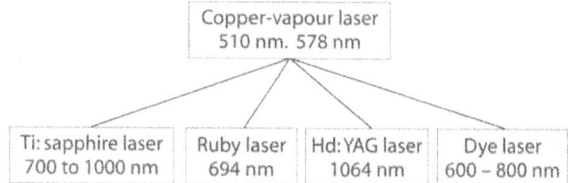

**Fig. 2.** The copper-vapour laser as pumping light source for various lasers. Here the copper-vapour laser pumps a crystal module, which may contain up to three different solid-state lasers. Alternatively, a dye laser can be attached. Its emission spectrum depends on the dye employed.

**Table 2.** Wavelength and output power of the copper-vapour laser and its modules

| Laser-active material | Wavelength (nm) | Output power (W) |
|---|---|---|
| Copper | 510 and/or 578 | 8 and/or 12 |
| Dyes | 600 – 800 | 2 |
| Ruby | 694 | 1 |
| Ti: sapphire | 700 – 1000 | 1.5 |
| Nd:YAG | 1064 | 1.5 |

**Table 3.** Applications for different wavelengths

| Wavelength (nm) | Application |
|---|---|
| 511 | Keratosis, warts lentigo, pigment spots |
| 578 | Teleangiectasia, rosacea, spider naevi |
| 600 – 700 | Tattoos, photodynamic therapy |
| 1064 | Coagulation, closing of small vessels |

Obviously, this system offers all wavelengths relevant in dermatology. The demarcation towards surgical systems is obvious; merely that the integrated Nd-YAG laser may perhaps be applied in microsurgery.

Typical applications for the available wavelengths are listed in Table 3.

## Summary

Metal-vapour lasers have undergone significant changes in the past years. Complete therapeutic systems have been developed, combining various lasers within one instrument. The therapist has a tool at his disposal with which he can carry out most applications suitable for laser treatment.

The largest component of such a system is the copper-vapour laser. Within a few years, significant reductions in the size and complexity of such systems can be expected, because the same wavelengths (510, 578 nm) and the same output powers can be achieved by a copper-bromide laser. These lasers are technically simple, very much smaller and less expensive. Recently, they became available on the German market (LASER 2000, Wessling).

# I-3
# Action Mechanisms of Laser Radiation in Biological Tissues

# I-3.1
# Properties of Biological Tissues

A. Roggan, U. Bindig, W. Wäsche and F. Zgoda

## Contents

## Introduction

If laser radiation strikes biological tissue, various effects can be observed. All are caused primarily by the interaction of photons with the molecules and the molecular compounds of the tissue. The resulting actions depend mainly on the application parameters, which, in turn, can be characterized by the wavelength of the applied laser, the exposure time and the power density. The action mechanisms may be divided approximately into photochemical actions, thermal processes and non-linear responses. The actions are dominated:

- For low power densities and long exposure times by photochemical processes.
- For average power densities and exposure times by thermal responses.
- For power densities exceeding $10^7$ W cm$^{-2}$ and ultrashort exposure times, in the ns range, by non-linear actions (see Chapter I-3.4).

Besides the application parameters, which can be freely chosen – within certain limits – by the user, the tissue properties have a decisive influence on the laser actions. These tissue properties can be subdivided roughly into optical tissue properties and thermal tissue properties. While the first set of properties determines the primary distribution of the laser light inside the therapeutic volume, the second deals with the conversion of light energy into heat, as well as the heat transport via thermal conduction.

## Optical Properties of Tissues

If a laser beam strikes biological tissue, a specific distribution of the laser light inside the irradiated volume is observed. Part of the radiation is absorbed by the tissue and, consequently, may be therapeutically active. Depending on the layer thickness, another part will be transmitted either directly at or after multiple scattering. A further fraction of the photons will be scattered in the tissue and leave it as reemission (diffusely reflected) radiation. For certain wavelengths in

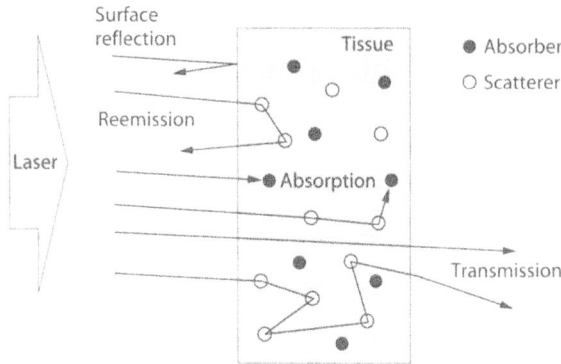

**Fig. 1.** Optical behaviour of a tissue layer during irradiation with laser light

the near-infrared, this contribution can reach 60%. Only a small percentage will be directly reflected at the surface, due to a step in the value of the refractive index.

The measurable quantities reemission, reflection and transmission are often called macroscopic optical properties; they depend on wavelength, tissue type and layer thickness. The distribution of the laser light inside the tissue and the resulting macroscopic optical properties are determined by three processes:

• Absorption
• Scattering
• Refraction.

If the tissue structure changes during laser therapy (bleaching by coagulation, blackening by carbonization), then the optical properties may vary considerably.

To find new applications for the laser, it is essential to have comprehensive knowledge of the various tissues' optical properties. On this basis, it is possible to predict for a given tissue type the efficiency of existing laser systems and even the applicability of wavelengths presently not utilized. The wavelength range of interest, available with medical lasers, can be subdivided into ultraviolet (UV: < 400 nm), visible (vis: 400 < 780 nm), near-infrared (NIR: 780 nm < µm 2.5) mean infrared (MIR: 2.5 µm < 25 µm), and far infrared (N: > 25 µm) [6]. The photon energy is directly proportional to the wave number $\alpha$ (cm$^{-1}$). This quantity is widely used in spectroscopy, especially in the infrared range. A simple relationship holds:

$$\lambda\ [nm] = \frac{10^7}{\gamma [cm^{-1}]}$$

## Absorption

The absorption of photons can, in principle, take place in all molecular components of the tissue. In the UV and vis regions these components will, in general, be tissue-specific chromophores like porphyrin, haemoglobin, melanin, flavin, retinol, ribonucleic acid, NADH (reduced nicotinamide-ademine-dinucleotide) etc., where electronic transitions are excited. This leads to discrete and intense absorption bands. In contrast, proteins display mainly barely structured and broad absorption bands, in general. In biological tissues the chromophores are predominantly coupled to cellular components (e.g. to membranes), or they are present in the form of a cellular substructure like the nucleic bases of DNS/RNS. In the NIR and MIR, tissue absorption is dominated mainly by the absorption in water. Absorption by molecules, e.g. glucose, plays an insignificant role. Besides low-energy electronic transitions, mainly higher harmonics and combination vibrations will be excited. The combination bands are composed of hydrogen tensile vibrations contributed mainly by molecular groups like N–H (amino), O–H (hydroxy) and C–H (hydrocarbon).

The absorption coefficient $\mu_a$ characterizes the absorption. It is the product of absorber concentration $c_a$ (cm$^{-3}$) and absorption cross-section $\sigma_a$ [cm$^2$]. It is measured in cm$^{-1}$. In general, different absorbers contribute at one wavelength, consequently $\mu_a$ must be considered the sum of all substances involved:

$$\mu_a = \Sigma\ c_a\ \sigma_a\ .$$

Biological tissues typically display absorption coefficients in the range 0.01 cm$^{-1}$ < $\mu_a$ < 100 cm$^{-1}$ [4, 14]. Note, that in tissue optics the absorption coefficient is often given in mm$^{-1}$ (1 mm$^{-1}$ = 10 cm$^{-1}$!).

For exclusively absorbing materials (except for the transparent components of the eye, tissues do not belong to this category), the intensity of the part $I_T$ transmitted through a layer of thickness d (cm) can be easily calculated according to Bouger-Lambert law. If $I_0$ is the intensity on entry into the substrate, then

$$I_T = I_0\ exp(-\ \mu_a\ d)\ .$$

To calculate the beam distribution in tissue for wavelength ranges which include large contributions from scattering, complex radiation models are used such as Kubelka-Munk theory or numerical solutions of the radiation transport equation (see Chapter I-3.1.2).

Frequently, especially in chemistry, the following terms are employed: absorber concentration c (mol dm$^{-1}$) and decimal molar extinction coefficient $\varepsilon$ (dm mol$^{-1}$ cm$^{-1}$). The latter term is defined by the Bouger-Lambert law and utilizes the decimal logarithm:

$$I_T = I_0 \times 10^{-\varepsilon \cdot c \cdot d} = I_0 \times 10^{-A} .$$

The exponent is called extinction coefficient or absorbance A of a substance, and mostly refers to a layer thickness d of 1 cm. The following relation holds:

$$\mu_a = 2.303 \; c \; \varepsilon$$

Figure 2 indicates the absorption coefficient of pure water and Figure 3 the absorption coefficient of human oxyhaemoglobin. Both substances are important absorbers in biological tissues.

## Scattering

The scattering of photons in biological tissues occurs at sites of inhomogeneity of the refractive index, e.g. at membranes, cell nuclei, mitochondria, lipids etc. The scattering coefficient $\mu_s$ characterizes the above process. It is defined, in analogy with the absorption coefficient, as the product of the concentration $c_s$ (cm$^{-3}$) of the scattering centres and the scatter-

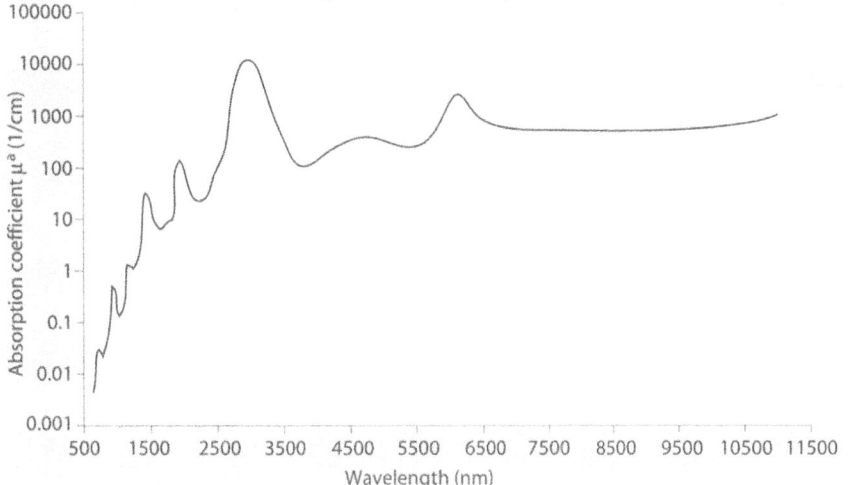

**Fig. 2.** Absorption coefficient of pure water [8]

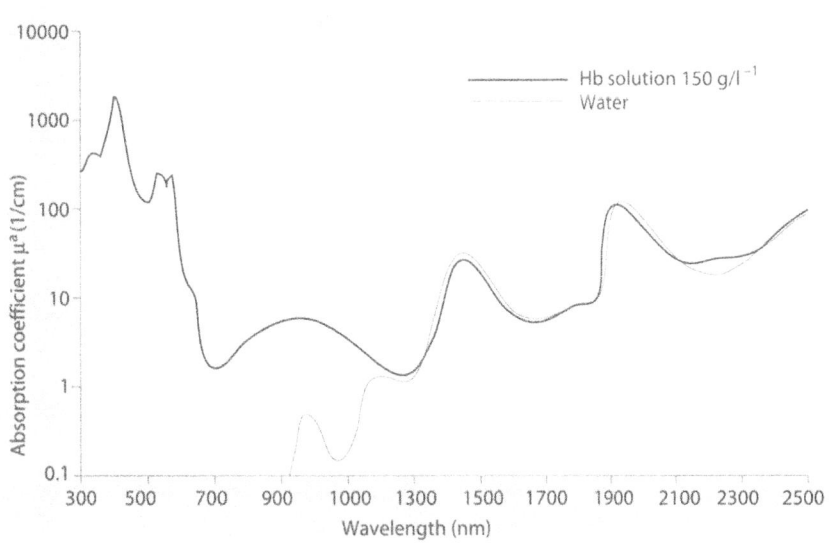

**Fig. 3.** Absorption coefficient of oxygenated human haemoglobin without cellular components. The physiological concentration is 150 g/l (uv/vis spectrometer and transmission measurements with a double Ulbricht sphere system. LMTB). For comparison, the absorption coefficient of water is also displayed

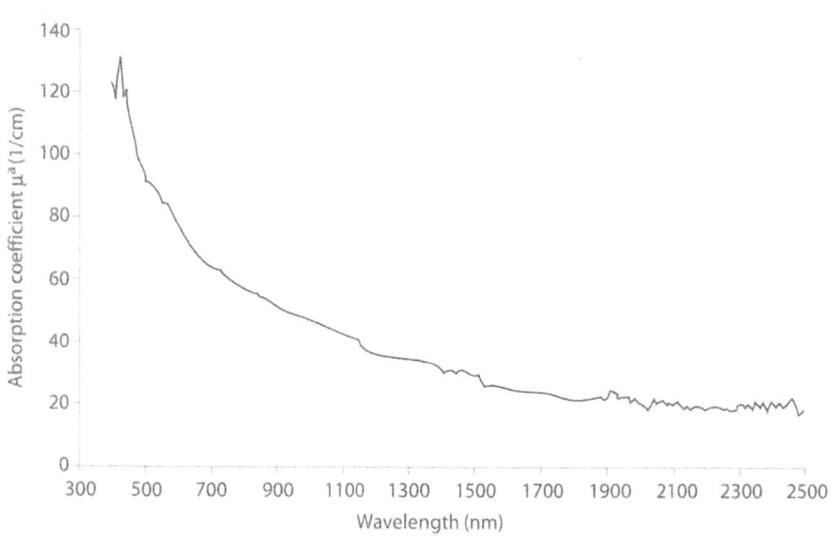

**Fig. 4.** Scattering coefficient of unstained porcine liver.

ing cross-section $\sigma_s$ (cm$^2$). It is summed over the various types of scattering centres:

$$\mu_s = \Sigma\, c_s\, \sigma_s\,.$$

Biological tissues typically display scattering coefficients in the range 10 cm$^{-1}$ < $\mu_s$ < 1000 cm$^{-1}$ [4, 14]. In tissue optics the scattering coefficient is frequently expressed in mm$^{-1}$. $\mu_s$ is a direct measure for the number of scattering events per unit path length covered by a photon. Over a wide range of wavelengths the scattering coefficient is significantly larger than the absorption coefficient. Also, $\mu_s$ depends only to a small degree on the wavelength. To first approximation many tissues display a $1/\lambda$ dependency, i.e. scattering decreases with increasing wavelength (Fig. 4).

The mathematical treatment of the real scattering events is difficult, because the geometrical structures involved in the scattering process are very numerous. It is only possible to describe mathematically exact single scattering at spherical and homogeneous scattering centres by solving the Maxwell equations[11]. Theoretical considerations show that it is impossible to describe the scattering properties of biological tissues completely by simply utilizing the scattering coefficients. In addition, the angular distribution which describes the directional change of the scattered photons must be known. It is called the scattering phase function and there are various mathematical approaches to formulate it for tissue optics (e.g. Henyey-Greenstein phase function, see chapter I-3.1.2).

Because it is difficult to determine scattering phase functions, their information content about the angular distribution of single scattering events is reduced to giving the so-called *anisotropy factor* g. This factor presents the cosine of the scattering angle $\theta$ (Fig. 5)

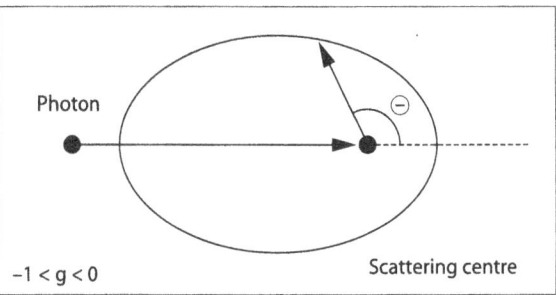

Backward scattering

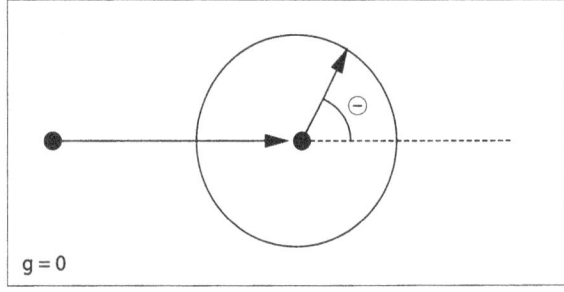

Isotropic scattering

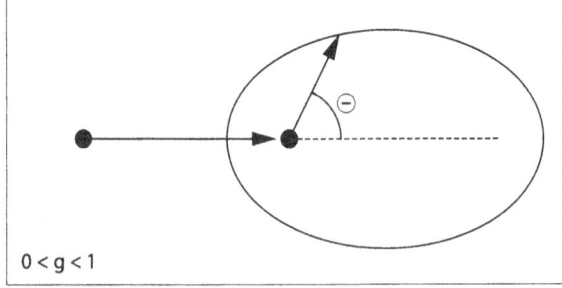

Forward scattering

**Fig. 5.** Definition of the anisotropy factor *g*

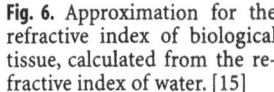

**Fig. 6.** Approximation for the refractive index of biological tissue, calculated from the refractive index of water. [15]

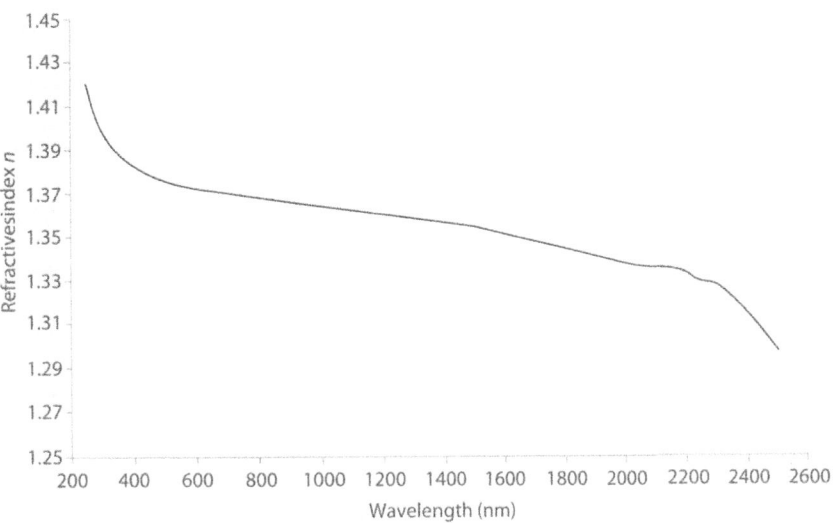

averaged over many scattering events. The range of values for *g* reaches from –1 (backward scattering) via O (isotropic scattering) to +1 (forward scattering). As a rule, biological tissues show pronounced forward scattering. For visible light and in the NIR g is in the range 0.8 to 0.99 [4, 14].

## Refractive Index

If a laser beam proceeds from an optically less dense medium (e.g. air) into an optically denser medium (e.g. water or tissue), it will be partially reflected at the boundary. This partial reflection depends on the wavelength and differs from reemission. The transmitted fraction of the laser beam changes its direction of propagation. Both effects depend on the refractive indices *n* of both media. The refractive index of many biological materials is correlated with that of water because for many tissues this is the main component (see. Fig. 6).

## Measurement of Scattered Light

In general, optical properties like absorption, scattering and the anisotropy factor of biological tissues cannot be measured directly. Rather measurements are restricted to macroscopic properties like backscattering and transmission by a tissue sample. In order to approximate or to calculate the beam distribution inside the therapeutic volume, it is necessary to know the microscopic optical parameters.

Very complex measuring and evaluation processes are needed to determine $\mu_a$, $\mu_s$ and *g*. So-called mephelometry is employed as the standard method. Here,

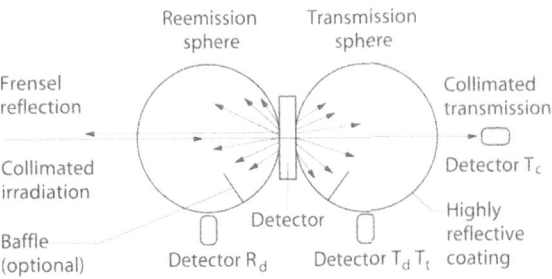

**Fig. 7.** Principle layout to measure scattered light, designed to determine optical properties of tissues. [13]

the tissue sample to be measured is placed between two intergrating spheres (see. Fig. 7; [14]). At the same time, the sample is illuminated by a collimated light beam of the wavelength of interest. The integrating spheres are covered on the inside by a highly reflective coat (e.g. $BaSO_4$). All of the diffusely reemitted and totally transmitted radiation of the probe will be recorded ($R_d$ and $T_t$). A photodetector, attached to each integrating sphere, will deliver an electronic signal proportional to the intensity of the respective radiation field. This signal may be further evaluated. Three independent measurements are needed to determine the three independent parameters $\mu_a$, $\mu_s$ and *g*. For this reason, the collimated, i.e. non-scattered transmission, $T_c$ of the tissue layer serves as a third measurement; for details see below. The measuring system is calibrated by reflection and transmission standards (Labsphere Corp.).

Figure 8 indicates the technical arrangement for such a double integrating sphere system [13–15]. A Hg or a Xe high-pressure lamp with attached monochromator serves as light source (LMTB, Berlin). The

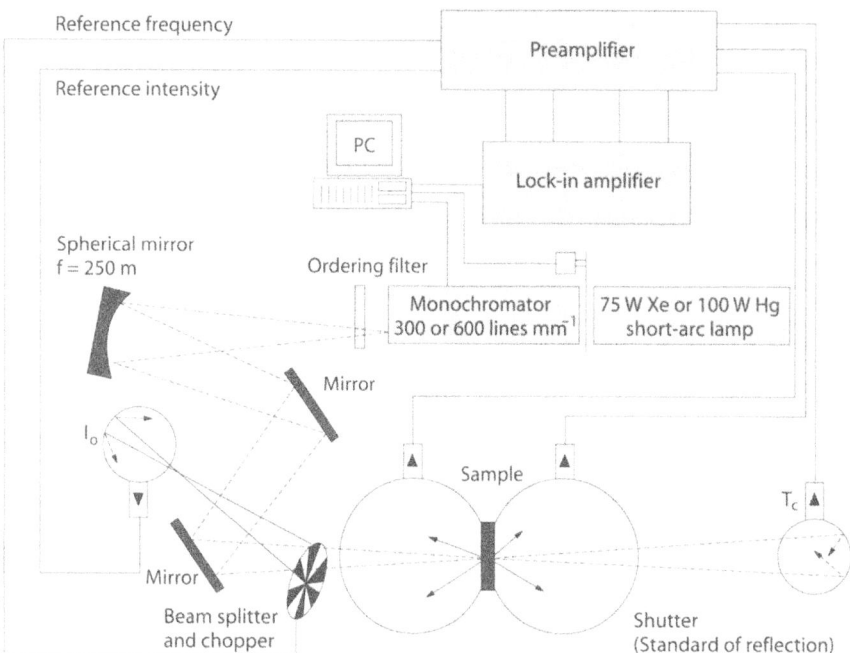

**Fig. 8.** Double integrating sphere system to measure optical parameters. [13–15]

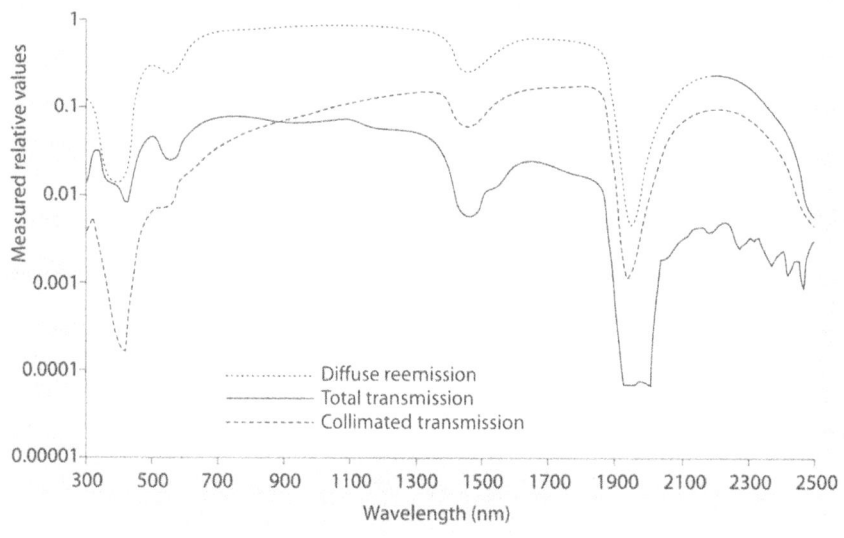

**Fig. 9.** Microscopic optical properties of liver (pig) measured with a double integrating sphere system (d = 500 μm). [15]

measuring range covers 300–2500 nm, the resolution is 8–16 nm (FWHM). The detectors are silicon photodiodes (for 300–1100 μm), Si-photodiodes or lead-sulphide photodiodes (for 1100 –2500 nm). Detection is via the lock-in technique and the entire arrangement is computer-controlled.

Of special importance for the measuring methods presented is the preparation of the samples, since their optical properties should not be altered at all. To prevent desiccation of the biological preparations during measurement, the tissue samples are fixed into cylindrical cuvettes made of quartz. Typical sample thicknesses range from 100 μm to 1 mm. The samples can be produced from body structures (e.g. skin), freeze-dried sections, direct sections, sawn sections (hard tissues) or from a homogenization procedure. Liquid substances showing a tendency towards sedimentation or clustering (e.g. blood) can be measured in flow cells. Figure 9 indicates the macroscopic quantities of a 500 μm thick layer of liver (pig) measured by a double Ulbricht sphere system.

The extraction of the optical parameters $\mu_a$, $\mu_s$ and g from the measured values is a complex task, because no analytical mathematical models of sufficient precision are available.

The method of choice, therefore, is Monte Carlo simulation. This statistical method computes the trajectories of a large number of photons and yields the reemission and transmission properties of a sample having specified optical parameters [22]. To determine for a given problem the unknown optical parameters from the measured macroscopic properties, the Monte Carlo procedure is started, using a set of estimated initial parameters for $\mu_a$, $\mu_s$ and g and its results are compared with the measurements. If the deviations exceed a predetermined error limit, a gradient matrix will be calculated yielding a new set of values for $\mu_a$, $\mu_s$ and g. This procedure will be repeated until it presents the measured values correctly. The

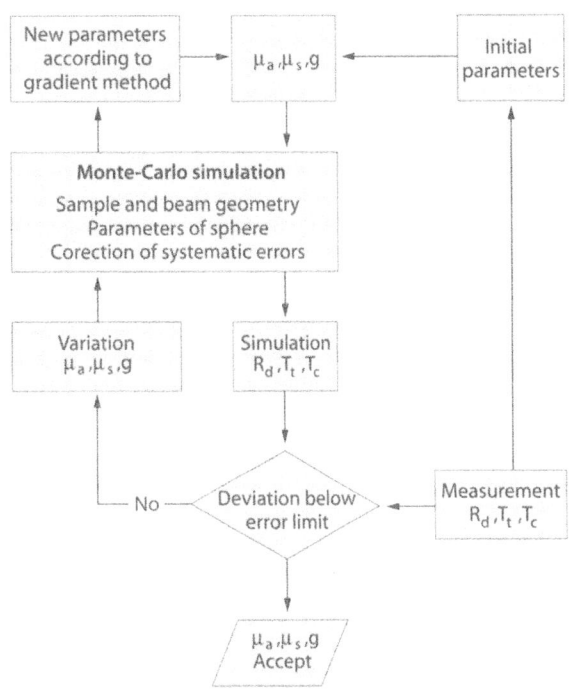

**Fig. 10.** Inverse Monte Carlo simulation to calculate the optical parameters from measured data. [13–15]

appropriate set of optical parameters is deemed acceptable.

Besides high computational accuracy, this method allows unambiguous compensation for systematic errors like lateral radiation losses and incomplete separation of the radiation fields. The potentially long computational times are significantly reduced by utilizing the presently available fast computers. Figures 11–15 indicate the optical properties determined for some tissues using double integration sphere systems and inverse Monte Carlo simulation.

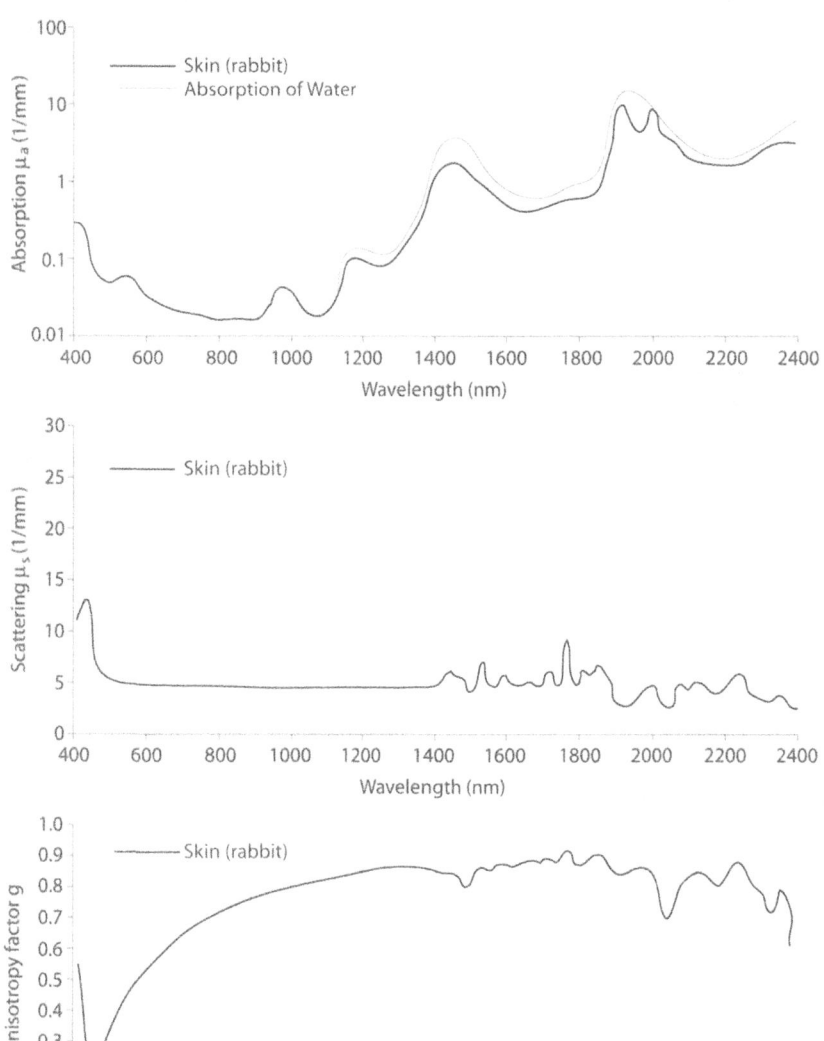

**Fig. 11.** Optical properties of rabbit skin. Mean values from three measurements, average thickness 1.4 mm. [15]

**Fig. 12.** Optical properties of rat skin (ear). Mean values from three measurements. Mean thickness 0.5 mm. [15]

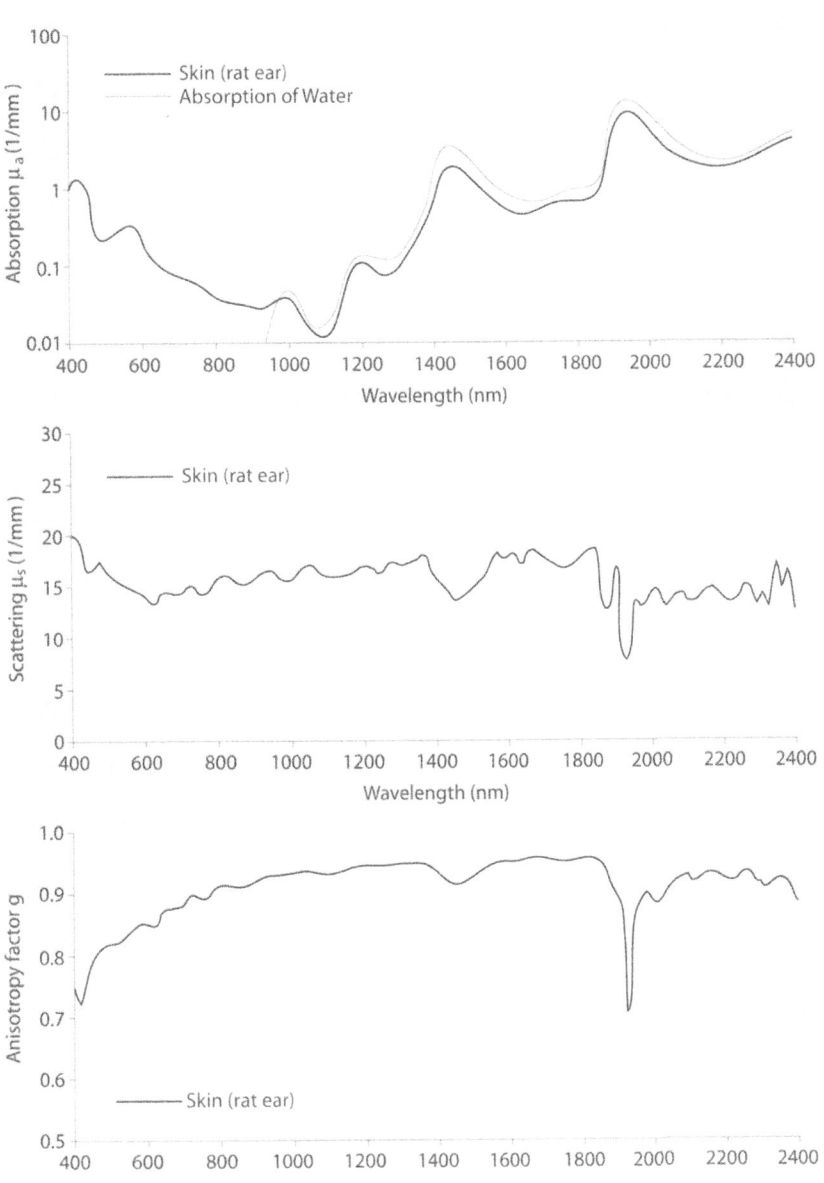

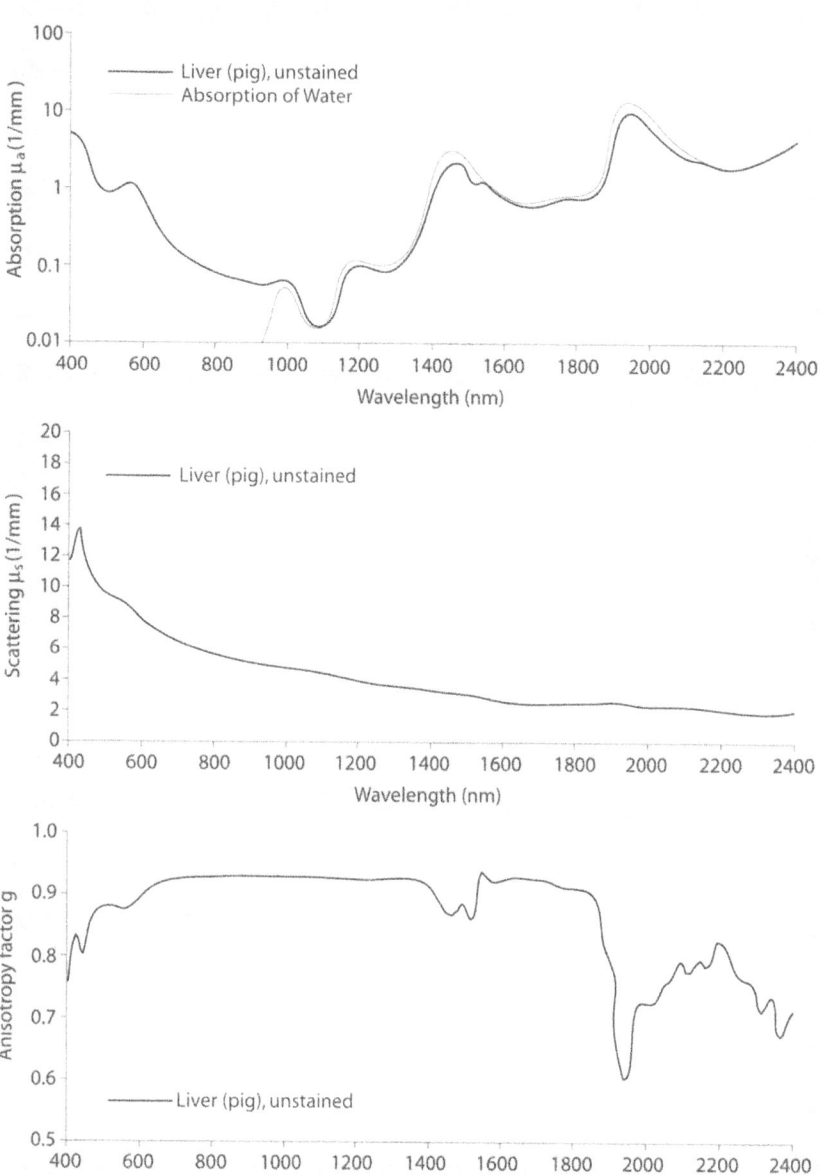

**Fig. 13.** Optical properties of unstained pig liver. Mean values from three measurements, mean thickness 0.5 mm. [15]

**Fig. 14.** Optical properties of coagulated pig liver (80 °C, 10 min). Mean values from three measurements, mean thickness 0.1 mm.

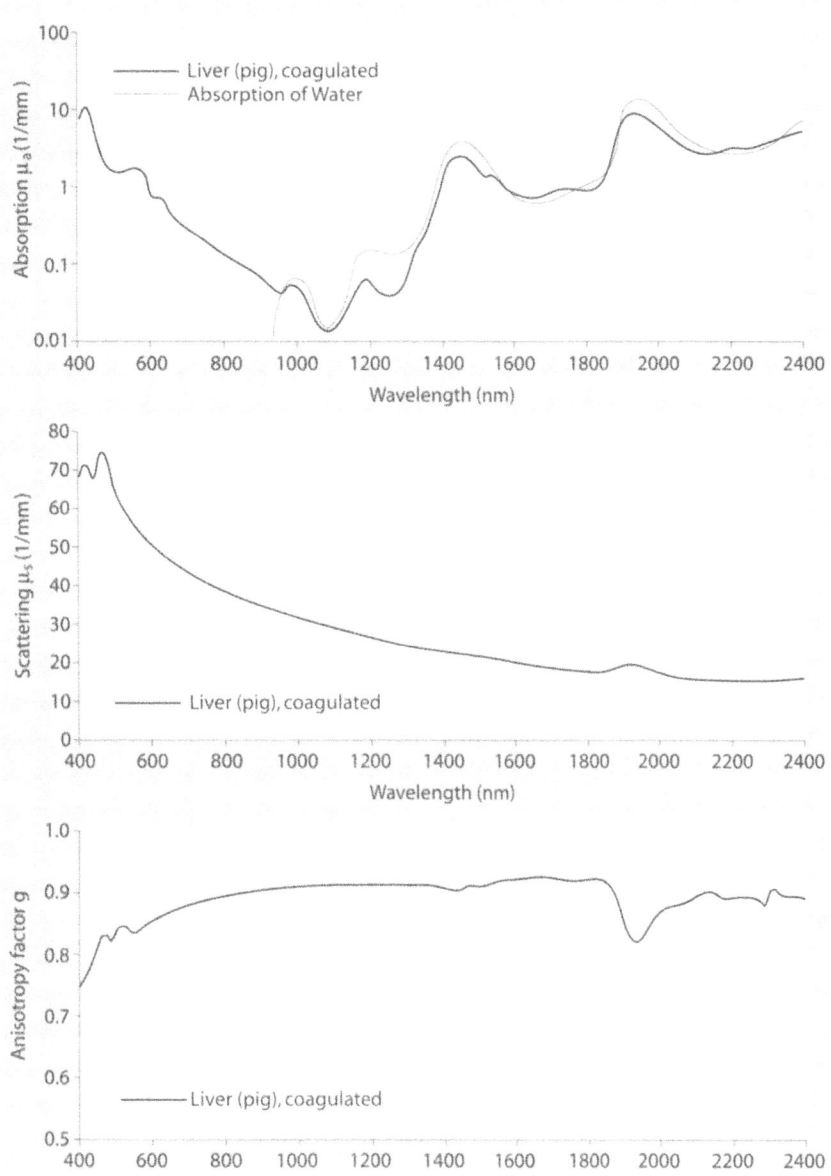

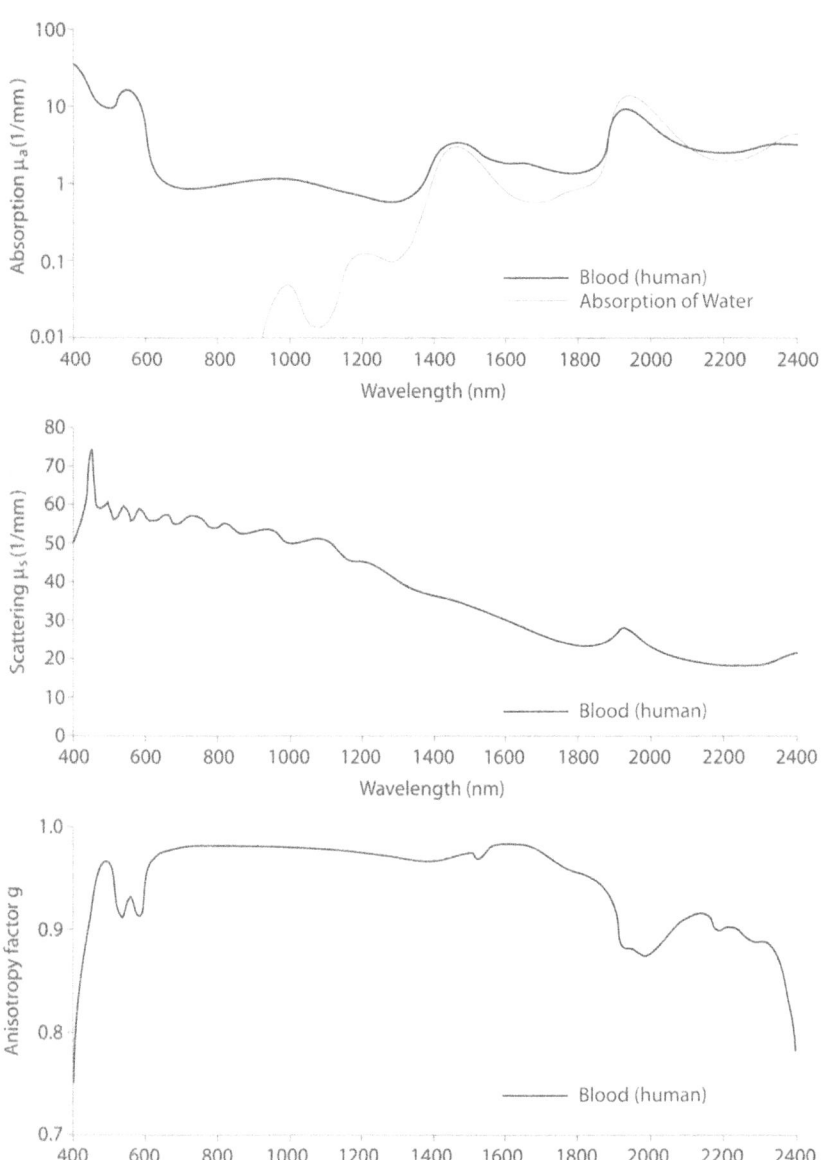

**Fig. 15.** Optical properties of human blood (phosphate buffer: Hkt = 44%, pH = 7.4; π = 0.3 mol 1). Mean values from three measurements, thickness of optic cell 97 μm. [15]

## Spectroscopic Methods

Besides the very involved determination of the microscopic optical parameters, there are various spectroscopic techniques to determine the transmission, sometimes also the re-emission, of biological tissues over a wide range of wavelengths. It must be pointed out that these results depend on the thickness of the specimen, whereas the microscopic parameters $\mu_a$, $\mu_s$ and $g$ are parameters independent of the layer thickness.

A transmission spectrum allows a relative comparison of the interaction of various wavelengths with an actual tissue type. However, it is not possible to estimate the interaction efficiency of one wavelength with various tissue types by comparing the various transmission spectra.

The transmission in the UV, VIS and NIR can be measured by a universal microscope spectrophotometer (e.g. UMSP 80, Zeiss Oberkochen). Mercury and Xenon short-arc lamps serve as light sources, and wavelength selection is performed by a monochromator. Figures 16–25 present typical spectra for various tissues in the wavelength range from 250 to 750 nm or up to 2000 nm.

The standard method to record vibrational spectra in the near and mid-infrared range (NIR 10 000–4000 cm$^{-1}$, MIR 4000–800 cm$^{-1}$) is Fourier transformation infrared spectroscopy (e.g. FTIR-spectrometer system 2000, Perkin-Elmer). The investigation of biological molecule / biological materials by vibrational spectroscopy offers a number of advantages as compared with other analytical techniques. FTIR spectrometers assure an optimum signal-to-noise ratio, very good wavelength reproducibility and short uptake times in the ms range. In the FTIR spectrometer the Michelson interferometer arrangement replaces the monochromator.

Figures 26–30 indicate NIR spectra and Figs. 31–45 indicate MIR spectra for various sample materials. The biological tissue samples were stored at room temperature and with light excluded inside a desiccator above a dehydrating agent. The soft tissue slices were prepared by a cryomicrotome, hard tissue slices by an inner-hole saw. Standard preparations of biologically active compounds in water solutions were mounted onto calcium fluoride specimen slides, without extra purification, dried by an airflow and measured as a film. As an additional standard method, KBr moulds were prepared and measured. Some of the IR spectra presented indicate an absorption at 2350 cm$^{-1}$. However, this IR band is caused by the atmospheric carbon dioxide content. For normalization, the IR spectra were in general set at an intensity of 0.3 absorption units and presented in transmission. Additional experimental conditions were: wave number range: 4000–800 cm$^{-1}$, resolution: 4 cm$^{-1}$, apodization: Filler, accumulation: 64 scans, interval: 2. Aperture: 100 μm$^{-2}$.

Of increasing importance to infrared spectroscopy is FTIR microspectroscopy. This technique makes it possible to gain spectral information about very small areas with the help of an IR microscope, see Fig. 45. This method can be applied to the analysis of materials, both as a diagnostic tool for hard and soft tissue slices and for blood [9]. In general, the graphic image of the spectral information is called an IR map.

From the various areas of an unstained human tissue slice displayed in a map (see Fig. 45B) it is possible to decode a number of discrete areas within the tissue composition. After the IR measurement, the above tissue slice underwent the usual standard procedure for HE staining. The spectral differences in the IR spectrum of these regions are caused by the quantitative composition of a few main components and their varying quantitative proportions, e.g. by protein/lipid/DNA and other components [1]. For example, HE staining discriminates in tumorous tissues between areas with increased and normal cell nucleus densities. In this context the phosphate vibration bands in the region around 1000 cm$^{-1}$ can serve as

markers for tissue-specific changes. These differences are presented in the IR map by distinct colour shades.

This technique permits high-resolution differentiation within an apparently homogeneous single-tis-

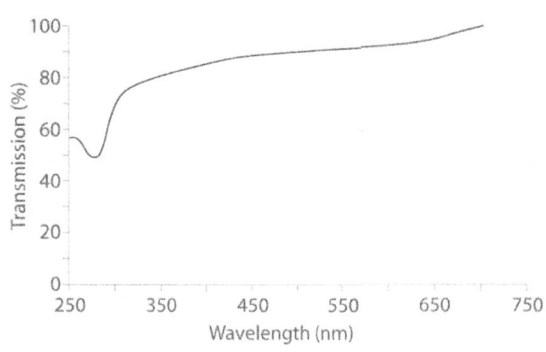

**Fig. 16.** UV/VI spectrum. Human patella tissue (post mortem) UMSP 80, freeze-dried slice, slice thickness 20 μm, measuring spot 10 μm

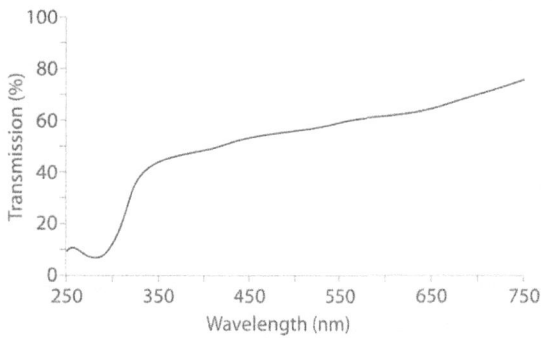

**Fig. 17.** UV/VIS spectrum. Human meniscus tissue (post mortem) UMSP 80, freeze-dried slice, slice thickness 20 μm, measuring spot 10 μm

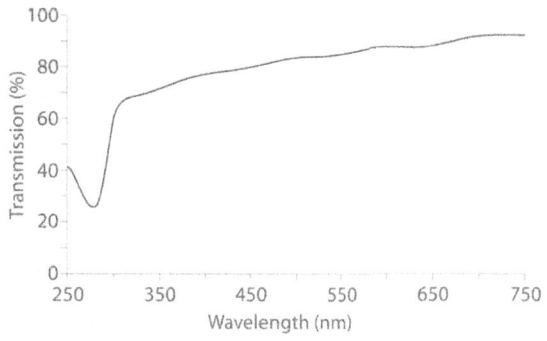

**Fig. 18.** UV/VIS spectrum. Muscle tissue (pig, post mortem) UMSP 80, freeze-dried slice, slice thickness 20 μm, measuring spot 20 μm

sue component (e.g. stroma). Figures 46–52 indicate typical IR transmission spectra for areas selected from Fig. 45. The spectroscopic differences in the MIR regions of these normalized spectra are detectable only within the finger-print region (below 1500 cm⁻¹ / 6.66 μm). Since the differences in the presentation of

the total absorption are sometimes very small, as a rule the higher mathematical derivatives of IR spectra are used. The result of such a procedure can again be presented as an IR map. Figure 45B indicates such an IR map for a discrete wavenumber (1221 cm⁻¹).

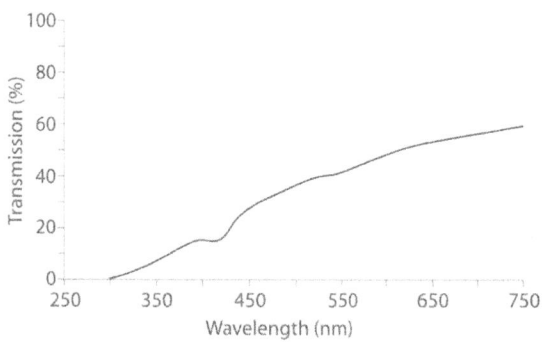

**Fig. 19.** UV/VIS spectrum. Cornea (pig, post mortem) UMSP 80, freeze-dried slice, slice thickness 1 mm, measuring spot size 40 μm

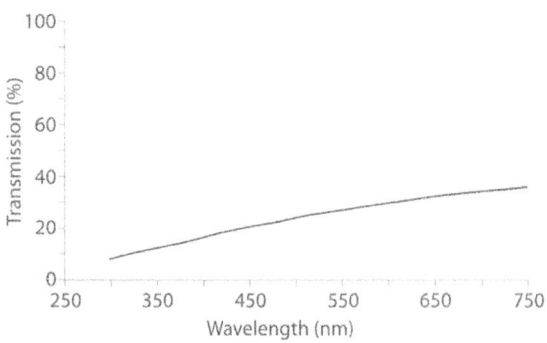

**Fig. 22.** UV/VIS spectrum. Human saliva stone (in vitro) USMP, hard-tissue slice, slice thickness 50 μm, measuring spot size 4 μm

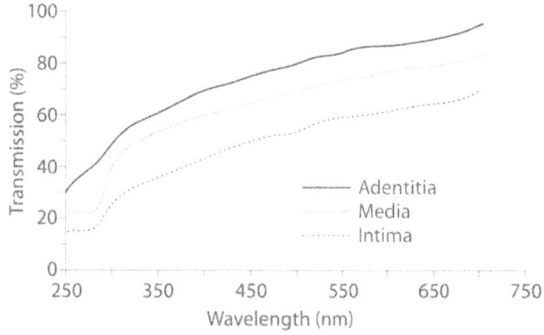

**Fig. 20.** UV/VIS spectrum. Human coronary artery (post mortem) UMSP 80, freeze-dried slice, slice thickness 10 μm, measuring spot size 20 μm

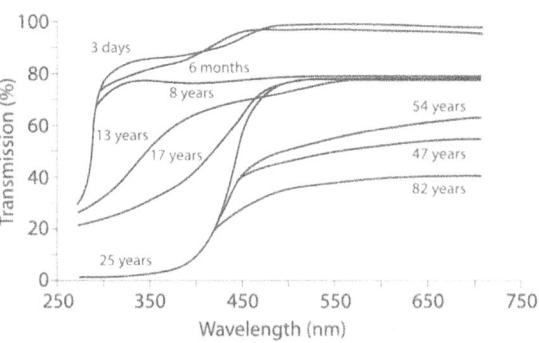

**Fig. 23.** Changes in the transmission of an ageing human lens, (after [10])

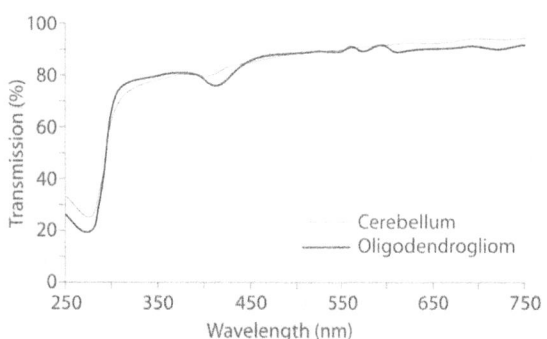

**Fig. 21.** UV/VIS spectrum. Human cerebellum: marrow support and obligodendrogliom (post mortem) UMSP 80, freeze-dried slice, slice thickness 20 μm, measuring spot size 80 μm

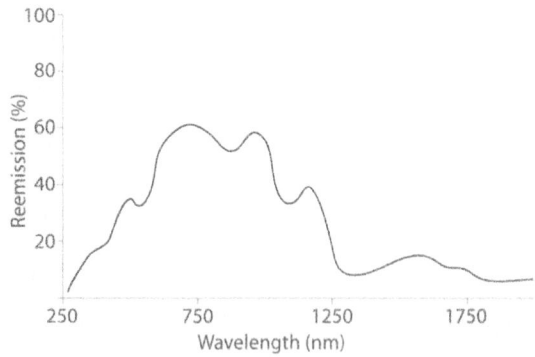

**Fig. 24.** Reemission spectrum. Human hand (in vivo, mean values from 27 light-skinned probands, reemission measuring system with Ulbricht sphere

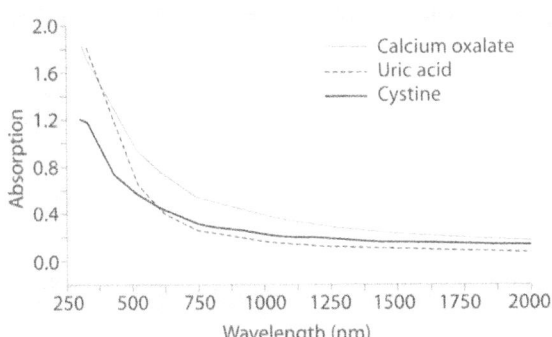

**Fig. 25.** Human kidney stones (in vitro), UMSP 80, thin sections, thickness 60 μm, measuring spot size 40 μm

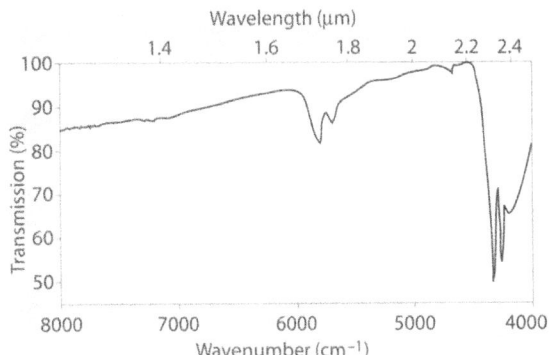

**Fig. 28.** NIR spectrum. Fatty tissue (pig, in vitro), freeze-dried slice, slice thickness 30 μm

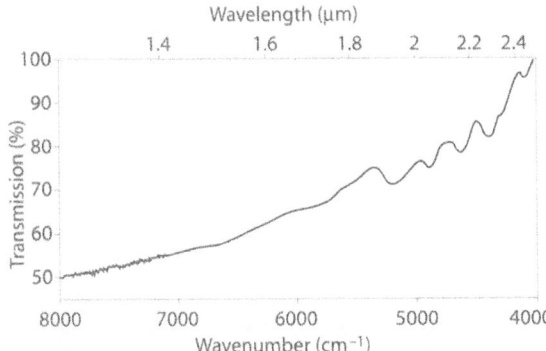

**Fig. 26.** NIR spectrum. Muscle tissue (pig. In vitro), freeze-dried slice, slice thickness 20 μm

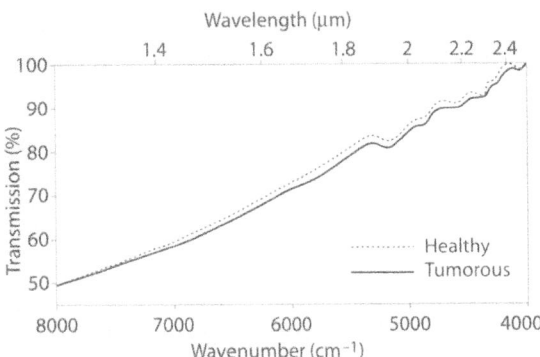

**Fig. 29.** NIR spectrum. Human tissue (colon), healthy and tumorous (post mortem), freeze-dried slices, each slice thickness 50 μm

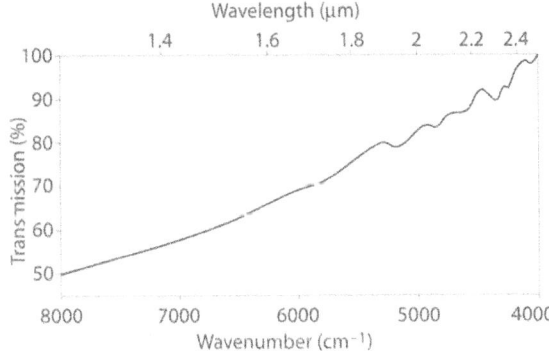

**Fig. 27.** NIR spectrum. Liver tissue (pig, in vitro), freeze-dried slice, slice thickness 20 μm

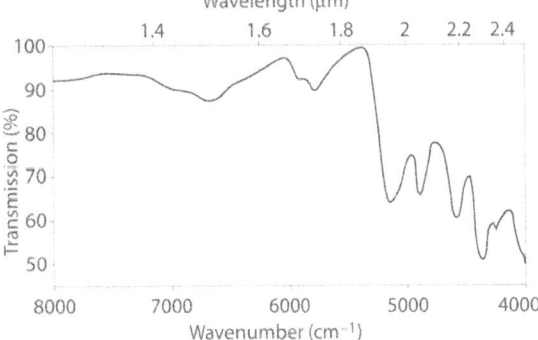

**Fig. 30.** NIR spectrum. Human abdomen skin (epithelium and top layer, post mortem), freeze-dried slice, slice thickness 100 μm

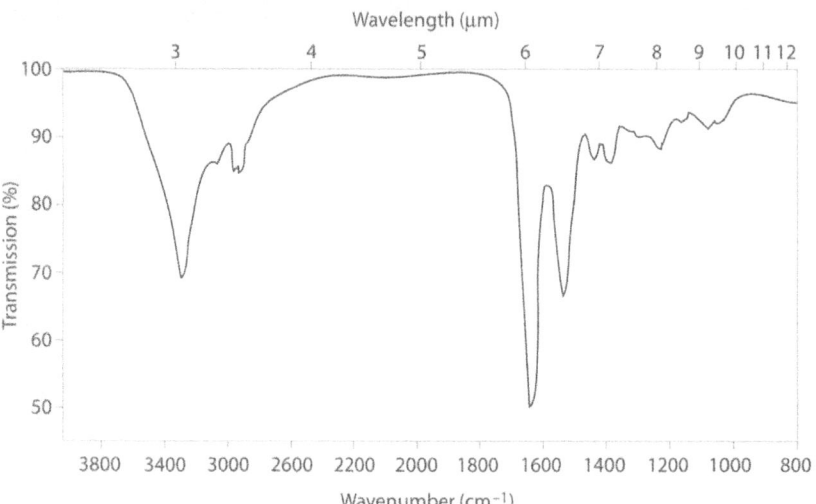

**Fig. 31.** NIR spectrum. Muscle tissue (pig, in vitro), freeze-dried slice, slice thickness 10 μm

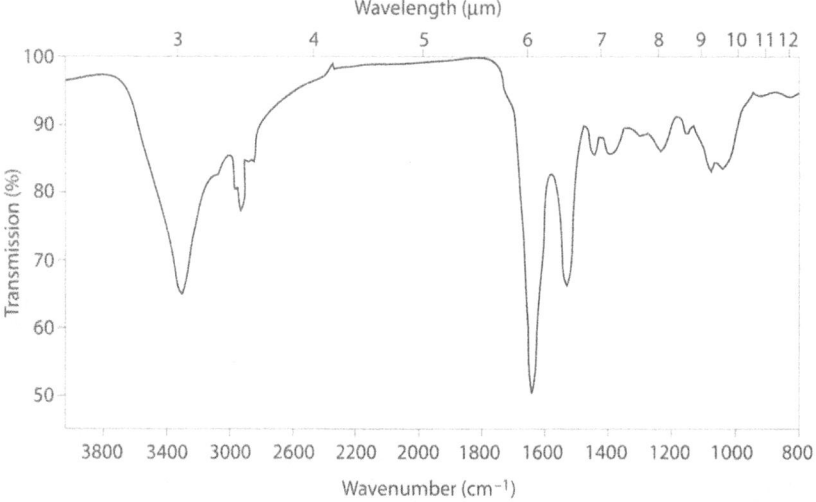

**Fig. 32.** IR spectrum. Liver tissue (pig, in vitro), freeze-dried slice, slice thickness 10 μm

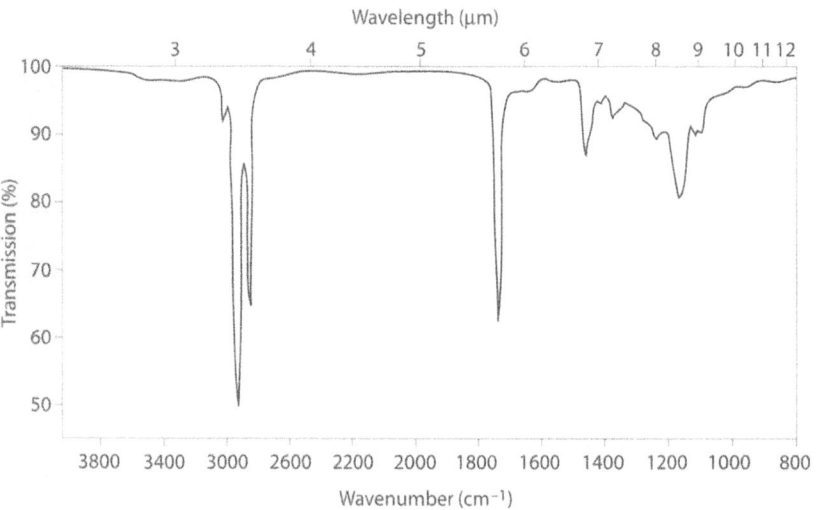

**Fig. 33.** IR spectrum. Fatty tissue (pig, in vitro), freeze-dried slice, slice thickness 10 μm

**Fig. 34.** IR spectrum. Human bone tissue (compacta), KBr mould (mixing ratio 1/250)

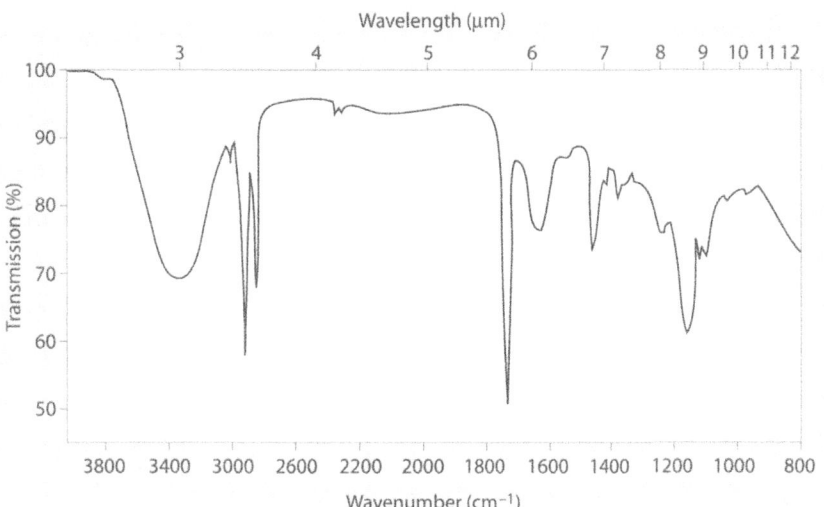

**Fig. 35.** IR spectrum. Human bone tissue (compacta), mortared and de-greased as KBr mould (mixing ratio 1/250)

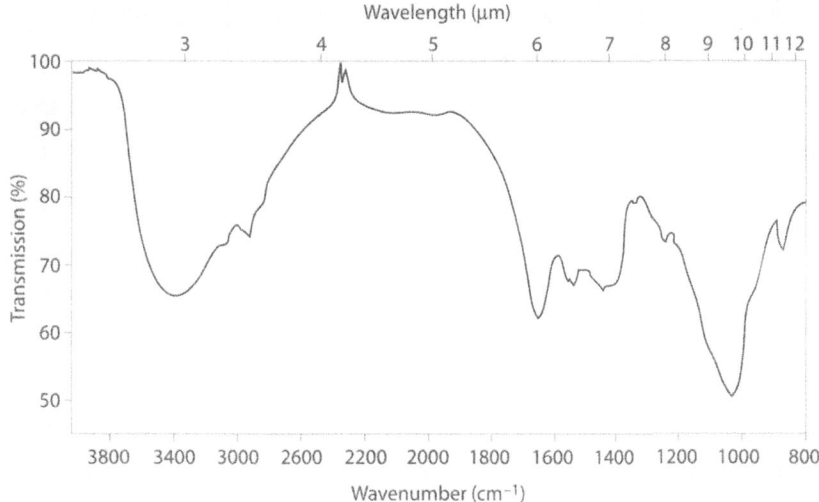

**Fig. 36.** IR spectrum. Human kidney stone (calcium-oxalate/phosphate, in vitro), KBr mould

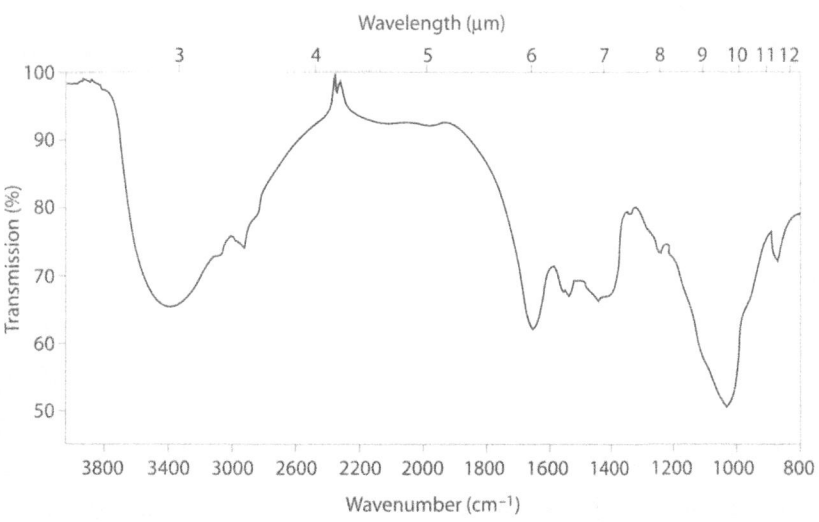

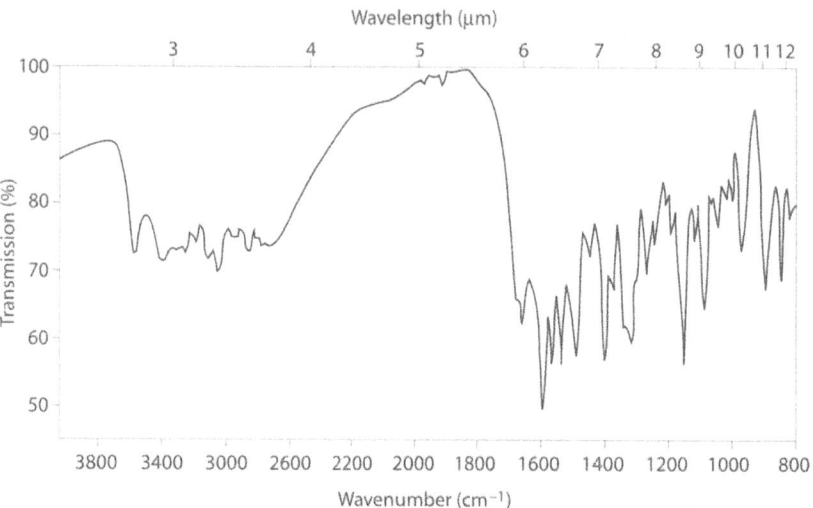

**Fig. 37.** IR spectrum. Human bladder stone (cystine, in vitro), KBr mould

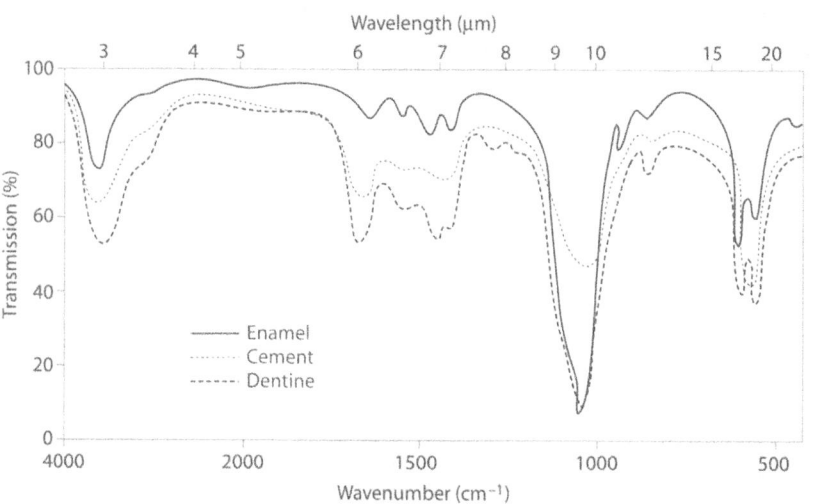

**Fig. 38.** IR spectrum. Human dental tissue (After [12])

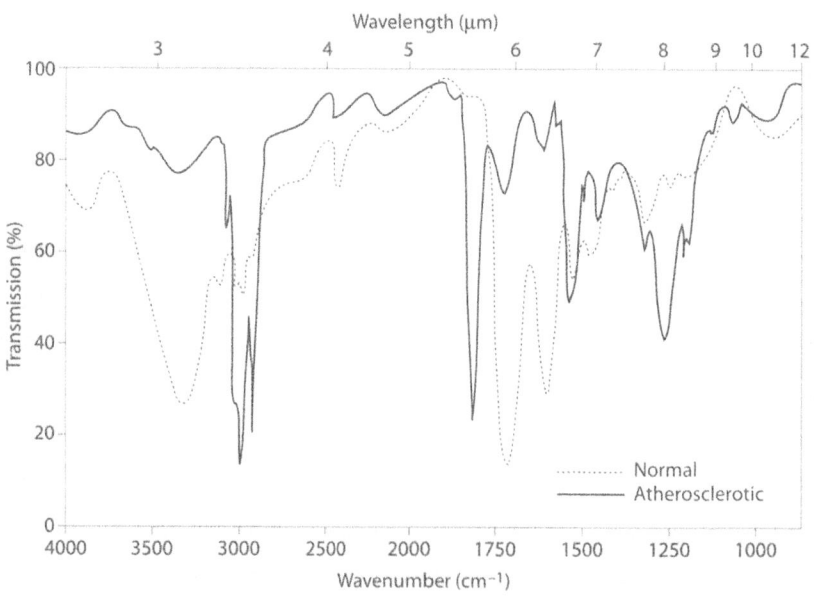

**Fig. 39.** IR spectrum. Human aorta (A. descendens, post mortem), FTIR microscope (Spectra Scope, Spectra-Tech. Inc), slice thickness 7 µm, measuring spot 600 µm

**Fig. 40.** IR spectrum, BSA [bovine serum albumine (Sigma)], film on CaF$_2$ object mount

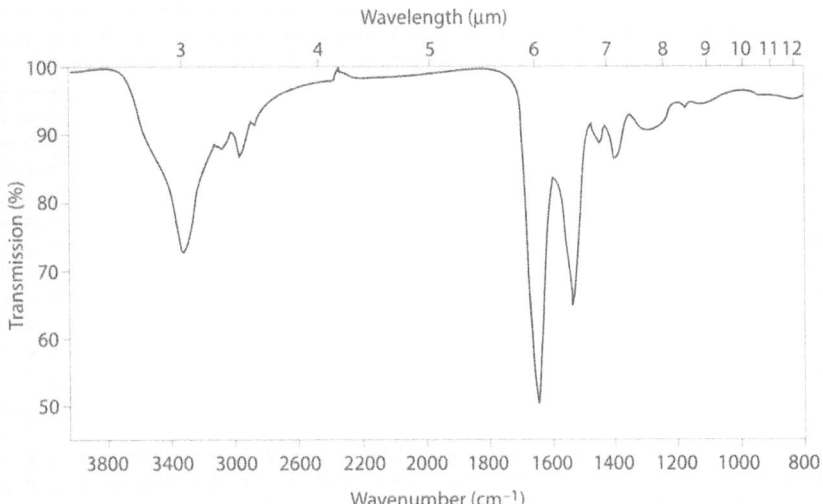

**Fig. 41.** IR spectrum. Collagene type I (calf skin), film on CaF$_2$ object mount

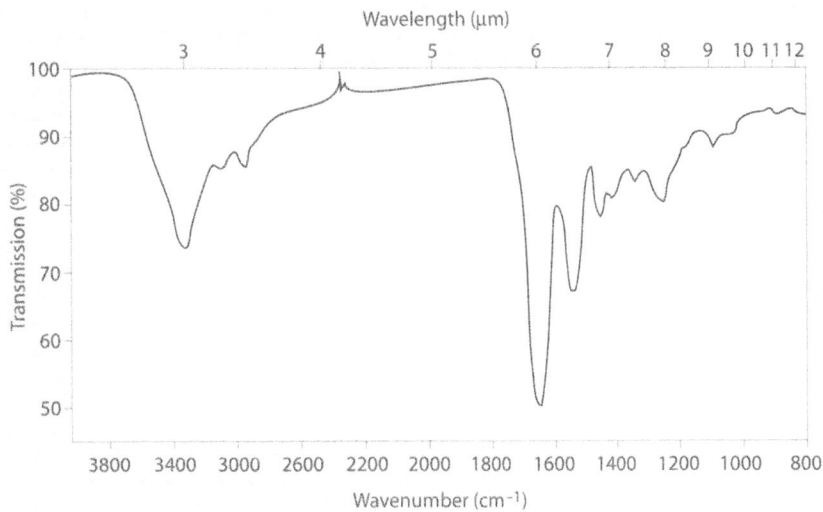

**Fig. 42.** IR spectrum. Calf thymus DNA (Sigma), film on CaF$_2$ object mount

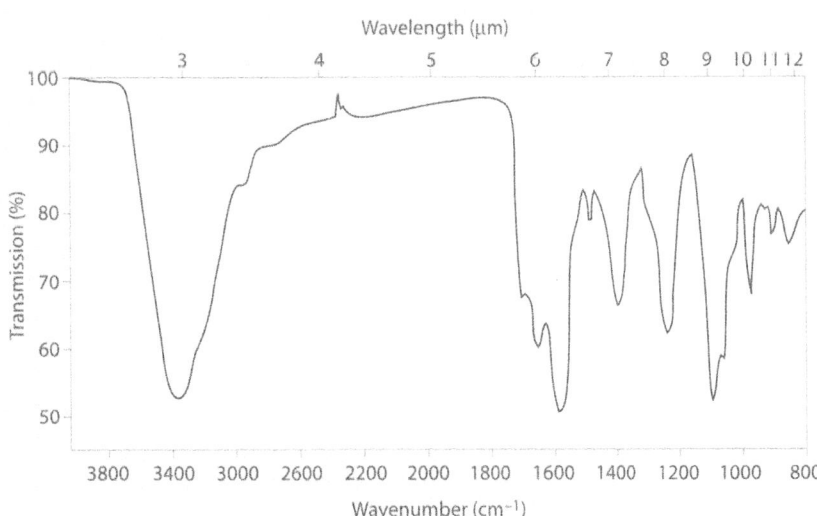

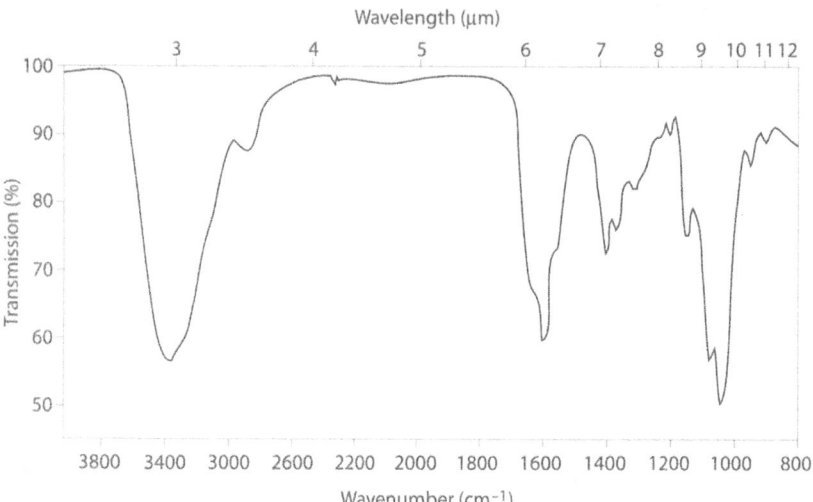

**Fig. 43.** IR spectrum. Hyaluronic acid from human umbilical cord (ICN), film on $CaF_2$ object mount

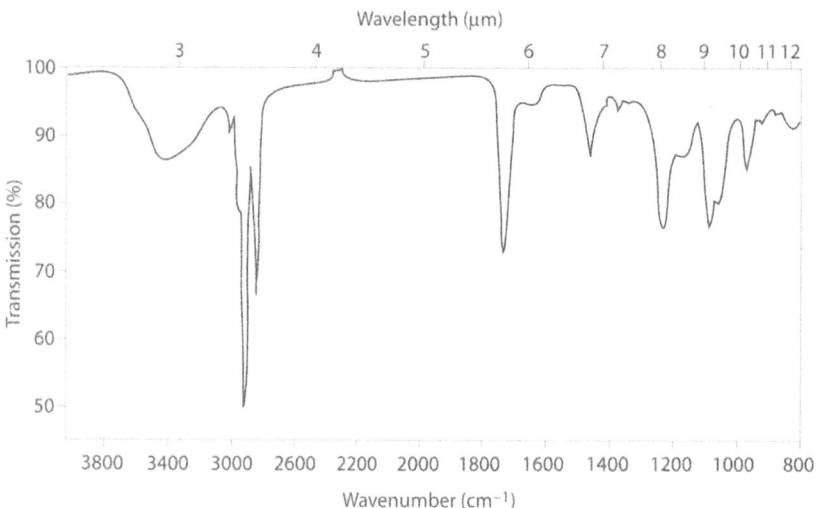

**Fig. 44.** IR spectrum. Lecithin (Merck), film on $CaF_2$ object mount

**Fig. 45.** **a** Carcinoma of the human colon, native freeze-dried slice, slice thickness 10 μm. **b** IR map in transmission, second derivative of the intensity at 1221 cm⁻¹. **c** With HE (haematoxyline/eosin) after-dried freeze-dried slice (see **a**)

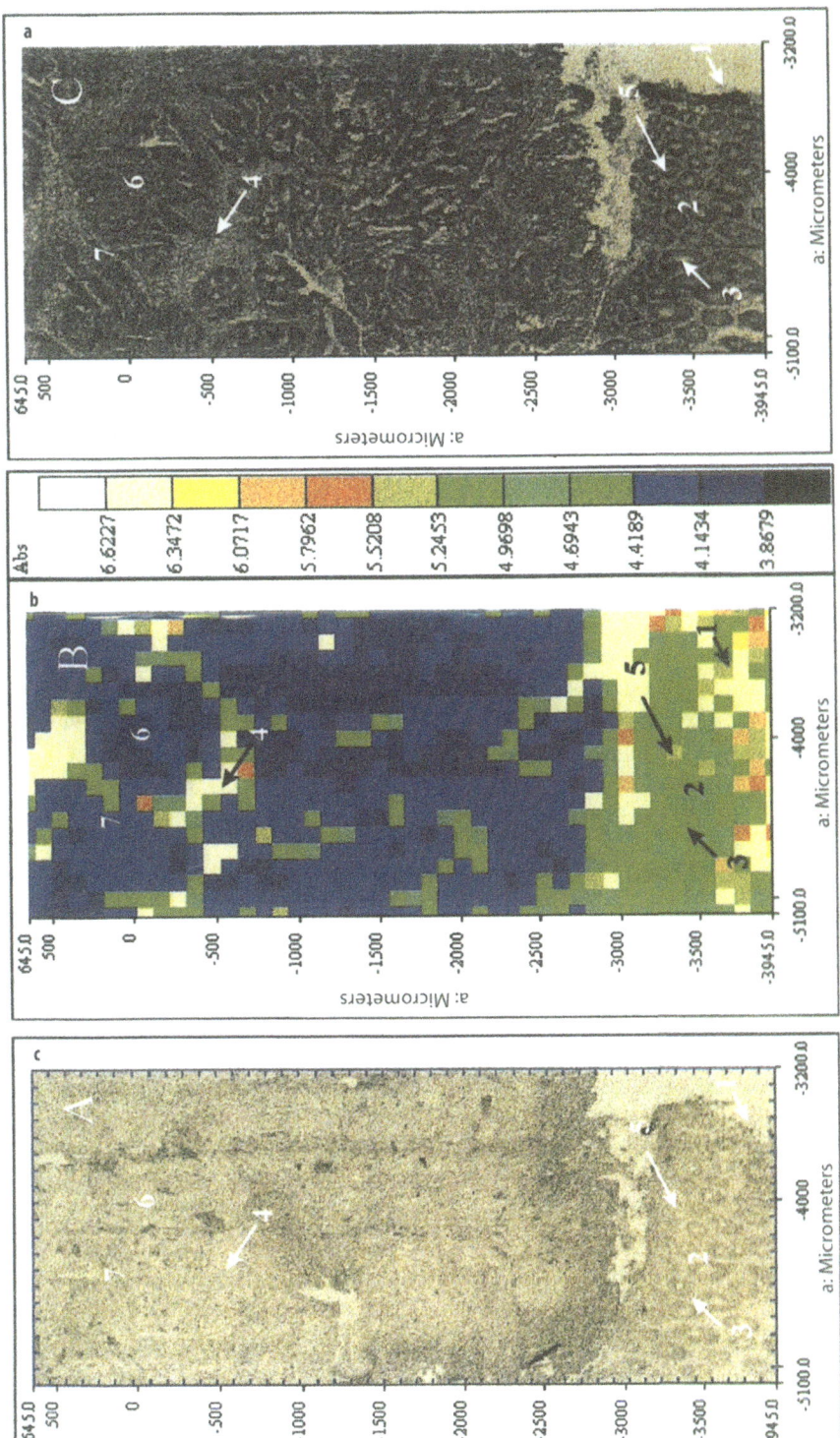

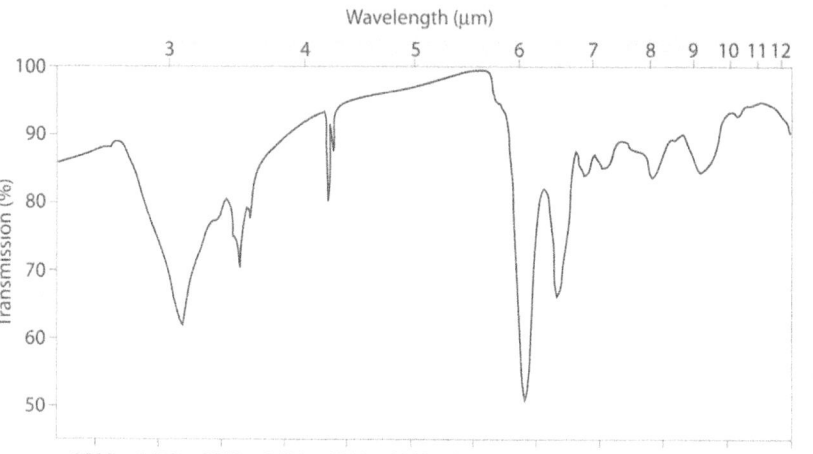

**Fig. 46.** (1) IR spectrum of the epithelium (see Fig. 45)

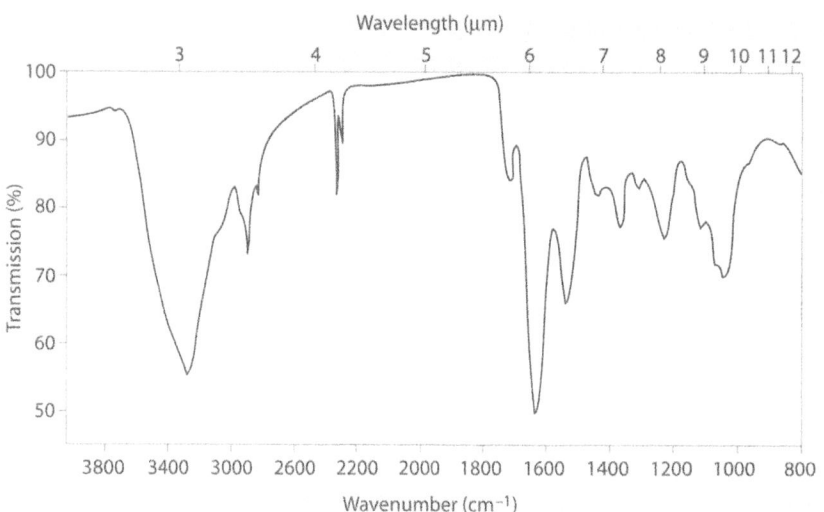

**Fig. 47.** (2) IR spectrum of crypts from colon tissue (see Fig. 45)

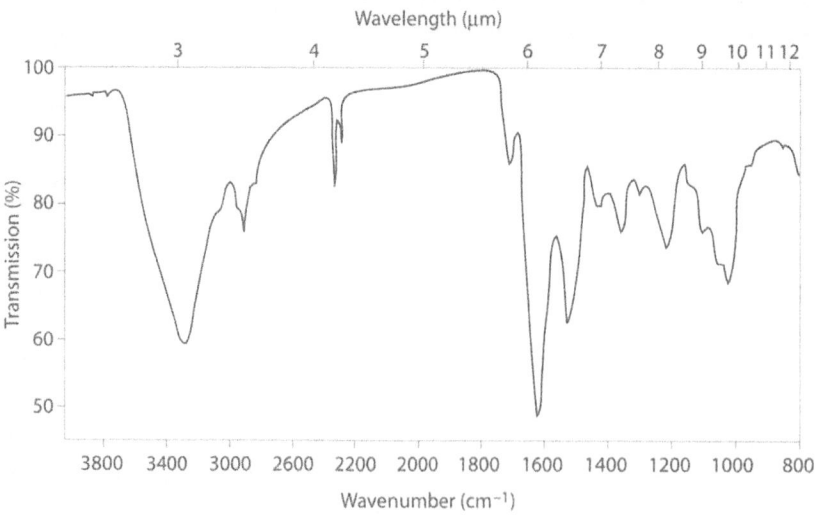

**Fig. 48.** (3) IR spectrum of the inner cryptal regions (mucus) (see Fig. 45)

**Fig. 49.** (4) IR spectrum of the stroma (see Fig. 45)

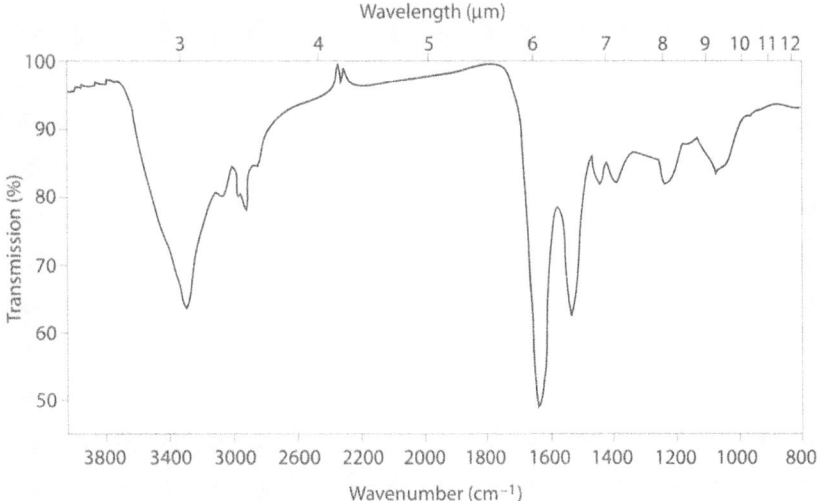

**Fig. 50.** (5) IR spectrum of the lamina propria (see Fig. 45)

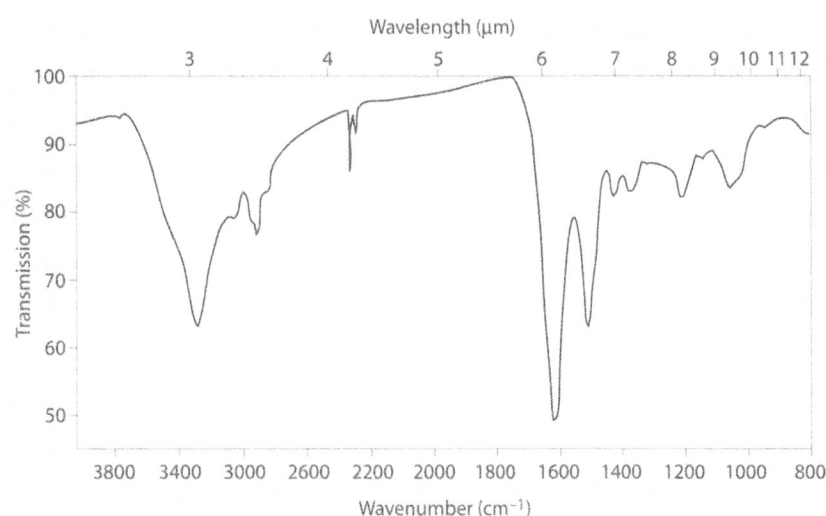

**Fig. 51.** (6) IR spectrum, of the tumorous region (see Fig. 45)

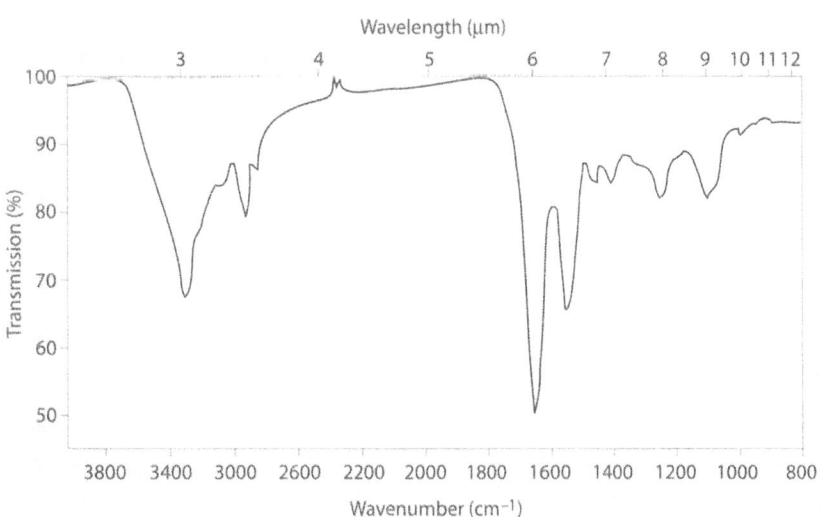

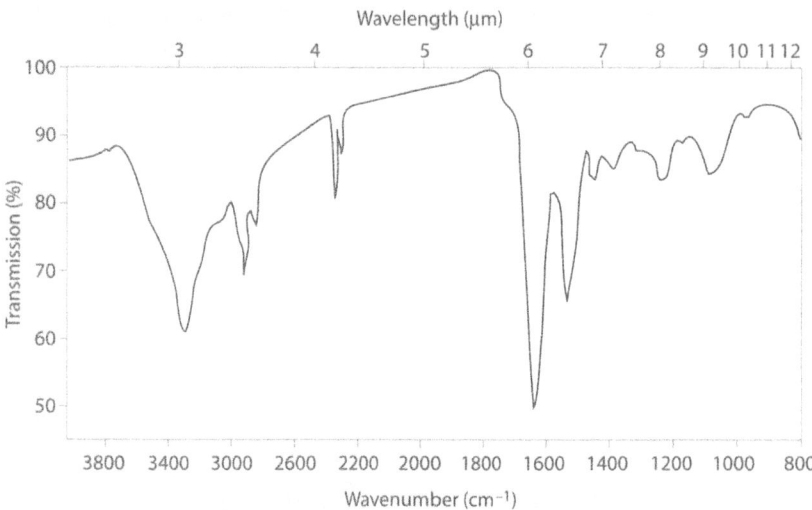

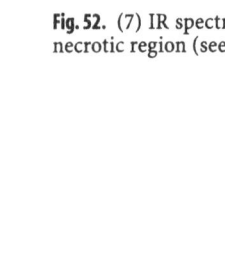

**Fig. 52.** (7) IR spectrum of the necrotic region (see Fig. 45)

## Consequences for the Application of Lasers in Medicine

Several factors, like absorption, scattering and refraction, determine the spread of laser light inside tissue, while its action is determined only by the absorption of the radiation energy by the tissue. Thermal action is characterized by the conversion of radiation energy into thermal energy, which, in turn, causes temperature increase and, consequently, thermal changes in the tissue. The irradiated tissue may remain in situ, as for coagulation, or it may be removed totally or in part from the overall tissue structure during abrasion by thermomechanical processes.

The tissue volume absorbing the radiation depends, for a given spot size, on the penetration depth of the laser radiation. The penetration depth is characterized approximately by a 1/e decrease in its intensity below the irradiated tissue surface. It also depends on the laser's wavelength and on the type of tissue involved. As discussed above, it is not straightforward to calculate the penetration depth by applying the Bouger-Lamert law of radiation transport, if scattering dominates. This is especially true for wavelengths in the region of the optical window (800–1100 nm). Here, scattering exceeds absorption by 1 to 2 orders of magnitude. This explains, for example, why radiation of the Nd:YAG laser (1064 nm) will penetrate only a few mm although its absorption in tissue is minimal (see Table 1). The influence of scattering can be neglected only for wavelengths showing extreme absorption ($CO_2$ lasers, Er:YAG lasers). In these cases, the penetration depth can be approximately estimated from the absorption in water.

Table 1 indicates typical values for the penetration depth into selected tissues for the most popular laser

**Table 1.** Depth of penetration (1/e intensity) into selected tissue types for various medical lasers. The numbers indicate only the correct order of magnitude since the published values differ greatly. [4]

| Laser | $\lambda$ (mm) | Depth of penetration (mm) | | | |
|---|---|---|---|---|---|
| | | Skin | Liver | Muscle | Blood |
| $CO_2$,[a] | 10600 | 0.01 | 0.01 | 0.01 | 0.01 |
| Er:YAG[a] | 2940 | 0.001 | 0.001 | 0.001 | 0.001 |
| Nd:YAG | 1064 | 4 | 5 | 4 | 0.8 |
| $Ar^+$ | 514 | 2 | 0.5 | 1 | 0.3 |

[a] For $CO_2$ and Er:YAG lasers the numbers represent the depth of penetration in water.

radiations used in medicine ($CO_2$, Nd:YAG, $Ar^+$ and XeCl lasers). The large differences are due to the different and wavelength-dependent absorption by compounds within the tissue. Water is the most important absorber in the mid- and near infrared region. Except for fatty tissues it comprises ca. 70% of the tissue mass. This explains the importance of the $CO_2$ lasers as well as the increasing application of Er:YAG lasers.

## Thermal Properties of Tissues

The action of the laser beam in surgery, as a tool for cutting or coagulation, is based on the conversion of electromagnetic energy into thermal energy. This conversion of radiation energy into heat will occur only if the laser radiation interacts with specific tissue chromophores, i.e. if it is absorbed. (There is no need for an absorber to be visible as a colour! For example, water is a specific absorber for $CO_2$ laser radiation ($\lambda$ = 10.6 μm). It follows, that the heat source density $q$

(W m$^{-3}$) inside an irradiated tissue volume is a function of the absorption coefficient $\mu_a$ and the total radiation density $L$. $L$ is composed of the directly contributing part of the collimated laser beam ($L_c$) and of contributions scattered in from the surrounding tissues ($L_s$):

$$q(\mathbf{r}, t) = \mu_a \, [L_C(\mathbf{r}, t) + L_S(\mathbf{r}, t)]$$
where $\mathbf{r}$ = position vector, and t = time .

The light energy, converted into heat, causes a local temperature increase inside the irradiated volume. As long as there are no phase transitions (transition of solid components into liquids or gases, vaporization of liquids), the temperature $T$ will increase proportionally the heat source density $q$. Part of the heat will diffuse by thermal conduction into the cooler surroundings: this process depends on the temperature gradient. For a given beam intensity this will limit the achievable maximum temperature of the irradiated region. Thus, a given beam intensity is associated with a certain maximum temperature rise. On the other hand, there is for each tissue, a specific intensity threshold which must be exceeded in order to reach a desired local temperature.

Since part of the energy is transported by thermal conduction and other processes into adjacent regions, not only the irradiated volume is heated, but also its surroundings. In vivo the local blood circulation will also transport heat away from the irradiated tissue. The thermal properties of living tissues are dominated by:
1. Heat conductivity,
2. Heat storage,
3. Heat dissipation via the vascular system.

## Heat Conductivity

Heat always flows from the warmer tissue regions to the colder. The heat flow dQ/dt is directly proportional to the temperature gradient. For a one-dimensional ideally homogeneous tissue sample of length $s$ and uniform cross-sectional area $A$, the heat energy $\Delta Q$ is conducted within time $\Delta t$ from a position of higher temperature T1 to a position with lower temperature T2, according to the following equation:

$$\frac{\Delta Q}{\Delta t} = \lambda A \frac{T_1 - T_2}{\sigma} \; .$$

The factor $\lambda$ is called thermal conductivity [W (m$^{-1}$K$^{-1}$)]. In general, it is low for gases and highest for metals. Table 2 summarizes typical values. The heat conductivity of liquids and solids is practically temperature-independent. For example, it increases from 0.62 W (m$^{-1}$K$^{-1}$) to = 0.64 W (m$^{-1}$K$^{-1}$) if the tem-

**Table 2.** Thermal conductivity of various materials and substances measured in W (m$^{-1}$K$^{-1}$) und standard conditions. [16, 20]

| Material | $\lambda$ (W m$^{-1}$ K$^{-1}$) |
| --- | --- |
| Air | 0.02 |
| Ethanol | 0.16 |
| Fatty tissue | $\approx 0.3$ |
| Watery tissue | $\approx 0.5$ |
| Water | 0.58 |
| Blood | 0.62 |
| Steel | 46.02 |
| Cooper | 418.00 |

perature of water increases from 37° to 57 °C. For tissues $\lambda$ ranges between 0.3 and 0.5 W (m$^{-1}$K$^{-1}$), depending on water content. To a good approximation, the following formula holds [19, 21]:

$$\lambda = (0.06 + 0.57 \, w/\varrho) \; [w \, (m^{-1}K^{-1})]$$
where $\rho$ = tissue density in kgm$^{-3}$ and
w = water content of the tissue in kgm$^{-3}$

With conversion of light energy into thermal energy, the statistical and random motion of the atoms and molecules (Brownian motion) will be accelerated. The energy transport by thermal conduction occurs towards lower temperature regions since the faster molecules in the warmer region will transfer kinetic energy by collisions with the slower molecules in the cooler tissue regions. During the collision processes the warmer molecules lose energy until finally equilibrium is established.

## Heat Storage

The specific heat c characterizes the ability of a tissue to store heat. This material constant describes the amount of heat needed to increase the temperature of a mass unit by 1 K. Typical values are summarized in Table 3. A good approximation is:

$$c = (1.55 + 2.8 \, w/\varrho) \quad [kJ \, (kg^{-1}K^{-1})] \; .$$

For temperatures leading to phase transitions (melting, vaporizing), the thermal motion increases so much that mutual attraction is insufficient to keep the atoms and molecules together. Solids lose their internal ordering in favour of the free mobility of the particles of a liquid. In the gaseous phase all particles move with high speed without influencing each all particles move with high speed and without mutual interaction. The thermal energy utilized to overcome the intermolecular forces is unavailable for a further temperature increase. Thus the temperature of the investigated volume remains constant until the phase

**Table 3.** Specific heat $c$ of some materials and substances (for gases $c_p$). [7]

| Material | kJ (kg$^{-1}$ K$^{-1}$) |
|---|---|
| Copper | 0.385 |
| Steel (V2A) | 0.477 |
| Air | 1.005 |
| Fat | 1.930 |
| Ethanol | 2.430 |
| Blood | 3.22 |
| Water | 4.183 |

transition is completed, even if heat is continuously supplied.

The heat-penetration coefficient b combines thermal conductivity and specific heat. It is a measure of the amount of heat penetrating, during a given duration, into a tissue volume after a sudden increase of the surface temperature:

$$b = \sqrt{\lambda \varrho c} \ .$$

It is also possible to characterize the temperature behaviour of a tissue sample by:

$$\chi = \frac{\lambda}{\varrho c} \ .$$

This temperature conductivity $\chi$ as defined by the above equation, has the same value for most tissues (about $1.2 \times 10^{-7}$ m$^2$ s$^{-1}$) [18]. The reason is that a decreasing thermal conductivity due to a lower water content in general is accompanied, and thus compensated for, by a decrease of the specific heat.

The general heat-conduction equation describes the entire space and time-dependent behaviour of the temperature distribution within an irradiated tissue volume:

$$\frac{dT}{dt} = \frac{q}{\varrho c} + \frac{\lambda}{\varrho c} \ \nabla^2 T,$$

$\nabla^2$ is the Laplace operator:

$$\nabla^2 = \left( \frac{\delta^2}{\delta x^2} + \frac{\delta^2}{\delta y^2} + \frac{\delta^2}{\delta z^2} \right) \ .$$

The first term describes the temperature change caused by the absorbed radiant power in the volume of interest. The second term deals with the temperature change due to heat dissipation into the surroundings. For real situations an analytical solution of the equation can be very difficult. Example: the one-dimensional temperature distribution generated if at time $t'$ an amount of heat $Q$ has been deposited at position $x'$ (the so-called pulse response [7]) is:

$$T(x,t) = \frac{Q}{2\varrho c \sqrt{\pi \lambda (t-t')}} \exp \left[ -\frac{(x-x')^2}{4\lambda(t-t')} \right] \ .$$

For a radiation field varying in space and time, the resulting heat source density $q(x', t')$ is [7]:

$$T(x,t) = \frac{l}{\varrho c \sqrt{\pi \lambda}} \int_0^t \int_{-\infty}^{+\infty} \frac{q(x', t')}{\sqrt{t-t'}} \exp \left( -\frac{(x-x')^2}{4\lambda(t-t')} \right) dx' \ dt' \ .$$

In practice, the thermal relaxation time $\tau$ can serve as a suitable parameter to estimate the time dependence of the propagation of local heating:

$$\tau = \frac{d^2 \varrho c}{\lambda} = \frac{d^2}{\chi} \ .$$

The significance of $\tau$ can be graphically interpreted:

• A cube-shaped piece of tissue having a side of length $d$ and exceeding the temperature of its surroundings by $\Delta T$, will cool to ambient temperature after the time $\tau$. The ambient temperature is thus slightly increased, i.e. the temperature gradient has disappeared in favour of a temperature increase of the surrounding tissue.

• If a short heat pulse strikes a tissue surface, a time $\tau$ will pass before a significant warming can be noticed at depth $d = \sqrt{\chi \tau}$.

## Heat Dissipation by Blood Flow and Other Mechanisms

Thermal energy dissipates from an irradiated tissue region not only by thermal conduction but also via the vascular system. As a good approximation it can be assumed that blood, having normal arterial temperature, enters the irradiated volume and is immediately warmed to local temperature inside the capillary region. Blood which leaves the venous part of the region will transport thermal energy which was stored according to the specific heat capacity of the blood components.

Whether or not the influence of the vascular region on the temperature distribution can be neglected for a given case can simply be estimated by multiplying the inverse of the product of the rate of blood throughput $v_B$ (Table 4) and the $\varrho$ density of the tissue. The resulting perfusion time $t_B$ represents the time during which the entire blood of a unit volume of tissue is exchanged. The influence of the vascular region must be taken into account if the irradiation time is of the same order of magnitude, or larger, than $t_B$ [18].

Especially during continuous irradiation, it is possible that heat transport via blood flow will be the dominating factor in maintaining a stationary temperature distribution. Without any blood flow the temperature during thermal equilibrium will decrease inversely with distance from the local heat

**Table 4.** Rate of blood throughput $v_B$ for various human organs, measured in ml $(m^{-1}g^{-1})$. [17]

| Tissue | $v_B$ (ml min$^{-1}$ g$^{-1}$) | |
|---|---|---|
| Fatty tissue | 0.012 ... | 0.015 |
| Muscle (arm) | 0.02 ... | 0.07 |
| Skin | 0.15 ... | 0.5 |
| Brain | 0.46 ... | 1.0 |
| Kidney | 3.4 | |

source $(T \propto 1/r)$. The cooling by blood flow introduces an additional factor presenting an exponential decrease. In this case, the temperature profile can be described by

$$T \propto \frac{1}{r} \exp\left( \frac{-d_{th}}{r} \right) .$$

The heat penetration depth $d_{th}$ is defined as $d_{th} = \sqrt{\chi t_B}$. Consequently it depends on the relation between heat conductivity and perfusion time [18]. The influence of blood throughput on the stationary temperature distribution is relevant only if the extent of the irradiated region is larger than the thermal penetration depth. If, on the other hand, the irradiated region is clearly smaller than $d_{th}$, then the heat transport via thermal conduction will dominate.

Furthermore, the irradiated volume can lose heat by metabolic processes, by water evaporation at the surface and by convection. Due to the relatively large time scales involved, these processes are of importance only for cw laser radiation [2, 3, 16].

# References

[1] BENEDETTI E, BRAMANTI E, PAPINESCHI F, ROSSI L, BENDETTI E (1997) Determination of the relative amount of nucleic acids and proteins in leukemic and normal lymphocytes by means of Fourier transform Infrared microspectroscopy. Appl Spectrosc 51: 792–796

[2] BIRNGRUBER R (1980) Thermal modeling in biologic tissues. In: HILLENKAMP S, PRATHESI R, SACCHI CA (eds) Lasers in biology and medicine. Plenum Press, New York

[3] CARSLAW HS, JAEGER JC (1973) Conduction of heat in solids. Clarendon Press, Oxford

[4] CHEONG WF, PRAHL SA, WELCH AJ (1990) A review of the optical properties of biological tissues. IEEE J Quant El 26: 2166–2185

[5] DEMPSEY RJ, DAVIS DG, BUICE Jr RG, LODDER RA (1996) Biological and medical applications of near-infrared spectrometry. Appl Spectrosc 50: 18A–34A

[6] GÜNZLER H (1996) IR-Spektroskopie. VCH-Verlag, Weinheim

[7] JACQUES SL, PRAHL SA (1987) Modeling optics for a slab geometry in the diffusion approximation. Lasers Surg Med 6: 494–503

[8] KOU L, LABRIE D, CHYLEK P (1993) Refractive indices of water and ice in the 0.65- to 2.5-µm spectral range. Appl Opt 32:3531–3538

[9] KUENSTNER JT, NORRIS K, KALASINSKY VF (1997) Spectrophotometry for human hemoglobin in the midinfrared region. Biospectrosc 3:225–232

[10] LERMAN S (1987) Effect of sunlight on the eye. In: BEN-HUR E, ROSENTHAL I (eds) Photomedicine 1. CRC Press, Boca Raton, Florida, pp 79–121

[11] MIE G (1908) Pioneering mathematical description of scattering by spheres. Ann Phys 25: 337

[12] NAGASAWA A (1983) Research and development of laser in dental and oral surgery. In: New frontiers in laser medicine and surgery. Excerta Medica, Amsterdam, pp 233–241

[13] ROGGAN A, MINET O, SCHRÖDER C, MÜLLER G (1993) Measurements of optical properties of tissue using integrating sphere technique. In: MÜLLER G et al (eds) Medical optical tomography: functional imaging and monitoring. SPIE-Press, Bellingham, 149–165

[14] ROGGAN A, DÖRSCHEL K, MINET O, WOLFF D, MÜLLER G (1995) The optical properties of biological tissue in the near infrared wavelength range – review and measurements. In: MÜLLER G, ROGGAN A (eds) Laser-induced interstitial thermotherapy. SPIE-Press, Bellingham, 10–44

[15] ROGGAN A (1997) Dosimetrie thermischer Laseranwendungen in der Medizin – Untersuchung der optischen Gewebeeigenschaften und physikalisch-mathematische Modellentwicklung. In: BERLIN H-P, MÜLLER G (eds) Fortschritte der Lasermedizin. ecomed, Landsberg

[16] SVAASAND LO (1982) Properties of thermal waves in vascular media: application to blood flow measurement. Med Phys 9: 711–714

[17] SVAASAND LO, BOERSLID T, OEVERAASEN M (1985) Thermal and optical properties of living tissue. Lasers Surg Med 5: 589–602

[18] SVAASAND LO, GOMER CJ, WELCH AJ (1989) Thermotics of tissue. In: MÜLLER GJ, SLINEY DH (eds) Dosimetry of laser radiation in medicine and biology. SPIE-Press, Bellingham, 133–145

[19] TAKATA AN, ZANEVELD L, RICHTER W (1977) Laser-induced thermal damage in skin. USAF School Aerospace Med, Rep SAM-TR-77-38, Broks AFB, TX.

[20] WEAST RC, ASTLE MJ (eds) (1982) Handbook of chemistry and physics. CRC Press, Boca Raton, Florida

[21] WELCH AJ (1984) The thermal response of laser-irridiated tissue. IEEE J Quant Elec QE-20: 1471–1481

[22] WILSON BC, ADAM G (1983) A Monte Carlo model for the absorption and flux distributions of light in tissue. Med Phys 10: 824–830

# I-3.2
# Interactions of Laser Radiation with Biological Tissue

R. Steiner

## Contents

## Introduction

The interactions of the laser light with biological matter are determined by its irradiation parameters, such as wavelength, power/energy and pulse length. However, they also depend on the „optical" properties of the biological matter such as, for example, absorption, scattering and reflection of the laser radiation. The absorption determines the energy transfer of the laser radiation into heat. The scattering of the light within the tissue defines the reaction volume. The exposure time of the laser radiation and the laser power also determine the laser-tissue reactions. Those reactions can be classified into non-thermal, thermal (coagulation and vaporization of the tissue, respectively), ablative and optomechanical reactions. Lasers affect soft as well as hard tissues. Figure 1 summarizes the therapeutic and diagnostic applications of lasers.

**Fig. 1.** Therapeutic and diagnostic applications of lasers in medicine. The applications are ordered according to the laser actions.

## Optical Properties of Tissues

It is necessary to understand the optical properties of tissues in order to understand the thermal reactions of laser radiation in tissue. These reactions depend on the wavelength and the penetration depth.

### Light Absorption and Scattering

To initiate a thermal reaction, the photons must be absorbed. Beer's law states that the light intensity in tissues is attenuated exponentially with depth, as described by the absorption coefficient $\mu_a$ (cm$^{-1}$). The inverse value $1/\mu_a$ is called the mean free path of the photons. Typical values in muscle tissue as a function of wavelengths are summarized in Table 1. Table 2 presents the appropriate laser types.

In the visible spectral range the photons will be repeatedly scattered before absorption. In this case, the scattering will not be isotropic but show a distinctly forward component. This anisotropy is expressed by the anisotropy factor $g$. having values between 0 and 1. The value 0 indicates isotropic scattering, all scattering angles are of equal probability; 1 indicates forward scattering. For soft tissues the values for $g$ range from 0.7 to 0.98. For theoretical considerations the reduced scattering coefficient, defined as $\mu_s = (1 - g)\,\mu_s$, describes the influence of anisotropy. Sometimes, to characterize the domination of the scattering or absorption, the ratio of scattering coefficient to the sum of scattering and absorption coefficients (albedo) is utilized. Photons are scattered, for example, inside the dermis on average 100 times before they are absorbed. The optical constants completely characterizing a tissue are summarized and listed below (Table 3).

It is possible to measure the light distribution inside tissue by introducing a miniaturized probe into the tissue via a hyperdermic syringe. The measurements confirm the theoretically derived phenomenon, that the light intensity directly below the tissue surface is enhanced by a factor of 2 to 4 as compared with the intensity of the incident beam [5].

The increased intensity is caused by backscattered photons overlapping with the incident photons. Another observation is that, due to the scattering effect, the penetration depth depends on the irradiated area. Consequently, the penetration depth will double if, for the same irradiance, the beam diameter increases from 1 to 5 mm. For dermatological applications this effect must be taken into account.

Since tissues consist mainly of water, the energy coupling of laser light into them is dominated by the absorption in water. This is especially the case in the infrared spectral range. In the visible wavelength range, absorption by blood and other pigments predominates. Figure 2 indicates the absorption spectrum.

The penetration depth of laser light into tissues is greatest in the wavelength range of 700–900 nm. Argon lasers and dye lasers utilize as their therapeutic effects (coagulation, vaporization) absorption by blood. For this reason, it is possible to treat vascular changes selectively.

The Nd:YAG laser has several emission lines, the most important being at 1064 nm. At this wavelength the penetration depth is great, consequently large tissue volumes will be heated, i.e. coagulated or vaporized. An application for this laser type is the recanalization of tumours within the gastro-intestinal tract and concurrent coagulation of the blood vessels in the border regions. Due to high absorption in water

**Table 1.** Penetration depth of laser light as a function of the absorption coefficients $\mu_a$ and $\mu_{eff} = \mu_a + \mu_s$, the sum of absorption and scattering coefficients. [1]

| Wavelength (nm) | Pentration depth $1\,\mu_a$ (µm) | Eff. opt. penetration $1\,\mu_{eff}$ (µm) |
|---|---|---|
| 193 | ca. 10 | ca. 1 |
| 309 | 50 | 6 |
| 532 | 830 | 240 |
| 1 064 | 2500 | 1900 |
| 2 060 | 286 | 250 |
| 2 940 | 3 | 3 |
| 10 600 | 17 | 17 |

**Table 2.** Common laser types for application in therapeutic medicine

| Laser type | Wavelength (nm) | Mode | Power (W) | Tissue reaction coagulation/vap. |
|---|---|---|---|---|
| COµ | 10 600 | cw/pulsed | 0–100 | (+)/++ |
| Nd:YAG | 1064/1320 | cw | 0–120 | ++/+ |
|  | 1064/1440 | Pulsed | 0– 40 (pulsed) | +/ablation |
| Argon$^+$ | 488/514 | cw | 0– 20 | +/+ |
| Dye | 490–790 | Pulsed 1–10 Hz | 0– 4 (mean) | +/–– |
| Holmium | 2100 | Pulsed, 1–10 Hz | 0– 20 (mean) | +/ablation |
| Er:YAG | 2940 | Pulsed, 1–30 Hz | 0– 15 (mean) | +/ablation |
| Excimer | 308 | Pulsed |  |  |

**Fig. 2.** Absorption spectra of water and of haemoglobin

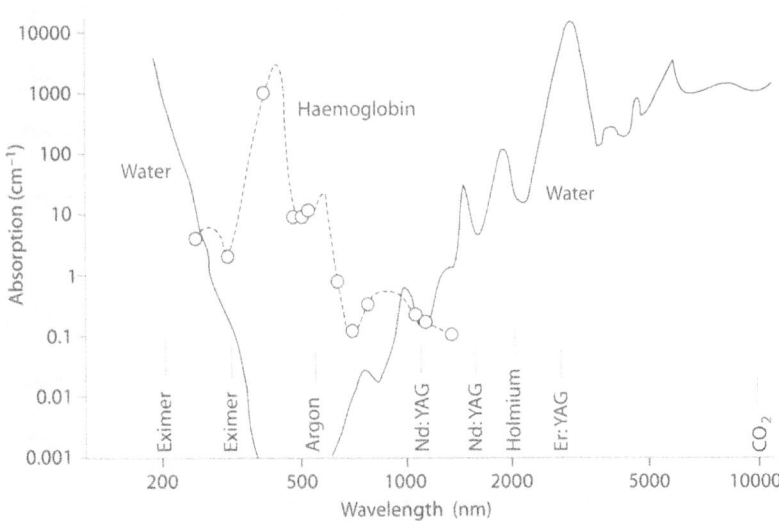

**Table 3.** List of optical constants

| Absorption coefficient | $\mu_a$ (cm$^{-1}$) |
|---|---|
| Scattering coefficient | $\mu_s$ (cm$^{-1}$) |
| Anisotropy factor | $g$ |
| Reduced scattering coeff. | $\mu_{s'} = (1-g)\,\mu_s$ |
| Effective attenuation | $\mu_{aff} = (\mu_a + \mu_s)$ |
| Albedo | $q = \mu s/(\mu_s + \mu_a)$ |

($\mu_a = 12\,000$ cm$^{-1}$), the radiation of the Er:YAG laser penetrates only a few µm. Consequently, this laser is well suited for the abrasion of soft and hard tissues without causing thermal necrosis. The $CO_2$ laser radiation at 10.6 µm wavelength is also strongly absorbed and thus well suited as a cutting tool.

## Interactions of Laser Light with Tissues

As indicated in Fig. 3, the exposure time together with the laser power determine the particular reaction mechanisms.

During continuous irradiation (> 0.25 s) of 30 J/ cm$^{-2}$ in total 1; tissues are easily coagulated. After shortening the interaction time to $10^{-3}$ s, the heat produced cannot diffuse into the tissue, so local overheating occurs and the tissue vaporizes at temperatures > 300° C. If the exposure time is reduced to several hundred microseconds then the time is insufficient even to vaporize the tissue and it will ablate explosively into fragments. If the time is shortened into the ns range ($10^{-9}$ s), optical breakdown will occur, i.e. within the laser focus a plasma will be created.

**Fig. 3.** Interactions of laser light depending on the exposure time

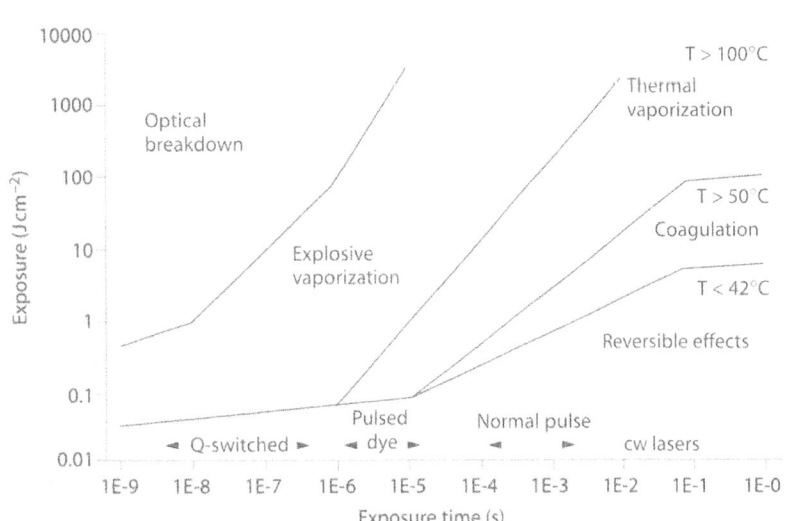

This plasma will expand and produce a cavitation bubble which subsequently collapses. Caused by this rapid thermal expansion, a shock wave will be generated, leading to optomechanical actions which will fracture concrements inside the body but also induce undesirable side effects. These reaction mechanisms will be explained in more detail.

## Nonthermal Reactions

During nonthermal reactions the quantum energy of the laser photons is utilized to excite certain specific molecules. This mechanism is the basis of photodynamic laser therapy. So-called sensitizers, i.e. dye molecules which predominantly accumulate in tumorous tissues, will be excited by the absorption of the laser light. After an internal energy transfer, this excitation energy will be transmitted to oxygen molecules. Singlet-oxygens of radicals are formed which are a cell toxin. In this case, a purely phototoxic effect is utilized.

## Thermal reactions

All who have experience with medical laser systems will utilize them for cutting or coagulating tissues. During coagulation the tissues pass through various stages of thermal damage until vaporization occurs at several hundred °C. The various thermal reactions of tissues are summarized in Table 4.

As already mentioned, thermal laser reactions are always accompanied by thermal necrosis, which depends on three parameters [4]:
1. The effective and wavelength-dependent penetration depth of the laser light.
2. The thermal diffusion inside the tissues, about $1.2 \times 10^{-7}$ $m_2/s^{-1}$ and
3. The exposure time $\tau$ of the laser light, its pulse length and consequently its power density $W/cm^2$.

For high power densities (irradiance) the necrotic zone will decrease but it still depends on the light penetration depth. Recently this has been carried to extremes by the Silk Touch, a scanner which moves the focused beam of a $CO_2$ laser very rapidly over the tissues in order to reduce the exposure time. This instrument is utilized to abrade the skin and in this way finally to smooth it via the healing process.

The other extreme would be laser – induced thermotherapy (LITT). In this case, the laser light – having lower power (6–12 W) – would slowly warm the tissue and coagulate it by a progressing thermal necrosis. The laser light would be guided into the tissue by an interstitially applied catheter.

Applications are in tumour therapy (brain tumours, metastases of the liver) and benign prostate hyperplasia (BPH).

As shown in Figs. 4 and 5 the coagulated regions are well presented in histological preparations.

During vaporization, a carbonized edge will be generated; behind this is a vascular zone. This, in turn, is followed by a region with irreversible thermal damage. The necrotic zone (see Fig. 5) and conse-

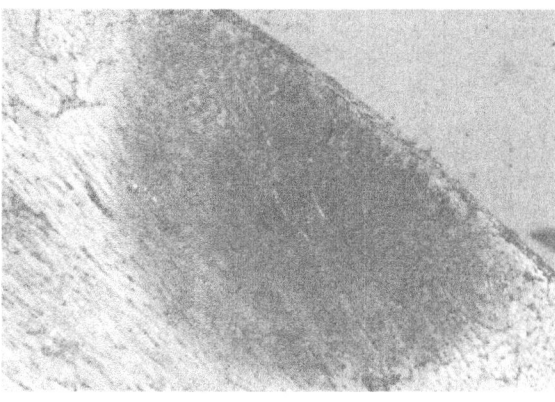

**Fig. 4.** Coagulation of muscle tissue (polarization microscope image)

**Fig. 5.** Vaporization of muscle tissue by the $CO_2$ laser, including necrotic zone

**Table 4.** Thermal effects at different temperatures during laser irradiation

| <40 °C | 40–50 °C | 60–65 °C | 80 °C | 90–100 °C | > 300 °C |
|---|---|---|---|---|---|
| Nonthermal effects | Enzymatic changes | Protein denaturation | Collagen denaturation | Tissue Vaporization | Carbonization |

quently the blood coagulation effect decreases with increasing irradiance of the $CO_2$ laser from several hundred µm thickness to below 100 µm.

## Ablative Effects of Laser Light

A prerequisite for the ablation of soft and hard tissues is high absorption of the laser light in the tissue, in order to reach the ablation threshold with the initial fraction of the pulse energy alone. The shorter the pulse, and consequently the higher the power density of the laser beam, the lower the ablation threshold. The ablation process and the occurrence of tissue fragmentation starts after 2 µs (see Fig. 6) and continues for free-running lasers with a pulse length of up to 500 µs, until the pulse energy drops below the ablation threshold.

The remaining energy will heat the tissue in the region of the penetration depth of the laser light (a few µm) and can even propagate by thermal diffusion [2]. For this reason, the necrotic zone is wider than the theoretical penetration depth of the light. Since during vertical laser irradiation the tissue fragments are ejected towards the laser beam, the radiation may be significantly absorbed by the ablation products. This is indicated in the irradiance/ablation plot by a nonlinear relationship and not, as expected, by a linear relation.

Due to the high absorption of water, tissues can be ablated principally by UV light (excimer lasers at 193 and 308 nm), by visible laser light (in the presence of absorbing pigments) and by radiation in the IR spectral region. The high precision of laser ablation allows its application in ophthalmology, surgery of the middle ear (auditory bones), dentistry (Fig. 7) and dermatology.

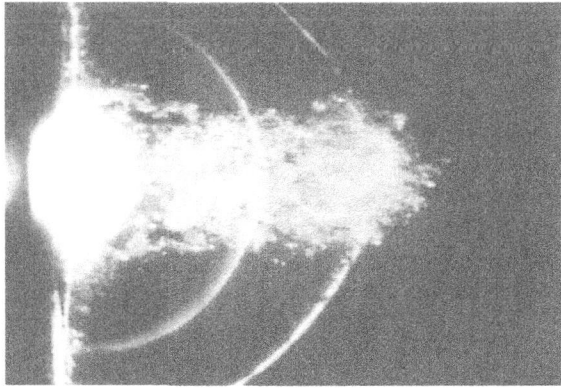

**Fig. 6.** Ablation process 2 µg and 5 µg after beginning of pulse (image from Schlieren microscope). The expansion of the shock wave is clearly visible. The tissue is ablated with ultrasonic speed.

**Fig. 7.** Ablation of hard dental tissue with the Er:YAG laser at 2.94 µm.

## Opto-Mechanical Laser Actions

For laser pulses in the ns region (Q-switched Nd:YAG lasers) and shorter, the electric field strength in the beam focus is so high that electrons are extracted from the atoms. A plasma develops which will be further heated by the laser pulse. The thermal expansion will create a cavitation bubble reaching its maximum diameter (up to 1 cm, depending on the pulse energy: see Fig. 8) after about 400 µs.

The dynamics of this process, initiated at the surface of body concrements, will fragment the stones. The shock wave itself contains only a small fraction of the pulse energy (about 4 to 8% [3]. A similar reaction can be achieved by the absorption of short laser pulses at the surface of stones (Alexandrite lasers, 750 nm, 150 ns – 1 µs; flash lamp-activated dye lasers, 1–50 µs). In this case, cavitation bubbles will develop following the same time schedule as for ns pulses.

The difference lies in the initiation of the processes, either without the absorption condition or with the necessary surface absorption. The cavitation bubble in Fig. 8 has a diameter of 6 mm and was created by a dye laser pulse having a length of a few µs. Up to now, laser lithotripsy has been unable to get established. One of the reasons is that it is a semi-invasive procedure (it is necessary for the probe to approach the stone via endoscopy). Another is the price of the laser instrumentation; it cannot compete with mechanical vibration devices.

In the meantime, understanding of the reaction mechanisms of the various laser types with tissue has grown significantly and optimum parameters for medical applications have been established. Nevertheless, the physician needs plenty of experience and even more patience to apply the laser properly.

**Fig. 8.** Cavitation bubble at the surface of a gall stone. [3]

## References

[1]  BOULNOUS, J-L (1986) Photophysical processes in recent medical laser developments: a review. Lasers Med. Sci: 47–66

[2]  DÖRSCHEL, K (1993) Thermische und nichtthermische Gewebewirkungen. In: BERLIEN H-P, MÜLLER, G (eds) Angewandte Lasermedizin. ecomed, Landsberg

[3]  IHLER, B (1992) LaserLithotripsie-Untersuchung der in-vitro-Fragmentierung mit Mikrosekunden-Impulsen. Dissertation Universität Karlsruhe

[4]  STEINER, R (1994) Thermal and Non-Thermal-Laser. Dissection. End Surg 2: 214–220

[5]  STEINER, R, MELNIK, IS, KIENLE, A (1993) Light penetration in human skin: in-vivo measurements using isotropic detectors. SPIE 1881: 222–230

# I-3.2.1
# Photochemical Effects

R. Senz

## Contents

## Introduction

Almost all living beings need light. Since the formation of the first single cell organisms, they have utilized light to initiate chemical reactions. Even today, the human body contains rudiments of these light-sensitive systems (mitochondria). These systems will be illustrated by three examples. The idea of stimulating (catalysing) photochemical reactions with laser light is nothing new, in principle. It repeats some of the processes of evolution.

**Fig. 1.** Survey of photochemical applications in medicine

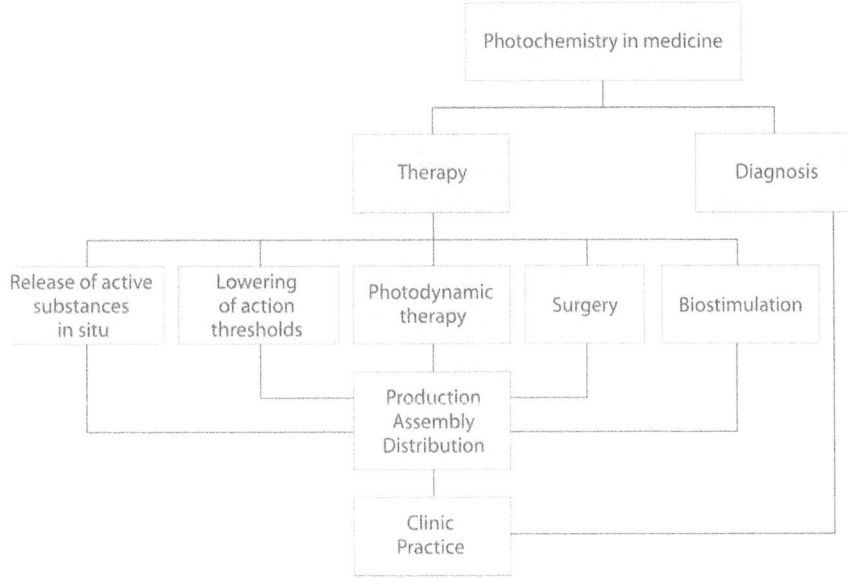

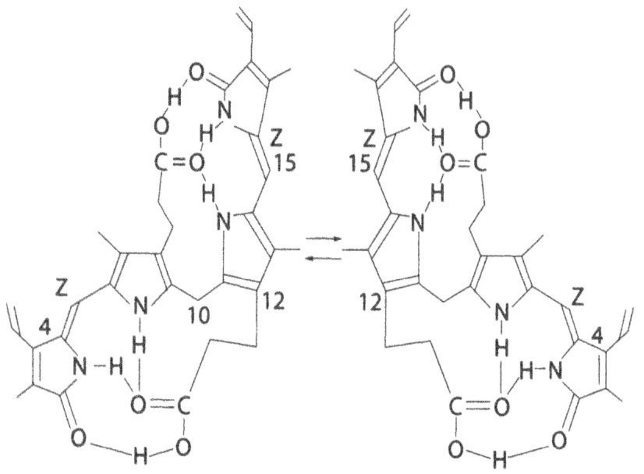

**Fig. 2.** Photochemical isomerization of bilirubin

## Photoinduced Isomerization

One example for this process is the conformational change of bilirubin by light. It is applied to treat hyperbilirubinaemia occurring, for example, in neonates. This is the currently used, so-called blue light therapy. Such an irradiation changes the conformation of the bilirubin in such a way that normal metabolistic processes are able to excrete this product (Fig. 2).

## Photoinduced Dissociation

For example, haematoporphyrine derivatives ($H_pD$) form singlet oxygen ($O_2$ at a high energy level). This will consequently destroy cells by forming hydroxy radicals which react with organic molecules (Fig. 3). However, the exact mechanism is still not fully understood. Such an action is employed in tumour therapy. By utilizing a photosensitizer, such as, e.g. di-haematophorphyrinether and/or -ester, which will accumulate predominantly in the region of malignant cells, it is possible to destroy such cells selectively. This will occur after appropriate time needed for bleaching the healthy tissue. The healthy tissue will be largely spared.

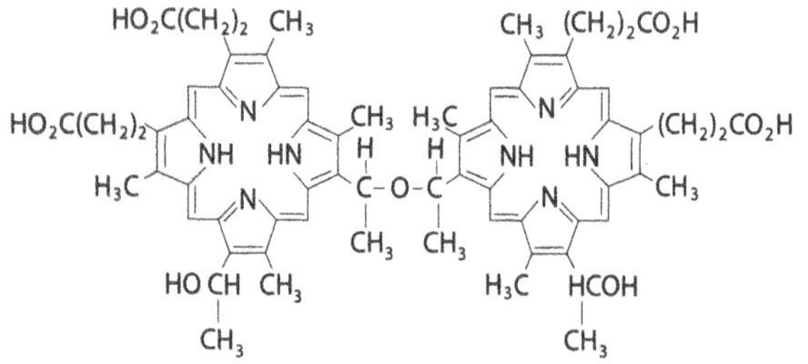

**Fig. 3.** Chemical structure of di-haematoporphyrinether

## Photoinduced Synthesis

A typical example is the photosynthesis in plants. UV-induced melanin production is also worth mentioning.

## Biostimulation

There are many publications dealing with biostimulation, i.e. with nonthermal, photochemical effects. However, very few publications are concerned with concrete ideas about the relevant biochemical mechanisms. The majority of the publications about biostimulation are purely speculative.

The condition for photochemically induced stimulation is that the laser radiation is absorbed within the tissue. To start a reaction it is necessary to influence a sufficient number of cells simultaneously. Up to now it is still unknown which wavelength is optimal for biochemical reactions in the biostimulating region. Wavelengths in the range of 442–1064 nm have been investigated.

One of the first questions arising in this connection regards the differences in the biochemical reactions initiated by laser radiation and by light from conventional sources. Since practically all biological systems display wide absorption bands, it should not be of primary importance to work with narrow spectral linewidths. In addition, any existing collimation – and polarization – will disappear to a large extent after passing through a few cell layers. Nevertheless, a few hypotheses assume that the special properties of laser radiation are necessary to initiate biostimulation. Laser radiation is not only monochromatic but also coherent and of high intensity. Each of these characteristics could be achieved by a conventional light source, but not all of them together in one system. Consequently, it is necessary to compare all the results achieved by a laser with those of a conventional light source having limited bandwidth and high intensity. One hypothesis to explain laser action is that the photon statistics of the laser radiation cause the biostimulation. The laser emits photons continuously due to the principle of the feedback amplifier. Other thermal light sources have different photon statistics (Bose-Einstein distribution). For example, the photons will not be emitted continuously but in „packets". Possibly, the time interval during which a biological system can be activated is very short, thus calling for a continuously emitted photon flux.

In vitro studies report on laser-specific reactions. There are very many studies in the field of practical applications of biostimulation, some of which are controversial. Some authors describe the treatment of various cell cultures with laser light, mostly helium-neon (633 nm) or gallium arsenide (904 nm) or a combination of both. Some apply continuous-wave (cw) lasers, others pulsed systems.

A plausible explanation of why these wavelengths are best suited is that the penetration depth into tissues is a maximum. However, there are no systematic studies indicating a correlation between wavelength and biological effect. In addition, quite different energy densities are applied to the tissues. This also explains the very different results achieved by various working groups.

Biostimulation lasers are employed for very different indications. Cure rates should be judged sceptically, because in general only a few patients have been treated. In addition, the spontaneous cure rate for these ailments reaches 50%. Up to now, no published studies could prove, in a controlled and randomized double-blind investigation, any clinical effectiveness surpassing a placebo effect. Current knowledge indicates that laser stimulation is best compared with other physical therapies such as stimulation by electric currents, magnetic field therapy or X-ray stimulation therapy.

## Photodynamic Therapy

For many years now there have been research activities in the field of photodynamic therapy (PDT), especially with haematoporphyrin ($H_pD$) as a photosensitizer. $H_pD$ ist a dye produced from blood, which accumulates in carcinomas (see. Fig. 3). Clinical studies on patients have already been performed, especially in the US. The potential for photodynamic therapy is enormous. Recent publications regarding photodynamic therapy are concerned with:

- Better understanding of the photodynamic processes involved: vascular or tumour cell hyperthermia and development of dosimetry tables for photodynamic therapy.
- New drugs: endogenous sensitizers (haemoglobin, melanin, flavin, carotenoid, porphyrin), exogenous sensitizers (porphyrin, porphin, chorin, diazos, melanin, carbocyanin, merocyanin 540, phthalocyanin).
- Drugs in new manifestations: microemulsions, liposomes, lipoproteins, special antibodies, drug combinations (e.g. vascular modified and immune modified).
- New light sources: pulsed metal-vapour lasers (gold/copper vapours), photo flash-pumped dye lasers, photo flash-pumped YAG lasers, diode lasers, high-energy arc lamps.
- New transfer systems: new light guides, e.g. for $CO_2$ lasers.

- New fields of indication: viral illnesses, psoriasis lesions, extracorporal blood cell analysis, cell biology, stone shattering, recognition of arteriosclerotic plaques.
- In ophthalmology: $H_pD$ and argon dye lasers to treat intraocular and periocular tumours, tumours of the conjunctiva.
- Combination therapy: hyperthermia/PDT, chemotherapy/PDT.

The following list indicates a summary of the advantages and disadvantages of photodynamic therapy on patients, according to knowledge current and probable findings in the near future.

### Photodynamic Tumour Therapy

#### *Advantages*

- High selectivity
- Non-toxicity of drugs employed
- Useful for both diagnosis and therapy
- May be repeated.

#### *Disadvantages*

- Photosensitizing of the skin
- No systematic applications possible (e.g. for leukaemia)
- Penetration depth of the light is limited
- Complexity of the laser systems.

It is possible that other HPD derivatives, or some synthetically produced phthalocyanines, are better suited. Synthetic drugs have a very high degree of purity and thus are better than HPD, which is extracted from a natural source (predominantly from cattle). They can be more effective and much more specific; in addition, they show less side effects. It would also be very useful to develop drugs with an absorption maximum at wavelengths longer than 630 nm, because at longer wavelengths the transmission through tissues improves. The greatest problem with the presently available substances is that the patient must remain inside a totally darkened room. This is due to the fact that the substances are not sufficiently specific to accumulate only in tumour cells. If the patient were exposed directly to sun-light or to light from fluorescent lamps, the surface of his entire body would be damaged. The concentrations of the photosensitizers in normal tissues will decay sufficiently after 2–3 days for the patient to have direct contact with sunlight. This necessary isolation is most likely to be one of the greatest problems for tumour patients.

# I-3.2.2
# Thermal Effects

J. Helfmann

## Contents

## Introduction

The thermal effects of laser radiation in medicine are in essence vaporization (cutting) and coagulation of tissues. This is achieved by various lasers with energy densities from a few $W/cm^{-2}$ to $10^6 W/cm^{-2}$ and irradiation times from milliseconds to several seconds (see. Fig. 1).

The various thermal effects of laser radiation can be described according to the temperature achieved and the duration of its action. Depending on the optical parameters of the tissue at the applied wavelength, these tissue temperatures can be achieved by different powers and irradiation durations.

**Fig. 1.** The power density – pulse length diagram presents all the various laser actions on a diagonal. Thermal action finds its place, with respect to pulse length as well as necessary power density, between the photochemical reactions and the nonlinear effects.

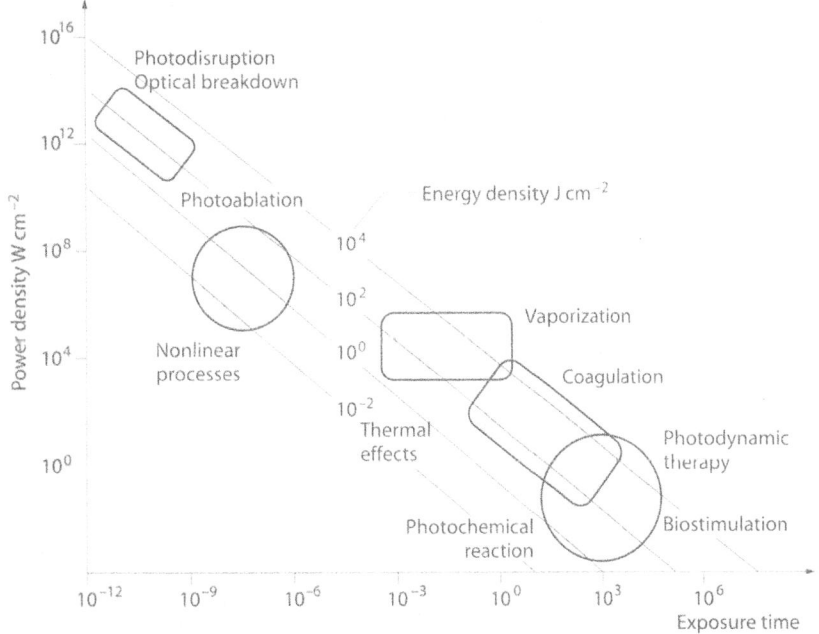

## Action of Laser Radiation

The thermal action of laser radiation in tissues is based on the absorption of the radiation and on the conversion of the laser energy into heat. The absorbency depends on tissue type and the lasers wavelength (see chapt. I-3.1 and I-2.4.1-6).

In the UV, the VIS and the NIR spectral ranges, the laser light will be mainly absorbed by the optically active electrons of the atoms. After that it will be converted into heat by collisions and nonradiative relaxation. In the IR range the radiation will be absorbed by excitation of rotational and vibrational states of the molecules. By atomic and molecular absorption and following relaxation of the excited particles, the optical energy will be converted into thermal energy. Depending on the material's heat capacity, the thermal energy will cause a certain temperature rise.

The amount of radiation energy absorbed decreases with the depth inside the tissue (Lambert-Beer law. See Chapt. I-3.1), i.e. the thermal energy and consequently the temperature, will decrease with depth. A fraction of the light will be scattered out of the forward direction and absorbed or reemitted lateral to the side of the beam path.

At the same time, heat will be dissipated by thermal conduction and by the blood flow (see Chapt. I-3.1; [2]). The above processes achieve temperature gradients in the forward and lateral directions. The decisive role in generating a given tissue temperature by laser radiation is played by the optical and thermal properties of the tissues involved.

## Temperature Dependence of Tissue Reactions

Figure 2 indicates a classification of the thermal reactions as a function of temperature. Accordingly, for up to 45 °C, no irreversible tissue damage is to be expected, for coagulation starts at 60 °C. Above 300 °C tissue can vaporize and can thus be cut.

However, it must be realized that tissue reactions will also depend on the length of time during which the temperature acts on it. This is demonstrated in Fig. 3 for irreversible tissue destruction. Accordingly, tissue destruction will be the same after heating for a short time (1s) to 70 °C as for a – heating duration of (10s) to 58 °C.

In addition, it must be taken into account that the optical, thermal and mechanical properties of tissues will change during laser irradiation, i.e. while the tissues are being heated. Figure 4 indicates this in detail for the various thermal effects.

For example, carbonization is important for radiation absorption because it causes an increased absorption of the laser radiation and thus leads to a

| Temp. (°C) | Tissue effects |
|---|---|
| 37° | No irreversible tissue damage |
| 40 – 45° | Enzyme inductions<br>Oedema formation<br>Membrane disintegration<br>and time-dependent cell death |
| 60° | Protein denaturation<br>Coagulation and necrosis |
| 80° | Collagen denaturation<br>Membrane defects |
| 100° | Desiccation |
| above 150° | Carbonization |
| 300° > | Evaporation, vaporization |

**Fig. 2.** Laser actions as a function of temperature

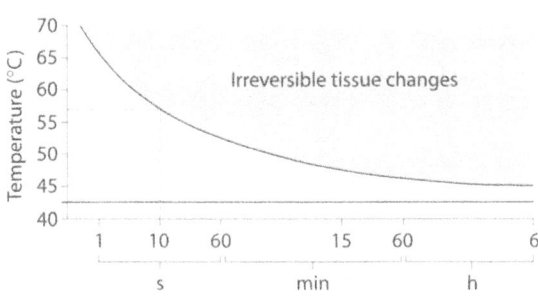

**Fig. 3.** Influence of temperature and exposure time on the irreversible destruction of tissues. (After Moritz and Henrique, published by Krizek, Hoopes and Steenburg)

rapid rise to high temperatures. The same effect is achieved by drying of the tissue. This decreases heat conductivity and leads to local thermal energy accumulation.

The various thermal effects of laser radiation and tissues never occur separately but, as during cutting, take place all at the same time.

If radiation enters homogeneous tissues, its power will decrease with increasing depth. It will be partly scattered and absorbed; accordingly a temperature gradient will develop inside the tissue. In the region where the temperature exceeds 300 °C the tissue will vaporize. This will be followed by a zone with a temperature exceeding 150 °C where the tissue will be carbonized; the next tissue layer will be coagulated. Even further out there will be a region with tissue very slightly warmed, hence suffering no irreversible damages.

Figure 5 indicates schematically the thermal effects caused by decreasing temperatures as a function of depth and lateral extension.

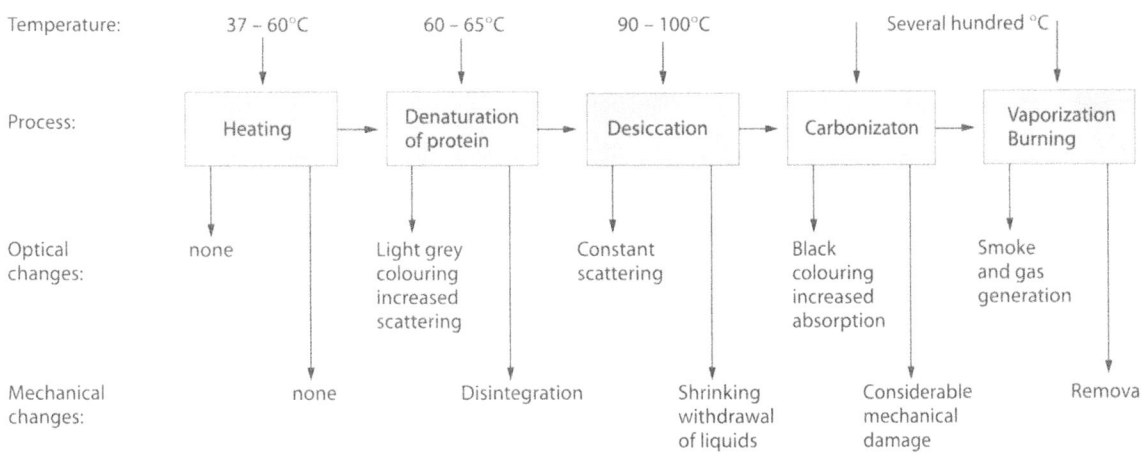

**Fig. 4.** Changes of the optical, thermal and mechanical properties of tissues during laser irradiation

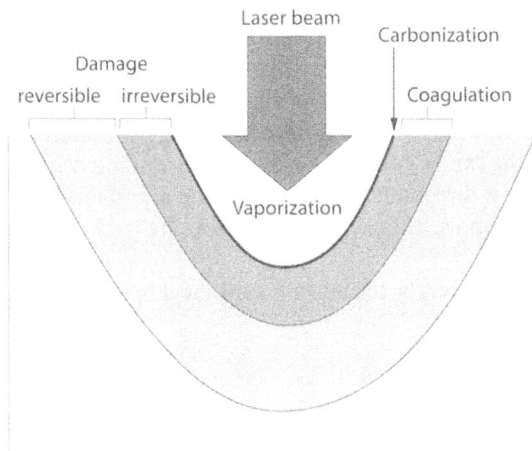

**Fig. 5.** The various action zones during the cutting of tissues

Of course, the extent of the individual zones, like cutting width, carbonization margin, coagulation margin and zone of reversible tissue changes, depend very much on the penetration depth of the laser wavelength utilized; besides many other parameters. Figure 6 presents these parameters in detail [1].

During cutting (vaporization) of tissues by the $CO_2$ laser, for example, the residual products play an important role in setting the width of the carbonization and coagulation margins. If the residual products are removed from the cutting channel by a gas jet, then the carbonization margin will be reduced. This will achieve a very narrow coagulation margin of about 100 μm as compared with more than 1000 μm without a gas jet (see. Fig. 7).

Likewise, the thermal effects will be influenced by the condition of the tissue surface: whether it is dry,

**Fig. 6.** The various parameters influencing the tissue reactions during laser irradiation

1. Coupling of the laser beam to the tissue surface (absorption inside tissue, absorption inside vapour)
2. Tissue conversion by the incident radiation (heating, vaporization, ionization, mechanical destruction)
3. Ejection of tissue materials (convection, heat conduction, condensation)

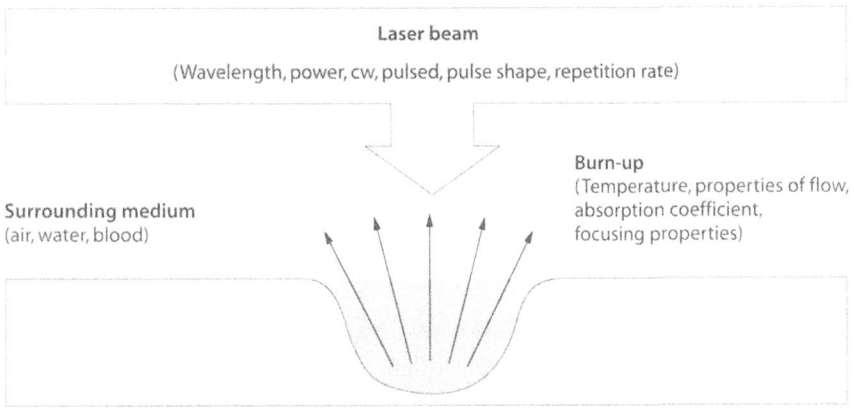

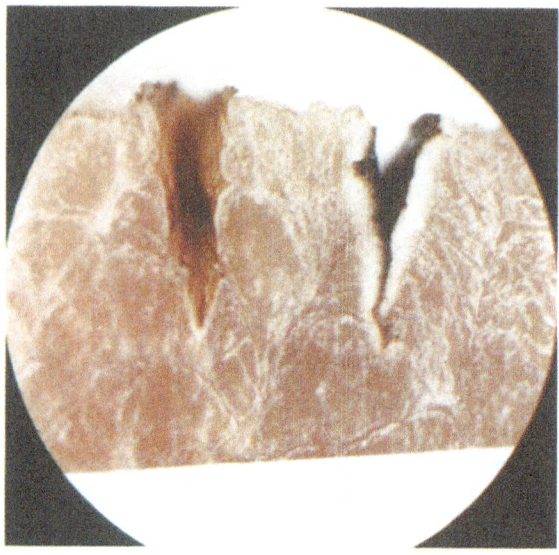

**Fig. 7.** The actions of a cw CO₂ laser on muscle tissue. Left: Including flushing with gas; right: no flushing with gas. Without flushing the section shows considerable carbonization and a coagulated border with a width of about 1 mm. In the left part of the section a strong jet was guided into the cutting channel. Consequently, the cutting power increased and the necessary laser power could be reduced by half. Carbonization is minimal and the coagulation border is about 0.1 mm wide.

moist or with blood; also by the type of tissue, each presenting different optical and thermal properties. Of further importance is the power output of the laser and the time regime of the laser radiation: cw or pulsed with various pulse durations.

Equation (1) presents the cutting depths achievable by vaporization.

Starting with the basic assumption that the energy per unit volume needed to vaporize tissue is independent of the tissue parameters, then the cutting depth achievable by a thermal laser should be proportional to the irradiation time and the applied power density (intensity)

$$T(x,y) = \int I_{(t)}^{(x,y)} \, dt \, , \qquad (1)$$

where $T(x,y)$ is the cutting depth and $I(x,y)$ the power density at the position $(x,y)$ and for the irradiation time $t$ (Fig. 8).

During drilling (lateral cutting speed $v = 0$) and for constant power density the following relation holds:

$$T(x,y) = I(x,y) \cdot t \, . \qquad (2)$$

In this case, the depth of the drilled hole reflects the power density of the laser beam, as presented in Fig. 8. If the diameter of the focal spot, created by a lens having the focal length f, is doubled by doubling

the focal length, then the power density will be reduced to a quarter of its initial value. Such a modified laser beam can be expected to produce a hole with a quarter of the depth but of double diameter.

In general, the relation between focal spot diameter, power density and drilling depth is:

$$T \sim I \sim 1/d^2 \, . \qquad (3)$$

However, during cutting $(v > 0)$ the power density $I(x,y)$ at a tissue position $(x,y)$ will be a function of time, even if the power density of the laser beam remains constant (see Fig. 8).

Replacing $dt = 1/v \, dx$ in Eq. (1) and for $v = $ constant $> 0$ one obtains:

$$T(y) = \tfrac{1}{v} \int I_{(x)}^{(y)} \, dx \, . \qquad (4)$$

Any spreadout of the intensity I in x direction will not influence the penetration depth because the integral over the x direction will be summed again.

Changing the focal spot size of a laser beam as described above by using lenses having various focal lengths can be conceived as spreading out intensity in x and y directions. According to the arguments presented above, this will lead to spreading only in the y direction.

Consequently, the cutting depth will be:

$$T \sim \sqrt{I} \sim \tfrac{1}{d} \, . \qquad (5)$$

In contrast to Eq. (2), the cutting depth is inversely proportional to the cutting width and not proportional to the power density.

Figure 9 indicates the experimental confirmation of Eq. (4). The cutting depth depends linearly on $1/d^2$. It can be estimated from the measurements that about 4–6 J of laser energy are needed to abrase 1 mm³ of tissue.

Looking at the achievable cutting depth in practice, it must be appreciated that tissue is a biological substance. It varies in its structure and will distort and change its shape during laser action.

Especially for very narrow and deep cuts, the theoretically expected cutting depth will not be achieved. The distortion of the tissue will cause additional substance to be abrased from the walls of the cut. In addition, this effect will be enhanced by any trembling movements of the hand.

Below are presented the essential effects caused by the currently most important thermal lasers.

- CO₂ laser $\lambda = 10.6 \, \mu m/9.6 \, \mu m)$
- Nd:YAG laser $\lambda = 1.06 \, \mu m)$
- Argon laser $\lambda = 488 \, nm/514 \, nm)$
- Dye laser $\lambda = 490–790 \, nm)$

**Fig. 8.** Schematic view of drilling and cutting of tissues

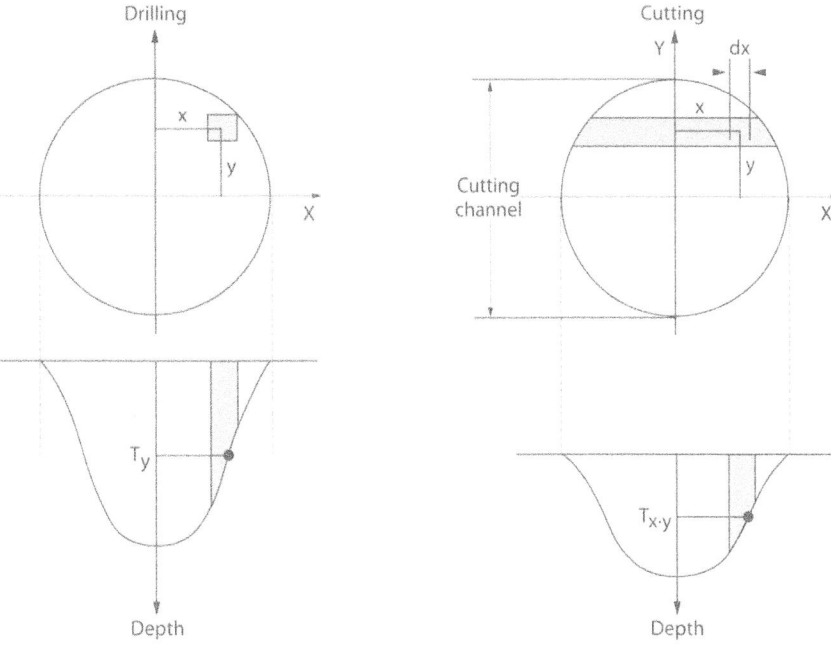

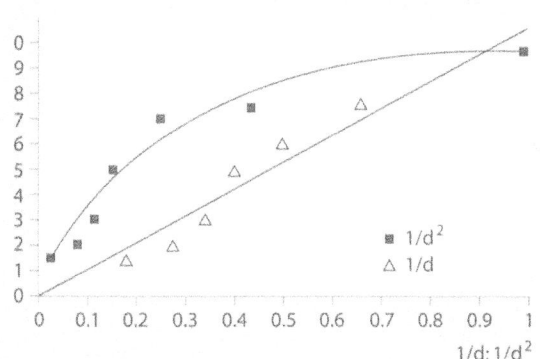

**Fig. 9.** The experimentally determined cutting depth by a laser is proportional to 1/2 d ($\triangle$) and not to $1/d^2$ and to the power density of the beam

Basically, the action of a particular laser is determined by the absorption at its actual wavelength or the inverse of its wavelength, i.e. by its penetration depth. Typical values for various biological substances are presented in Chapter I-3.1.

## CO₂ Lasers

The cutting action of this laser depends on its absorption in water. The penetration depth for the radiation of $CO_2$ laser into the tissue is less than 1/10 mm. The light absorbed by this thin layer will be converted into heat. This will rapidly raise the temperatures above 300 °C and thus vaporize the tissue.

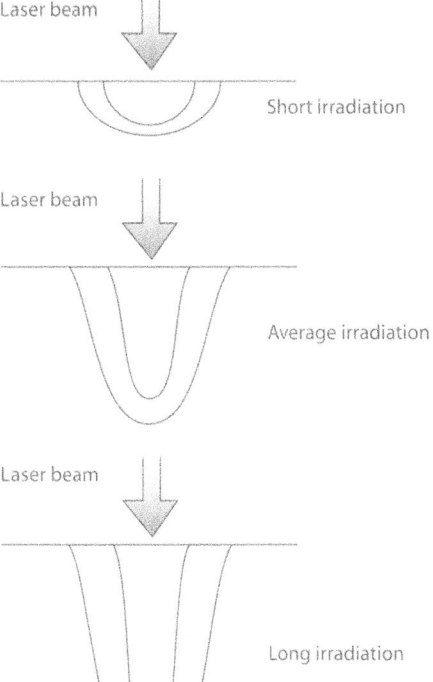

**Fig. 10.** Schematic display of the $CO_2$ laser actions

This special feature allows its application as a cutting laser successively abrasing one thin layer after the other. The cutting speed depends on the power density of the $CO_2$ laser.

The $CO_2$ laser coagulates tissues poorly, so that only vessels having a diameter of up to 0.5 mm can be closed safely. The following diagrams present the features of this laser regarding skin and muscle tissues.

## Nd:YAG Lasers

Compared with $CO_2$ lasers, the radiation of Nd:YAG lasers penetrates more deeply into the tissues, i.e. the associated effects differ considerably. Due to poor absorption and strong scattering, it is impossible to vaporize tissues instantly by applying average powers (30 W).

For this reason, only tissue coagulation can be observed. Extending the irradiation time, the tissue temperature will be raised to 100 °C and it will dry out. This drying out causes the heat conductivity to decrease and the temperature will climb further. If the irradiated surface becomes desiccated and starts to carbonize, the absorbency of the tissue will change. This means that the laser beam will be completely absorbed inside a thinner tissue layer, consequently the tissue will suddenly start to vaporize.

Even long exposure times by a low-power Nd:YAG laser will not achieve vaporization (cutting), because the energy will be completely transferred into the surrounding tissue. In this case, the tissue will merely be coagulated.

Due to the great penetration depth into tissue, a Nd:YAG laser beam can close vessels with diameters of up to 5 mm by coagulation and shrinkage.

Employing the usual hand-held attachments, efficient cutting will start (at low cutting speeds) for power outputs exceeding 70 W.

## Argon Lasers

The penetration depth of argon laser radiation is between 0.5 and 2.5 mm, this will place it between the $CO_2$ and the Nd:YAG laser.

## Dye Lasers

Using a dye laser it is possible to vary the wavelength and consequently adjust its penetration depth between a certain maximum or minimum (see Fig. 12).

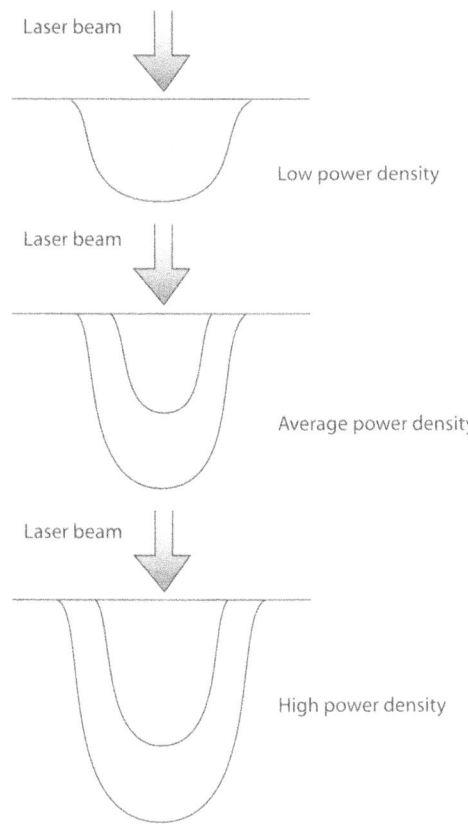

**Fig. 11.** Schematic display of the Nd:YAG laser actions

## References

[1]  BOULNOIS, IL (1986) Photophysical processes in recent medical laser developments: A review. Lasers in Medical Science, Vol. 1
[2]  CARRUTH JAS, McKenzie AL (1986) Medical lasers. Science and clinical practice. Bristol and Boston

## Further Reading

[1]  McKENZIE AL How far does thermal damage extend beneath the surface of $CO_2$ laser incisions? Phys. Med. Biol. Vol. 28, 905–912
[2]  STERN J, ENDERS S, FRANK F (1998) Laser in der Chirurgie – Grundlagen und Perspektiven. Biologische Wirkung des thermischen Lasers. Chirurg 59: 61–67
[3]  SVAASAND LO, BOERLID T, OEVERAASEN M (1985) Thermal and optical properties of living tissue: Application to laser-induced hyperthermia. Lasers in Surg Med 5: 589–602

**Fig. 12.** Absorption spectra of haemoglobin, melanin and xanthophyll of the chromophore

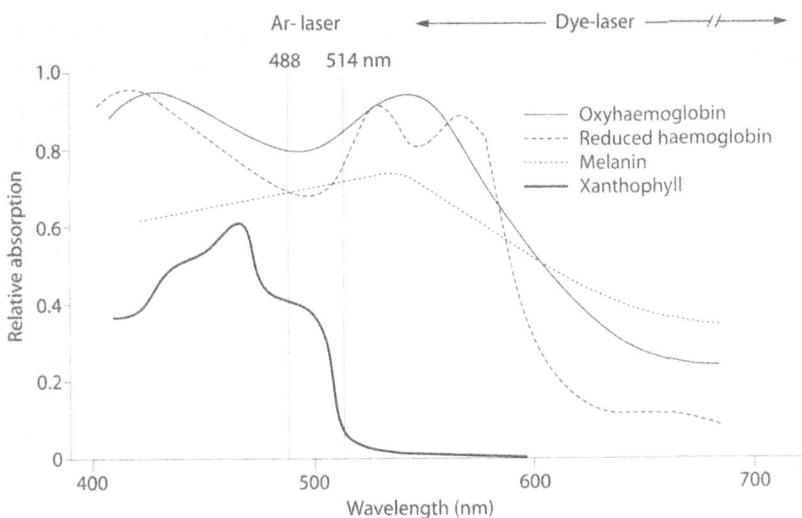

# I-3.2.3
# "Thermal" and "Non-Thermal" Effects on Tissues

K. Dörschel

**Contents**

## Introduction

Photoablation and photovaporization are tissue-abrading effects. The much-used differentiation, according to the pattern, photoablation is non-thermal and vaporization is thermal, is in fact very superficial and can rapidly lead to false conclusions. Both processes are ultimately of a thermal nature; they are distinguished only by their time dependence. Photoablation proceeds much faster than vaporization.

The terms thermal and non-thermal characterize the thermal influence in the border zones of the non-abraded tissue and, in general, are not a description of the abrasion process as such. In this sense, non-thermal means that thermal effects on the border zone are negligible. Whether effects in the border zone can be assumed to be negligible depends very much on the application. For the ophthalmologist abrading the cornea, a 20 µm thermally altered border zone will certainly be considered thermal, while during laser angioplasty in certain cases, 100 µm may be considered non-thermal.

Figures 1 and 7 in Chapter I-3.2.2 demonstrate that in certain cases the effects of photoablation on tissues are really very thermal, while vaporizing and cutting using a cw laser cause very little thermal stress.

Figure 1 displays the cutting of holes into discs of bone using a pulsed excimer laser (248 nm, 20 ns) and a pulsed $CO_2$ laser (10,6 µm, 300 ns) [1]. For a repetition frequency of 10 Hz, the walls of the holes show only small thermal changes, although both lasers emit extremely different wavelengths. However, for a repetition rate of 100 Hz while keeping the other laser parameters the same, the results look definitely thermal. Apparently, the wavelength influences not the outcome, but only the repetition frequency.

On the other hand, using a cw $CO_2$ laser, the results of cutting are normally thermal. By injecting a cold gas jet ($N_2$, Ar, compressed air) onto the cutting zone, the hot products of pyrolysis will be cooled and ejected rapidly out of the cutting zone. The result is a much smaller thermally changed border zone (see Fig. 7, Chap. I-3.2.2). The actual cut edge displays al-

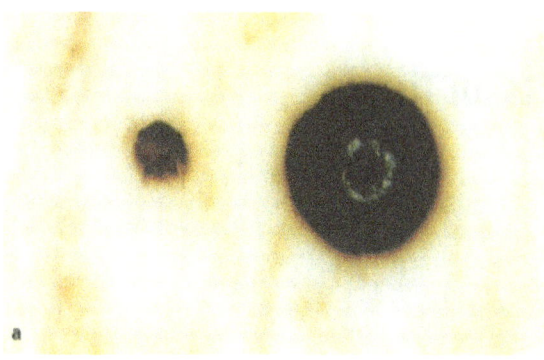

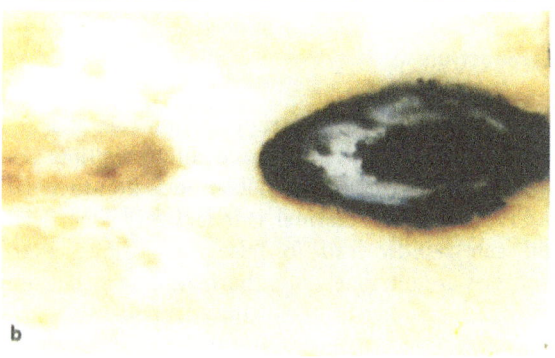

**Fig. 1a, b.** Holes cut into human bone discs display, independent of the laser type, narrow, thermally altered border zones, as long as the repetition rate remains low (10 Hz, left). Otherwise very strong thermal effects are visible (100 Hz, right). **a** Excimer laser (248 cm, 20 ns, 80 mJ). **b** $CO_2$ laser (10.6 µm, 300 ns, 100 mJ)

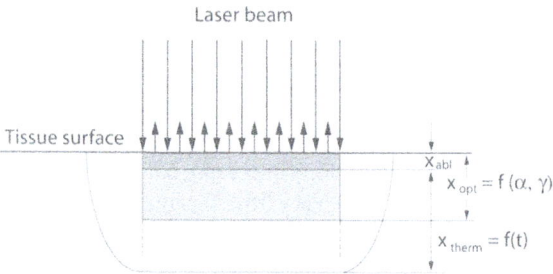

**Fig. 2.** The absorbed laser radiation ablates the tissue heated to several hundred °C ( ▭ ). However, radiation penetrates further into the tissue. The absorbed radiation energy, although insufficient to ablate, will coagulate a layer approximately corresponding to the optical penetration depth ( ▭ ). In addition to direct heating by the radiation energy, the surroundings may be heated by thermal conduction ( ▭ )

most no visible thermally changed border zone and could be termed non-thermal.

Besides answering the question of how much tissue will be ablated, it is equally important to estimate which tissue border zones will not be removed, but will be heated enough to initiate coagulation. In doing so, one can conclude that both processes create a minimally thermally altered border zone which cannot be

reduced. The size of the thermal border zones does not depend a priori on the ablation process but exclusively on its energy absorption.

For the energy absorption inside the border zone two processes are primarily responsible. The border zone can be heated either by optical radiation reaching it directly, or by thermal conduction (Fig. 2).

## Optically Deposited Energy

A fraction of the laser radiation will also penetrate into the non-ablated regions. There, the laser beam intensity will decrease exponentially with depth and the absorbed energy will heat this region. Since the energy needed to coagulate tissue is less than 10% of the vaporization energy, this heating is sufficient to produce a coagulated zone with a thickness ($x_{opt}$) of about the optical penetration depth ($1\alpha$) of the laser radiation. This thickness always presents a lower limit for the expected thermally altered border zone. For this border thickness the following relation holds:

$$x_{opt} = 1\alpha$$

Table 1 presents a review of the absorption coefficients (1) for normal soft tissues and the penetration depths ($x_{opt}$) of radiation having various wavelengths. It must be taken into account that tissues are strongly scattering substances. The resulting intensity distribution is thus determined not only by absorption, but also significantly by scattering. In Table 1 scattering is taken into account by a so-called effective absorption coefficient ($\alpha$). This leads to the effective penetration depth ($x_{opt}^*$). Scattering reduces drastically, up to a

**Table 1.** Normal ($\alpha$) [6] and effective (with scattering correction) ($\alpha^*$) absorption coefficients [7], with their respective optical penetration depths in soft tissues for the most common laser wavelengths

| Wave-length | Absorption coefficient | | Optical penetration depth | |
|---|---|---|---|---|
| $\lambda$ (nm) | $\alpha$ (cm-1) | $\alpha^*$ | $1/\alpha$ (µm) | $1/\alpha^*$ (µm) |
| 193 | > 400 | > 5000 | < 25 | < 2 |
| 248 | 600 | 5000 | 17 | 2 |
| 308 | 200 | 1670 | 50 | 6 |
| 351 | 40 | 170 | 250 | 60 |
| 532 | 12 | 42 | 830 | 240 |
| 1064 | 4 | 5 | 2500 | 1900 |
| 1320 | 8 | | 1250 | |
| 2060 | 35 | | 286 | |
| 2700 | 1000 | | 10 | |
| 2940 | > 2700 | | < 4 | |
| 9600 | 700 | | 14 | |
| 10600 | 600 | | 17 | |

factor of 10, the effective penetration depths for ultraviolet radiation.

Figure 3 indicates that the minimal thermal border zone depends on the optical penetration depth. There, the thermally altered border zone of an unstained cornea ablated by an excimer laser (308 nm, 20 ns) has a thickness of 70 μm, while this thickness is reduced to about 5 μm for a stained cornea [2]. The experimentally determined values of 70 and 5 μm correspond closely to the respective values of the optical penetration depths in unstained and stained corneas. The larger value attached to the unstained cornea, in comparison with normal soft tissues (Table 1), is caused by the lack of scattering in the cornea. Consequently, the effective penetration depth is not reduced.

The experimentally determined thermal border zones caused by photoablation with very different wavelengths, from UV to IR, also correspond very well to the effective optical penetration depths calculated from ablation rates ([3]; Fig. 4).

## Thermally Deposited Energy

In addition to the optically deposited energy, the region directly irradiated by laser radiation can dissipate energy by thermal conduction into the surroundings. The thermal conduction in tissues, besides depending on thermal diffusivity ($\alpha = 1.2 \times 10^{-7}$ m$^2$ s$^{-1}$), also depends on the time (t) during which the hot regions are in direct contact with their cold surroundings. The vaporized material streaming out of the irradiated region will also make contact with the surroundings.

In general, the laser pulse length can be considered the contact time, because tissue vaporized by short laser pulses will evaporate at very high speed. However, the actual contact time can be much longer if the hot gases must leave via narrow channels. In this case, the heat flow rate will be determined not only by the thermal diffusivity, but also by the heat transmitted by the outgoing gas into the wall.

Using a one-dimensional heat conductivity model it is possible to calculate the range ($x_{therm}$) of direct heat conduction into the surroundings. The penetration depth ($x_{therm}$) of an isothermal (temperature ($T_m$) between the thermostatically controlled hot surface ($T_w$) and the cold tissue ($T_k$) can be calculated as a function of time (Fig. 5):

$$x_{therm} = k_{Tm} \sqrt{4 \cdot \chi \cdot t} \,. \qquad (1)$$

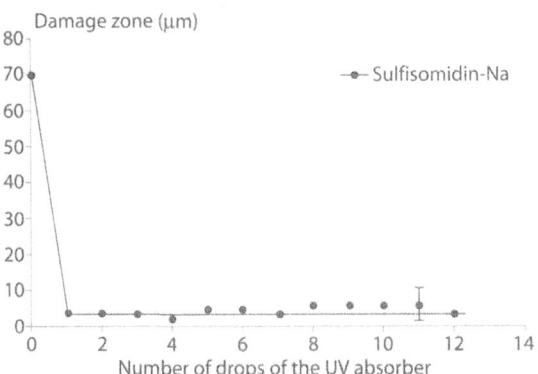

**Fig. 3.** It is possible to reduce the thermal border zone created in the cornea drastically, by an excimer laser (308 nm, 7.2 mJ 10 Hz) through staining this tissue with a dye. [2]. The width of the border zone corresponds to the appropriate optical penetration depth of the laser radiation

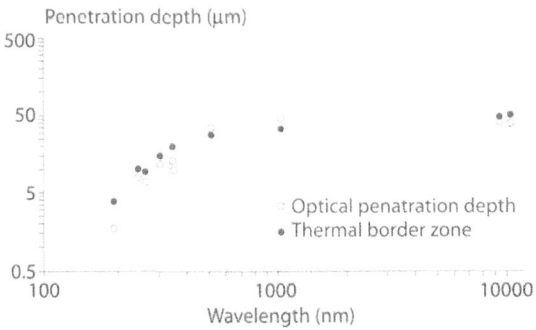

**Fig. 4.** Calculated effective penetration depths for ablation of aorta tissue employing short laser pulses and various wavelengths. The results based on ablation threshold and ablation rate agree well with the experimentally determined values for the thermally altered border zones. [3]

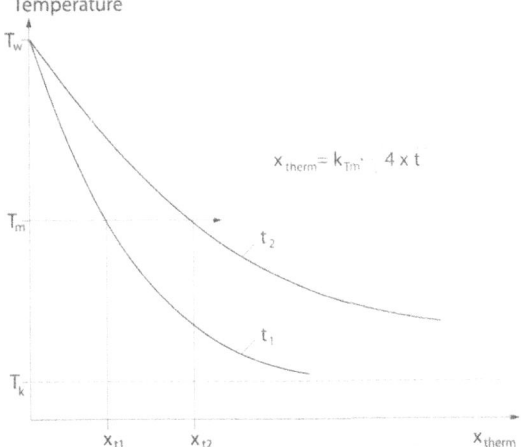

**Fig. 5.** Due to thermal conduction an isothermal ($T_m$) penetrates deeper and deeper ($X_{therm}$) into the tissue. At time t = 0 the tissue surface displayed a constant temperature ($T_k$). Thereafter, it was in contact with a constant temperature source ($T_w$)

**Table 2.** Relation between contact time (t) and range of thermal conduction ($x_{therm}$) from a one-dimensional thermal conduction model.

| Contact time $t$ | Thermal range $x_{therm} = \sqrt{4 \times t}$ |
|---|---|
| 1 µs | 0.7 µm |
| 10 µs | 2.3 µm |
| 100 µs | 7.2 µm |
| 1 ms | 23 µm |
| 10 ms | 72 µm |
| 100 ms | 0.23 mm |
| 1 s | 0.72 mm |

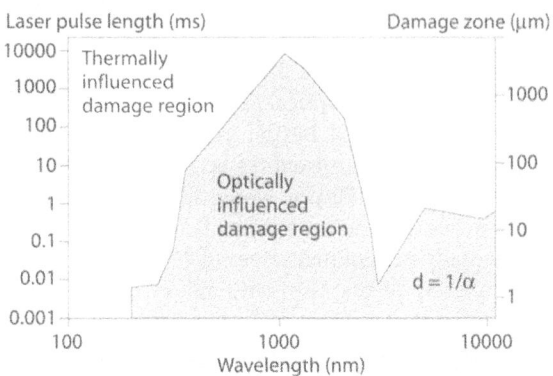

**Fig. 6.** Depending on wavelength and laser pulse length, it is possible to recognize two regions separated by the critical time. The thermally altered border zones of the regions are determined either directly by laser radiation or by thermal conduction

For temperatures (T,), approximately centred between $T_w$ and $T_k$, the constant ($k_{Tm}$) is about 1.

Table 2 presents a review of the relation between a range of heat conduction and contact times. Reliable guidelines are: heat penetrates into tissue about 1 µm deep within 1 µs and into a thickness of about 1 mm within 1 s.

Summarizing the results, it is possible to determine a critical time ($t_{crit}$) when the optical penetration depth exactly equals the range of the heat conduction ($x_{opt} = x_{therm}$) [5]:

$$t_{crit} = 4 \cdot \chi \cdot x_{opt}^{2}. \tag{2}$$

For irradiation times less than $t_{crit}$, the border zone is determined only by the optical penetration depth. It is independent of the laser pulse length. For irradiation times exceeding $t_{crit}$ the coagulation zone will be enlarged by heat conduction. Figure 6 graphically presents the relationship between $t_{crit}$, $x_{opt}$ and $x_{therm}$ for various wavelengths. The values for the effective optical penetration depths in normal soft tissues are taken from Table 1.

For most applications a thermally altered border zone of 20 µm could be considered non-thermal. According to Fig. 6, this is in general achievable for wavelengths in the UV range (< 350 nm) and in the IR range (> 1.8 µm) if the laser pulse length is below approximately 1 ms. Only for the extremely small optical penetration depths at 193 nm, 248 nm and 2.9 µm is it worthwhile to reduce the pulse length to 1 µs in order to achieve the smallest possible border zone of a few µm. However, reducing the pulse length even further will not provide additional improvement.

In the region of large penetration depths, between 500 nm and 1.5 µm, pulse lengths in the range of seconds can be utilized, practically corresponding to a cw regime. If in this wavelength range ns pulses are used, it must be realized that the large effective penetration depths due to non-linear effects are strongly reduced by the laser pulse itself. Consequently, in these cases, the width of the purely thermal border zone may be reduced to about 50 µm. This thermal

damage, however, will be superimposed upon the mechanical damage, which extends much further.

Up to now all considerations were based on the assumption that the ablated materials separated from the border zone once the laser pulse ended. Most likely this is true only for laser irradiation at a free surface. Especially while ablating inside narrow channels, the contact time can be lengthened considerably. Consequently, a widened border zone must be taken into account. This effect will be maximized if the irradiation takes place inside a closed vessel. In this case, the entire laser energy contributes to the heating of the border zone. However, the widening of the border zone depends very much on special circumstances.

Furthermore, the considerations have so far neglected repeated applications of the laser. Repetition can also widen the border zone, an effect which increases with increasing repetition rate.

These additional cases of widening of the border zone can be roughly approximated by the simple assumption that the response is proportional to the entire energy present. The entire available energy is made up of the average power ($N_m$) and the application time (t). An additional factor (k) will take into account the special form of application as described above. Of course, it is difficult to formulate the value of k quantitatively. Nevertheless, the influence of different average powers applied under the same circumstances, e.g. for the same k, should become apparent. For the entire thermal border zone ($x_{bord}$) we obtain:

$$x_{bord} = x_{opt} + x_{therm} + k \cdot N_m \cdot t \tag{3}$$

High average power applied over a longer time period at the same locality will contribute considerably to the border zone. For low average power this contribution will be negligible.

To achieve ablative action by using a cw laser or a pulsed laser with low pulse energy (below the ablation threshold) and a high repetition rate, the applied laser radiation must strike the tissue surface with a relatively high power density.

In contrast to the above, it is possible to achieve photoablation as long as the power density of the laser pulse exceeds a minimum value. In this case, the average power is the product of the energy of a single pulse and the repetition rate, i.e. in contrast to cw lasers "non-thermal" action can be achieved with low average power. It should not be overlooked, that "non-thermal" action is always accompanied by a relatively small ablation volume. For large repetition rates the average photoablative power increases. This will generate a "thermal" border zone, as already demonstrated in Fig. 1.

## References

[1] SCHOLZ C (1992). Neue Verfahren der Bearbeitung von Hartgewebe in der Medizin mit dem Laser. ecomed, Landsberg

[2] MÜLLER-STOLZENBURG N, BUCHWALD H, MÜLLER G, KAR H, DÖRSCHEL K. (1989) In-vitro-Untersuchungen zur refraktiven Hornhautchirurgie mit dem Excimer-Laser über Glasfaser. Fortschr Ophtalmol 86: 592–596

[3] KAR H, RINGELHAN H. (1992) Grundlagen und Technik der Photoablation ecomed, Landsberg

[4] WALSH J, FLOTTE T, DEUTSCH T (1989) Er:YAG laser ablation of Tissue: effect of pulse duration and tissue type on thermal damage. Lasers Surg Med 9: 314–326

[5] FURZIKOW N. (1987) Different lasers for angioplasty: thermooptical comparison. EEE u J. (1986), Quant Electr QE–23: 1751–1735

[6] ESTEROWITZ L, HOFFMAN C, TRAN D, LEVIN K, STORM M, BONNER R, SMITH P, LEON M. Angioplasty with a laser and fiber optics at 2,94 μm. In: Optical and Laser Technology in Medicine. SPIE, 605: 32–36

[7] ANDERSON R, PARRISCH J. (1981) The optics of human skin. J Invest Dermatol 77: 13–19

# I-3.2.4
# Non-Linear Processes

J. Helfmann

Contents

## Introduction

The range of short pulse durations and, consequently, high power densities, produces a new class of processes, which clearly differ from the purely thermal or photochemical effects of laser radiation on matter. These processes are called non-linear (Fig. 1) and extend to include the photoablation (photodecomposition) process, which SRINIVASAN investigated for the first time in 1982 using UV laser radiation and which is characterized by precision ablation of material at minimum thermal stress on the surrounding tissue. Photoablation is applied, e.g. in laser angioplasty or in cornea form correction. It requires power densities ranging between 0.1 and 10 J/cm$^{-2}$ and laser pulse durations in the ns and μs ranges.

Power densities higher than this (approx. $10^{10}$ W/cm$^{-2}$) cause still another process called optical breakdown. Here matter is ionized by the extremely high field strength of the laser radiation resulting in the formation of a plasma and mechanical shock waves (photodisruption). As this process needs no absorption by matter, it is observed also in transparent media such as air.

Photodisruption is applied in ophthalmology for the destruction of the secondary cataract membrane (destruction of the rear crystalline capsule membrane straight through the cornea) after an artificial lens implantation. Another field of application is laser lithotripsy, i.e. crushing of body concrements.

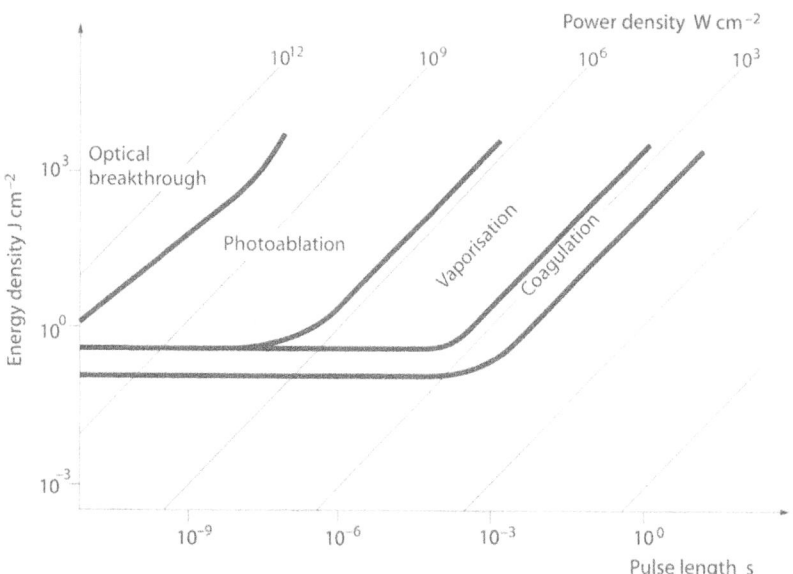

**Fig. 1.** Regions of energy density and pulse width for different laser effects on tissue. (Energy density, Pulse width, Optical breakdown, Photoablation, Vaporization, Coagulation)

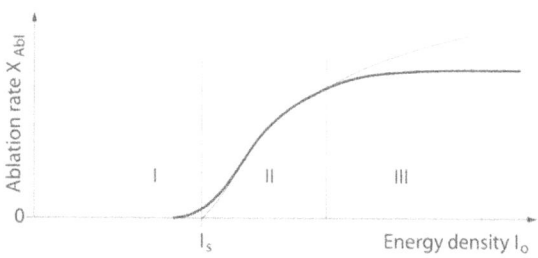

**Fig. 2.** Principle course of photoablation. The ablated thickness a single laser pulse versus the energy density is shown ($I_s$ ablation threshold)

## Photoablation

Figure 2 is a schematic view of the tissue ablation. The typical threshold behaviour is characterized by the sharp rise of the ablation rate from a tissue-specific threshold energy density (ablation threshold).

In the case of low energy densities within zone I (Fig. 2), the irradiated laser beam heats the tissue only slightly. This zone is followed by a transition zone with some higher energy densities where the heating-up causes a tissue vaporization. This effect is comparable to the vaporization of tissue by means of continuous wave laser systems. Once the ablation threshold ($I_s$) is reached, the ablation rate sharply rises, i.e. the irradiated tissue volume is blown off as in an explosion. This zone II is called photoablation. Then the curve flattens because the laser radiation is no longer fully effective. This saturation (zone III) is caused by a plasma or evaporated tissue which develops upon the surface and absorbs part of the radiation. Conse-

quently, that part does not effect an ablation on the tissue. The ablation is characterized mainly by the ablation threshold and the ablation rate increase. These values depend on the absorption coefficient of the tissue for the laser wavelength used.

## Optical Breakdown and Plasma Formation

If the energy densities are higher than required for photoablation, another phenomenon occurs which can be used for laser therapy. This phenomenon is called optical breakdown. In the ns range the threshold for the optical breakdown is approximately 10 GWcm$^{-2}$, depending on the tissue and the wavelength.

This process, which takes place in both absorbing and transparent tissue, has two stages. In the first stage, single free electrons are generated by multiphoton ionization or thermionic emission. In the second stage, each of these free electrons is accelerated by inverse bremsstrahlung in the electric field of the laser radiation. The electrons transfer the kinetic energy to the electrons still bound by collisions. If the energy transferred by collision is higher than the ionization energy, another free electron is produced, which is accelerated in the electric field of the laser radiation like the first free electron and can release other electrons by collision. The whole process causes an avalanche-type increase in the number of free electrons and is, therefore, called avalanche ionization. Consequently, the density of the free electrons rises exponentially. The state of matter with many free electrons and ion bodies is called plasma. At a molecule density of approximately $3 \cdot 10^{23}$ cm$^{-3}$ in tissue, the tissue properties are distinctly changed from a density of free electrons of approximately $10^{18}$ cm$^{-3}$.

The absorption of tissue in plasma state is very large and may lead to a penetration depth of merely 20 μm for 1064 nm laser radiation at an electron density of $10^{20}$ cm$^{-3}$ even if the medium, such as the eye lens, had been transparent before. We talk of plasma shielding if laser radiation is used for removing tissue e.g. by photoablation, but due to the development of an unwanted plasma the radiation is already absorbed in the plasma and does not penetrate to the tissue surface any more.

Due to the ionization energy, such plasma has a very large energy content, equivalently its temperature is very high ranging within some thousand Kelvin, and so is its pressure that ranges within some kbar. This is why the plasma expands rapidly; it is, so to speak an explosion in the ns range which, in turn, generates shock waves and cavitation bubbles.

The formation of the plasma depends on the pulse duration, pulse energy and wavelength of the laser. The tissue absorption is of minor importance only (influence on the formation of first free electrons), in contrast to the major importance of absorption for thermal laser effects.

## Laser-Induced Shock Waves and Cavitation

From a small plasma within the focus of a laser beam or at the end of an optical fibre the hot tissue volume expands at speeds ranging within several thousands of m/s pushing against the surrounding tissue or fluid. This causes a shock wave radially from the plasma. Due to the high pressure and the high temperature the expansion by far exceeds the acoustic velocity of the different tissues. The pressure wave, therefore, expands not as a sound wave, but as a shock wave. This means that the pressure front is characterized by a sharp rise and, contrary to a sound wave, changes its shape while expanding.

While the shock wave expands (in a homogenous medium as a spherical wave), the surface of the wave front increases, which, in turn, reduces the pressure jump. The shock wave becomes a sound wave, which now expands at acoustic speed. Due to the expansion, the hot plasma also cools, the pressure is reduced and the plasma surface expands more slowly and falls behind the shock wave, which continues as a pressure pulse. However, the plasma bubble continues expanding to some hundreds of μs. Its internal pressure falls below the ambient pressure due to its large volume and cooling off. Once the bubble wall is stagnant, the direction of motion reverses, which makes the bubble collapse. This process, i.e. the creation of a cavity by expansion and collapse, is called a cavitation and the bubble is called cavitation bubble.

Due to the collapse of the cavitation bubble, its volume is strongly compressed and heated, which causes the generation of another shock wave and another cavitation bubble. This periodic repetion of the same processes is called cavitation bubble oscillation. As energy is always released into the surroundings in the form of shock waves, heat and light, the oscillation decreases steadily.

While such cavitation is excellently visible in a fluid (e.g. when performing a laser lithotripsy), this type of bubble oscillation is, however, impeded in tissues that have structural strength.

# I-4
# Technical Basics of Medical Laser Systems

## I-4.1
## Basic Device

P. Greve

### Contents

### Introduction

In general, medical laser systems consist of a source of laser light, a shutter and control units, a beam-handling system and the optical end unit (Fig. 1).

For power lasers, the power supply, the cooling system and perhaps also the laser head are arranged in a so-called laser console. This will also contain the first shutter and the control units, as well as the input coupler for the beam-handling system. The beam-handling systems for the far IR and sometimes also for the UV region could be articulated arms. Light guides are used in the near-UV-region, in the visible and the near-IR regions (see Table 1 in Chap. I-4.2). The basic devices for lasers are usually unmodified pieces of equipment, except for the user interfaces. Beam-handling systems and optical end units must be selected for the particular medical application (Fig. 2).

In compact laser systems the laser head can be coupled directly to the optical end unit. In this case, the beam-handling system is reduced to a direct coupling and beam adaptation is via optical systems. Beam-handling systems are treated in Chapter I-4.2 and optical end units in Chapter I-4.3. The following is exclusively concerned with the basic laser system consisting of laser head, power supply, cooling unit, and, if necessary, the measurement and control of the laser parameters.

### Laser Head

The laser head contains the laser medium, the excitation mechanism for the laser medium, and resonance mirrors for the feedback and uncoupling of the electromagnetic radiation. If needed, the laser head can also include active and passive optical elements to shape the beam in space and time. A pilot light, typically an HeNe or a laser-diode, must overlap the beam of the active laser if the latter's wavelength is outside the visible spectral range.

In general, the laser medium determines the name of the laser system. Generally speaking, there are

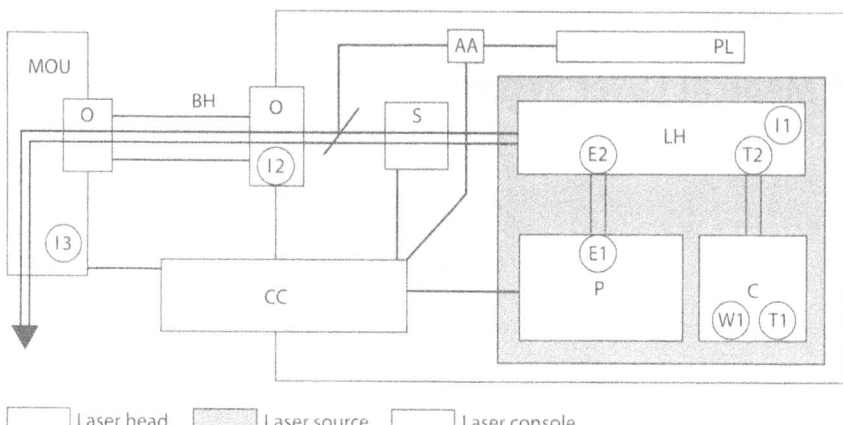

**Fig. 1.** Diagram of a medical laser

Laser head   Laser source   Laser console

C Cooling unit; P power supply; PL pilot laser; S shutter; AA attenuation of aiming light;
O optic adaption; BH beam handling; MOU medico-optical end unit; CC control console;
I1,I2,I3 intensity sensors; T1,T2 temperature sensors; W water flow sensors; E1,E2 electric
current sensors

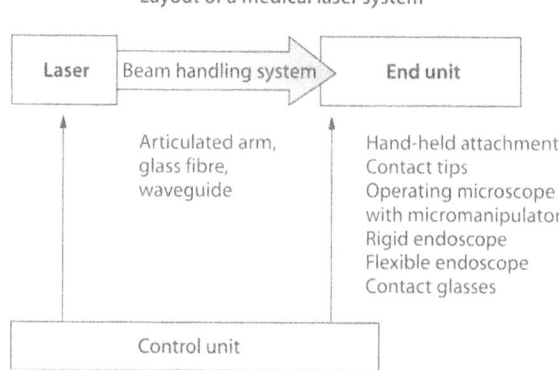

**Fig. 2.** Layout of a medical laser system

solid-state, gas and liquid-state lasers. Table 1 summarizes typical laser systems relevant to medicine.

In solid- and liquid-state lasers, the laser medium is excited by optical pumping. Flash lamps and arc lamps are utilized or other suitable light sources like laser diodes. In gas lasers an electric discharge directly through the laser medium will excite the atoms and the molecules via electron impact. To keep the optical losses low and to achieve optimum feedback and decoupling, the resonator mirror must be well coated and uncoated, respectively. Optical feedback is possible only if the resonator mirrors are aligned in such a way that the electromagnetic wave is reflected into itself. This condition calls for a precise, highly stabilized mechano-optical design of the laser head; highly stable under all temperature and pressure conditions.

The electric power supplies for the pumping lamps and the gas discharge devices are very compact units.

This compactness should not jeopardize the safety requirements for medical instrumentation. Cooling of the laser head is by convection or by circulating distilled water around the critical components. In turn, the cooling water will be cooled by a water/air or a water/water heat exchanger.

For safe operation of a medical laser system, all its essential functions must be monitored exhaustively to ensure that a device will switch off as soon as any malfunction occurs. The essential quantities are the laser output power (measured in W) and the output energy (measured in J) for pulsed lasers. Typically, the power and the energy will be monitored at three positions: directly at the laser head (Fig. 1: I1), distal to the shutter and control system, proximal to the beam-handling system (Fig. 1: I2) and inside the optical end unit as close as possible to the laser exit (Fig. 1: I3). A comparison of these signals will expose malfunctions in the regions of the shutter and the beam-handling system. The operating currents are monitored inside the power supply and the laser head (measuring stations (Fig. 1: E1, E2). Comparing the intensity signals (Fig. 1: I1 and I2 with E1 and E2) will allow checking the adjustment of the laser and/or the ageing process caused by the limited lifetime of individual components. If set limits are exceeded, an error signal will be generated. Temperature and flow rate of the cooling medium can be measured inside the cooling unit, in addition to inside the laser head, in order to avoid overheating. If a pilot laser has been installed, its intensity must also be controlled and checked.

All control and measuring signals are sent to the control console. There, the incoming information will be logically interconnected. All relevant data for the user are displayed. In general, these are the set power, the applied laser power (for pulsed lasers it is the laser

**Table 1.** Lasers applied in medicine

| Laser medium | Wavelength (nm) | Type | Power/(pulse energy) | |
|---|---|---|---|---|
| | | | (W) | (mJ) |
| Argon ions | Gas | 488.0 514.5 | cw | 1 – 20 |
| Helium, neon | Gas | 632.8 | cw | $1 - 5 \times 10^{-3}$ |
| Krypton ions | Gas | 647.1 | cw | 0.5 – 3 |
| Neodym:YAG | Solid state | 1064 | cw | 1 – 100 |
| Neodym: YAG | Solid state | 1064 | Pulsed | 5 – 50 |
| $CO_2$ | Gas | 10 600 | cw | 1–100 |
| Diode | Semiconductor | 650–1000 | 0.1–10 | |

energy) and the set effective time; possibly also the set repetition rate and the pulse-gap repetition frequency. If necessary, the laser beam diameter – controlled directly by the optical end unit – can also be displayed. All measured quantities are checked against stored criteria. Mismatches are indicated and then directly converted into signals, which switch off the device. Figure 3 shows a typical display.

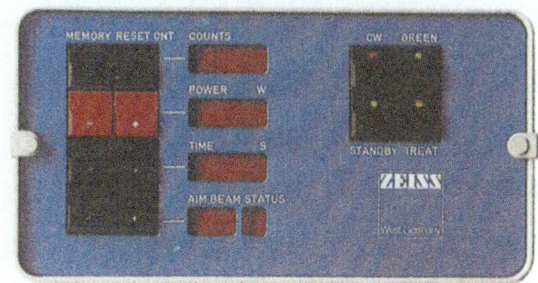

**Fig. 3.** Typical control console of a medical laser (Zeiss Visulas Argon)

# I-4.1.1
# Dose-Effect-Controlled Nd:
# YAG Laser System for the Cutting of Tissues

F. Frank and G. Hauptmann

The various interaction mechanisms between Nd:YAG laser radiation and biological tissue during non-contact applications are well known. The extinction of the laser radiation in biological tissues depends on the laser parameters (power density, irradiation time and wavelength), as well as the type of tissue. The range is typically a few centimetres [7]. Example: a laser output power of 10 W, transferred through an optical fibre having a diameter of 365 μm and a divergence of 19°, will achieve a power density of about 10 W/cm$^{-2}$ at a distance of 30 cm. This is sufficient to coagulate tissues. Bringing the fibre into direct contact with the tissue, the power density will increase 800-fold. Such a high power density will vigorously heat the tissue, causing carbonization, followed by vaporization. In this case, in contrast to non-contact application, a much smaller volume of tissue will be involved. This feature makes it possible to cut tissue in the contact mode with the Nd:YAG laser. But it is difficult to dissect tissues using a bare optical fibre in connection with a common laser device delivering a fixed, adjustable output power, because the fibre tip will be contaminated by tissue residues. On the one hand, this will improve the efficiency of the cutting process via increased absorption; on the other, it will destroy the fibre as soon as the tissue contact is interrupted. This is especially critical during endoscopy. Due to its very limited field of view, the user is often unable to keep the bare fibre in constant contact with the tissue during the entire cutting process. The various sapphire-tipped fibres try to overcome this problem. However, this method is technically very demanding and needs extensive training of the user. The relatively large diameters of tips limit their applicability. In addition, there is a danger of gas embolism caused by the need to cool the sapphire tips.

A dose-effect-controlled Nd:YAG laser system has been developed to achieve a constant temperature at the fibre-tissue interface. This leads to a constant cutting speed and allows the destructive thermal load on the tip of the fibre to be regulated. During cutting of biological tissues with the Nd:YAG laser system, visible light will be created. This is due to the burning and carbonization processes at the fibre-tissue interface. Light intensity and temperature at the terminal face of the fibre are closely correlated. This correlation allows utilization of the light intensity at the fibre-tissue interface as a control parameter. Thus it is possible to have a surgical laser system, based on an Nd:YAG laser, which allows optimum cutting while remaining in contact with the tissue.

The light emitted during the combustion process is returned to the laser device via the therapy fibre. A beam splitter removes the light from the beam path of the Nd:YAG laser into a detector, where it is converted into an electrical signal. A control unit, with an integrated microprocessor system, picks up this feedback signal and utilizes it as a control parameter for the laser output power. In this way, it is possible during the cutting process to produce visible light of constant intensity at the terminal face of the fibre (Fig. 1).

Control is achieved by comparison with a reference value. This value can be adjusted by the user according to the specific tissue (Fig. 2).

It is possible to obtain different cutting parameters by varying the light intensity and the maximum achievable laser power.

Such a dose-effect-controlled system presents the following advantages:

- The optical fibre is protected from thermal destruction by limiting the temperature at the distal fibre tip.
- The laser energy needed for the cutting process is optimized with respect to the cutting speed without destroying the optical fibre.
- As soon as the contact between fibre and tissue ceases, the laser power will be automatically reduced within a fraction of a second, thus preventing the destruction of the fibre.
- There is no need to cool the fibre tip.
- The well-defined, narrow cutting width generates very little smoke.
- A uniform coagulation necrosis is formed at the border of the vaporization zone. This is a consequence of the homogeneous distribution of the Nd:YAG laser radiation inside the tissue [2], but is also attributable to thermal conduction processes caused by rapid carbonization.

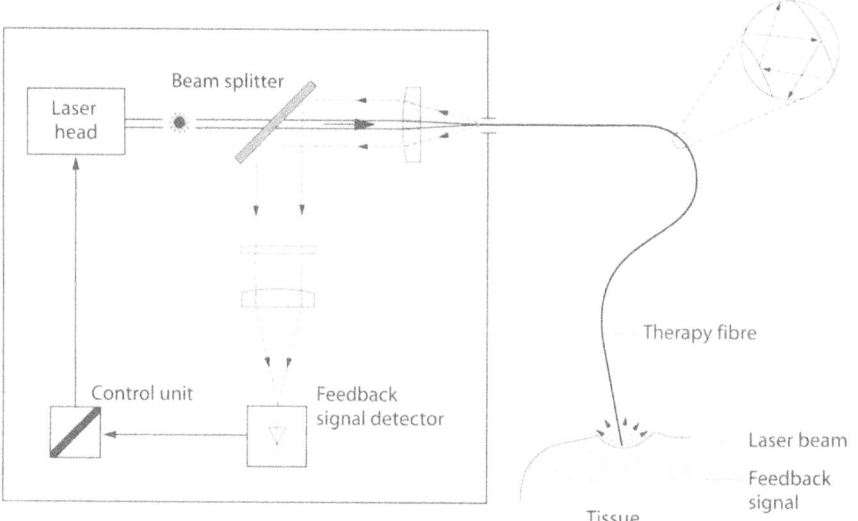

**Fig. 1.** Dose-effect-controlled Nd:YAG laser including therapy fibre for contact surgery. The feedback signal is transferred by the therapy fibre.

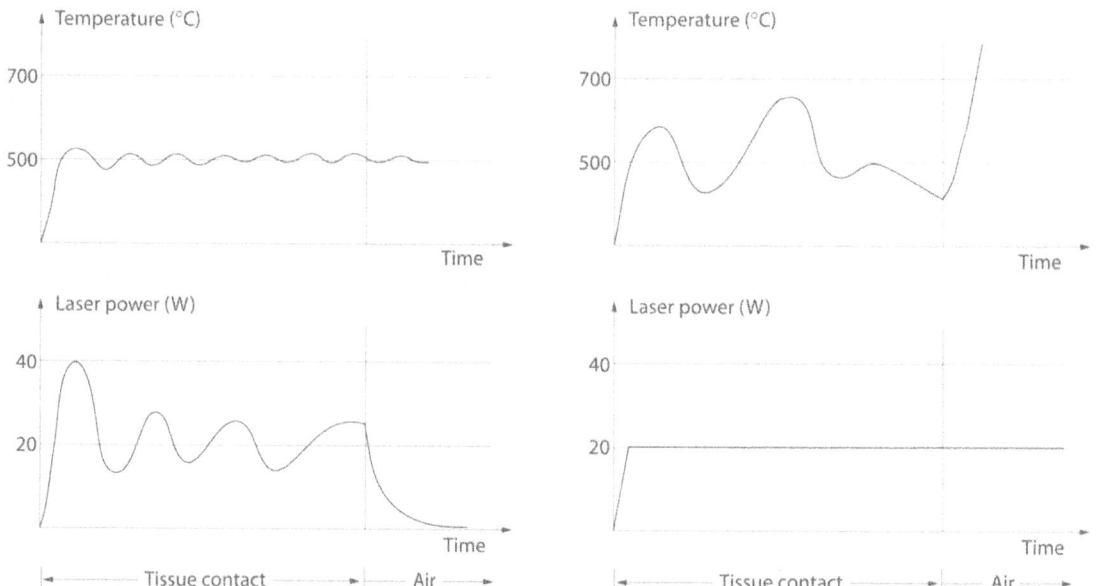

**Fig. 2.** Controlled temperature at the distal, contaminated fibre tip (left) compared with an uncontrolled system (right)

During the cutting process, some spots at the fibre tip will, for very short periods of time, reach surface temperatures sufficient to reduce the viscosity of quartz glass to such a degree that small carbonization and tissue particles will be melted and adhere to the surface of the fibre tip. Due to this "burn in" process, the fibre will be contaminated to such an extent that these particles will absorb part of the laser energy, which will then be converted into heat. The contaminated fibre will reach significantly higher temperatures compared to an uncontaminated one, leading directly to carbonization and vaporization of the tissue, neces- sary for a high cutting efficiency. Contamination will also alter the emission characteristics of the fibre. This will reduce the power density in the axial direc- tion, causing a homogeneous coagulation necrosis along the entire cutting edge (Fig. 3).

The laser device mediLas 4060 fibertom (produced by Dornier Medizin Laser GmbH) represents such a controlled system. It makes it possible to adjust the temperature at the distal laser tip in three steps (at about 400° C, 500° C and 600° C). This range of varia- tion is sufficient to adjust the cutting ability of the system to the various tissues. In addition, it allows to

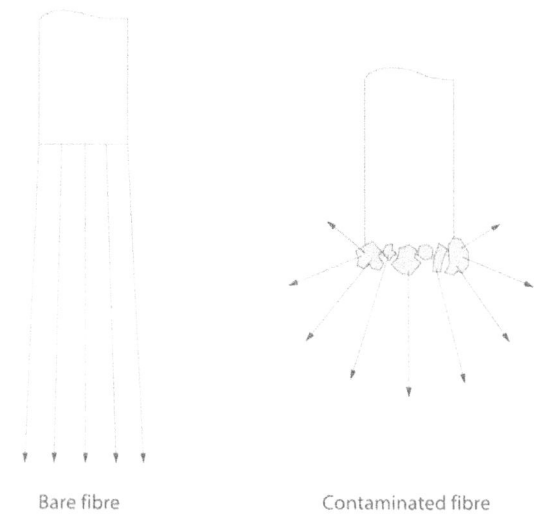

Bare fibre                    Contaminated fibre

**Fig. 3.** Emission characteristics of a standard bare fibre for non-contact application and a contaminated fibre for surgery

control the vaporization depth and coagulation border of the lesions supposed boundaries for appropriate therapy. In vivo investigations of uterus samples (pig) using a constant cutting speed of 2.5 mm s$^{-1}$, depending on the temperature setting of the fibre tip, lead to vaporization depths between 0.4 mm and 0.9 mm, and coagulation borders from 0.2 to 0.6 mm (Fig. 4a, b).

### References

BULNOIS J. (1986) Photophysical process in recent medical laser developments: a review. Lasers in Medical Science 1: 47
KEIDITSCH E, HOFSTETTER A, ZIMMERMANN I, STERN J, FRANK F, BABARYKA I. (1985) Histological investigations to substantiate the therapy of bladder tumors with the Neodymium-YAG-laser. Laser Med Chirurgie 19

### Further Reading

HESSEL ST, HAUPTMANN C Patentanmeldung: Chirurgisches Laserinstrument P 3934 647, 1–35
MÜLLER G, BRODZINSKI TH Patentanmeldung: Verfahren und Vorrichtung zur Leistung- bzw. Pulsenergiesteuerung bei der Lasermaterialbearbeitung organischer Materialien P 3934 646. 3–34

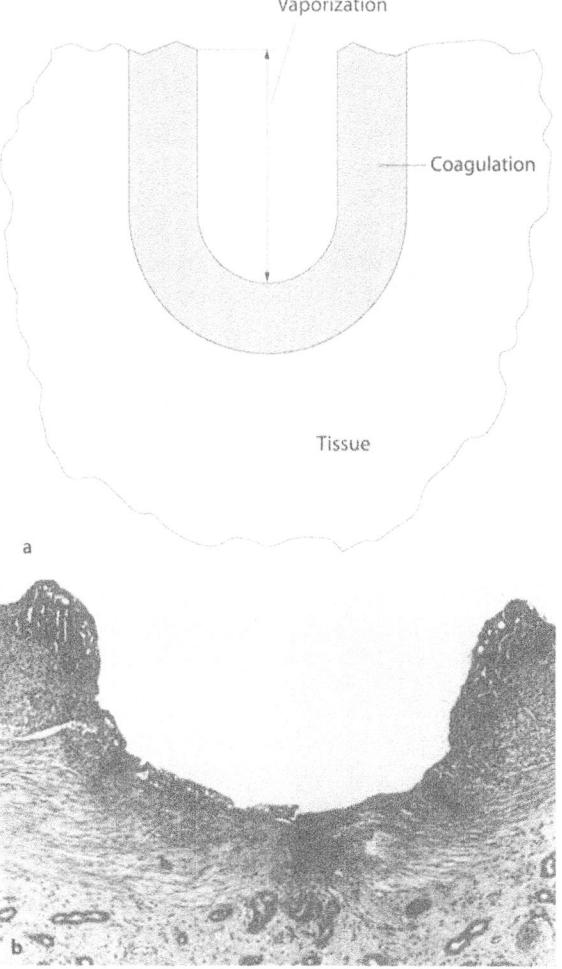

**Fig. 4 a,b.** Vaporization zone and coagulation zone in a tissue section generated by dose – effect controlled Nd:YAG Laser system. **a** Diagrammatic view, **b** Histological findings on a porcine uterine model

# I-4.2.1
# Beam-Handling Systems

P. Greve

**Contents**

## Introduction

Table 1 presents a summary of feasible beam-handling systems positioned between laser and medico-optical devices. Since direct coupling is used only in very compact laser systems, it will not be considered for the time being.

In general, articulated arms and light guides are considered to be beam-handling systems. They will be investigated in the following two chapters.

## Articulated Arms

Currently, there are no light guides available for many wavelength ranges, such as, e.g. in the IR region around 10 μm. In these cases, the beam transport from laser to hand pieces, or to an operating microscope or endoscope takes place via an articulated arm equipped with mirrors. An optimum beam-handling system requires at least five degrees of freedom, three for spatial positioning and two for the angular directions. Due to the rotational symmetry of the light beam there is, in general, no need for an additional sixth degree of freedom. Figure 1 presents the possible positional and directional changes of a laser beam.

For an arbitrary spatial alignment of the beam, it is necessary to place two rotating or tilting mirrors one after the other. This arrangement can be replaced by one mirror, able to tilt and to rotate around two axes. The latter arrangement needs high precision joints to bisect the deflecting angle. For this reason, it is not used in beam-handling systems. The series connection of two rotating mirrors, each having a deflection angle of 90°, is popular. This places no restrictions on the rotational angle (see Fig. 1f). In addition, the constant deflection angle at the rotating mirrors allows the use of mirrors optimal for a fixed angle.

Consequently, linear arrangements and pivot joints are widely employed as moving elements in articulated arms. To ensure that each element presents at least one mechanical degree of freedom, two linear arrangements or rotational axes cannot be aligned in

**Table 1.** Beam-handling systems

|  | Advantages/disadvantages | Wavelength range |
|---|---|---|
| Direct coupling | Good beam quality, conservation of coherence and polarization<br>Only for compact laser systems | Arbitrary |
| Articulated arm | Good beam quality, conservation of coherence and polarization<br>Inflexible, expensive, difficult to adjust | Arbitrary, used in UV and 10 µm range |
| Light guide | Flexible, inexpensive<br>Restricted for coherence, polarization and focus control | Near-UV-VIS and near-IR |

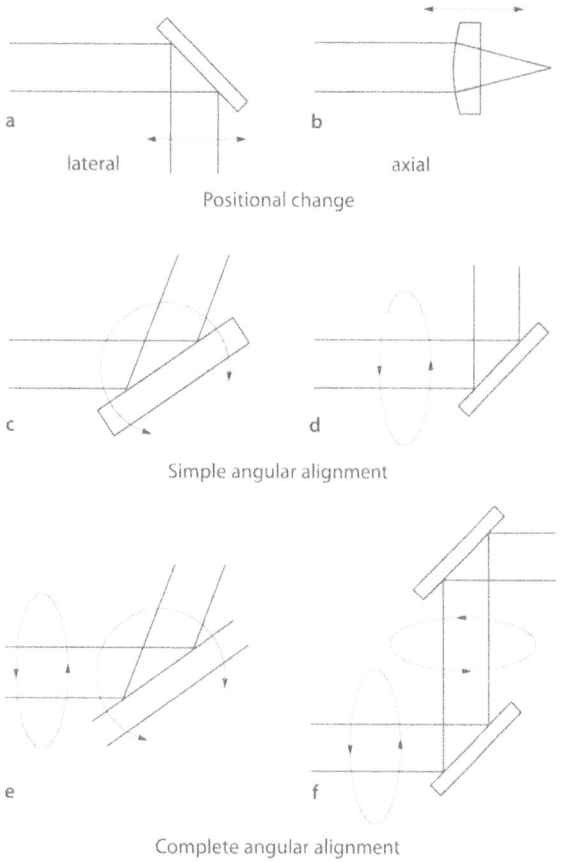

a    lateral            b    axial

Positional change

Simple angular alignment

c            d

Complete angular alignment

e            f

**Fig. 1.** Possible arrangements for changing position and angle of a laser beam

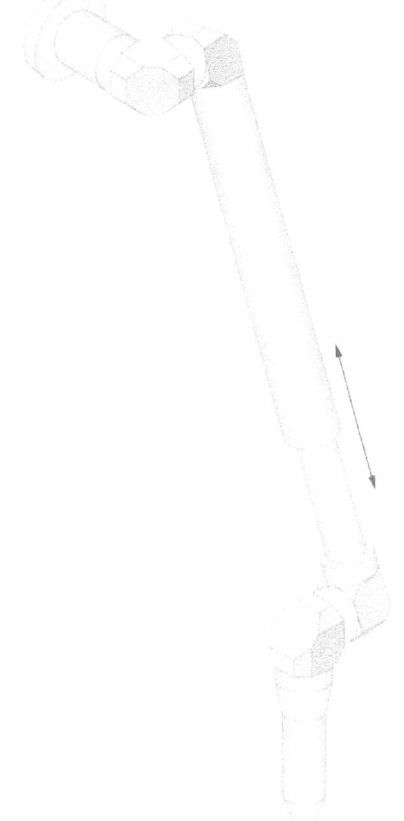

**Fig. 2.** Beam-handling system using a telescope

parallel within one system. Two consecutive linear arrangements or pivots are always separated by one deflection unit. There are a number of possibilities to combine linear arrangements and pivot joints resulting in a versatile beam-handling system covering all the needed degrees of freedom. A modular design is indicated in Fig. 3. Figure 2 presents one possible construction utilizing a telescope.

Figure 3 indicates a beam-handling system employing a simple hinge-like movement.

The quality of an articulated arm is characterized by its weight and the positional and angular accuracy of the exiting beam for all positions. Positional and angular accuracy are determined by the stiffness of the joints and the degree of bending of the straight tubes. In particular, the pivotal joints and the interconnecting tubes „straights" are critical factors. Their

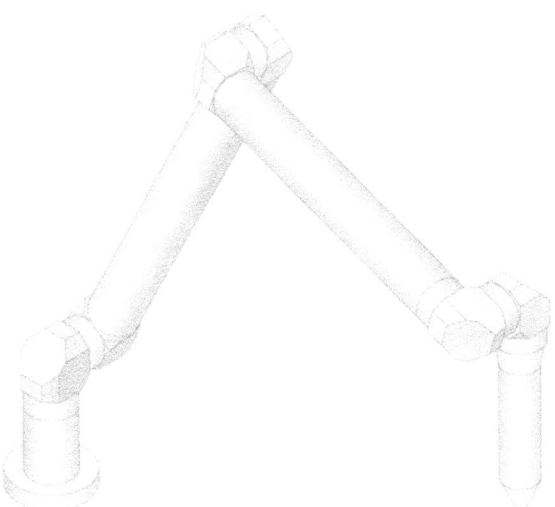

**Fig. 3.** Beam-handling system using a simple hinge-like joint

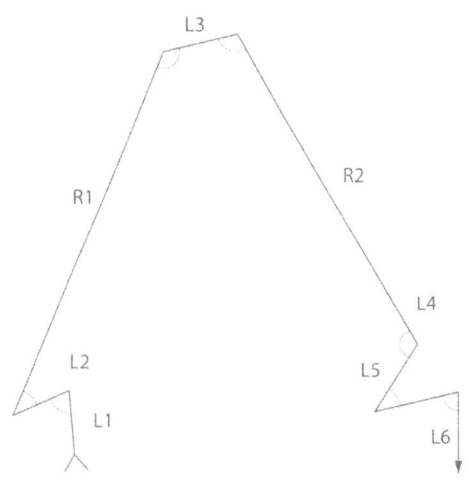

**Fig. 4.** Diagram of an articulated arm with six joints

axial as well as their radial dimensions prevent a stiff construction for the joints. Depending on their planned applications, and especially for medical systems, the relative advantages of the various design must be carefully considered in order to find the optimum.

Table 2 presents a list of the allowable influence of each system component on the lateral beam movement of an articulated arm with six joints (see Figs. 3 and 4).

Obviously, the first joints have to be the most accurate, due to their large distance from the beam exit. For a common four-point bearing an allowable tilt of 0.33' combined with a typical stiffness of 3–6 Nm/' at the joint. At a distance of about 1 m, measured from the beam exit, and not for a fully extended arm, the first joint should not carry a load effectively exceed-

ing 1 N. In general, this force is already present due to the net mass of the beam-handling system, thus requiring compensation by a counterweight.

A complete weight balancing is possible by employing counterweights (see Fig. 5 a, b) or by arranging a complex system of tension springs (see Fig. 5c).

These constructions are complicated; in addition, they will considerably increase the total weight of the system. For medical applications, in general there is no need for a complete weight balancing: the medico-optical devices (hand pieces, operation microscopes, etc.) are predominantly applied in a few positions only. In these cases, specially designed gas-powered springs are sufficient to achieve the necessary weight balance for the relevant regions. Figure 6 shows such a system.

**Table 2.** Uniform distribution of acceptable beam deviations due to individual system components. According to Fig. 4, the sum equals the total beam directional error at the beam exit for beam-handling systems

| System component | Axial distance to beam exit | Share of the lateral beam deviation | Acceptable pivoting per system component (mrad) |
|---|---|---|---|
| Joint 1 | 1305 | 0.125 | 0.10 (0.33') |
| Joint 2 | 1235 | 0.125 | 0.10 (0.33') |
| Straight 1 | 960 | 0.125 | 0.13 (0.47') |
| Joint 3 | 675 | 0.125 | 0.19 (0.65') |
| Straight 2 | 400 | 0.125 | 0.31 (1.06') |
| Joint 4 | 165 | 0.125 | 0.76 (2.61') |
| Joint 5 | 115 | 0.125 | 1.09 (3.76') |
| Joint 6 | 65 | 0.125 | 1.92 (6.62') |
| Total | 4920 | 1.000 | 4.60 (15.8') |

Complete weight balancing using counterweights

Weight balancing using counterweights

Weight balancing using tension springs

**Fig. 5.** Various methods for weight balancing

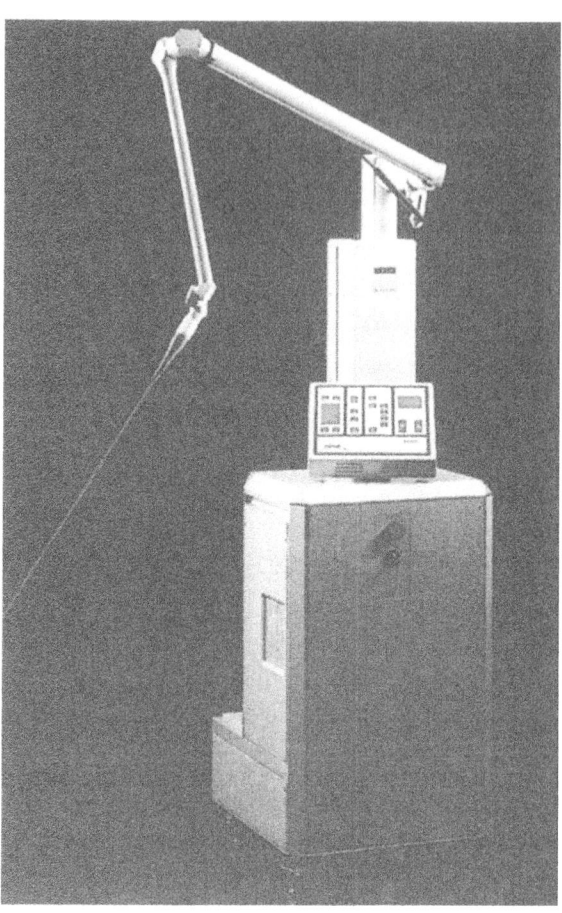

**Fig. 6.** Medical beam-handling system. Its weight is partially balanced by gas-loaded springs (ZEISS opmiLas $CO_2$)

# I-4.2.2
# Light Guides

K.-H. Schönborn

## Contents

## Introduction

The light guide, an optical fibre, has established itself in medicine as a simple light guide, i.e. a random bundle of fibres employed for illumination, and as an ordered fibre bundle for the transfer of images. These "coherent" bundles of light-guiding fibres are the optical backbone of every flexible endoscope.

While fibre bundles are well suited for the transfer of diffuse and incoherent light, single fibres offer the possibility to transfer the concentrated light of the laser and to conserve its special quality to a large degree. Even after passing through a fibre, the extremely high power density emitted by a laser is conserved. Thus, the large mechanical flexibility of the fibre is combined with the laser's properties, significantly increasing the therapeutic applicability of the laser beam.

The therapeutic applications may be divided into two categories: the first includes all applications where the optical fibre merely eases handling of the laser beam. The second kind of laser application would be impossible without a light guide.

Example for the first group of applications: retina coagulation by argon laser and slit lamp. In this case, the design can be simplified significantly by employing an optical fibre. The noisy and vibrating cooling unit of the laser system may be separated from the laser head. Due to the excellent transmission properties of quartz glass fibres, it is possible to remove the laser device completely from the treatment room.

Light guides are essential for all laser therapies requiring the laser energy to be guided into openings or cavities of the human body. Endoscopic methods are employed in urology, internal medicine, gynaecology, pulmology etc. It is also possible, by utilizing a fibre, to guide laser energy into the blood stream, e.g. laser angioplasty, which employs special laser fibre catheters. In these instances, optical fibres considerably extend the applicability of the laser. Beam-handling systems employing fibre optics belong to an ever-expanding group of therapeutic devices.

At present, all light guides used in clinical treatments are exclusively made of high-purity quartz glass. One of the reasons is the rapid development of information technology employing fibre optics. This leads to a highly efficient production of quartz fibres. Information technology requires extremely low light losses, very precisely adjusted optical properties and mechanical strength; qualities also very important in medical applications.

The transmission of quartz glass fibres (see below) is sufficient in the wavelength range between about 200 and 2000 nm. Outside this interval, the transmission losses increase considerably.

Medically important lasers operating inside this wavelength range are Nd:YAG lasers (1064, 1320 nm) as well as argon lasers (488, 514 nm). For these, quartz glass fibres are commercially available and may be used without causing any fibre-specific problems. Quartz glass fibres can also be employed with excimer lasers; however, only its wavelengths 308 and 351 nm will be transmitted. Solid-state lasers based on alexandrite and Ho:YAG also use quartz glass fibres for special applications; although with a Ho:YAG laser they are suitable for relatively short distances only.

Fibres developed for the most important surgical cutting laser, the $CO_2$ gas laser (wavelength 10.6 μm), are trailing technically far behind the quartz glass fibres. Considerable efforts at developing are being made. However, at present, no medically and technically satisfying fibre is available. There are also other interesting lasers, like the Er:YAG laser (2.94 μm), without suitable optical fibre.

For the above reasons, the next sections will deal exclusively with quartz fibres. At the same time, these sections will outline the essential facts of optical light transmission in addition to additional facts which must be taken into account while handling an optical fibre. After this other fibre materials will be considered. Evaluation and outlook will close the chapter.

## Propagation of Light Inside a Light Guide

In general, an optical fibre can be considered to be a dielectric waveguide. There are other types of light guides, working similarly to a waveguide in high-frequency technology, i.e. reflections at metallic boundaries, at electrically conducting layers (see below).

Every dielectric light guide [1] is made up of at least a central core transmitting the light and a surrounding optical cladding (mantle) keeping the light inside. The light guide is based on the physical principle of total reflection.

Total reflection at an interface of two transparent optical media occurs if the light ray approaches from the direction of the optically denser medium (Fig. 1) and if the angle of incidence exceeds the critical angle $\alpha_{tot}$ (angle of total reflection):

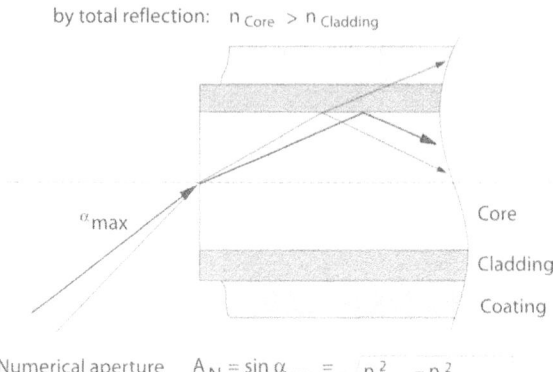

Light inside a light guide by total reflection: $n_{Core} > n_{Cladding}$

$\alpha_{max}$

Core

Cladding

Coating

Numerical aperture    $A_N = \sin \alpha_{max} = \sqrt{n^2_{Core} - n^2_{Cladding}}$

**Fig. 1.** Light propagation inside a dielectric waveguide

$$\sin \alpha_{tot} = \frac{n_2}{n_2}. \tag{1}$$

In this case, a light ray will not pass the interface; instead, the total optical power will be reflected. In principle, total reflection is loss-free. However, the light penetrates a few wavelengths into the optically less dense medium. This indicates that the totally reflecting medium in order to act, must have a certain, wavelength-dependent, minimum thickness.

To utilize total reflection, the central parts refractive index must be larger than that of the cladding.

$$n_{centre} > n_{cladding}.$$

In addition, the thickness of the cladding should be 10 to 20 times the wavelength of the light.

The critical angle of total reflection limits the rays transferred by the light guide: only rays entering the fibre core below a critical entrance angle, measured with respect to the axis of the fibre, will undergo total reflection. All other rays, while being reflected, will lose a fraction of their light to the cladding and will consequently be strongly attenuated. Rays of a power laser entering incorrectly can destruct the fibre (mismatched coupling). Taking into account the refraction of light while entering the fibre, the critical angle for total reflection is:

$$\sin \alpha_{tot} = \sqrt{n^2_{centre} - n^2_{cladding}} \tag{2}$$

This essential quantity characterizing a light guide is called numerical aperture $A_N$.

$$A_N = \sin \alpha_{max}$$

$A_N$ determines not only the coupling efficiency for light from a diffuse source but also the coupling between laser and optical fibre as well as the light losses inside bent fibres; the so-called bending losses (see below).

Although an ideal light guide is loss-free in principle, real light guides display light losses proportional to their lengths. In general, this attenuation of light is expressed in logarithmic form because the laser power P decreases exponentially with fibre length l:

$$P\,(l) = P_0 \cdot \exp\left(-\frac{a}{10 \cdot \lg e}\right) \cdot l \qquad (3)$$

The inverse of this formula presents the attenuation a. expressed in the unit dB per unit of length.

$$a = 10 \cdot \frac{1}{l} \cdot \lg \frac{P_0}{P\,(l)}\,. \qquad (4)$$

The basic unit of length for telecommunication fibres is in general 1 km, consequently attenuation is given in dB/km. For strongly attenuating fibres, e.g. those that are developed for the wavelength 10.6 μm, attenuation is expressed in d/B/m .

Loss mechanisms inside optical fibres are [l]:

- Basic attenuation of a perfect fibre.
- Additional attenuation caused by imperfections in the material like impurities, pores, bubbles, cracks and inclusions.
- Additional attenuation by geometric-mechanical deviations from a perfect fibre, like variations of the core dimensions with length, micro- and marcrobending of the fibre, as well as mechanical tensions.

There are also length-independent losses at the entrance and at the exit. Fresnel reflection losses at not dereflected fibre terminals and the loss of light at soiled or insufficiently prepared fibre terminals. For quartz fibres the reflected fraction of the laser power per terminal face passed is about 3.5%. In general, this is of no significance. There are other fibre materials, e.g. those utilized in the IR spectral range, having much higher reflection losses due to their larger index of refraction (see below).

## Self-Damping of Quartz Glass Fibres

Self-damping of a quartz glass fibre [2] in the UV and in the visible spectral range is predominantly determined by Rayleigh scattering. This scattering mechanism depends on the glass structure and is difficult to influence. It is strongly wavelength-dependent, decreasing proportionally with its fourth power. In the near-infrared, fibre damping increases rapidly with wavelength due to an absorption band in oxygen. These two damping mechanisms limit mainly the spectral range usable with quartz glass fibres.

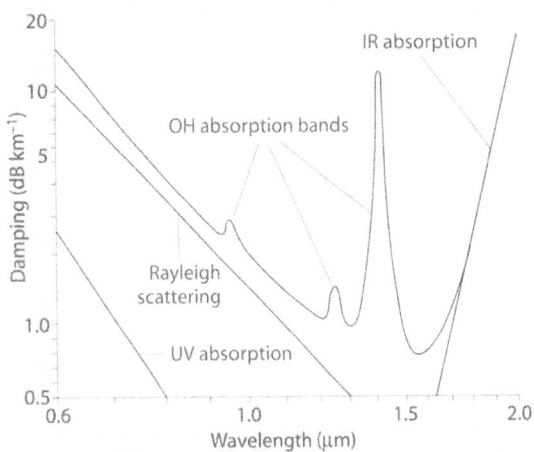

**Fig. 2.** Damping curves for a quartz glass fibre, semilog plot. (after [1])

Figure 2 shows a typical damping graph. The semilog plot allows a clear identification of both damping mechanisms.

The overlapping damping peaks are caused by trapped OH groups (water peaks) and by structural defects introduced during the manufacturing process (e.g. at wavelength 930 nm). Although the addition of $H_2O$ during manufacturing increases these water peaks, it reduces the losses in the vis and UV range by reducing Rayleigh scattering (water containing quartz glass fibres).

The extremely transparent quartz glasses currently used for optical fibres can no longer be melted from raw materials. They are manufactured by a vapour phase reaction. Silicate glasses are the product of the following reaction:

$$SiCl_4 + O_2 \rightarrow SiO_2 + 2\,Cl_2\,.$$

Using high-purity chemical materials, it is possible to reduce the concentrations of the critical contaminants to below a few ppm (ppm = parts per million).

There are two types of quartz glass fibres: fibres having a core of quartz glass and an optical cladding made of plastic and those with core and cladding made of quartz glass. In the latter case, both regions of the fibre cross-section have different doping concentrations.

The first fibre type is made by drawing out a quartz glass rod to a fibre and subsequently cladding it with a low refraction and transparent plastic (plastic-clad silica fibres = PCS). If the cladding is made of an especially hard and well-adhering plastic the fibres are called hard clad silica fibres (HCS). However, sometimes this name is also used to designate a fibre made solely of quartz glass (all-silica fibre).

The second fibre type is made from a so-called pre-form, that is a quartz glass rod with a diameter of about 15 to 150 mm that already has the desired distribution of the refractive index between core and cladding. This preform is then drawn out to a fibre.

There are a considerable number of technologies to manufacture a preform (see e.g. [1] and [3]). All are based on glasses made in vapour phase reactions. This method allows doping elements to be quantitatively added to the reacting gases in order to influence the refractive index of the manufactured glass. The elements germanium, phosphorus and aluminium increase the refractive index, while fluorine decreases it. Depending on the desired distribution of the refractive index, core, cladding or both can be doped accordingly. The concentrations of the doping elements are limited by technological, chemical and thermo-mechanical considerations. That is the reason why quartz glass fibres have a limited aperture (maximum 0.3 to 0.4).

The very complex production methods for the preforms offer, besides extreme purity of the glass, the possibility to continuously vary the refractive index across the fibre diameter. The result is a so-called graded-index fibre having an index of refraction which parabolically decreases toward the outside (Fig. 3).

These fibres are able to support much higher data-transfer rates than step-index fibres. In addition, it is possible to produce fibres having an extremely thin diameter. For core diameters of a few µm the fibre turns monomode, i.e. only one light wave will be transferred. There are also monomode, polarizing fibres as well as polarization-conserving fibres. All these special fibres have been developed for information technology or for fibre sensors. At present they are not utilized in medicine.

Nevertheless, it is possible that the special properties of these fibres will in future become of interest in the medical field [4, 5].

Monomode fibres allow the transfer of radiation from lasers working in the transverse principle mode $TEM_{OO}$ without loss of radiation quality. In this way, the excellent focus control of these laser beams is conserved. It is also possible that some special fibre sensors – at present under development – will be of interest in medicine because they do not require electrical connections, see e.g. [6].

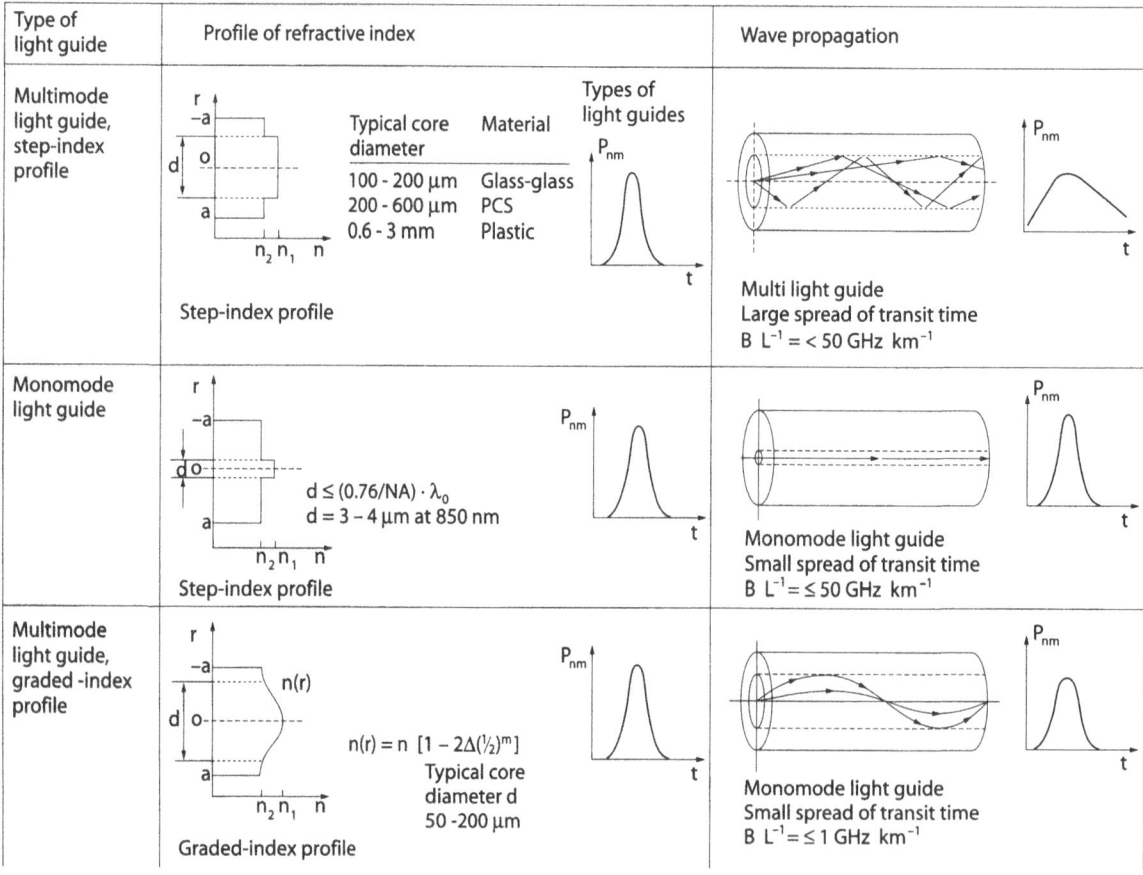

**Fig. 3.** Types of light guides. (after [11])

## Bending Losses

In addition to losses caused by absorption and scattering, a bent light guide experiences so-called bending losses [7, 8, 9]. In information technology the dominating additional losses are caused by so-called microbending. This may be important, for example, in cables containing fibres rounded about sinusoidally, if the lengths of the fibres exceed the length of the cable. Characteristic features: the radius of curvature is large compared with the diameter of the fibre core, and a large number of bends is distributed over a considerable length. Bending losses are noticeable only in long fibres. The additional damping is about 1–2 dB/km$^{-1}$ (i.e. losses of about 20–40% km$^{-1}$) and thus irrelevant for fibres applied in laser therapy. Depending on the kind of application, e.g. cardio-vascular laser angioplasty (CVLA) requires bending radii of under 20 mm, strong local bending is often required, but the usual fibre length is only about 1 m. Summary: laser therapy needs only a few but tight bends while information technology requires many weaker ones.

There are a number of theories dealing with losses due to microbending: however, their application to macrobending is difficult.

Some general statements can be made regarding interesting cases of tight bending in laser therapy [10]:

Bending losses do not increase continuously with decreasing bending radius. The losses start at a fibre-specific value and increase rapidly from there on (see Fig. 4).

For a given bending radius, a fibre having a large numerical aperture is advantageous, because it is less sensitive to bends.

A thin fibre is preferable to a thicker one (see Fig. 4).

If the aperture of a fibre is not fully utilized, a kind of "reserve" against bending losses is created. This reserve will by utilized in fibre bends.

Bending losses are undesirable in medicine not only because they decrease the efficiency of the laser beam transmission. The decoupled laser power may possibly endanger doctor or patient and/or destroy the fibre [11]. Locally emerging power losses can perforate a vessel wall during intravascular fibre application or separate a terminal fibre end inside the catheter. It may be impossible to remove this separated end without opening the vessel.

The above mentioned problems must be solved in the developmental phase of the manufacturing process. In any case, appropriate precautionary procedures must be developed for critical applications of fibres.

As long as other requirements are satisfied, light guides having a large aperture and a small core diameter should be applied if tight bending is unavoidable. The selection of materials for fibre cladding and catheter may be important. These points must be considered in each individual case.

## Power Limits of Quartz Glass Fibres

Previously, laser therapy procedures utilized thermal effects mainly, calling for low average laser power. By now, pulsed lasers exploiting non-thermal effects have become more and more important. Transferring such pulsed laser light via fibres leads to a dilemma: non-linear effects (like plasma creation) are desirable

**Fig. 4.** Bending losses in optical fibres: transmitted power of three different fibres vs. bending radius for a 180° bend. (Presented are core and cladding diameters of the fibres)

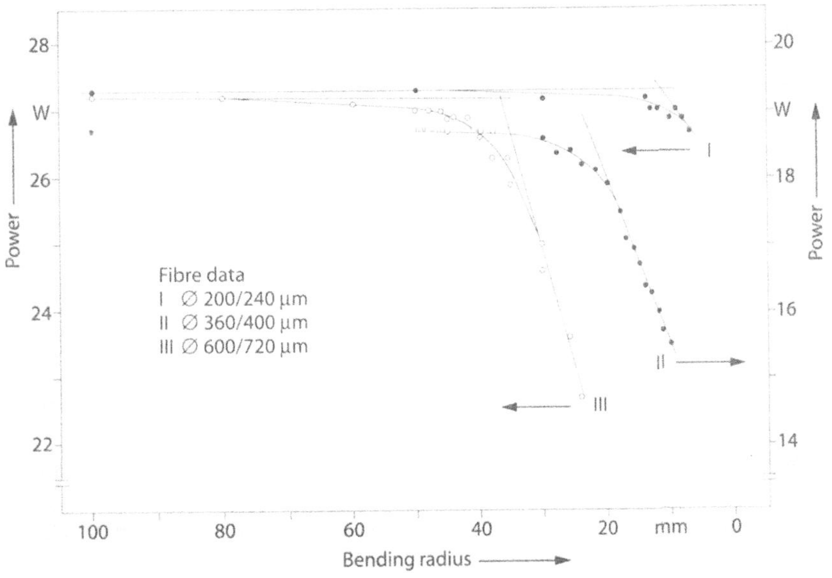

at the target site, non-linear effects inside the fibre or at the fibre surface are undesirable because they jeopardize the fibre. Sometimes, complex techniques are called for, e.g. optical devices or spherical surfaces at the fibre end, optoacoustical converters and concentrators for laser lithotripsy using Nd:YAG lasers [12].

In these cases, the limits for the transfer of light by light guides are encountered; they are summarized below:

Various physical effects limit the maximum laser power transferable by light guides [13, 15]:

- The fibre will be heated by absorbing some of the laser power. The cladding or the fibre itself may melt, decay or be destroyed by thermally induced mechanical stresses. Local overheating at lattice impurities must also be taken into account. The direct cause of fibre destruction is the average transported laser power.
- For quartz glass fibres, the effects mentioned above are not limiting factors, because of their exceptionally low absorption losses, their homogeneity and high resistance to thermal damage. Thermal damage can only be caused by mishandling, like inaccurate coupling of the laser beam, poor preparation of the terminal surfaces, or bending losses due to extremely small radii of curvature.
- For other fibre materials, e.g. those to be used for $CO_2$ lasers at 10.6 µm, thermal destruction is the critical type of damage (see below).
- During the transfer of pulsed laser radiation, very high-power densities and field strengths occur at the fibres terminal surfaces and inside the fibre. If they exceed critical values, the fibre will suddenly be destroyed locally. In many instances, this damage occurs at the fibre's terminal surfaces. The degree of cleanness of these surfaces and their method of preparation clearly determine the damage thresholds. An electrical breakdown in the atmosphere in front of the fibres terminal end is also possible, destroying the fibre. In comparison, the destruction thresholds for causes situated inside the fibre are about one order of magnitude larger. Consequently, there is a clear distinction between the threshold values for the destruction of surfaces and volumes.

The destruction threshold of light-guiding fibres caused by pulsed radiation depends on a multitude of factors. It is difficult to apply numerical values taken from the literature to concrete situations [15]. Well-established factors are:

- Quality of terminal fibre ends:
  Freshly broken fibres and faces are superior to fire-polished ones, which in turn are better than mechanically polished ends. The polishing materials used and their carriers exert a noticeable influence [14].

- Pulse duration [16]:
  The destruction threshold for a fibre surface, expressed as intensity $I_{max}$ is about inversely proportional with the square root of the pulse duration t.

$$I_{max} \sim 1/\sqrt{t}$$

If expressed as fluence F, it is proportional to the square root of the pulse length.

$$F \sim I_{max} \sqrt{t}.$$

- The above was checked for pulse lengths between ps and a few ns [16].
- Laser wavelength:
  Except for possible absorption maxima, the destruction threshold decreases with the wavelength. This is due to its higher quantum energy [17, 18].
- Number of pulses:
  An increased number of pulses will reduce the destruction threshold of quartz glass significantly, as compared with a single pulse. The literature indicates a reduction by a factor of two [19].
- Focusing:
  The type of focusing and the coupling of the laser radiation are very important. Inaccurate optical coupling (focusing into the fibre) or mismatching can lead to internal destruction distal of the fibres entrance surface [20, 21, 22].

Due to the uncertainties mentioned above, only a few reference values are given for the critical power densities of quartz glass fibres [18, 23, 24, 25]:

For Nd:YAG lasers (wavelength 1.06 µm) and pulse lengths between 5 and 20 ns, the destruction threshold of the fibre ranges from 10 to 40 J cm$^{-2}$. This corresponds with intensity values of about 0.5 to 5 GW cm$^{-2}$. Increasing the pulse length to about 100–150 ns decreases the fibre destruction threshold to about 100 MW cm$^{-2}$. For continuously emitting Nd:YAG lasers the destruction limit of a quartz glass fibre is approximately 5 MW cm$^{-2}$.

For excimer laser radiation (308 and 351 nm) with a pulse length of about 15 ns, the critical fluence is 5–10 J cm$^{-2}$. This value increases to 15–30 J cm$^{-2}$ for pulse lengths of 150–250 ns.

Other factors limiting the transfer of pulsed laser light through light guides are non linear effects like induced Raman scattering [26, 29]. These factors are insignificant for fibres and lasers used in medical therapy.

Excimer laser radiation may produce colour centres in quartz glass fibres. The causes are not fully understood. Colour centres will decrease the transmission of a light guide, but without destroying it, proportional to the number of transferred pulses [26, 27,

30]. The question of whether this effect is caused by material impurities or by defects in the glass structure is still awaiting detailed investigation.

## Mechanical Strength of Quartz Glass Fibres

The mechanical strength of optical fibres [31, 32] is of great importance during medical applications: lack of strength can jeopardize the operational safety of therapeutic laser devices. During intracorporal applications breakage of an uncoated fibre will put the patient in danger. During cardio-vascular laser angioplasty, for example, a broken fibre may cause perforation of a coronary artery. Also, it could be impossible to remove a broken-off piece of fibre without surgical intervention.

However, everybody who has used quartz glass fibres will realize that there are no problems regarding a fibre's mechanical strength. Even a very thin quartz glass fibre (e.g. having a standard diameter of 125 µm) cannot be torn by hand. This fibre will not break while negotiating very tight bends.

The extraordinarily convenient properties are the consequences of purposeful design and a production technology that has adapted well to this brittle material. If the mechanical load on a light guide is increased steadily, it will finally break. The break originates at a microscopically small defect, e.g. a microcrack or a pore. The size of such a microdefect determines the resistance of the fibre to fractionation. Consequently, the fibre's strength is solely a material property. It depends on the manufacturing process and the history of the individual fibre, because these determine the size and distribution of microdefects. For these reasons, any predictions regarding the strength of a fibre are of a statistical nature: what kind of microdefect exists at the position of highest load along the light guide will determine the probability of breakage. This fact is taken into account during design and manufacturing of light guides [1]:

- While drawing a light guide, it will be covered by plastic cladding immediately. This will conserve the virgin glass surface.
- The fibres will be mechanically tested immediately, i.e. every length of fibre will be subjected to a preset traction force acting along its axis. Positions weakened by surface defects will break. Every length of fibre passing this procedure will be free of critical microdefects (for the applied test load). Typical test loads for quartz glass fibres are about 200 to 1000 N mm$^{-2}$.

Another important problem related to the mechanical strength of quartz glass fibres is static fatigue strength. During long-term use of a fibre, its load limit decreases continuously. This effect is dominated by stress corrosion cracking: atmospheric water vapour will diffuse the cladding and stimulate growth of subcritical microcracks in the glass surface until they exceed a critical size and the fibre breaks. The speed of this process is controlled by water diffusion. It is possible to increase static fatigue strength by employing so-called hermetic cladding, especially developed for long-term applications.

The influence of static fatigue strength must always be taken into account for fibres stored for a long time, in particular if the fibres are stored in a coiled shape. Appropriate test methods must be found especially for fibre optic components to be sterilized by superheated steam.

Possibly, one-way fibres should be used exclusively for vital therapeutic procedures.

Until now, standardized test procedures are available only for light guides employed in information technology. The special requirements for light guides and components employed in medical therapy call for novel test procedures. Institutions dealing with medical laser applications are being challenged to develop appropriate standards.

## Fibres for Infrared Lasers

Quartz glass fibres, as well as fluoride fibres (see below), have been developed for use in information technology. Other fibres, still under development, can be used exclusively with $CO_2$ lasers ($\lambda = 9.6$ to 10.6 µm) for therapeutic applications [26]. In addition, fibres able to accommodate mean and far-infrared radiation are useful to carry information from sensors, e.g. the results of pyrometer measurements, at relatively low temperatures.

## Fibres for the $CO_2$ Lasers

The $CO_2$ laser offers the following, very specific advantages:
- Very good cutting abilities.
- High efficiency.
- Good radiation quality (operating often in the transverse principle mode $TEM_{00}$).
- Widely distributed as a medical appliance.

Articulated arms with mirrors are used as light-guiding systems for extracorporal applications or employed in open surgery. Appropriate optics allow the $CO_2$ laser beam to be spread or to be focused along the centre line of a rigid endoscope. In both cases, a flexible fibre guide would offer advantages as compared with an articulated arm.

**Table 1.** Fibre materials for $CO_2$ lasers (wavelength 10.6 μm)

| Material | Typical optimum damping values[a] (in dB/m⁻¹) | Losses per unit of length[a] (in % m⁻¹) | Index of refraction[b] | Reflection losses at a surface[c] (in db) | (in %) | Remarks |
|---|---|---|---|---|---|---|
| Chalocogenide glass (Ge, Sb, Se, As, Te) | 0.5 ···· 10 | 10 ···· 90 | 2.5 ···· 1.4 | 0.9 ···· 1.4 | 18 ···· 27 | Toxic |
| Thallium-bromo-iodide (KRS-5, crystalline) | 0.5 ···· 10 | 10 ···· 90 | 2.37 | 0.8 | 16.5 | Toxic |
| AgBr-AgCl (crystalline) | ca. 0.1 | ca. 2.0 | 1.98 | 0.5 | 11 | Water-soluble, plastic |
| Hollow fibre (dielectric) | 1 ···· 10 | 20 ···· 95 | – | – | – | Very sensitive to bending |
| Hollow fibre (metallic reflection coating) | 2 ···· 10 | 5 ···· 90 | – | – | – | Very sensitive to bending |

[a] Conversion: losses in % = 100 × [1–10 (– damping in dB .... 10)]
[b] According to [42]
[c] Reflectivity = reflected/incident radiant power (normal incidence)

$$R = \left(\frac{n_2 - n_1}{n_2 + n_1}\right)^2 = \text{refractive index taken from column 4}$$

For comparison: quartz glass at 1 μm wavelength: R = 3.6% ≙ 0.16 dBI-4.2.2 Figures

Intracorporal applications employing flexible endoscopes or catheters definitely need flexible optical fibres.

Besides some exotic and most likely unfeasible concepts, the developmental efforts concentrate on fibre materials listed in Table 1. They will be discussed below.

A number of compounds composed of the chalkonides Se, As and Te and the elements Ge and/or Sb solidify in a glassy state. They are transparent in the wavelength range of interest. Suitable combinations of these chalkonide glasses may be used to manufacture optical fibres. Possible processes are:

- Drawing out of a rod clad with optical polymers (e.g. urethane-acrylate).
- Rod-tube process.
- Drawing of a homogeneous fibre without optical cladding. The surrounding air will act as cladding. In the latter case, touching the surface of the fibre will cause additional losses.

The literature indicates damping loss values for chalkogenide fibres of about 0.5 to 10 dB/m, i.e. this corresponds to intensity losses of 10 to 90%/m [33, 34, 35]. Commercially available chalkogenide glass fibres have damping values of 5 to 10 db/m. Not only the relatively large losses are a disadvantage, but in addition, chalkogenide glass fibres are toxic – a serious problem during intracorporal applications.

Similar data and problems apply to fibres made of crystalline KRS-5 (thallium-bromo-iodide) ([33]; see Table 1).

More promising are crystalline fibres made of the components silver bromide and silver chloride. Changing the relative contributions of bromide and chloride, it is possible to control the index of refraction. Thus, it is possible to design a core-cladding structure. There are also fibres with a polymer cladding. At present AgBr-AgCl fibres with reasonable damping (about 0.1 dB/m, i.e. loss of about 2%/m⁻¹) are available. Crystalline fibres, in contrast to glass fibres, display plastic properties [43]. During deformation they behave approximately like a copper wire. Most likely this will cause problems if such a fibre is used inside a curved catheter.

Because AgBr-AgCl fibres are non-toxic, they hold great promise for possible intracorporal applications of the $CO_2$ laser. However, their plasticity may make them an interim solution only, with little prospect of their replacing the articulated arm.

Another light guide concept is the hollow waveguide [37]. There are two basically different designs: the dielectric hollow fibre [38] and the hollow fibre made electrically conductive by a metal coating of its inside [39].

A dielectric waveguide is a thin tube. The refractive index of its wall, made of special glasses, is below 1 (for the laser radiation to be transmitted). Consequently, the wall displays anomalous dispersion. The air-filled interior acts as fibre core. Transmission values of 1 to 20 dB/m (i.e. 20 to 95% loss/m) have been reported. The apertures of these fibres are relatively small and are very sensitive to bending.

Guiding fibres based on metallic reflection present relatively high damping values (2 to 10 dB/m, i.e. losses of 5 to 90%/m. Even surfaces of very high reflectivity, e.g. gold or silver coatings, lead to poor damping values for thin fibres, due to multiple reflections.

Overall, the concept of the hollow light guide presents no advantages when compared with the other fibres mentioned above.

## Fibres for the Mean Infrared Range (MIR)

The mean infrared range is dominated by a solid-state laser, the Er:YAG laser. The emitted wavelength of 2.94 µm is strongly absorbed in water and therefore also in tissues. For this reason, Er:YAG lasers are well suited as cutting devices. They are commercially available.

Optical fibres made of heavy metal fluoride glasses are transparent for this lasers radiation. The main components of the fibre are zirconium fluoride and barium fluoride. The theoretically achievable damping values for the wavelength range between 2 and 4 µm are below 0.01 dB/km [40].

Such low damping values have not yet been achieved. Very small impurities, especially transition metals, but also traces of oxygen or OH groups will increase the damping values tremendously. The damping values for these fibres are also increased by bubbles and devitrification caused by the usual casting of the preforms for core and cladding. The interface core cladding is also far from ideal. Commercially available fibres achieve, at $\lambda = 2.94$ µm, damping values of about 0.5 to 1 dB/m, i.e. 10 to 20%/m [33, 41].

Since fluoride fibres are currently under intense investigation, further improvements of the damping values can be expected.

## Summary

From the point of view of handling, mechanical strength, non-toxicity and state of development, quartz glass fibres dominate all other fibre types. For the transfer of high laser power, there is no competitor in the wavelength ranges near UV, VIS and NIR. Up to now, the development has been mainly aimed at optimizing and matching the components of the total therapy system, consisting of laser, fibre, connectors, applicators and special digital terminal devices.

An exception is the UV below 300 nm. Here fibre-specific problems are still to be solved, especially regarding pulsed lasers (excimer lasers).

The development of suitable fibres for special cutting lasers, like $CO_2$ lasers and Er:YAG lasers is by no means completed.

Great promise is held by AgB-AgCl fibres for $CO_2$ lasers and heavy metal fluoride fibres for Er:YAG lasers. In the first system the fibre is developed fur-

thest, in the second the laser. An interesting competition can be expected.

Er:YAG lasers (wavelength 2.94 µm) coupled with suitable heavy metal fluoride glass fibres and $CO_2$ lasers also utilizing suitable fibres, e.g. made of AgBr-AgCl, would be compatible systems. Both systems would be well suited as intracorporally cutting lasers.

The Ho:YAG laser (wavelength 2.08 µm) perhaps offers a compromise with respect to minimizing the transmission losses in the fibre and maximizing the absorption in the tissue to be abraded:

The proven quartz glass fibres (hydrophobic quality) are sufficiently transparent for short fibre lengths. Interesting and intracorporal applications can be expected in the near future in connection with the pulsed radiation emitted by the Ho:YAG laser and its high absorption in tissues.

## References

[1] LUTZKE D (1986) Lichtwellenleitertechnik. München
[2] KECK DB, MAURER RD, SCHULZ PC (1973), On the ultimate lower limit of attenuation in glass optical waveguides. Appl Phys Lett 22: 307
[3] ROSENBERGER D (1982) Optische Informationsübertragung mit Lichtwellenleitern. Grafenau/Württ.
[4] SCHONBORN KH, KERSTEN RTH, KORAYSAHI N (1987).Faseroptische Systeme für nicht-nachrichtentechnische Anwendungen. Proc MIOP'87, vol. 3, UV-1
[5] SCHONBORN KH (1987) Einsatz von Lichtwellenleitern: Auswahl und Konfektionierung. Electro J 22: 62
[6] Fiber optic and laser sensors V: Proc SPIE 985 (1989). Fiber optic and laser sensors VI: Proc SPIE 1169 (1990). Chemical, biochemical and environmental fiber sensors: Proc. SPIE 1172 (1989)
[7] GLOGE D (1972) Bending loss in multimode fibers with graded and ungraded core index. Appl Opt 11: 2506
[8] MARCUS D (1976) Curvature loss formula for optical fibers. J Opt Soc Am 66: 216
[9] WINKLER C, LOVE JD, GHATAK AK (1973) Loss calculations in bent multimode optical waveguides. Opt Quant Electro 11: 173
[10] SCHONBORNKH, BADER HCHR (1988) Advantages of using thin fibers. Proc. SPIE 906: 238
[11] SCHONBORN KH, WODRICH W (1988) Handling and safety aspects of fiber optic beam delivery systems. Proc SPIE 906: 244
[12] WONDRAZEK F, FRANK F (1988) Devices for laser-induced shock-wave lithotripsy (LISL) In Steiner (ed.) Laser-lithotripsie. Berlin p. 123
[13] SCHONBORN KH, KERSTEN RTH, BADER HCHR (1982) Faseroptische Lichtwellenleiter für die industrielle Fertigung. Laser Mag 3: 19
[14] HACK H, NEUROTH N (1982) Resistance of optical and coloured glasses to 3 nsec laser pulses. Appl Opt 21: 3239
[15] SCHONBORN KH, KERSTEN RTH (1986) Transmission of high-power laser light through optical fibers. Proc EFOC/LAN86: 227
[16] MILAM D, LOWDERMILK WH, RAINER F (1981) Pulse-duration dependence of 1064 nm laser damage thresholds of bare surfaces and optical thin films. Appl Opt 20: 169
[17] MERKLE LD, KOUMVAKALIS N, BASS M (1984) Laser induced bulk damage in $S_1O_2$ at 1.064, 0.532 µm. J Appl Phys 55: 772
[18] TAYLOR RS, LEOPOLD KE, BRIMACOMBE RK, MIHAILOV S (1988) Dependence of damage and transmission properties of fused silica fibers on the excimer laser wavelength. Appl Opt 27: 3124

[19] MERKLE LD, BASS M, SWIMM T (1983) Multiple pulse laser-induced bulk damage in crystalline and fused quartz at 1.064 and 0.532 µm. Optic Engin 22: 405

[20] WONDRAZEK F, FRANK F (1988) Lichtleitereinkopplung für Hochleistungslichtimpulse. Laser Optoelektro 20: 62

[21] KAR H, HELFMANN J, MÜLLER G, RINGELHAN H, SCHALDACH B (1989) Optimization of the coupling of excimer laser radiation (308 nm) into Q-Q fibres ranging from 200 to 600 µm core diameter. Proc SPIE 1067: 223

[22] SCHONBORN KH, KERSTEN RTH, KOBAYASHI N (1986) Highpower laser beam delivery systems in surgery: the technical aspect. Proc SPIE 1067: 223

[23] WEBER HP, HODEL W (1986) High power light-transmission in optical waveguides. Proc SPIE 650: 102

[24] TAYLOR RS, LEOPOLD KE, MIHAILOV S, BRIMASCOMBE RK (1987) Damage measurements of fused silica fibres using long optical pulse XeCl lasers. Opt Comm 63: 26

[25] SOWADA U, KAHLERT HJ, BASTING D (1988) Excimerlaser-Strahlung durch Quarzglasfasern – Grenzen und Möglichkeiten. Laser Optoelektron 20: 32

[26] RINGELHAN H, KAR H, HELFMANN J, DOERSCHEL K, MÜLLER G (1988) Lichtwellenleiter für die Medizin. Laser Optoelektron 20: 44

[27] NEVIS EA (1985) Alteration of the transmission characteristics of fused silica optical fibers by pulsed UV-radiation. Proc SPIE 540: 421

[28] VILHELMSSON K (1986) Simultaneous forward and backward Raman scattering in low-attenuation single-mode fibres. J Lightwave Technol 4: 400

[29] AOKI Y, TAJIMA K, MITO I (1988) Input power limits of single mode optical fibers due to stimulated Brillouin scattering. J Lightwave Technol 6: 710

[30] MÜLLER G, KAR H, DÖRSCHEL K, RINGELHAN H (1988) Transmission of short pulsed high-power UV laser radiation through fibres... Proc SPIE 906: 231

[31] MURATA H (1988) Handbook of optical fibers and cable. New York

[32] KURKIJAN CR, KRAUSE JT, MATTHEWSON MJ (1989) Strength and fatigue of silica optical fibers. J Lightwave Technol 7: 1360

[33] KLOCEK P, SIGEL GH JR (1989) Infrared fiber optics. Bellingham. WA USA

[34] RUSCONI DM, SIGEL GH JR (1988) Optical power propagation in chalkogenide fibers at 10.6 µm. Proc SPIE 929: 119

[35] NISHII J, YAMASHITA T, YAMAGISHI T (1989) Chalkogenide glass fibers with a core-cladding structure. Appl opt 28: 5122

[36] GAL D, KATZIR A (1987) Silver halide optical fibers for medical applications. J Quant Electron QE 23: 227

[37] NATTERMANN K, HOFFMANN H-J, NEUROTH N (1988) Which parameters are required for hollow fibers transmitting high-power IR-radiation? Proc SPIE 929: 124

[38] FALCIAL R, GIRENSI G, SCHEGGI AM (1984) Oxide glass hollow fiber for $CO_2$-laser radiation. Proc SPIE 494: 84

[39] DROR J, MENDLOVIC D, GOLDENBERG E, GOITROU N (1987) Hollow plastic waveguides with metal and dielectric films. Proc SPIE 843: 88

[40] SHIBATA S et al (1981) Prediction of loss minima in infrared optical fibres. Electron Lett 17: 775

[41] TRAN DC, LEVIN KH, MOSSADEGH M (1988) Surgical applications of heavy metal fluoride glass fibers. Proc SPIE 929: 115

[42] DRISCOLL WG, VAUGHAM W (ed) (1978) Handbook of optics. New York

[43] SAITO M, TAKIZAWA M, MIYAGI M (1988) Optical and mechanical properties of infrared fibers. J Lightwave Techno 6: 233

# I-4.3
# Optical Applicators

F. Frank, S. Hessel and C. Krampe

## Contecnts

## Introduction

Interaction between laser radiation and biological material will result in various tissue reactions, depending on wavelength and method of application. Examples are denaturation, abrasion and cutting of tissue, but also fusion of tissues, called tissue welding. The therapeutic application of laser radiation can be clinically useful only if the appropriate instrumentation is available [1].

Laser therapy needs transmission systems and applicators to transfer the radiation safely, and with losses as low as possible, to the point of application. In addition, the laser beam geometry is important for achieving the planned effect. Transmission systems are optical fibres, bending light guides and articulated arms. They transport the radiation from the laser's resonator to the point of application. Light guides are made up of flexible, transparent fibres. They are mostly composed of quartz glass, usable with laser radiation in the visible range as well as in the near-IR and UV. Articulated arms are employed with the infrared radiation of for example $CO_2$ lasers. Mirrors guide the radiation through interconnecting tubes.

Applicator systems are concerned with the desired beam geometry at the point of application. In addition, they allow easy handling of therapeutically applied laser radiation.

Since the laser has been introduced into medicine, a number of different applicators have been developed for day-to-day clinical use. This number is expanding with the introduction of new fields of application. For application methods to be accepted readily, new applicator devices must be developed, keeping in mind that they should be easily adaptable to well-known and standard instrumentation. Otherwise, the user will encounter problems using laser technology.

## Applicators Emitting Divergent Beams

Fibre light guides transmitting laser radiation have high mechanical stability and flexibility. This makes it possible to manufacture very thin and flexible applicator systems, usable with flexible endoscopes, for application in body cavities and narrow operating channels, operation microscopes and special probes, e.g. for intra-ocular use. A great variety of systems is available, especially with Nd:YAG and Ar lasers. Their transmission systems incorporate the transmitting device itself, the light-guiding fibre. In these arrangements, the light-guiding fibre determines the radiation divergence of the applicator. The divergence is defined by the coupler between laser and fibre at the laser head and by the numerical aperture of the fibre itself.

## Coupling and Radiation Divergence

To couple laser and fibre, the laser beam leaving the resonator is focused on the terminal end of the fibre. Laser radiation is transferred along the fibre by total internal reflection. The radiation leaving the distal end is divergent (see Fig. 1).

The focal length f of the coupler lens and the diameter d of the laser beam determine the coupler angle $\varepsilon_\sigma$. The radiation enters the light guide within the angle $\varepsilon_\sigma$.

$$\tan \varepsilon_\sigma = d_0/2f$$

$\varepsilon_\sigma$ is limited by the aperture angle of the fibre; it is indicated as numerical aperture (sine of the aperture angle) of the fibre.

The aperture angle – the divergence of radiation leaving the fibre end – for an ideal straight cylindrical fibre is equal to the coupler angle. In reality, the ever-present bending of the fibre will increase the divergence ($\varepsilon_1 \geq \varepsilon_0$). Very tight bending will even cause power losses.

Example. For a typical Nd:YAG laser, having a beam diameter of d = 4 mm and a coupler lens with f = 28 mm, the coupler angle can be calculated to be 4.1°. Consequently, the full angle of divergence for the exiting beam will be at least 8.2°. Figure 2 indicates measuring arrangement and a typical intensity distribution of such a light guide emitting divergent radiation. The light guide has a core diameter of 600 μm and the divergence is measured by a detector rotating around the fibre end at a distance of 10 cm.

The aperture angle of a fibre radiating in water is reduced by about 25% due to the different indices of refraction for air and water.

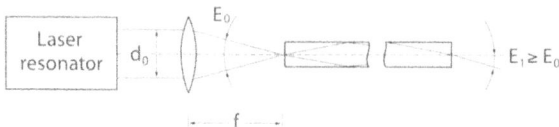

**Fig. 1.** Optical design of the coupling and transmitting of laser radiation by a glass fibre

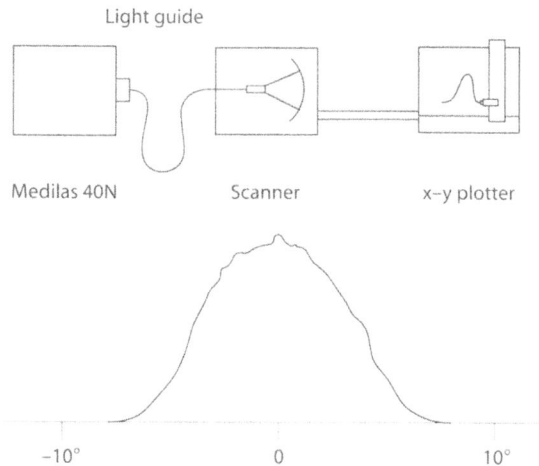

**Fig. 2.** Measuring arrangement and emission profile for measurement of the emission characteristics of a light guide with a core diameter of 600 μm

## Coupling into Thin Fibres

A prerequisite to coupling into a thin light guide is the ability to achieve a very small spot diameter by focusing. This is possible because a parallel beam leaves the resonator. However, although laser radiation is the most collimated form of electromagnetic energy, it will always have a very small divergence, caused by the configuration of the resonator and the modal structure. In addition, the divergence of solid-state lasers (e.g. Nd:YAG lasers) will be influenced by thermo-optical properties of the laser rod and by the emitted laser power.

The full angle of divergence α of the laser radiation leaving the resonator and the focal length f of the focusing lens, will determine a minimum focal diameter $d_{focus}$, which defines the minimum fibre diameter where coupling is possible.

$$d_{focus} = f \cdot \alpha$$

This implies that for a given divergence of the laser radiation, the focal length of the coupler lens must be reduced if the fibre's diameter decreases. Consequently, the coupler angle will increase, leading to an enlarged emission divergence for thinner fibres. For medical laser applications, the most popular light guiding fibres have core diameters between 100 and 600 μm.

## Power Density

To calculate power density, which determines the effect on tissue, at the point of application the divergence of the radiation leaving the light guide must be taken into account:

Power density (W cm$^{-2}$) = power (W) / irradiated area (cm$^2$)

The diameter D of the irradiated area (beam diameter) can be calculated from 1, the distance between fibre and tissue, the exit angle $\varepsilon_1$ of the emitted radiation and the core diameter $d_{fibre}$ of the fibre:

$$D = d_{fibre} + 2 \cdot l \cdot \tan \varepsilon_1 .$$

Consequently, the irradiated area F is:

$$F = (D/2)^2 \cdot \pi .$$

Assuming a Gaussian intensity distribution, 86.5% of the radiated power P arrives within this area. In general, the power density PD is given by a simplified formula, yielding to a good approximation:

$$PD \ (W \ cm^{-2}) = P \ (W) \cdot 100 \ / \ D \ (mm^2)$$

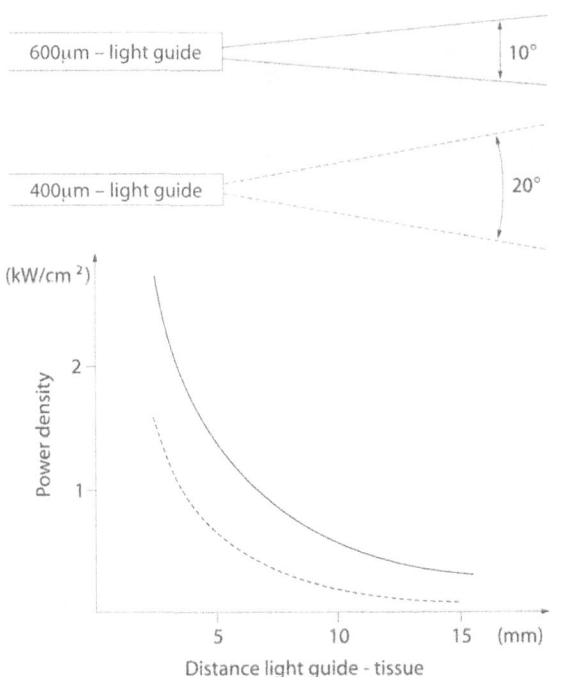

**Fig. 3.** Beam geometries for 600 μm and 400 μm light guides and pertinent power densities (30 W laser power)

Figure 3 indicates beam geometries and power densities of light guides having core diameters of 400 and 600 μm and emitting radiation into air.

## Flushing

Radiation with defined emission characteristics can be emitted with low losses only if the terminal fibre ends are free from scratches, cracks and contaminants. A damaged surface will distort the radiations scattering pattern and consequently heat neighbouring metal parts, e.g. applicators. Contamination will absorb laser radiation directly at the exit area and may cause an immediate charring of the fibre.

Applicators emitting divergent radiation must be placed very close to the tissue in order to achieve the necessary power densities. This poses the danger of contaminating the tip of the light guide with tissue or blood, especially by the vapours created during tissue vaporization employing Nd:YAG lasers. Flushing the light guide with gases or liquids serves two purposes: keeping the laser tip free from contamination and removing blood and blood coagulants. In this way, greater penetration depths can be achieved and the view of the operation field is kept clear.

Depending on circumstances, flushing agents are $CO_2$, $N_2$, compressed air, distilled water or solutions of NaCl. Liquid flushing achieves, in principle, better cooling of the tissue surface and thus will protect the surface to a certain degree.

In this way, deep coagulation may occur before carbonization effects commence [2, 3].

## Gas-Flushed Endoscopy Light Guides

Fiber light guides with integrated flushing are available for Nd:YAG lasers employed in endoscopy procedures in gastroenterology and pulmonology. The flushing fluid enters at the hand-held attachment of the light guide. Teflon tubing will conduct the fluid to the distal fibre end; it will leave through a nozzle. The total diameter of such a light guide is between 1.9 and

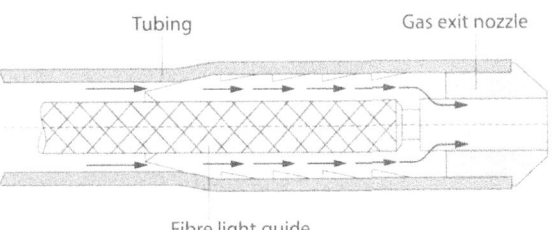

**Fig. 4.** Diagram of the gas nozzle at the distal end of a light guide flushed with gas

2.6 mm. All common endoscopes can be employed without modifications (see Fig. 4).

Although, in general gas is employed as flushing medium, the latest concentrically flushed light guides may be used with liquid flushing also. This may yield special tissue interactions effected by surface cooling [4].

Gastroenterology in general employs light guides with outer diameters of 2.6 mm. This will assure a sufficiently large flow of flushing gas, about 500 ml/min, to protect the light guide and to improve observation. The core diameter of the glass fibres is 600 μm. The light guides are fed into the working channels of flexible endoscopes and, for therapy advanced as far as possible (see Fig. 5).

At the disposal of pulmonologists are light guides having outer diameters of 2.1 mm and enclosing 600-μm fibres as well as 1.9-mm light guides with 400-μm fibres. It is also possible to feed these into the working channels of rigid and flexible bronchoscopes. There are special rigid bronchoscopes equipped with a mechanism which allows perpendicular irradiation of the tracheal wall. A separate channel will extract the vapour from tumour vaporization. With a diameter of 13 mm, its applications are limited (see Fig. 6).

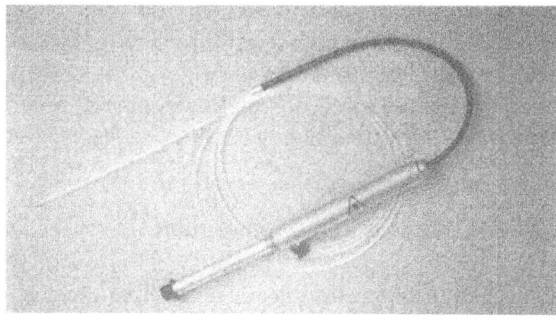

**Fig. 5.** Gas-flushed light guide for applications in endoscopy

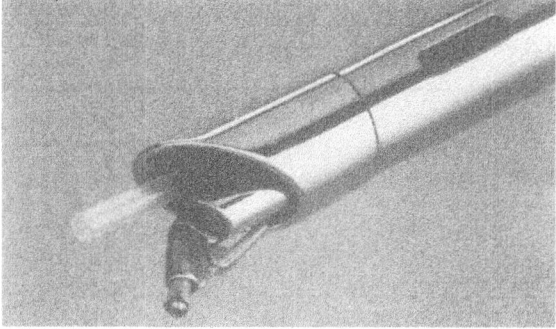

**Fig. 6.** Details of a rigid bronchoscope with suction and breathing channels including a working channel for the laser's light guide

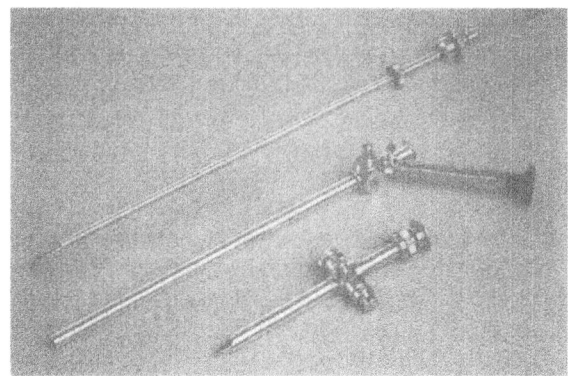

**Fig. 7.** Laparoscopy set, laser insert with bending mechanism

Gynaecology employs gas-flushed light guides in laparoscopy. The 11 mm diameter laparoscope uses a laser device with a fibre-bending mechanism ([5]; see Fig. 7).

## Liquid-Flushed Endoscopy Light Guides

Nd:YAG laser therapy utilizes liquid-flushed light guides in urology, neurosurgery and hysteroscopy procedures in gynaecology [6–12]. In these cases, flushing is not integrated with the light guide, but is separately attached to the endoscope. Only its distal part, the bare fibre, with an outer diameter of about 1 mm, is introduced into the working channel, and only its end will be flushed by the liquid (see Fig. 8).

A specially designed and spring-loaded nut connects light guide and endoscope. The spring-loaded device makes it possible to move the fibre, thus controlling the distance between fibre end and tissue. The free length of the fibre can be adjusted to fit the endoscope actually used (see Fig. 9).

The increasing multitude of endoscope models calls for further development of these fibres, to adjust them to the various types. The protective tubing of a universal light guide (see Fig. 10) moves in a telescopic fashion in order to adjust for the length of the endoscope.

A standard cystoscope (19–20 Charr), with a special laser attachment, uses a liquid-flushed light guide for the coagulation of bladder tumours. An Albarran lever makes it possible to bend the light-guiding fibre through angles up to 80°. This makes feasible the vertical irradiation of almost any region at the bladder wall. The circulating flow of flushing fluid will maintain the bladder's volume. For retrograde irradiation there is available a highly flexible fibre (400 μm core diameter) having a specially designed laser adapter which allows bending of the fibre through angles up to 130° (see Fig. 11).

**Fig. 8.** Diagram of a liquid-flushed light guide inside an endoscope

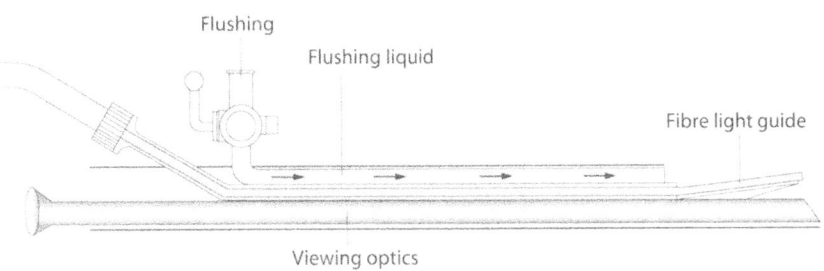

Flushing

Flushing liquid

Fibre light guide

Viewing optics

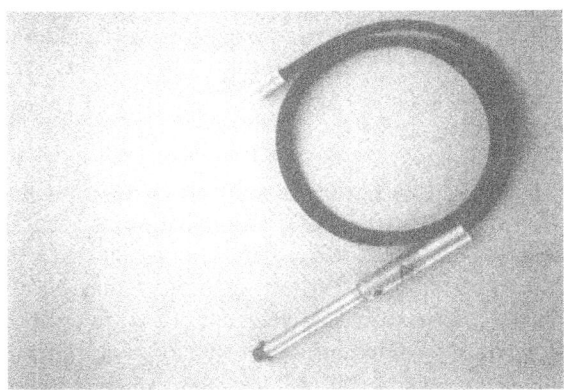

**Fig. 9.** Liquid-flushed light guide having a core diameter of 600 μm for endoscopic applications

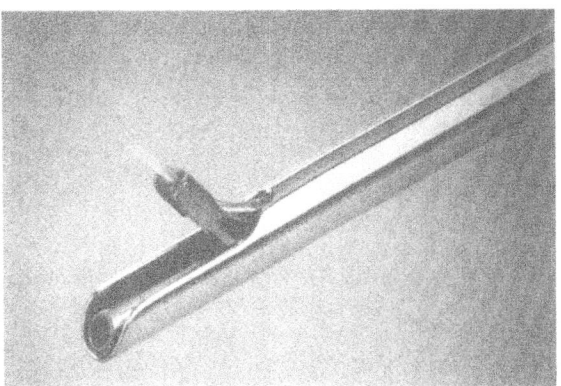

**Fig. 11.** Cystoscope: laser guiding channel with Albarran lever

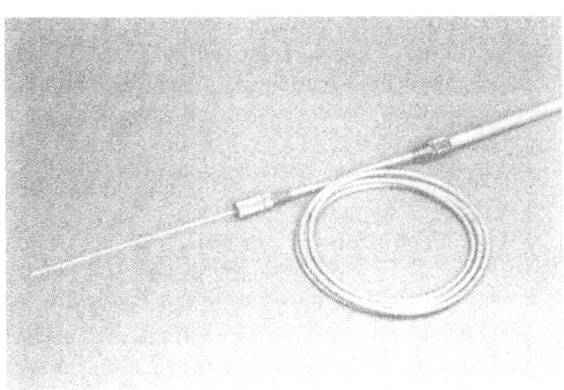

**Fig. 10.** Length-adjustable, liquid-flushed light guide

Similar instrumentation is utilized for hysteroscopy procedures. The intra-uterine pressure can also be controlled by the inflow of the flushing liquid.

Endoscopy Nd:YAG laser therapy is also employed in neurosurgery. A special laser encephaloscope is used, having the same type of flushing and light guide movement [13].

## Gas- and Liquid-Flushed Hand-Held Applicators

Hand-held applicators are used for laser applications in open surgery. Due to their slim construction, they hardly encroach upon the field of view of the operation microscope and do not impede the work in narrow operating channels. They contain a light guide (600 μm core diameter) ending at the distal tip of the applicator and radiating a divergent pattern.

Hand-held and gas-flushed applicators receive the flushing gas through the couplers grip of the light guide; the gas is led concentrically to the exit nozzle at the distal end. Hand-held applicators with an outer diameter of 4 mm are available in various shapes, adjusted to the different requirements of diverse situations during the surgical intervention (see Fig. 12).

The flushing liquid enters directly at the hand-held attachment; it flows around the light guide inside the distal tube. Various shapes are available and the outside diameter is 3 mm. For regions difficult to access, there are hand-held applicators with a flexible tube. This tube can be adjusted for all situations which may occur during an operation ([13]; see Fig. 13).

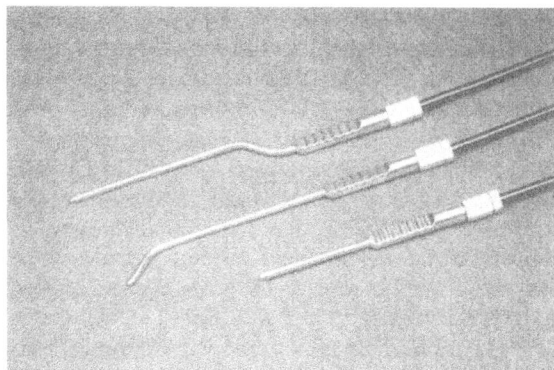

**Fig. 12.** Gas-flushed hand-held applicator

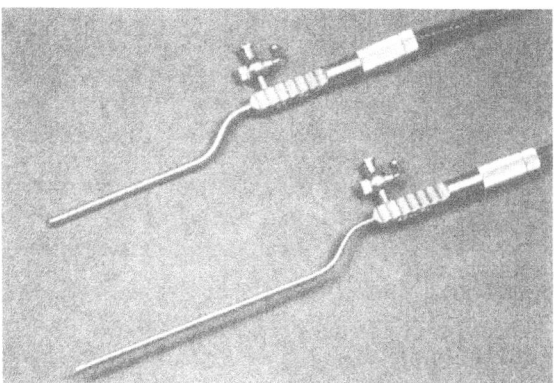

**Fig. 13.** Liquid-flushed hand-held applicators

## Focusing Applicators

Laser surgery, i.e. cutting and vaporizing of tissues, requires high power densities. These are achievable by concentrating the laser radiation into a small focal spot. The beam diameter determines the cutting width or the width of the vaporization zone, respectively. A smaller focal spot permits more delicate procedures to be performed, in addition to allowing reduced laser power. A hand – held attachment, placed at the end of the transmission system, houses the focusing optics, consisting of one or more lenses. The focal spot of the invisible laser radiation will be marked by a concentrically located pilot laser (e.g. HeNe laser).

Very precise and delicate work must proceed in the focal spot, because the width of the beam increases distally and proximally. To employ a wider beam, it is only necessary to increase the separation between fibre and tissue. Decreasing this distance will place the focal spot inside the tissue, causing uncontrolled tissue damage.

Cutting and vaporizing tissue produces much smoke. Besides providing efficient extraction, flushing the focusing applicator is important in order to clear the view of the operation area and to prevent contamination of the optics. In addition, this will cool the optics, a necessity if very high power densities are employed.

## CO₂ Laser Applicators

Due to wavelength-specific tissue effect, the $CO_2$ laser is a first-rate cutting instrument; the damage to surrounding tissues remains insignificant. This advantage is limited to focused beams. If unfocused, the beam will abrade the tissue layer by layer. Consequently, all $CO_2$ laser applicators are focusing systems. They achieve larger beam diameters by changing the distance between applicator and tissue, or by employing special defocusing devices.

Currently, no fibres are available to transmit the 10.6-μm wavelength of $CO_2$ laser radiation in a clinical setting. To transmit this radiation, an articulated arm must be employed. Various applicators may be attached to it: hand-held attachments, rigid endoscopes or operation microscopes enclosing focusing optics.

The articulated arm will transmit $CO_2$ laser radiation leaving the resonator with a virtually unchanged beam geometry (see Fig. 14).

Resonator configuration and mode structure determine the beam divergence; this influences the ability to focus the beam. The achievable focal diameter D is proportional to the full angle of divergence α of the beam leaving the resonator and to the focal length f of the focusing lens:

$$D = f \cdot \alpha$$

Lasers operating in the transverse principal mode $TEM_{00}$ have a Gaussian intensity profile and display the least divergence. Typically, a lens with a focal length of 200 mm will achieve a focal diameter of 350 μm. For a focal length of 50 mm, the focal spot will have a diameter of 100 μm. The power densities will increase appropriately. Choice of the lens for a given application depends on the minimum required beam diameter, the circumstances of the operation, and focal length required by the applicator.

To permit cutting, a power density of at least 1000 W cm$^{-2}$ is required. Table 1 indicates focal spot diameters and power densities for various focal lengths and working distances of a laser set at 30 W output power.

A standard focusing hand-held attachment, with a working distance of 125 mm is available for external applications and hand-held surgery employing a $CO_2$ laser. It is fixed to an articulated arm [14]. An attachable spacer eases the difficulty of working in focus and, at the same time prevents an unintended shift of the focal spot into the tissue. Hand-held attachments

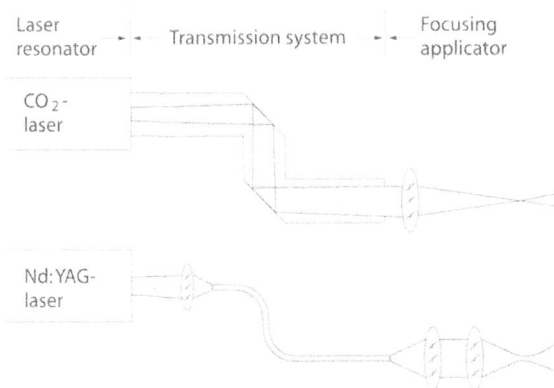

**Fig. 14.** Diagram of optical transmission systems and focusing applicators for a $CO_2$ and an Nd:YAG laser

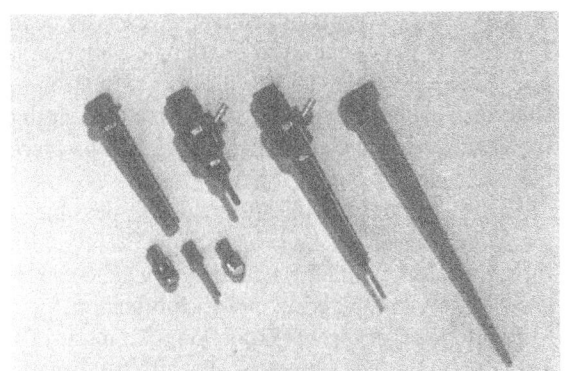

**Fig. 15.** Hand-held applicators for $CO_2$ lasers

**Table 1.** Typical beam diameters and power densities as a function of working distances (example: $CO_2$ laser)

| Working distance f (mm) | Beam diameter D (mm) | Power density at 30 W PD (W cm$^{-2}$) |
|---|---|---|
| 50 | 0.1 | 300 000 |
| 125 | 0.2 | 75 000 |
| 200 | 0.35 | 24 500 |
| 250 | 0.57 | 9 200 |
| 350 | 0.75 | 5 300 |
| 450 | 1.20 | 2 100 |

[a] PD (W cm$_{-2}$) = P (W) · 100/D (mm$^2$)

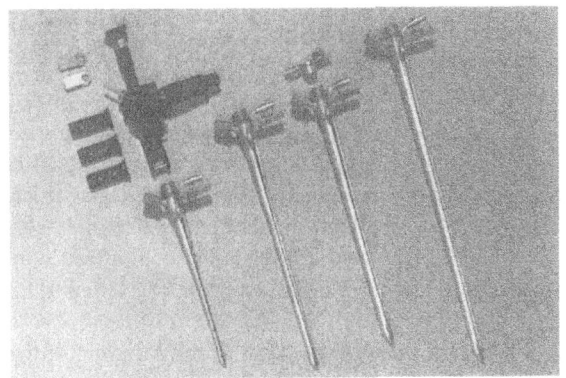

**Fig. 16.** $CO_2$ laser bronchoscopy set

with focal lengths of 50 and 200 mm are available. Attachable mirrors allow bending of the beam by 90° or 120° in order to reach into regions hidden from direct view. Nitrogen flushing is integrated into the hand-held attachment to clear the operation area of vapours during tissue vaporization (see Fig. 15).

For endoscopic treatments special laser endoscopes are available [15–21]. A coupler connects the articulated arm of the $CO_2$ laser in such a way that therapy and pilot beams are focused inside the field of view of the rigid endoscope. A built-in and movable mirror allows manipulation of the beam within the field of view. This calls for rigid endoscopes having a relatively large volume.

Figure 16 shows a laser bronchoscopy set. A swivelling rapid coupler connects endoscope and bronchoscope. To exchange instruments, the components can easily be decoupled and moved sideways. Via an eyepiece (4 x) the coupler allows direct and concentric view of the operation field. Depending on the length of the shaft used, lenses of 250, 350 and 450 mm focal lengths can be exchanged by a revolving coupler.

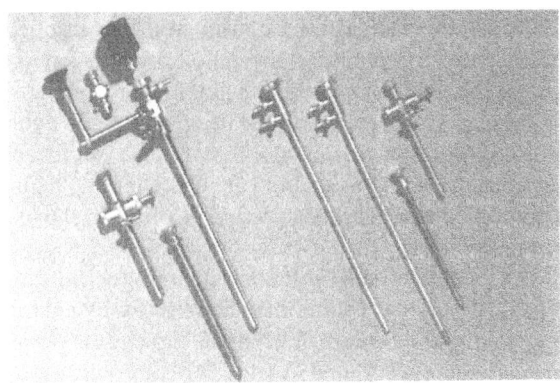

**Fig. 17.** $CO_2$ laser laparoscopy set

For rectoscopy there is a corresponding set available with an endoscopy coupler having focal lengths of 150 and 250 mm.

For laparoscopy (single or double-puncture technique) a very small and compact coupler connects articulated arm and endoscope. It contains only the manipulator to allow exact positioning of the laser beam. The coupler may be utilized in a single-puncture laparoscope with a parallel viewing port or in an 8-mm shaft for double punctures. According to the length of the shaft used, a 200- or 300-mm lens will be placed into the coupler (see Fig. 17).

For endoscopic applications, it is possible to forgo flushing of the laser applicator. However, it is recommended to employ a suction device to remove vapours developing during tissue vaporization.

In microsurgery an operation microscope and $CO_2$ laser can be coupled concentrically with the direction of view. A movable mirror guides the laser beam in the microscope's field of view. A lens focuses the beam into the focal plane of the microscope.

## Focusing Hand-Held Attachments for the Nd:YAG-Laser

Focusing hand-held attachments are employed for external point like applications using an Nd:YAG laser, e.g. in dermatology; and, in addition for precise tissue vaporization using high-power densities in open surgery. These hand-held attachments form a unit consisting of the light-guide transmission system and focusing optics.

The beam divergence distal to the focusing optics exceeds the divergence of a $CO_2$ laser by about 10°, due to coupling into the light guide fibre. Consequently, it is optically more difficult to focus radiation of a Nd:YAG laser into a limited focal spot (see Fig. 14).

To achieve a large power density in the focal spot it is necessary to minimize aberration. In addition, for easy handling the applicator should not be too long. A three-lens system, imaging the beam exit at the light-guiding fibre on a 1:1 scale, fulfils these conditions optimally. For a typical fibre core diameter of 600 μm this results in a focal spot diameter of about 0.6–0.7 mm (see Fig. 18).

An optical system, with a focal length (or working distance) of 50 or 70 mm may be selected. By enlarging the tissues distance, it is possible to change power density and focal spot diameter. For comparison, Fig. 19 indicates both quantities as a function of distance

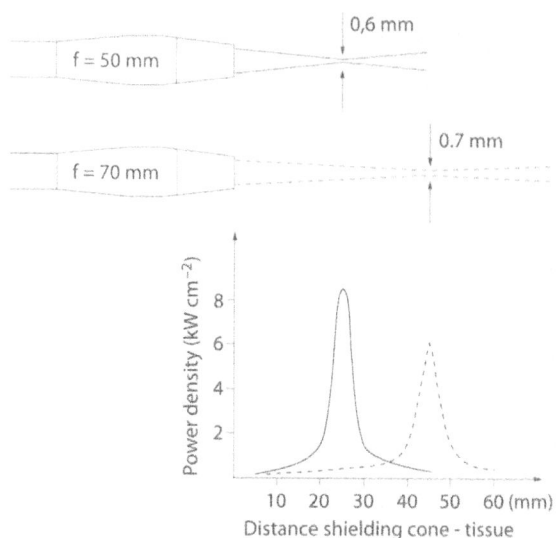

**Fig. 19.** Beam geometry of a focusing hand-held attachment with 50-mm and 70-mm focal lengths. The graph presents the pertinent power densities (30 W laser power)

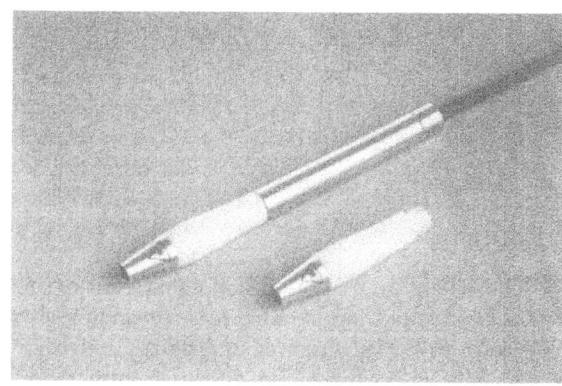

**Fig. 20.** Focusing hand-held attachment with interchangeable optics

for focusing hand-held attachments having a local length of 50 and 70 mm.

To protect the optical system against contamination during tissue vaporization, a metal jacket is attached in such a way that it reduces the free working distance by about 25 mm. In addition, focusing hand-

**Fig. 18.** Diagram of the optics of a hand-held focusing applicator for Nd:YAG lasers

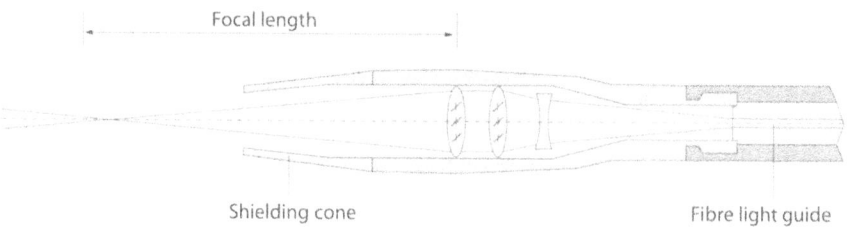

held attachments with integrated concentric gas flushing are available in order to keep the optics free from contamination and the operation field free from smoke (see Fig. 20).

## Contact Applicators

Various laser applicators have been developed recently to improve working in direct contact with tissues. The reason for this development is improvement in methodology, better control of laser power density and improved knowledge of the laser radiation – tissue interaction. All this has led to a considerable extension of the application range for lasers, e.g. in gastroenterology and gynaecology, as well as opening up entire new fields such as recanalization of blood vessels.

### Contact Tips

Employing contact tips yields the same wavelength-specific tissue effects as without direct tissue contact. However, the regions affected are considerably smaller. Contact tips are used mainly in connection with Nd:YAG lasers. They can also be utilized with other wavelengths, for example with the radiation of an Ar laser [22].

The contact tip is transparent for the wavelength applied. Its efficiency is determined, in principle by concentrating the incident radiation in a very small volume of tissue. With appropriate applicators and using a low power setting – about 10–25 W for a Nd:YAG laser – very high power densities can be achieved [23].

If the contact tip emits very low power over an extended period of time, it will cause deep coagulation, comparable with laser radiation applied without contact. As soon as the power density is sufficient for carbonization, the tissue's properties change and vaporization commences. Thereafter, the carbonized tissue particles at the contact tip will facilitate direct absorption at the surface and thus the conversion of the applied energy into vaporization of tissue.

**Table 2.** Material data for sapphire and quartz glass [24]

|  | Sapphire | Quartz glass |
| --- | --- | --- |
| Chemical composition | $Al_2O_3$ | $SiO_2$ |
| Index of refraction (1064 nm) | 1.75 | 1.45 |
| Mohs hardness | 9 | 7 |
| Melting temperature (°C) | 1800–2050 | 1300–1500 |
| Specific heat (cal $g^{-1}$ $K^{-1}$) | 0.18 | 0.177 cal $g^{-1}$ $k^{-1}$ |
| Thermal conductivity | 8.6 | 0.35 [cal $s^{-1}$ m K] |

Consequently, Nd:YAG lasers may be utilized to vaporize very small tissue volumes by direct contact. Thus, they achieve a certain cutting ability. This, together with their good coagulating features, allows almost bloodless cutting and dissecting.

In general, contact tips are made of synthetic single-crystal sapphires. Sapphire, $Al_2O_3$, is highly transparent over a wide wavelength range. It is physiologically inert, resistant to acids and bases, has a very high melting temperature (about 2000 °C), and, for a non-metal, a very large thermal conductivity. Due to its hardness it can only be worked by diamonds (see Table 2).

Contact tips are attached to the distal end of the transmission system with the help of a metal connector. There is a small gap between fibre and tip (see Fig. 21).

Because reflections of the laser radiation at the optical interfaces will heat the connector, the gap must be cooled continuously. The coolant flows concentrically with the light-guiding fibre and leaves through openings in the connector. The coolant may be $CO_2$, $N_2$, compressed air, water or a solution of NaC1. Since the absorption losses are higher in the case of a liquid coolant, this will require a higher setting of the power output.

Different shapes of contact tips will cause different emission characteristics, according to the laws of geo metrical optics and consequently different tissue effects.

**Fig. 21.** Diagram of a light guide with a contact tip for endoscopic applications

Tubing    Quartz fibre   Coolant        Metal mounting    Sapphire

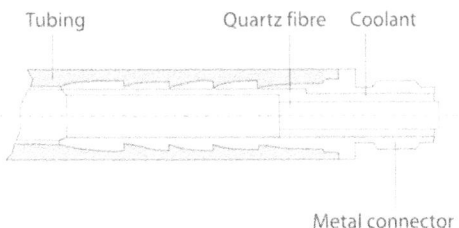
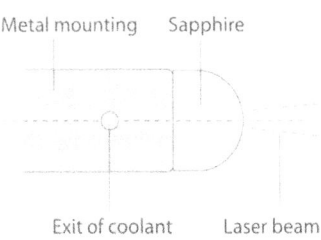

Metal connector        Exit of coolant    Laser beam

**Fig. 22.** Optical design of a cutting tip

**Fig. 23.** Optical design of a vaporization tip

## Cutting Tips

Conically shaped contact tips are used for fine cutting of tissues. Radiation is guided inside the sapphire tip to its apex by repeated total internal reflection at the interface with the surrounding and optically less dense materials (air, water or tissue). The radiation emerges with a large angle of divergence (see Fig. 22).

For a tip diameter of 0.4 mm and a power of 10 W, the power density at the tip reaches about 7500 W/cm$^2$. This corresponds to a power setting of 125 W and a distance of 5 mm if one of the common applicators emitting divergent radiation is employed. A tissue cut made by a conically shaped contact tip is insignificantly wider than the tip's diameter. A narrow coagulation zone surrounds the cut. The tip will become worn after extensive use, even if applied properly. Emission characteristics change, cutting abilities deteriorate and cuts will become coarser [25]. Cutting tips are applied if very delicate preparatory work is called for, e.g. in gynaecology for conization, in breast carcinoma recidives or vulva carcinoma, in microneurosurgery, in ear, nose and throat diseases, or in dermatology to remove melanomas. This cutting probe can also be utilized for laparoscopic lysis of adhesions, or for intratauterine anomalies [26].

## Vaporizing Tips

A hemisperical shape of the vaporizing tip acts as a focusing lens for laser beams in air. If laser radiation is emitted into surrounding water or tissue it will diverge slightly, due to small differences in the refractive indices (see. Fig.23).

A low divergence of the emitted radiation is observed for a "bullet probe". Its geometry is a combination of conical and hemisperical shapes. However, it is possible that transmission losses will occur, caused by uncontrolled total reflections at the surfaces of the beam's exit area [27]. These contact tips are well suited to vaporize small tissue volumes. They are applied in gastroenterology to treat tumour growth on endoprostheses or to relieve total intestinal stenoses. Lately, the main applications have shifted towards angioplasty. Vaporizing contact tips are employed for the recanalization of peripheral vessels which are not blocked by calcified stenoses [28].

## Coagulation Tips

The emission characteristics of cylindrical coagulation probes are like those of light-guiding fibres. They are recommended for the therapy of excess bleeding in gastroenterology. Contact pressure on the tissues exerted by the coagulation tip supports the blood-staunching effect during laser coagulation. If the beam power setting is too high, vaporization at the position of the beam's exit may be initiated. The temperature at the contact position causes large strains inside the crystal; cracks develop and the tip will eventually be destroyed.

There are other differently shaped crystal tips available. Wedge-shaped contact probes are employed for shallow removal of tissues and to combine cutting with vaporization. For interstitial irradiation and hyperthermia, contact probes with a rough surface are available. Diffuse laser radiation is emitted throughout the entire rough surface.

For endoscopic use, the light guides are flushed concentrically. The probes have a threaded nozzle – the connector – at their distal end. The contact tips are fastened to this connector. To calibrate the light guide, or for non – contact applications, a jacket can be attached. It protects the fibre end and thread against damage and contamination. Most manufacturers produce contact tips and light guides with diameters of 1.8 and 2.2 mm, suitable for endoscopes with 2 or 2.4 mm diameter working channels (see Fig. 24).

Special hand-held applicators are available for microsurgery and all open surgical applications. Practically all of them are used for cutting and dissection, i.e. with conical contact probes. Various lengths and shapes for different applications are on offer, as well as various systems, including one-way applicators. Some of the 2.2-mm diameter contact probes will also fit hand-held applicators. Figure 25 illustrates a hand-held applicator with an outside diameter of 3 mm. The correspondingly larger probes are more easily handled in open surgery and for deeper cuts. By screwing on a protective cover, they can also be used in non-contact laser therapy.

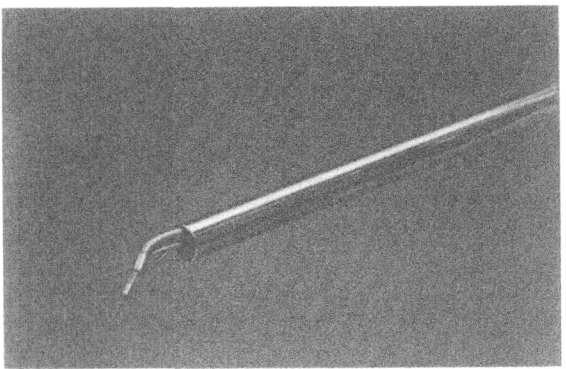

**Fig. 24.** Light guide with cutting probe inside a laparoscope

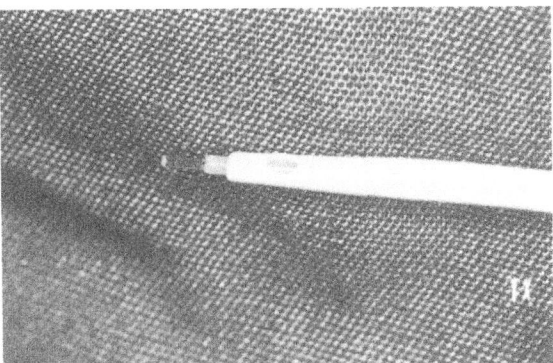

**Fig. 26.** Catheter with bare fibre for vessel recanalization (prototype)

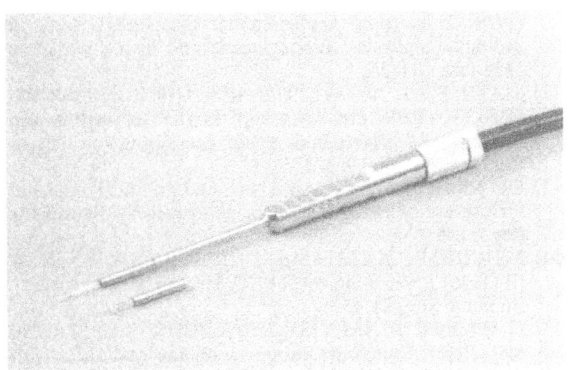

**Fig. 25.** Hand-held applicator with contact tip

### Bare Fibres

The bare fibre method employs a light-guiding fibre without its cladding. The quartz glass core vaporizes tissue in direct contact with it. In this case the transmission system is part of the applicator system. Laser radiation leaves a fibre with 600 µm core diameter at the full angle of divergence (8–10°). For thinner fibres, especially 400 µm fibres, the angle of divergence is larger.

Using a Nd:YAG laser, the action is similar to that of a sapphire probe. However, since the melting point of quartz glass is significantly lower (see Table 2), this method can induce only lower tissue temperatures. Time and power parameters of laser applications must be carefully controlled to prevent destruction of the fibre (see Chap. I-4.1.1.). Experience indicates that this method offers interesting [29] possibilities in percutaneous transluminal laser angioplasty. The critical warming-up and cooling-off periods of the bare fibre, because of its smaller volume, is very short (ca. 0.1 s) compared with that of a sapphire probe (cooling time ca. 2 s). During this phase vaporization does not occur; nevertheless, vessel walls may be thermally damaged [30].

Nonlinear interactions are mostly based on non thermal mechanisms; this can be observed, for example for excimer laser radiation (see Chap. I-3.2.4).

Special applicators are employed for laser angioplasty (see Fig. 26). The development of the therapy systems mentioned above requires close cooperation between manufacturers of catheters and laser systems because the light guides are an integral part of the catheters.

### Hot Tips

Hot tips are contact applicators transferring the energy to the tissue via a metal probe. The simplest design is an olive-shaped, hollow metal cap attached to the plastic lining of a light-guiding fibre. The end of the fibre protrudes into the tip. The metal absorbs the laser radiation and converts its energy into heat (see Fig. 27). The heated probe transmits only thermal energy, no direct radiation energy at all, to the tissue. Openings at the sides of the probe allow the escape of gases heated inside the probe.

The effect of the hot tip on tissue is independent of the wavelength of the laser radiation heating the probe. A hot tip can be used together with an Nd:YAG or an Ar laser. The Nd:YAG laser emits 10–12 W. The metal probe needs about 3–5 s to reach its working temperature of 400–600 °C [31]. Its cooling period is equally as long. During these phases undesirable thermal tissue damage may occur.

For vessel recanalization [32] various hot tip designs (1.5 to 2.5 mm diameters) are offered, together with the appropriate catheters in sterile single-usage packages. Some metal probes have a wire attached parallel to the light guide. This guiding wire is intended to ease manipulation inside the vessel. A short wire, connecting the metal probe and light guide, acts as a safety device in case the probe disconnects from the fibre during therapy. A 0.3-mm diameter channel

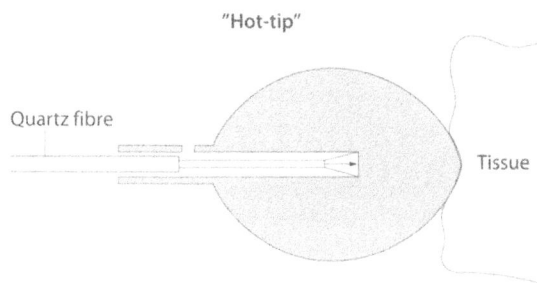

**Fig. 27.** Diagram of a hot tip

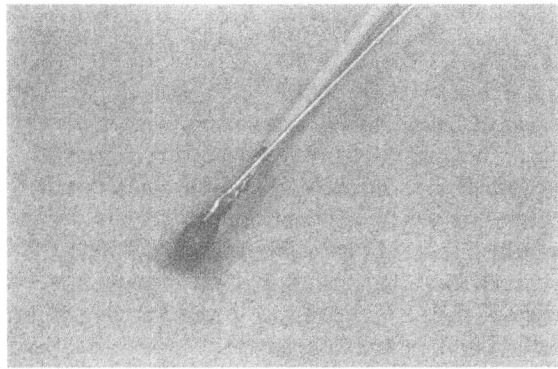

**Fig. 28.** Hot tip for vessel recanalization

in the metal probe facilitates the attachment to a guiding wire, previously positioned (see Fig. 28).

Hot tips with a diameter of 2.5 mm are offered for therapy of gastrointestinal bleeding, and coagulation and vaporization of obstructive tumour tissue. Due to the purely thermal effects, they are supposed to act on the tissues surface only.

So-called spectraprobes combine the actions of thermal and laser energy. These metal probes contain a sapphire window at their tips, thus allowing depending on design, 40 to 80% of the laser radiation to leave within an angle of 24° or 60°. Their applications include vessel recanalization, ablation and vaporization of tumour tissues.

### References

[1] FRANK F (1986) Biophysical basis and technical prerequisites for the endoscopic and surgical use of the neodymium-YAG laser. Laser 2: 124–132
[2] KROY W, HALLDORSSON T, LANGERHOLC J (1980) Laser coagulation: practical advice from a theoretical viewpoint. Appl Optics 19: 6
[3] ROTHENBERGER K, PENSEL J, HOFSTETTER A, KEIDITSCH E, STERN J (1981) Dosierung der Neodym-YAG-Laserstrahlung zur endovesikalen Anwendung bei Blasentumoren. Urologe A 20: 310–314
[4] SANDER R, POESL H, SPUHLER A, STROBEL M (1987) Nd:YAG laser with waster jet stream – a new transmission system. Abstr Laser Surg Med 7: 111

[5] LOMANO JM (1985) Photocoagulation of early pelvic endometriosis with the Nd:YAG laser through the laparoscope. Reproduct 30: 77
[6] HOFSTETTER A, FRANK F (1979) Ein neues Laser-Endoskop zur Bestrahlung von Blasentumoren. Fortschr Med 97: 232
[7] FRANK F, HOFSTETTER A, BOWERING R, KREIDITSCH E (1979) Endoscopic application of the Nd:YAG laser in urology. Biophysical fundamentals and instrumentation. SPIE Proc 211: 36
[8] ROTHENBERGER K, PENSEL J, HOFSTETTER A, KEIDITSCH E, FRANK F (1983) Transurethral laser coagulation for treatment of urinary bladder tumors. Lasers Surg Med 2: 255
[9] FRANK F, BAILER P, BECK OJ, BOWERING R, HOFSTETTER A (1983) Instrumentation for the surgical application of the Nd:YAG laser. SPIE Proc 405: 105
[10] LOFFER FD (1987) Hysteroscopic endomerial ablation with the Nd:YAG laser using a nontouch technique. Obstet Gynecol 69: 679
[11] FRANK F, HOFSTETTER A, BOVERING R, KEIDISCH E (1979) Endoscopic Application of the Nd:YAG laser in urology, biophysical fundamentals and instrumentation. SPIE Proc 211: 36
[12] STAEHLER G, HOFSTETTER A; FRANK F, HALLDORSSON TH (1978) Ein Zystoskop für die Applikation von Neodym-YAG-Laserstrahlung zur Zerstörung von Blasentumoren. Akt Urol 9: 271
[13] FRANK F (1984) Biophysical basis and technical requisites for the use of Nd:YAG laser in neurosurgery. Neurosurg. Rev 7: 145
[14] PENNINO R, LANZAFAME RJ, HERRERA HR, HINSHAW JR (1986) Applications of the $CO_2$ laser in general surgery. Contemp Surg 28: 13
[15] SIMPSON GT, SHAPSHAY SM; VAUGHN ChW, STRONG MS (1982) Rhinologic surgery with the carbon dioxide laser. Laryngosc 92: 412
[16] OSSOFF RH, KARLAN MS, SISSON GA (1983) Endoscopic laser Arytenoidectomy. Laser Surg Med 2: 293
[17] OSSOFF RH, DUNCAVAGE JA, GLUCKMAN JL et al (1985) Universal endoscopic coupler for bronchoscopic $CO_2$ laser surgery: a multi-institutional clinical trial. Otolaryngol – Head and Neck Surg 93: 824
[18] BAGGISH MS (1983) Laser endoscopy in obstetrics and gynecology. Clin Obstet Gynecol 26: 366
[19] TADIR Y, KAPLAN I, ZUCKMERAN Z, EDELSTEIN T, OVADIA J (1984) New instrumentation and technique for laparoscopic carbon dioxide laser operations: a preliminary report. Obstet Gynecol 63: 582
[20] MARTIN DC (1985) Laser techniques for pelvic adhesions. In: MS BAGGISH ed. Basis and advanced laser surgery in gynecology. Appleton-Century-Crafts, Norwalk, pp 331
[21] DANIELL JF, HERBERT CM (1984) Laparoscopic salpingostomy utilizing the $CO_2$ laser. Fertil Steril 41: 558
[22] FASANO VA; PEIRONE SM, FISCELLA B et al (1986), Preliminary experiences with contact Nd:YAG and argon laser in neurosurgery. Proc 4th General and Scientific Meeting of LANSI
[23] DAIKUZONO N, JOFFE SN (1985) Artificial sapphire probe for contact photocoagulation and tissue vaporization with the Nd:YAG laser. Med Instrument 19: 173
[24] Optics guide 3. Melles Griot 1985
[25] WALLWIENER D, POLLMAN D, MORAWSKI A, BASTERT G, KRAMPE C (1989) Die Nd:YAG-Laser Kontakttechnik mit Saphir-Schneidespitzen. Laser Med Surg Suppl 1: 11
[26] WALLWIENER D, MORAWSKI A, PLANTERER G, BASTERT G (1987) Ist die Adhäsiolyse mittels Nd:YAG-Laser eine Alternative zur $CO_2$-Laser-Adhäsiolyse? Laser Med Surg 2: 142
[27] VERDAASDONK RM, CROSS FW, BORST C (1987) Physical properties of sapphire fibre tips for laser angioplasty. Lasers Med 2: 183
[28] LAMMER J, PILGER E, KLEINERT (1988) Laser angioplasty by sapphire contract probe. Experimental and clinical results. Intervent Radiol 3: 1

[29] KLEPZIG M, NEUBAUR T, STRAUER BE, RICHTER EL, ZEITLER E (1987) Transfemorale periphere Laserangioplastie. Dt Med Wschr 112: 324

[30] HESSEL S, FRANK F, ISCHINGER T, HEINTZEN M (1987) Possibilities for the use of Nd:YAG laser in vascular recanalization. Vortrag, 1st German Symposium on Laser Angioplasty. Berlin

[31] VERDAASDONK RM, BORST C, BOULANGER LHMA, van GEMERT MJC (1987) Laser angioplasty with a metal laser probe (hot tip): probe temperature in blood. Lasers Med 2: 153

[32] CUMBERLAND DC, SANBORN TA, TAYLER DI, et al. (1986) Percutaneous laser thermal angioplasty: initial clinical results with a laser probe in total peripheral artery occlusions. Lancer 1: 1457

# I-4.3.1
# Laser Applicators for Interstitial Coagulation

A. Roggan

## Contents

## Introduction

Interstitial coagulation by laser energy is being used – and has been used for some time – to treat pathological tissue changes in various regions of the body. This contact procedure is applied to benign changes, e.g. to hereditary and acquired vessel anomalies [6, 48, 49], or to benign prostate hyperplasia (BPH) [24, 25, 27, 35–37, 39–42, 44, 45] as well as the destruction of malignant tumours or metastases. In the latter case, it is mainly a palliative treatment [1–5, 7, 8, 12, 14, 16, 18, 20, 22, 25, 29–34, 38, 45, 56, 60–67]. The aim is to congenital massive coagulation of the affected regions by raising the tissue temperatures to 45–100 °C. This procedure may succeed either by closure of the supporting vessel system or by destruction of the cells due to thermal damage [10]. The temperature distribution depends on optical properties of the tissue and additional factors like thermal conductivity and local blood perfusion [43, 52–54, 69].

Lasers emitting radiation in the near-IR are used for interstitial tissue coagulation. The field is dominated by Nd:YAG lasers (1064 nm) and by diode lasers (800–980 nm), more recently introduced [2, 15, 19, 26, 43, 68]. Lasers emitting radiation in the near-IR operate inside the so-called optical window, consequently their radiation penetrates deeply into biological tissues (typically 2–10 mm). Therapy of large volumes at moderate temperature gradients is thus possible [54].

As opposed to many other laser applications, interstitial tissue coagulation calls for relatively long irradiation times (2–20 min), low laser power (3–30 W) and a continuous-wave (cw) irradiation mode. In this way, coagulation can be achieved for volumes with diameters of up to 5 cm.

Transmission of the laser energy into the target region is by optical fibres with small diameters. These can transfer visible light and near-IR radiation over long distances. They are flexible and have low losses. Depending on the application, special applicator systems can be attached to the distal end of the light guide in order to optimize the achievable coagulation volumes for local conditions and to reach a high coagulation efficiency. The puncture of the distal end of

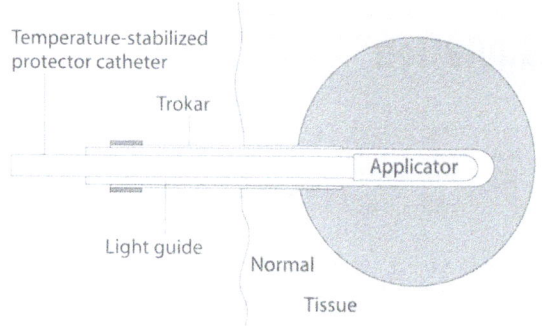

**Fig. 1.** General presentation of a percutaneous interstitial laser application

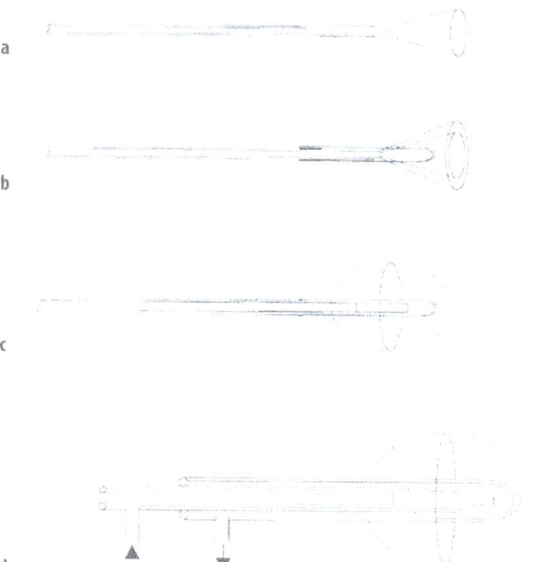

**Fig. 2.a-d** Schematic presentation of various applicators for interstitial laser applications. **a** Bare fibre. **b** DORNIER ring-mode applicator. **c** Scattering applicator. **d** Cooled scattering applicator

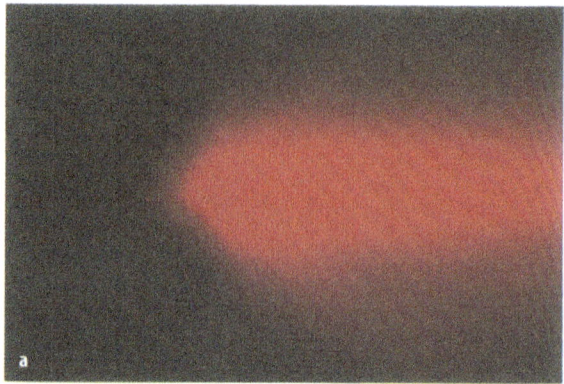

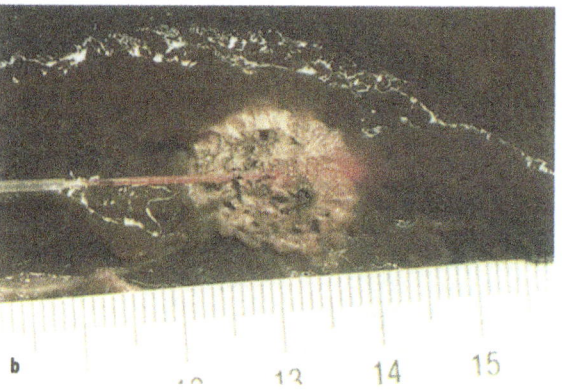

**Fig. 3a, b.** Bare fibre (400 μm core diameter. NA 0.37). **a** Emission characteristic inside a weakly scattering liquid. **b** Interstitial coagulation (in vitro), porcine muscle, Nd:YAG laser, 3 W, 5 min)

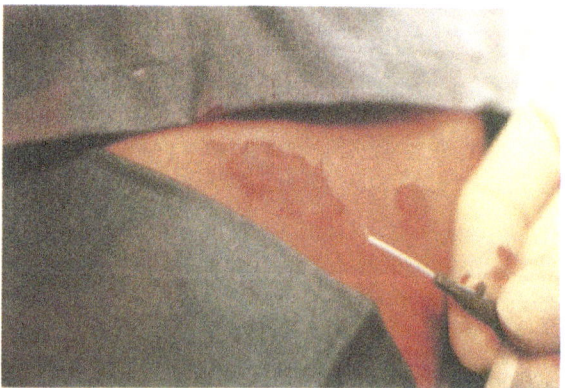

**Fig. 4.** Bare fibre application for therapy of a haemangioma. (Photo Prof. Berlien)

the light guide into the treatment region is monitored by palpation or under ultrasound, computer or magnetic resonance tomography. Figure 2 gives an overview of various applicators which are described in detail below.

## Bare Fibres

Initially, interstitial laser applications were performed by the so-called bare-fibre technique, see Figs. 3 and 4.

The terminal end of a optical fibre is moved directly into the target volume. This technique was optimized, and documented in detail, during more than 2000 applications, especially treating congenital malformations (CVD) [6, 48, 49].

In principle, the application of bare fibres is restricted to small lesions (d < 1 cm) due to their prograde emission characteristics. This is a consequence of the relatively high power densities at the terminal faces of the fibres even for low laser powers. At the terminal face of an optical fibre generally used in hospitals (600-μm core diameter) and with a laser power of 3 W. The power density is about 1 kW cm$^{-2}$. This output will raise the temperature after only 1 to 2 min. above a 100 °C threshold, initiating vaporization which, in the case of subcutaneous localization is clearly perceptible as crepitation. This crepitation may be used during therapy as a criterion for the necessary repositioning of the fibre or for switching off the laser.

If the laser beam is not switched off after commencement of vaporization, the temperature will exceed 300 °C and carbonization of the perifocal tissue layers begins. Carbonization will increase the absorption coefficient dramatically. The laser beam can no longer penetrate deeply into the tissue and the temperature will be raised even further. These processes are the reason for restricted power and time regimes during bare-fibre applications [2, 9, 22, 46, 68, 69].

Due to the strong scattering of laser radiation in biological tissues, the prograde emission has practically no influence on the shape of the coagulation volume. In the neighbourhood of the fibre's end , the beam intensity distribution decreases radially, leading to spherical lesions, as long as the influence of local inhomogeneous blood perfusion (local cooling) is neglected.

The advantage of the bare-fibre technique is the very small fibre diameter, thus making possible an interstitial positioning of the fibres with the assistance of thin catheters. This reduces stress on the patient and decreases the operation risk. The disadvantage of limited lesion sizes can be compensated for by multiple application tracks or retraction of the fibre in the access channel; however, this will increase duration of the treatment. Combined with a beam-splitter, bare fibres can also be applied to palliative tumour therapy [4, 5, 13, 18, 25, 31, 34, 38, 46, 56, 60–62, 67].

The aim of further developments, especially for therapy of large volumes (palliative therapy of tumours and metastases, treatment of prostate adenoma), is the design of distal applicators having an outer diameter as small as possible while transferring higher laser powers over longer periods of time. This would increase the efficiency for coagulation. It can be achieved only by reducing power density at the contact surface of the applicator to the tissue. Inevitably, this calls for applicators with increased outer diameters.

## Ring-Mode Applicators

The first of the special applicators for interstitial applications was the so-called ring-mode applicator (MBB/Dornier, see Fig. 5) [15, 19].

By introducing the radiation into the fibre at a given angle to the fibre's axis, selected modes are excited inside the light-guiding fibre and the laser radi-

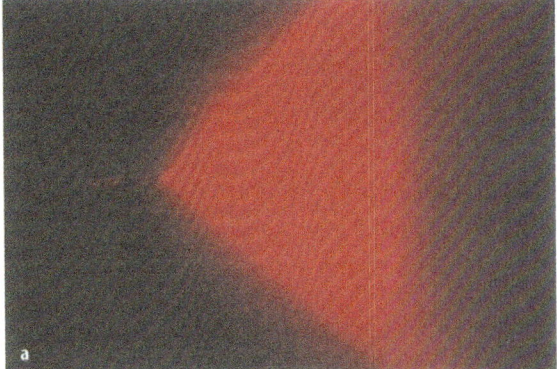
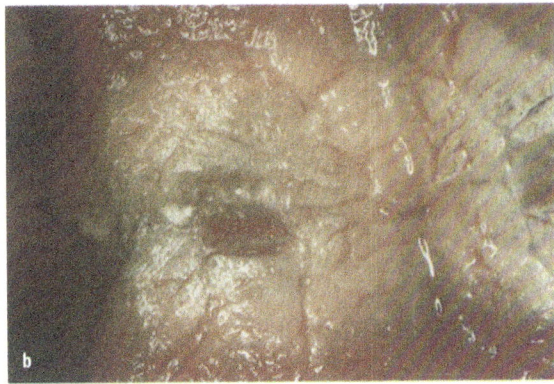

**Fig. 5a, b** . Ring-mode applicator (MBB/DORNIER) **a** Emission characteristic inside a weakly scattering liquid. b Interstitial coagulation (in vitro, porcine muscle, Nd:YAG laser, 5 W, 10 min)

ation will display a conical emission characteristic when leaving the fibre. Employing a glass dome, the radiative power will be spread over a large radial surface and the power density as compared with that of a bare fibre will be reduced.

To use a ring-mode applicator a special laser adaptor is needed to ensure the correct coupling conditions. Combining ring-mode applicators and Nd:YAG lasers, it is possible to apply a tissue-dependent power of 5–7 W over a time span of 10 to 15 min. Various applicators are available, varying in length and diameter of the glass dome (standard dome: L = 20 mm, d= 1.9 mm). For a standard dome and an output power of 6 W, the maximum power density will be about 20 W cm$^{-2}$ at the surface of the glass dome. Even using a glass dome, it is nevertheless necessary to avoid carbonization of the tissue, otherwise the thermal load on the glass dome will be excessive and it will stick to the tissue. This would considerably impede the ease of removal of the applicator from the channel. Some laser systems offer a safeguarding feature, which will switch off the laser as soon as carbonization occurs at the applicator's shaft (lightguide protection system, LPS, for lasers of the DORNIER Fibertom series). The protection system utilizes the pyrolysis glow emitted

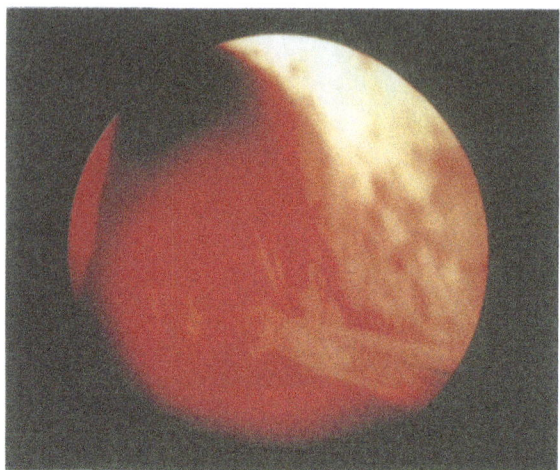

**Fig. 6.** Transurethral positioning of a ring-mode applicator in the side lobes of a prostate for treatment of BPH. (Photo Dr. Muschter)

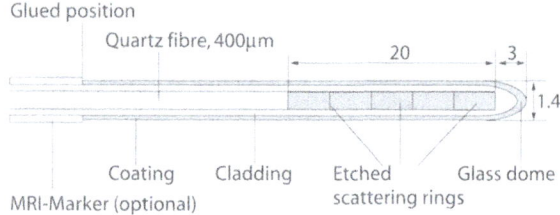

**Fig. 7.** Schematic design of a scattering applicator following an etching technique (LMTB)

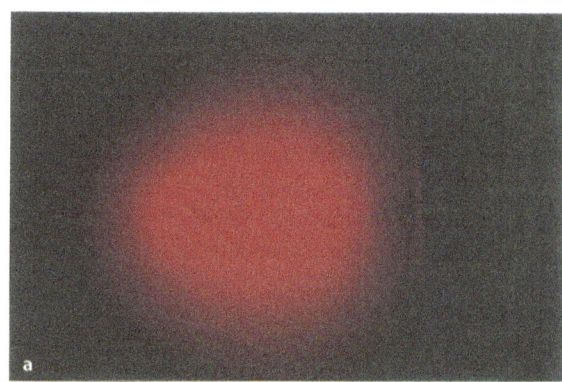

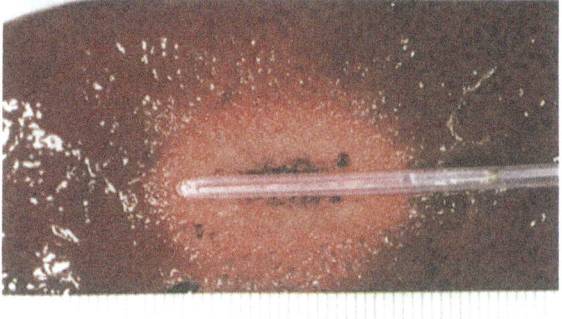

**Fig. 8 a, b.** Scattering applicator (LMTB). **a** Emission characteristic in a weakly scattering liquid. **b** Lesion of an interstitial coagulation (in vitro, porcine muscle, Nd:YAG laser, 6 W, 10 min)

in parallel with carbonization. This glow is returned to the laser through the therapy fibre (see also Chap. I-4.1.1).

Besides the treatment of tumours, ring-mode applicators are mainly employed for transurethral, interstitial laser coagulation of BPH (Fig. 6; [23, 24, 39–44]). By shaping the tip of the glass dome appropriately and under cystoscopic visual control, it is possible to puncture the prostate lobes directly. Multiple punctures may produce up to seven lesions, making it possible to destroy the periurethral adenoma tissue. During surgery it is possible to place a ring-mode applicator directly into accessible tumours or metastases for palliative therapy [7, 8, 29, 30].

## Scattering Applicators

Further developments lead to the so-called scattering applicator. This is a special preparation of the distal fibre end, causing diffuse emission of the laser radiation over an extended length [20–22, 51, 55, 57–59]. In general, this applicator also carries a protective glass dome over its fibre end. Various techniques are available to prepare the distal fibre end. Coating and cladding of the light guiding fibre are always removed over a length of 20–40 mm.

To achieve the intended scattering properties, the bare end of the fibre core can be mechanically roughened or supplied with a matt surface by using an appropriate etching solution. The actual length emitting diffuse laser radiation is often called the active length of the applicator (Fig. 7).

Emission characteristics and tissue reaction are indicated in Fig. 8.

Figure 9 shows the etched end of a fibre (scanning electron microscope image).

The advantage, as compared with a ring-mode applicator, is a smaller outside diameter. The geometry will further reduce the power density at the surface of the glass dome. For an active length of 20 mm, 1.4 mm diameter and 6 W power output, the power density will be only 7 W cm$^{-2}$. It is also necessary to prevent carbonization. Only then will a (tissue-dependent) power range of 5–8 W be available for a time span of 10 to 20 min., assuming an active length of 20 mm.

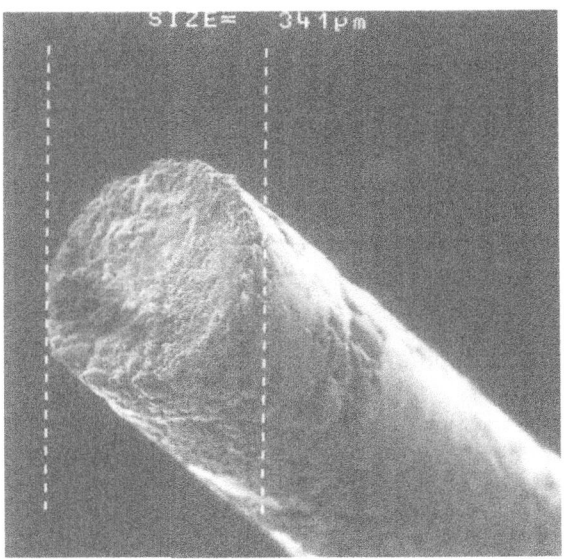

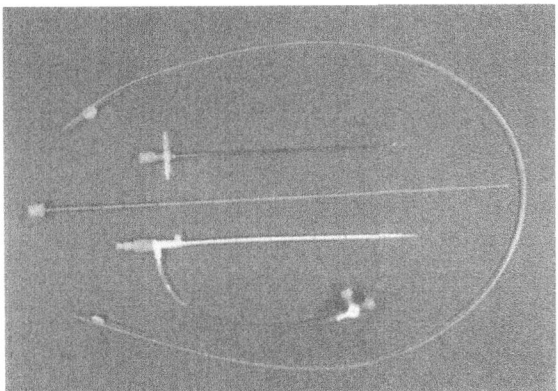

**Fig. 10.** Applicator system for percutaneous positioning and access of laser applicators (SOMATEX)

**Fig. 9.** Scanning electron microscope image of an etched fibre end for diffuse emission (LMTB)

Placing an NMR marker (iron oxide, Fig. 7), on the applicator's shaft will ease tracing of the applicator during NMR-controlled interventions.

To treat deep-seated processes percutaneously requires an additional access system, adapted to the applicator. Such a system, developed especially for scattering applicators, has become available (SOMATEX). Besides a punctation needle, guiding wire, mandrel and trocar (7F), it includes a protective catheter (6F) which is temperature-stabilized, transparent and distally closed (Figs. 1 and 10); [51, 55].

The advantages of a protective catheter are: no direct contact between laser applicator and tissue, protection of the glass dome against mechanical damage and easy repositioning of the applicator inside the protective catheter during therapy. In addition, it is possible, to place several approaches to a large tumour for laser therapy under NMR control (see Fig. 11).

After a final power check and positioning of the patient, it is sufficient to load the various approaches with applicators at the beginning of laser therapy. Temperature probes may be placed in unoccupied approach channels in order to gain additional information during therapy. After the end of therapy, fibrin adhesive may be injected through the trocar as a precaution against bleeding caused by the access channel. This procedure has been proven useful, especially for treatment of liver metastases and recurrences in the head and neck region [16, 63–66]. NMR was employed for the first time in monitoring of interstitial laser therapy of brain tumours. Besides precise positioning of the applicators, this allowed an on-line check of the actual temperature distribution inside the target volume [5, 7, 8, 11, 17, 28–30, 56, 6].

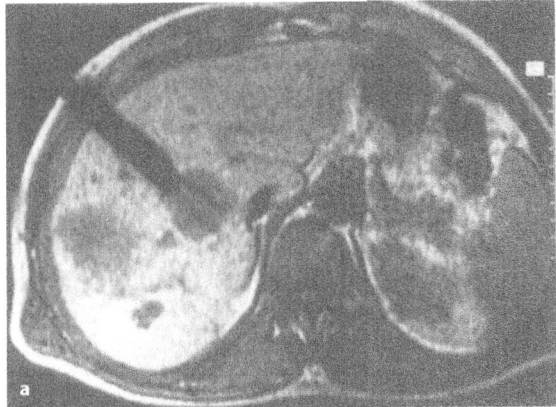

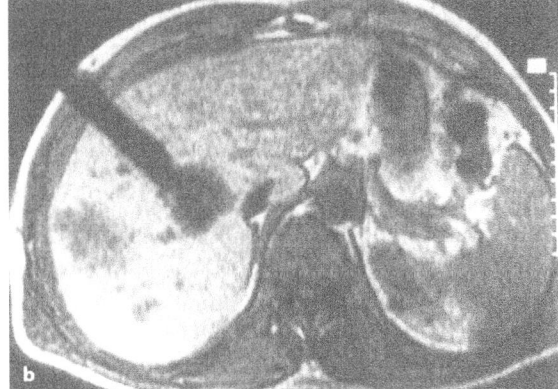

**Fig. 11.a, b  a** NMR tomography of a percutaneous application. Positioning of two applicators in a liver metastasis. **b** Hypointense presentation during laser irradiation. (Photo Prof. Vogl)

An alternative method to etching the free end of the light-guiding fibre is coating it with a plastic, achieving the desired diffuse emission by embedded scattering particles. Such applicators are on offer for the treatment of benign prostate hyperplasia by diode lasers (INDIGO-Medical; Fig. 12).

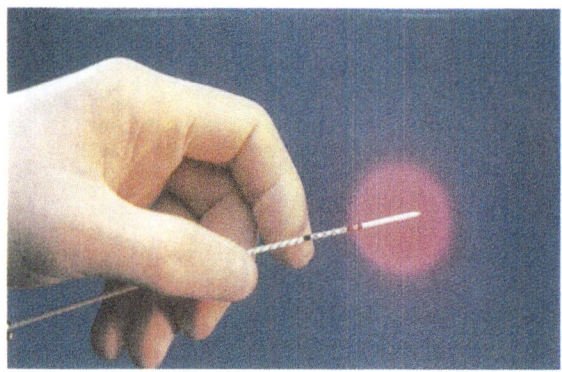

**Fig. 12.** Scattering applicator with integrated temperature sensor (INDIGO-Medical, Münster)

A special feature of these applicators is integrated temperature measurement using a distal temperature probe. This makes it possible to control laser power automatically, so that a maximum contact temperature of 100 °C cannot be exceeded. In this way, it is possible, to reproduce coagulation volumes, even if the optico-thermal tissue properties vary or if blood perfusion is ishomogeneous. The system is available with radiating (active); lengths of 5–20 mm and its outer diameter is 1.8 mm.

Although the power density is low at the contact surface of tissue and applicator, power and time ranges are limited for scattering applicators together with protective catheters. The coagulation volumes do not exceed a maximum diameter of 3 cm. However, there are indications that the power range may be extended if the active zone is lengthened significantly or if the applicator's surface is actively cooled.

## Zebra Applicators

Further research yielded applicators having active lengths of up to 40 mm [55]. It is impossible to make these by simply etching the fibre's core, because the result would be a non-uniform intensity distribution. Instead, radiation is distributed over a long distance by alternating active and passive regions. To achieve a very uniform temperature profile, despite intermittent radiation emission, tissue scattering and thermal conduction are again exploited (Fig. 13).

Since the outer diameters are the same, these applicators can be used with the same access channel system as scattering applicators having a short length. Corresponding to the larger surface, it is possible to almost double the maximum parameters, i.e. to 10–12 W over a time period from 10 to 20 min, depending on the tissue.

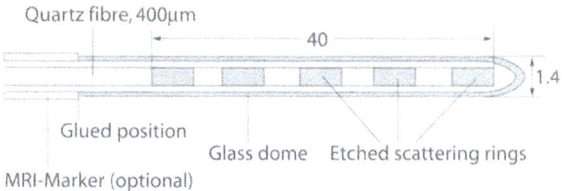

**Fig. 13.** Zebra applicator (L = 40 mm, d = 1.4 mm, LMTB)

## Cooled Applicators

One way to increase the maximum power tolerated by the applicator is to increase its surface area. Another is to cool the applicators surface actively; however, this is indicated only if the wavelength employed has a substantial optical penetration depth. In this way an appropriate and precooled liquid (e.g. water) moves in a counter current around the active part of the applicator which is, in general, the hottest and is cooled by heat conduction. Due to the large penetration depth of the radiation the region of maximum temperature can be found at a radial distance from the applicator's surface of about 2–4 mm [51, 53, 55, 57–59].

Figure 14 presents results of calculations of the temperature distribution for a cooled and an uncooled scattering applicator [52, 53]. It includes for comparison the temperature distribution of a so-called hot tip.

Power output was adjusted in such a way that the same maximum temperature was reached at the end of emission. Obviously, the cooled applicator achieved the largest coagulation radius and the lowest contact temperature. Compared with the cooled applicator, the hot tip produces a very large temperature gradient and only half the coagulation radius. It should be kept in mind that the coagulation volume – a significant parameter in every comparison – increases with the third power of the radius. Consequently, doubling the radius leads to an eightfold increase in volume, indicating an eightfold increase in efficiency. The efficiency of the uncooled scattering applicator is thus placed between the two other systems.

Figure 15 presents the design of a cooled prototype and the achievable tissue reactions.

The necessary inflow and outflow enlarges the outer diameter of the system as compared to an uncooled one (diameter about 4–5 mm). Coagulation volumes of up to 25 cm$^3$ are achievable: the maximum power is about 30 W over a time period of 20 min.

The large outer diameter of cooled applicators restrict their range of applications. The system may become important for transurethral therapy of BPH and for very large tumours in liver and abdomen.

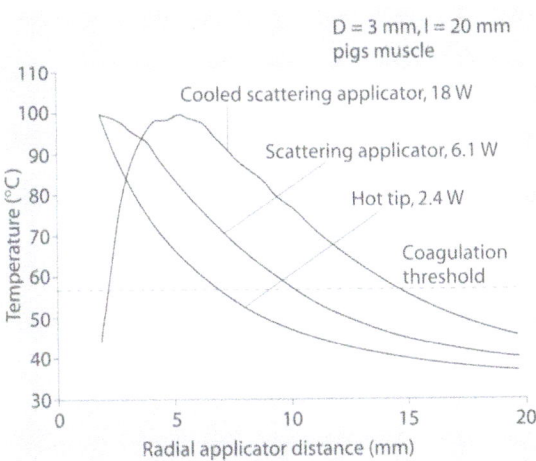

**Fig. 14.** Calculated radial temperature distribution after 10 min application of a hot tip, using a scattering applicator and a cooled scattering applicator (d = 3 mm, L = 20 mm, d = 5 mm, muscle) [50, 52, 53]

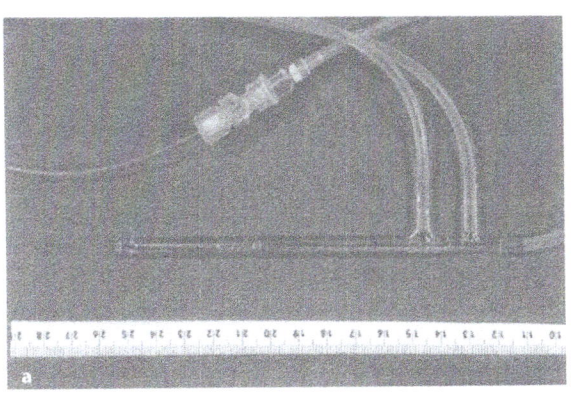

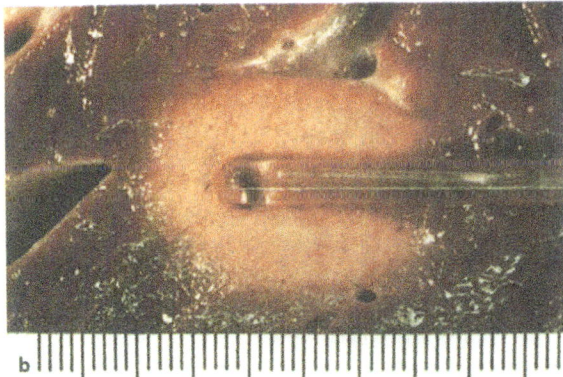

**Fig. 15.a,b a** Prototype of a cooled applicator (L = 20 mm, d = 5 mm, LMTB). **b** Interstitial coagulation (in vitro, porcine liver, Nd:YAG laser, 25 W, 10 min)

A further option for the application of cooled applicators is the targeted shaping of the coagulation region. Experiments trying to form asymmetric coagulation regions by modifying the emission characteris-

tic of normal applicators (e.g. by absorption at selected solid angles) were unsuccessful. This can be explained by a combination of strong tissue scattering and simultaneous (homogeneous) thermal conductivity. This combination had already presented an almost spherical form for a lesion produced by a bare fibre, although this fibre has an extremely prograde emission characteristic (Fig. 3). The situation differs for cooled applicators. There it is possible to influence the shape of the lesion by predetermined and selective absorption, because cooling prevents thermal balancing by heat transfer into the unexposed regions [59].

In summary, it can be said that, for the time being, cooled applicators are the most effective method to generate coagulation lesions. Applicators with a small diameter are available if the larger applicators are not suitable in a given clinical situation. Their efficiencies can be increased by the following appropriate procedures.

## Powermode

The simplest way to increase efficiency is the so-called powermode, offered by some laser systems, especially for interstitial applications. This method should not be confused with a pulsed mode; in powermode the output power of the laser is modified according to a predetermined regime (Fig. 16).

The aim is to apply increased power during the initial minutes of laser therapy when the temperature at the applicator is still low, and to decrease power as therapy proceeds. In this way, the power output is adjusted as the tissue's temperatures are raised, thus avoiding overheating of the applicator system.

To demonstrate this, Fig. 17 indicates the calculated temperature directly at the applicators surface and

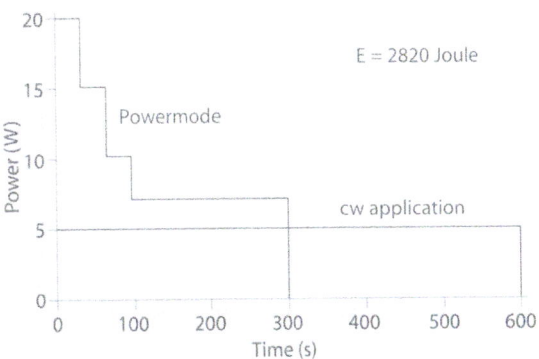

**Fig. 16.** Power vs. Time diagram for interstitial coagulation in powermode (DORNIER, Fibertom series). A continuous application with the same total power is presented for comparison

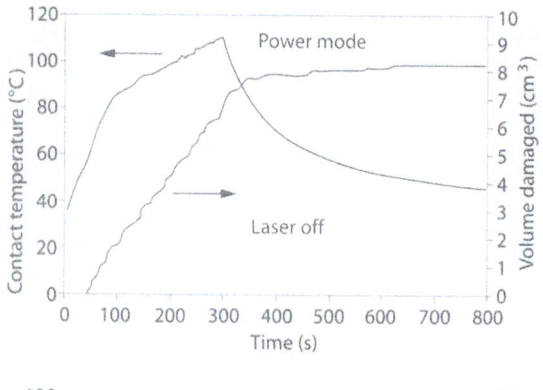

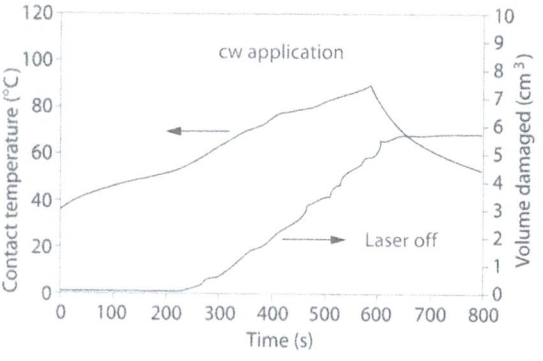

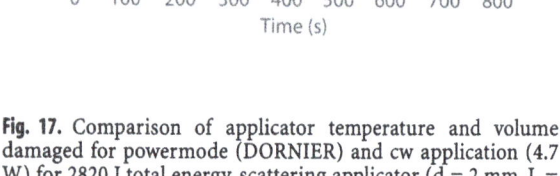

**Fig. 17.** Comparison of applicator temperature and volume damaged for powermode (DORNIER) and cw application (4.7 W) for 2820 J total energy, scattering applicator (d = 2 mm, L = 20 mm), simulation [53]

**Fig. 18.a, b** a Prototype of a four-channel beam splitter for ring-mode applicators (DORNIER). **b** Lesion caused by an interstitial coagulation with four ring-mode applicators (in vitro, porcine liver, separation 15 mm, 6 W, 10 min)

the measured coagulation volume. The temperature variation is shown for powermode and for cw application [53]. Obviously, efficiency is improved in the powermode and thus 5 min of applicator time are saved.

## Beam Splitter

Another option to increase efficiency and to adapt the coagulation lesion optimally to the actual shape of the pathological process is the employment of a beam-splitter. Such a system allows simultaneous connection of up to four applicators or bare fibres with the laser system [13, 15, 19, 63, 70]. By appropriate super-positioning of the individual temperature fields, large-sized lesions can be achieved (d > 5 cm!).

Complex beam splitters consist of lens systems and dielectric mirrors, each subtracting a fraction of the laser radiation from the primary beam (Fig. 18).

Designed appropriately, such systems allow to vary the power in the individual channel. Simplified systems consist of so-called fibre-optical couplers: two light-guiding fibres are fused into the shape of an Y:

Combining these, it is possible to achieve four exit channels; however, individual adjustment of the output power is impossible. Advantages are robustness to vibrations and very good durability.

## Power Measurements

As already indicated in previous sections, employing interstitial applicators poses the danger of exceeding the limits for carbonization of tissues close to the applicator. In addition, the final size of the coagulated region depends significantly on the power output. For these reasons, a measurement of the power directly at the applicator system should always precede every interstitial coagulation. For bare-fibre applications, power may be adjusted via the calibrated input of the laser device (sterile plug-in) or by using an external power meter having a planar measuring area. Due to their emission characteristics, this is not possible for ring-mode or scattering applicators. In these cases, a special power-meter is employed which utilizes the principle of the integrating sphere and thus is able to integrate the entire emitted power (Fig. 19).

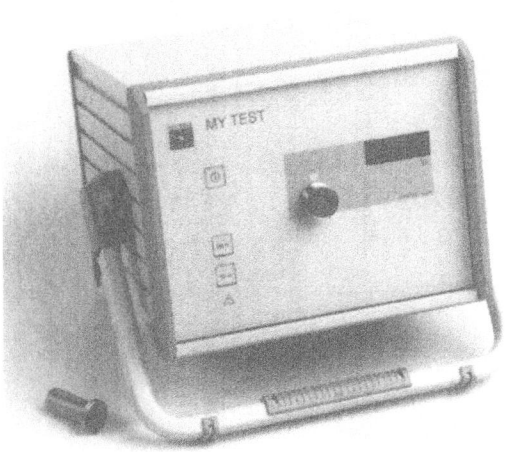

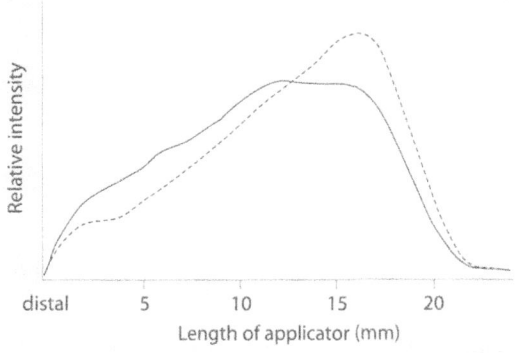

**Fig. 20.** Axial intensity distribution of a scattering applicator optimized for Nd:YAG lasers. Connected to an appropriate Nd:YAG laser (–) and diode laser (----) having the same power output. An ideal, uniformly emitting applicator would display a rectangular profile having a width of about 20 mm. Measurement by Integrating sphere [55]

**Fig. 19.** Power meter for diffuse emission applicators, wavelength range 400–1100 nm, measuring range 0–10 W and 0–100 W (HÜTTINGER, MY test)

## Temperature Distribution

Employing special sensors (thermopiles) allows their use over an extended wavelength range. Nd:YAG lasers can be calibrated as well as diode lasers emitting in the near infrared.

## Various Laser Systems

It should be mentioned that in general laser applicators for interstitial applications cannot be connected to any laser system. For example, numerical aperture (NA) and diameter of the light-guiding fibre are essential parameters. Nd:YAG lasers can employ almost any fibre having a core diameter larger than 300 μm and an NA exceeding 0.2. On the other hand, for diode lasers it is important to keep in mind the coupling conditions while selecting a fibre. For many diode lasers the minimum allowable NA is 0.37, making quartz/quartz fibres unsuitable. Some instruments require fibres with a minimum core diameter of 600 μm.

Coupling conditions also influence the applicators emission characteristics; this is another critical factor. Ring-mode applicators definitely require special adaptors; scattering applicators are optimized with respect to certain coupling parameters. Utilizing an Nd:YAG laser scattering applicator together with a diode laser can produce a significantly changed emission characteristic. Hot spots may occur and point – like tissue carbonizations cannot be excluded (Fig. 20). Consequently, applicators optimized for diode lasers are in general unsuitable combined with Nd:YAG lasers.

To evaluate an applicator system for interstitial coagulation, the time-dependent temperature distribution created in tissue is of the utmost importance. This influences the safety margins as well as the achievable effects on tissues. It is possible to measure, using a thermocouple, the temperatures at certain positions during irradiation. However, information is restricted to a limited number of points. In addition, absorption by the thermoelement makes any measurements close to the applicator impossible [47]. Using an infrared-sensitive camera, a two-dimensional temperature distribution in the plane of the applicator can be measured for an in vitro model [55]. To do this, the model tissue is cut in halves and the applicator to be investigated is placed between both parts. A special holder allows one to rapidly flip open the tissue and to register a thermogram in the applicator's plane (Fig. 21). The opening time period of the holder is very short (< 2 s), allowing one to neglect, in a first approximation, tissue cooling by convection. Figures 22 and 23 indicate sequential images, spaced at 2-min intervals, for an uncooled and a cooled scattering applicator.

The image for the cooled applicator clearly indicates a shift of the hottest region into the tissue. In addition, the extension of the thermal region is significantly larger than the corresponding one for the uncooled applicator.

In addition to in vitro investigations, an important role in designing optimal laser applicators for interstitial applications is played by simulation calculations. In this way, it is possible to determine the temperature profiles and variations caused by different emission characteristics even before applicators are employed. Increasingly, computer simulations are uti-

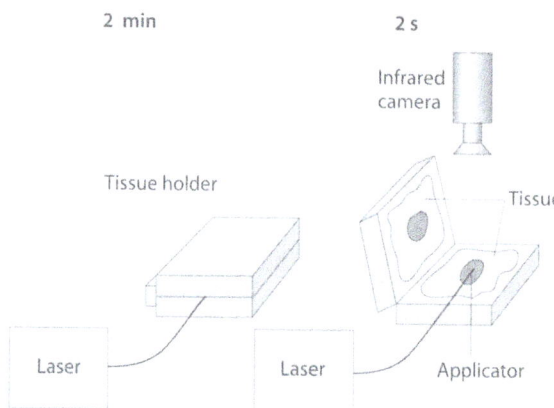

**Fig. 21.** Measuring arrangement to determine interstitial temperature fields [55] by an IR camera

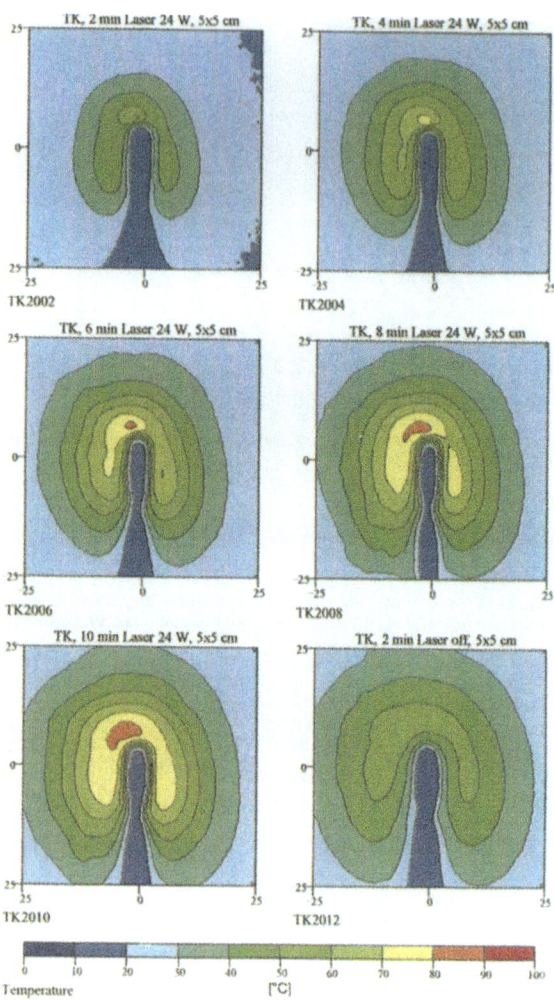

**Fig. 23.** Temperature distribution caused by a scattering applicator (LMTB, L = 20 mm, d = 5 mm, T = 10 °C) for 24 W from an Nd:YAG laser (in vitro, porcine muscle, T = 20 °C)

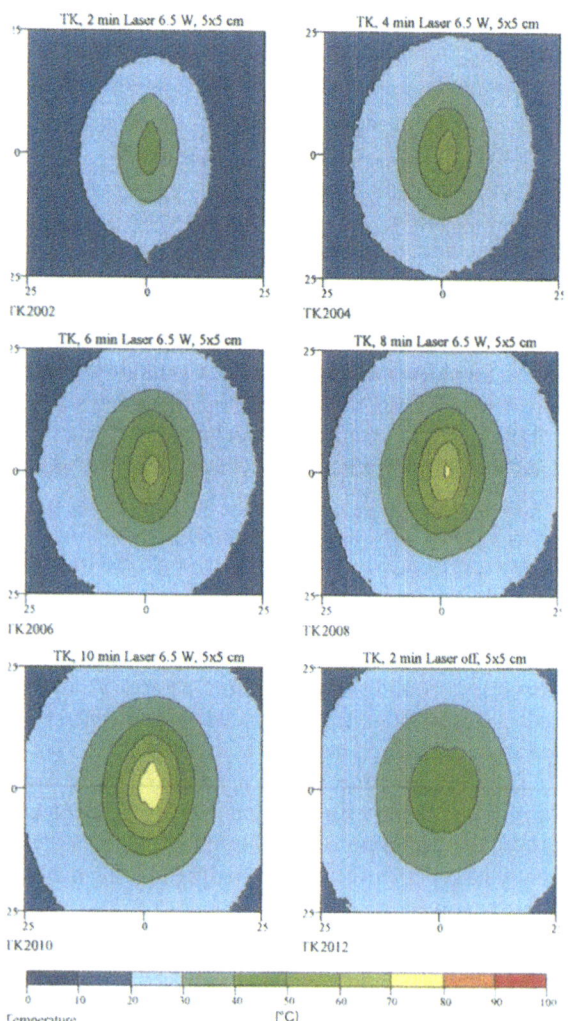

**Fig. 22.** Temperature distribution caused by a scattering applicator (LMTB, L = 20 mm, d = 1.4 mm) for 6.5 W from an Nd:YAG laser (in vitro) porcine muscle, T = 20 °C)

lized for the planning of interstitial therapy. Thus, application parameters can be optimized, leading to high efficiency and increased patient safety [50, 52, 53].

## References

[1] AMIN Z, BOWN SG, LEES WR (1993) Local treatment of colorectal liver metastases: a comparison of interstitial photocoagulation (ILP) and percutaneous alcohol injection (PAI). Clin Radiol 48: 166–171

[2] AMIN Z, BUANACCORSI G, MILLS T, HARRIES SA, LEES WR, BOWN SG (1993) Interstitial laser photocoagulation: Evaluation of a 1320 nm Nd:YAG laser and 805 nm Diode laser: the significance of charring and the value of pre-charring the fiber tip. Las Med Sci 8: 113–120

[3] AMIN Z, THURREL W, SPENCER GM, HARRIES SA, GRANT WE, BOWN SG, LEES WR (1992) Computed tomography-pathology assessment of laser-induced necrosis in rat liver. Invest Radiol 38: 1148–1154

[4] AMIN Z, DONALD JJ, MASTERS A, KANT R, STREGER AC, BOWN SG, LEES WR (1993) Hepatic metastases: inter-

stitial laser photocoagulation with real-time US monitoring and dynamic CT evaluation of treatment. Radiology 187: 339–347

[5] ASCHER PW (1990) Interstitial thermal therapy of brain tumours with Nd:YAG laser under real-time MRI control. Proc SPIE 1200: 242–245

[6] BERLIEN HP, PHILIPP C, ENGEL-MURKE F, FUCHS B (1993) Laseranwendungen in der Gefäßchirurgie. Zentralblatt für Chirurgie 118: 383

[7] BETTAG M, ULRICH F, SCHOBER R, SABEL M, BOCK WJ (1991) Stereotactic interstitial laser irradiation of brain tumours – a new therapeutic approach. Las Med Sci 87

[8] BETTAG M, ULRICH F, SCHOBER R, SABEL M, KAHN T, BOCK WJ (1992) Laser-induced interstitial thermotherapy in malignant gliomas. Adv Neurosurg 22: 253–257

[9] BEUTHAN J, MÜLLER G, SCHALDACH B, ZUR CH (1991) Fiber design for interstitial laser treatment. Proc SPIE 1420: 234–241

[10] BEUTHAN J, GEWIESE B, WOLFF KJ, MÜLLER G (1992) Die Laserinduzierte Thermotherapie (LITT) – biologische Aspekte ihrer Anwendung. Minimal Invasive Medizin 3: 102–106

[11] BEUTHAN J, GEWIESE B, FOBBE F, GERMER CT, AL-BRECHT D, BOESE-LANDGRAF J, ROGGAN A, MÜLLER G (1993) Investigations of MRI-controlled Laser-Induced Interstitial Thermo-Therapy (LITT). Med Tech 4 (4): 27–30

[12] BOWN SG (1983) Phototherapy of tumours. Word J Surg 7: 700–709

[13] DAVIS M, CAPADIA P, DOWDEN JM (1989) A model for laser treatment of tumours using four fibers. Las Med Sci 4: 41–53

[14] FAN M, ASCHER P, SCHRÖTTNER O, EBNER F, GERMANN RH, KLEINERT R (1992) Interstitial 1.06 Nd:YAG laser thermotherapy for brain tumours under real-time monitoring of MRI: experimental study and phase I clinical trial. J Clin Las Med Surg 10 (5): 355–361

[15] FRANK F, HESSEL S (1990) Technische Voraussetzungen für die interstitielle Thermotherapie mit dem Nd:YAG-Laser. Lasermedizin 10: 36–40

[16] GERMER CH, ALBRECHT D, BOESE-LANDGRAF J, ROGGAN A, ISBERT C (1995) Laser-induced thermotherapy (LITT) in the treatment of colorectal liver metastases – A clinical pilot study. In: MÜLLER G, ROGGAN A (eds) Laser-induced interstitial thermotherapy. PSIE Press, Bellingham, pp 393–399

[17] GEWIESE B, BEUTHAN J, FOBBE F, STILLER D, MÜLLER G, BOESE-LANDGRAF J, WOLF KJ, DEIMLING M (1994) Magnetic resonance imaging-controlled laser-induced interstitial thermotherapy. Invest Radiol 29 (3): 345–351

[18] HARRIES SAG, AMIN Z, SMITH ME, LEES WR, COOKE J, COOK MG, SCURR JH, KISSIN MW, BOWN SG (1994) Interstitial laser photocoagulation as a treatment of breast cancer. Br J Surg 81: 1617–1619

[19] HESSEL S, FRANK F (1990) Technical prerequisites for the interstitial thermotherapy using the Nd:YAG laser. Proc SPIE 1201: 233–238

[20] HILLEGERSBERG R, VAN STAVEREN HJ KORTH WJ, ZONDERVAN PE, TERPSTRA OT (1994) Interstitial Nd:YAG laser coagulation with a cylindrical diffusing fiber tip in experimental liver metastases. Las Surg Med 14: 124–138

[21] HILLEGERSBERG R (1995) Cylindrical diffusing fiber-tip for interstitial coagulation. In: MÜLLER G, ROGGAN A (eds) Laser-induced interstitial thermotherapy. SPIE Press, Bellingham, pp 195–211

[22] HILLEGERSBERG R, VAN STAVEREN HJ, ROGGAN A, MÜLLER G, IJZERMANS JNM (1995) Interstitial laser coagulation as a treatment of breast cancer. Br J Surg 82 (6): 856

[23] HENKEL TH, GRESCHNER M, LUPPOLD T, ALKEN P (1995) Transurethral and transperineal interstitial laser therapy of BPH. In: MÜLLER G, ROGGAN A (eds) Laser-induced interstitial thermotherapy. SPIE Press, Bellingham, pp 416–425

[24] HOFSTETTER A, MUSCHTER R, SCHNEEDE P (1993) Surgical treatment of BPH – LITT: State of the art in BPH. Med Tech 4 (2): 12–14

[25] HIELE M, PENNINCKY F, GEVERS AM, VAN EYKEN P, GEBOES K, NI Y, MARCHAL G, VANTRAPPEN G, FEVERY J, FRANK F, HESSEL S, RUTGEERTS P (1993) Interstitial thermotherapy for liver tumours: studies of different fibres and radiation characteristics. Las Med Sci 8: 121–125

[26] JACQUES SL, RASTEGAR S, MOTAMEDI M, THOMSON S, SCHWARTZ J, TORRES J, MANNONEN I (1992) Liver photocoagulation with diode laser (805 nm) versus Nd:YAG laser (1064 nm) Proc SPIE 1646: 107–112

[27] JOHNSON DE, PRICE RE, CROMEENS DM (1992) Pathologic changes occurring in the prostate following transurethral laser prostatectomy. Las Surg Med 12: 254–263

[28] JOLESZ FA, BLEIER AR, JAKAB P, RUENZEL PW, HUTTL K, JAKO GJ (1988) MR imaging of laser tissue interaction. Radiology 168: 853–857

[29] KAHN TH, BETTAG M, ULRICH F, SCHWARZMAIER HJ, SCHOBER R, FÜRST G, MÖDDER U (1992) MRI-guided laser-induced interstitial thermotherapy of cerebral neoplasms. J Comp Assist Tomogr 18: 519–532

[30] KAHN TH, BETTAG M, ULRICH F, SCHWARZMAIER HJ, HARTH TH, MÖDDER U (1995) MRI-guidance of laser-induced interstitial thermotherapy of brain tumours – three years experience. In: MÜLLER G, ROGGAN A (eds) Laser-induced interstitial thermotherapy. SPIE Press, Bellingham, pp 325–339

[31] MATSUMOTO R, SELIG M, COLLUCCI VM, JOLESZ FA (1992) Interstitial Nd:YAG laser ablation in normal rabbit liver: trial to maximize the size of laser-induced lesions. Las Surg Med 12: 650–658

[32] MASTERS A, STEGER AC, BOWN SG (1991) Role of interstitial therapy in the treatment of liver cancer. Br J Surg 78: 518–523

[33] MASTERS A, BOWN SG (1992) Interstitial laser hyperthermia. Semin Surg Onc 8: 242–249

[34] MASTERS A, STEGER AC, LEES WR, WALMSLEY KM, BOWN SG (1992) Interstitial laser hyperthermia: a new approach for treating liver metastases. Br J Canc 66: 518–522

[35] McNICHOLAS TA, STEGER AC, BOWN SG, O'DONOGHUE N (1991) Interstitial laser coagulation of the prostate: experimental study. In: WATSON GM, STEINER RW, PIETRAFITTA JJ (eds) Lasers in urology, laparoscopy and general surgery (Series Editor Katzir A). The Society of Photo-Optical Instrument Engineers, Bellingham, pp 30–35

[36] McNICHOLAS TA, STEGER AC, BOWN SG (1993) Interstitial laser coagulation of the prostate. An experimental study. Br J Urol 71 (4): 439–444

[37] McNICHOLAS TA, ASLAM M, LYNCH MJ, O'DONOGHUE N (1993) Interstitial laser coagulation for the treatment of urinary outflow obstruction. J Urol 149: 465A

[38] MUMTAZ H, HARRIES SA, HALL CRAGGS MA, DAVIDSON T, LEES WR, BOWN SG (1995) The potential of interstitial laser photocoagulation in the treatment of breast cancer. In: MÜLLER G, ROGGAN A (eds) Laser-induced interstitial thermotherapy. SPIE Press, Bellingham, pp 426–433

[39] MUSCHTER R, HESSEL S, HOFSTETTER A, KEIDITSCH E, ROTHENBERGER KH, SCHNEEDE P, FRANK F (1993) Die interstitielle Laserkoagulation der benignen Laserhyperplasie. Urologe [A] 32: 273–281

[40] MUSCHTER R, HESSEL S, HOFSTETTER A, ZELLNER M, SCHNEEDE P, OBERNEDER R (1993) One year experience in interstitial laser coagulation for benign prostatic hyperplasia. J Urol 149: 466A

[41] MUSCHTER R, HOFSTETTER A (1994) Erfahrungen mit der interstitiellen Laserkoagulation in der Therapie der benignen Prostatahyperplasie. Lasermed 10: 133–139

[42] MUSCHTER R, EHSAN A, STEPPE HG, HOFSTETTER A (1995) Clinical results of LITT in the treatment of benign prostatic hyperplasia. In: MÜLLER G, ROGGAN A (eds)

Laser-induced interstitial thermotherapy. SPIE Press, Bellingham, pp 434–442

[43] MUSCHTER R, HESSEL S, JAHNEN P, YALAVAC H, HOFSTETTER A (1995) Evaluation of different laser wavelength and application systems for LITT. In: MÜLLER G, ROGGAN A (eds) Laser-induced interstitial thermotherapy. SPIE Press, Bellingham, pp 212–223

[44] MUSCHTER R, HOFSTETTER A (1995) Technique and results of interstitial laser coagulation. World J Urol 13: 109–114

[45] MÜLLER G, WOLF J, FOBBE F, BÖSE-LANDGRAF J, GERMER C, BEUTHAN J, ROGGAN A (1992) Interstitial laser hyperthermia - a new method to treat tumors. In: SPINELLI P, DAL FANTE M, MARCHESINI R (eds) Photodynamic therapy and biomedical lasers. Excerpta Medica, New York, pp 406

[46] NOLSOE C, TORP-PEDERSEN S, OLLDAG E, HOLM HH (1992) Bare fiber low power Nd:YAG laser interstitial hyperthermia. Comparison between diffuser tip and non-modified tip, an in vitro study. Laser Med Sci 7: 1–7

[47] PHILIPP C, SHALTOUT J, ZGADO F, BERLIEN HP (1992) Zur Problematik von Temperaturmessungen mit Thermoelementen während Laserbestrahlung in streuenden Medien. Lasermed 8: 188–195

[48] PHILIPP C, BERLIEN HP, POTHKE M, WALDSCHMIDT J (1994) Ten years of laser treatment of congenital vascular disorders - technical techniques and results. Proc SPIE 2327: 44–53

[49] PHILIPP C, RHODE E, BERLIEN HP (1995) Treatment of congenital vascular disorders (CVD) with laser-induced interstitial thermotherapy (LITT). In: MÜLLER G, ROGGAN A (eds) Laser-induced interstitial thermotherapy. SPIE Press, Bellingham, pp 443–458

[50] ROGGAN A, MÜLLER G (1993) Computer simulations for the irradiation planning of LITT. Med Tech 4 (2): 18–24

[51] ROGGAN A, ALBRECHT D, BERLIEN HP, BEUTHAN J, GERMER C, KOCH H, WODRICH W, MÜLLER G (1994) Development of an application set for intraoperative and percutaneous laser-induced interstitial thermotherapy (LITT). SPIE 2327: 2503–2620

[52] ROGGAN A, MÜLLER G (1994) Development of a computer model for the irradiation planning of laser-induced interstitial thermotherapy (LITT). Proc SPIE 2100: 69–82

[53] ROGGAN A, MÜLLER G (1995) Dosimetry and computer based irradiation planning for laser-induced interstitial thermotherapy (LITT). In: MÜLLER G, ROGGAN A (eds) Laser-induced interstitial thermotherapy. SPIE Press, Bellingham, pp 114–157

[54] ROGGAN A, DÖRSCHEL K. MINET O, WOLFF D, MÜLLER G (1995) The optical properties of biological tissue in the near infrared wavelength rang - review and measurements. In: MÜLLER G, ROGGAN A (eds) Laser-induced interstitial thermotherapy. SPIE Press, Bellingham, pp 10–44

[55] ROGGAN A, ALBRECHT D, BERLIEN HP, BEUTHAN J, FUCHS B, GERMER C, MESECKE V, RHEINBABEN I, RYGIEL R, SCHRÜNDER S, MÜLLER G (1995) Application equipment for intraoperative and percutaneous laser-induced interstitial thermotherapy. In: MÜLLER G, ROGGAN A (eds) Laser-induced interstitial thermotherapy. SPIE Press, Bellingham, pp 224–248

[56] ROUX FX, MERIENNE L, LERICHE B, LUCERNA S, TURAK B, DEVEAUX BC, CHODKIEWICZ JP (1992) Laser interstitial thermotherapy in stereotactical neurosurgery. Laser Med Sci 7: 121–126

[57] SCHWARZMAIER HJ, GOLDBACH T, ULRICH F, SCHOBER R, KAHN T, KAUFMANN, R. WOLBARSCHT ML (1994) Improved laser applicators for interstitial thermotherapy of brain structure. Proceedings SPIE Vol 2132: 4–12

[58] SCHWARZMAIER HJ, GOLDBACH T, KAUFMANN R, ULRICH F, BETTAG M, KAHN T (1994) New applicators for the laser induced interstitial thermotherapy. Min Invasive Med 5: 32–35

[59] SCHWARZMAIER HJ, KAUFMANN R, KAHN TH, ULRICH F (1995) Applicators for the laser-induced interstitial thermotherapy - basic considerations and new developments. In: MÜLLER G, ROGGAN A (eds) Laser-induced interstitial thermotherapy. SPIE Press, Bellingham, pp 249–262

[60] STEGER AC, SHORVON P, WALMLEY K, CHRISHOLM R, BOWN SG, LEES WR (1993) Ultrasound features of low power interstitial laser hyperthermia. Clin Radiol 46: 88–93

[61] TRACZ RA, WYMAN DR, LITTLE PB, TOWNER RA, STEWARD WA, SCHATZ SW, WILSON BC, PENNOK PW, JANZEN EG (1993) Comparison of magnetic resonance images and the histopathological findings of lesions induced by interstitial laser photocoagulation in brain. Laser Surg Med 13: 45–54

[62] TRANBERG KG, MÖLLER PH (1995) Interstitial laser treatment: preliminary experience in patients. In: MÜLLER G, ROGGAN A (eds) Laser-induced interstitial thermotherapy. SPIE Press, Bellingham, pp 468–476

[63] VOGL TH, MÜLLER P, WEINHOLD N, HAMMERSTINGL R, MACK M, BÖTTCHER H, PHILIPP C, FELIX R (1995) MR-guided laser-induced interstitial thermotherapy (LITT) of liver metastases. In: MÜLLER G, ROGGAN A (eds) Laser-induced interstitial thermotherapy. SPIE Press, Bellingham, pp 477–492

[64] VOGL TH, MACK M, MÜLLER P, PHILIPP C, BÖTTCHER H, ROGGAN A, DEIMLING M, KNÖBBER D, FELIX R (1995) MR-guided laser-induced thermotherapy of head and neck tumours. In: MÜLLER G, ROGGAN A (eds) Laser-induced interstitial thermotherapy. SPIE Press, Bellingham, pp 493–504

[65] VOGL T, MÜLLER PK, HAMMERSTINGL R, WEINHOLD N, MACK MG, PHILIPP C, DEIMLING M, BEUTHAN J, PEGIOS W, RIESS H, LEMMENS HP, FELIX R (1995) Malignant liver tumours treated with MR imaging-guided laser-induced thermotherapy: technique and prospective results. Radiology 196: 257–265

[66] VOGL T, MACK MG, MÜLLER PK, PHILIPP C, BÖTTCHER H, ROGGAN A, JUERGENS M, DEIMLING M, KNÖBBER D, WUST P, FELIX R (1995) Recurrent nasopharyngeal tumours preliminary clinical results with interventional MR imaging-controlled laser-induced thermotherapy. Radiology 196: 725–733

[67] WALLWIENER D, KUREK R, POLLMANN D, KAUFMANN M, SCHMID H, BASTERT G, FRANK F (1994) Palliative therapy of gynecological malignancies by laser-induced interstitial thermotherapy. Lasermed 10: 44–51

[68] WYMAN DR, WHELAN WM, WILSON BC (1992) Interstitial laser photocoagulation Nd:YAG 1064 nm optical fiber source compared to point heat source. Las Surg Med 12: 659: 664

[69] WYMAN D, WILSON B, ADAMS K (1993) Dependance of laser photocoagulation on interstitial delivery parameters. Laser Surg Med 14: 659–664

[70] WYMAN DR (1993) Selecting source locations in multifiber interstitial laser photocoagulation. Laser Surg Med 13: 656–663

# I-4.4
# Laser Medicine – Technology and Dosimetry

## Present and Future

G. Müller

### Contents

## Range of Applications for Medical Lasers

The number of applications for medical lasers is growing continuously. There are many laser procedures already well established in clinical practice, e.g. retina coagulation in ophthalmology. All the time, new possibilities and procedures are being tested, e.g. laser-induced thermotherapy and laser angioplasty.

Often enough, lasers are in direct competition with high-frequency procedures in general surgery, especially in minimally invasive surgery. Lasers offer definite advantages for the potential-free application of energy (e.g. intracardial applications), high power density in restricted localities (lithotripsy of salivarycalculi, transmyocardial revascularization) or for differentiated, tissue-specific effects. The complexity of the ever-expanding applications constantly demands new solutions to problems arising in dosimetry.

## Dosimetry

In medicine the term dose is used as a description for a biologically effective quantity of drugs or as the amount of biologically effectie ionizing radiation administered. The following contribution will treat dosimetry as a method to determine the (laser) radiation parameters necessary for achieving a predetermined action, i.e. it determines laser power, power density and energy of a suitable wavelength to achieve the desired response.

## Action Principles

Historically, laser applications in medicine are closely associated with lasers as cutting instruments. Besides this conventional preparatory method, relying on the evaporation of tissues, an ever-widening spectrum of further interactions of different laser types with various tissue types is being utilized.

Laser action is subject to strongly wavelength-dependent absorption, different penetration depths and distinctive tissue selectivity. However, the possibility

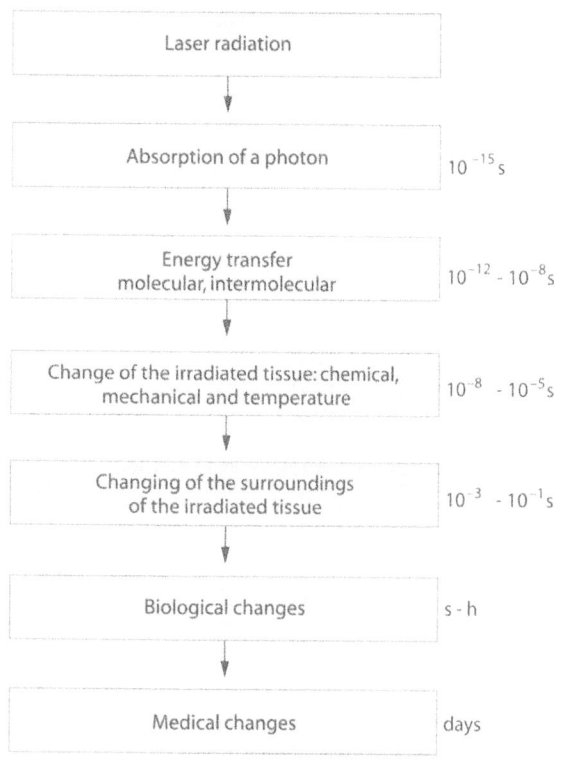

**Fig. 1.** Time scales during laser treatment [6]

of applying radiation energy over various time spans – resulting in very different power densities – leads to very different tissue reactions. It is necessary to distinguish between the direct physical results of irradiation and the desired medical-biological consequences, often manifesting themselves long after the end of irradiation. Depending on the type of treatment, various time periods must be taken into account. Figure 1 gives a breakdown of the processes and times involved.

Figure 2 shows the variation in the most important application parameters – power density and interaction time – and the multitude of actions achievable by choosing between these different combinations of laser parameters.

In addition, Table 1 gives an overview of further action principles based on different physical tissue reactions during medical application of lasers.

The pronounced dependency on the treatment parameters indicates that dosimetry is very important for the success of therapy. This is especially so, because the dose-effect relationship is by no means a uniquely determined function but depends on many other influences, individual parameters and fluctuations. To achieve success in medical treatment, the relevant parameters must be chosen appropriately. The treatment dose must be predetermined such that the long-term medical aim is achieved after the end of the biological reactions.

**Fig. 2.** Laser action for various application parameters

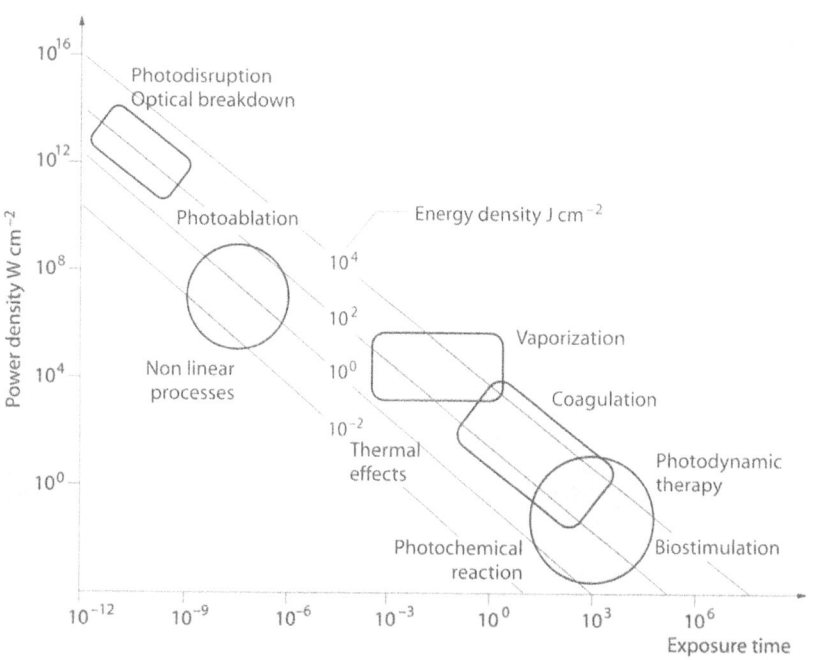

**Table 1.** Classes of action principles

| Photochemical effects | Action principle |
| --- | --- |
| Photoinduction | Biostimulation |
| Photoactivation of drugs Photoirradiation | POD |
| Photochemotherapy | Photodynamic therapy (PDT) Black light therapy (PUVA) |
| Photoresonance | |
| *Photothermal effects* | |
| Photohyperthermia | 37 °C–43 °C: Reversible damage of normal tissue. 45–60 °C: Celloedema, tissue welding; protein coagulation |
| Photothermolysis | Thermodynamical effects microscopically small sites of overheating |
| Photocoagulationsis | 60–100°C: Coagulation, necro- |
| Photocarbonization | 100–300°C: Drying out, vaporization of water, carbonization |
| Pyrolysis | > 300°C |
| *Photoionization or photofission* | |
| Photoablation | Rapid thermal explosion (angioplasty) |
| Photodisruption | Optical breakdown |
| Photofragmentation | Mechanical shockwave (lithotripsy) |

Laser Nd:YAG (1064 nm)
Applicator: bare fibre (200 - 600μm)
non-contacting, in air

Coagulation
narrow coagulation margin (1 - 2 mm)

| | |
| --- | --- |
| Leistung: | 30 W |
| Distance: | 2 mm |
| Duration of exposure: | 0.5 s |
| Interval: | ≥ 0,2s |

Directions:
- Decontaminate fibre
- Avoid carbonization
- Multiple exposures, if necessary
- Keep moist, if necessary
- Stop as soon as bleaching occurs

Laser Nd:YAG (1064 nm)
Applicator: bare fibre (200 - 600μm)
non-contacting, in air

Vaporization

| | |
| --- | --- |
| Power: | 15 - 50 W |
| Duration of exposure: | 0.1 - 0.3 s |
| Interval: | ≥ 0.1 s |

Directions:
- Gas cooling for P > 30 W
- Blackening of fibre
- Coagulation fringe between 3 and 5 mm, depending on relation time/power
- Do not drill, place fibre softly
- Fibre end must be visible at all times

**Fig. 3.** Therapeutic guidelines, extract from [1]

## Methods in Dosimetry

Historically, various dosimetry methods have been developed. The starting point, in many instances an important treatment tool, is the systematic analysis of experience gained regarding the empirical parameters chosen for treatment. The listing of guidelines for therapy was an early step in this direction.

## Therapy Guidelines

To apply such guidelines with confidence, it is necessary that the parameters important in dosimetry – power density and interaction time – can be reproduced precisely. This is outlined in the safety regulations for medical lasers.

Figure 3 is an example of such a summary of experience.

For applications in medical practice, these guidelines should not be followed to the letter. The doctor in charge intuitively needs to take into account the influence of many different parameters.

To assist in this task, thus making laser applications safer, research in laser applications tries to ob-

tain a deeper understanding of the dose – effect relationships and to convert this knowledge into suitable medical resources. An automatic response is not being asked for, but information supporting the physician in the appropriate application or by reducing the risk making application possible at all.

From the wealth of information available a few characteristic examples will be presented.

## Computer Simulation

Models are employed to investigate cause and effect relationships for the interaction between biological tissue and laser radiation. This modelling emphasizes the spread of radiation energy, absorption properties of tissue and the resulting temperature distribution.

Analysing the models behaviour regarding the spread of photons, as well as the dynamic changes of absorption and the resulting thermal, chemical and mechanical effects in biological tissue will allow useful information to be gained about the efficiency of new therapeutic and diagnostic methods.

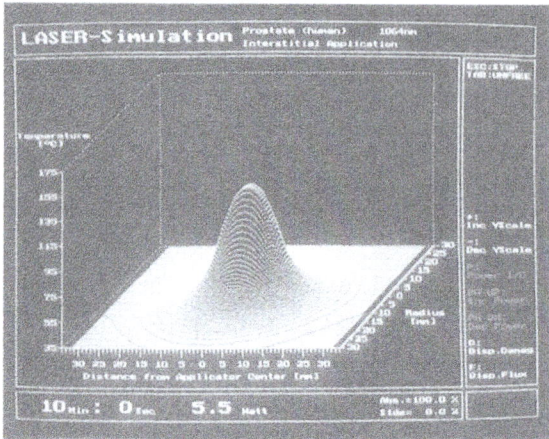

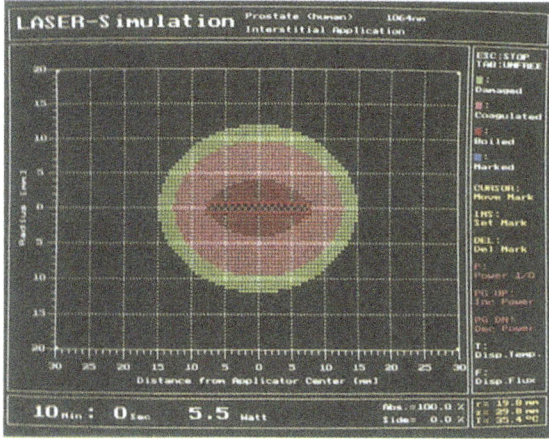

**Fig. 4.** Comparison of temperature distribution for LITT, computer simulation and experiment

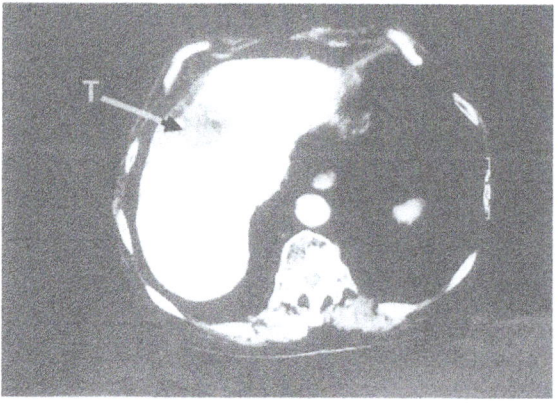

**Fig. 5.** MRT-controlled LITT of liver tumour (left), (CT of image of liver tumour 2 months after LITT)

Such simulations are already available in "real time", e.g. for laser-induced thermotherapy (LITT). They are being developed into an instrument for therapy control. Figure 4 shows a comparison between measured and simulated temperature distributions in tissue; a flushed applicator, developed for interstitial LITT, serves as an example [5].

The further development of models and measuring techniques will increase application of such computer-assisted procedures, especially for the predetermination and on-line control of irradiations. A typical example is the measurement of temperature distributions by magnetic resonance tomography (MRT), see Fig. 5 [3].

## Power Control of Lasers

The individual variability of tissue parameters limits the usefulness of such simulation for the dosimetry of laser radiation. This problem may be overcome by measuring and analysing physical tissue parameters with the help of suitable sensor systems.

Tissue displays a characteristic colour change during coagulation, a feature widely used as a visual feedback while applying a surgical laser. It is a consequence of the change in the optical properties of the tissue. Sensors are able to measure the onset of coagulation precisely and the extent of the damaged region. This information may be used to control the laser power (Fig. 6).

By employing such sensors, it is possible to prevent tissue carbonization. This is an essential safety feature if lasers are to be utilized in surroundings prone to explosions.

## Pyrometry

The preparatory surgical removal of tissue is still most important in clinical praxis.

It is necessary to distinguish between two main applications: (1) Using a hand-held focusing applicator and achieving a uniform cut, independent of variaions of the moving speed of the laser spot at the tissue surface, and (2) employing a fibre for the same task. In this case, the end of the fibre must be well protected against undesirable burnup.

Contact-free measurement of the temperature at the action spot allows determination of the relevant abrasion parameters. Increased surface temperatures can be expected if the surroundings of the affected tissue begin to dry without further tissue vaporization. In general, this will be the case if radiation is applied over a relatively long period of time. Contact-free temperature measurements at the ablation spot are made by pyrometer sensors. Figure 7 presents a possible design. Besides the infrared radiation detector IR-D.1, it contains a second detector (IR-D.2) to compensate for radiation from the surroundings.

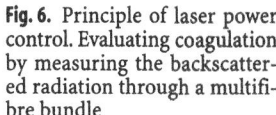

**Fig. 6.** Principle of laser power control. Evaluating coagulation by measuring the backscattered radiation through a multifibre bundle

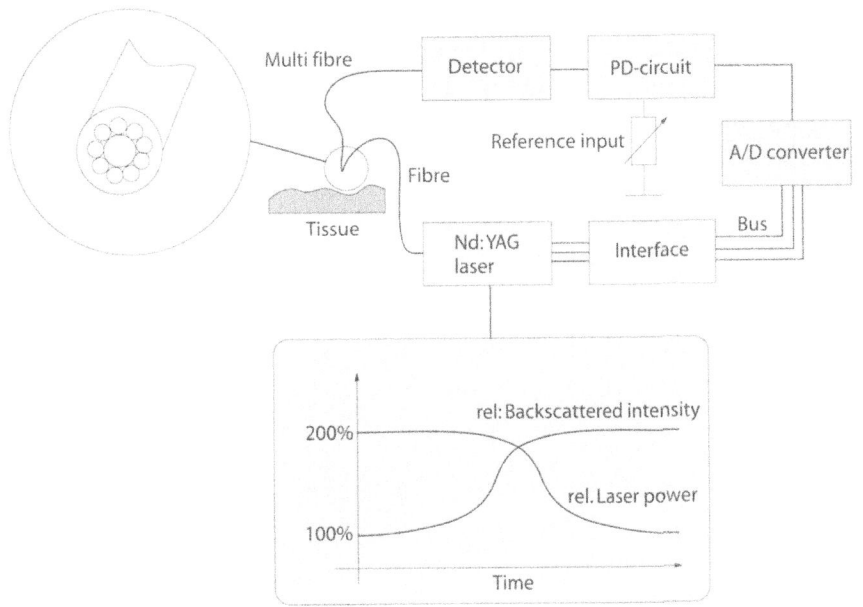

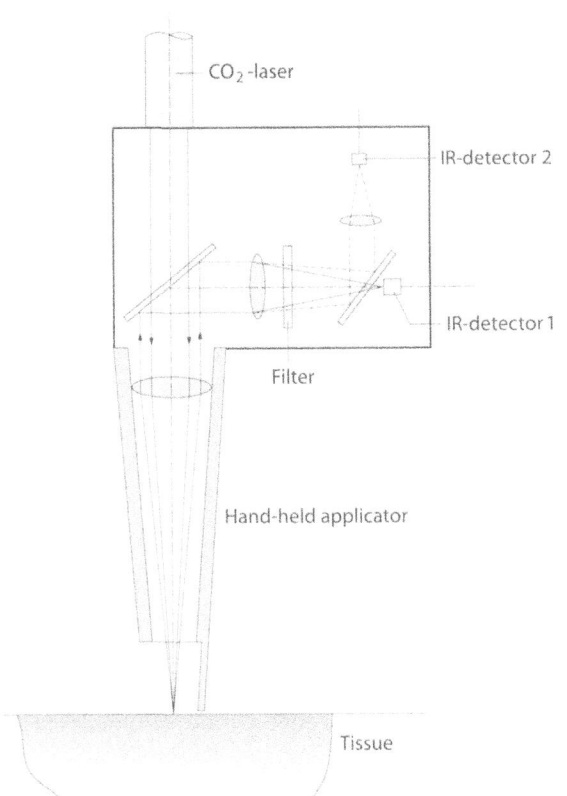

**Fig. 7.** Basic design of a hand-held temperature sensor for $CO_2$ Laser treatment

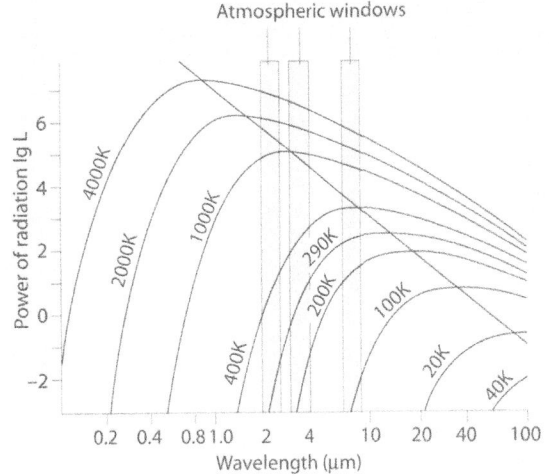

**Fig. 8.** Experimental investigation of integrating a pyrometer into a hand-held fibre applicator

For fibre application, e.g. Nd:YAG laser radiation, the above-described measuring principle is of limited use, because currently available pyrometer sensors are too large to be integrated into an applicator. Suitable fibres are available for the transmission of thermal radiation (maximum between 2 and 10 µm; see Fig. 9) achieving a temperature range of 300–600 K. Initial experimental investigations (see Fig. 8) confirm the possibility of detecting, with uncooled radiation sensors, thermal radiation transmitted by fibres. The agreement with commercially available pyrometers is good [8].

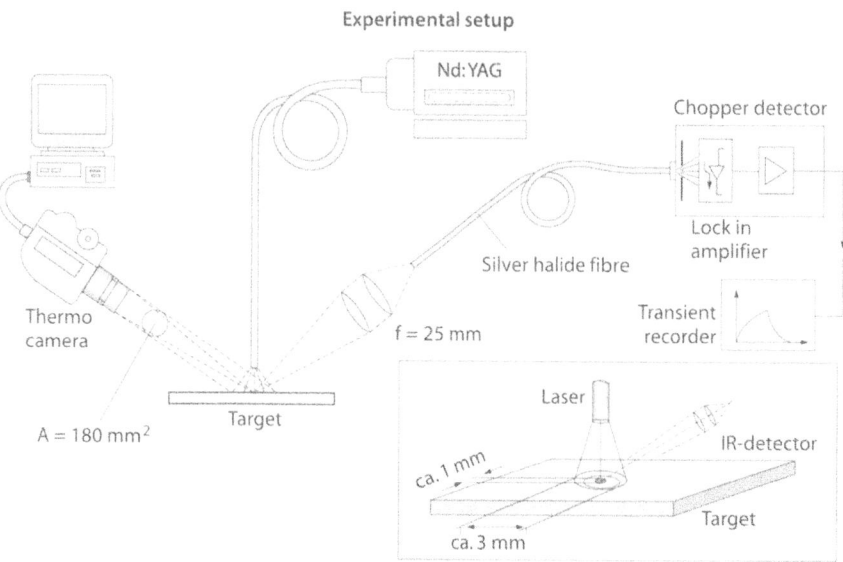

**Fig. 9.** Distribution of heat radiation

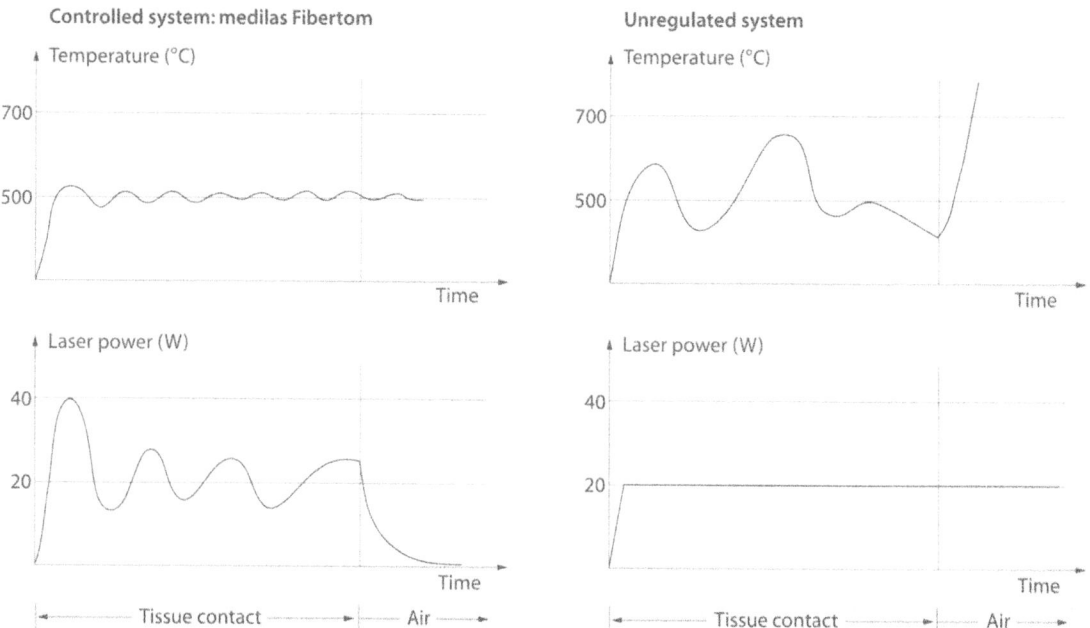

**Fig. 10.** Principle of the mediLas fibertom

Besides measuring the temperature at the point of radiative interaction, it is also possible to reduce burnup by monitoring the temperature at the end of the fibre. If the fibre is in contact with tissue, then heat conduction will transfer the absorbed radiant power into the surroundings. If, during non-contact application, fibre overheating threatens to occur – it is recognizable by an increased pyrolysis glow of attached particles – fibre destruction can be prevented by reducing the laser power. In this way, the fibre can be conserved for an extended period of time, even for large average ablative laser powers (Fig. 10).

## Lithotripsy

Another example of the effective use of feedback systems in medical laser applications is lithotripsy. Extracorporal stone destruction is widely used; however, for atypical stone types and restricted ac-

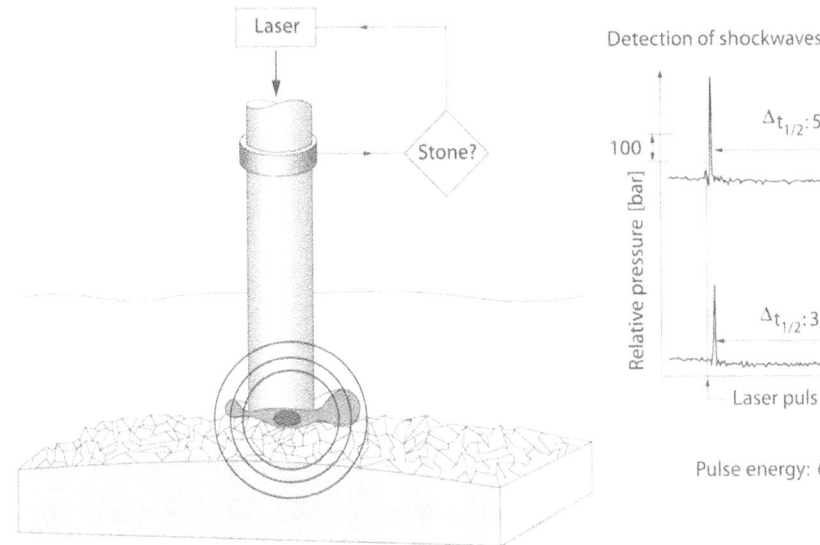
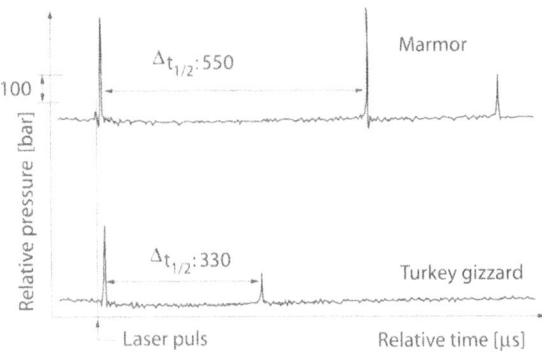

**Fig. 11.** Stone recognition by sound detector

cess, laser lithotripsy acts as a complementary therapy.

The action is as follows: a train of laser pulses is guided by a fibre to the application site and ignites a plasma at the stone's surface. The breakdown of the plasma creates a shockwave which detaches some fragments. After many repetitions, the stone will be destroyed. Tiny stone fragments suspended in the body fluid will impair the view through the endoscope. To ensure safe application, the stone should be abraded only slowly. Besides intense flushing, various other methods to monitor this application have been developed.

The evaluation of plasma glow is well known; its spectral distribution indicates the type of material involved [4]. In this way, unintentional irradiation of surrounding tissues will be avoided. However, this method calls for expensive investment in fibres and sensors.

Another method to control applications is to investigate sound waves transmitted by the laser fibre. Their time delay depends significantly on their point of origin, either from a stone or from inadvertent exposure of soft tissue (Fig. 11; [7]).

## Summary

It is impossible to include a complete review of the application of lasers in medicine, a list of examples could be extended virtually indefinitely. The fields of application mentioned are merely representative of trends in laser therapy. These can be characterized by:
- better mathematical understanding of the physical and biological processes,
- modelling the multitude of action principles involved,
- integrating sensors to measure process parameters,
- ensuring therapy success by compensating for variations of individual parameters.

The examples presented demonstrate that the development of procedures controlling therapy in laser medicine is still actively being pursued. On the one hand, new laser applications in medicine are calling for new therapy control procedures, for example the range of the various dental lasers or laser applications in cardiology. On the other hand, optical diagnostic procedures will control not only medical laser applications but all kinds of other therapies which apply energy as well.

## References

[1]  BERLIEN H-P, MÜLLER G (eds) (1989ff.) Angewandte Lasermedizin, Lehr- und Handbuch für Praxis und Klinik. Ecomed, Landsberg

[2]  DÖRSCHEL K, MÜLLER GJ (1991) Photoablation; future trends in biomedical application of lasers, SPIE Proc 1525: 253

[3]  GEWIESE B, BEUTHAN J, FOBBL F, et al (1994) Magnetic resonance imaging – controlled laser-induced interstitial thermotherapy. Invest Radiol 3: 345

[4]  KAR H, RINGELHAN H (1992) Grundlagen und Technik der Photoablation. Adv Laser Med 6 . ecomed Landsberg

[5]  MÜLLER GJ, ROGGAN A (eds) (1995) Laser-induced interstitial thermotherapy. SPIE Press PM 25

[6]  SCHOLZ C (1992) Neue Verfahren in der Bearbeitung von Hartgewebe in der Medizin mit dem Laser. Adv Laser Med. 7. Ecomed, Landsberg

[7]  TSCHEPE J (1994) Übertragung und Detektion von laserinduzierten Stoßwellen über Quarzglasfasern. Med Tech 1/95, Shaker (1995)

[8]  WÄSCHE W, et al (1994) IR-fiber-radiometry for high-temperature measurements during laser-tissue-interaction. SPIE Proc 2131: 155

# Part II
# Principles of Laser Application in Medicine

# II-1
# The Laser's Position in Medicine

B. Fuchs, H.-P. Berlien and C.M. Philipp

## Contents

## Introduction

Lasers are employed in diagnostic and therapeutic procedures, analogous to the applications of ultrasound. This similarity extends even further: both modalities must distinguish between pulsed and cw systems operating at various frequencies. With respect to the clinical application of lasers, the following has been proven: it is not sufficient to take into account merely the purely scientific explanation of the laser processes, i.e. the "how?", nor to distinguish only between the fields of application, the "where". It is most important to ask what is the laser's position, i.e. the "why" and "for what purpose".

Lasers in therapy:
- surgical tool
- central therapeutical procedure.

Lasers in diagnostics:
- monitoring metabolism
- imaging procedures.

## Basics

Laser radiation is characterized by the relations between its most important elementary parameters, such as: power and peak pulse, power, cw and single pulse duration, shape and repetition rate and beam spot diameter. Quantities like power density, totally absorbed energy, energy density and pulse energy are useful only as long as the more elementary parameters are known.

Laser radiation is characterized by:
- power and peak pulse power
- cw and single pulse duration
- pulse shape
- pulse repetition rate
- beam spot diameter.

Light is electromagnetic radiation; its optical penetration depth is determined by absorption and scattering processes.

Penetration depth depends on:
- absorption
- scattering.

Scattering displays two main components: backscattering from the issue surface ad forward scattering into the tissue. If the tissue layer is not too thick, photons may leave the tissue's distal boundary and thus will not be absorbed. The more often a photon is scattered within the tissue, the higher the probability that it will be absorbed (Fig. 1).

Absorption is determined by photochemical processes in the body's chromophores, like haemoglobin, myoglobin and melanin, and in the ultraviolet region mainly by proteins. Absorption by water exceeds specific absorption by the body's chromophores in the regions of shortwave ultraviolet and middle infrared. Scattering decreases with increasing wavelength. This creates a so-called optical window. In it scattering is low, specific absorption by the body's chromophores is negligible and absorption by water has not yet started (Fig. 2).

This optical window is situated in the wavelength range between 850 and 1100 nm. Due to non specific absorption in the near-infrared, it is possible to control the action depth (0.1–10 mm) via exposure time. Besides the wavelength-specific tissue properties, other quantities will influence the tissue's reaction. These include thermal conductivity, heat capacity and mechanical characteristics, e.g. elasticity.

## Laser Techniques

### Types of Application

There are various ways to transfer laser radiation to the tissue. In practice, only two beam-handling systems are used. Articulated arms are utilized where transmission by fibre is not possible, either due to the wavelength range, e.g. far-ultraviolet or mean infrared, or because the transmitted pulse power is so large that it would destroy the fibre.

Transmission of laser radiation:
- mirrored articulated arm
- fibre.

Whenever possible, fibre transmission is at present utilized because of the high flexibility offered by the various fibre diameters (between 0.2 and 0.6 mm) (Fig. 3).

In the simplest case, laser radiation leaving the light guide is applied directly to the tissue. However, it is possible to employ a focusing handpiece or a micromanipulator attached to an operation microscope, to focus the radiation into a very small spot diameter. This conventional non contact technique can be supplemented by contact techniques as long as the laser's radiation can be transmitted by fibre. In this case, the fibre is brought directly in contact with the tissue surface or guided into the tissue. Depending on the task, different tissue reactions can be initiated in this way.

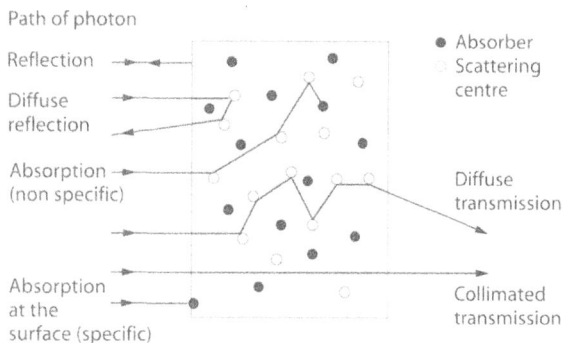

**Fig. 1.** Path of a photon through a volume of tissue

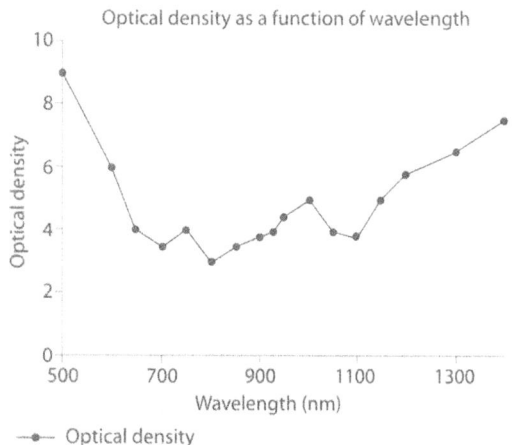

**Fig. 2.** Optical density and optical window

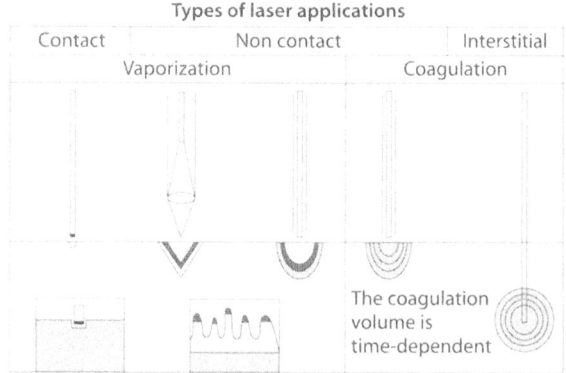

**Fig. 3.** Various applications of laser radiation

**Fig. 4.** Various ways to protect the surface by cooling

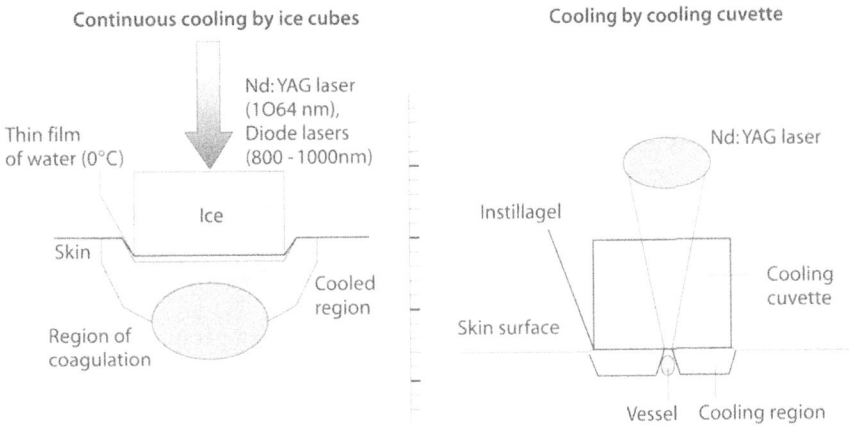

High-power radiation causes rapid carbonization at the fibre-tissue interface. All emitted radiation will be absorbed immediately and converted into vaporization energy. At low powers this does not happen, and the fibre tip can be placed into the tissue. In this way, it is possible to irradiate large tissue volumes without putting a load on the surface.

Types of application:
- non contact
- contact
- interstitial.

To protect the tissue surface during non contact applications, cooling may be applied (Fig. 4).

Using ice cubes, a tissue thickness of 1.5 to 2 mm can be cooled satisfactorily, so that even during transcutaneous irradiation no thermal damage will occur. Cooling by a flowthrough cuvette is also possible.

Protection of issue surface
- cooling by ice cubes
- cooling by cuvette.

The flowthrough cuvette has a flexible, transparent membrane and is in direct contact with the treated skin area. It consist of a cylindrical cooling body the upper part of which is a glass plate and the lower part a 0.04 mm-thick latex membrane, transparent in the visible (VIS) and in the near-infrared range (NIR). The total transmission of the system is 70% for NIR (1064 nm). The cooling medium is a 40% propyleneglycol solution with an initial temperature of –18 °C. The temperature of the interface skin-membrane area is between 0 and -10 °C. The cuvette is positioned without pressure, allowing of its contact area to be matched with the shape of the body's surface, without compressing any vessels close to the surface. A large proportion of the heat delivered by scattering radiation into the epidermis, and heat conducted from deeper layers, will be taken up by the cooling liquid. In this way, the skin layers close to the surface, and especially the area surrounding the beam's entrance, will be spared.

When using ice cubes, it is important that the ice is free of enclosures, especially air bubbles, because this would cause scattering and reflection of the applied radiation. Enclosure-free ice cubes (made by a special process) are split into halves and used in this form. These icicle halves are placed directly onto the skin, partly compressing the angioma. The icicles are penetrated by the radiation. Two effects dominate: compressing the cross-section of the vessels close to the surface by the pressure of the ice will reduce their diameter. Increasing the pressure causes compression deeper down, so that a greater volume of blood will be driven into greater depths, at least to 1 cm. Since the blood concentration in the upper regions is reduced, absorption decreases, as will thermal damage of the skin. In addition, heat produced in the top skin layers is conducted away to such a degree that the skin temperature will not exceed 45 °C. Beginning at a depth of about 1.5 mm, cooling becomes insufficient, due to the limited thermal conductivity of the skin, and temperatures may exceed 60 °C, resulting in vessel coagulation. For effective cooling, the pieces of ice should be exchanged often and shifted around during irradiation. Also, close contact between skin surface and ice cubes is essential. The radiation should pass only through intact cubes. This ice cube cooling technique is applied to voluminous and subcutaneous haemangiomas or to vascular malformations. Continuous cooling by ice is suitable for easily compressible and poorly perfused vessels in order to protect the skin against irreversible thermal damage. At the same time, tissue at depths of about 4 mm will be coagulated. Applied to large areas, it is possible that secondary trophic defects may develop. In this case, and for cases extending to a considerable depth as well as for vessels having a large throughput, the upper skin layers will be protected by using percutaneous interstitial or intravascular application. This is especially the case of the cutis is not affected.

## Laser Procedures

It is possible to differentiate between three laser effects:
- photochemical,
- photothermal and
- photomechanical.

These three basic laser effects make possible the following laser procedures:
- removing/cutting of tissue
- thermotherapy
- photochemistry and
- optical imaging.

### Removing/Cutting

The effect most often employed clinically is photovaporization. If a very narrow coagulation edge is called for, as in microsurgical preparations, a $CO_2$ laser will be employed. However, using an Nd:YAG laser together with bare-fibre contact technique, fine cuts with a small coagulation seam can also be achieved. For cuts having a wide coagulation seam, as required in haemostasis, an Nd:YAG laser using a focusing handpiece and non contact technique is required. Vaporization of extended areas is bet achieved by using a $CO_2$ laser together with a micromanipulator, or a scanner system, or a laser emitting short pulses, e.g. Er:YAG or Ho:YAG lasers. Photoablation by excimer laser is essential only in ophthalmology (for keratotomy) and laser angioplasty. Photomechanical effects leading to photodisruption are predominantly employed, besides in treatment of cataract membranes, in laser lithotripsy and for treatment of naevus flammeus. In this case, the diameter of the capillaries should not be too large, otherwise their elasticity increases and rupture only becomes possible using disproportionately high energies causing secondary heat damage.

Removing/cutting
- photovaporization,
- photoablation,
- photodisruption and
- photofragmentation.

### Thermotherapy

Laser-induced thermotherapy causes the following tissue reactions:

Hyperthermia is not a therapy by itself because no cells are damaged. As opposed to microwave hyperthermia, laser hyperthermia is locally limited to tissue heating.

Thermotherapy:
- laser-induced hyperthermia (LIHT),
- thermodynamic therapy (TDT) – leading to apoptosis,
- laser-induced coagulation (LIC) = necrosis.

Temperatures between 40 and 60 °C result in disorder of cellular metabolism and membrane function. The result is a thermodynamic reaction which may result in apoptosis of the affected regions. Definite coagulation necroses occur between 60 and 100 °C.

Thermic dynamic reactions are dominated by inflammatory reactions, leading via necrobiosis with reparative conversions to fibrillar structural changes This is of special advantage for non-neoplastic proliferations and for treatment of congenital vessel diseases.

Laser-induced coagulation in situ leads to tissue destruction by heating to temperatures above 60 °C. A special case is interstitial thermotherapy. Besides special applicators having an enlarged emission area, the most important procedure in this case is using the bare-fibre technique (Fig. 5).

After skin puncture, the bare fibre is guided through a Teflon or a metal hypodermic syringe channel, monitored, e.g. by ultrasound, to the lesion, which is 10 minutes irradiated at a low power of about 5 W. The diameter of the coagulation zone can be controlled by irradiation time per area; however, this time should not exceed 120 s. If it is not possible to monitor the procedure directly, it can be checked by colour duplex sonography. This is a very effective and simple method to localize the lesion and to control fibre position and procedure (Table 1).

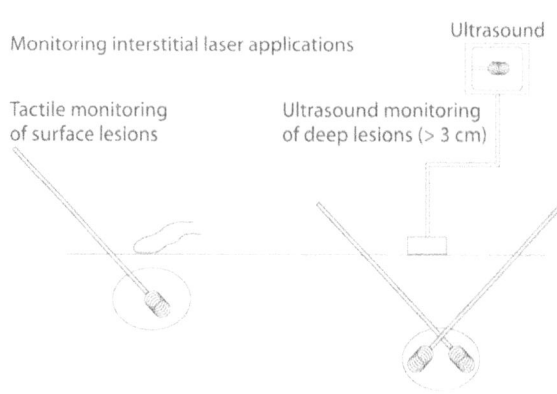

**Fig. 5.** Tactile and ultrasonic monitoring during LITT

**Table 1.** Monitoring LITT

Clinical
- Visual monitoring
  Position
  Blood content
- Digital monitoring
  Position
  Temperature
  Gas escape

FKDS
- B-image
  Position
  Structural change
- Colour coding
  Procedure monitoring

MRT
  Position
  Temperature
  Structural change

**Table 2.** Substances for PDT

- Pophyrin
- Phthalo-/naphthalocyanin
- Pheophorbide
- Ala
- Combined with monoclonal AK

### Photodynamic Therapy

Photodynamic therapy is based on the fact that there are, in effect, totally inert molecules inside the body, which can be induced to produce oxygen radicals if irradiated. Systematically applied, these molecules either accumulate selectively in specific tissues or are selectively eliminated. If irradiated at a specific time, radicals will be produced only within these tissues. Cytotoxic reactions will be the result, ending in apoptosis. Besides porphyrin derivatives, synthetic pheophorbide and phthalocyanin are also currently employed for this purpose, as is the delta-aminolaevulin acid (ALA)-induced protoporphyrin-IX synthesis (Table 2).

In addition, it is possible to attach synthetic substances like pheophorbide, phthalocyanin and naphthalocyanin to liposomes or monoclonal antibodies. This will significantly increase tissue selectivity. These procedures still carry the risk of skin sensitizing, causing phototoxic reactions. Appropriate protective measures should be taken.

### Laser-Induced Fluorescence

The selective accumulation of the above-mentioned substances can also be used for fluorescence diagnosis. In this case, radiation power – at a value below that used for photodynamic therapy – does not trigger creation of oxygen radicals, but leads to luminescence. This luminescence can be detected and used for diagnostic purposes. This is called xenofluorescence.

Fluorescence:
- autofluorescence
- xenofluorescence.

It is also possible to excite fluorescence in chromophores of the body. Here, the enzymes of the respiratory chain or the citris acid cycle are essential, especially NADH and FAD (auto). ALA-induced protoporphyrin-IX production can also be used for this purpose (xemo).

### Diaphanoscopy

Scattered and transmitted photons can be detected and consequently utilized for optical tomography or diaphanoscopy. White light diaphanoscopy is an old investigative tool, now, however, practically replaced by ultrasound, because of poor imaging qualities and very limited transmission depth. Lasers emitting in the near-infrared allow penetration of thicker tissue layers and fibres can bring the light source into the body's interior. Improved infrared cameras will deliver high-resolution images. Thus, intraoperative differentiation between cystic and solid structures is possible during endoscopic and especially during laparoscopic operations.

## Fields of Application

### Plastic Surgery

Lasers are well established in a number of fields. These will be presented below without any claim to completeness.

If it is inadvisable to remove tattoos surgically or by derma-abrasion, lasers offer an alternative treatment procedure. Laser treatment is the method of choice especially for line tattoos, residual after surgical intervention and for tattoos on hands, face and eyelids. $CO_2$ laser vaporization using a field technique or pulse ruby alexandrite is preferred. Nd:YAG lasers did not live up to expectations. In addition, there is a danger of heavy metal reactions initiated by the disruption process. Thermoabrasion by a $CO_2$ laser was used for a long time but, due to poor beam guidance, has been abandoned. $CO_2$ lasers with rotating scanning systems have been developed to treat necrotic burns. They are at present advertized by companies specializing in skin resurfacing. It remains to be seen whether or not this scanning procedure will be suitable for this purpose. Benign intraepithelial lesions are a domain of $CO_2$ laser application. Especially virus-induced lesions, like molluscum contagiosum and condyloma aciminata, can be vaporized very well.

Various laser types may be utilized to treat hypertrophic scares or acne scars. Coagulation of visible single vessels in hypercapillarization, which is combined with redness and skin irritation, is the aim of Argon laser treatment. Dye lasers can treat diffuse redness caused by invisible single vessels (new lesion). Nd:YAG lasers are well suited for treatment of lesions dominated by fibrosis (old lesion). The aim is to reduce the fibrotic mass of the keloid. Under local anaesthesia, short exposure times will lead to interstitial vaporization. The surroundings of larger keloids are exposed to radiation for 0.3 s. Since histology-based diagnosis is not possible, the treatment of pigmented naevi must be considered very carefully and should be performed only on the benign type. Treatment of lentigo is well suited for the application of Argon lasers. For erythrosis facialis and spider veins, coagulation by Argon laser radiation is the safest and most efficient method of treatment. It is superior to any other technique including other lasers like, e.g. dye lasers.

## Angiology

### Congenital Vessel Diseases

Before selecting an appropriate laser treatment technique, congenital tumour and vascular malformations must be clearly differentiated (Table 3).

Naevus flammeus (portwine stain) is a malformation and should preferably be treated by dye lasers. For pronounced tuberous formations, often observed in adults, treatment by an Argon laser is the method of choice. Other forms of congenital vessel diseases, such as haemangioma, extratruncular venous, arterio-venous and lymphatic malformations, are treated by Nd:YAG lasers. The treatment methods are either direct transcutaneous irradiation, the transcutaneous ice-cube technique, interstitial or intraluminal irradiation. Truncular vascular malformations, e.g. large AV fistulas or malformations of the basic venous system, should be embolized or treated by surgery.

### Treatment by Argon Laser

Small planar haemangiomas and their prodromal stages in young infants are well suited for treatment by Argon lasers. A pulsed laser beam is directed through the skin directly at the vessels. The radiation penetrates the epidermis without much absorption, but it is well absorbed by haemoglobin and skin pigments. Argon laser radiation is able to coagulate ves-

**Table 3.** Congenital vessel diseases

Haemangioma
– Planecapillary
  Tuberous

– Vascular malformations
– Capillary (naevus flammeus)
– Venous
– Lymphatic
– Arterio-venous
– Mixed

sels with a diameter of up to 1 mm, as long as the exposure time exceeds the vessel's relaxation time. To avoid coagulation damage to the skin, short exposure times, low beam power and the polka-dot technique are indicated. (This is a coagulation technique, irradiating point by point with an appropriate spot size. The spots do not overlap.)

### Treatment by Flashlight-Pumped-Dye Laser (FDL)

Using an FDL, specially developed to treat naevi flammei, in combination with optimal wavelength and an exposure time adapted to the heat conductivity of the tissue, a very high treatment specifity for the smallest vessels is achieved. Due to the very high power density of the FDL, small vessels will rupture. After a delay of about 20 – 60 s, a reddish coloring of the skin appears. Within a few minutes, the colour intensity deepens to a livid blue (purpura). At its centre a small, possibly itching, blister may appear. Purpura may remain, depending on the individual, for up to 2 weeks; it will display the distinct colour changes of a haematoma. Larger vessels will be coagulated. Capillary-sized vessels are much less affected due to the short time available for heating by thermal conduction. Strong absorption of the radiation in the proximal lumen of larger vessels will shield its distal components.

Multiple exposures of the same area should be avoided even if no primary reaction is observed. On the other hand, since the beam profile of a widened beam is inhomogeneous, the individual fields must overlap by about 20%.

Surface cooling of the skin will considerably reduce perceptible pain. Cooling with ice cubes is not suitable for treatment of naevus flammeus, since the target organ would be directly in the cooled region.

FDLs appear to be a much more efficient tool compared to an Argon laser, i.e. they achieve better bleaching of the naevus after fewer treatments. This is especially so for juvenile and very uniform naevi flammei. In addition, FDLs are well suited for treatment of the prodromal stage of haemangioma occurring as naevus flammeus like skin redness, spider naevus-like lesions or as teleangiectatic vessel budding. Treatment is similar to the treatment of naevus flammeus.

### Treatment by Nd:YAG Laser

Nd:YAG lasers are well suited for treatment of larger haemangioma and vascular malformations, with the exception of naevus flammeus. They emit radiation in the infrared (1064 nm), which is significantly better absorbed in blood than in surrounding tissues. The influence of scattering and long exposure times induces relatively uniform tissue reactions. The observed reactions in the endothelial structures are stronger than in surrounding fatty and connective tissues. Depending on tissue and irradiation parameters, by heat conduction it is possible to reach a penetration depth of a few millimetres. Distributing the energy over a large volume leads to uniform heating and coagulation at increased temperatures. Depending on localisation and extension of the disease in the vessel, perfusion volumes and lumens of the involved vessels, various irradiation parameters and application procedures are selected.

- Direct transcutaneous Nd:YAG laser application
  Intracutaneous parts of congenital vessel disease can be coagulated directly using an Nd:YAG laser beam. A polka-dot technique is applied. Multiple exposures and overlapping of individual fields should be avoided. The reaction depth is about 1–1.5 mm, and damaging the epidermis can often be avoided. This technique is employed mainly in teleangiectactis prodromal stages, with a strong central vessel and at the fast-growing offshoots of larger haemangioma, or at their margins, to prevent further spread.
  Undesirable coagulations and burning of regenerative skin layers may occur in cases of extended lesions, interlaced applications and especially during irradiation of deep-seated vessel malformations. To avoid this, another procedure is of value:
- Transcutaneous Nd:YAG laser application with continuous cooling of the surface.
  Transparent ice cubes, free of air bubbles, are well suited for cooling skin layers (thickness about 5 mm) close to the surface.
- Interstitial bare-fibre technique.
  Direct skin puncture is performed with a Teflon hypodermic syringe (Abocath G16) acting as a catheter. After introducing a freshly broken light guide (bare fibre) into the catheter, it is pulled back about 5 mm, the fibre remaining in place. The positioning procedure is often assisted by a red Ne laser pilot beam. Inside a slightly darkened room it is possible to recognize the position of the fibre and by observing the pilot beam radiation passing

through the tissue. This works for a depth of up to about 20 mm.

For interstitial applications two different procedures are available:
1. Keeping the fibre in position and determining its end point.
2. Withdrawing the fibre at $1 mm/s^{-1}$ while the laser beam remains switched on.

If possible, the first procedure should be applied because it is more precise and the extent of the coagulation can be controlled by the application time. The coagulation volume will also spread retrogradely along the fibre. During interstitial laser application, one or more channels is punctured by the guiding catheter loaded with a bare fibre in such a way that coagulation creates a fan like pattern. The bare fibre should not be withdrawn all the way into the catheter or the syringe, acting as a catheter. This will avoid damaging the catheter and so leaving foreign material within the tissue or vessel.

The spread of heat in the area under irradiation is monitored by digital palpation. This will avoid coagulation necrosis of the top skin layers. Compressing the skin must be avoided to prevent pressing skin directly against the end of the fibre, when the skin would be burned. Non invasive systems measuring skin temperature during laser irradiation with sufficient precision are unavailable at present.

- Intraluminal bare-fibre technique
  Intraluminal lasers are applied to treat large ectactic vessels, observable manly in venous malformations, or arterio-venous malformations. To do this, the end of the fibre must be flushed continuously, otherwise coagulated blood will accumulate and carbonize at the fibre tip. A physiologically neutral medium like 0.9% NaCl solution is well suited for the flushing. There is no need for flushing during treatment of predominantly lymphatic malformations. In this case, laser radiation is barely absorbed by lymphatic fluids and the vessel walls are directly irradiated. For the same reason, draining of the lymphatic fluid before laser treatment is not recommended. Monitoring skin temperature is again by continuous digital palpation. To prevent unchecked skin coagulation in the neighbourhood of the puncture spot, multiple laser application through a puncture channel should be avoided. To achieve a sufficiently long channel, the spot selected for skin puncture should be separated by at least 1 cm from the diseased vessel. In the case of afterbleeding, intermittent laser irradiation (duration about 0.5 s) will stop the blood flow inside the puncture channel. Irradiation of the cutis should be avoided. The use of a compression bandage ap-

pears to be helpful. This will reduce the thrombotic component and support coagulation within the vessel walls.

Selection of laser:

for superficial, planar, smooth surfaces
- argon, FDL (Nd:YAG)
for tubal, organ-specific, voluminous
- Nd:YAG
transcutaneous (+/- cooling)
interstitial/intraluminal.

### Acquired Vessel Diseases

Varicosis (Table 4)

Insufficiency of trunk for branch veins cannot be treated by lasers. However, treatment by percutaneous laser coagulation of isolated insufficiency of perforating veins, or residues after vein stripping is an alternative to endoscopic perforans dissection. It is a less traumatic treatment and can be performed on an ambulant patient under local anaesthesia. Laser treatment will perform epifascial coagulation of the convoluted veins; it will not generate a thrombus as in sclerosing. It leads to perivasculitis combined with shrinkage of the convolute and scarring of the fascia gap as a site of insufficiency. The coagulation generated must be resorbed, so the patient will feel a lump at the treated position. Treatment of a full vein allows better selection of the pathologically affected perforating veins. There are three main differences as compared with sclerotherapy:

- Sclerosing can be performed through a central puncture site; the laser must always be aimed directly at the target tissue.
- The effect of laser treatment is always local and sclerosing always regional.
- Side effects of laser treatment are locally limited. Sclerotherapy presents – besides regional side effects – a risk of severe systemic side effects in addition.

The success rate of laser therapy is comparable to that of sclerotherapy, but without its risk of valve damage and systemic side effects. It can be applied straightaway to high-risk patients, e.g. patients with repeated lung embolisms.

Laser treatment of spider vein varices should never be the first choice, but only for recurrence after un-

successful sclerosing. Treatment results for recurrences in the legs using Argon or pulsed dye lasers are unconvincing. As compared with teleangiectasis in the face, which responds well to Argon laser treatment, the diameter of the leg vessels is too large and the overlying skin is too thick. Nd-YAG lasers should be suitable, in pulsed operation, as their depth of action can be reduced, depending on pulse duration, to 1 mm. However, during direct transcutaneous irradiation employing relatively high power and longer irradiation times, coagulation of the skin may occur. This can be prevented by surface cooling employing a flowthrough cuvette.

### Arterial occlusion disease (Table 5)

Laser angioplasty is indicated if primary balloon dilatation is not possible, e.g. the guiding wire cannot be advanced due to total occlusion or if the stenosis is too long. This procedure is called laser-assisted balloon dilation. Only an excimer laser should be employed. Any other type of laser will cause significant thermal or mechanical damage of the vessel wall or its surroundings. The introduction of multifibre catheters with individually controlled fibres has significantly increased flexibility of the system and the efficiency of ablation. These systems are also used in coronary angioplasty. Laser angioplasty offers advantages compared with conventional balloon angioplasty, especially for long occlusions and obliteration in vessels below the knee. However, one problem remains: the rate of restenosis. Even this new procedure cannot reduce it significantly, at least for the time being.

**Table 5.** Advantages and disadvantages of laser angioplasty

Advantages:
- Unresticted repeatability
- Ablation of calcified matter
- High flexibility of the system
- Applicable in small and peripheral arteries

Problems:
- Low ablation rate
- Perforation
- Dissection
- Vasopasmus
- High costs for system

### Organ Resection

To perform microsurgical resections presenting a narrow coagulation edge, $CO_2$ lasers are the optimum tool; but this laser is not well suited for resection of parachymatose organs or plethoric tumours. This is the field of application of the Nd:YAG lasers using non-contact techniques and having high output

**Table 4.** Laser indications for varicosis

- No cross-insufficiency and trunk varicosis
- Branch varicosis after unsuccessful sclerotherapy
- Perforans varicosis for isolated veins and residuals after OP
- Spider veins only after unseccessful sclerotherapy

power. Using a small focal spot diameter and 60-W continuous output power, vaporization resulting in a wide coagulation edge is achieved after initial carbonization.

The coagulation edge will close capillaries and small vessels. By tissue spreading, especially during partial liver resection, veins and arteries can be identified. This will reduce the risk of unintentional vaporization and opening of these vessels. The procedure resembles that of using an ultrasound aspirator, but it has the additional advantage of presenting coagulated resection areas. In this way, risk of immediate or delayed bleeding is reduced, as well as the formation of a biliary fistula. In the case of bleeding, the blood can be removed using an NaCl solution and sucking dry. The thin film of water will not absorb Nd:YAG radiation and the bleeding vessels can be coagulated through the liquid film. The same procedure may be used for in situ coagulation, e.g. removal of tumour remains leaving a narrow coagulation edge.

It is also possible to achieve a small coagulation edge by using lower exit powers (of about 30 W), larger focal diameters and shorter exposure times (0.2–0.5 s). If needed, the same position can be exposed several times until blanching begins. To achieve deep coagulation, but avoiding carbonization at the same time, the surface must be moistened with an NaCl solution. To achieve deeper coagulation, longer exposure times are used because coagulation width depends to a large degree on exposure time.

To coagulate larger haemorrhages, 60-W laser power, large focal diameter ad continuous rinsing with NaCl solution are utilized during irradiation. The NaCl solution will wash away the blood, which otherwise would absorb a large percentage of the radiation and induce carbonization. In addition, the surface will be cooled for deeper coagulation:

Organ resection:
- lung
- liver
- spleen
- pancreas
- kidney
- tumours.

## Endoscopic Surgery

A wide variety of lasers is utilized in endoscopic surgery (Table 6). The $CO_2$ laser is the method of choice for surgery of the larynx and treatment of epithelial dysplasia in the oral cavity and perineal region.

For laparoscopic and thoracoscopic operations, Nd:YAG lasers with the bare fibre technique should be preferred. The non contact technique allows controlled coagulation via exposure time variation. Using

**Table 6.** Aims of endoscopic surgery

- Application of standard instruments
- No limits in miniaturization
- Possibility of precise preparations
- Possibility of controlled coagulation
- Management of complications
- Multifunctional tool
- Patient safety

the same instrument and the same parameters in contact technique, vaporization will occur. Advantages compared with high-frequency surgery are: the same instrument is able to coagulate and to cut, thus avoiding an interchange between bipolar and monopolar mode. In addition, the risk involved by applying monopolar currents with leakage currents does not exist. Operations under water do not need electrolyte-free solutions, which would pose a significant risk of water intoxication. Application of the so-called argon beamer, i.e. monopolar HF technique, cannot be recommended because during all endoscopic procedures it poses a very high risk of gas embolus. For treatment of malignant stenoses in the bronchial and gastrointestinal tracts, the combination of non contact-contact-bare-fibre techniques, is also well established. In this case, the visible tumour parts are coagulated first by non contact technique and under endoscopic monitoring.

Then the stenosing tumour parts are removed with the same fibre using the contact technique. The advantages as compared with primary non contact vaporization are: smoke development is reduced significantly, especially for bronchial stenoses, and unchecked deep coagulations, carrying the risk of perforation or via falsa, can be avoided.

Recanalization of tumours and benign stenoses:
- non contact coagulation
- contact abrasion.

During abrasion of suspected colon polyps, a sample for histological investigation can be removed first using an HF loop followed by Nd:YAG laser coagulation of the polyp's base. Laser treatment of acute gastrointestinal bleeding is inferior to injection procedures. However, bleeding in Osler-Rendu-Weber-disease and other angiodysplasia respond very well to laser treatment. Lasers with bare-fibre contact technique are also well suited for treatment of extreme anal and rectal stenoses in in Crohn's disease or obliteration of persistent ano-rectal fistula.

Laser lithotripsy using pulsed-dye lasers or frequency-doubled Nd:YAG lasers is indicated especially for impacted choledochal stones inaccessible to simple extraction after papilotomy. This method of extracorporal shock wave lithotripsy (ESWL) is also

**Table 7.** Lasers in orthopaedics and arthroscopy I

| Laser/wavelength (nm) | Fibre transmission | Hyaline cartilage | Fibrous cartilage | Osteotomy |
|---|---|---|---|---|
| Excimer/308 | + (+) | + | + | + |
| Nd:YAG/1320 | ++ | + | + | |
| Holmium/2100 | + (+) | (+) | ++ | (+) |
| Erbium/2940 | (+) – | | + (+) | ++ |
| CO2 pulsed/10600 | (+) – | | + | ∠ |

**Table 8.** Lasers in orthopaedics and arthroscopy II

| Laser/wavelength (nm) | PDT | IR diaphanoscopy | Coagulation | Cutting (soft tissues) | Cutting (hard tissues) |
|---|---|---|---|---|---|
| Nd:YAG/1064 | | ++ | ++ | + | |
| Nd:YAG/1320 | | | + | + (+) | |
| $CO_2$/10600 | | | – | + | |
| Pulsed Systems | | | | ++ | + (+) |
| Diodes/780–900 | + | ++ | + | + | |
| Dye (cw)/633 | + | | | | |

preferable for pancreatic stones since it significantly reduces mechanical trauma to the pancreas.

Indications for laser lithotripsy:
- contra indication for ESWL
- failure of ESWL.

Holmium lasers are preferred for discectomy and arthroscopic operations. They are able to remove tissue almost without thermal damage. During arthroscopic operations using mechanical instruments, especially for resection of the posterior horn of the meniscus, significant trauma to the knee joint will be the result. The laser fibre is able to puncture the skin at any position via a thin syringe (Tables 7 and 8).

Due to the delicate laser instrumentation, pressure and shear forces acting on other regions of the cartilage are significantly reduced as compared with mechanical shavers.

Arthroscopy:
- meniscectomy
- synovectomy.

Percutaneous laser disc decompression has sidelined chemonucleolysis almost completely. Compared with mechanical procedures, it results in much reduced trauma.

Discs:
- disc vaporization
- disc decompression.

This is indicated if conservative treatment fails, where there is no compelling reason for an operation, e.g. a prolapse.

Access:
- endoscopic
- percutaneous.

## Tumour Therapy

Laser therapy of tumours is not limited to the previously mentioned methods of tumour resection and endoscopic recanalization of tumour resection and endoscopic recanalization of tumour stenoses. It is also worthwhile to mention that there is interstitial laser coagulation of liver metastases, colorectal and mammary carcinoma, thoracoscopic coagulation of lung metastases and interstitial in situ coagulation of skin metastases and local recurrence of breast carcinoma, melanomas and of Kaposi's sarcoma. As opposed to these thermal methods of destruction, photodynamic therapy is indicated for advanced dysplasia which has not yet developed into an infiltrating tumour, also for micro-invasive tumours and tumours spreading on the surface. However, it must be stated that all these so-called destruction procedures, i.e. in situ coagulation, interstitial coagulation and photodynamic therapy, at our present state of knowledge, are all merely palliative procedures and should be applied only if all other methods have failed (Table 9). Time will tell whether or not the currently very encouraging results will lead to procedures for primary curative therapy.

**Table 9.** Laser procedures in tumour therapy

- Tumour resection
- Recanalization of tumour stenoses
- Tumour destruction
  In situ coagulation
  Interstitial
  PDT

## Further Reading

[1] BERLIEN HP, MÜLLER G, WALDSCHMIDT J (1986) Correct selection of different types of laser treatment of surface and deep located vessel anomalies. 3rd Congress European Laser Association, Amsterdam
[2] BERLIEN HP, WALDSCHMIDT J, MÜLLER G (1987) Laser treatment of cutan and deep vessel anomalies. Laser optoelectronics in medicine. Springer, Berlin Heidelberg New York, pp 526–528
[3] BERLIEN HP, MÜLLER G, WALDSCHMIDT J (1990) Lasers in pediatric surgery. Prog Ped Surg 25: 5–22
[4] BERLIEN HP, PHILIPP C. Engel-Murke F, Fuchs B (1993) Laseranwendungen in der Gefäßchirurgie. Zentralbl Chir 118: 383–389
[5] BERLIEN HP, CREMER H, DJAWARI D, GRANTZOW R, GUBISCH W (1993/94) Leitlinien zur Behandlung angeborener Gefäßerkrankungen. Pädiat Prax 46: 87–92
[6] FUCHS B, BERLIEN HP, PHILIPP C, MÜLLER G (1989) Laserapplikationen in der Medizin. Angew Lasermed, ecomed, Landsberg
[7] FUCHS B, PHILIPP C, ENGEL-MURKE F, SHALTOUT J, BERLIEN HP (1993) Techniques for endoscopic and non-endoscopic intracorporeal laser applications. Endoscopic surgery and allied technologies 1: 217–223
[8] GIERING K, BERLIEN HP, YAACOF ZB, PHILIPP C (1995) Risiken der Laser- und Hochfrequenzchirurgie. Lasermed 11: 97
[9] KUHLS R, BIER J, BERLIEN HP (1995) Behandlung von hypertrophen Narben bzw. Keloiden durch zentrale Exzision mit anschließender post-operativer, prophylaktischer Argon-Laser-Bestrahlung – Ergebnisse einer präliminaren Studie. Lasermed 11: 67–72
[10] LANDTHALER M, HOHENLEUTNER U, ABD EL RAHEEM T (1995) Therapie vaskulärer Fehlbildungen. medwelt 46: 357–359
[11] MÜLLER U, PHILIPP C, FUCHS B, BERLIEN HP (1994) Laser in der Perforansbehandlung. Lasermedizin 10, Gustav Fischer, Stuttgart: 150
[12] PHILIPP C, ROHDE E, WALDSCHMIDT J, BERLIEN HP (1994) Laser induced thermotherapy of benign and malign tumors controlled by color coded duplex sonography. SPIE Proc 2327: 262–268
[13] PHILIPP C, SHALTOUT J, BERLIEN HP (1995) Die kontinuierliche Eiskühlung der Haut bei der Nd:YAG-Laserbehandlung von CVD. Lasermed 11: 123
[14] PHILIPP C, GIERING K, BERLIEN HP (1994) Einsatzmöglichkeiten verschiedener Laser in der Traumatologie. Unfallchir 249: 495–498
[15] PHILIPP C, BERLIEN HP, KOWALSKI D, DASKALAKI A, DRESSLER C (1995) Comparison of different dyes used already in medicine for their photodynamic potential. XIV Annual Meeting American Society Lasers in Medicine and Surgery, San Diego, Lasers Med Surg, Suppl 7: 77
[16] PHILIPP C, ROHDE E, BERLIEN HP (1995) Nd:YAG laser procedures in tumor treatment. Sem Surg Oncol 11: 290–298
[17] SOKOLL C, PHILIPP C, BERLIEN HP (1995) Behandlung von Besenreisern mit einer neuen Kühlküvette. Lasermedizin 11, Gustav Fischer, Stuttgart, p 204
[18] TORSTEN U, PHILIPP C, ROHDE E, WEITZEL H, BERLIEN HP (1994) Laser-induced thermotherapy (LITT) for the treatment of local recurrences in patients with breast cancer (Poster). XIVth FIGO World congress of Gynaecology and Obstetrics in Montreal.

# II-2
# Fundamentals of Nd:YAG Laser Applications

B. Fuchs, H.-P. Berlien, C.M. Philipp and G. Müller

**Contents**

## Feasibility of Nd:YAG Laser Applications

Nd:YAG lasers are some of the most frequently utilized laser systems. Due to the types of application (contact, non contact) and the resulting different tissue reactions (hyperthermia, coagulation, vaporization), Nd:YAG applications are described here in detail. (Some applications have already been discussed in Chapter II-1, in a different context, however).

The tables referred to below are presented in Chapter II-3, Therapy Guidelines.

Contact or non contact techniques can be utilized because radiation in the near-infrared is transmitted by glass fibres. This is illustrated in Fig. 3 of Chapter II-1.

Non contact applications can be realized using a fibre emitting a divergent beam or employing a focusing hand-held applicator. In the latter case, it is possible to utilize either the focal point of the laser beam, or a defocused beam. Depending on the irradiation technique employed – coagulation or vaporization – beam power density at the tissue varies. Every non contact vaporization generates a 3- to 5-mm-wide coagulation seam with good haemostatic characteristics. Veins with diameters up to 3 mm and arteries with diameters of up to 1.5 mm are securely closed. To achieve good cutting ability or a small coagulation edge, bare fibres are utilized in contact application. Contact tips, e.g. made of sapphire and very popular in the past are no longer recommended. Reflections at the contact area between fibre and sapphire may lead to absorption of radiation and consequently to heating. This requires supplementary cooling of the contact area, leading to a larger diameter of the device due to the metal connector (Fig. 1).

Cooling gases pose the additional danger of a gas embolism, especially during endoscopic or interstitial procedures. Because of the high absorption, the sapphire tip works more like a hot wire than a precision cutting tool. Figure 2 presents a comparison between bare fibres, sapphire tips and the so-called heater probes.

The main effect of cutting with a bare fibre in contact technique is an interface phenomenon between

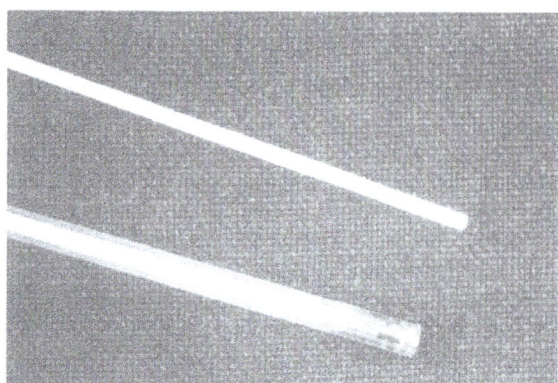

**Fig. 1.** Comparison of the diameters for a 600-μm bare fibre and a fibre with a rinsing tube

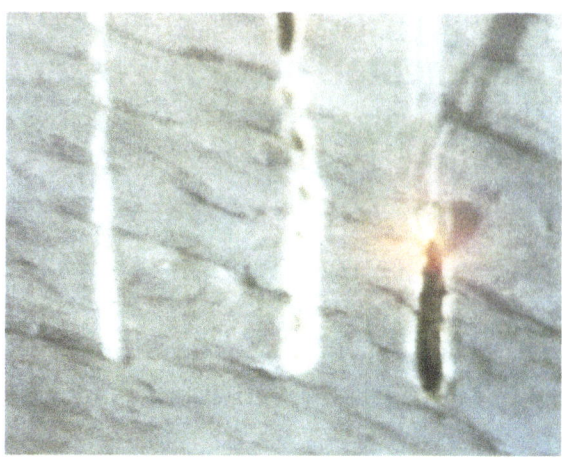

**Fig. 3.** Blackening the fibre and changing the laser-tissue interaction from coagulation to cutting

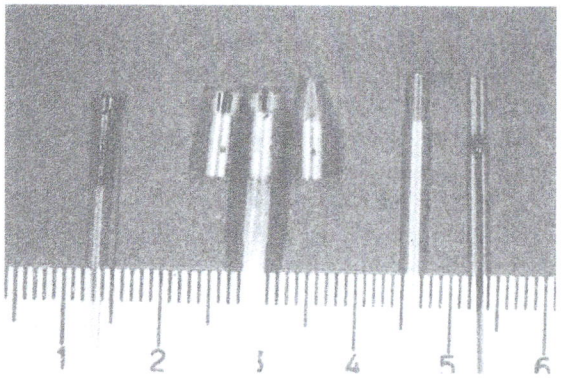

**Fig. 2.** Heater probes, sapphire tips and bare fibres for the Nd:YAG laser

the fibre's terminal face and tissue (Table 3A, Therapy Guidelines).

Carbonization, occurring at the boundary between the fibre's terminal face and tissue, will absorb most of the Nd:YAG radiation. Consequently, almost no photons penetrate into the tissue – as opposed to the „typical" interaction between Nd:YAG lasers and tissues – because most of the photons are absorbed by the carbonization layer. This kind of vaporization is similar to that caused by $CO_2$ lasers. The width of the coagulation edge depends to a large degree on exposure time and on the tissue's thermal conductivity. Figure 3 presents the results of an in vitro experiment.

Low laser power does not produce sufficient heat to induce carbonization. For this reason, only a coagulation seam will develop, limited by short exposure times. However, if a carbonization layer develops, a large proportion of the subsequent radiation will be absorbed by this layer and cannot penetrate more deeply into the tissue. For this reason, the width of the

coagulation edge is limited from 0.1 to 1 mm, depending on the power/time relation (Table 3B).

An additional parameter to control cutting or the coagulation width is the appropriate selection of the fibre's diameter (Table 3C).

To achieve precision cutting without danger of bleeding, a fibre with the smallest possible diameter should be selected. This fibre will achieve the highest power density, so that there is instant carbonization and optimal vaporization. However, a small fibre can achieve only poorly developed non contact coagulation. For proper coagulation – realizing the danger of bleeding – a fibre with a larger diameter should be preferred.

For a maximum and pulsed output power of 50 W the fibre will not be destroyed in a surrounding liquid, or if it is cooled by gas or liquid. Cleaning the fibre during an operation at the appropriate time will also prevent its destruction. For efficient cutting it is important that the fibre's terminal face remains carbonized. I fit is necessary to break the fibre or to prepare a new end, a new carbonization layer must be generated. It must be realized that the coagulation width in this case will increase as compared with the previous exposure.

A bare fibre makes it possible to alternate between cutting using contact techniques and coagulating by employing non contact methods (Table 3D).

To cut immediately, the fibre must be blackened before treatment. This can be achieved by a sterile cork, a wooden spatula or by single, short exposures with the tissue to be removed. If during a non contact procedure it becomes necessary to coagulate, e.g. to close a larger vessel initially, the carbonized layer can be removed from the fibre's terminal face by pyroly-

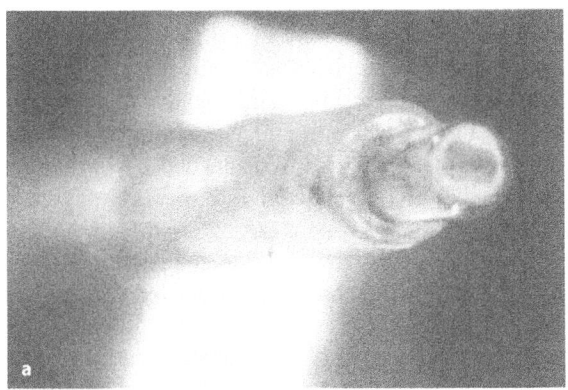

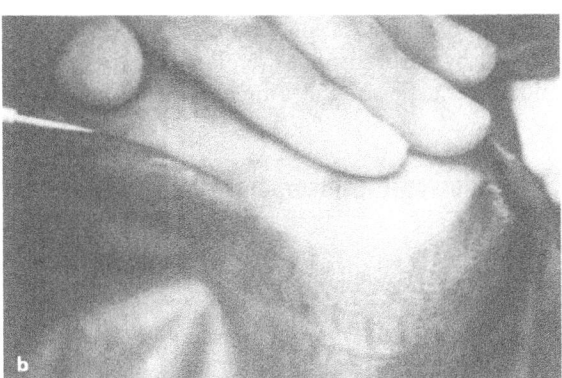

**Fig. 4a,b. a** A freshly broken bare fibre. **b** Interstitial application of a bare fibre through a syringe which punctures a vein

sis. For this, the fibre will be moved away to a safe distance and laser operation will be switched into the pulsed mode. Most of the radiation will then clean again at the fibre's terminal face. Thus, there is no need to break the fibre anew during a procedure for non contact application. Especially during endoscopic surgery it is an advantage to use only one instrument and, by merely changing its type of application (contact, non contact), to interchange between cutting and coagulating.

It is also possible to perform interstitial laser coagulation with a simple bare fibre. In this case, a freshly broken bare fibre without any trace of carbonization is utilized (Fig. 4a).

The advantage of a bare fibre, as compared with special ITT applicators, is its smaller diameter, which permits the use of commercial vessel puncture systems (Fig. 4b) to puncture tissue and so to insert the laser fibre. This will also allow multiple puncture and the shaping of the coagulation region.

## Parameters for Nd:YAG Laser Applications (Therapy Guidelines)

It is possible to present general information regarding the usage of Nd:YAG lasers having different applicators. This information is independent of disease and medical discipline. Although special parameters must be adjusted to suit the special conditions of operation, the guidelines will lead to appropriate results combined with low risk of side effects. Increasing experience will allow the user of laser systems to vary the presented parameters to achieve a higher efficiency.

## Focusing handpiece

Using a focusing handpiece opens a wide field of various tissue effects, e.g. a narrow coagulation edge for microsurgical preparations, or a wide coagulation edge for excellent, haemostasis. It is possible to cut effectively as well as coagulate subcutaneously without destroying the overlying tissue layer. For microsurgical preparations during an operation, a hand-held focusing applicator producing a small focal spot diameter is needed (Table 3.31).

Exposure times between 0.1 and 0.5 s (depending on the width of the coagulation edge desired and 30-W beam power will lead to narrow coagulation points for tissue preparation. For better handling, it is sometimes advisable to stretch the tissue slightly. If larger vessels are identified, these parameter values are insufficient for safe preliminary closure by coagulation. Defocusing the laser beam and increasing the exposure time to 1 to 2 s will close these vessels safely without danger of vaporization or bleeding.

To resect parenchymatous organs (Table 3.32) or very well-vascularized tumours, larger beam powers are needed. For a small focal diameter, 60 W and cw mode, carbonization will be achieved followed by vaporization, creating a wide coagulation edge.

In the case of bleeding, the blood can be removed using NaCl solution and suction. The thin film of liquid does not absorb Nd:YAG laser radiation and bleeding vessels may be coagulated through this film.

For further details, especially regarding in situ coagulation (Table 3.29), very deep coagulation (Table 3.26) and subcutaneous coagulation (Table 3.30) see Chapter II-1.

## Bare-Fibre Applications in Air

The bare-fibre technique can also be used for non contact applications similar to the focusing handpiece. The main region of application is endoscopic surgery, but also open surgery. For a procedure dominated by coagulation it is best to employ a 600-µm fibre. However, if microsurgical contact vaporization is also called for, a 200-µm fibre may be used. For a 600-µm fibre there is no need for cooling if it is employed for short exposure times and in a single pulse mode (Table 3.21).

To avoid destruction of the fibre's terminal face, a 200-µm fibre should be cooled. A wide coagulation edge can be achieved by longer exposure times (Table 3.23).

For these longer exposure times, a 400-µm or even better, a 600-µm fibre is most appropriate. More extensive bleeding, as well as open applications, requires continuous flushing with NaCl solution to remove the blood (Table 3.26). These cases require 50 to 60 W and exposure times of about 0.3 s per pulse and continuous radiation.

When applying exposure times longer than 0.5 s but without flushing with a NaCl solution, additional gas cooling is necessary. Vaporization with a wide coagulation edge can be achieved only if carbonization has been induced previously (Table 3.16). This is contrary to procedures for homogeneous coagulation and haemostasis, where carbonization must be avoided.

This procedure is used especially in endoscopic tumor removal. A disadvantage of the method is its extensive smoke production, so sufficient smoke elimination is required. Another way to remove tumours without encountering these problems is homogeneous non contact vaporization followed by contact vaporization. A great advantage of using a bare fibre is that it is possible to utilize the same fibre and the same parameters for both precise microsurgical preparation and cutting (Table 3.15).

To achieve effective vaporization with a narrow coagulation edge, a very thin fibre should be used together with short exposure times. Employing more than 30 W, specially if using a 200- or 400-µm fibre, additional gas cooling is needed. Gas cooling is not needed for an exit beam power of less than 35 W, a 600-µm fibre and exposure times below 0.3 s. Note, that the bare fibre must not be handled as a mechanical scalpel drilling into the tissue: only the tip of the fibre has contact with tissue. Thus, the end of the fibre must always be visible in the operation field.

## Contact Surgery Under Water

As compared with mono- and bipolar HF applications, lasers have the great advantage that they can be applied within liquid media. Basically, they can be applied in the same way as in air, but there are two exceptions: due to cooling in liquid media, non contact vaporization is impossible and larger-output beam powers and longer exposure times are necessary.

However, the cooling effect leads to a narrower coagulation edge for precision microsurgical vaporization as compared to application in air. For this reason, continuous NaCl flushing is utilized during open surgery to imitate working under water. To vaporize, a power output of 25 to 60 W is needed, depending on the applied fibre diameter and the desired effect (Table 3.13).

To achieve a narrow or wide coagulation edge, a higher beam power than in air is needed (Tables 3.22 and 3.23).

There is no danger of carbonization. As previously described for procedures in air or in open surgery, it is also possible to interchange between contact cutting and non contact coagulation. Fibres can also be cleaned by pyrolysis; however, larger output powers and longer exposure times are required.

## Non-endoscopic Intracorporal Laser Applications (NEILA)

Intracorporal laser application is possible not only via endoscopes but also by direct puncturing of vessels, hollow organs and tissues.

Endoscopic laser applications are performed under direct visual control, while puncturing must be indirectly monitored. Here, the most important clinical methods monitor visually, or by using a finger. For lesions at depths of less than 3 cm, it is possible to localize the exact position of the fibre by its crepitation and palpation. This is possible because laser radiation releases gases into the tissue. It also allows checking of the skin temperature, to prevent overheating and undesired coagulation. The possibility of observing the pilot beam under the surface depends on tissue properties. Vessels can be identified by direct visualization. A change in the surface structure indicates reactions inside the tissue. For deeper-lying lesions or in combination with endoscopic procedures, MRT or colour-coded Doppler ultrasound methods must be employed. MRT yields more precise information about the actual temperature, but colour-coded Doppler ultrasound investigation is, in general, more advantageous. On line monitoring is very useful during puncture, and position monitoring is equally useful during irradiation. Tissue reactions indicate loss of perfusion by coagulation. In addition, there are no problems due to the presence of metallic objects and it is a less expensive system than MRT.

Laser-induced hyperthermia (LIHT) requires an output power of 2 W and long exposure times. Contrary to microwave hyperthermia, it allows hyperthermia of a precisely defined volume of tissue.

Laser-induced coagulation (LIC) yields tissue destruction by coagulation necrosis, contrary to LIHT.

With a freshly broken fibre (600 or 800 µm) not more than 5 W should be used. The width of the coagulation edge should be controlled by exposure time.

An advantage of the bare fibre, as compared with special ITT applicators, is its low price and its ease of handling. Employing 5 W, there is no danger of fibre destruction. Since its diameter is smaller than that of the ITT applicator, readily available puncturing sets can be used. This is especially helpful during treatment of colorectal metastases. In addition, it is possible to crate precisely well-defined coagulation volumes – 5 to 15 mm in diameter, depending on exposure time – especially in regions which are accessible only with difficulty. To achieve larger necrotic volumes, multiple punctures are necessary. Due o easy handling of the system, the total exposure time of larger volumes does not exceed those achieved by special ITT applicators, or by multiple-fibre systems.

Intravascular coagulation, employed for treatment of varices or congenital venectasia, requires an increase in the power output to 8 to 10 W. The end of the fibre must be flushed to prevent adherence of blood. Procedure monitoring is similar to that of interstitial laser coagulation. The aim of this procedure is to avoid immediate vessel closure by inducing thrombosis. Instead, it should induce vasculitis followed by fibrosis of the vessel wall and perivascular tissue. After about 4 weeks, this will lead to occlusion of the vessel.

Treatment of congenital or acquired fistulas by the Nd:YAG laser is based on the same concept. Flushing during irradiation is not employed for these cases. The aim is the total destruction of the fistula's mucous membrane. The subsequent oedema closes the fistula and induces an intramural and perifistular fibrosis with subsequent shrinkage. This therapeutic method may be used for fistulas with diameters of up to 2 mm. Larger-diameter fistulas need an additional fibrin adhesive.

Preblackened fibres allow the removal of tissue by interstitial vaporization; 30-W beam power and exposure times between 0.1 and 0.3 s are applied. Control is the same as for interstitial laser coagulation. In contrast to interstitial coagulation, antegrade or retrograde procedures are possible.

In combination with endoscopic laser applications, the range of minimum invasive surgery (MIS) is extended, thus allowing a partial reduction of open surgery.

# II-3
# Additional Equipment

B. Fuchs, C. Philipp, P. Urban, H.-P. Berlien

## Contents

Not only the clinical laser effect can be determined by the selection of the suitable application equipment, but also by further additional measures. Two basic principle stand in the foreground:
- amplification of the laser effect
- decrease/prevention of side effects.

For the first aspect an additional homogenization of the photon distribution is responsible, but also the specific absorption. For the second aspect, particularly compressing and cooling procedures can be named. These procedures lead to an amplification of the laser effect in the target tissue.

## Amplification of the laser effect

At a given power and/or energy density can be amplified at and in the tissue with two principles:

### 1. Conditioning of the tissue and/or increasing of the absorption.

Particularly photosensitizer are used for conditioning. With laser powers, that not once for a clinically relevant temperature rise, not to mention for the thermal effect suffices, photochemical reactions can be caused as during the Photodynamic Therapy (PDT). But also fluorescencing dyes appertain that. Dyes, where the specific absorption does not suffice, can be used, however, also to increase the absorption. Because the complete use of the photons for the transformation of energy can so at low power and/or energy densities are caused thermal or really photomechanical effects in tissues that would be almost transparent for the corresponding wavelengths. On the border between photodynamic singelett oxygen production and thermal effect is indocyanine green (ICG) which is beside PDT and fluorescence diagnosis in use for thermal tissue welding. Ultra-violet absorbers are an example for the cornea surgery or angioplasty. Pigments are even more effective for that as they were used for the lithotripsy and epilation. Finally to use the same principal without supply of external absorbers double-pulse-laser, at which a weaker laser, however with high specific absorption, a plasma ignites and the second pulse that has a specific absorption for the respective tissue, takes up in the plasma completely and amplifies the effect.

An amplification of the laser effect can be reached not only by better photon utilisation, but also by a more effective photon distribution. Described we already under application techniques reflection and remission can be prevented, by contact and/or interstitial application that the whole laser distributes in the tissue. Next to the use of scattering applicators the photon distribution can be increased by scattering solutions, as Intralipid® solutionsparticularly in hollow organs. Is the specific absorber blood, the perfusion can by reduced by compression, that no specific absorption occurs. The compression reduces the thickness of the tissue, that the depth of distribution of the photons increase.

## 2. Decrease/prevention of side effects

If the photon absorption is limited on the target tissue, no unwished side effects can also occur in the environment. This is the main effect of compression and rinsing procedures. Is there no prevention of absorption and heating the tissue, however, the heat must be dissipated. Also unwished heat conduction into the environment must be prevented. The already mentioned rinsing solutions can be used for this procedure. The heat capacity of such common salt solutions – only physiological solutions come into consideration for that – is limited that a complete protection is not guaranteed to the surface except cooled air. Also the continuous gas cooling achieves only a pain reduction, however, not a surface protection. Spray cooling with liquids with low vaporisation temperature can cool the surface effectively, offers however some disadvantages. Because of the heat capacity of these liquids is very low, the cooling effect is based exclusively by vaporisation cold. With that only very superficial layers can be cooled by less 1 mm. Also a high liquid flow is necessary. If no laser application takes place, however, it can result easily in a frostbite as at the well-known etchloride. Therefore extensive control methods are necessary. Further the liquids drain the skin, it can result also in a ice crystallisation so that the scattering and reflection of the laser light at the surface increases.

## Compressing and cooling procedures

### Glass spattle compression

Due to the high specific absorption of haemoglobin superficial vessels can take up the whole laser radiation so that on the one hand unwished coagulation and even vaporisation can arise form the high power densities, on the other hand the underlaying target organ is protected. To prevent this effect the blood must be squeezed out from the superficial vessels. Most simply this occurs to a standard glass slide under which normal salime solution is brought as an index-matching liquid. Next to the compression steerable by the pressure another 2 more effects are achieved:
1. Due to the salt solution wetting of the horned surface scattering effects are avoided and the wet horn lamellas do not tend to the vaporisation.
2. The modification of the refractive index of the glass sheet/water film combination leads to a so called index matching, which represents to a better visualisation of capillaries, particularly in a focus shift under the skin, so that at the surface higher power densities can be avoided with simultaneous effectiveness in the depth.

If the simple compression does not suffice, however, to the protection of the surface, additional cooling procedures must be used.

### The cooling chamber

For protection of the epidermis a cooling chamber is used as a passage cooling system. A 40%-propyleneglycol solution is used as cooling medium, that onto an initial temperature of –18 °C is cooled. A good skin contact is reached by an elastic latex membrane also without pressure at rough faces. A contact gel (Instillagel®) as index-matching liquid on the skin allows a better heat transfer.

Because of the compression the reduce of the profile by the approximation of the vessel underside leads in the field of the penetration depth to a better illumination of the vessels that is also the underside can be heated up primarily from the light pulse. Because of reducing the flow off of the cooling chamber a compression onto larger superficially sited vessels with reduce of the diameter can be performed. On the other hand an easy low air pressure which arches the membrane after inside is made by restriction of the rush of the refrigerant. With good adhesion onto the skin this can leads to a lumen extending to the cutaneous vessels and to be reached to an increase of the local absorption by stronger filling in the smallest vessels. The negative effects of a raised pulsed energy onto the epidermis can be compensated by the cooling up to boundary value (cooling capacity).

### Ice cube cooling

In this case 2 effects stand in the foreground. Through the compression of vessel lumen by the supporting pressure of the icecube, the lumen of superficial vessels is reduced in the diameter. A depth of 1cm is reachable. In addition occurs in the upper parts because of the reduced blood contents a lower absorption of radiation, so that thermal damages are reduced by that in the skin. On the other hand the heat arising from absorption in the upper skin layers is dissipated, that in the skin niveau no temperatures arise about 45 °C. From a depth of approx. 1,5 mm the cooling is not anymore effective because of the limited heat conductivity of the skin. with that temperatures arises of 60 °C in which result in a coagulation of the vessels.

An often change of the ice cube and a movement and turning of the ice cube during the irradiation are necessary for the maintaining of an effective cooling. Furthermore it is to be ensure that during the irradiation the direct contact is guaranteed continuously between the skin surface and the ice cube and is irradiated only through intact ice cube.

The perfusion can be disturbed at large-area applications however so strongly that secondary skin necrosis are possible. The confluent treated area should not be extended therefore about 5 cm², however, many areas can be treated in one session, providing that these are confluent.

The advantage of the ice cube cooling is in the high heat capacity in the changing of frozen in the fluid phase of water. By the melting of the ice a continuous contact temperature is guaranteed by 0 °C.

### Monitoring procedures

Indeed not laser equipment in the actual sense since the following procedures are independent diagnosis methods they are supposed to be treated however here. The procedures are often helpful increasing the laser effect directly and/or to detect only with short chronological delay and to control the laser application. It can subdivided in following aspects from which some systems unite several aspects in themselves:

- Measurement of the temperature
- Imaging procedures
- Measurement of the perfusion
- Detection of the metabolism

## Measurement of the temperature

### Thermoprobes

The simplest kind of measurement of temperature represents the thermometer. The reaction time both from mercury as also standard electronic thermometers is, however, so small that a monitoring is not possible in the current sense.

But thermoelements with reaction times of few milliseconds permit a direct temperature control. Into preformed body cavities or into hollow organs as the rectum at the interstitial coagulation of the prostate, they can be introduced without problems. But into tissue they have to be pierced with which they do not fulfil the criterion of "non-invasive". A second aspect is added. These thermoelements are based on metal alloys which seem as "black bodies" also for optical radiation. That is, that they absorb the laser radiation where appropriate more strongly than the surrounding tissue. This causes two different measurement errors:

One the one hand they reach a higher temperature than the surrounding tissue and on the other hand heat up the surrounding tissue in a second way like a heating element. For a correct measurement they must always localized therefore outside of the penetration depth of the laser radiation. This problem can be avoided with optical fiber probes which may also in the direct radiation area.

### Thermocamera

A non invasive measurement is the thermocamera which leads to a copy of the measurement area with in colour marked temperature gradients. It detects the heat radiation in the field of 3 – 5 µm. So it is a pure measurement procedures for surfaces, because the radiation in these fields of wavelengths is completely absorbed by water and so that no measurements of temperature from deeper tissue is possible. The actual systems permit a picture build-up of less than 0.1 °C. The wavelength of detection lies outside of the at this time medically used wavelengths so that all laser reaction can be detected at the body surface. It is also usable in case of interstitial laser application, because the heat conduction can be measured, however the delay in time is to thought of that.

## Imaging Procedures

### Magnet-Resonance-Imaging (MRI)

MRI initially was developed as an analytical procedure of the spectroscopy, is next to the imaging procedures also for measurement of the temperature change suitable. The relaxation time T1 is dependent on the tissue temperature so that also intracorporal measurements of temperature are possible. The gauging accuracy conducts only ± 2 °C and is quite interference-prone. It importance has the MRI as a morphological procedure for the representation of the coagulation zones. As primary metabolism-descriptive procedure the MRI indicates grade of vitality of laser treated tissue. Thus a direct representation is on principal possible for the damage zone. Since for a sufficient precise representation as a rule only one picture is per minute is built up, a "on-line"-control is not possible with this procedure. Additionally no metals and also metal accessories may be in the examination field and the examinations are very expansive.

### Ultrasound

At the ultrasound is to be differ between B-picture with the morphological representation and the colour-coded-duplex scan that projects the measurement of perfusion into the b-picture including the vessel density and diameter and flow rate and direction. In contrast to the MRI, with which indeed a position control, but not puncture control is possible, also the "on-line"-control is feasible in the ultrasound. Considering is that one in the contrast to the MRI worse morphological representation. In the duplex-mode can be represented additionally the perfusion, which at hyper- and/or hypoperfused lesions an additional support of discrimination is. In the case of an obliteration of an afferent or efferent vessels or a decrease of the microcirculation can give an indication of the laser destruction next to the density change in the b-picture. At interstitial laser coagulation can be represented the heating of the tissue up to 55 °C in the duplex-mode and the $CO_2$-gasification of tissue as a "colour bruit". Also irregular patterns show themselves in the coagulated tissue until the temperature of environment was reached again. Under investigation is the measurement of temperature – depending on speed of ultrasound for a direct ultrasound measurement.

### Measurements of the perfusion

Also the already mentioned ultrasound duplex appertains to that of course. It is used as an independent examination procedure only in the phlebology, it has it great importance by the possibilities of the colour coded duplex sonography receive.

### Laser Doppler

Also the laser doppler is based on the same doppler principal of the frequency shift of the ultrasound. Instead of acoustic waves the reflection of light waves occurs at erythrocytes in motion. Because of the shorter wavelength a by far higher solution is possible. The He:Ne-laser with 633nm and diode laser in the near infrared with 780nm is used. Because of the limited penetration depth of the light only superficial layer are registrable. With infrared radiation a perfusion can be represented up to a depth of 3mm because of the increasing scattering from the depth the representation becomes increasingly more indistinct. With fiber led point dopplers even an absolute measurement of flow is possible. To be used clinically 2-dimsional 2-wavelength-dopplers, one allows cartographic representation of the microcirculation and with red and near infrared scan is able to distimguish between different layers.

### Monitoring of metabolism

Next to the laser doppler can represent the metabolism indirectly about the microcirculation also the metabolism of energy can be represented with the laser induced fluorescence (LIF). The most important enzyme for that is NADH. With fiber led point measurement even an absolute measurement is possible for the NADH-concentration, but comparable dates which allow a monitoring from the kinetics are still missing. At this time examinations in the perinatale medicine, where instead of the micro-scalp electrodes pO2- and the pH-measurement the NADH-concentration is supposed to be measured. Physiological background is that the NADH already shows modifications before it results in the permanent dysfunction

### O$_2$-electrodes

The central role of the oxygen in the photodynamic therapy (PDT) lets appear continuous monitoring of the concentration of the $O_2$-concentration in the tissue as a most important parameter. Because the measurement of the $O_2$-concentration in air is being a simple and reliable, a direct measurements of the dissolved oxygen in the tissue is only difficult or not possible. Membrane-contact-oxymeters change even the local $O_2$-concentration that they permit a snapshot, micro-glass-capillaries must be punctured into the tissue so that they lead to a modification of $O_2$-concentration about the trauma. These not invasive methods are based on the infrared spectroscopy so that they can represent only superficial layers because of the emission wavelength in the maximum of the water absorption.

## References

1. B. Fuchs, C. Philipp, H. P. Berlien Laser procedures for tumour ablation and destruction In: Min Invas Ther & Allied Technol 1998:7/6:489-494
2. M. Poetke, C. Philipp, H.-P. Berlin Ten Years of Laser treatment of Hemangiomas and Vascular Malformations : Techniques and Results. In: Schmittbecher, P.P. Laser Surgery in Children, Berlin: Springer Verlag 1997:82-91
3. V. Prapavat Laser in der Diagnostik In: Berlien/Müller: Angewnadte Lasermedizin Landsberg:ecomed Verlag 1997:1-16
4. J. Beuthan et al: Investigatins concerning the determinations of NADH-concentration using optical biopsy. In: Alfano, R.R., Advances in Laser and Light Spectroscopy to diagnose Cancer and other Diseases. Proc. SPIE 2135 (1994): 147-156
5. P. Urban, C. Philipp, l. Weinberg, H. P. Berlien: Bildgebende Verfahren in der Beurteilung vaskulärer Läsionen. Proceedings 13th International Congress Lesermed München (1997); 5
6. P. Urban et al. Monitoring von ALA-PDT Effekten durch Infrarot Thermgraphie, Laser-Doppler Flussmessung und zwei Fluoreszenz-Detektions-Systeme, Laser Medizin 14 (3): (1999): 78
7. P. Urban, B. Algermissen, C. Philipp, L. Weinberg, H. P. Berlien: Monitoring of ALA-PDT effects using infrared thermography, laser doppler perfusion imaging an d a fluorescence detection systems (SIMAS). Proc. 7th Biennial Congress, International Photodynamic Association, (1998) Nante-France

# II-4
# Therapeutic Guidelines

B. Fuchs, H.-P. Berlien, C.M. Philipp

**Table 1.1.** Lasers inducing photothermal reactions

| Laser | Wave-length | Temporal modalities/ pulse width | Exposure time | Power (W) | Property of emitted irradiation | Tissue interaction | Applicators/ transmission systems accessories |
|---|---|---|---|---|---|---|---|
| Ar⁺ | 488/514 | cw | 0.01 s , cw | 0.3 – 5 (–15) | Specific absorption by haemoglobin and melanin | Specific coagulation | Fibre/focusing hand piece |
| Kr⁺ | 350 – 667 | cw | 0.1 s , cw | 2 – 5 | Specific absorption by xenogenic chromophores | Specific coagulation/ photodynamic therapy | Fibre/focusing hand piece |
| KTP | 532 | cw | 0.01 s , cw | 1 – 30 | Specific absorption | Specific coagulation, vaporization | Fibre/focusing hand piece, bare fibre contact |
| Diode | 780 810 980 | cw | 0.1 s , cw | 1 – 60 | Unspecific volume absorption | Unspecific coagulation (vaporization) | Fibre/bare fibre contact, side fire, diffuser tip, focusing hand piece |
| Nd:YAG | 1064 1032 | cw | 0.01 s , cw | 1 – 120 1 – 40 | Unspecific volume absorption (max in vessels) | Unspecific coagulation, vaporization | Fibre/bare fibre contact, side fire, diffuser tip, focusing hand piece |
| CO₂ | 10600 | cw | 0.05 s , cw | 0.03 – 80 | High absorption by water | Vaporization, coagulation | Articulated arm, micro-manipulator, hollow wave guide, scanner, focusing hand piece |

**Table 1.2.** Lasers used for photodynamic therapy

| Laser | Wave-length (nm) | Temporal modalities/ pulse width | Exposure time | Power (W) | Property of emitted irradiation | Tissue interaction | Applicators/ transmission systems accessories |
|---|---|---|---|---|---|---|---|
| KTP/ Ar⁺ Dye | 532/514 633 (–650) | cw | 0.02 , cw | 0.2 – 7 | Specific absorption by natural and synthetic chromophores (photosensitizer) | Photodynamic therapy | Fibre, side fire, micro-lense, spherical or cylindrical diffuser tip |
| FPDL | 633 | Pulsed | 80/200 µs | 400 mJ/ pulse | Specific absorption by natural and synthetic chromophores (photosensitizer) | Photodynamic therapy | Fibre, side fire, |
| Diode | 635 650 675 690 | cw | 0.01 , cw | 3 – 5 – 25 | Specific absorption by xenogenic chromophores (photosensitizer) | Photodynamic therapy | Fibre, side fire, micro-lense, spherical or cylindrical diffuser tip |

**Table 1.3.** Lasers with ablative effect

| Laser | Wave-length (nm) | Pulse duration | Typical pulse energy/cm$^2$ | Frequency of repe-tition | Property of emitted irradiation | Indication | Applicators/ trans-mission systems accessories |
|---|---|---|---|---|---|---|---|
| Excimer | 193 | 10 ns | 180 mJ | 20 Hz | Ionization, absorption by water | Cornea surgery | Directly/slit lamp |
| Excimer | 308 | 100 – 250 ns | 5 – 200 mJ | 1 – 200 Hz | Absorption by water | Ophthalmol ogyangioplasty | Fibre/ multifibre-catheter |
| Nd:YAG pulsed | 1064 | 100 – 200 µs | 100 – 200 mJ | 15 – 30 Hz | Unspecific absorption | Dental medicine | Fibre, hand piece |
| Ho:YAG | 2100 | 1 – 2 ms | 0.8 – 4.5 J | 2 – 20 Hz | Absorption by water | Surgery/ orthopaedic surgery/ urology | Fibre/ bare fibre contact, side fire, hand piece |
| Er:YAG | 2940 | 0.1 – 1 ms | 200 mJ – 1.5 J | 1 – 20 Hz | Absorption by water | Surgery/ dental/ plastic surgery | Articulated arm, hollow wave guide, sapphire fibre |
| CO$_2$/ pulsed | 106000 | < 950 µs | 1 – 500 mJ | 1 – 10 Hz | Absorption by water | Plastic surgery | Articulated arm, hollow wave guide, scanner |

**Table 1.4.** Lasers inducing photodisruptive effects

| Laser | Wave-length (nm) | Pulse duration | Typical pulse energy/cm$^2$ | Frequency of repe-tition | Property of emitted irradiation | Indication | Applicators/ trans-mission systems accessories |
|---|---|---|---|---|---|---|---|
| Flashlamp-pumped dye (FPDL) | 504+ 595 | 1.4 – 2.5 µs | 50 – 120 mJ | 1 – 10 Hz | Absorption by chromophores | Lithotripsy | Fibre |
| Flashlamp-pumped dye (FPDL) | 585 | 200 – 500 – 800 – 1500 µs | 4 – 9 J | 2 Hz | Absorption by blood vessels/ haemoglobin | Plastic surgery/ dermatology | Fibre/focusing hand piece |
| Ruby (Q-switch) | 694 | 10 – 30 ns (Q-switch) 120 µs (non-Q-switch) | 5 – 20 (–50) J | 1 – 2 Hz | Absorption by pigment | Plastic surgery/ dermatology | Fibre/focusing hand piece |
| Alexandrite | 755 | 300 – 500 – 700 ns | 10 – 150 mJ | 1 – 20 Hz | Absorption by chromophores | Lithotripsy | Fibre |
| Nd:YAG (Q-switch) | 1064 | 20 – 25 ns | 10 – 100 mJ | 1 – 20 Hz | Optical breakdown (mechanical) | Secondary cataract membrane, dental medicine | Fibre/slit lamp |
| Double-pulse Nd:YAG/ KTP | 1064/.532 | 1 – 1.2 – 1.4 µs | 65 – 80 (–120) mJ 8 – 16 mJ | 10 Hz | Photoacoustic disruption (mechanical) | Lithotripsy | Fibre |

**Table 1.5.** Laser systems and tissue interaction

| Laser | Fluorescence | Photo-dynamic | Photo-coagulation | Photo-vaporization | Photo-ablation | Photo-disruption |
|---|---|---|---|---|---|---|
| Excimer (308) | | | | | ++ | |
| Nitrogen (337) | ++ | | | | | |
| Titanium sopphire (frequency – doubled) | ++ | | | | | |
| Krypton (350/800) | ++ | | ++ | | | |
| FPDL (504) | | | | | + | ++ |
| Argon/KTP (488/814) | | ++ | ++ | + | | |
| KTP (frequency – doubled) (532) | | | ++ | | | |
| Copper vapour (519/578) | | + | | | | ++ |
| FPDL (585) | | | | | | ++ |
| FPDL (633) | | ++ | | | | |

(+) = minor effect  + = good effect  ++ = excellent effect

**Table 1.6.** Laser systems and tissue interaction

| Laser | Fluorescence | Photo-dynamic | Photo-coagulation | Photo-vaporization | Photo-ablation | Photo-disruption |
|---|---|---|---|---|---|---|
| Diode (635) | + | ++ | | | | |
| Gold-vapor (628) | | ++ | | | | |
| Diode (650) | | ++ | | | | |
| Diode (675) | | ++ | | | | |
| Diode (690) | | ++ | | | | |
| Ruby (694) | | | | | | ++ |
| Alexandrite (755) | | | | | | ++ |
| Titan sapphire (tumable/700-1000) | | | ++ | | | |
| Diode (810) | | | ++ | | | |
| Diode (980) | | | ++ | + | | |

**Table 1.7.** Laser systems and tissue interaction

| Laser | Fluorescence | Photo-dynamic | Photo-coagulation | Photo-vaporization | Photo-ablation | Photo-disruption |
|---|---|---|---|---|---|---|
| Nd:YAG (1064/cw) | | | ++ | + | | |
| Nd:YAG (1064/pulsed) | | | | | ++ | |
| Nd:YAG (1064/Q-switch) | | | | | | ++ |
| Nd:YAG (1320/cw) | | | + | ++ | | |
| Ho:YAG (2100/pulsed) | | | | ++ | ++ | + |
| Er:YAG | | | | | ++ | |
| $CO_2$ (10.000) | | | (+) | ++ | | |
| $CO_2$ pulsed (10.000) | | | | | ++ | |

(+) = minor effect  + = good effect  ++ = excellent effect

**Table 1.8A.** Laser systems and tissue interaction

| Tissue interaction | Exc.[a] | Ar/KTP[b] | Nitrogen | Krypton | KTP/Ar-Dye | FPDL[c] (504/595 nm) | FPDL (585 nm) |
|---|---|---|---|---|---|---|---|
| Fluorescence | | | ++ | ++ | | | |
| Photodynamic-therapy | | ++ | | | ++ | | |
| Photocoagulation | | ++ | | ++ | ++ | | |
| Photovaporization | | + | | | | | |
| Photoablation | ++ | | | | | + | |
| Photodisruption | | | | | | ++ | ++ |

**Table 1.8B.** Laser systems and tissue interaction

| Tissue interaction | Diode (635) | Nd:YAG n.c.[f] | Nd:YAG c.[g] | Nd:YAG pulsed | Ho:YAG[d] | Er:YAG[e] | CO₂-pulsed | CO₂ |
|---|---|---|---|---|---|---|---|---|
| Fluorescence | + | | | | | | | |
| Photodynamic-therapy | ++ | | | | | | | |
| Photo-coagulation | | ++ | ++ | | + | | | (+) |
| Photo-vaporization | | + | ++ | | (+) | | | ++ |
| Photo-ablation | | | | ++ | + | ++ | ++ | |
| Photo-disruption | | | | | + | | | |

| | |
|---|---|
| (+) = minor effect | [c]FPDL = Flashlamp pumped Dye Laser |
| + = good effect | [d]Ho:YAG = Holmium:YAG |
| ++ = excellent effect | [e]Er:YAG = Erbium:YAG |
| [a]Exc. = Excimer | [g]c. = contact |
| [b]Ar = Argon | [f]n.c. = non-contact |

**Table 2.** Scores

| Indication | Realization |
|---|---|
| 0 = Contraindication | 0 = Not possible |
| 1 = Not recommended | 1 = Difficult to perform |
| 2 = Usable, when other techniques fail | 2 = Difficult to perform, usable when no other alternatives available |
| 3 = Usable as an alternative to other techniques | 3 = Possible |
| 4 = Appropriate indication | 4 = Simple to perform |
| 5 = Strict indication, superior to other techniques | 5 = Very easy to perform |

**Table 2.1.** Brain and spinal cord

| | Nd:YAG non-contact | Nd:YAG contact | CO$_2$ | Photodynamic therapy | Fluorescence diagnostics (monitoring) |
|---|---|---|---|---|---|
| Intracranial | | | | | |
| Preparation and resection of tumours | 4/4* | 4/4* | 4/4** | | |
| In situ destruction of tumours | 4/5 | | | 3/3 | |
| Interstitial thermotherapy (ITT) of tumours | | 3/4 | | | |
| Intraoperative monitoring | | | | | 4/4 |
| Intraspinal | | | | | |
| Malformations | | 5/5 | | | |
| Lipoma | | 4/4 | 4/4 | | |

\* Alternating techniques.
\*\* Operation microscope (OPMI) with micromanipulator.

**Table 2.2.** Vertebral column (arthroscopy)

| | Nd:YAG non-contact | Nd:YAG contact | Ho:YAG | Nd:YAG (1320 nm) |
|---|---|---|---|---|
| Percutaneous laser discectomy (PLD) | | 3/5 | 4/5 | 4/5 |
| Removal of plica | | 3/4 | 4/5 | 4/5 |
| Synovectomy | 4/4 | | 4/4 | 4/4 |
| Meniscectomy | | | 4/5 | 4/5 |
| Lateral release | | | 4/5 | |
| Tightening of capsule | | | 4/5 | |

**Table 2.3.** Face, oral cavity and pharynx

| | Argon/KTP | Nd:YAG contact | Nd:YAG non-contact | $CO_2$ | Nd:YAG – KTP (double-pulse)/ FDL (<100 μs) | PDT | Fluorescence diagnostics |
|---|---|---|---|---|---|---|---|
| Angioma of the lip (venous lake) | 4/5 | | 5/5 | | | | |
| Basalioma | | 2/5 | | 2/5 | | 2/5 | 4/5 |
| Epithelial dysplasia of the lip | | | | | | | |
| – Cheilitis actinica | | | | 5/5 | | 3/5 | 4/5 |
| – Leukoplakia | | | | 3/5 | | 3/5 | 4/5 |
| – Bowen's disease | | | | 4/5 | | 4/5 | 5/5 |
| – Carcinoma in situ | | | | 4/5 | | 3/5 | 5/5 |
| Epithelial dysplasia of the oral cavity/ pharynx | | | | | | | |
| – Leucoplakia | | | | 4/5 | | 3/5 | 3/5 |
| – Carcinoma in situ | | | | 2/5 | | 3/5 | 5/5 |
| Carcinoma | | | | | | | |
| – Lip | | | | 3/4 | | 2/4 | 3/4 |
| – Tongue | | 2/4 (ITT) | | 5/5[b] | | 2/4 | 4/5 |
| Adenoids | | 3/4 | | | | | |
| Hyperplasia of gingiva | | | | 4/5 | | | |
| LAUP[a] | | 3/4 | | 4/5 | | | |
| Tonsils | | | | | | | |
| – Tonsillotomy | | 3/4 | | 3/4 | | | |
| – Tonsillectomy | | 2/3 | | 2/3 | | | |
| Sialolithiasis (lithotripsy) | | | | | 4/3 | | |

[a]   LAUP Laser-assisted uvula-palatoplasty.
[b]   Depending on localization and size.
[*]   Depending on depth.

**Table 2.4.** Upper airways

| | Ar/KTP | Nd:YAG non-contact | Nd:YAG contact | $CO_2$ |
|---|---|---|---|---|
| Recurrent nasal polyps | | 2/3 | 4/5 | 3/3 |
| Hypertrophic inferior nasal turbinates | 3/4 | 2/4 | 4/5 | 3/3 |
| Nasal bleeding (hereditary telangiectasia) | 2/3 | 5/5 | | |
| Papillomatosis | | 4/4 | 4/5 | 4/5 |
| Laryngeal polyp | 3/5 | | 3/5 | 3/5 |
| Laryngeal carcinoma | | | 3/3[a] | 4/5[a] |
| Laryngeal stenoses | | | | |
| – Congenital | | | 4/5[b] | 3/3[b] |
| – Scarred | | | 4/5[b] | 3/4[b] |

[a]   Age-related.
[b]   Depending on localization and size.

**Table 2.5.** Lower airways

| | Nd:YAG non-contact | Nd:YAG contact | CO$_2$ |
|---|---|---|---|
| Tracheal and bronchial fistula | | 4/4[b] | |
| Tracheal and bronchial stenoses (benign) | | | |
| – Congenital | | 4/5 | 2/4 |
| – Scarred | | 4/5 | 3/4 |
| – Congenital vascular disorders (CVD) | 5/5 | 2/5 | |
| – Papilloma | | 3/4 | 3/4 |
| – Granuloma | | 3/4 | 3/4 |
| Tracheal and bronchial stenoses (malignant) | 4/4[a] | 4/4[a] | |

[a]Alternating techniques.
[b]Smaller 2 mm diameter.

**Table 2.6.** Thoracic wall and pectoral cavity

| | Nd:YAG non-contact | Nd:YAG contact | ND:YAG (1320 nm) | CO$_2$ | PDT | Fluorescence diagnostics |
|---|---|---|---|---|---|---|
| Thoracic wall | | | | | | |
| – Mastectomy | | 2/3 | | 3/4 | | |
| – Thoracic wall tumours | 3/4 | 3/4 | | | 4/4 | 4/4 |
| – Metastasis | 4/4 | 4/4 (ITT) | | | 4/4 | 4/4 |
| Pectoral cavity | | | | | | |
| – Pulmonary parenchyma | | | 4/4 | 3/3 | | |
| – Fistulas | 3/5 | | | | | |
| – Decortication | | 3/5 | | 3/5 | | |
| Thoracoscopy | | | | | | |
| – Decortication | | 4/5 | | 3/3 | | |
| – Pleurodesis (recurrent pneumothorax) | 4/4 | | | | | |
| – Sympathectomy | 5/4 | 4/5 | | 2/2 | | |
| – Wedge resection | 3/4 | 4/5 | 5/4 | 2/2 | | |

**Table 2.7.** Gastrointestinal tract

| | Nd:YAG non-contact | Nd:YAG contact | PDT |
|---|---|---|---|
| Esophagotracheal fistula | | 4/3[b] | |
| Esophagus stenosis | | | |
| – Tumorous | 5/4[a] | 5/4[a] | 5/4 |
| – Erosive trauma | | 4/4 | |
| – Barrett's ulcers | 4/4 | | 4/4 |
| – Congenital | | | |
| Bleeding | | | |
| – Angiodysplasia | 5/4 | | |
| – Ulcer | 2/3 | | |
| Polyps | | | |
| – Polyposis coli | | 3/5 | |
| – Villose adenoma | 3/5 | | |
| Colon/rectum carcinoma | | | |
| Palliative mass reduction | 5/4[a] | 5/4[a] | |
| Recanalization | 5/4 | 5/4 | 3/3 |

[a]Alternating techniques.
[b]Smaller than 2 mm diameter.

**Table 2.8.** Abdominal cavity (laparoscopy)

| | Nd:YAG non-contact | Nd:YAG contact | PDT |
|---|---|---|---|
| Cholecystectomy | 3/5* | 3/5* | |
| Appendectomy | 3/5* | 3/5* | |
| Adhesiotomy | | | |
| – Congenital | | 4/5 | |
| – Inflammatory | | 4/5 | |
| Tumour destruction | 4/5 | | |
| – LITT | | 5/5 | |
| Lymphadenectomy | | 4/5 | |
| Orchidolysis | 4/4[a] | 4/4[a] | |
| Vagotomy | | 3/4 | |

[a] Alternating techniques.

**Table 2.9.** Abdominal cavity (open access)

| | Nd:YAG non-contact | Nd:YAG contact | PDT |
|---|---|---|---|
| Tumour | | | |
| – Resection | 3/4 | 3/4 | |
| – Preparation | 3/4 | 3/4 | |
| Parenchymatous organs | | | |
| – Rupture | 4/4 | | |
| – Resection | 4/4 | 3/4 | |
| – Destruction | 4/5 | | |
| – LITT | | 5/5 | |
| Adhesiotomy | | | |
| – Congenital | | 4/5 | |
| – Inflammatory | | 4/5 | |
| Lymphadenectomy | | 4/5 | |

**Table 2.10.** Urogenital tract (gynecology)

| | Nd:YAG non-contact | Nd:YAG contact | $CO_2$ | PDT | Fluorescence diagnostics |
|---|---|---|---|---|---|
| Acuminate condyloma | | | | | |
| – Epithelium, keratinous | 3/5 | | 5/5 | 2/3 | |
| – Epithelium, non-keratinous (mucosa) | 5/5 | 4/5 | | 2/3 | |
| Epithelial dysplasia | | | | | |
| – Paget's disease | | | 2/5 | 2/5 | 4/5 |
| – Leucoplakia | | | 4/5 | 3/5 | 4/5 |
| – Bowen's disease | 2/5 | | 4/5 | 4/5 | 5/5 |
| – Cervical dyslasia | | | | | 4/4 |
| Carcinoma in situ | | | | | |
| – Vulva | | | 4/5 | 3/5 | 4/5 |
| – Vagina | | | 3/3 | 4/4 | 4/4 |
| – Cervix of uterus | | | 0/5 | | |
| Conization | | | 4/4 | | |
| Hysteroscopy | | | | | |
| – Benign endometrial polyp | | 3/5 | | | |
| – Metromenorrhagia | 4/5[a] | | | | |
| – Submucosal fibroma | | 4/4 | | | |
| – Uterine septum | | 4/5 | | | |
| Laparoscopy | | | | | |
| – Endometriosis | 4/4[b] | 4/4[b] | 4/5 | | |
| – Adhesions, peritubar | 4/5[b] | 4/5[b] | 4/5 | | |
| – Polycystic ovary | 4/5[b] | 4/5[b] | 3/3 | | |
| – Ectopic pregnancy | | 4/5 | | | |

[a] Side fire.
[b] Alternating techniques.

**Table 2.11.** Urogenital tract (urology)

| | Nd:YAG non-contact | Nd:YAG contact | $CO_2$ | Nd:YAG – KTP (double pulse)/ FDL ($<100\ \mu s$) | PDT | Fluorescece diagnostics |
|---|---|---|---|---|---|---|
| Acuminate condyloma | | | | | | |
| – Epithelium, keratinous | 3/5 | | 5/5 | | 2/3 | |
| – Epithelium, non-keratinous (mucosa) | 5/5 | 4/5 | | | 2/3 | |
| Epithelial dysplasia | | | | | | |
| – Querat's disease | 4/4 | | 4/4 | | 3/5 | 4/5 |
| – Paget's disease | | | 2/5 | | 2/5 | 4/5 |
| – Leucoplakia | | | 4/5 | | 3/5 | 4/5 |
| – Bowen's disease | ?/5 | | 4/5 | | 4/5 | 5/5 |
| Urinary bladder carcinoma | | | | | | |
| – Carcinoma in situ | 3/5 | | | | 3/5 | 4/5 |
| – Invasive carcinoma ($T_1$) | 3/5 | | | | 3/5 | 3/5 |
| Ureterostenosis/ urethrostenosis | | | | | | |
| – Congenital | | 4/5 | | | | |
| – Acquired | | | | | | |
| – Benign | | 3/5 | | | | |
| – Malignant | | 3/5 | | | | |
| Prostrate | | | | | | |
| – BPH: Transurethral | | | | | | |
| – Side fire vaporization | 4/4 | | | | | |
| – Interstitial | | 4/5(ITT) | | | | |
| – Prostatic carcinoma | 2/4 | 2/4 | | | | |
| – Laparoscopic lymphadenectomy | | 4/5 | | | | |
| Carcinoma of the penis | 3/5 | 3/4 | 3/3 | | 3/4 | |
| Heminephrectomy | 4/4 | 4/4 | | | | |
| Urolithiasis | | | | 4/3 | | |

**Table 2.12.** Proctology

| | Nd:YAG non-contact | Nd:YAG contact | CO$_2$ | PDT | Fluorescence diagnostics |
|---|---|---|---|---|---|
| Acuminate condyloma | | | | | |
| – Epithelium, keratinous | 3/5 | | 5/5 | 2/3 | |
| – Epithelium, non-keratinous (mucosa) | 5/5 | 4/5 | | 2/3 | |
| Epithelial dysplasia | | | | | |
| – Paget's disease | | | 2/5 | 2/5 | 4/5 |
| – Leucoplakia | | | 4/5 | 3/5 | 4/5 |
| – Bowen's disease | 2/5 | | 4/5 | 4/5 | 5/5 |
| Anal fissure | | | | | |
| – Excision | | 4/5 | 4/4 | | |
| – Coagulation | 4/5 | | | | |
| Anal mariske | | 4/5 | 4/4 | | |
| Haemorrhoidectomy | | 4/5 | | | |
| Fistula | | | | | |
| – Congenital | | 3/5[a] | | | |
| – Inflammatory | | 3/5[a] | | | |
| – Tumorous | | 4/5[a] | | | |
| Anal stenosis | | 4/5 | | | |
| Pilonidal sinus | | 4/4 | 4/5 | | |
| Polyps | | | | | |
| – Polyposis coli | | 3/5 | | | |
| – Villous adenoma | 3/5 | | | | |

[a] Depending on size.

**Table 2.13.** Skin and appendages of the skin

| | Ar/ KTP | Nd:YAG n.-c. | Nd:YAG contact | Nd:YAG-pulsed | Er:YAG | CO$_2$ | Ruby | Alexan-drite pulsed | FDL (>100 µsec) | PDT | Fluores-cence diagnostics |
|---|---|---|---|---|---|---|---|---|---|---|---|
| Acuminate condyloma | | | | | | | | | | | |
| – Epithelium, keratinous | | 3/5 | | | | 5/5 | | | | 2/3 | |
| – Epithelium, non-keratinous (mucosa) | | 5/5 | 4/5 | | | | | | | 2/3 | |
| Epithelial dysplasia | | | | | | | | | | | |
| – Leukoplakia | | | | | | 4/5 | | | | 3/5 | 4/5 |
| – Bowen's disease | | 2/5 | | | | 4/5 | | | | 4/5 | 5/5 |
| – Actinic keratosis | | | | | | 3/5 | | | | 5/5 | 5/5 |
| Molluscum contagiosum | 3/5 | 3/4 | | | | 2/4 | | | | | |
| Verruca | | | | | | | | | | | |
| – juvenilis | | 4/5 | | | | | | | | | |
| – palmaris | | | | | | | | | | | |
| – ungual, nail fold | | 3/4 | | | | 4/5 | | | 3/5 | | |
| – plantaris | | 3/4 | | | | 4/5 | | | 3/5 | | |
| Benign skin tumour | 4/5 | 2/4 | | | 3/5 | 3/5 | | | | | |
| Cutaneous/sub-cutaneous metastasis (inoperable) | | 3/5 | | | | 2/3 | | | | 4/5[a] | |
| Basalioma | | 2/5 | | | | 2/5 | | | | 3/5[a] | |
| Chromatopathy | 4/5 | | | | 3/5 | 3/5 | 4/5 | | | | |
| Tattoo | | | | | 2/2 | 2/2[a] | 5/5[a] | | | | |
| Hair removal | | | | 4/5 | | | 3/4 | 4/5 | | | |

[a]    Depending on depth.
[b]    Depending on localization and size.

**Table 2.14.** Scars and keloids of the skin

| | Ar/KTP | Nd:YAG non-contact | Nd:YAG contact | Er:YAG | CO$_2$ | FDL (>100 µs) |
|---|---|---|---|---|---|---|
| Scars | | | | | | |
| – Keloid | | | | | | |
| – Telangiectatic | 3/5 | | | | | |
| – Active | | 4/5 | | | | 4/5 |
| – Fibrous | | 4/5[a] | 3/5 (ITT)[a] | | | |
| – Hypertrophic scar | | 3/5 (ITT) | | 3/5 | 3/5 | 2/5 |
| – Atrophic scar | | | | 3/5[b] | 3/5[b] | 3/5 |
| – Acne scar | | | | | 5/5 | |
| – Reddish stria (stretch marks) | | | | | | 4/5 |

[a] Depending on depth.
[b] Flattening of edges.

**Table 2.15.** Vascular system

| | Ar | FDL (>100 µs) | Nd:YAG non-contact | Nd:YAG contact | Nd:YAG long-pulse | Excimer |
|---|---|---|---|---|---|---|
| Acquired | | | | | | |
| Angioplasty | | | | | | 3/3 |
| Varicose perforating veins | | | | 4/4[a] (ITT) | | |
| Subcutaneous varicose veins (AV-shunts) | | | | 4/4[a] (ITT) | | |
| Spider veins of the legs | | 2/4 | 3/3[a] | | 3/5[a] | |
| Spider nevus (spider angioma) | 5/5 | | 2/4 | | 3/5[a] | |
| Couperose | 5/5 | 2/4 | | | | |
| Congenital | | | | | | |
| Vascular malformation | | | | | | |
| – Portwine stain (flammeous nevus) | 2/4 | 5/5 | | | 3/5 | |
| – Tuberous transformation | 4/4 | 3/3 | 5/5 | | | |
| – Venous, lymphatic, arterial, mixed | | | 5/5[b/a] | 5/5 (ITT) | | |
| Hereditary telangiectasia (Osler's disease) | 3/5 | | 5/5 | | 3/5** | |
| Hemangioma | | | | | | |
| – Prodromal stage | 4/5 | 5/5 | 3/4 | | | |
| – Cutaneous | 3/4 | 3/5 | 5/4[b] | | | |
| – Subcutaneous/ visceral | | | 5/5[b] | 4/5 (ITT) | | |

[a] With ice cube cooling.
[b] With cooling of the surface, (e.g. cooling chamber).

**Table 2.16.** Oncology

| | Nd:YAG non-contact | Nd:YAG contact | CO$_2$ | PDT | Fluorescence diagnostics |
|---|---|---|---|---|---|
| Curative | | | | | |
| Dysplasia | | | 4/5 | 3/5 | 4/5 |
| Palliative | | | | | |
| Metastasis | 4/4 | | | | |
| Local recurrence | | 4/4 (ITT) | | 3/5 | 4/5 |
| **Recanalization** | 5/5* | 5/4[a] | | 3/5 | |

[a] Alternating techniques.

**Table 3.A.** Basics of bare fibre contact – surgery I (cutting)

- Boundary face phenomenon
- Induction of carbonized fibre end
- Carbonization leads to a high absorption of laser irradiation

- Guided by the fastest process
- Process with the highest absorption is the fastest
- Fastest process is the cutting process

- Effective vaporization reduces coagulation
- Tissue effect is depending on power/time ratio

Compare non-contact surgery technique with the $CO_2$ laser (Tables III.6, III.7)

**Table 3.B.** Basics of bare-fibre contact – surgery II (cutting)

- Exposure time influences the width of coagulation seam
  – long exposure time leads to wide coagulation seams
- Power influences the vaporization efficiency
  – high power leads to a high level of vaporization efficiency
- Fibre diameter influences the width of coagulation seam and of vaporization efficiency
  – small fibre diameter leads to narrow coagulation seams and a high level of vaporization efficiency

In a tissue with a high grade of vascularization, the coagulation seam should be wider; use repeated exposition if necessary

**Table 3.C.** Basics of bare fibre contact – surgery III (cutting)

- For more cutting applications with low risk of bleeding use
  the smallest possible fibre
- For indications with high risk of bleeding use the thickest possible fibre – more coagulation effects
- Your fibre is not a scalpel – do not press
- Your fibre is not a drill – fibre end has to be always visible
- Max. power:
  – with cooling or in water:   60 W
  – without cooling in air:       30 W
- Chopped mode:
  – max. exposure time pulse: 0.5 s
  – exposure time interval:      1:1 – 2:1

Also during endoscopic laser treatment, make sure that the fibre end is always visible, otherwise stop laser immediately

**Table 3.D.** Basics of bare-fibre contact – surgery IV (Preparation of the fibre)

| Preblackening for contact cutting | Precleaning for non-contact coagulation |
|---|---|
| - In air | - Freshly broken or |
| - Tissue contact or | - Non-contact pyrolysis |
| - Sterile cork or wooden spatulas | - Fibre flashes up shortly |
| - Power:        30 W | - Min. 5 mm distance to the underlying tissue |
| - Exposure time: 0.5 s | - Power:        30 W |
| - Single pulses until blackening | - Exposure time:  0.5 s |

The technique of non-contact pyrolysis is very helpful in endoscopic procedures; here it is possible to change from cutting to coagulation without taking the fibre out for new trimming of fibre tip

**Table 3.1.** Laser: Er:YAG (2940 nm) Applicator: focusing hand piece

| Ablation of soft tissue | |
|---|---|
| Energy: | 500 – 1000 mJ |
| Spot diameter: | 4 – 5 mm |
| Pulse duration: | 300 – 500 μs |
| Repetition rate: | 1 – 2 Hz |
| Remarks: | – Sufficient particle suction |
| | – Bleeding diathesis |
| | – Haemostyptic agents if necessary |
| | – Ablation of multiple layers if necessary |
| | – Ablation until capillary bleeding occurs (for skin resurfacing) |
| | – Ablation of hard tissue: 2 J$^{-1}$ pulse necessary |

The ablation of skin should be stopped when a capillary bleeding occurs, otherwise the risk of scarring is increased. For a homogenous superficial ablation, scanner systems can be used

**Table 3.2.** Laser: $CO_2$ (10 600 nm) Applicator: focusing hand piece, or operation microscope (OPMI) with micromanipulator

| Superficial vaporization | |
|---|---|
| Power: | 5 – 10 W |
| Spot diameter: | 0.5 – 1.5 mm |
| Exposure time: | 0.1 s |
| Interval: | 0.1 s |
| Remarks: | – Ablation depth depends on power density |
| | – Multiple exposures, if necessary |
| | – Plume suction |

Indications: acuminate condyloma, verruca, epithelial dysplasia, benign skin or mucosal tumors, recanalization as palliative cancer treatment. If massive bleeding occurs, see Tables 3.26, 3.28. For more intensive treatment see Table 3.5

**Table 3.3.** Laser: $CO_2$ (10 600 nm) Applicator: rotation or high-speed scanner

| Superficial vaporization | |
|---|---|
| Power: | 10 – 20 W |
| | or 5 – 7 W superpulse |
| Spot diameter: | 0.5 – 1 mm |
| Exposure time: | scan cycle |
| Remarks: | – Ablation depth depends on duration of scan cycle |
| | – Multiple exposures, if necessary |
| | – Plume suction |

Indications: skin resurfacing, homogenous superficial removal of flat benign skin tumours, in special indications superficial basaliomas, scars, tattoos. Be aware of increased risk for scars with high power or multiple exposures. (Tables 3.2, 3.4, 3.47)

**Table 3.4** Laser: $CO_2$ (10 600 nm) Applicator: cw – scanner system or short pulsed

| Scars | |
|---|---|
| Power: | 10 – 15 W |
| Spot diameter: | 0,5 mm |
| Pulse duration: | 400 – 980 μs (scan cycle) |
| Remarks: | – Treatment in 2 – 3 passages, depending on localization (forehead 2, cheek 3 passages) |
| | – Fill up deep scars with gel (e.g. Instillagel) |
| | – Postoperative treatment with hydrophilic cream and alternating several times daily moist compress with black tea |

Indications: acne scars, hypertrophic scars, and to flatten the skin relief (edges) in atrophic scars. For the treatment of keloids see Table 3.40. Be aware of increased risk for additional scars using high-power or multiple exposures. Often multiple treatments are necessary after a period of several months

**Table 3.5.** Laser: CO$_2$ (10 600 nm) Applicator: focusing hand piece or operation microscope (OPMI) with micromanipulator

Vaporization

| | |
|---|---|
| Power: | 8 – 10 – (20) W |
| Spot diameter: | 0.5 – 2 – (3) mm |
| Exposure time: | 0.1 – 0.2 s |
| Interval: | 0.1 – 0.2 s |
| Remarks: | – multiple exposures, if necessary<br>– plume suction |

Indications: extended acuminate condyloma, verruca, epithelial dysplasia, benign skin or mucosal tumours, acquired or congenital stenosis of the airways, recanalization as palliative cancer treatment. In the case of massive bleeding see Tables 3.26, 3.28. For more superficial treatment see Table 3.2. For endoscopic treatment (OPMI), e.g. in the larynx be aware of inflammable materials (tube).

**Table 3.6** Laser: CO$_2$ (10 600 nm) Applicator: focusing hand piece or operation microscope (OPMI) with micromanipulator

Preparation

| | |
|---|---|
| Power: | 5 – 10 W |
| Spot diameter: | 0.5 mm |
| Exposure time: | 0.1 – 0.2 s |
| Interval: | 0.1 – 0.2 s |
| Remarks: | – Tissue should be spread<br>– Depending on the intended coagulation change exposure time or flush incision channel with gas<br>– Maximal coagulation seam of 1.5 mm<br>– Plume suction |

Indications: surgery of benign or malignant tumours, preparation of organs before resection. These parameters are also suitable for vaporization of small structures (e.g. laryngeal polyp). Be aware of larger vessels, try to prevent bleeding

**Table 3.7.** Laser: CO$_2$ (10 600 nm) Applicator: focusing hand pieceor operation microscope (OPMI) with micromanipulator

Resection

| | |
|---|---|
| Power: | 15 – 25 W |
| Spot diameter: | 0.5 – 1 mm |
| Exposure time: | cw |
| Remarks: | – If necessary, remove carbonized layer<br>– Maximal coagulation seam of 2 mm (heat conduction)<br>– Plume suction |

Indications: surgery of epithelial dysplasia, benign or malignant tumours of the skin, airways, gastro intestinal tract, oral cavity, nervous system, adhesion/scar formations. In cases of extended tumours, a high-power resection technique with a power up to 50 W is possible. Be aware of limited coagulation effect; the coagulation ability can be increased with defocused beam. See Table 3.6

**Table 3.8.** Laser: Ho:YAG (2100 nm) Applicator: bare fibre

Ablation of soft tissue (arthroscopy/intervertebral disc surgery)

| | |
|---|---|
| Energy : | 0.8 – 1,2 (–4) J |
| Spot diameter: | 0.5 – 1 mm |
| Pulse duration: | 450 – 500 µs |
| Repetition rate: | 4 – 15 Hz |
| Remarks: | – High repetition rate may lead to heat accumulation<br>– Microcavities in the tissue possible (rupture of tissue)<br>– Treatment of hard tissue possible (calcaneal spur) |

A supervision of laser effects with an endoscope is recommended whenever possible

**Table 3.9.** Laser: Nd:YAG (1320 nm) Applicator: focusing hand piece

| Resection | |
| --- | --- |
| Power: | 25 – 40 W |
| Spot diameter: | 0.5 – 1.5 mm |
| Exposure time: | cw |
| Speed: | ca. 1 mm s$^{-1}$ |
| Remarks: | – Induce carbonization<br>– Tissue should be spread lightly<br>– Larger vessels have to be ligated preliminarily<br>– Plume suction<br>– In the case of blood in the incision channel: continuous suction and intermittent rinsing with saline solution during resection |

Useful technique for tumour resection or resection of parenchymatous organs: lung surgery. For resection of hard tissue see Table 3.11

**Table 3.10.** Laser: Nd:YAG (1320 nm) Applicator: bare fibre, contact

| PLDD/ P(E)LD<br>Percutaneous laser disc decompression/Percutaneous (endoscopic) laser discectomy | |
| --- | --- |
| Power: | 45 W |
| Spot diameter: | 0.5 – 1.0 mm |
| Exposure time: | 1 s |
| Intervall: | 10 s |
| Remarks: | – Induce carbonization<br>– Preblackening of the fibre tip |

Useful technique for percutaneous disc surgery; if possible, use endoscopic control

**Table 3.11.** Laser: Nd:YAG (1320 nm) Applicator: bare fibre, contact

| Meniscectomy | |
| --- | --- |
| Power: | 40 W |
| Spot diameter: | 0.5 – 1.0 mm |
| Exposure time: | cw |
| Remarks: | – Induce carbonization<br>– Preblackening of the fibre tip<br>– Plume suction |

Useful technique for arthroscopic surgery; if possible, use endoscopic control

**Table 3.12.** Laser: Nd:YAG (1320 nm) Applicator: bare fibre, non-contact

| Synovectomy | |
| --- | --- |
| Power: | 40 W |
| Spot diameter: | 0.5 – 1.0 mm |
| Exposure time: | cw |
| Remarks: | – Clean fibre, freshly broken<br>– Avoid carbonization, additional rinsing if necessary<br>– Avoid exsiccation<br>– Repeat exposure, if necessary<br>– Stop if you see blanching starting |

Useful technique for arthroscopic surgery; if possible, use endoscopic control

**Table 3.13.** Laser: Nd:YAG (1064 nm) Applicator: bare fibre (200 – 600 μm), contact, in water

| Vaporization small coagulation seam (0,5 – 2 mm) | |
| --- | --- |
| Power: | 25 – 60 W |
| Exposure time: | 0.1 – 0.3 s |
| Interval: | ≥ 0.1 s |
| Remarks: | – Preblackening of the fibre<br>– Coagulation seam between 0.5 and 2 mm depends on power/time ratio<br>– do not press, handle fibre without force<br>– fibre tip must be visible |

**Table 3.14.** Laser: Nd:YAG (1064 nm) Applicator: bare fibre (400 or 600 μm), contact

| Interstitial vaporisation | |
|---|---|
| Power: | 30 W |
| Exposure time: | 0.1 – 0.3 s |
| Interval: | > 0.1 s |
| Speed: | 1 mm s$^{-1}$ |
| Remarks: | – Preblackening of the fibre<br>– Insert fibre through a 16 G Teflon or metal cannula<br>– Position visualized by pilot laser (HeNe), ultrasound or MRI<br>– Prograde or retrograde application possible<br>– Process control also possible by CCDS or MRI-temperature measurements<br>– Extension of coagulation seam depends on exposure time |

Indications: treatment of thick keloids, see Table 3.40. Be aware of sufficient surface temperature measurement

**Table 3.15.** Laser: Nd:YAG (1064 nm) Applicator: bare fibre (600 μm), contact

| Interstitial coagulation impression technique | |
|---|---|
| Power: | 5 – 8 W |
| Exposure time: | cw |
| Remarks: | – Clean fibre, freshly broken<br>– Extension of coagulation seam depends on the exposure time<br>– Impression on the tissue surface with little force<br>– Only small coagulation on the surface, volume coagulation beneath the surface |

Indications: treatment useful for vascular malformations, especially for endoscopic treatment of lymphatic malformations

**Table 3.16.** Laser: Nd:YAG (1064 nm) Applicator: bare fibre (600 μm), non contact, in air

| Vaporization broad coagulation seam (3 – 5 mm) | |
|---|---|
| Power: | 50 – 60 W |
| Spot diameter: | 2 – 5 mm |
| Exposure time: | 0.5 s – cw |
| Remarks: | – Use additional gas cooling<br>– Clean fibre, freshly broken<br>– Induce carbonization first<br>– Sufficient plume suction<br>– Underlying coagulation seam depends on exposure time |

Indications: in the case of tumour recanalization, an additional coagulation and removal with forceps may be helpful, see Table 3.25

**Table 3.17.** Laser: Nd:YAG (1064 nm) Applicator: bare fibre (600 μm), contact

| Interstitial coagulation Laser-induced coagulation (LIC) | |
|---|---|
| Power: | 5 W |
| Exposure time: | cw |
| Remarks: | – clean fibre, freshly broken<br>– 16 G teflon or metal cannula with an excess length of 8–10 mm to prevent overheating and melting<br>– position of fibre visualized by pilot laser (HeNe), ultrasound or MRI<br>– only retrograde procedure<br>– optimal reaction is signalized by crepitation, palpable through the skin or by fibre vibration<br>– process control also possible by CCDS or MRI-temperature measurements<br>– extension of coagulation seam depends on the exposure time |

Indications: treatment of subcutaneous vascular malformations, extended haemangiomas, varicose perforating veins, (be aware of sufficient surface temperature measurement), interstitial treatment of malignant tumours or metastases

**Table 3.18.** Laser: Nd:YAG (1064 nm) Applicator: bare fibre (600 μm), contact

| | |
|---|---|
| Interstitial/intravascular coagulation Laser-induced coagulation (LIC with flushing) | |
| Power: | 8 – 10 W |
| Exposure time: | cw |
| Remarks: | – Clean fibre, freshly broken, rinsing with 100 mm $H_2O$ and 2 ml min$^{-1}$ |
| | – 16G teflon or metal cannula |
| | – position of fibre visualized by pilot laser (HeNe), ultrasound or MRI |
| | – only retrograde procedure |
| | – optimal reaction is signalized by crepitation, palpable through the skin or by fibre vibration |
| | – process control also possible by CCDS or MRI-temperature measurements |
| | – power and retraction speed depend on vessel radius and flushing volume |

Indications: intravascular treatment of pathologically altered vessels, e.g. vascular malformations

**Table 3.19.** Laser: Nd:YAG (1064 nm) Applicator: Somatex-LITT, contact

| | |
|---|---|
| Interstitial coagulation Laser-induced thermotherapy (LITT) | |
| Power: | According to applicator length |
| Exposure time: | cw (20 min) |
| Remarks: | – A 7 F Applicator without rinsing: power according to active length of applicator : 2.5 – 3 W cm$^{-1}$ |
| | – B 9 F Applicator, systems with cooling (rinsing system) |
| | 12 W cm$^{-1}$ of active Applicator length |
| | – Position of applicator by ultrasound, CT or MRI |
| | – Process control also possible by CCDS or MRI-temperature measurements |
| | – extension of coagulation seam depends on the exposure time |

Indications: liver metastases of operated colorectal cancer, depending on localization, size and number, in special cases metastases of other cancers (e.g. mamma carcinoma) or hepatocellular carcinoma.

**Table 3.20.** Laser: Nd:YAG (1064 nm) Applicator: bare fibre (200 – 600 μm), contact, in air

| | |
|---|---|
| Vaporization small coagulation seam (0,5 – 2 mm) | |
| Power: | 15 – 50 W |
| Exposure time: | 0.1 – 0.3 s |
| Interval: | ≥ 0.1 s |
| Remarks: | – If more than 30 W, use gas cooling |
| | – Preblackening of the fibre tip |
| | – Coagulation seam between 0.5 – 2 mm depends on power/time ratio |
| | – Do not press, handle fibre without force |
| | – Fibre tip must be visible |

This parameters can be used also for precise endoscopic resection of scars or tumours in the airways, digestive tract or during laparoscopic treatment, especially in children

**Table 3.21.** Laser: Nd:YAG (1064 nm) Applicator: bare fibre (200 – 600 μm), non contact, in air

| | |
|---|---|
| Coagulation small coagulation seam (1 – 2 mm) | |
| Power: | 30 W |
| Spot diameter: | 2 mm |
| Exposure time: | 0.2 – 0.5 s |
| Interval: | ≥ 0.2 s |
| Remarks: | – Clean fibre, freshly broken |
| | – Avoid carbonization, additional rinsing if necessary |
| | – Avoid exsiccation |
| | – Repeat exposure, if necessary |
| | – Stop if you see blanching |

Indications: endoscopic treatment of hereditary telangiectasia (M. Osler) or vascular malformations. Avoid popcorn effect, use repeated exposure

**Table 3.22.** Laser: Nd:YAG (1064 nm) Applicator: bare fibre (200 – 600 μm), non-contact, in water

| Coagulation small coagulation seam (1 – 2 mm) | |
|---|---|
| Power: | 30 – 40 W |
| Spot diameter: | 2 mm |
| Exposure time: | 0.2 – 0.5 s |
| Interval: | ≥ 0.2 s |
| Remarks: | – Clean fibre, freshly broken<br>– Multiple exposures, if necessary<br>– Stop if blanching begins |

Indications: polyps in the urinary bladder, acuminate condyloma inside the urethra

**Table 3.23.** Laser: Nd:YAG (1064 nm) Applicator: bare fibre (200 – 600 μm), non contact, in water

| Coagulation broad coagulation seam (3 – 5 mm) | |
|---|---|
| Power: | 40 – 60 W |
| Spot diameter: | 2 – 5 mm |
| Exposure time: | cw |
| Remarks: | – Clean fibre, freshly broken<br>– Avoid carbonization<br>– Continuous exposure over a few seconds<br>– Avoid popcorn effect, interruption if necessary |

**Table 3.24.** Laser: Nd:YAG (1064 nm) Applicator: bare fibre (600 μm)

| Fistula shrinking | |
|---|---|
| Power: | 4 – 5 W |
| Exposure time: | cw |
| Retraction speed: | 0.5 – 1 mm s$^{-1}$ |
| Remarks: | – Clean fibre, freshly broken<br>– Fibre visualized by pilot laser (HeNe)<br>– Only retrograde procedure<br>– Use additional fibrin glue > 2 mm diameter<br>– Maximal diameter 5 mm<br>– Avoid vaporization |

Indications: fistulas of the airways, or digestive tract. Including postoperative, congenital, acquired fistulas (e.g. anal fistula)

**Table 3.25.** Laser: Nd:YAG (1064 nm) Applicator: bare fibre (400 or 600 μm), non-contact, in air

| Coagulation, minor bleeding broad coagulation seam (2 – 3 mm) | |
|---|---|
| Power: | 30 W |
| Spot diameter: | 2 – 5 mm |
| Exposure time: | 1 – 2 s pulse$^{-1}$ |
| Remarks: | – Clean fibre, freshly broken<br>– Avoid carbonization<br>– Flush out blood with saline, or suction if necessary<br>– Repeat exposure, if necessary |

In cases with major bleeding, see Table 3.26

**Table 3.26.** Laser: Nd:YAG (1064 nm) Applicator: bare fibre (600 μm), non-contact, in air

| Coagulation, major bleeding broad coagulation seam (3 – 5 mm) | |
|---|---|
| Power: | 30 – 60 W |
| Spot diameter: | 2 – 5 mm |
| Exposure time: | cw |
| Remarks: | – Use additional gas cooling<br>– Clean fibre, freshly broken<br>– Avoid carbonization<br>– Continuous saline flushing during laser procedure<br>– Continuous exposure over a few seconds<br>– But avoid popcorn effect, interruption if necessary |

Indications: bleeding from ulcer, tumour, vascular malformations

**Table 3.27.** Laser: Nd:YAG (1064 nm) Applicator: focusing hand piece

| Coagulation, minor bleeding broad coagulation seam (2 – 3 mm) | |
|---|---|
| Power: | 30 W |
| Spot diameter: | 2 – 5 mm |
| Exposure time: | 1 – 2 s pulse$^{-1}$ |
| Remarks: | – Avoid carbonization<br>– Flush out blood with saline, if necessary<br>– Repeat exposure, if necessary |

Useful in a tissue with a high grade of vascularization

**Table 3.28.** Laser: Nd:YAG (1064 nm) Applicator: focusing hand piece

| Coagulation, major bleeding broad coagulation seam (3 – 5 mm) | |
|---|---|
| Power: | 30 – 60 W |
| Spot diameter: | 2 – 5 mm |
| Exposure time: | cw |
| Remarks: | – Avoid carbonization<br>– Continuous saline flushing during laser procedure<br>– Continuous exposure over a few seconds<br>– But avoid popcorn effect |

**Table 3.29.** Laser: Nd:YAG (1064 nm) Applicator: focusing hand piece

| In situ coagulation small coagulation seam (1 – 2 mm) | |
|---|---|
| Power: | 20 – 30 W |
| Spot diameter: | 1 – 2 mm |
| Exposure time: | 0.2 – 0.5 s |
| Interval: | 0.2 – 0.5 s |
| Remarks: | – Avoid carbonization<br>– Keep moist, if necessary<br>– Repeat exposure, if necessary |

Indications: acuminate condyloma, verruca or the central vessel of spider naevus/spider angioma. In the treatment of acuminate condyloma or verruca start laser irradiation at the base until blanching occurs/in vascular structures repeated exposure can be necessary.

**Table 3.30.** Laser: Nd:YAG (1064 nm) Applicator: focusing hand piece

| Subcutaneous coagulation | |
|---|---|
| Power: | 25 – 50 W |
| Spot diameter: | 5 mm |
| Exposure time: | cw |
| Remarks: | – Sufficient cooling of skin with ice cubes free of air bubbles<br>– good contact between ice cubes and skin<br>– laser through ice; with continuous movement of the ice cube<br>– in case of blanching stop laser but continue cooling of the skin |

Indications: Subcutaneous parts of vascular malformations, haemangiomas, keloids, small metastases. Cold water or a cold microscopic slide is not sufficient for skin protection.

**Table 3.31.** Laser: Nd:YAG (1064 nm) Applicator: focusing hand piece

| Preparation small coagulation seam (1 – 2 mm) | |
|---|---|
| Power: | 30 W |
| Spot diameter: | 0.5 mm |
| Exposure time: | 0.1 – 0.5 s |
| Interval: | > 0.1 s |
| Remarks: | – Induce carbonization |
| | – Tissue should be spread |
| | – Coagulation of vessels with defocused beam or ligature |

In a tissue with a high grade of vascularization a repeated exposure to laser irradiation can be necessary; do not increase exposure time (risk of vaporization – popcorn effect)

**Table 3.32.** Laser: Nd:YAG (1064 nm) Applicator: focusing hand piece

| Resection broad coagulation seam (3 – 5 mm) | |
|---|---|
| Power: | 60 – 120 W |
| Spot diameter: | 0.5 – 1.5 mm |
| Exposure time: | cw |
| Speed: | ca. 1 mm s$^{-1}$ |
| Remarks: | – Induce carbonization |
| | – Tissue should be spread lightly |
| | – Vessels larger than 5 mm have to be ligated preliminarily |
| | – In the case of blood in the incision channel: continuous suction and intermittent rinsing with saline solution during resection |

Useful technique for tumour resection out of hard tissue or resection of parenchymatous organs, lung (see Table 3.9, for preparation techniques see Table 3.31)

**Table 3.33.** Laser: Nd:YAG (1064 nm), long-pulsed Applicator: focusing hand piece

| Spider veins (leg), telangiectasia | |
|---|---|
| Power: | 80 – 145 J cm$^{-2}$ |
| Spot diameter: | 2.5 mm |
| Exposure time: | 3 – 8 ms |
| Repetition rate: | 2 – 3 Hz |
| Remarks: | – Parameters depending on vessel diameter and skin type (pigmentation of the skin) |
| | – No overlap (single – spot technique) |
| | – Cooling of the skin with gel or pre/postcooling by ice cube |

First investigate the deep veins of the legs. In large vessels prove the possibility of operative treatment or sclerotherapy first. Use sufficient skin cooling to prevent scars

**Table 3.34.** Laser: argon (514 nm)/KTP (532 nm) Applicator: focusing hand piece

| Couperose | |
|---|---|
| Power: | 2 – 3 W |
| Spot diameter: | 0.5 – 1 mm |
| Exposure time: | 0.02 – 0.1 s |
| Interval: | 0.1 s |
| Remarks: | – Compress area with a glass spatula until vessel is slightly visible |
| | – Induce an initial blanching |
| | – Treat spot by spot, avoid overlapping |
| | – Avoid confluent coagulation |

The vessels should be visible for the treatment with argon/KTP laser. In cases of homogenous reddish skin, other laser systems (e.g. FPDL) should be used. (see Table 3.39)

**Table 3.35.** Laser: argon (514 nm)/ KTP (532 nm) Applicator: focusing hand piece

| Superficial vaporization | |
| --- | --- |
| Power: | 5 W |
| Spot diameter: | 0.5 mm |
| Exposure time: | 0.01 – 0.02 s |
| Interval: | 0.01 – 0.02 s |
| Remarks: | – Induce carbonization |
| | – Repeat exposure, if necessary |

For treatment of flat benign (pigmented) lesions, e. g. chloasma

**Table 3.36.** Laser: argon (488/514 nm) Applicator: focusing hand piece

| Coagulation minimal blanching | |
| --- | --- |
| Power: | 5 W |
| Spot diameter: | 1 – 2 mm |
| Exposure time: | 0.1 - 0.2 s |
| Interval: | 0.1 s |
| Remarks: | – Avoid carbonization |
| | – Irradiate until blanching occurs |
| | – Multiple treatments |

For more intense treatment see Table 3.37. Prefer multiple treatment, prolonged exposure time increases the risk of scarring

**Table 3.37.** Laser: argon (488/514 nm) Applicator: focusing hand piece

| Coagulation | |
| --- | --- |
| Power: | 5 W |
| Spot diameter: | 1 – 5 mm |
| Exposure time: | 0.2 s |
| Interval: | 0.1 s |
| Remarks: | – Induce confluent coagulation (circular/tangential irradiation) |
| | – Multiple treatments |
| | – Local anaesthesia, if necessary |
| | – Risk of scarring |

Indications: benign tumours of the skin, molluscum contagiosum. A repeated treatment can be useful/ be aware of higher risk of scarring when exposure time is increased. Treatment until popcorn effect occurs

**Table 3.38.** Laser: argon (514 nm)/KTP (532 nm) Applicator: focusing hand piece

| Tuberous transformation of portwine stain Small venous angioma (e.g. small venous lake of the lip) | |
| --- | --- |
| Power: | 2 W |
| Spot diameter: | 2 mm |
| Exposure time: | 0.1 – 0.2 s |
| Interval: | 0.1 s |
| Remarks: | – Compression with a glass spatula |
| | – Induce an initial blanching |
| | – Polka-dot technique |
| | – Avoid confluent coagulation |

For treatment of combined portwine stain and tuberous transformations see Table 3.39. In the case of more voluminous venous angiomas of the lip, other laser systems should be used (e.g. Nd:YAG); Table 3.30. Be aware of immediate blanching (risk of scarring)

**Table 3.39.** Laser: flashlamp-pumped dye laser [FPDL] (585 nm), Applicator: focusing hand piece

| Portwine stain | |
|---|---|
| Energy density: | 4 – 8 J cm$^{-2}$ |
| Spot diameter: | 5 mm |
| Pulse duration: | 100 – 500 µs |
| Repetition rate: | 1 – 2 Hz |
| Remarks: | – Treat with an overlap of approximately 25%<br>– Avoid coagulation, but induce purpura<br>– Avoid multiple exposures even if no primary reaction is visible<br>– Darker lesions require lower power density |

The treatment of children or sensitive patients may require a local anaesthesia (e.g. anaesthetic cream) or cooling techniques. In cases of tuberous transformations, other laser systems (argon/ KTP) should be used in addition (Table 3.38)

**Table 3.40.** Laser: flashlamp-pumped dye laser (585 nm) Applicator: focusing hand piece

| Keloid | |
|---|---|
| Energy density: | 7 – 9 J cm$^{-2}$ |
| Spot diameter: | 5 mm |
| Pulse duration: | 200 – 500 µs |
| Repetition rate: | 1 – 2 Hz |
| Remarks: | – Appropriate for erythematous keloids only<br>– Use cooling chamber<br>– Treat with an overlap of approximately 30%<br>– Treatment in two passages: compression/decompression<br>– use combination with argon laser in teleangiectatic forms<br>– in cases of voluminous keloids, use interstitial vaporization |

In the treatment of keloids a combined use of different laser systems is often necessary. In keloids with an diameter of more than 1 cm, an additional interstitial vaporization is recommended (Table 3.14)

**Table 3.41.** Laser: flashlamp-pumped dye laser (504/595 nm) Applicator: bare fibre (200 – 280 – 320 µm), contact

| Lithotripsy | |
|---|---|
| Energy: | 50 – 120 mJ |
| Pulse duration: | 1.4 – 2.5 µs |
| Repetition rate: | 1 – 10 Hz |
| Remarks: | – Use only in liquid medium<br>– Ideally with endoscopic control<br>– Ideally with a stone recognition system<br>– Use the lowest effective energy and repetition rate (depending on the type of the stone) |

The laser parameters for lithotripsy differ in a wide range, depending on fibre diameter, wave-length, localization and absorption of concrement. Endoscopic control of laser effect should be used. Other laser systems for lithotripsy see Table 3.42, 3.43

**Table 3.42.** Laser: Nd:YAG/KTP (1064/532 nm) Applicator: bare fibre (280 µm), contact

| Lithotripsy | |
|---|---|
| Energy: | 65 – 80 (–120) mJ/8 – 16 mJ |
| Pulse duration: | 1 – 1.4 µs |
| Repetition rate: | 5 – 10 Hz |
| Remarks: | – Use only in liquid medium<br>– Ideally with endoscopic control<br>– Ideally with a stone recognition system<br>– Use the lowest effective energy and repetition rate (depending on the type of the stone) |

The laser parameters for lithotripsy differ depending on localization and absorption of concrement. This laser system works with a frequency-doubled double-pulse technique. The green light produces an absorption-induced plasma on the surface of the concrement, the Nd:YAG laser pulse serves to 'pump' the ignited plasma to increase the plasma-induced local shock wave. Other laser systems for lithotripsy see Tables 3.41, 3.43

**Table 3.43.** Laser: alexandrite (755 nm) Applicator: bare fibre (200 μm), contact

| Lithotripsy | |
|---|---|
| Energy: | 10 – 50 – 80 (–150) mJ |
| Pulse duration: | 300 – 500 ns |
| Repetition rate: | 1 – 20 Hz |
| Remarks: | – Use only in liquid medium |
| | – Ideally with endoscopic control |
| | – Ideally with a stone recognition system |
| | – Use the lowest effective energy and repetition rate (depending on the type of the stone) |

The alexandrite laser is rarely used, because of high burndown of the fibre tip. Other laser systems for lithotripsy see Table 3.41, 3.42. Endoscopic control of laser effect should be carried out

**Table 3.44.** Laser: alexandrite (755 nm) Applicator: focusing hand piece

| Hair removal/epilation | |
|---|---|
| Energy density: | 10 – 20 J cm$^{-2}$ |
| Spot diameter: | 10 mm |
| Pulse duration: | 7 – 20 ms |
| Repetition rate: | Up to 3 Hz |
| Remarks: | – Advantage: large spot diameter |
| | – High penetration depth |
| | – Cooling with gel, or precooling with ice |
| | – Avoid multiple exposures |
| | – Pretreat with shaving |
| | – Appropriate for light brown to black hair |

Compare other laser systems for epilation, Tables 3.45, 3.46. Repeated treatments are necessary. Cooling techniques are useful to reduce the risk of skin damage. Be aware of higher risk for side effects in more pigmented skin (skin type)

**Table 3.45.** Laser: Nd:YAG long-pulsed (1064 nm), diode Applicator: focusing hand piece

| Hair removal/epilation | |
|---|---|
| Energy density: | 28 – 40 (–60) J cm$^{-2}$ |
| Spot diameter: | 4 mm |
| Pulse duration: | 10 ms |
| Repetition rate: | 2 – 3 Hz |
| Remarks: | – High penetration depth |
| | – Cooling with gel |
| | – Avoid multiple exposures (atrophic scars) |
| | – Pretreat with shaving |
| | – Appropriate for light brown to black hair |

Compare other laser systems for epilation, Table 3.44, 3.46. Repeated treatments are necessary. Cooling techniques are useful to reduce the risk of skin damage. An advantage of the Nd:YAG and diode laser system for epilation is the increased penetration depth

**Table 3.46.** Laser: ruby (694 nm), long-pulsed Applicator: focusing hand piece

| Hair removal/epilation | |
|---|---|
| Energy density: | 10 – 60 J cm$^{-2}$ |
| Spot diameter: | 3 – 12 mm |
| Pulse duration: | 2 – 12 ms |
| Repetition rate: | 1 Hz |
| Remarks: | – limited penetration depth |
| | – avoid multiple exposures |
| | – only appropriated for dark brown and black hairs |

The long-pulsed ruby laser is one of the best-investigated laser systems for epilation. The parameters used differ over a wide range. Repeated treatments are necessary (Tables 3.44, 3.45)

**Table 3.47.** Laser: ruby (694 nm), short-pulsed Applicator: focusing hand piece

| Tattoo | |
|---|---|
| Energy density: | 6 – 20 (–50) J cm$^{-2}$ |
| Spot diameter: | 2 – 4 mm |
| Pulse duration: | 10 ns |
| Repetition rate: | 1 – 2 Hz |
| Remarks: | – Better response for Indian ink (e.g. Sriptol) than for ink, response depends on colour of tattoo and laser system used<br>– Colours containing pigments out of heavy metal may provoke local or systemic (allergic) reactions<br>– Change of colour from red to brown possible<br>– If no sufficient reaction to ruby laser occurs, use other wave-length, or dermablation with Er:YAG or $CO_2$-laser is possible |

The result of laser treatment of a tattoo is dependent on several factors: the colour of pigment and its often unknown ingredients, the density, distribution and the layer of the skin where the pigment is placed (Table 3.3). Increase the energy density stepwise when necessary

**Table 3.48.** Laser: diode (635 – 690 nm) Applicator: fibre with microlense, isotropic applicator

| Photodynamic therapy (PDT) | |
|---|---|
| Power density: | 50 – 150 mW cm$^{-2}$ |
| Total dose: | 50 – 100 J cm$^{-2}$ |
| Spot diameter: | 3 – 10 cm |
| Pulse duration: | cw |
| Exposure time: | 700 – 1000 s |
| Remarks: | – Use pain-reducing techniques in topical PDT<br>– Increased power density leads to increased penetration depth<br>– Cover the margins with aluminium foil and zinc<br>– Be aware of additional light exposure (ceiling lighting)<br>– Use precise dosimetry |

For topical PDT combined pain-reducing techniques are helpful. Often the combination of local anaesthesia and cooling systems is suitable (See Tables 3.49, 3.50)

**Table 3.49.** Laser: FPDL (633 nm) Applicator: handpiece

| Photodynamic therapy (PDT) | |
|---|---|
| Power density: | 100 – 600 mW cm$^{-2}$ |
| Total dose: | 50 – 100 J cm$^{-2}$ |
| Spot diameter: | 3 – 7 cm |
| Pulse duration: | 80 µs |
| Repetition rate: | 4 – 10 Hz |
| Exposure time: | 700 – 1000s |
| Remarks: | – Use pain-reducing techniques in topical PDT<br>– Increased power density leads to increased penetration depth<br>– Cover the margins with aluminium foil and zinc<br>– Be aware of additional light exposure (ceiling lighting)<br>– Use precise dosimetry |

For topical PDT combined pain-reducing techniques are helpful. Often the combination of local anaesthesia and cooling systems is suitable (see Tables 3.48, 3.50)

**Table 3.50.** Laser: KTP pumped dye laser (635 nm) Applicator: handpiece

| Photodynamic therapy (PDT) | |
|---|---|
| Power density: | 50 – 150 mW cm$^{-2}$ |
| Total dose: | 50 – 1 00 J cm$^{-2}$ |
| Spot diameter: | 3 – 15 cm |
| Pulse duration: | cw |
| Exposure time: | 700 – 1000 s |
| Remarks: | – Use pain-reducing techniques in topical PDT<br>– Increased power density leads to increased penetration depth<br>– Cover the margins with aluminium foil and zinc<br>– Be aware of additional light exposure (ceiling lighting)<br>– Use precise dosimetry |

For topical PDT combined pain-reducing techniques are helpful. Often the combination of local anaesthesia and cooling systems is suitable (see Table 3.48, 3.49)

# II-5
# Laser-Induced Thermotherapy (LITT), Basics

Carsten M. Philipp and H.-Peter Berlien

**Contents**

## Introduction

Because of the physical properties of light and the physical properties of mammalian tissues, light of the very near-infrared (NIR) penetrates most deeply of all laser wavelengths into the soft tissues of the human body. This results from the relatively weak absorption in water or chromophores, even if visually darker structures such as melanin or carbonized tissues lead to a higher absorption compared to fibrous tissue or fat. The lack of absorption increases the influence of scattering in beam propagation within the tissues. As a result, a wide distribution of the photons is possible and the Nd:YAG laser, emitting in a continuous-wave mode at a wavelength of 1064 mm, was used first for coagulation purposes. During the early 1980s, when the techniques referred to in this chapter were developed, diode lasers emitting in the infrared were not available for medical use. As the emission characteristics of today's NIR-diode lasers may differ compared to the Nd:YAG laser, resulting tissue effects may either vary or be comparable for individual tissues. Tissue-dependent variations of absorption and scattering coefficients at a given wavelength, as well as thermal diffusion parameters and the influence of local perfusion, are responsible for the observed quantitative and qualitative differences.

## Definition

Laser-induced thermotherapy (LITT) covers the treatment method of laser-induced hyperthermia (LIHT) for temperatures from 42 to 47 °C, thermic dynamic reaction (TDR) at temperatures between 48 and 60 °C, as well as high-temperature treatment of laser-induced coagulation (LIC) for temperatures above 60 °C. According to the indication and localization of the diseased area, either an interstitial or transcutaneous approach with surface cooling, or an intraperative access is used. The effect results in immediate (at temperatures > 60 °C) and delayed (at temperatures < 60 °C) thermal destruction of the viable tissues. Apoptosis and delayed cell death due to membrane changes are discussed as the mechanism of TDR. The influence of the LIHT is moderate, in combination with chemotherapy it may enhance the efficiency. It is assumed that the effect is comparable to other hyperthermia applications as generated by focused ultrasound or radiofrequency, but laser hyperthermia is far more limited regarding the tissue volume.

During LITT, heat is generated by photon absorption and propagated by thermal diffusion. This results in a gradually decreasing temperature from the centre of the incident beam and light-absorbing tissues towards the rim of the thermally altered volume. All zones of LITT will be found in an irradiated volume (Table 1, Fig. 1).

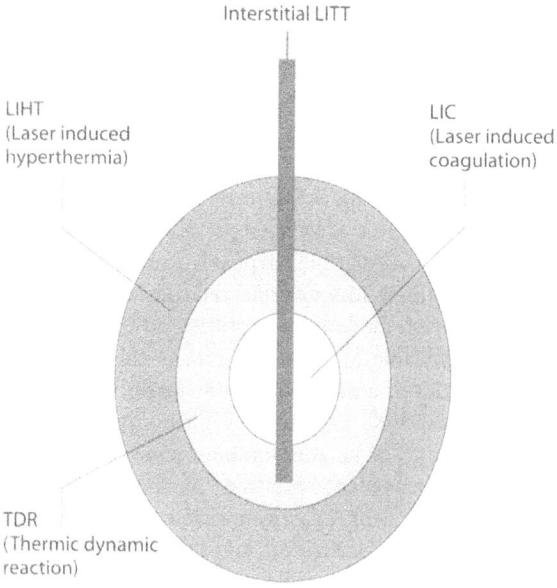

**Fig. 1.** Two-dimensional illustration of volumes of laser-induced coagulation (LIC), thermic dynamic reaction (TDR) and laser-induced hyperthermia (LIHT) during interstitial LITT

**Table 1.** Definition of laser-induced thermotherapy

| | |
|---|---|
| LITT– LIHT (laser-induced hyperthermia) | 42 – 47 °C |
| LITT– TDR (thermic dynamic reaction) | 48 – 60 °C |
| LITT– LIC (laser-induced coagulation) | > 60°C |

Interstitial application
– Percutaneous, endoscopic or intraoperative access
Superficial application
– With surface cooling

Effect
Immediate and delayed tissue destruction by heat
– Due to photon absorption and thermal diffusion

## Historical Development of Interstitial LITT

As penetration of light into tissue is somewhat limited, medical applications of lasers were exclusively surface application until the historical invention of fibre transmission of power irradiation from Nd:YAG lasers by G. Naht [1]. With flexible laser fibres the endoscopic use became a routine procedure in gastroenterology [2] and urology [3, 4]. But until the early 1980s only non-contact applications were performed, as experience showed that the fibre lost their transmission at the fibre tip when brought into contact with the tissue. Due to carbonization of adjacent tissues, which occurs faster with higher power densities, the optical properties of the surface of the fibre tip were destroyed, leading to a change in tissue interaction. The strong absorption at the carbonized layer, which heats the tip of the fibre to more than 300 °C, causes vaporization of the tissue being in contact with the fibre tip. As a result, a cutting effect as with a $CO_2$ laser is observed, but contrary to the $CO_2$ laser cut, it usually comes with a wide coagulation seam of one or more millimetres. This technique, known as the bare-fibre contact technique has become a basic surgical treatment method for cutting endoscopically, or to combine safe coagulation with cutting efficiency in open surgery. Today, special computer-controlled lasers with fibre protection systems reduce the emission, if the tissue contact of the fibre is interrupted and the fibre tip could get damaged. With endoscopic vision, this situation can be controlled by the surgeon, but the same feature can be helpful while using fibres interstitially.

As thermal conduction is the cause for variations in the relatively small coagulation seam of the $CO_2$ lasers, the local exposure time (e.g. the time a beam is irradiating a defined volume, heating it and leading to a thermal conduction into the surrounding tissue) was observed to be one factor in the development of a wide coagulation. Additionally to thermal conduction, a second factor must influence the width of the coagulation seam. Comparisons between heater probes and Nd:YAG bare fibre later showed the superiority of coagulation width for the laser [5]. In Europe surgeons had already used high-power Nd:YAG lasers for resections of parenchymatous organs because of their pronounced coagulation quality. It was known that besides the absorption of photons in the carbonization layer, a portion of distorted and widely scattered laser irradiation would pass into the tissue and lead to the pronounced coagulation seam mainly influenced by optical tissue properties, perfusion and thermal conduction [6].

During a meeting of LANSI (Laser Association of Neurosurgeons International) in Fuschl, Bavaria, in 1984, some discussion occurred as to whether this observation could lead to a new type of therapy. At this meeting, P.W. Ascher and D.S.J. Choy discussed the possibility of inserting such a fibre into the centre of a diseased area (intervertebral discs, brain tumours), using the coagulation effect from the inside toward the boundaries of the volume, with minimal access leaving the surrounding nearly untouched. The idea of the Interstitial Laser-Induced Thermotherapy was born.

In parallel, the first applications of interstitial LITT were reported by St. S. Bown [7], who used an Nd:YAG laser in combination with a bare fibre in the liver, and from P.W. Ascher [8, 9], who treated brain tumours with the same method. In the following years, several indications were defined. D.S.J. Choy and P.W. Ascher introduced the percutaneous laser disc decompression (PLDD) technique for the treatment of herniated intervertebral discs in 1986 and reported it in 1987 [10]. The technique of interstitial coagulation as a treatment method for vascular disorders and haemangiomas was first described by H.-P. Berlien and J. Waldschmidt [11, 12] in 1987. Next indications introduced to the clinic were the interstitial Coagulation and prostate (ILP) for benign prostatic hypertrophy [13] and the interstitial coagulation of intramural myomas in otherwise fertile women [14]. Head and neck tumours were identified as a possible indication [15] and many other indications followed, such as the interstitial vaporization of keloids, The interstitial thermotherapy of breast tumours is still under discussion [16, 17, 18], but some of these methods are now standard procedures. ILP and PLDD have been modified using other laser, e.g. diode lasers of various wavelengths (805–980 nm) or the holmium:YAG laser (2100 nm) in order to achieve a more ablative effect.

More recently in the mid 1990s, the treatment of secondary liver tumours of colorectal carcinoma again became a major topic in research. Meanwhile, the conditions and prerequisites had been investigated to achieve a maximum of coagulated volume and to predict its size with regard to the different tissue characteristics and their dynamic optical and thermal properties [19–26]. The application techniques for the Nd:YAG laser during interstitial LITT with the bare fibre had to be modified widely. Whereas in the earlier phases rather higher powers were used, as was common with endoscopic or superficial applications, the power levels now used are low (approximately 5–10 W), with the aim of avoiding critical carbonization temperatures for each tissue.

The bare fibre with its forward emission beam characteristic develops an egg-like coagulation volume, due to the influence of scattering (Figs. 2, 3). Its use is common; some advantages are obvious: access is minimal, larger volumes can be coagulated by multiple punctures, the volume can be shaped to anatom-

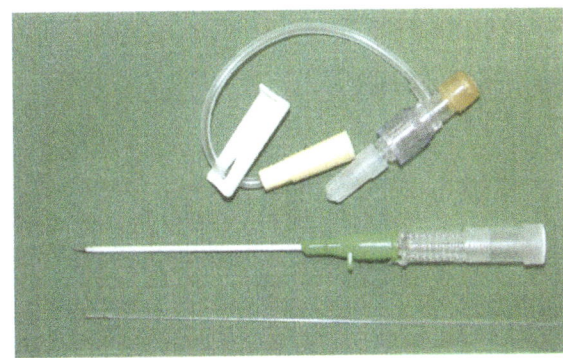

**Fig. 2.** Bare fibre and application set, disassembled

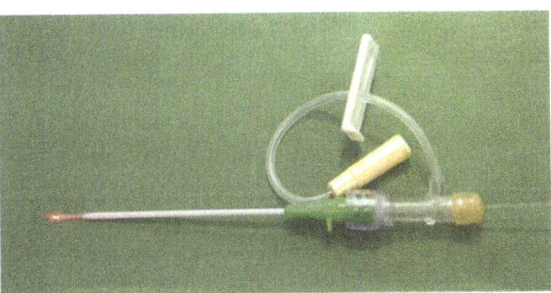

**Fig. 3.** An Abbocath G18 is used as catheter to sheath the fibre. The position of the fibre is predefined by the connector (Abbott Venisystems Extension Set with T Adapter SL). Its additional tube connector enables possibly produced gas or vapour to leave the system or add some saline flushing if needed. The system is mounted and the length of the fibre protruding from the catheter is positioned in advance, for the puncture it is dissembled again and the fibre is replaced by the needle. The connector stays in position with the fibre, keeping it in a defined length with its tip protruding from the catheter) after reinsertion

ical needs, intravascular application is possible and the bare fibre can be used for interstitial ablation of tissues by vaporization. Disadvantages are the low power settings possible for coagulation purposes and the limited coagulation volume at one position (Figs. 2, 3).

Carbonization was identified as negatively influencing the technique twofold. First, it leads to an increase in absorption at the carbonized layer, consequently the amount of photons transmitted into the tissue is lowered. Second, the temperature at the direct surrounding of the fibre tip shows a marked increase leading to vaporization, and significant amounts of gas are generated. The gas expands into the surrounding tissue and along the puncture channel, and may cause coagulation at unwanted regions [27]. Although, this effect is not preferred in the liver, there are indications where a controlled vaporization is the typical tissue interaction of the interstitial application, e.g. in the treatment of keloids and in PLDD when using a holmium:YAG laser.

## Technical Development

### *Fibres and Applicators*

Two ways to reduce the contact temperature at the fibre tip were identified: an increase in the emitting applicator's surface and flushing with physiological solutions around the fibre tip. Flushing in an open system obviously had to be limited to benign tumours, and is today the method of choice for the subcutaneous coagulation of vessels with interstitial or intraluminal approach. As the relation between power (W, J/s) and contact area (cm$^2$) are the parameters that most influence the boundary temperature at the fibre tip, the power density (W/cm$^{-2}$) had to be reduced to prevent carbonization. Engineers developed basically two technical solutions to increase the surface of the fibre tip:

1. The ring-mode applicator, which emits the laser beam to the circumferencial sides in a still rather small ring-like window.
2. The isotropic dome applicator, which emits the laser beam from its cylindrical sides at its full length.

Both were no longer just fibre tips, but became interstitial laser applicators. Avoiding carbonization, these applicators emit the light into a distinct volume of the tissue, according to the wavelength used. Photons are absorbed and heat is generated. The larger the primary volume of light distribution, the larger are the primary coagulation and its surface. Further from this surface, heat conduction alone will protrude coagulation into a larger volume, as long as energy is delivered and a sufficient power density (W cm$^{-2}$) of thermal energy at the given surface can be reached without overheating the centre [28–32].

Clinically, the isotropic dome applicators have found a wider distribution, due to their potential for further modification. Applicators with various lengths and diameters became available, two technical solutions were developed.

The frosted fibre applicator, where the coating and cladding of a fibre is removed at a certain length and the surface of the silica core is altered to a rougher surface to allow photons to be emitted in all directions. Both the ring dome and the frosted fibre applicator need special boundary conditions which permit the photons to be emitted sideways, e.g. with air as surrounding media, so both types are covered with a closed glass dome which is melted to the silica fibre.

Alternatively, some types of plastic fibres have been developed more recently which are doped with scattering elements at various lengths and may be completed with an integrated mirror at the very end. As this fibre has to be connected to the transmission

**Fig. 4.** Flexible interstitial isotropic scattering fibre in a plastic tube, bent to a radius of 2 cm

fibre (to form the emitting applicator), some problems occur at higher power densities with pulsed lasers. The connection site is heated due to transmission losses as they are observed at every optical boundary. The advantages of these fibres are their flexibility, better mechanical resistance as there is no need for a glass dome and the greater available lengths compared to the all-silica fibres. The performance with continuously emitting Nd:YAG lasers is adequate. This type of fibre is also used for endoscopic and endoluminal photodynamic therapy (Fig. 4).

### *Procedure*

These technical developments were accompanied by modifications of the procedure of applicator placement. Using the bare fibre, a puncture with an Abbocath or a biopsy needle (e.g. Turner needle) are sufficient, the fibre is inserted and the needle retracted for a certain distance. Similarly, or by direct puncture with the applicator in endoscopic procedures (e.g. transurethral approach for interstitial prostate coagulation), the ring-mode applicator is placed in the tissue. Otherwise, due to the mechanical instability of all-silica and the flexibility of the scattering plastic fibre a guiding and placement system had to be developed for the puncture. It consists of a sequence of needle, dilatators and port to insert a tube in Seldinger's technique (Figs. 5, 6).

The tube is made of thermoresistant and transparent teflon material and closed at the patient's side. Within this tube the applicator is placed at any predefined localization and the surrounding tissue is irradiated through the tube, comparable with afterloading therapy. As the contact surface with the tissue is defined by the diameter of the tube and the length of the applicator, the power density can be kept in a low range. At this low fluence carbonization is avoided, but the coagulated volume is on average around 3 cm$^3$ and perfusion effects such as cooling around vessels have severe influence on local temperature develop-

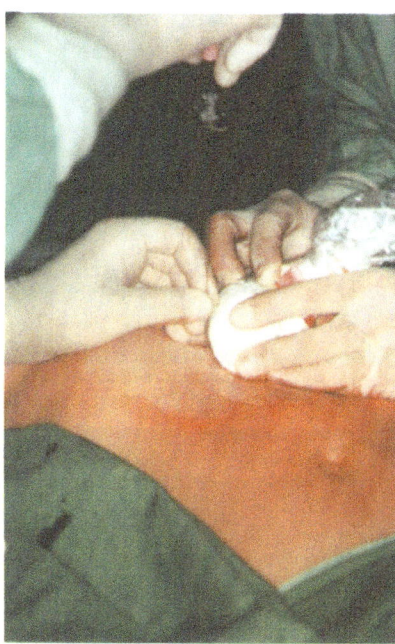

**Fig. 5.** Puncture with needle (17.5 gauge), after positioning of the tip at the desired position the port (12 F) is placed in Seldinger's technique over a guidewire

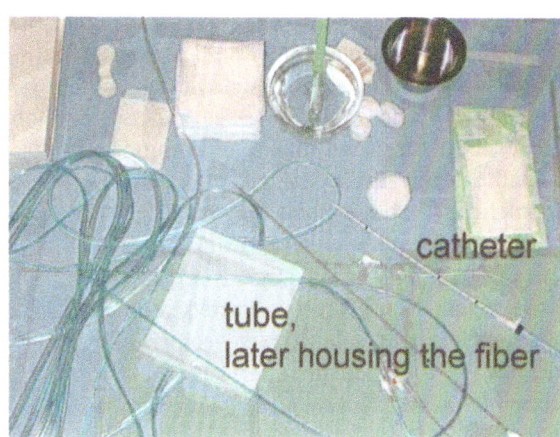

**Fig. 7.** Somatex interstitial Power-LITT set

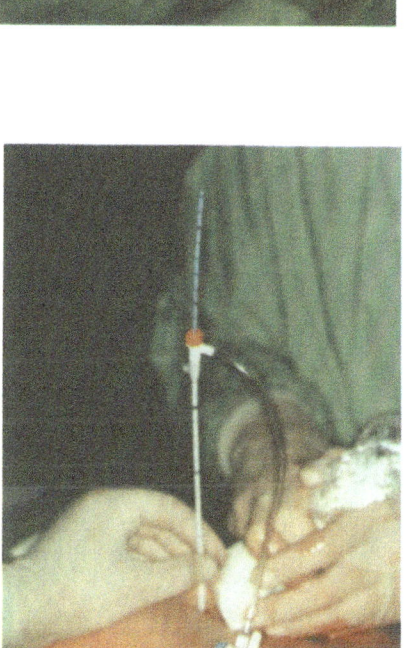

**Fig. 6.** Through the port a plastic tube (9 F, made of Teflon, closed at the end) is inserted and positioned with its tip protruding from the white port catheter at least the length of the applicator and an additional 1.5 cm. Then the laser fibre is inserted into the tube, retracted from the very end for 5 mm and fixed with tape to the end of the tube

means, the local tissue temperatures at the surface of the tube are kept low even with higher power densities and with a subsequent increase of coagulated volume at a given time (Somatex Interstitial Power-LITT set) (Fig. 7); other brands will be available soon for transcutaneous or intraoperative access (TRUMPF Microflexx and the surgical C-LITT Applicator). The influence of local perfusion is also lowered at higher power densities as it reduces flow and leads to an increase in coagulation volume. Vessel walls may become coagulated, which is desirable in small vessels [33]. Endogenous cooling by blood perfusion in larger vessels may prevent their coagulation, but leads to an increased risk of damaging the vessel walls, which must be taken care of during fibre placement.

Most LITT applications are transcutaneous procedures, as a minimized trauma during access is one of the major advantages of this method; but the method is also helpful in many operative situations. The open-access interstitial LITT during an open procedure enables the use of additional measures such as temporarily lowering the amount of perfusion of an organ (Prengel manoeuvre) [34].

Interstitial LITT may be applied under local, regional or general anaesthesia. The use of local anaesthesia is limited to smaller lesions and lesions where only the puncture route is pain-sensitive. Laser irradiation of soft tissues, bone and vessels is painful and the amount of applicable anaesthetic is limited by side effects at higher doses.

### Monitoring

Parallel to the development of the technical prerequisites for a safe application in hepatic secondaries, the route control of the puncture, the localization of the active portion of the fibre and the attempts to measure the occurring temperatures were in the focus of

ment. Clinically, a multifibre approach was necessary, where three or more fibres were placed into or around one tumour. In this way, the coagulation volume was increased to 15 cm$^3$ on average, but the procedures were time-consuming and needed improvement.

Improvement came by the introduction of a flushing system where the laser applicator is placed into an open inner tube which is hosted by a closed outer tube, allowing controlled power densities. By this research. A.C. Steger and St.S. Bown concluded that the visualization would not be sufficient for an online monitoring. Since then, the quality of ultrasound imaging has dramatically increased and the development of colour-coded duplex ultrasound and ultrasound contrast media have improved, thus leading to a redefinition of this monitoring method [37].

Other attempts were made using CT scans to visualize the affected tissues, but this remained clinically not relevant. Neurosurgeons started to utilize MRI to give images that would visualize temperature changes during the course of the intervention and, by using contrast images, the damaged volume after the procedure [38–41].

### Ultrasound and CCDS

The use of B-mode ultrasound for puncture control during liver biopsies has become a gold standard (Fig. 8). The puncture route and the localization of the needle tip can be precisely visualized in many body regions. By use of an CCDS, additional information about the vessels can be obtained. Applicator placement is safer and more precise, as the anatomy of the vessels is displayed with high precision and local bleeding at the puncture site can be prevented. A high blood content of tissues (e.g. a haematoma) leads to increased absorption and heating. On the „lee" side of the haematoma – with regard to the photons' propagation – a relative lack of photons can be assumed. By scattering, a part of this effect may be gradually reduced, but in the case of a larger „shadowing" zone, it may affect coagulation safety, even with higher power densities. Both effects increase the risks of the intervention; the first may lead to a local overheating and destruction of the tube, the second may influence procedure effectiveness. CCDS may also facilitate the identification of the needle tip if its position is questioned. By minimal movements, an artificial colour signal is produced, predominantly at the tip of the needle or catheter.

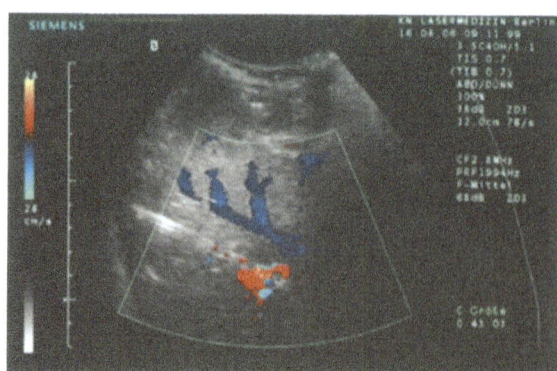

**Fig. 9.**

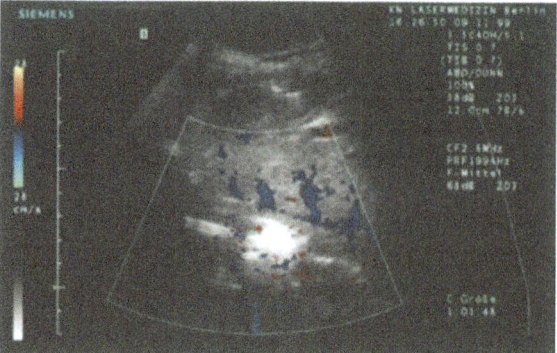

**Fig. 10.**

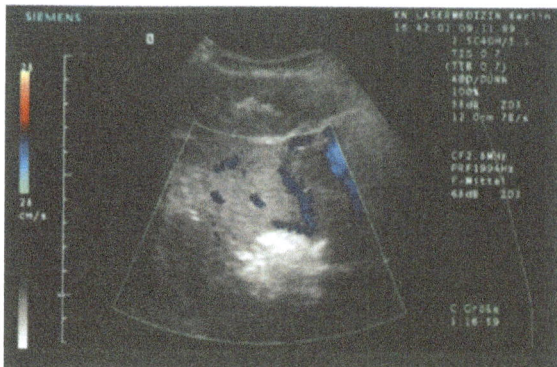

**Fig. 11.**

**Figs. 9, 10, 11.** The tip of the application system is indicated by colour signals when gently moving the needle. The course of the coagulation is visualized by CCDS images of the colour bruit; it starts at the applicator (9), later surrounding the centre of the coagulation. The intensity and size increases during the laser application until no more viable tissue can be reached by the coagulation front that extends perpendicularly from the applicator. 10 shows the colour bruit below the coagulation zone at a late stage of the intervention; it decreases when the largest achievable coagulation size is reached. About 20 min after the intervention the extent of the coagulation appears in the B-scan after the local gas is dissolved by the surrounding tissues (11)

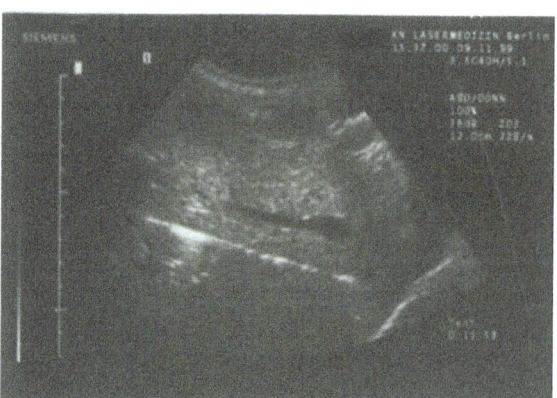

**Fig. 8.** B-scan image during liver puncture

During the interstitial LITT of subcutaneous lesions, a manual control of the skin temperature is used for protection; sometimes cooling of the skin is required. Together with the temperature increase, a crepitation caused by vapour and a degasification of the $CO_2$ in tissue becomes palpable. This reaction can also be monitored as an artefact in colour-coded duplex sonography (CCDS) in the form of an indeterminate random mixtures of red and blue signals. This colour bruit presents itself quantitatively differently according to the intensity of tissue reaction which is related to the actual heat development and transport in tissue [42, 43].

The typical course of images during an interstitial laser coagulation with a scattering dome applicator of 2 cm length is displayed in Fig. 9, 10 and 11.

After 10–30 s, some hyperechoic zones around the active portion of the applicator become visible, followed by the typical colour-bruit signals. In the course of procedure the hyperechoic zone above the applicator gains in intensity and subsequently limits the visualization of tissues below the applicator. In contrast, the colour-bruit signal increases to a certain extent and is displayed more intensely below the applicator. In the case of applicator damage, the signal increases its intensity remarkably. After the end of irradiation the colour-bruit signal decreases and stops. Temperature measurements could verify the relation between temperature and signal intensity [44]. As important additional monitoring information, changes in the perfusion of surrounding vessels can be monitored.

In bare-fibre LITT ultrasound guidance is often required. Due to the relatively small coagulation volume, multiple punctures or applications are frequent, control of the puncture route and the hot fibre tip are mandatory. Under most conditions, the position of the fibre tip outside the catheter can also be visualized by ultrasound or small colour signals when the fibre is moved up and down gently.

After start of the laser irradiation an increase in echogenity occurs in the B-scan, followed by a colour signal starting at the fibre tip and enlarging. Some colour signals along the puncture channel, interstitial spaces and nearby vessels indicate the propagation of the thermal front in the tissue by detecting the tissue movements resulting from degasification. After a certain time, depending on tissue variations, the increase in the colour signal stops and the intensity decreases. This marks the endpoint of the local application; the fibre may be replaced into a nearby volume or retracted.

In the follow-up the coagulation is detected as hyperechogenic; often the puncture route an be visualized for some days. In some procedures a certain

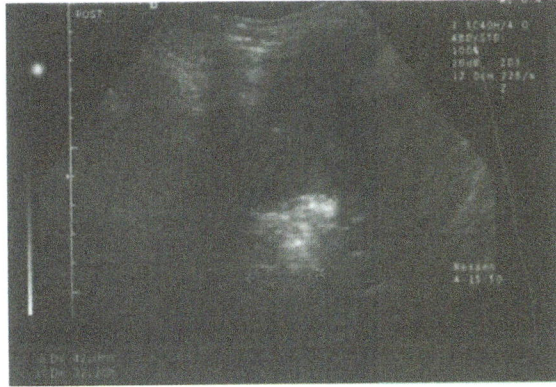

**Fig. 12.** The coagulation is depictable over the following days, the hypoechoic rim increases at day 2 to 3

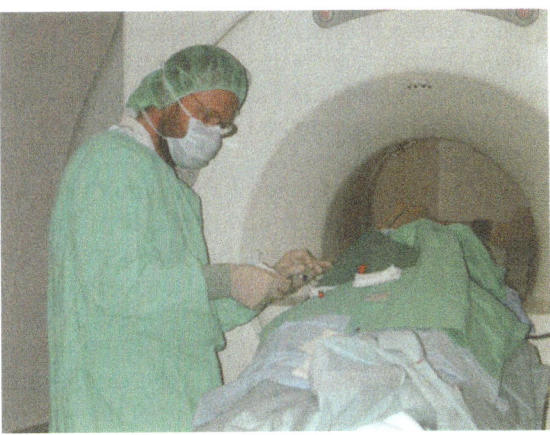

**Fig. 13.** After CT puncture the patient has to be transferred to the MR for thermal imaging. A control of the position is required

amount of carbonization cannot be avoided, especially in dark tissues and highly vascularized volumes. The resulting B-scan equivalent is a sharply demarcated echogenity at the centre of the coagulation image.

Some systematic limitations apply to this technique. The coagulation zone is not visualized exactly during the course of coagulation, as the increase of gas in the tissue reduces the resolution and shields the transmission of the primary ultrasound signal and its reflections. After the irradiation has been stopped according to the temperature of the tissue, dissolved gas will become resolved and the coagulation becomes visible as a hyperechoic region with a hypoechoic rim after 20 min (Fig. 12). Comparisons between size measurements in the B-mode ultrasound and contrast-enhanced MRI images display comparable results during follow-up [45]. Side effects and complications such as subcapsular bleeding of the liver or a pulmonary oedema are also depictable.

The greatest advantage of ultrasound monitoring is that the complete procedure can be performed in a sterile situation, using only one method. MR monitoring usually requires transport from the puncture room (CT or ultrasound) to the laser application room, the MR and additional position control or repositioning of the tubes (Fig. 13).

### MR Imaging for Interstitial LITT Control

Since 1993, MR imaging has been used alternatively by several groups [46, 47] for monitoring interstitial LITT. Using thermosensitive T1-weighted MR sequences, temperature rise around the applicator can be monitored during the course of irradiation. The loss of signal correlates with temperature rise and reverts during cooling (Fig. 14).

The loss of perfusion – an early equivalent of coagulation necrosis – can be displayed by using gadolinium (Gd-DTPA)-enhanced dynamic and static imaging directly after therapy and for later control examinations (Fig. 15).

The monitoring method of choice may vary according to the indications and localizations that are targeted with interstitial LITT. For many purposes ultrasound will deliver satisfactory image controls. In the brain and other regions which are not acccessible with ultrasound MRI control is mandatory. Today's open MR technique allow puncture and treatment to be performed without the interruption of transport of

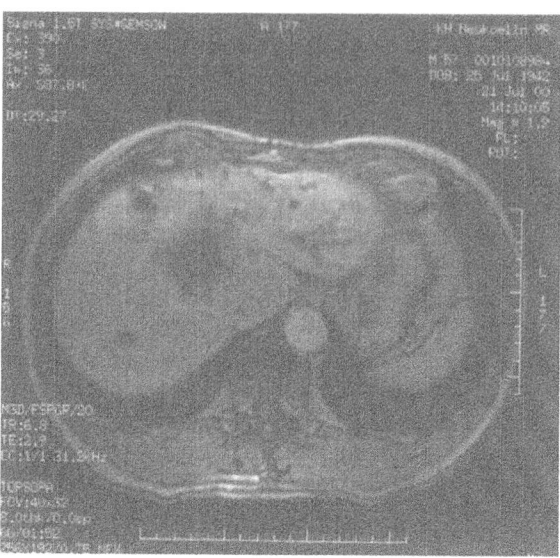

**Fig. 15.** Gadolinium (Gd-DTPA)-enhanced imaging after LITT. The loss of perfusion marks the coagulated area

the patient. With the wider use of these systems, it will become the preferred method of the coming years.

### Simulation

As in radiotherapy, the prediction of the damaged tissue volumes is crucial for interstitial LITT applications. With growing understanding of light transport and thermal diffusion in tissue, its dependency on boundary conditions, absorption and scattering, several authors have developed mathematical models to describe and predict the LITT effect in tissues [19–26]. A simulation program called LITCIT, developed by A. Roggan [48–50] became clinically relevant. It contains a wide variety of preselectable tissues properties, irradiation and heating modes (also RF probes) and takes care of the gradual decrease in perfusion and change in optical properties during the coagulation process (Fig. 16). In the 3-D version, vessels or multiple tissues can be placed virtually in a predefined volume, each with its own optical and physical properties. The prediction of the damaged zone is influenced by a factor that represents the viability of cells at certain temperatures. The strong correlation of the predicted thermal alterations were proven in clinical trials [51].

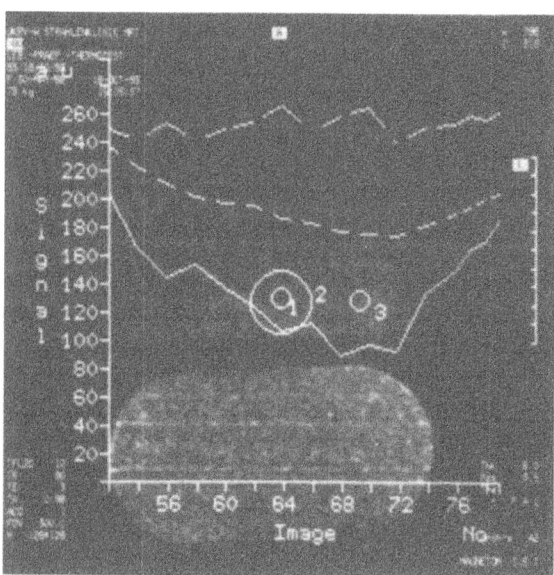

**Fig. 14.** A liver (ex vivo, pig) is placed on the water phantom, three ROI are marked, 1 at the applicator site, 2 in the proposed coagulation volume (slice) and 3 outside the heated volume. The decrease of the signal 1 is most pronounced, followed by the signal 2, 3 is not altered during the laser application. 2 and 3 return to base values after the laser is topped

## Interstitial LITT Application Techniques

The application technique and instrumental setup used may vary according to the various indications. For LITT of hepatic secondaries or malignant and non-malignant but solid tumours at other body re-

**Fig. 16.** The 3-D simulation of interstitial laser coagulation with LITCIT allows to predict the effect of multiple heat sources with regard to the changing tissue properties during the course of interstitial LITT

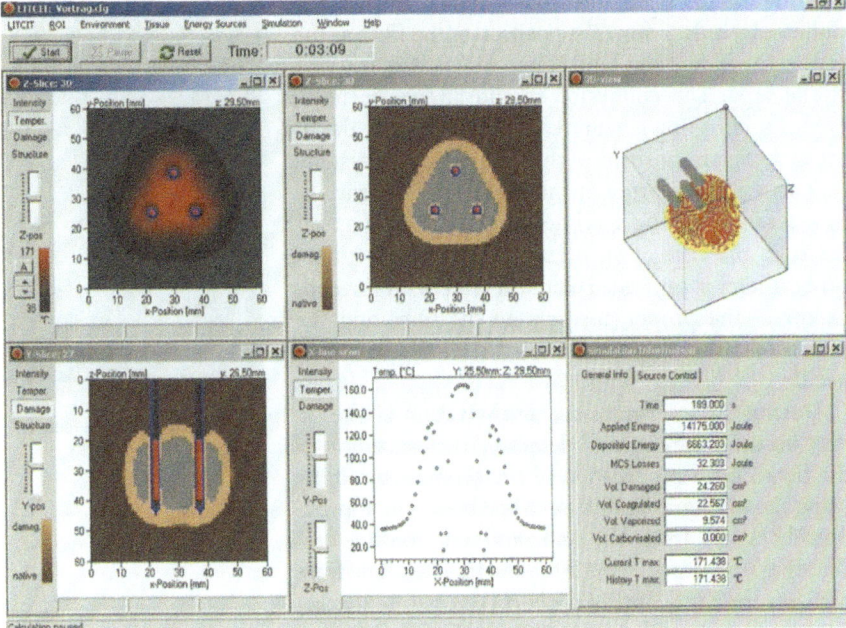

**Table 2.** Technical prerequisites for bare-fibre LITT

Nd:YAG laser 1064 nm

B-scan ultrasound (CCDS if available)

Bare fibre (600 µm, for most applications)

fibre preparation set
  (cutter, stripper, silicon plate)

Puncture system
  (Abbocath G14, G16, G18; Turner Biopsy Needle;
  SmartGuide Needle)

Connector
  (Abbott Venisystems Extension set with T Adapter SL)

Syringe

NaCI 0.9%

Pressure-controlled infusion pump (optional)

Wound tape

Ice cubes in sterile glove

gions including the brain, isotropic applicators with or without flushing are standard [33]. Ring-mode applicators (Dornier Medical Technologies) are used mainly for interstitial coagulation of benign prostatic hyperplasia, as they can be advanced into the tissue endoscopically without additional punctures or the need for an additional tube [13]. In the treatment of benign vascular tumours and malformations, the bare fibre is the commonly used applicator today [52, 53] but in selected cases also the Power-LITT Set was evaluated [54].

## Bare-Fibre Applications

Due to the minimal access trauma of the bare fibre, only an 18G access is required for most cases, and with shorter laser application times, a number of volumes can be treated with multiple punctures. This is a benefit in the LITT of small subcutaneous secondaries or irregularly shaped tumour volumes. Vessels can be targeted intra- and extravasally with precision. Due to the high power density at the tip also an interstitial vaporization can be achieved to ablate subsurface tissues. Since 1987, we have used bare-fibre LITT for various indications.

### Bare-Fibre LITT Instrumentation

The following list of technical prerequisites represents the currently used instrumentation for inter-

stitial LITT using bare fibres in our department (Table 2).

### Preoperative Preparation of the Set

Interstitial LITT is a sterile procedure. All preoperative preparations have to be performed under strictly sterile conditions. First the fibre is checked, to ensure that it is completely intact and no carbonization at the fibre tip is present. Second, the fibre is inserted to the required puncture set through the flushing connector, extending the end of the catheter tip for 10 mm. If flushing is intended, the fibre should exceed the catheter by only 5 mm. The position of the fibre at the connector is marked with water-resistant marker or sterile tape. The connector should always be used, even if no flushing is required, as it helps to fix the fibre at a certain position within the puncture system. It is also used as proof of intravascular localizations by aspiration, taking specimens of liquids, and as gas outlet in the case of vaporization. Finally, the puncture catheter is separated from the fibre with the connector at the proper position on the fibre and the puncture catheter is reassembled with the trocar or needle.

During the interventions, carbonization of the tip may occur. In any case of detecting carbonization of the fibre end, a preparation of the tip is required. By using the fibre preparation set, the fibre is cut approximately 2 cm from the end, and the coating and cladding is stripped 5 mm from the tip itself. The fibre now has the same optical properties as before its carbonization and can be used again for coagulation purposes. Otherwise the carbonization of the fibre will lead to immediate vaporization, as in the bare-fibre contact cutting mode, and the tissue effect is changed from coagulation to vaporization within the first seconds of laser emission.

## Puncture Technique

After identification of the lesion, the puncture route has to be defined. The puncture site should always be chosen laterally to the targeted volume to gain a puncture channel that can be used for compression if needed and to achieve the best imaging conditions. If using ultrasound guidance, it is advisable to target the most distant volume first, as interstitial gas between the transducer and the fibre may negatively influence visualization. Larger lesions are preferably coagulated in either a continuous or intermittent pull-back technique; if necessary multiple punctures are placed in a fan-like fashion or in parallel.

After positioning the puncture system, the trocar or needle is removed and replaced by the fibre. By connecting the needle to the connector, the intended distance of the fibre tip from the catheter tip is ensured, using the marking at the fibre as additional guidance. Under most conditions, the position of the fibre outside the catheter can also be visualized by ultrasound; this additional information should be used if available (Fig. 17).

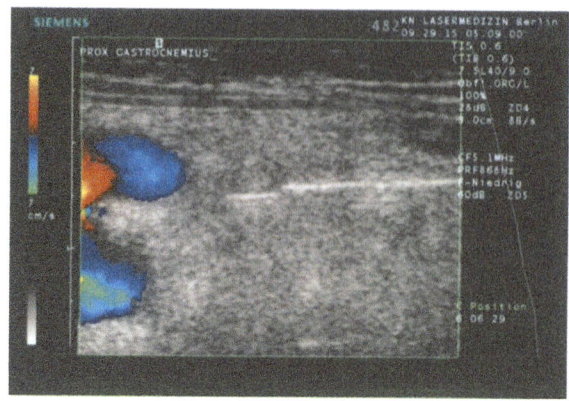

**Fig. 17.** Interstitial placement of catheter and bare fibre, visualized with ultrasound

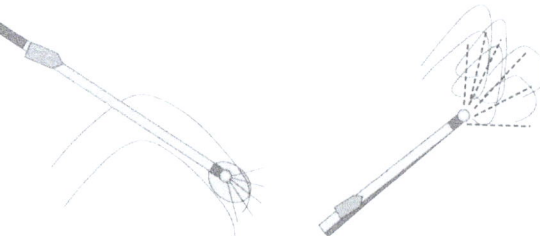

Puncture route and loss of perfusion controlled by CCDS

**Fig. 18.** Intravascular and interstitial treatment of vascular lesions (schematic drawing). Intravascular application requires saline flushing of the fibre tip to avoid fibre damage and clotting. Interstitial application in vascular lesions may be applied to a larger number of smaller vessels at one time; usually no flushing is required

## Flushing

In the case of intraluminal applications in blood vessels (low-flow ectasias and malformations), a flushing of the fibre end with 20 ml min of NaCl 0.9% is required, to prevent blood clotting at the fibre tip. If possible, the puncture route into the vessel should be directed against the flow direction. This enhances dilution of the blood around the fibre tip and facilitates the direct irradiation of the vessel wall (Fig. 18).

In the case of extraluminal position of the fibre, it may be helpful to install a few $cm^2$ NaCl 0.9% to generate a space between the targeted vessel and interstitial tissues, for better light penetration and a more selective coagulation. In the treatment of malignancies the open flushing system should not be used [55, 56].

### Irradiation

Clinically advisable are power settings between 5 and 12 W for different coagulation needs according to the indication: the parameters listed in Table 3 should be used as a guidance when using 600 μm fibres; some intraoperative adjustments may be necessary.

For coagulation of one vessel or small tumour volume, one application may be effective, but in larger lesions repeated applications will be necessary. Two different pull-back techniques are differentiated. The continuous retraction of the whole system consisting of fibre, catheter and connector for approximately 0.5–1 mm/s during continuous laser emission (preferably used in benign lesions, e.g. haemangiomas) and the intermittent retraction of approximately 10 mm after a localized application of the laser light for an effective time, with times varying according to the indication and the desired width of the coagulation. When using the intermittent technique a single application requires approximately 30 s until detectable changes in the tissue can be monitored by CCDS and approximately 3 min to achieve a maximum coagulation width. In immediately following applications at related sites, the onset of coagulation and thermal propagation may be faster due to preheating the tissue, and the required exposure times may be shortened to 120 s.

In benign lesions the application times can be lowered; with regard to the observed tissue reaction the application may be stopped prior to the achievement of the maximum volume. It is important to realize that the coagulation will take place around the end of the fibre, not only in front of it. It will propagate about 2/3 of its volume in front of the fibre, but due to backscattering and thermal diffusion 1/3 of the coagulation will develop behind the fibre tip. With the given laser parameter (Table 3) the maximum coagulation width is approximately 10 mm measured horizontally and 15 mm longitudinally to the fibre. In the centre of the coagulation a zone of vaporization surrounded by a carbonized layer may occur during extended applications.

After the application of laser light into the diseased volume, the whole system is retracted slowly. In order to close the puncture channel, some small coagulations may be produced by 0.5 s exposures repeatedly during the retraction along the puncture route, taking care that the skin is spared from coagulation. Keeping the catheter still in the puncture channel, the fibre is removed first. In the case of bleeding, a reinsertion of the fibre and a further coagulation may be necessary.

**Table 3.** Power settings for interstitial bare-fibre use

| | |
|---|---|
| LITT without flushing in highly absorbing tissues (e.g. haemangiomas). | 5 W |
| LITT with intraluminal flushing in blood vessels (low flow): | 8–10 W |
| LITT without flushing in low absorbing tissues (e.g. lymphangiomas): | 8 W |
| LITT with flushing in low absorbing tissues (e.g. lymphangiomas): | 12 W |

In the more likely case of no bleeding, the catheter is removed and a sterile wound tape is fixed at the puncture site. In the treatment of vascular malformations the wound care can be completed with additional compression of the lasered volume.

### Monitoring

During laser emission time, a complete monitoring of the tissue reaction is necessary. In superficial applications the pilot beam can help to monitor the fibre's position; a digital palpation of the surface is used for temperature and process control (Fig. 19). After the onset of coagulation a crepitation by dissolving gas becomes palpable. Strong crepitation or popcorn effect-like gas production indicates that a carbonization of tissue has already occurred or is very likely to occur within the next seconds. If this is not intended, the application catheter should be positioned into a new volume after retracting the fibre and replacing it with the needle or trocar. A fibre preparation may be required to continue the coagulation procedure.

The manual control of the surface temperature has shown its superiority compared to infrared thermographic readings regarding the response time of the cameras and of the surgeons. Thermographic controls may be used for documentation purposes. In benign subcutaneous lesions the emission from the laser must be stopped in the case of superficial temperature rise. Unwanted thermal damage of the skin can be limited by cooling with ice-cubes placed in a sterile glove above the centre of the pilot beam emission from the tissues. In very superficial applications (less than 1 cm depth) a continuous surface cooling during laser emission is mandatory; pre- and after-cooling may be helpful.

In malignancies such as thoracic secondaries of breast carcinoma, skin usually cannot be preserved. The coagulation must reach healthy tissue to destroy the tumour safely, followed by secondary healing.

Ultrasound should be used very carefully in superficial lesions, the skin (and the transducer) could easily be thermally damaged when the tissues are compressed by the transducer. The use of large amounts

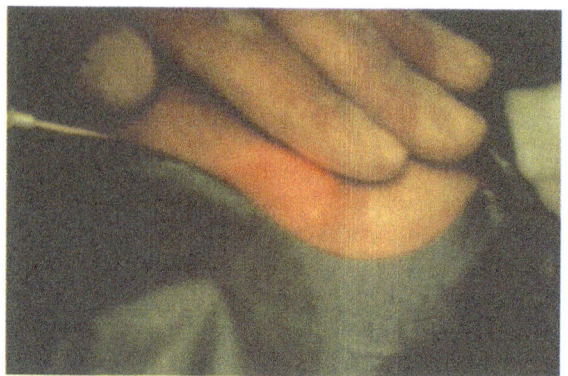

**Fig. 19.** Monitoring of subcutaneous LITT with transillumination and manual control of temperature

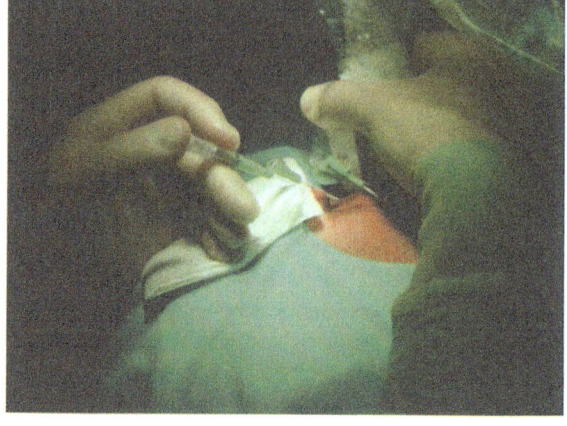

of ultrasound gel and a non-compression approach with the transducer are mandatory.

In deep-seated applications where the crepitation is not palpable, the use of CCDS is required to detect the dissolving gas and the subsequent tissue movements as colour bruit; but nevertheless, direct temperature control by palpation is strongly recommended additionally also for deeper applications. Interstitial gas may protrude to the surface regions along the puncture channel, through vessels or interstitially, causing unwanted skin reactions off-site.

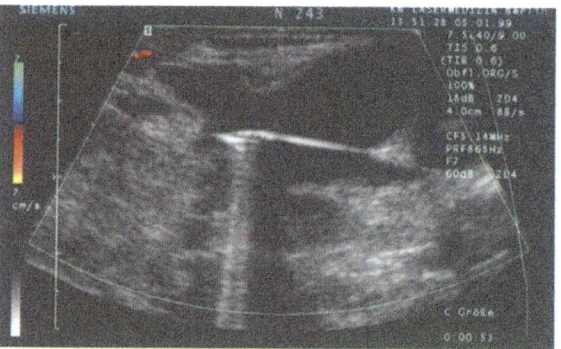

**Figs. 20, 21.** Puncture of a lymphatic cyst (20) and display of needle in ultrasound image (22)

## Special Techniques

### Vascular Lesions

#### LITT of Lymphangioma

In the treatment of lymphangiomas with larger cystic portions, the procedure is modified. After puncture of the cyst a specimen is taken from the liquid for laboratory analysis. If it contains old blood, repeated installation and washing with saline are advisable. When the aspirated fluid has little or no blood content, the intracystic liquid is removed partially. The laser emission is directed towards the wall of the cyst, with careful continuous movements, preferably in the near-contact mode. In multicystic lymphangiomas, a multipunture approach is used, where all catheters are placed before the first laser application (Fig. 20).

With the laser the endothelium of the cysts is destroyed, and after aspiration of the residual fluid the walls become adherent. Thus, followed by phases of inflammation and reparation the cysts are destroyed, resulting in an interstitial fibrosis. In some case, the additional injection of scleroting agents or fibrin glue into laser-pretreated cysts is required for satisfactory

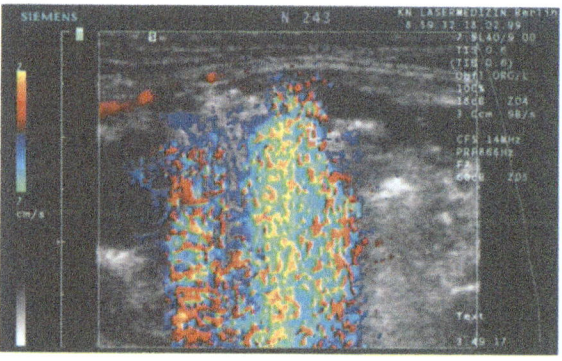

**Figs. 22.** colour bruit during the course of laser application to the cyst wall (21)

results. Care has to be taken to protect intraseptal or capsule-related blood vessels from leakage by puncture or vaporization. A gentle compression may help to secure the permanent occlusion of the cysts where applicable (Figs. 21, 22).

### LITT of Predominantly Venous VM

Vascular malformations appear at all types of vessels; mostly veins are involved. The intraluminal approach is an effective and less invasive way to deliver the laser energy to low flow lesions. Theoretically, the fibre could be advanced in the lumen also to distant places. Practically, a multipuncture approach is more useful, as unwanted perforations of the vessel by the advanced fibre and extravasating blood could lead to unwanted side effects. During the application of the laser energy a small volume of saline is continuously flushed using the connector as port for the fibre and the saline. As whole blood shows only little transmission for Nd:YAG irradiation, a dilution is needed to achieve a direct irradiation of the vessel wall. Earlier investigations showed a minimal perfusion of about 20 ml min$^{-1}$ to be sufficient to also protect the blood from clotting at the fibre tip and dilute whole blood locally. After insertion of the fibre the system is either placed locally for 60–180 s in large vessels and retracted step by step or continuously retracted with a speed of 1 mm 10$^{-1}$ s in small vessels until the vessel is occluded. The puncture should allow a small subcutaneous channel until entering the vein for compression purposes after retraction of the system. It is helpful to retract the fibre first, leaving the catheter in the vessel for proof of occlusion.

The occlusion of single venous vessel (e.g. perforating veins) may also be performed in an interstitial mode. After localization of the vessel in the ultrasound or with Doppler technique and local anaesthesia, two (three with larger diameters) application catheters are placed from the side with the tip directed towards the vein and superior to the muscle fascia. The vein should not be perforated; 3–5 ml saline (or Xylocain 0.5% if additional anaesthesia is required) are injected through the catheters to achieve a higher transparency in the subcutaneous tissue while the vein contains the highly absorbing blood. After positioning of the fibres an interstitial coagulation aiming at the outer vessel wall is performed until occlusion of the vein is depictable by CCDS or indicated by Doppler ultrasound.

### LITT of Haemangiomas

In large haemangiomas an interstitial approach may be valuable: nevertheless, in most cases it is not needed. The transcutaneous irradiation is sufficient in most cases and severe side effects such as nerve lesions have not been observed with this method. Nevertheless, in cases where a resection is planned during the course of treatment, interstitial LITT offers an effective way to reduce perfusion and facilitate and optimize the surgical procedure.

The puncture route should be precisely planned, in order to irradiate the most distant region from the puncture site first. The system is inserted subcutaneously aside from the haemangioma, forming a short channel in the subcutis for better wound closure at the end of the procedure. By pulling the whole system (application catheter, fibre and connector) slowly back a cylindrical volume is exposed and coagulated, sparing the channel and the haemangioma region close to the puncture. The fibre is then removed with the connector and the fibre tip checked; in the case of carbonization the fibre should be prepared. While the laser nurse is doing so, the needle or trocar is placed again in the catheter and a new puncture is done approximately 1 cm parallel to the first site. Step by step the complete volume of the haemangioma can be reached in a fan-like fashion, sometimes in multiple layers. Pathways of nerves must not be crossed while using this technique; as the bare fibre coagulation does not discriminate between tissue types, serious nerve damage could be the result.

### Interstitial Vaporization of Keloids

Several modifications of preparations and laser parameters are necessary for the purpose of interstitial vaporization (e.g. in the treatment of selected keloids). Most important is the use of a preblackened fibre. For this purpose the fibre is brought into contact with a wooden spatula during short laser emissions, 0.1 to 0.2 s of 30 W, until an immediate vaporization of the wood occurs. This is the same procedure as is used for preparation of the fibre for the contact-cutting technique. The fibre is then cleaned with a wet gauze pad and inserted into the puncture catheter already placed in the tissue. The application also requires 30 W and repeated exposures of 0.1 to 0.3 s during slow retraction. The catheter should be larger than in interstitial coagulation to allow the significant amounts of gas produced to leave the tissue in a controlled fashion. The irradiation must be interrupted repeatedly as surface temperature increase should be limited. As this is a very superficial application, great experience is necessary to spare the surface from thermal damage and to chieve best results.

## Future Trends

### Applicators

The high-power density applicators with cooled systems are already replacing simpler systems for large volume coagulation. In combination with flexible plastic fibres they allow curved punctures with special needles (Smart-Guide) and precisely shaped co-agulations. Using adjusted cooling by variations in flushing temperatures and volumes, the shapes of coagulation zones could be modified. Frosted fibre applicators and ring-mode applicators will be limited to special applications. The bare fibre will continue to be valuable, because of ist favourable rate between access trauma, effort and coagulation size, its ma-noeuverability and versatility.

Alternative energy sources using microwaves, HF current and bipolar applicators are in concurrence with laser-based systems today. Current clinical re-sults show the superiority of the laser techniques re-garding precision in application and clinical outcome, but the race is on.

## Monitoring

A significant improvement for the ultrasound visuali-zation of unperfused (coagulated) volumes in the liver can be expected by the use of ultrasound con-trast media such as Levovist or Optison. Similarly to the MR contrast-enhanced Gd-DTPA images, neither an early arterial nor a portovenous enhancement is seen in the treated volumes, while the liver tissue dis-plays a bright enhancement. This technique is not commonly used today, but as it is applicable during the procedures and is faster at lower cost compared to MR imaging, it has the potency to replace current standards after its evaluation. Application research is currently targeting the direct display of temperatures using ultrasound for online monitoring of LITT:

MRI imaging will improve in spatial resolution and the calculation time of the images will decrease. The open-MRI technique already allows direct access to the patient as with ultrasound, real-time image fusion of simulation and simultaneously acquired images during the procedure will be added to optimize process control.

Perfect positioning is crucial to achieve a homoge-neous coagulation. Technical and electronic aids will be developed in order to improve puncture precision and planning.

## Simulation

Preoperative simulations will attain the precision of radiotherapy planning as optical and physical proper-ties of tissues are investigated for all laser wave-lengths. Multiple applicators are needed for coagula-tion of larger volumes. To obtain optimal results, the position, power, total energy and local tissue proper-ties of the surrounding for each applicator will be in-dividually simulated and the laser emission will be adjusted to match with on-line acquired images by smart monitoring systems.

## References

[1] NAHT G, GORISCH W, KIEFHABER P (1973) First laser endoscopy via a fiber optic transmission system. En-doscopy 5: 2028

[2] KIEFHABER P, KIEFHABER K Therapeutic Nd: YAG-Laser Application in the Gastrointestinal Tract. In: van Maerke (eds) Stomach deseases - current status: Proceedings of the 13 th International congress on stomach dieases, Antwerpen, 199

[3] HOFSTETTER A, STACHLER G, KREIDITSCH E, FRANK F (1978) Lokale Laserbestrahlung eines Peniskarzinoms. Fortschr Med 96: 369

[4] HOFSTETTER A, BÖWERING R, FRANK F, KREIDITSCH E, PENSEL J, ROTHENBERGER KH, STAEHLER G (1980) Laserbehandlung von Blasentumoren. DMW 105: 1442

[5] WYMAN DR, WHELAN WM, WILSON BC (1992) Intersti-tial laser photocoagulation: Nd:YAG 1064 nm optical fiber source compared to a point heat source. Las Surg Med 12: 659–664

[6] BERLIEN H-P (1987) Resection of parenchymatous organs and bloody tumors with the laser I: lasers in Medicine. Proceedings of the 1st European Workshop on Lasers in Medicine, Crete 1987, p 243

[7] BOWN SG (1983) Phototherapy of tumors. World J Surg 7: 700–709

[8] ASCHER PW (1983) Verhandlungsbericht der DGLM-Tagung. Graz

[9] ASCHER PW (1981) Application of the laser in neuro-surgery. Las Surg Med 2: 91–97

[10] CHOY DJS, CASE RB, FIELDING W (1987) Percutaneous laser nucleolysis of lumbar disc. N Engl J Med 317: 771–772

[11] BERLIEN HP, WALDSCHMIDT J, MÜLLER G (1987) Laser treatment of cutan and deep vessel anomalies. In: WAIDELICH W, WAIDELICH R (eds) laser 87 – Optoelec-tronics in medicine. Springer, Berlin Heidelberg New York

[12] WALDSCHMIDT J, BERLIEN HP, HAUCK GW, EL-DESSOUKY ML (1988) Auswahl verschiedener Lasertypen bei der Behandlung von oberflächlichen und tiefen Gefäßanomalien. Kinderchir 43: 6–10

[13] MUSCHTER R, HOFSTETTER A (1995) Interstitial laser therapy outcomes in benign prostativ hyperplasia. J En-dourol 9 (2): 129–135

[14] CHAPMAN R (1998) New therapeutic technique for treat-ment of uterine leiomyomas using laser-induced intersti-tial thermotherapy (LITT) by a minimally invasive method. Las Surg Med 22 (3): 171–178

[15] PAIVA MB, SAXTON RE, VANDERWERF QM, BELL T, ESHRAGHI AA, GRAEBER IP, FEYH J, CASTRO DJ (1997) Cisplatinum and interstitial laser therapy for advanced head and neck cancer. A preclinical study. Las Surg Med 21: 433–431

[16] PANJEHPOUR M, WILKE AV, FRAZIER DL, OVERHOLT BF (1991) Nd:YAG laser hyperthermia treatment of rat mammary adenocarcinoma in conjunction with surface cooling. Las Surg Med 11 (4): 356–362

[17] HARRIES SA, AMIN Z, SMITH MEF, LEES WR (1994) In-terstitial laser photocoagulation as a treatment for breast cancer. Br J Surg 81: 1617–1619

[18] BASU S, RAVI B, KANT R (1999) Interstitial laser hyper-thermia, a new method in the management of fibroade-noma of the breast: a pilot study. Las Surg Med 25: 148–152

[19] PRIEBE LA, WELCH AJ (1979) Dimensionless model for the calculation of temperature increase in biologic tissues exposed to nonionizing radiation. Trans Biomed Eng, vol BME 26 (4): 244–250

[20] WELCH AJ, WISSLER EH, PRIEBE LA (1980) Significance of blood flow in calculations of temperature in laser-irra-diated tissue. IEEE Trans Biomed, vol BME 27 (3): 164–166

[21] JAQUES SL, PRAHL SA (1987) Modeling optical and thermal distributions in tissue during laser irradiation. Las Surg Med 6: 494–503

[22] MOTAMEDI M, RASTEGAR S, LECARPENTIER G, WELCH AJ (1989) Light and temperature distribution in laser-irradiated tissue: the influence of anisotropic scattering and refractive index. Appl Opt 28 (12): 2230–2237

[23] SVAASAND LO, GOMER CJ, WELCH AJ (1989) Thermotics of tissue. In: MÜLLER GJ, SLINEY DH (eds) Dosimetry in laser radiation in medicine and biology. SPIE Inst Ser IS 5: 133–145

[24] VAN GEMERT MJC, WELCH AJ (1989) Time constants in thermal laser medicine. Las Surg Med 9: 405–421

[25] PATTERSON MS, WILSON BC, WYMAN DR (1991) The propagation of optical radiation in tissue I. Models of radiation transport and their application. Las Med Sci 6: 155–168

[26] ESSENPREIS M, VAN DER ZEE P, MILLS TN (1992) Monte Carlo modelling of light transport in tissue: the effect of laser coagulation on light distributions. In: Bioptics: optics in biomedicine and environmental sciences. SPIEE, LA, PP 1524–1540

[27] FUCHS B, PHILIPP G, ENGEL-MURKE F, SHALTOUT J, BERLIEN HP (1993). Techniques for endoscopic and non-endoscopic intracorporal laser applications. Endo Surg Allied Technol 4 (1): 217–223

[28] DOWDEN J, JORDAN T, KAPADIA P (1988) Temperature distribution produced by a cylindrical etched fibre tip in laser treatment of tumors by local hyperthermia. Las Med Sci 3: 47–54

[29] VOGL TJ, WEINHOLD N, MULLER P, MACK M, SCHOLZ W, PHILIPP C, ROGGAN A, FELIX R (1996) MR-controlled laser-induced thermotherapy (LITT) of liver metastases: clinical evaluation. Roentgenpraxis 49: 161–168

[30] HEISTERKAMP J, VAN HILLEGERSBERG R, SINOFSKY E, IJZERMANS JN (1997) Heat-resistant cylindrical diffuser for interstitial laser coagulation: comparison with the bare-tip fiber in a porcine liver model. Las Surg Med 20: 304–309

[31] ORTH K, RUSS D, DUERR J, HIBST R, STEINER R, BEGER HG (1997) Thermo-controlled device for inducing deep coagulation in the liver with the Nd:YAG laser. Las Surg Med 20: 149–156

[32] IVARSSON K, OLSRUD J, STURESSON C, MOLLER PH, PERSSON BR, TRANSBERG KG (1998) Feedback interstitial diode laser (805 nm) thermotherapy system: ex vivo evaluation and mathematical modeling with one and four fibers. Las Surg Med 22: 86–96

[33] VOGL TJ, MACK MG, ROGGAN A, STRAUB R, EICHLER KC, MULLER PK, KNAPPE V, FELIX R (1998) Internally cooled power laser for MR-guided interstitial laser-induced thermotherapy of liver lesions: initial clinical results. Radiology 209: 381–385

[34] ALBRECHT D, GERMER CT, ISBERT C, RITZ JP, ROGGAN A, MULLER G, BUHR HJ (1998) Interstitial laser coagulation: evaluation of the effect of normal liver blood perfusion and the application mode on lesion size. Las Surg Med 23: 40–47

[35] STEGER AC, SHORVON P, WALMSLEY K, CHISHOLM R, BOWN SG, LEES WR (1992) Ultrasound features of low-power interstitial laser hyperthermia. Clin Radiol 46: 88–93

[36] AMIN Z, DONALD JJ, MASTERS A, KANT R, STEGER AC, BOWN SG, LEES WR (1993) hepatic metastases: interstitial laser photocoagulation with real-time US monitoring and dynamic CT evaluation of treatment. Radiology 187: 339–347

[37] PHILIPP C, BOLLOW M, KRASICKA-ROHDE E, FOBBE F, BERLIEN HP (1994) Color-coded duplex sonography as a new method for monitoring of laser-induced thermotherapy. SPIE Proc 2132: 287–294

[38] LEWA CJ, MAJEWSKA Z (1980) Temperature relationships of proton spin-lattice relaxation time T1 in biological tissue. Bull Cancer 67 (5): 525–530

[39] PARKER DL (1984) Applications of NMR imaging in hyperthermia: an evaluation of the potential for localized tissue heating and noninvasive temperature monitoring. IEEE Trans Biomed Eng Vol. BMW 31 (1): 161–167

[40] DICKINSON RJ, HALL AS, HIND AJ, YOUNG IR (1986) Measurement of changes in tissue temperature using MR imaging. J Comp-Assisted Tomogr 10 (3): 468–472

[41] JOLESZ FA, BLEIER AR, JAKAB P (1988) MR imaging of laser-tissue interactions. Radiology 168: 249–253

[42] PHILIPP C, BOLLOW M, ROHDE E, BERLIEN H-P (1994) Color Doppler as a new method for the monitoring of laser-induced thermotherapy. Las Med Surg Suppl 6:3

[43] PHILIPP CM, ROHDE E, BERLIEN HP (1995) Nd:YAG laser procedures in tumor treatment. Sem Surg Oncol 11 (4): 290–298

[44] ROHDE E, PHILIPP C, BERLIEN HP (1996) Untersuchungen zur Farbkodierten Duplexsonographie (FKDS)-Kontrolle der Interstitiellen laserinduzierten Thermotherapie (LITT). Lasermedizin 12: 121–129

[45] Philipp CM, Urban P, Berlien HP (unpublished data, currently continuing observations) B-mode ultrasound and contrast-enhanced MRI images display comparable results during follow-up.

[46] TRACZ RA, WYMAN DR, LITTLE PB, TOWNER RA, STEWART WA, SCHATZ SW, WILSON BC, PENNOCK BW, JANZEN EG (1993) Comparison of magnetic resonance images and the histopathological findings of lesions induced by interstitial laser photocoagulation in brain. Las Surg Med 13: 45–54

[47] VOGL TJ, ASSAL J, PHILIPP C, ROHDE E, FISCHER P, BÖTTCHER H, GREMMLER M, ROGGAN A, DEIMLING K, FELIX R (1994) Nichtinvasives Temperaturmonitoring für die interventionelle laserinduzierte Thermotherapie von Lebermetastasen. Tagber 75. Deutscher Röntgenkongreß Wiesbaden, 11.–14. Mai 1994, Zentrbl Radiol Bd. 150, Heft 1–3, Nr. 160

[48] ROGGAN A, MINET O, SCHRÖDER C, MÜLLER G (1993) Measurements of optical tissue properties using integrating sphere technique. Inst Adv Med Opt Tomogr IS 11: 149–165

[49] ROGGAN A, MÜLLER G (1993) Computer simulations for the irradiation planning of LITT. Min Invas Med 4 (2): 18–24

[50] ROGGAN A, RITZ JP, KNAPPE V, GERMER CT, ISBERT C, SCHÄDEL D, MÜLLER G (2001) Radiation planning for thermal laser treatment. Med Laser Appl 16: 65–72

[51] Handke A, Roggan A, Müller G, Miller K (1995 Laser-induced interstitial thermotherapy (LITT) of benign prostatic hyperplasia (BPH) – basic investigations and first clinical results. In: Müller G, Roggan A (eds) Laser Induced Interstitial Thermotherapy. SPIE-Press, Bellingham 403–415

[52] REICHMANN A, PHILIPP C, URBAN P (1999) Interstitielle Lasertherapie von vaskulären Tumoren und Malformationen. Ultraschall Suppl 20, p 80

[53] POETKE M, PHILIPP CM, URBAN P, BERLIEN HP (2001) Interstitial laser treatment of venous malformations. Med Laser Appl 16: 111–119

[54] CHOLEWA D, WACKER F, ROGGAN A, PHILIPP CM, WALDSCHMIDT j (2001) Interstitial laser therapy of congenital vascular malformations with MRI guidance. Med Las Appl 16: 103–110

[55] BERLIEN HP, PHILIPP C, ENGEL-MURKE F, FUCHS B (1993) Laseranwendung in der Gefäßchirurgie. Zentralbl Chir 118: 383–389

[56] PHILIPP C, BERLIEN HP, POETKE M, WALDSCHMIDT J (1994) Ten years of laser treatment of congenital vascular disorders – techniques and results. SPIE vol. 2327 Medical Applications of Lasers II, 44–53

# II-6
# Basics of Photodynamic Therapy (PDT)

Basil Jamil and H,-Peter Berlien

**Contents**

## Introduction

### The Basis of PDT

Photodynamic therapy (PDT) is a term encompassing a collection of both curative and palliative modalities, traditionally described as treating precancerous lesions and superficial tumours, using light. The list of medical fields in which PDT has managed to find a place as an accepted option for specific problems includes gastroenterology, dermatology, gynaecology, ophthalmology and ENT. Its increasing importance is underlined by a comparison with both traditional chemotherapy and radiotherapy, which can often significantly compromise patients health. The therapeutic use of these well-established therapies, is accordingly, limited by their toxicity. In contrast, PDT not only shows a distinct degree of tumour specificity, but can be repeatedly applied without apparently damaging the health of the patient.

Historically, PDT looks back on a long empirical evolution. Its beginnings are to be found in antiquity. However, a satisfactory theory as to how it actually works could be developed only during the past few decades. In principle, substances, called photosensitizers, absorb optical energy and transmit this to oxygen molecules, thereby forming highly aggressive forms of oxygen-based molecules. The latter are often denoted by the term "reactive oxygen species" (ROS). PDT's ancient roots are explicable because of the natural occurrence of photosensitizers with a significant capacity for forming ROS. These include, for example, bilirubine and the flavines. Following the uptake of photosensitizer by individual cells or cell groups, the absorption of light with consequent generation of ROS results in their death. Essentially, the therapeutic value arises from a more or less selective uptake of certain photosensitizers by precancerous or cancerous tissue.

Nowadays, we know that light is only one of three basic factors involved. As pointed out, PDT relies on three factors:
- light,
- oxygen
- photosensitizer.

An understanding of PDT demands at least an elementary description of the physics lying behind these factors, and this will be the subject of the first section given here.

Light cannot be fully characterized without taking account of its generation, transmission and absorption. One needs to understand the nature of matter for this. Modern physics, in the form of the quantum theory - with its wave/particle dualism - has blurred the classical boundaries between light and matter. In this framework, light is described in the form of both waves and particles - so-called photons.

Generally, with no extra help, oxygen reacts with organic substances very slowly - combustion is rarely spontaneous. It can be activated, however, in various ways, that is, brought into a form where such reactions progress rapidly. The substances become oxidized. The extraordinary biological role of oxygen, which is closely connected with this question of activation, can be understood only with the help of quantum physics.

The list of activated forms includes
- ozone
- superoxide radicals
- hydroxyl and peroxide radicals and
- so-called singlet oxygen.

For example, ozone ($O_3$) is generated in the atmosphere following the formation of atomic oxygen (O) by ultraviolet light together with a subsequent reaction with molecular oxygen ($O_2$). As photosensitizers do not generate atomic oxygen, ozone does not play a role in PDT, and the term ROS will be used here to denote only the other forms.

During PDT, activated photosensitizers can mediate either an energy or electron transfer to oxygen. As a mediator, it itself is not used up. Under the same physical conditions, different photosensitizers can vary widely with regard to both the absolute amount of activated oxygen they can generate and also the relative amounts of the different species involved. The targets of the therapy are cell components that exist in a range of very different environments, including the cell nucleus, the membranes with their apolar fatty acid interior and polar exterior layers, and the aqueous cytosol. The different physical qualities of these ROS, such as their oxidation potential, solubility and polarity, may therefore well open up the possibility of differing biological effects and therapeutic efficacies from different photosensitizers under otherwise identical clinical conditions. The photosensitizers themselves also have different pharmacological and pharmacokinetic properties. This all means that the best choice of photosensitizer depends on the case at hand, and clinical trials should take account of such fundamental differences between the various photosensitizers.

Because of their primary importance, photophysical mechanisms in PDT will be dealt with first in order to better understand the clinical aspects.

A sketch of the ideas underlying the classical physical formalism will be given to help make the modern and less intuitive quantum description more understandable. Just as the modern radiotherapist is expected to know about the interaction of ionizing radiation with matter, including knowledge of Compton scattering and pair production, and the modern radiologist about the magnetic properties of nuclei, pulse sequences and gradient fields, so the modern laser therapist should know more than the most basic aspects of the tools with which he works – first and foremost, oxygen, but also the interaction of light with matter. Thus, a good deal of attention will have to be given to the question of matter, not just light. The level of detail given in the section on the physical background should be viewed in this regard.

The perhaps very personal view lying behind the presentation given here is very much influenced by the fact that a great deal of effort is being put into both qualitative and quantitative theoretical work in the fields of photobiology, photochemistry, photophysics and consequently PDT as well; but to understand PDT is not simply to calculate light doses and distributions at various depths in natural or highly unnatural media. PDT is also not just a sophisiticated light therapy. We hope that the presentation given here, though more conceptual than some, is broad yet involved enough to be of help to those interested in looking into this very special and highly involved field in even more detail. If the reader needs a better grasp of the technical and physical aspects of PDT, including the generation, absorption, reflection and transport of light, there is a growing choice of interesting and well-written works that are available.

## The Physical Background

The emission and absorption of light energy corresponds to changes in the energy states of the electrons of the affected molecules. The theoretical characterization of electron states requires quantum theory, which gives us a basis for the description of the reactions involving light (photons), photosensitizer and oxygen.

For a better grasp of the abstract mathematical solutions, geometrical models that limit themselves strictly to either wave or particle qualities are used. Although any model by itself cannot do justice to the dualism of modern physics, together they help to successively clarify the underlying physics. This dualism has more or less unconsciously already become established in medicine. The higher the quantum energy of an individual photon, the more likely it is that one speaks in terms of energy. On the other hand, the lower the energy, the more one talks about the wavelength. For example, in radiotherapy one talks in terms of MV or MeV. In PDT, with visible light, wavelength is what is considered biologically important.

## The Physics of Light

Classical physics succeeded in describing Nature in the form of either waves or particles. These represent spatial extremes, waves being extended but particles strictly localized at a point. These counterpoles have the advantage of being graphic. Matter was represented in the form of particles, light as a form of energy with wave qualities.

The light intensity at a point corresponds to the incident energy per second per unit area, corresponding to a power density. The penetration depth of light in a material is defined as the average depth at which the light intensity, through both absorption and scattering, declines to about 1/3 (technically 1/e, where e is the base of the natural logarithms) of its incident value.

### Light as a Wave

Classical physics characterizes light on the basis of the three parameters of wave like motion

- frequency
- speed and
- wavelength,

symbolized conventionally through $\nu$, c and $\lambda$, respectively, whereby

$$\nu = c/\lambda \tag{1}$$

Because c is a constant factor, wavelaength and frequency are equivalent as parameters, so either one determines the other one unambignously. In other words, $\nu$ is inversely proportional to $\lambda$: the higher the frequency, the shorter the wavelength.

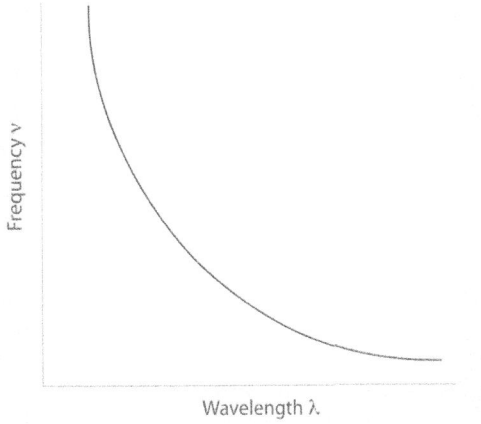

**Fig. 1.** The inverse relationship between frequency and wavelength for light

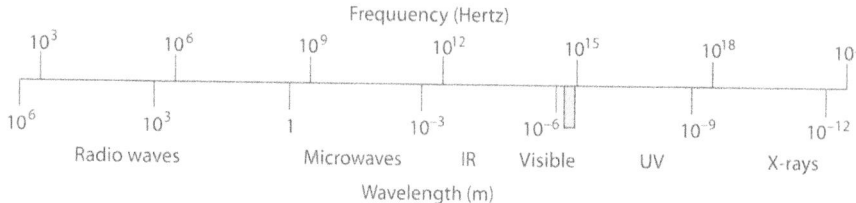

**Fig. 2.** Electromagnetic spectrum (frequency as function of wavelength)

## Rutherford's Model of the Atom

Rutherford and his employees developed a model in which the negatively charged electrons circle in orbits around a small, positive nucleus. Following the analogy of a mass in the Earth's gravity field, the height of a path, that is its distance from the nucleus, corresponded to the "energy" of the electron. Higher paths corresponded to states of higher energy. The absorption of energy could be represented as the transition of an electron from a lower into a higher level. The emission of optical energy, i.e. the loss of energy from the atom, corresponded to a drop to a lower level. If a photon has enough energy, its absorption can cause an electron to be completely ejected from the atom. Radiation with this sort of energy is called ionizing radiation, as an atom which has gained or lost electrons, and which is thus no longer electrically neutral, is called an ion.

Spectroscopic experiments showed that the emission of light from any specific element does not occur with wavelengths spread continuously in bands. Instead, one observes line spectra, each characteristic for that particular element. This phenomenon could not be explained by Rutherford's model. The same applies for the absorption spectra. Another and independent problem consisted in the stability of atoms: according to the rules of classical physics, the electrons should drop out of their orbits and collide with the nucleus, to which they are electrostatically strongly attracted. Quantum physics boldly declared the problems to be conceptual ones – based on notions that were simply wrong. It solved these problems by firstly accepting the facts, then developing essentially new concepts appropriate to these.

## Light as Particles

Max Planck was the first in the modern history of physics to describe the emission of light from matter as the ejection of energy packets that he christened quanta. He regarded quantization as a basic property of the emitting body, that is, he saw it as an attribute of matter. His idea that light energy is exchanged only in multiples of the light frequency $\nu$ – that the energy E involved had to equal a constant of Nature which he

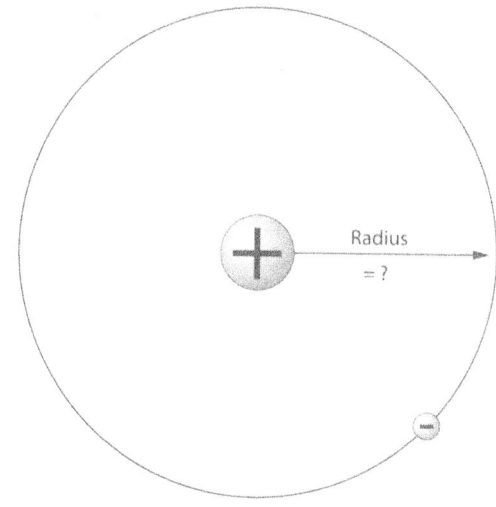

**Fig. 3.** Rutherford's hydrogen atom

denoted by h, times $\nu$ – was a new concept that fitted the facts. Thus he arrived at:

$$E = h\nu \qquad\qquad (2)$$
$$\text{i.e. } E = hc/\lambda \qquad \text{[using Equation (1)]}$$

This constant h is not simply a number, but a physical quantity, a unit of angular momentum. Angular momentum will be explained in more detail in a section on Bohr's atomic theory.

Einstein took Planck's ideas considerably further with his explanation of the photoelectric effect, in which he described light itself as existing in the form of quanta, called photons. He hereby complemented rather than overturned the classical description of light. Light did not simply lose its previous characterization, as one can see through the fact that the identification of light using wavelength never lost its validity. Instead, light gained particle qualities, such as momentum. The photon momentum p is related to its energy E and speed c by the equation

$$E = pc$$
$$\text{i.e. } p = h/\lambda \ .$$

**Fig. 4.** Electromagnetic spectrum (energy as a function of the wavelength)

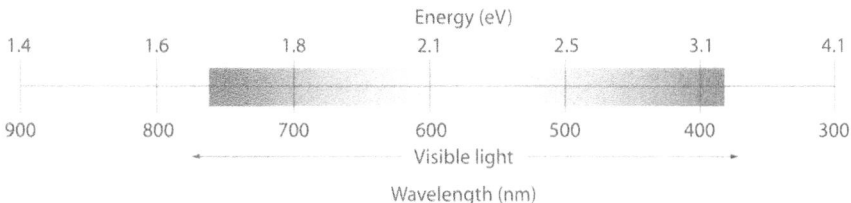

The concepts involved can be illustrated by the resonances produced by a piano string, a picture that will be used again later in connection with modern developments concerning the synthesis of photosensitizers. Each string has a lowest frequency, called the fundamental, and integer multiples thereof, which are the harmonics. It is as if Planck focused on the material resonating string, corresponding to an emitting atom, and Einstein on the resonance itself, that is, the energy. The conceptual boundaries between matter and energy were disappearing.

During the early years of quantum physics, Bohr developed a model, explained in more detail below, that was free from the problems of the classical one given above. He took the quantization of angular momentum in Planck's units of h as his starting point and deduced the quantization of energy. As h represented the smallest angular momentum in Bohr's model, he deduced that a lowest energy level had to exist. He thus provided a conceptual framework – an explanation – for the stability of atoms. The line nature of the emission and absorption spectra followed. Indeed, one of the most important tasks during this phase was the deciphering of electronic states from the variety of the spectroscopically observed transitions; but a unified description of electron states in atoms and molecules and the effects of light was obtained only after quantum mechanics had evolved further, as represented for example, by the Schroedinger equation. Its solutions are so-called wave functions, and provide the basis for the characterization of matter. Their particle qualities are to be found in their energies and momenta, which are quantized – as in the piano string frequencies in the analogy above.

According to (2), photons with a higher energy have a higher frequency, that is a shorter wavelength than those with a lower energy. For example, photons from the infrared region have less energy than those of UV light and X-rays.

## The Physics of Matter

Following this sketch of the physics of light, the physics of matter will be presented in more detail, starting with atoms. This should lead to an understanding of the nature of the special patterns that arise when electrons are grouped together in one system. The aim is to achieve a grasp of the molecular nature of matter, so that the reader can both appreciate the special nature of oxygen and have more insight into how photosensitizers are distinguished from simple dyes. This will greatly help to better comprehend what lies behind the development of new photosensitizers.

### *Angular Momentum and Bohr's Atomic Model*

In classical physics, a particle in orbit around a point has an orbital angular momentum that is related to its speed and distance from that point. An automobile passenger clearly senses centrifugal forces when being driven in a curve. These correspond to flying out of the curve. What the passenger has is an angular momentum in the curve, related to the inertia tending to keep the passenger going straight on rather than stay in the curve. Directly, he or she will notice the centripetal forces, acting against the tendency to fly out of the car. These will be provided by friction with the seat or, ultimately, the door of the car. The faster the automobile, or the tighter the curve, the larger the forces involved will be. Whilst these forces are not the same as the passenger angular momentum, they stand in direct relationship to it. The angular momentum has a magnitude, corresponding to both the passenger speed and size of the bend.

Bohr demanded that the orbital angular momentum of an electron equal an integer n times Plancks h divided by $2\pi$:

$$L = nh/2\pi \quad \text{with } n = 1, 2, 3 \dots . \tag{3}$$

The corresponding numbers L are called the angular momentum quantum numbers. By considering the forces of attraction between the electrons and the nucleus, Bohr was able to derive the quantization of the path radii and energy states: the electron orbits are stable, so the tendency of the electrons to fly off due to their motion is exactly balanced by the electrostatic attraction. The quantization of angular momentum had to be reflected in the quantization of these forces and the energy levels. Thus, Bohr's atomic model was based largely on classical concepts.

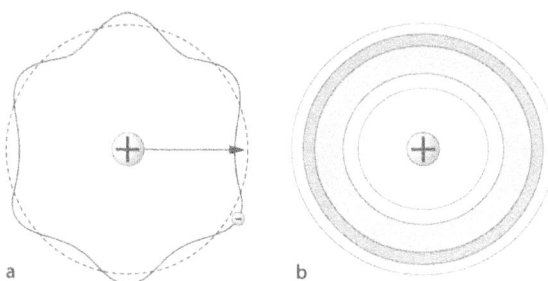

**Fig. 5.** Modified Bohr atom with electron wave. Quantum mechanical hydrogen atom with electron wave and with electron probability density

### Matter as Waves

De Broglie stood the new dualistic ideas concerning light on their head. He asked what the consequences would be if particles, such as electrons, also had wave qualities. He postulated that, just as $p = h/\lambda$ expresses the momentum of a photon in terms of its wavelength, $\lambda = h/p$ (de Broglie relationship) should define the "wavelength" of a particle in terms of its momentum.

One way of illustrating such electrons, in a modification of Bohr's model, represents them as waves in a ring, like vibrating circular strings. The later quantum mechanics, which is statistically oriented, replaced a path by the probability density for an electron, a surrogate for its concentration, which led to the picture of "electron clouds". This concept will be met again later, where it will help depict the role of photosensitizer molecule rings.

In the semiclassical model, an electron path lies in a plane, so that an orientation for the orbital angular momentum can be assigned – conventionally through the axis of rotation, like the axis of a spinning top.

However, states exist whose quantum mechanical probability densities are spherically symmetric, that is, with a complete rotational symmetry around the nucleus. These are not associated with any orbital angular momentum, alien to the classical picture. That is, motion can be associated with the electrons in such states, but no preferred direction and no plane at all. These very important states are the s-orbitals, and will be examined in more detail later.

As mentioned, the angular momentum has a certain magnitude, and, with the exception of s-states, an orientation. Such a quantity is designated as a vector and graphically represented as an arrow, the length of which corresponds to its size.

A classical orbit lies in a plane, and a convenient way to represent the orientation of a plane is as an arrow perpendicular to that plane. This is the background to the conventional way of representing an angular momentum.

### Energy Quantization of Electron States

Every spectral line corresponds to the exchange of light with a definite wavelength, and therefore to the exchange of a photon with unambiguous energy. One speaks of a discrete as opposed to a continuous emission or absorption spectrum. Before the development of quantum mechanics, an empirically based mathematical description of the spectral lines of the hydrogen atom had already been developed that organized these in different series, named respectively after their first describer (Fig. 7). It was discovered that the corresponding frequencies – and therefore energies – could be represented in the form $(1/n^2 - 1/m^2)$.

Starting with the quantization of the electron angular momentum, Bohr now devived the quantization of the corresponding force and thus energy. He could thus derive a formula for the energy states of the hydrogen atom:

$$E_n = -(13.6 / n^2) \text{ eV},$$

where n is like the so-called main quantum number.

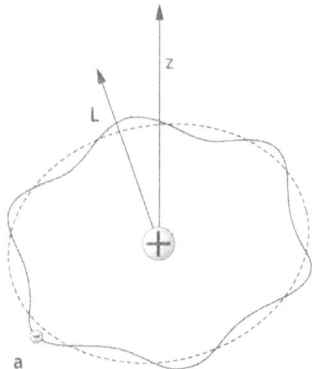

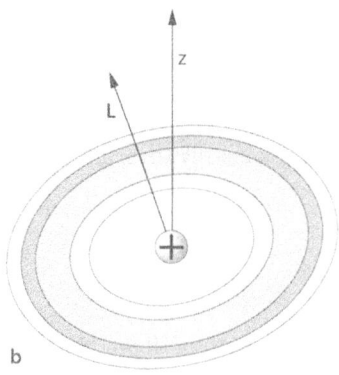

**Fig. 6** Geometrical representations of the (orbital) angular momentum L

## Electron Orbits

The quantization of the frequency and wavelength of the photon that corresponds to a transition is now a simple consequence of the unambiguous energy difference between two definite states of an electron. The line spectrum results from this quantization. Interpreted geometrically, the paths which represent the electronic states lie at fixed and unique distances to the nucleus. They can thus be grouped into shells or orbits, which are denoted by the main quantum number. No electron can exist in the no-man's land between these spatially separate shells. For the hydrogen atom, the energy difference between shells with main quantum numbers p and n is, accordingly

$$\Delta E = 13.6 \left( 1/n^2 - 1/p^2 \right) eV$$

in very good agreement with the empirical spectroscopic results. The deepest lying orbit is given a main quantum number of 1. The zero of energy is given by the formula for $E_n$ above. Setting n equal to infinity, corresponding to an ionized atom, E equals 0. Stability is associated with lower energy, and so bound states conventionally have a negative energy.

Bohr's formula agrees with the spectroscopically determined values very well. The Lyman series, for example, represents transitions ending on the lowest energy state, the ground state.

The importance of the ground state lies in its quality as the most stable state. Substances which are not somehow activated (e.g. by heat or light) should be assumed to be in this state.

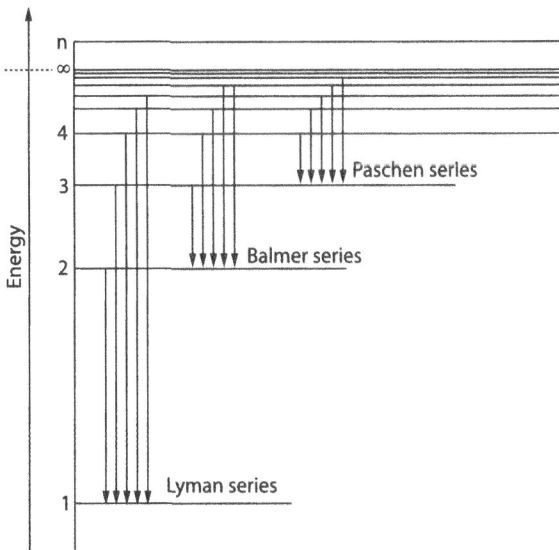

**Fig. 7.** Energy level scheme of the hydrogen atom ($\Delta E$ = constant × $1/n^2$)

## Electron Orbitals

For any given energy, several probability density distributions and directions and angular momentum orientations can exist. Expressed classically, for every energy level there are several possible orbits with different curvatures and at different angles. These spatially completely characterized paths are the electron subshells or orbitals. The electron orbitals are classified by main quantum numbers, corresponding to their respective orbit, with other quantum numbers corresponding to the other physical qualities such as curvature and orientation.

## Atomic Absorption and Emission Spectra

Hydrogen serves as the prototype for calculation of atomic spectra, because it consists of one proton and one electron, whose charges have a known a magnitude, conventionally denoted by e. Together with the known masses of the electron and proton, and Planck's constant, all the parameters necessary for the spectral calculation are given. It is a different matter for the larger multielectron atoms, as the high degree of complication inherent in the mutual interactions of the electron charges has resulted in the requirement of highly complicated models to calculate the spectra; but the underlying ideas are the same, and one can often achieve some insight by singling out the electron with the highest energy, and bundling the rest of the electrons together with the atomic nucleus, treating the rest of the atom as a "hydrogen-like" atom.

## Spin and Spin Orbitals of Electrons

Bohr's simple model is principally limited to a discussion of energy levels. It is incapable of explaining the pattern of occupation of the ground state electron orbitals of the elements, ultimately reflected in the Periodic System of elements of Mendeleev and Meyer. For this, criteria other than the energy must be considered.

An experiment of Stern and Gerlach in 1922 showed that the line spectrum of an element with a single electron, for example sodium or hydrogen, is split in two under the influence of a magnetic field. This leads to the conclusion that

- the energy state of an electron depends on the magnetic field, and
- there are two possible states per electron.

In a magnetic field, an electron thus behaves like a simple magnet with a south and north pole. One speaks of a magnetic dipole. This is represented for the sake of simplicity as an arrow.

According to quantum mechanical rules, these dipoles must be oriented strictly parallel or antiparal-

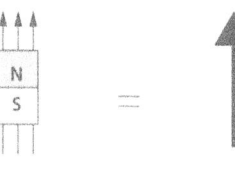

**Fig. 8.** Arrow representation of a dipole

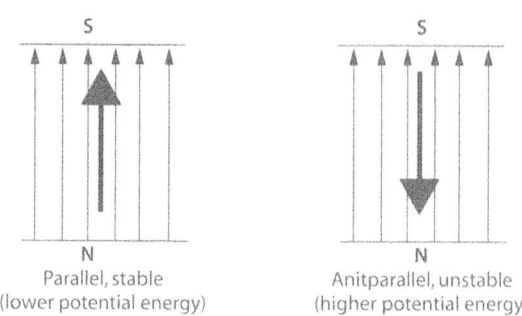

Parallel, stable
(lower potential energy)

Anitparallel, unstable
(higher potential energy)

**Fig. 9.** The two possible energy states of a magnetic dipole (doublet)

lel to an external magnetic field. There are therefore two possible energy states: those with a parallel orientation have a lower potential energy, i.e. are more stable, than those with an antiparallel one.

The consequence is a potential separation between the energy states of parallel (positive spin components) and antiparallel (negative spin components) spin states, with a degree of separation depending linearly on the strength of the applied magnetic field (Fig. 10). As states with parallel spin lie energetically lower, they are more stable. The parallel states are occupied preferentially.

A magnetic field can be generated by charges in motion. The classical case of an electromagnet coil illustrates this phenomenon: the electron current circles many times around the coil's axis. The circles lie parallel to each other, so that the microscopic field becomes amplified and macroscopic.

In analogy to this, a representation of an electron's own magnetic field as originating from some type of movement is natural. However, this field is independent both of external influences and of the electron's position, so the idea developed of an electron as moving on the spot – turning like a spinning top – and this developed into the concept of spin. Because of the independence from outside influences, spin is viewed as an intrinsic parameter. This proposal was first published by Uhlenbeck and Goudsmit. However, it should be remembered such a form of movement has no place in quantum mechanics, since the concept of a point is classical. Dirac finally managed to provide a sound theoretical basis for the phenomenon of spin by uniting Einstein's special theory of relativity with quantum mechanics, which underlines how removed this phenomenon is from the classical framework.

The complete expression for the state of an individual atomic or molecular electron is the simple product of its spin state and spatial orbital function, which is an extrinsic parameter. The product state is then designated a spin orbital.

### States with Several Electrons: Electrons as Indistinguishable Particles

In groups of electrons, such as found in an atom or molecule, no distinct particle states can be ascribed to the individual electrons within the framework of quantum mechanics. This is related to the quantum mechanical replacement of the classical point-parti-

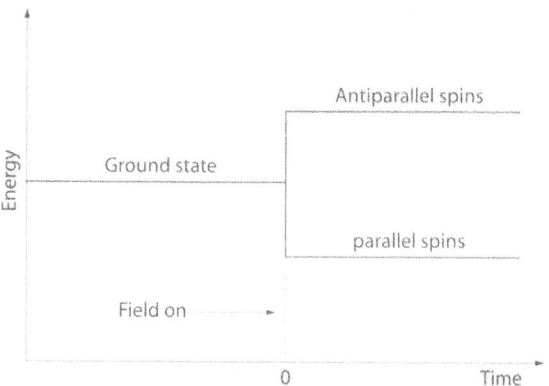

**Fig. 10.** Energy level splitting following application of an external magnetic field (case of a doublet state)

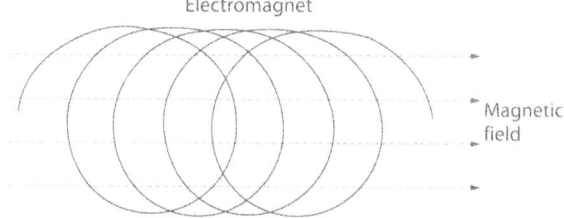

**Fig. 11.** Solenoid: magnetic field generation by electron motion

cle representation of matter by a wave-mechanical one. Electrons are characterized by probability densities, which reflect the chances of finding an electron at a particular point in space. No two electrons in the same atom can really be meaningfully spatially separated. This situation is often depicted by electron clouds. Heisenberg's uncertainty principle can be invoked here: there is no physically valid mathematical expression for the state of a solitary electron from the group, but only one for the group as a whole. A valid expression in terms of products of the individual

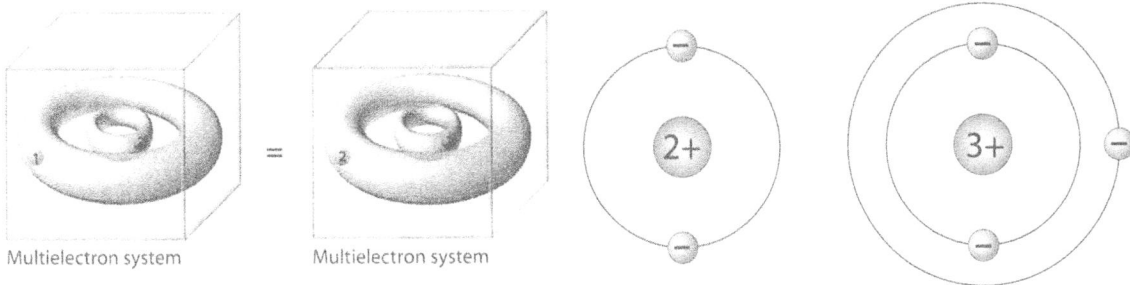

**Fig. 12.** Indistinguishability of electron states in a system

**Fig. 13.** The elements: helium and lithium

spinorbitals can, however, be written. One obtains a well-defined spatial state, defined for the group as a whole, and a collective spin state.

### The Pauli Exclusion Principle and closed Electron Orbitals

The elementary quantum of angular momentum is h, as Planck discovered a century ago. But his observations were based on the emission and absorption of light, i.e. changes in electron states. It may thus be no surprise, that it is not an angular momentum itself that has to equal an integer multiple of h, but changes of angular momentum. The surprise comes in the fact that there are two classifications of elementary particles in nature, which are divided into particles with integer spin quantum number (0, 1, 2 etc.) called bosons, and those with "$\frac{1}{2}$ spin quantum number", $1\frac{1}{2}$, $2\frac{1}{2}$ and so forth, called fermions. Electrons and also protons have a spin of $\frac{1}{2}$h, i.e. they are fermions. Their spin state can change from $-\frac{1}{2}$h to $+\frac{1}{2}$h or vice versa. This may appear bizarre at first, but mathematically it actually relates to the simplest possibility for a change of state to correspond to a difference in angular momentum of h.

The principle of indistinguishability is valid for all elementary particles. An extremely important restriction on the product states for fermions was first formulated by Pauli. His exclusion principle says that two electrons cannot exist in the same state. The parameters involved are angular momentum and energy. Two electrons in the same electron orbital have identical spatial functions and energies by definition. They must therefore differ in their spins. Since there are only two possibilities for the dipole spin orientation, this means that

- there can be at most two electrons in the same orbital, and
- two electrons in the same electron orbital must have antiparallel spins.

### The Elements and the Aufbau Principle

An orbital with two electrons is closed. This pairing is one principle behind the Periodic System of the elements. By definition, all the elements of nature are atoms.

All atoms can be pictured as arising from the prototype atom, hydrogen. This is the simplest, with a proton as its nucleus and one electron. The next heaviest element is helium, which has two protons in its nucleus, and has two electrons. Both these electrons fit into the same orbital. A third electron has no place in this orbital, it has to go into a different one, generally an energetically higher one. We can imagine going from helium, with two protons in its nucleus and two electrons, to the next element in the Periodic System, lithium, with three protons and three electrons. The idea is repeated to arrive at all the other elements one by one.

The pairing of electrons also plays a key role in the combination of atoms to form molecules. It is fundamental to the stability of matter. Helium is the prototype: it is a very stable element, as are all so-called rare gases, which include xenon, neon and argon. These simply cannot react with other elements, as they have no single electrons that can easily be paired with electrons from other elements.

### Light Energy and its Relationship with Chemical Reactivity

As the electron orbitals "fill up", the additional electrons lie on average at a greater distance to the nucleus than the previous ones. They are shielded from the attraction of the positively charged nucleus by the lower electrons. They are thus less firmly bound to the atom, and can be raised through the absorption of energy into a higher and as yet still vacant electron orbit. Changes in the configurations of the lower-level electrons demand energies much higher than those the heat or light can supply, and correspond to X-ray energies.

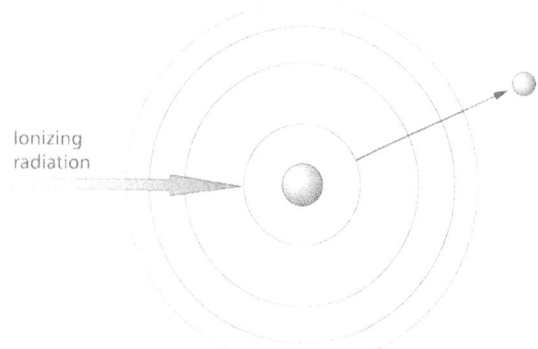

**Fig. 14.** Ionizing radiation ejects tightly bound low-lying electrons from atoms

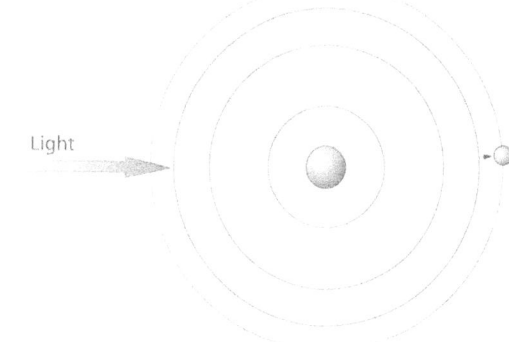

**Fig. 15.** Light absorption causes transitions of weakly bound higher-level electrons

In contrast, changes regarding the higher-lying electrons involve lower energies, those in the range of visible light.

Because the electrons involved are the outermost ones, they also determine the interactions with other atoms, that is, the reactivity of an atom. The energies involved in chemical reactions are thus in the optical range: photophysics is intertwined with chemistry.

The higher energy states have a greater number of possible electron orbitals than the lower ones: the further the orbitals are from the nucleus, the more space there is for them. According to quantum mechanics, the $n^{th}$ shell has $n^2$ electron orbitals, and can therefore take up $2n^2$ electrons.

### Interatomic Bonds and Atomic Orbitals

The step from atoms to molecules will involve interatomic bonds, and these can be understood on the basis of atomic orbitals. There is a large and complicated hierarchy of possible atomic orbitals, but we will need

an acquaintance with only the two simplest types for the purpose of understanding the molecules we will encounter.

The simplest atomic orbitals are the spherically symmetric s orbitals. These form the basis of the simplest interatomic bonds, as explained in more detail below.

The next higher ones are the *p* orbitals. These are not spherically symmetric, but only axially symmetric, and represent the basic component of the molecular structures that are involved in the absorption of light. There are $p_x$, $p_y$ and $p_z$ orbitals, named according to the respective axis of symmetry. The first energy shell has one electron orbital (the 1s orbital), and can accommodate two electrons. The second shell has four electron orbitals (the 2s, $2p_x$, $2 p_y$ and $2p_z$ orbitals), and can accommodate eight electrons ( $2n^2 = 2 \times 2^2 = 8$). Additional orbitals, with increasing geometric complexity, exist for higher shells, but as far as molecular bonding is concerned, they are of no concern to us here.

### Groups of Electrons

Both the indistinguishability of electrons and the Pauli principle must be considered in order to gain a physically valid mathematical expression for a group of electrons. Mathematically, an expression is needed for the impossibility of a constellation with electrons in the same total state. In the language of statistically oriented quantum mechanics, this is expressed as a probability of zero. The consequence is most easily seen by considering the case of just two electrons, and the concept of "electron pairs" is ubiquitous in physics.

An electron orbital can contain up to two electrons without violating the Pauli principle. Two electrons in the same orbital must have the same spatial function, whose product is therefore symmetrical with regard to a spatial exchange. The individual electron functions must differ in their spins. This isn't true however for two electrons in different orbitals and, as shown in more detail later, exactly this is the case for the two electrons most fundamental to the chemistry of oxygen.

The Pauli principle finds an expression within the framework of the formalism of symmetry and antisymmetry. To appreciate these terms, one must have some idea of the physical meaning of the wavefunction.

### Symmetry and Antisymmetry in Quantum Mechanics

The physical interpretation of the wavefunction involves a function known as the probability density of the particle state. By way of illustration, the familiar

**Fig. 16.** s- and $p_x$-atomic electron orbitals

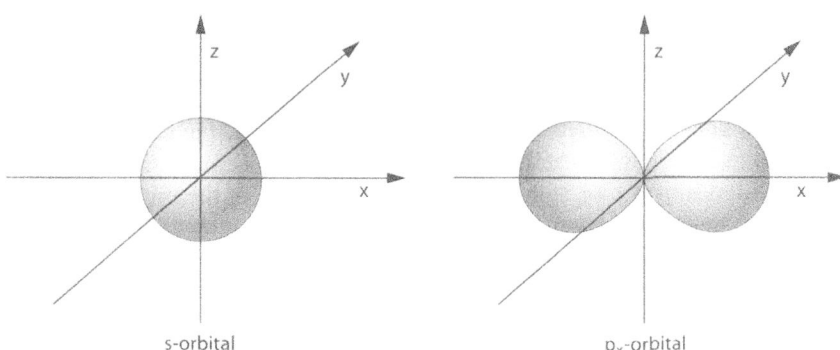

s-orbital                    $p_x$-orbital

**Table 1.** Scheme of orbitals for the "first" ten elements (electron = $\downarrow$ or $\uparrow$)

| Electron orbital | H Hydrogen | He Helium | Li Lithium | Be Beryllium | B Boron | C Carbon | N Nirogon | O Oxygen | F Fluorine | Ne Neon |
|---|---|---|---|---|---|---|---|---|---|---|
| 1s | $\downarrow$ | $\downarrow\uparrow$ | $\downarrow\uparrow$ | $\downarrow\uparrow$ | $\downarrow\uparrow$ | $\downarrow\uparrow$ | $\downarrow\uparrow$ | $\downarrow\uparrow$ | $\downarrow\uparrow$ | $\downarrow\uparrow$ |
| 2s | | | $\downarrow$ | $\downarrow\uparrow$ | $\downarrow\uparrow$ | $\downarrow\uparrow$ | $\downarrow\uparrow$ | $\downarrow\uparrow$ | $\downarrow\uparrow$ | $\downarrow\uparrow$ |
| $2p_x$ | | | | | $\downarrow$ | $\downarrow$ | $\uparrow$ | $\downarrow\uparrow$ | $\downarrow\uparrow$ | $\downarrow\uparrow$ |
| $2p_y$ | | | | | | $\downarrow$ | $\downarrow$ | $\downarrow$ | $\downarrow\uparrow$ | $\downarrow\uparrow$ |
| $2p_z$ | | | | | | | $\downarrow$ | $\downarrow$ | $\downarrow$ | $\downarrow\uparrow$ |

"area under the curve" (AUC) for a pharmaceutical is a measure of the quantity that appears in the blood. It corresponds to the integration of the concentration of the substance over the time – simply all the pharmaceutical that was in the blood.

In quantum mechanics, the probability density is analogous to a concentration. A central question regarding the interpretation of the wavefunction is clarified at this point: what is the meaning of a negative wavefunction? There is no such thing as a negative concentration, and similarly, $\Psi$ cannot be equated to the probability density. This problem is illustrated by the simple example of a sine-wave oscillation:

In quantum mechanics, one solves this problem by taking the square $\Psi^2$ of the of the wavefunction to represent the probability density of the state. As a consequence, both $\Psi$ and $-\Psi$ correspond to the same physical state. The probability that an electron is in a specific element in space is calculated by integrating $(\Psi)^2$ over this element.

Although one cannot physically distinguish the individual electrons, one must start with a mathematical expression for the single electrons in order to formulate a physically valid representation of the group afterwards. The problem of indistinguishability is then tackled by demanding that swapping the "individual electrons", i.e. the mathematical parameters, of an electron pair must lead to the same state. The fact

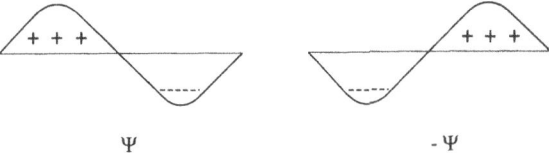

**Fig. 17.** Probability density of a state (i)

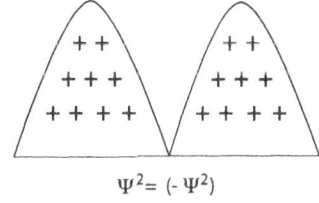

**Fig. 18.** Probability density of a state (ii)

$$\Psi^2 = (-\Psi^2)$$

that $\Psi$ and $-\Psi$ correspond to the same physical state must also be taken into account.

Symmetry and antisymmetry for the state corresponding to the pair describe the behaviour of the function (when one exchanges the parameters (e.g. positions) of the individual electrons in the pair. If $\Psi$ is unaltered, the state is symmetrical. If $\Psi$ is changed to $-\Psi$ the state is antisymmetric. As explained above, the total wavefunction for a system of electrons is antisymmetric with respect to an exchange of two electrons.

As explained above, two electrons in the same spatial system must have same spatial (orbital) wavefunction. They would then be distinguished through their spins. The physical subdivision in extrinsic (= spatial) and intrinsic (= spin) qualities for the electrons into independent systems is the justification for the analysis of the function for an electron pair $\Psi(1,2)$ can in terms of a product of an orbital component $\Phi(1,2)$ and a spin component $\Sigma(1,2)$, so that $\Psi(1,2) = \Phi(1,2)$ times $\Sigma(1,2)$, written $\Phi(1,2)\Sigma(1,2)$.

A completely antisymmetric function can be written as the product of an antisymmetric spin function $\Sigma$ with a symmetrical spatial function $\Phi$ or vice versa. Symbolically, $\Phi \times (-\Sigma) = (-\Phi) \times \Sigma$, and both = $-(\Phi \times \Sigma)$.

The separation of the spinorbital functions into orbital and spin states now allows one to take the analysis further.

### Spin as a Binary Quantity: Dipoles as Digits

Borrowing from concepts common to information technology, one can characterize spin as a bit (binary digit) of information. It is not only free of the geometrical associations of classical physics, but reflects the truly binary nature of the electron spin. Instead of the usual depiction in terms of spin components of $+\frac{1}{2}h$ and/or. $-\frac{1}{2}h$, the spin states can be abstractly represented by 0 and 1.

There are then four possible pair states: 00, 01, 10 and 11.

### Triplet and Singlet States as Products of Spin States

The condition of indistinguishability can now be understood as meaning that one cannot distinguish between elecrons in pair states by means of the order of the individual bits. The symmetrical pairs 00 and 11 fulfill this condition automatically. However, with the pairs 01 and 10 one must form combinations:

- (01 – 10),
- (10 – 01) and
- (01 + 10), which is identical to (10 + 01).

These are then combined into symmetrical and antisymmetric groups. One must take into account the fact that the wave functions $\Psi$ and $-\Psi$, here in particular (01 – 10) and (10 – 01), correspond to the same physics. We therefore really have only two combinations.

The result is a triplet of the symmetrical spin states $\Sigma_s$

- 00
- 11

- (01 + 10)

and the antisymmetric singlet $\Sigma_A$

- (01 – 10), which is the same as (10 – 01).

In the classical representation one describes the two spins of the triplet as parallel (to each other) because of the symmetry, since the exchange of the individual states does not alter the product state. The total spin is said to equal h, and one envisages the spins of $\frac{1}{2}h$ each as simply adding up. In the case of the singlet, the individual spins are described as antiparallel because of the antisymmetry, with a total spin of 0.

### The Total-Product States

The three symmetrical spin states can now be combined with an antisymmetric spatial state and the antisymmetric spin function can be combined with a symmetric spatial state. The complete state function for the electron pair now satisfies the Pauli exclusion principle. A symmetrical spin state $\Sigma_S$, has to be combined with an antisymmetric spatial state $\Phi_A$ or vice versa, in order to guarantee the antisymmetry of the entire wave function – the requirement of the Pauli exclusion principle. We thus arrive at the possibilities $\Sigma_A \times \Phi_S$ and $\Sigma_S \times \Phi_A$ for the antisymmetric electronic total-product states.

### The Spectroscopic Splitting of Product States in a Magnetic Field

The classification of states as singlet, doublet or triplet states originated in the field of optical spectroscopy, where their existence was first discovered.

The spatial structures of the electron orbitals affect principally the electrostatic energy of the product state, i.e. the energy relating to the attraction and repulsion due to the charges of the electrons and nuclei. In contrast, as already seen for a single electron, spin is an essentially magnetic parameter, so the state's spin affects the magnetic qualities of the state.

The orientation of the spin components of the product state behaves differently to that of a single electron. This can perhaps best be grasped by remembering the quantum principle on which Bohr based his model of the atom: angular momentum and its components change in units of h. A state spin of magnitude 1h parallel to a magnetic field is said to have a component of 1h parallel to the field. If it were antiparallel, the component would equal –1h. If these were the only two "orientations", as is the case with the spin of a single electron, the quantum principle would forbid the state from changing from an orientation with –1h to one with 1h. A state with component 0h does exist, so the changes can occur.

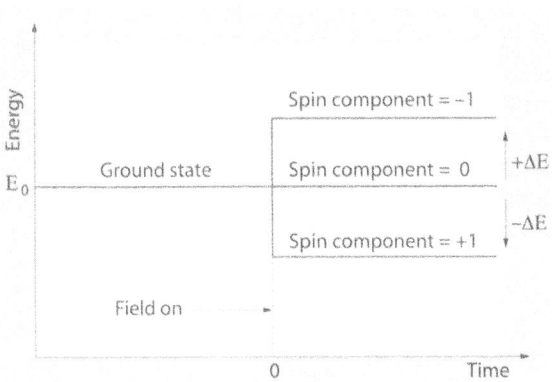

**Fig. 19.** Energy splitting of a triplet state in an external magnetic field

Using the classical terminology, the spin component corresponding to the state 11 is +h, described as parallel (to the magnetic field), graphically signified as ↑↑.

That for 00 is –h (antiparallel), signified by ↓↓.

The spin components of the individual electrons, $+\frac{1}{2}$h for the 1 individual state and $-\frac{1}{2}$h for the 0 individual state, are simply added. This can be done for the triplet's (01+10) state, which is thus seen to have a resulting component of zero. This state is described as being perpendicular to and thus independent of, the magnetic field, and – because of the indistinguishability of the particles – would have to be written as a combination of the symbolic pairs ↓↑ and ↑↓. This is the reason for the use of the algebraic system above, as the nature of an arithmetic combination is algebraic.

The singlet (01+10) state has only one possible component, which equals 0.

The consequences for the energy levels of singlet and triplet states are illustrated (Fig. 19) and should be compared to the case for the electron spin (doublet) given above.

Just as in the elementary case of the hydrogen atom, the energy levels of the compound states can be spectroscopically determined through the emission or absorption of light. With no applied magnetic field, a triplet state has a spectrum with a single line, corresponding to the energy $E_0$. The application of the field splits this level into three, giving levels at $E_0 - \Delta E$, $E_0$ and $E_0 + \Delta E$. The split $\Delta E$ is proportional to the strength of magnetic field.

This is analogous to a compass needle seeking the Earth's north pole. The parallel orientation is the most stable, i.e. the lowest energy state, the north pole of the needle pointing north. This corresponds to the state with spin component +1 and energy $E_0 - \Delta E$ in the diagram above. A singlet state always has only a one-line spectrum.

## Molecules, Molecular Orbitals, Chemical Bonds and Radicals

### Molecular Basics

The structural basis of a molecule can be regarded as a skeleton made up of all the nuclei from the atoms involved. The nuclei are positively charged, so they repel each other. The stability of the molecule is to be found in the bridging influence of the negatively charged electrons that are found between the nuclei. The interatomic molecular bonds thus depend on electrons. According to quantum mechanical calculations, e.g. that of Heitler and London, these so-called covalent bonds arise through the pairing of electrons, with each atom involved contributing one electron to this pair. The term covalence expresses the common contribution of a valence electron from the outermost shell of each atom.

Here too, the spins of these electrons play a major part in the structure of the bond: the spins of the two electrons in a bonding pair must be antiparallel. In this respect, the bond strongly resembles a closed electron orbit. The formation of a bond results in the release of energy. A double bond corresponds to two simple ones, with two electron pairs involved. Covalent bonds can be broken through the absorption of sufficient energy.

The term radical came from observations that the breakup of a covalently bound molecule could lead to two fragments, each with one electron from the original bonding pair. The fragments were called radicals of the molecule (radix = root). The term radical has since been widened to describe all molecules or molecule parts with an unpaired electron.

A bond is represented by a single line – , or by two dots: representing the two electrons.

Unpaired electrons are responsible for the high reactivity of radicals because of the generally strong tendency of the electrons to pair and thereby release energy.

A radical molecule M is denoted as M· the point representing the unpaired electron.

The electrons move partly between the atoms and partly around them, and this distinction forms a basis for the definition of molecular orbitals. In one widely used model, the LCAO model (linear combination of atomic orbitals), these are mathematically defined by combining the functions originally corresponding to the atomic orbitals.

The simplest case of a molecular orbital is the σ-type orbital (= σ-bond), named in analogy to the simplest atomic orbital, the s orbital. The σ-orbitals are defined as those in which the electrons lie between the atomic nuclei, i.e. in the plane of the molecule if one exists. It is these electrons that hold the nu-

$$A\text{-}B = (A : B) \rightarrow A. + B. \qquad A - B \rightarrow A^+ + B^-$$

Homolytic separation          Heterolytic separation
Formation of readicals        Formation of ions

**Fig. 20.** Homolytic and heterolytic separations of molecular bonds

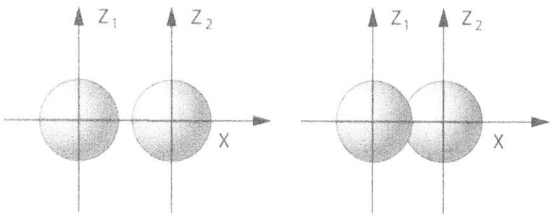

**Fig. 21.** Two atomic s-orbitals combined into one σ-orbital

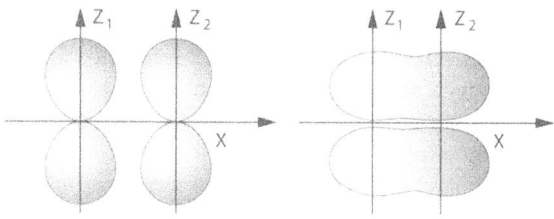

**Fig. 22.** Two $p_z$-orbitals combined into one π-orbital

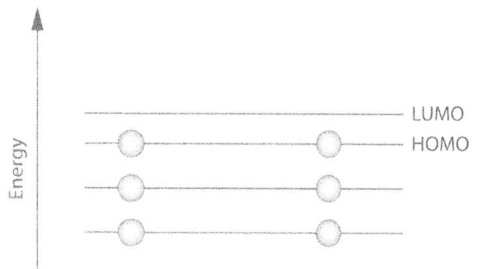

**Fig. 23.** Energy diagram: representation of HOMO and LUMO

clei together and are thus responsible for the basic form of the molecule. Their construction out of atomic s-orbitals is simple to envisage, but they can also be made on the basis of $p_x$-orbitals, where the nuclei concerned lie on the x-axis.

The so-called molecular π-type bonds correspond to the electrons that move relatively freely above or below the plane of the molecule. If one imagines that the molecule lies in a plane parallel to the x-axis, they are definable on the basis of the atomic $p_y$ and $p_z$ orbitals. They occur in the double and triple bonds found between carbon atoms in many organic molecules, and are central to the special properties of the ring structures at the heart of photosensitizers.

Just as for an atom, the molecule's energetically highest electrons are those that determine its chemical reactivity. The most important orbital in this respect, with the most weakly bound electrons, is labelled the HOMO (highest occupied molecular orbital). The absorption of energy causes an electron to jump into the next higher orbital, the LUMO (lowest unoccupied molecular orbital).

The representations presented here correspond to a systemization known as hybridization, which can take on a variety of forms. Basic to all is the mathematical construction of various types of molecular orbitals from unifomly defined atomic orbitals.

If the HOMO one is closed, i.e. occupied by two electrons, one speaks of a singlet state of the molecule because there is only a single spin state with antiparallel spins (Pauli exclusion principle). The molecular orbitals of the ground state of almost every substance are closed, so substances are generally in a singlet state. This is true in particular for photosensitizer and substrate molecules in cells (see also below).

### Molecular Spectra

The theories of molecular spectra go beyond those of multielectron atoms, as one now has to take into account the movement of the nuclei, both absolute and relative to each other. This means that rotational and vibrational effects on the energy spectra have to be considered. Because the nuclei are very much heavier than the electrons, one can separate the motions of the nuclei from those of the electrons, leading to an analysis of molecular energies in terms of vibrational, rotational and electronic spectra. This physical principle lies behind the Born-Oppenheimer approximation, as explained in more detail later.

Essentially, the transitions between the various molecular states relevant to PDT are electronic in nature, and the concepts encountered with multielectron atoms provide a part of their theoretical basis. Simplified, the transitions are represented by spectral peaks, analogous to those of atomic spectra, but highly broadened by the molecular motions.

## Conditions Regarding Transitions Between States Relevant to PDT

### Selection Rules and Spin-Forbidden Transitions

The absorption or emission of light by matter involves more than simply the energy. In quantum mechanics, the functions which describe the states of the electrons and photons have spatial symmetry qualities linking the initial and final states. These symmetries determine the possibility of an interaction between electrons and photons, and result in a series of so-called selection rules for the transitions.

A rule which is fundamental to light emission and absorption – and thus to PDT – says that there can be no change in the spin multiplicity of a state.

A transition from the singlet ground state of a molecule to a triplet is spin-forbidden. We can assume that every molecule in its ground state is singlet – except oxygen.

### Spin-Orbit Coupling and its Role

The concept of allowed and forbidden transitions is an idealization. In fact, prohibited transitions do occur – not because the laws of the quantum mechanics break down, but because the strict division in singlet and triplet states is an idealization.

From the viewpoint of a nucleus in the molecule it is the electrons that move, and this is the somewhat Copernican point of view one is most familiar with. However, from the Ptolemaic viewpoint of an electron it is the nuclei that move. These are positively charged, and represent a (positive) current, that, just as with an electromagnet, generates a small magnetic field at the location of the electron. Since the spin itself corresponds to a magnet, there is not only the well-known electrostatic but also a magnetic interaction between the nuclei and electrons. In other words, there is a coupling between an electron's motion and its own spin. This is known as the spin-orbit coupling. As a result, neither a singlet nor a triplet state can be pure, the combination between spatial and spin functions mixes triplet and singlet states.

A transition can therefore occur from the triplet part of the nominally singlet ground state to a higher (nominally) triplet state. Because the degree of mixing is usually small, such transitions are generally relatively rare. The molecules can be conceptually subdivided into the two groups of singlet and triplet states. These groups are designated systems, and a transition from a singlet to a triplet state or vice versa is called intersystem crossing (ISC).

The generation of photosensitizer in the triplet state of central importance to PDT, as will now be explained in the section on oxygen.

## The Multi-Electron Ground State

### The Influence of Symmetry and Electron Charge

The ground state of an atom or a molecule is determined by the spatial distribution of the electrons. The nearer the electrons are on average to a nucleus, the tighter they are bound. On the other hand, because of their negative charge, the electrons suffer a mutual repulsion that also increases with proximity, and the nearer they are to a nucleus, the closer they are to each other.

Charge thus has a direct influence on the spatial orbital functions. Classically seen, the orbitals of the ground state are those in which the electrons are as far away from each other for as long as possible. Quantum mechanics gives a similar result. The consequence is that the electrons of the ground state will tend to be in different orbitals, as far as this is energetically allowed (see Table 1 for an illustration of these principles for the atomic orbitals). In other words, the spatial antisymmetry of the state is energetically favoured. Through arguments based on Pauli's principle, this has an indirect influence on the spin function, which will accordingly be as symmetric as possible. These facts correspond to the empirical Hund's rules, which are applicable to both atoms and molecules.

The orbitals of most molecules, including the HOMO, are closed, i.e. the electrons paired and their spins parallel. The molecules are in the singlet state.

In oxygen, however, the uppermost orbitals are not closed. The $O_2$ molecule has a total of 16 electrons, 6 of which are in the uppermost π-type orbitals. Exceptionally, however, but in agreement with Hund's rules, the two energetically highest electrons are in the same energy state, in different molecular orbitals and with parallel spins. The spin state is therefore symmetric so the spatial orbital function must be antisymmetric. The most important consequence is that the ground state of oxygen is a triplet. As both these two electrons are unpaired, oxygen can be designated a biradical.

At the heart of PDT lies the generation of reactive oxygen species (ROS), generated through the action of photosensitizer molecules on normal molecular oxygen. The toxic effects of ROS on cells are responsible for the effectiveness of the therapy. The hydroxyl and superoxide radicals are two such species.

Radicals arise from electron transfer processes, such as can occur during the activation of the sensitizers. One highly reactive form of oxygen that does not arise from electron transfer but purely from energy transfer can also be generated, namely singlet oxygen. This denotation arose initially because of the behaviour of its characteristic spectral line under the

**Table 2.** Occupation of diatomic molecular electron orbitals (LCAO model)

| Electron orbit | H₂ Hydrogen | L₂ Lithium | Be₂ Beryllium | B₂ Boron | C₂ Carbon | N₂ Nitrogen | O₂ Oxygen | F₂ Fluorine |
|---|---|---|---|---|---|---|---|---|
| $\sigma(1s)$ | ↓↑ | ↓↑ | ↓↑ | ↓↑ | ↓↑ | ↓↑ | ↓↑ | ↓↑ |
| $\sigma^*(1s)$ | | ↓↑ | ↓↑ | ↓↑ | ↓↑ | ↓↑ | ↓↑ | ↓↑ |
| $\sigma(2s)$ | | ↓↑ | ↓↑ | ↓↑ | ↓↑ | ↓↑ | ↓↑ | ↓↑ |
| $\sigma^*(2s)$ | | | ↓↑ | ↓↑ | ↓↑ | ↓↑ | ↓↑ | ↓↑ |
| $\pi(2p_y)$ | | | | ↓ | ↓↑ | ↓↑ | ↓↑ | ↓↑ |
| $\pi(2p_z)$ | | | | ↓ | ↓↑ | ↓↑ | ↓↑ | ↓↑ |
| $\sigma(2p_x)$ | | | | | | ↓↑ | ↓↑ | ↓↑ |
| $\pi^*(2p_y)$ | | | | | | | ↓ | ↓↑ |
| $\pi^*(2p_z)$ | | | | | | | ↓ | ↓↑ |
| $\sigma^*(2p_x)$ | | | | | | | | |

influence of an external magnetic field – the line does not split. According to the two modes of activation of oxygen, one talks of type-I reactions (electron transfer) and type-II (energy transfer) reactions. Radicals are accordingly type-I products, singlet oxygen a type-II product.

The production of radicals originating from cellular molecules, in the form of either final or intermediate products of a reaction involving any form of activated oxygen, is fundamental to the biological effectiveness of PDT.

### Inactivity of Ground State Triplet Oxygen

The title for this section may be somewhat misleading – an answer to the basic question of the "activity" of oxygen cannot be sought simply in the properties of oxygen itself. Its scope must include the properties of the substances that oxygen can interact with.

Many intracellular metabolic pathways are involved in the physiological utilization of oxygen. Some of the most important are found in the respiratory chain, in the tricarboxylic (citric) acid cycle and in (-oxidation. In all of these, a broad variety of enzymes and co-enzymes comes into play, including the oxidases with their corresponding haem groups, NAD(P)+, the flavines and ubiquinone. The range of mechanisms involved accentuates the essential nonreactivity of ground-state oxygen. The reason for this inertness is explicable only with the help of the quantum mechanical selection rules mentioned. The elec-

tron configuration of the oxygen ground state is a triplet. This leads to a suppression of oxygen reactivity with organic molecules, which we can generally assume to be in a singlet state.

In a pair of molecules where one is in a singlet state and one in a triplet state, there is exactly one electron orbital with parallel spins. Chemical reactions involve electron transfer, and a pure transfer does not lead to a spin flip, that is, a reversal of orientation of one of the spins, which would involve a magnetic interaction. The product of a reaction of a ground (singlet) state organic molecule with ground (triplet) state oxygen would accordingly still have one electron orbital with "parallel" spins. It would therefore also be in a triplet state. This corresponds to an energized – an excited – state of the product. This can occur only after the input of additional energy.

Such an activation can occur physically, for example through a direct transformation of a triplet into a singlet with the help of a photosensitizer. However, a biological activation can occur through electron transfer, that is through the reduction of the oxygen molecule to the superoxide anion $O_2^{\bullet-}$, such as occurs with xanthine oxidase. The production of metal-oxygen complexes in the reactions of cellular metal enzymes with molecular-oxygen, for example as with the oxygenases and oxidoreductases in the liver, is of a different nature, and is due to the special configurations of the transition metals' electron shells.

## Oxidation and Reduction

### Redox Reactions

The term oxidation originally meant the reaction of a substance with oxygen, and reduction meant the loss of oxygen. The archetype was the burning of organic material, which leads to the generation of water as a by-product. This is also effectively what occurs in the respiratory chain, where hydrogen reduces oxygen in the mitochondria, leading to the production of water, $H_2O$. From the point of view that these reactions lead to the production of water, a loss of O is equivalent to a gain of H, so reduction was also taken to mean the addition of hydrogen. The modern standpoint, which aims at viewing these mechanisms from the perspective of their chemistry, defines them in terms of electron processes, as the participation of electrons is central to all chemical reactions.

Oxygen atoms attract electrons more strongly than hydrogen or carbon atoms, a quality formulated as the relatively high electronegativity of oxygen. This aspect has led to the more abstract formulation of oxidation and reduction in terms of electron transfers, reduction being the gain of electrons and oxidation the loss of electrons from an atom or molecule. Together, oxygen and reduction are designated redox reactions. For example, electrons are more strongly attracted to oxygen than to hydrogen, so that oxygen has a more positive electronegativity than hydrogen. Effectively,

$$\tfrac{1}{2}\,O_2 + 2\,H \rightarrow H_2O\,.$$

Each of the pairs $O_2^{2-}\,/\,O_2$ and $2H\,/\,2H^+$ is called a redox pair, and the two pairs together are called the redox partners. The redox potential of the redox partners expresses the tendency for the reaction to occur. By taking one redox pair to be the hydrogen pair $2H\,/\,2H^+$, one can draw up a list of standard potentials for all the elements. This expresses the strength of attraction of a particular substance for electrons, a more positive value corresponding to a stronger attraction, and is related to the electronegativity of an element. It depends on the concentrations of the oxidized and reduced forms of the redox pair and the temperature.

Nernst's equation:

$$E = E^0 + \ln\frac{R.T}{z.F}\frac{c_{ox}}{c_{red}}$$

corrects for these dependencies. E expresses the potential of the redox pair relative zo the standard element hydrogen, $E^0$ the potential under standard conditions, $c_{ox}$ and $c_{red}$ the concentrations of the oxidized and reduced form of the substance, z the number of electrons involved in the redox reaction, T the tem-perature (in degrees Kelvin), R the general gas constant and F the Faraday constant.

### Photoinduced Electron Transfer

Radicals are often generated in photochemical processes. The absorption of light leads to the transition of an electron into a higher orbit, i.e. to a change in the electron configuration. The presence of a molecule with a different electronegativity in the proximity means an electron transfer should then occur more easily.

In this way, the excitation of a photosensitizer molecule can lead to an electron transfer. The first step is purely energetic, written:

$$PS + h\nu \rightarrow PS^*\,.$$

When a molecule Q with higher electronegativity is near the excited photosensitizer molecule PS*, it can capture an electron from it. The process is called oxidative quenching, as Q oxidizes PS. A radical ion pair is formed:

$$PS^* + Q \rightarrow PS^{+\bullet} + Q^{-\bullet}\,.$$

In exactly this way, oxygen can be reduced.

Analogous to this is reductive quenching. If Q has a lower electronegativity than PS, it can act as an electron donor:

$$PS^* + Q \rightarrow PS^{-\bullet} + Q^{+\bullet}\,.$$

### Oxygen Radicals

The respiratory chain, which is anchored in the mitochondria and linked with both the tricarboxylic acid cycle and (-oxidation of fatty acids, is a biological system for the reduction of oxygen. In it, the transfer of four electrons to an oxygen molecule takes place, and results in the production of two molecules of water and energy in the form of ATP. Other enzyme systems, such as the oxidases and peroxidases, are able to transfer either one or two electrons to a molecule of oxygen. The ubiquitousness of the oxidoreductases and their co-enzymes leads to the continuous generation and destruction of superoxide and hydrogen peroxide in the cells. In a healthy cell, there is a steady-state balance between the enzymatic generation of these reactive oxygen species and their annihilation through combined protection mechanisms, including coupled systems of enzymatic and simple molecular antioxidants. With PDT, the influence of the additional burden through the action of photosensitizers can disturb this balance so much that the elimination

$$O_2 \xrightarrow[(+H^+)]{+e^-} O_2^- \cdot / (OH \cdot) \xrightarrow[(+2H^+)]{+e^-} H_2O_2 \xrightarrow[+H^+]{+e^-} H_2O + OH \cdot \xrightarrow[+]{+H^+} 2H_2O \cdot$$

**Fig. 24.** The reduction of oxygen

of the highly toxic molecules is no longer adequate. The precise type of product generated by the photosensitizer will also be important in this respect, because the various ROS have different physical properties and oxidizing potentials.

Hydroxyl radicals are extremely reactive. Enzymatic mechanisms cannot fully protect the cells against them, although antioxidants, as for example tocopheroles, flavonoids, phenoles, ascorbic acid, uric acid and glutathion, are of great importance in reducing their impact. On the other hand, the cells, under normal conditions, are able to reduce the rate of primary generation by enzymatically suppressing the generation of hydrogen peroxide and superoxide, themselves sources of hydroxyl radicals, as explained in more detail below. Singlet oxygen shows a similar oxidative capacity.

One can schematically trace the reduction of oxygen from its initial molecular state right through to the final product in the form water under the aspect of four single electron transfers. Products with a charge of −3 or −4 do not occur in nature, since the bond between the two oxygen atoms dissociates during the reduction of $H_2O_2$, leading to a molecule of $H_2O$ and an OH-radical. The products are the superoxide anion or the hydroperoxyl radical, hydrogen peroxide, the hydroxyl radical and finally water.

### The Superoxide Anion and Hydroperoxyl Radicals

The first reduction, that leads to the generation of the superoxide anion $O_2^-\bullet$, is cited as having a standard potential in an aqueous environment of -0.33 V. This means that the transfer of an electron to an oxygen molecule, i.e. the reduction of molecular oxygen to the superoxide anion, needs energy. Because this potential is negative, the superoxide anion can act as a reducing agent. However, it is also an oxidizng agent in the cell, and the explanation for this lies in its capability to react with protons and form the hydroperoxylradical $HO_2$. In a biological environment, there is a high availability of protons, particularly in the direct proximity of phosphatide membranes (Gouy-Chapman-Stern layer), where the pH can locally be as much as 3 lower than that generally found in the cytosol. Energy is gained by the protonation – the potential is quoted in the literature as being around +1 V. This more than compensates for the −0.33 volt

required for the reduction of $O_2$. $HO_2\bullet$ is correspondingly a more important biological oxidizing agent than superoxide, not only because of its oxidizing strength, approximately 1 V higher than that of $O_2^-\bullet$, but also because of his higher concentration in the neighbourhood of membranes. Whether superoxide acts as an oxidizing or reducing agent naturally depends on the coreactant. If this shows a higher electronegativity than superoxide, and is therefore a stronger oxidizing agent, superoxide will be able to reduce it.

A good example of the reductive capability of $O_2^-\bullet$ is given by the reaction with cytochrome c, frequently used as a proof for the presence of superoxide:

$$O_2^-\bullet + Fe^{(III)}\text{-cytochrome c} \rightarrow$$
$$Fe^{(II)}\text{-cytochrome c} + O_2 \,.$$

Superoxide can oxidize biologically relevant molecules, for example adrenaline, hydroxylamines and hydroquinones.

The generation of superoxide in PDT requires the presence of oxygen and electron donors which generate photosensitizer anions during the irradiation. The photosensitizer anions then act themselves as electron donors to reduce oxygen. Its relevancy lies in its ability to form oxidants such as hydrogen peroxide and the hydroxyl radical.

### Hydrogen Peroxide

$H_2O_2$ is regarded as a stable product from the reduction of oxygen. There are several indirect ways to generate it from oxygen. For example, hydrogen peroxide can arise through a simple dismutation of superoxide:

$$O_2^-\bullet + 2H^+ \rightarrow H_2O_2 + O_2 \,.$$

This reaction is considerably accelerated by superoxide dismutase (SOD). The biological cell protection system depends primarily not on the suppression of hydrogen peroxide formation, but, paradoxically on its generation. The cellular protective effect of SOD is rooted in the higher cell toxicity of the superoxide and hydroperoxyl radicals as compared to hydrogen peroxide. The subsequent rapid removal of hydrogen peroxide is very effectively realized tby the enzyme catalase and by different peroxidases.

$H_2O_2$ can also arise through the reduction of oxygen by hydroquinones:

$$QH_2 + O_2 \rightarrow H_2O_2 + Q \,.$$

A further possibility consists in the interaction of semiquinones and oxygen with the help of the superoxide radical as a mediator:

$$O_2^{-\bullet} + QH_2 \rightarrow Q^{-\bullet} + H_2O_2$$

and subsequent regeneration of the superoxide anion,

$$Q^{-\bullet} + O_2 \rightarrow Q + O_2^{-\bullet}.$$

This can be summarized as

$$QH_2 + O_2 \rightarrow H_2O_2.$$

The quinones, together with their simply reduced form the semiquinones and the doubly reduced hydroquinones, are of considerable importance as subsystems in biological redox systems. They are co-enzymes of the oxidoreductases. In this sense, they function as prosthetic groups of the flavoproteins, for example flavoadenine dinucleotide (FAD) or flavine mononucleotide (FMN). These proteins are components of the respiratory chain, localized in the mitochondrial membranes. Some photosensitizers accumulate strongly in these membranes, and the consequent particular vulnerability of the mitochondria is certainly one of the most important aspects regarding the efficacy of PDT. This compounds the previously mentioned formation of superoxide and hydroperoxyl radicals in the proximity of these membranes.

One can recognize four classes of oxidoreductases in the chain: the flavine coenzymes FMN and FAD, the ubiquinone-hydroquinone system, haem groups in the form of cytochrome c, and the pyridine nucleotides $NAD^+$ or $NADP^+$.

Another form of reaction catalysed by flavines is manifested by xanthinoxidase. This enzyme reduces oxygen, thereby generating superoxide. A flavine bound peroxide rest (HOO-) is an intermediate. Peroxides are very toxic, since they are capable of forming highly reactive peroxyl radicals.

Catalase catalyses the production of water and molecular oxygen from $H_2O_2$:

$$2H_2O_2 \rightarrow 2H_2O + O_2.$$

Peroxidases induce the reduction of $H_2O_2$ through $SH_2$-groups, as present for example in glutathion, thereby generating water:

$$H_2O_2 + SH_2 \rightarrow S + 2H_2O.$$

The monovalent reduction of hydrogen peroxide is of great importance, since it leads to the formation of aggressive hydroxyl radicals.

### The Hydroxyl Radical

In $O_2$, the two oxygen atoms are linked together with two electron pairs. $H_2O_2$ has two more electrons, which now fill two orbitals with the help of two electrons from one of the original bonds. The two oxygen atoms are linked through only one electron pair. A further reduction step leads to the occupation of a so-called antibonding orbital, the dissolution of the bond and the generation of a water molecule and a hydroxyl radical $OH^\bullet$.

An important pathway to OH-radical generation is opened up by the formation of both superoxide and hydrogen peroxide. The superoxide anion can then act as an electron donor for the hydrogen peroxide. This is the Haber-Weiss reaction:

$$O_2^{-\bullet} + H_2O_2 \rightarrow O_2 + OH^- + OH^\bullet \qquad (9)$$

The uncatalysed reaction is slow. Various metals can vastly accelerate it:

$$Metal^{n-1} + H_2O_2 \rightarrow Metal^n + OH^- + OH^\bullet \qquad (10)$$
$$Metal^n + O_2^{-\bullet} \rightarrow Metal^{n-1} + O_2 \qquad (11)$$

The first step, the reduction of hydrogen peroxide, is the Fenton reaction. It is very familiar in the form of $Fe^{III}$ as $Metal^n$.

These various routes to OH-radical generation, based on the initial generation of superoxide radicals, can explain much of the effects of PDT.

### Molecular Vibrations and the Born-Oppenheimer Approximation

As is the case for the electrons themselves, the energy levels for the molecule as a whole are quantized. There are several types of energy involved, the most important being

- the vibrations of the nuclei, representing the vibrational states of the whole molecule
- the energy corresponding to a specific configuration of the molecular electrons.

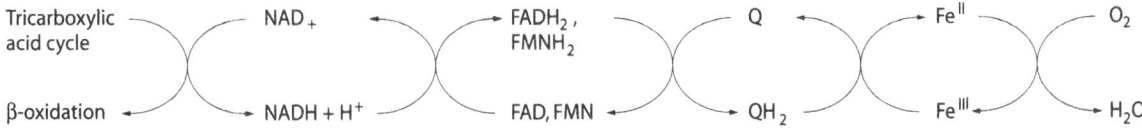

**Fig. 25.** The respiratory chain

If a molecule is excited with infrared light (i.e. heat) of the correct wavelength, its increased vibrational energy leads to a greater average size for the molecule. Eventually, if enough energy is absorbed, the molecule will dissociate.

The amplitude of vibration for the nuclei of a molecule as a function of energy. Level 0 represents the lowest energy state (non-zero oscillation corresponding to the finite zero-point energy). For energy level 4, the vibrational amplitude corresponds to an average movement represented by the distance between A and B. The steepness of the curve for low vibrational amplitudes corresponds to the sharply rising repulsional energy of the nuclei as they approach each other.

The electron configuration does not change, however, as this would involve a much higher energy. If the molecule, however, were excited with visible or UV light of the correct wavelength, this could suffice to put it into a qualitatively different state, where the electron configuration (the order of the electrons within the molecular orbitals) is changed.

Briefly, low-energy, infrared light translates into nuclear vibrations. The higher energies of visible or UV light translate into changes in the electronic configuration. This splitting of a molecule's motion into electronic and nuclear components is embodied in the Born-Oppenheimer approximation.

### The Frank-Condon Principle

In the vibrational ground state, a molecule's nuclei will be found, on average, at the mid-point between the limits of the oscillation. With higher energy states, the situation is the opposite – the nuclei will spend more time around these maxima than in the middle. This is analogous to the situation of a pendulum in motion, which, because it is slowest at the points of reversal of motion, spends more time on average around these positions than in the middle, where it is fastest.

If an electron of the molecule is excited through light and raised into a higher electronic state, This happens rapidly, as it is relatively light. The nuclei are relatively heavy, their positions cannot adapt rapidly to the new electronic configuration. Effectively, the nuclei stay where they are. They thus form a framework of fixed size in which the electronic transitions occur. The splitting of the motions of the heavy nuclei and the much lighter electrons is the content of the Born-Oppenheimer approximation mentioned above. However, the nuclear positions now correspond to a higher vibrational state of the excited molecule (see Fig. 27). This extra energy represents a part of the energy delivered by the photon. This means that the nuclei will pass from their vibrational ground state (i.e. from the middle of the "resting oscillation") into a higher vibrational state. This is represented by the arrow 1 in Fig. 27.

Before the – now larger – molecule yields light again, a part of this energy can be given up in the form of heat. The light photon that is then finally emitted (arrow 2) has a lower energy, that is a longer wavelength, then than the absorbed light had. This is what lies behind the phenomenon of fluorescence seen in photodynamic diagnostics, PDD. Blue light is absorbed by the tissue involved, and red light is re-emitted.

These transitions can be more clearly represented if one leaves out the dimension of the spatial extension, and considers only the energy.

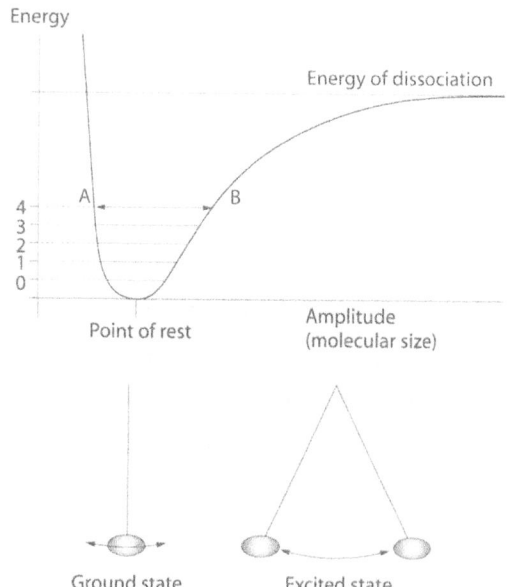

**Fig. 26.** Molecular vibrational states

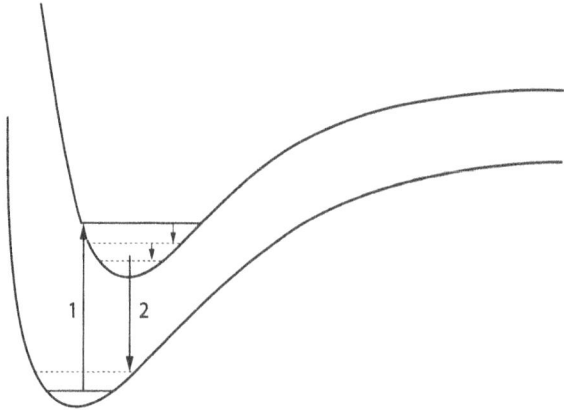

**Fig. 27.** Frank-Condon principle: electronic transition from the ground state to a higher vibrational state of the excited, larger molecule (1), and relaxation back (2)

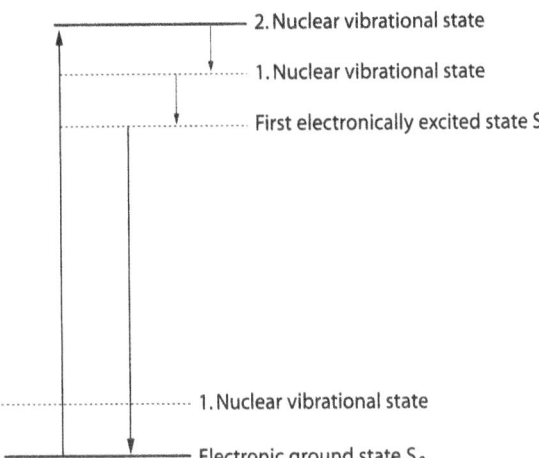

**Fig. 28.** Energy states and transitions

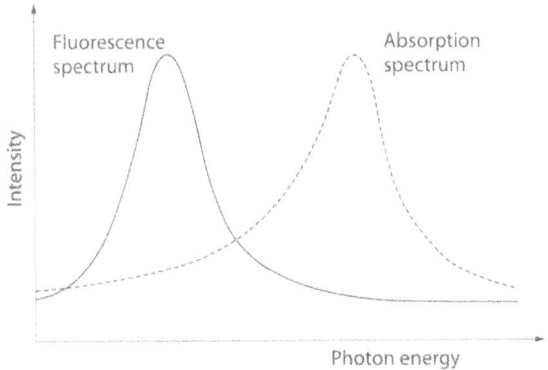

**Fig. 29.** Wavelength shift between absorption and fluorescence spectrum

These transitions can be more clearly represented if one leaves out the dimension of the spatial extension, and considers only the energy (Fig. 28).

One should bear these properties in mind when considering the fluorescence and absorption spectra of a photosensitizer. The two seem are generally very similar in shape, but the fluorescence spectrum will be shifted to longer wavelengths relative to the absorption spectrum. This is shown schematically in Figure 29.

### Photoinduced Energy Transfer

The redox reactions above correspond to the chemical activation of oxygen, i.e. they involve electron transfer. In contrast to this, a physical activation can also occur. In this case, the absorbtion of light in the visible range leads to a change in the electronic structure of the molecule, with an electron in the HOMO

being raised into the next higher level, i.e. the LUMO. This state has a limited lifetime, as the electron can fall back into the HOMO and give up the absorbed energy.

There are three ways in which this can happen:
1. light is rapidly re-emitted with minor energy loss (fluorescence)
2. light is re-emitted with a delay and with major loss (phosphorescence)
3. the energy is transferred to other molecules, for example oxygen.

The phenomena of fluorescence and phosphorescence therefore represent processes that compete with the excitation of oxygen. This provides a fundamental distinction between the diagnostic and therapeutic uses of photosensitizers. Protoporphyrin IX can be employed to serve both causes, but the mechanisms involved are rival ones, as is explained in detail below.

The normal configuration of the ground state of a photosensitizer molecule, just as for most organic molecules, corresponds to a singlet state. That is, all electron orbitals are closed. The ground state is designated by $S_0$. The first excited state is also a singlet, denoted by $S_1$. Transitions to the triplet state are spin forbidden, that is, the probability of such a transition depends on the strength of the spin-orbit coupling. For most organic molecules this is weak, and is increased by the presence of heavy atoms, e.g. metals. The degree of "mixing" of triplet in the ground state of an organic molecule is therefore, typically, small.

Phosphorescence also corresponds to a forbidden transition, representing the reverse case, from the triplet to the singlet state with the emission of light.
1. Fluorescence is seen when the absorbed light energy is re-emitted from the $S_1$ singlet state. Conceptually, this is the simplest process that enables the photosensitizer to return to the ground state. The process is not forbidden, so there is no significant average delay.
2. Phosphorescence occurs when the triplet $T_1$ state is occupied. The $T_1$ to $S_0$ transition is also forbidden, so there is a significant average time delay between absorption and emission. This is the phenomenon seen with watches that glow in the dark.

### Photodynamic Diagnostics (PDD)

Figure 30 illustrates the phenomena of fluorescence and phosphorescence. For example, photodynamic diagnostics (PDD) with protoporphyrin IX is finding increasing use in urology, where urothel carcinomas can be very clearly delineated, and neurosurgery, with promising trials aimed at providing a clear distinction between gliomas and normal brain tissue during an operation to exstirpate the tumours.

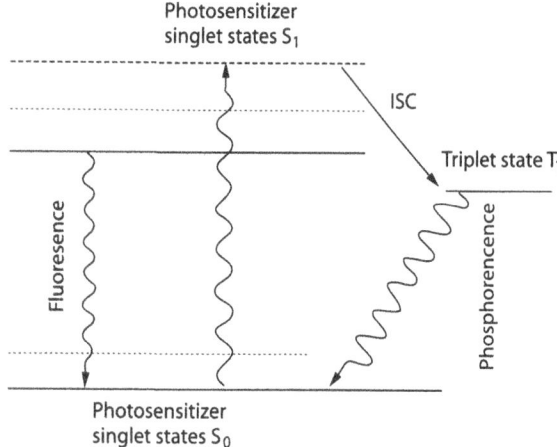

**Fig. 30.** Absorption and re-emission of light energy

As is generally true for the porphyrins, the absorption spectrum of PPIX shows a strong absorption band at the blue end of the UV-visible spectrum in the region around 380 nm. Blue light centred around this band is used to illuminate the tissue. A porphyrin molecule in its $S_0$ ground state can absorb a photon and is thereby excited to an $S_1$ state. The molecule, that is its nuclei, thereby starts to vibrate, and this $S_1$ state has an excess of vibrational energy that is quickly lost from the molecule, before the molecule re-emits a photon. This thus has a lower energy than the absorbed photon, which explains why the re-emitted light is red.

The common factor linking PDD and PDT is the selective uptake or retention of photosensitizer in tumour cells, resulting in a much higher yield
- of red light in PDD and
- of singlet oxygen in PDT
from tumorous rather than from normal tissue.

### The Quantum Yield

Given a photochemical (or photophysical) reaction X + hν → Y, the ratio

$$\frac{\text{Number of molecules of Y generated per second and per unit volume}}{\text{Number of photons absorbed per second and per unit volume volume by X}}$$

is designated the quantum yield ($\Phi_Y$ of this reaction.

It is expedient to subdivide ( into three categories that characterize the reaction:

- $\Phi = 1$. One absorbed photon leads to the transformation of one molecule of X to one of Y. The efficiency is 100%.
- $\Phi < 1$. The efficiency of the transformation in Y is less than 100 %. This means that there are alternative reaction paths for X*.
- $\Phi > 1$. The excitation initiates a chain reaction.

c) The transfer of energy from photosensitizer to oxygen is a special case of energy transfer between don(at)or and acceptor molecules.

It is conventional to write the rate at which a molecule undergoes a particular reaction R by $k_R$. This "rate constant" is in essence a measure of the probability of this reaction occurring. If $k_F$ represents the rate at which excited molecules fluoresce, $k_{ISC}$ the rate at which molecules undergo intersystem crossing, then the fraction of molecules fluorescing can be written as

$$\frac{k_F}{k_F + k_{ISC}},$$

ignoring the possibility of "internal conversion", whereby excited molecules lose energy by processes such as collisions with neighbouring particles. This is equivalent to the quantum yield $\Phi_F$.

The existence of acceptor molecules opens up a new channel for the loss of energy. If the transfer involved is represented by the rate constant $k_{TRAN}$, then the quantum yield for fluorescence is reduced to

$$\Phi_F = \frac{k_F}{k_F + k_{ISC} + k_{TRAN}},$$

That for the energy transfer can be written as

$$\Phi_{TRAN} = \frac{k_{TRAN}}{k_F + k_{ISC} + k_{TRAN}}.$$

On the assumption that the presence of acceptor molecules causes negligible changes to the fluorescence rate and rate of ISC (the so-called weak-coupling case), a theoretical treatment – Förster theory – can be used to calculate the rate of transfer $k_{TRAN}$. It is based on the reradiation of the light energy from the donator to the acceptor. The results depend on the distance between the two molecules, their relative orientation, the refractive index of the solution and on the overlap between the emission spectrum of the donator and the absorption spectrum of the acceptor. This last factor corresponds to a resonance condition. This weak coupling theory is plausible in situations where the spectra of the donators are not significantly altered by the presence of acceptors and vice versa.

The transfer of energy from the photosensitizer to oxygen is theprocess that leads to the generation of the highly toxic non-radical singlet oxygen. The rate-limiting step in this process is the occupation of the

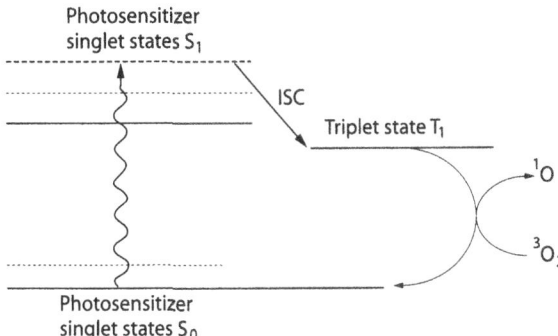

**Fig. 31.** PDT – $^1O_2$ generation

photosensitizer $T_1$, state. As the $S_0$-$T_1$ transition is forbidden, the probability of occupation of the $T_1$ state depends essentially on that of the $S_1$ state, so the photosensitizer's absorption coefficient is an important parameter regarding singlet oxygen generation.

The allowed reaction between two triplet states – photosensitizer in the excited $T_1$-state ($^3PS$) and oxygen in the ground state ($^3O_2$) – leads to the generation of two singlet states: photosensitizer in the ground state ($^1PS$) and activated singlet oxygen ($^1O_2$).

$$^3PS \ (= T_1 \text{ triplet state}) + {}^3O_2 \rightarrow$$
$$^1PS \ (= S_0 \text{ singlet state}) + {}^1O_2 \,.$$

### Dyes and Photosensitizers

An effective photosensitizer must be a good dye; but $^1O_2$-production also depends on the transition probability from the $S_1$ to the $T_1$ state, and on the lifetime of the photosensitizer molecule's triplet state, both of which are independent of the absorption. A potent photosensitizer is distinguished by having a high probability for the $S_1$-$T_1$ transition and a low one for the $T_1$-$S_0$ transition.

Energy transfer to oxygen can occur only during a collision between the photosensitizer in the T1 state and oxygen molecules. A short triplet lifetime reduces the chances of such a collision occurring before the energy is lost in the form of light emission (phosphorescence), and is thus is equivalent to a low probability for the activation of oxygen. The collision frequency is highly dependent on the availability of the oxygen, so the higher the oxygen partial pressure, the larger the probability that energy will be channelled into oxygen activation rather than being lost through phosphorescence.

### Type-I Reactions: Superoxide Radical Formation, Radical Chain Reaction and Autoxidation

According to the above scheme of photoinduced electron transfer, excited photosensitizer molecules can capture electrons from substrate molecules, being thereby converted into radical anions:

$$PS^* + X \rightarrow PS^{-\bullet} + X^{+\bullet} \,.$$

Oxygen can react with both radical ions. In either case, the final product is oxidized substrate. The results of the reaction of oxygen with the photosensitizer radical anion are, however, different from those of the more direct reaction with the substrate radical cation.

In the first case, superoxide radical anions are generated:

$$PS^{-\bullet} + O_2 \rightarrow PS + O_2^{-\bullet} \,.$$

The superoxide can subsequently oxidize a substrate molecule:

$$O_2^{-\bullet} + X^{+\bullet} \rightarrow XO_2 \,.$$

This completes the reaction.

In the second case, the reaction progresses without the generation of superoxide:

$$O_2 + X^{+\bullet} \rightarrow XO_2^{+\bullet} \,.$$

This raises the possibility of a chain reaction:

$$XO_2^{+\bullet} + X \rightarrow XO_2^{+\bullet} + X^{+\bullet} \,,$$

So that $X^{+\bullet}$ is regenerated. A condition for this to occur is that the standard potential of the radical cation $XO_2^{+\bullet}$ is more positive than that of the substrate.

Highly simplified, the reactions can both be written as

$$X + O_2 \rightarrow XO_2 \,.$$

However, this completely masks the radical reaction involved.

Electron donors can interrupt the chain reaction:

$$e^- \ X^{+\bullet} \rightarrow X \,.$$

The radicals themselves can also act as electron donors.

The hallmark of a chain reaction is a measured quantum yield $\Phi_{XO2}$ greater than 1. The possibility of an oxidative chain reaction is of extreme biological

Initiation:        $RH \longrightarrow R^{\bullet} + H^{\bullet}$

Propagation:    $R^{\bullet} + O_2 \longrightarrow ROO^{\bullet}$

                      $ROO^{\bullet} + RH \longrightarrow R^{\bullet} + ROOH$

Termination:    Electron donor $+ ROO^{\bullet} \longrightarrow$ Non-radicals

**Fig. 32.** The autoxidation of unsaturated lipids

$$R_1 \overset{\frown}{\quad} CH_2 \overset{\frown}{\quad} R_2 + ROO^{\bullet} \longrightarrow R_1 \overset{\frown}{\quad} \overset{\bullet}{CH} \overset{\frown}{\quad} R_2 + ROOH$$

**Fig. 33.** Autoxidation of linoleic, linolenic and arachidonic acids

significance, not only because of the greater local degree of destruction, but also because of the possibility that this damage can be spread over a far larger range.

The action of oxygen radicals and singlet oxygen on the lipids of the cellular membrane structures leads to the formation of alkoxyl and peroxyl radicals and subsequent damage to neighbouring lipid molecules.

Radical chain reactions can be analysed in terms of initiation, propagation and termination.

Long-chain fatty acids are widespread in various forms in cells, combined for example with glycerole in the form of triglycerides, or in membranes as a component of the phosphatides. Fatty acid residues with 18 carbon atoms, such as linoleic, linolenic and arachidonic acids are very common. The latter ones have a dialkene structure, that is, two double bonds. The carbon atom between these bonds is particularly vulnerable to a radical attack.

### Type-II Activation: Intersystem Crossing (ISC) and Singlet Oxygen – The Influence of the Environment on Oxygen Activation

Neither the absorption coefficient nor fluorescence nor phosphorescence suffices to determine the efficacy of a photosensitizer in generating singlet oxygen. The environmental factors that determine whether the energetic activation of oxygen is an important process in PDT, independent of the photophysical factors mentioned above, concern the availability of the three determining factors photosensitizer, oxygen and light. The question of the relative importance of each of the competing processes poses significant theoretical problems.

For example, regarding the generation of singlet oxygen, the oxygen solubility in lipids is significantly higher than that in water because oxygen is apolar, so

the solubility of photosensitizer in lipids will be important in determining the damage to membranes.

As far as radical formation is concerned, i.e. electron transfer, as water molecules are polar they interact with charged ions, forming a water cage around each ion. Overall, such a system then becomes more ordered, corresponding to a configuration with lower energy. An aqueous environment thus stabilizes the ionic product state that results from electron transfer, making it more likely that the reaction will occur.

In the end, it seems that only direct measurements of $^1O_2$ or radical generation in the situation of interest can answer the need to quantify the relative significance of any such process.

## Photophysics of the Photosensitizers

### Classification of Photosensitizers

PDT is an intrinsically selective therapy, in the sense that some photoactive substances accumulate more in degenerate than in normal tissue. Such an accumulation, together with a precise application of light, determine how discriminating PDT can be in distinguishing between degenerate and healthy tissue with respect to inflicting damage. It has to be remembered that the selective enrichment in the target tissue depends not only on the photosensitizer, but also on differences in cell composition and compartmentalization. In contrast, the extent to which optical energy is channelled into oxygen activation is a physical quantity that, in analogy to the absorption spectrum, is characteristic for each individual substance and depends on the particular optical wavelength in question.

The photosensitizers that are of interest regarding current and future clinical applications can be classified and exploited according to their different chemical, physical and pharmacokinetic qualities. The same major aim, the selective generation of singlet oxygen in the tumour cells and their destruction, is fundamentally valid for all photosensitizers.

The substances most frequently used in systemic PDT are the porphyrins, which belong to the first generation photosensitizers. The chlorins, which form one of the groups of the second generation photosensitizers, and are in essence reduced porphyrins, are finding increasing use. Bacteriochlorophyll derivates and pheophorbides, stable derivates of chlorophyll varieties that are found in bacteria, are related to the porphyrins and are simple to produce. Phthalocyanins and naphthalocyanins, which are azaporphyrin derivates, show high tumour selectivity, are simple to synthesize and are very stable. They are also associated with a high yield of reactive oxygen species.

As far as topical PDT is concerned, prominent photosensitizers include the thiazin dyes methylene blue and toluidine blue, and the protoporphyrin IX precursor δ-aminolaevulic acid.

## Dyes

The thiazine dyes contain a triple ring system and represent a widely used class of photosensitizer which are topically applied on mucous membranes, e.g. in dentistry, where the destruction of bacteria via PDT without significant side effects is currently being researched clinically. They are also used to treat superficial lesions of mucous membranes in the field of HNO, e.g. leucoplakia, where long remission rates have been achieved.

Methylene blue is soluble in water, and has a broad absorption band in the red area of the spectrum with a maximum, depending to some extent on the solvent, around 665 nm. Toluidine blue is very similar, but has a maximum that corresponds to somewhat shorter wavelengths, around 625 nm. Being water-soluble, it cannot pass easily through biological membranes, and finds use in topical but not in systemic PDT.

## Porphyrins

### The Place of Porphyrins in PDT

Porphyrins are the most frequently used photosensitizers in PDT, and their well established effectiveness makes them the standard for the newer substances.

In 1913, Meyer-Betz conducted the first documented clinical trial in the modern history of the PDT in the form of a self experiment with i.v. injection of haematoporphyrin. A wealth of experience has since been gathered with porphyrins, and because of their efficacy and low intrinsic toxicity, they make up most of the photosensitizers in current clinical use. Ubiquitous in nature, the decades since 1913 have seen improvements with respect to the chemical and photophysical qualities. The aims of present developments include progress regarding the

- selectivity, i.e. specificity, through better retention in degenerate than in healthy tissue,
- clinical response,
- side effects,
- depth of the tumour tissue which can be handled efficiently,
- monitoring and control of the therapy.

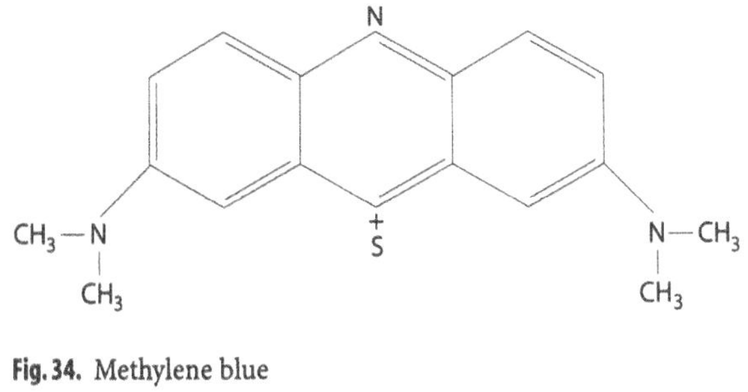

### Basic Structures of the Porphyrins and Their Derivatives

The tetrapyrrole ring represents the basic hydrophobic frame of the porphyrins, consisting of carbon, nitrogen and hydrogen atoms. The apolar tetrapyrrole ring is hardly soluble in water. Porphyrin derivatives are distinguished by their side chain groups and by the number of monomers subunits involved. For example, dimers and oligomers arise through linking with ester or ether bonds. In addition, loose aggregates of monomers arise when a solution is prepared, higher concentrations reinforcing their tendency to form. The solubility in water is determined primarily by the sidechains, and is increased by the presence of polar or charged side groups, for example those of acid residues.

The aims of present research and development efforts include improvements in the biochemical and pharmacological qualities, including an increase in tumour selectivity, one of the most important criteria for their successful employment. This should lead to a simultaneous reduction in side effects such as skin photosensitization. Furthermore, the proven high efficacy of these natural substances in converting optical energy into the activation of oxygen remains essential.

The most important substances that have found a therapeutic application belong to the family of the porphyrins, mainly in the form of the naturally occurring haematoporphyrins and their derivates. They also form a foundation for many of the current developments. Several groups of highly promising photosensitizers are based on expansion of the porphyrin ring system and modifications to the side chains. These so-called second generation substances show improvements in their pharmacokinetics and a shift in the absorption spectra maxima to longer wavelengths. The pharmacokinetics are altered mainly via modifications to the side chains. In contrast, the photophysical qualities depend fundamentally on the ring system.

Although most new compounds are still only at the stage of preclinical evaluation or phase-I clinical studies, a considerable palette of promising substances has arisen. In this respect, a scientifically

**Fig. 34.** Methylene blue

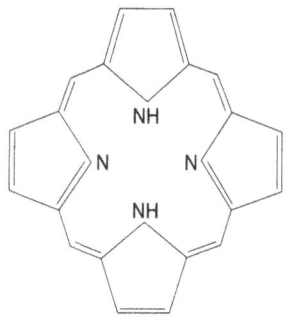

Tetrarapyrrol ring

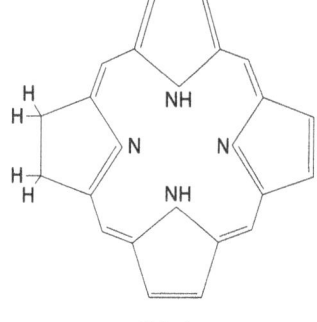

Chlorine

**Fig. 35.** Structure diagrams of some representative photosensitizers

m-THPC

Metallophthalocyanine

based understanding of the mechanisms involved is needed if one is to prepare an optimal therapy plan.

Benzoporphyrin derivative (BPD) is a modern chlorine at an advanced stage of development and undergoing clinical trials. Bacteriochlorophylls and pheophorbides are derivates of chlorophyll varieties occurring in bacteria, closely related to the porphyrins. Phthalocyanins and naphthalocyanins are azaporphyrin derivates. An essential difference to the porphyrins consists of the existence of eight nitrogen atoms in the inner ring of the molecule, instead of four. Theoretical considerations of the effect of these extra nitrogen atoms on the molecular electronic structure and the absorption properties of these substances are in accordance with experimental observations. They have a high absorption and show a high yield of reactive oxygen, a good tumour selectivity, are simple to manufacture, are very stable and have a low toxicity.

## The Physical Importance of the Size of the Photosensitizer Molecule

### Wave Representation Corresponding to Orbitals

In the case of a linear molecule, the electron wavefunction can be envisaged as a wave that spans the length of the molecule, as with a vibrating string. The ends are fixed, the probability density at these points must be zero. This wave corresponds to an orbital, and thus represents up to two electrons.

**Fig. 36.** "Standing wave"

### Ring Systems: Delocalization and the Analogy to Bohr's Atom

An analogy between molecular ring systems and the atomic ring of Bohr's hydrogen atom should help to illustrate some aspects of light absorption by photosensitizer molecules.

The benzene molecule of organic chemistry, $C_6H_6$, is regarded as the archetype of the ring system. It has a ring based on six carbon atoms.

The bonds that are responsible for the spatial form of the molecule will essentially be due to σ-orbitals, whose electrons lie between the nuclei. In contrast to these rigidly bound electrons, the π-electrons, which lie above and below the ring system, are relatively free. Benzene has 6 π-electrons, that lie in polycentric molecule electron orbits. This expresses the fact that these electrons cannot be strictly associated with particular carbon atoms. In this molecule, both the carbon atom nuclei and the electrons form separate but parallel ring systems.

The ring system of π-electrons can be treated quantum mechanically in analogy to the single electron of Bohr's atomic model: the electrons can be treated as waves whose extension corresponds to the entire ring. They are delocalized. Remembering the relation $E = hc/\lambda$ , which expresses an inverse relationship between wavelength and energy for light, and which, according to de Broglie's ideas, is also valid for a particle; higher energy levels correspond to shorter wavelengths. One is familiar with this property in the case of light.

In the free electron model of a π-electron system, higher orbitals correspond to waves with shorter wavelengths running around the molecule ring. As these waves circle round the ring, it would be inaccurate to represent the wavefunction by a simple standing wave. Such a representation would have points of permanent maxima and minima, appropriate for a linear system with fixed ends, not a ring. A ring of free, delocalized electrons cannot have privileged points of minimum or maximum probability.

However, the combination of two waves, 180° out of phase, can deal with this problem. By combining the resulting wavefunctions, no point in the ring is privileged over any other. Figure 38 represents the ring cut and rolled out as a line, so that one should envisage the two endpoints of a diagram joined and representing one point on the ring. The broken line represents an arbitrary point of the ring, the appropriate wavefunction is found by combining those of the two individual waves at that point.

One striking consequence is that linear molecules such as ethene are represented in the free electron model with half-integer waves, as the two points of zero amplitude – corresponding to the ends of the

**Fig. 37.** The benzene molecule

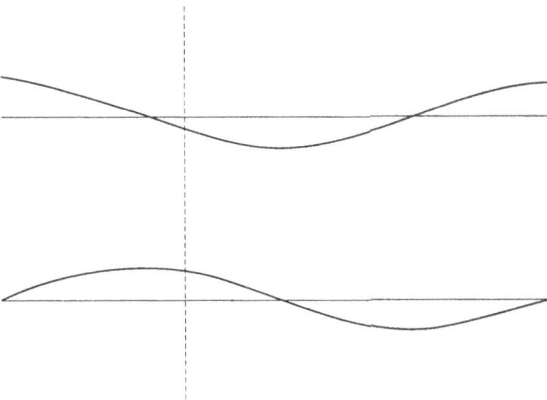

**Fig. 38.** Combination of two standing waves to represent a running wave

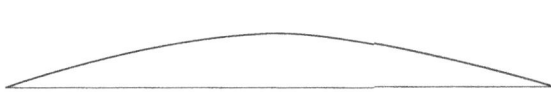

**Fig. 39.** The "fundamental" electron function of a linear molecule (standing wave)

molecule – can be perfectly represented using half-waves:

For a ring system, this does not apply, and the waves must be given the full integer wavelengths. As far as the size of the molecule is concerned, the most fundamental factor is the number of electrons contributing to the ring's π-system, as this determines the wavelengths of the electrons in the HOMO and LUMO.

### The Role of the HOMO and LUMO in the Theory of the Absorption Spectrum

Following the absorption of a photon, an electron from the HOMO is excited into the next higher orbital, the LUMO, and its standing wave corresponds to a higher energy.

Looking at the benzene molecule in the ground state, the six π-electrons will occupy three molecular orbitals. The lowest will correspond to a constant amplitude. The next highest level will correspond to a wave whose wavelength equals the circumference of

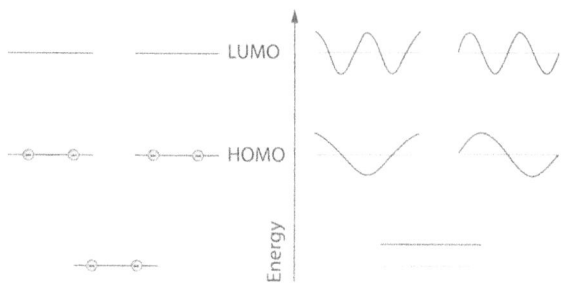

**Fig. 40.** The six π-electrons of the benzene molecule in orbitals and as waves

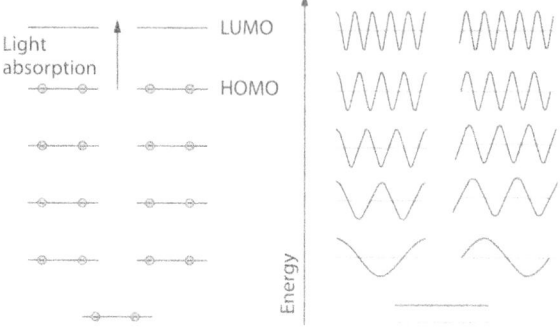

**Fig. 41.** The 18 π-electrons of the metalloporphyrin / metalloph-thalocyanin ring in the free electron model as orbitals and as waves

the ring. The wave for the next level will have a wavelength half of that.

The two lowest-lying electrons fill one orbital, the next four fill two orbitals at the same energy level. The principles of the free electron model mean that each energy level corresponds to two orbitals.

Quantum mechanics offers a method of calculating the energies of the electrons in this model, as can be seen by looking at the fundamental quantum equations

$$p = h/\lambda, \text{ and } E = p^2 / 2m$$

for a particle of mass m. Knowing λ and knowing m means knowing p and thus E. The bond length between the carbon atoms in the benzene molecule is known to be 140 nm, so the ring circumference is 840 nm.

This will be λ for the HOMO, as illustrated on the right of Fig. 40. As it represents the next level, in analogy to the next harmonic, the LUMO corresponds to electrons with a wavelength of 420 nm. A calculation of the HOMO-LUMO gap then follows from this.

The most important part of these molecules, as far as light absorption is concerned, is the inner ring. This can be treated analogously to the benzene ring, with the major difference that it is constituted of 16 and not just 6 atoms. Just as with benzene, the bonds alternate (single then double, repeated eight times), and the π-electrons can be viewed, precisely as in the case of benzene, as delocalized over the whole ring.

In the case of a metal ion such as $Cu^2+$ or $Zn^{2+}$ being at the centre of the ring, the metal contributes another two electrons to the ring system, making 18 in total. Thus, nine orbitals are now occupied in the ground state. Yet after taking this into account, the basic arguments for calculating the energies corresponding to the excitation from the HOMO to the LUMO can be developed along precisely the same lines just given for the case of benzene.

Just as with benzene, the lowest two molecules occupy one orbital. The other electrons occupy the higher orbitals with each energy level corresponding to four electrons.

If the circumference of the ring is denoted by L, then the wavelength

- of the second level     = L
- of the third level     = L/2
- of the fourth level     = L/3
- of the fifth level     = L/4
  (with this system, the HOMO)
- of the sixth level     = L/5          (the LUMO),
  corresponding to ever higher energies.

This allows an evaluation of the energy levels of the HOMO and LUMO using the fundamental quantum equations for p and E in terms of λ. The energy difference between these two levels is precisely that corresponding to light absorption. The resulting theoretical description reflects some important experimentally confirmed aspects, including the basic forms of the spectra, quite well. An aspect not be treated here is that of the influence of the side chains and side rings on the optical and photophysical properties of the photosensitizers.

Broadly speaking, the larger the molecule, the greater the shift in the bands of the absorption spectra to longer wavelengths.

## Biological Background

As mentioned right at the beginning, PDT works through the interaction of the three components: light, oxygen and photosensitizer. Attention will now finally be turned to the influence of the biological environment on this interaction, and, reciprocally, to the effects of PDT on this environment. This will necessitate consideration of the transport and distribution of these three components in tissue.

### The Cell

The direct destruction of the target cells can be due to one of two mechanisms: necrosis or the induction of apoptosis. Both processes are due to damage at the level of individual molecules (e.g. enzymes), molecular complexes (present in the cytosol, membranes or nucleus), or intracellular structures such as organelles. Which of these processes is or are significant will fundamentally depend on which oxygen species, which are chemically very different, are formed, on their concentration and localization. In turn, these factors are contingent primarily on the distributions of photosensitizer and oxygen in the cell, and therefore on physical qualities, for example their lipophilic or hydrophilic nature, their size and charge.

The consequence of therapy is a defect in the targeted tissue in the form of a necrosis, or its involution following the induction of apoptosis. Indirect damage to tumours is caused by the inevitable but also to some degree desired effects of PDT on endothelium cells in and around the tumour, leading to the thrombosis of blood vessels. Furthermore, the induction of various cytokines, for example TNF-$\alpha$, and consequent modulatory effects involving the immune system have been described and are presently the subject of some discussion.

### The Membrane System of Cells

Cell membranes are taken to be one of the main targets for the reactive oxygen species that PDT generates, particularly because these have a high content of readily oxidizable unsaturated fatty acids. Arachidonic acid, a common component of membranes, forms the substrate for inflammatory mediators, such as prostaglandins, prostacyclins and leucotrienes (Henderson and Donovan 1989). Cells are separated from their external environment through a plasma membrane, and are internally structurally organized into various compartments and cytoplasma with the help of specialized membranes and cell organelles. The membranes are responsible for functional coordination between the organelles, for example the exchange of specific signals and molecules. Differences in their functional specialization, evidenced, for example, in the endoplasmatic reticulum and the mitochondrial membrane, are reflected in differences in composition. Other possible targets include the cytoskeleton, which serves both as both a mechanical frame and motor and guides intracellular transportation processes. All these highly specialized structures can suffer damage through singlet oxygen, the extent of which can range from some degree of functional loss up to destruction.

Because of the extreme reactivity of singlet oxygen and radicals their lifetime is very short, in the range of microseconds. Consequently, primary damage will occur only in the neighbourhood of their generation. This leads to the assumption that the enrichment of photosensitizer molecules in specific cell structures would lead to characteristic types of damage and patterns in the functional failure of the structures concerned. The mitochondria are an important case in point, as their destruction through PDT is taken to be one of the main reasons for its effectiveness (Salet and Moreno 1990).

### Membrane Composition

Membranes consist of a fluid lipid bilayer in which proteins and carbohydrates are imbedded. Their exact composition depends on the respective organelles, cells and tissue concerned. In addition, they are polar, i.e. the composition of the outer side of a membrane is fundamentally different from that of the inner side, which explains their resulting functionality.

Lipids can be subdivided into three classes. The class of phosphatides represents the main components of a membrane. Phosphatides have a hydrophilic phosphate head and a hydrophobic tail. The second class, cholesterol, regulates membrane fluidity. Glycolipids are generally found on the outside of a cell membrane, and this class includes many molecules with a receptor function.

The action of oxygen radicals and singlet oxygen on the lipids leads to the phenomenon of autoxidation, where a type of oxidative chain reaction between neighbouring lipid molecules is triggered. Depending amongst other things on the oxygen partial pressure in the membrane, damage can spread far.

Proteins can either be integral components of the membrane, or, depending on their amino acid composition, be bound only peripherally to the membrane. Fundamentally, hydrophilic amino acid residues are to be found in the outer layers, hydrophobic ones within the lipid bilayer. Their oxidative damage can lead to changes in transportation processes or signal transduction through the cell membrane.

### Cell Organelles

The largest cell organelle is the nucleus, which harbours all the DNA of the cell – other than that of the mitochondria. In vitro experiments indicate that singlet oxygen can generate at most single-strand damage to the DNA. In contrast, other reactive species such as the hydroxyl radicals can cause double strand breaks, considered significantly more mutagenic.

The nucleus membrane is physically connected with the nominally closed system of the endoplasmatic reticulum. The rough endoplasmatic reticulum bears the ribosomes. These synthesize proteins including the enzymes essential for the cell function. The smooth endoplasmatic reticulum is responsible mainly for the synthesis of various lipids. The Golgi apparatus represents a further organelle, consisting of a series of compartments with membrane hulls. Its main responsibility is for the secretion of various substances packaged in granules. Endosomes and exosomes are mobile membrane vesicles. They provide for the exchange of cellular material with the cell's environment through its external membrane.

Lysosomes are used for the disassembly of the cell's own organelles and also of material which the cell took in through phagocytosis or endocytosis. Mitochondria are of central importance for supplying the cell with energy.

The cytoplasma is the central reaction compartment of the cell. In it, most of the disassembly of the imported material and the biosynthesis of fatty acids and proteins occur. Hydrophilic photosensitizers are generally more highly concentrated in the cytoplasma than in the membrane system.

### Mitochondria

Mitochondria are bacterium-like organelles (approximately $1 \times 2$ μm). Typically, there are thousands in a normal cell, and together they can occupy up to a quarter of the cell volume. They are enclosed by two membranes each of which fulfills different tasks. The external one is smooth, and equipped with pores which allow small molecules to pass through it. In contrast, the inner membrane is strongly folded, with a correspondingly large surface area. It houses the enzymes of the respiratory chain and the synthesis of ATP, which supply the energy for the cell. Furthermore, unique and essential metabolic pathways are found in the mitochondria, for example the synthesis of haem groups and the tricarboxylic acid cycle. The effect of singlet oxygen on these membranes can then lead to the loss of cell viability.

### Photosensitizer Distribution in Tissue

Hematoporphyrin derivates can accumulate strongly in the spleen and liver, in the stroma of the tumour, in the serosa and submucosa of the stomach and in the skin. This enrichment is most marked in the reticuloendothelial system, for example in the Kupffer cells of the liver, the macrophages in connective tissue, in the tumour, and in the spleen.

The underlying mechanism of the differences in photosensitizer distribution in the various types of tissue and organs is not completely clear. The trans-port processes in the blood certainly play a role, for example through the binding of photosensitizer molecules to the lipoproteins in the plasma or to albumin. Lipoprotein binding seems to favour the uptake in tumour cells, possibly with the aid of LDL receptors, while binding to albumin would seem to lead to a preferential enrichment in the tumour stroma.

Two other factors may well come into play in tumours. They are generally more hypoxic than normal tissue, and thus have a lower average intracellular pH (see the section below on anaerobic glycolysis), and may thus be targeted by electrically charged molecules attracted to an acid environment. Certain photosensitizers can have this property. Tumours also generally have a less well developed system of blood vessels. The latter means that the vessels, including the capillaries, can be more easily penetrated by relatively large molecules, such as photosensitizer molecules, which would otherwise remain in the intravascular compartment.

### Metabolism

#### Tricarboxylic (Citric) Acid Cycle

This is considered to be the central system in the intermediary metabolism, and has both essential catabolic and anabolic functions. It has a catabolic role in the preliminary stages of the process of oxidative phosphorylation. In the process known as glycolysis (see below), the breakdown of glucose supplies ATP, $NADH + H^+$ and pyruvate:

Pyruvate can then be decarboxylated to give acetyl-CoA:

Fat can also supply acetyl-CoA directly through the process of β-oxidation, i.e. the breakdown of fatty acids.

Acetyl-CoA is oxidized to $CO_2$ in the tricarboxylic acid cycle through links with the respiratory chain. This is tantamount to the reduction of $NAD^+$ to $NADH^+$ and $H^+$, or, equivalently, of ubiquinone to ubihydroquinone.

$$\text{Glucose} + 2\,NAD^+ + 2\,ADP \longrightarrow 2\,NADH + H^+ + 2\,ATP + 2\,\text{pyruvate}$$

**Fig. 42.** Generation of two molecules of pyruvate from one of glucose

$$\text{Pyruvate} + NAD^+ \longrightarrow CO_2 + NADH + H^+ + \text{acetyl-CoA}$$

**Fig. 43.** Generation of acetyl-CoA from pyruvate

The anabolic functions of the tricarboxylic acid cycle include the supply of important precursors for the biosynthesis of:

- glucose, through uptake of malate,
- fatty acids, through uptake of acetyl-CoA (synthesis instead of breakdown),
- porphyrins, through use of succinyl-CoA,
- most amino acids, through use of malate and
- 2-oxoglutarate.

All the relevant enzyme systems are to be found in the mitochondria.

### Respiratory Chain

The second subprocess of the oxidative phosphorylation is made possible only through the special double membrane structure of the mitochondria. Oxygen is reduced by $NADH + H^+$ and ubihydroquinone by means of enzyme complexes which are found in the inner mitochondrial membrane (Fig. 25). The content of haem groups and cytochrome c in these enzyme complexes, reflecting the requirements of the cell for haem, seems central to the mechanisms with which protoporphyrin IX can accumulate in tissue.

These redox reactions supply energy which is used in order to shift protons out of the matrix spatial between the two membranes. This leads to the build-up of an electrochemical gradient, as the protons, being charged particles, cannot diffuse back through the membrane. This biological system can be graphically likened to a hydroelectric power station – the electrochemical gradient is the water level, the inner mitochondrial membrane the dam. The protons can stream back into the mitochondrial matrix only through the ATP-synthetase enzymes, which are special transportation proteins in the membrane, viewed as the "turbines of the dam". The phosphorylation is driven by the supplied energy to change ADP to ATP. In effect, the energy carrier ATP is regenerated by oxygen by the "combustion" of reduction equivalents. The linking of the respiratory chain to the tricarboxylic acid cycle supplies 12 molecules of ATP per molecule of acetyl-CoA.

### Glycolysis

This is a catabolic process which takes place in the cytoplasma. One speaks of an aerobic as well as an anaerobic glycolysis, depending on whether or not oxygen is involved.

### Anaerobic Glycolysis

The respiratory chain, which oxidizes $NADH + H^+$ to $NAD^+$, cannot run without oxygen. Without NAD+, or equivalently, via negative feedback, with too much

$$Pyruvate + NADH + H^+ \longrightarrow Lactic\ acid + NAD^+$$

**Fig. 44.** Regeneration of NAD+ from pyruvate with generation of lactic acid

NADH, the tricarboxylic acid cycle comes to a standstill. However, glucose can be converted to pyruvate by means of NADH + H+ (Fig. 39). This process uses no oxygen, generates some ATP and regenerates the NAD+ needed in the aerobic production of ATP. Without oxygen, this is the only possibility of synthesizing ATP out of ADP and inorganic phosphate. The energy balance is poor – from each glucose molecule only two molecules of ATP are obtained, along with two molecules of NADH + H+ (from NAD+), which are then linked into the respiratory chain as reduction equivalents, together with two pyruvate molecules. The $NAD^+$ is regenerated outside of the mitochondria with the help of the enzyme lactate dehydrogenase:

This is the process responsible for the drop in pH in cells functioning under anaerobic conditions.

### Aerobic Glycolysis

Pyruvate can be decarboxylated to acetyl-CoA in the mitochondria through the action of pyruvate dehydrogenase (Fig. 39). Acetyl-CoA serves as a substrate in both the synthesis of fatty acids and in the tricarboxylic acid cycle. In the latter, the generation of ATP is markedly increased by the involvement of oxygen – in total, 26 molecules of ATP can be generated for each molecule of glucose. The NAD+ is then regenerated in the respiratory chain (Fig. 25).

### Tumour Metabolism

Known aspects of degenerate cells are an abnormally high rate of cell division, a relatively large nucleus, an increased uptake of nutrients needed for cell division and the loss of regulatory forces. In the framework of PDT, it is the enrichment of photosensitizer in tumour tissue that is of particular importance. A reason for this accumulation is seen in changes in metabolic processes, especially those that run in the mitochondria. In particular, the growth in protoporphyrin IX (PPIX) levels following the application of δ-aminolaevulic acid δ-ALA) can be partly explained on this basis.

Tumour cells often display a reduction in the number of mitochondria of up to 50% when compared with normal cells. Furthermore, the mitochondria often are smaller and have fewer folds, so that there is less surface area to accommodate the components of the respiratory chain. In addition, tumour cells are often forced to exist under hypoxic conditions due to an

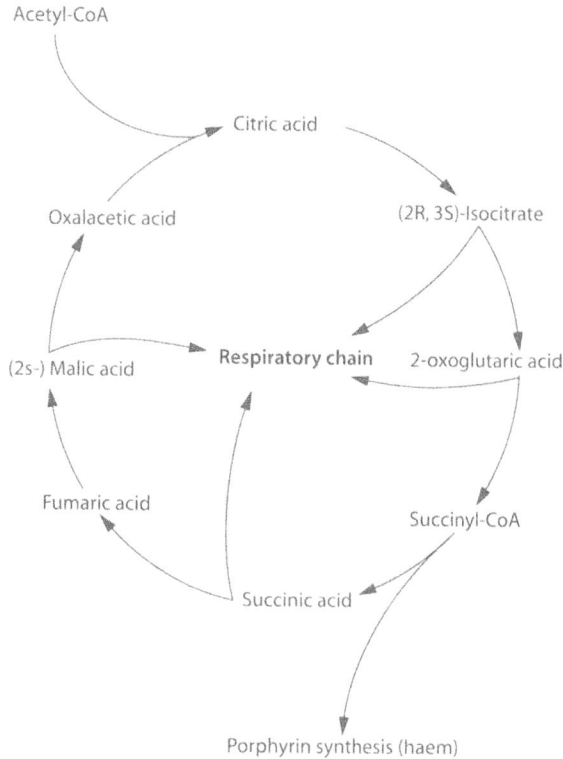

**Fig. 45.** Haem biosynthesis I (tricarboxylic acid cycle)

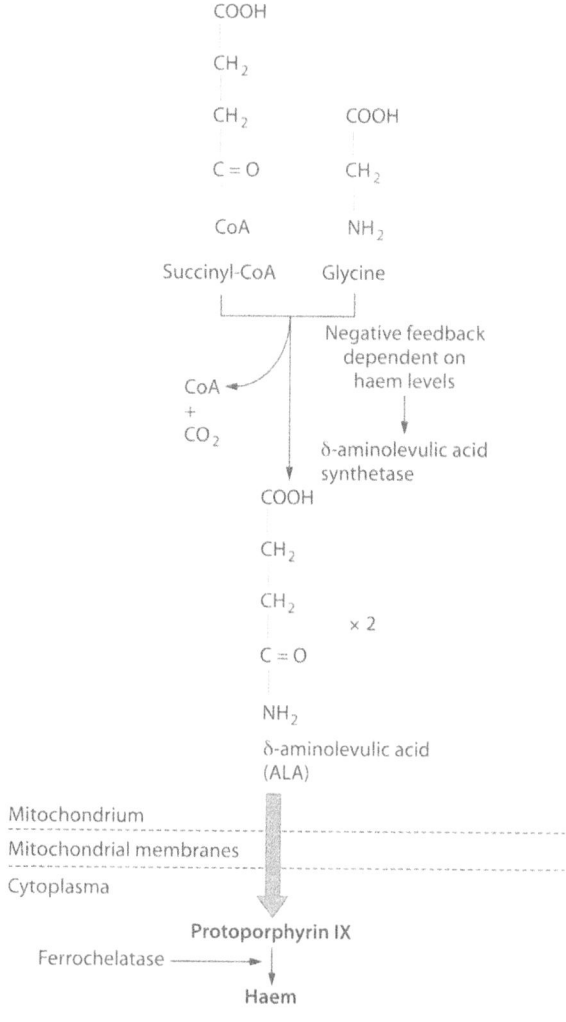

**Fig. 46.** Haem biosynthesis II

inadequate blood supply. The tumour cells must then satisfy their considerable demand for energy through an increased rate of anaerobic glycolysis.

The strong reduction in the respiratory chain activity leads to a correspondingly decreased requirement of cytochrome and haem groups. Furthermore, the slowing of the tricarboxylic acid cycle simultaneously slows the production of succinyl-CoA, required for the endogenous production of haem groups. Via negative feedback, the generation of $\delta$-ALA is reduced. Precisely this provides an explanation for the observed consequence of the application of exogenous ALA. The production of haem runs at high speed up to the point where the ferrochelatase should manufacture protoporphyrin IX. Here, the process slows dramatically, and the levels of PPIX in the tumour cells rise sharply.

## Clinical Aspects

### Pharmacokinetics

Knowledge of the pharmacokinetics of a photosensitizer is of essential importance for an optimal therapy planning. This includes determining the best time to

irradiate and adopting the correct strategy regarding the most important side effects, for example the photosensibilsing of the skin or the retina, in their anticipation. This will of course be based to a large extent on field experience, but a good idea of the dynamics of the transportation and distribution of the photosensitizer between the various compartments, including the blood and liquor as well as the individual organs, is indispensable. Most of our knowledge about the tissue distributions of photosensitizers comes from animal experiments, though the analysis of sera from patients also provides essential information on transportation phenomena. In vitro studies on the binding of photosensitizers to albumin and other proteins in blood, and the exchange of substances between the various vehicles involved, e.g. between albumin and lipoproteins, also serve to shed light on these matters.

## Photosensitizer Delivery and Transportation

It appears that hydrophilic substances, including haematoporphyrin and porphyrins substituted with hydrophilic side groups, are transported generally with albumin, the hydrophobic ones in lipoproteins, including LDLs. The selective uptake of certain of these photosensitizers could then be partly explained by an increased expression of LDL receptors by tumour cells, which can depend on proliferation. However, the distribution will depend on the dose used, as well as on the homogeneity of the photosensitizer. HpD, for example, corresponds to a mixture of oligomers and, when applied intravenously, a significant portion forms aggregates in blood, even though most becomes bound to albumin. The distribution profile can be complicated by transfers from albumin-bound photosensitizer to lipoproteins.

## Tissue Uptake: The Problem of Targeting

The accumulation in a tissue occurs both through passive mechanisms, such as an intrinsically high membrane affinity and leakiness in the tumour vasculature. Active ones, such as the selective uptake of photosensitizer in the cells through carrier systems, can also play a role. For example, experiments involving $^{14}$C-labelled hematoporphyrin derivative (Bellnier et al. 1989), revealed triexponential kinetics for the clearance from the blood, the excretion of the majority of the substance in the faeces and the accumulation in organs of the reticuloendothelial system, e.g. the liver and spleen. The last fact was interpreted as indicating the possible involvement of macrophages in the process of tissue uptake.

Apart from the possibility of a direct uptake of photosensitizer, precursors in the sense of prodrugs can be given, for example δ-aminolaevulic acid, that are taken into the cell and then converted to the active form, which would be protoporphyrin IX in the given example. Photosensitizers (such as phthalocyanins) can be bound to carrier molecules such as lipoproteins or antibodies before injection. This offers the possibility of increasing the selective uptake by the targeted tissue and, for hydrophobic substances, of allowing an intravenous application in the first place (Chan et al. 1988).

The distribution at the cellular level itself is an additional and crucial factor that needs to be taken into account in trying to understand the patterns of damage that PDT can inflict. Thus, hydrophilic photosensitizers are found to be concentrated primarily in lysosomes and in the cytoplasma, hydrophobic ones on the other hand, in the various membrane systems, for example those of the mitochondria, the endoplasmatic reticulum and the cell membrane itself (Moan et al. 1989). One can very well imagine for the sake of argument, for instance, that the destructive effects of a high rate of ROS generation restricted to the lysomes of a cell may be nowhere as deleterious as a significantly lower rate located in the nucleus or the mitochondrial membranes.

It should therefore be clear that the proper and comprehensive concept of a target should not simply be taken to mean an organ, a cell or some part of a cell. The aim is ultimately not simply selective uptake but rather selective cell death.

Passive accumulation mechanisms can include the delayed breakdown of the photosensitizer into photochemically ineffective metabolites and its delayed clearance from the tissue.

## Elimination

Photosensitizer elimination occurs via the liver and kidneys. Measurements on the elimination in faeces and urine with animals point to haematoporphyrin as being essentially hepatically eliminated. Interestingly, certain phthalocyanins appear to be eliminated to a very significant extent in the urine. The ratio of excreted photosensitizer concentration in urine to that in the faeces is not constant, probably reflecting shifts in the distribution between the various blood and tissue compartments.

In discussing the elimination, the plasma clearance also has to be considered. This will generally not show a simple exponential fall in plasma concentration, but rather a multicomponent behaviour with components having half-lives of between a few days and over a month.

## Vascular Damage and Hypoxia

In PDT, effects that are not specific to degenerate cells can also play an important role in tumour control. Almost all photosensitizers employed at this time, including the haematoporphyrins, have a high affinity to the endothelium of blood vessels. Therefore, as could be expected, one of the regular consequences of PDT is destruction of blood vessels. In fact, the effects on the vasculature are essentially the very first ones seen. Inflammatory reactions, with edema, vasoconstrictive and vasodilatative phases and haemorrhagic necrosis (Nelson et al. 1988). This can have a considerable impact on the overall results. Such an effect could well be disadvantageous near diffusely infiltrating tumours because of the risk of extensive necrosis and excessive soft tissue destruction. Since an adequate tumour selectivity in this respect is not given, the effects of the PDT on the healthy tissue would have to be minimized by an exact planning of the irradiation field. However, this unselective process can also have certain advantages:

- tumour regions with a poor microcirculation, and which therefore would not be capable of taking up sufficient photosensitizer for a selective effect, may also be controlled, and
- even an adequate perfusion in the irradiated volume may be reduced so much, that tumour cells which may have been only reversibly damaged through the PDT nevertheless die.

A very clear disadvantage of the resulting acute hypoxia is the loss of the key component oxygen to the process of PDT itself. This line of thought leads to the possibility that fractionation is at least as important to PDT as it is to radiotherapy, on condition that the hypoxia induced both by vascular occlusion and the activation of oxygen by photosensitizer is reversible.

As far as the damage to the vasculature through PDT using, amongst others, hematoporphyrin derivative seems to depend strongly on the level of photosensitizer in the blood (Fingar and Henderson 1987).

### Choice of Photosensitizer and Light Source

In the spectral range from UV light to the deep infrared, the penetration depth of light in tissue grows with increasing wavelength. On this basis alone, the excitation of a photosensitizer at the longest possible wavelength would generally be advantageous. Limitations are imposed by the discrete (line) nature of the photosensitizer absorption spectra. With haematoporphyrin derivates, currently the most frequently employed photosensitizers, an optical wavelength of approximately 630 nm is generally used. This corresponds to the absorption peak lying furthest in the red. In in vivo experiments, measurements show that the penetration depth of light at such a wavelength is approximately 1 cm. Clinical results, however, show that damage to the targeted tissue can occur to a depth of several centimetres. Nevertheless, this re-

striction regarding the penetration means that the use of PDT is limited to that of superficial lesions, for example superficial precancerous lesions or tumours, or the palliative treatment of cutaneous metastases. Deeper-lying tissue would have to be treated by means of an interstitial PDT. A central aim of current research is the development of photosensitizers which can be usefully employed at longer wavelengths.

There are specific indications, for instance the treatment of a Barrett oesophagus or bladder cancer, for which light with a shorter wavelength, such as provided by the argon laser, is employed. Its wavelength of 514 nm also corresponds to one of the absorption peaks of haematoporphyrin derivates. This, however, represents the exceptional deliberate use of a restriction in the light penetration. The idea lying behind this method is to avoid damage to deeper lying structures. Nevertheless, it demonstrates how PDT can be modulated through the tailored choice of the wavelength used.

### Side Effects

The photosensitization of the skin, the most important side effect of PDT with porphyrins, corresponds to that known from experience with other photoactive substances such as antibiotics, phytotherapeutic agents and cosmetics. Depending on the sensitizer used and the degree of exposure to light, this sensitization can take on an appearance ranging from a simple polymorphous light dermatosis (sun allergy) up to a serious epidermolysis or necrosis such as seen with congenital erythropoetic porphyria. The basic mechanism of this porphyria corresponds exactly to photodynamic therapy. As a rule, such skin irritations are short-lived, and permanent scars occur only in cases of absolute disregard of protective measures.

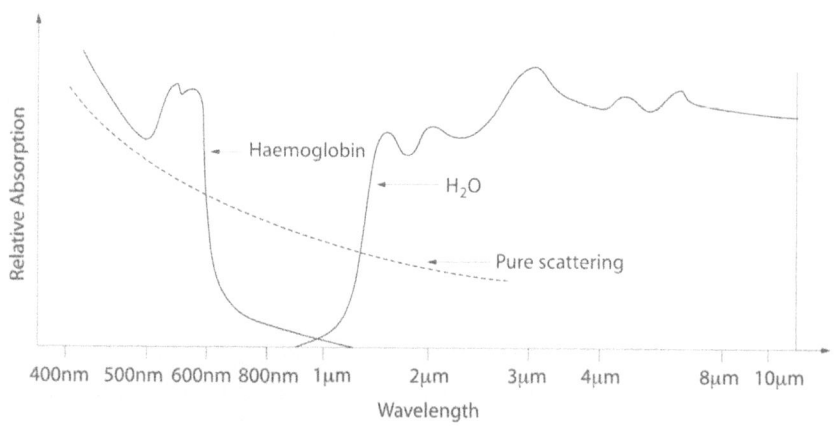

**Fig. 47.** Relative absorption of light by water and the pigment haemoglobin as a function of wavelength

## Metabolic Disorders

The major part of the metabolism and elimination of haematoporphyrin derivates takes place in the liver. However, a poor liver function does not automatically represent a counterindication for the application of these substances, since their use displays a large therapeutic width, due to their low intrinsic toxicity. In patients with liver malfunctions, the possibility of a considerably lengthened and intensified photosensitization should be taken into account.

## Immune Reactions

The allergic potential of a photosensitizer depends on its basic chemical structure. The haematoporphyrins generally have such a potential, which, however, is very dependent on the method of preparation and purification. Modern preparations, such as dihaematoporphyrin ether (DHE: Photofrin II) or haematoporphyrin derivate (HpD), have such a high degree of purity that reports about allergic reactions represent the exception. HpD has, for instance, been added in significant concentrations to several geriatric compounds for many years. This naturally represents a potential risk of photosensitization that has hardly been considered, but there has as yet been no notable number of accounts concerning allergic incidents.

## Further Reading

### Background Literature

ATKINS RW (1970) Molecular quantum mechanics. Oxford University Press Oxford
ATKINS RW (1974) Quanta. Oxford University Press, Oxford
BARLTROP JA, COYLE JD (1975) Excited states in organic chemistry. Wiley, London
BELLAMY LJ (1980) The infrared spectra of complex Molecules. Vol 2, Advances in infrared group frequencies. Chapman and Hall, London
BLUM FH (1964) Photodynamic action and diseases caused by light. Hafner, New York
BOCKHOFF FJ (1969) Elements of quantum theory. Addison-Wesley, Reading, Massachusetts
CALVERT JG, PITTS JN Jr (1966) Photochemistry. Wiley, New York
CANDLER C. (1964) Atomic spectra. Hilger and Watts, London
CARRINGTON A (1974) Microwave spectroscopy of free radicals. Academic Press, New York
CONDON EV, SHORTLEY GH (1953) The theory of atomic spectra. Cambridge University Press, London
COULSON CA, MCWEENEY R (1979) Coulsons valence. Oxford University Press, Oxford
COXON JM, HALTON B (1987) Organic photochemistry. Cambridge University Press, Cambridge
COYLE JD (1986) Introduction to organic photochemistry. Wiley, Chichester
COYLE JD, HILL RR, ROBERTS DR (eds) (1982) Light, chemical change and life Open University, Milton Keynes

DOLPHIN D (ed) (1987) The porphyrins, vol 3. Academic Press, New York
FEYMNAN RP, LEIGHTON RB, SANDS M (1965). The Feynman lectures on physics. Addison-Wesley, Reading, Massachusetts
FRENCH AP, EBISON MG (1986) Introduction to classical mechanics. Van Nostrand Reinhold, London
FRENCH AP, TAYLOR EF (1979) An introduction to quantum physics. Van Nostrand Reinhold, London
GANS R (1971) Vibrating molecules. Chapman and Hall, London
GERLOCH M (1986) Orbitals, terms and states, Wiley, Chichester
GILBERT A, BAGGOTT J, (1991) Essentials of molecular photochemistry. Blackwell, Oxford
GORDY W, COOK, RL (1984) Microwave Molecular Spectra, 3rd edn. Wiley-Interscience, New York
HERZBERG G (1944) Atomic spectra and atomic structure. Dover, New York
HERZBERG G (1945) Infrared and raman speetra. Van Nostrand, New York
HERZBERG G (1950) Spectra of diatomic molecules. Van Nostrand, New York
HERZBERG G (1966) Electronic spectra of polyatomic molecules. Van Nostrand, New York
HOLLAS JM (1996) Modern Spectroscopy. Wiley, Chichester
HORSPOOL WM, ARMESTO D (1992) Organic photochemistry: a comprehensive treatment, Ellis Horwood, New York
HUBER KP HERZBERG G (1979) Constants of diatomic molecules, Van Nostrand Reinhold, New York
JAMMER M (1974) The Philosophy of quantum mechanics. Wiley, Chichester
JORGENSEN R, ODDERSHEDE J (1983) Problems in quantum chemistry, Addison-Wesley, Reading, Massachusetts
KAUZRNANN W (1957) Quantum chemistry. Academic Press, New York
KETTLE SFA (1985) Symmetry and structure. Wiley, London
KING GW (1964) Spectroscopy and molecular structure. Holt, Rinehart and Winston, New York
KROTO HW (1975) Molecular rotation spectra. Wiley, London; (1992) Dover, New York
KUHN H, FÖRSTERLING H-D (2000) Principles of physical chemistry. Wiley Chichester
KUHN HG (1969) Atomic spectra. Longman, London
LANDAU LD, LIFSHITZ EM (1959) Quantum mechanics. Pergamon Press, Oxford
LONG DA (1977) Raman spectroscopy. McGraw-Hill, London
McWeeny R (1979) Coulsones valence, 3rd edn. Oxford University Press, Oxford
MERZBACHER E (1970) Quantum mechanics, 2nd edn. Wiley, Chichester
MOSER JG Ed. (1998) Photodynamic tumor therapy – 2nd and 3rd generation photosensitizers. Harwood Academic Publishers, Chur, Switzerland
MURRELL JN, KETTLE S, TEDDER JM (1978) The chemical bond. Wiley, London
ROSEN B (ed) (1970) Spectroscopic data relative to diatomic molecules, Pergamon, Oxford
MURRELL JN, KETTLE S, TEDDER JM (1965) Valence theory. Wiley, London
OKABE H (1978) Photochemistry of small molecules. Wiley, New York
PAULING L, WILSON EB (1935) Introduction to quantum mechanics. McGraw-Hill, New York
RICHARDS WG, SCOTT PR (1985) Structure and spectra of molecules. Wiley, Chichester
SCAIANO JC (ed) (1989) CRC handbook of organic photochemistry, vols I, II. CRC Press, Boca Raton, Florida
SCHINKE R (1993) Photodissociation dynamics. Cambridge University Press, Cambridge
SCHUTTE CJH (1968) The wave mechanics of atoms. Molecules and Ions, Arnold, London
SOFTLEY TR (1994) Atomic spectra. Oxford University Press, Oxford

STEINFELD J (1974) Molecules and radiation. Harper and Row, New York

STUART B (1996) Modern infrared spectroscopy. Wiley, Chichester

SUGDEN TM, KENNEY CN (1965) Microwave spectroscopy of gases. Van Nostrand, London

SUPPAN P (1994) Chemistry and light. Royal Society of Chemistry, Cambridge

TEDDER JM, NECHVATAL A (1985) Pictorial orbital theory. Pitman, London

TOWNES CH, SCHAWLOW AL (1955) Microwave spectroscopy. McGraw-Hill, New York

TURRO NJ (1991) Modern molecular photochemistry. University Science Books, Mill Valley, Ca

WALSH AD (1953) J Chem Soc, 2260-2317

WAYNE CE, Wayne RP (1996) Photochemistry. Oxford University Press, Oxford

WAYNE R P (1988) Principles and applications of photochemistry. Oxford University Press, Oxford

WILSON EB, DECIUS, JC, CROSS PC (1955) Molecular vibrations. McGraw-Hill, New York

WOLLRAB JE (1967) Rotational spectra and molecular structure. Academic Press, New York

WOODWARD LA (1972) Introduction to the theory of molecular vibrations and vibrational spectroscopy. Oxford University Press, Oxford

ZIMMERMAN HE (1975) Quantum mechanics for organic chemists. Academic Press, London

## Fundamentals of transitions and energy exchange

ALLEN HC JR, CROSS RC (1963) Molecular vib-rotors. Wiley, New York

ATKINS PM (1983) Molecular quantum mechanics, 2nd edn. Oxford University Press, Oxford

BIRKS JB (1970) Photophysics of aromatic molecules. Wiley-Interscience, Chichester

COTTON FA (1971) Chemical applications of group theory, 2nd edn. Wiley, Chichester

DAWBER PG (1987) Vectors and vector operators. Adam Hilger, Bristol

FEYNMAN RP, LEIGHTON RB, SANDS M (1965) The Feynman lectures on Physics, Vol 111. Addison-Wesley, Wokingham

HECHT E, ZAJAC A (1974) Optics. Addison-Wesley, Wokingham

RICHARDS W, SCOTT PR (1985) Structure and spectra of molecules, Wiley, Chichester

WELTNER K, GROSJEAN J, SCHUSTER P, WEBER WJ (1986) Mathematics for engineers and scientists. Stanley Thornes, Cheltenham

## Photophysics of multielectron systems and molecules

BRUINEN JS (1968) J Chem Phys 49: 586

C V SHANK (1986) Science 233: 1276

EI-SAYED MA (1962) J Chem Phys 36: 573

GORDON JP, ZIEGER HJ, TOWNES CH (1954) Phys Rev 75: 282

HATCHARD CG, PARKER CA (1956) Proc Roy Soc A 235: 518

HORROCKS AR, MEDINGER T, WILKINSON F (1967) Photochem Photobiol 6:21

HORROCKS AR, KEARVELL A, TICKLE K, WILKINSON F (1966) Trans Faraday Soc 62:3393

KASHA M (1950) Discuss. Faraday Soc 9:14

LAMOLA AA, HAMMOND GS (1965) J. Chem. Phys., 43, 2129

MEDINGER T, WILKINSON F (1965) Trans Faraday Soc 61: 620

O'CONNOR DV, PHILLIPS D (1984) Time-correlated single photon counting. Academic Press, London

PARKER CA (1968) The photoluminescence of solutions. Elsevier, Amsterdam

PARKER CA, HATCHARD CG (1959) J Phys Chem 63: 22

PORTER G, WEST MA (1974) In: Hammes GG (ed)Techniques of chemistry. VIB, Wiley-Interscience, Chichester

SIEBRAND W (1967) The triplet state. Cambridge University Press, Cambridge

SMOLUCHOWSKI M (1917) Z Phys Chem 92: 129

STERN O, VOLMER M (1919) Z Physik 20: 183

STRICKLER SJ, BERG RA (1962) J Chem Phys 37: 814

## Electronic Energy Transfer and Photochemistry

ABRAHAM JR, SMITH KM (1983) J Am Chem Soc 105: 5734

ALMOND MJ, DOWNS AJ (1989) Spectroscopy of matrix-isolated species. In: Clark ((Initial??)), Hesler ((Initial??)) (eds) Advances in spectroscopy. 17: 1

BIRKS JB, BRAGA, CL, LUMB MD (1965) Proc Soc A 283: 83

BLAUNSTEIN R.P, CHRISTOPHOROU LG (1971) Radiat Res Rev 3: 69

BUSCH GE, APPLEBURY ML, LAMOLA AA, RENTZEPIS P (1972) Proc Natl Acad Sci USA, 69: 2802

CHAPMAN L, MCLNTOSH CL,J PACANSKY (1973) J Am Chem Soc 95: 614

CLOSS GL, KATZ JJ, PENNINGTON FC, THOMAS MR, STRAIN HH (1963) J Am Chem Soc 85: 3809

CLOSS GL, DOUBLEDAY CE (1973) J Am Chem Soc 95:2735

CREED D, DEMARCO DC, MELTON LA, OHTA H, WILDE PH (1980) J Am Chem Soc 102: 2369

DAVYDOV AS (1951) Theory of light absorption by crystalline benzene. J Exp Theor Phys 21: 673

DEXTER DL (1953) J Chem Phys 21: 836

FLEMING L (1976) Frontier orbitals and organic chemical reactions. Wiley-Interscience, Chichester

FÖRSTER TH (1946) Energieumwandlung und Fluoreszenz. Naturwissenschaften 6: 166

FÖRSTER TH (1951) Fluoreszenz organischer Verbindungen. Vandenhoek & Ruprecht, Göttingen

FÖRSTER TH (1959) Discuss. Faraday Soc 27: 7

FRENKEL YI (1931) Excitons Phys Rev 37: 17

GILCHRIST TL, STORR RC (1979) Organic reactions and orbital symmetry, 2nd edn. Cambridge University Press

HUI M-H, WARE WR (1976) J Am Chem Soc 98: 4718

HUNTER CA, SANDERS KM (1990) J Am Chem Soc 112: 5525

LIU RSH, BROWNE DT (1986) Acc Chem Res 19: 42

MAEDA K, FISCHER E (1979) Mol Photochem 9: 309

MARCUS RA (1956) J Chem Phys 24: 966

MARCUS RA (1960) Faraday Discuss Chem Soc 29: 2l

MATAGA N, OKADA T, EZURNI K (1966) Mol Phys 10: 203

MCLAUCHLAN KA (1972) Magnetic resonance. Clarendon Press, Oxford

MILLER JR, CALCATERRA LT, CLOSS GL (1984) J Am Chem Soc 106: 3047

MIZUNO K, PAC C, SAKURAI H (1974) J Am Chem Soc 96: 2993

MOORE JW, PEARSON RG (1981) Kinetics and mechanism, 3rd edn. Wiley, New York

O'CONNOR DV, WARE WR (1979) J Am Chem Soc 101: 121

ORBACH N, OTTOLENGHI M (1975) Chem Phys Lett 35: 175

POTASHNIK R, GOLDSCHMIDT CR, OTTOLENGHI M, WELLER A (1971) J Chem Phys 55:5344

REHM D, WELLER A (1970) Z Phys Chem NF 69:183

REHM D, WELLER A (1974) Isr J Chem 8: 259

REHM D (1970) Z Naturforsch 25a: 1442

SALTIEL J, DABESTANI R, SCHANZE KS, TROJAN D, TOWNSEND DE, GOEDKEN VL (1986) J Am Chem Soc 108: 2674

SALTIEL J, D'AGOSTINO JT, CHAPMAN 0L, LURA RD (1971) J Am Chem Soc 93: 2804

STEVENS B (1962) Spectrochim Acta 18: 439

TERENIN AN, ERMOLAEV VL (1956) Trans. Faraday Soc 52: 1042

WASIELEWSKI MR, NIEMCZYK MP, SVEC WA, PEWITT EB (1985) J Am Chem Soc 107: 1080

WOODWARD RR, HOFFMANN, R (1969) Angew Chem Int Ed 8:781

## Photochemistry of Chromophores

AKHTAR M (1964) In: Noves WA, Hammond GS, Pitts LN (eds) Adv Photochem 2: 263

BINKLEY RW, FLECHTNER TW (1984) Synthetic organic photochemistry. Horspool, Ed, Plenum, New York

COX A (1982) In: Coyle ((Initial??)), Hill ((Initial??)), Roberts ((Initial??)) (eds) Light, chemical change and life. Open University Press, Buckingham

COYLE JD (1974) Chem Soc Rev 4, No. 4, 523

COYLE JD (1985) Tetrahedron 41: 5393

CRISTOL SJ, BINDEL TH (1983) In: Padwa A (ed) Organic photochemistry. Marcel Dekker, New York, 6

FOX MA (1986) In: Volman, DH, Hammond GS, Gollnick K (eds) Adv Photochem 13: 237

GASSMAN PG (1988) In: Fox MA, Chanon ((Initial??)) (eds) Photo-induced electron transfer part C: organic substrates. Elsevier, Amsterdam

GIVENS, RS. (1981) In: Padwa A (ed) Organic photochemistry. Marcel Dekker, New York, 5, 228

Griffin G W, Padwa A (1976) In: Buchhardt ((Initial??)) (ed) Photochemistry of heterocyclic compounds. Wiley-Inter-science, New York

Hixon S S (1979) In: Padwa A (ed) Organic photochemistry. Marcel Dekker, New York, 4, 192

KROPP PJ (1984) Acc Chem Res 17: 131

MAJERJR, SIMONS JP (1964) In: Noves WA, Hammond GS, Pitts LN (eds). Adv Photochem 2: 137

MANRING LE, PETERS KS (1984) J Am Chem Soc 106: 8077

MARIANO PS, Stavinoha LL (1984) In: Horspool WM (ed) Synthetic organic photochemistry. Plenum, New York

MATTAY J (1989) Synthesis 4: 233

NASTASI M, STREITH J (1980) In: de Mayo ((Initial??)) (ed.) Rearrangements in ground and excited states Vol 42-3. Academic Press, New York

PAC C (1986) Pure Appl Chem 58: 1249

RAMAMURTHY V (1985) In: Padwa A (ed) Organic photochemistry. Marcel Dekker, New York, 7, 23

SRINIVASAN R, ORS JA (1978) J Am Chem Soc 100: 7089

STEER RP, RAMAMURTHY V (1988) Acc Chem Res 21: 380

STILL IWJ (1985) Stud Org Chem 19: 596

VON SONNTAG C, SCHUCHMANN HP (1977) In: Pitts LN,Hammond GS, Gollnick (eds) Adv Photochem10: 59

### Electron Transfer, Oxygen Radical Production and Photooxidation

BAUMSTARK AL (1985) In: Frimer (ed) Singlet $O_2$, Vol 2, Part 1, CRC Press. Boca Raton, Florida

BELLUS D (1979) Adv Photochem. Pitts, Hammond, Grosjean (eds) 11: 105

Bloodworth, AJ, Eggelte, JJ (1983) in Singlet $O_2$, Frimer, Ed., Vol 2, Part 1, CRC Press, Boca Raton, Florida

COYLE JD (1985) Tetrahedron, 41, 5393

FOOTE CS (1985) Tetrahedron, 41, 2221

FOX MA (1986) Adv. Photochem., Volman, Hammond, and Gollnick, Eds, 13, 237

FRIMER AA (1979) Chem. Rev., 79, 359

FRIMER AA (ed) (1985) Singlet $O_2$ vols 1-4, CRC Press, Boca Raton, Florida

FRIMER AA, Stephenson LM (1985) In: Singlet $O_2$, Vol. 2, Part 1, CRC Press, Boca Raton, Florida

GEORGE MV, Bhat V (1979) Chem Rev, 79, 447

GOLLNICK, K, Griesbeck, A (1985) Tetrahedron, 41, 2057

GOLLNICK K, KUHN HJ (1979) in Organic chemistry, Vol. 40, Singlet Oxygen, WASSERMAN, MURRAY (eds) Academic Press, New York

GORMAN AA, RODGERS MAJ (1981) Chem Soc Rev, 10, 205

HURST JR, WILSON SL, SCHUSTER GB (1985) Tetrahedron, 41, 2191

JEFFORD CW, BOUKOUVALAS J, KOHMOTO S, BERNARDINELLI G (1985) Tetrahedron, 41, 2081

KAVARNOS GJ (1993) Fundamentals of photoinduced electron transfer, VCH Publ. Inc.

KAVARNOS GJ, TURRO NJ (1986) Chem Rev, 86, 401

KEARNS DR (1969) J Am Chem Soc 91, 6554

LEWIS FD In: Photo-induced electron transfer part C: photo-induced electron transfer reactions: organic substrates, Fox and Chanon, Eds, Elsevier, Amsterdam

MARCUS RA (1956) J Chem Phys 24, 966

MARCUS RA (1963) J Phys Chem 67, 853

MARCUS RA (1965) J Chem Phys 43, 679

MARCUS RA (1993) Angew Chem 105, 1161

MATSUMOTO M (1985) In: Singlet $O_2$, Frimer, Ed., Vol. 2, Part 1, CRC Press, Boca Raton, Florida

MATSUMOTO M, DOBASHI S, KURODA K, KONDO K (1985) Tetrahedron, 41, 2147

MATTES SL, FARID S (1986) J. Am Chem Soc 108, 7356

MATTES SL, FARID S (1988) Organic Photochem, Padwa, Ed., Marcel Dekker, New York, 1983, 6, 233

MCCAPRA F (1970) Pure Appl Chem, 24, 611

MELLOR J (1982) In: Light, Chemical Change, and Life, Coyle, Hill, Roberts, Eds, Open University Press

MIZUNO K, KAMIYAMA N, ICHINOSE N, OTSUKI Y (1985) Tetrahedron, 41, 2207

NAKAGAWA M, YOKOYAMA Y, KATO S, HINO T (1985) Tetrahedron, 41, 2125

RAMAMURTHY V (1985) Organic Photochem, Padwa, Ed., Marcel Dekker, New York, 7, 231

RIGAUDY JR (1968) Pure Appl Chem, 16, 169

SCHAAP AP, SIDDIQUI S, PRASAD G, PALOMINO E, SANDISON M (1985) Tetrahedron, 41, 2229

TURRO NJ (1985) Tetrahedron, 41, 2089

WASSERMAN, MURRAY (eds) (1979) Organic Chemistry, Vol. 40, Singlet Oxygen, Academic Press, New York

## Biochemistry and Physiology

ALBERTS B, BRAY D, JOHNSON A, LEWIS J, RAFF M, ROBERTS K, WALTER P (1998) Essential cell biology, Garland

EVANS WH, GRAHAM JM (1989) Membrane structure and function. Oxford University Press, Oxford

JAIN MK, WAGNER RC (1980) Introduction to biological membranes. Wiley, New York

LAMB JF INGRAM CG, JOHNSON LA, PITMAN RM (1991) Essentials of physiology, 3rd edn. Blackwell, Oxford

LEHNINGER AL, NELSON DL, COX MM (2000) Principles of Biochemistry

MURRAY RK, GRANNER DK, MAYES PA, RODWELL VW (2000) Harpers biochemistry, 25th edn. Appleton & Lange, New York

STRYER L (1995) Biochemistry, 4th edn. Freeman, San Francisco

## Photosensitizer Pharmakokinetics and Tissue Distribution

BELLNIER D, HO K, PANDEY RK, MISSERT J, DOUGHERTY TJ (1989) Distribution and elucidation of the tumor-localizing component of hematoporphyrin derivative in mice. Photochem Photobiol 50, 221–228

CHAN W-S, MARSHALL JF, LAM GYF, HART IR (1988) Tissue uptake, distribution and potency of the photoactivatable dye chloraluminium sulfonated phthalocyanine in mice bearing transplantable tumors. Cancer Res 48, 3040–3044

DOIRON DR, GOMERS CJ (1984) Porphyrin localization and treatment of tumors. Alan R Liss, New York

GINEVRA F, BIFFANTI S, PAGNAN A, BIOLO R, REDDI E, JORI G (1990) Delivery of the tumour photosensitizer zinc(II)-phthalocyanine to serum proteins by different liposomes: Studies in vitro and in vivo. Cancer Lett, 49, 59–65

GOMER CJ, FERRARIO A (1990) Tissue distribution and photosensitizing properties of mono-L-aspartyl chlorin e6 in a mouse tumor model. Cancer Res. 50, 3985–3990

HENDERSON BW, DONOVAN JM (1989) Release of prostaglandin E$_2$ from cells by photodynamic treatment in vitro. Cancer Res 49, 6896–6900

JORI G (1989) In: Photosensitizing Compounds: their chemistry, biology and clinical use. Wiley, Chicester, pp 78–86

KESSEL D (1986) Porphyrin-lipoprotein association as a factor in porphyrin localization. Cancer Lett 33, 183–188

KONGSHAUG M, MOAN J, BROWN SB (1989) The distribution of porphyrins with different tumor localizing ability among human plasma proteins. Br J Cancer 59, 184–188

NELSON JS, LIAV LH, ORENSTEIN A, ROBERTS WG, BERNS MV (1988) Mechanism of tumor destruction following photodynamic therapy with hematoporphyrin derivative, chlorin and phthalocyanine J Natl Cancer Inst 80, 1599–1605

RICHTER AM, CERRITO-SOLA S, STERNBERG ED, DOLPHIN D, LEVY JG (1990) Biodistribution of tritiated benzoporphyrin derivative ($^3$H-BPD-MA), a new potent photosensitizer in normal and tumor-bearing mice. J Photochem Photobiol, B, 5, 231–244

SALET C, MORENO G (1990) New trends in photobiology. Photosensitizazion of mitochondria. Molecular and cellular aspects. J Photochem Photobiol B, 5, 133–150

# Part III
# Areas of Application

# III-1
# Laser Stereotaxy

F. Ulrich, H.-J. Schwarzmaier and W. von Tempelhoff

## Contents

## Introduction

Conventional treatment of malignant gliomas includes surgery and radiation. Other methods, such as chemotherapy, may be effective in palliation and improved median survival. The cumulative toxicity of radiotherapy and development of chemotherapyhyphon – resistant tumour cell lines can limit the efficacy of these treatments. Hyperthermia, on the other hand, does shows no cumulative toxicity. However, hyperthermia is used infrequently for the treatment of brain tumours because of the physical constraints of heating tissue of the skull.

Laser-induced interstitial thermotherapy (LITT) is a minimally invasive therapeutic approach in the treatment of local tumours within solid organs and was first described by Bown in 1983 [1].

In vivo animal LITT experiments on normal rat brains and on F98 transplantation tumours have been reported previously [2]. Laser irradiation of the basal ganglia with an output power of 2–4 W and exposure times of 30–90 s resulted in temperatures of up to 45 °C in a 5 mm distance, and in ellipsoid lesions 4–8 mm in diameter. A central zone of coagulative necrosis was surrounded by a sharply demarcated rim of edema. Lesion size increased with increasing energy application and varied in different tissues. Similar observations were made in transplanted tumours. The necrosis became more evident after 3–4 days. After 1 week it was surrounded by a wall of granulation tissue and the resorption progressed towards the lesion centre.

Tumors are destroyed thermally by directing low power laser energy into the target volume through a laser light guide. Thermal diffusion that is preceded by the absorption of the optical energy results in tissue necrosis with a zonal architecture [2].

LITT is performed within the MR unit with the patient fully awake. MRI is used for monitoring the temperature-affected zone and to localize and determine the volume of the induced irreversible lesion [10, 20]. The safety of the procedure could be further improved by integrating information derived from functional MRI to avoid damage of cortical areas [7]. All masses had an approximately spherical configuration. In all subjects the tumor diameter determined by MRI was <35 mm due to the limited maximum diameter of the laser-induced lesion. Neoplasms with intratumoral bleeding and mainly cystic components were excluded. In all instances LITT was performed immediately following stereotactic biopsy and histological analysis [6].

Single lesions were induced in 17 patients, and in 1 patient LITT was performed in two foci along different tracts. Two patients underwent LITT twice due to tumours with two different localizations. An Nd:YAG laser (Medilas 4060N, 1–064 nm wavelength, continuous wave; Dornier Medizintechnik, Germering, Germany) was used in combination with an ITT light guide (Dornier, Germany) with an applicator length of 20 mm and a diameter of 1.9 mm. The laser energy was emitted circumferentially from the cylindrical surface of the tip, resulting in ellipsoid-shaped necrosis. The tissue was heated slowly over a longer exposure time (10–20 min) at low laser power levels (3.9–5 W) to avoid carbonization at the tip of the light guide.

The amount of energy deposited was tailored to the size of the neoplasm. LITT was terminated when the diameter of the peripheral zone determined on images during LITT was similar to the size of the neoplasm.

MRI was performed on a 1.5 T Magneton SP (Siemens, Erlangen, Germany) with a circularly polarized head coil. The temperature sensitivity inherent to the MR experiment provides the basis for using MR imaging to monitor thermosensitive therapies. Various forms of MR temperature mapping have been demonstrated in standard 1.5 T systems [3, 4, 5, 8, 9, 17]. The availability of 0.5 T „open-configuration" systems in conjunction with different tracking systems for real time instrument guidance in an MR environment allows for easy placement and interactive positional adjustment of laser fibres in various organs. For clinically effective monitoring, MR temperature maps need to be acquired in near real time with sufficient spatial and contrast resolution. Furthermore, reliable temperature control must be tissue-independent and provide quantitative data.

## Exemplary Material

A 30-year-old male with newly diagnosed intrinsic brain tumour of the right parietal operculum was treated by MR-guided laser therapy (Fig. 1). The otherwise healthy patient had experienced a focal and secondarily generalized epileptic seizure 3 weeks before therapy, and he was neurologically unremarkable. The MRI examination 2 weeks prior to laser

therapy revealed an oval mass in the right parietal operculum. The centre of the tumour was displayed with low signal intensity on the T1-weighted images. After i.v. administration of Gadolinium-DTPA (Magnevist, Schering, Berlin, Germany) in a dosage of 0.1 mmol kg body weight, a central enhancement was evident. The T2-weighted images showed high signal intensity of the lesion and the surrounding white matter edema

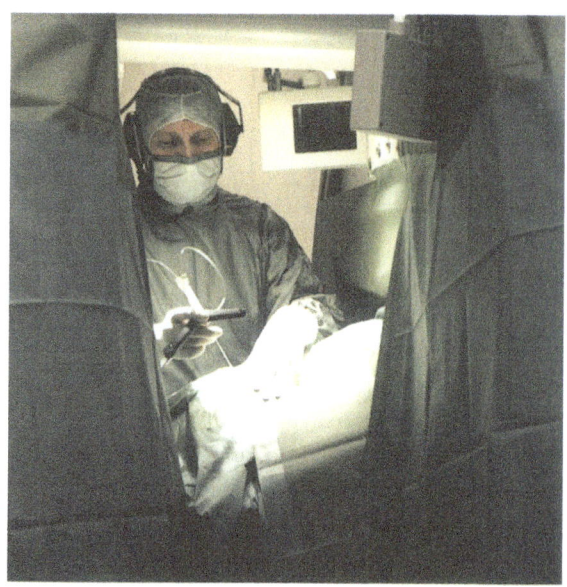

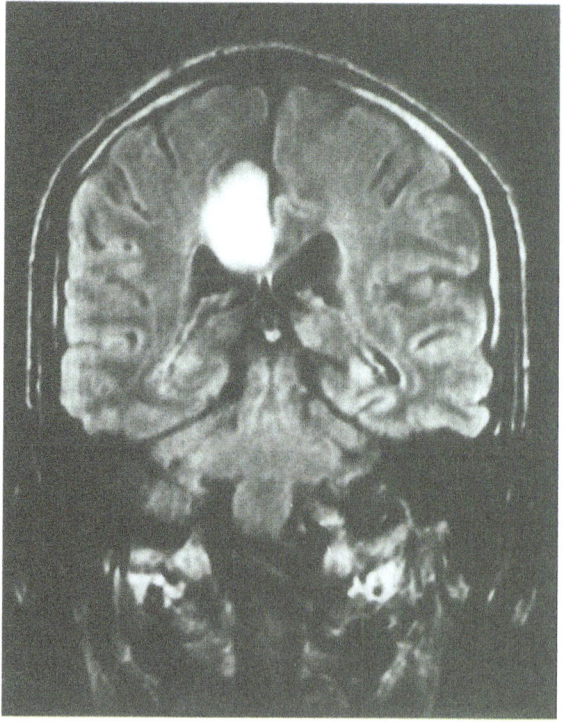

**Fig. 1.** MR-guided laser-induced thermotherapy

**Fig. 2.** Interventional MRI unit

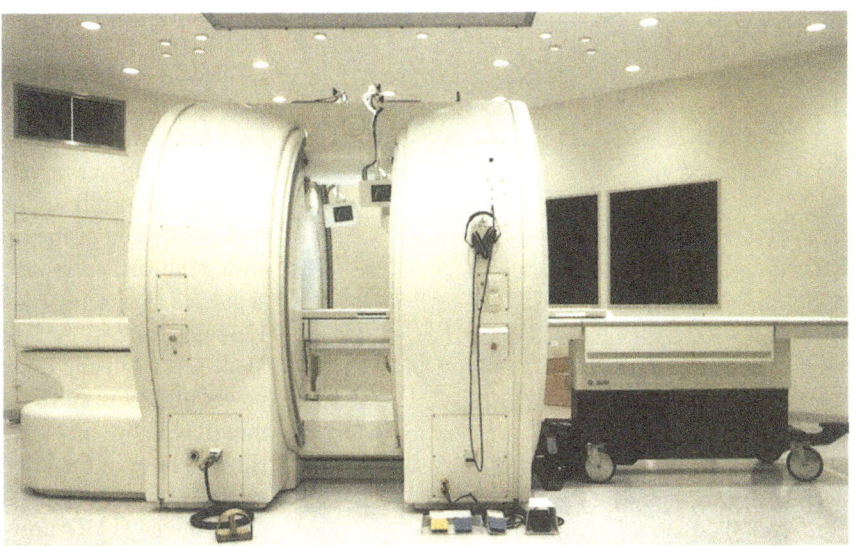

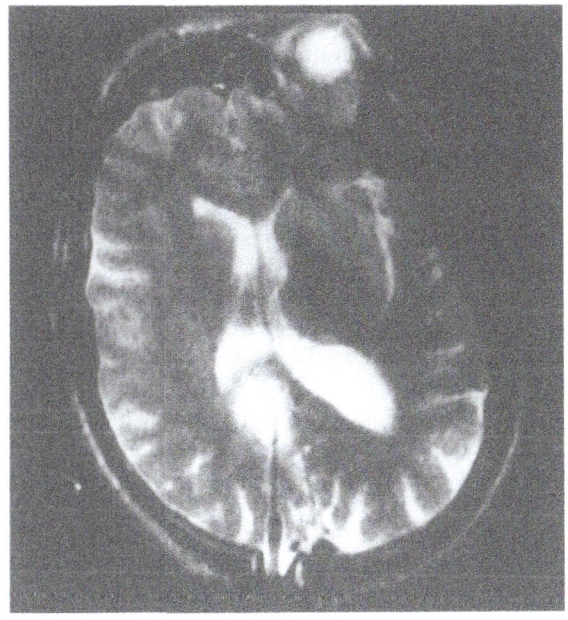

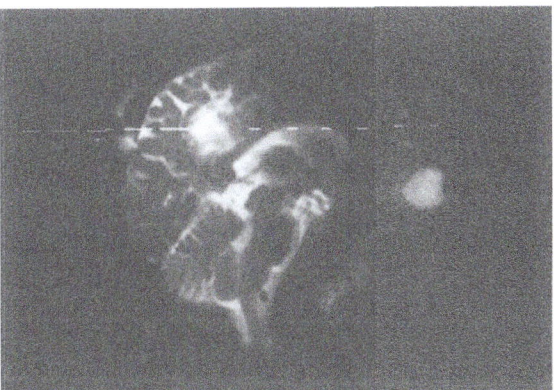

**Fig. 4.** Intraoperative MR imaging multiplanar trajectory of a cerebral glioma

(Fig. 1). The patient was intubated in the open 0.5 Tesla Magnet Signa SP, GE (Fig. 2). First of all an MR-compatible Mayfield clamp was placed on the skull of the patient. In a second step the coordinates of the tumour centre as well as the coordinates for best location for a burrhole were determined. A burrhole-mounted device has been constructed (Snapper Stereoguide, Magnetic Vision GmbH, Rüti, Switzerland) which allow the fixation of a biopsy needle or laser fibre with an angulation in a certain range. MR-guided biopsy of the tumour was carried out (Fig. 3).

Neuropathological examination was done both intraoperatively and in several small biopsy pitches embedded in paraffin. They showed a glioma of predominantly astrocytic differentiation. The final neuropathological diagnosis was astrocytoma, probably WHO grade II (Fig. 3). The stereotactic biopsy was followed by MR-guided implantation of an applicator probe (Somatex).

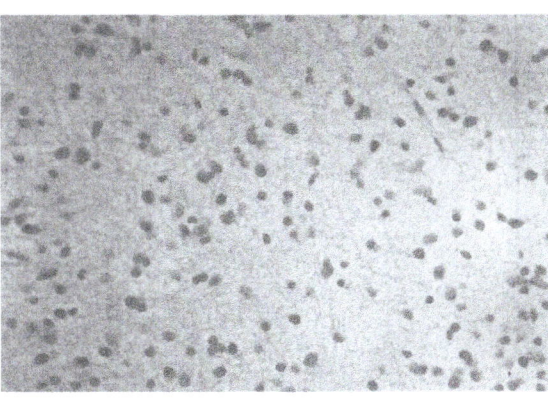

**Fig. 3.** MR-guided stereotactical biopsy of a cerebral glioma (WHO II)

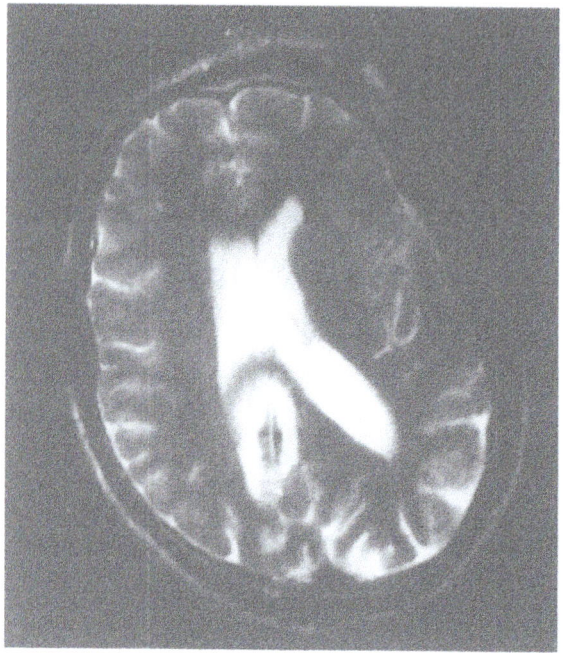

**Fig. 5.** Interstitial laser-induced thermotherapy of a cerebral glioma using a Nd:YAG laser (5 W, 18 min)

The very sensitive MRI, however, can only be performed in specially equipped rooms, where no laser unit is installed. As a consequence a 13 m extension light guide was connected with the ITT fibre. The ITT light guide tip was inserted into the tumour centre via the stereotactically applied applicator probe.

The position of the tip of the guide was controlled with multiplanar reconstructions of a T1-weighted sequence (Fig. 4). For laser intervention light from a 1064-nm Nd:YAG cw laser (MediLas 4060 N, Dornier Med Tec, Munich, Germany) was applied using a power of 5 W over 18 min (Fig. 5). Laser output was monitored by an integrated powermeter of the Nd:YAG laser in the laser head during treatment. The system was calibrated before treatment by a separate powermeter (LMG universal). Transmission loss of the system was 20%, which was measured before fibre implantation. For calculation of the temperature maps, a reference phase map was acquired at the beginning of the treatment and subtracted from the continuously updated phase maps.

The actual temperature distribution could be displayed as colour-coded images as well as time course of the hot spot temperature (Fig. 6 a-d).

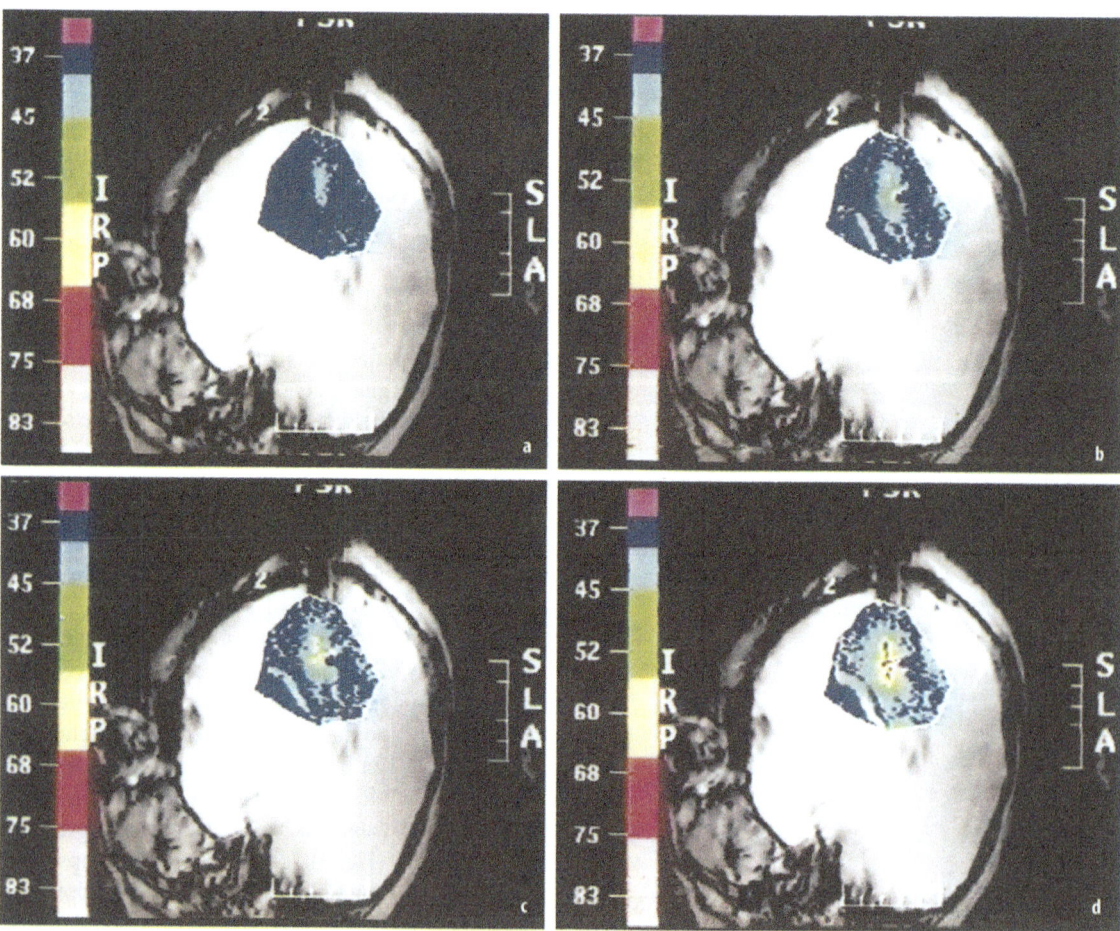

**Fig. 6 a-d.** Temperature measurement during laser therapy of a cerebral glioma (WHO II) using magnet resonance imaging

The zonal architecture of the laser lesions consisted of a central and peripheral zone. The size of the central zone was determined by segmenting a relatively low signal intensity zone in T2-weighted images adjacent to the light guide track. The size of the peripheral zone was calculated by subtracting the size of the central zone from the total lesion size.

## Discussion

On T1-weighted images the signal intensity of the central zone decreased and the signal intensity of the peripheral zone increased, resulting in a more homogeneous lesion without differentiation into zones. Zonal architecture was evident within a period of 8–200 days (mean 60 days) post-LITT. The enhancing rim after Gd-DTPA – parallel to the shrinkage of the lesion during follow-up – showed a continuous reduction of diameter and enhancement. However, the residual lesion and a spot-like enhancement after Gd-DTPA were visible in all late controls, even nearly 4 years after LITT. In two cases, variations with a different development of the zonal architecture were observed on T1-weighted images. In 16 lesions the central and peripheral zone could still be separated during subsequent follow-up, whereas in 1 case the central zone grew until it covered the whole irreversible lesion 13 days after LITT and hence a peripheral zone was no longer evident. Peripheral Gd-DTPA enhancement was still present. In a second case the peripheral zone could not be clearly delineated on plain T1-weighted images either immediately after therapy or during further follow-up. Slight peripheral enhancement was present on post-Gd-DTPA T1-weighted images. These variations were not apparent on T2-weighted images.

The initial lesion size varied from 171 to 581 mm$^2$ (mean 380 mm$^2$). The total size of the laser-induced lesion increased by 0–45% (mean 23%) within 1–40 days (mean 10 days). In most patients this was caused by an expansion of the peripheral zone. The maximum diameter of the laser-induced lesion in which cells were killed was 16–30 mm (mean 25 mm). After the initial increase, total lesion size decreased exponentially. Total lesion size decreased to 50% of the initial lesion size within a mean of 93 days after LITT. Recurrences occurred in 7 of 18 patients, indicated by tumour growth outside the laser-induced lesion.

In 16 patients the follow-up studies showed a slight to severe increase of perifocal odema on T2-weighted images. In no case was an increase of perifocal edema noted immediately after LITT. The onset of edema was 1–3 days after therapy with maximum extent after 4–27 days (mean 6 days). The regression of edema lasted 15–45 days [mean 23 days).

In the central zone there was a generalized damage of cellular and subcellular membranes observed and the intravascular red blood cells appeared empty. The high signal intensity of the central zone most likely represents heat-induced methaemoglobin conversion from deoxyhaemoglobin, which normally takes >72 h in cerebral haematoma without heat induction [10, 11]. On the other hand, high-protein content fluid collections may also contribute to the high signal intensity of the central zone. Similar findings in patients who underwent MRI-guided radiofrequency ablation of brain tumours were reported by Anzai et al. [12].

In that study, high signal intensity foci were seen only in patients for whom biopsy was performed immediately before radiofrequency ablation, which presumably caused a small amount of local bleeding [12]. The histological examination of the peripheral zone in rat brain revealed no membrane disruptions but an edema with generalized swelling and empty appearance of nerve cell processes and astrocytic foot processes [2]. The MR appearance with low signal intensity on T1-weighted and high signal intensity on T2-weighted images corresponds to these histopathological observations. Similar histological findings with an inner zone of coagulation necrosis and a peripheral zone of scattered cellular damage have been reported by Tracz et al. [13,14].

The peripheral zone has been labelled necrotizing edema and „delayed liquefaction necrosis", indicating the irreversible damage within this zone [2, 15, 16]. The low signal intensity rim on T2-weighted images most likely represents deoxyhaemoglobin in the acute stage and haemosiderin deposition in later follow-up studies. This zone could also be visualized in radiofrequency lesions [12]. After Gd-DTPA, an enhancement in this zone was evident in all patients independently of a blood-brain barrier disruption prior to LITT. It has to be considered that the central and peripheral zone together form the total lesion size. Follow-up studies with exponential reduction of lesion size and accompanying diameter reduction of the enhancing rim after Gd-DTPA support the assumption that the enhancing rim represents the outer border of the irreversibly damaged lesion. Similar results were reported form Tracz et al. in cat brain with a good correlation between histological lesion size and diameter of an enhancing rim after Gd-DTPA [13, 14].

However, within several days after LITT, a slight to moderate increase of the diameter of the enhancing rim was observed. As there is no sharp border of damaged and viable tissue in interstitial laser therapy, a spatial gradient of cellular destruction with some microscopic delayed cell damage may occur outside the enhancing rim visible immediately after LITT. The delayed cell damage may contribute to the increase in the diameter of the enhancing rim during the first days after LITT. In tumours with Gd-DTPA enhancement prior to LITT due to a disrupted blood-brain barrier, there was no enhancement within the laser induced lesion after LITT. There was no apparent relation between the size of the laser-induced lesion and the laser energy deposited. However, in vitro studies with pig brain revealed that, in comparison to the histological results, the outer border of the lesion corresponds to the 60 °C isotherm. The perifocal edema is not apparent immediately after therapy. It evolves 1–3 days after LITT and regresses completely within 15–45 days, indicating that no persistent damage occurs within this zone.

In the majority of the induced lesions, the long-term development is uniform with variations in the zonal architecture in two lesions. In all patients the decrease of lesion size followed an exponential pattern with a half-life period of 93 days. The shrinkage of the lesion is accompanied by a corresponding reduction of the size of the neoplasm. As the laser-induced enhancing rim persists over a long period, difficulties may arise in differentiating the residual laser lesion from recurrent tumour. The sequential time course of the laser-induced enhancing rim with an ongoing reduction of size may be used to exclude a recurrent tumour that appears as increasing volume and new onset of mass effect [17]. There is no apparent relation between tumour histology and response to therapy.

The advantage of therapeutic control by online monitoring of the awake patient makes LITT particularly suitable in brain tumours that are located in areas of functional relevance. Thirteen patients had brain tumours located in or close to sensorimotor cortex fields.

As the majority of these tumours were benign or semibenign usually slowly growing WHO II astrocytomas, it is particularly critical to avoid posttherapeutic neurologic deficits. Using functional MRI in seven patients, the localization of the cortical motor hand area could be determined [7]. This is important, because the localization of the cortical motor hand area may vary, especially in the presence of space-occupying lesions nearby or within, due to dislocation or reorganization.

In these patients, the border of the evolving irreversible lesion facing the motor hand area was monitored and LITT was terminated when the distance was less than 8 to 12 mm. Currently available light guides for LITT are limited in respect to the maximum lesion size.

Newer developments are aimed at increasing inducible lesion size. Besides the enlargement of the fibre tip, this can be achieved by cooling the tissue areas surrounding the optical fibre tips. This additional cooling of the regions offers in principle the induction of asymmetrical tissue lesions.

The ultimate role of LITT as a therapeutic alternative for brain tumours still has to be defined. Experimental studies after stereotactic laser induced interstitial thermotherapy [SLITT) have shown that, in contrast to interstitial radiotherapy, reparative processes undoubtedly remain preserved. In chronic preparations after stereotactic laser induced interstitial thermotherapy (SLITT), not even cell necrosis or signs of demyelinization could be detected in the relative small rim of edema [18].

There are only a few preliminary reports published about the clinical applications of LITT in brain tumours. LITT was of low intraoperative morbidity and no mortality. Only one out of 16 patients who suffered a persistent deficit [19]. In this case, the laser-induced lesion exceeded the tumour margin. Transient neurological deficits occurred in the early postoperative period due to vasogenic perifocal edema. In all patients concerned, recovery was obtained within 3 to 5 weeks. These results suggest that LITT is a safe therapy if the laser-induced lesion is confined to the tumor margins. In all patients, a marked tumour reduction could be obtained. Five of nine patients with low-grade gliomas (WHO II astrocytoma) showed a neurological improvement with a decrease in severity and frequency of their seizures [19].

At present, local hyperthermia for treating tumours is performed in different ways (ultra-, micro-, radiowaves), but a number of biophysical problems arise which have not been solved satisfactorily.

Laser energy combines several advantages which lead in summary to a superiority of the laser over conventional heat sources (radiofrequency, hot tips).

The major goal of interstitial thermal therapy is the heating of a given tissue volume without adverse effects such as charring or shock waves due to vaporization at the probe tip. Consequently, the peak temperature at the probe tip must not exceed 100 °C. This limits the peak power energy per given time and probe surface area. On the other hand a minimal radiant power density is required to achieve a sufficient temperature rise for tissue destruction. In systems predominantly based on thermal probes the complete energy is thermalized at the probe/tissue. Consequently, all these systems are inflicted with a high peak power density at the probe tip.

The penetration depth of laser light into tissue, on the other hand, allows a direct energy thermalization in a tissue volume. The penetration depth is dependent on the optical tissue properties (absorption, scattering, blood content) at the wavelength used.

The result of this procedure is a lower temperature at the laser fibre compared to thermal probes of the same surface area using the same power input for both systems. Therefore, higher energies can be coupled into the same tissue volumes at identical temperatures at the fibre tissue interface. Thus, in principle, the laser is capable of heating greater tissue volumes with the same or less side effects.

No known available systems for neuronavigation in neurosurgical procedures can provide the feature of intraoperative, dynamic adaptation to an eventual shift of the target point during surgery.

This situation is common in cystic lesions, soft necrotic tumour tissue or as an effect of CSF loss or brain swelling. Only the open MR in its real-time/ online mode allows the surgeon to update his target point in its new position after shifting and displacement. A 0.5-T SIGNA SP OPEN MRI (General Electric Medical Systems) is used as a fully equipped neurosurgical operation theatre. Frameless stereotaxis, craniotomies and LITT of tumours are performed routinely. The online/ real-time-mode provides the option of frameless stereotaxis by flashpoint-tracking. This is helpful in the puncture of cystic lesions or in positioning the laser-fibre after biopsy for LITT in tumours (so that it becomes a one-time and one-place procedure). The surgeon can visualize and react to dynamic changes during surgery, related to loss of cyst fluid, CSF or necrotic regions of a glioma/tumour.

In the T1-mode with contrast medium (Gadolinium) small lesions can be targeted and identified easily, choosing an optimal approach over a minimized craniotomy. The intraoperative MR imaging is a welcome improvement on the classic, static neuronavigation, adding the important feature of a dynamic online control of surgical procedures. The latest updates and developments of the MR software allow first fusions of preoperative and actual intraoperative MR data, even in 3-D mode.

The feasibility of quantitative temperature mapping is demonstrated in an open 0.5 Tesla interventional scanner (GE, Signa SP). For calculation of temperature maps, a reference phase map is acquired at the beginning of the treatment and subtracted from the continuously updated phase maps. With the presented technique, it was possible to monitor spatial and temporal temperature changes during laser irradiation in various tissues under in vivo conditions. Since the technique presented is based on a subtraction method, one limitation is respiratory motion. While its effects can be limited by adhering to a breath holding scheme, the use of navigator echoes could further improve the reliability of this form of temperature mapping.

Nevertheless, in malignant brain tumours MR-guided laser therapy can be combined with conventional radio- and/or chemo- and/or photodynamic therapy [20]. In addition to tumour therapy, MRI-guided laser-induced interstitial thermotherapy can also be applied in eliminating therapy resistant-

epileptic foci [21]. In general, the method of stereotactic laser-induced thermotherapy (SLITT) opens up a new dimension for treatment of focal cerebral dysfunctions. An electrophysiological control during laser irradiation is possible using electrodes that are mounted on the tip of the light guides [22].

## References

[1] BOWN SG (1983) Phototherapy of tumours. World J Surg 7: 700–709
[2] SCHOBER R, BETTAG M, SABEL M, ULRICH F, HESSEL S (1993) Fine structure of zonal changes in experimental Nd:YAG laser-induced interstitial hyperthermia. Lasers Surg Med 13: 234–241
[3] KAHN T, BETTAG M, ULRICH F, ET AL (1994) MRI-guided laser-induced interstitial thermotherapy of cerebral neoplasms. J Comput Assist Tomogr18: 519–532
[4] KAHN T, BETTAG M, ULRICH F, SCHWARZMAIER HJ, HARTH T, MÖDDER U (1995) MRI-guidance of laser-induced interstitial thermotherapy of brain tumours-three-year experience. SPIE 25: 325–339
[5] BETTAG M, KUNESCH E, KAHN T, ULRICH F, SCHMITZ F, BOCK WJ (1995) Neurological and functional changes after laser-induced interstitial thermotherapy (LITT) of brain tumours. SPIE 25: 382–392
[6] KAHN T, BETTAG M, HARTH T, SCHWABE B, SCHWARZ-MAIER HJ, ULRICH F(1997) Clinical applications of laser-induced interstitial thermotherapy in the brain. In: De Salles A, Lufkin R (eds) Minimally invasive therapy of the brain. Thieme, New York, 116–129
[7] KAHN T, SCHWABE B, HARTH T, ET AL (1996) Mapping of the cortical motor hand area with functional MR imaging and MR imaging-guided laser-induced interstitial thermotherapy of brain tumours. Radiology 200: 149–157
[8] HARTH T, KAHN T, RASSEK M, SCHWABE B (1995) Temperature monitoring using fast T-1 measurement. Proceedings of the 3rd Annual Meeting, Society of Magnetic Resonance (SMR) and the 12th Annual Meeting, European Society for Magnetic Resonance in Medicine and Biology (ESMRMB), Nice Acropolis, Nice, France 2: 1170
[9] HARTH T, KAHN T, SCHWARZMAIER HJ, RASSEK M, SCHWABE B (1996) Determination of temperature distributions using echo-shifted turboflash. Proceedings of the 4th Scientific Meeting of the International Society of Magnetic Resonance in Medicine (ISMRM), New York, NY, USA, 1: 46
[10] JOLESZ FA, BLEIER AR, JAKAB P, RUENZEL PW, HUTTL K, JAKO GJ (1988) MR imaging of laser-tissue interactions. Radiology 168: 249–253.
[11] GOMORI JM, GROSSMAN RI, GOLDBERG HI, ET AL (1985) Intracranial hematomas: imaging by high field MR. Radiology 157: 87–93
[12] ANZAI Y, LUFKIN R, DESALLES A, HAMILTON DR, FARAHANI K, BLACK KL (1995) Preliminary experience with a technique for MR-guided thermal ablation of brain tumours. AJNR 16: 39–48
[13] TRACZ RA, WYMAN DR, LITTLE PB, ET AL (1992) Magnetic resonance imaging of interstitial laser photocoagulation in brain. Lasers Surg Med 12: 165–173
[14] TRACZ RA, WYMAN DR, LITTLE PB, ET AL (1993) Comparison of magnetic resonance images and the histopathological findings of lesions induced by interstitial laser photocoagulation in the brain. Lasers Surg Med 13: 45–54
[15] KIESSLING M, HERCHENHAN E, EGGERT HR (1990) Cerebrovascular and metabolic effects on the rat brain of focal Nd:YAG laser irradiation. J Neurosurg 73: 909–917
[16] EGGERT HR, KIESSLING M, KLEIHUES P (1985) Time course and spatial distribution of neodymium: yttrium-aluminium-garnet (Nd:YAG) laser-induced lesions in the rat brain. Neurosurgery 16: 443–448

[17] KAHN T, BETTAG M, HARTH T, SCHWABE B, SCHWARZ-MAIER HJ, MÖDDER U (1996) Laser-induced interstitial thermotherapy of brain tumours – value of MRI guidance. Radiologe 36: 713–721

[18] SCHOBER R, BETTAG M, SABEL M, ULRICH F, HESSEL S (1991) Neuropathological aspects of stereotactic interstitial laser therapy (Abstract). J Neuropathol Exp Neurol 50

[19] SCHWABE B, KAHN T, HARTH T, ULRICH F, SCHWARZ-MAIER H-J (1997) Laser-induced thermal lesions in the human brain: short- and long-term appearance on MRI. J Comput Assist Tomogr, vol 21, No 5

[20] ULRICH F (1991) Stereotactic interstitial laser irradiation of brain tumours: International Symposium on MR-Guided Laser Interventions, Harvard Medical School, Boston, Massachusetts, USA, Jan 28–29

[21] ULRICH F, BETTAG M (1990) Stereotactic laser therapy in symptomatic epilepsy. Second International Cleveland, Clinic Epilepsy Symposium, Epilepsy Surgery, Cleveland, Ohio, USA, June 19–23

[22] HESSEL ST, FRANK F (1990) Technical prerequisites for the interstitial thermotherapy with the Nd:YAG laser. 5th International Congress of ELA, Graz, Austria, Nov 8–10

# III-2
# Laser Applications in Ophthalmology

Michael Mrochen and Theo Seiler

## Contents

## Introduction

If one thinks about therapy by means of light, the eye comes immediately to mind. The anatomy of the eye is constructed specifically to let light into the organ and, therefore, most of the structures inside the mammalian eye can be hit by light.

The first approach to use light in ophthalmic therapy dates back to 1945 when Gerhard Meyer-Schwickerath started coagulations of the retina using focused sunlight in Hamburg. Because sunny days are rare in Hamburg (this was one of the jokes of Meyer-Schwickerath) not much later carbon arcs, Beck arcs, and the high pressure xenon light-coagulators were developed which were in clinical use until the 1970s. In the 1960s, lasers were invented and already in 1963 the ruby laser was used to create retinal coagulations, followed by the argon laser that is still used clinically all over the world. In 1980, Franz Fankhauser started a new era of laser application in ophthalmology when he performed the first intraocular photodisruption using the Q-switched Nd:YAG laser without opening the eye, a technique that we still use for capsulotomy. Not much later, in 1983, Steven Trokel introduced photoablation by means of an excimer laser into ophthalmology.

Today, we classify the therapeutic applications of laser light according to the laser-tissue interaction in photodynamic therapy, photocoagulation, photoevaporization, photoablation and photodisruption. In some cases, two or several of these interaction mechanisms may take place simultaneously (for example photoevaporization and photocoagulation), however; always a main process can be identified. Throughout this chapter we will, therefore, follow this classification. Other classifications include laser applications at different tissue structures and with different indications, for example adnexa, anterior segments, posterior segments and glaucoma which are more clinically oriented.

We do not cover in this review the topics of laser stimulation and laser acupuncture, not because we do not believe in such applications, but simply because we are not aware of well-designed well-performed and well-documented studies that have been published in peer-reviewed journals. Also, limited space in this book did not allow the discussion of diagnostic laser applications. In addition, this field is experiencing very rapid changes and information is very short-lived.

Ophthalmic laser application is today a substantial part of ophthalmic therapy and in this chapter we describe both well-accepted therapies like retinal photocoagulation and new applications like resurfacing of periocular skins or Er:YAG laser vitrectomy.

Laser therapy is still a rapidly evolving subdiscipline, and we are sure that within the next few years with new technology also new applications will be developed and others will disappear. Many of these new applications, however, will again take place in ophthalmology, because the eye is the organ par excellence for lasers.

## Photocoagulation

Photocoagulation bases on the transformation of the electromagnetic energy of the light beam into heat at the target. This change is accomplished by absorption of the light by atoms and molecules that pass this energy onto their environment, in most cases as kinetic energy which means that the temperature in the molecular environment of absorbers increases. Biologic tissue is (optically) an inhomogenous matter because it consists of many types of organic and inorganic components with different absorption properties. Molecules that are capable of absorbing light of a specific wavelength range (more strongly than other wavelengths) are called chromophores. Some of the chromophores demonstrate strongly wavelength-dependent absorption (such as haemoglobin) and, therefore, appear colourful; others absorb light in broad bands and therefore, appear dark (melanin) or even transparent (cornea) and white (sclera). For the application of photocoagulation in the eye the following chromophores are important and will be discussed in detail: xanthophyll (yellow dye of the retina), haemoglobin (blood dye), melanin (dye of the uvea and retinal pigment epithelium) and water (absorption in the mid-infrared).

Before we discuss the absorption spectra of these chromophores, however, a brief description of the anatomy of the human eye (Fig. 1) and the optical properties of the structures of the eye is helpful for understanding treatment strategies.

Approximately 70 to 80% of the input of the human brain originates from vision in the form of electrically coded information. The transfer of the image of the environment into electrical signals occurs in the retina, a part of the brain that forms a photosensi-

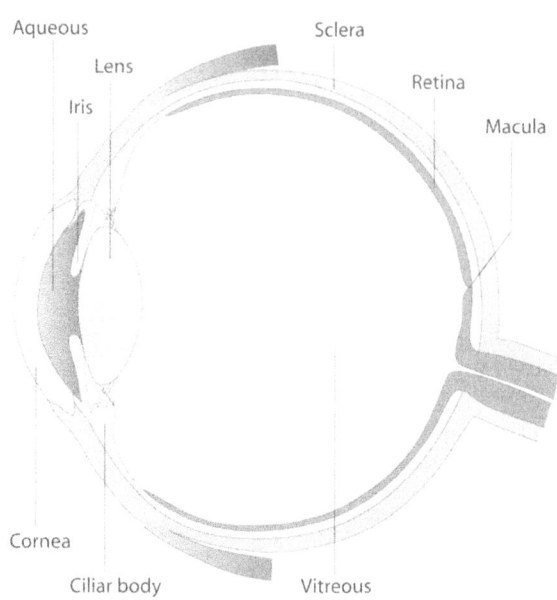

**Fig. 1.** Schematic cross-section of the human eye. The retina covers the inner wall of the eyeball

tive layer at the inner side of the eyeball. The retina is nourished by two parallel pathways: the retinal vasculature that is supporting the inner third of the retina, and the powerful complex choroid/pigment epithelium, that supports the outer layers of the retina where a much higher cellular energy turnover than in the inner retina takes place because of the photoreceptors and their metabolism. The rest of the eye structure's duty is to protect the retina, to establish constant mechanical and chemical conditions and to form constant and good optics. The optical apparatus that projects the image onto the retina consists of the cornea ($\frac{2}{3}$ of the optical power), the lens ($\frac{1}{3}$ of the optical power) and the iris that builds the pupil.

Light entering the eye has to pass the cornea, aqueous, lens, vitreous and inner retina to be transformed into an electrical signal in the outer retina. The optical properties of vitreous and aqueous are practically identical to those of water, and even the cornea's spectrum is similar to water absorption (Fig. 2). The cornea is transparent in the near ultraviolet (UV), optical range and near infrared (IR), but absorbs strongly in the far UV and mid-IR, where photoablation (excimer laser) and photocoagulation (Ho:YAG laser, laser diodes) of the cornea find clinical applications. The transmission of the sclera important for transscleral applications is more complicated. Here, scattering and absorption contribute to a reduced transmission of the light. Both processes, however, decrease with increasing wavelength and as a consequence, the transmission increases from 10% at

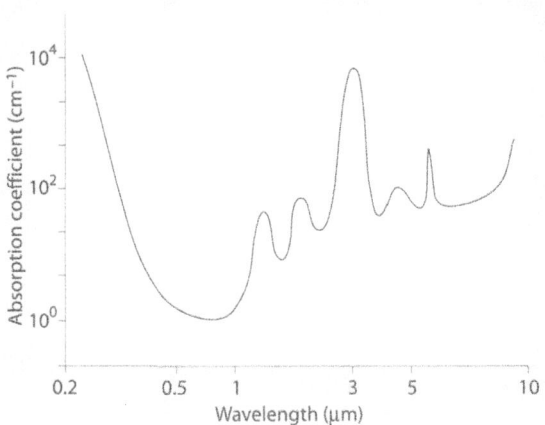

**Fig. 2.** Absorption spectrum of the cornea as a function of the wavelength. The cornea is transparent from near-UV to near-IR

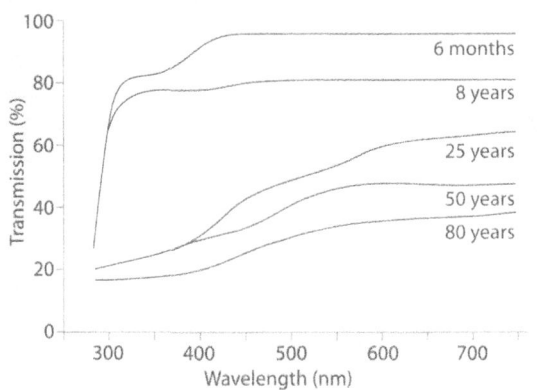

**Fig. 3.** Absorption of the lens as a function of the wavelength and the age of the patient. Only young children have clear lenses

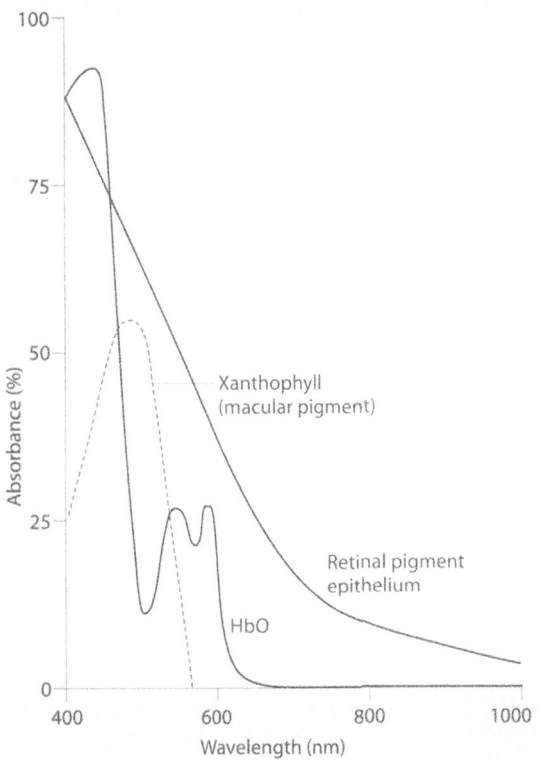

**Fig. 4.** Absorption spectra of different biological dyes present in the human eye. Xanthophyll appears yellow because it absorbs mainly in the blue range. Oxygenized and desoxygenized haemoglobin is red because of the absorption in the yellow/green

500 nm to approximately 53% at 1200 nm [1]. The absorption spectrum of the lens is strongly age-related (Fig. 3). A clear lens (fully transparent) can be expected only in very young children and already the juvenile lens shows the typical absorption that is stronger in the blue/UV range than at longer wavelengths. This absorption steadily increases throughout our live.

In Fig. 4, the absorption spectra of the three dyes that occur in the posterior part of the human eye are displayed. The xanthophyll is located in the inner retina concentrated towards the posterior pole of the eye (macula lutea). Whether it really plays an (unwanted) role during photocoagulation of the retina is not yet clear because of its relatively low concentration in the retina. Nevertheless, during clinical applications today blue laser light is avoided for central coagulations (for example argon all line). Haemoglobin

has to be divided in oxygenized (arterial blood) and deoxygenized (venous blood) haemoglobin because the absorption spectra are slightly different. Both types of haemoglobin absorb strongly in the blue/UV range with a relative minimum at approximatively 500 nm (green/blue) and a relative maximum at 570 to 590 nm (yellow). Towards the red and infrared (wavelengths longer than 620 nm), the haemoglobin appears more or less transparent. Melanin demonstrates good absorption within the optical range, however, clearly absorbs less towards longer wavelengths. This is an important aspect because longer wavelengths obviously can penetrate deeper into melanin-containing tissue (such as retinal pigment epithelium, choroid and inner sclera).

The increase in temperature induced by laser light in the centre of an irradiated area is dependent on geometrical and physical parameters of the beam and tissue. Parameters to be adjusted by the surgeon are spot size, beam power and application time. Physical models to explain the temperature increase dependent on these parameters exist, but it is beyond the scope of this chapter to describe these. In Fig. 5 the most important parameter, application time and its

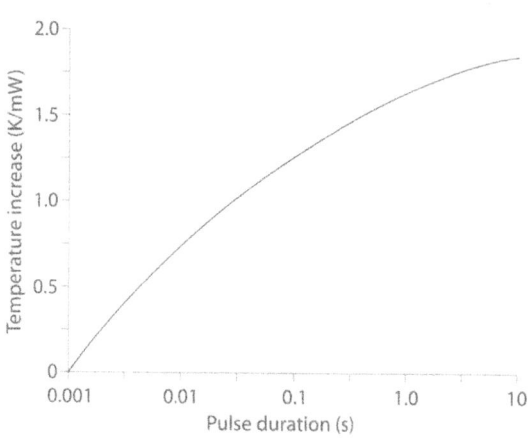

**Fig. 5.** Temperature increase in the centre of a 20-μm spot as a function of the exposure time

**Table 1.** Application modes of photocoagulation in ophthalmology

| Application | Contact | Non-contact |
| --- | --- | --- |
| • posterior segment (Argon, krypton 2 × Nd:YAG, dye, laser diode) | No | Slit lamp delivery Indirect ophthalmoscopy fibre/hand piece |
| • Glaucoma (Argon, dye, cw Nd:YAG) | Fibre/hand piece for transscleral appl. | Slit lamp delivery |
| • Refractive (Ho:YAG, laser diode) | Fibre/hand piece | Slit lamp delivery |

influence on the temperature increase are depicted. Note the logarithmic time scale; typical treatment times are 0.1 to 1 s.

As a last aspect in this section, the biological reaction to photocoagulation may be discussed. During photocoagulation the temperature produced exceeds 100° C and denaturation of the proteins is intended to destroy tissue. The consecutive inflammation processes induce the removal of such denaturated proteins by phagocytosis and, in a second stage, replacement by scar tissue. On the other hand, also proliferation of less damaged adjacent tissue may occur. At the retina during photocoagulation diffusion barriers (pigment epithelium/choroid) are disrupted and, therefore, facilitation of diffusion of oxygen and glucose is achieved as a desired side effect.

## Technical Aspects

The following lasers are currently in clinical use for photocoagulation in ophthalmology, ordered by occurrence: (1) argon laser with wavelength selection (514 nm), (2) laser diodes with wavelengths ranging from 650 to 1900 nm, (3) cw frequency-doubled Nd:YAG laser (532 nm) replacing the argon laser (514 nm) step by step, (4) dye laser, (5) krypton laser with two wavelengths in yellow and red, (6) cw Nd:YAG laser (1064 nm) for cyclodestruction and (7) Ho:YAG (2100 nm) laser for coagulation of the cornea.

Ophthalmic lasers are defined by special laser parameters and application modes (Table 1). Except for the Ho:YAG laser, virtually all lasers are run in the cw mode with exposure times ranging from 0.05 to several seconds (typically 0.1 to 1.0 sec). The maximal cw power applied is less than 2 W. The spot size at the target ranges from 50 μm (laser trabeculoplasty) to

2 mm (coagulation of a choroidal tumour). The application is regularly performed via the slit lamp, the operation microscope or fibres directly. The clinical use of the slit lamp for photocoagulation needs a special adapter that couples optically the laser light transmitted from the laser by means of a quartz fibre into the viewing path. The incomplete mirror usually is connected to a micromanipulator that permits small adjustments. In special cases, the fibre is used directly with special hand pieces, for example during endophotocoagulation in vitrectomy and cyclodestruction for glaucoma. Also, photocoagulation may be accomplished by indirect ophthalmoscopy, where the fibre is coupled into the headlight of the surgeon, which is beneficial in treating children and during retinal surgery.

## Posterior Segment of the Eye

The main chromophore for photocoagulation at the posterior segment is melanin which absorbs over the whole optical spectrum; however, penetration increases with longer wavelengths (krypton, laser diode). The site of coagulation depends on the presence of melanin within the complex retinal pigment epithelium/inner choroid, however, longer wavelengths may shift the coagulation towards the sclera and shorter wavelengths towards retina. The strength of the coagulation is adjusted under view and a whitish effect should become visible immediately after exposure.

Since photocoagulation implies destruction of tissue, it is important to understand the strategy of such a treatment. There are three main effects that may be achieved: (1) to destroy tissue or tissue layers; (2) to enhance diffusion through Bruch's membrane; (3) to induce scarring. The corresponding pathologies to be treated by photocoagulation are microangiopathy of the retina (diabetic retinopathy, venous occlusion), tumours of retina and choroid, prophylaxis of retinal detachment and age-related macular degeneration

(AMD). The strategic goals of photocoagulation in microangiopathy are threefold: to destroy consumers to compensate malnutrition, which means photoreceptors and pigment epithelium must be destroyed but unfortunately the underlying choriocapillaris is also damaged. Nevertheless, it was shown that oxygen in the vitreous next to the coagulations increased, indicating an improved diffusion from the choroid. Since many cases need additional ocular surgery, such as vitrectomy, also a fixation of the retina by scarring is of benefit. Regarding tumours, the goal is clearly destruction with minimal side effects. The prophylaxis of retinal detachment is scarring around lesions known to develop retinal detachment such as some forms of equatorial degenerations. Finally, in AMD the photocoagulation intends to destroy subretinal neovascularizations and to improve diffusion from the choroid.

A detailed review of the clinical benefits and risks of photocoagulation of the posterior segment may be found in Bloom and Brucker [2] summarizing the results obtained until 1990. Meanwhile, some technological aspects may have changed due the introduction of new lasers such as the frequency-doubled cw Nd:YAG laser and various laser diodes; however, the indications and contraindications of photocoagulation have not changed much.

Diabetic retinopathy (DR) is still the most frequent indication for retinal photocoagulation. Using the clinical classification in background DR, preproliferative DR, proliferative DR and diabetic maculopathy, photocoagulation has been proved beneficial only in proliferative DR and diabetic maculopathy. The optimal technique for proliferative DR is panretinal scatter laser coagulation with 2000 spots and more per eye (0.25 mm to 0.5 mm in diameter) sparing the macula. In cases of preexisting macular edema, panretinal photocoagulation appeared to aggravate diabetic maculopathy [3] and, therefore, macular edema should be focally treated prior to initiating panretinal photocoagulation. In a controlled study [4] the risk of severe visual loss was 6% in eyes treated with laser photocoagulation compared to 16% in control eyes within 2 years of follow-up. Later studies confirmed this significant benefit of panretinal photocoagulation in proliferative diabetic retinopathy.

Diabetic maculopathy may be treated either by focal photocoagulation or grid photocoagulation depending on fluorescein angiography and clinical estimation of the macula edema. The ETDR study (early treatment diabetic retinopathy study) found in patients with clinically significant macular edema that photocoagulation decreased the risk of severe visual loss by about 50% [5]. It was concluded that photocoagulation is the only treatment that has been proved to be beneficial for diabetic macular edema.

Retinal vein occlusion may lead to loss of visio due to macular odema and to retinal neovascularization with the risk of severe secondary glaucoma. Regarding macular odema, grid photocoagulation led to significant better visual results compared to untreated control eyes [6]. Early panretinal photocoagulation appeared to decrease the risk of developing neovascularization and of vitreous haemorrhage. Rubeosis iridis and neovascular glaucoma were significantly reduced.

Age related macular degeneration may induce subretinal neovascularization, leading to severe visual loss due to macular edema and subretinal haemorrhage. In cases of extrafoveal and juxtafoveal neovascular membranes, photocoagulation by means of krypton, argon (514 nm) and dye lasers (approximately 600 nm) was recommended by several studies. Laser treatment of subfoveal lesions, however, yielded only questionable results, and photodynamic therapy may be the better alternative.

Malignant melanoma of the uvea is the most common primary intraocular malignancy; however, only 5 to 10% of melanomas are treatable with laser surgery because only small posterior pole tumors seem to show adequate response (diameter < 10 mm, height < 3 mm). Multiple treatment sessions are employed. No prospective data are currently available but clinical cure was reported in as much as 71% [7]. On the other hand, many histopathologic studies have shown tumor to be present even when the patient was clinically „cured".

Retinal breaks may be considered as a risk factor for retinal detachment only if they are symptomatic (associated with photopsia and/or entopsia), and prophylactic treatment should be considered in cases of retinal breaks caused by vitreoretinal traction. The goal of treatment is to seal down the retina surrounding a retinal break with a chorioretinal scar.

Aside from these clearly defined pathologies, laser photocoagulation has been found beneficial in other disorders such as toxoplasmosis, acute retinal necrosis syndrome, cytomegaly retinitis, miscellaneous retinal neovascularization and optic nerve head pit.

### Anterior Segment of the Eye

Nearly all clinical applications of laser photocoagulation of structures of the anterior segment are focused on the treatment of glaucoma. Three types of glaucoma treatment are worth discussion: (1) laser trabeculoplasty, (2) laser iridectomy and (3) cyclophotocoagulation. Besides these, only laser thermokeratoplasty for hyperopia correction is another type of photocoaculation of the anterior segment to be cited.

During laser trabeculoplasty, small coagulations are created in the trabecular meshwork by means of a gonioscope. Argon lasers are mostly used for this application. It is believed that scarring consecutive to the coagulation and contraction of the scars may widen the pores of the trabecular meshwork, thus facilitating outflow of the aqueous. Typical laser parameters are a spot size of 50 μm, exposition time 0.2 s., and power setting 0.5 to 1.0 W. The efficacy of laser trabeculoplasty is limited: depending on the type of glaucoma 50 to 90% of the eyes treated show a significant reduction of intraocular pressure after 6 weeks. At three years after laser trabeculoplasty, 40 to 70% of the glaucomas were still regulated however, the effect is decreasing further with time. Although this treatment seems not to be a final solution of the glaucoma, in the majority of the cases it has gained some clinical value by delaying invasive glaucoma surgery for years.

In eyes with a narrow chamber angle, glaucoma may occur because of forward bulging of the iris, thus blocking the outflow of the aqueous. The therapy of this so-called narrow-angle glaucoma is an iridectomy to equalize the hydrostatic pressure in the posterior and anterior chamber by means of a hole in the peripheral iris. Such a hole can be created surgically, by strong photocoagulation (rather photoevaporation) and by photodisruption with the pulsed Nd:YAG laser.

Surgical iridectomy has the disadvantage of an invasive and eye-opening intervention such as infection and inflammation. The iridotomies performed with the Nd:YAG laser are small and tend to close. In order to prevent such healing, the iridotomy may be enlarged by photocoagulation spots around the laser iridotomy. On the other hand, in eyes with a too narrow angle, the iris is very close to the cornea, and by the combined laser treatment the corneal endothelium may be damaged.

By destruction of the ciliary body the formation of aqueous can be reduced and, therefore, the intraocular pressure lowered. Although the locus generis of the glaucoma is located in an increased outflow resistance of the aqueous in the trabecular meshwork in some cases, for example secondary glaucoma, cyclodestruction represents the only alternative. To perform transscleral cyclocoagulation by means of light, the radiation has to penetrate conjunctiva and sclera and the appropriate wavelength range is the near-IR. So far, cyclophotocoagulation has been achieved by means of laser diodes (810 nm and more) and the cw Nd:YAG laser (1064 nm). Typical laser parameters for the Nd:YAG laser are up to 0.6 mm spot size, exposure time 0.2 s. and power up to 20 W. Up to 40 coagulation spots can be set during one session. In many cases, several treatments are necessary for a continued ef-

fect. Even in refractory glaucomas this approach (using the diode) has proved to be a safe and effective method [8].

Another application of photocoagulation at the anterior segment is the laser thermokeratoplasty, where shrinkage of the peripheral corneal collagen induces steepening of the central cornea. This procedure is considered safe and effective for corrections of hyperopia up to + 2.0 D. Sixteen coagulations are positioned in two concentric rings with diameters of 6.0 mm and more. The procedure may be performed by a contact technique with special hand pieces or in the non-contact mode via slit lamp delivery. It is a simple and short operation; however, regression of the effect takes place during the first year post-op [9]. An initial overcorrection of 100% and more is intended. The major disadvantage is changing refraction for months and a relatively small residual effect. This procedure was successfully used to compensate for overcorrections after myopic PRK [10].

## Photodynamic Therapy (PDT)

Photodynamic therapy has been in clinical use for more than 10 years, however, it has gained clinical attention in ophthalmology only during the past 5 years because of beneficial effects in treating some forms of age-related macular degeneration. This technique involves intravenous injection of a photosensitizer that accumulates in neovascular and tumour tissue. The photosensitized tissue is then irradiated by light at the absorption maximum of the photosensitizer dye leading to cytotoxicity. In the past, various porphyrin derivatives were tested, but none of the dyes created damage selective enough to treat intraocular pathologies. With the development of verteporfin, a benzoporphyrin derivative complexed with low-density lipoproteins or formulated as a liposomal preparation, the disappointing situation changed dramatically. The results of the prospective trials created a sensation in 1999, leading to a socio-economic discussion because the treatment is very expensive. Verteporfin is a safe drug for human use and has been used in clinical trials in dermatology [1].

Age-related macular degeneration (AMD) is the most frequent cause of legal blindness in the elderly in the western world. In a fair percentage, the loss of vision occurs due to subretinal neovascularization that originates from the choroid. Natural barriers like Bruch's membrane and the retinal pigment epithelium are disrupted and the barrier functions diminish, creating a macular edema with consecutive decrease in vision. In a later stage, such neovascularizations may bleed and the subretinal haemorrhage leads to further deterioration of visual acuity close to legal blindness. All this happens in the majority of the cases at the posterior pole of the eye close or at the fovea centralis, the centre of vision. Approximately 10% to 20% of the cases of neovascular AMD are eligible for laser treatment where the neovascularization area is destroyed with thermal photocoagulation. In detail, photocoagulation applies only in small neovascular membranes that are located at least 400 μm away from the fovea. Alternative treatments, such as γ- or β-irradiation and medical treatment, have not been effective or safe. Surgical interventions, such as membranectomy or macula rotation are currently being investigated, however, the results reported are not very encouraging.

In animal experiments, optimal treatment parameters were identified as leading to an absence of angiographic leakage and histologic occlusion of the neovascularizations [2]. At the parameters required for closure of the neovascular membrane, one sees in normal structures adjacent to the membrane choriocapillaris occlusion, retinal pigment epithelium cellular necrosis and mild pyknosis of the outer nuclear layer of the retina; at 4 to 7 weeks after PDT these lesions show generalized recovery. The retinal pigment epithelium repopulates the treated area, even when larger areas (up to 4 mm in diameter) are treated. These experiments led to the conclusion that PDT using verteporfin is selective and effective enough to enroll human trials.

Three multicentre clinical trials were published in 1999 to evaluate the safety and efficacy of PDT with verteporfin in patients with choroidal neovascularization (CNV) secondary to age-related macular degeneration. The laser used was a diode laser emitting at a wavelength of 690 nm. The beam was placed over the lesion at the fundus and irradiation began 15 minutes after the start of the infusion. The delivered irradiance was kept constant at 600 mW cm$^{-2}$. In one study [3] including 97 patients, 5 dosage regimens were tested to evaluate adverse effects and efficacy in terms of short-term cessation of fluorescein leakage. Except for non-perfusion of neurosensory retinal vessels at a light dose of 150 J cm$^{-2}$, no other adverse events were noticed. In the second trial, retreatments 2 to 4 weeks after the initial treatment were performed [4]. The rate and severity of ocular and systemic adverse events were not increased by multiple applications. The TAP Study Group consisting of 22 ophthalmology practices in Europe and North America presented the results of verteporfin PDT treatments in 609 patients [5] (TAP Report 1). Inclusion criteria included CNV lesions of 5.4 mm or less in diameter and a minimal visual acuity of 0.1 (20/200 or better). The study was planned as a multicentre, double-masked, placebo-controlled, randomized trial. The placebo group received laser treatment identical

to the verum group; however, instead of verteporfin, dextrose was administered intravenously. Fifteen min after the start of the infusion, a laser light at 689 nm (diode laser with slit lamp delivery system) delivered 50 J cm$^{-2}$ over 83 s using a spot size with a diameter 1 mm larger than the greatest linear dimension of the CNV lesion. At any post-op examination up to 12 months after treatment, visual acuity, contrast sensitivity and fluorescein angiographic outcomes were better in the verteporfin-treated than in the placebo-treated eyes. At 12 months post-op the visual acuity benefit (< 15 letters lost) of verteporfin therapy was clearly demonstrated (67 vs. 39%, p < 0.001). Although this result appears not overwhelming it is important because no alternative treatment is available. Injection-site adverse events (13% vs. 3%) and transient photosensitivity reactions (3 vs. 0%) were of concern. The study group recommended verteporfin therapy for treatment of patients with predominantly classic CNV from AMD.

Besides these encouraging results, also in 1999 another trial including only 16 eyes demonstrated similar results using simple laser diode treatment only (810 nm) of a much higher light dosage (up 100 times more) [6].

The clinical value of the PDT treatment of subfoveolar neovascularization may further be optimized; however, already in the currently used mode it is appreciated because of the lack of alternatives. Long-term follow-up of these treatments and trials using other photosensitizes is eagerly awaited.

## Photoevaporization

During the past few years, lasers emitting at 3 μm have created an increasing interest as operation instruments for ophthalmic surgeons. With wavelengths in this spectral range, a high precision and good ablation efficiency combined with a relatively small lateral thermal damage can be achieved in soft and hard tissue ablation due to the high absorption coefficient of the 3 μm radiation in water (absorption coefficient $\alpha \approx 10\ 000$ cm$^{-1}$). Especially the erbium:YAG laser (Er:YAG) emitting at a wavelength of $\lambda = 2.94$ μm and transmittable through optical fibers (such as zirconium-fluoride or sapphire) was proved to be a useful tool for cutting and drilling water containing-tissue [1].

### Laser Vitrectomy

The Er:YAG laser was tested in different fields of ophthalmic application such as corneal ablation for refractive surgery or sclerostomy for glaucoma treatment [2, 3]. However, none of these techniques has gained widespread clinical acceptance. The main reasons for this seem to be higher wound-healing effects, longer treatment times and, unfortunately, higher purchase and maintenance costs compared to standard treatment techniques. However, there are other ophthalmic procedures such as phacoemulsification and vitrectomy that involve cutting and removing tissue from inside the eye. Each of these surgery techniques involves the insertion of a probe into the patient's eye, where tissue may be photoevaporized and then removed by an aspiration unit.

The major contribution of underwater tissue photovaporization with the Er:YAG laser is the expansion of laser induced vapour bubbles. Hereby, the erbium-laser radiation is capable of ablating tissue in a non-contact mode under water. The laser radiation is transmitted through a water-vapour channel initiated around the fibre tip by the early portion of the laser pulse. This is possible because steam, in contrast to liquid water, has a 1000 times lower density and, in addition, a lower absorption coefficient [4]. The formation and the collapse of this vapour bubble, the so-called Moses effect, are both accompanied by pressure transients and demonstrate complex physical dynamics. It has been shown that the formation of the vapour bubbles can be effectively controlled by choosing adequate fibre-tip and hand piece geometries as well as optimized laser parameters [5–7]. Beside this, the induced pressure transients are spherically shaped, and this results in a drastic decrease of the pressure amplitudes with distance (pressure amplitude ~ 1/r).

The removal of the gel-like vitreous from within the eye while still maintaining a normal intraocular pressure, the so-called vitrectomy, is one of the most delicate surgical problems today. The surgical situation for vitrectomy is shown in Fig. 6. The operating site is visualized though the cornea and the lens with the assistance of an operating microscope, while the implements for operating (vitreous cutter, optical fibres for illumination, infusion etc.) are inserted through pressure-tight holes in the pars plana, a region in which the retina is both firmly attached and bloodless. The normal internal pressure of the eye is maintained and is required for control of bleeding.

Mechanical cutting systems have become standard to remove altered vitreous and to fragment intravitreal structures. Most of these devices are based on the guillotine principle developed and improved in the early 1970s [8]. The mechanical cutting procedure necessarily includes traction and shear forces that act on the fragile retinal and vitreal structures and as a result intraoperative complications may occur. Ideal tissue cutting is accomplished with low-suction, high-cutting velocity and frequency and large aspiration ports [9].

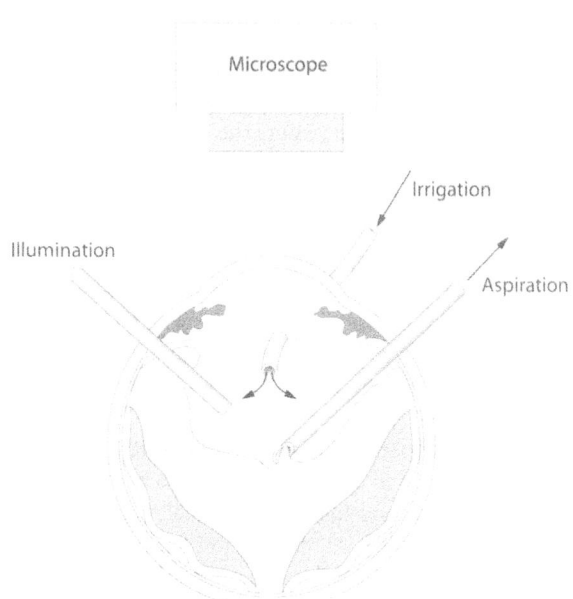

**Fig. 6.** Surgical situation during pars plana vitrectomy. The whole surgery is performed within the vitreous under microscopic control

Unfortunately, standard mechanical cutting systems are limited to a cutting frequency of less than 20 Hz, due to their mechanical nature. Beside this, the cutting velocity is limited to few ms and, as a result, standard suction forces of more than 200 mmHg are routinely used during surgery. To overcome such high suction forces, one may consider that the same aspiration flow can be achieved by increasing the cutting rate and cutting velocity. The use of modern Er:YAG lasers enables cutting rates (repetition rates) up to 200 Hz.

An appropriate laser hand piece for pars plana vitrectomy consists of a quartz rod (core diameter 320 µm) mounted inside an interchangeable microsurgical probe (Fig. 7). The microsurgical probe (length 23 mm) has an outer diameter of 0.9 mm and the inner diameter of the aspiration port is 0.6 mm. The hand piece consists of stainless steel and is sterilizable by standard methods. The radiation of the Er:YAG laser can be transmitted by mid-infrared optical fibres such as zirconium fluoride and is optically coupled into the quartz rod.

In vitro experiments as well as clinical studies have demonstrated a high efficiency for vitreous liquefaction and fragmentation [10–12]. Furthermore, it has been shown that the high cutting rate available with modern Er:YAG laser systems results in a reduced suction force during vitrectomy. Petersen et al. [11] have demonstrated that the averaged suction force was significantly smaller in a group of patients treated with an Er:YAG laser (40 mmHg) compared to

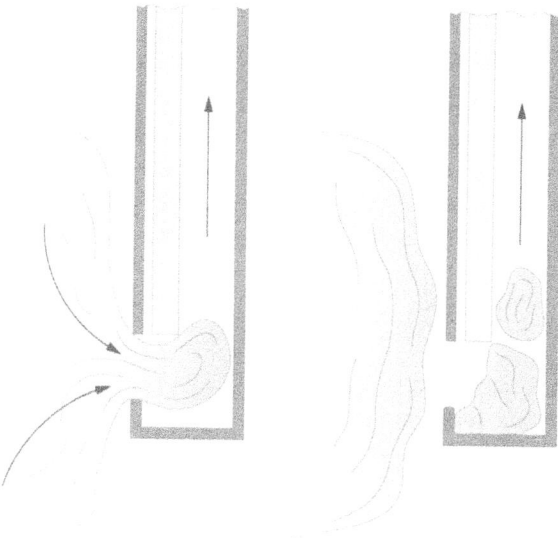

**Fig. 7.** Principle of a microsurgery probe used for erbium:YAG laser vitrectomy. The vitreous is sucked into the aspiration port of the microsurgery probe by the suction forces (left). The laser-induced evaporation bubble is capable of cutting the vitreous within the microsurgery probe (right).

a control group treated with a standard mechanical vitrectomy unit (180 mmHg). This reduction in suction force did not increase the vitrectomy surgery time.

The Er:YAG laser vitrectomy is currently under clinical investigation at different clinical sites. Binder et. al. [12] as well as Petersen et. al. [11] have reported that no laser associated complications occured within an observation time of more than 6 months on using cutting rates up to 70 Hz. Thus, the Er:YAG laser seems to be an appropriate tool to decrease the suction forces and decrease the risk of intraoperative complications caused by mechanical stress.

Besides this application, some authors have demonstrated the feasibility of cutting vitreoretinal structures and membranes [13–15]. The advantages and limitations of this technology in contemporary vitreoretinal surgery were explored in a clinical study that was followed by a multicentre study. The suitability of this laser for precise tissue cutting and ablation in vitreoretinal surgical manoeuvres was accompanied by a high degree of safety. However, a clinically acceptable Er:YAG laser vitrectomy system should be able to remove vitreous structures and cut strains and membranes. In future laser vitrectomy units, intraoperative photocoagulation should also be incorporated.

## Periocular Laser Surgery

Common uses of lasers in the periocular region may be divided into incisional (blepharoplasty) and resurfacing applications. For laser-assisted blepharoplasty and ptosis procedures, the $CO_2$ laser has been used since the early 1980s, but there is still some doubt whether laser application is advantageous compared to conventional surgery. Several reports suggest some relative advantages including decreased operation time, reduction of intraoperative bleeding and decreased postoperative pain and ecchymosis with faster recovery. Long-term results of laser-assisted procedures, however, seem not to differ significantly from those of cold-steel procedures [16]. On the other hand, increasing public awareness of laser-assisted surgery may represent a certain advantage to the surgeon who chooses to use a $CO_2$ or Er:YAG laser in the highly marketing-sensitive field of cosmetic eyelid surgery [16]. This argument is even more important in the performance of relatively new non-incisional procedures such a cutaneous laser resurfacing for the treatment of rhytides, wrinkles, photoaging, acne scars and other skin lesions in the periocular region. Any laser surgery for resurfacing must compete with conventional modalities including mechanical dermaabrasion and chemical peeling techniques that share a number of potential limitations: (1) the inability to control the depth, (2) variability and unpredictability in the wound healing process, (3) potential laryngeal edema and cardiac complications with deep chemical peels and (4) significant postoperative hyper- and hypopigmentation.

With the $CO_2$ laser (wavelength 10.6 μm) as well as the Er:YAG laser (wavelength 2.94 μm) the target of laser-tissue interaction is intra- and extracellular tissue water because both lasers emit at wavelengths of strong absorbance in water. The penetration depth is in the order of 10 to 30 μm in skin. The laser beam induces a sudden temperature rise in tissue water far beyond the boiling point, inducing its vaporization. However, a variable amount of heat is conducted to surrounding tissue, causing coagulative necrosis and other lateral thermal damage. This lateral thermal damage is responsible for the majority of unwanted effects secondary to laser treatment and, therefore, must be minimized by selection of appropriate laser parameters. The thermal relaxation time for skin has been determined at approximately 700 μs [17], but often ranges from 300 μs to 1 ms. Clearly, the pulse duration of the ideal laser should be less than 500 μs with a relatively constant energy level within the pulse. The estimated energy density (fluence) per pulse necessary to vaporize tissue is about 5 J cm$^{-2}$ [18]. These conditions are met by ultrapulsed $CO_2$

lasers (better than superpulsed $CO_2$ lasers) and by free-running and Q-switched Er:YAG lasers. The Er:YAG laser is more efficiently absorbed by water and collagen than the $CO_2$ laser, inducing less thermal damage. However, due to a lesser induction of heat, the Er:YAG laser lacks haemostatic effects and induces less collagen shrinkage [19].

Histologic studies of laser resurfacing demonstrated that ultra- and superpulsed $CO_2$ lasers removed skin precisely and bloodlessly without char formation and with thermal damage zones of 70 to 80 µm, one-tenth the amount of thermal damage reported with cw $CO_2$ lasers.

Also, there were no deleterious effects in wound healing in any of the studies [20–22]. A subepidermal collagen repair unique in laser therapy may correspond to the improvement in clinical appearance noted after laser resurfacing as compared to chemical peel treatment.

One of the results of clinical studies on laser resurfacing was that the outcome, especially postoperative hypo- and hyperpigmentation, is closely related with the skin type as classified by history [23] (Table 2). Hyperpigmentation is most commonly seen in darker skin types (types III to VI) and in patients with a history of postinflammatory pigment alteration. In such patients, pretreatment with bleaching agents 2 to 8 weeks preoperatively is recommended. Although long-term follow-up is limited, a number of clinical studies describe short-term results and complications associated with cutaneous laser resurfacing. Fitzpatrick et al., using ultrapulsed $CO_2$ laser to treat 73 perioral and 38 periorbital cases, reported a good response with improvement of 45–50% [23] . Twenty-seven % of patients with skin type III and more experienced hyperpigmentation that resolved over the ensuing 2 to 4 months of appropriate topical treatment, and 31% of the patients had erythema that persisted at 90 days post operation. Alster [24] found marked improvement in all patients and all patients were able to return to their regular activities within 1 week of treatment (with camouflage makeup). Similar results were reported by Waldorf et al. [25], who categorized the outcome as excellent for the periorbital and perioral areas treated.

Eythema is the most frequent sequel after laser resurfacing and may persist for 6 to 12 weeks with the $CO_2$ laser and for 4.2 ± 1.5 weeks with the Er:YAG laser [26]. Excessive erithema occurs in 5 to 8% of the cases and is a risk factor for hypertrophic scarring and atrophic textural changes. A recent survey in the USA revealed that 6.5% of patients developed hyperpigmentation after laser resurfacing whereas hyperpigmentation was present in only 3.2% of the patients [27]. Bacterial (6.5%) and viral infections (1.7%) may follow periocular laser surgery and may need appropriate systemic and topical medical treatment [27]. Superficial fungal infection seems to be rare. Hypertrophic scarring, a severe complication with disastrous consequence may occur in 0.9% of patients and the incidence of corneal injury was 0.3% [27].

In conclusion, recent advances in $CO_2$ laser technology and the introduction of Er:YAG lasers have facilitated the development of new applications for rejuvenation of facial skin and have increased the popularity of periorbital laser surgery over the past few years. However, potentially severe complications in the perioocular region may follow these aesthetic procedures. An appropriate selection of patients, an adequate pre-operative treatment, proper training of the surgeon and a cautionus approach are strongly recommended to minimize hazardous outcomes [28].

## Photoablation

More than 15 years after the discovery that corneal tissue can be micromachined in a non thermal fashion with extraordinary precision, the application of the ArF-excimer laser on the cornea for myopic, hyperopic and astigmatic corrections has been gradually accepted by the ophthalmic community [1]. Excimer lasers are gas lasers in which the lasing medium inside the resonator consists of a gas mixture that is excited by special pretreatments. Depending on the gas composition used, an excimer laser is able to emit laser light at various wavelengths, although only the 193 nm wavelength has gained clinical attention in ophthalmology. This far-ultraviolet light is obtained by means of a mixture of argon (Ar) and fluorine (F) gas inside the laser tube. The pulse duration of such excimer lasers used for refractive surgery is in the order of 20 ns.

In the year 1983, Trokel et al. [2] realized that intense excimer laser light can be used not only for etching plastics, but also for corneal tissue. Since then, many publications have reported the physics of this „photoablative decomposition" or simple „photoablation" laser-tissue interaction process. However,

**Table 2.** Fitzpatrick skin classification

| Skin type | Skin colour | History |
|---|---|---|
| I | White | Always burns, never tans |
| II | White | Usually burns, tans less than average |
| III | White | Sometimes mildly burns, regularly tans |
| IV | White | Rarely burns, tans more than average |
| V | Brown | Rarely burns, tans profusely |
| VI | Black | Never burns, deeply pigmented |

there is still confusion in the literature whether this process is photochemical or photothermal in nature, although the best model probably employs a combination of these processes. Basically, the intense 193-nm radiation is mainly absorbed by the collagen macromolecules of the cornea at the beginning of the laser pulse. Here, the high photon energy of the 193 nm wavelength is capable to disrupt the chemical bonds of the molecules such as C-C (photochemical process). However, the later portion of the laser pulse is more strongly absorbed by the tissue than is the early part. The major part of the incoming energy is then converted into heat by intersystemic energy transfer. The increase in temperature results in a breaking of the hydrogen bonds of tissue water and, as a consequence, the water of the cornea becomes a strong absorber at a wavelength of 193 nm (photothermal process). The material undergoes a phase transition into gas that is heated to several hundred degrees Celsius during the photoablation process. This hot gas, including the protein fragments, bursts out due to increased pressure. Gas chromatography and mass spectroscopy of the expelled products have revealed molecules typical of thermal changes in the protein fraction. In contrast, the remaining tissue shows only minimal signs of thermal processing since the photoablation process occurs within a few nanoseconds. Thus, the photoablation process was termed cold laser ablation.

The major advantage of ArF-excimer lasers for photorefractive surgery is the high precision obtained by corneal photoablation. A common method of studying photoablation is to irradiate a sample with series of laser pulses and then to measure the resultant etch crater depth. A collection of reported corneal etch data for the 193 nm laser radiation is shown in Fig. 8. The ablation threshold, the minimum radiant exposure for tissue removal, has been measured at approximately 50 mJ cm$^{-2}$. Above this threshold, the etch depth per pulse, called ablation rate, increases in a logarithmic fashion with the radiant exposure. Since currently used clinical excimer lasers work with a radiant exposure between 120 and 250 mJ cm$^{-2}$, the resultant ablation rate ranges from be 0.2–0.5 µm pulse (Fig. 8).

The use of intensive ultraviolet laser light in photorefractive surgery is combined with minimal thermal, mechanical and actinic damage of the remaining corneal tissue. As mentioned before, the ablation products that are expelled with ultrasonic speed are in the physical state of a hot gas, and may have temperatures of more than 500 K [3]. This hot gas condenses and creates some heat as well as a thin layer at the „cold" edges of the remaining tissue. This „pseudomembrane", however, disappears within a few days after surgery, due to the wound healing. Under

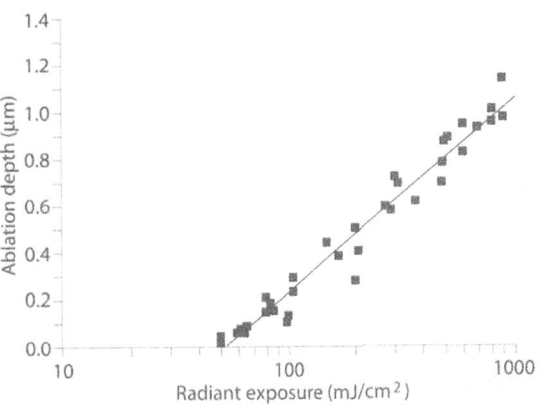

**Fig. 8.** Ablation depth per ArF-excimer laser pulse in corneal tissue as a function of the radiant exposure. The intersection of the logarithmic fit (R$^2$ = 0.94; p < 0.001) yields to an ablation threshold of 50 mJ cm$^{-2}$. The radiant exposure of commercially available medical excimer lasers ranges from 120 mJ cm$^{-2}$ to 250 mJ cm$^{-2}$

standard surgical conditions, the maximal averaged temperature increase at the edge of the irradiated tissue ranges from 5–10° C, which is considered insignificant [4]. In summary, thermal side effects during excimer laser surgery are minimal and should not induce any kind of inflammation of the cornea.

During laser refractive surgery mechanical damage to deeper layers of the cornea may originate from acoustic stress waves produced during the ablation process. Kermani and Lubatschowski [5] found pressure waves with an amplitude of 80 bar at a distance of 3 mm behind the cornea when an excimer laser beam with a diameter of 4 mm and a radiant exposure of 200 mJ cm$^{-2}$ was applied to the corneal surface. These waves travel with sonic velocity through the eye and, in contrast to the shock waves generated by Nd:YAG laser photodisruption, the amplitude decreases slowly with distance. However, stress waves of less than 100 bar probably do not induce damage in cellular ocular structures except the corneal endothelium, but might contribute to development of postoperative subretinal haemorrages that have been reported on rare occasions after laser refractive surgery.

Ultraviolet light is known to have the potential to generate actinic damage. Light with wavelengths of 300 nm and less is considered to be cytotoxic and mutagenic because in this wavelength range DNA shows strong absorption. ArF-excimer laser light, however, penetrates tissue thickness by less than 1 µm, so this radiation is unlikely to reach the nucleus of human cells (cytoplasmatic shielding). Cytobiological experiments showed that the initial 193 nm light bears little risk of causing mutagenic changes in mammalian cells [6]. However, during photoablation of the cornea, a faint bluish light is observed, the so-called

secondary radiation or fluorescence, which includes
wavelengths longer than 193 nm and with components in the dangerous 250–300-nm range. It is this
secondary radiation that accounts for the mutagenic
cellular damage that has been detected by very sensitive assays. The radiant exposure, however during
dinical photoablation, has been determined to be less
than 5 µJ cm$^{-2}$ , well below the estimated mutagenic
threshold of 10 µJ cm$^{-2}$ [7,8]. The mutagenic potential
of 193 nm laser light is thus considered to be insignificant, in accordance with the clinical absence of neoplasms after photorefractive surgery in millions of
cases.

Tissue can be removed from a large area in two
ways. Either the cornea can be irradiated with a number of laser pulses, each of which has a varied energy
distribution, or the eye can be irradiated with a series
of laser pulses of uniform irradiance but varying
geometry. In Fig. 9a, the centre of the laser beam contains the highest concentration of energy, which decreases towards the beam periphery. This beam will
remove more tissue centrally than peripherally with
each pulse, which flattens the cornea for myopic correction. If the irradiance were made greater in the periphery, more tissue would be removed from the
edges than from the centre, which would steepen the
cornea – hyperopic correction. In principle, any contour can be transferred to the cornea by controlling
the energy distribution with the laser beam. The same
effect can be achieved by exposing onto the corneal
surface a series of circular laser beams of uniform energy density but increasing the diameter up to 7 mm.
The result of this is, for example, that the centre of the
cornea receives more laser pulses, and has more tissue
removed.

Use of a small diameter spot or narrow slit beam
scanned across the ablation area allows for wide-area
ablations by means of a laser not generating the high
energies required of lasers that use a stationary large
beam. The clinical device can, therefore, be smaller
and less expensive. Recently, small ArF-excimer lasers
have been used to perform such scanning-spot photorefractive surgery. To complete the treatment within
a reasonably short period of time, the small diameter
spot must be rapidly scanned across the cornea, and
the repetition rate must be higher than with large-di-
ameter spot excimer lasers. However, small inevitable
eye movements that have nearly no impact on large-
diameter refractive surgery must be considered during scanning-spot corneal laser surgery for a precise
positioning of each single laser spot to omit an increased surface roughness, which may result in a
greater stromal wound healing response and haze.
Reasonably fast optical eye-tracking systems have
been established to overcome this problem.

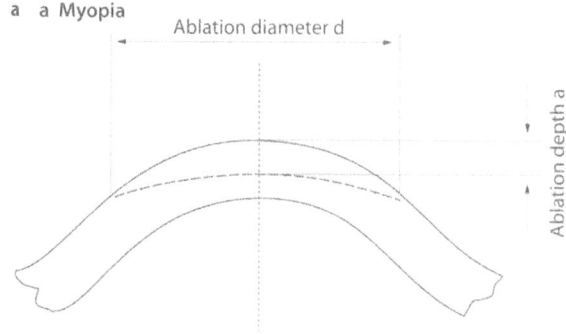

**a** Myopia

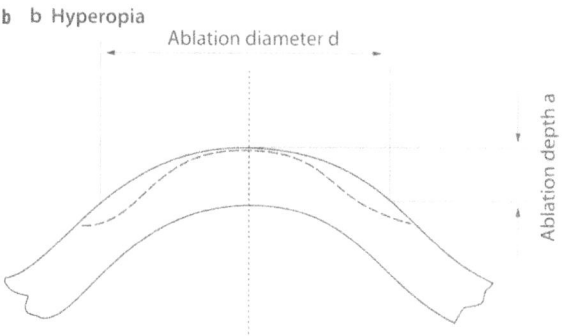

**b** Hyperopia

**Fig. 9a,b.** Schematic patterns of tissue removal during myopic
and hyperopic photorefractive keratectomy. The ablation diameter $d$ ranges from 6 to 7 mm and the central ablation $a_0$ results
from the amount of spherical correction

Currently, excimer laser systems with large-beam
diameters of more than 6 mm and those that scan
with a circle-beam diameter of 1 to 2 mm as well as
with a narrow slit beam are in clinical use. Some systems combine scanning elements with large-area
beams in an attempt to develop maximum flexibility
in the computer-controlled scanning algorithms.

## Photorefractive Keratectomy (PRK)

Photorefractive keratectomy (PRK), as well as the
photorefractive astigmatic keratectomy have become
clinical standard techniques to correct moderate myopia and astigmatism. During PRK for myopia correction a direct flattening is achieved by removal of a
convex-concave lenticule of tissue from the outer surface of the central cornea (Fig. 9). The central depth
($a_0$) of the keratectomy is determined by the intended
change of refraction, but is even more dependent on
the diameter of the ablation zone [9]. The following
approximation for $a_0$ (in microns) allows rapid estimation of this central ablation depth

$$a_0 = \frac{1}{3}\Delta D \cdot d^2 , \tag{1}$$

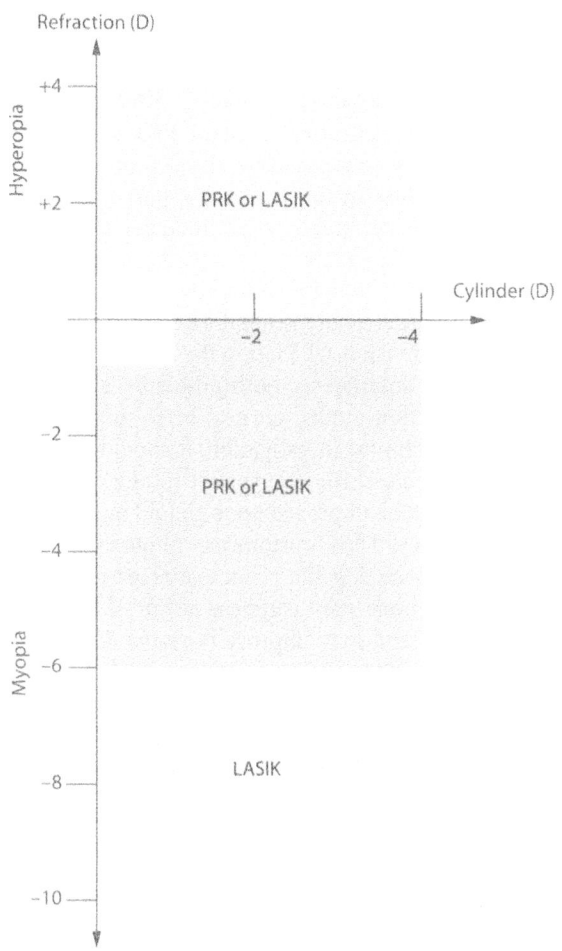

**Fig. 10.** Currently proceeding refractive corrections using photorefractive keratectomy or laser in situ keratomileusis (LASIK)

where ΔD represents the refractive change in diopters, and *d* the diameter of the ablation zone in millimeters. For example, a myopic correction of 6 D with a typical diameter of 6.0 mm of the ablation zone results in a central ablation depth of 72 μm.

By convention, data regarding efficacy of the refractive correction after photorefractive keratectomy generally report two overall measures: the percentage of eyes that achieve a postoperative refraction within 0.5 D of emmetropia and the percentage of eyes that achieve 20/40 or better uncorrected visual acuity. Clinical studies determined refractive success rates between 80–95% for corrections up to -6D of myopia and the range of patients achieving 20/40 or better distance acuity without correction ranges from 80 to 100% [1]. The overall incidence of vision-threatening complications such as a loss of best-corrected visual acuity and decreased contrast sensitivity was established to be in the order of 1–2% for corrections of moderate myopia with < -6 diopters. Higher myopic

correction leads to a significant higher regression and to a higher risk of scarring during wound healing. Thus, photorefractive keratectomy has been considered as an effective and safe surgery technique for myopic corrections up to -6 D (Fig. 10).

It is important to relate these success rates to a defined time after surgery, because of the wound healing. Epithelial as well as stromal wound healing has been documented to occur over a period of months after PRK and, thus, we choose a follow-up time of 12 months. Also, the success rates are helpful for assessing efficacy and predictability of a procedure but are not absolute measures of the „refractive success". Patients with preoperative refractive errors close to emmetropia may continue to complain of the need for spectacles or contact lens correction, at least for a part of the day. In contrast, patients with a residual myopia of, for example, –1.5 D, may consider the procedure successful if the eye was highly myopic prior to the surgery. In addition, glare and halos around sources of light, or loss of contrast sensitivity may not be detected under conditions used typically for measuring postoperative acuity. Because pupil diameter, and variation in pupil diameter affect acuity, measuring visual acuities with a single ambient light will not necessarily reflect the qualitiy of vision experienced by the patient.

## Laser in Situ Keratomileusis (LASIK)

The older literature on excimer laser photorefractive keratectomy for myopia describes ablation beginning at the corneal surface and extending into the deeper structures of the cornea such as Bowman's layer and the stroma. A modification of this technique involves the initial creation of a lamellar flap (average thickness 120–160 μm) of anterior corneal stroma, followed by refractive ablation of the exposed stromal bed. The flap is then repositioned onto the exposed stroma, and usually good adhesion is obtained without need for sutures (Fig. 11). Termed laser in situ keratomileusis (LASIK), the procedure was particularly investigated for high myopic corrections of more than -6 D, for which the precision and stability of the refractive outcome after photorefractive keratectomy have been somewhat disappointing. Advantages of this application include rapid visual recovery, because a central epithelial defect is not created, and a relative decrease in corneal haze compared to surface ablations of similar magnitude. In addition, refractive success rates seem to be identical after LASIK and PRK.

A potential disadvantage of the technique is the high amount of tissue removed during surgery and laser treatment, leading to a biomechanically effective

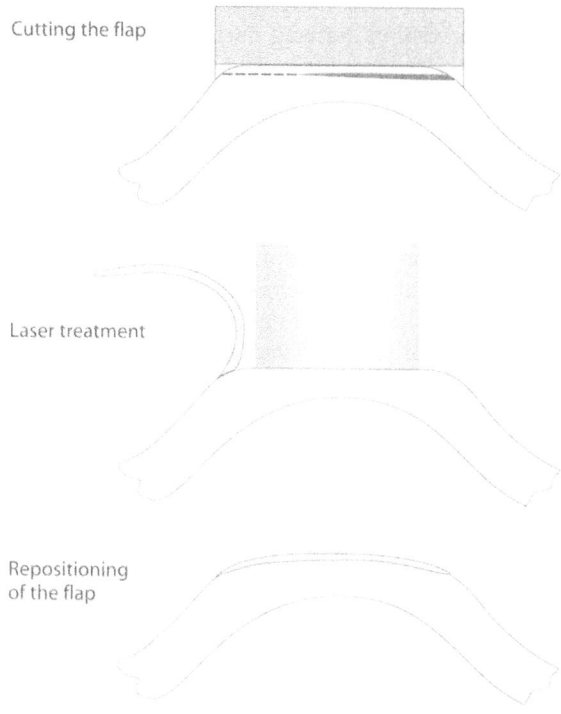

Cutting the flap

Laser treatment

Repositioning
of the flap

**Fig. 11.** Scheme of the laser in situ keratomileusis (LASIK) procedure

central thickness of a „normal" cornea (520 µm) to less than 300 µm, as is common in LASIK. Applying biomechanical data, the maximal myopic corrections (6 mm ablation zone, 160 µm flap thickness) would be limited to −3 D in a thin cornea (500 µm) and −10 D in a thick cornea (600 µm). Thus, the maximum myopic correction with LASIK is somewhere in the order of −10 D (Fig. 10). Besides the risk of creating a residual corneal thickness of less than 250–270 µm, the reported intra- and postoperative complications in LASIK are in the order of 1–2%, comparable to the complication rate that occurs after photorefractive keratectomy [10].

## Wavefront-Guided LASIK

Monochromatic optical aberrations of the human eye were discovered in the 19th century [11, 12, 13], but were not considered to play a clinically relevant role. Indeed, up to now wavefront errors of higher order could be corrected by optical means only in cases of irregularities of the outer corneal surface by hard contact lenses. Ocular aberrations, however, originate in normal eyes also from small decentrations of all refractive surfaces and their shapes. Since the introduction of excimer lasers to correct refractive errors, and in detail appropriate scanning-spot lasers, this situa-

tion has to be revised because now also optical errors of higher order may be corrected by selective photoablation of the cornea.

A reduced "visual performance" after refractive surgery as reported after standard PRK and LASIK procedures may be defined by visual problems such as halos and loss in low-contrast visual acuity, and was currently attributed to an increase in optical aberrations of the eye due to the operation. Also, optical aberrations or imaging of higher order may be the reason why visual acuity generally does not reach the retinal limit of vision, which is in the order of 20/10 or better [14]. A solution for both problems is the concept of wavefront-guided corneal laser surgery currently under clinical investigation. It should be mentioned that monochromatic spatial resolution of the human eye can be improved up to sixfold by means of adaptive optics. Thus, customized photoablation of the cornea correcting the refractive error plus individual aberrations may compensate for side effects such as halos and may improve postoperative visual acuity.

The optical wavefront deviations can be measured by means of wavefront analyzers such as an aberrometer of the Tscherning type or by a Hartmann-Shack sensor. From the obtained wavefront deviation maps, a corresponding ablation pattern can be calculated within a circular area with a diameter from 6.0 to 7.0 mm that differs from the standard ablation patterns commonly used in PRK and LASIK. The ablation pattern can be transferred to the excimer laser and used for corneal laser surgery. First patients treated with these technique in July 1999 achieved a postoperative uncorrected visual acuity up to 20/10 [15]. Ocular aberrations, however, are not constant during life, and increase with age and may change during accommodation. Although the healing response after LASIK is small compared to after PRK, epithelial healing may level out some of the corrections achieved during wavefront-guided LASIK. Longer follow-up and a greater database are necessary to estimate such influences. Recently in 2002, the FDA approved wavefront-guided ablation in the USA.

## Photodisruption

Photodisruption in ocular media generated with high-intensity laser pulses operating at non absorbing wavelengths has become a well-established tool for non-invasive microsurgery since the 1970s. During this time, Q-switched or mode-locked lasers such as the Neodymium:YAG laser (yttrium aluminum garnet) with pulse durations within the nano- or picosecond time scale have become a standard tool for producing well-defined cavitation bubbles for photodisruption. In ophthalmology, the most common application of photodisruption is the Nd:YAG laser capsulotomy after cataract surgery. Here, laser induced cavitation bubbles are used to open turbid posterior capsules in the presence of lens implants.

The behaviour of such cavitation bubbles in liquids has been widely studied since the late 1800's when propeller-driven ships were systematically developed. It is remarkable, that one of the earliest theoretical models of the collapse of such a spherically shaped cavitation bubble was solved by Lord Rayleigh [1]. Today, it is well known that during the collapse of such a spherically shaped cavitation bubble, acoustic transients (shock waves) with amplitudes of several hundred bars are induced [2]. Numerous theoretical and experimental studies were carried out to investigate the cavitation bubble dynamics, especially during their collapse and the building of water jet, with the goal of understanding the primary fragmentation mechanism. Nevertheless, the use of high-intensity lasers for clinical photodisruption requires a precise knowledge regarding the formation of laser-induced cavitation bubbles to avoid or reduce possible risk factors leading to postoperative complications [3].

Today, commercially available photodisruptors for capsulotomy consist mainly of a Q-switched Neodymium:YAG laser that emits invisible infrared radiation at a wavelength of 1064 nm with single pulse energies in the range of 0.1 to 30 mJ and pulse durations of 3 to 10 ns. There are two mechanisms that protect the retina from irradiation. First, a major fraction of the laser energy is absorbed or scattered in the region of the laser-induced plasma and, as a result, only a small portion of the laser pulse energy (ca. 5%) reaches the retina. In addition, the divergence of the beam posterior to the focal point of the laser beam protects the retina. For example, an incident beam with an cone angle of 20° and a focal diameter of 20 μm becomes approximately 4 mm in diameter at the retina. Thus, it can be estimated that if the creation of the plasma does not occur, the radiant exposure at the retina is still far below the damage threshold.

The laser-induced plasma that results from the high intensity of the laser radiation in the focal spot (up to $10^{12}$ W cm$^{-2}$) increases drastically in volume within a few ns and a velocity far above the speed of sound in water or tissue. This volume increase leads to a cavitation bubble due to the incompressible characteristics of the water. Figure 12 represents the time-dependent radius of the cavitation bubble as well as the corresponding pressure amplitude. The first pressure peak corresponds to the pressure wave induced by the optical breakdown (the plasma formation in water). The resulting cavitation bubble expands to its maximum radius $R_{max}$ of up to 1.5 mm within 100 μs. Due to the pressure of the liquid surrounding the

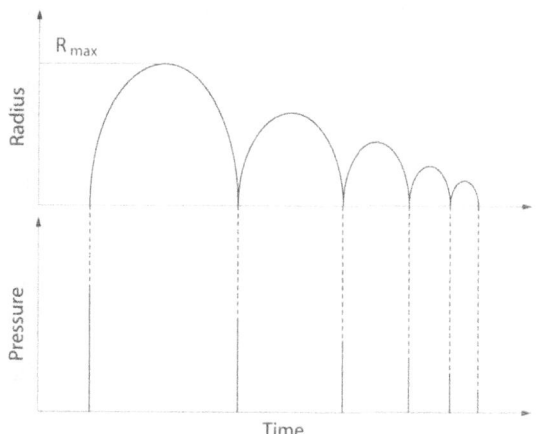

**Fig. 12.** Radius-time curve (top) and pressure-time signal (bottom) of an oscillating cavitation bubble in water. $R_{max}$ represents the maximum radius of the cavitation bubble during the first expansion of the bubble

cavition bubble, the bubble reduces its size and finally collapses after approximately 200 μs. A second pressure peak can be observed during the collapse of the cavitation bubble. The gas within the bubble that is compressed during the collapse leads to an re-expansion of the cavitation bubble with a smaller maximum radius and, in addition, this bubble also collapses. Further collapses, however, occur only at higher laser pulse energies in the order of several 10 mJ.

Rink and coworkers [4] have demonstrated in good agreement with the results presented by Vogel and Lauterborn [3], that ns-pulse fragmentation occurs mainly due to shock waves induced by the plasma expansion and the bubble collapse. The mechanical strength of this process is related to the laser pulse energy and duration. Besides this, the shock waves are of spherical character and, as a consequence, the intensity of these mechanical stress waves reduces drastically with its spatial radius ~1/r (r – spatial radius of the shock front). This leads to a very small region of approximately 1 mm in diameter where the shock waves are harmful and effective for ocular tissue structures.

The clinical use of photodisruption with the neodymium:YAG-laser for capsulotomy has been widely accepted since the early works of Franz Fankhauser [5] and Daniele Aron-Rosa [6]. The incidence of posterior capsular opacification after intraocular lens implantation is time-related and increases with longer follow-up after surgery time. An incidence rate of 18% to 50% is reported in patients who undergo extracapsular cataract extraction. Capsulotomy using the Nd:YAG laser has become the predominant approach for treating visually significant posterior capsular opacity.

The degree of visual loss necessary to justify a laser capsulotomy is difficult to estimate; however, a decrease in visual acuity to 20/40 or worse should be present before a treatment might be considered. Also, near-distance vision should be evaluated, since in some eyes the distance acuity remains relatively unaffected by an opacified capsule, while the reading vision is impaired. Furthermore, slit-lamp investigations should be carried out to evaluate other factors which make treatment more difficult, such as a lack of corneal clarity. On the other hand, the relative contribution of an opacified posterior capsule to visual loss is somewhat difficult to assess and is often a subjective decision.

Javitt and coworkers [7] used an immense database in their evaluation of retinal complications of Nd:YAG laser capsulotomy. A cohort of 46 000 individuals undergoing cataract surgery were followed for a minimum of 1 year, and a between three-fold and four-fold increased risk of retinal detachment was demonstrated in eyes in which a capsulotomy was performed. Thus, the increased risk of retinal complications that is associated with laser capsulotomy should be considered by physicians and their patientens in choosing whether to use the procedure. In detail, younger and myopic patients are at particular risk.

Aside from the retinal complications, after the laser capsulotomy increase in the intraocular pressure over a short period of time my occur. Damages of the corneal endothelium and cystoid macular odema have been reported [8].

# References

## 1 Photocoagulation

[1] VOGEL A, DLUGOS C, NUFFER R, BIRNGRUBER R (1991) Optical properties of human sclera, and their consequences for transscleral laser applications. Laser Surg Med 11: 331–340
[2] BLOOM SM, BRUCKER AJ (1991) Laser surgery of the posterior segment. Lippincott Company, Philadelphia
[3] FERRIS FL, ET AL (1987) Macula odema in diabetic retinopathy study patients. Diabetic retinopathy study report number 12. Ophthalmology 94: 754–760
[4] Diabetic Retinopathy Study Research Group (1981) Photocoagulation treatment of proliferative diabetic retinopathy: clinical application of DRS findings, DRS report number 8. Ophthalmology 88: 583–600
[5] Early Treatment Diabetic Retinopathy Research Group (1987) Photocoagulation for diabetic macular odema: ETDR report number 4. Int Ophthalmol Clin 27: 265–272
[6] Branch Vein Occlusion Study Group (1985) Argon laser photocoagulation for macular edema in branch vein occlusion. Am J Ophthalmol 99: 218–219
[7] JALKH AE, ET AL (1988) Treatment of small choroidal melanomas with photocoagulation. Ophthal Surg 19: 738–742
[8] SPENCER AF, VERNON SL (1999) „Cyclodiode": results of a standard protocol. Br J Ophthalmol 83: 311–316
[9] TASSIGNON MJ, TRAU R, MATHYS B (1997) Treatment of hypermetropia using the Holmium laser. Bull Soc Belge Ophtalmol 266: 75–83

[10] GOGGIN M, LAVERY F (1997) Holmium laser thermokeratoplasty for the reversal of hyperopia after myopic photorefractive keratectomy. Br J Ophthalmol 81: 541–543

## Photodynamic Therapy

[1] LUI J, HRUSA L, KOLLIAS, ET AL (1993) Photodynamic therapy of malignant skin tumors with bensoparphyrin derivate: preliminary investigations. Proc Lasers Otolaryngol Dermatol Tissue Weld 1876: 147–151
[2] HUSAIN D, KRAMER M, KENNY AG, ET AL (1999) Effects of photodynamic therapy using verteporfin on experimental choroidal neovascularisation and normal retina and choroid up to 7 weeks after treatment. Invest Ophthalmol Vis Sci 40: 2322–2331
[3] MILLER JW, SCHMIDT-ERFURT U, SICKENBERG M, ET AL (1999) Photodynamic therapy with verteporfin for choroidal neovascularisation caused by age-related macular degeneration. Results of a single treatment in a phase 1 and 2 study. Arch Ophthalmol 117: 1161–1173
[4] SCHMIDT-ERFURT U, MILLER JW, SICKENBERG M, ET AL (1999) Photodynamic therapy with verteporfin for choroidal neovascularisation caused by age-related macular degeneration. Results of treatments in a phase 1 and 2 study. Arch Ophthalmol 117: 1177–1187
[5] TAP Study Group (1999) Photodynamic therapy with verteporfin for choroidal neovascularisation caused by age-related macular degeneration. One-year results of two randomized clinical trials – TAP report 1. Arch Ophthalmol 117: 1329–1345
[6] REICHEL E, BERROCAL AM, IP M, ET AL (1999) Transpupillary thermotherapy of occult subfoveal choroidal neovascularisation in patients with age-related macular degeneration. Ophthalmology 106: 1908–1914

## Photoevaporization

[1] HIBST R (1997) Technik. Wirkungsweise und medizinische Anwendungen von Holmium- und Erbium-Lasern. Ecomed, Landsberg
[2] WENZEL W, OTTO R, FALKENSTEIN W, SCHMIDT-ERFURTH U, BIRNGRUBER R (1995) Development of a new Er:YAG laser conception for laser sclerostomy ab externo: experimental and first clinical results. German J Ophthalmol 4: 283–288
[3] MROCHEN M, SEMCHISHEN V, FUNK RHW, SEILER T (2000) Limitations of Erbium:YAG laser ($\lambda$ = 2.94 μm) photorefractive keratectomy. J Refract Surg 2000; 16: 51–59
[4] Vodopyanov KL (1990) Bleaching of water by intense light at the maximum of the $\lambda$ = 3 μm absorption band. Sov. Phys JEPT 70: 114–121
[5] MROCHEN M, DONITZKY C, WENIG M, RIEDEL P, REINDL M, SEILER T (1999) Investigations on the vapor bubble formation during erbium:YAG laser vitrectomy SPIE Proceed 3591: 171–181
[6] ITH M, PRATISTO H, ALTERMATT HJ, FRENZ M, WEBER HP (1994) Dynamics of laser-induced channel formation in water and influence of pulse duration on the ablation of biotissue under water with pulsed erbium-laser radiation. Appl Phys B 59: 621–629
[7] JANSEN DE, VAN LEEUWEN TG, MOTAMEDI M, BORST C, WELCH AJ (1995) Partial vaporization model for pulsed mid-infrared laser ablation of water. J Appl Phys 78: 564–571
[8] MACHEMER R (1995) Reminiscences after 25 years of pars plana vitrectomy Am J Ophthalomol 119: 505–510
[9] CHARLES S (1994) Principles and techniques of vitreous surgery. In: Ryan JR (ed) Retina, vol 3. Mosby Year Book, St Louis, Missouri, pp 2063–2067
[10] MROCHEN M, PETERSEN H, WÜLLNER C, SEILER T (1998) Experimentelle Ergebnisse zur Erbium:YAG Laser-vitrektomie Klin Monatsbl Augenheilkd 212; 50–55

[11] PETERSEN H, MROCHEN M, SEILER T (2000) Comparision of Erbium:YAG-laser vitrectomy and mechanical vitrectomy – a clinical study. Ophthalmology 2000; 107: 1389–1392

[12] BINDER S, STOLBA U, KREBS I, KELLNER L (1999) Erste Erfahrungen mit dem Erbium:YAG Laser Vitrektomie System in der Vitrektomie. Spektrum Augenheilkd 13: 105–108

[13] D'AMICO DJ, BLUMENKRANZ MS, LAVIN MJ, QUIROZ-MERCADO H, PALLIKARIS IO, MARCELLINO GR, GRACE RM (1996) Multicenter clinical experience using an Erbium:YAG laser for vitreoretinal surgery. Ophthalmology 103: 1575–1585.

[14] BRAZITIKOS PD, D'AMICO DJ, BERNAL MT, WALSH AW (1995) Erbium:YAG laser surgery of the vitreous and retina. Ophthalmology 102: 278–290

[15] MARGOLIS TI, FARNATH DA, DESTRO M, PULIFATIO CA (1989) Erbium:YAG laser surgery on experimental vitreous membranes. Arch Ophthalmol 107: 424–428

[16] GOLDBERG RA (1996) The carbon dioxide laser in oculoplastic surgery and sliced bread Arch Ophthalmol 114: 1131–1133

[17] WALSH JT, FLOTTE FJ, ANDERSON RR, ET AL (1988) Pulsed $CO_2$ tissue ablation: effect of tissue type and pulse duration on thermal damage. Lasers Surg Med 8: 108–118

[18] GOLDBAUM AM, WOOG JJ (1997) The $CO_2$ laser in oculoplastic surgery. Surv Ophthalmol 1997: 42: 255–167

[19] ZIERING CL (1997) Cutaneous laser surfacing with the Er:YAG laser and the char-free carbon dioxide laser: a clinical comparison of 100 patients. Int J Aesthet Restorat Surg 5: 29–37

[20] CHIN-CHIANG Y, CHEE-YIN C (1995) Animal study of skin resurfacing using ultrapulse carbon dioxide laser. Am Plast Surg 35: 154–158

[21] FITZPATRICK RE, TROPE WD, GOLDMAN MP, ET AL (1996) Pulsed carbon dioxide laser, trichloroacetic acid Baker-Gordon phenol and dermabrasion: a comparative clinical and histologic study of cutaneous resurfacing in a porcine model. Arch Dermatol 132: 469–471

[22] GREEN HA, BURD E, NISHIOKA N, ET AL (1992) Mid-dermal would healing. A comparison between dermatomal excision and pulsed carbon dioxide laser ablation. Arch Dermatol 128: 639–645

[23] ITZPATRICK RE, GOLDMANN MP, SATUR NM, TROPE WD (1996) Pulsed carbon dioxide laser resurfacing of photoaged facial skin. Arch Derm 132: 395–402

[24] ALSTER TS (1996) Comparison of two high-energy carbon dioxide lasers in the treatment of periorbital rhytids. Dermatol Surg 22: 541–545

[25] WALDORF HA, KANVAR AN, GERONEMUS RG (1995) Skin resurfacing of fine deep rhytids using a char-free carbon dioxid laser in 47 patients. Dermatol Surg U: 940–946

[26] BASS LS (1998) Er:YAG laser skin resurfacing: prelimany clinical evaluation. Ann Plast Surg 40: 328–334

[27] APFELBERG DB (1998) Summary of the 1997 ASAPS/ASPRS laser task force survey on laser resurfacing and laser blepharoplasty. Plast Reconstruct Surg 101: 511–518

[28] BLANCO G, SOPARKAR CN, IORDAN DR, PATRINELY IR (1999) The ocular complications of periocular laser surgery: Curr Opin Ophthalmol 10: 264–269

## Photoablation

[1] SEILER T, MCDONNELL PJ (1995) Excimer laser photorefractive keratomy. Survey Ophthalmol 40: 89–118

[2] TROKEL SL, SRINIVASAN R, BRAREN B (1983) Excimer laser surgery of the cornea. Am J Ophthalmol 96: 710–715

[3] VALDERAME GL, FREDIN LG, BERRY MJ (1991) Temperature distribution in laser-irradiated tissue. SPIE Proc 1427: 200–213

[4] BENDE T, SEILER T, WOLLENSAK J (1988) Side effects in excimer corneal surgery: corneal thermal gradients. Graefes Arch Clin Exp Ophthalmol 226: 277–280

[5] KERMANI O, LUBATSCHOWSKI H (1991) Struktur und Dynamik photoakustischer Schockwellen bei der 193 nm Excimerlaser-photoablation. Fortschr Ophthalmol 88: 748–753

[6] KOCHEVAR IE (1989) Cytotoxicity and mutagenicity of excimer laser radiantion. Laser Surg Med 9: 440–445

[7] LUBATSCHOWSKI H, KERMANI O (1992) 193 nm Excimerlaserphotoablation der Hornhaut: Spektrum und Transmissionsverhalten von Sekundärstrahlung. Ophthalmologe 89: 134–138

[8] MACHETTE LS, WAYNANT RW, ROYSTON DD, ET AL (1989) Induction of lambda prophage near the site of focused UV laser radiation. Photochem Photobiol 49: 161–167

[9] MUNNERLYN CR, KOONS SJ, MARSHALL J (1988) Photorefractive keratectomy: a technique for laser refractive surgery. J Cataract Refract Surg 14: 46–52

[10] KNORZ MC, JENDRITZA B, HUGGER P, LIERMANN A (1999) Komplikationen der Laser-in situ Keratomileusis (LASIK). Ophthalmologe 96: 503–508

[11] TSCHERNING M (1894) Die monochromatischen Aberrationen des menschlichen Auges. Z Psychol Physiol Sinne 6: 456–471

[12] VOLKMANN AW (1846) Sehen. In: Wagner R (ed) Handwörterbuch der Physiologie III. Vieweg, Braunschweig, pp 289–293

[13] GULLSTRAND A (1909) Die Dioptrik des Auges, V. Die monochromatischen Aberrationen des Auges. In: von Helmholtz H (ed) Handbuch der Physiologischen Optik. 3rd edn. L Voss, Leipzig, pp 353–376

[14] HOLLADAY JT, WARING GO III (1992) Optics and topography of radial keratectomy. In: Waring GO III (ed) Refractive keratectomy for myopia and myopic astigmatism. Mosby-Year Book, St Louis, Missouri, p 37–141

[15] MROCHEN M, KAEMMERER M, SEILER T (2000) Wavefront-guided LASIK: early results in three eyes. J Refract Surg 2000; 16: 51–59

## Photodisruption

[1] RAYLEIGH J (1917) On the pressure development in a liquid during the collapse of a spherical cavity. Philos Mag 34: 94

[2] LAUTERBORN W (1974) Kavitation durch Laserlicht. Acustica 31: 51–78

[3] VOGEL A, HENTSCHEL W, HOLZFUSS J, LAUTERBORN W (1986) Kavitationsblasendynamik und Stoßwellenstrahlung bei der Augenchirurgie mit gepulsten Neodym:YAG-Lasern. Klin Monatsbl Augenheilkd 189: 308–316

[4] RINK K, DELACRETAZ G, SALATHE RP (1992) Fragmentation process induced by nanosecond laser pulses. Appl Phys Lett 61: 2644–2646

[5] FANKHAUSER F, LOERTSCHER HP, VAN DER ZYPEN E (1982) Clinical studies on high- and low-power laser irradiation upon some structures of the anterior and posterior segments of the eye. Int Ophthalmol 5: 15–32

[6] ARON-ROSA D, ARON JJ, GRIESMANN M, THYZEL R (1980) Use of the neodymium-YAG laser to open the posterior capsule after lens implant surgery. A preliminary report. J Am Intraocul Implant Soc 6: 352–354

[7] JAVITT JC, TIELSCH JM, CANNER JK, KOLB MM, SOMMER A, STEINBERG EP (1992) National outcomes of caract extraction – Increase risk of retinal complications associated with Nd:YAG laser capsulotomy. Ophthalmology 99: 1487–1498

[8] STEINER RF, PULIAFITO CA, KUMAR SR, ET AL (1991) Cystoid macular edema, retinal detachment, and glaucoma after Nd:YAG laser posterior capsulotomy. Am J Ophthalmol 112: 373–380

# III-3
# Dental Laser Applications:
# Periodontal Treatment and Intra-Oral Surgery

Petra Wilder-Smith

## Contents

## Extended Summary

One of the greatest advantages of laser use is a high rate of patient acceptance. Patients expect reduced intra-operative and postoperative discomfort and improved treatment speed and outcome from laser procedures. When used for suitable applications with appropriate techniques, lasers can indeed provide these advantages. Particularly in the field of soft tissue surgery, lasers can offer substantial advantages over conventional techniques for procedures such as frenectomy, gingivectomy, gingivoplasty, ablation of pathological lesions, incisional and excisional biopsy, soft tissue tuberosity reductions, operculum removal, coagulation of graft donor sites and crown-lengthening procedures. The strong absorption of radiation from the argon laser in pigmented tissues permits excellent haemostasis and selective destruction of lesions which have a large vascular component, such as haemangiomas, telangiectasias, epulides, haemorrhagic gingival hyperplasias and granulomas containing a large blood component, while sparing superficial non-pigmented structures from damage.

Use of lasers for periodontal and intra-oral soft tissue surgery is significantly restricted by two factors. The first is the extreme thermal sensitivity of intra-oral structures such as the dental pulp and bone, which precludes application of laser techniques in close proximity to these structures. The second is the still significant cost of laser systems. Although the cost/benefit ratio of such technology may be acceptable in multi-user or specialized practices, it may not be appropriate for surgeries devoting a lesser portion of their time to laser-related procedures.

## Indication

In approaching the concept of laser use for periodontal and intra-oral soft tissue surgery, laser efficacy for achieving any specific treatment goal, as well as the clinical acceptability of the side effects of the laser use must be considered.

In many respects, laser properties are ideally suited to soft tissue surgery in the oral cavity: characteristics such as surgical precision and efficacy, haemostasis, bacterial elimination, minimal postoperative scarring, reduced postoperative pain and swelling provide many advantages over conventional surgical techniques. The thin, flexible fibre or waveguide delivery systems currently available on most clinical lasers allow easy access to all areas of the mouth. Specially configured laser tips, ranging from very fine for excellent surgical precision, to wide for rapid ablation of large surfaces, and including curved or side-firing designs, are available for ensuring optimal surgical access, control and outcome.

Many areas of routine $CO_2$ and, to a lesser extent, argon or Nd:YAG laser applications for soft tissue surgery have developed over the past 30 years. Typical applications include haemostasis and treatment of patients with haemorrhagic disorders, ablation of pathological lesions, incisional and excisional biopsies, coagulation of graft donor sites, implant exposure, frenectomies, preprosthetic surgery, soft tissue tuberosity reductions, gingivectomies, gingivoplasties, crown-lengthening procedures and operculum removal.

In almost all intra-oral soft tissue surgery, the minimal penetration depth of $CO_2$ laser irradiation into the tissues proves very advantageous. Rapid, effective surgical performance with effective haemostasis and minimal involvement of adjacent and underlying structures is achieved. Typically, a short burst of $CO_2$ laser light is absorbed within the first 0.3 mm of soft tissue, and thermal effects in collateral soft tissues extend 15–300 µm, depending on the laser configuration used.

The strong absorption of radiation from the argon laser in pigmented tissues achieves excellent haemostasis and selective destruction of lesions which have a large vascular component, such as haemangiomas, telangiectasias, epulides, haemorrhagic gingival hyperplasias and granulomas containing a large blood component, while sparing superficial non-pigmented structures from damage. Irradiation from the argon laser is absorbed within the first 1–2 mm of soft tissue, and collateral tissue effects are slightly greater than those produced by the $CO_2$ laser.

Lasers emitting at wavelengths with lower tissue absorption and significant optical scattering, such as the Nd:YAG laser, or some of the semiconductor or diode systems now available, will tend to impact underlying and adjacent structures to a greater extent than $CO_2$ or argon lasers, although collateral effects can to a certain extent be mitigated by use of appropriate tips and cooling.

## Criteria for Exclusion

Because lasers used for soft tissue surgery achieve their effects mainly through thermal mechanisms, tissues lying under or adjacent to the surgical site will experience greater or lesser degrees of thermal alteration or damage. Thus, laser procedures must be performed with caution in proximity to heat-sensitive structures. This is particularly relevant in the oral cavity. The periodontal tissues overlying bony structures are extremely thin and tightly bound in some areas of the mouth. A temperature increase of as little as 10 °C can cause osseous necrosis. The cutting effects of surgical lasers are achieved at temperatures exceeding 100 °C, necessitating extreme caution when working with a laser in thin oral soft tissues with close proximity to bone. In the enclosed environment of the periodontal pocket, laser-generated heat is retained. Here, also, exists the potential for thermal damage to bony periodontal structures. Additionally, transmission of laser energy through the dentinal tubules directly to the pulpal tissues can threaten pulpal vitality. A temperature rise as small as 5.5 °C can cause pulpal necrosis.

The Nd:YAG and the Er:YAG laser have been proposed for calculus ablation and smear layer removal after conventional root planing procedures. Although root cleaning can be achieved, effects are usually patchy, with areas of inadequate cleansing and zones of thermal damage [1, 2]. Laser-induced char layers prevent connective tissue and/or epithelial reattachment [3]. These limitations, as well as concerns regarding the achievement of complete bacterial elimination, in order to avoid bacterial repopulation and re-infection of adjacent tissues, speak against the use of laser techniques as an alternative or adjunct to conventional root planing procedures [4–7]. Moreover, a post-irradiation decrease in protein/mineral ratio and surface alterations seems to render irradiated root surfaces potentially unfavourable for fibroblast attachment [8, 9].

Caution is necessary when working in proximity to metal dental restorations, due to the danger of oral or ocular damage from reflected laser light [10].

## Comparison with Alternative Methods

### Haemostasis

One of the great advantages of laser surgery is the establishment of excellent operative visibility due to a relatively blood-free surgical field. Moreover, coagulation of bleeding areas or of soft tissue graft donor sites is extremely effective. Postoperative bleeding from a donor site after soft tissue grafting poses a major challenge to dental clinicians. Once laser coagulation is achieved, chances for postoperative bleeding are minimal. For active bleeding sites, argon, Nd:YAG and Ho:YAG are the lasers of choice. In non-bleeding areas, to prevent subsequent bleeding, nearly all lasers are applicable. To coagulate an actively bleeding site with a $CO_2$ laser, bleeding must first be momentarily stopped, usually by pressure or administration of additional local anaesthetic.

## Surface Lesions

Removal of oral lesions by application of laser energy is readily achieved for surface exophytic and invasive lesions. Rapid removal of surface lesions is easily achieved using an unfocused beam of relatively large diameter (1 cm or more). Applications of this sort include ablation of a wide range of white lesions (once a biopsy has been taken and a diagnosis established), especially of the buccal mucosa, palate and floor of the mouth. In our experience, the $CO_2$ laser is well suited to these applications because of its ability to ablate tissue rapidly in the unfocused mode with minimal underlying tissue damage.

Symptomatic relief of aphthous ulcers and herpetic lesions can be achieved using the laser in the defocused mode. Here, goal of the laser therapy is to remove necrotic material from the lesion surface, disinfect the remaining wound and create a conservative laser wound which will heal more quickly, with less discomfort and fewer complications than the original lesion. As these lesions usually occur on the lip, potential laser effects on underlying bone are not a concern, and a wide range of lasers including argon, $CO_2$ and Nd:YAG have been successfully used for this application. It should be noted that appropriate precautions should be employed when applying laser treatment to herpetic lesions since the herpes virus and verruca virus may be transmitted via the laser plume.

## Excision and Biopsy

Precise biopsy or excision of exophytic, invasive or other lesions is achieved using a focused beam with a minimal diameter where the surgical technique is similar to that used with a scalpel. The depth of penetration of the incision made by the focused laser beam or contact tip will vary with the power density used and the speed of movement of the laser across the surgical site. In most instances, sutures are not necessary and the laser wound is left to heal by secondary intention.

Some authors have hypothesized minimization of seeding and metastasis from malignant lesions dur-

ing surgery by using laser techniques to seal blood vessels and lymphatics related to the lesion. Animal studies and a small number of clinical investigations appear to confirm this concept [11]. Conversely, recent investigations in animals have raised concerns about promotion of malignant transformation in potentially pre-malignant leukoplakic lesions by $CO_2$ laser incision [12].

On the tongue conventional biopsies or excisional procedures tend to cause immediate and copious bleeding, significantly hindering the surgical process. Lesions on the tongue can be removed almost bloodlessly using the $CO_2$ laser, and most often without the need for sutures.

## Implant Exposure

Exposure of single or multiple fixture implants is readily performed using the laser by simple vaporization of the tissue overlying the implant until the surgical healing cap is reached. This is usually best accomplished in the defocused mode, often eliminating the need for a flap and suturing, and reducing the level of postoperative discomfort. Where there is a need for apically positioning a flap, or for uncovering implants within osseous structures, or where regeneration or augmentation materials have been used, conventional flap procedures may be indicated.

## Frenectomy

Goal of frenectomy is to release attachment with minimal damage to surrounding tissues, and particularly the $CO_2$ and argon laser provide enormous advantages over conventional scalpel surgery, including very rapid, efficient tissue removal with a bloodless field. As with most laser procedures, sutures are usually not needed – which can be difficult and uncomfortable to place and remove in the mobile non-keratinized oral mucosa – and the patient experiences minimal discomfort during the healing period.

## Preprosthetic Surgery

Conventional surgical techniques for preprosthetic surgical needs such as removal of inflammatory hyperplasia, epulides, for tuberosity reduction or vestibular deepening procedures commonly cause extensive bleeding, limiting intra-operative visibility and requiring suturing. Significant postoperative pain results, and grafting may become necessary to cover the wound. Wound contraction after healing may compromise the surgical result. The laser provides good surgical access and precision, haemostasis during and after the procedure, speed, minimal scarring and wound contraction, minimal or absent postoperative pain as well as near-normal colour and texture of the wound. Suturing is rarely needed, so that the normal anatomical contour is preserved.

## Gingivectomy and Gingivoplasty

If thermal or direct damage to underlying bone and tooth substance is avoided by judicious case and wavelength selection, gingivectomy and gingivoplasty can advantageously be accomplished using the laser. Tissue overgrowth, pseudopockets or hyperplasias resulting from disease, drug therapy such as phenytoin, cyclosporine, nifedipine and others, orthodontic treatment or congenital conditions are precisely and rapidly removed. Intra-operative haemostasis and reduced postoperative pain are further advantages of laser use over conventional techniques for this type of application, where because of its minimal penetration depth in soft tissue (thus minimally impacting underlying bone and tooth structures) and its speed, the $CO_2$ laser is particularly advantageous. In some cases, tooth protection can be achieved by shielding with a device such as a periosteal elevator. Near the tooth margins, the beam should be used in the focused mode to precisely contour the gingival margin and to permit optimal control of laser beam impingement on target and adjacent tissues. Distant from the gingival margin, the beam is used in the defocused mode to ablate and vaporize excess tissue. Particularly for

mentally handicapped patients, the laser offers enormous advantages over conventional treatment options, due to the speed of treatment, and greatly reduced postoperative bleeding and pain.

### Patients with Haemorrhagic Disorders

Patients with disorders such as haemophilia, Sturge-Weber syndrome and idiopathic thrombocytopoenic purpura can be effectively treated for intra-oral lesions using the laser, as they lose minimal amounts of blood intra-operatively due to the coagulative effects of lasers. These patients should, however, still be cross-matched for blood to cover all contingencies.

### Effects on Collateral Structures and Wound Healing

Obviously, scalpels cause no collateral thermal damage. Lasers tend to cause less thermal damage in adjacent or underlying tissues than electrocautery, whereby the amount of collateral damage caused depends on the laser wavelength and laser pulse regimens used, as well as coolant and surgical technique. The amount of tissue destroyed adjacent to a laser incision is of considerable importance in determining the rate and quality of wound healing. Using a $CO_2$ laser with extremely short pulses and high peak powers, collateral thermal damage in oral soft tissues ranged from 15–170 µm (Figs. 1–3) [13]. The ability to predict and control collateral effects of laser use are of prime importance in ensuring good treatment out-

**Fig. 1.** Typical incision profile in oral soft tissue using a $CO_2$ laser at 9W average and peak power, continuous wave mode and 40 Hz. Incision is deep and wide, and collateral damage is relatively great

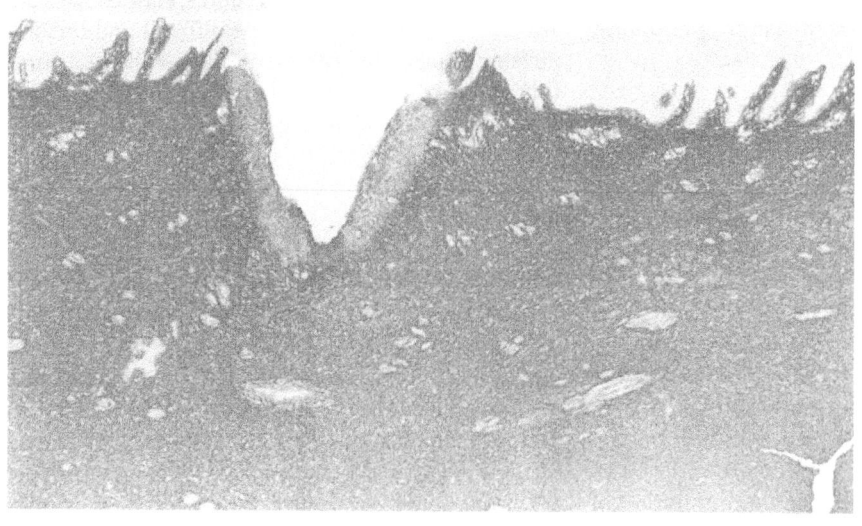

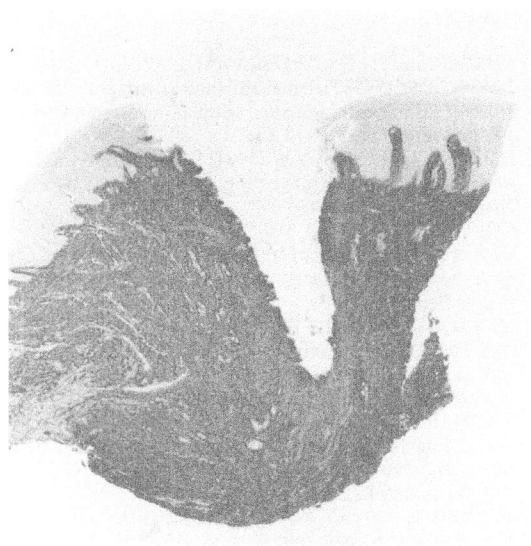

**Fig. 2.** Typical incision profile in oral soft tissue using a $CO_2$ laser at 3 W average power, 20 W peak power, 300 µs pulse duration and 1100 Hz. Incision is deep and narrow, and collateral damage is moderate

come and avoiding damage to adjacent structures. Moreover, a thick layer of thermal damage at the margins of a wound will tend to delay healing and weaken wound cohesive strength.

Laser wounds heal in a fashion similar to scalpel wounds. Although initial healing appears to be slightly delayed, wound contraction and scar formation are reduced, leading to an improved cosmetic result.

## Results

Accumulated clinical and research experience over the past 30 years by clinicians and researchers pertaining to laser use in dentistry has clearly demonstrated many advantages of laser surgery, as well as clear and imperative contraindications. Applications where laser use provides many advantages over conventional techniques include haemostasis and treatment of patients with haemorrhagic disorders, removal of soft tissue surface lesions, excision and biopsy, implant exposure, frenectomy, preprosthetic surgery, gingivectomy and gingivoplasty.

Because lasers used for soft tissue surgery achieve their effects mainly through thermal mechanisms, tissues lying under or adjacent to the surgical site will experience greater or lesser degrees of thermal alteration or damage. Thus, laser procedures must be performed with caution in proximity to heat-sensitive structures, specifically to avoid damage to osseous and dental/pulpal structures.

**Fig. 3.** Typical incision profile in oral soft tissue using a $CO_2$ laser at 1 W average power, 100 W peak power, 300 µs pulse duration and 10 Hz. Incision is shallow and narrow, and collateral damage is minimal

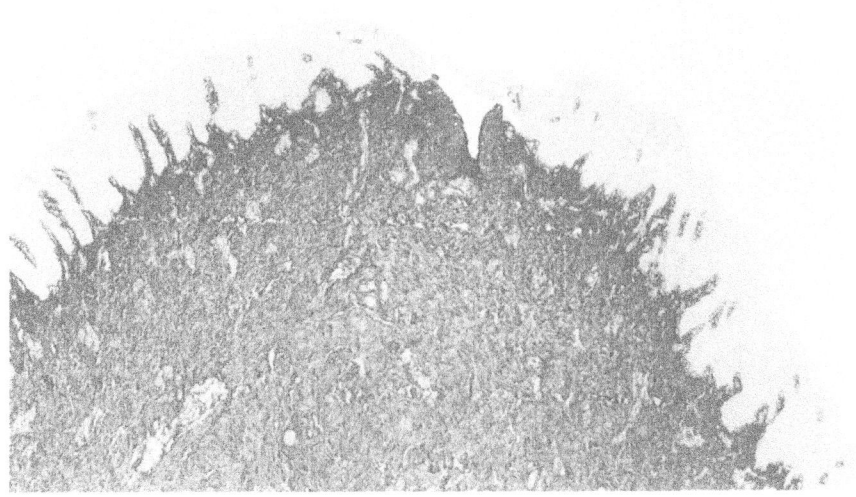

## Discussion

In approaching this topic, an unusual combination of excitement and reluctance is generated: excitement at the true potential that lasers have for improving on many conventional techniques in oral soft tissue surgery; reluctance because lasers need to be used in an informed and appropriate fashion to avoid the very real potential for negative side effects in our patients. Moreover, with a relatively costly, high-tech device such as a laser system, the temptation to convert this modality into a universal panacea is great – and inappropriate.

Thus, dental clinicians are encouraged to maximize the great, beneficial potential of laser soft tissue surgery for the welfare of their patients by developing a thorough understanding of basic laser mechanisms and by obtaining good, clinical instruction in all the laser techniques and systems they wish to use.

## References

[1] WILDER-SMITH P, ARRASTIA AM, LIAW L-H, GRILL G, BERNS MW (1995) Thermal and microstructural effects of Nd:YAG laser irradiation and root planing on root surface. J Periodontol 66: 1032–1039
[2] SCHILKE R, GEURTSEN W (1994) REM Analyse von Zahnoberflaechen nach Bearbeitung mit einem Er:YAG Laser. Dtsch Zahnarztl Z 49: 160–162
[3] GOPIN BW, COBB CM, RapGley JW, KILLOY WJ (1997) Histologic evaluation of soft tissue attachment to $CO_2$ laser-treated root surfaces: an in vivo study. Int J Periodont Restor Dent, in press
[4] MORLOCK BJ, PIPPIN DJ, COBB Cm, KILLOY WJ, RAPLEY JW (1992) The effect of Nd:YAG laser exposure on root surfaces when used as an adjunct to root planing: an in vitro study. J Periodontol 63: 637–641
[5] THOMAS D, COBB C, RAPLEY J, SPENCER P, KILLOY W (1994) Effects of the Nd:YAG laser and combined treatments on in vitro fibroblast attachment to root surfaces. J Clin Periodontol 21: 38–44
[6] TUCKER D, RAPLEY JW, COBB CM, KILLOY WJ (1996) Morphologic changes following in vitro $CO_2$ laser treatment of calculus laden root surfaces. Lasers Surg Med 18: 150–156
[7] COFFELT DW, COBB CM, MACNEILL S, RAPLEY JW, KILLOY WJ (1997) Determination of energy density threshold for laser ablation of bacteria: an in vitro study. J Clin Periodontol 24: 1–7
[8] SPENCER P, TRYLOVICH DJ, COBB CM (1992). Chemical characterization of lased root surface using Fourier transform infrared photoacoustic spectroscopy. J Periodontol 63: 633–636
[9] TRYLOVICH DJ, COBB CM, PIPPIN DJ, SPENCER P, KILLOY WJ (1992) The effects of the Nd:YAG laser on in vitro fibroblast attachment to endotoxin-treated root surfaces. J Periodontol 63: 626–632
[10] SIEVERS M, FRENTZEN M, KOORT HJ (1994) Reflexion infraroter $CO_2$ Laserstrahlung an Fuellungswerkstoffen. ZWR 103: 288–290
[11] PICK R, POGREL A, LOH HS (1995) Clinical applications of the $CO_2$ laser . In: MISERENDINO L, PICK R (eds) Lasers in dentistry. Quintessenz, Chicago, pp 145–160
[12] IIDA K, KATO M, YOSHIDA K, KuritGa K, TATEMATSU M (1999) Promotional effects of $CO_2$ laser on DMBA-induced buccal cheek pouch carcinogenesis. Laser Surg Med 24: 360–367
[13] WILDER-SMITH P, DANG J, KUROSAKI T (1997) Investigating the range of surgical effects on soft tissue produced by a carbon dioxide laser. JADA 128: 583–588

## III-4
## ENT

## III-4.1
## Endoscopic Laser Treatment of Laryngeal and Tracheal Diseases

J. A. Werner and W. Steiner

### Contents

### Introduction

The use of lasers for laryngeal surgery began in Boston in the early 1970s. It was thought then that a laser might offer an advantage over other conventional therapeutic modalities in treating cancer of the larynx. This driving force for the development of an instrument to treat cancer of the larynx started with the use of the argon laser to excise laryngeal tissue. It was soon found that the argon laser lacked the necessary precision for laryngeal surgery. As early excitement with the argon laser was fading, a new laser, the neodymium in glass laser, became available experimentally. Again, the researchers were unable to produce a precise cut. An experimental laser wavelength produced by American Optical became the next focus of interest. This was a gas laser using carbon dioxide. The carbon dioxide ($CO_2$) laser was found to allow precise injury to the canine vocal cord. This opened the door to clinical use of the $CO_2$ laser [6, 27].

### Safety Aspects in Laryngeal and Tracheal $CO_2$ Laser Surgery

The surgeon operating the laser is responsible for the correct implementation of the pertinent regulations and for the safety of patient, personnel and environment, as well as his own. A good knowledge of the hazards of laser radiation is therefore necessary. Laser radiation is hazardous for several reasons.

Laser beams of the visible or near-infrared part of the spectrum are focused by the refractive media of the eye. Very small powers as low as some mW are sufficient to cause damage to the retina. Lasers with wavelengths in the middle and far infrared spectrum damage the cornea. Some laser radiation is invisible for human eye. The aversion response of the eye, which otherwise protects against low power visible laser radiation hence does not function.

If the laser beam hits the endotracheal tube, it can ignite. The combustion of the endotracheal tube threatens the patient's life.

In order to reduce the hazards to a minimum, certain special safety measures should be adhered to in laser surgery of the larynx and trachea. Precautions should be taken against the toxic fumes or fires. Personal protective eye-wear should be worn by all personnel, when the laser is operable. The eyes of the patient should be shielded.

The main problems in laryngeal laser surgery are the potential effects of the laser beam on the tube and the narrow anatomic situation. A modification of regular anaesthesiological techniques is therefore necessary with respect to the physical properties of the tube, the ventilation technique, the anaesthetic gases and the relaxation [10].

Laser surgery in the upper aerodigestive tract necessitates protection against tube fires. There are different possibilities of achieving this goal. One can use regular PVC tubes, which are to be covered with wet gauze. Further possibilities are the uses of special, commercially available laser tubes, or the laser surgery in intermittant apnea, or during jet ventilation. By using special laser tubes, it has to be mentioned that some of them can produce problems with the airway resistance. Jeckström et al. [10] surprisingly found that this special problematic may be of great importance since many of the patients with laryngeal disorders also suffer from chronic respiratory diseases with an increased airway resistance which adds to the increased airway resistance of some types of laser tubes. High intrathoracic pressures may result, which can cause complications with ventilation and circulation.

## Benign Diseases of the Larynx

The operation begins with the transoral insertion of the laryngoscope. Prior to inserting the laryngoscope, a tooth guard should be placed in the upper jaw to prevent accidental damage to the teeth. It seems to be recommendable that the right-handed surgeon handles the laryngoscope from the right and pushes the endotracheal tube to the left, which can be done digitally without visual control (Fig.1). The surgeon should also be cautious that the tongue should not be caught between teeth and laryngoscope.

## Operation Technique

While with the first laser surgical operations [27] still developed a considerable ot of carbonization debris, current laser technology effects dissection with a focused laser beam, meaning dissection with hardly any carbonization. The laser beam is controlled via micromanipulator, in our departments we favour a micromanipulator which gives the focal diameter of 0.5 mm. If the laser is set at 6 W, we operate with 3.056 W cm$^{-2}$. At a laser power of 20 W, this equals a laser power density of 10.188 W per cm$^2$. At this range the dissecting and coagulating capacity of the laser is used to its maximum advantage [22], an effect due to the biophysical characteristics of the laser.

A general recommendation for laser surgery is to remove carbonization debris from the operating site by wiping it off repeatedly with a moist swab to facilitate dissection. The video camera which is attached to the operating microscope provides demonstration and documentation of the operation. Video technology enables anaesthesiologists, operating room staff, residents and visiting surgeons to follow the operation on the television screen. The attendant operating staff is able to follow the operation and anticipate which instrument may be needed next. Cooperation between the operating surgeon and the nurses is hereby facilitated. The whole operating team is able to comprehend the course of the operation. A further advantage of video technology is the recording of operations for education purposes.

## Polyps

In phonosurgical operations, for example in polyps or cysts, the $CO_2$ laser may be used with a small focus (e.g. 0.25 mm) and with low power settings (0.5–1 W).

Polyps are usually found unilaterally at the free edge of a vocal cord. Predisposing factors are usually voice abuse and chronic inflammation. As polyps do not respond to speech therapy or medication, surgery is usually required to remove these lesions. Additionally, it is wise to send a removed polyp as operation specimen for histologic examination. We therefore recommend an excision rather than a vaporization of a polyp. Although polyps have a typical clinical ap-

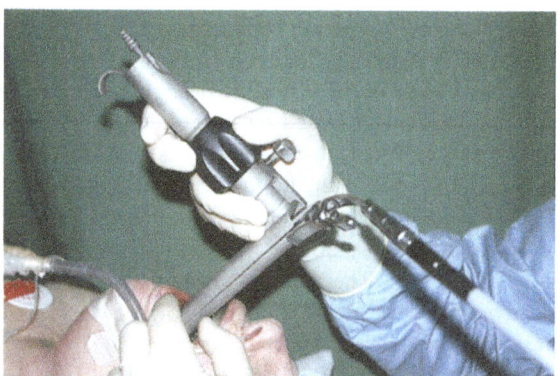

**Fig. 1.** The laryngoscope is inserted under digital control into the mouth, in order to avoid injuries particularly to the lip and tongue

pearance, rare tumours of the larynx or early glottic carcinomas must always be ruled out.

The glottis is exposed via a laryngoscope. To prevent accidental damage to distal mucosa by the laser beam, a moistening neurosurgical gauze is introduced subglottically. The polyp is then grasped gently with the cup forceps and retracted medially. This technique lifts the polyp from the vocal ligament, which can be spared from accidental trauma. The polyp is dissected from the underlying mucosa at its base. At the end of the operation, nerely a tiny epithelial defect above the vocal ligament is visible.

### Laryngoceles and Cysts

Internal laryngoceles are dilatations of the laryngeal ventricle which are usually lined with respiratory epithelium. Laryngoceles are filled with either air or mucous, and they are likely to spread into false cord or up to the aryepiglottic fold. The operation is again performed under general anaesthesia.

First, the supraglottic region is exposed with the distending laryngoscope. The mucosa above the laryngocele is incised in the area of the cord or the aryepiglottic fold. The lining of the laryngocele is then identified. The bloodless dissection of the laser facilitates the identification of the outerlining of the laryngocele. The cele can then be dissected along its lining, on else the cele can be incised and the air or the mucous can be suctioned out. The laryngocele can then be dissected down to its origin at the sinus of Morgagni where it should be excised. If the larynocele extends through the thyroid membrane into the soft tissues of the neck, we are dealing with a mixed laryngocele. Mixed laryngoceles should be followed into the soft tissues of the neck until the base of the laryngocele can be identified and excised.

The operation is performed under general anaesthesia through the laryngoscope. The mucosa which covers the cyst should be incised, and the sack of the cyst should be grasped with a fine cup forceps. The cyst can then be dissected step by step until its base is reached and can be excised.

External laryngoceles, however, should be managed by an external approach. The most common lesions are retention cysts. Retention cysts result from obstructed excretory ducts of submucous glands, the obstruction of the excretory duct is usually being due to inflammatory processes. Most common sites of retention cysts are the vocal cords, the false cords and the epiglottis. The complete excision of the cysts is mandatory to prevent recurrent disease.

### Granulomas

Intubation granulomas can be found unilaterally or bilaterally at the vocal process of the arytenoid cartilage. Intubation granulomas should not be removed directly after diagnosis because spontaneous healing can be observed frequently within 4 weeks after the intubation. However, if the granulomas are of a considerable size, meaning they cause symptoms like dyspnea or severe hoarseness, they may have to be removed before the 4 week period has passed. If the decision for surgery is made, repeat intubation should be avoided; the excision should rather be performed in apnea or with the jet ventilation technique. The glottic region is exposed as for typical microlaryngoscopy, the granuloma is then grasped with the small cup forceps and excised at its space. Care should be to that the arytenoid cartilage. However, recurrence of the granulomas is not infrequent and may be a sign of persistent inflammation of the perichondrium.

The excision of contact granulomas is only rarely indicated. The main therapy for this disease is speech therapy and psychotherapy. If for certain reasons surgery is considered, the technique is the same as for intubation granulomas.

### Recurrent Papillomatosis of the Larynx

Laryngeal papillomatosis is a benign tumour disease induced by HPV 6 or 11; as of today no curative therapy is available. New therapeutic concepts have been reported and discussed controversely, most recently treatment with interferon or acyclovir or photodynamic therapy, these methods lack long-term results. They are also non curative and have side effects. Evaluation of new methods is difficult because of the long course of this disease. The symptomatic removal of papillomas is so far the only method where long-term results are available.

Papillomavirus can be detected not only in the lesions but also in the intact mucous membranes all over the upper aerodigestive tract [13,15]. For this reason, surgery cannot be curative. The surgical intention is to remove as much possible of the papillomatosis without causing further damage.

Papillomas usually become evident in childhood between the ages of 1 and 3 years, but they may occur at any age. Apart from the typical clinical finding, e.g. stridor and hoarseness the diagnosis should be made by endoscopy. As the disease tends to spread along the whole respiratory tract, it seems to be recommendable to perform a full panendoscopy rather than only microlaryngoscopy. Once the focus of the papillomatosis has been identified, the laser surgical operation should be performed.

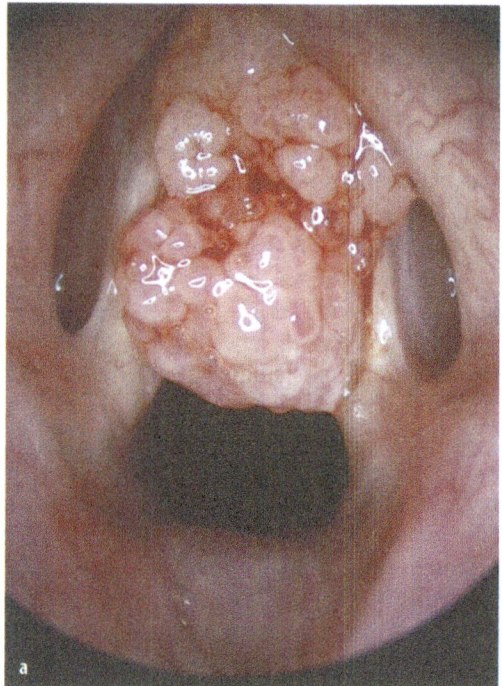

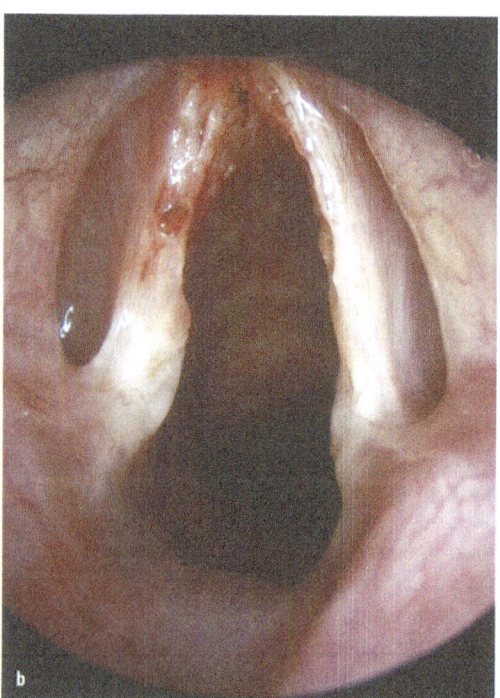

**Fig. 2a,b.** Extensive recurrent papillomatosis of the larynx. **a** Endoscopic aspects before and **b** immediately after transoral $CO_2$ laser removal

The operation can be performed using jet ventilation technique or endotracheal intubation combined with intermittent apnea which we prefer. We recommend the use of a slightly defocused laser beam and low laser power settings. Low laser power settings

permit an excision of the papillomas in the plane of the mucosa, thus avoiding damage to the deeper tissue layers. Islands of non-diseased mucosa should be left intact in the anterior part of one vocal cord to allow a reepithelialization of the laser lesions. The aim of the laser surgical procedure, however, should be the complete removal of the papillomas to achieve prolonged symptom-free intervals (Fig. 2). If both vocal cords are involved, special care has to be taken to spare the anterior commissure from trauma to prevent synechia. In this area a more conservative approach seems to be advisable, especially in children. If the anterior commissure is affected, small areas of papillomas should rather be left intact than risk the formation of webs in the anterior commissure. The risk of postoperative oedema is low, extubation is usually possible at the end of the operation. Additionally, it seems to be advisable to administer a single dose of steroids (e.g. 3–5 mg prednisolon – kg body weight) for oedema prophylaxis. In adult patients, a thorough follow up by indirect laryngoscopy is mandatory. Here, fibrinous membranes should be removed if they become evident in the anterior commissure. Fibrinous webs can be removed with a damp swab under surface anaesthesia in the surgeon's office.

In summary, the use of the laser did not reduce the rate of recurrence. It did, however, cut back the rate of complications. The reduced number of glottic webs may be a result of the less traumatic technique when using the laser. The immediate haemostasic effect of the laser allows a better exposure with a more precise removal of the papillomas and less unintensional damage to the functional structures of the larynx [12, 17, 28].

## Haemangiomas and Vascular Malformations of the Larynx

Laryngeal haemangiomas can be distinguished in the more capillary haemangioma of early childhood and the more cavernous haemangioma in the adult patient, which is synonymous with vascular malformation.

Capillary haemangiomas are present as well defined reddish tumours of the posterior subglottis or of the subglottic belly of the vocal cord. Horseshoe-shaped haemangiomas which extend from the subglottic belly of one vocal cord to the posterior commissure and the contralateral vocal cord are usually rare. Similar to the treatment concept of the papillomatosis, it seems to be advisable that not only the larynx is examined by endoscopy, but also the trachea and the main bronchi. Larynx, trachea and bronchi should examined with the 0°- and 30°-degree rigid endoscope at the beginning of the operation. For laryngeal hemangiomas the larynx is exposed via the laryngoscope. The haemangioma should then be excised step by step with a focused beam at a low power setting. The vessels of capillary haemangiomas are sealed instantly by the laser beam, and hardly any haemorrhage is evident. The endotracheal tube may be removed temporarily if it obstructs the view, and the operation can be performed under intermittent apnea. Alternatively, jet ventilation can be employed. Unilateral haemangiomas can be completely resected by this technique; however, care should be taken that the vocal folds are completely preserved.

In adult patients, mostly cavernous haemangiomas can be found. Resection with the $CO_2$ laser is possible if it is a pedunculated lesion, or if the haemangioma is confined to the circumscribed area of the supraglottis. We recommend frequent inspections by indirect laryngoscopy in the first postoperative days because secondary haemorrhages may be possible. Additionally, the Nd:YAG laser [23, 31] can be employed in cases of rare extensive vascular malformations which spread between larynx (Fig. 3) and hypopharynx sometimes even into the oropharynx.

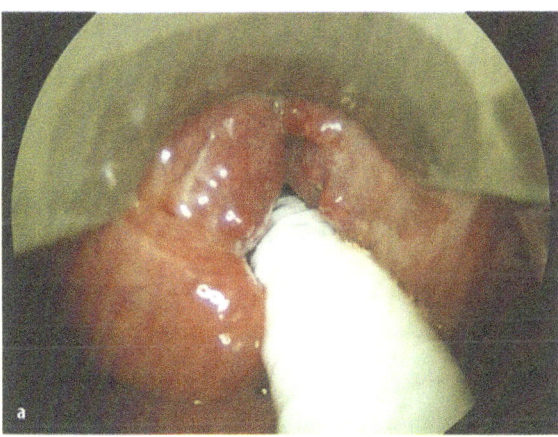

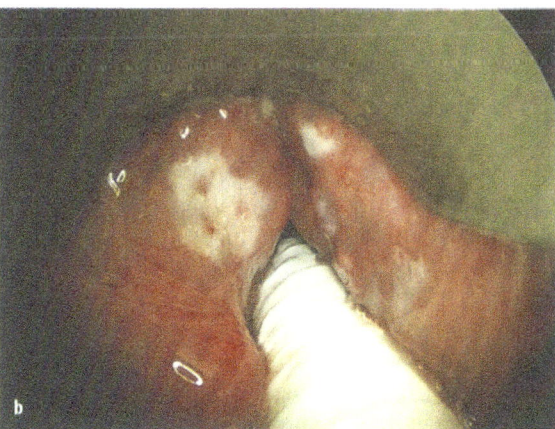

**Fig. 3a,b.** Endolaryngeal vascular malformation **a** just before und **b** just after Nd:YAG laser coagulation. Treatments are accomplished step by step to avoid extended endolaryngeal swelling

## Stenoses of the Larynx

Stenoses of the larynx may be classified according to their anatomic location. Stenoses can be found in the supraglottic, glottic and in the subglottic regions. Depending on their location, the prognosis is different.

### Supraglottic Stenosis

Supraglottic laryngeal stenosis mainly occurs after partial supraglottic laryngectomy. The formation of scar tissue can be performed easily with the $CO_2$ laser, and this procedure has usually a good prognosis. Wide excision of the scar tissue is, however, recommended.

### Glottic Stenosis

Webs in the glottic region may be divided in to primary and secondary webs: also the differentiation between webs of the anterior and the posterior glottis is of importance. Primary webs may be found as congenital lesions, whereas secondary membraneous webs are usually due to trauma, long-term intubation or open surgery. The surgical instrument of choice is the $CO_2$ laser. Membranous webs, on the other hand, can be simply excised. However, the structures of the vocal ligaments should be preserved carefully. The laser is set at a low power setting. Solid scar formations in the anterior commissure should be excised in a key-hole pattern onto the very edge of the thyroid cartilage to enable brushing of the fibrin exsudations and to prevent recurrence of stenosis. With the compliance of the patient, the anterior commissure should be brushed vigorously with a swab under local anaesthesia two to three times a week postoperatively. In some cases, especially in paediatric patients, repeat endoscopy may be necessary to perform this postoperative care. In cases of recurrent webbing of the anterior commissure, it may be necessary to insert a silicone foil, according to the technique of Friedrich [6]. This foil is used as a stent that facilitates re-epithelization of the anterior commissure. Posterior glottic stenoses, on the other hand, usually have a good prognosis, if only a fibrous bridge is evident. Interarytenoid fibrosis usually has a less favourable prognosis.

Subglottic stenoses of the larynx have a good prognosis, if they are of membranous consistence. Stenoses of this kind can be found after intubation. However repeated procedures are necessary to achieve a patent airway. The stenoses may be excised completely or incised in the form of a five-pointed star. Widening the stenoses with bougis of increasing size may be beneficial additionally. All other stenoses are not suitable for laser surgery alone – stenting and/or reconstructive surgery are mandatory.

## Bilateral Recurrent Nerve Palsy

The difficulty in the operative enlargement of glottis is to perform a sufficient widening of the airway by simultaneous preservation of the highest possible voice quality. For many years the most commonly performed procedure for a bilateral recurrent nerve palsy has been the unilateral arytenoidectomie. Numerous variations of this procedure exist, and simultaneous removal of other parts of the glottic area has been described.

In our own experience, a bilateral excision of the posterior part of the vocal cord in the area of the vocal process (posterior chordectomy) is a procedure which finds the best compromise between voice and sufficently widened airway [2]. In posterior chordectomy the glottis is widened only posteriorly. The main part of arytenoid cartilage and vibrating parts of the vocal cords remain intact, thus a good balance between voice and airway can be found. As the arytenoid cartilage remains intact, there are no problems with aspiration.

## Laryngomalacia

Laryngomalacia is the most common cause for stridor in pediatric patients. Several different types of it have been described and not all of them seem to be defined by immature laryngeal cartilage. The pathologic changes which may be observed are the following:

1. In cases of enlarged cuneiform cartilages the aryepiglottic folds may tend to collapse into the glottic space, thus leading to stridor.
2. A tubular-shaped epiglottis shows collapse towards the glottis on inspiration.
3. The arytenoid cartilages show collaptic movements towards the glottic space during respiration.
4. An immature epiglottic cartilage is aspirated during inspiration into the glottis.
5. Shortened aryepiglottic folds obstruct the glottis.
6. An extremely acute angle between epiglottis and glottis leads to obstruction of the laryngeal entrance.

These observations describe the most commonly observed pathologic changes, but combined forms may also be found.

The clinical picture of laryngomalacia may vary. Mild forms with occassional episodes of stridor may be found, which can resolve with maturation of the laryngeal cartilage. Several forms of laryngomalacia with continuous stridor and oxygen desaturation have frequently necessitated tracheotomy in the past. However, tracheotomy may be avoided, if endoscopic laser microsurgical options are considered. About 20% of laryngomalacia may require surgery to resolve deglutitions disorders, failure to thrive and last but not least, life-threatening phases of respiratory failure.

The assessment of stridor in paediatric patients always requires laryngotracheobronchoscopy to define the location of obstruction. Video monitoring of endoscopy is recommended for the following reasons:

1. It allows nursing staff and the anaesthesiologists to follow the course of the operation. Assisting personal can estimate which instruments are needed next, thus reducing operating time. Especially for operations which are performed in intermittent apnea this method has been found to be of great value.
2. Findings can be recorded on video tape for secondary observation and for interdisciplinary discussion.
3. Video tapes for educational purposes can be produced.

### Surgical Techniques

Shortened aryepiglottic folds should be divided by the laser beam at a power setting of 1 Watt. The underlying tissue is sheltered from accidental trauma with a so-called protector. After division of the mucosal folds, a wide gap relieves the epiglottis, thus widening the laryngeal entrance.

Supraglottoplasty is indicated for cases where excess supraglottic tissue is the cause of laryngeal obstruction. The hypermobile supraglottic tissue is resected with the laser beam with simultaneous preservation of functional structures.

Epiglottopexia (Fig. 4) is indicated for cases in which the epiglottis is aspirated on inspiration [32]. At the beginning of the operation, the base of the tongue is cleared of from mucosa in a horseshoe-shaped pattern. This can be done by a complete resection of the mucosal layer as well as by vaporization of the mucosa. A corresponding wound is created on the surface of the epiglottis. The laser power setting should be 1 W at a focal diameter of 0.25 mm. The next step of the operation is to lay three sutures into the tongue base and the epiglottis. The epiglottis is then elevated with these sutures towards the base of the tongue, thus giving wide access to the glottis. Once a satisfactory enlargement of the glottis has been achieved by this manoeuvre the sutures should be tied in tight knots. It has been found to be advantageous to use single stitch sutures at the perimeter and an inverse mattress suture at the centre; the use of resorbable sutures is recommended.

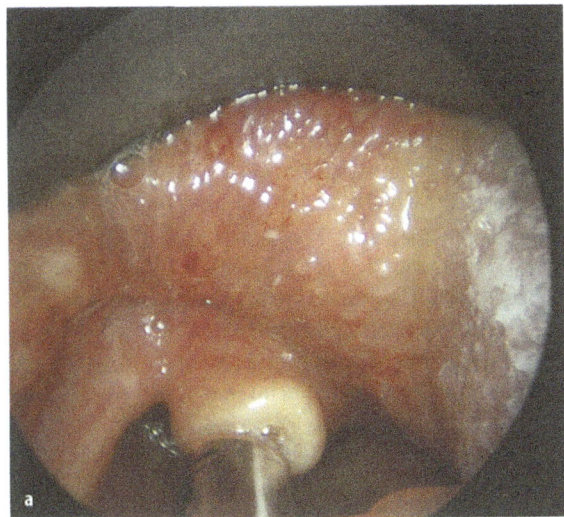

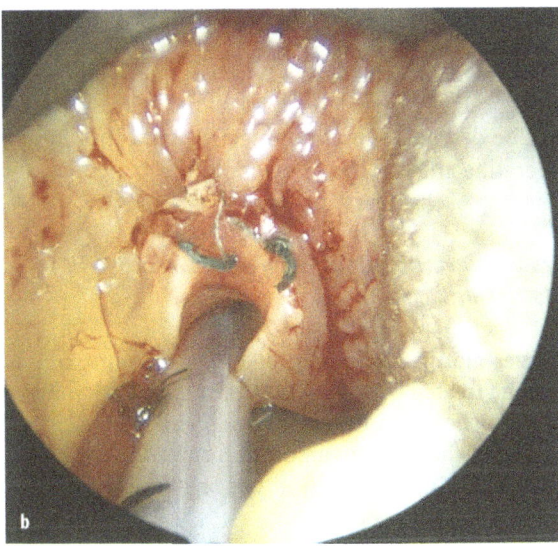

**Fig. 4a.** In cases of posterior caudal displacement of the epiglottis the mucosa of the lingual epiglottis is removed initially. A corresponding lesion is created at the tongue base, before the two areas are connected **b** by three single-stitch sutures

## Benign Diseases of the Trachea

Several benign diseases of the trachea can be treated very efficiently by endoscopy or microlaryngoscopy employing different kinds of laser systems ($CO_2$, argon, Nd:YAG). The laser beam is directed to the operating site via a rigid tracheoscope, with a ventilator connection or through a laryngoscope in intermittent apnoea or with jet ventilation. Under direct vision, lesions like papillomas, granulomas, fibromas, chondromas or stenoses can be excised. Membraneous sickle-shaped stenoses can be resected quite efficiently with the laser. However, solid circular stenoses of considerable length should rather be treated by conventional open surgery.

## Carcinoma of the Larynx, Hypopharynx and Trachea

Laser microsurgical resection of carcinomas follows the same oncological principles as conventional surgery [17]. The main concern is the complete resection of the carcinoma with simultaneous functional preservation. Endoscopic laser microsurgery fulfills these requirements and offers reduction of surgical morbidity at the same time. Usually surgical margins of between 1.5 to 5 mm depending on the location and size of the tumours should provide sufficient safety. In early glottic superficial lesions excisional biopsies have been found to be of sound oncological safety, offering good vocal function at the same time.

Generally, a $CO_2$ laser, mounted on an operating microscope is used for dissection as introduced by Steiner in the early eighties [16]. The magnification of the operating microscope allows superior assessment of tumour extension compared to the unguarded eye. Under microscopic control the tumour is resected with clinically safe margins and the operation specimen is mounted on cork with pins according to its location for histologic assessment. The reduction of carbonization at the edge of the line of dissection, which may obscure the histologic assessment, is achieved by the use of modern micromanipulators, like the Microspot or Accuspot. Further reduction of char formation may be achieved through the use of the laser at a low power setting or in superpulse mode.

Larger tumours may be removed by dissecting the tumour in several pieces [16]. At first sight this concept seems to be unconventional, for fear of the accidental spread of tumour cells. However, long-term observation of patients, who have been treated in such a fashion did not show increased numbers of local or distant spread of carcinomas. This clinical observation is consistent with experimental findings on the buccal mucosa of the rat, where a sealing effect of the laser beam onto lymphatic vessels could be demonstrated [33]. On the 15th postoperative day after $CO_2$ laser incisions in the buccal mucosa of the rat, no lymphatics could be demonstrated on indirect India ink lymphography. Following Nd:YAG laser incisions this effect was even more delayed. Superficial and profound lymphatics could only be found only after 36 days postop.

## Laryngeal Carcinoma

### Carcinoma in Situ, Microinvasive Carcinoma, Localized T1a Carcinoma of the Vocal Cord

Transoral $CO_2$ laser excision of carcinoma in situ and early T1 glottic cancer was introduced by Strong and Jako [27, 29] in the early 1970s. At that time, it was felt that resection of the arytenoid cartilage coupled with resection of the soft tissue of the true vocal cord was absolutely contraindicated, due to the high likelihood of postoperative aspiration. Steiner [19] reported encouraging results in stage II glottic cancer patients utilizing transoral $CO_2$ laser microsurgery. His approach has shown that the traditionally accepted limitations to transoral glottic cancer surgery earlier discussed were, in fact, too restrictive [3].

For limited glottic lesions, excisional laser biopsies may be indicated [4, 9, 24]. In cases with negative cytologic or histologic results, a diagnostic/therapeutic excision is indicated. This consists of a complete removal of all macroscopic disease in the form of an excisional biopsy. In cases of a positive cytology, or biopsy-proven carcinoma, similarly an excision biopsy is performed, this time with the advantage of hindsight. The entire lesion is excised with an appropriate resection margin. Account is taken of the individual circumstances of each patient, but the aim in all cases is to perform an adequate excision biopsy. The aim is to avoid a second general anaesthesia and the associated risks to an elderly patient, which would become necessary if the final histological result showed positive margins. Although this occurs in less than 5% of our patients, it is a factor that must be considered in the individual patient with a high risk factor for general anaesthesia.

The use of modern micromanipulators is strongly recommended, as these micromanipulators perform excellent dissection at low power settings thus reducing char formation at the edge of the cut. This applies of both benign and malignant lesions of the larynx. The reduction of carbonization in operation specimens facilitates the histologic assessment considerably. Recommended surgical margins should be 1.5 mm in diameter and towards the depth of the tumour. Vocal function usually is not impaired after these limited excisions. Sometimes it has been found to be advantageous for the less experienced surgeon to outline the zone of resection with a few laser spots prior to dissection. For the application of these dots the laser should be set at interval mode rather than at continuous mode. In interval mode the laser emits single impulses and the surgeon can thus outline his resection margin more easily.

### Large T1a and T1b Carcinomas of the Glottis

Resectability of larger glottic tumours depends first and foremost on the extension of the tumour and the suspected depth of infiltration. Again, it may be possible that resection of larger glottic tumours becomes possible through the blockwise resection of the neoplasia and stepwise resection. Division of the tumour is also recommended for estimating the depth of invasion clinically. An incision is placed in the middle of the carcinoma until a clinically safe margin is defined. The tumour may then be resected in two or seldom three pieces through anterior and posterior dissection of the neoplasia.

Up to an infiltration depth of approximately 2 mm, the resection is certainly feasible with sound oncological results.

Involvement of the anterior commissure without subglottic extension of the tumour (T1b) requires dissection along the thyroid cartilage towards the involved contralateral side. The lesion on the contralateral vocal cord is resected until clinically safe margins can be defined. The use of high magnification in the operating microscope is strongly recommended for this step of the operation.

### T2a Glottic Carcinomas Moderately Advanced Glottic Carcinomas

Larger glottic carcinomas, like T2a tumours, should also be treated by transoral laser microsurgery. There seems to be no limitation to this kind of treatment, regardless whether the tumour has spread to the supraglottic region, the subglottic region, the anterior commissure or both vocal cords.

Especially suited for laser surgery are superficially spreading carcinomas. These lesions may be removed through partial excision of the mucosa of the endolarynx, no matter how large the spread of superficial carcinoma is. However, the removal of the large superficial carcinomas is hardly feasible in one piece. Again, the removal of the lesions in several pieces is recommended. In such cases, it is of greatest importance to mount the specimen on cork plates matching the exact anatomic location in the larynx to give the pathologist a better orientation. The basal surface of the operation specimen which is of main interest for the surgeons resecting the tumour in several pieces may also be marked with blue ink, to avoid confusion.

Additionally to the video documentation of the relevant steps of the operation, it is very helpful to perform a schematic drawing and an exact description of the anatomic situation on the pathology request form and in the patient's charts. These drawings can be reviewed postoperatively with the definite pathology report, and it can then be judged whether the resection has been complete or whether control endoscopy with repeated tissue resection should be considered.

### T2b (Mobility-Impaired) and T3 Glottic Carcinoma

Resection of large, more advanced, carcinomas of the glottic region (T2b and T3) again become feasible through the division of the tumours into several smaller portions [25].

Laterally the margin of dissection is defined through the thyroid cartilage: inferiorly, the dissection should be carried out onto the superior surface of the cricoid cartilage. The line of dissection follows the extension of the tumour plus an additional rim of healthy mucosa. Towards the deep parts of the tumour, a dissection into the musculature is necessary and very often the partial or complete removal of the arytenoid cartilage. A safe margin in the depth can often be defined by the different reaction of the healthy tissue to laser light, as opposed to the effect which laser light has on tumorous tissue.

If the lesion extends towards the perichondrium of the thyroid cartilage the inner layer of the thyroid cartilage, should be included in the operation specimen. Sometimes, even the removal of entire chips of the thyroid cartilage is necessary in cases of suspected or evident involvement of the thyroid cartilage. In these cases, it is advisable to remove pieces of prelaryngeal soft tissue to document an entire resection of the tumour.

### Supraglottic Carcinoma

An external or transoral laser surgical supraglottic partial resection are two possible approaches in surgery for supraglottic carcinoma. Besides the oncological results of these procedures, the functional ones are extremely important for the patients. Aspiration of saliva and food particles is the most common complication after external supraglottic partial resection. The results after laser surgery show a much lower rate of aspiration pneumonia than for the group treated by conventional surgery [11].

The ability to stabilize the airway without tracheotomy in selected advanced supraglottic cancer patients by performing $CO_2$ laser endoscopic subtotal supraglottic resections during the initial biopsy procedure was reported in 1981 [5].

Steiner [19, 25] carried out supraglottic resections extensively into the pre-epiglottic space and has even included partial resections of the thyroid cartilage and removal of one arytenoid if oncologically necessary, i.e. where cancer spread has dictated this. Patients treated with histologically positive surgical margins had reresections to obtain clear margins. This approach has probably led to greater success with the more advanced cancer patients.

The extension of the operation also in the supraglottic region is basically defined through the intraoperative findings. The surgeon should follow the tumour through repeat readjustment of the operating microscope. By this approach extensive tumours may become resectable, if the surgical principles described above are applied [8,14,17]. Smaller, well defined, neoplasias of the supraglottis are rather rare findings. These tumours may be resected as one piece, similar by to small carcinomas of the glottic region [1].

For supraglottic carcinomas, the use of distending, bivalved laryngoscopes is recommended. The use of these laryngoscopes allows improved access to the operation site. Technically, the operation is relatively easy. In the supraglottic region the surgeon has to worry less about functional impairments, as even deglutition of fluids can be trained after the operation to avoid aspiration.

Additionally, the supraglottic area allows wider resections than the glottis area, safe margins can therefore be achieved more easily. The most problematic region in supraglottic carcinomas is the area around the petiole. Here, it is difficult to judge the depth of infiltration preoperatively. If the preepiglottic space is infiltrated by the malignancy, a seemingly small T1 carcinoma can easily turn into a T3 carcinoma. Tumours, which are localized in the infrathyroid area of the supraglottis should always be suspected of having infiltrated the preepiglottic space for adequate staging of the tumour and surgical treatment.

### Dissection Technique

After the insertion of the operating distending bivalved laryngoscope, the epiglottis is exposed. The epiglottis is dissected in the midline, the resection is then extended to the pharyngo-epiglottic fold and to the glosso-epiglottic fold. Sometimes electrocautery is necessary to achieve good haemostasis, especially in the highly vascularized area of the tongue base.

Readjustment of the bivalved laryngoscope should be performed repeatedly to achieve adequate exposure of the tumour. The plane of dissection reaches from the epiglottic cartilage to the pre epiglottic fat and the laryngeal surface of the infrathyroid epiglottis where the tumour is situated. The dissection is extended inferiorly; depending on the tumour extent, horizontal dissection is necessary in addition.

Larger supraglottic tumours can infiltrate through the epiglottic cartilage into the preepiglottic space, which may be permeated entirely with tumour cells. Despite an exact staging with fine-layer CT scan of the larynx preoperatively circumscript, involvement of the preepiglottic space may be an accidental intra-operative finding. Entire resection of the tumour should, however, always be the aim of the operation, with a safety margin of about 5 mm.

Functional impairment after anterior resection in the supraglottic region is usually minimal. Degluti-tion might be impaired if one of the arytenoid carti-lages has to be included into the operation specimen. However, suspected involvement of thyroid or ary-tenoid cartilage should always lead to extended sur-gery, sometimes even into the soft tissues of the neck. In this way, total laryngectomy can be avoided in many patients with advanced supraglottic tumours [25].

## Hypopharyngeal Carcinoma

The preoperative staging of hypopharyngeal carcino-mas comprises endoscopy of the entire upper aerodi-gestive mucous membranes under general anaesthe-sia, ultrasound scanning of the neck, and facultatively CT imaging of the neck and the thorax, of course.

The results of a meticulous clinical staging should then lead to an individually designed therapy. Hy-popharyngeal carcinomas may obscure their true ex-tent, because their submucosal spread is likely to in-vade the paraglottic and preepiglottic space, the area of the arytenoid cartilages, the thyroid cartilage and the soft tissues of the neck. The entire extension of the tumour may not be evident on endoscopy, imag-ing studies are therefore strongly recommended in most hypopharyngeal carcinomas.

If the patient complains about hoarseness, and indi-rect laryngoscopy reveals reduced mobility of the vocal cord usually the cricoarytenoid joint is infiltrated by the tumour. In such cases, most oncologic head and neck surgeons would stage the disease beyond re-sectability by partial pharyngolaryngeal resection.

As previously mentioned the concept of step wise resection also applies for hypopharyngeal carcino-mas. For hypopharyngeal carcinoma resection the ap-plication of a bivalved laryngo-tracheoscope is useful. It should be positioned in such a fashion that it allows to distinguish between tumour and a circular rim of healthy mucosa of about 5–10 mm width.

The operating microscope should be adjusted to the highest possible magnification The incision may then be performed circularly around the tumour, if one-step resection seems feasible, or in a mosaic pat-tern, if the tumour has to be resected in several pieces.

In the hypopharynx, distinction between healthy and carcinomatous tissue is usually mainly possible. Functional impairment after endoscopic resection of hypopharyngeal carcinomas is usually low if func-tionally important structures of the larynx can be preserved.

After microendoscopic laser surgical resection, the laser wounds usually epitheliaze within a time range of 4–6 weeks spontaneously. Grafting of the defect is not necessary and functional results are good to ex-cellent. Perioperative morbidity is also usually lower than compared to open partial laryngo-pharyngo-tomy [21, 26].

The oncological results are very encouraging, even better than after radical mutilating procedures [26].

## Malignant Tumours of the Trachea

Primary squamous cell carcinomas of the trachea are rare. Secondary involvement of the trachea in sub-glottic carcinomas may be found comparatively often. In these cases, it has been discussed whether the tu-mours can be resected completely via translaryngeal approach or combined with open surgery. The pre-requisite for this kind of surgery would be an excel-lent endoscopic exposure and the absence of signifi-cant infiltration of the neck.

The main indications for endoscopic laser surgery are palliative treatment of different kinds of stenotic malignant tumours or metastases of the trachea. One difficulty in the treatment of stenosing tumours on the trachea lies in respiratory management of these patients. Jet ventilation sometimes may lead to pneu-mothorax if the ventilating pressure is too high. How-ever, high ventilating pressure may be necessary to bypass stenosing tumours. In cases of significant nar-rowing of the tracheal lumen, it may be beneficial to perform laser surgical debulking of the tumour in in-termittent apnea.

The risks of this type of surgery are due to the lock at good exposure, inadequate smoke evacuation and the difficulties in controlling major haemorrhages, if they occur.

In cases of difficult exposure of tracheal carcino-mas, sometimes a combined approach or only a con-ventional surgical approach should be considered. For advanced resection involving removal of cartilage or soft tissue of the neck, inferior tracheostomy should always be performed to secure the airway.

Once control of the disease has been achieved, de-canulation can be performed after a recurrence-free interval.

## References

[1] AMBROSCH P, KRON M, STEINER W (1998) Carbon diox-ide laser microsurgery for early supraglottic carcinoma. Ann Otol Rhinol Laryngol 8: 680–688
[2] BIGENZAHN W, HÖFLER H (1996) Minimally invasive laser surgery for treatment of bilateral vocal cord paraly-sis. Laryngoscope 106: 791–793
[3] DAVIS, RK (1995) Transoral laser surgery for glottic can-cer. Adv Otorhinolaryngol 49: 254–258

[4] DAVIS RK (1997) Endoscopic surgical managemant of glottic laryngeal cancer. Otolaryngol Clin North Am 30: 79–86

[5] DAVIS RK, SHAPSAY SM, VAUGHAN CW, STRONG, MS (1981) Pretreatment airway management in obstructing carcinoma of the larynx. Otolaryngol Head Neck Surg 89: 209–214

[6] DUNCAVAGE JA, OSSOFF RH (1990) Laser surgery for benign laryngeal lesions. In: DAVIS RK (ed) Lasers in otolaryngology – head and neck surgery. Saunders, Philadelphia, pp 26–32

[7] FRIEDRICH G (1998) Externe Stimmlippenmedialisation – chirurgische Erfahrungen und Modifikationen. Laryngorhinootol 77: 7–17

[8] IRO H, WALDFAHRER F, ALTENDORF-HOFMANN A, WEIDENBECHER M, SAUER R, STEINER W (1998) Transoral laser surgery of supraglottic cancer: follow-up of 141 patients. Arch Otolaryngol Head Neck Surg 124:1245–1250

[9] JÄCKEL M, STEINER W (1998) Ear, nose and throat techniques in endoluminal surgery. Min Invas Ther Allied Technol 7/1: 9–14

[10] JECKSTRÖM, W, WAWERSIK J, HOFFMANN P, WERNER J A, LIPPERT B.M, CHRISTIANSEN B, PAUSTIAN R, SOWADA U (1995) Anesthesiological problems of endolaryngeal and endotracheal laser surgery. Adv Otorhinolaryngol 49: 15–19

[11] KÖLLISCH M, WERNER JA, LIPPERT BM, RUDERT H (1995) Functional results following partial supraglottic resection: Comparison of conventional surgery versus transoral laser microsurgery. Adv Otorhinolaryngol 49: 237–240

[12] MAHNKE C, WERNER JA, FRÖHLICH O, LIPPERT BM, HOFFMANN M, RUDERT H (1998) Klinische und molekularbiologische Untersuchungen zur respiratorischen Papillomatose. Laryngorhinootology 77: 27–33

[13] OGURA H, WATANABE S, FUKUSHIMA K, BABA Y, MASUDA Y, FUJIWARA T, YABE Y (1993) Persistence of human papillomavirus type 6e in adult multiple laryngeal papilloma and the counterpart false cord of an interferon-treated patient. Jpn J Clin Oncol 23: 130–133

[14] RUDERT H, WERNER JA, HÖFT S (1999) Transoral $CO_2$ laser resections of supraglottic carcinomas. Ann Otol Rhinol Laryngol 108: 819–827

[15] STEINBERG BM, TOPP WC, SCHNEIDER PS, ABRAMSON AL (1983) Laryngeal papillomavirus infection during clinical remission. N Engl J Med 308:1261–1264

[16] STEINER W (1984) Transoral microsurgical $CO_2$-laser resection of laryngeal carcinoma. In: WIGAND ME, STEINER W, STELL PM (eds) Functional partial laryngectomy. Springer, Berlin Heidelberg New York

[17] STEINER W (1988) Experience in endoscopic laser surgery of malignant tumours of the upper aero-digestive tract. Adv Otorhinolaryngol 39: 135–144

[18] STEINER W (1991) Die endoskopische Lasertherapie im oberen Aero-Digestivtrakt. In: BERLIEN H-P, MÜLLER G (eds) Angewandte Lasermedizin, ecomed, Landsberg, pp 1–5

[19] STEINER W (1993) Results of curative laser microsurgery of laryngeal carcinomas. Am J Otolaryngol 14: 116–121

[20] STEINER W (1997) Endoskopische Laserchirurgie der oberen Luft- und Speisewege. Schwerpunkt Tumorchirurgie. Thieme, Stuttgart

[21] STEINER W, AMBROSCH P (1996) Laser im Kopf- und Halsbereich. In: BERLIEN H-P, MÜLLER G (Hrsg): Angewandte Lasermedizin 13. Ecomed, Landsberg. S 127–136

[22] STEINER W, AMBROSCH P (2000) Endoscopic laser surgery of the upper aerodigestive tract, with special emphasis on tumor surgery. Thieme, Stuttgart

[23] STEINER W, WERNER J A (2000) Lasers in otorhinolaryngology, head and neck surgery. Endo-Press, Tuttlingen

[24] STEINER W, AURBACH G, AMBROSCH P (1991) Minimally invasive therapy in otorhinolaryngology, head and neck surgery. Min Invas Ther 1: 57–70

[25] STEINER W, AMBROSCH P, MARTIN A, LIEBMANN F, KRON M (1996) Results of transoral laser microsurgery of laryngeal cancer. 3rd European Congress of the European Federation of Oto-Rhino-Laryngological Societies "Eufos", Monduzzi Editore S.p.a. Bologna, pp 369–375

[26] STEINER W, AMBROSCH P, HESS CF, KRON M (2002) Organ preservation by transoral laser microsugery in piriform sinus carcinoma. Otolaryngol Head Neck Surg (in press)

[27] STRONG MS, JAKO GJ (1972) Laser surgery in the larynx; early clinical experience with continuous $CO_2$ laser. Ann Otol Rhinol Laryngol 81: 791

[28] STRONG MS, VAUGHAN CW, COOPERBAND SR, HEALY GB, CLEMENTE MA (1976) Recurrent respiratory papillomatosis: Management with the $CO_2$ laser. Ann Otol Rhinol Laryngol 85: 508

[29] VAUGHAN CW, STRONG MS, JAKO GJ (1978) Laryngeal carcinoma: transoral treatment utilizing the $CO_2$ laser. Am J Surg 136: 490–493

[30] WERNER JA, RUDERT H (1992) Der Einsatz des Nd:YAG-Lasers in der Hals-, Nasen-, Ohrenheilkunde HNO 40: 248–258

[31] WERNER JA, LIPPERT BM, GOTTSCHLICH S, FOLZ BJ, FLEINER B, HÖFT S, RUDERT H (1998) Ultrasound-guided interstitial Nd:YAG laser treatment of voluminous hemangiomas and vascular malformations in 92 patients. Laryngoscope 108: 463–470

[32] WERNER JA, LIPPERT BM, SCHÜNKE M, RUDERT H (1995) Tierexperimentelle Untersuchungen zur Laserwirkung auf Lymphgefäße. Laryngorhinootology 74: 748–755

[33] WERNER JA, LIPPERT BM, ANKERMANN T (2000) Transorale Behandlung der schweren Laryngomalazie. Übersicht und Beschreibung einer modifizierten Operationstechnik Laryngorhinootol 79: 416–422

# III-4.2.1
# Laser Stapedotomy

S. Jovanovic

## Contents

## Introduction

Although stapedotomy is a preferred operation in stapes surgery, mechanical instruments such as a drill or a perforator cannot create a precise, round perforation. In fact, in some situations, mechanical instruments can prove hazardous. For example, a partially fixed stapes is often accidentally mobilized by manipulations (floating footplate), and a thin footplate is not infrequently fractured. In obliterative otosclerosis, perforation of a thick obliterating footplate with the drill can result in significant inner ear trauma due to vibrations.

The aim of stapedotomy with the laser is to enable management of the stapes in such a way as to ensure the greatest possible protection of the inner ear and the avoidance of damage to residual middle ear structures. Advocates of the laser technique agree that non-contact laser vaporization of the bone covering the vestibule is less traumatic for the inner ear than manual instrumental extraction or perforation of the stapes footplate.

A few thermally acting lasers (argon, KTP-532 and $CO_2$ lasers) in continuous wave (cw) and super-pulse (sp) mode have thus far been used in stapes surgery. In 1980, Perkins and DiBartolomeo and Ellis used the argon laser for the first time [72] and in 1989 following the development of precision micromanipulators Lesinski (1989) [52] successfully used the $CO_2$ laser in stapes surgery. Nevertheless, their effectiveness and safety are still controversial [3, 11, 14–16, 18, 20, 23, 27–50, 52, 52–57, 60–62, 70, 71, 75, 80, 82, 85, 86, 89, 91–100]. This initially led to scepticism regarding their application in stapes surgery.

Since the publication of the author's experimental and clinical studies confirming that $CO_2$ laser is suitable for stapedotomy [29–39, 41, 44–46, 49], this wavelength in the far-infrared range has found greater acceptance and broader application in ear surgery. Particularly for revision stapedotomies, but also for primary surgeries, clinical studies demonstrate that the $CO_2$ laser achieves significantly better hearing results and less complication rates than conventional surgery [4, 20, 29, 30, 45, 46, 52, 55, 56, 82].

The argon and KTP lasers seem to prove their value in primary and revision cases also [27, 61, 68, 72, 78, 95, 100]. Here, the introduction of a fiber optical microhand piece (Endo-Otoprobe) [21], which, in comparison with laser application by help of micromanipulators attached to the microscope, has the advantage that, due to the strong laser beam divergence at the exit of the optical fibre, a rapid decrease of energy density in relation to the increase of distance will result [8, 19]. Thus, the penetration depth and temperature problem in the perilymph with possible damage to the inner ear will be reduced. Moreover, the use of the fibre optical microhand piece facilitates the vaporization especially also of the anterior crus, while reducing the technical equipment needed [21]. Nevertheless, the suitability of the argon and KTP lasers for stapedotomy is doubtful in view of the lower absorption coefficient of the stapes for the argon and KTP beam and the considerable influence which the degree of pigmentation of the irradiated medium exerts on its effect, with the resultant poor reproducibility of the perforation. The beam of the $CO_2$ laser is far better absorbed at the footplate than that of the argon and KTP lasers. This results in higher effectivity, lower thermic side effects and better reproducibility of the perforation diameter [44].

More recent investigations show that novel pulsed laser systems (excimer, holmium:YAG, erbium:YSGG, erbium:YAG), which can act almost athermically, may likewise prove to be efficient and safe for stapes management [2, 14, 15, 22, 25, 26, 29, 31, 33, 37, 40, 42, 43, 47, 48, 50, 51, 66, 67, 76, 77, 79, 80, 81, 83, 90, 101].

From among the group of pulsed laser systems, the Er:YAG laser at first seemed to possess the most suitable wavelength for middle ear surgery. Due to different wavelengths and the differing relation of irradiation and time, the impact and effectivity of the Er:YAG laser as against the $CO_2$ laser differ in the tissue. While the continuously radiating $CO_2$ laser is suitable for use on soft tissue as well as – if well focused – for vaporization of thin bone structures [29, 39, 44], the Er:YAG laser offers advantages mainly in the treatment of bone structures [9, 29, 40, 47, 67, 69, 74, 77]. However, as soon as bleeding occurs, the oligothermic Er:YAG laser radiation is completely absorbed by blood and no longer reaches the intended area; it is then ineffective.

Moreover, the measured sound level in Er:YAG laser therapy is higher and implies the risk of inner ear trauma and tinnitus [22, 29, 42, 50, 77]. Meanwhile, it is suspected that the pressure waves resulting from Er:YAG laser therapy may cause transitory or even permanent inner ear damage such as deteriorated hearing of high frequencies or tinnitus [5, 22, and own experience]. Thus, application safety in the Er:YAG laser is less than in the $CO_2$ laser. For the time

being, the erbium laser may therefore not be recommended for stapes surgery.

## $CO_2$ Laser and Application Systems

One of the great advantages of the $CO_2$ laser for stapedotomy is its high absorption of radiation in the perilymph with a resultant low penetration depth of only 0.01 mm. The $CO_2$ laser beam, which could not be adequately focused in the past due to the longer wavelength and the poor beam quality, can now be focused with high-precision micromanipulators to a spot diameter of about 180 μm at a focal length of 250 mm (Fig. 1). Thus, values are reached which permit the finest microsurgical work.

When applying laser irradiation with microprocessor-controlled rotating mirrors, the so-called scanner systems (SurgiTouch, ESC Sharplan Co., modified according to the author's recommendations), a spiral figure is traced within the defined pulse duration. This enables the $CO_2$ laser to achieve high power density with minimal side effects, even in large irradia-

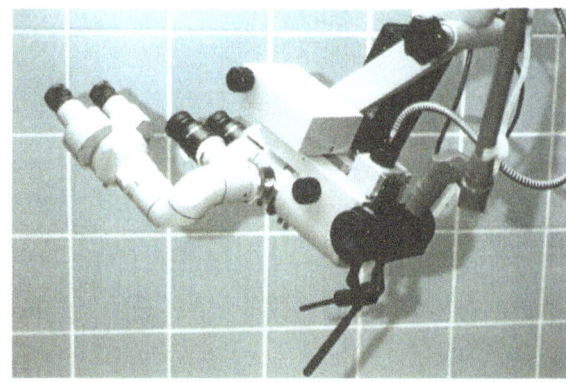

**Fig. 1.** High-precision micromanipulator Acuspot 712, ESC Sharplan Co., with variable working distances at 200 to 400 mm

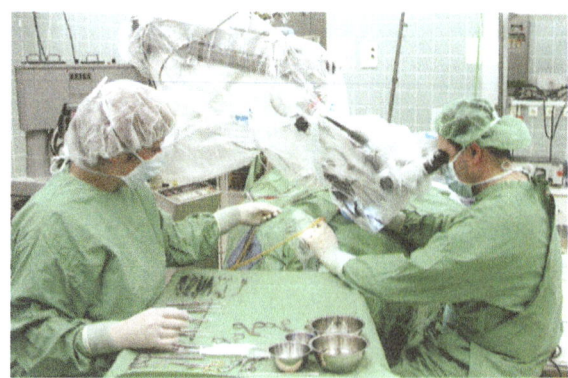

**Fig. 2.** Intraoperative setup for $CO_2$ laser stapedotomy

tion fields. At a working distance of 250 mm, the scanner can be set for irradiation fields of various sizes depending on the anatomical variations and preselected perforation diameter. Irradiation fields of 0.5, 0.6 and 0.7 mm are suitable for stapedotomy. A radius of less than 0.5 mm or greater than 0.7 mm can be achieved with SurgiTouch, but is rarely applied. Thus, a suitable selection of laser parameters basically enables a single-shot footplate perforation of preselected diameter. The laser beam is guided via a hinged mirror arm to a micromanipulator connected to the operating microscope and transmitted from there into the operating area; 250 mm proved to be the most favourable working distance (Fig. 2).

## Effective and Safe Laser Energy Parameters

Based on data obtained in petrous bone preparations and in a cochlea model [44], we determined effective parameters for stapedotomy with the $CO_2$ laser [Type 40C SurgiTouch (ESC Sharplan Co.) and micromanipulators Acuspot 712 (ESC Sharplan Co.)] (Table 1). The mode is continuous wave. A favourable pulse duration proved to be the shortest time of 0.03 to 0.05 s. The choice of powers ranged from 1 to 22 W (4000–88000 Wcm$^{-2}$). To reduce the thermic effects of $CO_2$ laser irradiation, perforation of the footplate is performed by several juxtapositioned single shots with low power, a short pulse duration and a small beam diameter or by a microprocessor-controlled rotating laser beam with scanning figures of different irradiation diameters (SurgiTouch, ESC Sharplan Co.).

$CO_2$ laser irradiation of high power density and low single-pulse energy is applied in this connection. Footplate perforations of defined diameter (0.5–0.7

mm) can usually be achieved by a single laser shot with rotating mirrors. If necessary, the perforation diameter can be increased by additional single shots without the rotating beam (Table 1). Applying a good beam profile enables optimal tissue results with minimal thermic side effects. Restricting the laser energy parameters accordingly obviates any risk to middle and inner-ear structures by thermic or acoustic stress.

### Laser Stapes Surgery

#### Primary Surgery

Injection of 1% xylocaine with 1:200 000 epinephrine and preparation of the tympanomeatal flap are followed by the opening of the middle ear and the removal of the meatus bone covering the oval niche with the sharp double-ended curette according to House or with the diamond drill while preserving the chorda tympani. As with the conventional technique, access to the oval niche is adequate when the pyramidal process and part of the tympanic segment of the facial nerve are clearly visible (Fig. 3). Application of the $CO_2$ laser is preceded by some test shots for instance at, a wooden spatula, in order to exclude a possible disadjustment between the HeNe pilot beam and the invisible infrared $CO_2$ laser beam. If the pilot beam is not in conformity with the surgical beam, a correction can be effected prior to lasering. Ablation of the suprastructures and perforation of the footplate is then performed in a non-contact manner with the $CO_2$ laser beam.

Figures 3 to 29 show the operation technique schematically and intraoperatively.

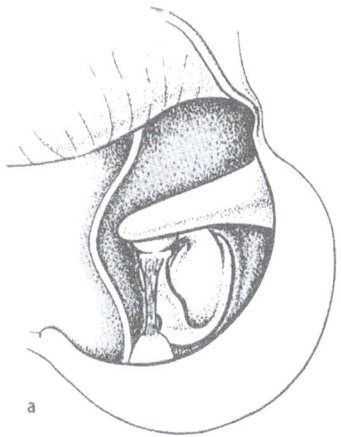
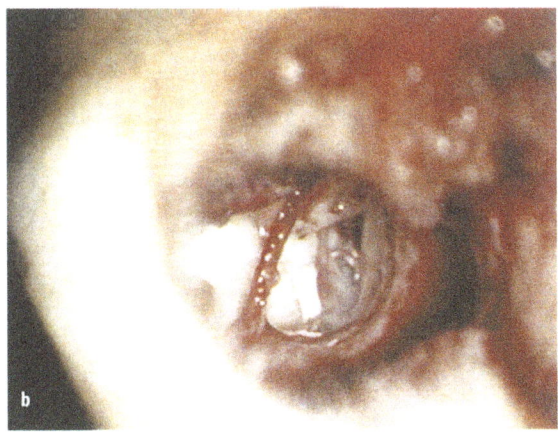

**Fig. 3a,b.** Middle ear after exposure of the stapes

## Vaporization of the Stapedius Tendon

The stapedius tendon is first vaporized with two to three single pulses at low power of 2 W (power density 8000 Wcm$^{-2}$) and a pulse duration of 0.05 s (Fig. 4). The developing smoke is removed by suction. In the case of a favourable anatomical situation, the stapedius tendon can be preserved.

## Separation of the Incudo-Stapedial Joint

The incudo-stapedial joint is then separated by vaporizing the head of the stapes applying 8 to 14 single pulses of the laser beam at 6 W (power density 24 000 Wcm$^{-2}$) and a pulse duration of 0.05 s (Fig. 5).

The fact that the $CO_2$ laser beam does not fall entirely perpendicular to the joint necessitates additional instrument-assisted checking and, if necessary,

severing of residual connections between the processus lenticularis and the caput stapedis.

## Vaporization of the Posterior Crus

The posterior crus of the stapes is severed near the footplate with for to eight pulses and the same power of 6 W (power density 24 000 Wcm$^{-2}$) and pulse duration of 0.05 s applied for the incudostapedial joint (Fig. 6).

While severing the joint and the posterior crus with this relatively high laser power, care must be taken that middle-ear structures situated in the beam direction (footplate, facial nerve canal, etc.) are not irradiated by mistake and damaged. Reliable protection is provided by filling the cavum tympani with saline or covering these structures with saline-tinctured gelatin sponge (Gelita, Spongostan, etc.) (Fig. 7).

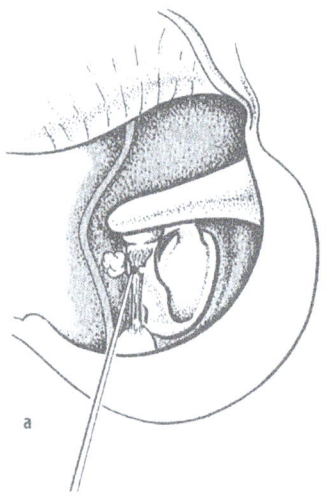
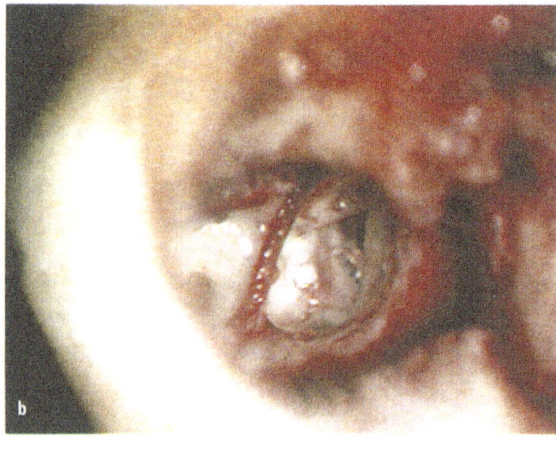

**Fig. 4a,b.** Severance of the stapedius tendon

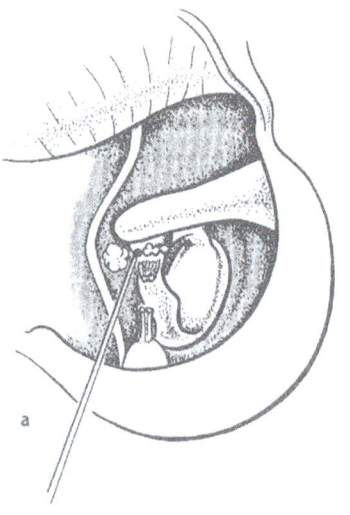
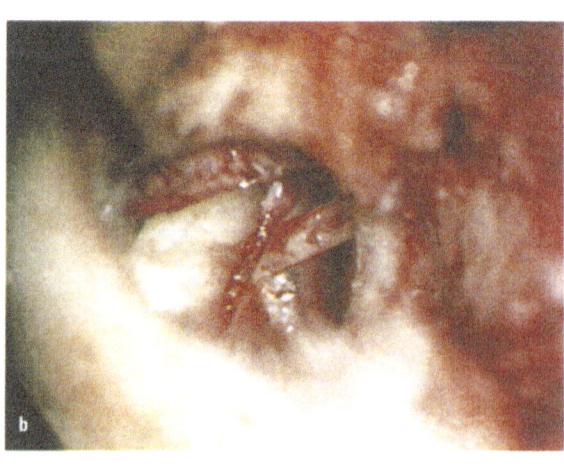

**Fig. 5a,b.** Severance of the incudostapedial joint

**Fig. 6a, b.** Severance of the posterior crus of stapes

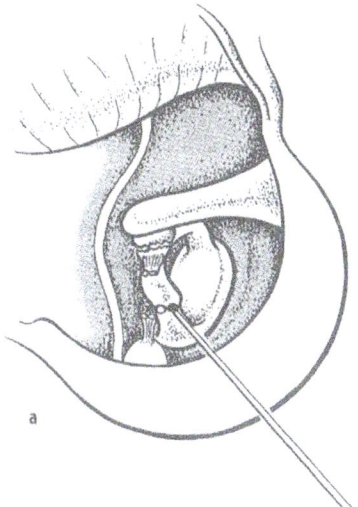
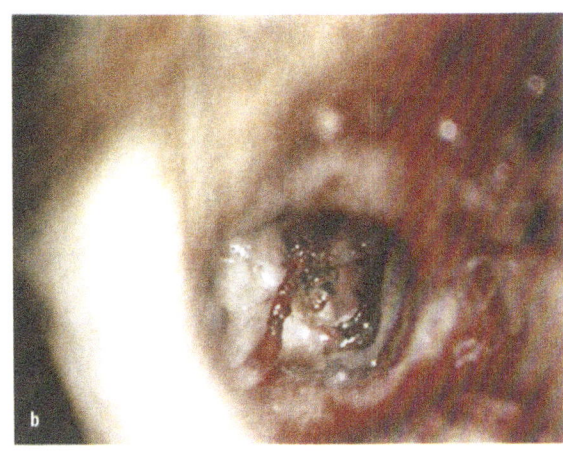

**Fig. 7.** Covering the oval niche with saline-tinctured gelatin sponge in order to protect the footplate and the facial nerve from inadvertently $CO_2$ laser irradiation (right ear)

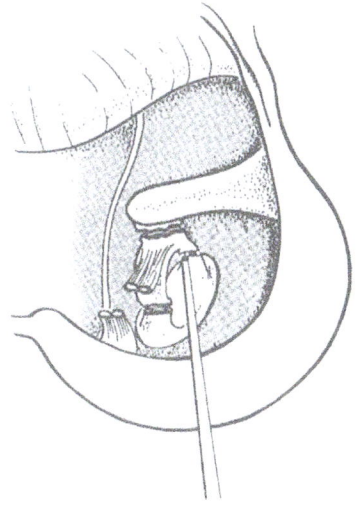

## Vaporization of the Anterior Crus

The anterior crus of the stapes is often not directly accessible to the laser beam; it is possible with the aid of a mirror to deflect the $CO_2$ laser beam in such a way

as to enable vaporization of the anterior crus under visual control. The mirror must meet high demands: it must reflect the $CO_2$ laser beam completely and without scattering, that is, it must transmit the laser energy without loss, if possible, and must still reflect well enough to ensure clear recognition of the anterior crus to be lasered and precise focusing of the HeNe pilot beam and thus the $CO_2$ laser beam. The mirrors available to us so far have not yet been optimal in this respect, so that we have preferred in most cases to fracture the anterior crus conventionally with the hook.

If it is intact or only partially visible, it is likewise vaporized with the $CO_2$ laser beam using the same parameters applied for the posterior crus (Fig. 8). Even in cases of incomplete severance, it can thus be submitted to controlled fracturing with the hook at the vaporized site (Fig. 9a, b). Footplate mobilisation or even partial or total footplate extraction is thus almost entirely excluded.

In experiments on petrous bone preparations, we found the same effective powers for severance of the anterior crus via a deflecting mirror as for direct irradiation. Here too, it is advisable to cover the sur-

**Table 1.** Effective laser energy parameters for stapedotomy (Sharplan 40c $CO_2$ lasers). Specified powers correspond to real powers at the end of the application system. Use of rotating application systems at the stapes footplate may require perforation enlargement by additional individual applications without a rotating laser beam (power: 6 W, pulse duration: 0.05 s)
Focal length: f = 250 mm; Focal size: 0.18 mm (Acuspot 712)

| Anatom. structure | Real power (W) | Power density (W cm$^{-2}$) | Pulse duration (s) | Mode | Diameter of irradiation (mm) | No. of pulse | Diameter of perforation (mm) |
|---|---|---|---|---|---|---|---|
| Stapedius tendon | 2 | 8 000 | 0.05 | cw | 0.18 | 2– 3 | |
| Incudostapedial joint | 6 | 24 000 | 0.05 | cw | 0.18 | 8–14 | |
| Crura | 6 | 24 000 | 0.05 | cw | 0.18 | 4– 8 | |
| Stapes footplate | 6 | 24 000 | 0.05 | cw | 0.18 | 6–12 | 0.5–0.7 |
| | or 20–22 | 80 000–88 000 | 0.03 or 0.05 | cw | ca. 0.5, 0.6 or 0.7 | 1 | 0.5–0.7 |

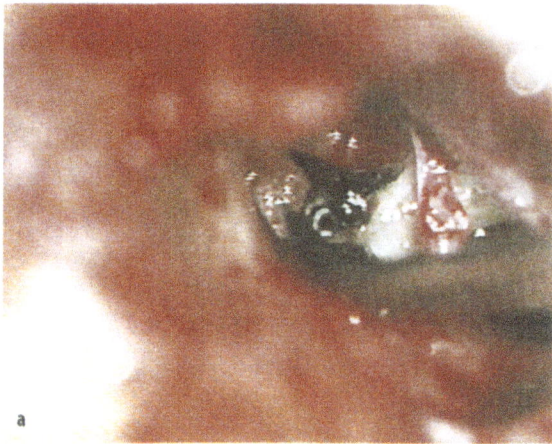

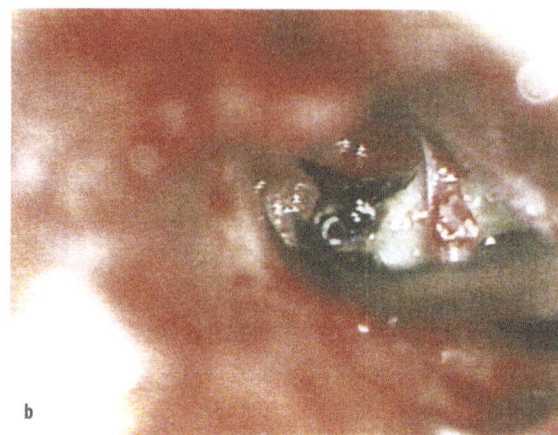

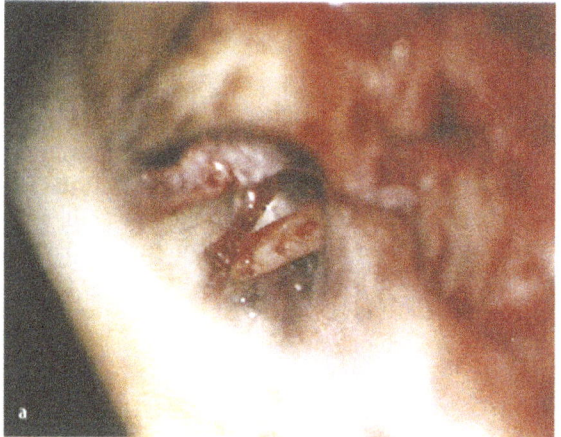

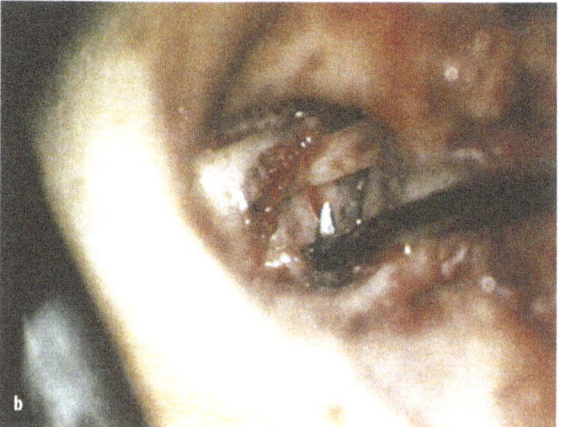

**Fig. 8a,b.** Direct vaporisation of the anterior crus with the $CO_2$ laser beam

**Fig. 9a,b. a** Incomplete vaporised anterior crus. **b** Controlled fracturing of the anterior crus at the vaporized site with the 90° hook

rounding area (footplate, facial nerve canal) with saline or moist gelatin sponge. The stapes suprastructure is extracted with the microforceps (Fig. 10).

## Perforation of the Footplate

Vaporization of the posterior part of the footplate can be performed after the suprastructure has been removed. The aim is to achieve an adequately large, nearly round, reproducible perforation of 0.5 to 0.7 mm with one-shot application (Fig. 11) or with a few juxtapositioned, slightly overlapping multiple applications (Figs. 12, 13).

## One-shot Application Technique

When applying laser irradiation with a scanner system (SurgiTouch, ESC Sharplan Co., modified according to our recommendations), a spiral figure is traced within the defined pulse duration. This enables the

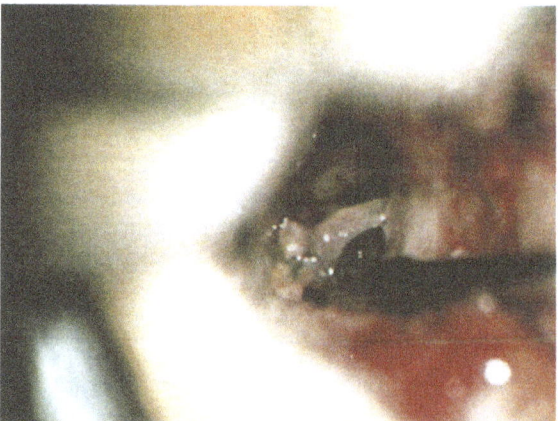

**Fig. 10.** Extraction of the stapes suprastructure with a microforceps

$CO_2$ laser to achieve high power densities and, thus, optimal results with minimal side effects even in large irradiation fields. At a working distance of 250 mm, the scanners can be set for irradiation fields of vari-

**Fig. 11a,b.** Vaporization of the stapes footplate with a scanner (SurgiTouch, irradiation diameter 0.6 mm) and achievement of a perforation 0.5 mm in diameter (right ear)

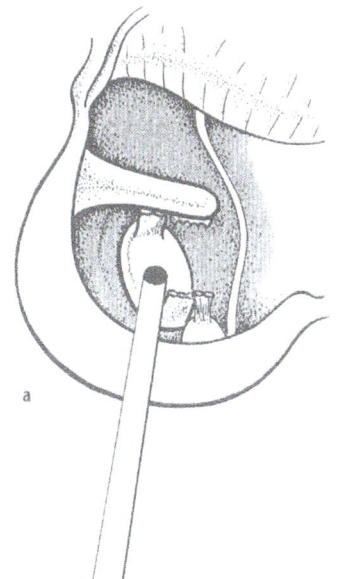
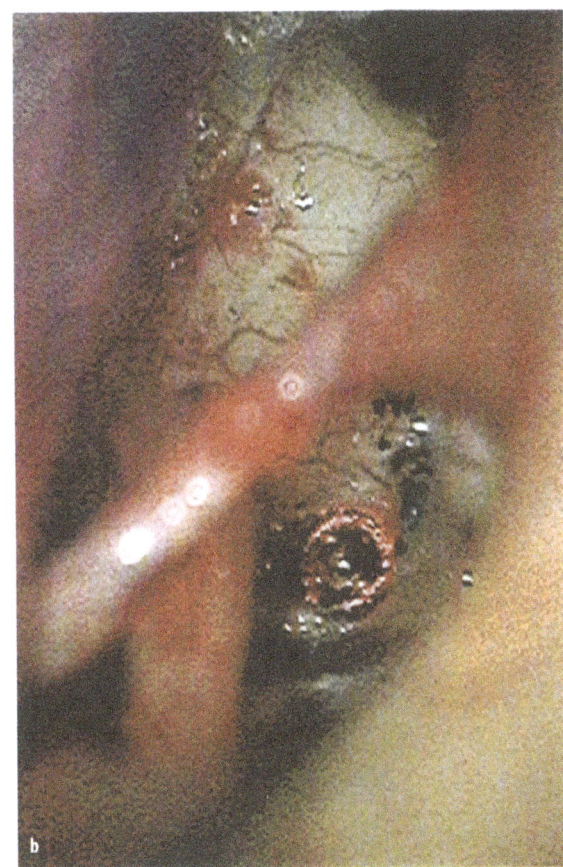

**Fig. 12a,b.** Perforation of the footplate with multiple juxta-positioned slightly overlapping $CO_2$ laser applications until the desired perforation diameter is achieved

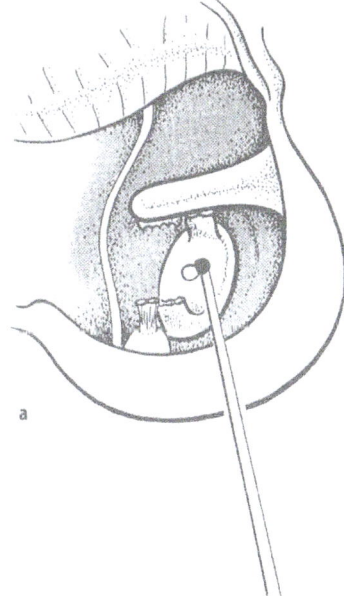
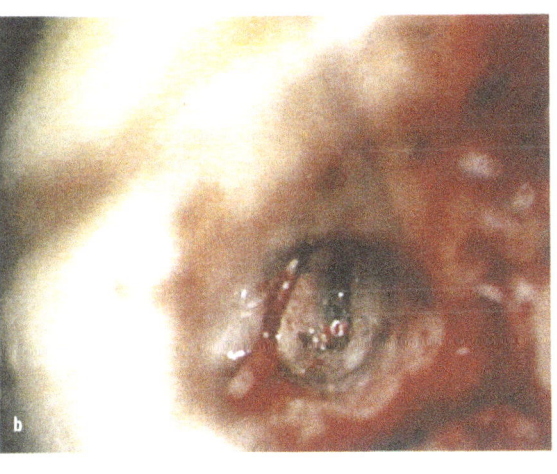

ous sizes depending on the anatomic conditions and defined perforation diameter. Irradiation fields of 0.5, 0.6 and 0.7 mm are suitable for stapedotomy. A radius of less than 0.5 mm (0.4 and 0.3 mm) or greater than 0.7 mm (0.8, 0.9 and 1 mm) can be achieved with SurgiTouch, but is rarely applied. Thus a suitable selection of laser parameters basically enables a single-shot footplate perforation of defined diameter. The selected powers for the one-shot application technique with rotating mirrors (SurgiTouch, ESC Sharplan Co.) are 20 to 22 W (power density 80 000–88.000 Wcm$^{-2}$). The pulse durations involved range between 0.03 and 0.05 s pulse. Depending on the thickness of the footplate and the irradiation diameter of the scanning figure applied, the perforation diameters with the one-shot stapedotomy technique with rotating mirrors range between 0.5 and 0.7 mm, in 90% of the cases. In those cases in which the desired perforation diameter is not achievable with one shot, the enlargement of the perforation is performed by additional laser applications without the scanner system.

## Multiple-shot Application Technique

When a scanner system is not available, the stapedotomy can be performed with several juxtapsitioned laser applications of the focused laser beam (beam diameter: 180 μm) in a circular manner. The selected power for the multiple-shot application technique is 6 W (power density 24 000 Wcm$^{-2}$) and the pulse duration 0.05 s. The number of pulses varies between 6 and 12 applications depending on the thickness of the footplate. During the enlargement of the perforation with additional laser shots until the desired perforation diameter of 0.5–0.7 mm is achieved, care must be taken that the vestibulum is filled with perilymph to ensure adequate protection of inner ear structures and prevent damage by direct laser irradiation. If the perilymph is inadvertently aspirated from the vestibulum, no further laser irradiation may be applied to the footplate.

A platinum Teflon piston 0.4 or 0.6 mm in diameter is then inserted in the perforation and fixed to the incus neck. Finally, the oval niche is sealed with connective tissue or a blood clot (Fig. 14).

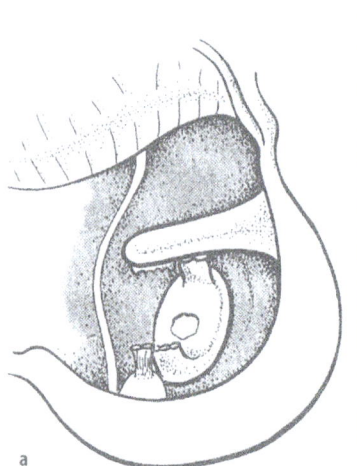
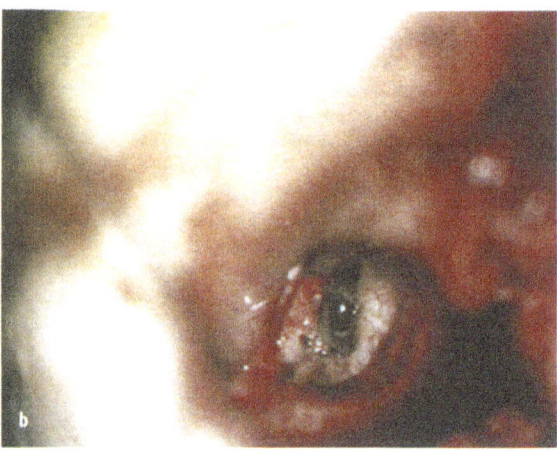

**Fig. 13a,b.** Final footplate perforation with a diameter of ca. 0.5 mm

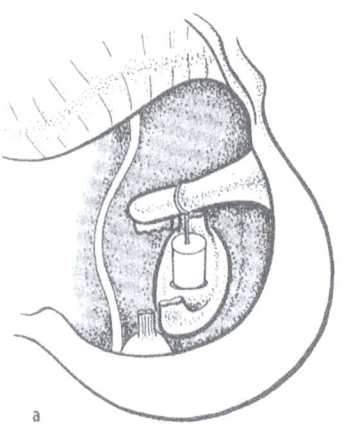
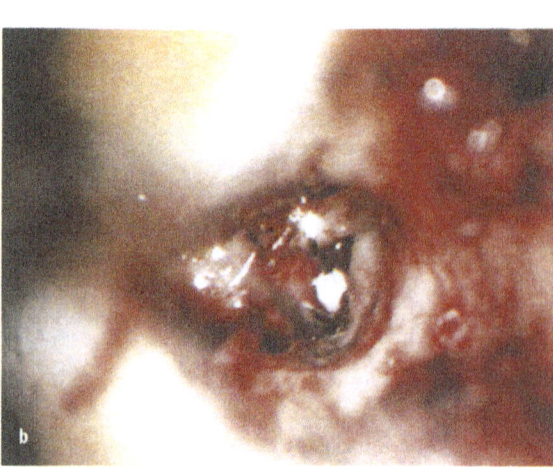

**Fig. 14a,b.** Site after insertion of a platinum Teflon piston 0.4 mm in diameter in the perforation and fixation to the long crus of the incus. The oval niche is sealed with a blood clot.

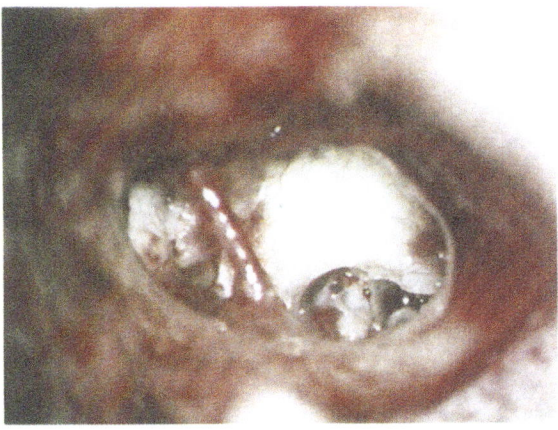

**Fig. 15.** Intraoperative finding of an obliterative otosclerosis a Otosclerotic foci completely filling the oval niche. b Situs after shifting chorda tympani caudally

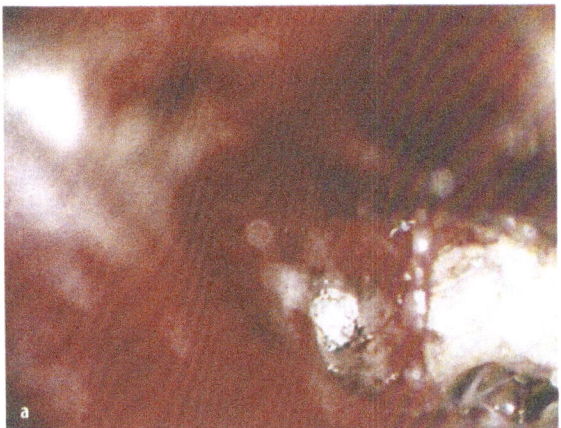

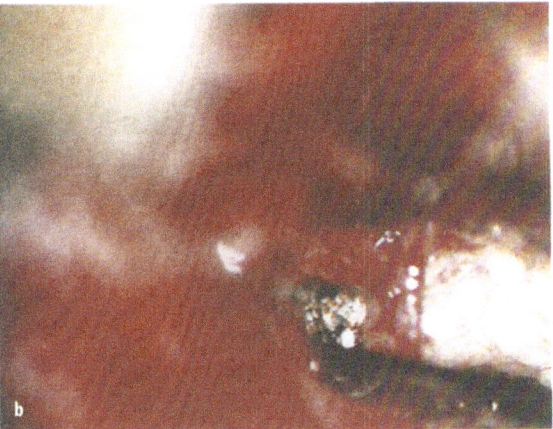

**Fig. 17a,b.** **a** Extensive vaporization of the otosclerotic foci in the oval niche with the SurgiTouch scanner. While ablating bone masses, crystallization product occurs, which leads to increased reflection and thus to ineffectiveness of the $CO_2$ laser irradiation. **b** Removal of the crystallzsation product with a needle

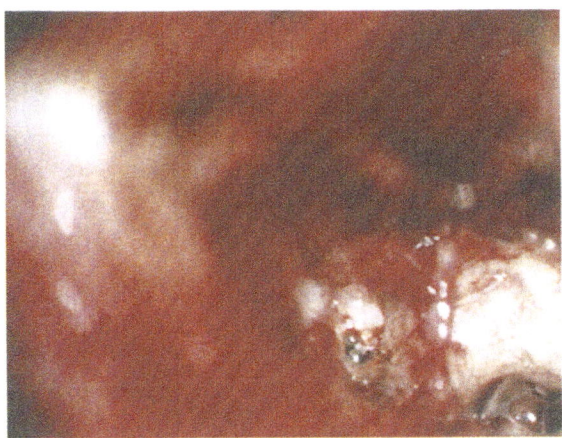

**Fig. 16.** Situs of an obliterative otosclerosis after removal of the suprastructure. Carbonization product (black) in the area of the vaporized posterior crus

## Special Cases

### Obliterative Otosclerosis

Using the drill to perforate a thick footplate obliterating the oval niche (Fig. 15) can cause a significant inner-ear trauma through vibrations. $CO_2$ laser stapedotomy, on the other hand, enables the ear surgeon to vaporize a perforation in the stapes footplate regardless of its thickness or degree of fixation without mechanically traumatizing the inner ear.

After removal of the suprastructure (Fig. 16), the otosclerotic foci obliterating the oval niche are extensively and symmetrically ablated by applying the laser irradiation with the SurgiTouch (power 20–22 W, 0.03–0.05 s, scanner diameter 0.5–0.7 mm) until the lateral borders of the oval window can be precisely identified (Fig. 17a). In part, lower powers are applied

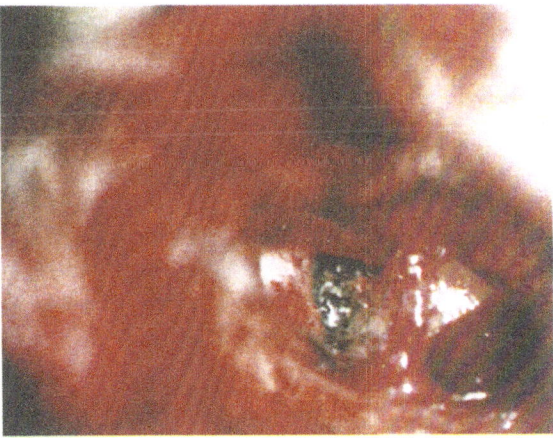

**Fig. 18.** Opening of the vestibulum at the centre of the oval niche

around the oval niche to avoid inadvertently opening the inner ear. Ablation of these bone masses yields large amounts of thermic products such as carboniza-

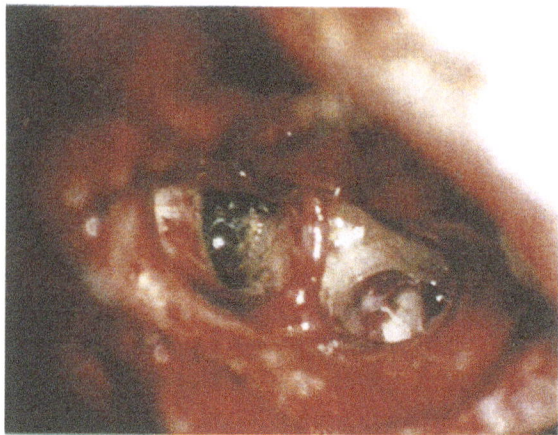

**Fig. 19.** Enlargement of the perforation until the desired diameter of ca. 0.7 mm is reached

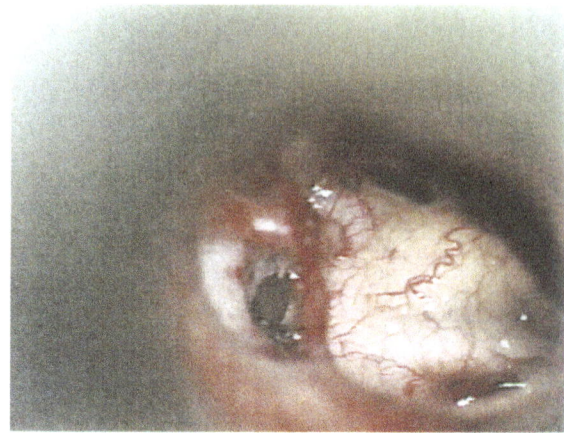

**Fig. 21.** Perforation of a floating footplate with the $CO_2$ laser

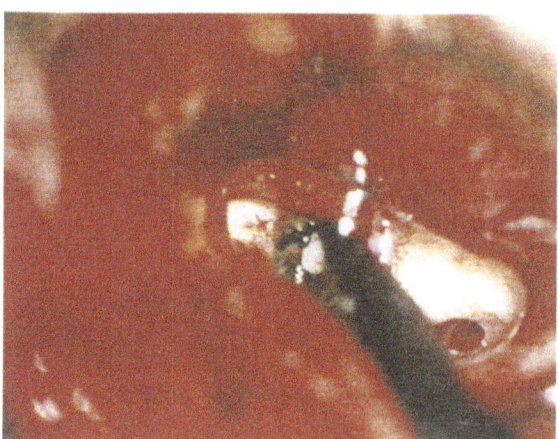

**Fig. 20.** Insertion of a platinum Teflon piston (4.5 mm long, 0.6 mm in diameter) in the perforation

tion and crystallization products. Since the crystallization product leads to increased reflection of the $CO_2$ laser irradiation, and thus to ineffectiveness and lower ablation, it must be removed with conventional instruments such as a curved needle or suction device (Fig. 17b). The vestibulum is opened at the centre of the oval niche (Fig. 18). The perforation is then concentrically enlarged to the desired diameter by further single applications (Fig. 19). The prosthesis is inserted in the typical manner (Fig. 20).

## Floating Footplate

Particularly with a partially fixed stapes, manipulation by conventional stapedotomy often leads to accidental mobilisation of the stapes and results in a so-called floating footplate. The operation is usually abandoned or total stapedectomy carried out, since the footplate can no longer be perforated with conventional instruments in many such cases. On the other hand, the $CO_2$ laser achieves a non-contact perforation of defined diameter even in a floating footplate (Fig. 21) without the associated trauma caused by conventional instruments. A platinum Teflon piston can then be inserted in such a perforation. However, the incidence of a floating footplate is extremely low in laser stapedotomy compared to conventional operations. In the authors' experience it is as low as 0.5%. A stapedectomy was not necessary in any of the cases.

## Revision Surgery

Successful restoration of hearing in revision stapedotomies comprises the precise identification and correction of the particular abnormality without damaging the inner ear.

In revision stapedectomy, conventional surgical procedures not infrequently lead to unsatisfactory hearing results and inner-ear damages. Studies by numerous ear surgeons have shown that a successful closure of the airbone gap of less than 10 dB could be achieved only in less than half the patients [10, 17, 58, 84, and others]. These studies indicate that 8 to 33% of the patients have poor hearing after revision operations. The incidence of a significant postoperative sensorineural hearing loss is 3 to 20% (up to 14% for profound sensorineural hearing losses).

Surgeons are warned particularly against damage to the inner ear through mechanical trauma due to excessive manipulations at the prosthesis and/or the connective tissue occluding the oval niche.

Histopathological studies on petrous-bone preparations from stapedectomized patients show that,

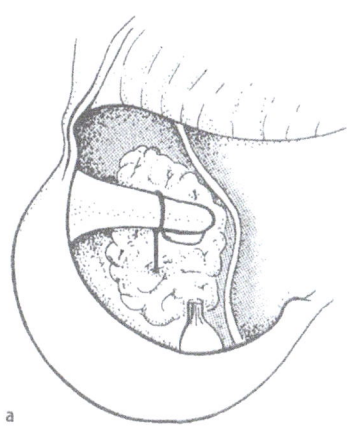

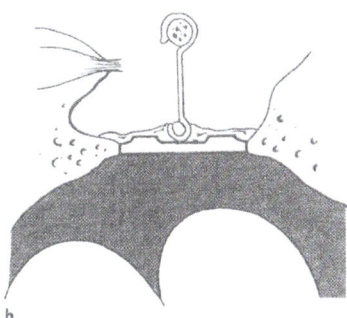

**Fig. 22a.** Oval niche filled with soft tissue. The lateral borders as well as the depth and position of the prosthesis cannot be precisely determined. **b** Fixed stapes footplate below a connective-tissue neomembrane

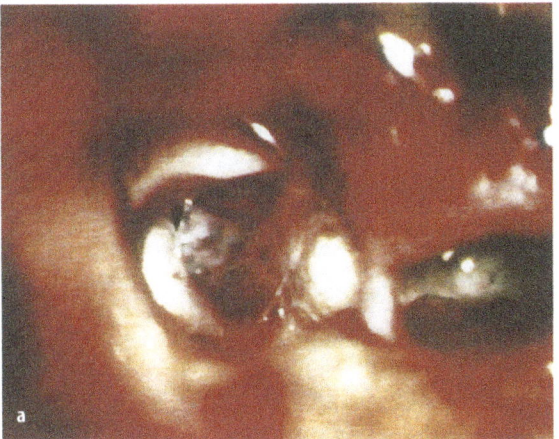

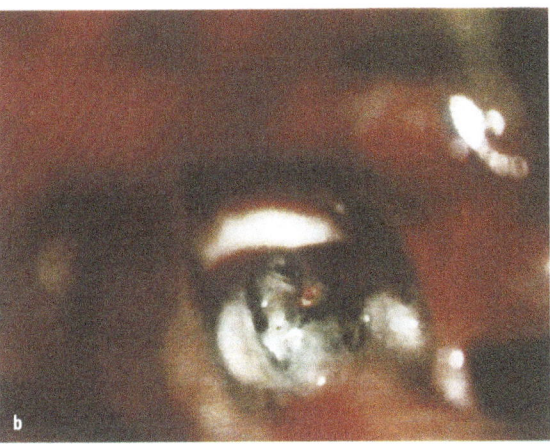

**Fig. 23a,b. a** Adhesions between the prosthesis and the middle ear mucosa. **b** Non-contact exposure of the prosthesis by vaporisation of the soft tissue surrounding it with the $CO_2$ laser beam

between the prosthesis and/or neomembrane of the oval window and the utricle and saccule, there are frequently adhesions [24, 58] which, through surgical manipulations, can lead to ruptures of these fine inner-ear structures with resultant vertigo and labyrinthine damages.

Use of laser for revision surgery offers distinct advantages over the conventional method. The complication rate is low and the, success rate is statistically significant improved, independent of the laser system used [18, 20, 28, 52, 56, 100 etc.]. Measured as closure of air-bone gap of 20 dB or less, the rate of success with the laser ranges from 70 to 92% for laser surgery, compared to 49 to 85% for that of the conventional technique.

While exploring the middle ear for failed stapedotomy, the ear surgeon is in a dilemma. In order to find reasons for the existing conductive hearing loss, the operator must test the mobility and integrity of the entire chain of auditory ossicles and precisely assess the status of the oval window and the prosthesis position relative to the entrance of the vestibule. In this connection, he is frequently unable to determine the depth and lateral border of the oval window or to see the structures behind the connective tissue covering the oval niche (Fig. 22a, b).

While confining the palpation of these structures to a minimum in order to keep the inner ear trauma as mild as possible, the surgeon inevitably runs the risk of not recognizing the exact cause of the existing conductive hearing loss (often more than one), and thus not initiating an adequate therapy of the hearing defect.

The old prosthesis should be removed very carefully; if vertigo occurs, it should be left in place in order to avoid permanent labyrinthine dysfunction or hearing loss.

If extraction of the prosthesis is nevertheless achieved without appreciable traumatization of the inner ear, the new prosthesis, which is often too short, is inserted in the presumed centre of the oval niche.

If the oval window is free of residual disease and the surgeon has not caused any inner ear damage with his manipulations, the patient's hearing will improve initially as a rule. The cause to which most failures after stapedotomy are ascribed, migration of the prosthesis, however, is frequently not eliminated by

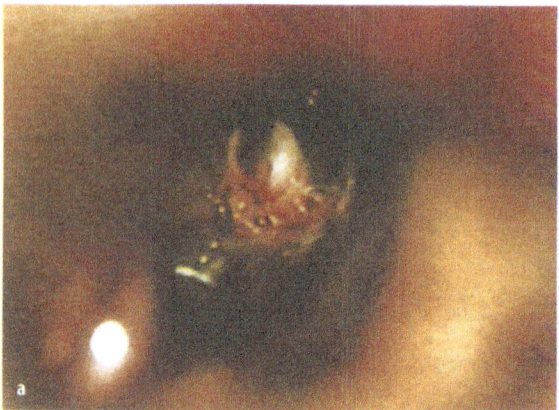

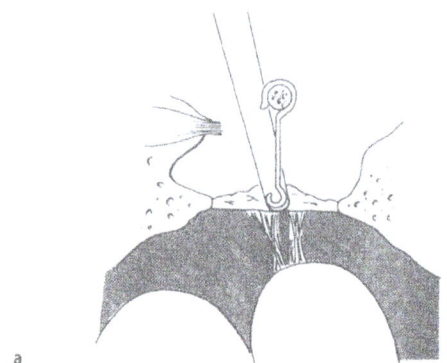

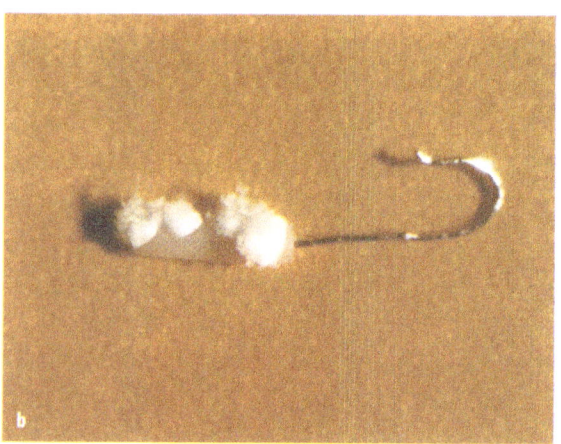

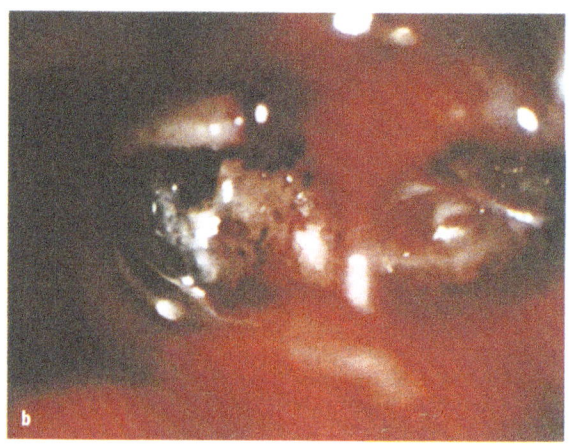

**Fig. 24a,b. a** Vaporization of soft tissue around the platinum Teflon piston after stapedotomy at primary surgery. **b** Platinum Teflon piston swollen mushroom-like after direct $CO_2$ laser irradiation

**Fig. 25a,b.** Extensive symmetrical vaporization of the soft tissue covering the oval niche and precise identification of the borders of the oval window

this. The new prosthesis can again migrate out of the oval niche. In view of these considerable difficulties in performing revision stapedectomies, the reported success rates of 30 to 50% are rather incomprehensible.

Following the elevation of the tympanomeatal flap, the middle ear is first inspected. The integrity and mobility of the malleus and incus are checked by palpation with the Rosen needle.

With the experimentally determined effective and safe laser energy parameters (Table 2) the frequently existing adhesions are first vaporized with the $CO_2$ laser. Using a beam diameter of 0.18 mm, low powers of 1 to 2 W at a pulse duration of 0.05 s are adequate for this purpose. If the SurgiTouch scanner system is used, powers of 4 to 8 W at a pulse duration of 0.03 to 0.05 s and variable scanner diameters (0.3 to 0.5 mm) are sufficient for the soft tissue treatment. With these parameters, the prosthesis is then exposed by vaporization of the soft tissue surrounding it (Fig. 23a, b).

In the case of a wire/connective-tissue prosthesis (e.g. made of platinum), even direct lasering of the prosthesis is harmless. For a piston with Teflon parts

**Fig. 26.** $CO_2$ laser vaporization of all soft-tissue connections to the distal end of the prosthesis permits extraction of the prosthesis without mechanically traumatizing the inner ear

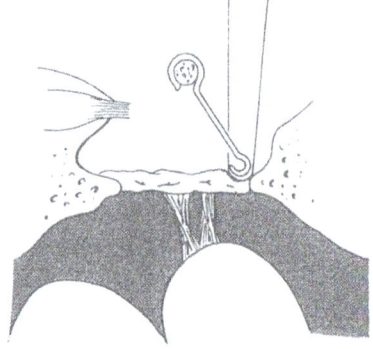

(e.g. platinum Teflon piston), direct irradiation of the prosthesis must be avoided, since the Teflon cannot withstand the high temperatures ($> 300°$ C) of the laser irradiation, and its surface swells up mushroom-like without disintegration or inflammation (Fig. 24a, b).

By non-contact vaporization of the connective-tissue bands, the prosthesis is exposed without mechanically traumatizing the inner ear. The soft tissue cov-

**Table 2.** Effective laser energy parameters for revision stapedotomy (Sharplan 40c $CO_2$ lasers). Specified powers correspond to real powers at the end of the application system. Use of rotating application systems at the stapes footplate may require perforation enlargement by additional individual applications without a rotating laser beam (power: 6 W, pulse duration: 0.05 s) Focal length: f = 250 mm; Focal size: 0.18 mm (Acuspot 712)

| Anatom. structure | Real power (W) | Power density (W cm$^{-2}$) | Pulse duration (s) | Mode | Diameter of irradiation (mm) | No. of pulse | Diameter of perforation (mm) |
|---|---|---|---|---|---|---|---|
| Soft tissue | 1–2 | 4 000– 8 000 | 0.05 | cw | 0.18 | | |
| Bony stapes footplate | 6 | 24 000 | 0.05 | cw | 0.18 | 6–12 | 0.5–0.7 |
| | or 20–22[a] | 80 000–88 000 | 0.03 or 0.05 | cw | ca. 0.5, 0.6 or 0.7 | 1 | 0.5–0.7 |
| Connective-tissue neomembrane | 1–2 | 4 000– 8 000 | 0.05 | cw | 0.18 | 6–12 | 0.5–0.7 |
| | or 4–8[a] | 16 000–32 000 | 0.03 or 0.05 | cw | ca. 0.5, 0.6 or 0.7 | 1 | 0.5–0.7 |

ering the oval niche is then symmetrically and extensively vaporized until the lateral borders of the oval window can be identified exactly (Fig. 25). If the prosthesis is still situated within this connective tissue, the vaporization is continued until it is completely freed from it. Only when all connective-tissue bands to its distal end have been separated (Fig. 26) is the prosthesis extracted from the site after disconnection from the incus with a 2-mm-long 90¡ hook. If vertigo occurs (during interventions under local anaesthesia), the surgeon must interrupt the manipulations

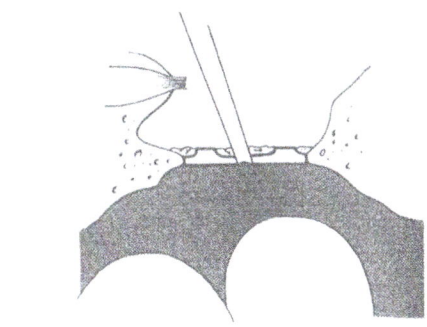

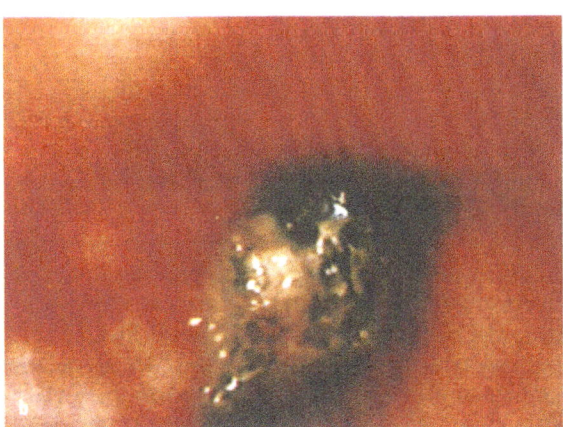

**Fig. 27a,b.** $CO_2$ laser stapedotomy of the connective-tissue neomembrane and/or bony stapes footplate

immediately and again check the distal end of the prosthesis for residual connective tissue bands.

In the centre of the oval window, a stapedotomy opening 0.5 to 0.7 mm in diameter is created by uniform vaporization of the tissue until the perilymph of the vestibulum is identified (Fig. 27). Depending on the findings (connective-tissue neomembrane and/or bony stapes footplate), powers of 1 to 6 W (0.05 s) are required with a beam diameter of 0.18 mm and multiple juxtapositioned, slightly overlapping applications (6 to 12 applications) (Fig. 28). Using the one-shot application technique of laser irradiation with rotating mirrors (SurgiTouch), powers of 4 to 22 W and pulse durations of 0.03 to 0.05 s are required.

The prosthesis length is determined precisely (usually 4.5 to 4.75 mm) by measuring the distance between the vestibulum and the lower surface of the incus and adding 0.2 mm. To reduce the risk of renewed prosthesis migration, the prosthesis should project 0.1 to 0.2 mm into the stapedotomy opening. The platinum Teflon piston is then inserted in the perforation and, if the incus is intact, fixed to the incus neck (Fig. 29a, b). In the case of a completely eroded incus, a malleo-vestibulopexy is performed to restore ossicular continuity.

Finally, the oval niche is sealed with connective tissue or a blood clot.

## Discussion

The advantages of stapedotomy compared to stapedectomy have already been confirmed by numerous authors [7, 13, 63, 64, 73, 87, 88]. All studies demonstrate that stapedotomy causes less inner-ear damage and less vertigo than conventional stapedectomy. Despite these advantages, it is nevertheless difficult to create a precise, round stapedotomy opening with mechanical instruments (drill, perforator). A partially fixed stapes is often accidentally mobilized by the manipulations (floating footplate) and a thin footplate

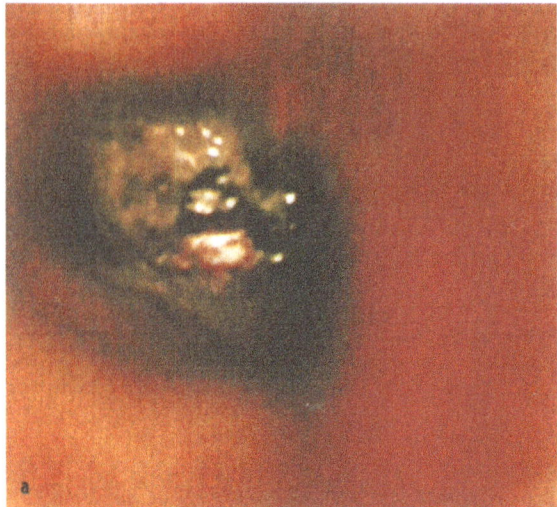

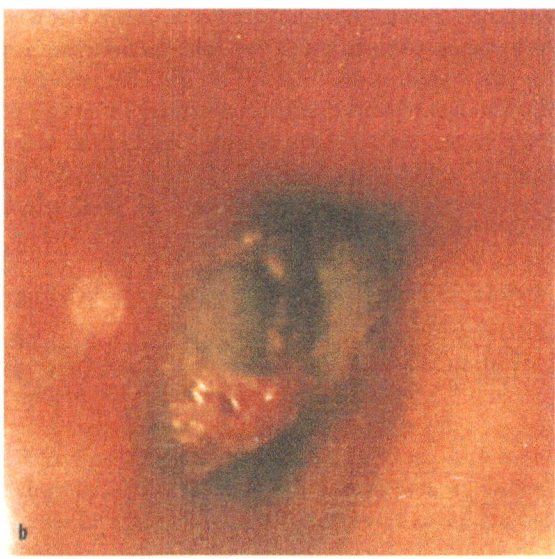

**Fig. 28a,b.** Creation of stapedotomy opening 0.5 to 0.6 mm in diameter with eight juxtapositioned, slightly overlapping applications of $CO_2$ laser irradiation. The perilymph of the vestibulum provides adequate protection for the inner-ear structures behind it.

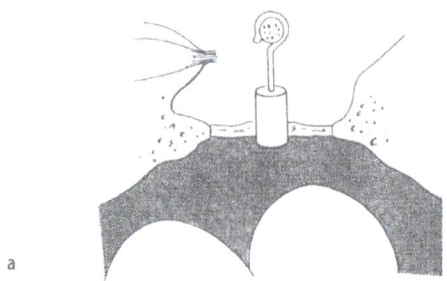

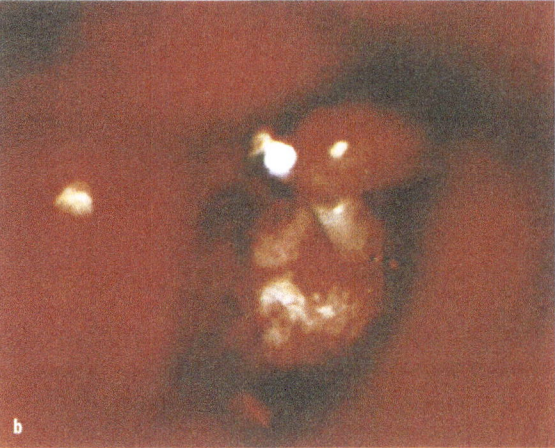

**Fig. 29a.** Prosthesis in situ. The platinum Teflon piston projects 0.1 to 0.2 mm into the stapedotomy opening. This stabilizes the prosthesis in the centre of the oval window and minimizes the risk of renewed prosthesis migration. b Oval niche sealed with connective tissue

not infrequently fractured. Perforation of a thick footplate obliterating the oval niche (obliterative otosclerosis) with the drill can cause significant inner ear trauma through vibrations.

Laser stapedotomy, on the other hand, when performed with the suitable wavelength and the effective and safe parameters, enables the ear surgeon to create, by precise and non-contact vaporization, a circular perforation in the stapes footplate regardless of its thickness or degree of fixation, without mechanically traumatizing the inner ear.

Prior to the clinical application of the laser, extensive laboratory studies had to be performed in order to determine the ideal wavelength of laser radiation for stapedotomy and to establish the effective and safe laser parameters for each type of laser.

The results of previous studies support the usage of both visible (argon and KTP) and invisible, far infrared ($CO_2$) laser systems for primary otosclerosis surgery [1, 3, 4, 6, 8, 12, 19, 23, 27, 52, 56, 61, 65, 70, 72, 78, 82, 85, 95].

The author has used the $CO_2$ laser for stapedotomy. The results so far clearly document that the incidence and severity of postoperative complications after $CO_2$ laser stapedotomy are lower than after conventional interventions [33, 35, 36, 44, 46].

In the author's experience of almost 300 stapedotomies and more than 50 revision surgeries with the $CO_2$ laser, there were no intra-operative complications. In primary surgery, postoperatively three patients (1%) showed significant sensorineural hearing loss, one patient (0.3%) a severe sensorineural hearing loss, which was probably caused by a granuloma and showed no improvement following the revision surgery. In revision surgery, one patient (1.8%) developed a severe sensorineural hearing loss. Late cases of deafness were not observed. The mean sensorineural hearing losses before and after $CO_2$ laser stapedotomy clearly show that there was no appreciable deterioration of inner ear function. A vestibular disturbance occurred in only eight cases because of the length of prosthesis. The postoperative airbone gap closure in primary laser stapedotomy is comparable to that following conventional surgery. The hearing results obtained after revision stapedotomy so far suggest an improvement of the air-bone gap and an elimination of a significant hearing loss. These results are similar to the results of Lesinski and Newrock [56] and Lesinski and Stein [55] in over 200 $CO_2$ laser stapedotomies and stapedectomy revisions.

In revision stapedotomy, the $CO_2$ laser provides the ear surgeon with three important advantages compared to the conventional technique: (1) improved diagnostic precision; (2) the possibility of better stabilization of the new prosthesis in the centre of the oval niche; and (3) reduction of the inner-ear trauma.

Thus, the $CO_2$ laser enables the ear surgeon to eliminate a sound conduction hearing loss recurring after stapedotomy with high precision and safety. Some of the revision operations could only be performed by using the laser.

For the operative procedure, this means:

1. $CO_2$ laser stapedotomy is already performed at first interventions in order to minimize the risk of prosthesis migration, which is by far the most frequent cause for recurrence of a sound conduction hearing loss after stapedotomy or stapedectomy.

2. In order to avoid erosions of the incus, which are frequently the result of prosthesis migration and fixation with the bone surrounding the oval niche, a revision should be performed soon after a significant sound conduction hearing loss has been detected.

3. A stapedotomy of the neomembrane of the oval window should always be performed. Three reasons argue in favour of this:

   a) The covering neomembrane frequently conceals a residually fixed stapes footplate.

   b) The depth of the oval niche can be precisely ascertained and the length of the prosthesis exactly determined.

   c) The stapedotomy opening stabilizes the new prosthesis in the centre of the oval window and minimizes the risk of renewed migration.

Thus, the $CO_2$ laser appears to be well suited for application in stapes surgery. It does not endanger the inner ear with the laser parameters limited as specified. The one-shot stapedotomy, which is able to achieve an adequately large (0.5 to 0.7 mm in diameter) circular footplate perforation with a single application of laser irradiation without appreciable thermic damage to

the surrounding area, is a great advantage in $CO_2$ laser stapedotomy. For this purpose, new modes of application have been developed, such as, for instance, application of laser radiation with rotating mirrors (Surgi-Touch, ESC Sharplan Co.), which in part show very promising results.

Application of the laser in stapes surgery contributes to the optimization of this high-precision intervention and shows promise of improving hearing results and reducing the incidence of inner-ear damages.

## References

[1] ANTONELLI PJ, GIANOLI GJ, LUNDY LB, LAROUERE MJ, KARTUSH JM (1998) Early post-laser stapedotomy hearing thresholds. Am J Otol 19 (4): 443–446
[2] ARNOLD W, NIEDERMEYER HP, ALTERMATT HJ, NEUBERT WJ (1996) Pathogenesis of otosclerosis. State of the art. HNO 44 (3): 121–129
[3] BARTELS LJ (1990) KTP laser stapedotomy: is it safe? Otolaryngol Head Neck Surg 103: 685–692
[4] BEATTY TW, HABERKAMP TJ, KHAFAGY YW, BRESEMANN JA (1997) Stapedectomy training with the carbon dioxide laser. Laryngoscope 107: 1441–1444
[5] BRETLAU P (1999) Argon laser stapedotomy vs. erbium laser stapedotomy. Otology 2000, XXII Annu Meet of the Politzer Society, Zürich
[6] BUCHMAN CA, FUCCI MJ, ROBERSON JB JR, DE LA CRUZ A (2000) Comparison of argon and $CO_2$ laser stapedotomy in primary otosclerosis surgery. Am J Otolaryngol 21 (4): 227–230
[7] CAUSSE JR, CAUSSE JB, BEL J (1985) Amélioration de l'audition en fonction du type de platinectomie ou de platinotomie effectué dans la chirurgie de l'otospongiose. Ann Oto-Laryngol (Paris) 102: 401–405
[8] CAUSSE JB, GHERINI S, HORN KL (1993) Surgical treatment of stapes fixation by fiberoptic argon laser stapedotomy with reconstruction of the annular ligament. Otolaryngol Clin North Am 26 (3): 395–416
[9] CHARLTON A, DICKINSON MR, KING TA, FREEMONT AJ (1990) Erbium:YAG and holmium:YAG laser ablation of bone. Lasers Med Sci 5: 365–373
[10] CRABTREE JA, BRITTON B, POWERS WH (1980) An evaluation of revision stapes surgery. Laryngoscope 90: 224–227
[11] DIBARTOLOMEO JR (1981) Argon and $CO_2$ lasers in otolaryngology: which one, when, and why? Laryngoscope 91 (Suppl 26): 1–16
[12] DIBARTOLOMEO JR, ELLIS M (1980) The argon laser in otology. Laryngoscope 90: 1786–1796
[13] FISCH U (1982) Stapedotomy versus stapedectomy. Am J Otol 4 (2): 112–117
[14] FISCHER R, SCHÖNFELD U, JOVANOVIC S, SCHOLZ C (1990) Experimenteller Vergleich zwischen kurzgepulsten und kontinuierlich strahlenden Lasern in der Stapeschirurgie – akustische und thermische Ergebnisse. Arch Otorhinolaryngol (Suppl) II: 224–227
[15] FISCHER R, SCHÖNFELD U, JOVANOVIC S, JAECKEL P (1992) Thermische Belastung des Innenohres durch verschiedene Lasertypen bei der Laser-Stapedotomie. Arch Otorhinolaryngol (Suppl) II: 251–253
[16] GANTZ BJ, JENKINS HA, KISHIMOTO S, FISCH U (1982) Argon laser stapedotomy. Ann Otol Rhinol Laryngol 92: 25–26
[17] GLASSCOCK ME (1987) Revision stapedectomy surgery. Otolaryngol Head Neck Surg 96: 141–148
[18] GHERINI SG, HORN KL, BOWMAN CA, GRIFFIN GM (1990) Small fenestra stapedotomy using a fiberoptic

hand-held argon laser in obliterative otosclerosis. Laryngoscope 100: 1276–1282

[19] GHERINI S, HORN KL, CAUSSE JB, MCARTHUR GR (1993) Fiberoptic argon laser stapedotomy: is it safe? Am J Otol 14 (3): 283–289

[20] HABERKAMP TJ, HARVEY SA, KHAFAGY Y (1996) Revision stapedectomy with and without the $CO_2$ laser: an analysis of results. Am J Otol 17: 225–229

[21] HÄUSLER R (2000) Fortschritte in der Stapeschirurgie. Laryngo-Rhino-Otol 79 (Suppl 2): 95–139

[22] Häusler R, SCHAR PJ, PRATISTO H, WEBER HP, FRENZ M (1999) Advantages and dangers of erbium laser application in stapedotomy. Acta Otolaryngol 119 (2): 207–213

[23] HODGSON RS, WILSON DF (1991) Argon laser stapedotomy. Laryngoscope 101: 230–233

[24] HOHMANN A (1962) Inner ear reactions to stapes surgery (animal experiments). In: SCHUKNECHT HF (ed) Otosclerosis. Little, Brown, Boston

[25] HOMMERICH CP, HESSEL S (1991) Untersuchungen mit dem Holmium:YAG-Laser an Amboß und Steigbügel. Eur Arch Otorhinolaryngol (Suppl) II: 280

[26] HOMMERICH CP, SCHMIDT-ELMENDORFF A (1993) Experimentelle $CO_2$-, Holmium:YAG- und Erbium:YAG-Laseranwendung an der Steigbügelfußplatte. Eur Arch Otorhinolaryngol (Suppl) II: 39–40

[27] HORN KL, GHERINI S, GRIFFIN GM (1990) Argon laser stapedectomy using an Endo-Otoprobe system. Otolaryngol Head Neck Surg 102: 193–198

[28] HORN KL, GHERINI S, FRANZ DC (1994) Argon laser revision stapedectomy. Am J Otol 15: 383–388

[29] JOVANOVIC S (1996a) Der Einsatz neuer Lasersysteme in der Stapeschirurgie. In: MÜLLER GJ, BERLIEN HP (eds) Fortschritte der Lasermedizin 14. Ecomed, Landsberg

[30] JOVANOVIC S, SCHÖNFELD U (1995g) Application of the $CO_2$ laser in stapedotomy. Adv Oto-Rhino-Laryngol 49: 95–100

[31] JOVANOVIC S, SCHOLZ C, BERGHAUS A, SCHÖNFELD U (1990) Experimenteller Vergleich zwischen kurzgepulsten und kontinuierlich strahlenden Lasern in der Stapeschirurgie – histologisch-morphologische Ergebnisse. Arch Otorhinolaryngol (Suppl) II: 72–73

[32] JOVANOVIC S, BERGHAUS A, SCHÖNFELD U, SCHERER H (1991) Bedeutung experimentell gewonnener Daten für den Klinischen Einsatz verschiedener Laser in der Stapeschirurgie. Eur Arch Otorhinolaryngol (Suppl) II: 278–280

[33] JOVANOVIC S, PRAPAVAT V, SCHÖNFELD U, BERGHAUS A, BEUTHAN J, SCHERER H, MÜLLER G (1992a) Experimentelle Untersuchung zur Optimierung der Parameter verschiedener Lasersysteme zur Stapedotomie. Lasermedizin 8: 174–181

[34] JOVANOVIC S, BERGHAUS A, SCHERER H, SCHÖNFELD U (1992b) Klinische Erfahrungen mit dem $CO_2$-Laser in der Stapeschirurgie. Eur Arch Otorhinolaryngol (Suppl) II: 249–250

[35] JOVANOVIC S, ANFT D, SCHÖNFELD U, TAUSCH-TREML R (1993a) Tierexperimentelle Untersuchungen zur Eignung verschiedener Lasersysteme für die Stapedotomie. Eur Arch Otorhinolaryngol (Suppl) II: 38–39

[36] JOVANOVIC S, SCHÖNFELD U, FISCHER R, SCHERER H (1993b) $CO_2$ laser in stapes surgery. Proc SPIE 1876: 17–27

[37] JOVANOVIC S, SCHÖNFELD U, FISCHER R, DÖRING M, PRAPAVAT V, MÜLLER G, SCHERER H (1995a) Temperaturmessungen im Innenohr-Modell bei Laserbestrahlung. Lasermedizin 11: 11–18

[38] JOVANOVIC S, ANFT D, SCHÖNFELD U, BERGHAUS A, SCHERER H (1995b) Tierexperimentelle Untersuchungen zur $CO_2$-Laser-Stapedotomie. Laryngo-Rhino-Otol 74: 26–32

[39] JOVANOVIC S, SCHÖNFELD U, PRAPAVAT V, BERGHAUS A, FISCHER R, SCHERER H, MÜLLER G (1995c) Die Bearbeitung der Steigbügelfußplatte mit verschiedenen Lasersystemen. Teil I: Kontinuierlich strahlende Laser. HNO 43: 149–158

[40] JOVANOVIC S, SCHÖNFELD U, PRAPAVAT V, BERGHAUS A, FISCHER R, SCHERER H, MÜLLER G (1995d) Die

Bearbeitung der Steigbügelfußplatte mit verschiedenen Lasersystemen. Teil II: Gepulste Laser. HNO 43: 223–233

[41] JOVANOVIC S, SCHÖNFELD U, FISCHER R, DÖRING M, PRAPAVAT V, MÜLLER G, SCHERER H (1995e) Thermische Belastung des Innenohres bei der Laser-Stapedotomie. Teil I: Kontinuierlich strahlende Laser. HNO 43: 702–709

[42] JOVANOVIC S, ANFT D, SCHÖNFELD U, BERGHAUS A, SCHERER H (1995f) Experimental studies on the suitability of the erbium laser for stapedotomy in an animal model. Eur Arch Otorhinolaryngol 252: 422–428

[43] JOVANOVIC S, SCHÖNFELD U, FISCHER R, DÖRING M, PRAPAVAT V, MÜLLER G, SCHERER H (1996b) Thermische Belastung des Innenohres bei der Laser-Stapedotomie. Teil II: Gepulste Laser. HNO 44: 6–13

[44] JOVANOVIC S, SCHÖNFELD U, PRAPAVAT V, BERGHAUS A, FISCHER R, SCHERER H, MÜLLER GJ (1996c) Effects of continuous wave laser systems on stapes footplate. Lasers Surg Med 19: 424–432

[45] JOVANOVIC S, SCHÖNFELD U, SCHERER H (1997a) $CO_2$ laser in revision stapes surgery. SPIE Proc 2970: 102–108

[46] JOVANOVIC S, SCHÖNFELD U, HENSEL H, SCHERER H (1997b) Clinical experiences with the $CO_2$ laser in revision stapes surgery. Lasermedizin 13: 37–40

[47] JOVANOVIC S, SCHÖNFELD U, PRAPAVAT V, BERGHAUS A, FISCHER R, SCHERER H, MÜLLER G (1997c) Effects of pulsed laser systems on stapes footplate. Lasers Surg Med 21: 341–350

[48] JOVANOVIC S, SCHÖNFELD U, FISCHER R, DÖRING M, PRAPAVAT V, MÜLLER G, SCHERER H (1998) Thermic effects in the "vestibule" during laser stapedotomy with pulsed laser systems. Lasers Surg Med 23: 7–17

[49] JOVANOVIC S, ANFT D, SCHÖNFELD U, BERGHAUS A, SCHERER H (1999) Influence of $CO_2$ laser application of the guinea-pig cochlea on compound action potentials. Am J Otol 20: 166–173

[50] JOVANOVIC S, JAMALI J, ANFT D, SCHÖNFELD U, SCHERER H, MÜLLER G (2000) Influence of pulsed lasers on the morphology and function of the guinea-pig cochlea. Hear Res 144: 97–108

[51] KAUTZKY M, A TRÖDHAN, SUSANI M, SCHENK P (1991) Infrared laser stapedotomy. Eur Arch Otorhinolaryngol 248: 449–451

[52] LESINSKI SG (1989) Lasers for otosclerosis. Laryngoscope 99 (Suppl 46): 1–24

[53] LESINSKI SG (1990a) Laser stapes surgery (letter). Laryngoscope 100: 106–107

[54] LESINSKI SG (1990b) Lasers for otosclerosis – which one, if any, and why. Lasers Surg Med 10: 448–457

[55] LESINSKI SG, Stein JA (1992) Lasers in revision stapes surgery. Oper Tech Otolyngol Head Neck Surg 3: 21–31

[56] LESINSKI SG, NEWROCK R (1993) Carbon dioxide lasers for otosclerosis. Otolaryngol Clin North Am 26: 417–441

[57] LIM RJ (1992) Safety of carbon dioxide laser for stapes surgery. Lasers Surg Med (4) 6

[58] LINTHICUM F (1971) Histologic evidence of the cause of failure in stapes surgery. Ann Otol Rhinol Laryngol 80: 67–77

[59] LIPPY WH (1980) Stapedectomy revision. Am J Otol 2: 67–77

[60] LYONS GD, WEBSTER DB, MOUNEY DF, LOUSTEAU RJ (1978) Anatomical consequences of $CO_2$ laser surgery of the guinea pig ear. Laryngoscope 88: 1749–1754

[61] MCGEE TM (1983) The argon laser in surgery for chronic ear disease and otosclerosis. Laryngoscope 93: 1177–1182

[62] MCGEE TM, KARTUSH JM (1990) Laser stapes surgery (letter). Laryngoscope 100: 106–107

[63] MARQUET J (1983) Otosclerosis: small-hole technique. J Laryngol Otol (Suppl) 8: 78–80

[64] MARQUET J (1985) Stapedotomy technique and results. Am J Otol 6: 65–67

[65] MOLONY TB (1993) $CO_2$ laser stapedotomy. J La State Med Soc 145 (9): 405–408

[66] NAGEL D (1996) Laser in der Ohrchirurgie. HNO 44: 553–554

[67] NAGEL D (1997) The Er:YAG laser in ear surgery: first clinical results. Lasers Med 21 (1): 79–87

[68] NISSEN RL (1989) Argon laser in difficult stapedotomy cases. Laryngoscope 108: 1669–1673

[69] NUSS RC, FABIAN RL, SARKAR R, PULIAFITO C (1988) Infrared laser bone ablation. Lasers Surg Med 8: 381–391

[70] PALVA T (1987) Argon laser in otosclerosis surgery. Acta Otolaryngol (Stockh) 104: 153–157

[71] PALVA T, KARJA J, PALVA A (1977) Otosclerosis surgery. Acta Otolaryngol 83: 328–335

[72] PERKINS RC (1980) Laser stapedotomy for otosclerosis. Laryngoscope 90: 228–241

[73] PERSSON P, HARDER H, MAGNUSON B (1997) Hearing results in otosclerosis surgery after partial stapedectomy, total stapedectomy and stapedotomy. Acta Otolaryngol (Stockh) 117: 94–99

[74] PFALZ R (1995) Eignung verschiedener Laser für Eingriffe vom Trommelfell bis zur Fußplatte (Er:YAG-, Argon-, $CO_2$-s.p.-, Ho:YAG-Laser). Laryngo-Rhino-Otol 74: 21–25

[75] PFALZ R, LINDENBERGER M, HIBST R (1991) Mechanische und thermische Nebenwirkungen des Argon-Lasers in der Mittelohrchirurgie (in vitro). Eur Arch Otorhinolaryngol (Suppl) II: 281–282

[76] PFALZ R, BALD N, HIBST R (1992) Eignung des Erbium:YAG Lasers für die Mittelohrchirurgie. Eur Arch Otorhinolaryngol (Suppl) II: 250–225

[77] PRATISTO H, FRENZ M, ITH M, ROMANO V, FELIX D, GROSSENBACHER R, ALTERMATT H, WEBER H (1996) Temperature and pressure effects during erbium laser stapedotomy. Lasers Surg Med 18: 100–108

[78] RAUCH SD, BARTLEY ML (1992) Argon laser stapedectomy: comparison to traditional fenestration techniques. Am J Otol 13 (6): 556–560

[79] SCHLENK E, PROFETA G, NELSON JS, ANDREW JJ, BERNS ML (1990) Laser assisted fixation of ear prosthesis after stapedectomy. Lasers Surg Med 10: 444–447

[80] SCHÖNFELD U, FISCHER R, JOVANOVIC S, SCHERER H (1994) "Lärmbelastung" während der Laser-Stapedotomie. Eur Arch Otolaryngol (Suppl II): 244–246

[81] SEGAS J, GEORGIADIS A, CHRISTODOULOU P, BIZAKIS J, HELIDONIS E (1991) Use of the excimer laser in stapes surgery and ossiculoplasty of middle ear ossicles: preliminary report of an experimental approach. Laryngoscope 101: 186–191

[82] SHABANA YK, ALLAM H, PEDERSEN CB (1999) Laser stapedotomy. J Laryngol Otol 113 (5): 413–416

[83] SHAH KU, POE DS, REBEIZ EE, PERRAULT DF, PANKRATOW MM, SHAPSHAY SM (1996) Erbium laser in middle ear surgery: in vitro and in vivo animal study. Laryngoscope 106: 418–422

[84] SHEEHY JL, NELSON RA, HOUSE HP (1981) Revision stapedectomy: a review of 258 cases. Laryngoscope 91: 43–51

[85] SILVERSTEIN H, ROSENBERG S, JONES R (1989) Small fenestra stapedotomics with and without KTP laser: a comparison. Laryngoscope 99: 485–488

[86] SILVERSTEIN H, BENDET E, ROSENBERG S, NICHOLS M (1994) Revision stapes surgery with and without laser: a comparison. Laryngoscope 104: 1431–1438

[87] SMYTH GDL, HASSARD TH (1978) Eighteen years experience in stapedectomy. The case for the small fenestra operation. Ann Otol Rhinol Laryngol (Suppl) 87: 3–36

[88] SOMERS T, GOVAERTS P, MARQUET T, OFFECIERS E (1994) Statistical analysis of otosclerosis surgery performed by Jean Marquet. Ann Otol Laryngol 103: 945–951

[89] STRUNK CL, QUINN FB, BAILEY BJ (1992) Stapedectomy techniques in residency training. Laryngoscope 102: 121–124

[90] STUBIG IM, ReDer PA, FACER GW, RYLANDER HG, Welch AJ (1993) Holmium:YAG laser stapedotomy: preliminary evaluation. Proc SPIE 1876: 10–19

[91] THOMA J, UNGER V, KASTENBAUER E (1981) Temperatur- und Druckmessungen im Innenohr bei der Anwendung des Argon-Lasers. Laryngo-Rhino-Otol 60: 587–590

[92] THOMA J, UNGER V, KASTENBAUER E (1982) Funktionelle Auswirkungen des Argon-Lasers am Hörorgan des Meerschweinchens. Laryngo-Rhino-Otol 61: 473–476

[93] THOMA J, MROWINSKI D, KASTENBAUER ER (1986) Experimental investigations on the suitability of the carbon dioxide laser for stapedotomy. Ann Otol Rhinol Laryngol 95: 126–131

[94] VERNICK DM (1990) Laser stapes surgery (letter). Laryngoscope 100: 106–107

[95] VERNICK DM (1996) A comparison of the results of KTP and $CO_2$ laser stapedotomy. Am J Otol 17: 221–224

[96] VOLLRATH M, SCHREINER C (1982a) Influence of argon laser stapedotomy on cochlear potentials. I. Alteration of cochlear microphonics (CM). Acta Otolaryngol (Stockh) (Suppl) 385: 1–31

[97] VOLLRATH M, SCHREINER C (1982b) The effects of the argon laser on temperature within the cochlea. Acta Otolaryngol (Stockh) 93: 341–348

[98] VOLLRATH M, SCHREINER C (1983a) Influence of argon laser stapedotomy on cochlear potentials. III. Extracochlear record DC potential. Acta Otolaryngol (Stockh) 96: 49–55

[99] VOLLRATH M, SCHREINER C (1983b) Influence of argon laser stapedotomy on inner ear function and temperature. Otolaryngol Head Neck Surg 91: 521–526

[100] WIET RJ, KUBEK DC, LEMBERG P, BYSKOSH AT (1997) A meta-analysis review of revision stapes surgery with argon laser: effectiveness and safety. Am J Otol 18 (2): 166–171

[101] ZRUNEK M, KAUTZKY M, HÜBSCH P (1993) Experimentelle Laserchirurgie bei ossifizierter Cochlea. Eur Arch Otorhinolaryngol (Suppl) II: 37–38

# III-4.2.2
# Laser Myringotomy

B. Sedlmaier and S. Jovanovic

**Contents**

## Introduction

Ear diseases due to impaired middle-ear ventilation, such as secretory otitis media (SOM), are the most common clinical pictures in otorhinolaryngology, head and neck surgery. After drug therapy fails, surgical ventilation of the tympanic cavity by myringotomy with or without tympanic drainage is the treatment of choice [1, 13]. It is now the surgical intervention most frequently performed by ENT specialists. Most of the patients are children. The incidence in children under the age of 7 comprises 5% [2].

A conventional incision myringotomy heals after 1 or 2 days, which is too short for a therapeutic effect in most cases. On the other hand, a ventilation tube (VT) has an average indwelling time of 4 to 6 months, which is too long. In general, a 3-week transtympanic ventilation period is regarded as adequate [1].

Laser perforations for transtympanic ventilation of the middle ear may provide a solution to this very relevant clinical problem if the appropriate laser system, mode of application, and laser parameters are chosen [8, 10, 19]. This also avoids rare undesirable side effects of tympanostomy tubes such as chronic otorrhea, permanent perforations and atrophic ear-drum scars as well as the development of tympanosclerosis or cholesteatomas [3, 6, 7].

Laser myringotomy is a proven method that has found increasing clinical application [15, 18, 19]. The ventilation time of the middle ear is largely determined by the diameter of the myringotomy perforation and, to a lesser extent, by the thermal effects of the laser in the tympanic membrane [10, 18, 19].

A myringotomy perforation of appropriate size should be created with a single laser application in the topically anaesthetized ear drum. $CO_2$ laser myringotomies generally heal without scar formation [18].

Alternative methods such as heat myringotomy and mono- or bipolar electrothermal myringotomy do not have the same precision and are less safe in their application. Further, their application is restricted when using a topical anaesthetic because they require a longer exposure time on the tympanic membrane [14, 20].

## Laser and Application Systems

The $CO_2$ laser (10 600 nm) and the Er:YAG laser (2940 nm) should be suitable for laser myringotomy due to their irradiation effects in tissue [8, 12]. Water molecules, which are a main component of biological tissue, are responsible for the strong absorption of both wavelengths in tissue. The continuous-wave, thermal $CO_2$ laser has a vaporizing effect in tissue using a focused beam. The oligothermal pulsed Er:YAG laser has a photo-ablative effect owing to its short pulse duration. In the past few years, both lasers have been applied in stapes and middle-ear surgery due to their controlled tissue penetration depth [9, 12].

The $CO_2$ laser has several advantages over the Er:YAG laser: computer-guided scanners can be used (SurgiTouch 780 A Office, ESC Sharplan Co., Tel Aviv, Israel) for selecting the size and shape of the scan area (Fig. 1). In addition, highly precise micromanipulators (e.g. Acuspot 712, ESC-Scharplan, Tel Aviv, Israel) and otoscopes (Fig. 2; Otoscan, ESC Sharplan Co., Tel Aviv, Israel) are available that can be used in combination with the scanners. With these systems one or only a few laser applications are required to create a laser myringotomy of sufficient size. Moreover, the $CO_2$ laser has an adequate haemostatic effect, thus the laser beam is not absorbed by blood in repeated applications. This laser has a very high application safety; even high power levels can be used in cases of marked thickening of the ear drum and tympanic effusion.

With the Er:YAG laser, an adequate perforation size is achieved by the application of several juxtapositioned focused pulses. The slight haemostatic effect makes repeated application difficult because of haem-

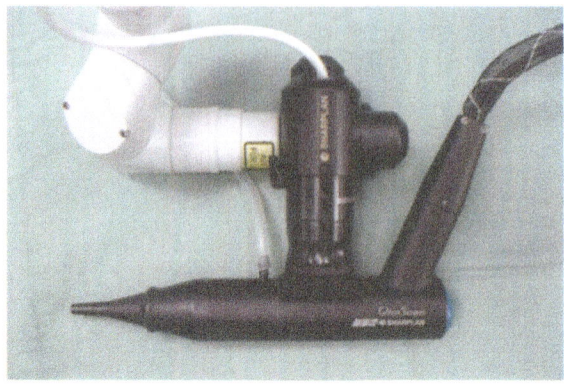

**Fig. 2.** Detail of the Otoscan otoscope: the otoscope consists of a mirror system (centre), a videocamera (left) and ear speculum attachments of various sizes (right). On top of the otoscope is the computer-guided scanner, to which the articulated mirror arm of the laser is connected

orrhaging at the perforation margin [17]. A laser energy of 100 mJ per pulse should not be exceeded because of the risk of acoustical trauma to the inner ear. Currently, no scanners are available that can enlarge the exposure area in the tympanic membrane.

Other lasers applied in clinical medicine, such as the argon and Nd:YAG lasers, are unsuitable for use in the tympanic membrane due to their tissue penetration depth.

The $CO_2$ laser is the laser of choice for tympanic membrane application.

### Micromanipulators

Micromanipulators applied in the tympanic membrane should have a precise beam profile and an exact beam focusing. A micromanipulator like the Acuspot 712 (ESC Sharplan Co., Tel Aviv, Israel) fulfills these requirements. A laser beam with a spot diameter of 200 µm yields a high power density and reduces thermal side effects. Micromanipulators must be connected to an operation microscope. Although this is more complicated technically, it provides a three-dimensional microscopic view and facilitates mechanical manipulation of the tympanic membrane.

### Otoscope

The $CO_2$ laser otoscope, Otoscan (Fig. 2; ESC Sharplan Co., Tel Aviv, Israel), consists of a mirror system with an integrated videocamera. Specula of varying length and diameter can be used, according to the patient's anatomy. The diameter of the focused laser beam is about 400 µm. The otoscope should be used in conjunction with a computer-guided scanner .

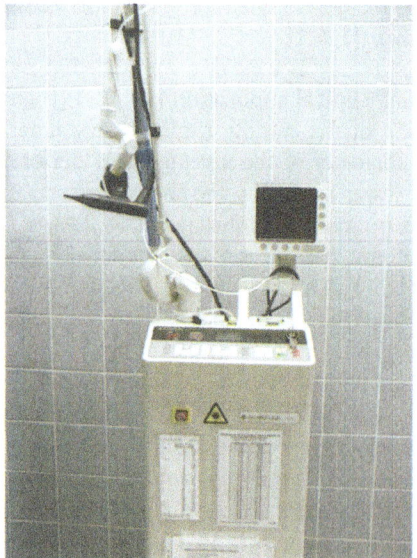

**Fig. 1.** 780 A Office scanner

## Scanners

Scanners suitable for $CO_2$ laser application in the tympanic membrane are those that move the focused laser beam over a defined area by means of computer-guided rotating mirrors. In the new systems, the laser beam describes a spiral figure that homogeneously irradiates the individual parts of the scanned area. The SurgiTouch 780 A Office system is incorporated in a 40C $CO_2$ laser (ESC Sharplan Co., Tel Aviv, Israel; Fig. 1) and can be combined with the Acuspot 712 micromanipulator as well as with the Otoscan otoscope. The diameter of the scanned area can be adjusted from 1 to 3 mm and the applied energy level can be set from 1 to 40 W. The pulse duration is automatically set by the system and is dependent on the diameter of the scanned area (30 to 300 ms).

## Surgical Technique

Prior to laser treatment, the auditory canal is cleansed and topical anaesthetic is applied to the tympanic membrane. This is a 16–32% tetracaine-based isopropanol solution that is left on the ear drum for 30 min. The solution should be carefully infused into the external auditory canal so that no air bubbles prevent its contact with the tympanic membrane. A small cotton pledget or a thin wick made of Merocel (Pope Ear Wick, Merocel Surgical Products, Mystic CT, USA) is placed on the tympanic membrane to prevent the anaesthetic from escaping the auditory canal through a motion of the head and to reduce the necessity of suctioning out the solution before the intervention.

The myringotomy perforation is typically made in the anterior-inferior quadrant, avoiding irradiation of the limbus and the manubrium. Irradiation of the posterior-inferior quadrant can occur if the auditory canal is narrow or if its anterior wall is prominent. The laser energy level should be reduced in cases of equivocal tympanic effusion. The laser beam should be optimally focused when using the otoscope as well as when applying the micromanipulator. The focal plane is adjusted by altering the distance between the target tissue and the application system. The best focal length is that which corresponds to maximal visual acuity. When using the operation microscope in conjunction with the micromanipulator, the largest magnification should be chosen at which to adjust the focal plane. The pilot beam circumscribes the selected scan area, whose diameter varies according to the indication. As a rule, tympanic membrane perforation is achieved in the first laser application using the parameters given in the next section. In the presence of marked thickening of the ear drum, several applications may be required. In cases of a secretion-filled

tympanic cavity, if the desired perforation diameter is not attained after the first laser application, the same site may be subjected to several pulses until the perforation size is adequate. In cases of an air-filled tympanic cavity, the perforation diameter can be enlarged by ablating the perforation margin without the scanner, or using the scanner only at its smallest scan diameter because otherwise the promontory may be accidentally irradiated. When the scanner is employed with the otoscope, the following settings should be selected: 10 W power level, 1 mm scan diameter, 50 ms pulse duration (with the Acuspot micromanipulator: 10 W/1 mm / 60 ms). When the scanner is not used, a low power level of 2 W and a short pulse duration of 50 ms should be applied.

Irradiation of the promontory at these parameters will not damage the vestibulocochlear organ but can lead to pain because the middle-ear mucosa is not anaesthetized. Smoke generated by the micromanipulator can be suctioned off between laser applications. The Otoscan otoscope has a built-in ventilator for removing charring products.

In principle, $CO_2$ laser myringotomy may be performed without scanners, without an otoscope, and with micromanipulators other than the ones mentioned here. The decisive parameter is the power density ($W\,cm^{-2}$) applied to the tissue. An effective power density of about 2000 $W\,cm^{-2}$ would be required for perforating a normal human tympanic membrane. The diameter of the focused laser beam is a characteristic size of a particular micromanipulator. When using a focused laser beam without a scanner that sets the pulse duration automatically, a short pulse duration (i.e. 50 ms) should be selected. To achieve the desired diameter of the laser myringotomy, several contiguous laser applications at the margin of the perforation are required.

Defocusing the laser beam to enlarge the diameter of the irradiated area leads to a reduction in the power density in the tissue proportional to the square of the radius of the irradiated area. Laser energy must be considerably increased to attain an effective power density. Defocusing to an area of 2 mm would necessitate a power level of about 60 W for an effective power density. Irradiation of the stapes footplate at this power level can lead to inner-ear damage. A power setting of 60 W at a pulse duration of 0.05 s yields an energy of 3 J per pulse ($J = W \times s$). As Jovanovic et al. [9] showed in laser stapedotomy experiments in animals, a total energy of 3 J and above can cause irreversible inner-ear disorders when applied to the basal spiral canal of the cochlea in guinea pigs.

In addition, defocusing makes the laser beam profile imprecise, which, in turn, reduces perforation capacity.

For these reasons, computer-guided scanners, which enlarge the irradiation area by moving the focused laser beam, should be given preference over defocused laser-beam application.

## Indications

### Secretory Otitis Media (SOM)

Childhood and adult SOM is a sequela of a eustachian tube (ET) dysfunction. Children under four very often have persistent middle-ear effusion following acute otitis media. If an 8-10-week drug therapy is unsuccessful, surgical ventilation of the middle ear is carried out with the $CO_2$ laser via the tympanic membrane.

In children, the intervention is very often performed under general anaesthesia, because hyperplastic adenoids are removed in the same session. If they are not enlarged, surgery on the tympanic membrane can be performed using a topical anaesthetic agent (see above). The perforation diameter created by the $CO_2$ laser should not be smaller than 2 mm, because otherwise the therapeutic ventilation time would be too short (Fig. 3). Myringotomy perforations of this diameter generally close after a mean 17 days and heal progressively without scar formation (Figs. 4, 5). This treatment time was shown to be adequate on the basis of previous experience [7, 9]. In our

hospital, the treatment schema for children with SOM comprises an adenotomy with $CO_2$ laser myringotomy as the primary intervention in the presence of impaired nasal breathing. The first recurrence of effusion can be treated by a $CO_2$ laser myringotomy under topical anaesthesia if nasal breathing is no longer im-

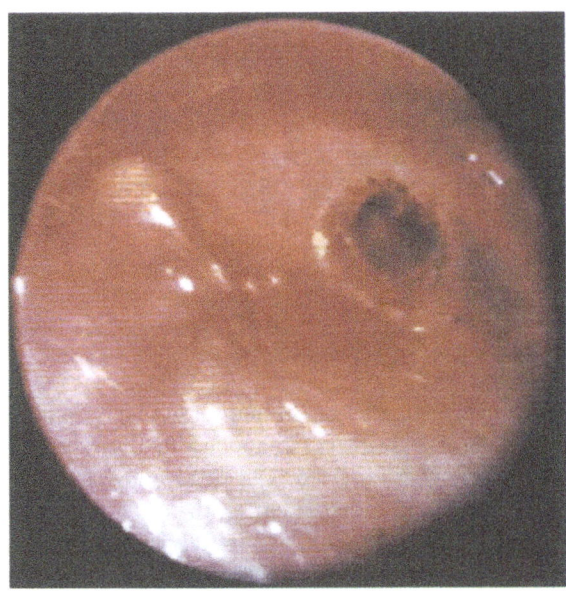

**Fig. 4.** Healing course of a $CO_2$ laser myringotomy 2 weeks postoperatively. The perforation is almost closed by a onion-skin-like membrane made of keratinised material (right ear)

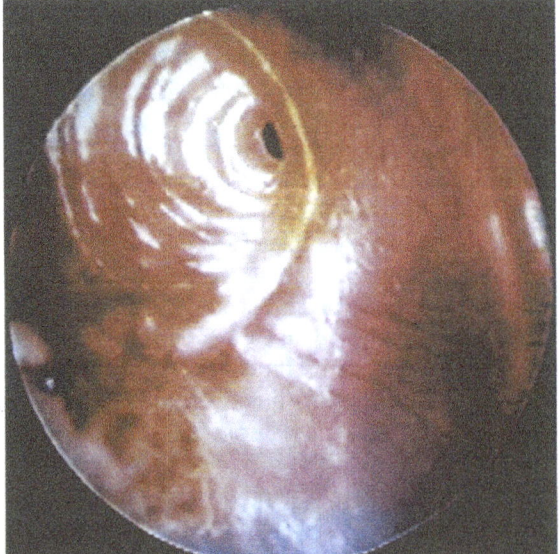

**Fig. 3.** Condition after performance of $CO_2$ laser myringotomy in the anterior-inferior quadrant (power level: 12 W, pulse duration: 180 ms, scan diameter: 2.2 mm, perforation diameter: ca. 2 mm). Coagulation traces are visible at the perforation margins

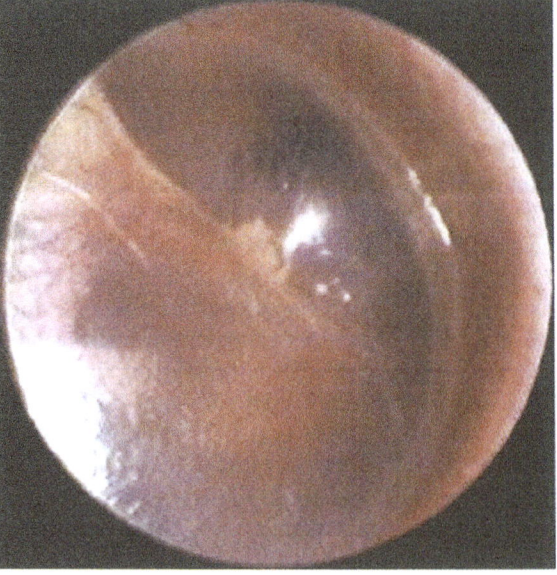

**Fig. 5.** Tympanic membrane 4 months after $CO_2$ laser myringotomy. An unremarkable, normal ear drum can be seen where laser myringotomy was previously performed (left ear)

peded. A recurrence of adenoids can be managed by renewed adenotomy under general anaesthesia and a renewed $CO_2$ laser myringotomy. A second relapse is then treated by ventilation tube insertion.

The myringotomy can be performed using the otoscope or the micromanipulator. Using the $CO_2$ laser otoscope, Otoscan, a power setting of 12 W and a scan diameter of 2.2 mm should be applied. A pulse duration of 180 ms is automatically set by the system. Using the Acuspot 712 micromanipulator connected to the operation microscope, a power level of 10 W and scan diameter of 2.2 mm should be selected. The system automatically sets a pulse duration of 260 ms.

In cases of unequivocal middle-ear effusion or thickening of the tympanic membrane, the power level of the Otoscan otoscope can be increased to 15 W and that of the Acuspot 712 micromanipulator to 13 W.

For a perforation diameter of 2 mm, the scan area would be 2.2 mm using either system.

## Acute Otitis Media (OMA) with Vestibulocochlear Complication

Acute otitis media is a bacterial infection that is generally secondary to a viral infection. In rare cases, vestibulocochlear complications can occur with hair-cell function impairment of the auditory and vestibular system. Toxic damage through bacterial products is assumed. A myringotomy is performed to treat this condition. More serious complications, such as acute inflammatory facial nerve paralysis, mastoiditis and endocranial extension of the bacterial infection via blood conduits or meninges, must be managed by mastoidectomy or, in some cases, ligature of the sigmoid sinus.

When acutely inflamed, the tympanic membrane is very often markedly altered, thickened, and covered with fluid-filled blisters (Fig. 6). This influences the laser's effect on the tympanic membrane. The perforation is made in the anterior-inferior quadrant, if possible (Figs. 7, 8). In most cases, a shorter ventilation time than that needed for secretory otitis media is adequate, thus a scan diameter of 1.6 mm is used. When using the Otoscan otoscope, the power level

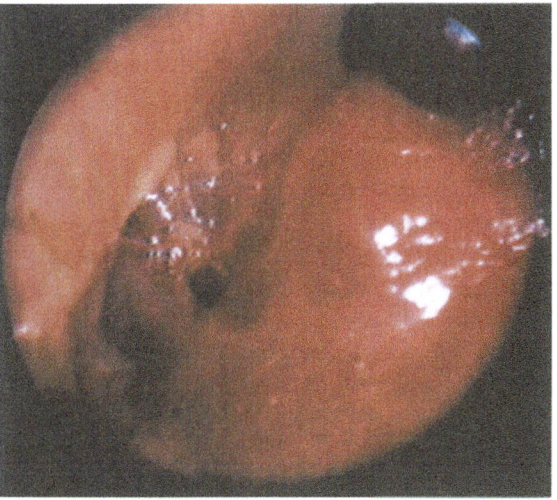

**Fig. 7.** Situation after a $CO_2$ laser myringotomy. The perforation margins show slight hemorrhaging

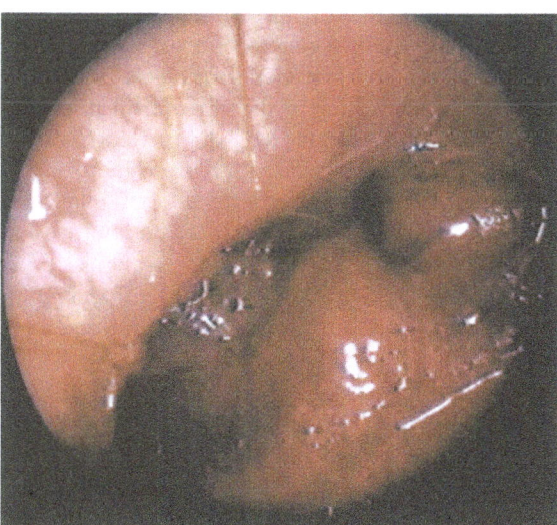

**Fig. 6.** Acute otitis media with redness and bulging of the ear drum and vascular injection

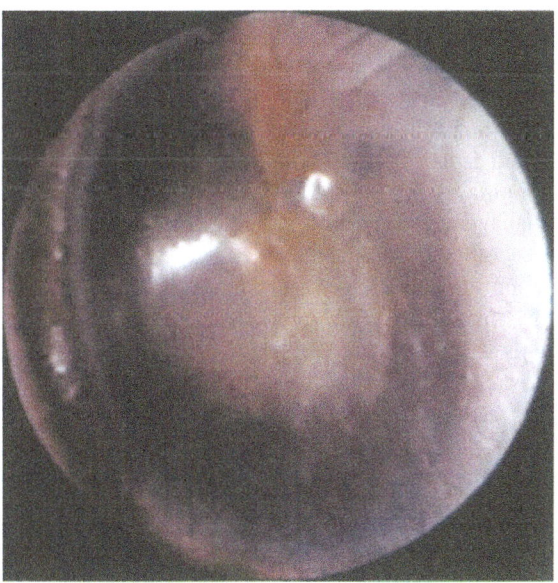

**Fig. 8.** Tympanic membrane 2 months after $CO_2$ laser myringotomy for OMA. The site of the former perforation is no longer identifiable

would be 20 W and pulse duration 80 ms. For the Acuspot micromanipulator, a power setting of 20 W and pulse duration of 110 ms is recommended. After creation of the laser perforation, the appropriate antibiotics therapy can be determined by a bacteriologic sample taken by swabbing the auditory canal, which was previously disinfected with alcohol while under topical anaesthesia.

Investigations are underway on the application of laser myringotomy as primary therapy for uncomplicated acute otitis media to prevent persistent tympanic effusions and to reduce multiple courses of antibiotics. The final results are not yet available.

## Acute Eustachian Tube (ET) Dysfunction

Acute ET dysfunction is predominantly caused by an acute inflammatory edema of the ET ostial and luminal mucosa. An acute viral infection or an allergy are possible causes of the inflammation. Prolonged low tympanic pressure leads to fluid accumulation. An immediate operative ventilation of the middle ear is rarely indicated; instead, drug therapy should first be tried. In cases of therapeutic resistance, or when a rapid improvement in hearing is desired, laser myringotomy is performed using the same parameters as those used in the treatment of secretory otitis media. The power level should be reduced when treating patients with equivocal effusion.

## Barotrauma

Barotrauma is the result of a sudden change in ambient pressure in conjunction with reduced ET function. It is an acute, painful increase or decrease of pressure in the middle ear with a protrusion or retraction of the tympanic membrane. Barotrauma may result from air travel, a quick altitude change in the mountains, or diving. The recommended conservative management is to reduce the mucosal swelling of the nose and to administer an analgenesic. The painful symptoms can be relieved immediately by $CO_2$ laser myringotomy under topical anaesthesia. This requires only a short-term ventilation of the middle ear. To create a perforation of 0.8 mm in diameter, the following settings are used with the Otoscan otoscope: 10 W power level, 1 mm scan diameter, 50 ms pulse duration. The parameters for the micromanipulator are 10 W power level and 60 ms pulse duration at the same scan size.

## Transtympanic Tympanoscopy

The indications for an inspection of the middle ear comprising laser myringotomy under local infiltration anaesthesia using 1.7-mm strong rigid angle optics (0°, 30°, 70°) or dedicated flexible microoptics, are sudden deafness, which may indicate window membrane rupture; renewed conductive deafness following stapedotomy, which could be caused by a dislocated prosthesis; and other equivocal middle-ear findings. Moreover, drugs can be applied locally to the round window membrane through a laser myringotomy. The tympanic tube orifice as well as the caudal parts of the sound-conducting apparatus can be inspected through a perforation in the anterior-inferior quadrant. The round window membrane can be seen in most cases through an opening in the posterior-inferior quadrant. The Otoscan otoscope is used at a power level of 10 W and pulse duration of 270 ms; the Acuspot micromanipulator is set at 10 W power level and a 360 ms pulse duration. A low power setting should be chosen because no secretion is present in the middle ear in conjunction with these indications. The scan diameter would be 2.6 mm using either system.

## Discussion

A ventilation dysfunction of the tympanic cavity is the cause of many acute and chronic middle-ear diseases. Particularly in childhood, ventilation problems of the middle ear are frequent due to recurrent mucosal infections and enlargement of the lymphoepithelial pharyngeal and palatine tonsils.

Persistence of this problem can lead to chronic inflammatory middle-ear diseases such as adult purulent otitis media involving the mucosa and the bone.

The most important therapeutic goal is to restore middle-ear ventilation. If drug treatment proves unsuccessful after a period of 8 to 10 weeks, surgical procedures are an option. Apart from removing the pharyngeal tonsils to eradicate the nasopharyngeal germ reservoir and, in some cases, to improve tube function, tympanic membrane interventions are performed that provide transtympanic ventilation of the middle ear. These are the most frequent operations performed by the ENT specialist [2].

Three to 4 weeks is considered adequate for therapeutic ventilation [1]. The simple incision in the tympanic membrane, the myringotomy, generally closes within 48 h. The ventilation tube, which may be inserted into the incision, has an average in-dwelling time of 4 to 6 months in the tympanic membrane. On the one hand, this time period is too long. On the other hand, rare complications of tympanic drainage can develop such as persistent tympanic membrane perforation, atrophic scarring, cholesteatoma or tympanosclerosis formation and persistent otorrhea [3, 6, 7]. Laser myringotomy facilitates ventilation of the middle ear via a perforation whose closure time determines the duration of tympanic ventilation. Closure time is primarily dependent on perforation diameter; to a lesser extent it is influenced by thermal effects on the tympanic membrane, such as carbonization and coagulation [8, 10, 19].

Thermal myringotomy using a hot needle described by Saito et al. [14] and its later development, monopolar or bipolar electrothermal paracentesis, seem to delay healing [14, 20]. These procedures do not possess the same precision as laser myringotomy and are painful due to the longer exposure time on the tympanic membrane. Therefore, their application under topical anaesthesia is restricted.

Compared with methods using conventional instrumentation, laser application on the tympanic membrane has the advantage of being a non-contact, relatively bloodless and very precise therapy [8, 10, 19]. As already discussed above, the Er:YAG and $CO_2$ lasers are suited for tympanic membrane application because of their laser beam tissue interaction. The greater application safety and available application systems of the $CO_2$ laser make it the laser of choice. Its irradiation can be applied to the tympanic membrane via either highly precise micromanipulators or dedicated laser otoscopes [4, 11, 16]. Using the available computer-guided scanners, the diameter of the irradiated area can be exactly preset according to the treatment indication. These systems enable tissue ablation without any noteworthy thermal effects. Treatment can be performed under topical anaesthesia in adults, in children and even in infants. Therefore, this may obviate the need for general anaesthesia in some interventions, particularly in young patients.

Other lasers employed in medicine such as the Nd:YAG (1064 nm) and argon (488 nm) lasers are unsuitable for tympanic membrane application due to their tissue penetration depth, their effects in tissue and their lack of resorption in blood-free fluids.

The recommended laser parameters for the specific indications listed earlier are oriented on the desired perforation size and closure time as well as on the presence of fluid behind the tympanic membrane. For SOM a therapeutic ventilation time of about 3 weeks is desirable. An average ventilation time of about 17 days is required for a perforation diameter of 2.0 mm at a scan size of 2.2 mm [18]. This ventilation time seems adequate on the basis of previous experience. Larger perforations are particularly difficult to create in infants because the tympanic membrane as a whole is smaller and there is the risk of irradiating the manubrium or the limbus. In a group of 81 patients (159 ears), 14.5% of the patients had recurrent effusion within a 6-month follow-up period. This relapse rate is no higher than that associated with conventional operative procedures for transtympanic ventilation, particularly after establishing tympanic drainage [5].

In Europe, greater importance is attributed to enlarged pharyngeal tonsils in the pathophysiology of SOM than in the USA. In the EU, recurrent acute ear inflammation and persistent tympanic effusion, even in the presence of non-severe impairment of nasal breathing, are generally managed by nasopharyngeal inspection with surgical removal of the pharyngeal tonsils to improve tube function and reduce the epipharyngeal germ reservoir. In the USA, establishment of tympanic drainage is frequently the only therapeutic measure. At our hospital, the indication for ventilation tube insertion is the second or multiple recurrence of SOM. The standard treatment schema for SOM is described in a previous section.

In acute otitis media, $CO_2$ laser myringotomy is always performed in the presence of vestibulocochlear complications, such as impaired inner-ear hearing or a vestibular disorder. The myringotomy perforation enables toxic inflammatory products from the inner ear to escape via the middle ear. The inflamed tympanic membrane is often very thickened and altered; thus perforation of the ear drum is not always accomplished with a single laser application. With increasing layer thickness and decreasing tissue pH, the topical anaesthesia of the tympanic membrane is not as sufficient as in cases of SOM or an unaffected tympanic membrane. Ongoing prospective and prospective-randomized studies in the USA and Germany are investigating the importance of laser myringotomy in the primary therapy of acute otitis media to prevent persistent tympanic effusion and avoid multiple courses of antibiotics. Final results are not yet available.

$CO_2$ laser myringotomy of the topically anaesthetised tympanic membrane can effect a rapid and immediate recession of the complaints in cases of acute tube dysfunction and barotrauma.

In the performance of endoscopic transtympanic tympanoscopy in adults, a bloodless perforation of 2.4 mm in diameter can generally be created using a scan area of 2.6 mm and a single laser application. A perforation of this size has a closure time of about 6 weeks.

Investigations are still underway on $CO_2$ laser application for the treatment of small atrophic scars and of small perforations of the tympanic membrane that do not heal spontaneously, which occurs, for example, in chronic purulent otitis media involving the mucosa. Preliminary results suggest that $CO_2$ laser application may be extended to this area.

$CO_2$ laser myringotomy is a modern method in the surgical treatment of ventilation disorders of the middle ear. This relatively painless outpatient procedure, which can be performed under topical anaesthesia even in children, frequently replaces the ventilation tube with a self-healing perforation that enables sufficient ventilation of the tympanic cavity. For long-term, chronically recurring functional disorders of the eustachian tube, the ventilation tube cannot be replaced completely.

Modern application systems such as the $CO_2$ laser otoscope, Otoscan, combined with a scanner facilitate the simple and fast performance of interventions in the topically anaesthetised tympanic membrane for various indications.

## References

[1] ARMSTRONG BW (1954) A new treatment for chronic secretory otitis media. Arch Otolaryngol 59: 653–654

[2] BLACK NA (1984) Surgery for glue ear – a modern epidemic. Lancet 1: 835–837

[3] BUCKINGHAM RA (1981) Cholesteatoma and chronic otitis media following middle-ear-intubation. Laryngoscope 91: 1450–1456

[4] DEROWE A, OPHIR D, KATZIR A (1994) Experimental study of $CO_2$ laser myringotomy with a hand-held otoscope and fiberoptic delivery system. Lasers Surg Med 15: 249–253

[5] GATES GA, AVERY CA, COOPER JC, PRIHODA TJ (1989) Chronic secretory otitis media: effects of surgical management. Ann Otol Rhinol Laryngol Suppl 138: 2–32

[6] GATES GA, AVERY C, PRIHODA TJ, HOLT GR (1998) Delayed onset post-tympanotomy otorrhea. Otolaryngol Head Neck Surg 98 (2): 111–115

[7] GOLZ A, NETZER A, JOACHIMS HZ, WESTERMAN ST, GILBERT LM (1999) Ventilation tubes and persiting tympanic membrane perforations. Otolaryngol Head Neck Surg 120 (4): 524–527

[8] GOODE RL (1982) $CO_2$ Laser myringotomy. Laryngoscope 92: 420–423

[9] JOVANOVIC S, ANFT D, SCHÖNFELD U, BERGHAUS A, SCHERER H (1993) Tierexperimentelle Untersuchungen zur $CO_2$-Laserstapedotomie. Laryngol Rhinol Otol 74: 26–32

[10] JOVANOVIC S, SEDLMAIER B, SCHÖNFELD U, SCHERER H, MÜLLER G (1995a) Die $CO_2$-Laser- Parazentese-tierexperimentelle und klinische Erfahrungen. Lasermedizin 11: 5–10

[11] JOVANOVIC S, SEDLMAIER B, SCHÖNFELD U, DESINGER K, SCHERER H (1995b) Ein neues Applikationsystem für die Laserparazentese – Erste Ergebnisse. Minim Invas Med 7 (2): 76–78

[12] PFALZ R (1995) Eignung verschiedener Laser für Eingriffe vom Trommelfell bis zur Fußplatte (Er:YAG-, Argon-, $CO_2$-, Ho:YAG-Laser). Laryngol Rhinol Otol 74: 26–32

[13] PULITZER A (1869) Diseases of the ear, 5th edn. (Translated by Ballin M J and Heller C L Lea and Febiger, Philadelphia, pp 145–155, 282–302

[14] SAITO H, MIYAMOTO K, ET AL(1978) Burn perforation as a method of middle-ear ventilation. Arch Otolaryngol 104: 79–81

[15] SEDLMAIER B, JOVANOVIC S, TÄGL P, SCHÖNFELD U (1998) Das neue $CO_2$- Laserotoskop und das Er:YAG-Laserotoskop-klinische Erfahrungen. HNO 46: 385.

[16] SEDLMAIER B, JOVANOVIC S, BLÖDOW A, SCHÖNFELD U (1998) Das $CO_2$-Laserotoskop-ein neues Applikationssystem für die Parazentese. HNO 46: 870–875

[17] SEDLMAIER B, TÄGL P, GUTZLER R, SCHÖNFELD U, JOVANOVIC S (2000) Experimentelle und klinische Erfahrungen mit dem Er:YAG-Laserotoskop. HNO 48 (11): 816–821

[18] SEDLMAIER B, JIVANJEE A, GUTZLER R, JOVANOVIC S (2001) Heilungsverlauf des Trommelfells und Dauer der Paukenbelüftung nach Lasermyringotomie mit dem $CO_2$-Laserotoskop Otoscan®. HNO 49/6: 447–453

[19] SILVERSTEIN H, KUHN J, CHOO D, KRESPI PY, ROSENBERG SI, ROWAN P (1996) Laser-assisted tympanostomy. Laryngoscope 106 (9): 1067–1074

[20] TOLSDORFF P (1998) Bipolare Thermoparazentese – Grundlagen und Klinik. HNO 4: 386

# III-4.3
# Laser-Assisted Surgery of the Soft Palate for Snoring and Obstructive Sleep Apnoea

Jürgen U.G. Hopf and Marietta Hopf

## Contents

## Introduction

Snoring – the common term for obstructive breathing while asleep – is an extremely often encountered disorder which may lead not only to severe medical problems. Predominantly, snoring is a socially disturbing factor that may shatter the best interpersonal relationships or at least put some significant borders on them.

Snoring is more common in men. Its prevalence ranges between 6.8 and 60% and increases with age and body weight. In the population aged between 30 and 35 years, snorers comprise 20% of the males and around 5% of the females. This percentage will increase at the age of 60 years; around 60% of the men and 40% of the women will snore habitually [7, 43, 15, 3, 69].

Nobody has a striking explanation for the preponderance of male snoring: there is an old legend applied to this puzzle: going back to the early days of mankind, primitive men could defend their wives at night when they were asleep only by making loud and terrifying noises, protecting them from attack by ferocious beasts of prey.

Compared to the rest of mammalian animals, snoring is found almost exclusively in human beings. One of the reasons for this fact may lay in the sleeping position. Wild animals either sleep standing upright or lie down in the ventral position or on the side so that the lower jaw is somehow sustained, preventing it from falling back. The problem with snoring began when primitive primate ancestors decided to turn to a new sleeping position – lying on their back! An exception to this observation are bulldogs and other brachycephalic dog races. They snore sometimes terribly, and some of them need surgical resections of the uvula and the soft palate to prevent them from strangling during sleep.

Snoring may be a minor annoyance in some households, but in others it can become disruptive to family life and partnership. It makes the snorer an object of ridicule, and commits other family members to sleepless nights filled with resentment. Hapless and frustrated bed partners become consumed with plotting strategies to get to sleep (sleeping pills, ear plugs, ear muffs, pushing the snorer out of the bed or fleeing the sleeping room themselves, etc.). A long list of complaints expressed by snorers and their wives gives poignant testimony to the hardships that snoring imposes on their everyday life. While some spouses can easily fall asleep in an noisy environment (including closed proximity to a snorer), others cannot. The noises of snoring are mostly hard to ignore because of their inherent irregularity. The "victims of noise" have been known to lie awake for hours just marvelling at the kaleidoscopic variety of sounds produced by the snorer, and some snorers make truly frightening sounds, suggesting that each breath may be their last.

Couples who have learned to cope with a snoring problem have usually established two separate sleeping rooms; but this is not done without some anguish, because some consider this an abandonment of their marital expectations. On the other hand, spouses are unwilling to leave their husbands alone during the night, so as to be ready to wake him up to breathe in periods of apnoea.

According to the Guiness Book of World Records, the highest measured sound level recorded by any chronic snorer is a peak of 93 dB A in distance of one-foot away from the head. It is easy to understand that loud snoring has been noted to be responsible for divorces; but it is also reported for a "significant sudden change in the pattern and the character of nightly snoring" normally heard by the neighbourhood. In one case, the husband just returning from an overnight business stay abroad was asked by his neighbour living in the apartment next to him: what was up with you last night. Your snoring was so completely different to what we have been used to for many years.

Charles Dickens already observed the association between obesity and daytime sleepiness in his description of Joe, the servant boy of Mr. Wardle, in the Posthumous Papers of the Pickwick Club. "His head was sunk upon his bosom; and perpetual snoring, with a partial choke occasionally; were the only audible indications of the great man's presence".

Snoring and obstructive sleep apnoea syndrome (OSAS) have become an issue in health care only since the past two decades. The conditions were considered to have minor importance until the social ills of loud snoring and the serious effects of OSAS were identified and adequately investigated. Snoring and OSAS represent a continuum of the same phenomenon, uncomplicated snoring being at one end and OSAS at the other.

## Pathophysiology of Snoring/OSAS

Snoring is a loud and recurrent breathing sound with variable intensity and frequency, that occurs upon inspiration during sleep. It is correlated with age and sex. The correlation to body weight is also often found but is still under controversial discussion.

Snoring is a symptom of a large variety of different disorders and etiopathologies [20]. Primarily, it is substantial to differentiate simple or habitual snoring (without additional symptoms of disease) from obstructive sleep apnoea (OSA) syndrome [56, 57]. Lugaresi [42, 44] judge snoring as the beginning and OSA as the most severe end of the sleep disturbance continuum.

Pathophysiological investigations of Issa and Sullivan [30, 31], Remmers et al. [49, 50] and Strohl et al. [55] confirm this hypothesis.

OSA is significantly characterized by periodic apnoeas and hypopnoeas that produce asphyxia and arousals from sleep. Leading symptoms of OSA are loud and irregular snoring, morning headache, daytime sleepiness, excessive daytime tendency to fall asleep, sudden drop of physical and mental performance, depression and libido problems.

Located in between the wide range from plain snoring to life-threatening severe OSA, there is the so-called upper airway resistance syndrome (UARS), also termed the "heavy snorer's disease". In literature it is also found as obstructive snoring, or "crescendo snoring". Guilleminault et al. [18, 19, 21, 22, 23] describe this syndrome, which is also characterized by excessive daytime sleepiness. During periods of loud snoring while asleep there is a partial or transient reduction of pharyngeal air flow, causing a pathological increase of the airway resistance. However in contrast to OSA, no oxygen desaturations are induced, because the patient is able to keep up an acceptable pulmonary air flow sufficient for an unrestricted gas metabolism by intensified breathing efforts.

The sounds of snoring originate in the collapsible part of the airway where there is no stiff and rigid substance creating anatomic subsegments. This area streches from the choanae to the hypopharynx and the epiglottis. Noisy vibrations can arise from the soft tissue structures in the pharynx, including the soft palate, uvula, tonsils, tonsillar pillars, base of the tongue and the posterior and lateral walls of the pharynx.

Generally speaking, the noisy vibrations occur because of airflow turbulence in the sleeper's pharynx, originating either in the nose due to restricted airflow or in the pharynx itself. Thus created turbulent airflow produces a flutter-valve effect in the collapsible pharyngeal tissues. OSA results, roughly speaking, from the collapse of the pharyngeal walls in response

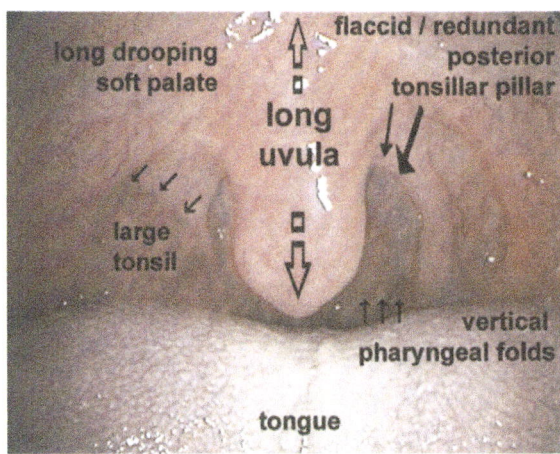

**Fig. 1.** Frequently found anatomical features in the oral cavity and the oropharynx of snorers: a long large uvula, a long, lax and drooping soft palate, a flaccid and redundant posterior tonsillar pillar, some vertical mucosal folds at the posterior (and lateral) pharyngeal wall as well as large tonsils

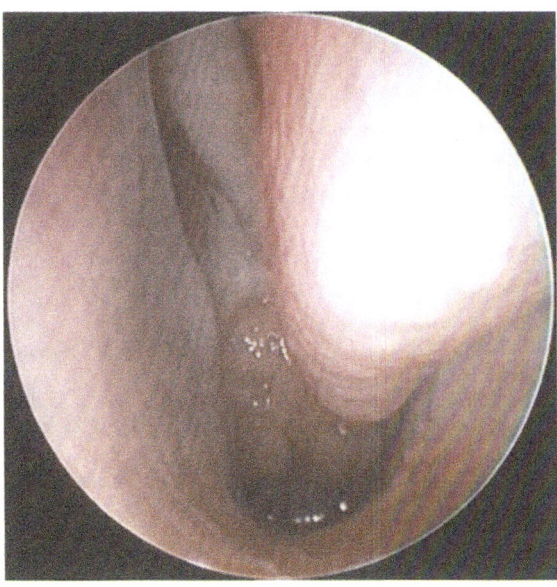

**Fig. 2.** Endoscopic view into the left nasal cavity of a 6-year-old boy with nightly snoring and impeded nasal breathing. At the rear end of the nasal cavity adenoids come into view

to negative inspiratory pressure in the upper airway [4]. Hypotonictiy of the pharyngeal musculature allows upper airway collapse even at more modest negative inspiratory pressures, leading to snoring or apnoea.

In detail the following factors, all single or in a combination, contribute to snoring/OSA.

## Soft Tissue Plus in the Epi-, Meso-, Hypopharynx (with Regular Skeletal Configuration)

The so-called "space-occupying masses" impinging on the lumen of the airway can contribute to snoring and OSA. This is very often the case in snoring children, where frequently enlarged tonsils and hypertrophied adenoids are found.

In 1965, eight snoring children with adenoid and tonsillar hypertrophy are reported having developed the full clinical picture of OSA. After adenoidectomy and tonsillectomy, the OSA pathology, including a cor pulmonary, was completely reversed. In children showing just a lymphatic tonsillar hypertrophy, modern $CO_2$ laser surgery allows a size reduction of the tonsils either by laser-assisted tonsillotomy or by vaporization techniques, provided there is definitely no history or sign of chronic tonsillitis in these children. This may help in many snoring children to preserve this immunorelevant organ during the childhood.

About one third of adult snorers also have tonsils large enough to contribute to this etiopathology. Tonsillectomy with or without laser-assisted reshaping of the soft palate will improve the situation.

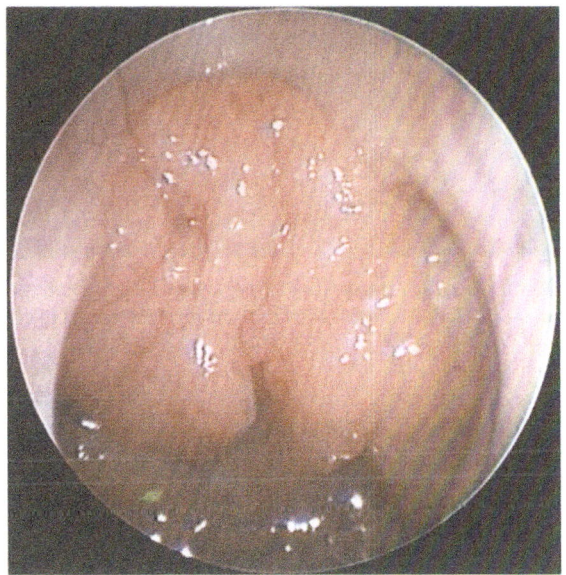

**Fig. 3.** Endoscopic close-up view of the adenoids with partial obstruction of the airway

Bulky pharyngeal tissues are notable in obese persons. Using MRI imaging of the neck in these patients, fat deposits in the parapharyngeal region are detected, narrowing the luminal situation as well as destabilizing the interior soft tissue integument. This prepares for soft tissue flutter excursions and faciti-tates the onset of snoring vibrations.

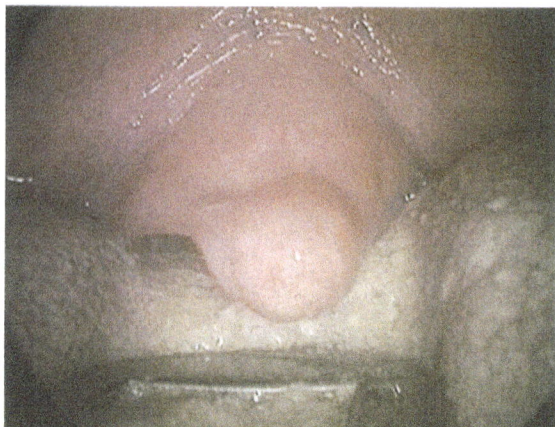

**Fig. 4.** Intraoral aspect of a huge elongated uvula and flaccid and drooping palatal arch

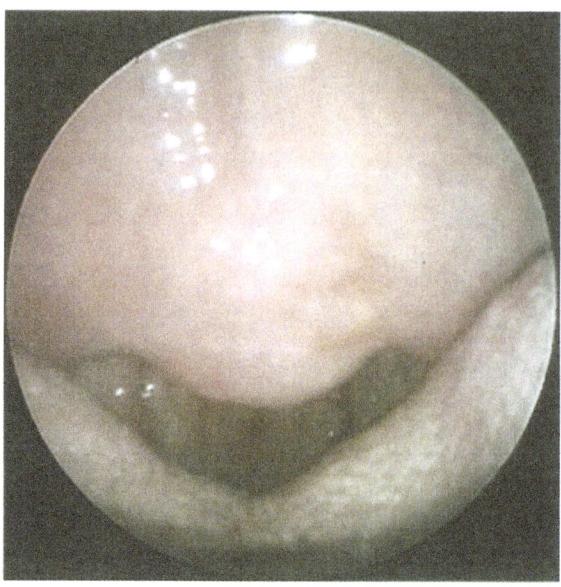

**Fig. 6.** The same patient (as in Fig. 4 and 5) 2 years after laser-assisted uvulopalatoplasty (LAUP) resting his reshaped, reinforced and stiffened soft palate in normal position; all the snoring problems have gone. The formerly large elongated uvula and the flaccid soft palate have been successfully treated by laser surgery. There were no side effects or complications.

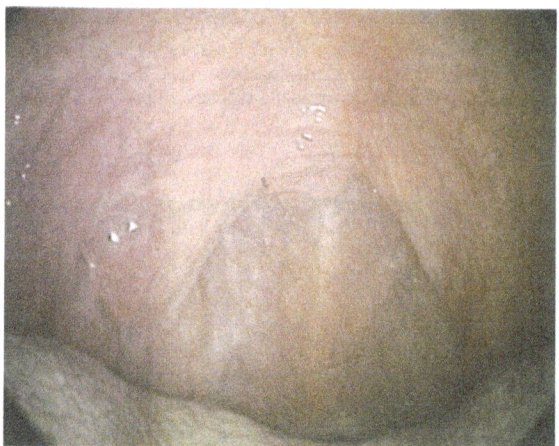

**Fig. 5.** The same patient (as in Fig. 4) while phonating "hiiiiiii", the large elongated uvula is nearly completely retracted in cranial direction by muscle activation.

## Vibration Trauma – Neurogenic Lesion

Either as a predisposing pathophysiological factor for primary snoring or as a sequela of snoring-induced vibration trauma, the excessive length of the uvula and the soft palate can dramatically narrow the nasopharyngeal aperture. The consequence is that the palate descends not only inferiorly in direction, but posteriorly as well. In many cases, snorers are encountered having no more than a slit-like lumen opening from the choanae to the lower portion of the pharynx. Thus, they breathe through a one-way valve. This can be found very easily when the patient is placed in supine position during examination.

There seems to be evidence that the development of OSA is preceded by the subocclusive stage of habitual snoring [55, 56, 57, 32]. Local neurogenic lesions in (oro-)pharyngeal tissues caused by the low-frequency

vibration of habitual snoring are believed to be the pathophysiological mechanism underlying progression to OSA. This theory is based on results from different biopsy studies. Friberg [8] found in the soft palatal mucosa an increased number of abnormally varicose afferent nerve endings in snorers and in OSA patients compared to healthy non-snorers. In a similar fashion, taking biopsies from the palatopharyngeal muscle, Friberg [9] and Friberg et al. [10, 11] showed typical morphologic abnormalities including neurogenic signs in snorers with and without OSA. Gutmann's et al [24] and Gutman's [25] muscle biopsy results showed a conspicuous share of atrophied muscle fibres just next to hypertrophied ones in snorers and OSA patients.

Low-frequency vibration has been found to disturb microcirculation in vibration-induced white fingers disease [58]. According to this, signs of disturbed nerval regulation of the microcirculation have been found in soft palate mucosa of habitual snorers and some patients with mild OSA [11]. In 1992, Larsson could demonstrate indications of local oropharyngeal neuropathy in OSA patients [42]. Moreover, muscle biopsy studies by Woodson et al. [68] and Edström et al. [5] reported typical lesions of peripheral motoneurons. Electronmicroscopically, they showed a high incidence of focal degenerative changes of myelinized nerve fibres and axons.

Together, these data suggest that a disturbance in the efferent and/or afferent nerve pathways involved

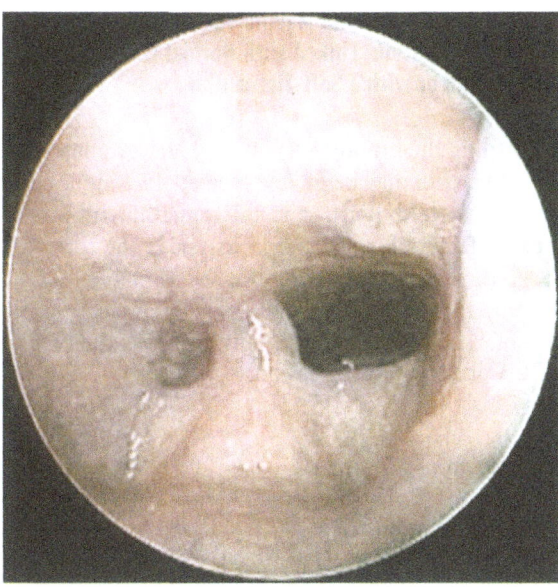

**Fig. 7.** Transnasal endoscopic view of the naso- and oropharynx exposing the pre-operative situation of a long and flaccid soft palate and a large elongated uvula creating a slit-like airway lumen and loud snoring sounds

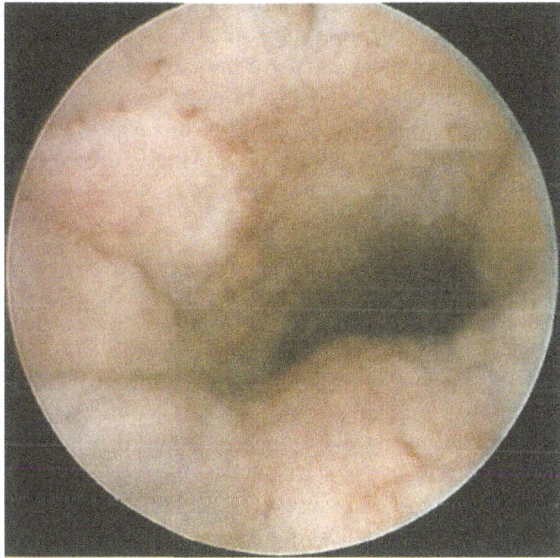

**Fig. 8.** Transnasal endoscopic view of the naso- and oropharynx showing the situation 18 months post-op. The soft palate is reshaped by LAUP. The lumen is enlarged also during sleep by stiffening of the tissues, the vertical folds in the area of the posterior pharyngeal wall are still partly present. Snoring is gone.

in the reflexogenic mechanism of the upper airways contributes to the pharyngeal collapsibility seen in patients with OSA. The correlation between snoring and the neurogenic lesions described does not prove causation. The possibility that neurogenic lesions could precede snoring cannot be completely precluded, but if neurogenic lesions were involved every

time in the development period of snoring, they would be present in every snorer; but the fact is that they are not consistently found. This finding lends support to the interpretation that snoring causes neurogenic lesions and not vice versa.

Taking a detailed history, specifically, the sudden onset of snoring symptoms may be an indicator of cystic lesions and tumours. These potentially life-threatening causes, as well as acute and chronic mucosal inflammations and edema formations, must be definitely excluded by an initial ENT examination, (video-)endoscopy, microscopy and contrast-enhanced CT scan or MRI imaging.

## Lacking or Insufficient Tone of the Lingual, Palatinal and Pharyngeal Musculature

Horner et al. [26], Horner et al. [27, 28] and Wheatley and White [64] and Wheatley et al. [65, 66, 67] found clear sign that the width of the upper airway is influenced or controlled by reflectory regulations via afferent and efferent nerval pathways. Induced by the above-discussed vibration trauma or caused by a different mechanism that is still unknown the tone of palatal, lingual and pharyngeal muscles in snorers and OSA patients is frequently found to be incompetent. This seems to be the predominant cause in adult-onset snoring. In deep-sleep stages, this musculature fails to participate properly in the respiratory cycle to keep the airway completely open during inspiration. Especially, the dilator effect of the pharyngeal muscles and the protrusive effect of the genioglossus muscle are not strong enough. Thus, the tongue falls backward into the airway and then vibrates against the flaccid soft palate structures, the uvula and the pharyngeal folds. This incompetent muscle tension mechanism is found in nearly every adult when the subject has consumed alcoholic beverages, sedatives, hypnotics, tranquilizers or sedating antihistamines before going to sleep. Hypothyroidism also contributes to poor muscle tone, snoring and apnoea, as do some neurological diseases and disorders such as myasthenia and muscular dystrophy, as well as cerebral palsy. Because inadequate muscle tone is often not very apparent on physical examination of an awake patient, this etiologic aspect is sometimes hard to detect. Sleep endoscopy with a flexible fibrescope [45, 48] will help out of this uncertainty and will give good impressions on sudden or persisting local limpness. A characteristic finding in these patients is redundant vertical folds in the tissue of the posterior and lateral wall of the pharynx, resembling in appearance more the interior of an intestine than an airway.

Smirne et al. [53] could demonstrate in a histomorphological analysis using muscle biopsies from pha-

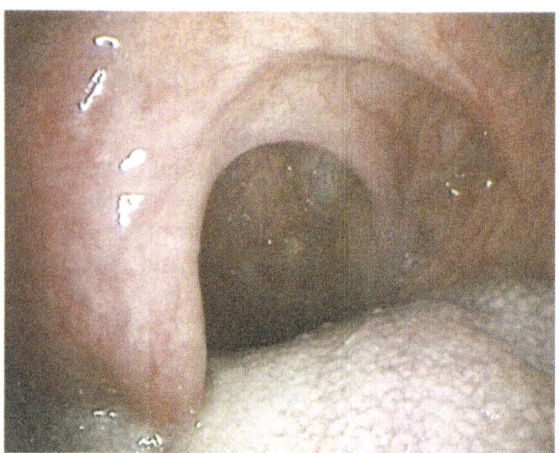

**Fig. 9.** Close-up look on the left side of the oral cavity and oropharynx in a heavy snorer. Typical aspect of webbing type palatal arch and bulky uvula. Ideal precondition for palatal vibrations and snoring.

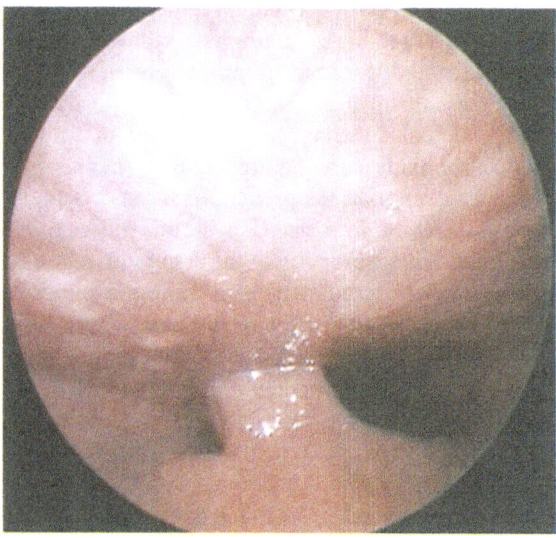

**Fig. 10.** The same patient (as in Fig. 8) in a videoendoscopic sleep examination exposing the backside of the bulky uvula, the drooping soft palate as the site of massive tissue vibrations with subtotal obstructions of the upper airway in the level of the naso-oropharynx.

ryngeal dilators that there were obviously fine structural changes of the muscle fibre composition. Snorers and OSA patients had a minor share of type-I and type-II b muscle fibres and a higher proportion of type-II a fibres.

## Maxillo-Mandibular and Maxillo-Facial Dysproportion (Skeletal Situation) Linked with Normal Local Soft Tissues and or with a Soft-Tissue Plus

Regarding the proportions of ideal airway anatomy, a receding chin may be unable to keep the tongue sufficiently forward. Retro- and micrognathia are predisposing factors for snoring in as far as the tongue is large in regard to the minor space available for it to occupy. In Down's syndrome, as well as in acromegaly, a remarkable total enlargement of the tongue is part of the phenotypic expression.

## Nasal Breathing Disorder – Recurrent or Persistent Blocked Nose

Severely impeded nasal airflow creates increased negative pressure during inspiration which draws together the flaccid tissues in the collapsible part of the airway, where they vibrate and cause snoring sounds. This correlation explains the very common observation that healthy non-snorers become snorers during periods of common cold or allergy attack. Nasal, septal and/or turbinate deformities, endonasal tumours, chronic and polypous rhinosinusitis, as well as nasal allergy or hyperreactivity, are therefore also possible etiologic factors for snoring. In every case, they have to be treated first before any kind of surgical procedure to the effector organs of snoring is drawn into consideration.

## Snoring and Obstructive Sleep Apnoea – Impact on and Consequence to Life

Snoring has long been ignored by most of the medical community and was addressed as a purely social problem primarily disruptive to family life. Snorers may be socially ostracized by roommates or even housemates; but it has also a medical implication and consequence. Chronic snorers very often also express cosymptoms like restless sleep, headaches or excessive fatigue in the morning. They may also suffer from daytime listlessness and hypersomnolence [46, 61, 62].

Heavy snorers run a higher risk of developing hypertension, and suffer from angina pectoris as well as having a higher probability of brain infarction and stroke. The most advanced stage of snoring is obstructive sleep apnoea, which causes profound cardiac, pulmonary and behavioural problems. Around 2% of adult women and 4% of adult men are supposed to be affected by this condition. Suffering from OSAS, the patients are not just simply snoring. A better description is "snoring" and "choking" as well, a frightening struggle to breathe while asleep. Whereas snoring is produced by partial obstruction of the airway, apnoea means total obstruction. It interrupts the loud snoring with episodes of silence during which the patient struggles with unsuccessful respiratory effort (obstructive apnoea). After a couple of seconds a deep loud "resuscitative snort" occurs as the patient partially or totally awakes, forces open the airways and resumes successful breathing. These moments are often characterized by gross and harsh body motions such as kicking or flailing of the arms or a body spasm. Sometimes the half-awake person rises up in bed or even falls out of it entirely. These snorts and compulsory body motions are quite sufficient to drive the bed partner to a different sleeping quarter.

Short occasional obstructive breathing events have no pathological implication and are quite common in the normal adult population, but it becomes critical to health when apnoea episodes last over 10 s each and occur seven to ten times per hour in average; 35% of habitual snorers are supposed to have significant apnoea.

However, not in every event of obstructive breathing while asleep is the airflow totally stopped. In those events in which at least some air squeaks through and such an airflow is reduced to 30% of the regular amount, the event is termed hypopnoea. An alternative definition is reduction of respiratory flow that is associated with marked oxygen desaturation and/or some degree of sleep arousal. To some extent, fast repetitive or continuing hypopnoea can be more dangerous than full apnoeic episodes, and because the consequences of hypopnoeas are rather the same as of apneas, the frequency of each should be added together in assessing the severity of the disorder.

Summing up the incidence of apnoea (A) and hypopnoea (H) per hour of regular sleep makes the apnoea-hypopnoea index (AHI or – to use another term – the respiratory disturbance index (RDI).

Augmented ventilatory effort in reaction to increased internal airway resistance results in repetitive arousals from sleep. Usually, these are partial arousals shifting the patient from deeper stages of sleep into lighter one, with the consequence that the amount of restorative sleep periods is not enough. This kind of deprivation keeps the patient sleepy during waking hours. Reduced mental and body fitness as well as decline of business performance are found. In addition, irritability and personality changes such as emotional depression are common consequences of sleep deprivation.

Whereas in adults symptoms like hypersomnolence with a reduced capability to take active part in their environment predominate, just the opposite can be found in children, who will make themselves un-

pleasantly conspicuous by hyperactivity and antisocial behaviour.

Chronic nocturnal hypoxaemia has some predictable cardiovascular consequences. Hypoventilation leads in every second patient to clinically relevant pulmonary hypertension, then to increased cardiac workload and to systemic hypertension. Blood oxygen desaturation and increased carbon dioxide levels may also frequently lead to cardiac arrhythmias and advance the risk of myocardial infarction.

Repetitive hypoxaemia in apnoeics also has an impact on general intellectual deterioration such as impairment of verbal fluency, attention, memory and executive functions.

## Diagnosis of Snoring and OSAS

The key point in diagnosing snoring and OSAS patients is the close interdisciplinary collaboration between:

- the "somnologist" (normally an internal medicine/cardio-pulmonary specialist),
- the otorhinolaryngologist, head and neck surgeon,
- the maxillofacial surgeon,
- the dentist and
- the neurologist.

In evaluating a patient with snoring or obstructive sleep apnoea (OSA), it is necessary to obtain the history from the patient, but also from his bed partner or family. Predisposing factors, e.g. sleeping on the back, alcohol, sedatives, hypnotics, daily eating and drinking habits, increase in body weight, active and/or passive cigarette smoking, nasal allergy and nasal obstructions should be inquired about.

The pathognomonic event is hidden from the patient's consciousness because it occurs while he or she is asleep. The patient himself is usually not aware of the loud snoring, irregular breathing patterns and other physical concomitant events such as thrashing about in bed. A detailed survey that explores the snorer's medical health condition, sleeping position, alcohol and sedative intake and weight changes is an important part of the history. Diagnosis of snoring is made primarily by history, much of which can be obtained from the patient's bed partner. The character and consistency of the snoring is reviewed to determine severity of the breathing disorder and the probability of an obstructive sleep apnoea.

Frequent episodes of breathing cessation followed by sudden arousals is a strong indicator of OSA. Diagnosis of OSA or UARS is confirmed by polysomnography. The extent of the disease in OSA can be classified according to the apnoea-hypopnoea index (AHI) and oxygen desaturation.

The Epworth score is a very simple and effective method of assessing sleepiness. It consists of eight questions related to the chance of dozing off during specific activity and situations, e.g. watching TV, sitting and reading. Each question is scored on a scale of 0, 1, 2 or 3: O indicates no dozing off, and 3 always dozing off. The sum of these eight questions can result in a maximum score of 24. A total score of less than 10 is considered not significant. The score of 10–12 is a borderline score, whereas a score of more than 12 is supposed to be significant. The Epworth scale correlates extremely well with daytime somnolence, which provides an important yardstick for advising aggressive management.

## Examination and Assessment of Snoring/OSAS

Physical examination should include complete evaluation of the nose, nasopharynx, oral cavity, oropharynx, hypopharynx and larynx.

There are no really specific physical findings that, if detected, give a clear-cut diagnosis of snoring or OSAS. Some common anatomical variations in the nose or the pharynx having minor pathological implication might create a composite picture of airway incompetence and arouse the examiner's suspicions. Among the physical findings and properties which characterize the pharynx in snoring and OSAS are increased collapsibility, increased compliance, increased resistance and decreased cross-sectional diameters.

The effectiveness of acting airway supporting muscles depends on their tonicity, the exact contractile coordination with the action of the diaphragm, the vector angles through which they operate and the linear distance through which they contract. Both of the latter two parameters are in close relation to the patient's individual anatomic situation. The amount of force required to keep the upper airway patent relates to the degree of negative pressure acting on the pharynx, the dimensions and the configuration of the pharynx, and to the degree of proper muscle action coordinated with the onset of inspiration. In particular, the following muscles have an active part in supporting pharyngeal airway patency: M. genioglossus, M. geniohyoideus, M. palatoglossus, M. palatopharyngeus, M. stylopharyngeus and M. tensor veli palatini, M. levator veli palatini.

Anatomic narrowing of the upper airway creates a situation whereby negative pharyngeal pressure generated in inspiration requires augmented activity from the dilating muscles to maintain patency of the upper airway.

When patients are awake, the contraction of this dilating musculature is usually adequate to cope with

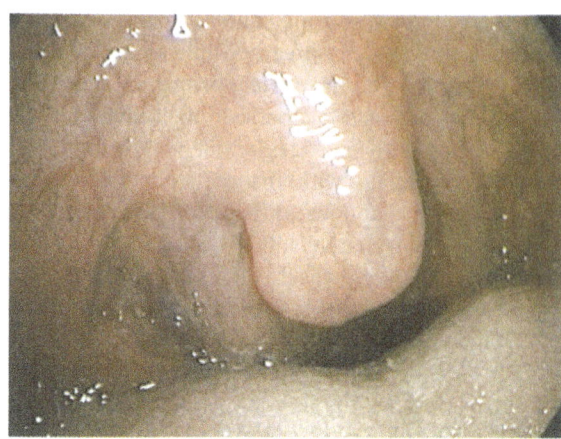

**Fig. 11.** Typical intraoral aspect of the soft palate in a male snorer: bulky uvula, redundant posterior tonsillar pillars.

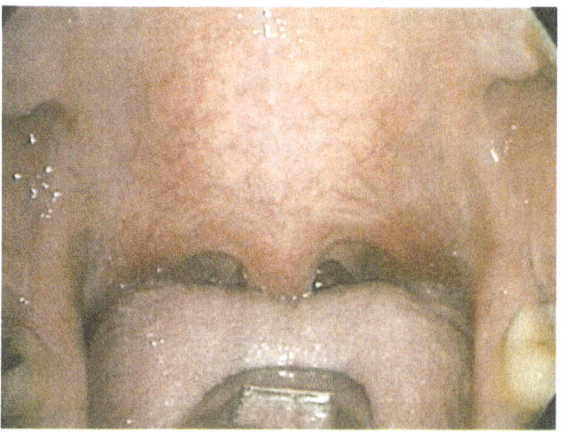

**Fig. 12.** Typical intraoral aspect in a male snorer: long partly invisible uvula, webbing and caudally drooping soft palate and posterior tonsillar pillars, relatively large tongue.

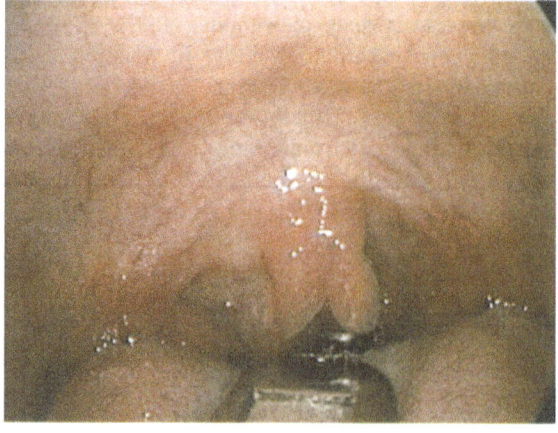

**Fig. 13.** In some cases, snoring is linked to anatomic variations of the oral cavity and the pharynx: uvula bifida in a 48-year-old heavy snorer.

any kind of anatomic narrowing, but during sleep, there is a normal decrease in the tonicity of these muscles and so snoring and sleep apnoea syndrome may arise [47]. An additional modulation of the muscular activity stage is known by the peripheral and central mechanisms of respiratory function. Therefore, not only neural control abnormalities of these muscles will interfere, but also alterations in neural regulation of breathing and reflex responses will lead to airway occlusion and hypopnoea or apnoea.

With regard to the underlying pathophysiology concerning muscle tone and neural regulations, physical examination must focus on the physical properties and spatial relationships of the pharyngeal airway, head and neck, as well as on the neuromuscular integrity of the airway and the mechanism of breathing control. Physical and functional narrowings from the nasal cavities down to the hypopharynx would definitely exacerbate the neurophysiologic tendency to aggravate plain snoring to an obstructive sleep apnoea syndrome in a patient.

Many specific pathologies – benign or malignant tumours as well as post-traumatic or inflammatory lesions and metabolic abnormalities – are proven either to induce or to intensify snoring and OSAS.

In children with no facial dysmorphia, craniofacial anomaly or neurologic abnormality, particularly hypertrophic tonsils and adenoids are frequently implicated as the underlying reason for snoring and OSAS (Fig. 2, 3).

In most of the adults, no specific focus of pathology can be identified. The most findings in "normal" individuals are multifactorial and more or less a combination of disproportionate anatomic relationships of the upper airway. Concerning the soft and hard tissue shape in head and neck there is a fluent transition from the individuals with normal structured dimensions and spatial relationships of musculature, mucosa and underlying craniofacial skeleton to well-defined craniofacial anomalies, resulting in snoring and OSAS.

Potential factors are micrognathia, maxillary hypoplasia, decreased size of the nasal capsule, choanal atresia and stenosis, decreased pharyngeal cicumference and distorted pharyngeal orientation due to abnormal angulation of the skull base. An abnormally acute angle of the skull base is regularly associated with a decreased anterior-posterior extension of the cranial base, a reduced anterior-posterior diameter of the bony pharynx, a diminished width of the bony pharynx (reduced interpterygoid distance) and a decreased anterior-posterior length of the nasal and oral airway. The associated soft tissue changes include diminished cross-sectional area and verticalization of the pharynx. In these cases, the palate is suspended straight down into the airway with the uvula extend-

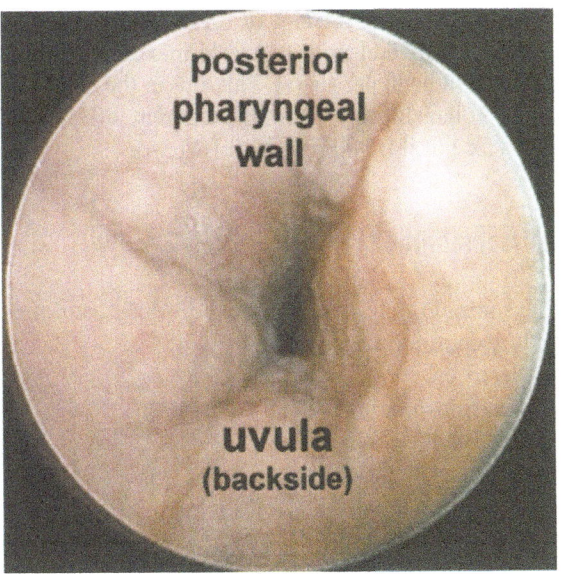

**Fig. 14.** Flexible fiberoptic naso-pharyngo-laryngoscopy of the patient performing the Mueller manoeuvre. Negative pressures are causing a pronounced pharyngeal narrowing of the airway – quite a good indication of the site of snoring.

ing to the inferior base of tongue. This anatomical situation causes an unusually long vertical portion of the posterior tongue, in apposition over an excessively long distance with the soft palate and posterior pharyngeal wall, predisposing to pharyngeal collapse.

If acute angulation of the skull base is linked with micrognathia, the long tubular lower pharynx would be even smaller in the anterior-posterior diameter, with an increased likelihood of pharyngeal collapse. In individuals where maxillary hypoplasia or choanal atresia are coexisting, the patient would be driven into the unfavourable dynamics of chronic mouth breathing, which would further jeopardize the already highly vulnerable airway. Add generalized muscular hypotonia, and the patient loses the one and only mechanism of protection against airway collapse while asleep. Besides clinical examination, the lateral X-ray image for cephalometry is helpful in these cases.

Special attention must be given also to retrognathia, macroglossia and obesity. There is evidence that the effect of obesity on snoring and OSA relates to local parapharyngeal fat deposits, but the interrelationship may be more complex.

Any diagnostic approach must take into account the specific craniofacial morphology, the contribution of superimposed pathologic space-occupying lesions, the neuromuscular status and the degree of obesity.

Flexible fiberoptic naso-pharyngo-laryngoscopy aids this examination as well as allows the performance of the Mueller manoeuvre. The Mueller manoeuvre consists of the attempt of breathing in against a closed mouth and nose to create maximal negative pressure in he upper airway. This aids in the detection of any collapsing site within the different levels of the pharynx, but it is no proof for it (Fig. 14).

## Polysomnography (PSG)

Polysomnography is a somnological-diagnostic procedure continuously recording
- electroencephalogram (EEG),
- electromyogram (EMG),
- electrooculogram (EOG),
- elctrocardiogram (ECG),
- nasal airflow,
- oral airflow,
- thoracic movement,
- abdominal movement,
- capillary blood oxygen saturation,
- body position,
- total sleep time and
  facultatively
- intra-oesophageal pressure,
- flexible sleep endoscopy

through a complete night of sleep.

By means of simultaneous analysing EEG, EMG and EOG, the cyclic repetition of the different sleep stages from level 1 to 4 and REM (rapid eye movement sleep) can be differentiated and monitored. Nasal and oral airflow and respiratory efforts (thoracic and abdominal movement) are used to check breathing activities.
- In cases of central apnoea [47], the recordings will show no airflow at nose or mouth and no respiratory effort,
- in cases with obstructive apnoea, no nasal or oral airflow will be detected despite large respiratory efforts.

Blood oxygen saturation monitors the degree of oxygen desaturation which possibly occurs during periods of hypopnoea or apnoea. Continuous ECG registration defines the cardiac arrhytmias corresponding to other parameters.

This procedure enables the examiner to differentiate between a plain snorer, a patient with UARS and a patient with OSA. These excellent sleep studies – but at least the screening test at the patient's home [54] – should be done for all patients prior to any kind of conservative or invasive treatment [63] and regularly in the follow-up period [60].

The following parameters gained by PSG indicate the severity of a pre-existing OSAS: Frequency of apnoeas (total lack of breathing) or hypopnoeas (significantly diminished nasal and oral airflow), duration of apnoeas and/or hypopnoeas, degree, duration and peak of oxygen desaturation, cardiac rate and rhythm abnormalities corresponding to apnoea/hypopnoea periods.

Other relevant aspects in assessing the patient by polysomnography are the degree of sleep architecture disruption caused by apnoea-induced arousals and the observations of the patient's behaviour during the night, thus providing a comprehensive analysis of a sleep study.

## Where Is the Site of Snoring, Where Is the Site of Upper Airway Collapse?

Today, the most widespread method to detect the level of snoring is flexible naso-pharyngo-laryngoscopy under sleep conditions and in different sleeping positions [63]. In cases where the patient is unable to fall asleep awaiting the flexible instrument to penetrate his nares, short acting narcotics such as propofol (1.5–2.5 mg $kg^{-1}$ BW) or midazolam are applied [45].

However, not all clinicians agree that this is a good method, particularly because it is not a natural sleep being induced. Quinn et al. [48] remark on the fact that hypnotics not only alter the central arousal level, but also interact with the reflectory processes of the upper airway patency, as well interacting with the peripheral muscle tone.

Another important aspect must be mentioned: the site where the snoring sounds originate may differ completely from the site of the upper airway collapse in OSA. Collapse, by the way, is a very silent process.

Acoustic frequency analysis (AFA) is been increasingly evaluated and it seems a promising method but yet not readily available or validated. Once OSAS is excluded and the source of the snoring accurately located, its management can be planned. The majority of the plain snorers will have the source producing snoring noise at palatal level. According to our last consecutive 200 patients with plain habitual snoring, we found in 39% mixed forms, in 47% the palatal area, in 9% the tongue or base of tongue and in 5% the epiglottis as the site where snoring sounds were produced. If once confirmed, and if the nasal airway is wide or surgically corrected, the goal of palatal surgery will be to reshape the anatomical situation, to reduce redundant soft tissue and to transform a floppy palate into a stiffer one, inducing scar formation and interstitial fibrosis of the soft palate so that it does not vibrate as much as before.

If an airway narrowing soft tissue plus different from the soft palate is to be surgically addressed,
- tissue-reducing procedures such as
  - laser-assisted lingual tonsillectomy,

- laser midline glossectomy and lingualplasty [13],
- excision of redundant aryepioglottic folds to
- partial epiglottectomy are performed.

If skeletal disproportions are in the main focus of surgical therapy,

- mandibular advancement with bilateral sagittal splitting osteotomy (in cases of mandibular retrognathia or micrognathia),
- maxillo-mandibular advancement with vertical osteotomy (in cases of maxillo-mandibular retrognathia),
- inferosagittal osteotomy of the mandible with hyoid myotomy suspension are performed.

## Classic Treatment

### Non-Surgical Management

The treatment of snoring classically starts with the elimination or the reduction of causative or exacerbating factors.

Various non-surgical methods have been utilized to alleviate snoring and/or OSA. There are today over 600 registered devices as cures for snoring commercially available. The first therapy for snoring and sleep apnoea may arguably be an elbow in the ribs to induce the bed partner to assume a lateral recumbent position. Patients and bed partners commonly describe greater snoring and more frequent apnoeas during sleep in the supine position. It was postulated that the supine position facilitated gravity-associated relapse of the tongue and the tongue base against the posterior pharyngeal wall. Some antisnoring devices are variations on the idea of taping a marble on the snorer's back so as to force time to sleep on his side. The lateral recumbent sleeping position can be accomplished, i.e. by attaching a "sleep ball" or a sleep sock to the back of the garment. Rolling onto the ball is so uncomfortable for the patient that he or she immediately turns to a lateral sleeping position in order to eliminate the pain or discomfort created by this hard thin pressure on the back. This is only one example, but the list of so-called anti-snoring devices seems to be endless. Chin and head straps, whiplash-type collars and oral devices have been promoted, as have a number of electrical devices which produce unpleasant stimuli when the patient starts snoring. None of these would-be remedies has gained wide acceptance for their efficiency. Another recommendation of postural treatment is to sleep on the stomach.

The avoidance of centrally sedating drugs and medicaments as well as alcoholic beverages prior to sleep, as well as weight reduction using strict dietary measures and daily exercise, can eliminate mild snoring. In particular, it is apparent that OSA and nocturnal haemoglobin oxygen desaturation in obese patients may be improved by weight reduction that normally leads to an increase of the lung volume.

Prosthetic and tongue-retaining devices may be effective in more than every second patient but have a poor compliance rate. Nasal allergy and/or hyperreactivity should be treated when present. Exposure to upper airway irritants such as smoke and fumes must be eliminated.

Bilevel positive airway pressure and continuous positive airway pressure (CPAP) require the patient to wear a bulky nasal device attached to a bedside positive pressure-generating machine that helps to maintain upper airway patency. However, this is the first and highly effective treatment of choice for OSA patients. It is non-invasive and can be continued into the postoperative period until wound healing is completely finished and a follow-up polysomnogram can be performed to evaluate the outcome after surgery.

Surgical procedures to address snoring entail certain risks and discomfort. It is therefore prudent to attempt medical intervention or postural and behavioural modification under appropriate circumstances. Since both medical and behavioural management requires prolonged follow-up and/or adherence to a restrictive lifestyle, not all patients are able to comply. Additionally, many patients do not respond to conservative treatment measures.

### Surgical Management

Every surgical treatment modality should be lacking in side effects, should be reversible and should be tailored to the individual findings and needs of the patient.

The treating doctor should critically discuss potential surgical procedures and the chances for positive surgical result and outcome with his patient, who mostly feels under tremendous pressure to do something against his nightly noise production.

When nasal symptoms are the primary complaint, success of nasal surgery alone may be predicted preoperatively by the nightly use of a long-acting nasal decongestant spray, e.g. xylometazoline, sometimes combined with an extranasal dilating device (i.e. Breathe Right®, AirPlus®) or an intranasal dilator (i.e. Nozovent).

All epipharyngeal pathologies also have to be diagnosed as well as treated prior to all the intra- and transoral procedures.

Focusing on the soft palate and the oropharynx, the anatomical region which is usually looked upon as the effector organ of snoring sounds, several different techniques have been used in the surgical management of snoring and or OSAS. The first surgical procedure was inaugurated by Ikematsu [29]. In 1964 he published his first large series of 152 patients with 82% subjective relief from snoring [14]. In 1981, Fujita et al. [12] introduced uvulopalatopharyngoplasty (UPPP), which represents a maximal removal of the soft palate and the tonsils including the entire uvula. Simmons et al. [51, 52] adapted, modified and popularized the technique in the US. Since first described, there have been a number of papers reporting the results and complications of the various methods [16, 17]. The published literature describes success rates in excess of 70% in relieving the symptoms of snoring. Potential complications of uvulopalatopharyngoplasty [2] may lead to reluctance on the part of both the patient and the surgeon to use this procedure for the treatment of plain snoring. General anaesthesia is usually required for the traditional technique. Intubation is often difficult and complicated in this patient population due to the fact that there is a large prevalence of short full necks and relatively thick tongues. In patients with OSA the incidence might be even higher. Postoperative complications sometimes include severe haemorrhage (2–5%), postoperative nasal regurgitation (20–60%), permanent velopharyngeal insufficiency (0.5%) and nasopharyngeal stenosis due to circumferential cicatrization [39].

A long-term minor complication can be voice or resonance changes (open rhinolalia) and a foreign body sensation. The latter is attributed to the loss of the uvula, which sweeps the posterior pharyngeal wall clear of mucosal secretion while swallowing. There also might be a problem in the articulation of the "rolling r" as well as "ch" speaking foreign languages (i.e. German, Spanish, Farsi, Russian and Arab).

Laser-assisted uvulopalatoplasty (LAUP) is a technique developed by Kamami in France in the late 1980s. It was introduced in the United States in 1992 as a treatment for snoring without apnoea. The procedure is designed to correct airway obstruction and soft tissue vibration at the level of the soft palate, by reducing and stiffening the tissues of velum and uvula.

## Laser-Assisted Uvulopalatoplasty (LAUP)

The operation of laser-assisted uvulopalatoplasty (LAUP) was developed as a modification and an alternative to traditional uvulopalatopharyngoplasty for the treatment of snoring due to palatal flutter. Various methods of performing the procedure as well as extending the indication for the treatment of mild, moderate and even severe OSA have been published [1, 59]. Kamami [33] first described the procedure in 1990 and called it laser vaporization of the palato-

pharynx (LVPP) performed under local anaesthesia, using a $CO_2$ laser via a designed hand piece. The subsequent fibrosis combined with the decrease in bulk of the soft palate has been shown to be effective in reducing the snoring or stopping it completely [34, 35, 36].

Laser assisted uvulopalatoplasty (LAUP), as this procedure was named when it was adopted in the US, has relatively few long-term complications when compared to the more extensive uvulopalatopharyngoplasty. Krespi et al. described a similar technique in 1994 [37]. They advocate the $CO_2$ laser surgery in patients with sleep-related breathing disorders [38], undertaken in sequential sessions just as Kamami proposed, to reach an adequate result with minimal morbidity such as velopharyngeal incompetence, nasal regurgitation or voice change.

Early in 1994, Ellis described a different technique, suggesting that snoring caused by palatal flutter could be controlled by stiffening the soft palate. The handheld Nd:YAG laser, having, due to its wavelength, a higher coagulation and scarring ratio was used to remove a longitudinal strip of mucosa from the soft palate [6].

## Contra-Indications

Absolute contra-indications to LAUP in an office setting are relatively few. They include: significant sleep apnoea (AHI greater than 30), uncontrolled hypertenison, trismus, cleft palate, pre-existing velopharyngeal insufficiency, imbalanced cardiopulmonary diseases, uncooperative patients, and an anatomical source of snoring other than the oropharynx, such as gross nasal or maxillo-facial problems.

Caution should be exercised in patients who use their voice professionally or play wind instruments. Linguistic constraints for certain languages that use of soft palate or the uvula extensively, such as Arabic, Russian, Hebrew and Farsi, may also be a consideration. Patients with allergies to local anaesthetics and a hyperactive gag reflex should be treated under general anaesthesia. In OSAS, it may be indicated if indeed the syndrome is in mild form. In severe OSAS, LAUP may help induce tolerance and reduction of pressure in CPAP management, but normally these patients are no ideal candidates for palatal surgery. In these cases, the decision for surgical treatment must be exactly discussed with the involved somnologist and the pneumological colleague who normally checks the n-CPAP device of this patient.

Doing palatal resections in OSA patients, the surgeon has to spare soft tissue. Otherwise, the n-CPAP air stream, while breathing in, will leave through the mouth at once without having penetrated the lungs.

## Choice of Laser Systems

Actually, the choice of the laser system for LAUP is linked to the surgeon's philosophy of how to change the morphology and the function of the fluttering soft palate, and his experience in the use of different laser systems and parameter settings.

Roughly speaking, almost every laser system can be used for LAUP.

Regarding the applicability of laser photons for this surgical procedure, one has to decide between free beam laser transmission (enhanced by hand pieces and scanner systems) and the fibre-guided laser systems. Another consideration would be the availability of a particular laser within the hospital or clinic setting and the feasibility of outpatient or day-case surgery.

In our hands, the $CO_2$ laser (10 600 nm) is the system of first choice, although it is not yet fibre-transmissible. Two modalities treating the fluttering soft palate are possible. The first is the resection technique, where the laser beam is predominantly used as a knife with a little coagulation zone around the cut. Here, the goal is to resect the redundant soft tissue and reshape the palate with best possible muscle preservation. The second modality is the vaporization technique, easy to use, where precise and rapid superficial debulking and vaporization of tissue, layer by layer, is best performed with a laser scanner system.

After the $CO_2$ laser reshaping of the lower parts of the palate is finished, additional stiffening of the palatal area is created by either an anterior rectangular mucosa resection cranial to the uvula base or interstitial laser or high-frequency coagulation of the soft palate.

A genuine alternative to the $CO_2$ laser is one of the fibre-transmissible laser systems, well known from functional endoscopic endonasal laser surgery (FEELS): diode laser (805–980 nm), Nd:YAG laser (1064 nm) or KTP laser (532 nm). They thus provide a useful tactile feedback. However, their application – although best performed in tissue-contact mode – may create a deeper coagulation zone and a period of slightly intensified postsurgical pain, and induces longer wound healing. The diode laser and the KTP laser are compact machines suitable for portability and trouble-free office procedures in diverse locations. The diode laser is marketed at a much lower price. Nd:YAG should be used only in contact mode with preblackened fibre, as the wavelength scatters widely within the tissue in free beam mode. Ho:YAG is a pulsed laser with splattering of the tissues. It requires dedicated combined suction-fibre applicator.

All the laser systems mentioned, including the $CO_2$ laser, seal off small blood vessels, offer a good intra-operative visibility, an excellent efficiency ratio and a very low rate of post-operative bleeding. The LAUP is thus also suited for selected patients in anticoagulation therapy. In these patients some authors prefer conventional surgery, radiofrequency-assisted surgery (the so-called RAUP) or electrosurgery (ES) to laser surgery.

## Surgical Procedure of LAUP Using a $CO_2$ Laser

The procedure is performed with the patient sitting in an upright or semirecumbent position in an outpatient operation facility or in a otorhinolaryngologic examination chair. Normally, the patient receives no sedatives or systemic analgetics before.

A topical anaesthetic such as Xylocaine is sprayed into the posterior oral cavity over the soft palate, the tonsils and the uvula. After some minutes, 4–10 ml Lidocaine 1–2% with an admixture of Epinephrine 1:100 000 or 1:200 000 is injected via four to six spots into the soft palate, the uvula and the pillars on both sides. If laser-assisted vaporization of the tonsils and the tonsillar pillars is to be performed, injection is increased into the junction of the anterior and posterior pillars bilaterally and to the lateral pharyngeal walls.

The laser surgical treatment itself is started after 15 min, to allow adequate anaesthetic effect and sufficient vasoconstriction.

The $CO_2$ laser is preferred due to its excellent laser-tissue interactions and technical and practical ease of use. Even though the carbon dioxide laser is not the best coagulating laser available, it has adequate coagulation for the diameter of vessels encountered in this

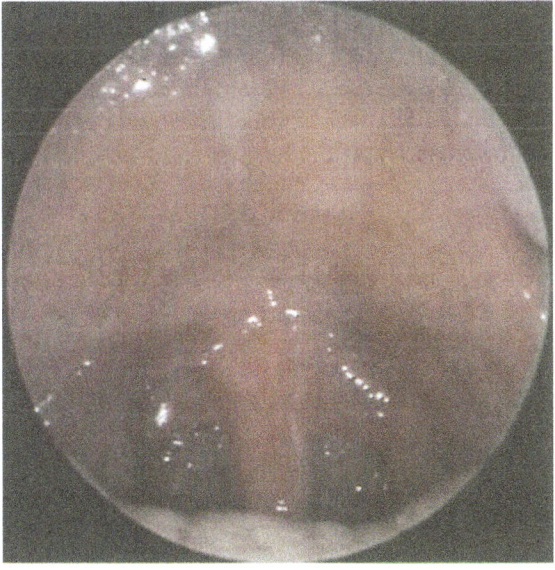

**Fig. 15.** Intra oral view of a pre-operative situs of a snorer: web type and drooping soft palate with long uvula.

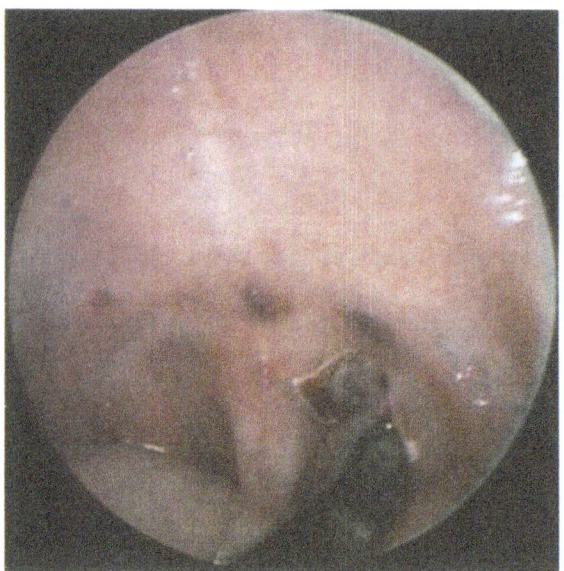

**Fig. 16.** LAUP – step 1: $CO_2$ laser vaporization of the drooping soft palatal arch with backstop hand piece and laser scanner system in local anaesthesia

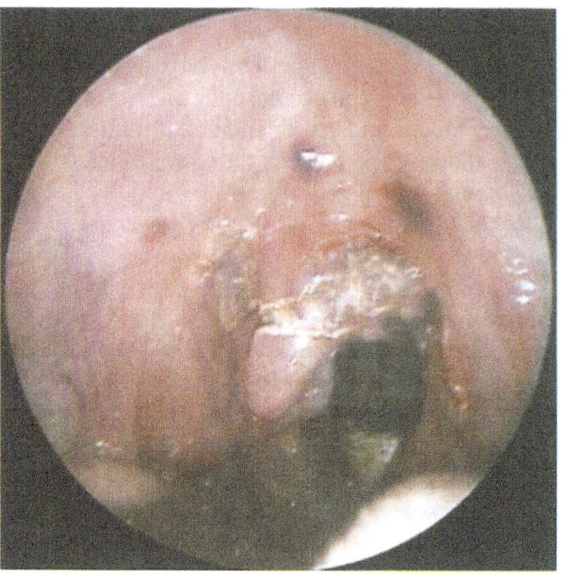

**Fig. 17.** LAUP – step 2: $CO_2$ laser-assisted size reduction and re-shaping of the uvula. If large soft tissue masses, such as a long slim uvula, have to be reduced, laser-assisted sharp resection with a focused laser beam can be done. If more superficially located redundant tissue masses are to be reduced, layer-by-layer vaporization with a scanner system is the best method.

procedure. The following description outlines LAUP performance with the $CO_2$ laser.

The patient and staff are equipped with protective goggles, and laser safety rules are followed. Using – as we do – a scanner system (i.e. Surgitouch or the Swift-lase flash scanner by the Lumenis Company (formerly ESC-Sharplan) attached to the $CO_2$ laser for careful vaporization of palatal soft tissues, a power setting of 16–24 W in the continuous mode is applied. The diameter of the scanner area is normally 2–3 mm. The tongue is retracted inferiorly with a specially designed tongue depressor, which is fitted with an integrated plume evacuation channel.

Now the laser-transmitting hand piece equipped with a backstop device is introduced and focused onto the palatal region where soft tissue resection and/or vaporization is to be done.

As described above, either sharp cut resection of an elongated palatal arch is performed with 5–15 W focused laser power or vaporization technique using the scanner system is applied to superficially debulking the surplus of palatal soft tissue.

Afterwards in most cases, through and through full thickness vertical trenches measuring 1.0–1.5 cm are fashioned on the free edge of the soft palate on either side of the uvula. These trenches are created using a focused beam and also the backstop hand piece. The patient is asked to inhale normally, and then the laser is activated during slow exhalation in order to avoid the patient ingesting the plume.

Shortening and thinning of the uvula are easily carried out with the backstop hand piece in combina-

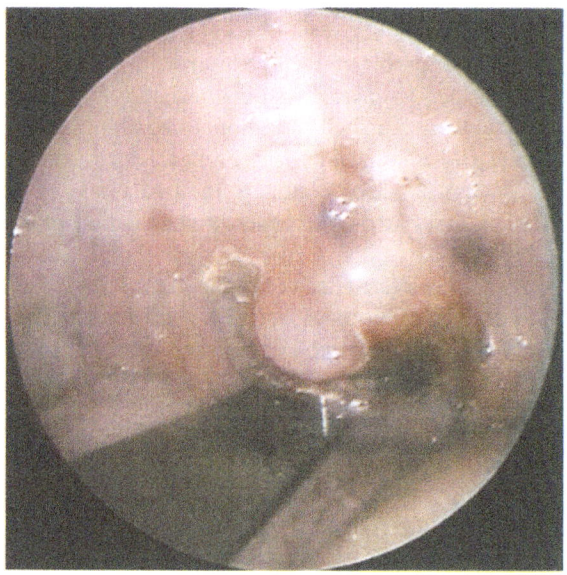

**Fig. 18.** LAUP – step 3: After the reshaping of the palatal arches and the gross volume reduction of the uvula is performed, $CO_2$ laser-assisted fine work at the uvula follows. In this anatomic area the surgeon should be aware that some blood vessels can be encountered which must be sealed either by defocused laser irradiation or electrocauterization.

tion with the Lumenis scanner. The uvula is reduced, if necessary, to around a half to a third of its original size by cutting it or carefully coring it in a cephalic direction. In every case, the surgeon has to keep in mind

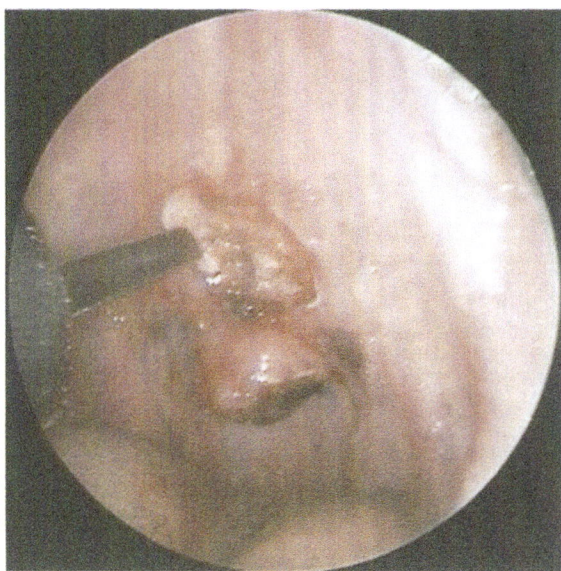

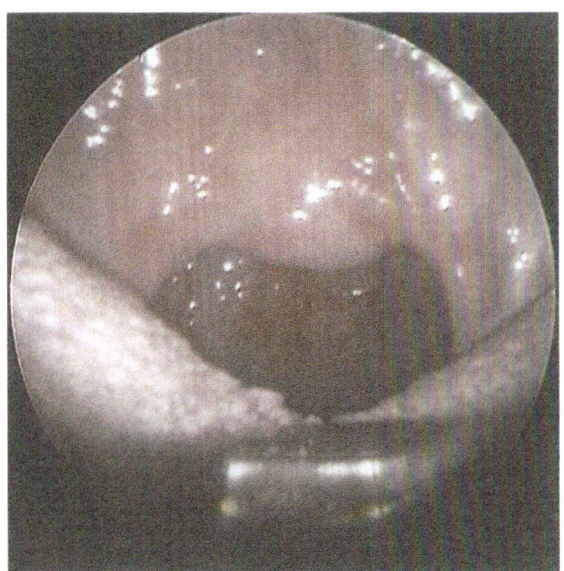

**Fig. 19.** LAUP – step 4: Having finished $CO_2$ laser-assisted fine work at the uvula, an additional vaporization of mucosa and submucosal layers of soft tissue right above the root of the uvula is performed best with the laser scanner system. Tissue removal ends when palatal muscles are visible. They should not be touched. Wound healing in this superficial resection area will lead to a stiffening and scarification, reducing vibration potential of the palate.

**Fig. 21.** Intra oral aspect of the situs 2 years after $CO_2$ laser-assisted LAUP. The patient reported that snoring was gone 6 to 8 weeks after LAUP. This is a routine follow-up 2 years after laser surgery. The volume reduction was moderate enough for the sweeping function of the uvula to be still preserved. Scar formation created a significantly reduced vibration potential of the palatal tissues.

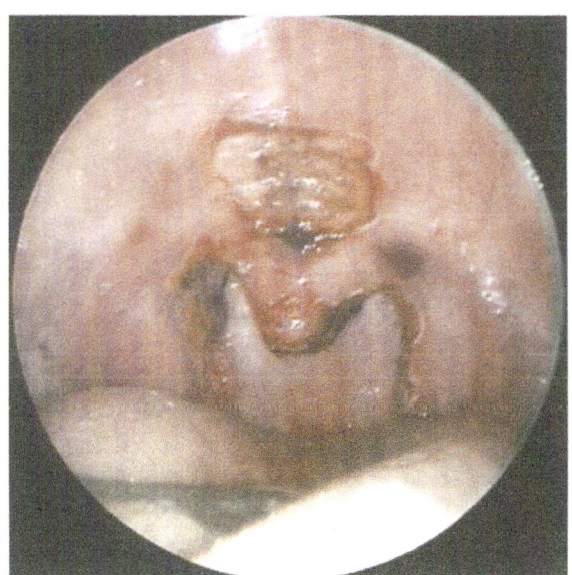

**Fig. 20.** Typical intraoral aspect of the situs at the end of the $CO_2$ laser-assisted LAUP: Uvula, soft palate and tonsillar pillars are size-reduced and reshaped. No bleeding occurred during and after surgery.

the position of the "dimple point", which must not even be reached or cranially exceeded in palatal resection. Overall, the goal is to reduce the length and reshape the soft palate and uvula. Care must be taken not to burn the mucosal covering of the soft palate

and the uvula excessively. The uvula is shortened by ablating the muscle from within, creating a "fish-mouth" appearance, while preserving the mucosa of the base of the uvula on the nasal and oral surfaces. Light bleeding during surgery can occur, which is easily controlled by applying bipolar coagulation forceps.

In addition, we vaporize with the $CO_2$ laser scanner system a rectangular mucosal area with a transverse diameter of 2–3 cm and a vertical height of 1–2 cm.

This strictly mucosal resection is median located around 0.5–1.5 cm cranial to the base of the uvula and leaves the palatal muscles untouched. After 6 weeks of scar formation, this additional procedure will help in stiffening the soft palate.

## Surgical Procedure of LAUP Using the Fibre-Guided Diode Laser

As a good alternative to the $CO_2$ laser, especially in those patients with a high laxity of the soft palate, we use the 940-nm diode laser (medilas D fibertom Skin-Pulse S, by Dornier Medizinlaser, Germany) for the LAUP procedure. Its laser tissue interactions allow a fibre-transmissible soft (and hard) tissue resection as well as thermically induced scar formation by suitable interstitial soft palate coagulation. The local anaesthesia preceding the treatment is exactly the same as

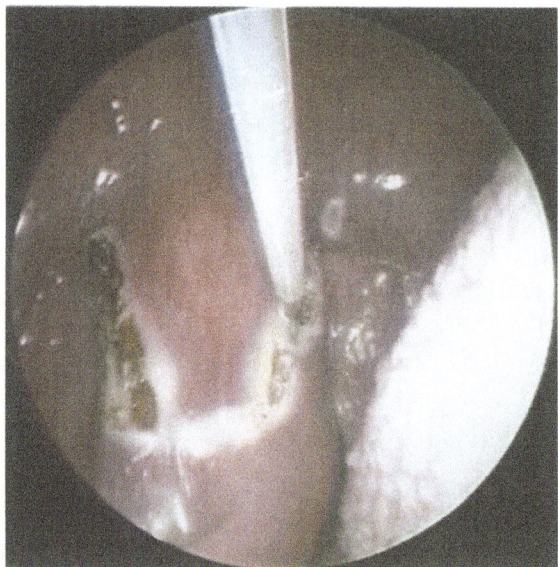

**Fig. 22.** LAUP with the fibre-guided Diode laser-step 1: Diode laser photons are transmitted with a flexible fibre onto the surgical target. The fibre is carried in an adjustable hand piece or an endoscopic device (known from FEELS procedures).

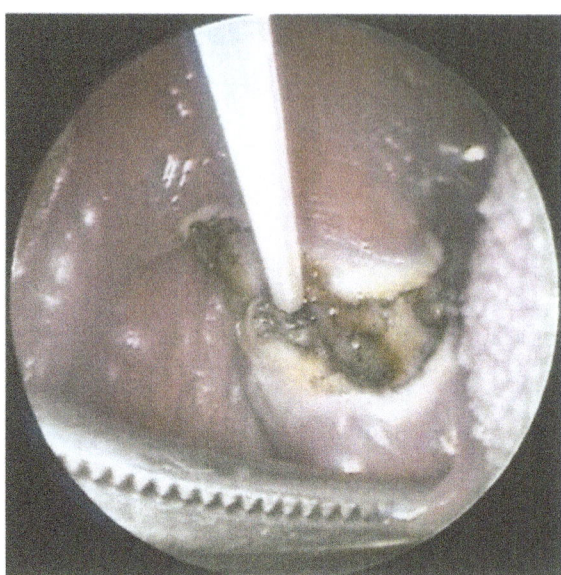

**Fig. 24.** LAUP with the fibre-guided diode laser-step 3: The volume and size reduction leaves behind a small uvula with moderately coagulated wound margins. The coagulation zone induces a scar formation which leads to stiffening and reinforcement of the soft tissue.

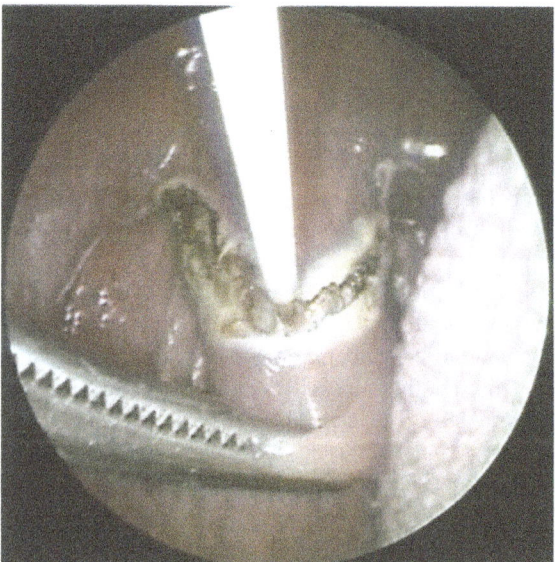

**Fig. 23.** LAUP with the fibre-guided Diode laser-step 2: After the markings, the diode laser-assisted resection of the uvula starts in a contact mode. With the help of a clamp the distal end is grasped.

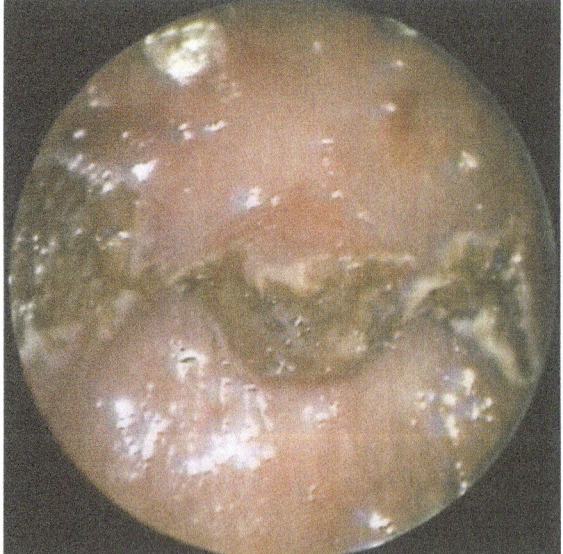

**Fig. 25.** Intra oral aspect of the situs at the end of the diode laser assisted LAUP: Uvula, soft palate and tonsillar pillars are size reduced and reshaped. The wound margins will re-epithelialize within 3 weeks.

when using the $CO_2$ laser. The freshly broken laser fibre with an outside diameter of 600 μm is introduced into a fibre handle in order to position it very accurately onto the surgical target.

The first step of this surgery to do symmetrical markings to the uvula and the soft palate by short laser pulses.

Using either the continuous mode (with or without fibertom mode) applying low laser powers (4–10 W) or the pulsed mode with high laser powers the surgeon is able to fit his surgical tissue effect of resecting with or without collateral coagulation as desired.

The volume and size reduction leaves behind a small uvula with moderately coagulated wound mar-

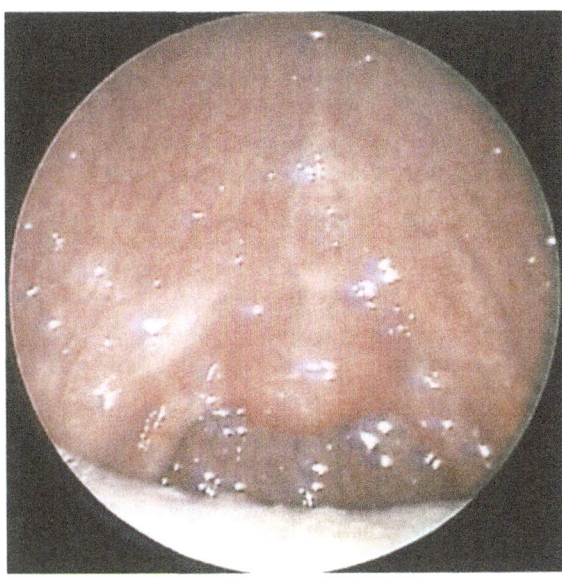

**Fig. 26.** Follow-up after diode laser-assisted LAUP: The postoperative situation 16 months after the surgery shows a re-inforced soft palate and a small uvula.

gins. The coagulation zone induces a scar formation which leads to stiffening and reinforcement of the soft tissue

The extent of resection may be somewhat lesser compared to with the $CO_2$ laser, because the scar formation and cranio-lateral wound edge retraction using the diode laser may be slightly more intensified.

For stiffening of the lax soft palate an interstitial coagulation is additionally performed in many cases. The fibre has to puncture the soft palate with the laser activated. Then, the laser fibre is pushed forward in the layer of the palatal muscles in cranio-caudal direction. Thus three to four stripe-like submucosal intramuscular suspensory scars are created by low power (3–5 W) diode laser application to help increase the stiffness of the soft palate.

### Postoperative Instruction

Patients may resume regular activities the day after surgery. A soft, bland diet, with avoidance of citrus products, is strongly recommended. Aggressive hydration, humidification and steam inhalation are emphasized. Mucous membrane dehydration and infection are thought to be important sources of postoperative pain. Systemic analgetics and antiphlogistics as well as xylocain gel are used to relieve pain. Gargling with sage tea and saline solution and/or non-alcoholic mouth washes is strongly recommended. The duration of the need for analgetics varies according to each patient's tolerance and the extent of the treat-

ment. Postoperative infection prophylaxis by systemic antibiotics is recommended for every patient.

Typically, LAUP procedure – if performed with caution – requires one to four treatments spaced at a minimum of 6–8 weeks apart. Elapsed time between the procedures allows proper healing of the soft palate mucosa. The endpoint of LAUP is when significant reduction or elimination of snoring has occurred. Confirmation is obtained by patient or partner history and/or by the inability of the patient to perform voluntary snoring.

### Outcome

Krespi [38] reviewed 280 patients who underwent LAUP in the office, with a 3-month to 2-year follow-up. He reported elimination of snoring in 84% of the patients, and reduction of snoring in 7%. Carenfelt [1] reported 85% total or near total elimination of snoring during a short-term follow-up of 60 patients. In a review of 741 patients, with a maximal follow-up of 5 years, Kamami (1994) reported 69.8% cure or significant reduction of snoring. Ellis (1994) published the results of laser-assisted palatoplasty with 3–6 month follow-up. The surgical technique described by Ellis was slightly different, in that only a central longitudinal strip of mucosa was removed from the surface of the soft palate. This resulted in 85% elimination or significant reduction of snoring.

Prolonged follow-up revealed more modest success for this operation. Walker et al. [60] and Michelson and Hollrah [46] presented long-term results of LAUP treatment for snoring. Kamami [34–36] found that, 8 years after treatment, 37% of patients initially cured had a recurrence of snoring. This was mainly related to weight gain or to a nasal obstruction (septal deviation, turbinate hypertrophy) which was not surgically treated prior to LAUP. In most cases, the appropriate treatment of the nasal obstruction cured or improved the symptom of snoring.

### Complications

Reported complications for LAUP are rare. Intra operative bleeding can occur in around 3% of the patients. Bleeding is usually from the apex of the palate trench incision and is usually stopped by bipolar coagulation.

Moderate to severe pain is the major side effect of the procedure. Pain intensity reaches its peak 4 to 7 days postoperatively, with complete relief of symptoms in 2 to 3 weeks. Pain is usually well controlled with hydration, anaesthetic gel and oral analgesics. Most patients report some degree of weight loss, typically less than 10 kg over the course of treatment. The superficial mucosal vaporization areas are completely covered after fibrinous tissue has fallen off after about 12 days. Definite healing of the palate occurs by scar formation in 3–6 weeks following the procedure. Krespi [38] reported two vasovagal episodes following injection of the local anaesthetic in a review of 280 patients.

Velopharyngeal insufficiency, either temporary or permanent, is a very rare complication, probably due to the graded surgical approach. Nasopharyngeal stenosis has been encountered only once in our department, because, by using the special hand piece with backstop to make the palatal incisions and vaporizations, the nasopharyngeal mucosa is protected from injury. Approximately 40% of patients may complain of "scratchy" or "dry mucous" sensation in the throat. This is usually self-limited and resolves within 2 months.

## Discussion

Laser-assisted uvulopalatoplasty (LAUP) is an effective method for treating patients with loud, habitual snoring, upper airway resistance syndrome and mild obstructive sleep apnoea. LAUP offers several advantages over the classical uvulopalatropharyngoplasty, including reduced cost, decreased intra- and postoperative morbidity, diminished postoperative pain and abbreviated reconvalescence period, as well as avoidance of general anaesthesia.

LAUP as an office or outpatient procedure performed under local anaesthesia has proved to be a safe and effective method of alleviating bothersome snoring. The surgery is undertaken in stages to allow titration of tissue removal and stiffening of the palate by scar induction with only minimal risk of overcorrection. Polysomnography prior to surgery is indicated in every patient who is potentially at risk for OSA, in regarding to his/her history.

Patient selection requires a careful review of the medical history and a thorough physical evaluation. The nose, tongue base and hypopharynx should be ruled out as the primary site of airway obstruction. Therefore, different non-surgical as well as surgical strategies have to be drawn into consideration until the single patient's differential diagnosis of snoring and OSA has found a final medical conclusion and the individually fitted management can be discussed with each single patient. LAUP, when performed in properly selected candidates, can result in excellent clinical outcome and patient satisfaction.

## References

[1] CARENFELT C (1991) Laser uvulopalatoplasty in treatment of habitual snoring. Ann Otol Rhinol Laryngol 100 (6): 451–454

[2] CARENFELT C, HARALDSSON P O (1993) Frequency of complications after uvulopalatopharyngoplasty. Lancet 13; 341 (8842): 437

[3] CIRGINOTTA F, D'ALESSANDRO R, PARTINEN M, ZUCCONI M, CRISTINA E, GERARDI R, CACCIATORE FM, LUGARESI E (1989) Prevalence of every-night snoring and obstructive sleep apnoeas among 30-69-year-old men in Bologna, Italy. Acta Neurol Scand 79 (5): 366–372

[4] CIRIGNOTTA F, LUGARESI E (1980) Some cineradiographic aspects of snoring and obstructive apnoeas. Sleep 3 (3–4): 225–226

[5] EDSTRÖM L, LARRSON H, LARSSON l (1992) Neurogenic effects on the palatopharyngeal muscle in patients with ostructive sleep apnoea: a muscle biopsy study. J Neurol Neurosurg Psych 55 (10): 916–920

[6] ELLIS PD (1994) laser palatoplasty for snoring due to palatal flutter: a further report. Clin Otolaryngol 19 (4): 350–351

[7] Ellis PD (1994) Snoring. Clin Otolaryngol 19 (4): 275–276

[8] FRIBERG D, GAZELIUS B, HOKFELT T, NORDLANDER B (1997) Abnormal afferent nerve endings in the soft palatal mucosa of sleep apnoics and habitual snorers. Regul Pept 23; 71 (1): 29–36

[9] FRIBERG D, GAZELIUS B (1998) Evaluation of the vascular reaction in pharyngeal mucosa. Acta Otolaryngol 118 (3): 413–418

[10] FRIBERG D, ANSVED T, BORG K, CARLSSON-NORDLANDER B, LARSSON H, SVANBORG E (1998) Histological indications of a progressive snorer's disease in an upper airway muscle. Am J Respir Crit Care Med 157 (2): 586–593

[11] FRIBERG D, GAZELIUS B, LINDBLAD LE, NORDLANDER B (1998) Habitual snorers and sleep apnoics have abnormal vascular reactions of the soft palatal mucosa on afferent nerve stimulation. Laryngoscope 108 (3): 431–436

[12] FUJITA S, CONWAY W, ZORICK F, ROTH T (1981) Surgical correction of anatomic abnormalities in obstructive sleep apnoea syndrome: uvulopalatorpharyngoplasty. Otolaryngol Head Neck Surg 89 (6): 923–934

[13] FUJITA S, WOODSON BT, CLARK JL, WITTIG R (1991) laser midline glossectomy as a treatment for obstructive sleep apnoea. Laryngoscope 101 (8): 805–809

[14] FUJITA S (1994) Pharyngeal surgery for obstructive sleep apnoea and snoring. In: FAIRBANKS F, FUJITA S (eds) Snoring and obstructive sleep apnoea, 2nd edn Raven Press, New York pp 77–96

[15] GISLASON T, ALMQVIST M, ERIKSSON G, TAUBE A, BOMAN G (1998) prevalence of sleep apnoea syndrome among Swedish men–an epidemiological study. J Clin Epidemiol 41 (6): 571–576

[16] GISLASON T, LINDHOLM CE, ALMQVIST M, BIRRING E, BOMAN G, ERIKSSON G, LARSSON SG, LIDELL C, SVANHOLM H (1988) Uvulopalatorpharyngoplasty in the sleep apnoea syndrome. Predictors of results. Arch Otolaryngol Head Neck Surg 114 (1): 45–51

[17] GUILLEMINAULT C, HAYES B, SMITH L, SIMMONS FB (1983) Palatopharyngoplasty and obstructive sleep apnoea syndrome. Bull Eur Physiopathol Respir 19 (6): 595–599

[18] GUILLEMINAULT C, STOOHS R, DUNCAN S (1991) Snoring (I). Daytime sleepiness in regular heavy snorers. Chest 99(1): 40–48

[19] GUILLEMINAULT C, STOOHS R, CLERK A, SIMMONS J, LABANOWSKI M (1992) From obstructive sleep apnoea syndrome to upper airway resistance syndrome: consistency of daytime sleepiness. Sleep 15 (6 Suppl): 13–16

[20] GUILLEMINAULT C, STOOHS R, QUERA-SALVA MA (1992) Sleep-related obstructive and nonobstructive apnoeas and neurologic disorders. Neurology 42 (7 Suppl 6): 53–60. Review

[21] GUILLEMINAULT C, PHILIP P (1992) Polygraphic investigation of respiration during sleep in infants and children. J Clin Neurophysiol 9 (1): 48–55. Review

[22] GUILLEMINAULT C, STOOHS R, CLERK A, CETEL M, MAISTROS P (1993) A cause of excessive daytime sleepiness. The upper airway resistance syndrome. Chest 104 (3): 781–787

[23] GUILLEMINAULT C, HILL MW, SIMMONS FB, DEMENT WC (1978) Obstructive sleep apnoea: electromyographic and fiberoptic studies. Exp Neurol 62 (1): 48–67

[24] GUTMANN L, GUTMANN L, SCHOCHET SS (1996) Neuromyotonia and type I myofiber predominance in amyloidosis. Muscle Nerve 19 (10): 1338–1341

[25] GUTMANN L (1996) AAEM minimonograph #46: neurogenic muscle hypertrophy. Muscle Nerve 19 (7): 811–818

[26] HORNER RL, INNES JA, HOLDEN HB, GUZ A (1991) Afferent pathway(s) for pharyngeal dilator reflex to negative pressure in man: a study using upper airway anaesthesia. J Physiol May; 436: 31–44

[27] HORNER RL, INNES JA, MURPHY K, GUZ A (1991) Evidence for reflex upper airway dilator muscle activation by sudden negative airway pressure in man. J Physiol 436: 15–29

[28] HORNER RL, GUZ A (1991) Some factors affecting the maintenance of upper airway patency in man. Respir Med 85 Suppl A: 27–30

[29] IKEMATSU T (1964) Study on snoring. J Jpn Otol Rhinol Laryngol Soc 64: 434–435

[30] ISSA FG, SULLIVAN CE (1984) Upper airway closing pressures in obstructive sleep apnoea. J Appl Physiol Aug; 57 (2): 520–527

[31] ISSA FG, SULLIVAN CE (1984) Upper airway closing pressures in snorers. J Appl Physiol 57 (2): 528–535

[32] JÄGHAGEN EL, BERGGREN D, ISBERG A (2000) Swallowing dysfunction related to snoring: a videoradiographic study. Acta Otolaryngol 120 (3): 438–443

[33] KAMAMI YV (1990) laser $CO_2$ for snoring. Preliminary results. Acta Otorhinolaryngol Belg 44 (4): 451–456

[34] KAMAMI YV (1994) Outpatient treatment of sleep apnoea syndrome with $CO_2$ laser. LAUP: laser-assisted UPPP results on 46 patients. J Clin Laser Med Surg 12 (4): 215–219

[35] KAMAMI YV (1994) Outpatient treatment of snoring with $CO_2$ laser: laser-assisted UPPP. J Otolaryngol 23 (6): 391–394

[36] KAMAMI YV (1994) Outpatient treatment of sleep apnoea syndrome with $CO_2$ laser: laser assisted UPPP. J Otolaryngol 23 (6): 395–398

[37] KRESPI YP, PEARLMAN SJ, KEIDAR A (1994) Laser-assisted uvula-palatoplasty for snoring. J Otolaryngol 23 (5): 328–334

[38] KRESPI YP (1998) The success of LAUP in select patients with sleep-related breathing disorders. Arch Otolaryngol Head Neck Surg 124 (6): 721

[39] KRESPI YP, KACKER A (2000) Management of nasopharyngeal stenosis after uvulapalatoplasty. Otolaryngol Head Neck Surg 123 (6): 692–695

[40] LARSSON SG, GISLASON T, LINDHOLM CE (1988) Computed tomography of the oropharynx in obstructive sleep apnoea. Acta Radiol 29 (4): 401–405

[41] LARSSON H, CARLSSON-NORDLANDER B, LINDBLAD LE, NORBECK O, SVANBORG E (1992) Temperature thresholds in the oropharynx of patients with obstructive sleep apnoea syndrome. Am Rev Respir Dis Nov: 146 (5 Pt 1): 1246–1249

[42] LUGARESI E (1975) Snoring. Electroencephalogr Clin Neurophysiol 39 (1): 59–64

[43] LUGARESI E, CIRIGNOTTA F, COCCAGNA G, PIANA C (1980) Some epidemiological data on snoring and cardiocirculatory disturbances. Sleep 3 (3–4): 221–224

[44] LUGARESI E, MONDINI S, ZUCCONI M, MONTAGNA P, CIRIGNOTTA F (1983) Staging of heavy snorer's disease. A proposal. Bull Eur Physiopathol Respir 19 (6): 590–4.83

[45] MARAIS J (1998) The value of sedation nasendoscopy: a comparison between snoring and non-snoring patients. Clin Otolaryngol 23 (1): 74–76

[46] MICHELSON E, HOLLRAH S (1999) Evaluation of the patient with shortness of breath: an evidence-based approach. Emerg Med Clin North Am 17 (1): 221–237

[47] OLSON LG, STROHL KP (1986) Pathophysiology and treatment of central sleep apnoea. Chest 90 (2): 154–155

[48] QUINN SJ, DALY N, ELLIS PD (1995) Observation of the mechanism of snoring using sleep nasendoscopy. Clin Otolaryngol 20 (4): 360–364

[49] REMMERS JE, DEGROOT WJ, SAUERLAND EK, ANCH AM (1978) pathogenesis of upper airway occlusion during sleep. J Appl Physiol 44 (6): 931–938

[50] REMMERS JE, YOUNES MK, BAKER JP (1978) Termination of inspiration through graded inhibition of inspiratory activity. Chest 73 (2 Suppl): 262–263

[51] SIMMONS FB, GUILLEMINAULT C, SILVESTRI R (1983) Snoring, and some obstructive sleep apnoea, can be cured by oropharyngeal surgery. Arch Otolaryngol 109 (8): 503–507

[52] SIMMONS FB, GUILLEMINAULT C, MILSE LE (1984) The palatopharyngoplasty operation for snoring and sleep apnoea: an interim report. Otolaryngol Head Neck Surg 92 (4): 375–380

[53] SMIRNE S, LANNACCONE S, FERINI-STRAMBI L, COMOLA M, COLOMBO E, NEMNI R (1991) Muscle fibre type and habitual snoring. Lancet 9; 337 (8741): 597–599

[54] STOOHS R, GUILLEMINAULT C (1992) MESAM 4: an ambulatory device for the detection of patients at risk for obstructive sleep apnoea syndrome (OSAS). Chest 101 (5): 1221–1227

[55] STROHL KP, CHERNIACK NS, GOTHE B (1986) Physiologic basis of therapy for sleep apnoea. Am Rev Respir Dis 134 (4): 791–802. Review

[56] STROHL KP (1996) Diabetes and sleep apnoea. Sleep Dec; 19 (10 Suppl): S 225–228. Review

[57] STROHL KP, REDLINE S (1996) Recognition of obstructive sleep apnoea. Am J Respir Crit Care Med Aug; 154 (2 Pt 1): 279–289

[58] TAKEUCHI T, FUTATSUKA M, IMANISHI H, YAMADA S (1986) Pathological changes observed in the finger biopsy of patients with vibration-induced white finger. Scand J Work Environ Health 12 (4 Spec No): 280–283

[59] WALKER RP, GRIGG-DAMBERG MM, GOPALSAMI C (1999) Laser-assisted uvulopalatoplasty for the treatment of mild, moderate, and severe obstructive sleep apnoea. Laryngoscope 109 (1): 79–85

[60] WALKER RP, GARRITY T, GOPALSAMI C (1999) Early polysomnographic findings and long-term subjective results in sleep apnoea patients treated with laser-assisted uvulopalatoplasty. Laryngoscope 109 (9): 1438–1441

[61] WEITZMAN ED (1979) The syndrome of hypersomnia and sleep-induced apnoea. Chest 75 (4): 414–415

[62] WEITZMAN ED, POLLAK CP (1979) Disorders of the circadian sleep-wake cycle. Med Times 107 (6): 83–88, 92–94

[63] WEITZMAN ED, KAHN E, POLLAK CP (1980) Quantitative analysis of sleep and sleep apnoea before and after tracheostomy in patients with the hypersomnia-sleep apnoea syndrome. Sleep 3 (3–4): 407–423

[64] WHEATLEY JR, WHITE DP (1993) The influence of sleep on pharyngeal reflexes. Sleep 16 (8 Suppl): 87–89

[65] WHEATLEY JR, TANGEL DJ, MEZZANOTTE WS, WHITE DP (1993) Influence of sleep on response to negative airway pressure of tensor palatini muscle and retropalatal airway. J Appl Phyiol 75 (5): 2117–2124

[66] WHEATLEY JR, MEZZANOTTE WS, TANGEL DJ, WHITE DP (1993) Influence of sleep on genioglossus muscle activation by negative pressure in normal men. Am Rev Respir Dis 148 (3): 597–605

[67] WHEATLEY JR, TANGEL DJ, MEZZANOTTE WS, WHITE DP (1993) Influence of sleep on alae nasi EMG and nasal resistance in normal men. J Appl Physiol 75 (2): 626–632

[68] WOODSON BT, GARANCIS JC, TOOHILL RJ (1991) Histopathologic changes in snoring and obstructive sleep apnoea syndrome. Laryngoscope 101 (12 Pt 1): 1318–1322

[69] YOUNG T (1993) Analytic epidemiology studies of sleep-disordered breathing–what explains the gender difference in sleep-disordered breathing? Sleep 16 (8 Suppl): 1–2

# III-4.4
# Functional Endoscopic Endonasal Laser Surgery – FEELS
# Laser Treatments of the Nose and the Paranasal Sinuses

Jürgen U. G. Hopf and Marietta Hopf

## Contents

## Extended Summary

Endoscopic surgery is a recognized procedure for the therapy of diseases of the main nasal cavities and the paranasal sinuses. Implementing suitable laser endoscopes, fibre-delivered laser radiation can be applied to the respective target tissue, using either the contact or non-contact method.

In our hands, the Diode laser and the Nd:YAG laser are normally used für this procedure; but in our department there are also specific indications for the Argon Ion laser as well as for the KTP laser and the $CO_2$ laser. Functional endoscopically controlled endonasal and transnasal laser Surgery (FEELS) provides the possibility of administering outpatient therapy under surface anaesthesia for a large variety of very common diseases.

"Blocked nose" is the leading symptom of morphological obstructions of the nasal cavity, which are mainly caused by benign hyperplasia of the lower and middle nasal turbinates, polypous rhinosinusitis and postoperative or postinflammatory septo-turbinal synechia and cicatricial stenosis. Some patients also exhibit small- to medium-sized septal spurs and ridges, which change the laminar flow of air inside the nasal cavity into a turbulent one. Nasal breathing is thus impeded and ventilation and drainage of the corresponding paranasal sinuses via their natural orifices are significantly reduced.

In these cases, the goal of laser-assisted endoscopic surgery is to reshape the ideal inner surface configuration of the nasal cavity in order to restore an unimpeded airway. This is a minimally invasive but maximally effective procedure performed under local anaesthesia on an outpatient basis. Using appropriate power and application parameters, laser photons transmitted via thin fibres are able either to vaporize or to coagulate and shrink soft and hard tissues without bleeding. Surgery is performed under video endoscopy to achieve excellent visualization and an enlarged representation of the site.

During the past 14 years, more than 6000 endoscopically controlled laser surgical operations have been carried out in our ENT department on different pathologies of the main nasal cavities, the paranasal sinuses and the nasopharynx, with mostly excellent clinical success. The various indications have also included massive polyposis nasi et sinuum, vomeroseptal soft-tissue hyperplasia, cysts, mucoceles, tumours, vascular haemorrhagic diatheses, hereditary haemorrhagic telangiectasis, choanal atresia and stenosis as well as hyperplasia of the lymphatic tissue of the nasopharynx.

Selecting the appropriate laser system and using correct laser parameters is decisive for the success of the treatment and for minimizing peri- and postoperative risks and complications and for avoiding damage being done to the surrounding healthy tissue. The technique used for endonasal laser surgery, the treatment protocol, the course of the therapy and the advantages of laser treatment will be demonstrated for the individual indications. These advantages include, in particular, the possibility of working under videoendoscopic monitoring without any bleeding, the ability to use regularly only surface anaesthesia with almost no pain involved, and the possibility of performing such operations on an outpatient basis. This means that hazard-free treatment is also possible for higher-risk patients who are unable to undergo general anaesthesia, as well as for patients with an increased or excessive bleeding tendency.

However, a good understanding of the biophysical basics is absolutely essential for any kind of successful laser therapy. The surgeon must be able to master these basics just as well as all the anatomical and pathophysiological aspects of the diseases involved, that is to say, the aspects which put him in a position to determine the therapeutic strategy in the case of conventional operations and to carry out the surgical intervention successfully after the diagnosis has been made and confirmed. In terms of laser therapy, this means that the surgeon must be able to choose the correct type of laser and to select safe and effective laser parameters in addition to the appropriate endoscope.

Sudden changes unexpectedly encountered during the course of the operation will require flexibility on the part of the surgeon. Accordingly, the surgeon should be sufficiently acquainted with the physical basics concerning laser–tissue interactions. The extent of the effect produced on the tissue depends on the laser wavelength, the associated optical penetration depth of the laser photons, the laser power parameters and the application technique used. Inappropriate selection of any one of these parameters would usually produce unsatisfactory therapeutic results and may possibly involve the risk of complications.

## Introduction

The pathological complex of obstructive nasal respiration with secondary symptoms such as olfactory disturbances, recurrent infections, watery or sticky mucous rhinorrhea and tubal ventilation disorders of the middle ear is very widespread. When conservative measures fail to bring about the desired alleviation of the symptomatology, surgical intervention frequently becomes necessary. The extent of the operative measure will then depend on the degree to which the disease has manifested itself, on the threatening secondary diseases such as chronic bronchitis, the sinu-bronchial syndrome or chronic otitis media and on the constitution of the patient. Where the secondary physical diseases are less evident, it is sometimes only a question of the patient's quality of life. Possible intra-operative and postoperative complications due to the traditional surgical treatment itself must similarly be taken into consideration. However, regarding the clinical picture of chronic polypous rhinosinusitis, obstructed nasal respiration is no longer a question of reduced comfort in life but an advanced indicator of a serious restriction of the paranasal ventilation and drainage. Since topical and systemic steroids can only intervene here in the manner of alleviating the symptoms and reducing the inflammatory mucosal swelling and oedema, in the large majority of cases, a traditional functional endoscopic or microscopic sinus surgery (FESS) with all the known, possibly also serious, side effects is advisable. According to experience, the intraoperative risk of a complication during the operation for recurrent polyposis increases with its frequency in the individual patient. On the one hand, this is based on the frequent lack of "anatomical landmarks", and on the other, on the lacking overview in the case of heavy bleeding.

Where the economic aspect of the health system is concerned, a reduction in the length of stay in hospital is being demanded nowadays more and more vehemently. It is normally also an issue for the patient not to stay away from work and from his family for too long. For this reason, attempts are being increasingly made to shift operative forms of treatment to the outpatient sector. Here, surgical strategies of a minimally invasive nature find a justified field of application. The endoscopic application of the laser is, in fact, predestined for the surgery of pathological changes in the nose, in the paranasal sinuses and the nasopharynx.

After the in-depth investigations carried out by Hosemann and Wigand as well as by Tos et al., the respiratory mucosa of the nose and the paranasal sinuses can no longer be considered as a uniform integument of homogenous structure.

The great histomorphological variability which characterizes this organ even in a normal and healthy functional state is considerably increased, when its manifold appearance in the pathologically altered condition of sinusitis is also taken into consideration. In this respect, the work of Messerklinger and Wigand has produced fundamental new insights into the physiology, pathophysiology and reparative behaviour of the nose and the paranasal sinuses. Based on the apparent reversibility of extremely pathologically changed respiratory epithelia in the nasal and paranasal anatomic area following the restoration of ventilation and drainage, these two authors have de-veloped new concepts for structure-preserving nasal and paranasal sinus surgery, which has increasingly replaced the radical-surgery principle of complete musoca removal.

This development rapidly led from endoscopically supported diagnosis to endoscopically controlled microsurgery.

With their activities, Messerklinger, Rudert, Draf, Wigand, Stammberger and Kennedy were some of the pioneers in this field.

The advantage of having a good overview of the entire surgical situs, which has become possible in the past decade thanks to significant improvements in endoscopic instrument technology, is marred, of course, if intraoperative bleeding destroys this overview. Moreover, the new knowledge gained on the regeneration capability of chronically inflamed musoca in this region requires the use of microsurgical techniques in order to keep collateral damage and the surgical trauma to healthy surroundings as low as possible.

In the past, the question was raised as to whether the implementation of the laser is even justified in endoscopic rhinosurgery, for at the beginning of the laser era in the 1980s, the attempt was made in all branches of surgery to carry out as many indications as possible with laser-surgical treatments. This aroused partly justified criticism on the part of the traditional otolaryngologists and head and neck surgeons.

The sense behind minimally invasive laser application for specific indications, however, could be proven by the excellent results of international studies and long-term observations. Thus, the initially widespread euphoria had to yield to a considerably narrower spectrum of indications, where the laser proved to have real advantages. The result of this controversy and the many long years of experience are now revealing a field of application that is becoming established in rhinosurgery worldwide. The method we favour, namely, fibre-delivered endoscopic laser surgery using preferably the Nd:YAG laser and the Diode laser and – for specific indications – the Argon laser and the $CO_2$ laser, thus represents a further development in many ways and, in certain cases, a sensible supplementation to the operative methods already established regarding both the pathophysiology and the restitution of mucosal function.

The targeted microsurgical application of laser radiation for the well-closed removal or thermal change of pathological soft and hard tissue presented visible surgical progress in this connection. However, a perfect understanding of the intended laser-tissue interactions is a necessary and indeed an absolute prerequisite for carrying out such forms of treatment.

Thanks to the progressive miniaturization in the field of optical fibre systems and their incorporation in miniature and microendoscopes in combination with (sometimes actively controllable) microcatheters, it was possible to develop so-called "laser endoscopes" specifically for surgery on the main nasal cavities, the paranasal sinuses and the nasopharynx, which enable therapeutic laser irradiation transmitted by means of a quartz glass fibre.

The intention of this chapter is to focus on a detailed overview of relevant biophysical fundamentals, the equipment technology, the systematics and the treatment protocol of endoscopic laser surgery used for pathological structures in the nasal cavities, the paranasal sinuses and the nasopharynx.

The indications and clinical results gathered over a period of 15 years are well presented and critically discussed. Along with a critical consideration of respective literature, precise recommendable laser parameter settings – safe and effective – are given for the suggested range of indications.

## Biophysical Basics of Laser-Tissue Interactions

As far as endonasal applications of the Diode laser, the Nd:YAG laser, the KTP laser and the argon laser are concerned, the endoscopically monitored fibre transmissions has proved to be by far the most suitable, safe and effective method. The infrared-fibre transmission of $CO_2$ laser photons has been technically realized, but is commercially not yet available from every laser company.

Fibre guided laser radiation can be transmitted to the target tissue by two different methods: On the one hand, there is the so-called contact application in which the tip of the fibre just touches the tissue surface without pressing it into deeper layers of the tissue. During laser exposure, the fibre must follow the blanching and shrinking tissue surface in such a manner that the emitted photons penetrate almost directly into the target tissue.

On the other hand, it is also possible to irradiate the tissue keeping the tip of the fibre constantly at a certain distance away from the surface of the tissue. Accordingly, this procedure is termed "non-contact application". To achieve predictable results, however, the latter application method requires a highly experienced laser surgeon having a good working knowledge of the laser-tissue interactions involved.

The non-contact procedure seems to be more sophisticated,
- since the laser power density which represents the determining factor for the surgical efficiency decreases reciprocally relative to the value of the

square of the distance from the tip of the bare laser fibre.
- since it is not possible, for anatomical as well as treatment-specific reasons, to keep the distance between the fibre tip and the target tissue absolutely constant throughout the entire period of irradiation, thus unwillingly producing unequal effects – and more or less unpredictable results at different areas of the target tissue and
- since the thermal tissue effects produced by the non-contact laser are normally not visible to their full extent in the very moment of treatment.

The laser-tissue interactions and the corresponding effects of the contact method for endo-and transnasal laser surgery are based on the following fundamentals of optical physics in biological tissues.

The "optical penetration depth" is a fixed physical parameter which is determined only by the wavelength, by the light absorption and scattering, as well as by the anisotropy of the tissue (i.e. scattering direction of photon). The "optical penetration depth" is largely independent of the irradiated laser power or energy, the laser power density and the method of application. The parameter "optical penetration depth" can be ascertained for each type of tissue by conducting an experimental analysis of the tissue-specific absorption coefficient µa, the scattering coefficient µs and the anisotropy factor g, implementing these three figures after averaging into a computer-assisted Monte Carlo simulation. It expresses the typical tissue-specific spatial penetration and distribution ratio of the irradiated photons characterized by one certain wavelength. The "optical penetration depth" parameter thus represents the spatial localization of an irradiated photon most likely to interact with the molecules of the tissue. The course of an irradiated photon can be accompanied by the physical phenomena of remission, transmission, scattering and absorption. In the latter process, the energy quantum of the irradiated photon will be totally transformed into thermal energy of the absorbing molecule. The optical penetration depth of the Nd:YAG laser photons ranges between 2 and 4 mm; the optical penetration depth of the 810–940-nm Diode laser photons ranges between 1 and 2 mm, depending on the structure and the resulting optical properties of the irradiated tissue.

The "optical penetration depth", however, must be clearly differentiated from the surgically "effective depth of the laser induced lesion".

The "effective depth of the laser-irradiated lesion" represents the acute therapy effect intended by the surgeon. This parameter is, therefore, an expression of the direct efficiency of the laser treatment.

The "effective depth" is determined, on the one hand, by the above-mentioned optical penetration depth of the monochromatic laser light and, on the other hand, by the mode of application, the exposure time, the power density, the total energy applied through laser radiation, plus by the specific thermodynamic parameters pertaining to the tissue under treatment. The thermodynamic effects are determined by the heat capacity and thermal conductivity as well as by intrinsic cooling (e.g. by blood vessels) or localized cooling measures taken by the surgeon during laser application (e.g. use of gas spray). However, the size of the irradiated surface is also determinant.

The following hypothetical model serves to illustrate the interactions between the laser and the tissue. The distinguished energy of a photon entering the tissue through laser irradiation is transferred only by absorption of the irradiated energy quantum into a distinguished atom or molecule. This absorption process causes the molecule to gain a higher energetic state. When laser photons are irradiated into relatively large target volumes, the absorption process taking place in the atoms and molecules causes heat to build up in the distribution volume of the tissue to temperatures between 45 and 90 °C, thus producing coagulative effects. Due to the optical properties and especially the high optical penetration depth involved, this is the case with most biological soft tissues when Nd:YAG, Diode or Argon laser radiation is used. The zones of irreversible and reversible damage are determined by the temperature gradient, which decreases at an increasing distance away from the point of irradiation of the target tissue. Temperatures of between 100 °C and far above 300 °C are produced at the immediate laser spot. Accordingly, the carbonization zone and the vaporization volume will develop in this area.

Well-vascularized areas possess an excellent cooling mechanism as a consequence of the high rate of blood flow, which is suitable for dispersing laser-induced thermal energy. As a rule, therefore, in such areas lesser effective coagulation depths are obtained although identical parameters of power and total energy are applied. Reversely, longer Nd:YAG and Diode laser exposure times are needed for treating these internally well-cooled anatomical areas in order to obtain exactly the same coagulation effect.

The following rules may be formulated as a guideline for fibre-delivered laser surgery:

At an identical total energy applied, a good vaporization effect can be obtained by increasing the power; a good coagulation effect can be achieved by increasing the exposure time.

A similar relationship also applies for the power density. High laser-power densities, i.e. small-fibre cross-sections with an identical laser power at the fibre tip, lead to higher vaporization volumes than large-fibre cross-sections when the contact method is used.

## Preblackening the Fibre Tip

The high optical penetration depth and the correspondingly large distribution volume of the Nd:YAG laser photons in biological tissues are an unchangeable parameter which, being only slightly dependent on output power, produces a relatively large coagulation volume at a simultaneously low vaporization volume. In order to be able to influence this unalterable biophysical behaviour and minimize undesirable large-volume thermal effects whilst at the same time maximizing the removal of tissue by genuine vaporization, we use a technique with the contact method called preblackening. This involves the prior blackening of the distal fibre end and, in particular, of the fibre cross-section being directly applied to the tissue. The carbonized matter needed is obtained by applying one or two short laser shots with high powers onto cork or wood or, possibly, the patient's own blood, which then adheres to the fibre tip; as a consequence, 90 to 95% of the irradiated energy is absorbed by this carbonizate, which heats up to temperatures between 300 and 600 °C in the laser process. This allows the local vaporization process to take place in the tissue area under the preblackened interface.

When applying the Diode laser and the Argon laser, it is not always necessary to preblacken the fibre-tip due to higher absorption on tissue chromophores like haemoglobin and myoglobin and the consecutively lesser optical penetration depth and the smaller distribution volume of their laser photons.

The choice of fibre diameter depends on the desired tissue effect and the size of the available working channel of the endoscope. The relationship between fibre-core diameter and tissue effect should be taken into consideration when choosing parameters.

The Nd:YAG laser, the Diode laser and the Argon laser are – in clear contrast to the Holmium:YAG laser – continuous wave systems. Therefore, two operating modes must be highlighted. The real continuous-wave application and the so-called chopped mode, that is characterized by a mechanical shutter-induced interruption of the laser beam. The irradiation parameters of the chopped mode thus contain the additional information about the exposure time (laser on) and the exposure break (laser out). This should not be confused with parameters of genuine pulsed lasers – like the Excimer laser and the Ho:YAG laser. In connection with the continuously emitting laser systems, the interrupted irradiation should be termed the "chopped mode".

If the clinical aim is to vaporize tissue, high power settings in the chopped mode are necessary. If the goal is to coagulate and shrink mucosal and submucosal soft tissue without bleeding, low power settings in the continuous-wave mode are effective.

## Biophysical Principles, Instrument Technology and Operational Procedure

### Selection of Laser Systems and Adequate Parameters

For endoscopically controlled endonasal and transnasal laser surgery, we routinely use the medilas D Diode laser manufactured by Dornier Med Tech (Germany) (940 nm, 80 W maximal output power), the Nd:YAG laser manufactured by Martin Medizintechnik (Germany) (1064 nm, 60 W maximal output power), the 6020 Diode laser from Lumenis (Israel 805 nm, 20 W maximal output power, an Argon ion laser made by Aesculap-Meditec (Germany, 488 and 514 nm, 5 W maximal output power) and the Aura KTP laser from Laserscope (USA) rated at 15 W maximal output power (532 nm).

The decision as to whether to use a contact or noncontact procedure with these systems is based on the desired effect and specific surgical strategy in each individual indication.

In all systems referred to above the laser beam is transferred to the pathological area with commercially available, flexible quartz glass fibres.

Considering the relative inaccessibility of pathological changes deep inside the nose and the nearly absolute inaccessibility of the paranasal sinuses and nasopharynx to the orthogonal free laser beam, the application of the free beam with a micromanipulator generally is not advisable. Moreover, the use of the free beam is only suitable to produce a focused noncontact laser effect and dramatically increases the potential risk of injuring anatomical structures located in the proximity of the treated area, such as the nasal vestibule.

For these reasons another instrument from our set of laser systems, the 40C $CO_2$ laser manufactured by Lumenis (10600 nm), has found only limited use in the treatment of rhinological diseases in our daily practice, despite the fact that its laser beam is easy to manipulate accurately under endoscopic and microscopic control with high-quality micromanipulators and handles or in combination with commercially available hollow-wave guides (the so-called "nasal probes"). However, these limitations will only be overcome once the technology is available to transmit $CO_2$ laser photons through water free IR fibres, which is an area of ongoing research.

## Nd:YAG Laser – Diode Laser

The Nd:YAG laser is a solid-state laser containing an yttrium-aluminium-garnet (YAG) crystal, i.e. a crystal of garnet structure doted with the rare-earth element, neodymium. Its light with a wavelength of 1064 nm is much more strongly absorbed by blood than by the surrounding tissue, though it is not specifically absorbed by haemoglobin. Rather, the beam is strongly scattered by cellular components effectively increasing the absorption by blood. Because of this scattering effect, the photons of the Nd:YAG laser undergo a relatively uniform volume distribution at prolonged exposure times with a large fraction of the energy being absorbed inside the vessels. Depending on the individual tissue and exposure parameters, thermal conduction may result in effective penetration depths of up to 8–10 mm. Progressive dehydration of tissues may be caused at prolonged exposure times or high-power or power-density settings. The ensuing changes in the physical parameters of the tissue lead to increased absorption of photons by the tissue, and eventually to carbonization, which, in turn, further increases absorption and finally brings about the vaporization of the tissue.

Diode lasers have the simplest and most trouble free laser technology. The photons are produced by an electric current. Several diodes are arranged in diode arrays to produce the output power required for medical application. However, it has been a true technological challenge to couple the laser beams in thin fibres. To keep the emitted wavelength constant, these systems require very delicate and stable temperature control. In the near-IR, i.e. at wavelengths between 800 and 980 nm, more than 50 W of output power can be achieved. Owing to their high efficiency, these lasers need no power units and cooling systems, which is very advantageous in that small, portable versions, barely larger than an HF surgery unit, can be manufactured.

Nd:YAG and Diode lasers can be used either in chopped mode or continuous-wave mode. The laser beam is produced continuously in chopped mode, but blocked out for a defined, present interval of time by means of a simple shutter mechanism integrated into the laser unit. Typical for this mode is that the exposure time laser on and interval between exposures laser off must be preselected. In contrast, the laser beam is guided from the laser unit to the target area with no interruptions in continuous-wave mode (cw mode). In both modes, flexible quartz fibres with core diameters of 600 or 400 μm are used to deliver the laser beam to the target area.

To reduce tissue mainly by means of vaporization the Nd:YAG lasers is applied at high power (35–50 W) in chopped mode with both exposure times and pauses in the range of 0.1 to 0.2. Under these conditions, a power density of 12 000 to 40 000 W cm$^{-2}$ is achieved.

For the same effects with the 6020 diode laser (Lumenis), we use a contact procedure at 15–20 W laser power in chopped mode. Tissues are exposed to the laser for 0.1–0.2 s with intervals of 0.2–0.4 between exposures.

The medilas D laser (Dornier Med Tech) produces much higher laser power, i.e. between 20 and 80 W in chopped mode. Laser exposure times are freely selectable between 0.01 and 0.2 interrupted by intervals of 0.2–0.8 s when vaporizing cartilaginous and bony tissues – such as septun, bullous concha and medial maxillary sinus wall – it is essential to apply this laser for short exposure times at high power settings with long intervals between exposures to minimize the coagulating effect on the neighbouring tissues.

If the surgical strategy calls for coagulation and thermal effects of the Nd:YAG laser in the target tissue aiming to promote scarring of tumefacient, and edematous tissues, the laser should be used in continuous mode at low laser power (3–8 W) in a contact procedure or at 15 to 20 W in a non-contact procedure. Under these conditions, power densities in contact versus non-contact procedures reach 1000–6400 W cm$^{-2}$ versus 5300–16000 W cm$^{-2}$ respectively.

Due to the slightly smaller optical penetration depth of the Diode laser, this device is used in continuous mode at lower laser power in a contact procedure (2–5 W) or non-contact procedure (2–15 W) to produce coagulating effects.

## Argon Ion Laser

The argon laser is a gas laser characterized by its strong selective absorption by melanin and haemoglobin at 488 nm (blue) and 514 nm (green), respectively. Its good optical features make this laser very suitable for medical applications. The laser can be used not only in continuous wave and chopped modes, but also in a short-pulse mode. The high spectral purity of the bands and advantageous beam characteristics of this laser are remarkable and the beam can be coupled into even very small fibre with a diameter of only 50 μm. This guarantees high power density and extremely precise technique. Unfortunately, broader acceptance of this laser in medicine is limited by its poor efficiency.

We prefer to use flexible quartz fibres with core diameters of 100, 200 or 400 μm to transmit the light of the Argon ion laser. The external diameters of these fibres are somewhat smaller than those used in Nd:YAG and diode laser surgery. Consequently, the Argon ion laser is easy to combine with miniature and microendoscopes. In addition, narrower fibres display slightly higher flexibility and resistance to fracture and, thus, can be bent more strongly in the tip of actively controlled endoscopes placed in the operation site.

In contact procedures, we use exposure times between 0.02 and 0.5 s in chopped mode and intervals of 0.1–0.5 s between exposures.

In contact procedures performed in continuous-wave mode, the laser power should be between 2 and 5 W.

## KTP Laser

In the KTP laser, the beam of the Nd:YAG laser, characterized by a wavelength in the near-IR (1064 nm), is guided through a postassium titanyl phosphate crystal. In the process, the frequency of the light doubles and the wavelength decreases by a factor of 2 to 532 nm, producing green light resembling that of the Argon laser. Some of the commercially available units are designed as multiple-wavelength lasers and produce light not only of the basic wavelength of the Nd:YAG laser (1064 nm), but also light of 532 nm. A remarkable feature is the high average power of the KTP crystal, which exceeds that offered by the Argon laser.

In functional endoscopic endonasal laser surgery (FEELS), this laser is used in contact procedures in continuous-wave mode at 2 to 15 W power (Fig. 1).

## Endoscopes

Depending on the indication and the patient-specific anatomical situation, we use the endoscope best suited to the therapy concerned. Prototypes of some of these endoscopes were developed by Hopf and Scherer in collaboration with the instrument manufacturer, Karl Storz (Tuttlingen, Germany) after the commercially available classical instruments of endonasal surgery were found to be inadequate for the specific requirements of this type of endoscopy. All these prototypes of endonasal and transnasal telescopes are now commercially available and can be attached to laser application sheaths. Though it is practically feasible to use a traditional rigid endoscope without laser application sheath by simply taping the laser fibre to the telescope and using a separate tube for suction, doing so is extremely inconvenient and gives the surgeon much less control over the procedure.

There are clear differences in use between rigid and flexible laser endoscopes.

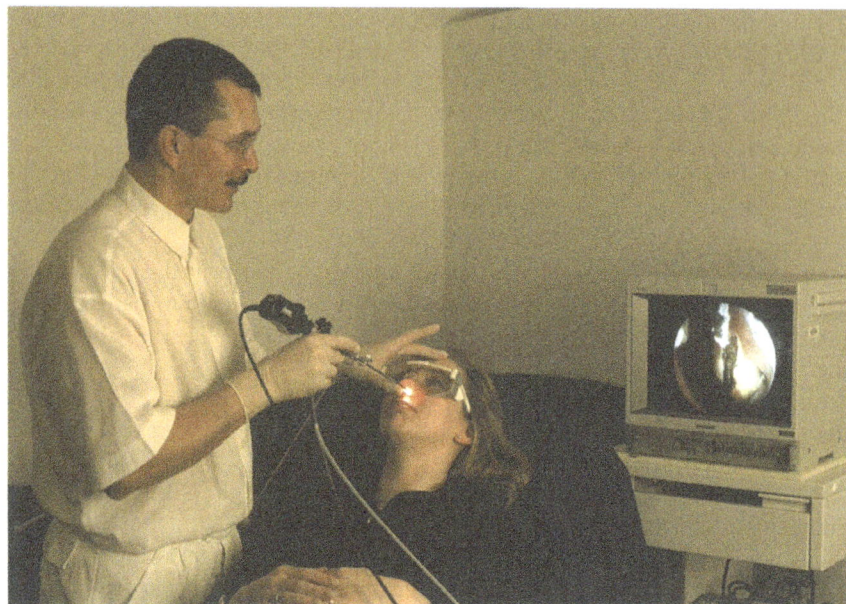

**Fig. 1.** Clinical situation of a FEELS procedure in the outpatient department.

## Rigid Endoscopes

Rigid endoscopes are fitted with high-quality rod lens optics and offer the highest resolution and, thus, afford the best detailed view, especially when working with a digital or analogous camera providing for endoscopic vision on the video monitor (videoendoscopy). In addition, rigid endoscopes are slightly easier to position inside the nose due to their better handling properties. The instruments rest securely (without dislocation) in the hand of the surgeon, who is usually familiar with the instrument from previous experience in conventional endoscopic sinus surgery. The ability to kink the fibre tip is a clear advantage and makes it easier to place the fibre tip on the surface of the target tissue. Thus, there is a reduced tendency of the fibre tip to slip from the target site in contact applications, which would result in inadvertent application of the laser energy in non-contact fashion. This reduces the vaporization and coagulation efficiency and negatively affects the immediate and long-term results of the procedure. In addition, the superior accuracy of laser application is a main factor in eliminating or minimizing traumatization of adjacent healthy tissues which should remain unaffected by the procedure. Thus, the highly sophisticated laser application system offered by Karl Storz (Fig. 2) is not only highly efficient, but also offers superior intraoperative safety. The detail view of the operation site provided by 0° wide-angle telescopes is lightly inferior, because a sharp image is produced only if the distance to the object is a least 2 mm or more. However, these telescopes allow for localized vi-

sualization of those areas selected for treatment, e.g., for coagulation of vascular convolutes, with no angular distortion, and give the surgeon superior control over the switch from contact to non-contact laser technique.

As in conventional paranasal sinus surgery, the surgeon uses the common anatomical structures as landmarks. Using endoscopes with smaller external diameter, the surgeon will find it easier to keep the treatment of the non-sedated patient under local anaesthesia free of pain for the entire course of the procedure. The narrower instruments can also be used in children from the age of 7 years. In addition, Karl Storz now equips the narrower laser endoscopes with improved tips (Fig. 3), which show a ball-shaped structure. This prevents iatrogenic mucosal damage during the introduction of the instrument.

As a disadvantage it must be mentioned that the smaller lumen channels of the narrower endoscopes do not evacuate smoke as efficiently. Since endonasal surgery is performed in a moisture-saturated environment, obstruction by wet smoke particles is certainly possible in these instruments. Smoke particles may also adhere to the telescope lenses, in which case the procedure must be interrupted to clean the lens, preferably with a cotton carrier soaked in detergent. The suction channel of the laser application sheath can be easily disassembled into an external and internal sheath. After disassembly the components should be soaked in a cleaning solution, and then cleaned with a set of small brushes. Later, both parts are subjected to liquid, gas or autoclave sterilization.

**Fig. 2.** Laser-endoscopic application sheath (Karl Storz Company, Tuttlingen, Germany) equipped with an Albarran lever for functional endoscopic endonasal laser surgery (FEELS).

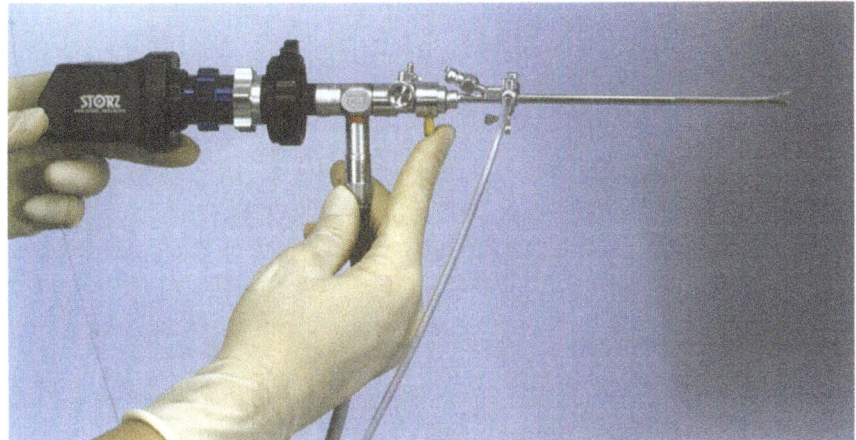

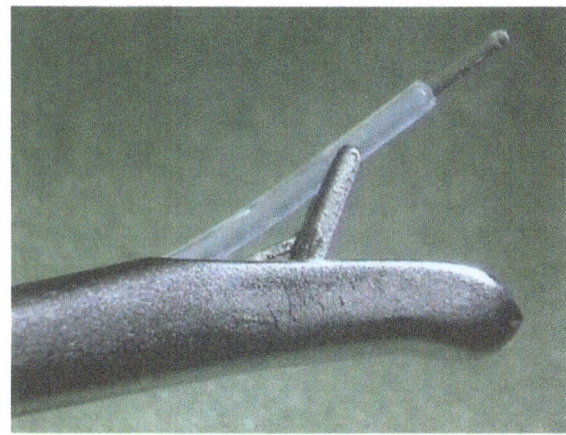

**Fig. 3.** Tip Segment of the laser-endoscopic application sheath (Karl Storz Company, Tuttlingen, Germany) with Albarran lever. This device enables the surgeon to point exactly onto the surgical target without moving the entire instrument in the narrow nasal cavity.

The rigid endoscopic surgery instruments are based on a Fess endoscope developed by Rudert which was later equipped with an additional channel to hold the laser fibre in laser surgery. Fitted with a smoke evacuation facility and integrated fibre channel, the handle is of oval shape and has a maximal transverse diameter of 5.6 mm. Handles can be attached to various angled endoscopes.

The fibre-guiding groove of the working channel is equipped with an Albarran lever that allows the fibre to be kinked by up to 50° to position the fibre. The relatively large external diameter of the assembled instrument sheath limits the application in confirmed anatomical areas. Consequently, smaller oval endoscope sheaths with a ball-shaped tip matching the internal lumen of the nasal cavity were developed by Karl Storz in collaboration with authers. In these instruments, the fibre telescope and suction channels are arranged in a concentric fashion. These telescopes are modular and exchangeable and provide for both suction and a laser fibre channel.

## Flexible Endoscopes

Flexible endoscopes are particularly recommended for laser surgery to be performed in the paranasal sinus area (Fig. 4). In cases of patients who have been pre-operated in a very gentle and structure-preserving manner, sometimes the anterior ethmoid and the nasofrontal recess, the posterior ethmoid and the sphenoid sinus can only be adequately accessed for laser surgery when a flexible endoscope is at hand. This particularly applies in cases where there is a large and extended middle concha which extensively shuts off the middle nasal meatus, or when, in the postoperative period after traditional FESS accompanied by moderate conchotomy, a cicatricial lateralization and scarification of the middle concha occurs. Access to such areas is also difficult when in FESS pre-operated patients too little surgical enlargement has been provided in the areas of the agger nasi and the uncinate process. If septum ridges and spurs are massive but the patient refuses to undergo the necessary surgical correction, it is often extremely difficult to perform an operation on the posterior parts of thenasal cavity, the posterior ethmoid, the sphenoid region and the nasopharynx, when using rigid instruments only.

We prefer to use the flexible, actively controlled nasopharyngoscopes and, for children, tracheobronchoscopes manufactured by Karl Storz, Germany. These instruments are equipped with a working channel and suction facility. As mentioned above, flexible endoscopes are somewhat more complicated to use due to the more difficult positioning in the target area.

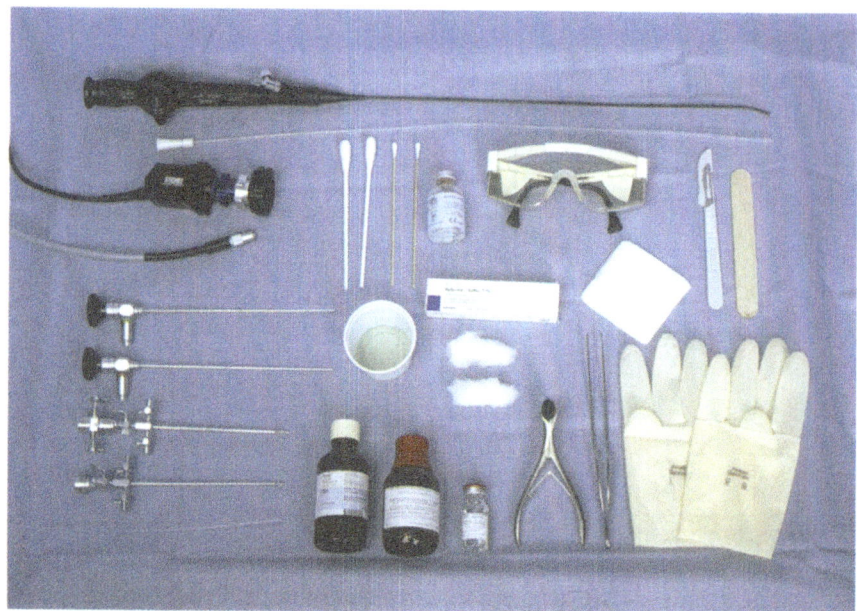

**Fig. 4.** Preparation of instruments and supplemental devices for functional endoscopic endonasal laser surgery (FEELS).

This disadvantage is related to the resilience of the endoscope tube in the absence of a firm sheath. Once the endoscope is properly positioned over the target site, each motion of the surgeon's hand in an endonasal approach is directly transmitted to the distal tip of the fibre endoscope. Consequently, the fibre may inadvertently move during the irradiation with laser light. In some clinical applications, it has proven beneficial to fix a part of the flexible endoscope with a rigid guiding handle serving as some kind of splint to fix the section of the endoscope spanning the main nasal cavity.

The detailed view and brilliance of the optical image are somewhat inferior, because of the wide-angle fibre telescope. The use of microendoscopes with gradient optics may be associated with a reduction of the image quality provided by the chip-camera/monitor system due to the presence of Moiré lines and appearance of spots related to the optical fibre. This problem is solved in the new digital video system manufactured by Karl Storz through a special integrated electronic videofilter: by filtering out Moiré lines, any interference with the optical image is eliminated.

Nevertheless, the following generalization must be kept in mind: the smaller the endoscope, the smaller is the total visualized area available to the surgeon to gain a quick overview of the operation site.

The introduction of flexible and actively controlled fibre telescopes with <2 mm external diameter and integrated working channel was a distinct step forward in our medical specially. These microendoscopes can be passed through into the paranasal sinuses using the natural or only slightly enlarged paranasal ostia. Second only to conventional flexible telescopes, these instruments are most commonly used in middle meatus operations in patients with previous ethmoidal cell, frontal and maxillary sinus surgery.

## Pretherapeutic Examinations and Treatment Protocol

After the detailed history and a complete oto-rhinolaryngologic status of the patient is established, decisions are taken concerning relevat pretherapeutic examinations dependent on the clinical requirements of the patient

- Rhinoresistometry and/or acoustic rhinometry
- Olfactometry
- Imaging procedures: plain film X-ray imaging, B-scan sonography, infrared-diaphanoscopy high-resolution CT scans of the paranasal sinuses in coronary and/or axial sections, MRI
- Endoscopy and micro-endoscopy
- Histologic proof of diagnosis by taking representative samples
- Allergologic diagnostics
- Haematological and clinical-chemical blood routine
- (Extended) analysis of the coagulation parameters
- Asthma diagnostics
- Cystic fibrosis diagnostics.

Endoscopically controlled laser surgery on pathological tissue of the main nasal cavities and the paranasal sinuses as well as the nasopharynx are carried out on an outpatient basis using local anaesthesia.

**Fig. 5.** Flexible endoscope (Karl Storz Company, Tuttlingen, Germany) with working channel for carrying the laser fibre penetrating the middle nasal meatus.

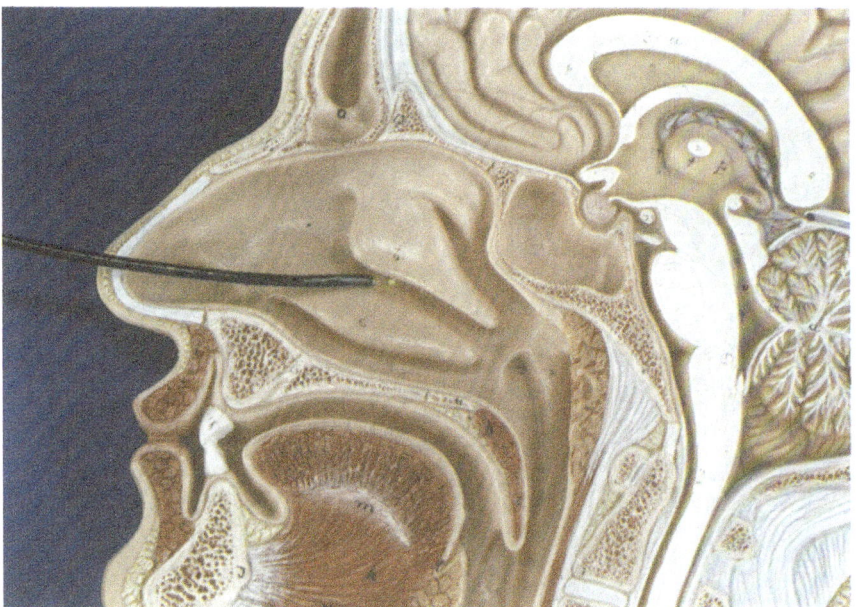

**Fig. 6.** For local anaesthesia and mucosal decongestion tretacaine/naphazoline-soaked cotton swabs are introduced for about 10 to 15 min prior to any FEELS procedure.

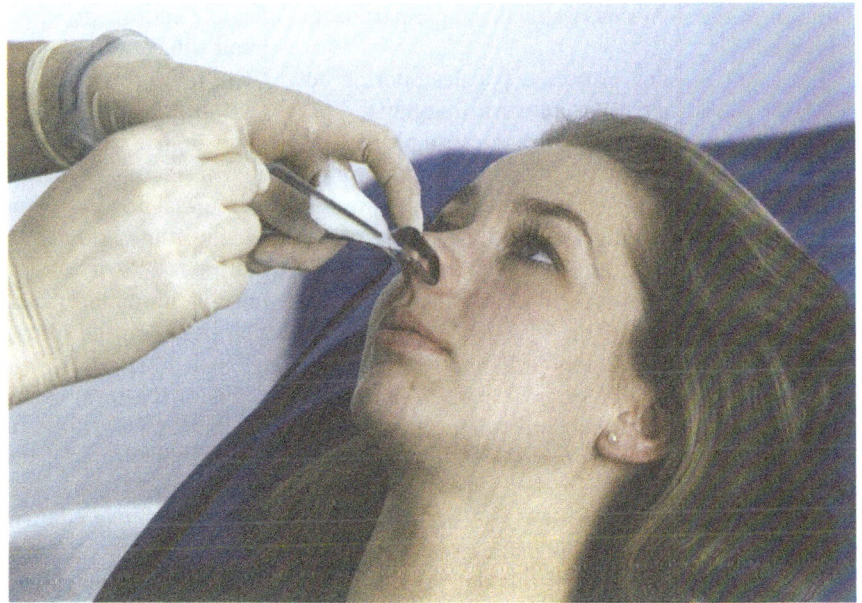

For this purpose, cotton wool swabs soaked in naphazoline and tetracaine (2%) are deposited in the main nasal cavity for 10 min (Fig. 6). An additional aimed local anaesthetic is then carried out with lidocaine gel (5%) by means of bent cotton-wool swab carriers, especially targeted at the areas to be irradiated. A local injection of 1–2% lidocaine solution with an epinephrine additive 1:200 000 is seldom necessary, only in the anterior part of the inferior concha. As a possible alternative, for use in a cases of highly pain-sensitive patients, we use a punctual application of cocaine-hydrochloride solution (of 4 to 10%). Normally, no sedation of the patients is required. Depending on the operating room facilities, the laser therapy can be undertaken directly in the ENT examination chair, with the patient in a semi-supine or supine position with the head slightly inclined. The supine position is, in our experience, preferred by many patients because of the sometime intense nervous tension and anticipatory attitude, especially when endonasal laser treatment is done for the first time. The ideal position of the surgeon is similar to that

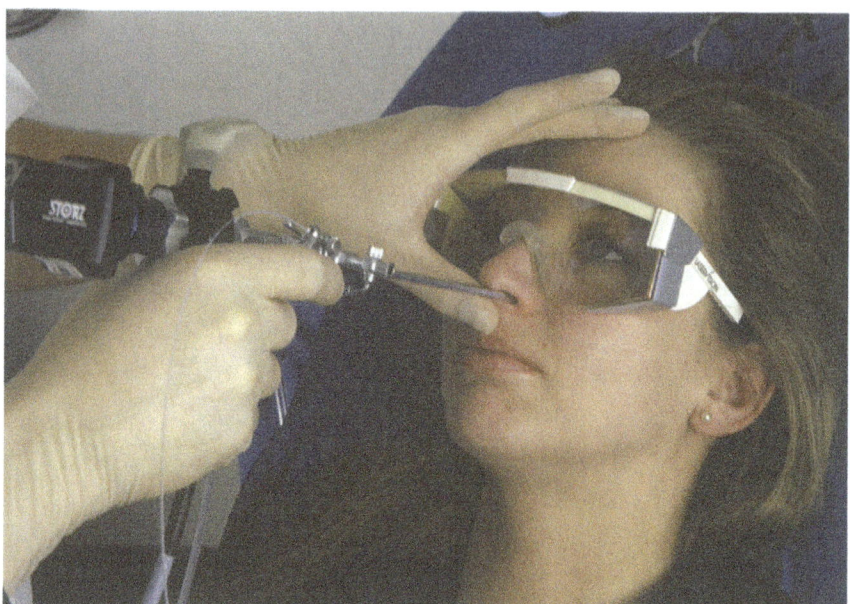

**Fig. 7.** To prevent the nasal cavity from iatrogenic lesions caused by sudden movements it is useful to put the instrument on the thumb of the left hand like a billiard queue.

adopted during conventional FESS procedures, i.e. standing at the right side, beside the upper part of the patient's body.

For safety reasons, protective goggles filtering out the laser wavelength must be worn by both the patient and the surgeon. After the endonasal positioning of the instruments and the ideal positioning of the laser fibre, the initial key to successful therapy is a calm, confident and targeted handling of the procedure. Since the patient is normally under only a local surface anaesthetic for the endonasal mucosal integument, the sensitivity of the external nasal pyramid area and the deep sensitivity of all the facial soft tissues are not completely blocked. A painful experience during insertion of the instrument into the nares, the nasal floor and the tip can lead to an immediate cramp-like reaction in the patient when uncontrolled movements with the endoscope are carried out. The patient then constantly anticipates a subsequent repeated painful experience of this nature during the laser therapy. Such an initial constellation of events will considerably reduce the theoretical treatment time available, as well as the compliance and stamina of the patient.

As far as the surgeon is concerned, it has proved beneficial to relax the arms, close to the body, keeping them close and bent at the elbow. Right-handers usually hold the endoscope in the left hand, thus gently supported by the facial soft tissues of the patient, if required. The right hand is exclusively responsible for being the active agent for the fine positioning and the forward movement or withdrawal of the laser fibre. Applying a rigid endoscope with shaft, the surgeon uses the right hand to operate the Albarran lever for

raising the tip of the laser fibre only. When using a flexible endoscope with a working channel, the right hand will control the lever for raising the entire distal tip of the endoscope.

If, in the case of large-volume vaporization treatment an increased development of fumes and incinerated residue is expected in advance – as in treating large nasal polyps – an additional evacuation catheter for the fumes should be inserted into the opposite main nasal cavity and should be passed far enough into the nasopharynx, so that the resulting incineration gases can be immediately evacuated out of the choanal area.

Following local surface anaesthesia and reducing the mucosal swelling, first the endonasal, paranasal and rhinopharyngeal inspection may be commenced with the endoscope. In our department, the enlarged endoscopic image shown on a monitor has completely replaced the direct from of observation through the ocular of the endoscope.

The endoscope is connected to a one-chip or three-chip camera which transmits the image simultaneously to a monitor. The pretherapeutic and posttherapeutic operating situs, as also the laser-therapeutic intervention itself, can be recorded for documentation purposes via a video recorder and video printer. The relevant findings can thus also be more easily archived, without any additional expenditure. According to our experience, demonstrating the findings and the surgical laser intervention procedure online or afterwards considerably enhances the patients' compliance.

On completing this pretherapeutic precision planning of the surgical procedure, the laser endoscope,

**Fig. 8.** Course of treatment and operational procedure reflected for orientation in a flow chart.

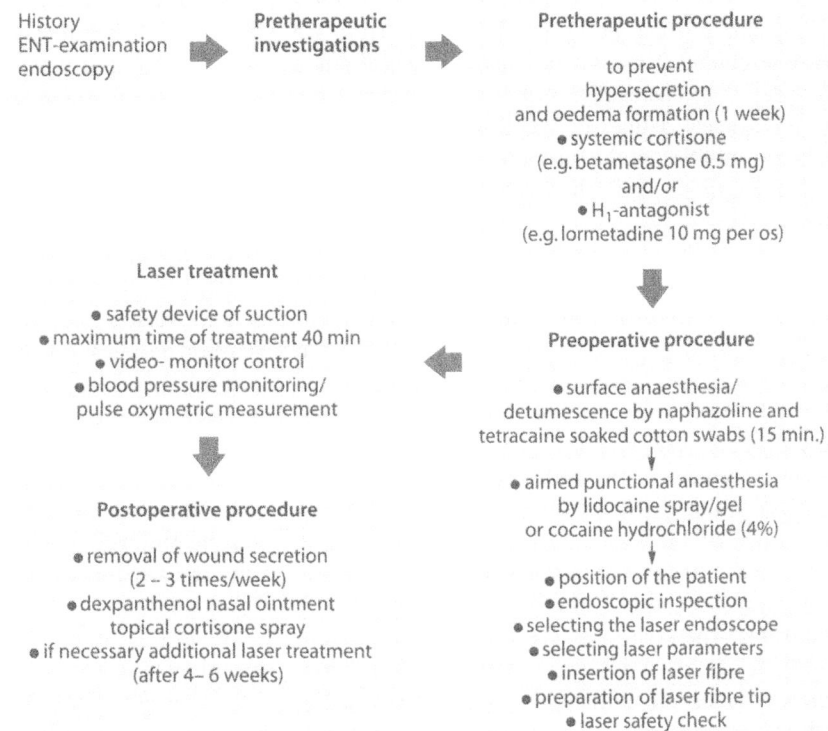

History
ENT-examination
endoscopy
→
Pretherapeutic investigations
→

Pretherapeutic procedure

to prevent hypersecretion and oedema formation (1 week)
• systemic cortisone
(e.g. betametasone 0.5 mg)
and/or
• H₁-antagonist
(e.g. lormetadine 10 mg per os)
↓

Laser treatment

• safety device of suction
• maximum time of treatment 40 min
• video- monitor control
• blood pressure monitoring/
pulse oxymetric measurement

Preoperative procedure

• surface anaesthesia/
detumescence by naphazoline and
tetracaine soaked cotton swabs (15 min.)
↓
• aimed punctional anaesthesia
by lidocaine spray/gel
or cocaine hydrochloride (4%)
↓

←

↓
Postoperative procedure

• removal of wound secretion
(2 – 3 times/week)
• dexpanthenol nasal ointment
topical cortisone spray
• if necessary additional laser treatment
(after 4– 6 weeks)

• position of the patient
• endoscopic inspection
• selecting the laser endoscope
• selecting laser parameters
• insertion of laser fibre
• preparation of laser fibre tip
• laser safety check

the laser system and the relevant laser parameters for the intended treatment are selected in accordance with the intended surgical strategy and the anatomic localization of the target region.

## Laser Treatment – Proper Parameter Selection Helps to Achieve the Intended Effect

The selection of a suitable laser application method is of paramount importance to ensure the successful outcome of laser surgery. This includes choosing between contact versus non-contact procedure, continuous versus chopped mode, clean versus preblackened fibre and selecting appropriate application parameters, such as power settings, exposure time and fibre diameter in order to carefully remove by vaporization or reduce by coagulation pathologically changed mucosal areas and hyperplastic tissue under optimal preservation of the neighbouring healthy tissue. These surgical steps are ultimately aimed at limiting the ability of affected areas to swell up.

Based on the intrinsic characteristics of the different laser beams (Nd:YAG, Diode, Argon Ion and KTP laser) and the wavelength-dependent absorption and scattering properties of tissues and ensuing photon distribution, for long years the Nd:YAG laser was the true workhorse for laser surgical interventions, in

which large volumes are removed with a moderate margin of coagulation, or thermal alterations are produced in relatively large volumes of tissue (this includes interstitial interventions).

Characterized by its (wavelength-dependent) slightly smaller optical penetration depth, strong absorption by haemoglobin and slightly smaller effective penetration depth for coagulation purposes, the Diode laser is the most suitable unit for volume reduction by vaporization and interstitial coagulation. Excellent results have been achieved with this laser mainly in the management of turbinal hyperplasia, of recurrent epistaxis, and in Osler-Weber-Rendu's disease. In addition, this laser appears well suited for surgical therapy of morphological obstructions on and inside the auditory tube. Our excellent experiences have shown this laser to be the prior choice in rhinosurgery and more than a promising alternative to the Nd:YAG laser.

The medilas D diode laser (Dornier Med Tech) offers unprecedented options for management of bony and cartilaginous tissues. The 940-nm photons emitted by this laser are strongly absorbed by water molecules. The laser can be used in chopped mode at high power with extremely short exposure times to vaporize bony and cartilaginous battens located at the nasal septum. Similarly to the Nd:YAG laser, extensive coagulation of the adjacent tissue is prevented by separating successive periods of laser exposure with long pauses. Under these conditions, septal perforation and ensuing necrosis as a potential complication are largely excluded. The laser can also be used to effectively vaporize obstructive soft tissue growth at the nasopharyngeal ostium of the Eustachian tube impairing ventilation of the tube and middle ear.

In our opinion, both the Argon Ion and the KTP laser are also quite suitable systems for finer detail work on smaller volumes. Both beams are strongly absorbed by haemoglobin, which is ideal for relatively well-vascularized tissues. Since these lasers have a relatively small coagulation depth and are excellent to combine with miniature and microendoscopes (the laser light can be transmitted through very thin fibres), they are very suitable for accurate management of the paranasal ostia and any work inside the paranasal sinus system as well as in the nasopharyngeal orifice of the Eustachian tube. Initially, highly vascularized tissue must not be exposed to the static laser beam, but the beam must be progressively moved over the area to be treated proceeding from healthy to diseased tissue in non-contact fashion. Strong exposure of individual spots of tissue to the laser beam would lead to the so-called popcorn effect, i.e. strong absorption of the laser light by haemoglobin would vaporize blood components and vascular walls in the irradiated area, causing rupture and intraoperative haemorrhage of the affected vessels. Only after gentle precoagulation of the target is it possible to treat the tissues in a contact procedure.

Upon removing the coating and flaming off the cladding, a nick is carefully made at the tip of the prepared laser fibre by means of a small preparation knife, either before or ideally after insertion of the fibre into the working channel. This is done in order to provide a breaking point where the fibre can be broken evenly with the thumb on a medium-solid silicon block. A flat broken cross-section of the fibre enables the photons to be radiated directly forward as homogenously as possible and without any wide scattering aperture. The result can be checked by observing the beam pattern of the red helium-neon pilot laser, which should ideally produce a circular, sharp-edged circle area. Preparing the fibre tip after insertion reduces the danger of damage being done to the fibre by the sharp edges of the components incorporated in the metallic endoscope shaft.

When flexible endoscopes are used, there is a danger of damage to the soft internal coating of the working channel during fibre insertion when the endoscope tube is not fully extended straight forward. During the next following sterilization of this flexible endoscope, any force exerted when rinsing the working channel may then cause stray fluid to seep into the vicinity of the optical and the light-guiding fibres, which would render the endoscope entirely unusable.

If preblackening is required, the fibre tip is now blackened by emitting several short laser pulses – with high output powers of between 25 and 50 W – onto a cork or wooden spatula surface.

Subsequently, the fibre is positioned inside the endoscope so that its distal end protrudes approximately 2 mm from the working channel – just enough for it to be seen clearly in the endoscopic image shown on the monitor.

Depending on the laser parameters selected – i.e. contact versus non-contact application, continuous-wave or pulsed mode, cleaned or preblackened fibre, output power, exposure time and fibre thickness – a targeted and well-dosed surgical operation can now be carried out on the pathological tissue, under conditions of maximum protection of the healthy mucosal, cartilagenous or bony environment: tissue can thus be removed by means of vaporization, or its volume and swelling capability can be reduced by means of coagulation.

If laser treatment on very small or punctuated pathologies is estimated not to exceed a treatment time of around 5 min, the patient is usually requested to breathe in through the mouth and out through the nose in order to avoid an ingestion of fumes into the lower respiratory tract. This constellation, however, is rather seldom. Whenever the laser treatment is applied over a longer period of time the additional use of a fumes evacuation device is absolutely necessary and mandatory. Such a facility is integrated into the endoscope; alternatively or additionally, evacuation is achieved by the insertion of a plastic suction tube through the opposite nasal cavity up to the choanal region. The endoscopic suction channel tends to become clogged from time to time, due to the relatively large size of the particles of the fumes developed and the high moisture content of the gases involved.

The average treatment time lies between 10 and 40 min, depending on the indication and the extent of the pathology and intended treatment as well as on patient's compliance.

During and after laser treatment, mediator substances may be released as a result of mechanical and thermal tissue irritation, whose appearance might be somewhat similar to a pseudo-allergic reaction. This can lead, more or less quickly, to hypersecretion and oedema formation. Such phenomena are quite often the limiting factor with respect to laser treatment time. These symptoms are particularly frequent in patients with polyposis nasi et sinuum. Such patients therefore receive betamethasone (0.5 mg) or loratadin (10 mg) per os, 1 week before the next treatment session. This conservative therapy is continued for 2 to 3 weeks following laser treatment, in order to reduce any delayed reaction in the tissue.

Subsequent treatment is then conducted as a rule after 4 to 6 weeks at the earliest. In the meantime, all crusts and fibrinous coatings have disappeared, the wound-healing process or the intended cicatrization is well underway, and the nasal passages are clear again. Depending on the reaction of the patient's mucosa and the extent of the laser irradiation, nasal care is required twice or three times each week during the initial postoperative 20 days.

Those patients admitted to our department for primary or recurrency-FESS because of polypous rhinosinusitis are given endoscopic postoperative examinations at 4-week intervals. In this way, small hyperplastic mucosal changes or beds or oedema developing at critical points can be detected at an early stage and can thus be removed by endoscopic laser treatment in order to support the healing process and counteract any early recurrence or relapse. By constantly maintaining ventilation and drainage, efforts are made to achieve the best possible restitution of chronically changed mucosal tissue on a minimally invasive basis.

## Safety Aspects of Laser Application

Current legal regulations require both patient and surgeon to wear safety goggles suitable to filter out the light produced by the laser. These regulations are undisputed as far as concerns the $CO_2$ laser, and affect all participants in the operating room, i.e. surgeon, patient and medical-technical support staff.

The situation may be somewhat different where Nd:YAG, Argon, KTP and Diode lasers are utilized under videoendoscopic control. Undoubtedly, protection of the patient's eyes with goggles suitable to filter out the applied laser light makes perfect sense. At institutions using several different kinds of lasers, it is necessary to confirm the suitability of the respective goggles by checking the label attached to the top or side (bow) of the goggles.

The worst-case scenario of laser surgery is that the fibre might break near the surgeon's or patient's eyes. Usually, this corresponds to the place at which the fibre enters the laser endoscope sheath. However, an accident of this kind poses a risk to the attending surgeon only if the or she has his or her eyes directly on the eyepiece of the laser endoscope. In videoendoscopic techniques, though, the surgeon gets no closer to this single potential source of fibre breakage than the length of his or her arm. Another potential source of radiation, i.e. back-scatter of laser light in endonasal and transnasal procedures, has little, if any, chance of escaping from the nasal cavity. In view of these considerations and the fact that the power density decreases reciprocally to the square of the distance from the source of radiation, it should be critically discussed whether surgeons should be obliged to wear eye protection. It must be emphasized, though, that these considerations are strictly limited to the endocavitary videoendoscopic application of laser systems emitting near-IR-light.

## Postsurgical Care

The frequency and intensity of postsurgical care was adjusted to the individual objective and subjective wound-healing conditions of the patient. As a rule, postsurgical care was carried out by sucking off secretion and removing fibrinous or crusted coating once or twice a week during the first 14 days. Nasal rinsing with saline solution as well as the application of dexpanthenol and vitamin A ointment was recommended. Vasoconstrictors and steroids should be avoided during the first wound-healing period.

In cases of a hyperreactive mucosa we occasionally found a strong tissue reaction with swelling and massively obstructed nasal respiration after laser-assisted cauterization of the turbinates. In these special cases, the application of a nasal ointment containing naphazoline as well as the administration of some doses of local or systemic steroids was inevitable.

## FEELS Is Part of a Therapeutic Master Plan for Nasal and Paranasal Sinus Disease

Functional endoscopic laser therapy is an integral component of our master plan for surgical management of nasal and paranasal sinus disease. For example, our patients with nasal or recurrent polyposis undergoing conventional functional endoscopic pansinus surgery as developed and described by Wigand, Messerklinger and Stammberger are re-evaluated by endoscopy 4 to 8 weeks after the first operation. Thus, small hyperplastic changes of the mucosa or edematous beds in critical areas are discovered early and can be surgically removed with the laser without delay, which promotes the healing process and prevents recurrence. The surgeon should aim to maintain ventilation and drainage in this minimally invasive procedure for the entire postoperative course in order to ensure optimal restoration of chronically altered tissues.

## Indications, Results and Literature Overview

### Preliminary Remarks

Impaired nasal breathing and its secondary symptoms, such as olfactory deficit, recurrent infection, aqueous or mucous rhinorrhaea and disturbed tubal ventilation, are very common phenomena and often require surgical intervention if conservative measures fail to remedy or mitigate the ailment. The extent of surgery then depends on the severity of manifest disease, possible sequelae, such as chronic bronchitis, sinubronchial syndrome or otitis media and the individual physical condition of the patient. If the physical sequelae are only mild, this may simply be a question of quality of life. Possible intra- and postoperative complications of a conventional surgical approach must certainly also be taken into consideration. However, impaired nasal breathing due to chronic polypous rhinosinusitis may serve as an example of a case in which this symptom is not merely a question of quality of life, but rather a serious indicator of a severe impairment of paranasal ventilation and drainage. As topical and systemic steroids only alleviate symptoms without curing the cause of disease, traditional functional endoscopic or microscopic surgery is indicated in the majority of cases, regardless of their well-known, and sometimes serious, potential adverse effects. The risk of intra-operative complications during surgical management of recurrent polyposis is known to increase with the number of recurrences. This is, in part, related to the absence of anatomical landmarks, and to some extent to the absence of an adequate overview to intraoperatively detect any serious haemorrhage.

The economic constraints becoming increasingly significant in health care demand that patients spend less time in the hospital. As a result, surgical procedures are gradually shifting to outpatient settings, i.e. to the domain of minimally invasive surgical techniques. Endoscopic laser applications are perfectly suited for surgical management of pathological changes in the nose, paranasal sinuses and nasopharynx.

In the early days of the laser era (early 1980s), at a time when surgeons of all subspecialties were still attempting to apply the laser in as many applications as possible, the question was raised as to whether the use of the laser was justified at all. Some very reasonable criticisms were expressed by traditional otorhinolaryngologists, head and neck surgeons.

However, the justification of the minimally invasive laser technique in certain specific applications has since been provided by the excellent results obtained in international studies and after long-term follow-up. With time, the overly enthusiastic broad interpretation of laser applications has given way to a much narrower spectrum of indications, in which the laser offers clear benefits. As a consequence of this debate and based on the long-standing experience with lasers, a set of well-defined rhinosurgical laser applications is emerging. The method of fibre-guided en-

doscopic laser surgery with Diode, Nd:YAG, KTP and Argon lasers, as described here, has developed from, and supplements, the established surgical techniques.

The biophysical principles, technical equipment and clinical procedures of laser surgery have been described above. The following section focuses on suitable indications for laser surgery of the nose, paranasal sinuses and nasopharynx, and is based on our 14-year experience with these procedures at the Department of Otorhinolaryngology, Head & Neck Surgery at the Universitätsklinikum Benjamin Franklin der Freien Universität Berlin (Germany). In addition to a critical discussion of the literature, reasonable laser parameters are spelled out for individual indications to serve as an aid for newcomers to this field.

## Indications and Results

In the past 14 years, we have performed over 6000 endoscopically controlled endonasal and transnasal laser interventions on ambulatory patients under local anaesthesia. The specific indications are discussed in the following.

## Turbinal Hyperplasia, Concha Bullosa, Chronic Recurrent Rhinosinusitis, Nasal and Paranasal Sinus Polyposis

Overall, more than 4500 laser interventions were performed in more than 3900 patients with nasal and paranasal sinus polyposis. These interventions were not limited to managing polyposis, but extended to surgical removal of turbinal hyperplasia, if manifest. This patient group included adolescents and children suffering from cystic fibrosis as the underlying disease.

In addition, we treated over 2000 patients with vasomotor, allergic or pseudoallergic rhinopathy with the laser.

In more than 500 chronic recurrent sinusitis maxillaris patients the minimally invasive laser technique was used to enlarge the middle meatus as a means of improving the morphology underlying sinusoidal ventilation and drainage.

In most of these cases, any manifest uni- or bilateral hyperplasia of the inferior turbinate was also treated with the laser. The size of the middle turbinate needed to be reduced with the laser in some cases, though without removing any bony structures from the medial turbinate bone.

The age of the patients ranged from 8 to 89 years with a mean of 46 years.

## Prerequisites

A current axial or coronary CT scan of the paranasal sinuses and indicative histological findings were prerequisites for the first application of the laser in nasal and paranasal sinus polyposis patients.

Exclusion criteria for primary or exclusive laser therapy included classical pansinus polyposis with complete obstruction of the main nasal cavity and complete opacity in the CT of all paranasal sinuses in patients not previously subjected to surgery as well as suspicion of ongoing sinugenic orbital, facial (soft tissues) and endocranial complications.

As an exception from these criteria, elderly or cardiopulmonary high-risk patients with moderate pansinusitis were treated by (usually) palliative application of the laser to improve symptoms. Occasionally, we accepted patients with multiple previous operations refusing to undergo another conventional intervention, even after receiving extensive information and advice.

In the ideal case, patients selected for endoscopic laser surgery had previously undergone conventional surgery to create a wide bony access to the affected sinus, which could be used to easily pass the endoscope.

Unless done previously, histological tests were performed before commencing laser therapy in order to exclude malignant disease. If the history revealed any evidence indicative of bronchial asthma or allergic disposition, a complete set of allergy diagnostic tests was performed and conservative treatment initiated. Conservative treatment was continued during the laser intervention (Fig. 9).

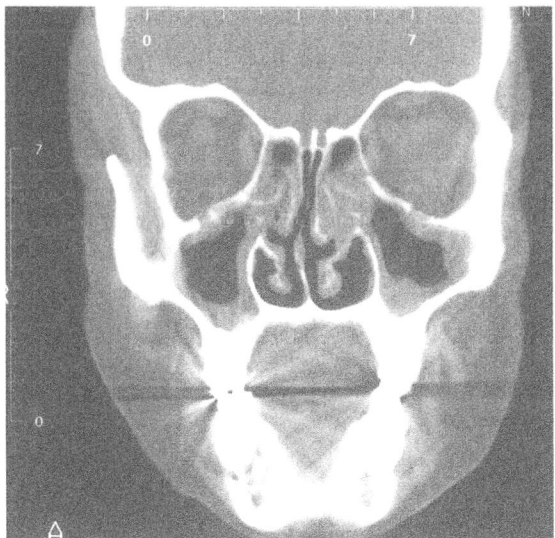

**Fig. 9.** Coronary section of a paranasal CT scan in a patient with chronic polypous rhinosinusitis. The patient had already undergone conventional paranasal sinus surgery six times when he asked for laser surgery in our outpatient department. This CT scan shows an ideal candidate for a FEELS procedure.

Qualitative olfactometry prior to the first session was obligatory in both symptomatic and asymptomatic patients. In addition, prior to receiving therapy, our patients were subjected to the standard examinations, rhinoresistometry and acoustic rhinometry (before and after decongestion with naphazoline solution) to obtain a quantitative description of the degree of nasal obstruction in each individual patient.

### Applying The Nd:YAG Laser

In the early phase of functional endoscopic laser surgery (FEELS), the Nd:YAG laser was employed in chopped mode in a fibre-assisted contact procedure. Under the selected parameters (35–40 W power, 0.1 or 0.2 s exposure, 0.1 or 0.2 s exposure pauses), this laser could be used to vaporize polypous tissue in places, in which immediate reduction with only a small penetration depth was sought (nasal cavity and maxilloethmoidal infundibulum). In the presence of massive polyposis in the main nasal cavity, we prefer to apply the Nd:YAG laser at low to moderate power (2 to 8 W) in continuous-wave mode in order to coagulate the obstructing polypous tissue without placing adjacent anatomical structures at risk. The treated tissue either shows intraoperative thermal demarcation and is subsequently removed by the surgeon in a haemorrhage-free procedure or is eventually (within several days postoperatively) rejected and removed in the course of normal cleaning of the nose. For laser ablation, the fibre is gently placed on the surface of the polypous tissue. Once the laser is activated, it is possible to observe the tissue shrinking towards the laser fibre tip. As an alternative, the laser can be used in continuous-wave mode at low power with slow advancement of the fibre to effect interstitial coagulation. The avital tissue masses thus created are rejected within up to 3 weeks after the intervention, depending on the degree of damage. The advantage of this type of irradiation procedure is that a relatively large target volume of polypous mucosal tissue can be thermally damaged in a reasonable amount of time such that there is progressive demarcation without risk of haemorrhage. As a consequence, free nasal passage of air is restored.

### Applying The Diode Laser

Added 3 1/2 years ago to the spectrum of laser instruments available at our hospital, the Diode laser is characterized by a slightly smaller coagulating penetration depth at similar efficiency. In contrast to the Nd:YAG laser, this laser can be used without preblackening of the fibre tip and its light is rapidly absorbed

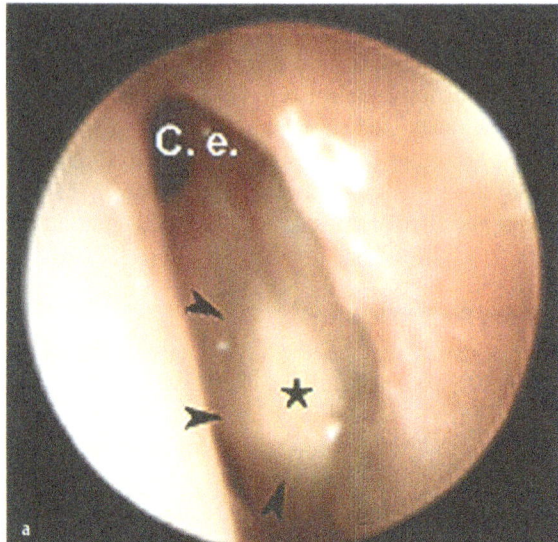
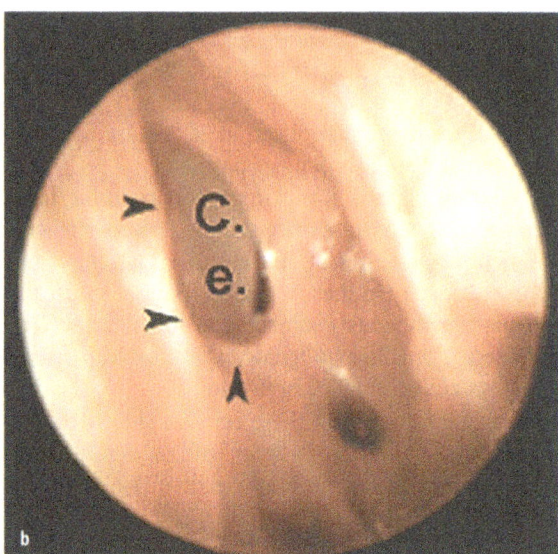

**Fig. 10. a** Pre-operated condition with recurrency of obstructed cellulae ethmoidales (C.e.) by hperplastic mucosa (➤) and polyps (★). **B** Wide open access to the ethmoidal cells, no hyperplastic mucosa (➤) 2 months after laser surgery

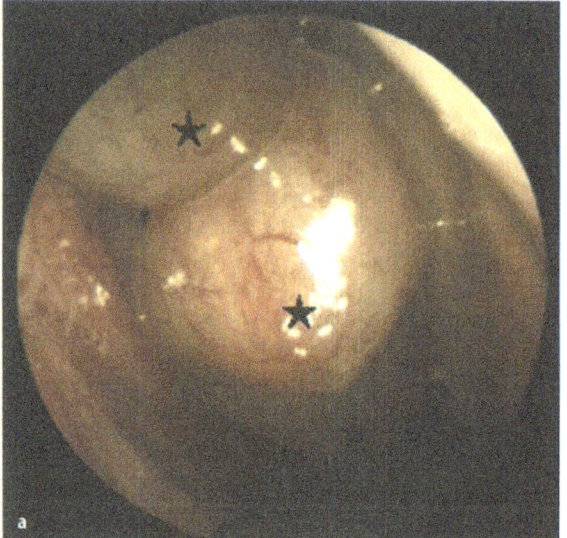
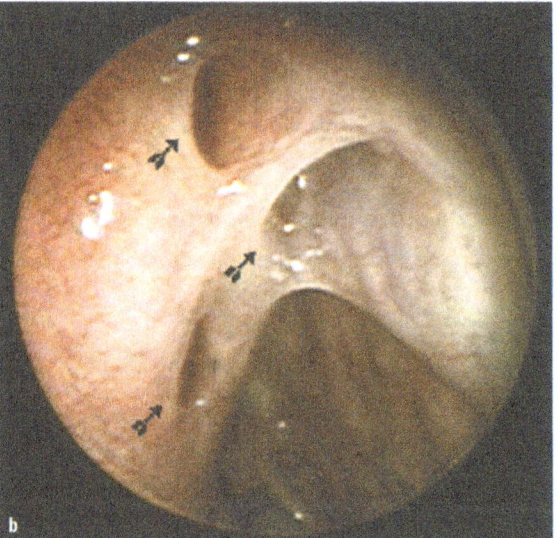

**Fig. 11. a** Recurrence of polyposis (★) in the middle nasal meatus, right side, prior to diode laser surgery. **b** Free ethmoidal cell system (➤) after successful FEELS procedure

at the surface of the target tissue. For this purpose, the laser is used at lower power settings. Consequently, a steeper temperature gradient is produced in the tissue and, in turn, in clinical routine applications our patients only rarely report sensing heat even during irradiation of very sensitive areas. The 940 nm Diode laser also permits a very gentle approach to sensitive areas. In chopped mode with extremely short exposure times and long pauses – this prevents heat accumulation in the tissue – this laser can be used for very precise work with only a minimal coagulation zone.

### Applying The Argon Laser – KTP Laser

The Argon ion laser and the KTP laser have been used to ablate tissue in the immediate vicinity of the skull base and olfactory epithelium utilizing the gentle effect of these lasers in anatomically high-risk areas.

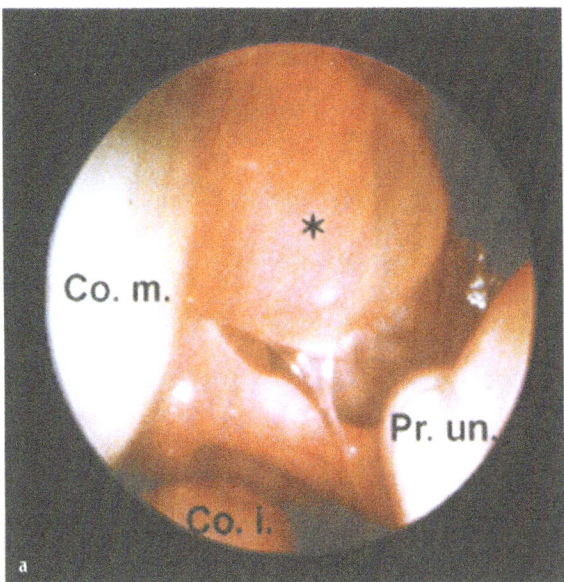
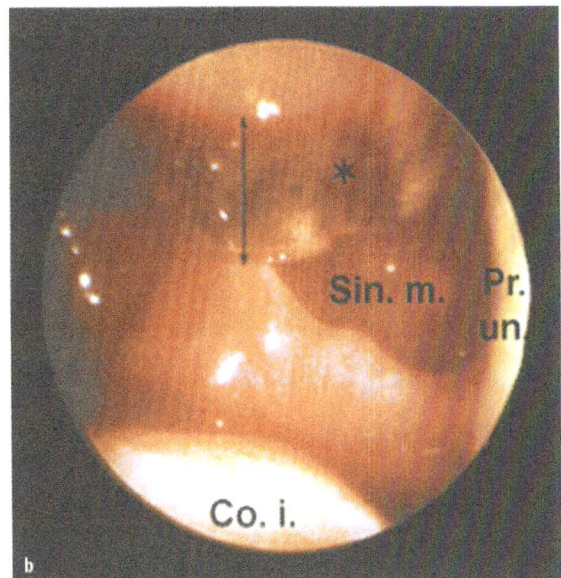
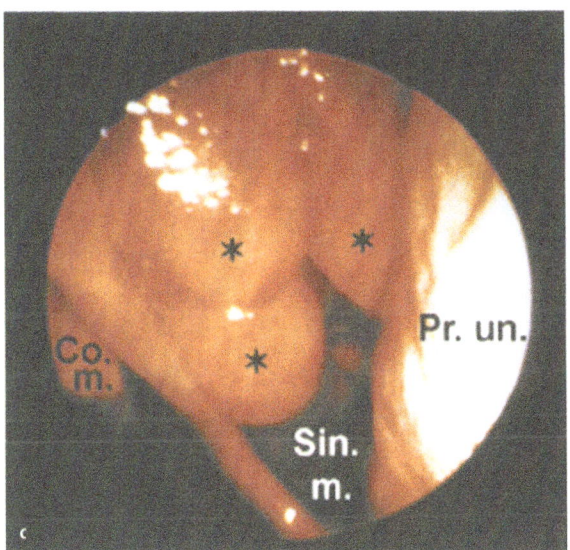
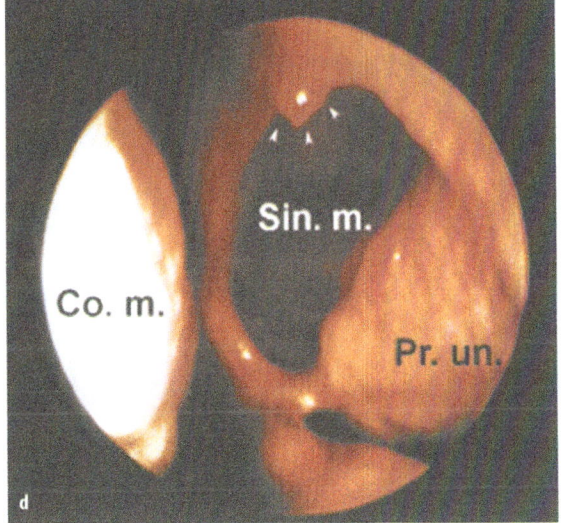

**Fig. 12. A** Polyposis nasi et sinuum, obstruction of the left maxillary ostium by polypous mucosa. **B** Situation after laser treatment by 3-W continuous mode. Clearly visible is the zone of irreversible damage (*arrow*). **C** Obstruction of the supraturbinal maxillary ostium by polyps in the maxillary sinus (★). **D** Supraturbinal ostium after endoscopic laser surgery with vaporization technique. The marking with arrows represents the wound margin immediately after polyp resection

This feature relates mainly to the much lower optical penetration and effective depth.

## Tactility Mediated by the Laser Fibre?

Unlike traditional functional endoscopic laser surgery, resecting instruments are not employed in laser surgery and suction tubes to remove secretions and blood are dispensable and, therefore, not used. Thus, the consistency of the tissue can be tested by gently touching with the laser fibre to distinguish between soft hyperplastic or polypous tissue versus scar tissue or bony structures coated by mucosa. This is an additional feature of the laser contributing to the preservation of periostal tissue and tissue in the immediate vicinity of bone to prevent inadvertent bone necrosis.

In nasal polyposis patients, we frequently observed hyperplasia of the inferior concha with thickened, "cauliflower-like" posterior edges and, in some cases, hyperplasia or polypous degeneration of the heads of the middle turbinates to obstruct ventilation of the infundibulum. No later than in the second session of polyp ablation, the affected nasal conchae were sub-

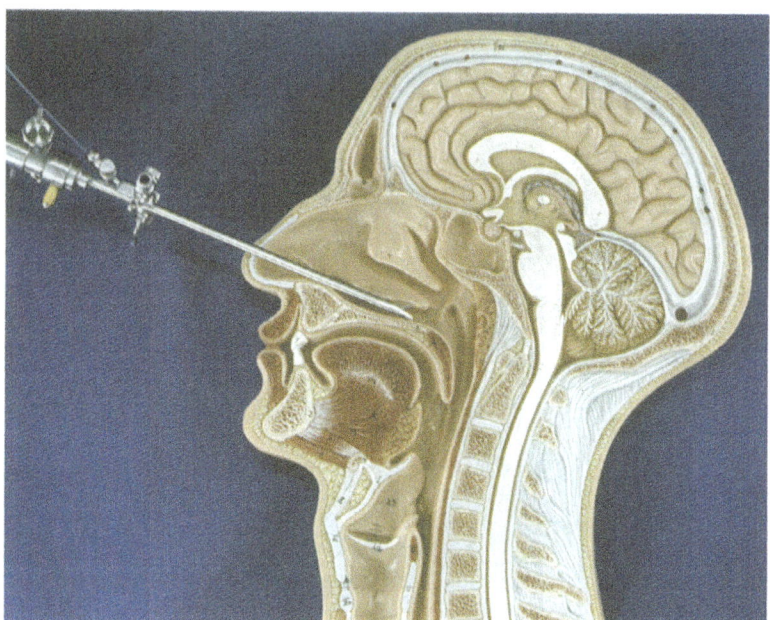

**Fig. 13.** Diagrammatic sketch shows the positioning of the laser fibre application sheath in the lower nasal meatus for laser surgery of turbinal hyperplasia

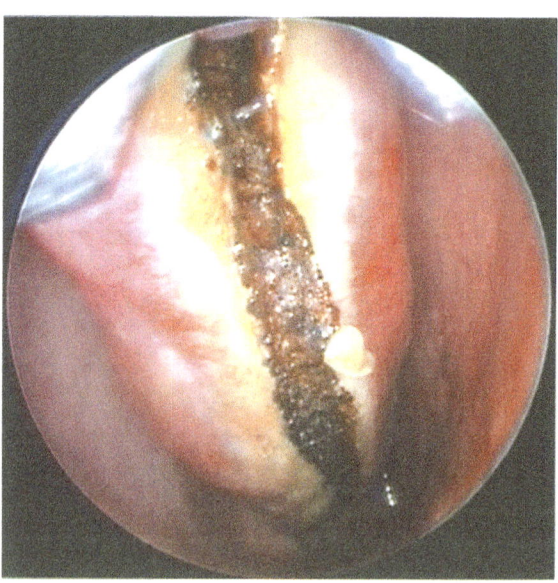

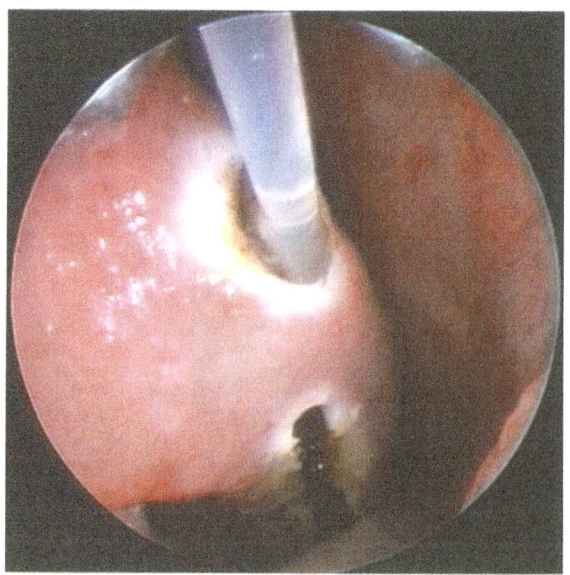

**Fig. 14.** Strip coagulation/vaporization for inferior turbinal hyperplasia

**Fig. 15.** Interstitial Diode laser coagulation at the anterior segment of a hyperplastic inferior turbinate

jected to laser cauterization. Especially the posterior edges of the inferior turbinates were thus much better visualized and more amenable to treatment after a first round of polyp ablation.

## Results

Since classical genuine polypous rhinosinusitis is an intrinsic mucosal disease, it was not expected that endoscopic laser surgery would increase the recurrence-free interval in patients with previous conventional surgery. However, not every recurrence is accompanied by symptoms necessitating another operation. In patients with previous conventional surgery presenting themselves at our institution for endoscopic follow-up every quarter, we succeeded in detecting polypous mucosal structures early and immediately commenced minimally invasive laser surgery to remove these structures under superficial anaesthesia. Com-

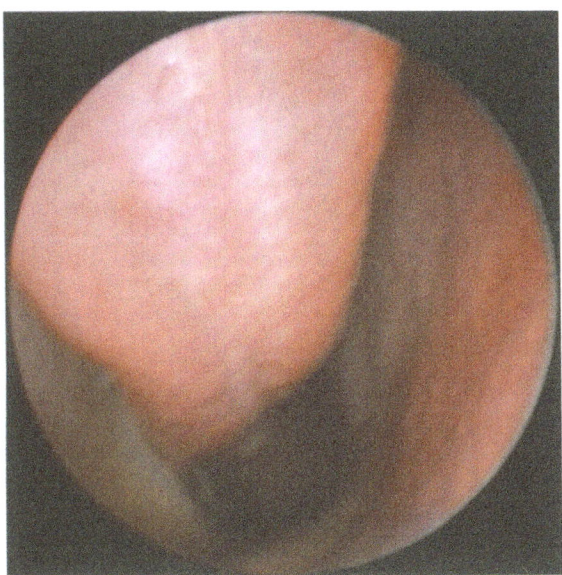

**Fig. 16.** The endoscopic inspection 12 weeks after laser treatment shows an excellent scarification of the formerly hyperplastic inferior turbinate and the really enlarged lower nasal meatus

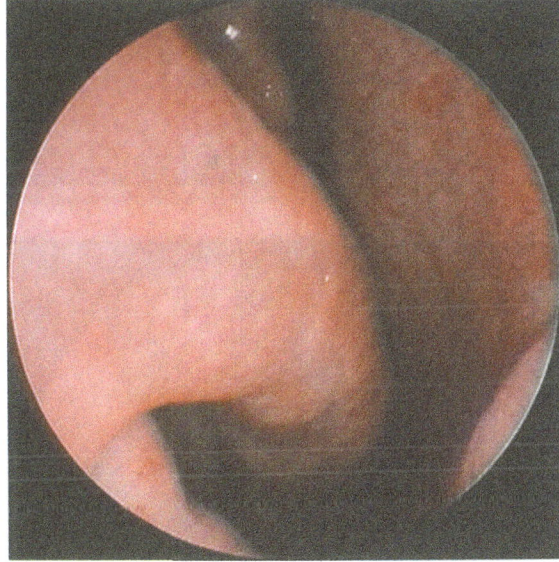

**Fig. 17.** The head portion of the formerly hyperplastic inferior turbinate has now become slim. A normal air passage is restored; 12 weeks after laser surgery

bined with microendoscopy-assisted diagnostic tests and therapy, early treatment of polypous structures managed to maintain the general well-being of the patients for a longer period of time than simple topical-inhalative steroid therapy.

The impairment of nasal breathing was sufficiently improved in all polyposis patients. Improvement of

symptoms was apparent immediately or within the first 4 weeks after the operation. A single session was required to remove solitary and choanal polyps, which cannot be removed with the loop without putting the patient at risk, as well as broad-based mucosal polyps or oedema beds located on critical orifices of the main nasal cavity and nasal ostia.

In recurrent polyposis with impaired nasal breathing, sufficient ventilation was restored after an average of three sessions of laser therapy scheduled at intervals of 4–6 weeks.

In many cases, we succeeded in remedying anosmia or hyposmia by carefully removing the polypous tissue obstructing the olfactory cleft under preservation of the olfactory epithelium (at low laser power, short pulses of irradiation). Olfactory function was restored either right away or within up to 3 to 6 weeks. There were no cases of olfactory dysfunction due to laser surgery.

In some cases, it was necessary to repeat laser therapy for the following reasons:

- Occasionally, hypersecretion, oedema and strong sneezing were observed, in some cases, reduced patient compliance. These patients were given topical and/or systemic steroids and antihistamines to prepare them for the next laser session.

- Coagulated tissue and carbonization in the main nasal cavity only rarely interfered with endoscopic vision of the anatomical details. In these patients, it proved sensible to wait for demarcation and rejection of the coagulated tissue that occurs as a late effect and depends on the penetration depth of the laser. It is not in all cases necessary to completely and immediately remove the diseased tissue. Rather, interstitial coagulation in the desired area is sufficient in many cases, even though the positive immediate effect to have restored nasal breathing is not achieved. Which laser power to select (3–10 W, continuous-wave mode) depends on the patient's individual sensitiveness to pain.

- There were limits as to how long non-sedated patients could be treated. This was determined more by personal anxiety and uncertainties than any objective inconvenience.

- After successful vaporization of main nasal cavity and middle meatal polyps with the preblackened fibre tip of the Nd:YAG laser, the reduced occlusion of the semilunar hiatus gave rise to the renewed formation of mucosal polyps from the more posterior paranasal sinus systems to relieve pressure.

- In some of the patients with turbinal hyperplasia (4%) we observed renewed impairment of nasal breathing no sooner than 12 months after primary management. This was observed mainly in patients with concurrent septal deviation affecting the aerodynamic condition of the nose. Patients with

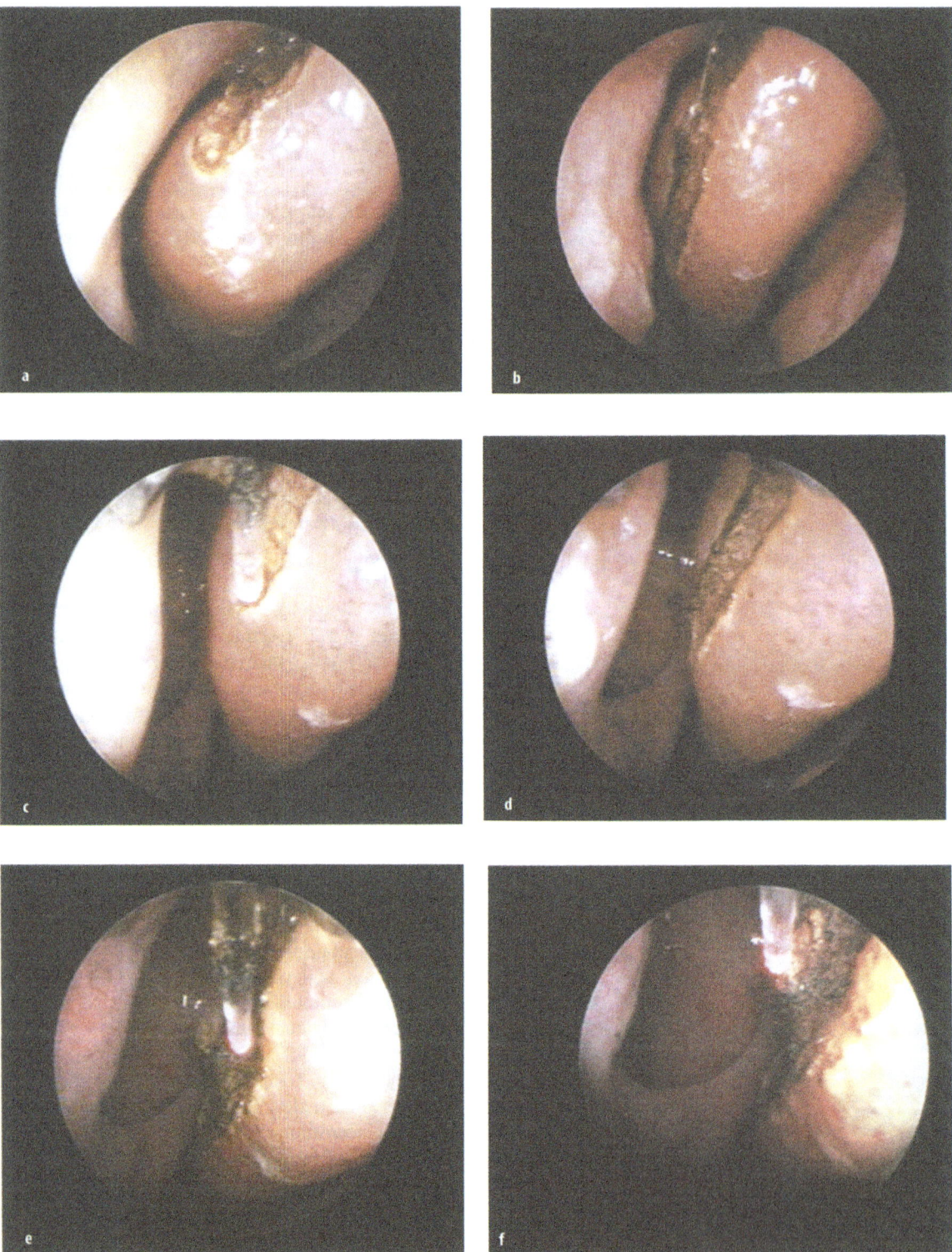

**Fig. 18 a-f.** Patient with recurrent Sinusitis maxilloethmoidalis. Extremely narrow infundibulum. Under videoendoscopic monitoring the middle turbinate (★) is reduced by Diode laser-assisted vaporization at its lateral face. This improves ventilation and drainage at this very keystone area of the paranasal system. After moderate vaporization of the caudolateral soft tissue hyperplasia of the middle turbinate the preoperated and enlarged ostium becomes visible again

recurrence of turbinal hyperplasia-related impaired breathing were again subjected to laser surgery; in some cases septal spurs were surgically removed.

Children with cystic fibrosis and recurrent polyposis (> 7 years of age) were successfully treated with the laser in general anaesthesia. The incidence of infections was reduced by maintaining the patency of nasal passage and improving ventilation and drainage. Due to the usually poor cardio-pulmonary status of these patient, it was essential to administer general anaesthesia in as few cases as possible. Many of these young patients had previously undergone one or several pansinus operations.

Pulmonary symptoms were reduced for moderate periods of time by improving the nasal passage in sin-ubronchial syndrome and bronchial asthma patients. With few exceptions, these patients were capable of resuming their normal activities directly after treatment, if feasible considering their general status.

We failed to effect superficial anaesthesia in only a few patients, in whom we then resorted to infiltration and nerve block anaesthesia.

Laser surgery successfully restored ventilation in all cases of polyps recurring in the middle meatus and natural and surgically enlarged ostia and chronic rhinosinusitis cases with narrowing of the middle and inferior meatus by hyperplastic tissue.

## Allergic Rhinitis – Pseudoallergic Rhinitis – Vasomotor Rhinopathy – Inferior Turbinal Hyperplasia

In patients with an endoscopic diagnosis of solitary turbinal hyperplasia confirmed by rhinoresistometry or acoustic rhinometry, the laser endoscope is advanced through the inferior nasal meatus to the dorsal end. Subsequently, the hyperplastic tissue is vaporized and coagulated proceeding from posterior to anterior. The thickened posterior edges are reduced first followed by Diode or Nd:YAG laser strip vaporization and coagulation (continuous wave mode, 3 to 6 W power) creating a line in the free caudal margin of the inferior turbinate proceeding from dorsal to ventral. In doing so, it is critical to only slowly withdraw the fibre tip to ensure that the turbinal vessels are adequately coagulated. Working under endoscopic vision, it is important not to proceed beyond the white coagulation zone (blanching) forming in the turbinal mucosa near the fibre tip. If these recommendations are followed, the procedure can be performed with either the Nd:YAG or Diode laser and no haemorrhage is caused, i.e. there is no need for a nasal packing.

Any haemorrhage caused by not exposing the tissue to the laser beam long enough to effect adequate coagulation or by an inadvertent movement with the endoscope can be staunched by simple resting the fibre on the respective location or with another cycle of laser exposure. Usually, the rather moderate vaporization leading to the desired immediate effect on nasal breathing is accompanied by good tissue coagulation. This, in turn, effects not only adequate closure of mucosal and submucosal vessels keeping the procedure free of haemorrhage, but also provides for good scarification of hyperplastic and hyperreactive swelling tissues due to the high effective penetration depth offered by both laser systems.

Aside from the strip coagulation described above, the surgeon should remove any hyperplastic tissue in aerodynamically significant areas. If the nose is very narrow, the surgical strategy must be changed and the procedure initiated in an anterior location progressing towards the back, as the endoscope can be advanced under these anatomical conditions only when the tissue is progressively reduced with the laser beam (Table 1).

The development of equipment for this type of laser surgery has continued and recently Storz endoscopes with smaller external diameters have become available as alternatives to the large laser endoscopicapplication sheaths.

**Table 1.** Parameter guidelines.
Hyperplasia of Concha nasalis inferior

| Diode laser ($\lambda$ = 940 nm; $\lambda$ = 810 nm) | |
| --- | --- |
| Therapeutical goal | Coagulation for permanent tissue shrinking |
| | Partial sealing of submucous venous plexus |
| Mode | Continuous application (continuous mode) |
| | Contact procedure |
| | Interstitial application |
| Power | 2–7 W |
| Special remarks | Preblackening of laser fibre tip not always necessary |

| Nd:YAG laser ($\lambda$ = 1064 nm) | |
| --- | --- |
| Therapeutical goal | Coagulation for permanent tissue shrinking |
| | Partial sealing of submucous venous plexus |
| Mode | Continuous application (continuous mode) |
| | Contact procedure |
| | Interstitial application |
| Power | 2–5 W |
| Special remarks | Preblackening of laser fibre tip |

**Table 2.** Parameter guidelines.
Enlargement of the middle nasal meatus

| Diode laser (λ = 940 nm) | |
| --- | --- |
| Therapeutical goal | Soft tissue resection by vaporization (without deep thermal impact) |
| Mode | Pulsed application (chopped mode) |
| | Contact procedure |
| Power | 20–60 W |
| Pulse duration (laser on) | 0.06–0.2 s |
| Interval (laser off) | 0.4–0.8 s |
| **Nd:YAG laser (λ = 1064 nm)** | |
| Therapeutical goal | Soft tissue resection by vaporization |
| Mode | Pulsed application (chopped mode) |
| | Contact procedure |
| Power | 35–60 W |
| Pulse duration (laser on) | 0.05–0.1 s |
| Interval (laser off) | 0.8–1.0 s |
| Special remarks | Preblackening of laser fibre tip |
| Caution | Thermal effects in close vicinity to orbita and rhinobase |
| **Diode laser (λ = 810 nm)** | |
| Therapeutical goal | Soft tissue resection by vaporization (without deep thermal impact) |
| Mode | Pulsed application (chopped mode) |
| | Contact procedure |
| Power | 20 W |
| Pulse duration (laser on) | 0.06–0.2 s |
| Interval (laser off) | 0.4–0.8 s |

## Laser-Assisted Surgical Enlargement of the Middle Meatus in Chronic Rhinosinusitis, Dissection and Partial Resection of the Media Concha Bullosa

### Laser Surgery of the Middle Meatal Soft Tissues

Laser-assisted surgical management of chronic rhinosinusitis, entails the excision of hyperplastic soft tissue at the lateral aspect of the middle turbinate in order to enlarge the transverse diameter of the middle meatus and facilitate ventilation of the maxilloethmoido-frontal compartment. In addition to patients with no prior surgery, this management may also be applied to patients with recurrent symptoms after previous sinus surgery (Table 2).

### Laser Surgery of the Concha Bullosa Media

However, the presence of a uni- or bilateral concha bullosa media is another factor contributing to chronic rhinosinusitis with ensuing massive narrow-

ing of the middle meatus and infundibulum. This situation requires not only the excision of soft tissues eo loco, but complete laser surgical resection of the lateral portion of the bullous concha. However, this intervention, if performed with conventional instruments, may well also destabilize the remaining medial middle portion. Inadvertent fractures due to mechanical trauma may loosen the fine bony insertion at the rhinobase. In contrast, endonasal laser surgery does not cause any more mechanical trauma than direct tactile contact with the laser fibre. A combination of the endoscopic laser rhinoscope sheath (Karl Storz, Tuttlingen, Germany) and the 940-nm Diode laser (Dornier), is ideally suited of this operation. By using the extremely short pulse mode which produces minimal heat, the area of thermal damage to the bone is limited and thus has a favourable effect on wound healing (Fig. 19).

The 810-nm Diode laser is also suitable for this type of procedure, although prolonged periods of exposure are required to vaporize bone with this laser in pulse mode with a maximal output power of only 20 W. This may in some cases lead to a pronounced area of thermal damage to the bone and, thus, somewhat prolonged healing periods.

Adequate partial resection of a medial bullous concha can also be achieved within acceptable periods of exposure using the Nd:YAG laser. Despite fibre tip preblackening and short pulses with long periods of laser pauses, pronounced thermal alteration to the adjacent bone may occur. Thus, the bone remains unepithelialized under a thick fibrin coat for several weeks until the mucosal layer is completely formed 6–8 weeks later. However, as with other laser systems, osteonecrosis is observed only in exceptional cases, provided this laser is used in compliance with our recommendations.

### Postoperative Care

The level of postoperative care is determined by the subjective well-being of the patient. Usually this includes nasal care with sucking off secretions and removal of fibrinous layers or scab once or twice a week in the first 30 days after the operation. It is recommended that the patient applies nasal ointments which contains dexpanthenol and/or vitamin A as well as continuously administrate (2–3 times daily) a nasal douche which contains physiological Nall solution. In contrast, vasoconstrictors or steroids should not be used during the acute wound-healing phase.

Following laser cauterization of the turbinates, we have occasionally observed that the hyper sensitive mucosa shows a strong tissue reaction with swelling and massively impeded nasal breathing. These excep-

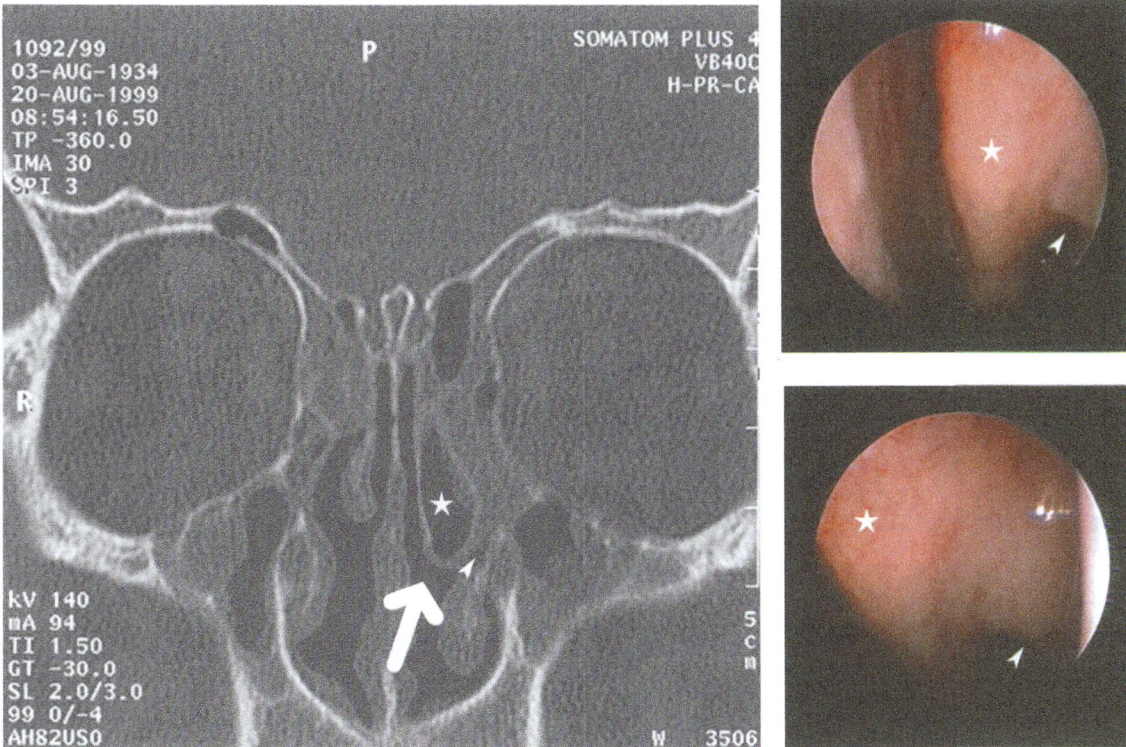

**Fig. 19.** Concha bullosa media left (★). The arrow (→) represents the direction of the endoscopic view to the infundibulum and the obstructed middle nasal meatus (➤). Patient with recurrent sinusitis

tional cases could be managed by administration of nasal ointment containing naphazoline to reduce the symptoms and, in extremely rare cases, systemic steroidal antiphlogistics.

However, laser-assisted surgical enlargement of the middle meatus must be founded more on physiology than on purely mechanical considerations. In using thermal techniques to remove or reduce tissue in this very sensitive area, the surgeon must keep in mind that this anatomical region is crucial to mucociliary clearance and also to the entire process of lymph drainage, especially in the ethmoidal and maxillary sinuses. Thus, one must take care not to coagulate the mucosal and submucosal tissue layers in the entire periphery of the ostium and previously operated neo-ostium and thus produce irreversible alterations and ensuing scar formation. No more than a semicircle should be thermally damaged by the laser-assisted incision. This is less significant for mucociliary clearance, which usually returns to normal within 3 months. In contrast, laser application-induced thermal damage of the periphery creates extreme problems for lymph drainage and ultimately leads to persistent edematous mucosal damage over a circumscribed are.

The use of flexible, steerable fiberscopes allows the surgeon to access even initially hidden pathological structures in patients with a constricted nasal cavity and narrow middle meatus. Miniaturized telescopes and microendoscopes may also be used for this purpose.

**Complications**

In terms of complications, we have encountered two cases of paranasal sinus infection with facial swelling. These were caused by additional swelling of the nasal mucosa due to the development of acute rhinitis after laser-assisted surgical polypectomy. This led to bacterial superinfection following acute exacerbation of the chronic sinusitis.

Moreover, there has been one case of an asthma attack following laser surgery. After laser cauterization of the turbinate, moderate haemorrhaging required nasal packing in three cases, whereas necrosis of the inferior turbinate bone was found in two other cases, but healed completely with routine therapy.

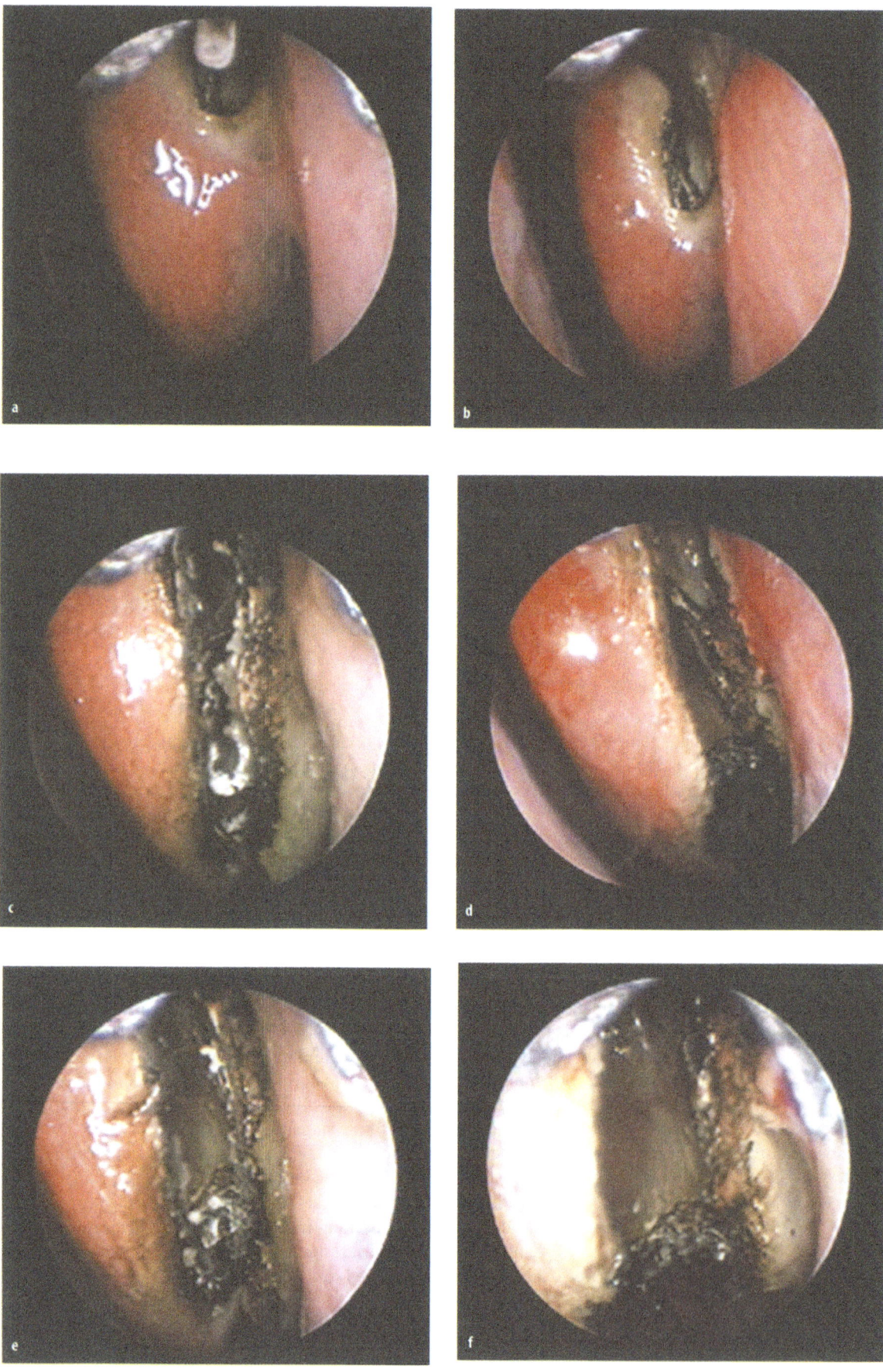

## Synechiae, Cicatricial Stenosis and Granulation (Figs. 21, 22)

Septoturbinal synechiae, i.e. postoperative or post-traumatic adhesions between the middle and lower turbinate and the septum, or between the middle turbinate and the medial maxillary sinus wall, as well as scar formations, stenosis and granulation tissue, are indications for laser surgery which can be managed endoscopically and resolve the problems on a long-term basis. Similarly, stenoses obstructing access to the maxillary sinus, ethmoidal cell system and the sphenoid sinus occurring after conventional surgery are also indications for laser application. For this purpose, the high power (35–40 W) fibre-guided Nd:YAG laser in pulse mode is used. The Diode laser and $CO_2$ laser have also been employed in this indication. It is recommended that the synechiae be transected and the adjacent tissue from which the synechiae originate then be removed by vaporization in order to create a separation in space and prevent the recurrence of adhesions. This procedure makes the use of packing or spacers unnecessary and the recurrence rate has been consistently low at less than 10%.

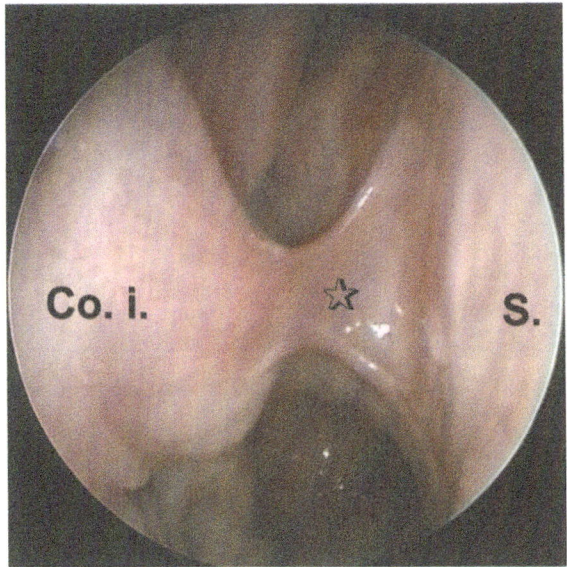

**Fig. 21.** Endoscopic view of the posterior third of the right main nasal cavity: there is a synechia (☆) between inferior turbinate *(Co. i)* and septum *(S)*

## Septal Spurs and Septal Crest

Small to moderate cartilaginous or partially bony septal crests and spurs can be excised with the Diode laser or the Nd:YAG laser in pulsed mode at high laser power and short exposure intervals (35–40 W, 0.1 s). However, it is important to limit the excision to one side of the septum only. We experienced one case of dorsal septal perforation without clinical symptoms after treating the patient in continuous mode at low laser power.

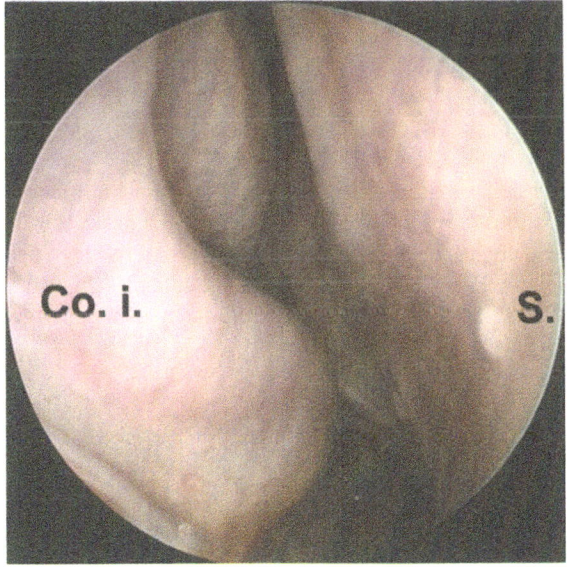

**Fig. 22.** Endoscopic view 4 weeks after fibre-assisted vaporization of the synechia. Normal mucosa at the septum *(S)* and at the inferior turbinate *(Co. i.)*

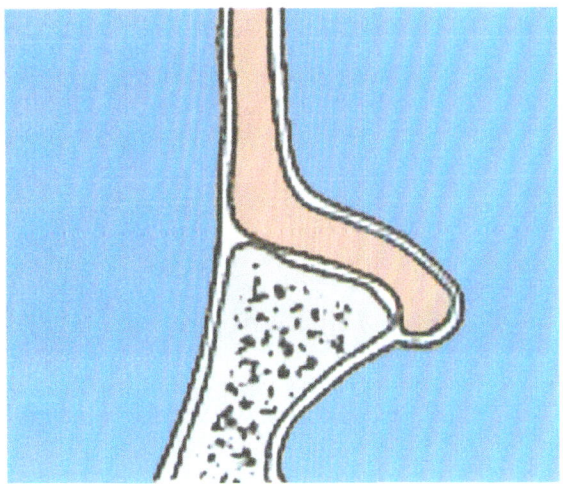

**Fig. 23.** Diagram showing the ideal situation for the laser resection of a septal spur/crest

**Fig. 20a-f.** With the short pulsed Diode laser medilas D (Dornier Med Teh, Germany) the lateral bony lamella of a concha bullosa media can be vaporized and resected with only minimal coagulation impact to the anatomic surroundings.

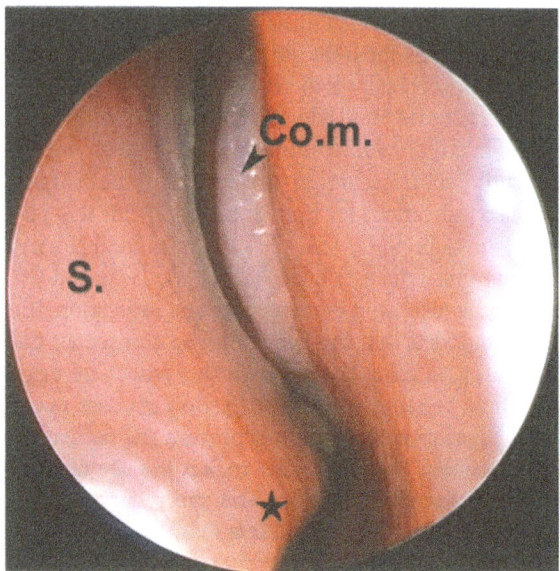

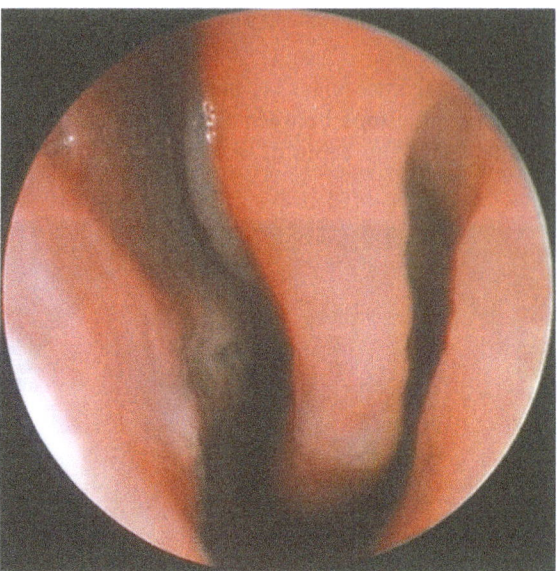

**Fig. 24.** 29-year-old patient with impeded nasal breathing. Endoscopic inspection of the left nasal cavity. The septum *(S)* is nearly median, but a large septal crest is narrowing the inferior nasal meatus. The inferior turbinate *(Co. i.)* is slim. The crest extends up to the posterior third of the nasal cavity. Condition pre-OP

**Fig. 26.** 12 weeks post-OP: the inferior nasal meatus is wide open again. Nasal breathing is normalized

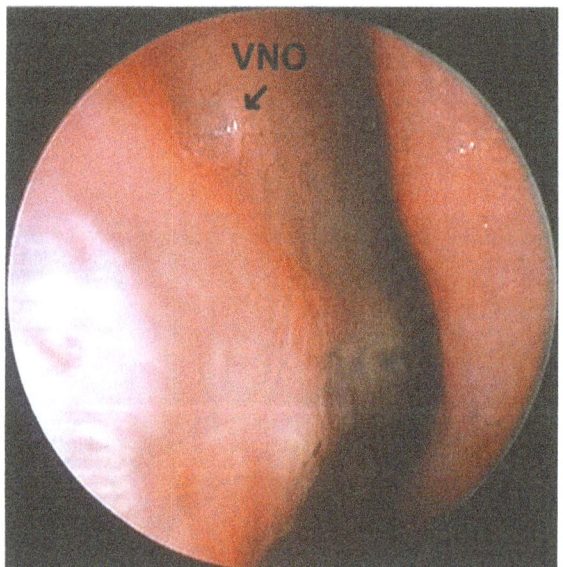

**Fig. 25.** 12 weeks post-OP: the re-epithelialization has completely closed the resection area. Organum vomeronasale *(VNO)*

As an alternative, the $CO_2$ laser can be applied with excellent success in this indication. These successes certainly do not mean that conventional septoplasty has become an obsolete technique in our hospital. S-shaped to wave-shaped septal deviations are a priori not amenable to laser therapy. The concave surface of these variations may suggest that the nasal cavity is very wide, whereas the convex side manifests narrowness. Ideally suitable are deviations which are found in the middle or lower meatus as sharply delineated crests contacting the turbinate (frequently with strictly unilateral occurrence), whereas the endoscopic view from the other main nasal cavity reveals the nasal septum to be located close to the middle (Figs. 23, 24, 25, 26).

## Soft Tissue Hyperplasia of the Nasal Septum

Aside from turbinal hyperplasia, a small number of patients presented with impairment of nasal breathing due to the formation of a large septal tissue hyperplasia. Permanent tissue shrinkage can be achieved in these patients, without perforation by using spot coagulation at low laser power.

## Mucoceles and Cysts

Postinfectious, posttraumatic or postoperative paranasal sinus and nasopharynx mucoceles confirmed by computed tomography can be permanently healed by appropriate vaporization of the wall of the mucocele, at the same time restoring the width and patency of the ostium. Similarly, cysts forming after previous surgery in the maxillary sinus area can be excised under fiberscopic guidance taking advantage of the enlarged ostia. This procedure has proven its value as a minimally traumatic technique which showed no recurrence over observation periods ranging from 2 to 5 years.

## Recurrent Epistaxis

### Laser Management of Vascular Haemorrhagic Diathesis, Hereditary Haemorrhagic Telangiectasia and Capillary Haemangioma

The use of the fibre-guided laser systems for surgical management of recurrent, non-tumorous nasal bleeding has become an internationally established method of treatment. For this purpose, continuously emitting laser systems, such as the Argon laser (Parkin and Dixon 1981, 1985; Haye 1991, 1992), KTP laser (Levine 1989) and Nd:YAG laser – which are widely used in Germany (Illum and Biewing 1988; Dobrovic and Hosch 1994; Werner 1997, 1999) – have been employed. The coagulative effect of the laser can be used to safely and definitively eliminate recurrent, circumscribed sources of bland haemorrhage in the nasal cavity. Similarly to many other laser-assisted techniques, the key to success is an endoscopic surgical procedure which allows operating with precision under excellent visual conditions and detailed recognition while at the same time optimally preserving the peripheral healthy anatomy. In contrast, if there are more extensive malformations of the vascular tissue structure, such as in Osler-Weber-Rendu disease and primary and secondary haemostasis deficiencies, the incidence of recurrent haemorrhage may be clearly reduced, but the effects of treatment are usually only temporary.

Due to its minimal optical depth of penetration and high absorption by haemoglobin, the light of the Argon Ion laser is very suitable for the treatment of superficial, richly vascularized alterations of the mucosa. The good endoscopic vision and optional optical magnification of the site displayed on the monitor allow for identification of capillaries which are hardly discernible with the naked eye.

If the structures are localized in the anterior part of the nose and easily identified, laser surgery is carried out with the bare fibre or the focusing handle at a focal spot diameter of 0.5 mm. Management usually involves several sessions scheduled at intervals of 4 weeks. In the no-touch or close-to-no-touch technique, the authors routinely use the following laser parameters: power setting at 2–5 W, period of exposure of 0.02–0.1 s and a repetition rate of up to 6 Hz.

The Nd:YAG laser and diode laser, used in a no-touch technique, are also suitable therapeutic options for this indication.

Reports regarding the coagulative management of vascular disease and malformations of the nasal mucosa with the Diode laser are rarely mentioned in the literature (Hopf). Depending on the type of Diode used, the wavelength emitted by this laser ranges from 810–980 nm. Thus, the depth of penetration (Hopf) ranges between the values of the Argon and KTP laser on the one hand, and that of the Nd:YAG on the other. Hence, the biophysical properties of the Diode laser are very promising for therapeutic interventions: the penetration depth in biological tissue is higher than that of Argon or KTP lasers, which permits coagulation not only of small-lumen vessels located directly at the mucosal surface, but also thermal penetration into subepithelial layers.

The site is endoscopically inspected on the video monitor only. Due to the excellent quality of the high-magnification images, the affected mucosal vessels can be visualized in detail. The laser endoscope is selected depending on the intended strategy and the target region. The extremely slim laser application sheaths with longitudinal-oval cross-sections (Karl Storz, Tuttlingen, Germany) have proven extremely useful for this application. These are equipped with a channel for guiding laser fibres of up to 1.1 mm outer diameter, which easily allows for the passage of fibres of 600 µm.

When the laser fibre (600 µm) exits the working channel, it is advanced so that it protrudes beyond the distal end of the telescope at a distance of approximately 3 mm, i.e. just enough to endoscopically see the image of the laser fibre tip on the video monitor.

The patients treated by the authors suffered from recurrent epistaxis of various etiologies. Approximately one third of the patients presented with recurrent local bland nasal bleeding. The causes included telangiectatic or pyogenic septal granuloma, capillary septal haemangioma, uni- or bilateral vascular ectasia in the Kiesselbach's plexus (Little's spot) and/or vascular convolution near the tip of the middle turbinate. In the remaining third of our patients recurrent systemic nasal bleeding due to hereditary haemorrhagic teleangiectasia (Osler-Weber-Rendu disease) was evident.

### Laser treatment

We routinely use the Diode laser in pulsed mode (940 nm) at a laser power between 10 and 35 W and exposure periods of 80, 100 or 200 ms with exposure breaks of 200 to 600 ms. The determination of parameters is based on the size and area which is to be treated with the laser. We always use the non-contact-technique for exposure and keep a working distance of approximately 1 mm from the tissue. The fibre tio

**Table 3.** Parameter Guidelines. Concha bullosa media – infundibulotomy

Diode laser (λ = 940 nm)

| 1st step – Preparation | Superficial mucosa coagulation to prevent from bleeding along the resection margins |
|---|---|
| Mode | Continuous application (continuous mode) |
| | Contact procedure (also: non-contact or interstitial application) |
| Power | 2–4 W |
| Special remarks | Preblackening of laser fibre tip not always necessary |
| Caution | Thermal effects in close vicinity to orbita and rhinobase |

Diode laser (λ = 940 nm)

| 2nd step – Therapy | Opening and partial resection of the bony concha bullosa by vaporization with minimal coagulation of the remaining medial bone lamella |
|---|---|
| Mode | Pulsed application (chopped mode) |
| | Contact procedure |
| Power | 20–80 W |
| Pulse duration (laser on) | 0.03–0.2 s |
| Interval (laser off) | 0.5–0.8 s |

Diode laser (λ = 810 nm)

| 1st step – Preparation | Superficial mucosa coagulation to prevent from bleeding along the resection margins |
|---|---|
| Mode | Continuous application (continuous mode) |
| | Contact procedure (also: non-contact or interstitial application) |
| Power | 2–4 W |
| Special remarks | Preblackening of laser fibre tip not necessary |
| Caution | Thermal effects in close vicinity of orbita and rhinobase |

Diode laser (λ = 810 nm)

| 2nd step – Therapy | Opening and partial resection of the bony concha bullosa by vaporization with minimal coagulation of the remaining medial bone lamella |
|---|---|
| Mode | Pulsed application (chopped mode) |
| | Contact procedure |
| Power | 15–20 W |
| Pulse duration (laser on) | 0.1–0.2 s |
| Interval (laser off) | 0.5–0.8 s |

Nd:YAG laser (λ = 1064 nm)

| 1st step – Preparation | Superficial mucosa coagulation to prevent from bleeding along the resection margins |
|---|---|
| Mode | Continuous application (continuous mode) |
| | Contact procedure (also: non-contact or interstitial application) |
| Power | 2–3 W |
| Special remarks | Preblackening of laser fibre tip |
| Caution | Thermal effects in close vicinity to orbita and rhinobase |

Nd:YAG laser (λ = 1064 nm)

| 2nd step – Therapy | Opening and partial resection of the bony concha bullosa by vaporization with minimal coagulation of the remaining medial bony lamella |
|---|---|
| Mode | Pulsed application (chopped mode) |
| | Contact procedure |
| Power | 15–60 W |
| Pulse duration (laser on) | 0.1–0.2 s |
| Interval (laser off) | 0.6–1.0 s |
| Special remarks | Preblackening of laser fibre tip |

we use for this purpose is clean, not preblackened. Assisted by the highly magnified monitor image, these parameters permit nearly selective treatment of the vessels.

Close to the septum we prefer to work at low power settings of 10–15 W and short periods of exposure (80 and 100 ms) and longer intermissions between the pulses (200 to 600 ms) to prevent any thermal necrosis of the cartilaginous septum. We increase the period of exposure to 200–400 ms, if required. During laser application in this area, we painstakingly avoid involving corresponding areas on the opposing side in the same session so as to prevent irreversible damage to the cartilaginous septum.

The laser power and the exposure intervals are slowly and gradually increased depending on individual requirements. This iterative approximation to the individual optimal parameter set is indispensable in all patients. Optimal parameter values are attained, once there is blanching of the tissue followed by complete disappearance of the vessels exposed to the laser light.

Carefully increasing laser power, power density and exposure period allows the surgeon to largely prevent the so-called popcorn effect, i.e. sudden bursting of the blood vessels by expanding gas bubbles formed early in exposure for short periods of time if the vascular walls are not coagulated with adequate restraint.

**Table 4.** Parameter guidelines. Telangiectasia – capillary haemangioma – M. Osler-Rendu

| Diode laser ($\lambda = 940$ nm; $\lambda = 810$ nm) | |
|---|---|
| Therapeutical goal | Coagulation for permanent vessel sealing |
| Start with | |
| Mode | Continuous application (continuous mode) |
| | Non-contact/quite-contact procedure |
| Power | 2–5 W |
| Special remarks | Starting from the healthy area circle around the pathological tissue creating a superficial coagulation volume |

| Diode laser ($\lambda = 940$ nm; $\lambda = 810$ nm) | |
|---|---|
| Therapeutical goal | Coagulation for permanent vessel sealing |
| Switch to aimed application | |
| Mode | Pulsed application (chopped mode) |
| | Non-contact/quite-contact procedure |
| Power | 10–30–50 W |
| Pulse duration (laser on) | 0.08–0.2 (–0.4) s |
| Interval (laser off) | 0.2–0.6 s |
| Special remarks | Beware of popcorn effect |

| Argon-Ion laser ($\lambda = 488/514$ nm) | |
|---|---|
| Therapeutical goal | Coagulation for permanent vessel sealing |
| Start with | |
| Mode | Continuous application (continuous mode) |
| | Non-contact/quite-contact procedure |
| Power | 2–5 W |
| Special remarks | Starting from the healthy area circle around the pathological tissue creating a superficial coagulation volume |

| Argon-Ion laser ($\lambda = 488/514$ nm) | |
|---|---|
| Therapeutical goal | Coagulation for permanent vessel sealing |
| Switch to aimed application | |
| Mode | Pulsed application (chopped mode) |
| | Non-contact/quite-contact procedure |
| Power | 2–5 W |
| Pulse duration (laser on) | 0.02–0.1 s |
| Interval (laser off) | 0.1–0.4 s |
| Special remarks | Beware of popcorn effect |
| Mode | Continuous application (continuous mode) |
| | Contact procedure (also: non-contact or interstitial application) |
| Power | 2–4 W |
| Special remarks | Preblackening of laser fibre tip not necessary |
| Caution | Thermal effects in close vicinity to orbita a,nd rhinobase |

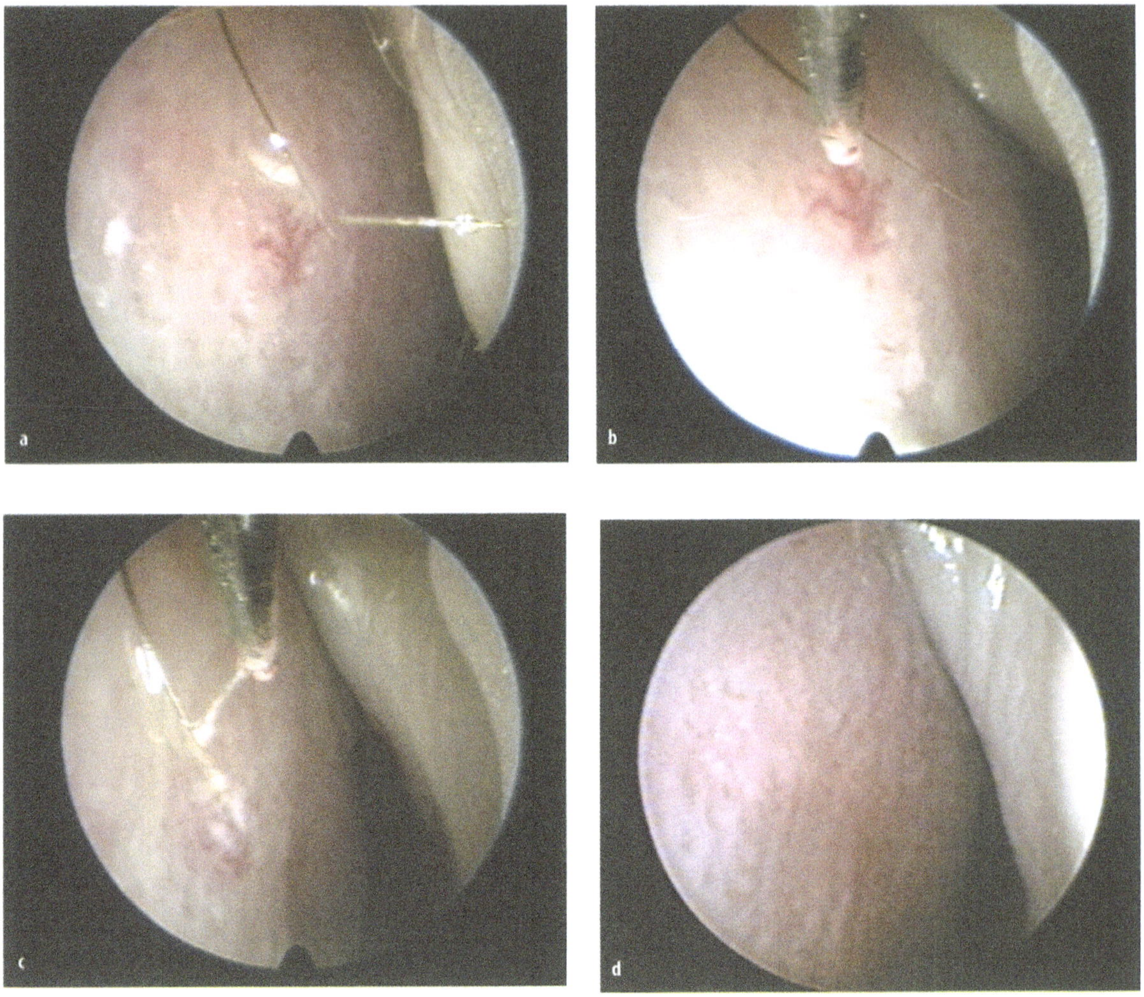

**Fig. 27a-d.** Laser treatment of vessel ectasias at Little's spot. **a, b, c** During treatment, **d** 6 weeks post-OP

Granulomas and voluminous haemangiomas are treated at increased laser power and longer exposure intervals in order to achieve a sound treatment effect even in deeper areas. In this application, the Diode laser again proves its good tissue absorption properties and coagulation capacity. Usually, granulomas are treated in just one session and any ensuing fibrous sclerotization of the area is evidenced by the tissue changing its colour to white and marked delineation from the adjacent tissue. The healing process depends on the coagulated volume and takes approximately 4 to 7 weeks.

Capillary haemangiomas often show a residual convolution, especially if located directly at the septum. To preserve the septal cartilage during the first session, it is recommended that the penetration depth of the laser be reduced and, apart from that, the laser effect on the tissues be first tested and deliberately abstain from complete excision at this stage. However, this requires a second session to be scheduled 6 weeks later (Table 4).

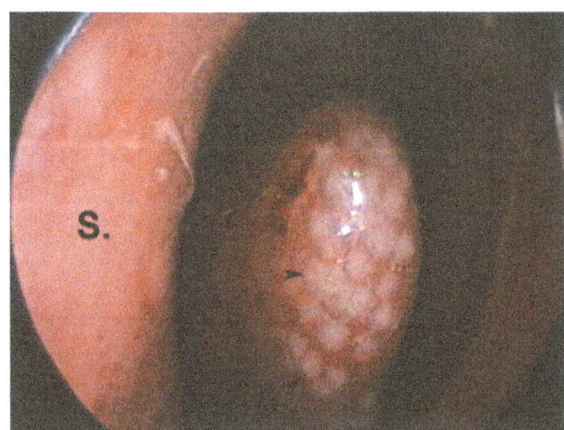

**Fig. 28.** Endoscopic appearance of the nasal mucosa with hereditary haemorrhagic telangiectasia (M. Osler-Rendu). Vessel convolutions at the septum *(S)*. The *middle turbinate* shows significant blanching phenomena (➤) with whitish appearance of the mucosa after argon laser therapy (3 W, 0.2 s, 6 Hz)

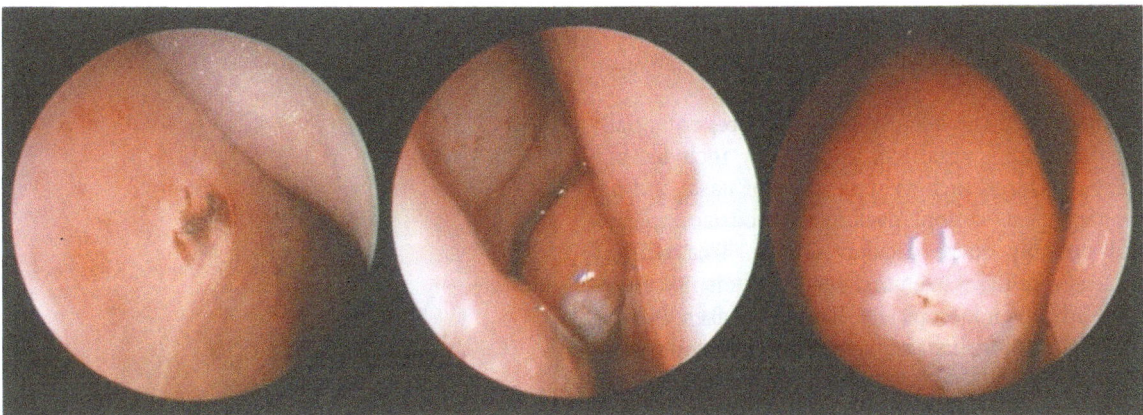

**Fig. 29.** Immediately after Argon laser therapy of an Osler's disease at the middle turbinate

The number of treatments which are to be applied to patients presenting with vascular ectasia at the locus Kiesselbachi (Little's area) depends on whether there is uni- or bilateral involvement of corresponding septal areas (Fig. 27). Poorly stabilized hypertonic patients frequently respond to laser application with iatrogenic (!) haemorrhage usually alio loco at the nasal mucosa. Haemorrhage should be immediately managed with continuous nasal packing for safety reasons. The nose of these patients needs to be prepared for the next session 1 week later with ample panthenol ointment for 2 days prior to laser treatment. As in bipolar diathermia treatment, isolated cases may show vascular ectasia at the Kiesselbach's area – adjacent to the pretreated area.

The majority of our patients presenting with hereditary haemorrhagic teleangiectasia show disseminated findings in both main nasal cavities (Fig. 28). Teleangiectasia can be found on the entire septum, inferior and middle turbinate and lateral wall of the nose. The sequence of treatment follows the symptoms, i.e. after thorough endoscopic examination those regions revealing the most extensive alterations are treated first, since they can be expected to be causative for the most recent epistaxis episodes. The surgeon must painstakingly avoid additional iatrogenic haemorrhage that might be caused by the tip of the instrument used (endoscope and laser fibre). Many of our patients required prior blood transfusions to manage the underlying disease.

All patients are afflicted by some additional mild episodes of nasal bleeding in the follow-up period, though at much reduced incidence and severity. Some of our patients may serve as examples: laser treatment succeeded in reducing the incidence of nasal bleeding from once daily to only once a week in one patient. Another patient with mild nasal bleeding prior to surgery experienced no nasal bleeding for 2–3 months.

Our patients subjectively reported their condition and quality of life to be much improved (Fig. 29).

The average duration of the treatment session depends on the indication and extent of pathological alterations and varies between 5 and 15 min.

Postoperative care depends on the individual patient's mucosal reaction and intensity of laser exposure and usually involves intensive application of panthenol ointment to the nose and lavage with solutions containing minerals (Sal Ems facticum) in the first 20 days after the operation. The mucus flow of these patients has a mildly bloody tinge for 1 to 4 days.

Any secondary surgery should not be scheduled until at least 6 weeks after the first treatment, since most scab and fibrin layers have disappeared by then and wound healing and the intended scar formation have progressed.

The vast majority of our patients subjectively described laser treatment as little or no stress. Only 2% of the patients reported some pain due to laser treatment. Invariably, in those cases, attempts to administer superficial anaesthesia had been discontinued in favour of infiltration anaesthesia. Especially the most anteriorly located area of the septum near the columella is pain-sensitive and thus a number of patients may require infiltration of a local anaesthetic.

No postoperative septal perforation, formation of synechiae or other complications were observed in our epistaxis patients treated with the argon or the diode laser.

## Discussion

Epistaxis is a very common symptom. It occurs sporadically or due to mechanical alterations once or several times in nearly every individual. The usually bland course requires no special treatment, since bleeding can be controlled with the usual counter-measures or ceases spontaneously within a few minutes.

Therapy of the acute phase of severe, potentially life-threatening nasal bleeding mainly includes primary measures, such as application of nasal and nasopharyngeal cotton packings, electrocoagulation and possibly open vascular ligation. Treatment of acute severe nasal haemorrhage usually is not an indication for endoscopically controlled laser therapy. Severe bleeding instantly obstructs the endoscopic view of the affected vessels and hampers targeted and carefully measured laser application.

Before working out a therapeutic strategy in cases of recurrent nasal bleeding, a thorough examination and subsequent diagnostic work-up has to be carried out to ascertain the underlying causes, thus allowing for differentiation between: local bland epistaxis and symptomatic epistaxis.

Amongst our patients, bleeding from haemangiomatous areas and malformation of the nasal vascular tissue were particularly common (Figs. 30, 31, 32).

The cause of local nasal bleeding is to be found in the nasal cavity itself, most often (approximately 90%) in Kiesselbach's area (Little's spot), a very dense superficial plexus of vessels in the anterior part of the nasal septum. Due to its close proximity to the nasal vault, this area is exposed to potentially intense mechanical trauma and epithelial dehydration as to flow physiology. They

very thin and easily injured mucosa of this area is firmly attached to the septal perichondrium.

Another related phenomenon is the haemorrhaging polyp which constitutes a teleangiectatic granuloma or haemangioma with a strong tendency to bleed even upon minor contact. In addition, benign tumours and highly vascularized malignomas may also show a strong haemorrhagic tendency and are amenable to palliative laser surgery.

In contrast, symptomatic nasal bleeding is caused by systemic factors and therefore indicative of a general disease. Laser surgery can be applied with success if morphological changes manifest in or at the vascular system of the nasal mucosa during the haemorrhage-free interval, e.g. Osler-Weber-Rendu disease (Teleangiectasia haemorrhagica hereditaria) (Osler 1901; Weber 1907; Rendu 1896) is characterized by a typical appearance of the mucosa. Convolution-forming arterio-venous shunts are visualized as discrete sites or, after dissemination, as thickened vessels. This autosomal-dominant congenital disease is characterized by generalized pinhead-sized darkred angiodysplasia of the skin and mucosa.

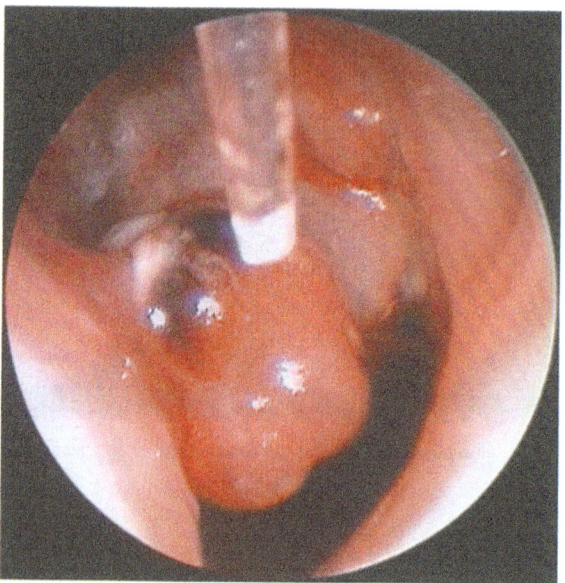

**Fig. 30.** Cavernous haemangioma in the left nasal cavity originating from the septum. Condition before laser treatment

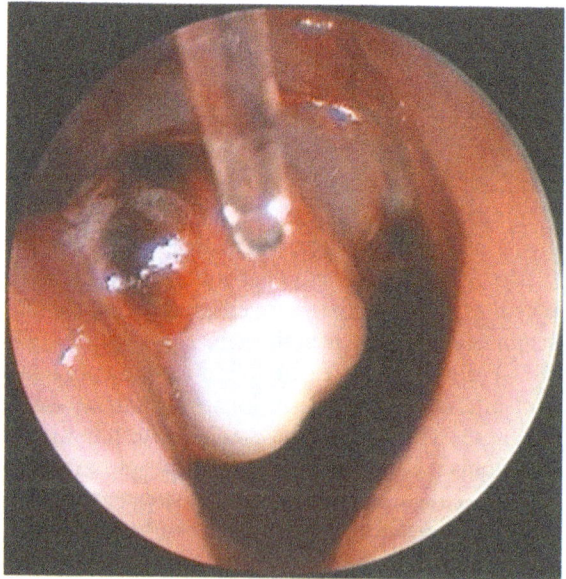

**Fig. 31.** Diode laser treatment of the haemangioma in the chopped mode with 50 W power, 200 ms exposure time and an interval of 400 ms

Laser surgery with non-contact technique is a valuable therapeutic measure for the management of recurrent epistaxis. The technique can be performed largely without causing iatrogenic haemorrhage and injury to healthy tissue, which helps not only to preserve adjacent anatomical structures, but also significantly reduces the operation time compared to other

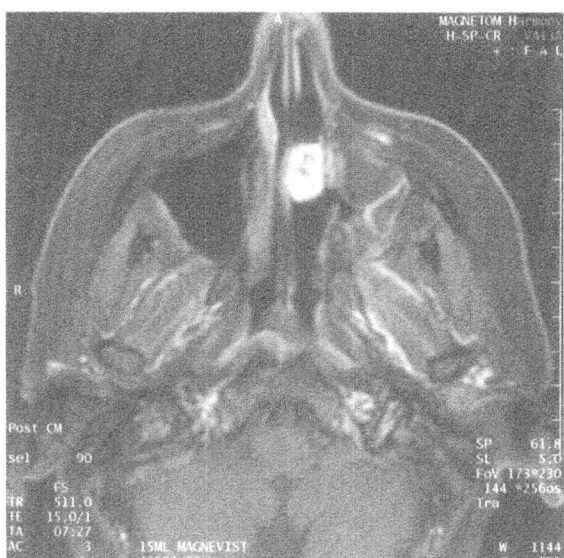

**Fig. 32.** Haemangioma of the left nasal cavity. MRI with contrast shows the high blood flow

forms of therapy. The risk of intraoperative haemorrhage or injury of healthy tissues resulting in poor vision can be clearly minimized under videoendoscopic guidance. We generally prefer to apply the laser light by means of the bare laser fibre in contrast to a laser hand piece even in the anterior areas of the nose, since it enables the surgeon to follow the high-magnification image displayed on the monitor in all situations.

As a matter of principle, all our laser surgery patients are treated under locall anaesthesia. The tendency of the mucosa to haemorrhage is substantially lower in patients having no general anaesthesia, as is known from functional endoscopically controlled nasal and sinus surgery. The intervention itself is short and can be performed on an outpatient basis.

Surveying the literature, we found that most of the authors have assimilated practical experience with endonasal Nd:YAG laser applications (Apfelberg et al. 1987; Dobrovic and Hosch 1994; Werner 1997, 1999; Hopf 1999; Shapshay and others 1984), Argon laser (Lenz 1984; Parkin and Dixon 1981, 1985; Haye and 1991; Lennox et al. 1997; Hopf 1999) and KTP laser (Levine 1989; Siegel 1991).

Since 1988, we have used various laser systems at our hospital to treat recurrent epistaxis of different aetiopathogenesis during the non-haemorrhaging interval. The majority of our data have been obtained with the argon and the diode laser.

With both lasers, strongly vascularized tissue should be exposed to the laser light using a non-contact technique, i.e. this involves general exposure of the area rather than spot-wise application of the laser

beam. Overly intense exposure of small spots of tissue would lead to the so-called popcorn effect (Jaques 1992; Hopf 1999), i.e. intense intravascular photon absorption resulting in vaporization and ensuing formation of gas bubbles, which cause the vessels to rupture and bleed. This risk is especially high in Osler foci because of the increased fragility of the vascular walls in these areas. Only after carefully precoagulating the target area with the laser beam is it realistic and reasonable to treat the tissue in a contact technique at low power with the laser in continuous mode.

Compared to the other laser systems referred to above, technological features render the beam emanating from the bare fibre of the Diode laser more strongly divergent, i.e. less focused. This might be regarded as a physical disadvantage of this system when using the no-touch technique, causing increased scatter to the areas adjacent to the region within the focus and the fibre. In clinical practice, though, this effect combined with the non-contact technique reduces the undesirable popcorn effect during exposure of the target tissue.

The exact values of laser parameters, such as power, power density, exposure period and pauses, to be used in laser treatment of recurrent epistaxis cannot be brought to a systematic form of fixed standards.

However, the following considerations should be included in the selection of laser parameters.

The fibre-guided laser beam can be applied to the target tissue in one of two ways. Firstly, there are contact techniques, in which the fibre tip just touches the surface of the target tissue at low pressure and follows the changes of the tissue surface induced by exposure to the laser beam, i.e. the photons penetrate the target tissue almost directly. However, this method is not suitable, as it may produce vascular lesions and intra operative haemorrhage and, therefore, poor vision during the operation.

In contrast, when using the non-contact technique, the laser beam is directed to the target tissue from some distance away. The critical factor for the efficiency of the laser, i.e. its power density, decreases with distance. For anatomical and intrinsically technological reasons, the distance between laser and target tissue cannot be kept constant during the entire intervention. Since some thermal effects on the tissue may become evident only at a later point in time, a second operation of the same anatomical area may occasionally be required. We observed this effect mainly in our haemangioma patients.

Between 1988 and 2001, approximately 250 patients with recurrent epistaxis of different aetiogenesis were treated at our institution with the Argon ion, the 940 nm Diode laser, and the Nd:YAG laser. Overall,

321 procedures were performed. Single laser application was usually sufficient in pyogenic granuloma cases and vascular ectasia of Kiesselbach's area, whereas haemangiomas required two sessions and hereditary haemorrhagic teleangiectasia needed repeated laser sessions.

## Rare Indications and Case Reports

### Squamous Cell Papilloma and Transitional Cell Papilloma

Squamous cell papillomas of the nasal vestibule are usually removed by conventional surgery rather than by laser technique. In contrast, we elected to manage three cases of nasopharyngeal papilloma confirmed by biopsy with the Nd:YAG laser under endoscopic control.

We also applied the laser to remove one inverted main nasal cavity papilloma in an 85-year-old patient with severe sinubronchial syndrome as a palliative measure to improve nasal breathing.

### Recurring Olfactory Neuroblastoma (Case Report)

In a 56-year-old female patient previously subjected to exhaustive radiation treatment, a strongly vascularized olfactory neuroblastoma extending from the anterior cranial fossa to the left main nasal cavity that responded to touch with immediate haemorrhage was treated in three laser sessions keeping 3-month intervals. It was intended to treat this neuroblastoma, which obstructed the main nasal cavity and gave rise to an extreme haemorrhagic tendency, palliatively to reduce the size of the tumour and tendency to bleed. For this purpose, the Nd:YAG laser was used in a non-contact procedure at 3 to 20 W (continuous-wave mode).

### Sinus Histocytosis (Case Report)

Multilocular histocytosis with nearly complete obstruction of the right main nasal cavity in a female patient, 62 years of age, caused recurrent infection of all paranasal sinuses. Endonasal histiocytic foci were successfully reduced in volume by application of the Nd:YAG laser to induce interstitial coagulation (2 to 5 W) and superficial vaporization. Laser treatment caused the target tissue to shrink and scarify. Having undergone three previous unsuccessful conventional interventions under general anaesthesia, three successive sessions of laser surgery succeeded in restoring nasal breathing, which has been maintained for 8 years of follow-up.

### Choanal Stenosis – Choanal Atresia

We successfully vaporized broad-based bony and cartilaginous and membranous choanal stenoses in children, in general as well as in adolescents and adults with the laser beam without administering infiltration anaesthesia, and experienced no restenosis. Placement of a spacer was dispensable in about 50 % of the patients. In a male patient 19 years of age, left-sided choanal atresia previously had been subjected to five conventional revision operations due to restenosis. The partially bony atresia was successfully incised with the laser beam and sufficiently enlarged in two subsequent sessions. The choana remained sufficiently wide for ventilation and drainage for the entire follow-up period of 5 years.

### Adenoids

We successfully removed adenoidal and hyperplastic mucosal tissue from the nasopharyngeal fold in patients from the age of 14 using an transnasal access. However, as an indispensable requirement, the absence of tumorous masses (e.g. juvenile nasopharyngeal angiofibroma, lymphoepithelial carcinoma) must be previously diagnosed and confirmed by pathohistology. The fact hat adenoids cannot be rendered sufficiently free of pain under local anaesthesia is a clear limitation, effectively excluding the application of this approach in children. Moreover, the tratment time with the laser ist too long

## Discussion

Over the past decade, functional endoscopic endonasal laser surgery (FEELS) has become an established method in the management of diseases characterized by the obstruction of nasal breathing. Especially vasomotor rhinopathy and allergic and pseudo-allergic turbinal hyperplasia are known to be curable with very good success rates or at least to permanently improve upon laser treatment. Other applications of lasers include minimally invasive postoperative care or second-line treatment of rhinosinusitis chronica et polyposa patients after conventional primary surgery as part of a medium- to long-term therapeutic plan. Obviously, the laser should not be regarded as a panacea of endo- and paranasal endoscopy that makes conventional functional endoscopic surgery (FEES) obsolete. Rather, it is a useful surgical instrument, which, like all other surgical instruments, should only be used with much thought and skill – it is nothing more and nothing less.

Since laser surgery involves the surgical-therapeutic application of coherent, collimated, monochromatic light, it is useful to brush up one's knowledge of the basic principles of optical physics that form the foundation of this elegant and effective therapeutic method.

Any attending surgeon must be familiar with the conditions of the target tissue, i.e. its absorption and scattering properties, and the anisotropy factor, all of which are only rarely mentioned in the literature, which focuses on clinical aspects. At least to some extent these factors depend on the wavelength of the incident light. On the basis of these optical parameters it is possible to calculate the probable distribution volume of photons in the tissue and, thus, determine the optical and effective depth of penetration in the specific tissue. Knowing the optical penetration depth of laser light of a certain wavelength, the surgeon is in a position to select adequate laser parameters for the application at hand and predict the probable long-term effect.

Unlike in laser surgery, the main focus in monopolar and bipolar high frequency surgery or application of the argon plasma beamer is to coagulate the tissue. These techniques clearly differ from modern laser surgery in that they are restricted to just one a single surgical effect, i.e. coagulation induced by thermal damage.

In contrast, the predictability of laser-tissue interactions and tissue effects under the selected laser parameters enables the surgeon to fine-tune the desired effect and choose between vaporization and/or coagulation.

Laser-tissue interactions are determined mainly by laser properties, such as incident wavelength, mode of application (continuously versus chopped pulsed), application procedure (contact versus non-contact procedure), power and energy density and the exposure time.

To vaporize tissue with the Diode or Nd:YAG laser it is essential to work at high power density using chopped mode in order to prevent excessive heating of the tissue (above 50 °C) and convective transport of the energy to neighbouring areas. In this application, the Diode laser in chopped mode with extremely short exposure times is more suitable than the Nd:YAG laser, even though it is possible to reduce the high optical penetration depth of the Nd:YAG laser by preblackening the fibre tip. The chopped mode guarantees that the briefly exposed tissue has ample time to recover and cool down after exposure.

If the main therapeutic effect of laser surgery desired by the surgeon is coagulation – seeking scarification at a later point in time – lower power-density settings are selected and the surgeon performs the procedure with the clean fibre tip in continuous mode. Under these conditions, application of the laser in non-contact mode effects coagulation almost ex-

clusively, whereas contact exposure produces an additional, low-level vaporization effect directly under the fibre tip.

Various laser systems with widely differing properties have been proposed for endo- and transnasal laser surgery of benign pathological changes. These systems may be subdivided in continuous-wave and pulsed lasers.

The most prominent member of the continuous-wave systems is the Nd:YAG laser, which has been the focus of our clinical experience in the beginning and was proposed for endonasal surgery by Werner, Krespi and Lippert. The Nd:YAG laser is a robust and reliable system, not prone to interferences, and characterized not only by its well-established laser-tissue interactions in routine work in the hospital or private office, but also by its affordability and low maintenance costs.

Another member of the continuous emission devices, the Argon Ion laser ismore expensive and much more difficult to transport within the facility. The vibration-sensitive technology of this laser strongly limits the mobility of the device and frequently makes extensive and time-consuming re-adjustment the system components necessary. Also, because of the lower maximal power of this system, the vaporization and coagulation efficiencies of the Argon Ion laser are lower than those of the Nd:YAG laser. Despite these drawbacks, the Argon Ion laser is the system favoured by Lenz for endonasal surgery.

The Anglo-American literature, especially Levine, reports good rates of success with the KTP laser in endonasal surgery. Due to frequency duplication in this device, light of a wavelength of 532 nm is emitted by the laser. Laser-tissue interactions and irradiation efficiency closely resemble those of the Argon Ion laser (488 and/or 514 nm).

The Diode laser is another continuously emitting laser system and is commercially available from various manufacturers with beams of 805, 810, 820 and up to 980 nm. The 805-nm beam is strongly absorbed by the haemoglobin in the blood regardless of its state of oxygenation. Photons with a wavelength between 940 and 980 nm are strongly absorbed by water. The optical penetration depth of the photons of this laser is substantially lower than that of the Nd:YAG laser. The Diode laser is manufactured at surprisingly small size and, thus, the device is easily portable. Due to its optical properties and efficiency, which are somewhere between those of the Argon and Nd:YAG lasers, this device proved to be very suitable for clinical applications.

The therapeutic options becoming available with this system are impressive. That the laser can be employed in both continuous-wave and chopped mode allows the surgeon to achieve excellent coagulation and vaporization of endonasal and paranasal tissues. However, the applicability of this laser is not limited to soft tissues. Even bony and cartilaginous shape variants of the nasal septum and ethmoid can be successfully and safely managed with this laser. Especially these systems with wavelengths between 940 and 980 nm, characterized by strong absorption by water molecules, are excellent to manipulate bony and cartilaginous tissues.

The $CO_2$ laser is a classical representative of the continuous-emission systems. The introduction of this gas laser into clinical use initiated the era of laser medicine in all sub-specialities. The photons emitted by this laser (10 600 nm) show a high affinity to water molecules and are strongly absorbed. Consequently, the photon distribution volume is only small and restricted to the surface, which explains the predominance of vaporization over other tissue effects of this laser. Usually, this laser light is applied to the target tissue in a microscope-assisted, free beam procedure using a micromanipulator and various spot diameters or a focusing handle. With the highly sophisticated scanner system available for this laser the surgeon can draw a two-dimensional figure with the microprocessor-controlled laser and homogeneously vaporize the target areas. In recent years, hollow-wave guides of varying diameters have become available, permitting application of this laser in the nose and at the ostoeomeatal unit. These tube-shaped applicator units contain a reflective internal coating and are capable of transporting the beam on a straight, curved and slightly bent path. Recently, fibres conducting the IR radiation emitted by the $CO_2$ laser have become commercially available. This is another option for clinical applications. Both hollow-wave guides and IR-conductive fibres are characterized by high power losses, which are proportional to the length of the applicator unit. The lost energy is converted into heat by absorption inside the applicator. Therefore, it is necessary to operate wave guides with gas coolingn to prevent from melting and burning the anatomic surroundings. However, power losses inside the hollow wave guide or fibre cannot be completely compensated for by increasing the laser power setting. Eventually, the internal coating is damaged at elevated laser power due to the limited heat resistance of the wave guides, which effectively destroys the applicator.

Hollow-wave guides and IR fibre applicators equipped with a diamond tip no longer have to be strictly protected against accidental tissue contact and operated at a safe distance from the tissue surface. Thus, contamination of the applicator tip by blood or secretions, which used to cause immediate destruction and melting of the unit, no longer poses a risk, provided the hollow wave guide is equipped with gas cooling. Due to the relatively poor coagulating

power of the $CO_2$ laser beam, insufflating the cooling gas in moderately to strongly vascularized areas may cause small blood droplets to splatter which, in turn, can impair the view of the operation site, especially in procedures performed under endoscopic control. In our opinion, its tissue interactions make the $CO_2$ laser less suited for laser-assisted reduction of the turbinates in vasomotor or allergic rhinopathy patients. Management of these patients mainly requires laser exposure of the turbinal submucosa to induce scarification and coagulate the large venous vessels. Routinely, this can be done on an outpatient basis. In addition, it is desirable to restore nasal breathing by vaporizing excess tissue without haemorrhage. However, as the $CO_2$ laser's specific interactions with superficial tissue layers predominate, the surgeon is forced to work with a defocused beam to keep the intervention largely free of haemorrhage. This, in turn, may be accompanied by broad epithelial irradiation with secondary formation of fibrin deposits. Since the inferior nasal meatus affected by pathology is only rarely wide enough to permit passage of the $CO_2$ laser wave guide, treatment must be initiated at the medial flank of the nasal concha resulting in an increased incidence of septo-turbinal synechiae formation. Occasionally, even on decongestants, it may not be possible to expose a hyperplastic rear end of an inferior turbinate to the laser since the confined spatial situation makes it difficult to place the laser in this remote anatomical area. Performing this intervention under endoscopic control as part of routine clinical practise has proved to be very time-consuming, but it is indispensable where the posterior portions of the main nasal cavity cannot be visualized under microscopic control. Compared to fibre-assisted laser applications, there is a significantly greater need to clean the endoscope optics, which balances out any time advantage the technique my possess.

The Holmium:YAG laser is a pulsed laser system and was first used in the early 1990s to surgically manage pathological changes of the nose and paranasal sinuses. The Holmium:YAG laser causes very rapid heating of the exposed area, which is in distinct contrast to the continuous-wave systems, whose laser-tissue interactions and penetration depths are determined to a large extent by processes of absorption and scattering and the photon-distribution volume. Rapid heating by the laser vaporizes some of the tissue, whereas the remainder is made to expand and eventually explode, splattering fragments of tissue in the direction of the incident beam. Mechanical effects on the underlying tissues, such as fissure formation, have also been observed. This process is oligothermal for the surrounding tissue, but not completely athermal (i.e. without heat effects); the lesion thus produced shows a small margin of coagulation in accordance with the following rule: the shorter the laser pulse, the smaller the extent of coagulation.

The introduction of laser systems with medium pulse duration, like the Ho:YAG laser, aimed to enable the surgeon to reduce cartilaginous and bony tissue down to the adjacent healthy anatomical structures without major thermal effects. This possibility had been demonstrated in the ground-breaking work of Shapshay. Although our own experimental and clinical experience also demonstrated the facility to reduce bony and cartilaginous structures with this laser both in vitro and in vivo. The use of this laser is limited not only by its high acquisition costs and operational expenses, but also by its limited efficiency and inconvenience in clinical applications, because the laser beam effects intraoperative haemorrhage. At low to moderate pulse energy of 300 to 500 mJ per pulse and a pulse frequency of 2 to 10 Hz the coagulating effect of the laser on the adjacent mucosal tissue was so poor that exposure to the laser light was accompanied by haemorrhage, mainly in the ethmoid, and ensuing obstruction of vision. In addition, some of our in vitro experiments showed that the use of this laser (mainly at the rhinobase) is accompanied with a substantial risk of causing iatrogenic injury due to perforation of the skull base with ensuing liquorrhoea. In our opinion, the Holmium:YAG laser is an efficient but expensive system suitable only for experienced rhinosurgeons working with endoscopic or microscopic techniques, who wish to use this laser as a supplemental instrument in functional sinus surgery.

Differently from the clinical procedure followed in functional endoscopic surgery of the nose and paranasal sinuses (FESS), laser surgery is exclusively performed under video-endoscopic vision at our hospital, observing the procedure on a monitor. This has four main advantages:

- The surgeon is not forced to peek through the keyhole-like eyepiece of the endoscope in a darkened environment, as used to be required in the early days of FESS. Rather, even the affordable one-chip camera transmits the image to a monitor and the magnified picture provides a good on-line view of the operation site.
- It is in principle no longer mandatory for the surgeon to wear laser safety goggles in strictly intracavitary fibre-delivered applications of the Nd:YAG, Diode, Argon and KTP lasers, in which the image is transmitted to a monitor through an endoscope – i.e. in numerous nasal, paranasal sinus and nasopharyngeal applications. In contrast, when the surgeon looks through the eyepiece of an endoscope (or microscope), his or her eyes must be protected from the laser beam with safety goggles or suitable filters attached to the eyepiece. The latter are quite expensive. The unprotected eye of the

surgeon is endangered mainly if the light hits the metallic edge of the fibre channel and is scattered back to the eyepiece. It is mandatory that the patient wears adequate (i.e. filtering out the wavelength of laser light) and well-fitting eye protection and closes his or her eyes during the laser procedure.

- Videoendoscopic vision allows the surgeon to perform the operation in a comfortable upright, standing position with no strain on back or extremities.
- Any relevant findings or procedural steps can be simultaneously documented by means of videoendoscopy. Allowing the patient to watch the intervention while it is ongoing may actually increase patient compliance in some cases.

Endoscopes have become well-established instruments in functional endonasal surgery and are indispensable tools in procedures of rhinosurgery. The advantages of the superior vision achieved with these instruments have been described in detail in the literature.

In contrast to in conventional FESS, haemorrhage and ensuing loss of vision during the endoscopy, which may occasionally necessitate suspension or termination of the operation, are uncommon in laser surgery with Diode and Nd:YAG lasers. The superior coagulating power of these systems resulting from the large penetration depth effectively prevents haemorrhage.

However, to maintain permanent visual control over all steps of the operation, it is indispensable to evacuate any smoke produced by the laser process. Volatile combustion products depositing at the lens due to insufficient smoke evacuation are the most common reason to briefly suspend laser surgery. Pauses for cleaning of the endoscope's optical parts can be substantially reduced using a suitable gas evacuation system.

A number of very useful endoscopes for endonasal laser surgery have become commercially available. Though these have, in general, been very successful in specific applications, a universally applicable endoscope has not been designed to date. Considering the multitude of laser surgical indications and new ways to approach regions which used to be completely inaccessible to an endoscope, there is a clear need that the industry should continue the vigorous development of new instruments.

In terms of technique, functional endoscopic endonasal laser surgery (FEELS) is very similar to the conventional endoscopic procedure described by Wigand, Messerklinger and Stammberger, and poses no difficulties to the experienced surgeon. Instead of using mechanical instruments for placing, grasping, punching and cutting, the surgeon has just to place the laser fibre accurately at the tissue to be treated. The only difference in operating technique to get used to is that the surgeon stands straight and views an enlarged monitor image of the operation site instead of peeking through the endoscope's eyepiece in a uncomfortable, crouched position.But this is easy

Performed as a minimally invasive technique and guided by endoscopic vision, laser surgery is an accepted and valuable method of modern medicine that can be performed on ambulatory patients under local anaesthesia. As such, the technique has found many applications in the management of a broad spectrum of rhinologic diseases at our institution.

Both the fibre-guided Nd:YAG laser and Diode laser have become established as very useful surgical tools.

In addition, there are several other fibre-guided laser systems in common use in surgery that offer different wavelengths, such as the Argon Ion laser, KTP laser, and Holmium:YAG laser.

As described above, there is a multitude of rhinosurgical indications for endoscopically controlled laser systems. In the past 14 years, lasers have been used most commonly to manage nasal and paranasal sinus polyposis and turbinal hyperplasia. As early as in 1984, Lenz, one of the pioneers of laser surgery of rhinologic disease, reported laser carbonization of small, firm nasal polyps with the Argon laser.

Ohyama focused on the application of the light emitted by the Nd:YAG laser to treat nasal polyposis in a superficial contact procedure. Ohyama placed a ceramic tip on the laser fibre to reduce the high optical penetration depth of the laser beam by scattering on tissue components, and at the same time emphasize the vaporizing effect inside the tissue. In our opinion, this very expensive technique offers no significant advantages over preblackening of the freshly produced bare fibre tip.

The study of Zhang addressed possible benefits of laser surgery in rhinologic disease and looked at whether the technique is superior to conventional surgery. In a prospective, randomized study in nasal and paranasal sinus polyposis patients, Zhang showed a delay in recurrence as compared to traditional polypectomy. Over the observation period of 18 to 30 months, the rate of recurrence in patients after Nd:YAG laser polypectomy was 46.6%, whereas the incidence in patients after conventional polypectomy was significantly higher (66.6%, p<0.01).

In our opinion, it is difficult to draw conclusions regarding a decrease in the rate of recurrence since there is a considerable interindividual variation in the pathogenesis and severity of disease. There are as many endogenous and exogenous factors as there are intra- and interindividual variations with respect to

localization, size and morphology of polypous mucosal areas and adjacent anatomical structures, just to name but a few contributing factors. Consequently, the analysis of our data obtained from patients did not produce statistical evidence corroborating that the rate of recurrence is lower than in conventional techniques.

However, the facility to rapidly detect any recurrence in the course of regular endoscopic postoperative follow-up examinations after both traditional and laser-assisted interventions of the nose and paranasal sinuses has proven very beneficial at our clinic, and routinely is taken as a reason to initiate early local laser management of the affected site. It is ideal to do this early, before polyp formation causes obstruction of ethmoidal areas and ensuing disturbance of ventilation and drainage of posterior paranasal areas. Hence, laser surgery of polyposis should be initiated before secondary symptoms become manifest. It is an advantage of the laser that even broad-based polypous mucosa covering a wide area can be treated gently and under preservation of the ventilation function of the paranasal sinus system, which is a prerequisite of functional restoration.

Quoting Selkin and Levin, Johnson listed some advantages of the KTP laser-assisted technique, such as accurate removal of tissue with little direct contact and facility to staunch haemorrhage by coagulation. This source also makes mention of a possible beneficial effect on the incidence of polypous rhinosinusitis recurrence.

Turbinal hyperplasia has been the second most common indication for laser surgery in our experience. By now, laser therapy has replaced submucous diathermia, conchotomy and submucous turbinectomy in many patients.

Conventional submucous turbinectomy and diathermia have been reported to be successful in providing relief from nasal breathing impairment in 38 and 88% of cases, respectively. A comparison of different literature sources is complicated by the fact that there are substantial differences between publications in the numbers of patients, observation periods and evaluation criteria.

In their prospective randomized study comparing laser cauterization and submucous diathermia, McCombe and Cook found the laser to produce superior results in therms of the lasting improvement of subjective symptoms.

It has been our own experience that the good visualization even of the posterior edges of the inferior turbinate constitutes a major intra operative benefit. In addition, the laser technique facilitates endoscopic inspection and management of the anterior head region of the middle turbinate and its lateral face, i.e. the medial margin of the middle meatus. In this area, hyperplastic tissue can be resected by means of vaporization. However, the main goal of allergic and pseudoallergic turbinal hyperplasia therapy is to moderately coagulate venous sinusoids and submucous glands. Submucous scarification under preservation of large areas of superficial epithelium leads to a lasting volume reduction after 2 to 6 months. The biophysical properties of the photons emitted by the Nd:YAG laser (1064 nm), including the high optical penetration depth (of up to 7 mm), render this laser system ideally suited to attain this therapeutic goal since the photon distribution volume is large and reaches great depths. This laser has been applied at out institution with great success for more than a decade. Similar experience has been reported by Lippert, Jovanovic, and Krespi. In an evaluation of randomly sampled cases of our hospital spanning follow-up periods after laser therapy between 18 and 24 months, 71% of a total of 198 interviewed patients stated that laser management resulted in "clear improvement" of nasal breathing, while 18% experienced "mild improvement", 9% "no improvement" and 1% actually reported some "deterioration" of this symptom. This patient population has been treated with Nd:YAG laser therapy for turbinal hyperplasia and/or polyposis.

The literature also propagates other laser systems to surgically reduce hyperplastic turbinates, including the Argon Ion laser, which is in use at out institution, $CO_2$ laser and KTP laser, which is quite popular in the US.

Introduced in the 1990s, the $CO_2$ laser has been employed in nasal cavity and paranasal sinus applications with good success, though it is not always easy to accurately place a hallow wave guide or a direct laser beam into the nasal cavity consisting of many small edges and orifices. Micromanipulator-assisted application of the laser, similar to the procedure followed in larynx surgery, does not allow the surgeon to access more than about the first two thirds of the main nasal cavity and ostium of the anterior ethmoid in patients not previously subjected to surgery. A treatment of more posterior sections of the main nasal cavity and paranasal sinuses is difficult with this laser.

Transmission of the laser beam with fibre guides has not yet become an established method at all institutions. An interim solution, the so-called hollow-wave guides, aiding the application of the $CO_2$ laser in anatomical areas that are difficult to access, has become commercially available, but these devices require extensive experience in endoscopic laser applications.

This is due firstly to the considerable power losses during transmission, and secondly to the relatively large external diameter of the applicator, i.e. this device competes with the operating endoscope for the available space inside the nose. Originally, the surgeon had to carefully avoid contaminating the applicator tip with secretions, as this would destroy the tip. The introduction of diamond-doted applicator tips has been a solution to this problem.

Fukutake presented a large study of laser surgery in the treatment of vasomotor rhinitis. In a total of 1000 patients receiving $CO_2$ laser therapy, he found excellent outcomes in 46%, good results in 34% and unchanged status in 20%.

Mittelman described the application of the $CO_2$ laser in patients with turbinal hyperplasia and synechiae. Since the $CO_2$ laser creates purely superficial defects, the surgeon obtains the benefit of having an immediate overview of the tissue effect of the laser beam. It is a disadvantage of this system in the opinion of the authors that the energy of the laser is emitted in a straight forward direction only.

Selkin focused mainly on clinical applications of the $CO_2$ laser in continuous-wave mode. Due to its poor coagulation features, the author finds this laser not very suitable for endonasal surgery.

In a comparative study, Lippert reported success rates of 85.7 and 77.1% (6 months and 5 years postoperatively, respectively) with his own $CO_2$ laser operating technique. The reported rates of success were clearly higher than those obtained in his work with the Nd:YAG laser (66.6 and 64.6%, respectively). It is our opinion that the observation periods in comparisons of the success of laser therapy must be at least 2 to 3 years to allow reasonable conclusions to be drawn in the literature. This closely corresponds to the approximate time after which a certain percentage of vasomotor rhinitis patients treated with conventional conchotomy or submucous turbinectomy presents again for recorrection surgery.

In 1990, Levine reported on 425 patients treated with the KTP laser of which 71.1% were asymptomatic and needed no vasoconstrictive nasal drops during the 2–4 years of follow-up. An additional 20.2% of patients were asymptomatic, but occasionally administered nasal drops, whereas no improvement was observed in 8.7%. Deterioration was observed in none of the patients.

Warnick-Brown reported high rates of complications in traditional submucous turbinectomy, including dryness of the nasal mucosa, epistaxis and cacosmia found in 43, 10 and 12% of patients, respectively. In contrast, except for one single case of postoperative haemorrhage after turbinal laser cauterization, we observed no complications in laser therapy of the inferior turbinate or the other indications listed. Haemorrhage was easily staunched by temporary nasal packing.

Extremely low rates of complications are reported in many papers in the literature investigating large numbers of patients.

In 1985, Selkin reported a study of the incidence of complication rates accompanying the use of the $CO_2$ laser. In a total of 250 patients with various indications, he found 11 cases with intra- or postoperative haemorrhage, 2 cases of septal perforation, and 1 case of rhinitis sicca. Recurrence of nasal breathing impairment was observed in 4 cases and burning of the nasal skin in two other patients.

Soh reported a low haemorrhage tendency during laser turbinectomy and cauterization, and as a beneficial effect, the fact that there is no need for nasal tamponade.

Applying the laser in recurrent polyposis in previously operated paranasal sinuses, even severe complications must be taken into consideration if inadequate laser power and irradiation time parameters are used. These include iatrogenic orbital penetration with ensuing risk of injuring the external eye muscles and nerve structures. Structures of the anterior cranial fossa may also be at risk, especially after previous conventional surgery or due to rarefying osteitis of the frontal base. Though this is an intrinsic risk of all laser systems, it is more pronounced in pulsed laser systems and the chopped mode of application.

Because of this complication, Ossof recommended to basically limit the use of the Nd:YAG and Holmium:YAG lasers to the septum and inferior turbinate.

However, our experience argues against this opinion that the chopped mode in continous emitting laser systems in hazardous. As described above, both the intended and potential side effects depend not only on the optical properties of the target tissue, but also - and very strongly - on the power and pulse energy applied in the individual case, total energy applied and application mode selected (i.e. chopped versus continuous-wave mode). It has been our experience that the endoorbital and endocranial bone particle spread described by Ossof can occur in applications of pulsed lasers only.

Undoubtedly, it is necessary to exercise particular care in patients with recurrent polyposis. The removal of the orbital lamina in previous FESS may have caused a local soft tissue prolapse and included some orbital material, which may rest directly under the polypous tissue. This issue can be resolved with a coronal CT scan prior to commencing laser therapy. The distance to the anterior cranial fossa is small in patients previously subjected to a complete ethmoidectomy. Under these circumstances, the selection of inadequate, i.e. extremely high, power settings may, at least in theory, lead to the coagulation of endocranial structures. In this region, it is advantageous

to apply the laser in chopped mode with short exposure times and long pauses between pulses. In our hands, there has not been one single case of the described complications.

The incidence of other complications in our patient population was small concerning to polyposis and turbinal hyperplasia.

Steiner found endoscopic laser surgery clearly superior to conventional techniques in removing residual peritubal adenoid tissue, scar stenosis, and cysts, and related this observation to the non-contact application of the laser and absence of intraoperative haemorrhage. These results were obtained with the Argon laser.

In 1984, Lenz recommended the Argon laser as another useful instrument for endonasal laser surgery in a report of his professional experience with the coagulation of the area of Kiesselbach in chronic recurrent nasal bleeding.

The Nd:YAG laser and KTP laser (532 nm) have also been applied with much success to the management of vascular disease. Soh and Ducic agree in that they consider coagulopathy and haemorrhagic diathesis suitable indications for these lasers.

Laser therapy of choanal atresia has been reported in the literature, mainly employing the $CO_2$ laser. Due to the confined anatomical conditions, it is difficult to treat children with lasers. The long-term results are quite similar to the outcome of conventional surgery. Metson, Woog, Ossoff and Soh reported another indication: Holmium:YAG or KTP laser-assisted endoscopic dacryocystorhinostomy, However, so far no long-term results have become available.

There are various literature reports and personal communications describing case reports, such as $CO_2$ laser treatment of sarcoidosis with granulomatous manifestation in the main nasal cavity. We treated one case with recurrent extreme haemorrhagic tendency due to olfactory nerve neuroblastoma. Even conventional surgery intervention with palliative intention would have put the patient at an extreme risk. In contrast, the Nd:YAG laser was applied in non-contact technique to gently coagulate the vessel and prevent the recurrence of epistaxis, which might have been fatal.

All interventions listed above can be performed on ambulatory patients under locall anaesthesia. With certain limitations this also holds true in paediatric applications. Thus, the method is not limited to therapeutic applications in clinical settings and involves relatively little inconvenience for the patient. Usually, patients can resume their regular habits directly after surgery.

The successful outcome of laser surgery depends on proper patient selection and choice of adequate application parameters. Based on our own experimental and clinical experience, we favour the Diode laser, the Nd:YAG laser and, under certain circumstances, the Argon laser. Due to the emitted wavelength and the tissue effects produced, the diode laser assumes an intermediary position between these two extremes and appears most suitable for use in a broad range of clinical indications. Because of the fact that early positive experience with this laser is confirmed, this laser system develops from being just a supplemental technique into a true work horse alternative to the Nd:YAG laser.

Under the massive budgetary restrictions on ENT specialists in both private office and clinical setting, it is not necessary to have available a whole set of different laser systems for all possible applications. A review of the literature clearly shows that many rhinological diseases can be treated with several laser systems. Since the mechanism of action of lasers depends on a number of physical principles, the skilled surgeon finds himself in a position to fine-tune equipment parameters to optimally suit a wide spectrum of indications.

Functional endoscopic laser surgery of the nose, paranasal sinuses and nasopharynx (FEELS) is a low-risk, painless, outpatient method that can be performed under local anaesthesia and provides excellent vision of the operation site. Laser application is characterized by little, if any, haemorrhage while the obstructing soft tissue structures are reduced in size. The technique is associated with a positive immediate effect on ventilation and drainage of the paranasal sinuses and, in some cases, immediate improvement of nasal breathing. In addition, this is a promising technique for patients excluded from surgery unter general anaesthesia or refusing to undergo risky revision surgery due to their advanced age or underlying severe cardiopulmonary disease. The laser can also be used in palliative management of tumours.

## Further Reading

APFELBERG DB, SMITH T, LASH H, WHITE DN, MASER MR (1987) Preliminary report on use of the neodymium-YAG laser in plastic surgery. Las Surg Med 7: 189–198

BERGLER W, GÖTTE K (1999) Hereditary hemorrhagic telangiectasias: a challenge for the clinician. Eur Arch Otorhinolaryngol 256: 10–15

COOK JA, MCCOMBE AW, JONES AS (1993) Laser treatment of rhinitis – one year follow-up. Clin Otolaryngol 18: 209–211

DOBROVIC M, HOSCH H (1994) Non-contact applications of Nd:YAG laser in nasal surgery. Rhinology 32: 71–73

DUCIC Y, BROWNRIGG P, LAUGHLIN S (1995) Treatment of haemorrhagic telangiectasia with the flashlamp-pulsed dye laser. J Otolaryngol 24 (5): 299–302

ELWANY S, HARRISON R (1990) Inferior turbinectomy: comparison of four techniques. J Laryngol Oto 194: 206–209

ELWANY S, ABDEL-MONHEIM MH (1997) Carbon dioxide laser turbinectomy. An electron microscopic study. J Laryngol Otol 111: 931–934

ENGLENDER M (1995) Nasal laser mucotomy (L-mucotomy) of the inferior turbinates. J Laryngol Otol 109: 296–299

FEYH J (1995) Endoscopic surgery of the nose and paranasal sinuses with the aid of the holmium:YAG laser. In: RUDERT H, WERNER JA (eds) Lasers in Otorhinolaryngology and in head and neck surgery. Adv Otolaryngol, Karger, Basel 49: 122–124

FREDERIKSEN LG, JORGENSEN K (1996) Sarcoidosis of the nose treated with laser surgery. Rhinology 34: 245–246

FUKUTAKE T (1993) $CO_2$ laser and turbinate dysfunction. Presented at the XII International Symposium on Infection and Allergy of the Nose (ISIAN), Seoul, Korea, October 8–11, 1993

Fukutake T, YAMASHITA T, TOMODA K, KUMAZAWA T (1986) Laser surgery for allergic rhinitis. Arch Otolaryngol Head Neck Surg 112: 1280–1282

Gleich LL, REBEIZ EE, Pankratov MM, Shapshay S (1995) The holmium:YAG laser-assisted otolaryngologic procedures. Arch Otolaryngol Head Neck Surg 121: 1162–1166

GUNDLACH P, SCHERER H, Hof J, LEEGE N, MÜLLER G, HIRST L, SCHOLZ C (1990) Die endoskopisch kontrollierte Laserlithotripsie von Speichelsteinen. In-vitro Unterschungen und erster klinischer Einsatz. HNO 38: 247–250

HAACKE VON NP, HARDCASTLE PF (1985) Submucosal diathermy of the inferior turbinate and the congested nose. ORL, J Otolaryngol Releat Spec 47: 189–193

HAYE R, AUSTAD J (1991) Hereditary haemorrhagic teleangiectasia – argon laser. Rhinology 19: 5–9

HAYE R, AUSTAD J (1992) Hereditary haemorrhagic teleangiectasia: unsuccessful treatment with the flashlamp-pulsed dye laser Rhinology 30: 134–137

HEALY GB, MCGill R, JAKO G (1978) Management of choanal atresia with the carbon dioxide laser. Ann Otol Rhinol Laryngol 87: 658–662

HOPF JUG, LINNARZ M, GUNDLACH P, SCHÄFER E, LEEGE N, SCHERER H, SCHOLZ C, MÜLLER G (1991) Die Mikroendoskopie der Eustachischen Röhre und des Mittelohres – Indikationen und klinischer Einsatz. Laryngol Rhinol Otol 70: 391–394

HOPF JUG, LINNARZ M, GUNDLACH P, SCHERER H, LUTZKOFFROTH C, LOERKE S, VÖGE K, TSCHEPE J, MÜLLER GJ (1992) Microendoscopy of the Eustachian tube and the middle ear. SPIE, Opt Fibers Med VII 1649: 264–268

HOPF JUG, HOPF M, GUNDLACH P, SCHERER H (1998) Miniature endoscopes in oto-rhino-laryngologic applications. Min Invas Ther Allied Technol 7/3: 209–218

HOPF JUG, REICHERT K, SCHILDHAUER S, SCHERER H-P (1999) Die endonasale und transnasale endoskopisch kontrollierte Laserchirurgie rhinologischer Erkrankungen. Teil 2: Indikationen, Ergebnisse und Literaturübersicht. In: BERLIN HP, MÜLLER G (eds) Angewandte Lasermedizin, Lehr- und Handbuch für Praxis und Klinik. 16. Ergänzungslieferung, III-3.4.3.2: 1–16, ecomed, Landsberg

HOPF JUG, HOPF M, SCHERER H, MÜLLER G, BERLIEN H-P (1999) Die endonasale und transnasale endoskopische kontrollierte Laserchirurgie rhinologischer Erkrankungen. Teil 1: Biophysikalische Grundlagen, Gerätetechnologie und Behandlungsablauf. In: BERLIN HP, MÜLLER G (eds) Angewandte Lasermedizin, Lehr- und Handbuch für Praxis und Klinik. 16. Ergänzungslieferung, III-3.4.3.1: 1–16, ecomed, Landsberg

HOPF JUG, JOPF M, KOFFROTH-BECKER C (1999) Minimal invasive Chirurgie obstruktiver Erkrankungen der Nase mit dem Diodenlaser. Las Med 14 (4): 106–115

HOPF JUG, HOPF M, ZIMMERMANN B, MERKER H (2000) Das Reparationsverhalten der Eustachischen Röhre in der Gewebekultur nach $CO_2$-Laserbehandlung. Las Med 15 (3): 44–71

HOPF JUG, HOPF M, ROHDE E, ROGGAN A, EICHWALD H, SCHERER H (2000) Die Behandlung der rezidivierenden Epistaxis mit dem Diodenlaser. Las Med 15 (3): 95–105

HOPF JUG, HOPF M, EICHWALD H, WOLTER H (2000) Das Training zur funktionell-endoskopischen endonasalen Laserchirurgie (FEELS) am Tiermodell Schaf. Las Med 15 (3): 123–138

HOPF JUG, HOPF M, ROGGAN A, HIRST L, BEUTHAN J, SCHERER H (2000) Optische Eigenschaften von Weich- und Hartgeweben der Eustachischen Röhre. Spektroskopische Untersuchungen. Las Med 15 (4): 189-196

HOSEMANN W, WIGAND ME (1985) Örtliche Unterschiede im Gewebebild der chronisch-hyperplastischen Nasennebenhöhlenschleimhaut. HNO 33: 311-315

Illum P, BJERRING P (1988) Hereditary hemorrhagic teleangiectasia treated by laser surgery. Rhinology 26: 19-24

JAKOBOWICZ M, FRECHE C, DELACOUR JF, DURAND JP (1990) Une nouvelle therapeutique endonasale: le laser YAG par voie endoscopique. Ann Oto-Laryng (Paris) 107: 21-25

JAQUES SL (1992) Laser-tissue interactions. Photochemical, photothermal and photomechanical. Surg Clin N Am 72: 531-558

JOHNSON LP (1990) Paranasal sinus applications of surgical lasers. Otolaryng Clin N Am 23 (1): 29-30

JOVANOVIC S, DOKIC D (1995) Nd:YAG-Laserchirurgie in der Behandlung der allergischen Rhinitis. Laryngol Rhinol Otol 74: 419-422

KAMAMI Y-V (1997) Laser-assisted outpatient septoplasty results of 120 patients. J Clin Las Med Surg 15: 123-129

KAUTZKY M, BIGENZAHN W, STEUER M, SUSANI M, SCHENK P (1992) Holmium:YAG-Laserchirurgie. Anwendungsmöglichkeiten bei entzündlichen Nasennebenhöhlenerkrankungen. HNO 40: 468-471

KENNEDY D (1985) Functional endoscopic sinus surgery. Arch Otolaryngol 111: 643-649

KOMISAR A, RUBEN R (1981) Use of carbon dioxide laser in pediatric otolaryngologic disease. New York State J Med 81: 1761-1764

KRESPI YP, SLATKINE M (1994) Nd:YAG fiber delivery system for submucosal interstitial coagulation of nasal turbinates. Las Surg Med 15: 217-248

LANDTHALER M, HOHENLEUTHNER U, TALAL ABD El RAHEEM (1995) Therapy of vascular lesions in the head and neck area by means of argon, Nd:YAG, CO₂ and flashlamppumped pulsed dye lasers. In: RUDERT H, WERNER JA (eds) Lasers in Otorhinolaryngology and in head and neck surgery. Adv Otorhinolaryngol, vol 49, Karger, Basel, pp 81-86

LENNOX PA, HARRIES M, LUND VJ, HOWARD DJ (1997) A retrospective study of the role of the argon laser in the management of epistaxis secondary to hereditary haemorrhagic telangiectasia. J Laryngol Otol 111: 34-37

LENZ H (1985) Acht Jahre Laserchirurgie an den unteren Nasenmuscheln bei Rhinopathia vasomotora in Form der Laserstrichkoagulation. HNO 33: 422-425

LENZ H, EICHLER J (1975) Wirkung des Argon-Lasers auf die Gefäße, Mikro- und Makrozirkulation der Schleimheut der Hamsterbackentasche. Eine intravitalmikroskopische Studie. Laryngol Rhinol Otol 54: 612-619

LENZ H, EICHLER J (1984) Endonasale chirurgische Technik mit dem Argon-Laser. Laryngol Rhinol Otol 63: 534-540

LENZ H, PREUßER H (1986) Histologische Veränderungen des respiratorischen Schleimhautepithels der unteren Nasenmuscheln nach Argon-Laserstrichkoagulation (Laser-Muschel-Kaustik) bei Rhinopathia vasomotora. Laryngol Rhinol Otol 65: 438-444

LENZ H, EICHLER J, KNOF J, SALK J, SCHÄFER G (1977) Endonasales Ar+-Laser-Strahlführungssystem und erste klinische Anwendungen bei der Rhinopathia vasomotora. Laryngol Rhinol Otol 56: 749-755

LEVINE HL (1989a) Endoscopy and the KTP 532 laser for nasal sinus disease. Ann Otol Rhinol Larnygol 98: 46-51

LEVINE HL (1989b) Lasers and endoscopic rhinologic surgery. Otolaryngol Clin N Am 22 (4): 739-748

LEVINE HL (1991) The potassium-titanyl phosphate laser for the treatment of turbinate dysfunction. Otolaryngol Head Neck Surg 104: 247-251

LEVINE HL (1997) Lasers in endonasal surgery Otolaryngol Clin N Am 30 (3): 451-455

LINNARZ M (1992) Microendoscopy of the nasal cavity and the paranasal sinuses via their naturla orifices. SPIE, Opt Fibers Med VII 1649: 273-276

Lippert BM, WERNER JA (1995) Reduction of hyperplastic turbinates with the CO₂ laser. In: RUDERT H, WERNER JA (eds) Lasers in otorhinolaryngology, head and neck surgery. Adv Otorhinolaryngo, Karger, Basel 49: 118-121

LIPPERT BM, Werner JA (1996) Nd:YAG-laserinduzierte Nasenmuschelreduktion. Laryngol Rhinol Otol 75; 523-528

LIPPERT BM, WERNER JA (1998) Long-term results after laser turbinectomy. Las Surg Med 22: 126-134

LIPPERT BM, WERNER JA, Hoffmann P, Rudert H (1992) CO₂- und Nd:YAG-Laser: Vergleich zweier Verfahren zur Nasenmuschelreduktion. Arch Otorhinolaryngol Supp II: 116-117

MCCOMBE AW, COOK JA, JONES AS (1992) A comparison of laser cautery and submucosal diathermy for rhinitis. Clin Otolaryngol 17: 297-299

MEHTA AC, LIVINGSTON DR, LEVINE HL (1987) Fiberoptic bronchoscope and Nd:YAG laser treatment of severe epistaxis from nasal hereditary hemmorhagic telangiectasia and hemangioma. Chest 91: 791-792

MESSERKLINGER W (1966) Über die Drainage der menschlichen Nasennebenhöhlen unter normalen und pathologischen Bedingungen. 1. Mitteilung Mschr Ohrenheilk 100: 56-68

MESSERKLINGER W (1979) Das Infundibulum ethmoidale und seine entzündlichen Erkrankungen. Arch Ortholaryngol 222: 11-22

MESSERKLINGER W (1987) Die Rolle der lateralen Nasenwand in der Pathogenese, Diagnostik und Therapie der rezidivierenden und chronischen Rhinosinusitis. Laryngol Rhinol Otol 66: 293-299

METSON R, WOOG JJ, Puliafito CA (1994) Endoscopic laser dacryocystorhinostomy. Laryngoscope 104: 269-274Min YG, KIM HS, Yun YS, MIN CS, JANG YJ, JUNG TG (1996) Contact laser turbinate surgery for treatment of diopathic rhinitis. Clin Otolaryngol 21: 533-536

MITTELMAN H (1982) CO₂ laser turbinectomies for chronic obstructive rhinitis. Las Surg Med 2: 29-36

MLADINA R, RISAVI R, SUBARIC M (1991) CO₂ laser anterior turbinectomy in the treatment of non-allergic vasomotor rhinopathia. A prospective study upon 78 patients. Rhinology 29: 267-272

NICKEL R, SCHUMMER A, SEIFERLE E (1987) Lehrbuch der Anatomie der Haustiere, vol II, 6th edn. Paul Parey, Berlin

OHYAMA M (1989) Laser polypectomy. Rhinology, Suppl 8: 35-43

OHYAMA M, YAMASHITA K, FURUTA S, NOBORI T, DAIKUZONO N (1988) Applications of the Nd:YAG laser in otorhinolaryngology. In: Joffe S N, Oguro Y (eds): Advances in Nd:YAG laser surgery. Springer, Berlin Heidelberg New York pp 156-178

OSLER W (1901) On family form of recurring epistaxis associated with multiple telangiectases of skin and mucous embranes. Bull Johns Hopkins Hosp 12: 333-337

OSSOFF RH, COLEMAN JA, COUREY MS, DUNCAVAGE SA, WERHAVEN JA, REINISCH L (1994) Clinical application of lasers in otolaryngology - head and neck surgery. Las Surg Med 15: 217-248

PARKIN JL, DIXON SA (1981) Laser photocoagulation in hereditary hemorrhagic telangiectasia. Otolaryngol Head Neck Surg 89: 204-208

PARKIN JL, DIXON SA (1985) Argon laser treatment of head and neck vascular lesions. Otolaryngol Head Neck Surg 93: 211-216

PLAUCHU H, DE CHADAREVUIAN JP, BEDEAZ A, ROBERT JM (1989) Age-related clinical profile of hereditary hemorrhagic telangiectasia in an epidemiologically recruited population. Am J Med genet 32: 291-297

RAILLY PJ, NOSTRANT TT (1984) Clinical manifestations of hereditary hemorrhagic telangiectasia. Am J Gastroent 79: 363-367

RATHFOOT CJ, DUNCAVAGE J, SHAPSHAY M (1996) Laser use in the paranasal sinuses. Otolarygol Clin N Am 29 (6): 943-948

RAUCHFUSS A (1990) Komplikationen der endonasalen Chirurgie der Nasennebenhöhlen. Spezielle Anatomie, Pathomechanismen, operative Versorgung. HNO 38: 309-316

RENDU HJLM (1896) Epistaxis reétéez chez un sujet porteur de petits angiomes cutanes et muqueux. Bull Soc Med Hop 13: 731–733

ROELLY R, ROGER G, BELLITY A, GRABENDIAN EN (1992) Choanal atresia: management and surgical treatment. Study of 50 cases. Ann Pediatr (Paris) 39: 479–483

ROGGAN A, BINDIG U, WÄSCHE W. ZGODA F (1997) Wirkungsmechanismen von Laserstrahlung im biologischen Gewebe, Eigenschaften von biologischen Gewebe. In: Müller G, berlien HP (eds), ecomed Landsberg, 13. Erg. Lfg., II-3.1:1–38

RUDER H (1988) Mikroskop- und endoskopgestützte Chirurgie der entzündlichen Nasennebenhöhlenerkrankungen. HNO 36: 475–482

SCHERER H, hopf JUG, linnarz M, gundlach P, vöge K (1992) New approaches in laser surgery of paranasal sinuses. SPIE Opt Fibers Med VII 1649: 269–272

SCHERER H, REICHERT K, SCHILDHAUER S (1999) Die Laserchirurgie des mittleren Nasenganges bei der rezidivierenden Sinusitis. Laryng-Rhino-Otol 78: 50-53

SELKIN SG (1985) Laser turbinectomy as an adjunct to rhinoseptoplasty. Arch Otolaryngol Head Neck Surg 111: 446–449

SHAPSHAY SM, REBEIZ EE, MICHAIL M, PANKRATOV MM (1992) Holmium:yttrium aluminium garnet laser-assisted endoscopic sinus surgery: clinical experience. Laryngoscope 101: 142–149 18

SHAPSHAY SM, OLIVER P (1984) Treatment of hereditary hemorrhagic telangiectasia by Nd:YAG laser photocoagulation. Laryngoscope 94: 1554–1556

SIEGEL MB, KEANE WM, ATKINS JP, ROSEN MR (1991) Control of epistaxis in patients with hereditary hemorrhagic telangiectasia. Otolaryngol Head Neck Surg 105: 675–679

SIMPSON JF, GROVES F (1958) Submucosal diathermy of the inferior turbinates. J Laryngol 58: 292–300

SLATKINE M, KRESPI YP (1994) Instrumentation for office laser surgery. Operative techniques in otolaryngology. Head Neck Surg 5: 211–212

SOH KBK (1996) Laser technology in research diagnosis and therapy in rhinology. Clin Otolaryngol 21: 102–110

STAMMBERGER H (1985) Unsere endoskopische Operationstechnik der lateralen Nasenwand – ein endoskopischchirurgisches Konzept zur Behandlung entzündlicher Nasennebenhöhlenerkrankungen. Laryngol Rhinol Otol 64: 559–566

STASCHE N, HÖRMANN K, CHRIST M, SCHMIDT H (1995) Carbon dioxide laser delivery systems in functional paranasal sinus surgery. In: RUDERT H, WERNER JA (eds),

Lasers in otorhinolaryngology and in head and neck surgery. Adv Otolaryngol. Karger, Basel, 49: 114–117

STEIN E, SEDLACEK T, FABIAN RL, NISHIOKA NS (1990) acute and chronic effects of bone ablation with a pulsed holmium laser. Las Surg Med 10: 384–388

STEINER W (1989) Die endoskopische Lasertherapie im oberen Aero-Digestiv-Trakt. In: MÜLLER G, BERLIEN HP (eds) Angewandte Lasermedizin, ecomed Landsberg, III – 3.4.1: 1–5

TOS M, MOGENSEN C, NOVOTNY Z (1978) Quantitative histology of the normal ethmoidal sinus. ORL 40: 172–180

WARNER M (1996) Light-tissue interaction. Lasers in facial and reconstructive surgery, vol. 4, 2: 223–229

WARNICK-BROWN NP, MARKS NJ (1987) Turbinate surgery: how effective is it? J Otorhinolaryngol Relat Spec 49: 314–320

WEBER EP (1907) Multiple hereditary developmental angiomata (telangiectasia) of the skin and mucous membranes associated with recurring hemorrhages. Lancet 2: 160–162

WERNER JA (1999) behandlungskonzept der rezidivierenden Epistaxis bei Patienten mit hereditärer hämorrhoagischer Teleangiektasie. HNO 47: 525–529

WERNER JA, RUDER H (1992) Der Einsatz des Nd:YAG-Lasers in der Hals-Nasen-, Ohrenheilkunde. HNO 40: 248–258

WERNER JA, GEISTHOFF UW, LIPPERT BM, RUDERT H (1974a) Behandlung der rezidivierenden Epistaxis beim Morbus Rendu-Osler-Weber. HNO 45: 673–681

WERNER JA, LIPPERT BM, GEISTHOFF UW, RUDERT H (1997b) Nd:YAG-Lasertherapie der rezidivierenden Epistaxis bei hereditärer hämorrhagischer Teleangiektasie. Laryngo-Rhino-Otol 76: 495–501

WIGAND ME (1989) Endoskopische Chirurgie der Nasennebenhöhlen und der vorderen Schädelbasis. Thieme, Stuttgart

WIGAND ME, STEINER W, JAUMANN MP (1978) Endonasal sinus surgery with endoscopical control: from radical operation to rehabilitation of the mucosa. Endoscopy 10: 255–260

WOLFSON S, WOLFSON LR, KAPLAN I (1996) $CO_2$ laser inferior turbinectomy: a new surgical approach. J Clin Las med Surg 14 (2): 81–83

WOOG JJ, METSON R, PULIAFITO CA (1993) Holmium:YAG endonasal laser dacryocystorhinostomy. Am J Ophthalmol 116: 1–6

ZHANG b (1993) Comparison of results of laser and routine surgery therapy in treatment of nasal polyps. Chin Med J 106 (9): 707–709

# III-5
# Laser Treatment in Gastroenterology

S. Khulusi and N. Krasner

## Contents

## Extended Summary for Non-Experts

- Laser is an acronym for light amplification by stimulated emission of radiation.
- In gastrointenstinal treatment lasers are used for their thermal properties or alternatively as a source of non-thermal light for photodynamic therapy (PDT)
- Thermal laser is employed for thermo-ablation of tumours to reduce bulk and to re-establish luminal patency where this has been compromised.
- The thermal effects of laser can be used for coagulation of a bleeding vessel or the bleeding surface of a tumour.
- Several laser types are used for thermal laser therapy, including the well-established and highly effective neodymium yttrium aluminium garnet (Nd:YAG) laser and the more recently developed semi-conductor diode laser. The diode laser has the added advantage of practicality, being compact and portable, and able to be operated directly from a standard electrical power point.
- Laser energy is delivered through light-transmitting fibres positioned close to the site to be treated (non-contact method).
- Alternatively, low-power energy delivered for prolonged periods of time through a probe placed on or into a lesion (interstitial laser therapy) can be effective in producing significant tissue ablation.
- Interstitial laser therapy does not produce the dramatic vaporization and smoke produced by high-power laser therapy and carries a reduced risk of perforation, but can be time-consuming.
- PDT can be used in the palliation of malignant disease in both the upper and lower intestinal tract.
- Malignant biliary obstruction resulting from cholangiocarcinoma, ampullary tumours, as well as metastatic disease, has been shown to be responsive to PDT.
- Tumours of the pancreas and those of the liver are also sensitive to PDT, although the accessibility of these solid organs to laser light poses a limitation.
- PDT is ideally suited to the superficial treatment of large mucosal surfaces, as seen in patients with

Barrett's oesophagus (columnar epithelium in the oesophagus).

- Definitive, non-surgical treatment of severe dysplasia or early-stage malignancy may in the future be most suitably undertaken by PDT.
- Ideally, the merits of different therapeutic modalities need to be considered for the task intended and the outcome required. Thermal and non-thermal modes of laser therapy have both advantages and limitations and should be considered both individually and in conjunction with other forms of therapy in multimodal treatment.

## Introduction

The devolopments of the flexible endoscope and of laser (light amplification by stimulated emission of radiation) both date back to the late 1950s. However, it was not until the mid 1970s that technological advances allowed the two to be used in conjunction. The main limiting factor was overcome with the development of a flexible light-diffusing fibre which could be inserted through the instrument channel of the endoscope, enabling focused laser light to be directed to the site in the alimentary tract where it was required.

Argon lasers were the first to be used, followed in the early 1980s by the neodymium yttrium aluminium garnet (Nd:YAG) laser and, more recently, semiconductor diode lasers (Fig. 1) and tunable dye lasers. The earliest use of laser therapy in the gastrointestinal tract was for management of upper gastrointestinal bleeding. Laser therapy has gradually developed both from a technical perspective and in terms of its uses. In recent years, lasers have become smaller, more portable, easier to use and more affordable. They are employed not only for effecting haemostasis, but also for ablation of tissue, both malignant and benign, from internal organs.

Photodynamic therapy (PDT) is a more recent application of laser energy in the gastrointestinal tract, although use of non-thermal light energy in medical treatment dates back to antiquity. PDT involves the initial sensitization of a tissue by photo-activated drugs, which are inert in their native state. Endoscopic delivery of monochromatic non-thermal laser light tuned to an approriate wavelength results in activation of the drug and the photochemical reaction that ensues produces tissue destruction. PDT has a number of unique features that make it ideal in many aspects of palliation and mucosal ablation.

This chapter provides an overview of the thermal uses of lasers in gastroenterology as well as their application as a non-thermal light source in photodynamic therapy.

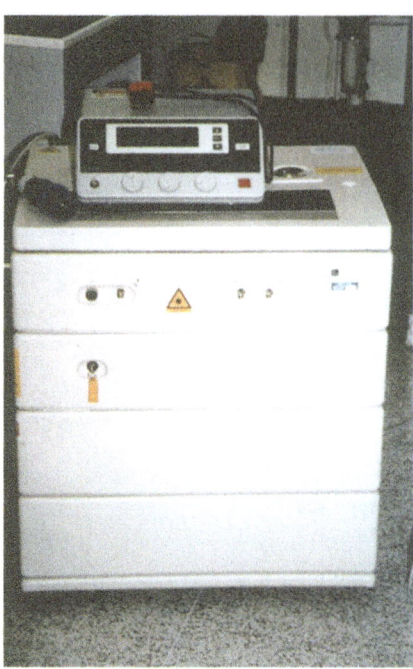

**Fig. 1.** Nd:YAG laser; GaALAs semiconductor diode laser

## Thermal Laser Therapy

### Types of Lasers

The argon laser was the first relatively high-powered continuous-wave laser produced. Its principal wavelengths are 488 and 514 nm in the blue and green regions of the visible spectrum. This enables selective absorption into vascular media and its 2–20 W power output makes it suited only for superficial tissue necrosis. Consequently, the argon laser has been superseded by a number of other more versatile and more powerful lasers in gastrointestinal therapy.

Nd:YAG lasers are some of the most powerful lasers used in gastroenterology today. High power levels are often required for precise and effective therapeutic results; 50–100 W lasers, which are the most commonly used systems, have greater efficiency than argon lasers, but require high levels of electrical input and consequently a large-volume water cooling system. At a setting of 60–80 W, a depth of penetration of 6–8 m can be produced by the Nd:YAG laser. The principal wavelength of 1064 nm enables absorption of energy equally between different tissue types, although there may be some increased absorption into vascular tissue.

Nd:YAG laser energy is delivered mainly by a noncontact method and light is scattered greatly within tissue, penetrating to several millimetres. The optimal diameter of fibre used with this type of laser is 600 μm, although occasionally fibres as small as

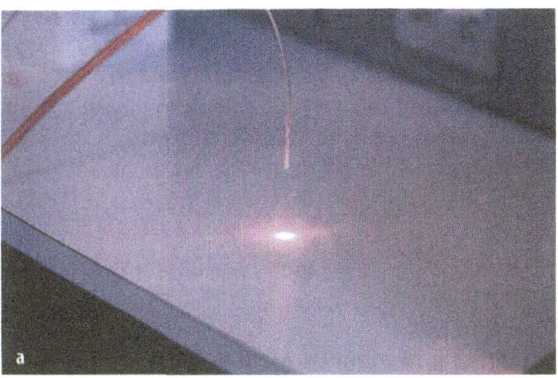

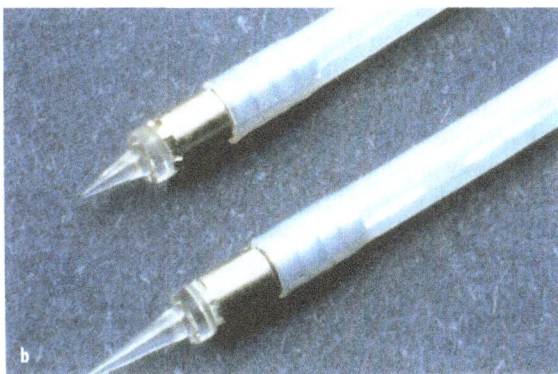

**Fig. 2.** (**a**) Quartz fibre transmitting laser beam. (**b**) Quartz fibre with sapphire probes

400 µm can be used (Fig. 2). The Nd:YAG beam is virtually undetectable by the naked eye as it is near the infrared end of the spectrum. An Nd:YAG laser therefore requires a helium-neon laser passed through the same quartz fibre and coupled into its path in order to provide an aiming beam. An alternative wavelength of 1320 nm has a different therapeutic spectrum. As a result of increased absorption into tissue, there is improved haemostasis at this wavelength. In addition, high-power laser energy applied at this wavelength to a limited area of tissue results in disintegration and vaporization locally and is useful as a surgical laser knife.

An alternative to the non-contact method is direct application of low-power Nd:YAG laser energy through a probe placed on or into a lesion (interstitial laser therapy). Interstitial therapy utilizes laser power that is low enough to produce necrosis but without the dramatic vaporization and pressure effects of higher-power laser ablation. Nd:YAG lasers set to outputs of 2 W for up to 1 000 s, can produce destruction of 16 mm diameter. A second contact method uses an artificial sapphire probe (Fig. 2). This is applied directly onto the surface to be treated, and using a power output of 15–20 W for 3–9 s produces controlled tissue destruction. These low-power methods can be more time-consuming, but carry a reduced risk of perfora-

tion and produce less smoke during the ablation process.

A more recent development is the gallium aluminium arsenide (Ga Al As) diode laser. This is a highly efficient semiconductor laser, which optimally operates at 805 nm. A diode laser at 25 W has tissue-destructive capacity similar to the Nd:YAG laser, used in the contact mode [1]. Although it is more usually applied at 25–45 W (maximum 60 W) using short pulses of 0.5–2 s or continuously in a "brush stroke" action. The added advantages of the diode laser are versatility and practicality compared to the Nd:YAG laser, as it does not require a water-cooling system, is compact and portable and operates directly from a standard electrical power point.

## Indications

Coagulation of a bleeding vessel can be produced by the thermal effects of laser. Rapid heating of tissue causes initial contraction and sealing of superficial vessels up to 1 mm in diameter. This is followed by thrombosis. Laser energy applied for 0.5 s produces this effect, while longer bursts of laser energy produce a wider and deeper necrosis of tissue. These features underpin the main uses of thermal laser therapy in the gastrointestinal tract, namely haemostasis and thermoablation.

Haemostasis may be required in cases of an acutely bleeding vessel within a peptic ulcer, vascular anomalies, or a bleeding tumour surface. The principal aims of laser haemostasis for bleeding ulcers are: definitive therapy by termination of bleeding in patients unfit for surgery, or the reduction of transfusion requirements and temporization prior to surgery in those who are surgical candidates. In the acute setting, the patients who tend to benefit most from laser therapy are those where active bleeding from an ulcer is detected or where a visible vessel is present. In cases of more chronic blood loss from angiodysplasia or haemorrhagic telangectasia, laser thermocoagulation again offers a very good therapeutic option, but needs to be repeated periodically as further vascular lesions arise.

The palliative role of laser in gastrointestinal malignancy includes not only thermo-ablation of the tumour to reduce its bulk and re-establish luminal patency where this has been compromised, but also coagulation of any bleeding surface, and tumour-eroded vessel, in order to reduce blood transfusion requirements. Ablative uses of laser are not limited to malignancy. Benign polyps may also be treated with laser under certain circumstances. A polyp too large for complete snare excison my initially be reduced in size with piecemeal snare resection, followed by laser therapy to the remaining adenomatous tissue. In addition, extensive small adenomas including "carpet adenomas" may also be treated with laser in preference to surgery or another modality, because of the location of the lesions in the gastrointestinal tract (e.g. low rectum) or because of a high operative risk.

The pulsed Nd:YAG thermal laser has been used in the management of concreted common bile duct stones [2]. A laser fibre is introduced through a duodenoscope or choledochoscope into the bile duct following a sphincterotomy. Thermal energy (10–200 J) is then directed to produce very high vaporization pressure within the stone, causing fragmentation. This method has not been widely practised because of the risk, however remote, of inadvertent thermal bile duct injury and the need for highly experienced operators. It has been somewhat superseded by the advent of endoscopic mechanical lithotripsy and extracorporeal shock wave lithotripsy.

## Patient Preparation

Initial correction of hypotension, anaemia and hypoxia prior to endoscopy is essential. Details of the therapy must be explained to the patient and, where appropriate, alternative forms of therapy or adjuvant therapy discussed. Risks of the procedure must then be explained and informed consent obtained. The patient needs to be fasted for at least 4–5 h and maintenance of adequate hydration (occasionally by intravenous infusion of crystalloids) during and after endoscopy is important. Judicious quantities of intravenous sedation should be administered. The sedated patient must be given oxygen and cutaneous oxygen saturation monitored. Prophylactic antibiotics are not routinely given unless there is a specific indication, such as a prosthetic heart valve.

## Patient Aftercare

Oversedation can result in respiratory depression and aspiration of gastric contents. Providing small "top-ups" of sedation during a prolonged procedure is preferable to a large initial dose. In spite of this careful titration, reversal of benzodiazepine sedation in postprocedure may be required in some patients. Monitoring of blood pressure post-therapy will allow early detection and enable correction of hypotension. Although sepsis is a recognized complication of laser therapy, we do not advocate routine administration of antibiotics post-procedure unless a complication is suspected.

Patients with gastro-oesophageal malignancy will quite often have lost confidence in swallowing solid food and occasionally in all forms of nutritional intake. It is therefore important, following palliative laser therapy, that they should be encouraged to take high calorie fluids initially and be gradually reintroduced to semisolid and ultimately, if possible, solid food. Carried out with dietetic supervision, this not only improves their nutritional status but also enhances their confidence in swallowing.

## Procedure Details

The procedure commences with insertion of the endoscope and a separate ventilation tube. The ventilation tube acts as an exhaust outlet for the products of laser vaporization that may otherwise obscure endoscopic views. It also allows evacuation of blood, which would reduce the effectiveness of the laser beam. Having identified the site of bleeding, it is important to wash away any loosely adherent clot by irrigation of the site with water. This then reveals the precise site of any visible vessel. Non-contact laser therapy is conventionally used for coagulation of bleeding points. The laser fibre is positioned approximately 1 cm away from the target. A power setting of 50–70 W at 0.5 s pulses is used to produce tissue shrinkage with six to eight laser bursts of energy aimed around the vessel. Direct energy into the vessel may accentuate the bleeding and should be avoided. Blanching of the tissue around the vessel can be usefully enhanced by the prior injection of 1 in 10 000 adrenaline in 1 ml aliquots into the four quadrants around the bleeding vessel.

An alternative method of vascular coagulation is with a contact sapphire probe attached to the tip of the laser fibre. At a power level of 8–10 W in 1–2 s pulses, adherence to a non-bleeding but visible vessel wall occurs. The thermal effect causes swelling and temporary occlusion of the vessel, which leads to a more permanent thrombosis. Continuous water irrigation through the laser fibre ensures that the tip is then released from the vessel.

There are two approaches to dealing with bleeding gastric tumours of tumours of the gastro-oesophageal region (Fig. 3). The first involves use of a high-power laser (80 W) in a non-contact mode to coagulate the surface of the tumour. While this method produces surface coagulation and reduces transfusion requirements, it also produces charring of the surface of the tumour and hampers further penetration of the laser beam. An alternative contact method involves the use of low-power interstitial laser photocoagulation. Low-power laser energy used for extended periods of time (200–1000 s) and repeated at several sites within the tumour produces ischaemic necrosis by occlusion of tumour vessels, resulting in reduction in the size of the tumour and improved haemostasis.

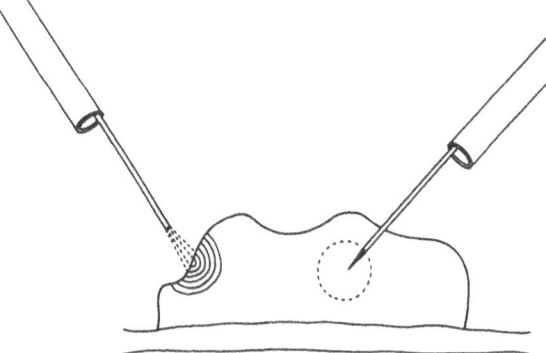

**Fig. 3.** Left Non-contact laser therapy producing surface haemostasis and deeper photcoagulation/necrosis; right interstitial laser therapy producing intra-tumour necrosis

Endoscopic laser therapy is suited to the task of debulking malignant tissue. Tumour vaporization can be produced by high-power lasers in a non-contact mode positioned approximately 1 cm from the surface of the tumour. Treatment commences from the distal end of the tumour in order to prevent the ensuing tissue from swelling, restricting access to the endoscope. Preliminary dilatation of the lumen may sometimes be required prior to laser therapy. However, this adds to the recognized risk of perforation associated with the procedure. An alternative laser method involves the use of a contact sapphire probe adapted to fit the end of the quartz delivery fibre and linked to the Nd:YAG system, delivering 10–15 W for up to 5 s, which can be repeated at different sites within the tumour. Laser ablation of malignant tissue has been used to good effect in cases of excessive tumour growth over the upper or the lower ends of a previously placed oesophageal stent. This role for laser thermoablation is useful in view of the exponential increase in popularity of oesophageal stents and the not uncommon "tumour overgrowth" problem encountered with the prolonged survival of patients [3].

The use of a beam splitter allows several fibres carrying the laser beam to increase the volume of tissue which can be thermally destroyed. This is particularly useful in the treatment of solid-organ tumours such as in the liver and pancreas, where the initial percutaneous placement of the laser fibres can be accurately guided by ultrasound or CT control.

## Criteria for Exclusion

There are few absolute contraindications to the use of laser therapy for haemostasis. A high initial haemostatic effect on bleeding oesophageal varices is unfortunately followed by an unacceptably high rebleeding rate. In a large uncontrolled study, initial haemostasis was recorded in 92% of cases of variceal bleeding but rebleeding occurred in 30% [4]. Injection sclerotherapy and endoscopic banding remain at present the accepted forms of endoscopic therapy for bleeding varices. Laser haemostasis for bleeding varices should be avoided, unless initial haemostasis is followed immediately by sclerotherapy to prevent rebleeding.

Laser therapy may not be the ideal palliative modality for malignancy under certain circumstances. A rapidly growing or long tortuous tumour will require multiple lengthy laser sessions. Laser should also be avoided with a submucosal lesion or a tumour arising in an adjacent organ and causing extrinsic compression, because of the added risk of perforation or posttherapy stricturing. A fistulating tumour treated with laser therapy may result in reduction in the tumour mass but enlargement of the fistulous tract. These are better treated by alternative methods such as insertion of a covered self-expanding metallic stent (SEMS). In addition, outside the relatively thick-walled gastro-oesophageal region or the rectum in its retroperitoneal position, laser poses a greater risk of perforation, and an alternative method may be considered more appropriate.

## Comparison with Alternative Methods

Alternative endoscopic methods of haemostasis include intramucosal injection of adrenaline or saline solution, or use of electrocautery, heater probe or argon plasma coagulation. Laser therapy is equally effective in promoting haemostasis. However, the alternative methods require less expensive equipment and consumables. Therefore on a cost basis, and in terms of the greater technical difficulty in using laser therapy, lasers may not be the first choice when considering the purchase of equipment intended predominantly for haemostatic therapy.

A preference for laser therapy over SEMS in the palliation of oesophageal cancer is usually relative. A higher oesophageal tumour where the patient may continuously sense a prosthesis and experience discomfort, or an exphytic tumour less than 5 cm in length, especially in the mid-oesophagus, where a short stent may dislodge, is more suitably treated with laser.

A completely obstructing oesophageal tumour poses particular difficulties (Fig. 4). Luminal patency may be re-established with antrograde laser therapy, but the risk of lasering into the mediastinum, peritoneum or pleural space is a not uncommon hazard, especially with an inexperienced operator. Such tumours are also likely to be rapidly growing and may completely restenose before the laser therapy can be repeated. An alternative method is to insert a SEMS. However, the obstructed oesophagus must first be

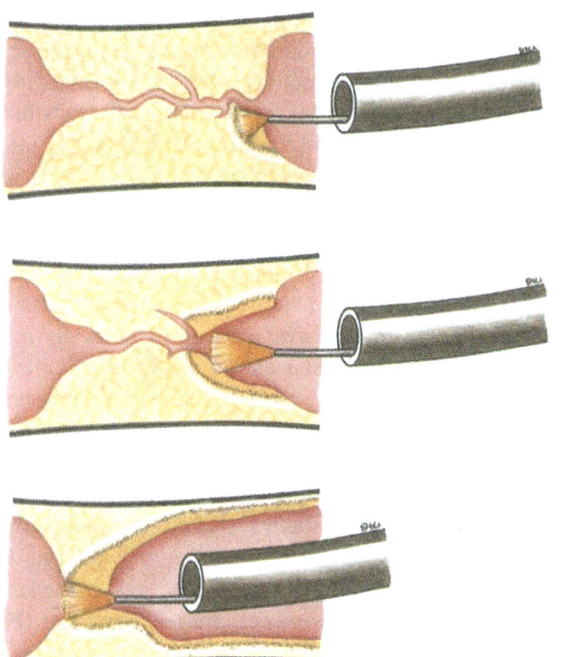

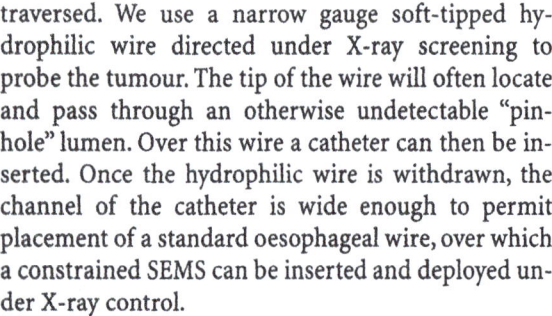

**Fig. 4.** Antrograde laser therapy to obstructed oesophageal carcinoma

**Fig. 5.** Argon plasma coagulator

traversed. We use a narrow gauge soft-tipped hydrophilic wire directed under X-ray screening to probe the tumour. The tip of the wire will often locate and pass through an otherwise undetectable "pinhole" lumen. Over this wire a catheter can then be inserted. Once the hydrophilic wire is withdrawn, the channel of the catheter is wide enough to permit placement of a standard oesophageal wire, over which a constrained SEMS can be inserted and deployed under X-ray control.

Apart from laser therapy, there are other methods by which heat may be used for ablation of malignant tissue. Monopolar electrical current applied through electrodes produces both coagulation and cutting, depending on the voltage used, and can be employed in snare resections to debulk a tumour prior to further therapy. Biopolar electrodes produce a smaller and more controlled depth of heat penetration, thus reducing the risk of perforation. Rigid BICAP tumour probes can reduce the bulk of circumferential low rectal and oesophageal cancers and have been shown to be cost-effective [5]. Haemorrhage and fistula formation, however, are significant when using this technique, thus limiting its wide-spread use.

At an affordable price, well below that of most, if not all, laser systems, is the more recent development, the argon plasma coagulator or APC (Fig. 5), which is rapidly growing in popularity. Ionized argon gas, otherwise known as argon plasma, acts as a vehicle for the delivery of monopolar electrical current. A flexible Teflon tube with a tungsten electrode contained in

a ceramic nozzle at its end (to prevent tissue adherence) is used to convey the argon plasma. The plasma remains ionized for 2–10 mm from the applicator tip which itself should not be in contact with the surface to be treated, otherwise the submucosa is inflated with argon gas. The current in the plasma travels to the nearest surface, penetrating uniformly to a maximum depth of 3 mm. APC is ideally suited to coagulation of large surface areas, especially if actively bleeding. Its limited penetration means that there is less risk of perforation than with laser. However, the energy delivered is less effective in debulking large tumours unless the same surface is treated repeatedly. An almost unique feature of APC is its ability to treat surfaces which are not in the axis of the applicator, as the ionized argon stream "bends" and seeks out the nearest surface. Lesions located behind mucosal folds are sometimes amenable to treatment with APC, where laser therapy may not be technically possible. In our Endoscopy Unit, the APC and various laser systems are located in close proximity and used in a complementary fashion.

## Results

The results of studies on laser therapy have provided us with a clearer understanding of its role as well as its limitations in haemostasis and tumour debulking [6]. The success rate of laser therapy in establishing haemostasis with bleeding ulcers ranges between 70

**Fig. 6.** Left Bleeding rectal cancer pre-treatment; right post-laser treatment showing haemostasis and reduction in tumour bulk

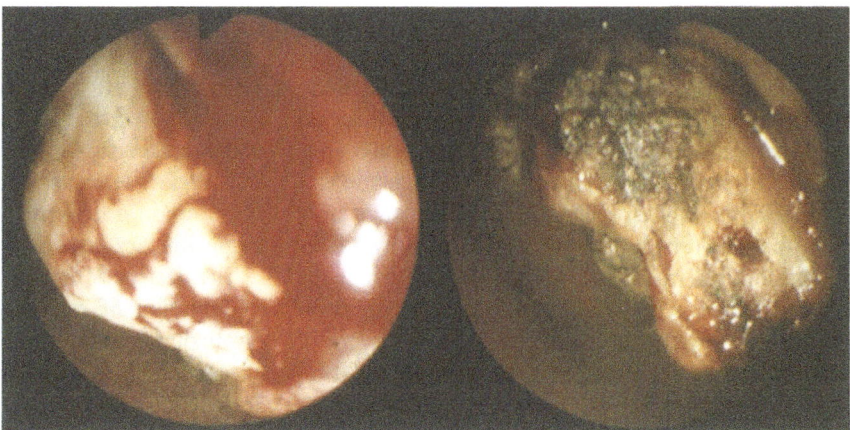

and 94%. Coagulation of a bleeding visible vessel or a friable vascular tumour can dramatically reduce the transfusion requirements of a patient. Unfortunately, on average in 9–15% of cases, bleeding will be aggravated or restarted by the effects of the laser [7].

Laser photo-ablation is well-tolerated, effective and associated with low complication rates in the palliation of both oesophageal and rectal tumours [8]. A significant improvement in dysphagia occurs in up to 90% of cancer patients treated with laser. However, complete relief from dysphagia and return to normal function is less common even with re-establishment of luminal patency, because of the persisting dysmotility caused by intramural tumour encasement. Often several laser sessions on a weekly basis are required to produce adequate debulking and functional improvement. Thermoablation can then be conducted monthly to prevent redevelopment or progression of dysphagia, or alternatively repeated at a frequency dictated by the patient's symptoms. A 1–2 month period of improved symptom control between laser sessions is common. However, this tends to be with disease affecting the mid and lower third of the oesophagus. The results of laser therapy for high oesophageal tumours are disappointing, dysphagia is only partially relieved and most subjects are unable to swallow solid food.

Laser therapy is the palliation of rectal cancer (Fig. 6) produces improvement in individual symptoms and an overall enhancement in the quality of life. Thermoablation of tumour and re-establishment of an adequate lumen often produces reduction in overflow diarrhoea and incontinence. Coagulation of the tumour surface also reduces bleeding and rectal discharge. Pelvic and perianal discomfort in rectal cancer may be due to tumour bulk and respond to thermoablation, although occasionally it is neuropathic, due to infiltration of the sacral plexus and therefore resistant to any laser destruction of luminal or mural elements of the cancer.

Laser can be used as a single-method therapy in the palliation of oesophageal malignancy. It is comparable in efficacy to endoluminal intubation with a SEMS, and may be associated with lower complication and mortality rates [9]. Alternatively, laser can be combined with another modality such as intubation of radiotherapy. The efficacy of the range of permutations and combinations possible is far from clear. The outcome of palliation of oesophageal malignancy can be enhanced by a combination of laser treatment with a stent prosthesis [10]. Similarly, when used in conjunction with radiotherapy for oesophageal malignancy [11] and in the treatment of rectal cancer [12] the combination of laser with external beam radiotherapy can reduce the number of laser sessions required and extend the length of time that a patient is symptom-free.

The main drawback to laser therapy in the palliation of malignancy, however, relates to the need for repeated and often regular relasering. This raises the costs in some patients to those of SEMS. In addition, the capital costs of acquiring a laser may be prohibitive to some smaller gastroenterology units, and proficiency in the use of laser requires a lengthy period of training.

**Complications**

Laser therapy when used for haemostasis by an experienced operator is a safe method. Perforation rates of 1% or less are often quoted. The main problem encountered is inadequate treatment to the bleeding vessel. Inaccessibility of the vessel because of its position will result in suboptimal tissue shrinkage and lack of secondary thrombosis. Similarly, a vessel over 1 mm in diameter, even if in an ideal position, will not be adequately thrombosed and may be left prone to further and possibly more brisk bleeding [13].

Laser palliation of gastro-oesophageal malignancy is associated with perforation rates of 5%. Both dilatation and the direct thermal effects of laser contribute to this figure. Conservative treatment of a perforation with analgesia, intravenous antibiotics and naso-jejunal feeding may suffice. Such patients, under the circumstances, are often reluctant to have further laser therapy, and a tube prosthesis may be the only palliative option left. In our practice we tend to favour insertion of a covered SEMS which seals the perforation and affords relief from dysphagia. Mortality rates of up to 4% can be attributed directly to laser palliation of gastro-oesophageal malignancy. Complication and mortality rates associated with laser palliation of colorectal cancer are variable but realistically are probably lower than those for upper gastrointestinal palliation. A small perforation through the rectal wall below the peritoneal reflection is rarely of any clinical consequence.

## Photodynamic Therapy (PDT)

### Mechanism of Action

The mechanism of action of PDT (Fig. 7) provides some insight into the potential applications of this therapy. The cell destruction produced by PDT is often referred to as being doubly selective. The first level of selection relates to the administration of a photosenistizer which is preferentially transported into the malignant cells. This is the result of a combination of activated cellular uptake, increased proliferation rate and "leaky" neovasculature within tumour tissue. Relatively low pH, poor lymphatic drainage and high lipid and collagen concentrations are be-

lieved to contribute to preferential retention of the photosensitizer drug within the tumour. Most compounds used are, however, only moderately selective, being concentrated to double or at most fourfold the concentration found in normal adjacent tissue. A second level of selection is provided by restricting light illumination, and therefore photosensitizer activation to specific targeted sites.

The destruction brought about by PDT is due to a combination of mechanisms. Direct malignant cell cytotoxicity is produced by the liberation of free oxygen radicals, which cause intracellular organelle destruction and protein denaturation, leading to apoptosis. Endothelial cells of tumour micovasculature are particularly sensitive to this effect, which is further compounded by their more avid uptake of photosensitizer drugs. The ensuing tumour vascular damage results in release of thromboxane A2 and other thrombogenic agents, which produce thrombosis and ischaemic necrosis of surrounding tumour tissue. Immune responses have been detected following PDT but their part in cell destruction has yet to be fully characterized. A role for immunity in retarding tumour growth in regions outside the treatment zone has been postulated.

### Photosensitizers

There are several photosensitizers currently used in PDT, although only porfimer sodium (photofrin) is at present licensed in the USA and various European countries specifically for this purpose. Photofrin is a complex purified mixture of porphyrin oligomers synthesized from haematoporphyrin dihydrochloride. Light at a variety of wavelenghts (300–700 nm)

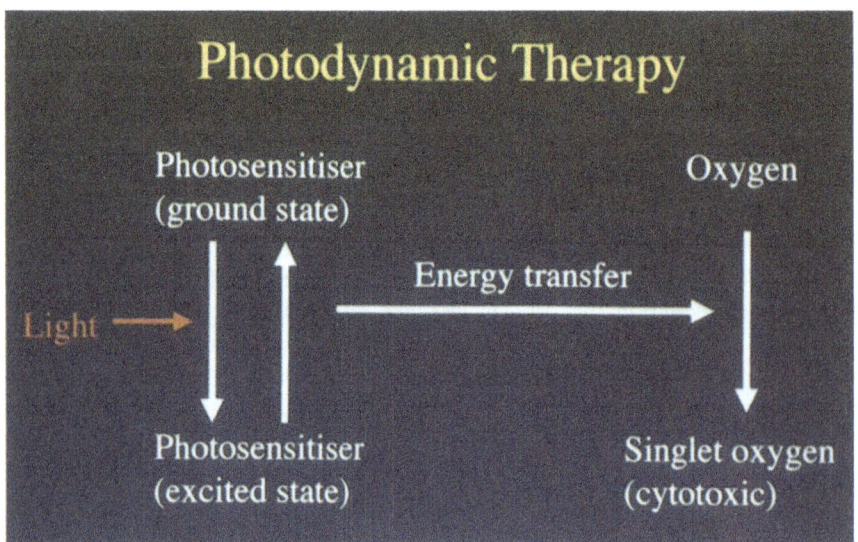

**Fig. 7.** Mode of action of photodynamic therapy (PDT)

can be used to activate Photofrin, although in a therapeutic setting red light at 630 nm is employed because of its superior transmission through tissue. Light activation is undertaken 40–50 h after administration of Photofrin, a time interval that allows for the drug to be cleared from a number of normal tissues and to be concentrated in malignant cells. Haematoporphyrin derivatives less well defined than Photofrin, which were previously used in gastrointestinal PDT, have to a greater extent been superseded. Newer photosensitizers including phthalocyanines, benzoporphyrins and chlorins, appear to have advantages over the older generation of haematoporphyrins in showing enhanced absorption shifted further into the red end of the light spectrum. Meso-tetra-hydroxyphenyl-chlorine (mTHPC), which has a higher activity than Photofrin, producing tissue necrosis to a depth of 1 cm, also produces a shorter period of skin photosensitization.

5ALA (5-aminolaevulinic acid) is a natural porphyrin precursor involved in haem synthesis. Exogenous administration of 5ALA results in transient conversion to the potent photosensitizer protoporphyrin IX (PpIX), before conversion to non-photoactive haem 5ALA is preferentially transported into mucosal cells (and less into the submucosa and muscle layers) and conversion produces accumulation of PpIX in the mucose [14]. Hence, 5ALA is suited principally to treating superficial structures, Barrett's epithelium and carcinoma in situ being notable examples. 5ALA has the added advantage of being conveniently administered by the oral route, and can be photoactivated within 4–10 h. Rapid conversion of PpIX to haem within a further 24 h diminishes the problems of cutaneous photosensitivity.

## Types of Lasers

The argon-ion pumped-dye laser is commonly used for PDT (Fig. 8). This laser produces red light tuned to a wavelength of 630 nm, and its beam can be focused

down a 200–400 µm quartz optical fibre. The fibre is either directed at the site to be treated or placed interstitially into the tumour. Circumferential treatment is possible with attachment of the fibre to a cylindrical diffuser or the recently developed balloon applicator. This is particularly useful in the treatment of oesophageal tumours.

Diode lasers have been developed to deliver light at specified wavelengths which match the absorption spectra of individual photosensitizers. The gold-vapour laser with a fixed output at 628 nm or the tunable copper-vapour system are alternative light sources for PDT. The KTP:YAG pumped-dye system employs a dye laser to convert the main laser output of 532 nm to the working wavelength of 630 nm and may also be used as a light source. Non-laser sources of light for PDT, at lower capital cost than laser, are also currently being developed [15].

## Indications

The utility of PDT is medical practice is well recognized. In gastroenterology, PDT has more commonly been used in the palliation of luminal malignant disease both in the upper and in the lower intestinal tract. In addition, malignant biliary obstruction resulting from cholangiocarcinoma [16].

Ampullary [17] tumours as well as mestatatic disease [18] hav been shown to be responsive to PDT. Tumours of the pancreas and those the liver are also sensitive to this form of therapy, although the accessibility of sites within solid organs to precise light illumination has posed a limitation and increased the risk of inadvertent damage to adjacent structures. In addition to the effectiveness of PDT in palliation, many practitioners believe that this form of therapy also has a role in the definitive treatment of early-stage malignancy in patients unsuitable for curative surgical resection. The algorhithm for management of dysplastic intestinal metaplasia of the lower oesophagus has yet to be universally agreed upon. PDT of-

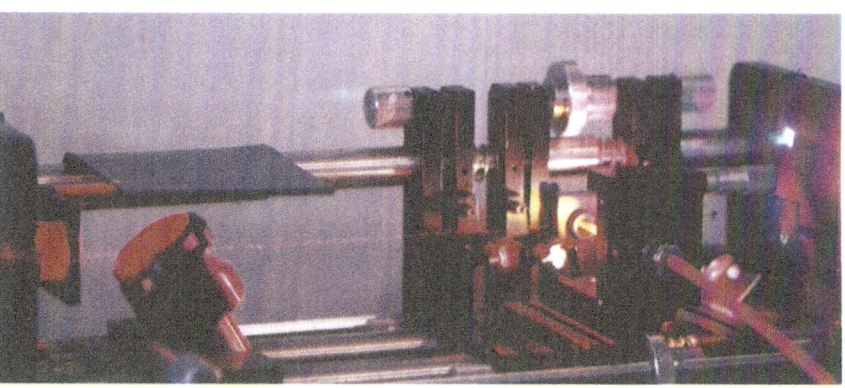

**Fig. 8.** Argon-ion pumped-dye laser. Argon laser is fired into jet of rhodamine B dye resulting in a red dye laser beam which is focused into a quartz fibre for delivery

fers a potentially highly efficacious from of non-surgical therapy as it is ideally suited to the superficial treatment of large mucosal surfaces, as seen in patients with long segment Barrett's oesophagus (columnar epithelium in the oesophagus).

## Limitations

There exist a number of limitations to the use of PDT. Een highly penetrative infrared light can penetrate no more than 1–2 cm into tissue. This relatively superficial effect means that bulky tumours may be only partially tretaed. Following necrosis of the treated superficial aspects of the tumour, deeper untreated malignant surfaces will be left exposed and prone to bleeding. In general, distant malignant tissues and infiltrated lymph nodes are unaffected by PDT. This feature restricts the present role of this therapy in malignant disease to that of palliation. The use of PDT as definitive treatment in early cancers, where surgery in contra-indicated, is a potential application which requires further assessment, as does use of PDT in the management of dysplastic conditions. PDT relies on an oxygen-dependent cytotoxicity in vitro and cytotoxicity is directly proportional to oxygen tension. Ischaemic areas may therefore be less sensitive to the destructive effects of PDT.

## Patient Preparation

In patients undergoing luminal PDT, initial preparation is similar to that prior to thermal laser therapy. Details of the treatment as well as the risks of the procedure must be explained in full and informed consent obtained. Prolonged skin photosensitivity is a particular hazard and its implications must be explained to the patient. The photosensitizer is systemically administered at an appropriate interval prior to the procedure and the patient needs to be fasted for at least 4–5 h prior to endoscopy. Sedation should be administered, as the procedure is often prolonged. The sedated patient must be managed in a way similar to that described for thermal laser therapy.

## Patient Aftercare

Following PDT patients are prone to sunburn on exposure to direct sunlight or high-powered artificial (Fig. 9) light. Simple protective measures for 6 weeks or so can prevent this. There is no need for the patient to remain in a darkened room. On the contrary, gradual light exposure and hence photochemical activation, promotes a reduction in the duration of photosensitivity. A light meter is a useful tool to enable graded exposure to light. Photosensitive eruptions are, in part, due to cutaneous histamine release, and the use of antihistamines in selected patients may be helpful in reducing both incidence and severity.

## Procedure Details

The initial process involves systemic administration of the appropriate photosensitizer. The dose often needs to be adjusted for body weight. At a set time interval after administration, monochromatic laser light is delivered to the tumour site by means of an

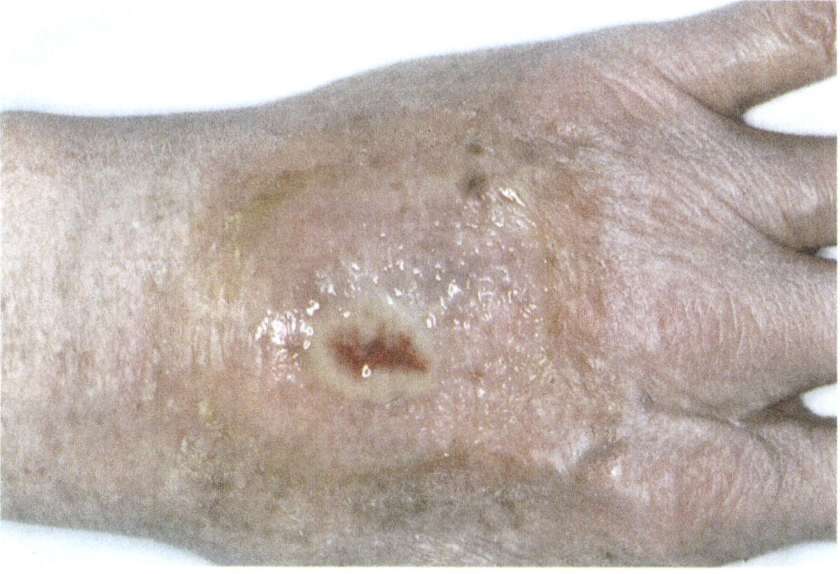

**Fig. 9.** Skin burn following exposure to bright light after PDT

optical fibre. This is inserted either endoscopically for luminal disease or percutaneously under computerized tomography (CT) guidance in the case of pancreatic cancer. Solid cancers are illuminated by insertion of the fibre directly into the tumour substance (interstitial therapy), while luminal cancers tend to be illuminated from above or around the tumour surface (intraluminal therapy). Several sites can be treated during the same procedure. The duration of light illumination, by the chosen method of delivery, must be calculated beforehand in order to provide the optimal light-activating energy. The calculations are best undertaken in collaboration between the gastroenterologist treating the patient and a suitable qualified light physicist. In luminal PDT, debridement of necrotic tumour is recommended by means of a further endoscopy 2–3 days post-therapy. This also allows for laser or other haemostatic therapy to any untreated and bleeding surfaces.

## Criteria for Exclusion

The general utility of PDT in palliation means that exclusion criteria are few and are relative rather than being absolute. The main decision for exclusion is whether the period of light sensitivity is an acceptable inconvenience when balanced against the survival advantage and symptomatic benefits afforded by PDT. Any severe photosensitivity reaction or other severe complication with previous PDT will also mean that alternative palliative methods may need to be considered in preference.

## Comparison with Alternative Methods

There are a number of advantages of PDT over other methods used for the palliation of gastrointestinal malignancy. Selective cytotoxicity of PDT, combined with the systemic non-toxic effects of the photoreactive drugs used, often enables repeated applications of PDT, with no limit on the number of courses, although the frequency will be determined by the duration of phototoxicity. This is in contrast to most if not all systemic chemotherapy regimes. As a result of PDT's mode of cytotoxicity, normal tissue inadvertently damaged in the course of tumour therapy, or mucosa at the site of previous tumour ablation, regenerates rather than healing by fibrosis and scarring. This results in a lower incidence of stricture formation than with radiotherapy. As light penetrates only 1–2 cm into tissue. PDT will treat superfically and is less likely than high-powered thermal methods to cause perforation of a treated hollow viscus.

## Results

A number of studies have demonstrated the efficacy of PDT in the palliation of advanced oesophageal cancer [19]. When compared with Nd:YAG thermo-ablation, a similar improvement in survival rate was noted with PDT. The improvement in dysphagia with either method was also similar, but significant complications were fewer with PDT [20]. There is evidence that treatment of upper oesophageal cancers, which can be technically difficult with thermal laser therapy, may be more effectively managed with PDT. In patients with colorectal cancer where surgery carries a high risk, PDT can be used for palliation to good effect [21]. In some instances in our practice we utilize both thermal laser and PDT sequentially. Thermal laser is used to debulk the tumour, and then PDT cleans up the persistent remnants.

The prolongation of survival from oesophageal cancer with use of SEMS has produced its own problems. Ingrowth of tumour through the mesh of an uncovered stent or overgrowth at the top of the stent can cause further dysphagia, and may be difficult to manage with thermal laser without damaging the stent. Insertion of a further SEMS or argon plasma thermo-ablation are safe alternatives. PDT may also relieve oesophageal obstruction under these circumstances without damage to the SEMS.

Barrett's oesophagus (Fig. 10) has been effectively ablated with a number of methods including thermal laser and APC. PDT is ideally suited to destroying Barrett's epithelium, including any dysplastic cells within it, and allowing regeneration back to squa-

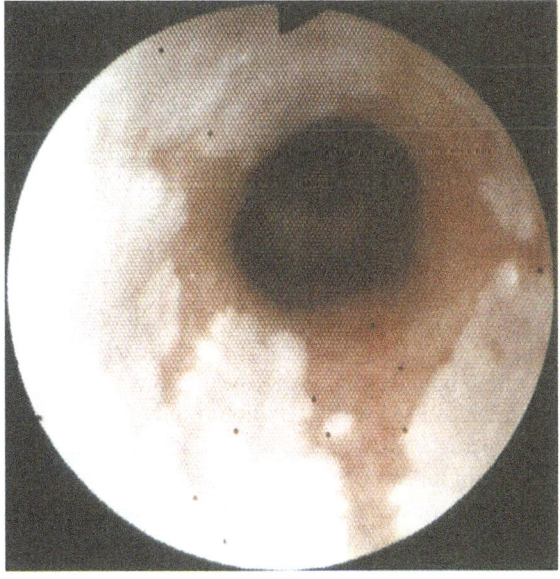

**Fig. 10.** Barrett's oesophagus showing tongues of columnar epithelium protruding into squamous epithelium

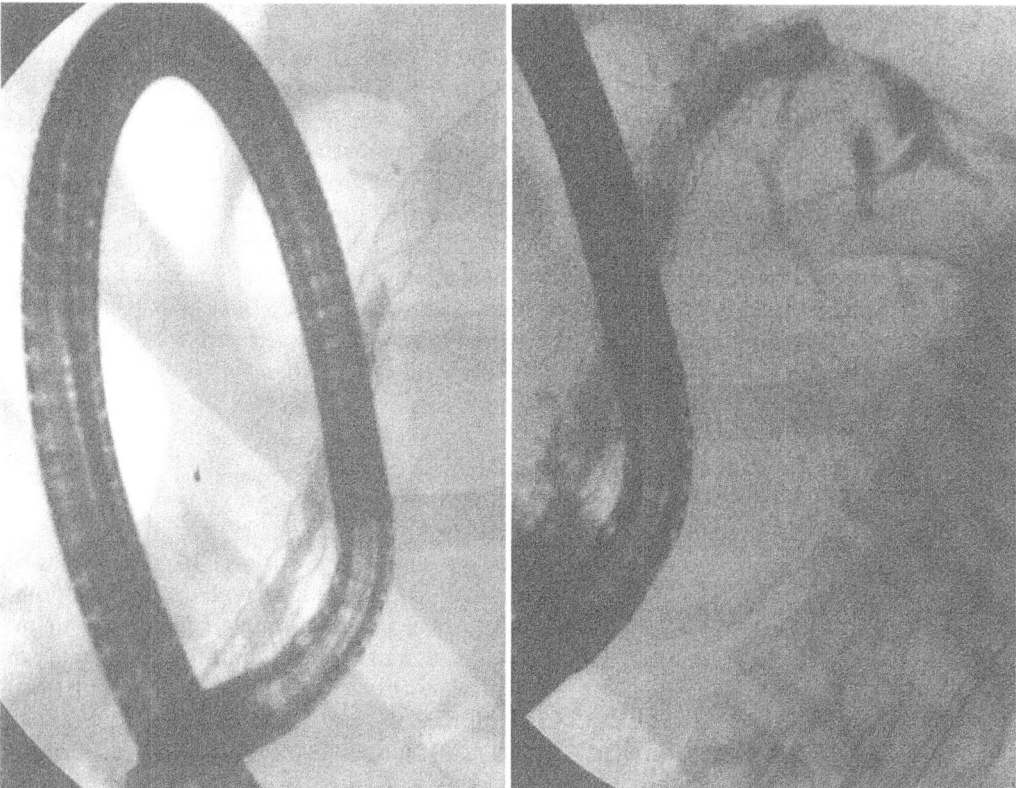

**Fig. 11.** Left cholangiocarcinoma overgrowth blocking metal mesh stent at ERCP; right post PDT therapy showing free flow of contrast into intrahepatic biliary tree

mous or non-dysplastic columnar epithelium. PDT has also been used in the definitive treatment of early-stage oesophageal cancers occurring within segments of Barrett's oesophagus [22, 23]. At present, this form of treatment is restricted to patients declining surgery or those in whom surgery carries an unacceptable risk. In a recent study, PDT achieved a complete response in 84% of patients with early carcinomas without recurrence after 2 years of follow-up [24]. Improvements in the accuracy of staging and future identification of tumours restricted to the mucosa and not infiltrating submucosal sites may improve the result of PDT when used for curative intention.

Cholangiocarcinoma has been shown to be responsive to PDT (Fig. 11). In one study PDT was applied using Photofrin in conjunction with argon dye laser light followed by insertion of a biliary stent [16]. PDT was repeated every 2 months in patients showing a response. A rapid decline in serum bilirubin level and relief of jaundice, as well as an increase in quality of life indices, was detected. This was in excess of that produced by stenting alone. In our practice, we have

also used PDT to good effect in unresectable Bismuth IV cholangiocarcinoma, and consider PDT to be an effective form of palliation in patients with proximal non-resectable cholangiocarcinomas in whom stenting has failed. As with oesophageal SEMS, metallic biliary stents can block with tissue ingrowth. In our unit and in others, PDT has been used to reduce ingrowth irrespective of its origin. Cholangiocarcinoma, hilar metastaes and even benign mucosal hyperplasia can be reduced in bulk and improvements in biliary drainage are produced with PDT.

Periampullary carcinoma is often amenable to curative surgery. However, in a proportion of patients where metastatic spread has occurred or where surgery is contra-indicated, PDT provides a suitable palliative method.. In a recent study, PDT was used every 3–6 months, producing a high rate of response in patients with biliary obstruction from inoperable periampullary malignancy [17]. The results of monotherapy with PDT alone and combination therapy with either snare resection or laser thermo-ablation plus PDT are similar.

Increased median survival of patients with pancreatic cancer has been possible with mTHPC-initiated PDT. Laser fibres are introduced percutaneously and positioned into the tumour by CT guidance. Monochromatic red light is then used to activate the mTHPC. Minor abdominal pain and transient ileus are not uncommon. Acute pancreatitis is rare and patients are frequently well enough to be discharged at 5–6 days post-therapy. The most serious problems encountered are duodenal necrosis and bleeding and duodenal stenosis [25].

## Complications

Complications associated with PDT tend to be minor and/or self-limiting. More serious side effects are relatively rare. In comparison with Nd:YAG thermo-ablation, the use of PDT in patients with oesophageal malignancy produces a greater frequency of pleural effusions, fevers and chest pain. These side effects tend to be transient and resolve spontaneously. Anaemia is also common and responds to blood transfusion. Oesophageal stricture formation results from circumferential illumination and damage deep to the smooth muscle layer of the oesophagus. PDT-induced strictures are usually amenable to courses of dilatation, although a few remain intractable. Unlike Photofrin and mTHPC, newer oral agents such as 5-ALA produce more superficial mucosal ablation and a reduced risk of stricturing. Formation of a tracheobroncheal fistula or prolonged photosensitivity can be more serious consequences of this therapy. Skin photosensitivity normally 2–4 weeks. However, in a small number of patients more severe reactions can occur. 5ALA has much shorter duration of activity and therefore reduced skin photosensitizing capacity.

## Conclusion

Laser therapy in gastroenterology has advanced considerably since its inception as a haemostatic technique for bleeding peptic ulcers. Both thermal laser and non-thermal laser light energy (employed through PDT) have established their individual therapeutic value. The more recent addition of diode lasers to the original range of hardware has been a significant development. These highly effective systems have the added advantages of being less costly, smaller and more portable, features that should popularize their use. Luminal and solid-organ tumours have been shown to be responsive to the effects of laser treatment, which now plays an important part in an ideal management algorithm for palliation of gastrointestinal malignancy. Laser therapy has many unique characteristics and needs to be considered on its own merits as well as in the context of multiple-method therapy. Unfortunately, a present there is scant information on the advantages of combining laser with other therapeutic methods.

Laser light energy has a number of effects on intestinal tissues. Vapourization and immediate necrosis are dramatic effects of high-power laser therapy often seen with tumour ablation. Recanalization of an obstructed viscus can be an immediate benefit of this property. Low-power laser and interstitial therapy produce little in the way of vaporization but significant necrotic effects, which can be more time-consuming to produce but more controllable and associated with fewer side effects. The tissue effects of PDT are potent and associated with a good safety profile. The safety aspect is linked to the selectivity of PDT for malignant tissue and to the mechanisms of cell destruction.

Barrett's mucosa is generally considered a premalignant condition and its management at present varies widely. A reliable non-surgical treatment that removes the abnormal mucosa and preserves the integrity of the organ would be an attractive option. Treatment of Barrett's epithelium with PDT has produced initial beneficial results. The mucosa is ablated and either non-dysplastic columnar epithelium or squamous epithelium replaces the surface. The potential problem remains of dysplastic islands of cells being left behind or dysplasia being covered with the regenerating columnar cells. However, this potential hazard can be encountered with any mucosal ablative therapy, and may even prove to be less of a problem with PDT. Further studies are in progress that will clarify the situation. Although surgery remains the preferred treatment for severe dysplasia and early oesophageal cancer, there is evidence that tumours restricted to the mucosa may be amenable to complete obliteration with PDT. However, this application for PDT requires excellence in imaging in order to accurately stage mucosal tumours. Alterations in the type of photsensitizer used and in the activating light energy employed can potentially enable treatment of tumours penetrating the intestinal wall to different depths. In this respect, the performance of newer photosensitizers will need further rigorous evaluation.

Additional therapeutic uses of laser and PDT are worthy of mention. Laser therapy has been used for the fragmentation of common bile duct stones, although the potential for bile duct injury and the increasing use of mechanical lithotripsy has relegated this application for laser. PDT has recently been found to be an effective means of destroying Helicobacter organisms in the stomach. These organisms tend to be located at superficial sites, on the cell surface of gastric cells and in the mucus gel layer, locations which are readily accessible to light illumination. It is, however, unlikely at present that PDT will displace current antibiotic-based treatment of this organism.

The benefits of laser therapy and its versatility in gastrointestinal treatment are undisputed. However, the practicality of laser therapy is a separate issue. Capital costs of purchasing laser equipment as well as the running costs, including operator training, nursing time, maintenance, administration and consumables, all need to be taken into consideration. Laser should complement, not replace, other palliative modalities and may be added to existing treatments to enhance the quality of cancer management services. Training, apprenticeship and experience of personnel in the uses of laser in general and in the particular application for which it is being used are important. It is equally important to audit a laser service at regular intervals in order to assess outcomes of treatment and impact on predetermined endpoints, Referrals to a laser service will be attracted by the quality of the service, and data on measurable benefits will add to the credibility of the service.

## References

[1] WYMAN A, DUFFY S, SWEETLAND HM, ET AL (1992) Preliminary evaluation of a new diode laser. Lasers Surg Med 12: 1–4

[2] ELL CH, LUX G, HOCHBERGER J, MULLER D, DEMLING L (1988) Laser lithotripsy of common bile duct stones. Gut 29: 746–751

[3] MADHOTRA R, RAOUF A, STURGESS R, KRASNER N (1999) Laser therapy in maintenance of the patency of expandable metal stents. Lasers Med Sci 14: 20–23

[4] KIEFHABER P, KIEFHABER K, HUBER F, NATH G (1983) Usefulness of Nd-AG laser applications in acute gastrointestinal haemorrhage. Laser Surg Med 3: 111

[5] JENSEN DM, MACHIADO G, RANDALL G, AN TUNG L, ENGLISH-ZYCH S (1988) Comparison of low-power YAG laser and Bicap tumour probe for palliation of oesophageal cancer strictures. Gastroenterology 94: 1263–1270

[6] KRASNER N (1988) Endoscopic application of lasers in gastrointestinal disease. Br J Hosp Med 40: 184–192

[7] BUCHI KM, BRUNETAUD JM (1987) Endoscopic gastrointestinal laser therapy. In: Dixon JA (ed) Surgical applications of lasers. Year Book Medical, Chicago pp 95–118

[8] DOHMOTO M, HUNERBEIN M, SCHLAG PM (1996) Palliative endoscopic therapy of the rectal mucosa. Eur J Cancer 32A: 25–29

[9] CLARKE G, DOLAN V, GOH J, ET AL (1997) A comparison of laser photoablation and mesh stenting in the palliation of oesophageal cancer. Gastrointest Endosc 45; AB66

[10] LOIZOU LA, GRIGG D, ATKINSON M, ROBERTSON C, BOWN SG (1991) A prospective comparison of laser therapy and intubation in endoscopic palliation of malignant dysphagia. Gastroenterology 100: 1303–1310

[11] SARGEANT IR, LOIZOU LA, TOBIAS JS, BLACKMANN G, THORPE S, BOWN SG (1992) Radiation enhancement of laser palliation for malignant dysphagia: a pilot study. Gut 33: 1597–1601

[12] SARGEANT IR, TOBIAS JS, BLACKMAN G, THORPE S, BOWN SG (1993) Radiation enhancement of laser palliation for advanced rectal and rectosigmoid cancer: a pilot study. Gut 34: 958–962

[13] SWAIN CP, STOREY DW, BOWN SG, ET AL (1986) Nature of the bleeding vessel in recurrently bleeding gastric ulcers. Gastroenterology 90: 595–608

[14] LOH CS, MAC ROBERT AJ, BUONACCORSI G, KRASNER N, BOWN SG (1996) Mucosal ablation using photodynamic therapy for the treatment of dysplasia: an experimental study in the normal rat stomach. Gut 38: 71–78

[15] ROCHE JVE, WHITEHURST C, WATT P, MOORE JV, KRASNER N (1988) Photodynamic therapy (PDT) of gastrointestinal tumours: a new light delivery system. Lasers Med Sci 13: 137–142

[16] ORTNER MEJ, LIEBETRUTH J, SCHREIBER S, HANFT M, WRUCK U, FUSCO V, MULLER JM, HORTNAGL H, LOCKS H (1998) Photodynamic therapy for nonresectable cholangiocarcinoma. Gastroenterology 114: 536–542

[17] ABULAFI AM, ALLARDICE JT, WILLIAMS NS, VAN SOMEREN N, SWAIN CP, AINLEY C (1995) Photodynamic therapy for malignant tumours of the ampulla of vater. Gut 36: 853–856

[18] ROCHE JVE, KRASNER N, STURGESS R (1998) Photodynamic therapy for occluded biliary metal stents. In: Progress in biomedical optics. Proc of Optical and Imaging Techniques for Biomonitoring IV. Europto Ser 3567: 36–39

[19] KRASNER N, CHATLANI PT, BARR H (1990) Photodynamic therapy of tumours in gastroenterology a a review. Lasers Med Sci 5: 233–239

[20] LIGHTDALE CJ, HEIER SK, MARCON NE, MCCAUGHAN JS JR, GERDES H, OVERHOLT BF, SIVAK MV JR, STEIGMANN GV, NAVA HR (1995) Photodynmaic therapy with profimer sodium versus thermal ablation with

Nd:YAG laser for palliation of oesophageal cancer: a multicerter trial. Gastrointest Endosc 42: 507–512

[21] BARR H, KRASNER N, BOULOS PB, CHATLANI P, BOWN SG (1990) Photodynamic therapy for colorectal cancer: a quantitative pilot study. Br J Surg 77: 93–96

[22] OVERHOLT BF, PANJEHPOUR M (1996) Photodynamic therapy for Barrett's esophagus: clinical update. Am J Gastroenterol 91: 1719–1723

[23] GOSSNER L, STOLTE M, STORKA R, RICK K, MAY A, HAHN EG, ELL C (1998) Photodynamic ablation of high-grade dysplasia and early cancer in Barrett's esophagus by means of 5-aminolaevulinic acid. Gastroenterology 114: 448–455

[24] SAVARY JF, GROSJEAN P, MONNIER P, ET AL (1998) Photodynamic therapy of early squamous cell carcinomas of the oesophagus: a review of 31 cases. Endoscopy 30: 258–265

[25] BOWN SG, LOVAT LB (2000) The biology of photodynamic therapy in the gastrointestinal tract. In: Gastrointestinal endoscopy clinics of North America (in press)

## Further Reading

KRASNER N (ed) (1991) Lasers in Gastroenterology, Chapman and Hall Medical Publication, London

# III-6
# Angiology

# III-6.1
# Laser Treatment in Haemangiomas and Vascular Malformations

M. Poetke

## Contents

## Introduction

A distinct classification of congenital vascular anomalies is necessary because there is a broad list of lesions with variability in signs, symptoms and clinical behaviour. Muliken and Glowacki [44] in 1982 proposed a biological classification that defines the cellular features of vascular birthmarks and correlates these with physical examination and natural history. There are two major types of vascular lesions: those exhibiting a history of rapid neonatal growth and slow involution, characterized by hypercellularity during the proliferative phase followed by diminishing hyperplasia and progressive interstitial fibrosis (haemangiomas); and those present at birth which grew commensurately with the child, and failed to regress, characterized by a normal rate of endothelial cell turnover and a normal mast cell count (malformations). The first type is classified as haemangioma, the second as vascular malformation. Vascular lesions can thus be classified by their clinical history and physical examination. (The clinical criteria are the most important in making a proper diagnosis.) These definitions are critical because haemangiomas and vascular malformations have completely different prognoses and treatment alternatives.

Benign vascular neoplasms are a distinctive disease of infancy and childhood, which can be misdiagnosed as a classic, involuting type of haemangioma of infancy. Infants with a benign vascular neoplasm exhibit a distinctly heterogeneous group of vascular tumours with different clinical and histological features and course, although some of them have few recent descriptions and may produce diagnostic difficulties, for example kaposiform haemangioendothelioma and tufted angioma.

## Haemangiomas

### Clinical Features and Morphology

Clinical Features. Haemangiomas are common benign vascular tumours that are present at birth in 2% to 3% of newborns and about 10% by the end of the first year of life [38]. There is a higher incidence in infants born prematurely, related to gestational age at birth, as for instance in up to 22% of preterm babies weighing less than 1000 g [3]. Therefore, there is speculation that intrauterine influences and/or extrauterine factors contribute to haemangiogenesis. Haemangiomas also occur more frequently in female than in male infants, with a female to male preponderance of 3 to 5:1 [44]. These tumours are also more common in monozygotic twins, where both twins can have a haemangioma, and have the same incidence of 2–3% in the case of dizygotic twins, as in every sibling. It is generally believed that haemangiomas are not familial; yet a positive family history can be elicited from the parents of 10% of affected infants.

Localization. The head and neck are the site of predilection for these tumours – 60 to 70% of all cases – but haemangiomas appear anywhere on the body surface. Often, haemangiomas may involve mucous membranes of oral and genital regions. In these cases, they may be associated with deeper internal lesions.

In most patients, a single lesion is present. However, in approximately 15 to 20% of infants with haemangiomas, the lesions are multiple [65]. In the latter cases in some instances, even other organs such as the lung or liver can be affected. This association has been called diffuse neonatal haemangiomatosis (DNH).

Histopathology. Histopathologic findings in infantile haemangiomas vary with the age and stage of the lesion. Haemangiomas in the early stage of development are highly cellular and characterized by plump endothelial cells that line vascular spaces with small inconspicuous lumina. Numerous mast cells are present in the intervening stroma. It has been suggested that these cells may play a role in the production of angiogentic factors that regulate the growth of the lesions. At this stage, the vascular nature of the lesion may not be readily apparent.

During the period of proliferative growth, plump endothelial cells and an increased number of mast cells are noted. As the lesion matures, the endothelium becomes flattened and lumina appear more obvious and larger. In this stage the dilation of the vessels within a haemangioma has a cavernous appearance that can be histopathologically confused with a venous malformation. Progressive interstitial fibrosis, fatty infiltration, diminshed cellularity and a normal mast cell count were present during the involuting phase [44], but the mechanism that stimulates this process is unknown.

Clinical Course. Infantile haemangiomas are benign but highly proliferative lesions involving aberrant localized growth of endothelium. The lesions generally progress unpredictably, initially appearing as a white or pink macule, a port-wine stain-like lesion (Fig. 1a, b) or a teleangiectasia with surrounding vasoconstriction (prodromal or early phase), enlarging rapidly during the first weeks of life (proliferative phase). A haemangioma is rarely fully grown at birth (prenatal matured haemangioma).

Capillary and cavernous haemangiomas have been the terms classically used to name the different variants of haemangiomas, but they are not the most appropriate denominations for these lesions. Most of the so-called cavernous haemangiomas are not haemangiomas at all, but venous malformations. Therefore,

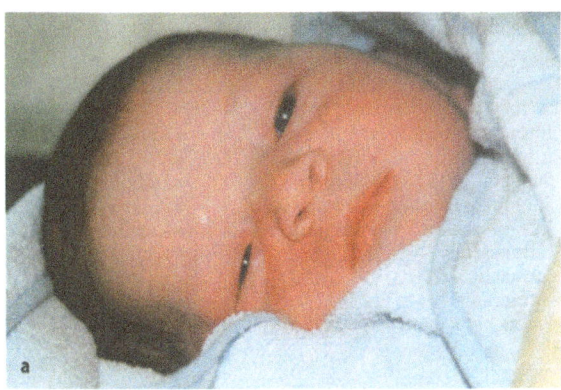

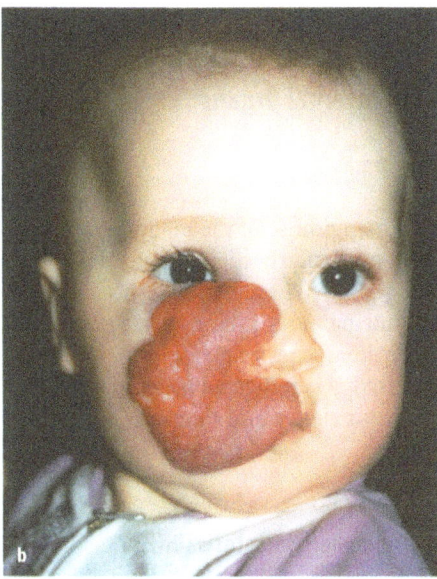

**Fig. 1. a** Newborn child with a pink macular stain of the cheek.
**b** By 3 months, the child has a large haemangioma of the right
face

the most important conceptual issue is that a particular haemangioma has its own histopathological pattern throughout the depth of the lesion. For this reason, haemangiomas arc classified as either cutaneous, subcutaneous or mixed. Their colour intensity depends on their depth and spread and the lumina of the vessels involved, but there may be fluctuation due to localization, state of excitement and temperature.

Haemangiomas are characterized by a proliferation growth phase followed by very slow inevitable regression (involutive phase) between 1 to 10 years of age. Herein, the endothelial component of the tumour is decreased and deposition of fibrous tissue ensues. Although haemangiomas resolve, the lesions persist in 35 to 50% of children who begin school. Even after spontaneous involution of the lesions, 15 % of children have residual skin changes, including de- or hyperpigmentation, teleangiectasia, atrophy and wrinkling of the

skin and cutaneous depression; in addition, the risk of scarring continues, especially at sites of previous ulcerations. If skin changes occur, these may correspond to the largest size of the haemangioma. However, spontaneous regression is no guarantee for a satisfactory cosmetic result, as is often presumed.

Differential Diagnosis. Usually, clinical history and physical examination are sufficient to establish the differential diagnosis between haemangioma and vascular malformation. In some instances, however, an unequivocal diagnosis between deep subcutaneous haemangioma and vascular malformation cannot be established accurately. In this situation, a second visit within a few days or weeks resolves the problem, because rapid growth within the first weeks of life favours a diagnosis of haemangioma. If necessary, colour-coded duplex sonography (CCDS) and magnetic resonance imaging can be used to help differentiate haemangioma from vascular malformation. A proliferating haemangioma demonstrates a well-circumscribed homogeneous density in the MRI and a well-circumscribed hyperperfusion with numerous vessels with spontaneous flow in the CCDS.

A vascular malformation, however shows a heterogenous density, occasionally with calcification and multilocular cystic spaces in the case of large ectatic venous or lymphatic vessels and in the CCDS diffuse hyposonic vascular spaces, wherein blood flow may be seen in the case of compression/decompression of the tissue.

## Different Stages of Haemangioma

A uniform and complete classification of haemangiomas can hardly be achieved and would be quite confusing. Several aspects have to be considered regarding a rational indication for therapy and therapy planning.

1. Localization. Cutaneous and mucosal haemangiomas with their different stages, growing forms and other biological characteristics have to be separated from those differently located, such as subcutaneous, submucosal, intramuscular, intraosseous and internal organs.

2. Biological activity. Here, a differentiation has to be made between haemangiomas with primary demarcation or diffusely infiltrative growth. Further, there is a difference between localized and multifocal or even disseminated haemangiomas with organic affection.

3. Growing stage. The most important criterium for the indication of therapy is the growing stage. In this context five different stages can be distinguished with respect to clinical, sonographic and histologic findings [79].

**Table 1.** Classification of congenital vascular anomalies (CVA)

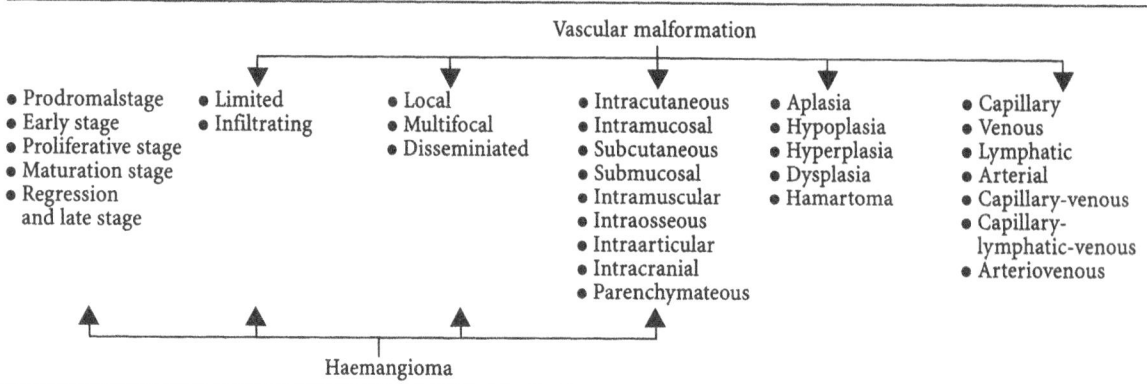

Although these stages can be well defined, there are the following problems.

1. Haemangiomas may skip over different stages, e.g. the tumour can start with a massive proliferation without passing through the prodromal or early stage. On the other hand, a prodromal or early stage can be directly followed by the maturation or even regression stage.
2. In the case of mulifocal localization, each tumour may develop differently. The division in different stages can be related to only one tumour.
3. Even within the tumour, different parts may have a different biological activity. It is possible that there is already regression of the cutaneous parts while proliferation is observed in the subcutis. Accordingly, in the centre of the haemangioma regressive changes can be seen, although a progressive infiltration of the periphery exists. Further, a tumour may show progression and infiltration while already being in the maturation or regression stage. This means that the growing properties of a haemangioma are not predictable and can vary in course from stoppage to foudroyant, exceeding the organ borders and destruction.

These growing properties can lead to several complications, the most common of which are ulceration and secondary infection, bleeding, disfigurement (especially with facial lesions) and ophthalmic problems related to periorbital lesions. In the last case, haemangiomas of the eyelids that block vision may result in amblyopia. Haemangiomas in the nose or mouth may interfere with vital functions of feeding or breathing, and haemangiomas on the ear may obstruct the external auditory canal and interfere with hearing. Further, an airway haemangioma may produce obstruction and respiratory failure.

## Involvement of Internal Organs

### Airway hameangiomas

Infants with haemangiomas of the mandibular region appear to be at particularly high risk of developing airway haemangiomas. These patients should be observed closely during early infancy for the development of stridor, hoarseness or respiratory distress. Tracheobronchoscopy is indicated because of the risk of respiratory insufficiency and to ensure early treatment (Fig. 2a, b).

### Diffuse Neonatal Haemangiomatosis (DNH)

DNH is a rare and life-threatening congenital disorder characterized by the presence of multiple cutaneous and visceral haemangiomas with thromboctopenia, haemorrhage and central nervous system involvement. Holden and Alexander [35] established three minimal diagnostic criteria for DNH: (1) onset in the neonatal period, (2) no evidence of malignancy and (3) the involvement of three or more organ systems. The skin, liver, lungs, intestines and central nervous systems are the most commonly affected organs; spleen, pancreas, adrenals, heart, kidney, bladder, thymus, thyroid, testes and bone are next. Complications of DNH include intestinal bleeding, obstructive jaundice, convulsions and central nervous system haemorrhage. Death can result from high-output cardiac failure as a result of arteriovenous shunting. The mortality rate is about 77% in untreated and 27% in treated patients [41]. Therefore, investigations by ultrasound and magnetic resonance imaging are recommended where indicated in any infant with multiple haemangiomas, to rule out visceral involvement. Since the clinical features and outcome of patients with only cutaneous and   hepatic haeman-

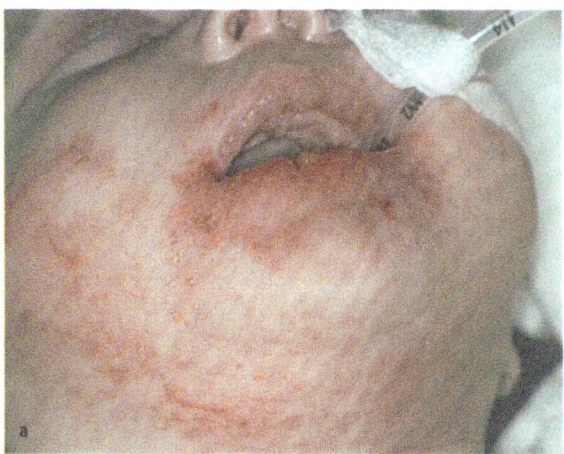

**Fig. 2. a** Four-week-old girl with a large haemangioma of the mandibular region and diffuse infiltration of the lower lip. **b** The tracheobronchoscopy showed a subglottic haemangioma, with a stenosis of 90%

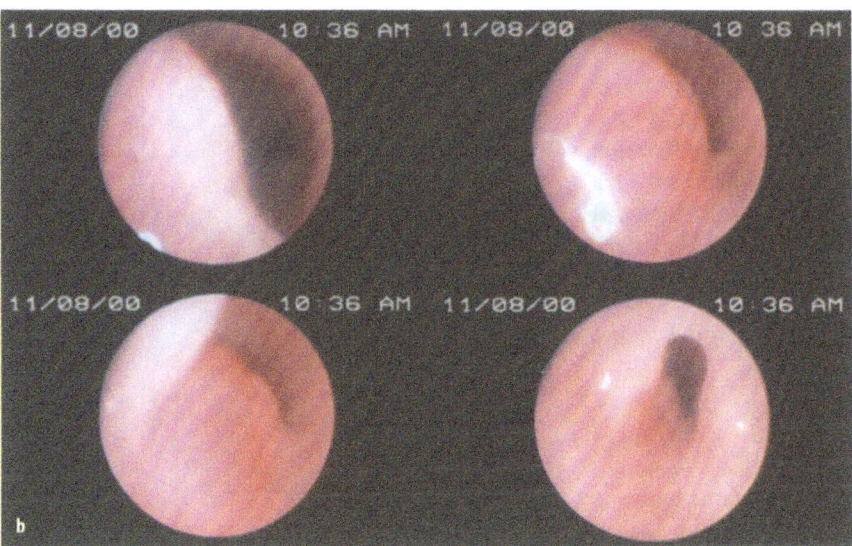

giomas are very similar to those of patients with DNH, Lopriore and Markhorst [41] recommend that the two groups be regarded as entity. Therefore they suggest a fourth criterion for DNH: (4) involvement of at least the skin and liver.

### Benign Neonatal Haemangiomatosis (BNH)

Multiple cutaneous haemangiomas without visceral involvement have been referred to as benign neonatal haemangiomatosis (BNH) because of the benign nature of this disease.

### Special Cases of haemangiomas

#### Prenatally Matured haemangioma

The already prenatally matured haemangioma represents a special form. It remains unclear why a tumour proliferation with further maturation takes place intrauterinely. Nevertheless, at birth there is a solid tumour, that can still show signs of hypervascularization according to the maturation stage. On the other hand, a complete fibromatous differentiation may also be seen.

New histopathologic observations showed that infants with prenatally matured haemangiomas do not have haemangioma of infancy, but tufted angiomas.

## Haemangiomas with Associated Anomalies

Haemangiomas are only incidentally seen with rare dysmorphic conditions. Even though several cases of large haemangiomas in association with different dysmorphic features and syndromes have previously been reported, only limited attention has been paid to haemangiomas and their relationship to other systemic abnormalities. However, there are three dysmorphic conditions that may accompany haemangioma.

### Haemangiomas in Association with Cardiac and Midline Ventral Defects

Facial haemangiomas can have associated cardiac abnormalities and midline ventral defects, particularly coarctation of the aorta. Schneeweiss and coworkers [68] reviewed 68 cases of coarctation and found 4 patients with haemangiomas. In addition to coarctation of the aorta, they found congenital aneurysm of a subclavian or innominate artery, and associated valvular aortic stenosis. However, all of these patients had haemangiomas of the face and neck, a higher incidence than might be predicted by chance.

Midline ventral developmental defects, including sternal clefting and supra-abdominal raphe, were also reported in association with facial haemangiomas. Igarashi et al. [37] documented one case with a supra umbilical midabdominal raphe and facial haemangioma and cite two cases of female infants with the same condition. Sternal cleft is another midline defect that could be coincidentally present with haemangioma. Hersh et al. [32] documented 2 cases of haemangiomas and presence of sternal cleft and reviewed 15 cases from the literature. Kaplan et al. [39] examined 50 cases of chest wall defects, divided into 30 cases of ectopia cordis, of which none was found to have associated haemangiomas, and 20 cases of sternal cleft, of which 8 had true haemangiomas.

### Sacral haemangiomas in Association with Other Dysmorphic Features

Sacral haemangioma can have associated dysmorphic features, including midline (sternal, abdominal) clefting and spinal and genito urinary defects [16, 28]. Goldberg et al. [28] previously proposed that infants with sacral haemangioma appear to have a consistent association with a variety of congenital abnormalities involving various organs. The majority of these are urinary tract malformations, including renal abnormalities, deformities of the external genitalia and imperforate anus associated with a fistula. Spinal defor-

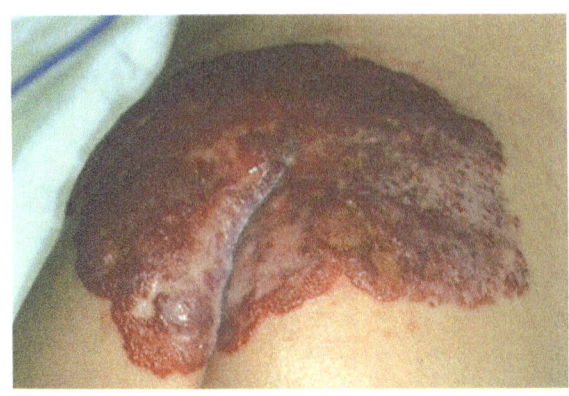

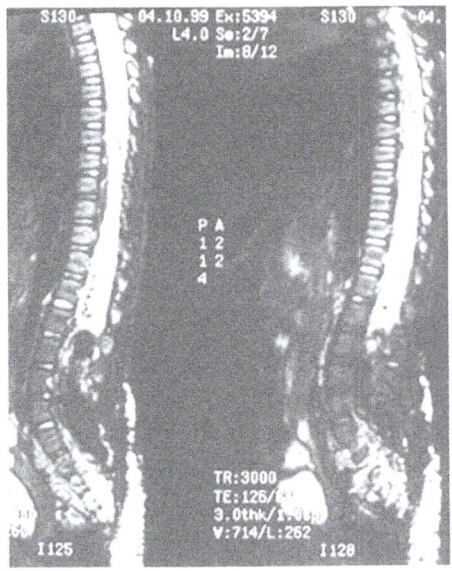

**Fig. 3. a** Infant with a large lumbosacral haemangioma. **b** Magnetic resonance imaging study demonstrated haemangioma of the conus medullaris, and sinus venosus defect

mities, particularly of the lumbar and sacral portions, are also commonly encountered in association with sacral haemangiomas. Of interest is that most of the patients had bony deformities of the sacrum and also had lipomeningomyelocele, which is considered to be a form of occult spinaldysraphism. Lipomeningiomyelocele can have associated neurologic deficits (e.g. lower extremity deformities, sphincteric malfunction, motor and sensory paralysis) that may be apparent at birth but usually develop somewhat later and have a deceptively slow progression (Fig. 3a, b).

Tethered cord syndrome is also commonly encountered in association with sacral haemangioma and lipomeningomyelocele. It refers to the association of neurologic deficits involving the lower extremities of sphincters and a low lying conus medullaris, and is mostly seen in patients with intraspinal lipomas, diastomatomyelia or fibrous bands. Albright and coworkers [2] postulated haemengioma as an indicator of tethered spinal cord. They reviewed the cases of seven consecutive children with lumbar haemangiomas. All of these children were neurologically normal, but were found to have tethered spinal cord.

## Large Facial Haemangiomas and their Association with Dandy-Walker-Syndrome and other Malformations of the Posterior Cranial Fossa. PHACE Syndrome

Large facial, particularly bilateral, haemangiomas, appear to have a consistent association with the Dandy-Walker syndrome (DWS) and other malformations of the posterior fossa of the brain [53, 56, 63].

The Dandy-Walker syndrome is characterized by a cystic expansion of the fourth ventricle, a dysgenesis of the vermis of the cerebellum and atresia of the foramen of Magendie. Dysgenesis or agenesis of the corpus callosum were also found in this association. The Dandy-Walker syndrom is estimated to have a total incidence of about 2–4%. About 70% of the cases are diagnosed before the patients are 1 year old, and only very few in adulthood. Macrocrania is by far the most frequent presenting symptom: it was found in about 80–90% of the cases. It is usually the consequence of hydrocephalus, but not always. Hirsch et al. [33] supposed that macrocrania preceded the development of hydrocephalus and was due to an enlarged posterior fossa. All the children sooner or later became hydrocephalic and therefore required treatment. However, hydrocephalus is common but not invariably present. Cerebellar symptoms such as ataxis, nystagmus or epileptic attacks are observed in only about 15% of the patients with Dandy-Walker syndrome.

Early diagnosis and treatment are essential conditions for the survival and optimum development of these patients.

Hypoplasia of the cerebellum and cerebellum vermis have also been found in these patients and are included as part of the spectrum of CNS abnormalities seen in the Dandy-Walker syndrome. Unilateral cerebellar hypoplasia does not seem to have any clinical significance from the neurological point of view. As reported by Erskine [20], unilateral agenesis of the cerebellum is asymptomatic and is normally an incidental finding. He suggested that the main part of the compensatory function in these cases is carried out by the cerebral cortex.

The aetiology of the Dandy-Walker syndrome is unknown. Benda [11] stated that the main pathology consisted in the formation of a cleft of the cerebellum associated with more or less severe malformations and with a meningocele-like sac in place of the posterior medullary velum. The association of facial and cardiovascular abnormalities favours the hypothesis that the onset of the malformation occurs between the formation and migration of the cells of the neural crest, between the 3rd and the 4th postovulatory week.

Because of these associations, neuro imaging has been recommended in infants with large facial haemangiomas, especially those which are plaque-like in nature and when the orbits were involved. MR imaging and MRI angiography is the most sensitive method for imaging CNS abnormalities. These children should also be examined for symptoms of the Dandy-Walker syndrome, e.g. hydrocephalus internus because a neurosurgical intervention (e.g. drainage operation) sometimes has to be performed. Signs that suggest the possible development of neurological problems should be given special attention.

The facial haemangiomas associated with DWS are large, and either unilateral or bilateral. They initially appeared as a macular redness a few days after birth, so that a port-wine stain was initially diagnosed in most cases. This misdiagnosis is understandable during the prodromal or initial phase of the haemangiomas, which can closely resemble port-wine stains. Most of the haemangiomas have a plaque like quality, covering at least one, mostly several dermatomes on the face, and in the case of bilateral distribution they were often butterfly like (Fig. 4a, b). Severe ophthalmic occlusion and ulcerations have been a feature of many of these facial haemangiomas. The mandibular region is often involved as well, and infants appear to be at particularly high risk of airway haemangiomas.

Numerous systemic anomalies, most of which were ophthalmologic abnormalities, structeral arterial anomalies and cardiac defects, have been found associated with large facial haemangiomas and the DWS. As published, around a third of patients with both the Dandy-Walker syndrome and facial haemangiomas had further anomalies of the eye (e.g. colobomas, optic nerve atrophy, microphthalmos, cryptophthalmos, exophthalmos, posterior embryotoxon, glaucoma). Retinal and choroidal haemangiomas have also been described in this connection. Right-sided coarctation of the aorta is also a typical association of large facial haemangiomas and DWS. Furthermore, several cardiac abnormalities (e.g. aortic and tricuspid atresia, dextrocardia, patent ductus arteriosus, ventricular septal defect, cor triatriatum) have also been reported. Frequent alterations of the intracranial vasculature, particularly in the region of the left carotid artery, were also present in some cases (e.g. persistent trigeminal artery, absence or hypoplasia of carotid or vertebral arteries, aneurysmal dilatation of the carotid artery, dilated cerebrovascular vessels and abberant left subclavian artery [48].

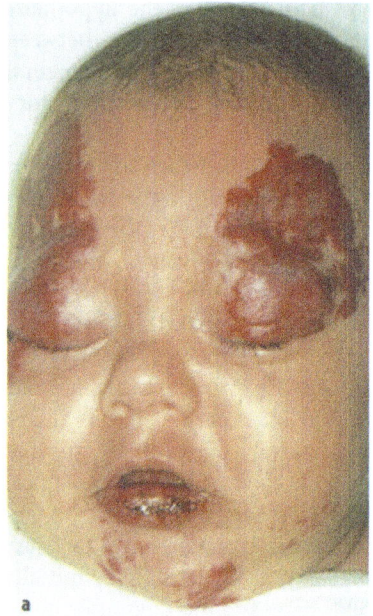

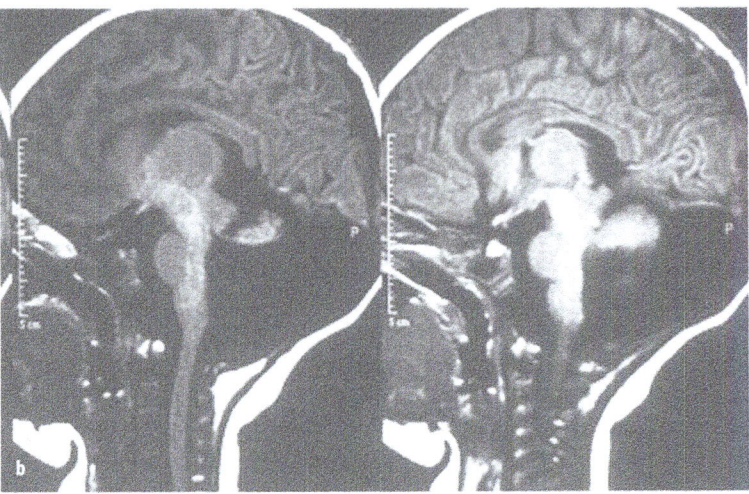

**Fig. 4. a** Bilateral haemangioma in an infant with Dandy-Walker malformation and coloboma of the right optic nerve. The right medial carotid artery was hypoplastic. **b** Transverse MRI scan shows right cerebellar hypoplasia, and communication between the fourth ventricle and the cystic mass

Because large facial haemangiomas may have a distinctive group of associated arterial, central nervous system and ophthalmologic abnormalities, Frieden and coworkers [23] proposed the acronym PHACE syndrome to emphasize the characteristic findings: posterior fossa malformations, haemangiomas, arterial anomalies, coarctation of the aorta, cardiac defects and eye abnormalities. When sternal clefting and/or supra umbilical abdominal raphe is present, they recommend calling the syndrome appropriately PHACES. Large plaque-like facial haemangiomas, covering at least on, or in most cases, several dermatomes on the face, in association with the Dandy-Walker syndrome and other central nervous system (CNS) abnormalities, are the main characteristic features.

Cases of PHACE syndrome were published earlier under various names, including 3C syndrome: cerebellar hypoplasia, cavernous haemangioma and coarctation of the aorta [27], and oculo-cerebro-acral syndrome that describes the association of facial haemangiomas with brain cysts, microphthalmia, ectrodactyly and several other malformations like cleft lip, malformed ears, cardiac and renal abnormalities. In 1991, Pascual-Castroviejo [48] et al. reported the association of cutaneous facial haemangiomas with anomalies affecting intracranial and extracranial arteries, the cerebellum, and, less frequently, the cerebral hemispheres and aortic arch. This association was called cutaneous haemangioma-vascular complex syndrome (CHVC syndrome). Although there is some difference of opinion regarding the correct ter-

**Table 2.** Distinguishing features of the PHACE syndrome and Sturge-Weber syndrome

| Feature | PHACE syndrome | Sturge-Weber syndrome |
| --- | --- | --- |
| Extracranial vascular anomalies | Haemangioma | Port-wine stain |
| Intracranial vascular anomalies | Arterial anomalies, intracranial haemangiomas | Leptomenigeal vascular malformation |
| Structural CNS defects | Dandy-Walkersyndrome, cerebellar hypoplasia, posterior fossa malformations | Calcification and atrophy of the cerebellum |
| Symptoms | Hydrocephalus internus, mental retardation in some cases | Epileptic attacks, mental retardation |
| Ophthalmologic anomalies | Colobomas, optic nerve atrophy, microphthalmos, cryptophthalmos, exophthalmos, glaucoma, embryotoxon, haemangiomas of the retina, conjunctiva | Glaucoma (buphthalmos), retinal vascularity |
| Cardiologic anomalies | Coarctation of the aorta, cardiac abnormalities | Usually none |

minology for this condition, PHACE syndrome has
been the preferred designation in the 4 years. However, the term PHACE describes the possible associated abnormalities, and requieres that the physician
look deeper for any underlying problem.

It is important to be aware of this association in infants with large facial haemangiomas with DWS and
ophthalmologic, cardiovascular and arterial abnormalities. Therefore, patients with large facial haemangiomas involving more than one dermatome, including those on only one side, should be examined for
symptoms of the Dandy-Walker syndrome, and
brain-imaging studies should be performed. Signs
that suggest the possible development of neurological
problems should be given special attention. In this
setting, neurological status and head circumference
should be monitored closely. A thorough ophthalmic
examination is also to be recommended, as some
cases of haemangiomas or associated anomalies, predominately anterior segment abnormalities, will thus
be exposed early on. Furthermore, it is noteworthy
that glaucoma has also been reported in the PHACE
syndrome, because some authors suggested that the
absence of glaucoma is an important means to help
distinguish the PHACE syndrome from the Sturge-
Weber syndrome (Table 2). Furthermore, cardiac abnormalities are important to consider in infants with
large facial haemangiomas. Careful cardiac examination and blood pressure measurements are necessary.
Goh and Lo [27] suggested that defining cardiac abnormality and anatomy of the head and neck vessels
is of vital importance in surgery to ensure the survival of these patients. Among other systemically associated anomalies, special attention should be given
to clefts. Finally, infants with large facial haemangiomas, especially in the mandible region, should be
observed closely for the development of respiratory
distress and be examined for subglottic haemangiomas.

Nevertheless, all the patients had a CNS abnormality associated with large facial extracranial haemangiomas. The Dandy-Walker syndrome is the most
common CNS abnormality reported in association
with the PHACE syndrome. Hypoplasia of the cerebellum and cerebellum vermis has also been reported
and is included as part of the spectrum of CNS abnormalities seen in the Dandy-Walker syndrome. Other
reported CNS abnormalities include arterial anomalies (e.g. persistent trigeminal artery, absence or hypoplasia of carotid or vertebral arteries, aneurysmal
dilatation of the carotid artery, dilated cerebrovascular vessels, abberant left subclavian artery), and these
anomalies have also been reported in several patients
with large facial haemangiomas and PHACE syndrome without posterior fossa malformations (Fig.
5a, b). Pascual-Castroviejo [48, 49], who routinely per-

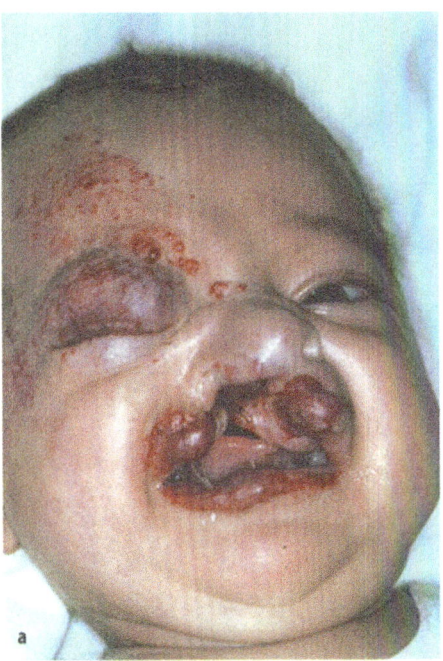

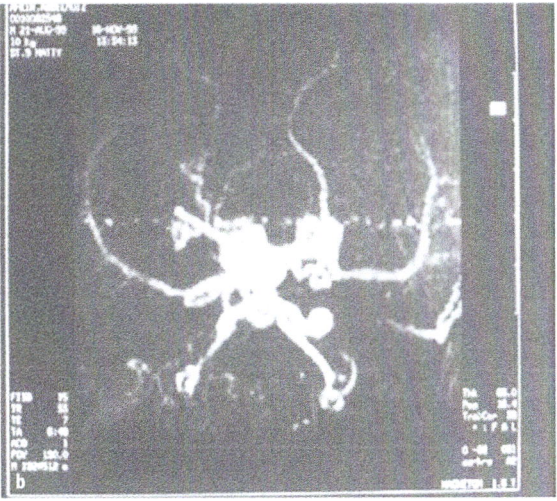

**Fig. 5. a** Bilateral haemangioma in an infant with intracranial
abnormalities and a complete cleft palate, VSD, ASD and periphere stenosis of the pulmonalis. **b** MRI angiography. Hypoplastic internal carotid artery and ectasia of the basilaris artery

forms carotid angiography on children with large facial haemangiomas, reports frequent alterations of
the intracranial vasculature. However, the incidence
of intracranial arterial anomalies is unknown, because most patients were not evaluated for it.

Torori et al. [77] emphasized a further CNS abnormality seen in association with large facial haemangiomas, namely intracranial haemangiomas. He described four cases of large facial haemangiomas with
associated intracranial, meningeal-based contrast-en-

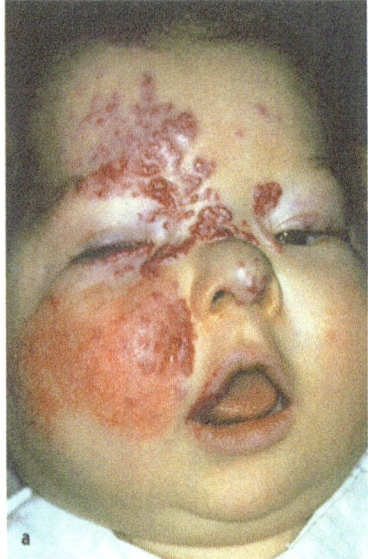
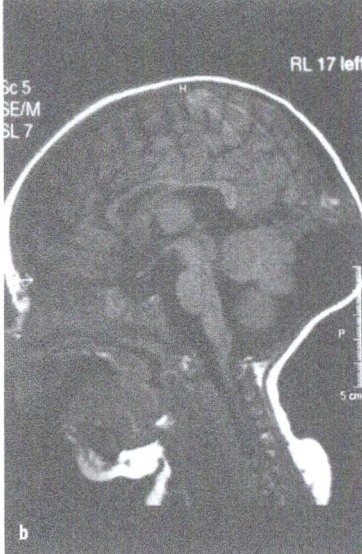
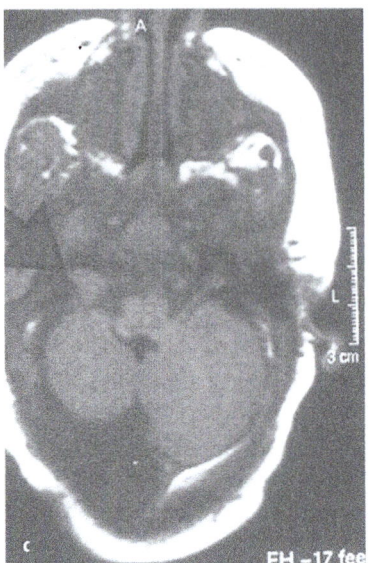
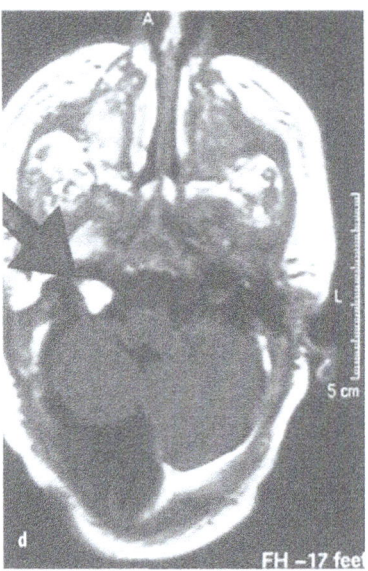

**Fig. 6. a** Infant at 10 weeks of age demonstrating large right orbitofacial haemangioma, a small haemangioma of the left face, and a deviation of the right eye into the adducted position. Medical history was notable for a right-sided aortic arch with mild aortic coarctation. **b** MRI scan revealed an ipsilateral hypoplastic cerebellar hemisphere, and an enlarged posterior fossa. The right internal carotid artery was hypoplastic. **c, d** Furthermore, there was a rounded enhancing mass at the right cerebellopontine angle (CPA) (**c**, arrow). This lesion was isointense on T1-weighted images and showed homogeneous contrast enhancing with contrast medium (**d**, arrow): intracranial haemangioma

hancing masses which they diagnosed as intracranial haemangiomas. The intracranial lesions were discovered with CT, MRI and DSA. There were three haemangiomas in the cerebellopontine angle, one in the unco-hippocampal region and one in the hypothalamus. Diffuse leptomeningeal enhancement was noted on the surface of the left cerebellar hemisphere in one patient. Furthermore, associated cerebellar atrophy was found in one patient. Follow-up of the patients showed that the intracranial haemangiomas behaved in most cases in parallel with the extracranial haemangiomas. The author proposed that intracranial haemangiomas could be a peculiar phenotype of the PHACE syndrome. This consideration is supported by the added association of a persistent trigeminal artery in one case, and cerebellar atrophy in another.

One case that we reported [61] has very similar CNS abnormalities: cerebellar hyoplasia as a part of the Dandy-Walker syndrome, hypoplasia of the right ICA and leptomeningeal enhancement at the cerebellar surface and two separate masses, one at the pituitary stalk and a second in the cerebellopontine angle (Fig. 6a–6d), Billson and Gilliam (1984) [14] reported a case with a large craniofacial haemangioma, and its association with Dandy-Walker malformation and intracranial haemangiomas of the pituitary stalk and the brain-stem meninges. Pascual-Castroviejo [49] reported a facial haemangioma associated with "angiomatous malformations" at the left carotid siphon and hypothalmic area.

The MRI features of intracranial haemangiomas showed the same signal characteristics as extracranial haemangiomas. The lesions were isointense on T1-weighted images and gave a high signal (hyperintense) on T2-weighted images, reflecting their content of unclotted blood. Typically, haemangiomas enhance markedly and homogenously with contrast medium on CT and MRI. The angiographic features are also quite characteristic. The tumours were fed by slightly enlarged branches of normal systemic arteries; usually, several vessels fan out into individual branches, supplying individual lobules of the lesion. In the venous phase, small venous branches seemed to drain each lobule, joining into large veins at the base of the mass. Such features are typical of extracranial haemangiomas, but were also seen in DAS in intracranial haemangiomas, as Tortori and other authors [77] reported.

In some cases, intracranial haemangiomas are part of the PHACE syndrome. We believe that this association, while rare, is important to consider in infants with PHACE syndrome. We therefore suggested that a seventh criterion should be added to the six minimal inclusion criteria for the PHACE syndrome. The inclusion criteria would then be arterial abnormalities and/or intracranial haemangiomas. The practical implication of this additional criterion is to ensure contrast-enhanced imaging to detect silent intracranial lesions. As extracranial haemangiomas are known to regress spontaneously or with systemic corticosteroids or interferon, surgery should be avoided unless follow-up studies demonstrate growth or, more importantly, the clinical picture deteriorates.

The PHACE syndrome is more frequent than previously supposed. Awareness of this entity and early recognition is important for appropriate therapy. Therefore, clinicans who treat children with large facial haemangiomas should consider PHACE syndrome in the differential diagnosis and carry out appropriate neuroimaging studies (MR imaging, MR angiography, and contrast-enhanced MR imaging), cardiac and ophthalmologic evaluation to detect other PHACE syndrome-associated anomalies. Besides these associated anomalies, intracranial haemangiomas are supposed to be a peculiar phenotype of PHACE syndrome.

### Haemangiomas in Association with other Syndromes

Reliable reports on haemangiomas associated with other syndromes are also found such as postnatal overgrowth and macroglossia in the case of the Wiedemann-Beckwith syndrome [3].

## Other Benign Vascular Neoplasms

### Kaposiform Haemangioendothelioma

Clinical Features. Kaposiform haemangioendothelioma (KHE) is a rare, locally aggressive vascular neoplasm of childhood that is clinically and histologically distinct from haemangioma of infancy. This disease generally manifests later than infantile haemangioma, usually during childhood or early adolescence. KHE can be present as multinodular soft tissue masses, purpuric macules and multiple teleangiectatic papules. Lesions are characterized by rapid growth and extension locally involving skin, soft tissue and even bone. It is associated with lymphangiomatosis and frequently complicated by the Kasabach-Merritt syndrome. Cutaneous kaposiform haemangioendothelioma lesions are non metastizing, and have no known association with Kaposi sarcoma, related to human immunodeficiency virus infection.

Localization. Kaposiform haemangioendothelioma are encountered in two forms. In the first and more common type, the skin is involved (at a average age of 43 months), whereby the second group is characterized by extracutaneous retroperitoneal presentation (average age 10 months) [80]. Complications of retroperitoneal and visceral KHE are serious and often lead to death, because of functional compromise of vital structures. Early recognition of the lesion is therefore crucial in order to prevent these complications, but if present retroperitoneally or viscerally, they may for the time being be asymptomatic and remain undetected. However, extracutaneous KHE may be almost impossible to differentiate from benign tumours without biopsies.

Vin Christian et al. [80] reported that cutaneous KHE manifests later than infantile haemangioma; they therefore recommended taking biopsies in children older than 3 months of age with a new onset of proliferating vascular lesions, rather than assuming that their lesions are simply late-onset infantile haemangiomas.

Histolopathology. Biopsies of these lesions exhibit a lobular or nodular growth pattern of densely packed spindle-shaped tumour cells closely associated with small slit like and sieve like blood vessels. Furthermore, a great tendency to form complete vascular spaces, and the presence of scattered fibrin thrombi and dense fibrosis is described. Each nodule consists of an admixture of small round capillaries and glomeruloids, solid nests of round and epithelioid endothelial cells that contain haemosiderin, hyaline globules of vacuoles as expression of primitive luminal differentiation. This multinodular pattern of kaposiform haemangioendotheliomas closely resembles that of tufted angioma, but nodules of KHE are larger and less circumscribed and they involve deep soft tissue and bone [65].

Spindle-shaped tumour cells stain only focally positive for CD34, CD31 and vimentin and are entirely negative for muscle-specific actin. Some authors [78] have reported immunoreactivity for factor VIII-related antigen and Ulex europaeus I lectin, whereas other authors found negativity for these endothelial markers [84].

Clinical Course. Kaposiform haemangioendotheliomas are associated with lymphangiomatosis and are frequently complicated by the Kasabach-Merritt syndrome (KMS), which consists of thrombozytopenia, microangiopathic haemolytic anemia and localized consumption coagulopathy. The ecchymotic appearance of the lesions was typical of the KMS (Fig. 7a, b). The Kasabach-Merritt syndrome most commonly occurs during the first weeks of life and has a high mortality rate, reported to be from 20 to 30%.

KMS is a variant of disseminated intravascular coagulation (DIC) in which platelets and clotting factors are locally consumed within a giant vascular neoplasm. The cause of KMS is not clear, but some authors [62] suggested that because the blood stays statically in the vessels, both platelets and contact factors may be activated by the abnormal endothelium.

Enjolras et al. [19] proposed that infants with Kasabach-Merritt syndrome do not have "true" classic, involuting type of infantile haemangioma. The clinical and biological features and course and further histopathologic aspects are different from infantile haemangioma. Vin Christian et al. [80], in the light of the relative frequency of the Kasabach-Merritt syndrome in KHE, suggested scrutinizing the diagnosis of infantile haemangioma in any child with a rapidly proliferating vascular tumour and aquired thrombocytopenia or other findings consistent with platelet trapping.

Differential Diagnosis. The so-called haemangioendotheliomas represent a heterogeneous collection of neoplastic and non-neoplastic vascular lesions. Several variants have been described, including infantile, spindle cell, retiform, epitheoliod and kapsiform types. Infantile haemangioendothelioma or benign haemangioendothelioma has been used synonymously for the common haemangioma of infancy. Because the term haemangioendothelioma connoted possible malignant behaviour, and this notion is mistakenly applied to the common haemangioma of infancy that regress spontaneously, it should be discarded. Spindle cell haemangioendothelioma is a proliferative but non-neoplastic lesion that develops in association with endogenous or iatrogenic vascular malformations. These lesions occur in the distal part of the extremities of children and young adults, as blue nodules of firm consistency. Retiform and ep-

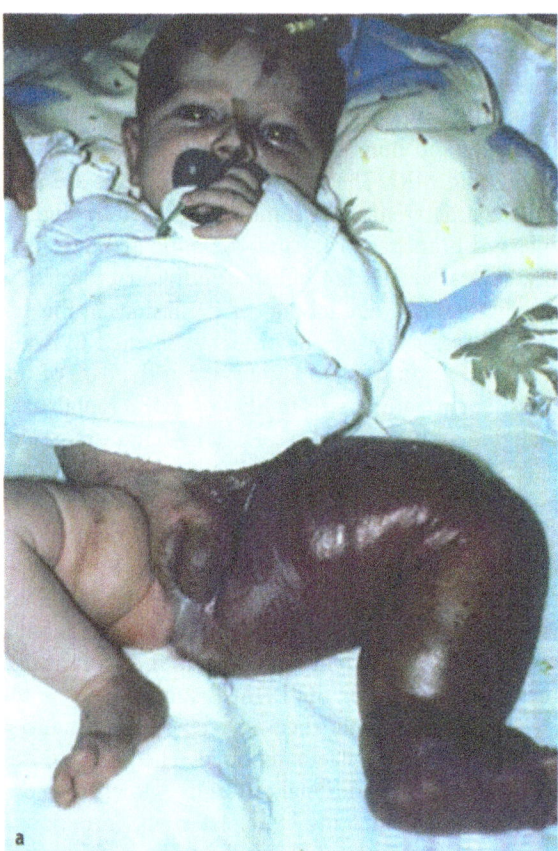

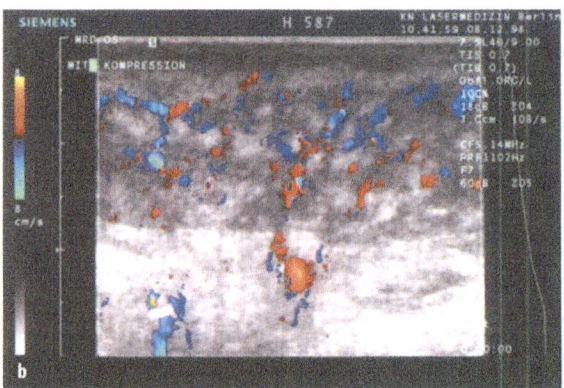

**Fig. 7. a** Infant with a congenital vascular neoplasm of the left thigh, that appeared shortly after birth concurrent with Kasabach-Merritt syndrome: haemangioendothelioma. **b** The colour-coded duplex sonography showed hyposonic nodular tufts between some pathological vessels

ithelioid haemangioendothelioma are low-grade forms of angiosarcoma that hold a small but real risk of metastasis for the patient. Kaposiform haemangioendothelioma is clinically and histologically distinct and has previously been called haemangioma with Kaposi's sarcoma-like features and Kaposi-like infantile haemangioendothelioma. It is a neoplastic proliferation that displays locally invasive growth but probably does not represent a form of angiosarcoma, as distant metastases have not been observed.

## Tufted Angioma

Clinical Features. Tufted angioma is a rare, recently described, vascular anomaly, that usually appears in infancy or early childhood but is rarely present at birth. It is named for it characteristic histological pattern of grouped dermal capillary tufts. The clinical appearance of the lesions is variable. In most cases, they appear as pink or red patches with a mottled appearance, and with clusters of superficial, small,

bright red angiomatous papules. Some of them are characterized by enlarging erythematous or brown macules or plaques, but other lesions may resemble granulomas or connective tissue abnormality. These lesions may be slightly warm, covered with lanugo hairs or exhibit hyperhydrosis.

Localization. Tufted angiomas are most commonly located on the neck, upper trunk, back and chest, although some cases on the head and extremities have been also described (Fig. 8a, b).

Histopathology. Biopsies of tufted angiomas exhibit multiple separated angiomatous cell lobules within the dermis and subcutaneous fat that are much larger and extend deeper than in infantile haemangiomas. Each lobule is composed of aggregates of endothelial cells within which small capillary lumina are seen, concentrically whorled along a pre existing vascular plexus. Cell-marker studies suggest that the cell lobules consist of closely packed blood capillary endothelial and perithelial cells, because they exhibit strong positivity for Ulex europaeus I lectin and EN4, as expression of their endothelial nature. Some cells show reactivity for smooth muscle cell actin and they are probably pericytes.

Clinical Course. Tufted angiomas grow slowly and insidiously, characterized by lateral extension for 5 months to 10 years. Instead of spontaneous involution, the lesions appears generally to persist indefinitely. In some rare cases, spontaneous regression has been reported.

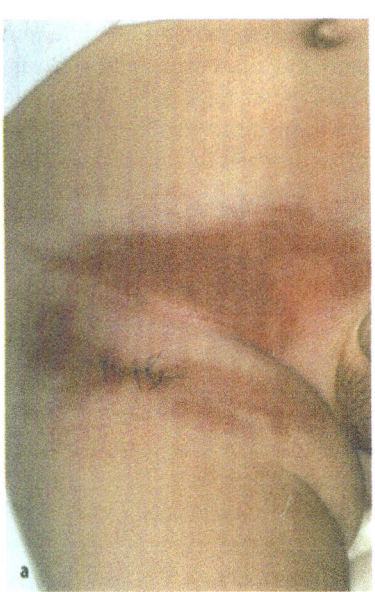

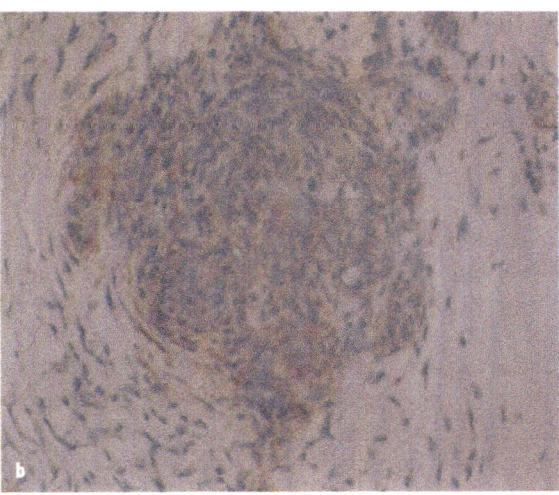

**Fig. 8. a** Infant with a pink patch, that first appeared at 3 months: tufted angioma. **b** Cell marker studies showed cannon ball patterns, consisting of perithelial and capillary endothelial cells

## Vascular Malformations

### Clinical Features and Morphology

Vascular malformations are structural abnormalities, errors of vascular morphogenesis, which can be localized in all parts of the vascular system. All vascular malformations, are by definition present at birth and grow proportionately with the child; the volume can change. In contrast to the haemangiomas, which proliferate only from the endothelial cells and where the division in stages is of clinical importance, vascular malformations are divided from the part of the vascular system which is affected. Vascular malformations may have any combination of capillary, venous, arterial and lymphatic components, with or without fistulae (Table 1). Their development is caused by changes in pressure and flow, ectasia, collateral formation, shunting and hormonal modulation.

It is critical to understand that these changes in vascular anomalies are not of proliferative nature such as those seen in haemangiomas. These changes tend to occur secondarily from haemodynamic causes peculiar to each vascular structure involved.

There are truncular and extratruncular forms [10], a combination of both is possible. The truncular form is fundamentally based on an obstruction, dilatation, arterivenous shunting or a combination of these forms. A malformation with an obstruction shows either agenesis, aplasia, hypoplasia or hyplasia of vessels. An irregular structure of the vessel wall, position anomaly as a result of an irregular origin and course or a persistent foetal vessel can be the reason for this. As with haemangiomas, a multifocal affection is relatively frequent, but there is also a broad variance and combination of different origins of vessels.

In some cases, the vascular malformation remains preformed and latent and grows as a result of a lesion, trauma or hormonal effect. This can happen during adolescence or even during adulthood.

Most vascular anomalies are well known with their eponymous conditions: Sturge-Weber, Klippel-Trenaunay, Rendu-Osler, Parkes-Weber, Blue-rubber-bleb-nevus syndromes are described in more details here.

## Capillary Malformations

### Port-Wine Stains

Clinical Features. A port-wine stain (PWS) is a congenital malformation of the superficial cutaneous vascular plexus, involving venules, capillaries and possibly perivenular nerves [9], that is present in 0.3 to 0.5% of children at birth [38]. It first appears as a pale pink macule that evolves with time and becomes dark red to purple. In 65% of these patches, nodularities and a cobblestone puffy pattern may develop, and severe hypertrophy of the soft tissue with facial asymmetry or deformity occurs by the fifth decade of life [25]. Particularly, eruptive angiomas (tiny bleb lesions) are a frequent complication of port-wine-stains.

Localization. Port-wine stains are usually unilateral and segmental, though they may be bilateral, and they most commonly occur on the face.

Histolopathology. Histologically, PWS is characterized by an increased number of thin-walled (capillary-venular size) blood vessels with normal ultrastructure confined predominantly to the 0.6 mm subepidermal zone [22].

Clinical Course. PWS in children are characterized histologically by a small vessel with few erythrocytes. With increasing age, there is progressive ectasia and associated colour shifts (pink to purple) because of an age-related increase in erythrocyte content of the vessels, mean vessel area and percentage of dermis occupied by the vessel [47]. Smoller [71] revealed a significant decrease in perivascular nerve density in port-wine stains; this may account for the decreased

neural modulation of vascular tone and subsequent dilatation of dermal vessels. Ashinoff [6] suggested that collagen degeneration that accompanies age can also weaken the supporting dermis and allow abnormal vessels to dilate.

Differential Diagnosis. Lesions that may be confused with port-wine stains are the following:
1. Large facial haemangiomas during the macular phase (prodromal phase) can closely resemble port-wine stains. The proliferation of the cutaneous lesions and increasing colour intensity, however, are not features of port-wine stains, and distinguish them from haemangiomas during the macular phase.
2. The transient macular stains so-called stork bite or salomon patch belong in a separate category from the permanent capillary malformations. The are usually located on the glabellar region, eyelids and nape. In contrast to port-wine stains, these lesions are paler and fade and disappear. Mulliken suggested that a small percentage may persist, although diminshed in intensity, and he would classify this subgroup as vascular malformations [16].

## Port-Wine Stains with Associated Vascular Malformation (Neurocutaneous and Dysplasia) Syndromes

Capillary or dermal vascular malformations are occasionally associated with deeper vascular anomalies and may be merely the marker which Mulliken called the red flag of other vascular anomalies. The key point of this fact is that these cutaneous signs permit early diagnosis, thus helping in further recognition of more complex syndromes. The Sturge-Weber syndrome is the best known vascular malformation complex associated with port-wine staining.

### Sturge-Weber Syndrome

Capillary malformations of the face may be associated with choroidal and leptomeningeal vascular malformations; well known is the eponymous condition, the Sturge-Weber syndrome. This syndrome is commonly found among those patients with port-wine stains involving the distribution of the first and second branch of the trigeminal nerve. It has a high incidence of congenital glaucoma of the ipsilateral eye, intracranial vascular malformation of the brain with calcification and atrophy of the underlying cerebrum, associated seizure disorders, and, in some cases, mental retardation and haemiparesis (Fig. 9).

The Jahnke syndrome describes a Sturge-Weber syndrome without ocular involvement.

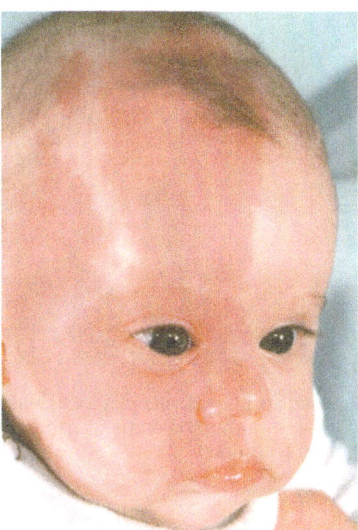

**Fig. 9.** Infant with diffuse vascular birthmark of the right face and scalp, buphthalmos of the right eye and epileptic attacks: capillary malformation (port-wine stain) as part of Sturge-Weber syndrome

In the case of Sturge-Weber syndrome with choroidal involvement but no glaucoma, the eponym Milles syndrome is used.

## Von Hippel-Lindau Syndrome

In the von Hippel-Lindau syndrome, facial port-wine stain is associated with retinal capillary malformation and non-calcified vascular malformation of the brain and the spinal cord. Polycystic tumours affecting other viscera, especially the kidneys, are also a diagnostic clinical feature.

## Klippel-Trenaunay Syndrome

Port-wine stains of the extremities may be associated with systemic abnormalities such as the Klippel-Trenaunay syndrome. The Klippel-Trenaunay syndrome (combined capillary-venous-lymphatic malformation with skeletal overgrowth) is characterized by a large port-wine stain of the extremities, malformation of the venous system associated with hypertrophy of the soft tissue and skeletal tissue. The lymphatic system could also be involved. It is also not uncommon to see a PWS overlying an arteriovenous malformation, therefore requiring the physican to look deeper for any underlying problem.

## Proteus Syndrome

Diffuse patchy port-wine stains also may be associated with partial gigantism of hands or feet, usually bilateral and asymmetrical, and haemihypertrophy. Numerous systemic abnormalities such as macrocephaly, skull exostoses, lipomas and sometimes café-au-lait patches are notable.

The Riley-Smith syndrome is probably part of the Proteus syndrome, that characteristically consists of capillary and/or venous malformations associated with lymphatic anomalies, macrodayctyly, macrocephaly, chylous cysts and pseudopapilledema.

Another association, partly subsumed by the term Proteus syndrome, is called Bannayan syndrome and consists of macrocephaly with subcutaneous lipomas and vascular malformations.

## Wyburn-Mason Syndrome/ Bonnet-Decaume-Blanc Syndrome

A port-wine facial stain may be the sign of unilateral arteriovenous malformation of the retina and intracranial optic pathway. This entity is known in the French literature by the syndromic term Bonnet-Decaume-Blanc, who first reported the association of retinal vascular malformation with ipsilateral cerebral arteriovenous malformations and facial port-wine stains in 1937; and after in the English literature Wyburn-Mason, who reviewed a large series in 1943.

Manifestations can be cerebral or ocular or both. Headaches, seizures or subarachnoid haemorrhage are the usual indicators of involvement of the central nervous system, although the specific symptoms and signs will vary depending on the location and extent of the arteriovenous malformations. The retinal lesions, generally unilateral, range from ophthalmoscopically, barely visible, vessels covering a substantial portion of the retina, and can cause cystic retinal degeneration between the dilated vessels and impair vision, to optic atrophy, enlargement of the optic foramen and occasionally exophthalmos. Some authors propose that retinal involvement is not essential for the diagnosis of the Wyburn-Mason syndrome.

## Cobb Syndrome

A port-wine stain of the posterior thorax, especially at the lumbar skin, may also indicate an underlying arteriovenous malformation of the spinal cord, called Cobb syndrome. In this syndrome, the vascular malformation is in the lumbar skin or vertebrae and underlying spinal meninges, causing neurological damage by the size of the malformation.

## Brushfield-Wyatt Syndrome

A capillary malformation in the trigeminal area may be the sign of an associated vascular malformation and calcificied cerebral cortex.

## Non-Fading Telangiectasias: Congenital and Acquired

Non-fading telangiectasias were subcategorized under capillary malformations. These present in a spectrum from the classical spider nevus, to the maculopapular punctate anomalies of the Osler-Rendu-Weber syndrome, and the characteristic reticulated marbling and cutaneous hypoplasia seen in cutis marmorata telangiectatica congenita (CMTC).

## Cutis Marmorata Telangiectatica Congenita (Van Lohuizen Syndrome)

Clinical Features. Cutis marmorata telangiectatica congenita is a rare skin disease in newborn or very young children. This is a pathological entity first described by van Lohuizen in 1922 and now known by the term cutis marmorata telangiectatica congenita. The characteristic lesion has a distinctive deep purple colour and is depressed in a serpiginous reticulated pattern (Fig. 10). In some cases of CMTC, associated deep venous anomalies, ulceration of the reticulated purple areas and hypotrophy of the involved limb and subcutaneous tissue have been reported.

Localization. These lesions occur in a localized, segmental or generalized distribution. The trunk and extremities are more commonly involved than the face and scalp.

Histolopathology. Biopsies of these lesions reveal dilated capillaries and veins in the dermis and subcutis that lack elastic fibres.

Clinical Course. Almost all infants affected with CMTC show steady improvement of their accentuated vascular pattern during the first year of life that continues into adolescence because of normal thickening and maturation of the skin. Nevertheless, the skin atrophy and deep vascular staining can persist into adulthood, along with diffuse ectasia of the veins in the involved extremitites . For these reasons, Muliken and coworkers placed the cutis marmorata telangiectatica congenita within the spectrum of capillary-sized vessel malformations.

Differential Diagnosis. The Adams-Oliver syndrome is a rare autosomal dominant neuroectodermal syndrome including strabismus convergens,

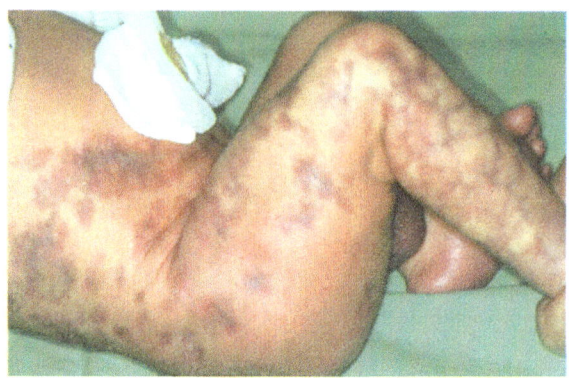

**Fig. 10.** Infant with a congenital net like vascular pattern of blue-violet colour localized at the extremity: cutis marmorata telangiectatica (van Lohuizen syndrome)

atrial septal defects, retro- and micrognathia, short toes with partially missing phalanges and nails, large vascular plaques on the scalp with atrophy and ulcerations, cutis marmorata and dilated veins on the trunk and extremities, which may be a maximal variant of van Lohuizen's syndrome.

In van-Bogaert-Divry syndrome cutis marmorata is associated with non-calcified cerebral vascular malformations, mental retardation and spasticity.

## Rendu-Osler-Weber Syndrome (Hereditary Haemorrhagic Telangiectasia)

Clinical Features. Rendu-Osler-Weber disease (hereditary haemorrhagic telangiectasia) is an autosomal dominant, systemic fibrovascular dysplasia that is classified in the extratruncular capillary malformations. This disease can appear as telangiectasiases in the skin and arteriovenous malformations widely distributed throughout the body.

Localization. Sites of predilection include the gingiva, lips, mucosa of the nose, face and fingers.

Histolopathology. In typical cases the patients developing discrete, spider-like, bright red maculopapules, usually 1 to 4 mm in diameter, which most commonly emerge after puberty. Some authors [43] have suggested that there is a formation of structurally weak vessels that become progressively dilated and elongated.

Clinical Course. Patients can suffer from a variety of serious clinical complications. These include recurrent bleeding, especially of the nasopharyngeal cavity and the gastro intestinal tract, secondary iron deficiency, hepatic portosystemic encephalopathy, embolic abscesses and a variety of neurological complications.

### Ataxia-Telangiectasia (Louis-Barsyndrome)

Clinical Features. Ataxia-telangiectasia is a rare syndrome, an autosomal recessive disorder consisting of cerebellar ataxia, ocular and cutaneous telangiectasis, frequent severe respiratory tract and sinus infections caused by immunological deficiency with diminshed levels of immunoglobulins.

Localization. Cutaneous and ocular teleangiectases occur at 3–6 years of age, first noted in the temporal and nasal area of the bulbar conjunctiva; later bright red cutaneous telangiectases appear on the eyelids, nasal bridge, cheeks, ears, neck, upper chest and flexor surfaces of the forearms.

Clinical Course. The course is foudroyant and lethal, and death occurs in the second decade, because of recurrent pulmonary infections and bronchiectasis, or from lymphoreticular malignancy.

### Generalized Essential Teleangiectasia (Angioma serpiginosum)

Clinical Features. The aquired, idiopathic vascular ectasia is characterized by multiple, minute, red to purple pin-sized vascular punctata, appearing in groups that extend over a period of months or years in serpiginous and gyrate patterns. Frequently, there is a background of diffuse erythema. The condition is asymptomatic and usually does not haemorrhage.

Localization. This may occur anywhere on the extremities, but the lower limbs are the sites most commonly affected. These lesions typically appear in females; the onset varies widely.

Histolopathology. Histologically, these lesions consist of clusters of dilated capillaries housed in dermal papillae and lined by thick walls with no signs of inflammation or hyperplasia.

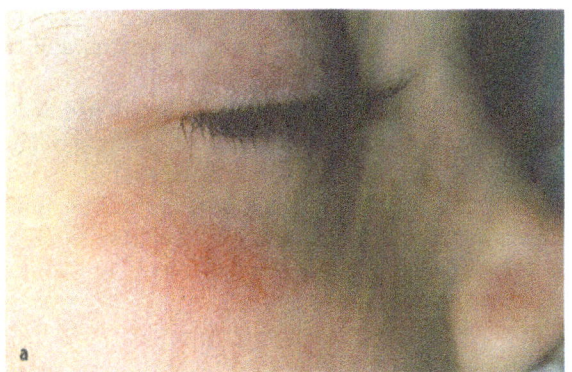

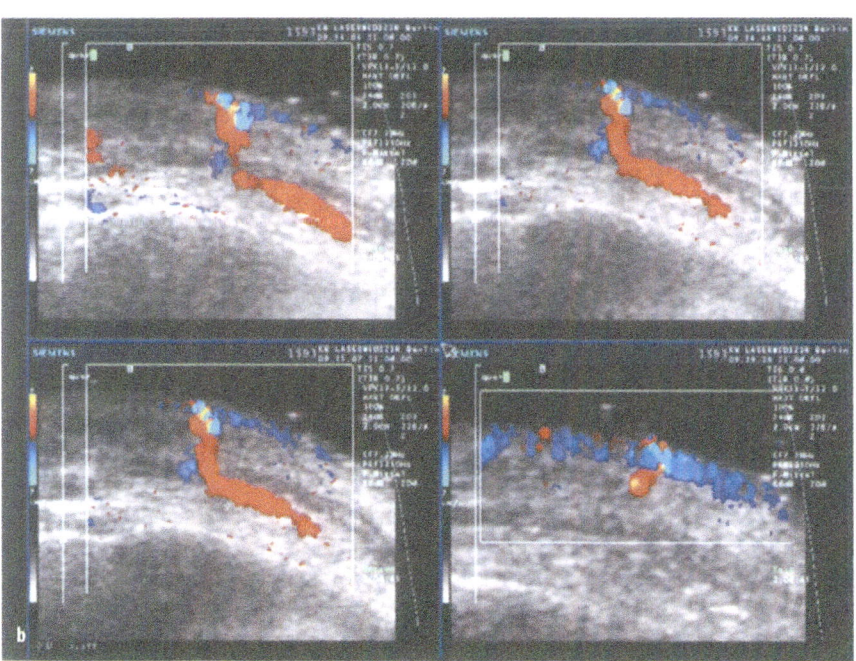

**Fig. 11. a** Infant with a vascular ectasia of the lower eye lid: spider nevus. **b** The colour-coded duplex sonography shows a central artery

Clinical Course. The lesions usually remain stable in adult life and sometimes partially regress but never completely.

### Spider Nevus (Nevus araneus)

Clinical Features. The aquired spider nevus is a extremely common lesion of the skin, that is present in up to 15% of children and young adults. These lesions occur more frequently between 7 and 10 years of age, with no significant difference between boys and girls, while the increased frequency of spider marks in pubertal females, more so than in males, suggests a hormonal mechnanism.

Localization. These lesions most commonly appear on a prominent portion of the face, especially at the lower eye lid and cheek; forearms then over the dorsum of the hands and fingers next are.

Histolopathology. These lesions usually consist of a central artery from which superficial vessels radiate (Fig. 11a, b). When a glass slide is pressed gently over the lesion (diascopy), the pulsations of the central artery will be observed, and blanching of the surrounding vessels. A centrifugal flush will occur when the pressure is released.

Clinical Course. Spontaneous disappearance has been seen in childhood, and does not occur after puberty.

### Capillary-Lymphatic Malformations

#### Angiokeratoma

Clinical Features. There are cases in which a vascular stain has a rough, hyperkeratotic surface. These lesions may have been labelled hypertrophic nevus flammeus or verrucous haemangioma in the past. Hyperkeratotic vascular stains are malformations, whereby formation of ectatic vascular vessels stays in focus. They are usually present at birth and remain throughout life: there is no involution. Therefore, the designation capillary-lymphatic malformation is accurate.

Capillary-lymphatic malformations may be light pink, bluish red to black in colour, and well demarcated with reactive epidermal hyperkeratosis and parakeratosis (Fig. 12).

Localization. They are most commonly located on the lower extremities, but are also often observed at the abdomen and upper extremeties.

The term angiokeratoma is usually remembered by its eponyms, with typically predilection of the lesions: Mibelli, lesions on the hands or feet, Fordyce, lesions on the scrotum and Fabry, lesions on the trunk or thighs.

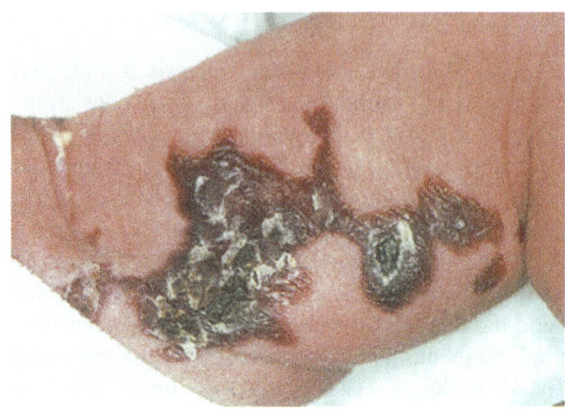

**Fig. 12.** Infant with firm, raised vascular lesion with hyperkeratotic surface of the forearm, present since birth: capillary-lymphatic malformation (angiokeratoma)

Histolopathology. Histologically, these lesions demonstrate dilated, capillary- to venular-sized vessels in the dermis and subcutaneous tissue [45]. In contrast to the PWS, the hyperkeratotic vascular stains can lead to an affection of lymphatic capillaries.

Clinical Course. With trauma, altered haemodynamics, or possibly secondary infection, these lesions become more keratotic and wart-like.

Differential Diagnosis. Capillary-lymphatic malformations are observed either in association with the Klippel-Trenaunay syndrome or alone.

### Lymphatic Malformations

#### Lymphangioma

Clinical Features. Lymphatic malformations manifest a wide clinical spectrum from localized (lymphangioma, Fig. 13) or diffuse swelling (lymphedema) and large cystic lesion (cystic hygroma) to vesicular excrescences on the skin or mucous membranes (lymphangioma circumscriptum). For these lesions of lymphatic origin, the term lymphangioma has been generically applied. Despite the long usage of lymphangioma, the term should be abandonned, because it is an inaccurate term usually used to describe congenital malformations of the cutaneous, subcutaneous or submucous lymphatic system. Therefore, the designation lymphatic malformation should be preferred.

Lymphatic malformations are usually obvious at birth, grow commensurately with the child and sometimes enlarge. They may be first present as a soft tissue swelling, most commonly involving the head and neck region, where they are usually called cystic hygromas. These malformed lymphatic cysts contain a

**Fig. 13.** Infant with cutaneous vesicles of the lower leg, present since birth. The colour is due to intralesional haemorrhage: localized lymphatic malformation

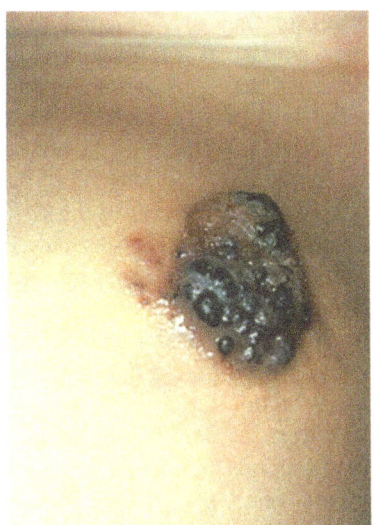

**Fig. 14.** Newborn child with swelling in right temple present since birth: lymphatic malformation (lymphangioma)

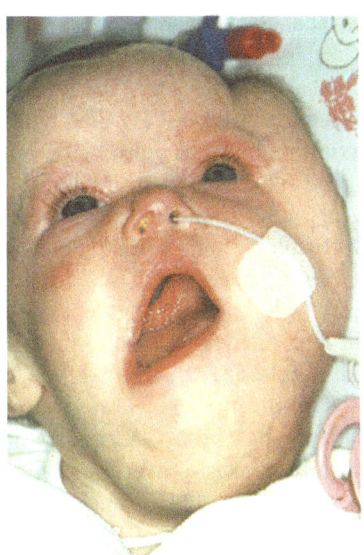

watery, sometimes lightly opaque fluid, occasionally discoloured by the presence of fresh or altered blood. The cysts may be single or multiple, interconnected or separate. They may communicate with adjacent lymphatics or veins. Progressive ectasia and dilatation of the lymphatic cysts corresponds to increase in volume. The overlying skin can appear to be quite normal, or there may be an associated dermal capillary malformation, especially at the extremities.

Localization. Deep-seated lymph cysts of the neck are recorded either in association with lymph cysts of the mouth, tongue and the oropharyngeal region or alone. They appear as clusters of thin-walled vesicles that look like frog spawn, usually filled with clear colourless fluid or bloody from surrounding capillaries.

Histopathology. Lymphatic malformations consist of irregular, variably sized channels with thick and thin walls, and cystic spaces surrounded by fibromuscular thickening. The channels contain pale eosinophilic material; lymphoid aggregates are also frequent.

Clinical Course. Lymphatic cysts of the oropharyngeal region may bleed, serving a portal of entry for infectious organisms, or may cause macroglossia with difficulty in feeding and speaking. In the newborn period, a large cervicofacial lymphatic malformation may present with rapid enlargement of the tongue or floor of the mouth, leading to respiratory obstruction. In these cases, tracheostomy may be necessary. Large cervicofacial malformations may also cause dysfunction of the upper alimentary tract, necessitating the placement of a feeding tube or gastrostomy (Fig. 14). Not infrequently, a large lymphatic malformation may cause cellulitis and swelling secondary to sepsis or in-

tralesional haemorrhage. Therefore, every episode of cervicofacial swelling should be promptly treated with antibiotics, preferably penicillin, on the presumption that there is cellulitis and that intra oral organisms are responsible. After the inflammation subsides, the involved area usually remains edematous and firm for a long period of time.

In addition to these lesional characteristics, which may cause an airway obstruction, bleeding, infection and physical deformity, there is a risk of lasting detrimental effects on a child's psychological and social interpersonal development.

Deep-seated lymph cysts in the trunk appear to have a consistent association with more extensive malformations involving all soft-tissue, mesentery, peritoneal cavity and bones, especially the vertebral bodies and the sternum with osteolysis due to intraosseous, usually venous malformations. Often they are combined with a chylothorax and chyloperi-cardium and a pleural adhesion, and are called the Gorham-Stout syndrome (or disappearing bone disease). In childhood the course is foudroyant and lethal.

### Lymphedema

The lymph vessels can also show structural abnormalities or predisposition in the sense of a vascular malformation. Lymphedema caused by a hypoplasia or hyperplasia of the lymph vessels is a well-known example.

In the second format, solitary or multiple cystic lymphatic malformations present as deep or superficial lesions.

## Lymphangioma circumscriptum

Clinical Features. Only the lymphatic capillaries are affected, this disease is called lymphangioma circumscriptum, but it can also be observed as an affliction of the subcutis. Often these lesions bleed easily and weep, either spontaneously or following trauma.

Histolopathology. Histologic examination demonstrates a collection of subcutaneous lymphatic cisterns with a thick muscle coat that communicates through dilated lymphatic channels with the superficial vesicles [81].

Clinical Course. Following surgical excision, many patients have subcutaneous vesicles or superficial lymphatic vesicles over these areas many years later in skin grafts applied onto abnormal subcutaneous tissue after removal of affected skin. Accordingly, excision of the skin vesicles without adequate excision of the feeding cisterns will not effect a cure, because the cisterns will produce new outpouchings that will appear on the skin as new vesicles.

Differential Diagnosis. A combination of venous and lymphatic malformations mostly affects the extremities. These combined lesions are often associated with skeletal elongation and hypertrophy. When there is also a port-wine stain, the eponym Klippel-Trenaunay syndrome is applicable.

## Venous Malformations

Clinical Features. Venous malformations are developmental abnormalities of veins, dysmorphic in configuration and structure. They usually occur in truncular or extratruncular form (described in more detail here), or they may be combined. Furthermore, the may coexist as capillary-venous or lymphatic-venous anomalies. A venous malformation is characteristically a soft vascular swelling, usually of a bluish colour, with skin normal overlying or bluish to purple-coloured (Fig. 15a). Combined capillary-venous malformations of the skin have a dark red to purple colour. Phleboliths may be palpated within the lesion and confirmed by plain-film radiography and sonography. These lesions characteristically expand with a Valsalva manoeuver, or when the involved area is placed in a dependent position. Particularly venous malformations are often soft, well compressible, and non-pulsatile. The combined capillary-venous malformations exhibit superficial lymphatic vesicles overlying deep venous anomalies (Fig. 16).

Venous malformation of the head and neck region present in a wide spectrum from circumscribed venous anomalies in which the venous lacunae are connected to the venous circulation by capillaries, through localized venous anomalies by connected veins to the venous circulation, and diffuse venous ectasias. Furthermore, multiple venous lesions tend to coexist with venous ectasias and deep vein anomalies.

Localization. Venous malformations are frequently found in the facial region, within the masseter, lips, tongue, pharynx, soft palate, forehead and eyelid, but can appear anywhere on the body surface. They may be localized or diffuse in their distribution.

Histolopathology. Venous malformations are histologically characterized by dilated large channels, varying from capillary to cavernous dimensions, and a lining of a flat, single endothelial layer. The larger channels show sparse smooth muscle cells and adven-

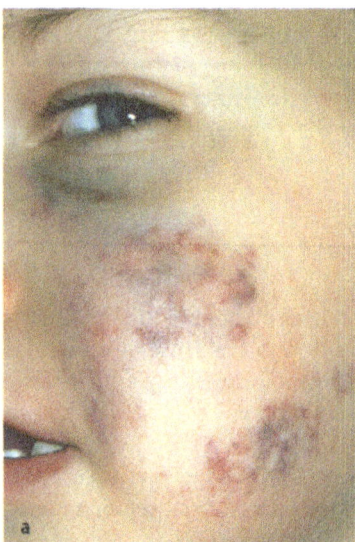
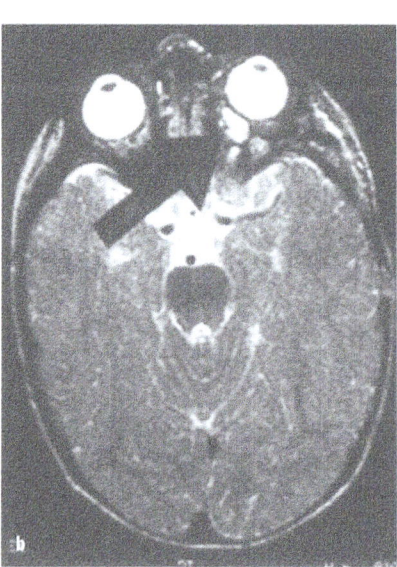

**Fig. 15. a** Two-year-old boy with vascular lesion of the face, which has grown commensurately since birth: venous malformation. **b** An additional retroorbital vascular anomaly (arrow)

**Fig. 16.** Five-year-old girl with vascular anomaly of lower extremity with soft tissue and bone enlargement. On the lateral side raised haemorrhagic vesicles within the portwine stained area: combined capillary, lymphatic and venous malformation with skeletal overgrowth (Klippel-Trenaunay syndrome)

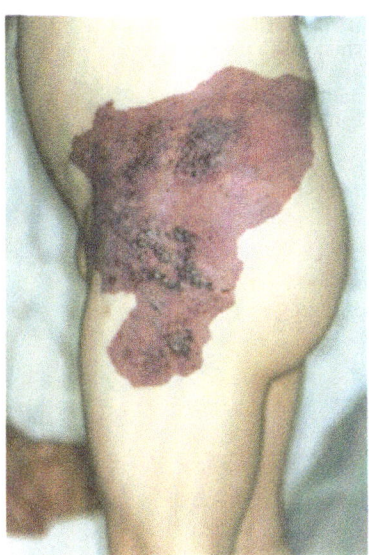

**Fig. 17a, b.** Extra-truncular venous malformation **a** with lacunar venectasias visible in the CCDS **b**, without connection to the deep venous system

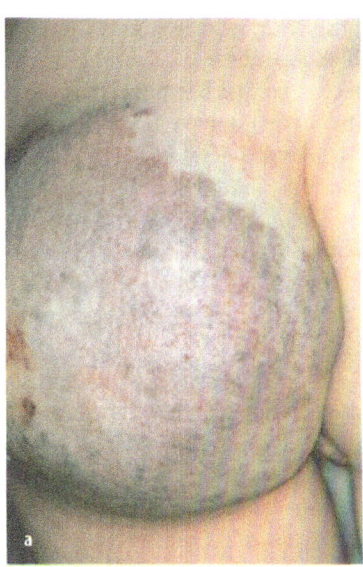

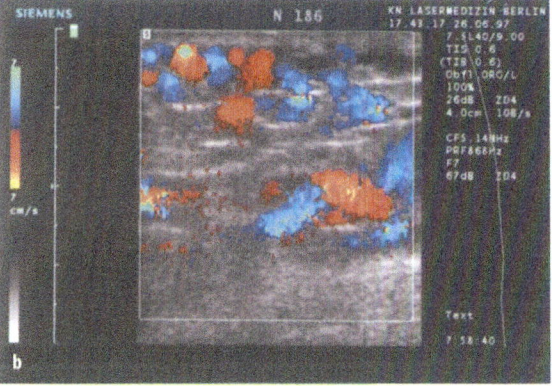

tial fibrosis; organizing thrombi and calcification are also frequent.

Previously, these lesions were often misinterpreted as cavernous haemangiomas because histologically large ectatic vessels of the subcutis are revealed. This term is, however, inaccurate, because the lesions are no true haemangiomas. Because the use of this term is confusing in this setting, it should be discarded.

Clinical Course. Like other malformations, they grow steadily with the patient and may expand with local trauma or progress to deformation of adjacent structures. The majority of venous malformations are isolated lesions and are physiologically benign; however, many are not. Lesions involving both the skin and soft tissue and other organ systems such as the pharynx can obstruct the airway, and may produce bleeding or impair vision. The vast majority of deeply placed venous malformations cause no symptoms, but with time, they will occasionally become symptomatic as a result of slow enlargement due to haemodynamic factors and simple stretching of the walls of the venous space, and create pressure on surrounding structures, especially nerves.

In the extremities, they are usually extensive and often associated with anomalies of the deep venous trunks, lymphatics and with bone disease in various ways. In the case of truncular venous malformations, hypo- or hypertrophy of bones and soft tissue is often present. When comparing the two sides of the body, one can see a difference in the length and circumference of the extremities.

Venous malformations are not easily identified with conventional regional arteriography. For this, magnetic resonance imaging and colour-coded duplex ultrasonography are the most sensitive method for imaging their extent (Fig. 17a,b). Truncular ve-

nous malformations must be demonstrated by phlebography and varicography.

In the case of large venous and venous/lymphatic malformations, coagulation abnormalities can occur, called localized intravascular coagulation (LIC). The coagulopathy is characterized primarily by low levels of plasma fibrinogen and coagulation factors, a high level of D-dimers (fibrin degradation products) and by elevation of fibrin split products, whereas the platelet count is normal or moderately low (at least 100 000 mm$^{-3}$). KMS must be distinguished from the chronic consumptive coagulopathy that occurs in association with extensive venous and lymphatic malformations on a lifelong basis. In contrast to KMS, the haemostatic profile is improved, but very slowly, by heparin treatment. Very low molecular weight heparins with prolonged kinetics are recommended as the most efficacious treatment [42].

Multiple venous malformations rarely associated with bleeding from similar vascular anomalies of the gastrointestinal tract, so-called blue rubber bleb nevus (Bean) syndrome. Even other organs such as the lung, liver, muscle, bones, kidneys, brain, spleen, gallbladder, adrenals, pleura and peritoneum can be affected, so that haemoptysis, haematuria, epistaxis and menorrhagia have been described in these patients. The lesions tend to enlarge with time and may be painful.

There are also cases of glomangiomas in which the lesions have been large enough to be raised, soft and compressible, and have been mistakenly diagnosed as blue rubber bleb nevus syndrome even though intestinal bleeding was absent.

Young [83] suggested that diffuse venous anomalies are rare, and whether the blue rubber bleb nevus syndrome truly exists as a separate disease is debatable.

In the case of diffuse venous malformation of the rectum and perirectal tissue the eponym Barker-Kausch syndrome is used.

In the Esau-Bensaude malformation diffuse venous malformation of rectum and perineum is associated with involvement of the ureterovesical mucosa.

A generalized intestinal involvement by venous-lymphatic malformations is called the Kaijser syndrome.

At the spinal cord is affected, with enlargement in calibre and thickness of the vessels and varicose postspinal vein, the disease is called the Foix-Alajouanine syndrome. Venous or venous-lymphatic type of vascular malformations associated with other lesions like enchondromas and dyschondroplasia of a limb or limbs is a pathological entity, first described by Maffucci in 1881 and now known by the term Maffucci syndrome (syn. Kast syndrome, Ollier-Klippel syndrome). Pleboliths and expansive bone lesions can be seen on plain films. Multiple organ involvement has been reported.

Differential Diagnosis: Different lesions that may be confused with venous malformations are the following:

1. Blue marks. The Mongolian blue spot (Fig. 18), the blue nevus and the nevus of Ota are characteristically purple-blue or grey patches usually involving the extremities; the lesion has a predilection for the sacrum and buttocks, and is melanocytic, not vascular.

2. Nodular lesions like glomus tumours. Another rare lesion that may be confused as a venous malformation is glomus tumours. These lesions are relatively uncommon neoplasm that arise from modified smooth muscle cells. Glomus tumours are thought to originate from their normal counterpart, the glomus cells; therefore they tend to occur most commonly in acral areas of the extremities, especially the nail beds of the fingers and toes. The diagnostic clinical feature is pain occurring when the lesion is pressed, but often also spontaneously. They characteristically occur as a small purple

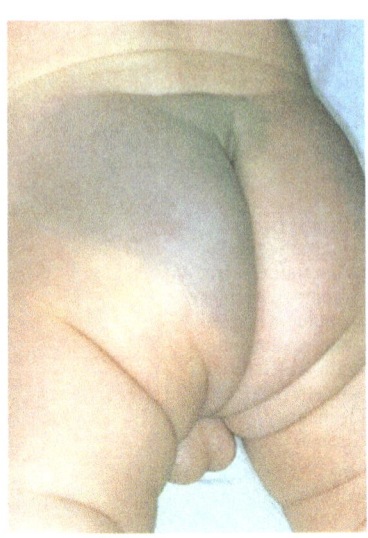

**Fig. 18.** Newborn child with a bluish grey patch on the buttocks: mongolian blue spot

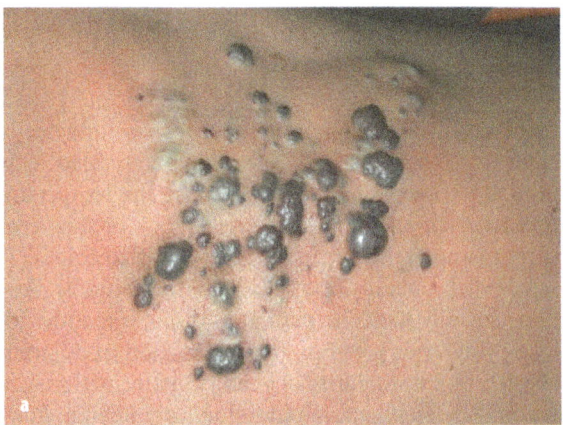

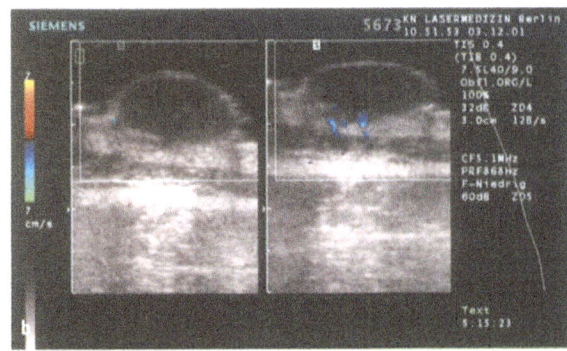

**Fig. 19. a** Fifty-year-old man with multiple, disseminated blue nodules of the shoulder: glomangiomas. **b** Sonographically, glomangiomas are hyposonic without pathological vessels

nodule that could appear solitary or scattered throughout a limb or diffusely. However, glomus tumours have also been described in extracutaneous sites such as bone, stomach, colon, trachea and mediastinum.

Multiple glomus tumours are much less common than solitary lesions, and they are named glomangiomas because of the angiomatous appearance of the lesion (Fig. 19 a, b). In contrast to solitary glomus tumours, glomangiomas often appear during childhood as small blue nodules, widely scattered over the skin surface. Glomangiomas have also been described grouped in a plaque on an anatomic region. Cases of large, raised, soft and compressible glomangiomas have been mistakenly diagnosed as venous malformations or as blue rubber bleb nevus syndrome. Glomangiomas are less likely to be painful than solitary glomus tumours.

Histopathologically, glomangiomas are less well-circumscribed lesions than solitary glomus tumours, and the number of glomus cells in the adjacent stroma is much smaller.

Histological examination is the only way to make a differential diagnosis between glomangiomas and venous malformations.

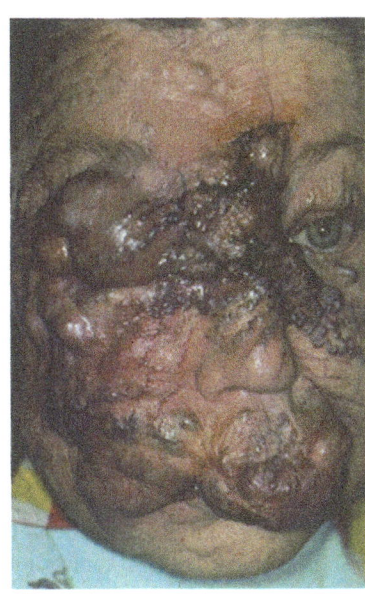

**Fig. 20.** Fifty three-old woman with vascular anomaly of the right and partially of the left face, present since birth, with large ectativ vessels, repeated arterial bleeding, and lost of vision of the right eye: large arteriovenous malformation of the face

## Arterial Malformations

Clinical Features. Arterial malformations include coarctation, ectasia, aneurysm, truncal arteriovenous fistulae, and arteriovenous malformation. These arterial anomalies are characterized by increased skin temperature, a bruit and thrill, in contrast to capillary, lymphatic, venous and combined malformations.

Localization. Arteriovenous malformations of the head and neck region are quite rare in contrast to venous malformations (Fig. 20) and found more frequently in the extremities. The term angioma racemosum (from Virchow) became synonymous with arteriovenous malformations, especially for the head and neck region, for lesions where the course of the lesion is progressive.

Histopathology. Biopsy of arteriovenous malformation characteristically shows close juxtaposition of medium-sized arteries and veins and vessels of indeterminate nature. Frequently, so-called arterialized veins, caused by intimal thickening of veins, suggest elevated pressure within the vasculature.

Clinical Course. Cervicofacial arteriovenous malformations are rarely noted in childhood. Usually, the may lie quiesent for years before the signs and symptoms become ominous. Typically, these lesions enlarge because of natural haemodynamic pressures, or there is rapid expansion following local trauma, or hormonal changes associated with puberty or pregnancy. The involved skin has an elevated temperature, and dermal staining is frequently obvious. Some lesions lie hidden within the masticatory muscle or beneath an innocent-appearing cutaneous stain. There may also be increased mobility of teeth, or expansion of the

buccal cortex. Repeated bleeding around the gingival necks of the teeth, gingiva or nasopharynx, frequently sudden and massive in nature, can be the first symptoms and may be very distressing and life-threatening for the patient. Unexpected haemorrhage also can follow tooth extraction or intervention for biopsy.

Arteriovenous malformations enlarge, not by cellular hyperplasia but by haemodynamic mechanisms. In the angiography, a network of abnormal vessels is seen, featuring one or several shunt zones (niddus) that shortcircuit the capillary bed. There are usually several dilated afferent arteries and multiple draining veins.

Arteriovenous malformations at the extremities are usually symptomatic at time of first presentation. Increased warmth of the limb, pain, paresthaesia and hyperhydrosis are common symptoms. They are frequently associated with bone distortion and overgrowth, in the past called Parkes-Weber syndrome. This syndrome exhibits bony overgrowth with capillary-lymphatic-venous malformations, but there is arteriovenous shunting.

Differential Diagnosis. The latter distinguishes it from the Klippel-Trenaunay syndrome. Arteriovenous malformations will show an inexorable progression and remain difficult, despite treatment.

## Laser and Laser-Tissue Interactions

### Argon Laser

The argon laser is extremely effective as a superficial photocoagulator since its wavelength of 514 nm penetrates the skin for photocoagulation of haemoglobin within the blood vessels. Skin-exposure times are determined either by mechanical shuttering of the beam or by the rate of movement of the hand-held treatment beam to another site. The shortest reliable exposure time that can be obtaine, 100 ms, results in coagulation necrosis of dermal connective tissue. Histological examinations showed that treatment caused general coagulation necrosis of the entire epidermis with injury to papillary dermis, thrombosis of vessels and preservations of dermal appendages [70] to a depth of 0.5 mm. This indicated thrombosis of pathological vessels as well as additional, rather extensive, non-specific damage to surrounding epidermis and dermis that was thought to be due to heat generated within the vessels by haemoglobin absorption of argon laser energy.

A minimal blanching of the tissue is intended, whereby a complete treatment of the whole area should be avoided. An application of the polka-dot technique is helpful in avoiding confluent coagulation of the skin. The distance between the single application spots should be as large as the focus diameter. Multiple exposures and overlapping of single irradiation areas must be avoided. By compressing the vessels through a glass plate, it is possible to simplify coagulation of all vessel parts and to minimize the shielding effect of absorption. With these procedures, there is a low risk of damaging the overlying epidermis.

## KTP Laser

The potassium titanyl phosphate (KTP) laser emits a green light at 532 nm, which has nearly the same effect on biological tissue as the argon laser. Regarding the tissue reaction the difference is caused by the fact that the clinically used KTP laser can be applied with short pulses and high peak power. Therefore, the clinical effect has an intermediate position between the argon and flashlamp-pumped pulsed dye laser. A pearl-string-like treatment following vessels or a polka-dot or scanning technique of larger areas can be performed, whereby double exposures should be avoided. Larger areas can be treated more homogenously using a scanner.

## Flashlamp-Pumped Pulsed Dye Laser (FDPL)

The flashlamp-pumped pulsed dye laser is the first laser especially designed for cutaneous vascular lesions. The development of this laser was based upon the Anderson and Parrish [4] theory of selective photothermolysis, which predicted that selective destruction of blood vessels is possible by matching the wavelength of light absorbed by haemoglobin into the vessels and by using an exposure time less than the calculated thermal relaxation time for the blood vessels [26]. The thermal relaxation time is the time required for the target tissue to lose 50% of its heat. An exposure time of 1 ms or less was advocated to confine that heat to the vessel to decrease the risk of heat diffusion, which can cause scarring.

The flashlamp-pumped pulsed dye laser at 577 nm was the first pulsed dye laser developed in the early 1980s to treat port-wine stains. Unlike its predecessor, the blue-green argon laser (488 and 514 nm), the dye laser's wavelength was tuned to the yellow at 577 nm, and its pulse duration was between 300 and 500 ms. The pulse duration was calculated to match the thermal relaxation time of cutaneous blood vessels and its wavelength to coincide with the third absorption spectral peak of oxyhaemoglobin. With these modified laser parameters, clearance of these malformations with normalization in texture and colour of the treated skin were seen.

The wavelength was then adjusted to 585 nm in the early 1990s to increase the depth of vascular injury. With the change of wavelengths from 577 to 585 nm for deeper penetration, the intravascular haemoglobin release may be greater because of the larger blood vessels and concentration of oxyhaemoglobin found in the deep vascular plexus of the dermis. This is in contrast to the argon laser, with which a coagulation effect of the epidermis and dermis is seen.

The pulsed dye laser, when compared with the argon laser, is the only one that produces intravascular thrombus formation without epidermal and dermal damage. Histological studies [46] immediately after treatment showed agglutinated red blood cell, fibrin and platelet thrombi confined to the blood vessels of the papillary and midreticular dermis, with little or no damage to the surrounding dermis and epidermis. Sequential biopsy specimens revealed destruction of the abnormally ectatic blood vessels in PWS with replacement by normal-appearing new vessels and little or no dermal scarring at 1 month [74]. Because treatment results in selective vascular injury, the area can be treated in a repetitive fashion with minimal risk of complications until the desired degree of lightening has been achieved.

The depth to which light penetrates tissue is also a critical limiting factor in laser treatment and is directly proportional to the wavelength of the light. For the pulsed dye laser adjusted to 585 nm, the depth of penetration for 50% of the energy is calculated to be 0.8 mm [26], a prediction confirmed histologically in laser-treated skin [46]. Because the dermis of facial skin is approximately 0.6 mm deep in children and approximately 0.9 mm deep in adults [73], the pulsed dye laser provides adequate penetration for cutaneous vascular lesions.

Although the flashlamp-pumped pulsed dye laser at 585 nm has been successfully used for the treatment of port-wine stains, there were still a number of patients whose port-wine stains remained incompletely cleared following pulsed dye laser treatment. The limitation of the flashlamp-pumped pulsed dye laser is generally regarded as its inability to penetrate to the level of the deep vessels. It was assumed that vessels within the incompletely cleared port-wine stain were those lying beyond the 1.5-mm penetration depth of the 585-nm laser beam.

Nowadays, response to the FDPL therapy is enhanced by tuning the laser to 595 nm to increase the depth of vascular injury. The pulse duration ranges from 200 μs to 2 ms, and the laser beam can transmit down by focus from a 5 to 12-mm spot beam.

Long-term results with 595 nm have not yet been reported, because treatment with this method began only at the turn of the century.

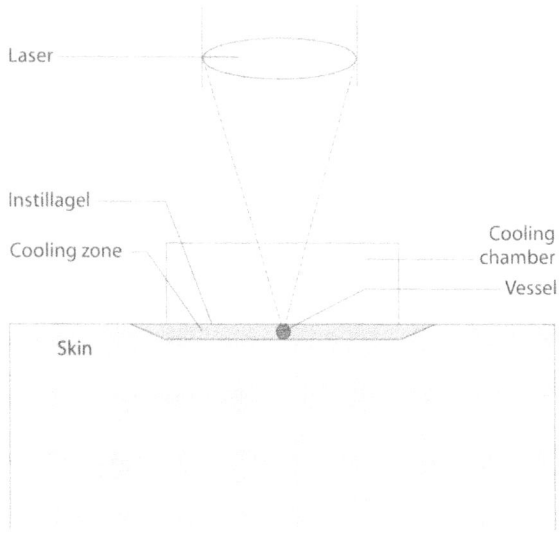

**Fig. 21.** Principle of the cooling chamber

### Nd:YAG Laser

The neodymium:yttrium aluminium garnet (Nd:YAG) laser is a solid-state laser. It emits light at 1064 nm, which is within the non-visible infrared part of the electromagnetic spectrum. The Nd:YAG laser wavelength is more highly absorbed in blood than in the surrounding tissue. As a result of scattering in the tissue, the laser irradiation is relatively uniformly distributed, and it can be observed that the endothelial structures are more prone to injury than the surrounding connective tissue. As a consequence of its deep penetration and wide scatter in tissue, the major effect of the Nd:YAG laser occurs below the surface.

Due to the possibility of transmitting the radiation through glass fibres, one can use either contact or non contact application. The non contact application can be performed with a fibre-divergent beam or with the focusing hand piece. Using both irradiation types, either coagulation or a vaporization is possible depending on the changes in power density.

### Transcutaneous Application

#### Direct Transcutaneous Application

Superficial cutaneous blood vessels that are most commonly found in a variety of cutaneous diseases respond quite easily to direct transcutaneous Nd:YAG laser application, to a depth of 1 mm. Here, the same concerns noted in the above discussion of the argon laser in terms of side effects are true, or higher, in the direct transcutaneous Nd:YAG laser application as well.

### Transcutaneous Laser Application with Cooling Chamber

The chamber is filled and rinsed with a cooling fluid, e.g. a 40% water-glycol solution having a temperature of -18 °C. The cooling effect reaches to about 1 mm of the tissue (Fig. 21).

Using the cooling chamber, the risk of side effects like scattering or heat conduction in the cutis can be reduced. Furthermore, the coagulation depth may be increased and a compression of decompression of the vascular anomalies is achieved by varying the filling volume of the chamber. Here, various parts of the vascular anomalies can be treated in different depths.

### Transcutaneous Laser Application with Continuous Ice-Cube Colling

With transcutaneous Nd:YAG laser application in continuous mode a coagulation depth of 8 mm can be reached. Less selective water absorption, reflection and remission of the emitted laser beam would damage the skin without protective cooling. Highly sufficient protection is provided by clear ice cubes placed directly on the skin. The defocused laser beam should be applied directly through these ice cubes. With the help of an ice cube pressing on the skin surface, the coagulation depth can be increased up to 10 mm.

Local skin cooling can reduce the risk of thermal epidermal damage for two reasons. First, due to the cooling effect of water at a constant temperature of 0 °C melting on the contact surface of the ice and skin, the heat transfer from the superficial layers into the water is very effective and the skin can be spared coagulation. At a depth of approximately 1.5 mm, where

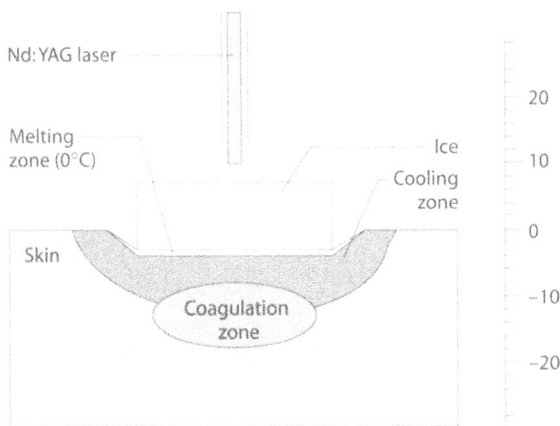

**Fig. 22.** Principle of continouous ice-cube-cooled irradiation

the cooling has no effect because of the limited heat conduction of the skin, the temperature rises to more than 60 °C, with a resulting coagulation zone. Second, by pressing the ice cube on to the skin surface, the epidermal thickness is reduced so that a lesser amount of laser irradiation is absorbed into this layer (Fig. 22).

When using this method, one must remember to change the ice cubes regularly, since they melt during laser irradiation. Generally, irradiation should be terminated when initial blanching becomes visible or after a 20-s irradiation period of the same region. In this way, thermal damage to the skin can be avoided.

## Percutaneous Laser Application

### Percutaneous Interstitial Bare-Fibre Technique

If a lesion is located deep in the subcutis or if there is a vascular anomaly with high-flow vessels, percutaneous application has proven to be very useful. Using this technique, direct puncturing of the lesion is performed by a Teflon cannula (Abocath) through which a newly broken bare fibre is introduced. One should ensure that there is a distance of about 5 mm between the end of the bare fibre and the end of the Teflon cannula, in order to avoid damage to the cannula. The red helium-neon pilot beam can often be used as a positioning aid. The correct position of the fibre end can be easily and clearly identified in a dimmed room by the pilot light up to a depth of approximately 2 cm or with ultrasound.

The bare fibre and the abocath are left in the same position during the irradiation of an are, but changed afterwards. A power of 5 W and an irradiation time of

180 s at maximum should be chosen because longer application time or higher power can cause carbonization of the tissue and the fibre. To avoid a coagulation necrosis of the superficial layers, the temperature of the irradiated skin region has to be controlled using two fingers placed on the skin surface of the treated area. However, one must be careful not to press the skin directly onto the fibre tip because the laser beam could then damage the skin.

Since 1992, colour-coded duplex sonography has been used for the online control of this procedure. This technique provides complete information, for example, for determination of the precise puncturing route, control of the fibre position, visualization of tissue changes during the procedure, as well as for depiction of the reduction of tumour perfusion and coagulated volume.

## Endoscopic Application

The main field of the non-contact technique are endoscopic procedures, but it can also be used in open applications. Often coagulation of tissue is intended, whereby the use of a 600-µm fibre is recommended. The 600-µm fibre is used with short exposure times of single chopped pulses (15 W, 0.2 s). For a deeper coagulation, longer exposure times are useful, whereby only the 400-µm or, better, 600-µm fibre should be used (15–20 W, 0.3–0.5 s).

An endoscopically controlled coagulation of vascular anomalies is achieved by placing the fibre directly on the tissue without any force. Hereby a lower power of 2–5 W and an exposure time of 0.5–5 s are recommended. By pushing the fibre gently into the lesion, an interstitial-like procedure can be performed. Thereby coagulation of deeper localized parts of the lesion is possible (e.g. lymphatic cysts of the mucosa).

## Laser Treatment of Vascular Lesions

Most vascular abnormalities can be successfully, and in fact preferentially, treated with lasers. The field is evolving rapidly and new laser systems are being commercialized. It is important that the appropriate laser and application form be used. Nowadays, we are treating lesions at a much earlier age to avoid complications and progression of the lesions and our expectations are much higher. Nevertheless, we have to carefully investigate the diagnosis and indications for treatment. However, lasers are expected to play an increasing role in the treatment of vascular abnormalities.

## Laser Treatment of Port-Wine Stains

Capillary vascular malformations are the most frequent and oldest indication for laser treatment in both children and adults.

A few lasers, mainly the carbon dioxide laser (10 600 nm), neodymium: YAG laser and copper vapour laser (578 nm) have been used to treat port-wine stains; cosmetic results have been poor. The previous treatment of choice for port-wine stains was the argon laser, which is noted to have an incidence of scarring ranging from 5 to 38%, and an incidence of permanent depigmentation of at least 20% [26]. Especially for port-wine stains in children there is a high risk of scarring with argon laser therapy, because of the poor vascular target in these characteristically light pink macular lesions. Thus, the argon laser (power 2 W, pulse duration 0.1 s, interval 0.1 s, spot size 0.05–1 mm) is only useful for telangiectatic port-wine stains, small eruptive angiomas and dark purple and nodular port-wine stains of adults, with a good response and low risk of scarring.

Port-wine stains should respond to KTP laser treatment as well. With longer exposure times of approximately 10 ms, the KTP laser shows effects (energy density 15–20 J cm$^2$, pulse duration 2–5 ms, spot size 1 mm, focusing hand piece or scanner system) similar to the argon laser; with 2 ms and a pulse energy above 15 J cm$^{-2}$ one can also induce a purpura beside thermal reactions. In port-wine stains in children, use of the KTP laser is also associated with a high risk of scarring and atrophy.

With the advent of the flashlamp-pumped pulsed dye laser, port-wine stains can be treated in infancy and early childhood; therefore the pulsed dye laser nowadays is generally accepted as the treatment of

choice for macular port-wine stains [7, 26, 29, 36, 59, 66, 75]. This represents a significant breakthrough, because, unlike previous lasers, scarring is a rare side effect of treatment.

The laser beam can transmit down to 1 mm fibre with the use of a convex lens and focus directly on the skin surface with a 5- to 12-mm spot beam. The energy fluence ranges from 4 to 10 J cm$^{-2}$ and is varied according to the age of the patient, the anatomic location and the colour of the lesion: 5.0 to 5.6 J cm$^{-2}$ in children less than 12 months of age, 5.6 to 6.4 J cm$^{-2}$ in children 12 months to 4 years of age, and approximately 7 J cm$^{-2}$ in children over 4 years of age. Energy fluency is frequently reduced over eyelids and hands. Treatments were started at the lowest energy density that shows a high degree of lightening. The lighter the lesion, the higher the energy density in the age range. The darker the lesion, the lower the energy density. Treatments are usually repeated at 6-week intervals until the desired degree of lightening is achieve. Repititive treatment sessions are required to obtain the maximum benefit of pulsed dye laser photocoagulation (Fig. 23 a, b). If the response was not adequate, the energy density is increased by 0.2 to 0.4 J cm$^{-2}$. Treatments are performed with a maximum spot overlap of 20%.

The high pulse peak power of the pulsed dye laser disrupts the vessels. Immediately after treatment, the area characteristically becomes blue-grey and turns purpuric in a few hours, with surrounding erythematous flare; this took 7–14 days to resolve. Some edema, especially in the periorbital area, is possible. If blanching or greying of the epidermis occurs during application, energy fluences should be reduced in order to avoid blistering of the epidermis. As the purpura disappears, the area lightens progressively for up

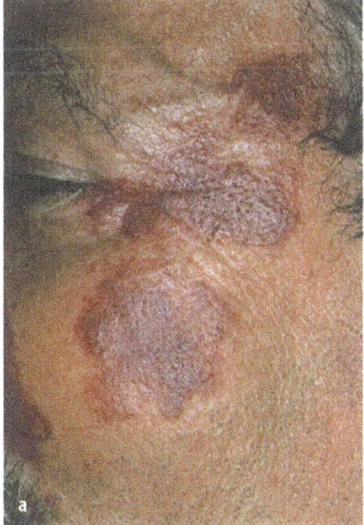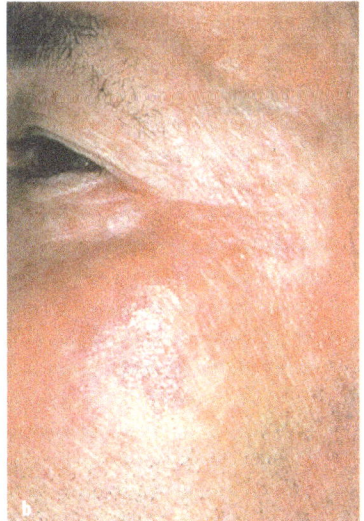

**Fig. 23a, b.** Port-wine stain of an adult **a** prior to and **b** after eight treatments with the pulsed dye laser

to 6 weeks. After the treatment, the treated areas were covered with panthenol ointment. In the case of blistering, the patients parents were instructed to cleanse the area with polyvidon-iodine solution, as well as if a crust formed. To avoid postoperative irritation of the treated areas, we instruct the parents to keep their children's fingernails short or that the children wear gloves to avoid trauma to the treated areas.

The laser pulses produce a mild to moderate degree of discomfort. Adult and teenage patients can often be treated without the need for any anaesthesia, although this is dependent upon the size and anatomic location of the lesion itself. General anaesthesia may be helpful for children with extensive lesions or in the case of central facial port-wine stains, because of eye protection. A significant reduction of pain as well as skin protection during laser treatment can be achieved by the cooling chamber.

The incidence of complete clearing is variable. In general, the red lesions require more treatments than the light pink lesions to achieve the same degree of lightening; the degree of response depends greatly upon the anatomical location of the port-wine stain. Studies [26, 64, 72] examining different anatomical sites on the head and neck revealed that PWS in centrofacial regions involving the medial portion of the cheek, upper lip and nose in both adults and children show a lesser degree of lightening than PWS of the other locations of the face (periorbital, forehead, temple, lateral aspect of the cheek), which respond quickly. Evaluation by dermatomal distribution reveal

that lesions involving dermatome $V_2$ respond less favourably than lesions involving the other dermatomes of the head and neck, which is consistent with the finding that centrofacial lesions respond less favourably than lesions of other regions, because dermatome $V_2$ primarily involves the central aspect of the face (Fig. 24). Furthermore, lesions on the hand and arm respond less well than lesions on the face, neck and torso.

The factors that account for the anatomical differences in both children and adults in response to treatment of PWS by FDPL are not clear. Perhaps, structural characteristics of the dermis such as orientation, proximity and density of nerves, vessels, density and depth of sebaceous follicles and adnexae to fibrous proteins are likely to play a role in treatment response. Accordingly, when treating PWS with the FPDL, the clinician may wish to consider using higher energy fluences in regions of the lesion that are more resistant to treatment.

Pulsed dye laser treatment of port-wine stains can be undertaken safely also in infancy (Fig. 25a, b). Several studies [7, 26, 36, 75] showed that patients younger than 4 years of age require fewer treatments to attain the same degree of response. Especially infants only a few weeks old have a much less evident lesion by the time they become psychologically aware. Several factors contribute to a greater response in younger patients and newborns. The lesions in children are smaller overall and the vessels are more superficial and of small diameter. Noe [47] showed his-

**Fig. 24a, b.** By anatomical subdivision of the head and neck into regions, the centrofacial region shows a poorer result than the other regions of the head and neck. By dermatomal distribution the dermatomes V1, V3, C2/C3 show a better result than the dermatome V2

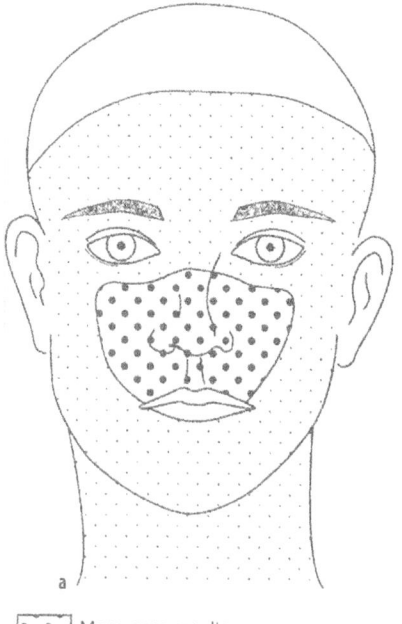
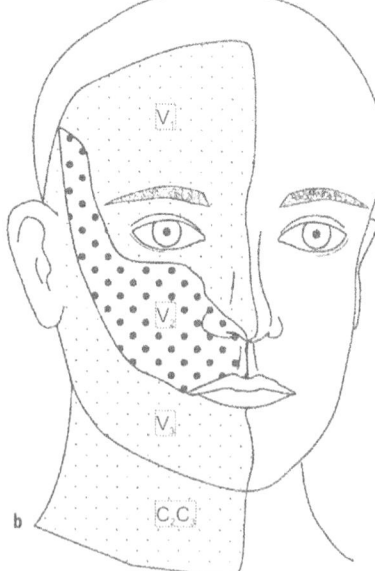

More pore results

Mostly excellent results

**Fig. 25a, b.** Port-wine stain of **a** newborn child and **b** after nine treatments with the pulsed dye laser

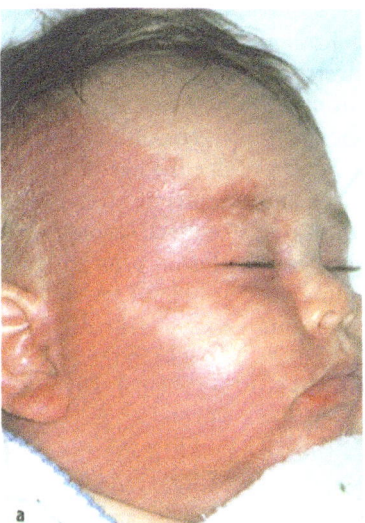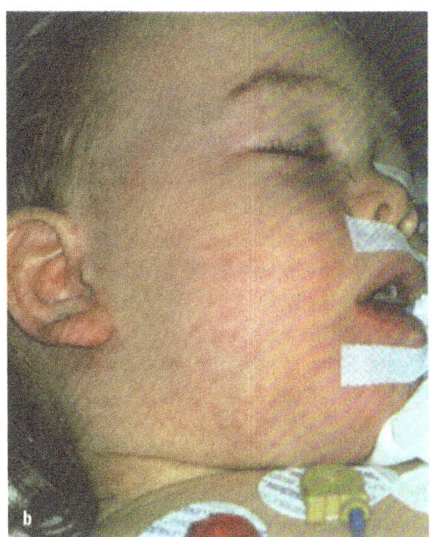

tologically a correlation between age, progression and an increase in the vascular area, mean vessel area and percentage of vessels containing erythrocytes. Furthermore, port-wine stains maintain a relative size throughout life, and, thus, are smaller in surface area in infancy and childhood. In addition, skin thickness increases linearly with age up to 20 years, so that laser treatment early in life may penetrate to the deeper component of the PWS [73]. For treatment during infancy, lower energy fluences would be appropriate with infants and young children (5–7 J cm$^{-2}$) as compared with adults (7–8 J cm$^{-2}$) with larger ectatic vessels within their port-wine stains. Generally, no textural changes after treatment, or damage to the surrounding dermis, were seen in treated skin. The incidence of scarring or pigmentary loss is extraordinarily low.

Patients with port-wine stains suffer a considerable degree of psychological stress because most port-wine stains occur on the face. Treatment of children at the earliest possible age, especially before formal schooling begins, could prevent considerable psychosocial impairment, and provide for a more complete response, because the lesions in children are smaller overall and the vessels more superficial and of small diameter, and while treatment appears to be more effective in pale pink and red port-wine stains in childhood. Early treatment of these lesions is expected to prevent the progression of the vessels in port-wine stains to the more ectatic structures that make the lesions dark purple, raised and nodular in many adults. It is hoped that hypertrophy of affected areas, which is a common complication of extensive port-wine stains, and permanent deformity associated with these lesions can be mitigated.

## Laser Treatment of Non-Fading Vascular Lesions

Spider vascular lesion can be obliterated by argon laser directed at the central artery under compression; however, the marks tend to recur unless the abnormal vessels are completely obliterated. In the case of recurrence, direct coagulation by Nd:YAG laser seems to be an alternative. Complete resolution of spider nevus after pulsed dye laser treatment has also been reported [26].

Other telangiectases, such as the angioma serpiginosum, also respond safely to this therapeutic approach with the argon laser. Widespread telangiectases as a component of other syndromes should also respond to pulsed dye laser treatment. Patients with diffuse telangiectases as a component of the Rothmund-Thomson syndrome or the Louis-Bar syndrome demonstrate dramatic clearing following treatment with the pulsed dye laser.

Other spider vascular lesions, as in the case of patients with Osler syndrome, show better results with Nd:YAG laser treatment. Because the central arteries are usually thick in diameter, patients generally show a higher therapeutic response on direct Nd:YAG laser coagulation.

## Laser Treatment of Haemangiomas

Because haemangiomas may involute spontaneously, allowing for spontaneous regression remains a viable therapeutic option. Attempted therapy may cause adverse systemic or cutaneous side effects, particulary scarring, so intervention has been reserved for patients with significant complications. Alternative treatments have included, electrosurgery, cryosurgery, surgical excision, sclerotherapy, embolization and drug therapy. In view of the potential adverse effects, these treatments have not generally been advised for patients with cutaneous haemangiomas. Thus, the quest for a therapy that eliminates haemangiomas before the development of complications and without systemic side effects has been difficult. Laser therapy has been demonstrated to treat cutaneous vascular lesions effectively and safely in children, while significantly minimizing any adverse cutaneous effects. Several clinical trials have been reported [6, 8, 24, 25, 54, 55, 58, 69]. Through laser treatment also an early and careful therapy of haemangiomas has become possible, so that haemangiomas can be treated in early or prodromal phases to avoid enlargement.

However, laser treatment is required in rapidly growing haemangiomas of the head, or when these lesions interfere with important functions (e.g. hands and feet), or when they endanger delicate structures because of their location (e.g. eye, anogential region). Treatment of large haemangiomas may also be desirable.

The diagnosis of haemangiomas is easily established by history and physical examination. The size, extent and perfusion of the haemangiomas are monitored by colour-coded duplex sonography. Subglottic haemangiomas are demonstrated by bronchoscopy, periorbital haemagiomas are measured by magnetic resonance imaging.

## Superficial Cutaneous Haemangiomas and Treatment with the Flashlamp-Pumped Pulsed Dye Laser

The flashlamp-pumped pulsed dye laser (FDPL) may also successfully prevent enlargement and promote involution of cutaneous haemangiomas (Fig. 26a–c) with minimal adverse effects [24, 26, 34, 40, 51, 55, 69]. Therapy should be initiated as early as possible, even in the first days or weeks of life, when the haemangioma is flat or superficial, to prevent enlargement, promote involution or eliminate these vascular lesions. The same spot size and overlap were used as with the treatment of port–wine stains. The energy fluence ranges from 5 to 7 J cm$^{-2}$ and has to be frequently reduced over eyelids, hands, scrotum and gluteal region. After FDPL treatment, the patients were then evaluated at 2 to 4 weeks, and, depending on the degree of response, the entire lesion is then treated again, usually 4 weeks after the first session. Treatment of cutaneous or superficial haemangiomas during the proliferative phase of growth is usually effective, utilizing frequent treatments (3–4 week intervals) at higher energies than would be used in the

treatment of port-wine stains to obtain a therapeutic response.

Because these lesions are usually small in diameter and few in number, children generally tolerate the treatment quite well without anaesthesia. Infants under the age of 1 year have topical anaesthetic cream (Eutectic Mixture Topical Anesthetics, Astra, Pharmaceutical, Oslow, Sweden) applied for $1\frac{1}{2}$ h under occlusion prior to laser treatment. General anaesthesia is helpful for children with extensive haemangiomas over a large dermatome and in the case of haemangiomas of the periorbital area, especially if the eyelids are treated. Therefore, an eyeshield is used to protect the globe.

FPDL for the treatment of haemangiomas should be reserved for use in carefully selected patients. Such patients are those with cutaneous and flat haemangiomas at sites of potential functional impairment (hands or feet, anogenital region), especially on the face. Furthermore, haemangiomas should be treated with the FDPL as soon as possible, even in the initial or early phase when possible, before the lesions have reached an exponential growth phase, and the limited depth of vascular injury of the FDPL is unlikely to effect any improvement in decreasing size of the lesion. Success depends on the depth of the haemangioma. Several studies have shown that the thicker the lesion at presentation, the less effective was FDPL treatment, which is not surprising given the limited degree of vascular injury caused by this laser. Flat cutaneous haemangiomas respond quite easily and effectively with the FDPL, while a more poorer result is noted in deeper haemangiomas. Although treatment should begin during the early phase of life while the haemangioma is beginning to proliferate, it is often difficult to predict whether or not there will be a superficial and deep component of the haemangioma or only a superficial component. In this case, a colour-coded duplex sonography (CCDS) is helpful. In addition, FDPL treatment of haemangiomas cannot prevent proliferation of the deep component, despite early intervention. Following treatment with the pulsed dye laser, the deep component can continue to proliferate, while the cutaneous component has resolved.

Subcutaneous or mixed haemangiomas do not benefit from pulsed dye laser treatment and one only loses time, because the efficacy of the FDPL is limited by its depth of vascular injury (1–2 mm) and mixed haemangiomas may extend far beyond this depth into the subcutaneous tissue. Treatment with the Nd:YAG laser with continuous ice-cube cooling is required for these types of haemangiomas. Once the underlying subcutaneous vessels have been treated, the superficial component may respond to the treatment with the pulsed dye laser much better.

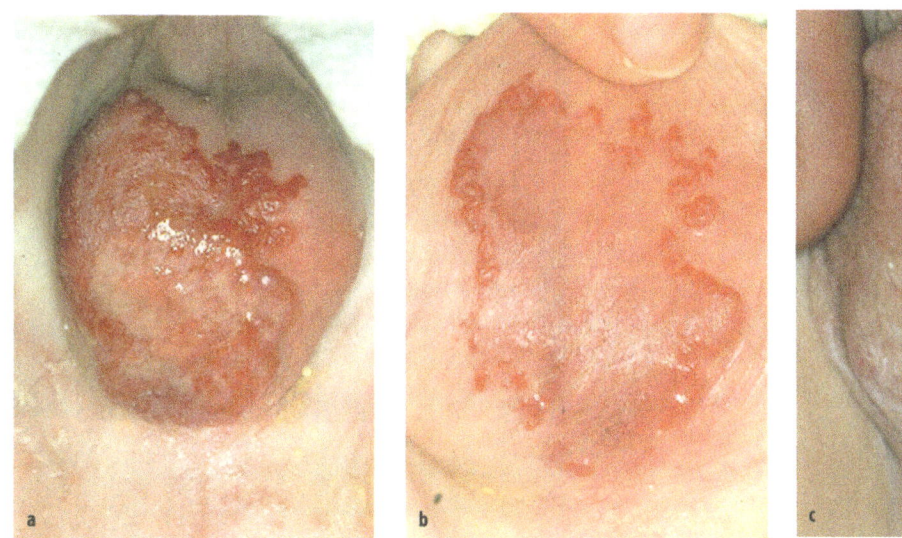
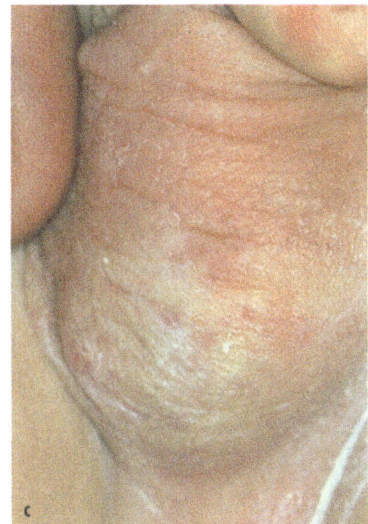

**Fig. 26a–c.**  a A cutaneous haemangioma at the scrotum, b after two treatments, and c after three treatments with the FDPL

Also cutaneous haemangiomas during the involution phase benefit from FPDL treatment. One may argue that superficial haemangiomas in the involution phase may not require treatment; however, many of these lesions are slow to resolve and can be present in a cosmetically or functionally prominent area, so that they have to be treated, especially in children of school age. These patients will benefit from treatment because their response is usually quick and simple (anaesthesia is usually not required). Deep haemangiomas during the involution phase are characterized by their thick appearance and bluish colour. These lesions usually do not benefit from pulsed dye laser treatment.

Using the cooling chamber, the risk of side effects can be reduced and a compression is possible. As with the treatment of port-wine stains, the incidence of side effects from pulsed dye laser treatment of haemangiomas is quite small, thus making the benefit-to-risk ratio high.

The argon laser and the KTP laser are also useful in the treatment of superficial telangiectasias and small, flat cutaneous haemangiomas. Superficial haemangiomas respond quite easily to the argon laser and KTP laser, while a more variable response is noted in deeper haemangiomas. Comparatively, for cutaneous haemangiomas, the treatment with the flashlamp-pumped pulsed dye laser is still more effective because of its greater focus, so that a larger area can be treated more easily. In addition, none of these lasers has an optimal selective combination of colour and pulse duration, which is necessary to ensure limitation of non specific thermal damage.

## Subcutaneous Haemangiomas and Treatment with the Nd:YAG Laser

Nd:YAG laser treatment is an important adjunctive therapy when haemangiomas threaten further compression of function and disfigurement (Fig. 27a–d). It is a well-known and effective method in the shrinkage and blanching of haemangiomas, and has been reported by many authors [1, 12, 30, 50, 51].

The laser beam can transmit down with the use of a focusing hand piece and focus directly on the skin surface with a 1–2-mm spot beam. Low-power energy, intermittent short exposure (20 W, 0.1/0.1 s repetitive mode) or higher energy with continuous surface cooling (20–46 W, cw, with ice cube cooling) accomplish photocoagulation with minimal damage to surrounding tissue. Especially Nd:YAG laser treatment with continuous surface cooling is safe, with a low incidence of scarring and pigmentary alteration. It is indicated for massive or deep subcutaneous haemangiomas. The power energy range from 20–46 W with lower energies used for treatment in small and dark red haemangiomas (20–34 W, with continuous ice cube cooling) and the progressively higher energies for patients with thicker and only subcutaneously localized haemangiomas (require up to approximately 46 W, with continuous ice cube cooling). Also haemangiomas of the eyelids and the conjunctiva can be treated safely (Fig. 28a, b), although protection of the ocular globe is required during treatment sessions.

Subcutaneous components of haemangiomas require repeated treatments with the Nd:YAG laser. Since they often grow extensively, one should treat them as soon as possible [13, 30]. Furthermore, prolif-

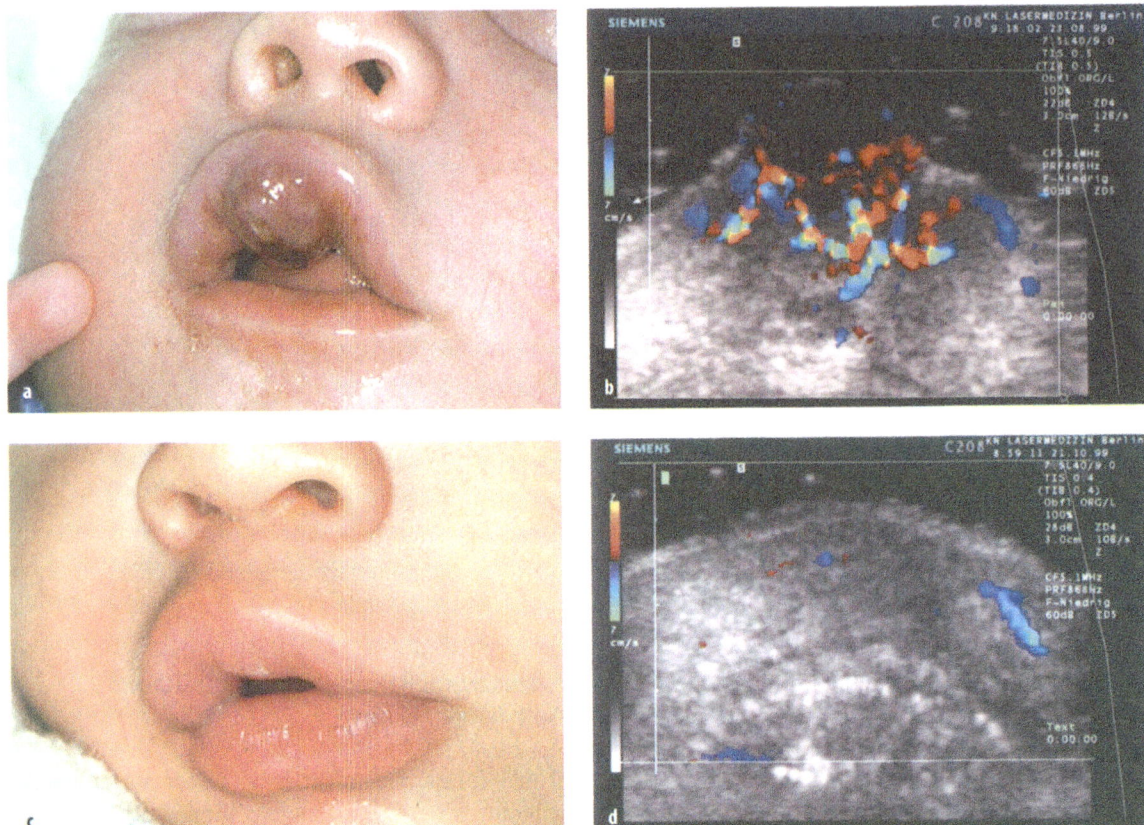

**Fig. 27 a–d.** Six-week-old boy with an ulcerated bleeding haemangioma of the upper lip **a** with pathologic hypervascularization **b** prior to, and after two Nd:YAG laser therapy treatments **c** with continuous ice-cube cooling. The decrease in coloration by perfusion in the colour-coded duplex ultrasound is notable **d**

erating haemangiomas may grow rapidly, and frequent treatments (approximately 4 to 6 weeks) will help counter the proliferative growth. Even if they no change in size, laser treatment can reduce the volume of haemangioma, initializing the process of regression.

For larger and deep-seated haemangiomas up to a depth of 2.5 cm, a percutaneous interstitial Nd:YAG laser treatment may be preferred because it may decrease possible cutaneous skin damage and more effectively reduce bulky, deep lesions (Fig. 29a–d).

Nd:YAG laser treatment is very painful, therefore general anaesthesia is usually required. Swelling and hardening of the haemangioma occur immediately following treatment. Blisters may appear and should not be opened because restitution of the epidermis should not be endangered. Minor crust formation is possible. The swelling slowly diminishes within 1 week after laser treatment. After 2 to 4 weeks, improvement with flattening of the lesion, less swelling, and improvement in the vascular red or bluish hue should be visible.

Response to photocoagulation with the Nd:YAG laser is immediate, with no bleeding and less destruction of surrounding tissue than with cryosurgery. The side effects following Nd:YAG laser treatment in children are low in incidence and usually of little clinical significance. Atrophic scarring can occur, which is usually a consequence of direct transcutaneous application with excessive energy. Permanent depigmentation is also seen, and this complication is mostly seen after spontaneous regression of large haemangiomas as well.

Another advantage of the Nd:YAG laser is its particularly useful fiberoptic transmission through available endoscopes. Thus, the Nd:YAG laser has been used primarily endoscopically for treatment of obstructing haemangiomas of the tracheobronchial tree [17] and gastrointestinal tract, and for haemangiomas of the bladder. Its properties of deep tissue penetration and scatter make it an ideal instrument for photocoagulation and haemostasis. Repeated applications are possible with minimal oedema and bleeding.

**Fig. 28.** **a** Thirteen-week-old girl with a rapidly growing haemangioma of the upper eyelid and conjunctiva. **b** The same infant as in a after three treatments with the Nd:YAG laser with continuous ice-cube cooling of the surface

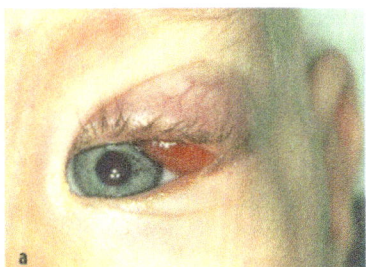
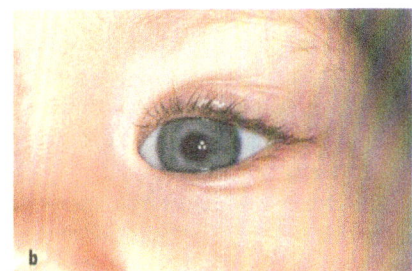

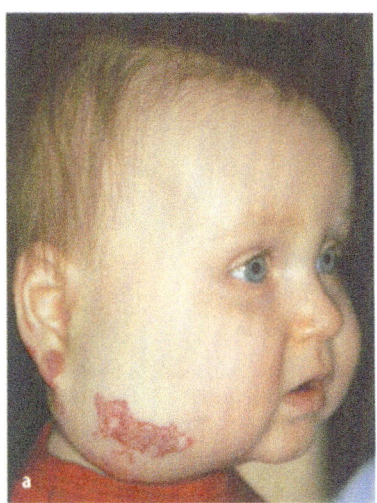
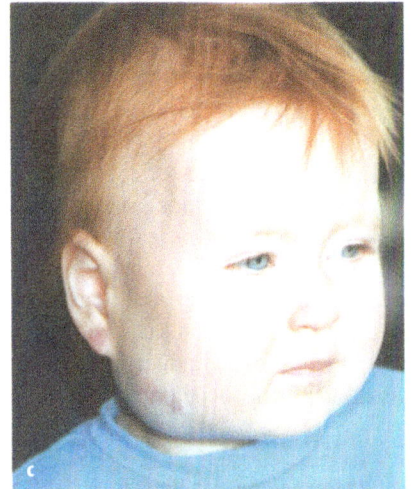

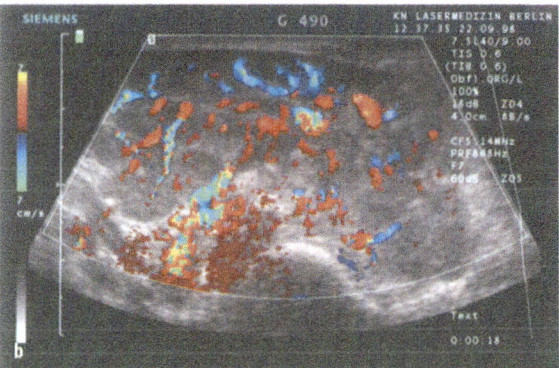
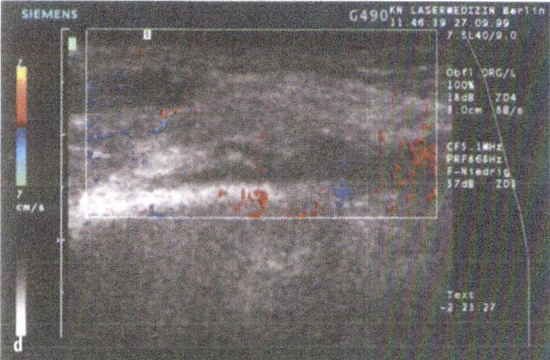

**Fig. 29a–d.** Twelve-week-old girl with a large haemangioma of the cheek **a** with pathological hypervascularization **b** prior to, and after three interstitial Nd:YAG laser treatments and three transcutaneous Nd:YAG laser therapies with continuous ice-cube cooling **c**. The decrease in coloration by perfusion in the colour-coded duplex ultrasound is notable **d**

## Combination of Laser and Drug Therapy

In the case of large haemangiomas and/or those impinging on vital structures or endangering life or vision (defined as lesions that invade or obstruct vital organs or tissue, such as the airway, oropharynx, orbit or adnexa, retroperitoneum and viscera [21]), the current guidelines [67] suggest that treatment should be started first with high oral doses of corticosteroids (5 mg kg$^{-1}$ day$^{-1}$ prednisone). After 2 weeks of treat-

ment the dose must be reduced gradually to 2.5 mg week$^{-1}$.

The key to therapy is the use of high doses of the drug for long periods of time. The side effects seem to be much rarer and better tolerated in children, and abate completely when the drug is stopped. In those patients in whom the drug was stopped, and the haemangioma began to grow again, children are treated a second time. With this regime there has been no emergence of drug resistance.

Congestive heart failure, consumptive coagulopathy and thrombocytopenia were also urgent indications for the institution of cortison therapy. Corticosteroids increase the survival time of platelets, reduce the vascular mass and restore the integrity of the clotting system.

If after 2–4 weeks no effect is seen and the clinical situation requires further treatment, interferon alfa is recommended [21]. Despite cortison or interferon therapy, laser treatment should be started at the same time to induce improvement of the haemangioma.

Although the Kasabach-Merritt syndrome is usually a self-limited condition that remits when the tufted angioma begins to involute, the mortality rate ranges between 30 to 60% without therapy [21]. In some cases, corticosteroid therapy has been sufficient to palliate the condition, resulting in elevation of the platelet count and shrinkage of the tufted angioma. Additional therapeutic options include laser therapy. Finally, interferon alfa is an effective treatment in infants with large tufted angioma and Kasabach-Merritt syndrome and/or those impinging on vital structures.

Operative treatment is indicated in cases with large residuals, especially before formal education begins, to prevent considerable psychosocial impairment.

## Laser Treatment of Haemangioendotheliomas and Tufted Angiomas

Although the response of haemangioendothelioma and tufted angiomas to different therapies has recently been described, there is no therapeutic approach that has constantly given good results. Lesions often recur after surgery and the use of the pulsed dye laser is ineffective, as confirmed in patients by Enjolras [19]. In the case of large kaposiform haemangioendotheliomas within mediastinum or retroperitoneum, these lesions are unresectable and may cause death by complication of consumption coagulopathy. In those cases, more satisfactory results have been reported with high-dose systemic steroids, and interferon alfa therapy. In most cases, residual vascular neoplasm was noted, while the haematological phenomenon was cured after a period of time. However, Nd:YAG laser treatment seems to be an interesting adjunct to drug therapy (Fig. 30a–c).

## Laser Treatment of Vascular Malformations

The attitude that vascular malformations should be radically excised like cancer has a high incidence of recurrence; the more accurate term is re-expansion because of expansion of remaining malformed channels. Unfortunately, experience shows that it is only

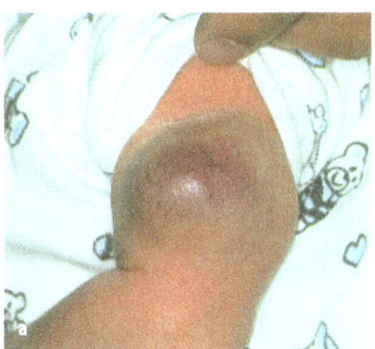

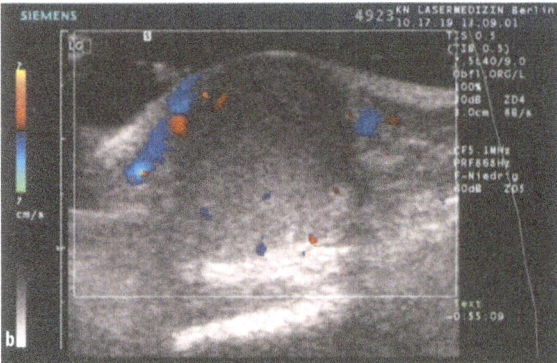

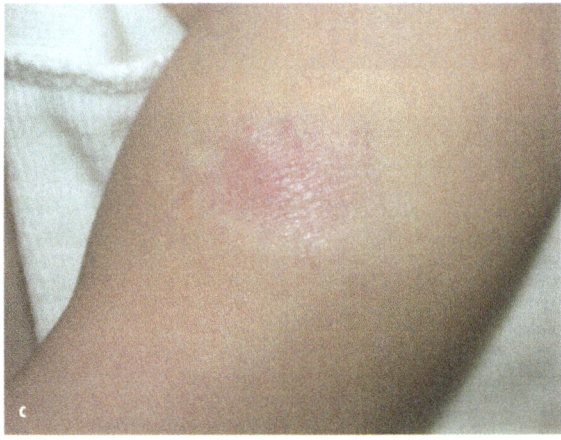

**Fig. 30. a** Tufted angioma in a newborn child. **b** Sonographically, tufted angiomas are hypersonic with some pathological vessels. **c** The same infant after two treatments with the Nd:YAG laser with continuous ice-cube cooling of the surface

very rarely that resection of large vascular malformation, either lymphatic, venous or combined forms, is feasible in its entirety. Furthermore, the extent of the lesion is a critical factor, because attempted removal of a large malformation is hazardous and there is always the danger of massive bleeding and the need for extensive resection, causing mutilation.

The wide variety of clinical presentations for these anomalies makes it difficult to outline specific management programs. Therefore, treatment of these difficult vascular lesions must be carefully individualized. Whereas laser treatment in capillary malformations is the first choice of therapy, in arteriovenous malformations, embolization, either alone or in conjunction with surgical excision, is the first choice. In venous and lymphatic malformations, laser therapy seems to be an interesting alternative to surgical excision and sclerosant therapy.

## Laser Treatment of Venous Malformations

Although total excision may be very successful in dealing with small to moderate-sized lesions, all too often only subtotal resection is possible. Angiography and embolization techniques are unsuitable for venous malformations, because embolic occlusion of the arterial inflow to a venous anomaly often causes necrosis of adjacent tissue and overlying skin [18]. A far more appealing concept is direct injection of the sclerosing solution into the centre of a venous malformation. Intralesional sclerotherapy by direct percutaneous injection is useful especially for large ectatic veins, while the arterial inflow and venous outflow are occluded. Capillary-venous malformations and isolated venous ectasias can be managed by laser treatment [50], in some cases following sclerosant therapy. Also, in the case of diffuse malformation, where only subtotal excision is possible in order no to sacrifice vital anatomic structures, laser treatment is possible (Fig. 31a, b). This type of laser treatment may diminish the vascular swelling and thrombose the malformation in order to minimize flow into the malformation. By this, the area of vascular anomaly decompresses to such a degree that later resection is not necessary. Also in extensive bulky malformations, laser

treatment helps surgery by providing a relative definition of the zones treated, thus simplifying later surgical access and minimizing the chance of re-expansion (Fig. 32a–d).

Superficial venous malformations may respond somewhat to treatment with the argon laser, KTP laser or pulsed dye laser. These lesions are often thicker in their involvement, and, thus, the potential for a therapeutic response from argon or pulsed dye laser is not as great as one would find with the Nd:YAG laser, that creates a greater degree of thermal injury. Furthermore, the deeper component cannot be reached with the argon, KTP or pulsed dye laser. The Nd:YAG laser has been used extensively and appears to be particularly effective for venous anomalies. Transcutaneous Nd:YAG laser treatment with continuous ice-cube cooling can usually be safely accomplished in selected cases of cutaneous(-mucous)/subcutaneous (-mucous) venous, lymphatic-venous and capillary-venous lesions, up to a depth of approximately 1.5 cm, for example a lesion that interferes with dental occlusion. For this, utilizing higher energies than would be used in the treatment of haemangiomas to obtain a therapeutic response is usually effective. Interstitial Nd:YAG laser coagulation is indicated to reduce bulk and improve contour and function in the case of large extensive lesions, thus minimizing the potential complications of possible future surgical excision. The Nd:YAG laser can also be used primarily endoscopically for treatment of venous anomalies of the pharynx and trachea. Excision with the $CO_2$ laser has been useful for small oral mucosal venous lesions [5].

Treatments are usually repeated at 6-week intervals until the desired degree of lightening and decompression is achieved. Repetitive treatments sessions are re-

**Fig. 31. a** Venous malformation in a thirty six-year-old female. **b** The same patient following three treatment sessions with the Nd:YAG laser with continuous ice-cube cooling

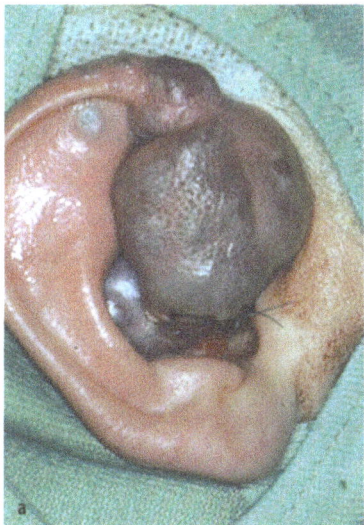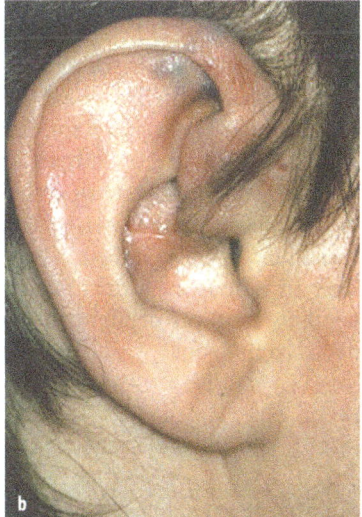

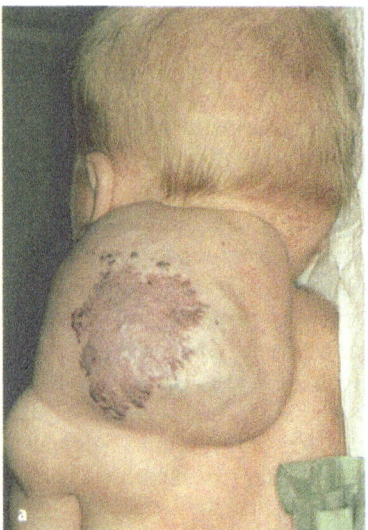

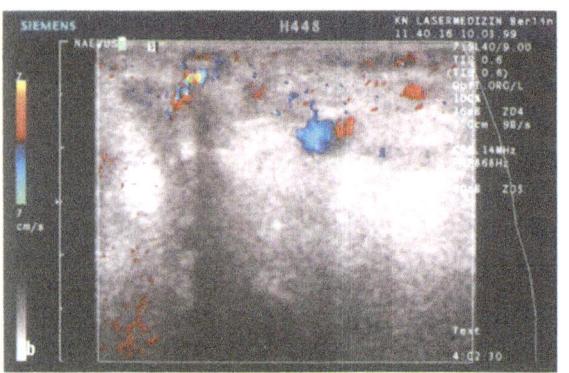

**Fig. 32. a** Two-month-old infant with large extensive venous malformation of the left shoulder, who also had associated coagulopathy, recurrent bleeding anaemia. **b** CCDS documenting large lacunar venectasias prior to and **c** after eight interstitial Nd:YAG laser treatments. The decrease in coloration by pathologic perfusion and fibrosis of the tissue is notable. **d** The same infant at 3 years of age after the first surgical excision by Prof. Grantzow of the von Haunersche Kinderspital in Munich

quired to obtain the maximum benefit of laser coagulation.

Truncular malformations such as the Klippel-Trenaunay syndrome with a persistent marginal vein, the Parks-Weber syndrome or av-fistulas do not respond to laser treatment. Treatment with excision and/or embolization are required for these types of malformations. Once the underlying vascular lesion has been treated, the superficial component may respond to treatment with the laser, in the case of the Klippel-Trenaunay syndrome treatment with the pulsed dye laser, in venous malformations the Nd:YAG laser, depending upon the depth of involvement of the cutaneous component.

## Laser Treatment of Glomangiomas

Multiple glomangiomas are a rare clinical occurrence. In the past, various treatments have been proposed for eradication of these tumours. In the case of single glomangiomas, excision is an effective approach; percutaneous sclerotherapy gives poor results. An effective therapy for multiple glomangiomas is treatment with the Nd:YAG laser with continuous surface cooling. Laser therapy was successful in relieving symptoms and improving the appearance of the glomus tumours (Fig. 33a, b).

## Laser Treatment of Lymphatic Malformations

Treatment of lymphatic malformations includes surgical resection, sclerosing injections and laser treatment. Bulky lymphatic malformations are amenable to surgical resection, with the proviso that, postoperatively, vesicles may bubble up in the scar or the adjacent remaining lymphatic anomaly may expand. Vesicular extension from retained lymphatic cysts frequently occurs in the scar after subtotal excision.

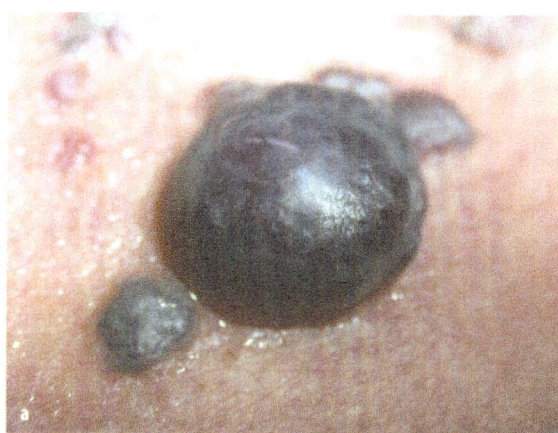

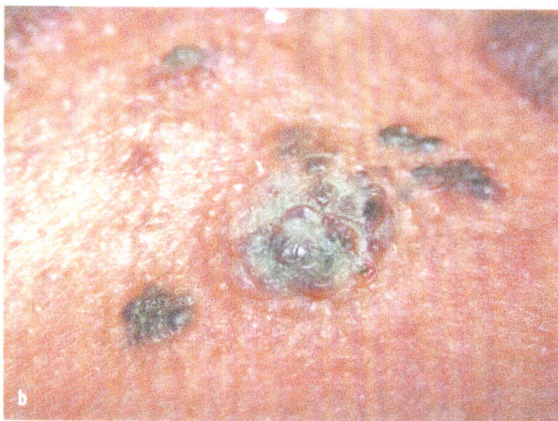

**Fig. 33. a** Glomangiomas in a fifty-year-old male with recurrent bleeding. **b** The same patient directly after the first Nd:YAG laser treatment with continuous ice-cube cooling. Shrinking and blanching of the glomangioma is visible

bose the malformation, thus simplifying later surgical access and minimizing the chance of re-expansion. An annoying complication of interstitial laser coagulation is the possibility of nerve damage that may result from postoperative swelling; transient facial nerve paralysis or weakness is an obvious concern, that will regress over several weeks. Nerve paralysis may also be the result of surgical resection, and is more likely with each procedure because dense scar tissue distorts the usual anatomic location of nerves, especially the facial.

Past experience with surgical resection and sclerosis has been unsuccessful in the management of lymphatic malformations of the oropharyngeal region. For this, endocopical Nd:YAG laser treatment allows elective coagulation of oropharyngeal lymph cysts that are narrowing the upper airway and may be causing bleeding. Nd:YAG laser treatment during infancy helps to prevent tracheostomy or free the airway. Herefore, repetitive treatment sessions are required (Fig. 34a, b).

Nd:YAG laser can also be used to coagulate oozing mucosal blebs in the mouth and the tongue that frequently appear in scars of the tongue and oral mucosa after previous surgical resection. In the case of macroglossia, operative reduction may be necessary to restore the tongue into the oral cavity. Herefore, excision with $CO_2$ laser or Nd:YAG laser has been useful in view of its simplicity, and the minimal bleeding.

Nd:YAG laser treatment may also be useful, particularly to control troublesome bleeding points or weeping lymphatic cysts in the case of lymphatic or capillary-lymphatic malformations.

Because surgical resection is limited in the therapy of lymphatic malformations, other authors reported more favourable results by using fibrosing agents. For this, the sclerosing agent is injected directly into the lymphatic malformation by percutaneous puncture. Best results have been seen in large lymphatic cysts, while in microcystic forms sclerotherapy resulted in failure [15]. Other products like alcohol have been used with variable results and long-term results which are not yet well known [82]. At the other end of the spectrum, large cystic lesions are more amenable to interstitial Nd:YAG laser coagulation and are less likely to deflate. As for venous malformations, treatment may diminish the vascular swelling and throm-

## Laser Treatment of Arteriovenous Malformations

At present, the first choice of therapy for the management of troublesome arteriovenous malformations is embolization, either alone or in conjunction with surgical excision. The therapeutic principle in embolization is to deliver the embolic material into the centre of the vascular anomaly (the niddus) in an attempt to block the smallest vessels first, from the inside out. For extensive lesions, interstitial Nd:YAG laser coagulation may help by obliterating all microfistulae in order to collapse the arteriovenous malformation permanently, so that collateral vessels can develop very slowly.

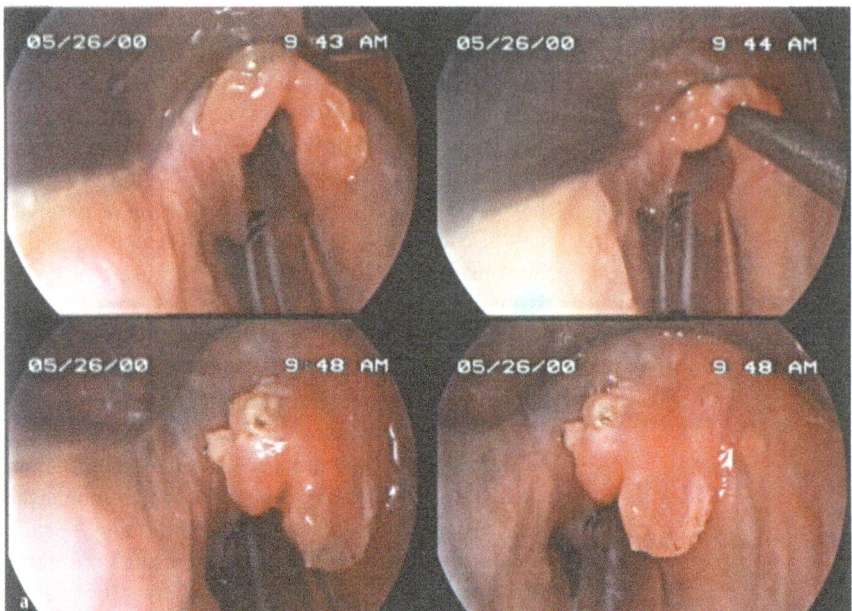

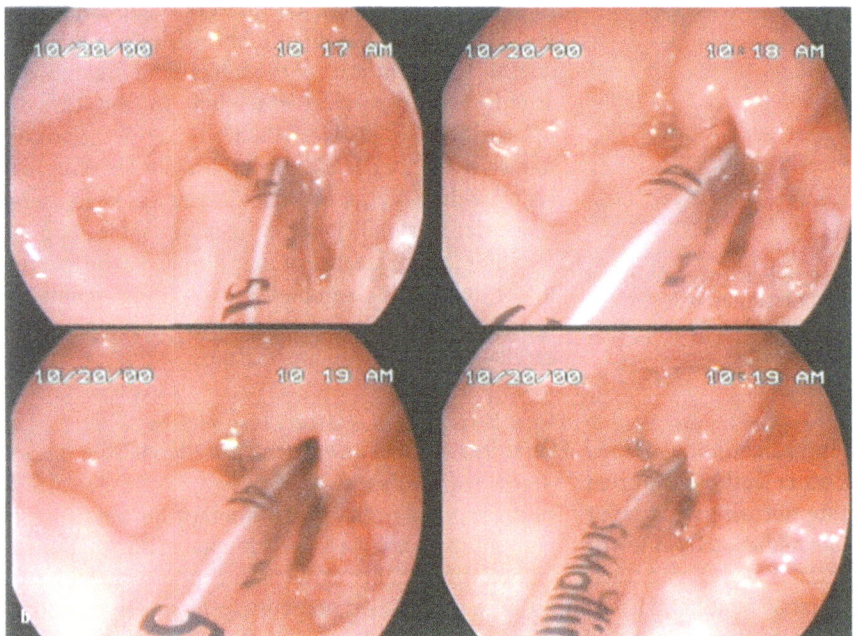

**Fig. 34a, b.** Lymphatic malformation of the pharynx **a** prior and **b** after one endoscopic direct transcutaneous Nd:YAG laser application

## References

[1] ACHAUER BM, VANDER KAM VM (1989) Capillary haemangiomas (strawberry mark) of infancy: comparison of argon and Nd:YAG laser treatment. Plast Reconstr Surg 84: 60–70

[2] ALBRIGHT A L, GARTNER J C, WIENER E S (1989) Lumbar cutaneous haemangiomas as indicators of tethered spinal cords. J Pediatr 88: 977–980

[3] AMIR J, METZKER A, KRIKLER R, REISNER SH (1986) Strawberry haemangiomas in preterm infants. Pediatr Dermatol 3: 331–332

[4] ANDERSON RR, PARISH JA (1981) The optics of human skin. J Invest Dermatol 77: 9–13

[5] APFELBERG DB, MASER MF, LASH H, WHITE DN (1985) Benefits of $CO_2$ laser in oral haemangioma excision. Plast Reconstr Surg 75: 46

[6] ASHINOFF R, GERONEMUS RG (1991) Capillary haemangiomas and treatment with the flash lamp-pumped pulsed dye laser. Arch Dermatol 127: 202–205

[7] ASHINOFF R, GERONEMUS RG (1991) Flashlamp-pumped pulsed dye laser for port-wine stains in infancy: Earlier versus later treatment. J Am Dermatol 24: 467–472

[8] ASHINOFF R, GERONEMUS RG (1993) Failure of the flashlamp-pumped pulsed dye laser to prevent progression to deep haemangioma. Pediatr Dermatol 10: 77–80

[9] BARSKY, SH, ROSEN S, GEER D, NOE JM (1980) The nature and evolution of port-wine stains: a computer-assisted study. J Invest Dermatol 74: 154–7

[10] BELOV ST (1989) Classification, terminology and nosology of congential vascular defects. In: Belov St, Loose D, Weber J (ed) Vascular Malformations. Periodica Angiologica 16, Einhorn Presse, pp 25–28

[11] BENDA CE (1954) The Dandy-Walker syndrome or the so-called atresia of the foramen Magendie. L Neuropathol Exp Neurol 13: 14–29

[12] BERLIEN HP, WALDSCHMIDT J, MÜLLER G (1988) Laser treatment of cutaneous and deep vessel anomalies. In: Waidelich W (ed) laser optoelectronics in medicine. Springer, Berlin Heidelberg New York, pp 526–528

[13] BERLIEN HP, CREMER H, DJAWARI D, GRANTZOW R, GUBISCH W (1993/94) leitlinien zur Behandlung angeborener Gefäßerkrankungen. Pädiatr Praxis 46: 87–92

[14] BILLSON V R, GILLIAM G l (1984) An unusual case of Sturge-Weber syndrome. Pathology 16 (4): 462–465

[15] BREVIÈRE GM, BONNEVALLE M, PRUVO JP, BESSON R, MARACHE P, HERBAUX P, DEBEUGNY P (1993) Use of ethibloc in the treatment of cystic and venous angiomas in children. 19 cases. Eur J Ped Surg 3: 166–70

[16] BURNS AJ, KAPLAN LC, MULLIKEN JB (1991) Is there an association between haemangioma and syndromes with dysmorphic features? Pediatrics 88: 1257–267

[17] CHOLEWA D, WALDSCHMIDT J (1998) Laser treatment of haemangiomas of the larynx and trachea. Lasers Surg Med 23: 221–232

[18] DEMUTH RJ, MILLER SH, KELLER F (1984) Complications of embolization treatment for problem cavernous haemangiomas. Ann Plast Surg 13: 135

[19] ENJOLRAS O, WASSEF M, MAZOYER E, FRIEDEN IJ, RIEU PN, DROUET L, TAIEB A, STALDER JF, ESCANADE JP (1997) Infants with Kasabach-Merritt syndrome do not have "true" haemangiomas. J Pediatr 130 (4): 631–639

[20] ERSEKINE CA (1950) Asymptomatic unilateral agenesis of the cerebellum. Psychiatr Neurol 119: 321–339

[21] ERZEKOWITZ RÁB, MULLIKEN JB, FOLKMAN J (1992) Interferon alfa-2a therapy for life-threatening haemangiomas of infancy. N Engl J Med 326 (22): 1456–463

[22] FINLEY JL, CLARK RAF, COLVIN RB (1982) Immunofluorescence staining with antibodies to factor VIII, fibronectin, and collagenase basement membrane proteins in normal skin and port-wine stains. Arch Dermatol 18: 971–975

[23] FRIEDEN IJ, REESE V, COHEN D (1996) PHACE syndrome. The association of posterior fossa brain malformations, haemangiomas, arterial anomalies, coarctation of the aorta and cardiac defects, and eye abnormalities. Arch Dermatol 132 (3): 307–311

[24] GARDEN JM, BAKUS AD, PALLER AS (1992) Treatment of cutaneous haemangiomas by the flashlamp-pumped pulsed dye laser: Prospective analysis. J Pediat 120: 555–560

[25] GERONEMUS RG, ASHINOFF R (1991) The medical necessity of evaluation and treatment of port-wine stains. J Dermatol Surg Oncol 17: 76–79

[26] GERONEMUS RG (1993) Pulsed dye laser treatment of vascular lesions in children. J Dermatol Surg Oncol 19: 303–310

[27] GOH, WSH, LO R (1993) A new 3C syndrome: cerebellar hypoplasia, cavernous haemangioma and coarctation of the aorta. Dev Med Child Neurol 35: 631–641

[28] GOLDBERG NS, HEBERT AA, ESTERLY NB (1986) Sacral haemangiomas and multiple congenital abnormalities. Arch Dermatol 122: 684–687

[29] GOLDMAN MP, FITZPATRICK RE, RUIZ-ESPARAZA J (1993) Treatment of port-wine stains (capillary malformation) with the flashlamp-pumped pulsed dye laser. J Pediatr 122: 71–77

[30] GRANTZOW R, SCHMITTENBECHER PP, SCHUSTER T (1995) Frühbehandlung von Hämangiomen: Lasertherapie. Monatsschr Kinderheilkd 143: 369–374

[31] HERMANN TE, MCALLISTER PW, DEHNER LP, SKINNER M (1997) Beckwith-Wiedemann syndrome and splenic haemangioma: report of a case. Pediatr Radiol 27: 350–352

[32] HERSH JH, WATERFILL D, RUTLEDGE J (1985) Sternal malformation vascular dysplasia association. Am J Med Genet 21: 177–186

[33] HIRSCH JF, KAHN AP, RENIER D, SAINT-ROSE C, HOPPE-HIRSCH E (1984) The Dandy-Walker malformation. A review of 40 cases. J Neurosurg 61: 515–522

[34] HOHENLEUTNER U, LANDTHALER M (1997) Die Behandlung der Säuglingshämangiome. Kinderarzt 28: 989–1000

[35] HOLDEN K, ALEXANDER F (1970) Diffuse neonatal haemangiomastosis. Pediatrics 46: 411–421

[36] HOLY A, GERONEMUS RG (1992) Treatment of periorbital port-wine stains with the flashlamp-pumped pulsed dye laser. Arch Ophthalmol 110: 793–797

[37] IGARASHI M, UCHIDA H, KAJAII T (1985) Supraumbical midabdomininal raphe and facial cavernous haemangiomas. Clin Genet 27 196–198

[38] JACOBS AH, WALTON RG (1976) The incidence of birtmarks in the neonate. Pediatrics 58: 218–222

[39] KAPLAN LC, MATSUOKA R, GILBERT EF, OPITZ JM, KURNIT DM (1985) Ectopia cordia and cleft sternum: evidence for mechanical teratogenesis following rupture of the chorion or yolk sac. Am J Med Genet 21: 187–199

[40] LANDTHALER M, HOHENLEUTNER U, TALAL AHMED ABD-EL RAHEEM A (1995) Therapie vaskulärer Fehlbildungen. Medwelt 46: 357–359

[41] LOPRIORE E. MARKHORST DG (1999) Difuse neonatal haemagniomatosis: new views on diagnostic criteria and prognosis. Acta Paediatr 88: 99–97

[42] MAZOYER E, ENJOLRAS O, MERLAND JJ, DROUET (1995) Differential coagulation abnormalities in venous vascular malformations and haemangiomas? Abstr for the Study of Vascular Anomalies (ISSVA) in Cirse (France)

[43] MENEFEE MG, FLESSA HC, GLUECK HI, HOGG SP (1985) Herediatary hemorrhagic telangiectasia (Osler-Weber-Rendu disease): an electron microscopy study of the vascular lesions before and after therapy with hormones. Arch Otolaryngol 101: 246–251

[44] MULLIKEN JB, GLOWACKI J (1982) Haemangiomas and vascular malformations in infants and children: a classification based on endothelial characteristics. Plast Reconstr Surg 69: 412–22

[45] MULLIKEN JB (1988) Capillary (port-wine) and other teleangiectatic stains. In: Mulliken JB, Young AE (eds) Vascular birthmarks. Saunders, Philadelphia, pp 170–195

[46] NAKAGAWA H, TAN OT, PARRISH JA (1985) Ultrastructureal changes in human skin after exposure to a pulsed laser. J Invest Dermatol 84: 396–400

[47] NOE JM, BARSKY SH, GEER DE, ROSEN S (1980) Port-wine stains and the response to argon laser therapy: successful treatment and the predictive role of colour, age and biopsy. Plast Reconstr Surg 65: 130–136

[48] PASCUAL-CASTROVIEJO I, VELEZ A, PASCUAL-PASCUAL SI, ROCHE MC, VILLAREJO F (1991) Dandy-Walker malformation: analysis of 38 cases. Childs nNerv Syst 7 (2): 88–97

[49] PASCUAL-CASTROVIEJO I, VIANO J, MORENO F, PALENCIA R, ET AL (1996) Haemangiomas of the head, neck and chest with associated vascular and brain anomalies: complex neurocutaneous syndrome. Am J Neuroadiol 17: 461–471

[50] POETKE M, PHILIPP C, BERLIEN HP (1997) Ten years of laser treatment of haemangiomas and vascular malformations: techniques and results. In: BERLIEN HP, SCHMITTENBECHER PP (eds) Laser surgery in children. Springer, Berlin Heidelberg New York, pp 82–91

[51] POETKE M, BÜLTMANN O, PHILIPP C, BERLIEN HP (1998) Hämangiome und vaskuläre Malformationen im Säuglings- und Kindesalter. 184: 40–47

[52] POETKE M, BÜLTMANN O, URBAN P, BERLIEN HP (1998) Vaskuläre Malformationen im Kindes- und Erwachsenenalter. Therapie mit dem Nd:YAG-Laser. Vasomed 10: 338–347

[53] POETKE M, BÜLTMANN O, BERLIEN HP (2000) Association of large facial haemangiomas with Dandy-Walker syndrome. Eur J Pediatr Surg 10: 125–129

[54] POETKE M, BERLIEN HB (2000) Treatment of a subungual haemangioma with flashlamp-pumped pulsed-dye laser. J Am Acad Dermatol 43: 1135–1136

[55] POETKE M, PHILIPP C, BERLIEN HP (2000) Flashlamp-pumped pulsed dye laser for haemangiomas in infancy. Arch Dermatol 136: 628–632

[56] POETKE M, BÜLTMANN O, BERLIEN HP (1999/2000) Ausgedehnte Hämangiome im Gesicht und ihr assoziiertes Auftreten mit einem Dandy-Walker-Syndrom und anderen Veränderungen der hinteren Schädelgrube. Pädiatr Praxis 57: 283–294

[57] POETKE M, JAMIL B, MÜLLER U, BERLIEN HP (2002) Diffuse neonatal haemangiomatosis associated with Simpson-Golabi-Behmel syndrome: A case report. Eur J Pediatr Surg (in press)

[58] POETKE M, PHILIPP C, BERLIEN HP (2001) Die Behandlung von Hämangiomen im Säuglings- und Kindesalter mit dem blitzlampengepumpten Farbstofflaser: Kutane versus gemischte kutan-subkutane Hämangiome. Hautarzt 52: 120–127

[59] POETKE M, PHILIPP C, GROßEWINEKLMANN A, URBAN P, BERLIEN HP (2001) Die Behandlung von Naevi flammei bei Säuglingen und Kleinkindern mit dem blitzlampengepumpten Farbstofflaser. Monatsschr Kinderheilkd 32: 405–415

[60] POETKE M, PHILIPP C, URBAN P, BERLIEN HP (2001) Interstitial laser treatment of venous malformations. Med Laser Appl 16: 111–119

[61] POETKE M, FROMMELD T, BERLIEN HP (2202) PHACE Syndrome. New views on diagnostic criterias. Eur J Pediatr Surg (in press)

[62] DE PROST Y, TEILLAC D, BODEMER C, ENJOLRAS O, NIHOUL-FEKETE C, DE PROST D (1991) Successful treatment of Kasabach-Merritt syndrome with pentoxifylline. J Am Acad Dermatol 25: 854–55

[63] REESE V, FRIEDEN I, PALLER A, ESTERLY N, FERRIERO D, LEVY M, LUCKY A, GELLIS S, SIEGFRIED E (1993) Association of facial haemangiomas with Dandy-Walker and other posterior fossa malformations. J Pediatr 122 (3): 379–384

[64] RENFRO L, GERONEMUS RG (1993) Anatomical differences of port-wine stains in response to treatment with the pulsed dye laser. Arch Dermatol 129: 182–188

[65] REQUENA L, SANGUEZA OP (1997) Cutaneous vascular proliferations. Part II. Hyperplasias and benign neoplasms. J Am Acad Dermatol 37: 887–919

[66] REYES BA, GERONEMUS RG (1990) Treatment of port-wine stains during childhood with the flashlamp-pumped pulsed dye laser. J Am Acad Dermatol 23: 1142–1148

[67] SADAN N, WOLACH B (1996) Treatment of haemangiomas of infants with high doses or prednisone. J Pediatr 128: 141–146

[68] SCHNEEWEISS A, BLIEDEN LC, SHEM-TOV A, MOTRO M, FEIGEL A, NEUFELD HN (1982) Coarctation of the face and neck and aeurysm or dilatation of a subclavian or innominate artery. A new syndrome? Chest 82 (2): 186–187

[69] SHERWOOD KA, TAN OT (1990) Treatment of a capillary haemangioma with the flashlamp-pumped dye laser. J Am Acad Dermatol 22: 136–137

[70] SOLOMON H, GOLDMAN L, HENDERSON B, RICHFIELD D, FRANZEN M (1968) Histopathology of the laser treatment of port-wine lesions: biopsy studies of treated areas observed up to three years after laser impacts. J Invest Dermatol 50: 141

[71] SMOLLER BR, ROSEN S (1986) Port-wine stains: a disease of altered neural modulation of blood vessels? Arch Dermatol 122: 177–179

[72] TALLMAN B, TAN OT, MORELLI JG, PIEPENBRINK BS, STAFFORD TJ, SHAWN TRAINOR, WESTON WL (1991) Location of port-wine stains and the likelihood of opthalmic and/or central nervous system complications. Pediatrics 87: 323–327

[73] TAN OT, STATHAM B, MARKS R, PAYNE PA (1982) Skin thickness measurement by pulsed ultrasound: Its reproducibility, validation and variability. Br J Dermatol 106: 657–667

[74] TAN OT, CARNEY JM, MATGOLIS R (1986) Histologic response of port-wine stains treated by argon, carbon dioxide, and tunable dye laser: a preliminary report. Arch Dermatol 122: 1016–1022

[75] TAN OT, SHERWOOD K, GILCHREST BA (1989) Treatment of children with port-wine stains using the flashlamp-pulsed tunable dye laser. New Engl J Med 320: 416–421

[76] TONG AKF, TAN OT, BOLL J, PARRISH JA, MURPHY GF (1987) Ultrastructure effects of melanin pigment on target specificity using a pulsed dye laser (577 nm). J Invest Dermatol 88: 747–752

[77] TORTORI-DONATI P, FONDELLI MP, ROSSI A, BAVA GL (1999) Intracranial contrast-enhancing masses in infants with capillary haemangioma of the head and neck: intracranial capillary haemangioma? Neuroradiology 41 (5): 369–375

[78] TSANG WYW, CHAN JKC, FLETCHER CDM (1991) Kaposi-like infantile haemangioendothelioma: a distinctive vascular neoplasm of the rtroperitoneum. Am J Surg Pathol 15: 982–929

[79] URBAN P, ALGERMISSEN B (1999) Stadieneinteilung kindlicher Hämangiome nach FKDS-Kriterien. Ultraschall Med 20: 36

[80] VIN-CHRISTIAN K, McCALMONT TH, FRIEDEN IJ (1997) Kaposiform haemangioendothelioma. An aggressive, locally invasive tumour that can mimic haemangioma of infancy. Arch Dermatol 133: 1573–1578

[81] WHIMSTER IW (1976) The pathology of lymphangioma circumscriptum. Br J Dermatol 94: 473

[82] YAKES WE (1989) Alcohol embolotherapy of vascular malformation. Sem Intervent Radiol 6: 146–161

[83] YOUNG AE (1988) Venous and arterial malformations. IN: MULLIKEN JB, YOUNG AE (eds) Vascular birthmarks. Saunders, Philadelphia, pp 196–214

[84] ZUCKERBERG LR, NICKOLOFF BJ, WEISS SW (1993) Kaposiform haemangioendothelioma of infancy and childhood: anaggressive neoplasm associated with Kasabach-Merritt syndrome and lymphangiomatosis. Am J Surg Pathol 17: 321–328

# III-6.2
# Excimer Laser-Assisted Recanalization
# of Chronic Peripheral Arterial Occlusions

D. Scheinert and G. Biamino

## Contents

## Background

According to the results of the Framingham study, symptomatic peripheral arterial obstructive disease (PAOD) is at least as frequent as angina, showing an annual incidence of about 26/10 000 in men and 12/10 000 in women. The high prevalence of PAOD in the age group over 65 years and the fact that within the next 30 years this population group in the Western Hemisphere will have a 100% increment (namely from 12.4% in 1993 to 22% in 2030), illustrates the sociopolitical relevance of this disease.

For many years, surgical bypass grafting has been the only standard method to treat patients with PAOD. The expansion of the simple Dotter concept [1] by transcutaneous balloon angioplasty has dramatically changed the conception of the management of obstructive arterial disease. Technological improvements of this method during the last few years have resulted in primary success rates of 95% in attempts to recanalize stenotic peripheral arteries.

One main limitation of arterial balloon angioplasty remains, however: the inability to cross total occlusion with a length of more than 5 cm in up to 50% of the cases. This relevant disadvantage of the technique is magnified by the Achilles heel of balloon angioplasty: the high degree of restenosis or reocclusions of approximately 50% in peripheral arteries [2, 3]. Analyzing these data, the idea was conceived that with debulking of obstructive material it may be possible to reduce the degree of stenosis or to transform an occlusion in a stenosis, so that the final balloon dilatation, which remains necessary in the majority of the cases, may provoke a limited damage of the arterial wall and, consequently, may reduce the stimuli activating the mechanism of restenosis. Subsequently, a variety of new techniques designed to remove obstructive vessel material has been developed [4].

In the early 1980s the innovative idea that laser energy might be also used to vaporize sclerotic material [5–7] convinced several groups to introduce laser angioplasty as a clinical modality very quickly [8–18]. Overly enthusiastic, maybe success-dictated, reports presaged the science fiction fantasy that by aiming a laser beam through an optical fibre at the obstructive material, blood flow might be restored without spasm, embolism, dissection or perforation. The clinical debacle of the first systems using continuous-wave laser irradiation was the consequence of an uncritical use of such a powerful, inadequate energy source for debulking sclerotic vessel material. Furthermore, the partial use of the acronym laser without regard to the specific tissue-related properties of different laser sources led to many simplifications and misunderstandings and may explain the present scepticism towards laser angioplasty by a large number of interventionalists.

The main goal of our cardiovascular laser angioplasty program since November 1985 has been the development of a technology of percutaneous application of laser energy, which would remove plaque rather than mechanically reshape the obstructive material. On January 25, 1989, we successfully attempted to recanalize the occluded superficial femoral artery of a 58-year-old woman with a 7 French multifibre catheter. Since then, we have performed laser-assisted angioplasty of peripheral vessels in more than 7000 cases. In this chapter we will report our experiences in laser-assisted recanalizations of long chronic iliac and femoropopliteal occlusions.

## Technical Aspects

During the last few years, it has been demonstrated that the majority of laser wavelengths of the electromagnetic spectrum can debulk vessel material [10, 12, 13, 17–19]. Two main parameters must be considered for laser–tissue interaction effects: the interaction time (between laser beam and tissue) and the tissue-specific absorption or effective energy density [20]. Electrical fields large enough to break chemical bonds can be produced only at very high energy densities with an extremely short interaction time. This effect, called optical breakdown or photoablation, can be obtained with the pulsed excimer laser with high-energy densities [21–24].

At lower energy densities and longer interaction time, tissue ablation is a consequence of local heating with resulting dessication and subsequent vaporization [25]. Examples of these thermal effects are the continuous-wave Nd:YAG or argon laser systems. Although the laser light of the continuous-wave laser sources has been transmitted via bare fibre or with metal tips [26, 27], quartz window [28] or sapphire probes [29, 30] at the distal end of the optical fibre, heat can be regarded as the main cause of tissue vaporization of all systems. Particularly when the recanalization speed is slower than 1.5 mm$^{-1}$ s, considerable transversal temperature rise is observed [25]

and thermal damage of adjacent tissue structures can hardly be avoided. The different transmitting systems of continuous-wave laser beam may be used with a relatively low risk of damage in non-calcified lesions when the probes are able to cross the obstruction very fast [8]. Because heavily calcified material is refractory to the ablation mechanism of thermal systems [31], a recanalization stop due to calcified obstacle will cause an unacceptably high risk of perforation [22]. Furthermore, the increase of the vessel-surrounding temperature to 60–80 °C may be deleterious with regard to the long-term results of those interventions.

In contrast to the vaporization caused by the continuous-wave lasers, the excimer laser is a pulsed system which induces photoablation during a so-called athermic process [22, 32]. This photoablation phenomenon was first described by Srinivasan in 1982 [33, 34] and is related only to energy densities below 1 J cm$^{-2}$ at a wave length of 193 nm. However, the medical systems introduced for angioplasty use a xenon chloride 308-nm excimer laser as the source of energy. At this wavelength the ablation of the irradiated tissue is no longer a consequence of a photochemical disruption of molecular chains. It is predominately a local, very fast micro-explosion provoked by an extremely high temperature rise of the irradiated volume with energy densities of about 3 to 6 J cm$^{-2}$. As a result of the excimer laser beams small penetration depth and the extremely short pulse duration in comparison to the pulse repetition rate used, the thermal damage induced by the excimer laser is minimal even when high-energy densities are used [19].

## Clinical Application

### Recanalization of Chronic Iliac Artery Occlusions

Percutaneous transluminal angioplasty has been proved to be an effective technique for the treatment of symptomatic iliac artery stenoses and short occlusions [35–37]. When combined with the use of adjunctive stent placement, the immediate technical success rate improved significantly by up to 95% [38–43]. Patency rates of 80 to 90% after 5 years, reported for short iliac stenoses, are comparable to surgical results [44–46].

Long total occlusions of iliac arteries have not yet been considered as a generally accepted indication for percutaneous treatment [47–49]. Results reported with balloon dilatations of iliac occlusions showed a technical failure rate of up to 20% [50] and less favourable long-term outcome when compared with iliac stenoses [50–51]. Primary stenting after percutaneous recanalization of long iliac occlusions may contribute to an improvement of the long-term results [52–54].

Because of the relatively high complication rate, due to dislodgement of atheromatous material and distal embolization, the use of thrombolytic therapy prior to balloon angioplasty has been advocated [48, 55]. Lysis therapy, however, has not been uniformly reported as effective for chronic iliac occlusions. Alternatively, the use of percutaneous debulking techniques prior to primary stenting may improve the acute and long-term results, especially after recanalization of chronic occlusions of >5 cm in length, which are normally not considered for percutaneous procedures. Pulsed laser systems with a short absorption depth, such as the 308-nm xenon chloride excimer laser, have been shown to permit controlled photoablation of sclerotic material [8, 19].

## Study Population

We evaluated the initial and long-term results of primary stent implantation after excimer laser-assisted recanalization of chronic iliac artery occlusions. The data of 212 consecutive patients with unilateral chronic iliac artery occlusions, who were percutaneously treated in our institution between March 1992 and December 1997, were analysed. Only patients with angiographically proven iliac occlusion and stable clinical symptoms of at least 3 months duration were included. The baseline clinical characteristics of the study population are given in Table 1.

**Table 1.** Baseline characteristics and pre-interventional clinical categories of 212 patients undergoing stent-supported recanalization of iliac artery occlusions

|  | Value | % |
| --- | --- | --- |
| Age (y) | | |
| Mean ± SD | 60 ± 10.6 | |
| Range | 38–89 | |
| Sex | | |
| Male | 166 | 78.3 |
| Female | 46 | 21.7 |
| Cardiovascular risk factors | | |
| Smoking | 186 | 87.7 |
| Arterial hypertension | 109 | 51.4 |
| Hyperlipoproteinemia | 86 | 40.6 |
| Diabetes mellitus | 28 | 13.2 |
| Family history | 47 | 22.2 |
| Pre-interventional clinical categories (Rutherford classification) | | |
| 1  Mild claudication | 9 | 4.2 |
| 2  Moderate claudication | 51 | 24.0 |
| 3  Severe claudication | 126 | 59.4 |
| 4  Rest pain | 17 | 8.0 |
| 5  Minor trophic changes | 9 | 4.2 |
| 6  Major trophic changes | – | – |

According to the guidelines of the American Heart Association (AHA), the Rutherford categories were used for clinical classification [48, 49, 56]. The majority of patients presented with markedly impaired walking capacity due to claudication. Accordingly, pre-interventional standardized treadmill tests (5 min at 2 mph on 12% incline) were completed by only 9 of the 212 patients (4.2%). On the side of the occlusion, the mean ankle brachial index (ABI) before and after exercise testing was 0.53 ± 0.16 and 0.41 ± 0.18, respectively. On the contralateral leg, ABI values of 0.88 ±.0.21 at rest and 0.71 ± 0.22 after treadmill tests were found.

Pre-interventional angiography showed occlusions that involved the common iliac artery (CIA) in 67, the external iliac artery (EIA) in 74, and both vessel segments in 71 cases. The mean length of the occlusion was 8.9 ± 3.9 cm (range 3 to 19 cm). In 24 cases, a contralateral ostial stenosis of the CIA was present. Based on the criteria of the Society of Cardiovascular and Interventional Radiology [49], lesions were graded class III (<5 cm) in 46 cases and class IV (>5 cm or bilateral) in 166. A poor runoff with additional occlusion of the ipsilateral superficial femoral artery was observed in 43 patients.

## Recanalization Procedure and Stent Implantation

In 193 cases (91.0%) the initial passage of the obstruction with the guide wire was performed by the cross-over technique, whereas the retrograde approach was used in only 19 patients (9.0%).

In the case of unilateral iliac occlusion (Fig. 1), after retrograde puncture and sheath placement into the contralateral common femoral artery, the occlusion was initially passed by cross-over technique with a 0.035" (0.89 mm) hydrophilic guide wire (length 260 cm, stiff type angled tip, Terumo, Tokyo, Japan) finally placed in the superficial femoral artery. Using the guide wire as a marker, the ipsilateral common femoral artery was punctured under fluoroscopic control and a second 8 French introducer sheath was positioned. Using an angled-shaped wire loop, introduced through the ipsilateral sheath, the tip of the guide wire in cross-over position was snared and retrieved from the sheath.

Using this technique, the cross-over position of the guide wire enables the manoeuvre of a 2.5-mm multifibre laser catheter through the occlusion, minimizing the risk of perforation and avoiding the possibility at tracking the laser catheter subintimally in the area of the aortic bifurcation.

In order to achieve an optimal debulking of the occluded vessel, several retrograde passes with the laser catheter were performed. Pulsed excimer laser sys-

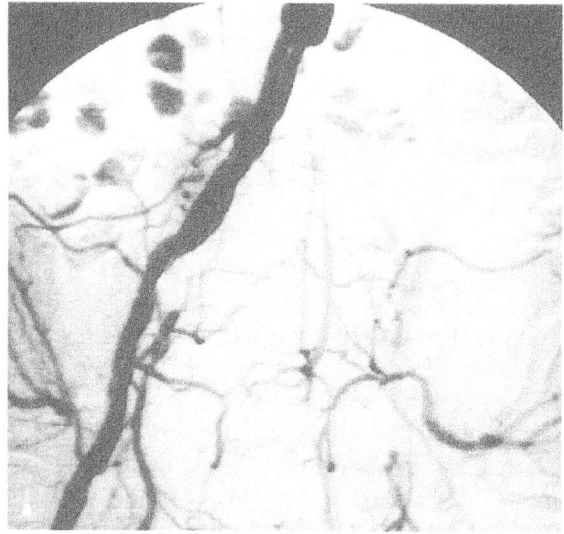

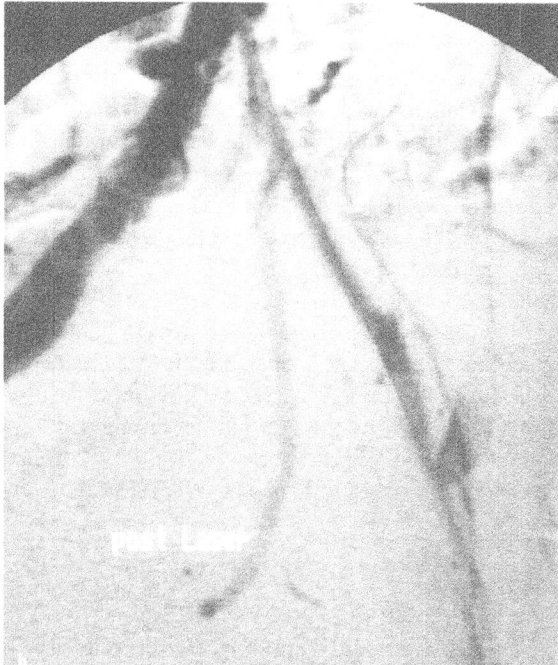

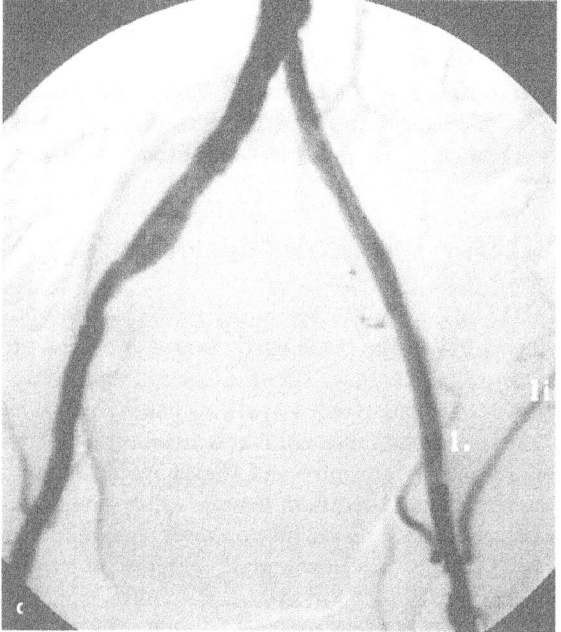

**Fig. 1a–c.** Total chronic occlusion of the left common and external iliac artery **a**. Primary laser canal after excimer laser-assisted recanalization with a 2.5-mm multifibre laser catheter **b**. Final results after implantation of three Palmaz stents (length proximal to distal: 7.6, 4 and 4; final balloon diameter: proximal 8 mm, distal 7 mm) **c**

tems, which work at a wavelength of 308 nm (LAIS DYMER 200+, pulse duration 200 ns, calibrated fluence 40–60 mJ mm$^{-2}$ or Spectranetics CVX 300, pulse duration 120 ns, fluence 45 mJ mm$^{-2}$) were used for the generation of the laser beam. In a considerable number of patients (n = 49; 70.3%) a primary laser canal with a diameter of more than 2 mm was achieved. Additional predilatations with an undersized balloon (mean diameter 6.3, range 5 to 7 mm) were performed in the majority of cases (n = 168; 79.6%) prior to stent implantation.

After excimer laser-assisted recanalization, a total number of 527 stents were implanted in 196 patients in order to stabilize the treated vessel segment.

According to the morphology and location of the lesions, different types of stents were implanted (Table 2). Palmaz stents were used in the majority of cases (132 patients), mainly for highly calcified and ostial lesions of the CIA. Self-expanding wallstents were placed to stabilize longer, less calcified vessel segments distal to the aortic bifurcation. In 31 patients, who showed major dissections after recanalization, Dacron-covered self-expanding nitinol stents were used to repair the vessel wall. In 53 patients a combination of two different stents was chosen.

With the intention of stabilizing the entire target lesion, implantation of a single stent was sufficient in only 42 patients (21.4%). In the remaining cases, mul-

**Table 2.** Endovascular stents used for stabilization of the recanalized iliac arteries

|  | No. of stents | |
|---|---|---|
|  | Value | % |
| Total number | 527 | |
| Palmaz stent (P 394)[a] | 346 | 65.7 |
| Wallstent[b] | 94 | 17.8 |
| Strecker Stent[c] | 38 | 7.2 |
| EndoPro System 1[d] / Passager[e] | 49 | 9.3 |

[a] Johnson and Johnson Interventional Systems, Warren, NJ, USA.
[b] Schneider Europe/Boston Scientific, Bülach, Switzerland.
[c] Medi-tech/Boston Scientific, Jyllinge, Denmark.
[d] Minte, Freeport, Bahama Islands.
[e] Meadox/Boston Scientific, Natick, MA, USA.

tiple stents (mean number 3.1, range 2 to 6) were implanted. The mean length of the stented segment was 10.2 ± 4.4 cm (range 3 to 22 cm). Diameters of the implanted devices ranged from 6 to 8 mm (mean 7.3 ± 0.61) mm) and were selected by comparison with the contralateral corresponding artery or with residual patent segments in order to avoid overdilatation.

Most of the stents were delivered using a retrograde approach ($n = 421$). In patients recanalized by a single cross-over approach, stents (Palmaz stent $n = 73$, Wallstent n = 26, Strecker stent $n = 7$) were implanted by the cross-over technique. In patients with proximal lesions of the CIA, which involved the aortic bifurcation and additional ostial stenosis of the contralateral CIA ($n = 24$), after initial recanalization of the occlusion the aortic bifurcation was reconstructed by bilateral simultaneous implantation of Palmaz stents by the "kissing balloon technique" [43]. In a further 32 patients the kissing balloon technique was used to protect the contralateral common iliac artery during stent implantation into the proximal CIA. In these cases, only low-pressure balloon inflations in the non-affected CIA without stent implantation were performed.

**Postprocedural Treatment**

During intervention, all patients received 10 000 units of unfractionated heparin intra-arterially, followed by intravenous heparin 1000–1200 units h$^{-1}$ for 24 h (aPTT 60–80 s). Anticoagulation was continued with 0.3 mg subcutaneous low molecular weight heparin (Fraxiparin, Sanofi Winthrop GmbH, Munich, Germany) twice daily for 2 weeks. Oral anticoagulation with phenprocoumon (INR >2.5) was given in 78 patients, which included those with low blood-flow conditions due to impaired peripheral runoff. The re-

maining patients ($n = 112$) received ticlopidine (250 mg BID for 4 weeks) concomitant to ASA (100 mg OD).

# Results

## Primary Technical Results

Out of 212 patients, the occlusion was successfully passed with a guide wire in 196 cases (92.5%). After excimer laser-assisted recanalization primary stent implantation was performed in these 196 patients. In three cases a patent vessel could not be achieved due to subintimal tracking, even after stent implantation. Furthermore, acute thrombosis immediately after recanalization and stent implantation, which could not be recanalized by local thrombolysis and PTA, occurred in two patients. In one case, an iliac vessel ruptured during stent implantation, which required emergency surgical intervention. Consequently, a primary technical success with a residual diameter stenosis <30% was achieved in 190 of the 212 patients (89.6%).

## Primary Clinical Results

A marked clinical improvement of ±3 or ±2 grades, according to the AHA guidelines, was achieved in 112 (52.8%) and 67 cases (31.6%), respectively. In 7 patients (3.3%) symptoms improved only slightly (±1 grade), and in 26 cases (12.2%), which includes the technical failures, no changes were observed. Nine of these 33 patients with no or minor clinical improvement had coexistent occlusions of the ipsilateral SFA, which were not considered for recanalization at the time. After successful recanalization, the mean ABI at rest increased from 0.53 ± 0.16 to 0.91 ± 0.13 ($p < 0.001$) and after exercise from 0.41 ± 0.18 to 0.84 ± 0.16 ($p < 0.001$).

## Complications

The overall rate of major procedure-related complications was 1.4% ($n = 3$). In addition to the patient with arterial rupture during stent implantation, one patient had embolic occlusion of the ipsilateral tibioperoneal trunk. Local rt-PA thrombolysis resulted in complete resolution of the thrombus. In another patient, who underwent stent implantation into the proximal CIA, peripheral embolization of the contralateral leg occurred. This was due to dislodgement of thrombotic material, which occurred during balloon inflation and stent placement at the aortic bifurcation. The contralateral peripheral occlusion was recanalized by PTA and concomitant application of a weight-adjusted bolus of ReoPro (absiximab, Lilly & Co. Ltd., Basingstoke, Hampshire, UK) of 0.25 mg kg$^{-1}$, followed by a standard ReoPro inusion of 0.125 g kg$^{-1}$ min$^{-1}$ for 12 h.

There were no cases of major bleeding or extensive haematoma. Minor puncture site complications occurred in 14 patients (6.6%),which included 8 haematomas (3.8%) and 6 false aneurysms (2.8%) that were successfully treated by ultrasound guided compression. Considering the fact that in 172 cases a bilateral percutaneous access was used, the total number of groins at risk was 384. Accordingly, the adjusted incidence of puncture site complications was 3.6%, which included haematomas in 2.1% and false aneurysms in 1.6%.

## Follow-Up Results

Cumulative primary, assisted primary and secondary patency rates were calculated using the Kaplan Meier life table method, which included those patients with primary interventional failure (Table 3).

In three patients with markedly impaired peripheral runoff, thrombotic re-occlusions of the target vessel occurred within the first 30 days after the intervention, which led to a 1-month primary patency rate of 88.2%. In two of these three cases, the re-occluded segment was successfully recanalized by rt-PA thrombolysis and PTA. Thus, the secondary patency rate at 1 month was 89.1%.

During the long-term follow-up 2 patients (0.9%) died due to acute cardiac events. In only 13 patients (6.1%) was deterioration of the initial clinical improvement, according to the AHA classification, observed. The mean ABI remained almost stable during the follow-up period and accounted for 0.89 ± 0.15 before and 0.81 to ± 0.16 after exercise testing at the last available follow-up (mean 27 months, range 3–67 months).

Acute or subacute re-occlusions of the stented segment were found in five patients, three of whom were successfully treated by excimer laser-assisted angioplasty. In one case the secondary intervention failed and the patient was referred for elective bypass surgery. One patient refused further intervention.

Re-stenosis (> 50%) of the stented segment was revealed by follow-up angiography in 17 cases. However, re-stenosis never involved the complete stented segment, but was located in the proximal portion in 9 patients and in the distal portion in 2 patients. In 6 patients with multiple stents, re-stenosis was located at the overlap site between the stents. In 13 of the 17 patients, re-stenosis could be successfully treated by secondary interventions.

Univariate analysis of angiographic, clinical and procedural variables was performed to identify factors that determined re-stenosis or re-occlusion. Length of the occluded vessel segment as well as length of the stented segment were the only variables associated with the occurrence of re-obstructions. There were no statistically significant differences with

**Table 3.** Cumulative patency rates (Kaplan-Meier life-table method) after stent-supported iliac recanalization

| Follow-up (months) | No. of patients | Lost to follow-up | Local events | Patency (%) | SD |
|---|---|---|---|---|---|
| Primary cumulative patency | | | | | |
| 0 | 212 | 0 | 22 | 89.6 | 2.1 |
| 6 | 176 | 16 | 4 | 87.7 | 2.3 |
| 12 | 145 | 21 | 7 | 83.9 | 2.6 |
| 24 | 110 | 34 | 4 | 81.2 | 2.8 |
| 36 | 83 | 19 | 4 | 77.9 | 3.2 |
| 48 | 54 | 28 | 2 | 75.7 | 3.5 |
| 60 | 15 | 35 | 4 | 66.1 | 5.8 |
| Assisted primary patency | | | | | |
| 0 | 212 | 0 | 22 | 89.6 | 2.1 |
| 6 | 176 | 16 | 4 | 87.7 | 2.3 |
| 12 | 149 | 21 | 3 | 86.1 | 2.4 |
| 24 | 113 | 38 | 2 | 84.7 | 2.5 |
| 36 | 83 | 24 | 1 | 83.8 | 2.7 |
| 48 | 54 | 29 | 1 | 82.4 | 3.0 |
| 60 | 15 | 38 | 1 | 77.5 | 5.5 |
| Secondary patency | | | | | |
| 0 | 212 | 0 | 22 | 89.6 | 2.1 |
| 6 | 178 | 16 | 2 | 88.7 | 2.2 |
| 12 | 153 | 21 | 1 | 88.2 | 2.2 |
| 24 | 114 | 42 | 1 | 87.5 | 2.3 |
| 36 | 83 | 25 | 1 | 86.5 | 2.5 |
| 48 | 54 | 29 | 1 | 85.1 | 2.8 |
| 60 | 15 | 38 | 1 | 80.0 | 5.6 |

regard to the distribution of baseline clinical characteristics (gender, cardiovascular risk factors, pre-interventional clinical categories) or other angiographic and procedural variables (vessel segment, stent type, number of stents, final balloon diameter, postinterventional drug treatment) for patients with and without re-stenosis.

## Discussion

After the first successful percutaneous recanalization procedure of a completely occluded iliac artery, reported by Tegtmeyer et al. in 1979 [57], several small studies have been published, all of which show a high rate of failure to pass the occlusion [47, 58]. Concomitantly, complication rates of up to 20% were reported, which raised the question as to whether an iliac artery occlusion is an appropriate indication for angioplasty [47, 59].

More recently, larger studies that investigated the initial and long-term success of percutaneous transluminal balloon angioplasty for the treatment of iliac occlusions have been published [50, 51, 60]. Johnston et al. Achieved a considerable primary recanalization rate of 82% (67 of 82 patients treated successfully); however, the long-term results were disappointing and showed a cumulative patency rate of 58% after a mean follow-up of 36 months [51].

There are three major challenges in the percutaneous treatment of chronic iliac occlusions, given in the following sections.

### Passage of the Occlusion with the Guide Wire

Safe advancement of a guide wire through the occluded segment must be achieved. Colapinto et al. Described the technical problems that can occur with mechanical traversal of an occluded artery and found that the chance of successful recanalization decreases with increased length of the occluded segment [50]. Vorwerk et al. reported that subintimal passage of the occlusion with an eccentric re-entry through the aortic wall occurred in 5.8% of the cases, which eventually may cause technical failure [52]. In contrast to the majority of authors, who prefer a primary retrograde approach, in our study the initial passage of the occlusion was performed by the cross-over technique, whenever possible (91.0%). Using this technique, we were successful in initially passing the lesion with a guide wire in 92.5% of the cases, which compares favourably with the results reported in the literature [44–47, 52]. Furthermore, in patients with occlusions of the proximal CIA, this approach may have contributed to the avoidance of major aortoiliac dissections.

### Debulking of Atherosclerotic and Thrombotic Material

A second aspect for successful recanalization of occluded iliac arteries is the fact that clot material usually extends both proximally and distally to the underlying stenotic lesion. Various techniques for removal of the occluding thrombotic material have been proposed. Several authors used local thrombolytic therapy as an adjunct to balloon angioplasty [59, 61, 62]. Blum et al. reported an excellent primary recanalization rate of 98% by the combined use of local fibrinolysis and PTA in patients with relatively new thrombotic iliac occlusions (0.1–12 months; mean 4.2 months between onset of symptoms and recanalization) [55]. Additional stent placement was performed in 18 of 47 patients (38.3%). In contrast, Hausegger et al. treated 42 patients for chronic iliac occlusions [63]. They achieved instant recanalization after local thrombolysis in only 12 patients (28%) and reduction of thrombotic mass in 5 cases (12%). The overall complication rate in both studies was approximately 15%, with a 7% rate of peripheral embolizations.

In our study, the removal of obstructing material was attempted by performing several passes with an excimer laser catheter prior to PTA and stent implantation. The excimer laser is a pulsed laser system, which induces photablation during a so-called athermic process. It is predominantly a local, very fast micro-explosion, which is provoked by an extremely high temperature rise of the irradiated volume with energy densities of about 3 to 6 J cm$^{-2}$. As a result of the excimer laser beams small penetration depth and the extremely short pulse duration, the thermal damage induced is minimal, even when high energy densities are used [8, 9]. In 70% of the cases an effective debulking with a primary laser canal of at least 2 mm was achieved, which facilitated subsequent balloon dilatation and stent implantation. As a result, the frequency of peripheral embolic events (0.9%) was considerably lower than in previously published studies [52, 55].

### Stabilization of the Recanalized Vessel Segment

Maintaining patency of the vessel lumen after transluminal angioplasty represents the third major challenge after achieving the initial technical success. Endoluminal stent placement in iliac artery stenoses has been proved to be efficient in eliminating residual stenoses and for providing a high long-term patency rate [38–40, 43]. Patency rates reported for stent placement after recanalization of iliac artery occlusions are somewhat lower than for treatment of iliac artery stenoses [52, 53, 55, 59]. Including interventional failures, we observed cumulative primary patency rates of 83.8% after one, 81.2% after 2 and 75.7% after 4 years. Vorwerk et al. [52] reported on 103 patients with chronic iliac occlusions and achieved a primary 2-year patency rate of 83% and a 4-year patency rate of 78%; however, primary interventional failures were excluded from their calculations and the mean occlusion length was only 5.1 cm. Analysing the influence of the lesion length on the long-term outcome, we found a significantly higher rate of re-occlusions and re-stenoses in the subgroup with a stented segment >10 cm in length. However, secondary interventions within the stent were successfully performed in almost all of the patients with re-obstructions, independent of the lesion length. As a result, the secondary patency rate showed no significant differences between the two subgroups.

## Recanalization of Long Chronic Occlusions of the Superficial Femoral Artery

Chronic atherosclerotic obstructions of the superficial femoral artery are a leading cause of life-style-limiting intermittent claudication, which occurs in about 6–10% of the population over 65 years of age [64] percutaneous transluminal angioplasty is normally recommended as the primary treatment for short-segment femoropopliteal stenoses and occlusions. In contrast, long chronic occlusions of the superficial femoral artery are still mainly considered for vascular surgery [48, 49].

Since bypass grafting is associated with a considerable procedure-related morbidity and mortality, surgical intervention is usually preserved for patients with ischemic rest pain or very advanced claudication [65]. Consequently, many patients with long chronic superficial femoral artery occlusions remain untreated, systematically underestimating the subjective discomfort induced by the disease for the single patient.

As transcutaneous revascularization techniques permit a lower threshold of intervention than has been traditionally practised for surgical procedures, the improvement of endovascular recanalization techniques for treatment of total occlusions is particularly desirable.

The pulsed excimer laser has been extensively evaluated in debulking atherosclerotic material in vitro and in vivo, demonstrating that the photo-ablative effect of laser light can be used to recanalize even lesions not amenable to conventional PTA [14, 24, 66–68].

## Study Population

In this study we analysed the data of 318 consecutive patients with occlusions of the superficial femoral arteries which were percutaneously treated by excimer laser-assisted angioplasty in our institution during a 1-year period between January 1 and December 31, 1996.

Only patients with long chronic occlusions of the superficial femoral artery, which normally would not be considered for perutaneous treatment, were included in this analysis. Accordingly, the following inclusion criteria applied:
- angiographically proven chronic occlusions of the superficial femoral artery at least 10 cm in length,
- history of claudication of at least 6 month's duration without deterioration of symptoms.

The baseline clinical characteristics of the study population are listed in Table 4. On average, there were 2.7 cardiovascular risk factors per individual, with smoking, arterial hypertension and hyperlipoproteinemia being the most frequent predisposing factors.

According to the guidelines of the American Heart Association (AHA), the Rutherford categories were used for clinical classification [48, 49, 56]. The majority of patients presented with markedly impaired walking capacity due to claudication. Accordingly, preinterventional standardized treadmill test (5 min at 2 mph on a 12% incline) was completed by only 21 of 318 patients (6.6%). Critical lower limb ischemia

**Table 4.** Baseline data of 318 patients with a total of 411 long SFA occlusions undergoing excimer laser-assisted recanalization

|  | $n$ | % |
|---|---|---|
| No. of patients | 318 | |
| Male | 207 | 65.1 |
| Female | 111 | 34.9 |
| Mean age (years) | 64.2 ± 10.7 (range 33–91) | |
| Cardiovascular risk factors | | |
| Smoking | 240 | 75.5 |
| Arterial hypertension | 222 | 69.8 |
| Hyperlipoproteinemia | 168 | 52.8 |
| Diabetes mellitus | 99 | 31.1 |
| Family history | 89 | 27.9 |
| Cerebrovascular disease | 36 | 11.3 |
| Clinical categories (Rutherford classification) | | |
| 1  Mild claudication | 21 | 6.6 |
| 2  Moderate claudication | 24 | 7.5 |
| 3  Severe claudication | 252 | 79.2 |
| 4  Rest pain | 6 | 1.9 |
| 5  Minor trophic changes | 15 | 4.7 |
| 6  Major trophic changes | – | – |
| Relative walking capacity | 102 ± 92.5 m | |
| Absolute walking capacity | 142 ± 90.0 m | |
| ABI at rest | 0.62 ± 0.15 | |
| ABI after exercise | 0.40 ± 0.18 | |

**Table 5.** Lesion characteristics of 411 total SFA occlusions

|  | $n$ | % |
|---|---|---|
| No. of lesions | 411 | |
| Right | 199 | 48.4 |
| Left | 212 | 51.6 |
| Bilateral lesions (pts.) | 93 | 29.2 |
| Involvement of vessels segment | | |
| Proximal SFA | 255 | 62.0 |
| Medial SFA | 381 | 92.7 |
| Distal SFA | 369 | 89.8 |
| Length of occlusion | 19.4 ± 6.0 cm (range 10–31 cm) | |
| Length of proximal SFA stump | 4.6 ± 5.5 cm (range 0–20 cm, median 2 cm) | |
| Runoff | | |
| One vessel | 97 | 23.6 |
| Two or three vessels | 314 | 76.4 |

with rest pain or minor tissue loss was present in 6 and 15 patients, respectively. The mean ankle brachial index (ABI) determined at the index leg before and after exercise was 0.62 ± 0.15 and 0.40 ± 0.18, respectively.

In total, 411 occluded vessels were interventionally treated in 318 patients (Table 5). In 93 cases bilateral occlusions of the SFA were recanalized during separate procedures. The mean length of the occlusion was 19.4 cm, almost always involving the medial and distal segment of the SFA. The length of the patent proximal stump was 4.6 cm on average; however, in 53 cases the artery stump was shorter than 1 cm.

### Recanalization Procedure

Laser-assisted recanalization was performed using pulsed XeCl-excimer laser systems, working at a wavelength of 308 nm: Spectranetics CVX 300, pulse duration 120 ns, fluence 45 mJ mm$^{-2}$ or LAIS DYMER 200+; pulse duration 200 ns calibrated fluence 40–60 mJ mm$^{-2}$.

In the majority of cases (89.7%) the primary recanalization procedure was attempted by contralateral approach, starting with retrograde femoral puncture and introduction of a F6-8 haemostatic sheath. After cross-over placement of a hydrophilic guide wire (Terumo 0.035 0.89 mm, stiff type, angled or straight tip, Terumo Inc., Tokyo, Japan) into the origin of the occlusion, two different techniques were used to cross the lesion: firstly, the guide-wire was navigated through the lesion, supported by a multipurpose guiding catheter (F4-5, Cordis, Miami Lakes, FL), to an angiographically proven intraluminal popliteal position. Then the activated laser catheter was ad-

vanced over the wire (over-the-wire technique). Alternatively, to enter the occlusion or to pass a segment resistant to guide-wire crossing, the activated laser catheter was stepwise advanced for a short (<5 mm) distance without wire guidance, followed by further crossing with the guide wire (step-by-step technique).

Fluoroscopic "road mapping" as used throughout to verify the alignment of guide wires and catheters to the vessel lumen. Particular attention was given to thoroughly flushing the vessel with saline before lasering to remove remaining contrast medium. In all cases, the advancement of the activated laser catheter was performed very slowly, not exceeding 1 mm$^{-1}$ s. After initial laser passage, the 0.035" (0.89 mm) hydrophilic guide wire was changed with a 0.018" (0.46 mm) guide wire (High-Torque, Mallinckrodt Medical Inc., St. Louis, MO) to allow distal saline flushing during a second or third passage of the obstructed area. Angiographic controls of the intervention site and the distal runoff were performed before the recanalization was completed by complementary dilatations with low-profile balloons of 5 to 7 mm diameter and 4 to 8 cm length (Smash, Schneider Europe, Bülach/CH) at 6 to 14 atm for 0.5 to 3 min. Finally, the result was angiographically controlled, including the lower limb runoff (Fig. 2). The haemostatic sheath was removed after completion of the intervention.

Alternatively to the cross-over technique, the antegrad approach was used in selected cases with very calcified lesions of the distal superficial femoral artery. A popliteal approach was performed in 15 patients as a primary technique to recanalize long-segment superficial femoral artery occlusions without patent proximal stump. However, this technique was mainly used as a secondary approach after failed cross-over recanalization. To achieve a safe puncture of the popliteal artery, an F4 catheter was first placed in the external iliac artery via ipsilateral or contralateral femoral access. This patients were then turned to a prone position. The popliteal puncture was performed with help of roadmap fluoroscopy after injection of contrast media through the femoral access. After introduction of an F6 sheath into the distal superficial femoral artery or the first segment of the popliteal artery, laser catheters of 1.7 to 2.0 mm diameter were used to retrogradely recanalize the occlusion in over-the-wire technique. Implantation of endovascular stents after recanalization of superficial femoral artery occlusions was considered in 30 patients (7.3%) with threatening acute reocclusion due to extensive dissections or marked wall instability. In total, 87 stents were placed. For cross-over implantation of Palmaz stents a 7 French guiding catheter Cordis, Miami Lakes, FL) was used to facilitate passage around the aortic bifurcation.

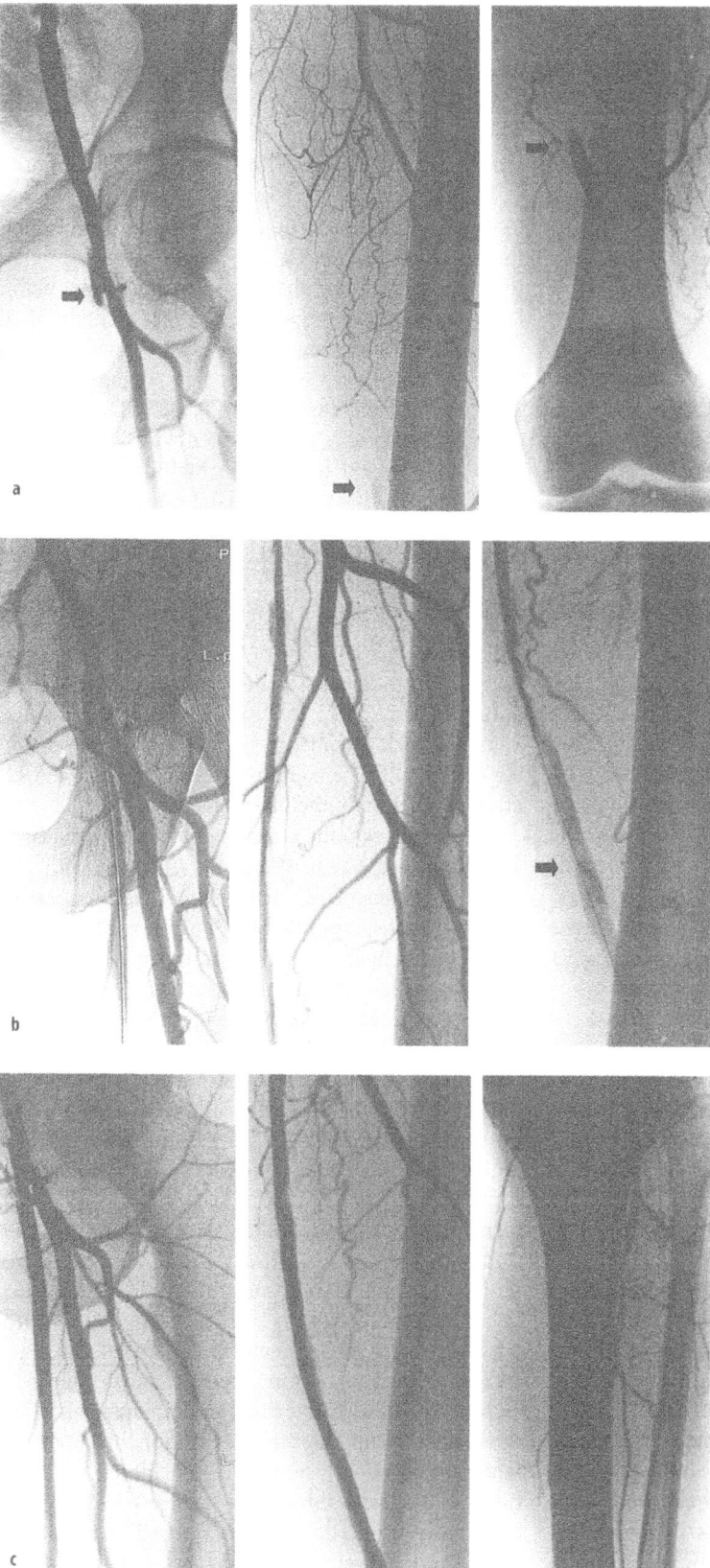

**Fig. 2a–c.** Long chronic occlusion of the left superficial femoral artery **a**. Angiographic image after successful recanalization with a 2.5-mm multifibre laser catheter (two passes) **b**. Final result after balloon dilatation (diameter 5 mm, length 80 mm, dilatation pressure 10 bar) **c**.

## Postprocedural Treatment

During intervention, all patients received 10 000 units of unfractionated heparin intra-arterially, followed by intravenous heparin 1000–1200 units$^{-1}$ for 24 h (aPTT 60–80 s): Anticoagulation was continued with subcutaneous low molecular weight heparin (Fraxiparin, Sanofi Winthrop GmbH, Munich, Germany) 0.3 mg BID for 2 weeks. Oral anticoagulation with phenprocoumon (INR 2.2–2.7) was given to 13 patients (4.1%) with markedly impaired distal runoff. All patients were put on ASA (100 mg OD) as long-term medication.

## Results

### Primary Technical Results

Table 6 details the technical results of the initial recanalization procedures. In 342 of 411 cases (83.2%) the primary approach to cross the occlusion was successful. In 44 of the 69 initially failed procedures (63.8%) a secondary procedure was performed, mainly utilizing the transpopliteal technique; 68.2% of secondary recanalization procedures were completed successfully. Thus, the total primary recanalization rate was 90.5%.

The average diameter of the laser catheter used for recanalization was 2.18 ± 16 mm. Balloon dilatations were performed in all 372 cases after successful initial laser passage, with a maximal balloon diameter of 5.8 ± 0.43 mm.

Implantation of endovascular stents into the SFA was performed in 30 cases with threatening re-occlusion due to extensive dissections or marked wall instability after excimer laser-assisted angioplasty. In total 87 stents were implanted with a mean length of the stented segment of 11.3 ± 3.3 cm.

## Complications

There were no serious procedure-related adverse events like death, amputation or acute surgical intervention (Table 7). Relevant complications involving the target lesion of the intervention or the distal runoff vessels were encountered in 29 cases (7.1%). Distal embolization or periinterventional thrombosis of the infragenouidal arteries occurred in 16 cases. Using adjunctive interventional techniques including mechanical recanalization of the occlusion and catheter-based local thrombolysis with rt-PA (10 mg bolus, followed by continuous infusion of 1 mg-$^1$ for 24 h) a complete resolution of the thrombus could be achieved in 11 cases. In the remaining 5 patients, pa-

**Table 6.** Interventional data of excimer laser-assisted recanalization of SFA occlusions in 411 cases

|  | $n$ | % |
| --- | --- | --- |
| Primary approach | | |
| Cross-over | 369 | 89.7 |
| Antegrad | 27 | 6.6 |
| Popliteal | 15 | 3.6 |
| Primary approach successful | 342 | 83.2 |
| Primary approach failed | 69 | 16.8 |
| Secondary approach | | |
| Secondary approach attempted | 44 | 63.8$^a$ |
| Cross-over | 2 | |
| Antegrad | 3 | |
| Popliteal | 39 | |
| Secondary approach sucessful | 30 | 68.2$^b$ |
| Cross-over | 1 | |
| Antegrad | 2 | |
| Popliteal | 27 | |
| No secondary approach | 25 | 36.2 |
| ELA success (total) | 372 | 90.5 |
| Size of laser catheter (mm) | | |
| 2.0 | 117 | 28.5 |
| 2.2 | 234 | 56.9 |
| 2.5 | 60 | 14.6 |
| Final balloon diameter (mm) | | |
| 5 | 78 | 21.0$^c$ |
| 6 | 288 | 77.4 |
| 7 | 6 | 1.6 |
| Implantation of stents | | |
| No. of cases | 30 | 7.3 |
| No. of stents | 87 | |
| Wallstents | 5 | |
| Palmaz stents | 81 | |
| Length of stented segment | 11.3 | |
| | (range 4–24 cm) | |

$^a$ % of primarily failed cases ($n = 69$).
$^b$ % of secondary attempts.
$^c$ PTA performed in 372 cases after successful laser passage.

tency of only one runoff vessel could be achieved; however, the patients were postinterventionally completely asymptomatic.

Minor puncture site complications occurred in 27 patients (6.5%), including 12 haematomas (2.9%) and 15 false aneurysms (3.6%), all of which could be successfully treated by ultrasound-guided compression. In none of the cases did a surgical intervention became necessary.

### Primary Clinical Results

The postinterventional clinical state according to the Rutherford scale is given in Table 8. A marked clinical improvement of +2 or more according to the limb status grading scale was observed in 247 patients. Minor clinical improvement (+1) occurred in 26 cases and in 45 patients no changes in clinical status were ob-

**Table 7.** Interventional complications and serious adverse events

|  | n | % |
|---|---|---|
| Death | 0 |  |
| Amputation | 0 |  |
| Acute surgical intervention | 0 |  |
| An target lesion |  |  |
| Acute reocclusion | 4 | 1.0 |
| Perforation | 9 | 2.2 |
| Embolisation / distal thrombosis | 16 | 3.9 |
| At puncture site |  |  |
| False aneurysm | 15 | 3.6 |
| Haematoma (reduced Hb > 3mg/dl$^{-1}$) | 12 | 2.9 |

**Table 8.** Postinterventional clinical data

|  | n | % |
|---|---|---|
| Clinical category |  |  |
| (Rutherford classification) |  |  |
| 0 | 219 | 68.8 |
| 1 | 53 | 16.6 |
| 2 | 26 | 8.2 |
| 3 | 17 | 5.3 |
| 4 | – | – |
| 5 | 3 | 0.9 |
| 6 | – | – |
| ABI at rest | 0.92 ± 0.15 |  |
| ABI after exercise | 0.87 ± 0.17 |  |

served. The group of patients with no or minor improvement included those with primary interventional failures ($n = 33$ including 6 cases with bilateral failure) as well as patients with limitations of the walking capacity due to contralateral ($n = 28$) iliac or femoral obstructions. The remaining 10 patients showed diffuse obstructions in the ipsilateral infragenouidal arteries which, despite treatment ($n = 6$), remained symptomatic.

A postinterventional treadmill test was completed by 272 patients (85.4%) without occurrence of claudication. As compared to the pre-interventional reference value, the ABI showed a significant increase from 0.62 ± 0.15 to 0.95 ± 0.15 (at rest) and 0.40 ± 0.18 to 0.87 ± 0.17 (after exercise).

## Follow-Up

Follow-up was continued over a 1-year period using clinical examination, standardized treadmill test and transcutaneous colour-coded duplex ultrasound. Cumulative primary, assisted primary and secondary patency rates were calculated, including those patients with primary interventional failures (Table 9).

In total, during 1 year 257 cases with re-obstructions were observed, with the majority of events being recorded during the second half of the observation period. Because of the close monitoring during follow-up, in a considerable number of resymptomatic patients re-stenotic lesions could be detected early. Thus, the majority of re-obstructed vessels ($n = 158$) could be successfully retreated before eventual re-occlusion occurred, resulting in an assisted primary patency rate of 64.6% after 1 year. Furthermore, in 40 cases with partial or total re-occlusions of the treated vessels segment, successful re-interventions could be performed, accounting for another 15.6% of all re-obstructions. In consequence, the secondary patency rate after 1 year was 75.1%.

## Discussion

After more than 20 years of experience with angioplastic procedures in the femoropopliteal segment, a number of factors that directly affect the success of the procedure have been identified. Short lesion length, minimal vascular disease elsewhere with good peripheral runoff, claudication as opposed to limb-threatening ischemia as the presenting symptom, stenosis rather than occlusion and absence of diabetes all correlate with improved patency [69, 70].

Unfortunately, in the femoropopliteal arteries, as opposed to the iliac system, occlusions predominate by a factor of at least 3. Furthermore, most femoropopliteal occlusions are long and coexistent multilevel atherosclerotic disease is frequent. In consequence, patients with long chronic occlusions of the superficial femoral artery, which have been investigated in this study, cannot be considered good candidates for percutaneous recanalization and surgical revascularization is proposed to be the treatment of choice.

According to a metanalysis by Dalman and Taylor, 4-year primary patency rates of surgical above-knee revascularization varied from 60 to 70%. For below-knee procedures, the 4-year secondary patency ranged from 70% for venous grafts to 40% for PTFE grafts [65].

Despite these good long-term results, even at the infra-inguinal level vascular surgery is associated with a substantial risk for the patient. Mortality for femoropopliteal bypasses varies from 3 to 12% in the literature. Accordingly, it is common practice that patients with long total SFA occlusions often do not undergo surgery before they have reached at least category 3 of chronic limb ischemia [48, 49]. As percutaneous procedures are generally associated with a significantly lower rate of general and local complications, effective transluminal recanalization tech-

**Table 9.** Follow-up data – excimer laser-assisted recanalization of long SFA occlusions

| Follow-up | Lesions at risk | Events | Lost to follow-up | Event-free survival | SD |
|---|---|---|---|---|---|
| Primary patency | | | | | |
| 0 | 411 | 39 | 2 | 0.905 | 0.014 |
| 3 | 370 | 34 | 18 | 0.822 | 0.019 |
| 6 | 318 | 66 | 20 | 0.651 | 0.024 |
| 9 | 232 | 109 | 8 | 0.345 | 0.025 |
| 12 | 115 | 48 | 67 | 0.201 | 0.021 |
| Assisted primary patency | | | | | |
| 0 | 411 | 39 | 2 | 0.905 | 0.014 |
| 3 | 370 | 31 | 18 | 0.829 | 0.018 |
| 6 | 321 | 36 | 22 | 0.736 | 0.022 |
| 9 | 263 | 28 | 12 | 0.658 | 0.024 |
| 12 | 223 | 8 | 219 | 0.646 | 0.024 |
| Secondary patency | | | | | |
| 0 | 411 | 39 | 2 | 0.905 | 0.014 |
| 3 | 370 | 19 | 18 | 0.859 | 0.017 |
| 6 | 333 | 20 | 22 | 0.807 | 0.020 |
| 9 | 291 | 16 | 12 | 0.763 | 0.021 |
| 12 | 263 | 4 | 259 | 0.751 | 0.022 |

niques would offer a treatment option for many patients with long SFA occlusions who currently remain untreated.

## Initial Recanalization Procedure

The first major challenge in percutaneous treatment of chronic superficial femoral artery occlusions is to achieve a safe initial passage of the occlusion. Using conventional recanalization techniques with guiding catheter-supported navigation of a guide wire through the occlusion followed by PTA, initial technical failure rates of 18 to 26% have been reported for chronic SFA occlusions [69, 70].

Since the first clinical feasibility study of percutaneous transluminal laser angioplasty of peripheral arteries using a continuous-wave laser was reported in 1984 [71], many different laser sources and catheter delivery systems have been developed. A number of randomized trials were performed in the early 1990s comparing these different laser sources and application systems with conventional PTA. Although none of the studies could demonstrate a statistically significant superiority of laser recanalization alone as compared to PTA, the combined treatment with guide wire and laser resulted in technical success rates of up to 91% [72, 73].

In the present study, a total recanalization rate of 90.5% could be achieved using the excimer laser-assisted recanalization technique. Considering the fact that with 19.4 cm vs. 7.8 cm the average occlusion length in this study was considerably longer than in the report by Lammer at al. [72], our technical success rate compares favourably with these results and in substantially higher than in previously published

studies without laser assistance [69, 70]. The following technical modifications, which may have contributed to the high success rate, have to be addressed:

1. In most of the cases the step-by-step-technique, which has been described in detail in the methods section, was applied to initially cross the occluded vessel segment. This technique was developed to safely navigate the guide wire through the occlusion using the guidance and ablative force of the laser catheter. The technique was particularly beneficial to enter flush occlusions without visible proximal stump or to pass a segment resistant to guide-wire crossing. Furthermore, the routine use of this technique may have contributed to the low rate of relevant arterial wall dissections with a resulting stent frequency of only 7.3%.

2. In 89.7% of the cases the cross-over technique was used as the primary approach to cross the occlusion. Using this technique, peri- and postinterventional blood flow reduction at the treated vessel segment due to obstruction of the ipsilateral common femoral artery by the introducer sheath or postinterventional compression bandage could be prevented. As a result, early re-occlusions of the recanalized vessel, which have been reported in the literature with a frequency up to 41% [74], were extremely rare in our study (1.0%). Furthermore, this approach allows the recanalization of occlusions extending to the origin of the superficial femoral artery, which would not be accessible using the antegrad technique.

3. Whereas the primary interventional approach was successful in 83.2% of the cases, the application of alternative secondary approaches contributed to an increment of the overall technical success rate by 7.3%. Particularly the popliteal access was

highly effective to recanalize occlusions which had not been able to pass using the antegrad way. Puncture of the popliteal artery under fluoroscopic control using the roadmap technique was safe with only minor puncture site complications occurring in four cases (10.2%). However, a popliteal approach should be attempted only if the distal superficial femoral artery as well as the popliteal artery is free of stenoses and if there is a sufficient distal runoff. The size of the introducer sheath should not exceed 6 French.

In general, the excimer laser-assisted angioplasty of long chronic superficial femoral artery occlusions has been proven to be a safe technique. There were no major adverse events.

Furthermore, all minor procedure-related complications could be treated without surgical intervention.

In almost all cases, a successful recanalization was associated with a marked improvement of the clinical symptoms. Postinterventionally, 85.4% of all patients referred for percutaneous recanalization of long superficial femoral artery occlusions were able to complete 5 min standardized treadmill test without claudication, which is equivalent to a pain-free walking distance of almost 300 m. In 68.8% of the patients, complete relief of symptoms with an unlimited walking capacity could be achieved.

## Long-Term Results

Maintaining long-term patency of the recanalized vessel is another major aspect for successful treatment of long chronic superficial femoral artery occlusions. By removing as much occluding atherosclerotic and thrombotic material as possible, a total occlusion can be converted into a stenosis, facilitating subsequent balloon dilatation and reducing the risk of thrombo-embolic events. Furthermore, it is the concept to limit the arterial wall stress during balloon inflation which subsequently may lead to a reduction of the re-stenosis rate. In our study, debulking of obstructing material was attempted by performing several passes with an excimer laser catheter prior to balloon dilatation. The excimer laser is a pulsed laser system which induces photo-ablation during a so-called athermic process. It is predominantly a local, very fast micro-explosion provoked by an extremely high temperature rise of the irradiated volume with energy densities of about 3 to J cm$^{-2}$. As a result of the excimer laser beam's small penetration depth and the extremely short pulse duration, the thermal damage to the vessel wall induced by the excimer laser is minimal even when high-energy densities are used. Accordingly, in the randomized comparison of excimer laser-assisted angioplasty, continuous-wave Nd:YAG laser angioplasty and conventional PTA, published by Lammer et al. [72], the 12-month angiographic patency rate after recanalization of long superficial femoral artery occlusions (>8 cm in length) was best in the excimer laser group (42% vs. 24% for Nd:YAG and 12% for PTA).

As the calculation of patency rates based on recurrence of clinical symptoms tends to overestimate patency rates, in our study long-term results were evaluated using a standardized treadmill test with calculation of the Doppler indices at rest and after exercise. Duplex-derived peak systolic velocity ratios have been shown to accurately reflect the anatomic status of the arterial tree [75, 76]. Accordingly, serial colour-flow duplex scanning was used for long-term surveillance after recanalization and angioplasty. In the case of suspected re-stenosis, angiography was performed to confirm the re-obstruction.

Based on a standard life-table survival analysis the primary patency rate after 1 year was 20.1%. Due to the fact that a considerable number of patients without clinical symptoms did not attend the follow-up examinations, the calculated primary patency rate may have been biased towards less favourable results. However, recurrent disease after femoropopliteal angioplasty is a relevant problem and restenoses seems to be even more frequent after recanalization of very long chronic superficial femoral artery occlusions.

By performing an aggressive surveillance program including functional clinical testing as well as colour-coded duplex ultrasound, re-stenoses can be detected early. Re-interventions on the target lesion could be performed in the majority of cases on an outpatient basis. As a result, eventual re-occlusion could be prevented in most of patients, reflected in an assisted primary patency rate of 64,6% after 1 year. The performance of repeat recanalization procedures on re-occluded arteries contributed to a further 10.5% improvement in vessel patency, resulting in 75,1% of patients being free of symptoms after 1 year.

## Conclusion

### Chronic Occlusions of Iliac Arteries

There are several problems associated with the recanalization of chronic iliac occlusions, which have to be solved to achieve a favourable result. In most of the cases, especially using the cross-over approach, it is possible to pass the occluded vessel segment successfully without creating major dissections. With reference to the literature, at least in some cases it seems to be possible to remove some thrombotic material by means of local thrombolysis. However, thrombolytic therapy is not universally efficient and exposes the patient to additional risks. In our experience, excimer laser recanalization offers an efficient debulking technique and substantially reduced the risk for the patient.

By comparing the patency rates achieved with stent-supported recanalization of chronic iliac occlusion with the results reported with PTA alone [50, 51], there seems to be clear evidence that primary stenting provides better long-term results after recanalization of iliac artery occlusions. Although there is a potential risk of re-stenosis, repeat recanalization of the re-obstructed segment is possible in the majority of cases. In consequence, the assisted primary and secondary patency rates are almost similar to the mid-term surgical results.

In our opinion, primary stent implantation after excimer laser-assisted recanalization of chronic iliac occlusions is a safe and effective technique and provides a real alternative to surgical aortofemoral bypasses. This procedure should be considered as a first-line treatment for unilateral iliac occlusions, before vascular surgery is performed.

## Long Femoropopliteal Occlusions

The combination of excimer laser technology with current interventional devices and advanced recanalization techniques allows recanalization of long chronic superficial femoral artery occlusions in more than 90% of the cases. The successful revascularization is associated with excellent immediate clinical results. A complete relief of symptoms or at least a significant improvement in functional capacity could be achieved in almost all patients.

The high frequency of re-stenoses remains the major limitation of this interventional technique. An aggressive surveillance program with standardized treadmill test and colour-coded duplex sonography is essential to detect re-stenoses. Re-interventions should be performed early to prevent eventual re-occlusion and to maintain the achieved clinical benefit.

## References

[1] DOTTER CT, JUDKINS MP (1964) Transluminal treatment of arteriosclerotic obstruction. Description of a new technique and the preliminary report of is application. Circulation 30: 654

[2] HEWES RC, WHITE RI, MURRAY RR, KAUFMAN SL, CHANG R, KADIR S, KINNISON ML, MITCHELL SE, AUSTER M (1986) Long-term results of superficial femoral artery angioplasty. Am J Roentgenol 146: 1025–1029

[3] COLAPINTO RF, HARRIES-JONES EP, JOHNSTON KW (1980) Percutaneous transluminal angioplasty of peripheral vascular disease: a two-year experience. Cadiovasc Intervent Radiol 3: 213–218

[4] WALLER BF, CRACKERS, BREAKERS, STRETCHERS, DRILLERS, SCRAPERS, SHAVERS, BURNERS, WELDERS, MELTERS (1989) The future treatment of atherosclerotic coronary artery disease? A clinical-morphologic assessment. J Am Coll Cardiol 13: 969–987

[5] CHOY DSJ (1988) History of lasers in medicine. Thorac Cardiovasc Surg 36: 114–117

[6] CHOY DSJ, STERTZER SH, MYLER RK, MARCO J, FOURNAIL G (1984) Human coronary laser recanalization. Clin Cardiol 7: 377–381

[7] FORRESTER JS, LITVACK F, GRUNDFEST WS (1986) Laser angioplasty and cardiovascular disease. Am J Cardiol 1986: 57: 990–992

[8] CUMBERLAND DC, TAYLER DI, WELSH CL, MOORE DJ, WELSH CL, GREENFIELD AJ, GUBEN JK, RYAN TJ (1986) Percutaneous laser thermal angioplasty: initial clinical results with a laser probe in total peripheral arterial occlusions. Lancet ((Band-Nr.??)): 1457–1459

[9] SANBORN TA, CUMBERLAND DC, GREENFIELD AJ, WELSH CL, GUBEN JK (1988) Percutaneous laser thermal angioplasty: initial results and 1-year follow-up in 129 femoropopliteal lesions. Radiology 168: 121–125

[10] NORDSTROM LA, CASTANEDA-ZUNIGA WR, LINDEKE CC, RASMUSSEN TM, BURNSIDE DK (1988) Laser angioplasty: controlled delivery of argon laser energy. Radiology 167: 463–465

[11] GINSBURG R, WEXLER L, MITCHELL RS, PROFITT D (1985) Percutaneous transluminal laser angioplasty for treatment of peripheral vascular disease: clinical experience with 16 patients. Radiology 156: 619–624

[12] SANBORN TA, FAXON DP, HAUDENSCHILD CC, RYAN TJ (1985) Experimental angioplasty circumferential distribution of laser thermal injury with a laser probe. J Am Coll Cardiol 5: 934–938

[13] WELCH AJ, BRADLEY AB, TORRES JH, MOTAMEDI M, GHIDONI JJ, PEAREE JA, HUSSEIN H, O'ROURKE RA (1987) Laser probe ablation of normal and atherosclerotic human aorta in vitro: a first thermographic and histologic analysis. Circulation 76: 1353–1363

[14] ABELA GS (1988) Laser arterial recanalization: a current perspective. J Am Coll Cardiol 12: 103–105

[15] CREA F, DAVIS G, MCKENNA W, PASHAZADE M, TAYLER K, MASERI A (1986) Percutaneous laser recanalization of coronary arteries. Lancet 1986 198: 214–215

[16] GINSBURG R (1988) Percutaneous laser angioplasty in the treatment of peripheral vascular disease. Thorac Cardiovasc Surg 36 (Suppl 2): 142–145

[17] ABELA GS, NORMANN S, FELDMAN RL, GEISER EA, COHEN D, CONTI CR (1982) Effects of carbon dioxide, Nd:YAG and argon laser radiation on coronary atheromatous plaques. Am J Cardiol 50: 1199–1205

[18] GESCHWIND HJ, BOUSSIGNAC G, TEISSEIRE B, BENHAIEM N, BITTOUN R, LAURENT D (1984) Conditions for effective Nd:YAG laser angioplasty. Br Heart J 52: 484–489

[19] BIAMINO G, DÖRSCHEL K, HARNOSS BM, KAR H, MÜLLER 1988 Experience in excimer laser photoablation of atherosclerotic plaques. In: BIAMINO G, MÜLLER GJ (eds) Advances in laser medicine I. 1st German Symp on Laser Angioplasty. ecomed, Berlin, pp 147–156

[20] BERLIEN HP, MÜLLER GJ 1988 Laser in medicine. In: BIAMINO G, MÜLLER GJ (eds) Advances in laser medicine I. 1st German Symp on Laser Angioplasty. ecomed, Berlin, pp 45–55

[21] SRINIVASAN R, BRAREN B, DREYFUS RW, HADEL L, SEEGER DE (1986) Mechanism of the ultraviolet laser ablation of polymethyl methacrylate at 193 an 248 nm: laser-induced fluorescence analysis, chemical analysis, and doping studies. J Opt Soc Am 3: 785–791

[22] BIAMINO G 1990 Coronary and peripheral laser angioplasty. In: B. Meier (ed) Interventional cardiology. Hogrefe & Huber, Göttingen, pp 243–260

[23] LINSKER R, SRINIVASAN R, WYNNE JJ, ALONSO DR (1984) Far ultraviolet laser ablation of atherosclerotic lesions. Lasers Surg Med 4: 201–206

[24] ISNER JM, DONALDSON RF, DECKLEBAUM LI, CLARKE RH, LALIBERTE M, UCCI AA, SALEM DN (1985) The excimer laser: gross, light microscopic and ultrastructural analysis of potential for use in laser therapy of cardiovascular disease. J Am Coll Cardiol 6: 1102–1109

[25] DÖRSCHEL K, BIAMINO G, BRODZINSKI T, AXEL T, MÜLLER G (1988) Comparison of the feasibility of laser angioplasty using heater probes, sapphire tips, and bare fibers. Eur Heart J 9 (Suppl): 331

[26] ABELA GS, FENECH A, CREA F, CONTI CR (1985) Hot tip: another method of laser recanalization. Lasers Surg Med 5: 327–335

[27] HUSSEIN H (1986) A novel fiberoptic laser probe for treatment of occlusive vessel disease. Opt Laser Technol Med 6: 59–66

[28] COTHREN RM, HAYES GB, KRAMER JR, SACKS B, KITRELL C, FELD MS (1986) A multifiber catheter with an optical shield for angiosurgery. Laser Life Sci 1: 1–12

[29] FOURRIER JL, BRUNETAUD JM, PRAT A, MARACHE P, LABLANCHE JM, BERTRAND ME (1987) Percutaneus angioplasty with a sapphire tip. Lancet 105

[30] GESCHWIND HJ, BLAIR JK, MONGOLSMAI D, KERN MJ, STERN J, DELINOGUL U, KENNEDY HL (1987) Development and experimental application of contact probe catheter for laser angioplasty. J Am Coll Cardiol 9: 101–107

[31] BIAMINO G, KAR H, HARNOSS BM, DÖRSCHEL K, MÜLLER G 1988 Feasibility of Nd:YAG laser angioplasty. In: BIAMINO G, MÜLLER GJ (eds) Advances in laser medicine I. 1st German Symp on Laser Angioplasty. ecomed, Berlin, pp 134–140

[32] PACALA TJ, MCDERMID IS, LAUDENSLAGER JB (1984) Ultranarrow linewidth, magnetically switched, lon pulse, xenon chloride laser. Appl Phys Lett 44: 658–660

[33] SRINIVASAN R, MAYNE-BAUTON R (1982) Self-developing photoetching of poly films by far ultraviolet excimer laser radiation. Appl Phys Lett 4: 576–578

[34] SRINIVASAN R, LEIGH W (1982) Ablative photodecompensation: action of far ultraviolet (193 nm) laser radiation on poly films. J Am Chem Soc 104: 6784–6785

[35] TEGTMEYER CJ, HARTWELL GD, SELBY JB, ROBERTSON R JR, KRON IL, TRIBBLE CJ (1991) Results and complications of angioplasty in aortoiliac disease. Circulation 83 (Suppl I) I-53–I-160

[36] KADIR S, WHITE RI, KAUFMAN SL, et al (1983) Long-term results of aortoiliac angioplasty. Surgery 94: 10–142

[37] BECKER GJ, KATZEN BT, DAKE MD (1989) Noncoronary angioplasty. Radiology 170: 921–940

[38] MURPHY KD, ENCARNACION CE, LE VA, PALMAZ JC (1995) Iliac artery stent placement with the Palmaz stent: follow-up study. J Vasc Intervent Radiol 6: 321–329

[39] HAUSEGGER KA, LAMMER J, HAGEN B, ET AL (1992) Iliac artery stenting: clinical experience with Palmaz stent, Wallstent, and Strecker stent. Acta Radiol 33: 292–296

[40] HENRY M, AMOR M, ETHEVENOT G, HENRY I, AMICABILE C, BERON R, MENTRE B, ALLAOUI M, TOUCHOT N (1995) Palmaz stent placement in iliac and femoropopliteal arteries: primary and secondary patency in 310 patients with 2–4 year follow-up. Radiology 197: 167–174

[41] SULLIVAN TM, CHILDS MB, BACHARACH JM, GRAY BH, PIEDMONTE MR (1997) Percutaneous transluminal angioplasty and primary stenting of the iliac arteries in 288 patients. J Vasc Surg 25: 829–838

[42] LABORDE JC, PALMAZ JC, RIVERA FJ, ENCARNACION CE, PICOT MC, DOUGHERTY SP (1995) Influence of anatomic distribution of atherosclerosis on the outcome of revascularization with iliac stent placement. J Vasc Intervent Radiol 6: 513–521

[43] SCHEINERT D, SCHRÖDER M, BALZER JO, STEINKAMP H, BIAMINO G (1999) Stent-supported reconstruction of the aorto-iliac bifurcation using the kissing balloon technique. Circulation 100 (Suppl II): II-295–II-300

[44] BROTHERS TE, GREEFIELD LJ (1990) Long-term results of aortoiliac reconstruction. J Vasc Intervent Radiol 1: 49–55

[45] KWASNIK EM, SIOUFFI SY, JAY ME, KHURI SF (1987) Comparative results of angioplasty and aortofemoral bypass in patients with symptomatic iliac disease. Arch Surg 122: 288–291

[46] DEVRIES SO, HUNINK MG (1997) Results of aortic bifurcation grafts for aortoiliac occlusive disease: a meta-analysis. J Vasc Surg 26: 558–569

[47] RING EJ, FREIMANN DB, MCLEAN GK, SCHWARZ W (1982) Percutaneous recanalization of common iliac artery occlusions: an unaccaptable complication rate? Am J Roentgenol 139: 587–589

[48] PENTECOST MJ, CRIQUI MH, DORROS G, GOLDSTONE J, JOHNSTON W, MARTIN EC, RING EJ, SPIES JB (1994) Guidelines for peripheral percutaneous transluminal angioplasty of the abdominal aorta and lower extremity vessels. Circulation 89: 511–531

[49] Standards of Practice Committee of the Society of Cardiovascular and Interventional radiology. Guidelines for percutaneous transluminal angioplasty. Radiology 177: 619–626

[50] COLAPINTO RF, STRONELL RD, JOHNSTON WK (1986) Transluminal angioplasty of complete iliac obstructions. Am J Roentgenol 146: 859–862

[51] JOHNSTON KW (1993) Iliac arteries: reanalysis of results of balloon angioplasty. Radiology 186: 207–212

[52] VORWERK D, GUENTHER RW, SCHÜRMANN K, WENDT G, PETERS I (1995) Primary stent placement for chronic iliac artery occlusions: follow-up results in 103 patients. Radiology 194: 745–749

[53] REYES R, MAYNAR M, LOPERA J, ET AL (1997) Treatment of chronic iliac artery occlusions with guide wire recanalization and primary stent placement. J Vasc Intervent Radiol 8: 1049–1055

[54] DYET JF, GAINES PA, NICHOLSON AA, CLEVELAND T, COOK AM, WILKINSON AR, GALLOWAY JM, BEARD J (1997) Treatment of chronic iliac occlusions by means of percutaneous endovascular stent placement. J Vasc Intervent Radiol 8: 349–353

[55] BLUM U, GABELMANN A, REDECKER M, NÖLDGE G, DORNBERG W, GROSSER G, HEISS W, LANGER M (1993) Percutaneous recanalization of iliac artery occlusions: results of a prospective study. Radiology 189: 536–540

[56] RUTHERFORD RB (1991) Standards for evaluating results of interventional therapy for peripheral vascular disease. Circulation 83 (Suppl I): I-6–I-11

[57] TEGTMEYER CJ, MOORE TS, CHANDLER JG, WELLONS HA, RUDOLF LE (1979) Percutaneous transluminal dilatation of a complete block in the right iliac artery. Am J Roentgenol 133: 532–535

[58] MOTARJEME A, KWEIFER JW, ZUSKA AJ (1980) Percutaneous transluminal angioplasty of the iliac arteries: 66 experiences. Am J Roentgenol 135: 937–944

[59] REES CR, PALMAZ JC, GARCIA O, ROEREN T, RICHTER GM, GARDINGER G JR, SCHWARTEN D, SCHATZ RA, ROOT HD, ROGERS W (1989) Angioplasty and stenting of completely occluded iliac arteries. Radiology 172: 953–959

[60] JOHNSTON KW, RAE M, HOG-JONSTON SA, COLAPINTO RF, WALKER PM, BAIRD RJ, SNIDERMAN KW, KALMAN P (1987) Five-year results of a prospective study of percutaneous transluminal angioplasty. Ann Surg 206: 403–412

[61] AUSTER M, KADIR S, MITCHELL SE, WILLIAM GM, PERLER BA, CHANG R, WHITE RI JR (1984) Iliac artery occlusions: management with intrathrombus streptokinase infusion and angioplasty. Radiology 153: 385–388

[62] KICHIKAWA K, UCHIDA H, YOSHIOKA T (1990) Iliac artery stenosis and occlusion: preliminary results of treatment with Gianturco expandable metallic stents. Radiology 177: 799–802

[63] HAUSEGGER KA, LAMMER J, KLEIN G, FLUCKIGER F, LAFER M, PILGER E, ASCHAUER M (1991) Perkutane Rekanalisation von Beckenarterienverschlüssen: Fibrinoyse, PTA, Stents. Fortschr Roentgenstr 155: 550–555

[64] KANNEL WB, MCGEE DL (1985) Update on some epidemiologic features of intermittent claudication: the Framingham study. J Am Geriatr Soc 33: 13

[65] DALMAN RL, TAYLOR LM (1990) Basic data related to infrainguinal revascularization procedures. Ann Vasc Surg 4: 309–312

[66] DECKELBAUM LI, ISNER JM, DONALDSON RF, CLARKE RH, LALIBERTE S, AHARON AS, BERNSTEIN JS (1985) Reduction of laser-induced pathologic tissue injury using pulsed energy delivery. Am J Cardiol 56: 662–667

[67] GRUNDFEST WS, LITVACK IF, GOLDENBERG T, et al (1985) Pulsed ultraviolet lasers and the potential for safe laser angioplasty. Am J Surg 150: 220–226

[68] GRUNDFEST WS, LITVACK F, FORRESTER JS, GOLDENBERG T, SWAN HJC, MORGENSTERN L, FISHBEIN M, MCDERMID W, RIDER DM, PACALA TJ, LAUDENSLAGER JB (1985) Laser ablation of human atherosclerotic plaque without adjacent tissue injury. JACC 5: 929–933

[69] CAPEK P, MCLEAN GK, BERKOWITZ HD (1991) Femoropopliteal angioplasty: factors influencing long-term success. Circulation (Supp I): I-70–I-80

[70] JOHNSTON KW, RAE M, HOG-JONSTON SA, COLAPINTO RF, WALKER PM, BAIRD RJ, SNIDERMAN KW, KALMAN P (1987) Five-year results of a prospective study of percutaneous transluminal angioplasty. Ann Surg 206: 403–412

[71] GESCHWIND J, BOUSSIGNAC G, TEISSEIRE B (1984) Percutaneous transluminal laser angioplasty in man. (letter) Lancet 844

[72] LAMMER J, PILGER E, DECRINIS M, QUEHENBERGER F, KLEIN GE, STARK G (1992) Pulsed excimer laser versus continous-wave Nd:YAG laser versus conventional angioplasty of peripheral arterial occlusions: prospective, controlled randomized trail. Lancet 340: 1183–1188

[73] BELLI AM, CUMBERLAND DC, PROCTER AE, WELSH CL (1991) Total peripheral artery occlusions: conventional versus laser thermal recanalization with a hybrid probe in percutaneous angioplasty: results of a randomized trail. Radiology 181: 57–60

[74] JORGENSEN B, MEISNER S. HOLSTEIN P, TONNESEN KH (1990) Early rethrombosis in femoropopliteal occlusions treated with percutaneous transluminal angioplasty. Eur J Vasc Surg 4: 149–152

[75] KÖHLER TR, NANCE DR, CRAMER MM, VANDERBRUHE N, STRANDNESS DE JR (1987) Duplex scanning for diagnosis of aortoiliac and femoropopliteal disease: a prospective study. Circulation 76: 1074–1080

[76] VROEGINDEWEIJ D, TIELBEEK AV, BUTH J, VOS LD, VAN DEN BOSCH HCM (1997) Patterns of recurrent disease after recanalization of femoropopliteal artery occlusions. Cardiovasc Intervent Radiol 20: 257–262

# III-7
# Laser Treatments in Plastic Surgery and Dermatology

## III-7.1
## Laser Skin Resurfacing

A. F. Osterhaus

### Contents

## Introduction

The goal of skin resurfacing is the rejuvenation of the skin by stimulation of its regenerative potential. The outer layer of epidermis is ablated in order to improve the quality of the skin. Sun damage is removed and age-related changes are improved in skin resurfacing. In general, there are several methods for resurfacing procedures: mechanical, known as dermabrasion, chemical peel, electrical, known as radiowave ablation and photoablation, normally referred to as laser skin resurfacing. The difference between these procedures is the varying results, the side effects, the difficulty in performance and reproducibility. Laser skin resurfacing is a well-controlled and precise procedure that shows defined, predictable results with fewer risks of complications [10, 15, 40]. There are laser wavelengths suitable for laser skin resurfacing. One is the $CO_2$ laser (10 600 nm) and the other is the Er:YAG (erbium:yttrium-aluminium-garnet) laser (2940 nm). The coefficient of absorption in water is more than 16 times higher in the Er:YAG wavelength. This results in a more complete absorption in a thinner layer of tissue with ablation of up to 30 μm and a thermal damage zone of less than 50 μm. In contrast, the $CO_2$ laser has an ablation depth up to 100 μm and a thermal damage zone of 150 μm. To understand the principles of skin resurfacing, we have to understand the ageing process of the skin, which is characterized by intrinsic and extrinsic factors. The intrinsic factor is thought to be a result of cell membrane damage and genetical determination. Extrinsic ageing, however, is mainly caused by photodamage. The layers of the most severe photoageing on the skin are the epidermis and upper dermis. The structural changes of the ageing skin are collapse of the elastic fibre system, reduction of skin moisture due to loss of water-binding capacity and structural changes of collagen fibre [4, 20, 35, 38]. Surgical treatment of the ageing skin is limited to the epidermis and upper dermis to replace the tissue without scar formation. A key role in the role regeneration of healthy skin is taken by the fibroblasts. Deep resurfacing causes the induction of fibroblasts followed by the production of new collagen, proteoglykanes, glykosaminoglykanes and hyaluronic acid. Water-binding capacity and skin turgour increase. Re-ep-ithelialization is related to the stimulation, prolifera-tion and migration of keratinocytes from the edges and adnexes influenced by fibronectin, type I collagen and blood-borne cytokines. Compared to dermabra-sion, re-epithelialization in deep $CO_2$ laser resurfacing is twice as slow, but the laser wound contains signifi-cantly more active fibroblasts, resulting in an impres-sive regeneration of photoageing effects in the skin. Very unique in thermal resurfacing (laser and ra-diowave) is the phenomenon of collagen shrinking due to the thermal effect on the collagen [37]. Similar histological results were obtained in deep phenol peels besides the effect on the melanocytes. Phenol stops the production of melanin permanently whereas even deep laser resurfacing has little long-term effect on melanocytes.

The healing process of the skin has a fixed pattern of regeneration according the damage. To achieve the same increase in new collagen formation and in inter-cellular substances, the depth of damage has to be the same, no matter what procedure of resurfacing is used. Superficial $CO_2$ laser resurfacing treatment pro-duces a wound depth of 60–100 µm (epidermis and upper papillary dermis), medium treatment a depth of 200–450 µm (papillary dermis and upper reticular dermis) and deep treatment a depth of 500–800 µm (midreticular dermis). Deeper skin damage to the midreticular dermis will result in scar formation. Laser skin resurfacing should affect a depth of 250–400 µm in the skin (papillary dermis and reticu-lar dermis). Because of the varying thickness of the skin in different areas, the energy applied to the tissue has to be adapted.

As the histological structure of the skin and the ad-nexes differs from the face to the neck, the capability of wound-healing changes. Apart from the face, resur-facing deeper than the upper reticular dermis most likely will result in scar formation. The average thick-ness of the epidermis in the face is 122 µm, while the epidermal thickness of the neck is 87 µm. The papillary dermis in the face is 113 µm, in the neck is average 81.3 µm; the reticular dermis of the face is 1200 µm, in the neck 1333 µm.

## Indications and Criteria for Exclusion

Indications for laser skin resurfacing are facial and neck wrinkling, photodamage of the face and neck, acne scars, traumatic, surgical and atrophic burn scars, actinic cheilitis, rhinophyma and benign growths. Contraindications are abnormal wound healing, decreased adnexal structures, infectious dis-eases, isomorphic diseases, medical and psychological conditions.

## Patient Selection

Candidates for laser resurfacing must be asked for their history and medication very carefully. Examina-tion of the skin has to consider skin types, pigmenta-tion and condition, zonal areas, scars, skin pathology, type of rhytids and, last but not least, patients' expec-tations. Patients who have undergoing an Accutane therapy should wait periods of 6 to 18 months prior to a resurfacing therapy. Accutane inhibits the collage-nase and suppresses the activity of the pilosebaceous apparatus. This interferes with post-operative wound healing and can result in scar formation. The inci-dence of herpes infection is increased by 3–6% even in patients without a history of herpes infection. Therefore, a herpes prophylaxis is recommended. Pa-tients with pigment disorders, hyper- or hypopigmen-tation, have a high risk of agravating the dyspigmen-tation. Patients with multiple scars from excoriatians, isomorphic diseases, abnormal wound healing or psy-chotic patients must be excluded. The patients' expec-tation, in contrast to the potency of the resurfacing procedure, is the most common reason for dissatisfac-tion in our patients. Deep dynamic wrinkles and an oily thick skin are unfortunate conditions for good re-sults. In contrast, the fine static wrinkles on a pre-treated moist skin are excellent for getting good re-sults. The postoperative healing process can take sev-eral weeks and redness can last up to 6 months. Pa-tients who are not willing to accept this long duration are not suitable for a resurfacing treatment.

## Skin Preparation

Pretreatment of the skin has been shown to have a beneficial effects on the resurfacing results. Retin-A, glycolic acid, hyaluronic acid, vitamin C., E, sun-screen, hydrochinon and kojic acid are used for a pe-riod of 2–4 weeks. Longer pretreatment periods up to several months are even more effective [5, 24]. Herpes prophylaxis is mandatory. The prevention of infec-tions must cover the three most critical types of infec-tion. One is cutaneous candidiasis, second stahylococ-cal aureus and the third pseudomonas aeroginosa. To prevent this type of infection, repeated skin cleaning prior to the resurfacing is recommended. In endan-gered patients, a prophylaxis with ciprofloxacin (500 mg twice a day) and fluconazole (400 mg single dose) on the day of operation is useful. Patients with acne should take minocycline for a longer period to treat and prevent inflammatory skin lesions.

## Preoperative Measures

The skin should be cleaned with non toxic microbicide, antiviral and antifungal agents. Flammable agents should be avoided immediately to or during the laser procedure. Eyes, hair and teeth should be protected by eye shields and wet gauze. The laser must be tested before starting the therapy. Evacuation must be checked. Laser security measures must be observed [6, 30].

## Anaesthesia

Facial resurfacing is highly painful and requires efficient anaesthesia. For a full-face procedure, at least 14 injections are necessary to perform all nerve blocks. The amount of local anaesthetics needed is critical. The use of tumescent anaesthesia results in swelling of the face, that interferes with the original findings of facial rythids. For most patients undergoing a full-face treatment, general anaesthesia is recommended. For partial resurfacing, local anaesthesia with additional analgosedation is advisable. Topical anaesthesia (EMLA) is not effective enough to tolerate several passes of Er:YAG or $CO_2$ laser resurfacing.

## Technique of Resurfacing

Most important in the resurfacing process is to apply enough energy per pulse to exceed the threshold for vaporization of epidermal cells with a pulse duration shorter than the dermal relaxation time. In this way, energy will be consumed for the vaporization process. With energy below the level of cell vaporization, the heat will be conducted to the surrounding tissue. Under these conditions, the pulse duration becomes a much more important factor. If the pulse duration exceeds the thermal relaxation time, a repetition rate of more than 5 Hz can build up enough heat to provoke thermal damage. Laser systems for resurfacing procedures must have a high absorption in water and a high-energy single pulse with short pulse durations. Overlapping of pulses by more than 20% should be avoided. If a scanner is used, avoid very high densities because of too much overlapping in high pulse frequencies (Fig. 1).

The different zonal areas of the face and neck have different thicknesses and different adnexal structures. For this reason, the laser parameters and the number of laser passes have to be adjusted to the area that is being treated. The central part of the face, the glabella region, nose and upper lip are least sensitive to laser damage. In contrast, the peripheral parts, the lids, lateral cheeks, lower perioral region and the neck are

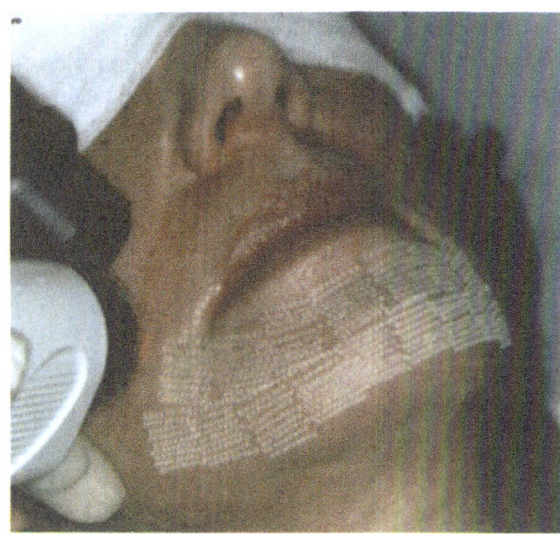

**Fig. 1.** Technique of laser skin resurfacing using a scanner. Patterns are set next to each other without overlapping. The upper lip has already been treated

much more vulnerable. This results form the different regenerative potential according to the histological structure of the skin. For example, the thin skin of the lids needs lower energy and less passes compared to the midface area. The lateral cheeks, lateral forehead and neck have less re-epithelization capacity, due to the pilosebaceous apparatus.

Performing the resurfacing with Er:YAG or $CO_2$ lasers on extended areas it is recommended to use a scanner. The first pass evaporates the superficial epidermal layer. Before the second pass the skin has to be cleaned with saline. The second pass should be perpendicular to the first to avoid visible linear patterns on the skin. The third pass is applied selectively to central face areas with deep wrinkles or severe photodamage. To avoid demarcation effects to the untreated skin, feathering of the borders is necessary. By lowering the energy and density of the laser, a harmonic transition can be created.

Using an Er:YAG laser the risk of ablating too deeply increases with each pass, because the absorption in the tissue is not self-limiting. In contrast, resurfacing with the $CO_2$ laser shows very clearly the limit of ablation and the beginning of thermal reaction of the dermal layers. For both system, Er:YAG or $CO_2$ there are signals of limitation. For the Er:YAG laser, papillary bleeding shows the critical depth. Continuing the procedure can end up in irregular skin texture and microscars. The endpoint in $CO_2$ laser resurfacing is the lack of further vaporization, further shrinking and yellow discoloration of the skin. The phenomenon of discoloration indicates the vasoconstriction of the reticular vessels dependent on thermal effects. Continuing the treatment beyond this point does not improve the final result, but increases the risk of side effects like hypopigmentation and scar formation and prolongs the time of erythema. In $CO_2$ laser resurfacing it is helpful to the wound healing to remove the zone of thermal necrosis. This can be done with a single pass of Er:YAG laser at the end of the treatment. Remaining keratoses or deep wrinkles could be treated isolated with Er:YAG as well.

## Postoperative Wound Healing and Care

Wound healing after resurfacing is divided into the inflammatory, proliferation and maturation phases (Fig. 2). All phases are controlled by fibroblasts. During the early phase (day 3–10 the macrophages are responsible for wound debridement. Abnormal or damaged collagen is degraded. The fibroblasts form new interstitial collagen and granulation tissue. Of some interest is the persistence of hyaluronic acid, that may have some influence on avoiding scar formation. Preloading the skin with hyaluronic acid precursors seems to be of some benefit for the wound healing. The next 10 to 14 days are dominated by the formation of procollagen and collagen in the fibroblasts and angiogenesis. The increase in the amount of collagen lasts for almost 3 weeks (Fig. 3), but remodelling of the new collagen continues for over 1 month. Intercellular substances like proteoglycanes are produced and enhance the water-binding capacity. This is important for the skin turgor and the disappearance of the wrinkles. Re-epithelialization starts from the wound edges and from the adnexes within the first 3 days and is completed after 5–7 days.

Numerous studies have shown that occlusive or semi occlusive dressings improve wound healing. The most common dressings after resurfacing are petroleum jelly or biosynthetic films like polyurethane, hy-

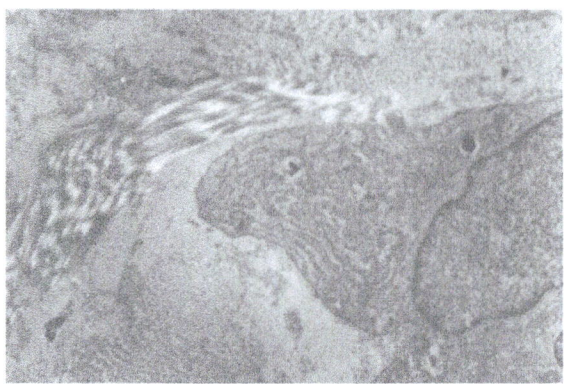

**Fig. 3.** Ultrastructural findings 3 weeks after $CO_2$ laser skin resurfacing. Active fibroblast with proliferated RER and Golgi apparatus producing a new matrix, collagen and intercellular substance (U. Westermann)

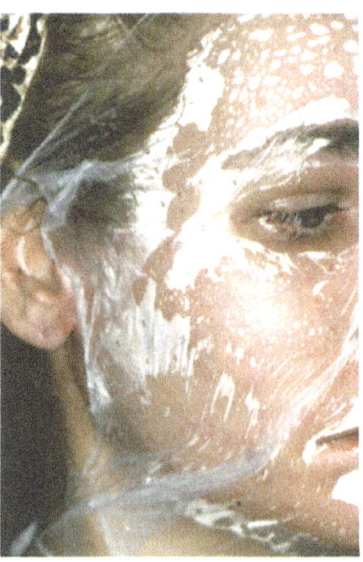

**Fig. 4.** Wound dressing with a semi occlusive membrane (TSR), applied immediately after the procedure (U. Westermann)

drocolloids, hydrogels and alginate foams (Fig. 4). The application should start immediately after resurfacing and be continued for 3–5 days. To prevent infections, the dressing should be changed every 24–48 h after cleaning the skin. After this period, petroleum cream can be used for another 5 days in a lighter preparation. For the following weeks moisturizers are applied. Some authors describe the benefit of a combination with vitamin C, E, essential fatty acids, hyaluronic acid and aloe vera to the moisturizer [3]. Close follow-up of the patient in the early phase of wound healing is essential to detect infections and abnormal wound healing. The oral application of antibiotics is widely recommended, whereas topical antibiotics are avoided because of the high incidence of allergic reactions and the local effect on skin regeneration. Steroids have some advatanges, perioperatively reducing the swelling and burning sensation.

**Fig. 2.** Immediate ultrastructural findings after $CO_2$ laser skin resurfacing. Disintegration and fragmentation of collage fibre into filaments next to intact but swollen cross-linked fibres (U. Westermann)

## Complications

There are several normal side effects accompanying resurfacing, like swelling, erythema, burning, itching and ecchymosis. They must be differentiated from complications such as dyspigmentation, infections, scar formation, milia and prolonged erythema. The most severe complication is scarring, which results from deep tissue injury due to extensive tissue vaporization (Er:YAG), from thermal injury ($CO_2$ laser resurfacing) or from postoperative infections. The early signs for the development of scars are erythema and pruritus. These should indicate the ruling out of infection and lead to early treatment with high-potency steroids to prevent scarring [34]. Milia are often the result of inappropriate postoperative care. Rich petrolium-based and occlusive ointments can be responsable for the milia. Frequent skin cleansing, lighter skin care and antibiotics will be beneficial. Postoperative hyperpigmentation is a very common complication in darker skin types (20–30% in Fitzpatrick type 3, over 80% in skin type 4). It is a result of the stimulation of melanocytes and will disappear spontaneously in most cases after 2–6 months [13, 27,

29]. Sun protection, hydrochinon, tretinoin, vitamin C and E, azelaic acid, kojic acid and glucosamine are helpful in treating hyperpigmentation. Hypopigmentation must be differentiated into two forms. Pseudo-hypopigmentation is a result of vaporizing sun-damaged pigmented epidermis. Permanent hypopigmentation relates strictly to the depth of skin damage due to vaporization or thermal injury (Figs. 5, 6). It becomes evident 6–12 months after the resurfacing in 8% of the patients with higher incidence in preoperatively severely photodamaged skin [7]. Treatment of hypopigmentation is unsatisfactory [8, 12, 31]. Blending the aesthetic areas or treating the whole face and neck can make this phenomenon less visible.

## Results

Review of the literature confirms excellent results to fine and moderate wrinkles in both $CO_2$ and Er:YAG laser treatment [22, 26, 36, 41, 42, 43, 44]. Deep wrinkles gave better results with $CO_2$ laser resurfacing

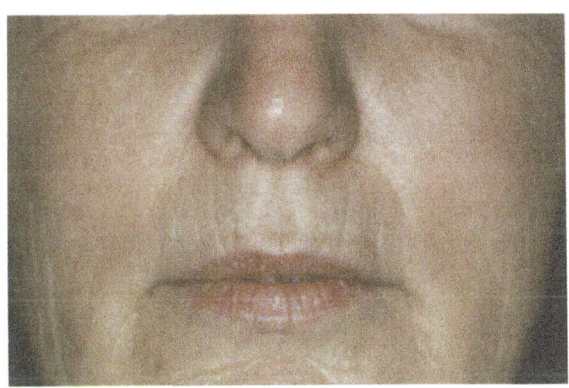

**Fig. 5.** Deep upper lip lines prior to $CO_2$ laser skin resurfacing

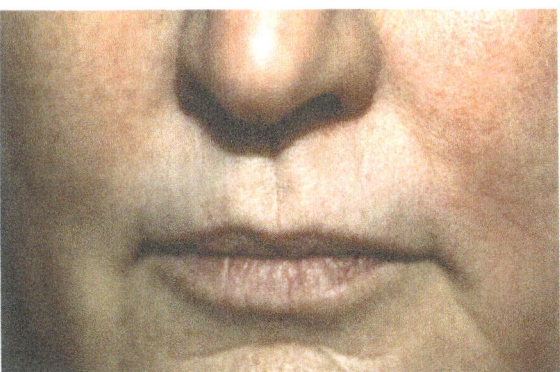

**Fig. 6.** Same patient as in Fig. 5 1 year posttreatment, long-term result on the upper lip lines, resistant moderate hypopigmentatation of the upper lip

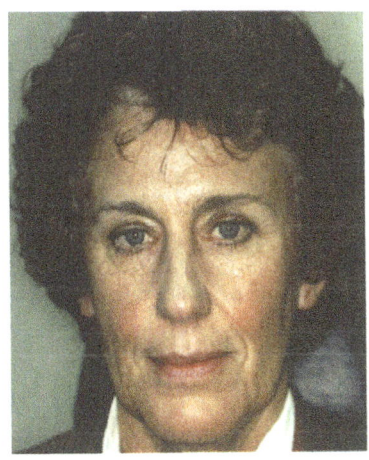

**Fig. 7.** 60 year-old female patient prior to full-face $CO_2$ laser skin resurfacing, severe photodamage, dynamic and static wrinkles

**Fig. 8.** Result 1 year after a full-face $CO_2$ laser skin resurfacing, disapearance of photodamages and fine lines, improvement of dynamic wrinkles, no complications

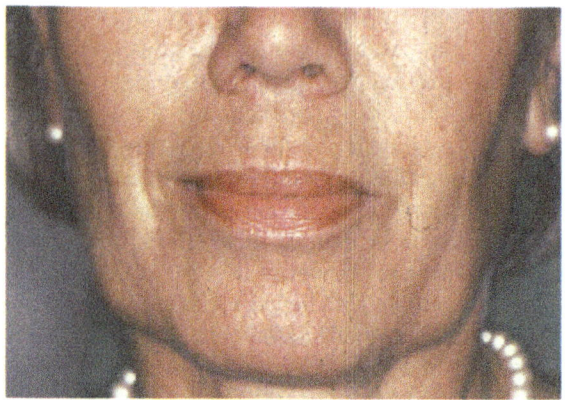

**Fig. 9.** Female patient prior to full-face $CO_2$ laser skin resurfacing, moderate photodamage, dynamic and static wrinkles

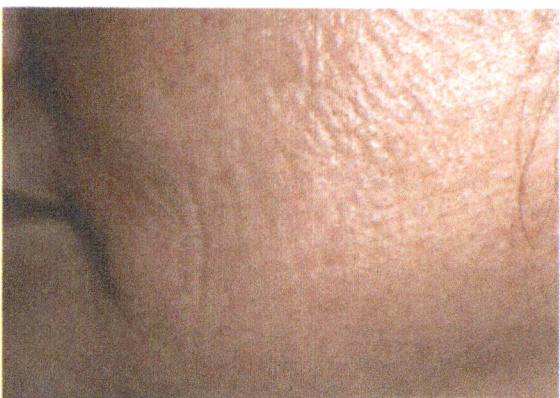

**Fig. 11.** Left cheek prior to a combined Er:YAG-$CO_2$ laser skin resurfacing of the cheeks, photoageing

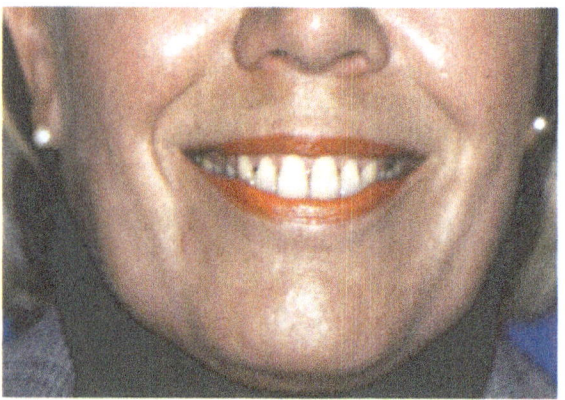

**Fig. 10.** Result 6 months after a full-face $CO_2$ laser skin resurfacing, complete disapearance of photodamage and fine lines, improvement of dynamic wrinkles, no complications

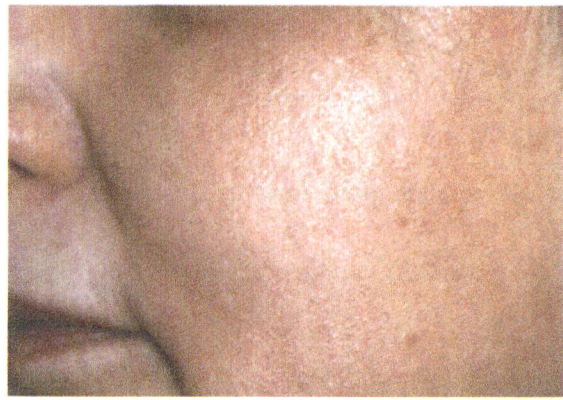

**Fig. 12.** 6 weeks after an isolated combined Er:YAG-$CO_2$ laser skin resurfacing of the cheeks, disapearance of photodamage and fine lines, rapid wound healing, no demarcation lines

(Fig. 7–10). As to reach the same depth of tissue needs three $CO_2$ passes (300 mJ) compared to 12–15 Er:YAG passes (5J cm$^{-2}$) the results can be similar in both systems [1, 2, 25]. This confirms the principals of biological reaction to adequate damage. There is evidence that the combination of both laser systems (Er:YAG and $CO_2$) have some advantages in resurfacing procedures (Fig. 11, 12). The Er:YAG laser can remove the necrotic tissue after the last pass of $CO_2$ resurfacing and sculpture isolated deep wrinkles. By enhancing the wound healing, this may lead to better results [39, 42]. The Er:YAG laser is indicated wherever ablation is the main subject. It can be useful in feathering the borders. For the treatment of epidermal lesions it is the laser of first choice. Resurfacing of the neck needs to be more superficial without thermal damage. Therefore only the Er:YAG laser can be recommended for treatment of the neck [26]. The number of complications following resurfacing procedures is highly correlated to the experience of the physician. More aggressive treatment and more passes do not induce

better results. Alternatively, to improve the results, a second treatment after 6–12 months can be considered.

## Comparison with Alternative Methods

Comparison of chemical peeling, dermabrasion and laser resurfacing shows the differences between these procedures [11]. The skin damage corresponds with consecutive re-epithelization and regeneration of the skin. The depth of the lesions, the zone of necrosis and the process of healing differ, while the final results can resemble each other regarding the penetration depth of the procedure. Clinically, the results of Er:YAG resurfacing can be compared to medium-depth peeling [28]. It can improve the epidermis and flatten superficial wrinkles. $CO_2$ laser resurfacing is comparable to deep peeling, but without injury to the deep reticular dermis (Figs. 13, 14). This reduces the risk of complications but produces improvement to

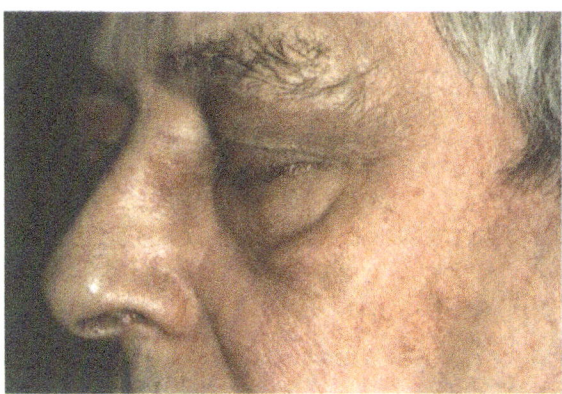

**Fig. 13.** Male patient with dermatochalasis of the lower lid, cheek pads and photodamage, prior to $CO_2$ laser skin resurfacing and transconjunctival blepharoplasty (U. Westermann)

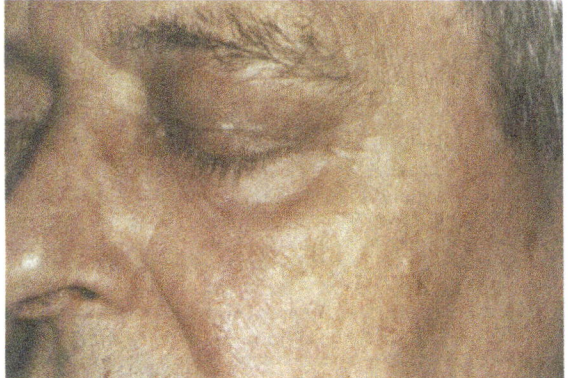

**Fig. 14.** Result 6 months after $CO_2$ laser skin resurfacing and transconjunctival blepharoplasty (U. Westermann)

distinct wrinkles. The re-epithelialization time of 35% TCA peeling is comparable to moderate energy $CO_2$ laser resurfacing (150–250 mJ, three passes). Laser treatment with higher energies (350–450 mJ, single pass) is comparable in re-epithelialization to dermabrasion [23]. Deep phenol peelings result in prolonged wound healing and re-epithelialization (3 weeks). The advantage of laser resurfacing is its reproducibility and control. Lasers vaporize and influence a predictable depth of tissue.

## Discussion

The difference between resurfacing systems of the $CO_2$ laser is the irradiance: This term describes the power delivered to the tissue per unit area in W cm$^{-2}$. For resurfacing, the irradiance has to be high enough to vaporize tissue fast with little heat conduction to surrounding tissue. All high-power ultrashort pulsed $CO_2$ laser deliver sufficient irradiance [14, 21, 32]. They vary in ablation depth and residual thermal

necrosis after several passes. The $CO_2$ laser with he highest irradiance has the smallest zone of thermal necrosis. In Er:YAG laser resurfacing, with its 16 times higher absorption coefficient in water and small zone of thermal damage, the tissue interaction differs. Er:YAG resurfacing produces no coagulation effect on the papillary vessels. Therefore, the papillary bleeding is clinically the limiting point or Er:YAG resurfacing. Physically there is no limit in ablation to the dermal tissue because of the high absorption in water and collagen fibres corresponding to the reduced water content of the deeper dermis. This phenomenon makes the Er:YAG an excellent tool for ablation, but there is no self-limiting factor in ablation as there is in $CO_2$ laser resurfacing. The depth of ablation in $CO_2$ lasers after three passes is 20 µm because of the reduced water content from the epidermis to the dermis. The second reason is the degree of reduced water content in deeper layers with each pass, due to the thermal conduction. Therefore $CO_2$ laser resurfacing is self-limiting or a depth up to 250–400 µm; but where are the danger signs in $CO_2$ or Er:YAG-resurfacing? An important factor for damage to the skin is heat. The threshold for irreversible thermal damage to human collagen is above 70 °C. Underlying the zone of thermal necrosis up to 150 µm is the zone of reversible thermal damage, showing a gradient between 70 and 37 °C body temperature. At 63 °C the collagen fibres shrink. This shrinkage is characteristic for $CO_2$ resurfacing and is partially responsible for the excellent cosmetic results. Another important factor is heat conduction. This is determined by the thermal conductivity of tissue, the temperature gradient and the heat loss in time. In vitro, a heated layer of dermal tissue 50 µm thick takes 70 ms to loose heat and reach the temperature of the underlying tissue. This clearly shows that pulse stacking, overlapping of pulses, number of passes, pulse width and repetition rate become important. Studies show that spots overlapping more than 35% using high repetition rates and multiple passes produce significantly higher rates of complications like scars, hypopigmentation and poor wound healing [33]. The critical repetition rate is about 10 Hz. With each pass after vaporization of the epidermis, the zone of thermal necrosis increases by about 30–50 µm because the laser has lost it target, the water. A reduction in energy per pulse is no guarantee to lower the thermal damage. On the contrary, there is even more heat conduction as a result of lacking heat loss from tissue vaporization. In $CO_2$ resurfacing there are three clinical visible end points of the treatment. The elimination of the wrinkle, scar or epidermal lesion, a yellow-brown discoloration and no further shrinking effect. It is safe to stop treatment if one of these signs occurs. Proceeding further than this increases the risk of irreversible complications.

Retreatment after 6-months period of healing is the appropriate answer to incomplete results. The combination of both laser wavelengths in one system or separately in one resurfacing procedure seems to have some benefits [9, 25]. The zone of thermal necrosis is reduced. This can have some advantages for the period of wound healing, especially the re-epithelialization time being shortened. The protracted period of wound healing, the incidence of pigmentary alteration and erythema is still inacceptable for many patients. Therefore, other laser systems are being investigated for their effects on collagen and fibroblasts. One is a 1320-nm modified Nd:YAG laser inducing thermal damage to the papillary dermis without damaging the epidermis by cooling the epidermis with cryogen. The intention is to stimulate fibroblasts to produce new collagen and intercellular substances. Studies have to prove the clinical safety and results of these new techniques.

## References

[1] ADRIAN RM (1999) Pulsed carbondioxide and erbium: YAG laser resurfacing: a comparative clinical and histologic study. J Cut Laser Ther 1: 29–35

[2] ADRIAN RM (1998) A clinical and histologic comparison of erbium: YAG and pulsed carbon dioxide lasers in the treatment of facial rhytids. Lasers Surg Med 10: 38

[3] ALSTER TS, WEST TB (1998) Effect of topical vitamin C on postoperative $CO_2$ laser resurfacing erythema. Dermatol Surg 24: 331–334

[4] ALSTER TS, KAUVAR AN, GERONEMUS RG (1996) Histology of high-energy pulsed $CO_2$ laser resurfacing. Semin Cutan Med Surg 15 (3): 189–193

[5] BEASLEY D, ET AL (1997) Effect of pretreated skin on laser resurfacing. Lasers Surg Med Suppl 9: 43

[6] APFELBERG DB (1997) Perioperative considerations in laser resurfacing. Int J Aesthet Restor Surg 5: 21–28

[7] BERNSTIEN LJ, ET AL (1997) The short and long-term side effects of carbondioxide laser resurfacing. Dermatol Surg 23: 519–525

[8] BOLOGNIA JL, PAWELEK JM (1988) Biology of hypopigmentation. J Am Acad Dermatol 19: 217

[9] GOLDMAN MP, MANUSKIATTI W (1999) Combined laser resurfacing with the 950-microsec pulsed $CO_2$ + Er:YAG lasers. Dermatol Surg Mar 25 (3): 160–163

[10] FITZPATRICK RE (1997) Laser resurfacing of rhytids. Dermatol Clin 15 (3): 431–447

[11] FITZPATRICK RE, ET AL (1995) Pulsed carbon dioxide laser, trichloroacetic acid, Gordon-Baker phenol, and dermabrasion: a comparative clinical and histologic study of cutaneous resurfacing in a porcine model. Arch Dermatol 132: 469–471

[12] FITZPATRICK RE , ET AL (eds) (1981) Biology and diseases of dermal pigmentation. University of Tokyo Press, Tokyo

[13] FULK CS (1984) Primary disorders of hyperpigmentation. J Am Acad Dermol 10: 1

[14] GROSS EA, ROGERS GS (1998) A side-by-side comparison of carbon dioxide resurfacing lasers for the treatment of rhytids. J Am Acad Dermatol 39: 547–553

[15] GROVER R, ET AL (1999) A quantitative method for the assessment of facial rejunivation: a prospective study investigating the carbon dioxide laser. Br J Plast Surg 51 (1): 8–13

[16] HIBST R, STOCK, KAUFMANN R (1997) Ablation and controlled heating of skin with the Er:YAG laser. Lasers Surg Med Suppl 9: 40

[17] HOHENLEUTNER U, ET AL (1997) Fast and effective skin ablation with an Er:YAG laser: Determinatin of ablation rates and thermal damage zones. Lasers Surg Med 20 (3): 242–237

[18] KAUFMANN R, HIBST R (1996) Pulsed erbium: YAG laser ablation in cutaneous surgery. Lasers Surg Med 19: 324–330

[19] KAUFMANN R, HIBST R (1990) Pulsed 2.94- microns Er:YAG laser skin ablation – experimental results and first clinical application. Clin Exp Dermatol 15 (5): 389–393

[20] KAUVAR AN, GERONEMUS RG (1997) Histology of laser resurfacing. Lasers Dermatol 15 (3): 459–467

[21] KAUVAR AN, WALDORF HA, GERONEMUS RG (1996) A histological comparison of „char-free" carbon dioxide lasers. Dermatol Surg 22: 343–348

[22] KHATRI K, ET AL (1997) Comparison of erbium-YAG and $CO_2$ lasers in skin resurfacing. Lasers Surg Med Suppl 9: 37

[23] KOSH MM, ET AL (1999) Safety and efficacy of high-fluence $CO_2$ laser skin resurfacing with a single pass. J Cut Laser Ther 1: 37–40

[24] LOWE NJ, ET AL (1995) Laser skin resurfacing. Pre- and posttreatment guidelines. Dermatol Surg 21 (12): 1025–1029

[25] MILLMAN AL, MANNOR GE (1999) Histologic and clinical evaluation of combined eyelid erbium: YAG and $CO_2$ laser resurfacing. Am J Ophthalmol 127 (5): 614–616

[26] MCDANIEL D, ET AL (1997) The erbium: YAG laser: a review and preliminary report on resurfacing of the face, neck, and hands. Aesthet Surg J 17: 157–164

[27] NANNI CA, ALSTER TS (1998) Complications of carbon dioxide laser resurfacing. An evaluation of 500 patients. Dermatol Surg 24: 315–320

[28] REED JT, ET AL (1997) Treatment of periorbital wrinkles. Silktouch carbondioxide laser compared with a medium-depth chemical peel. Dermatol Surg 23: 643–648

[29] ROBERTS TL, ET AL (1996) $CO_2$ laser resurfacing: recognizing and minimizing complications. Aesthet Surg Q 143–148

[30] ROHRICH RJ, ET AL (1997) $CO_2$ laser safety considerations in facial skin resurfacing. Plast Reconstr Surg 100 (5): 1285–1290

[31] ROSENBERG G, ET AL (1997) Treatment of post-laser resurfacing complications. Aesthet Surg J ((Band-Nr. fehlt)): 119–123

[32] ROSS E, ET AL (1997) Long-term results after $CO_2$ laser skin resurfacing: a comparison of scanned and pulsed systems. J Am Acad Dermatol 37: 709–718

[33] ROSS E, ET AL (1997) Effect ofoverlap and pass number in $CO_2$ laser skin resurfacing: preliminary results of residual thermal damage, cell depth and wound healing. Proc SPIE 2970: 395

[34] SCRIPRACYA-ANUNT S, ET AL (1997) Infections complicating pulsed carbon dioxide laser resurfacing for photoaged facial skin. Dermatol Surg 23: 527–536

[35] STUZIN JM, ET AL (1997) Histologic effects of the high-energy pulsed $CO_2$ laser on photoaged facial skin. Plast Reconstr Surg 99 (7): 2036–2050

[36] TEIKEMEIER G, GOLDBERG DJ (1997) Skin resurfacing with Er:YAG laser. Dermatol Surg 23: 685–687

[37] TOPE W, ET AL (1997) Ultrastructure of collagen thermnally denatured by microsecond-domain $CO_2$ laser. Lasers Surg Med Suppl 9: 43

[38] TRELLES MA, ET AL (1999) Electron microscopy comparison of $CO_2$ laser flash scanning and pulse technology 1 year after skin resurfacing. Int J Dermatol 38 (1): 58–64

[39] TRELLES MA et al. Skin resurfacing improved with a new dual wavelength Er:YAG/$CO_2$ laser system: a comparative study. J Clin Laser Med Surg 17 (3): 99–104

[40] TRIMAS SJ, ET AL (1997) The carbon dioxide laser. An alternative for the treatment of actinically damaged skin. Dermatol Surg 23: 885–889

[41] WEINSTEIN C (1999) Erbium laser resurfacing: current concepts. Plast Reconstr Surg 103 (2): 602–616

[42] WEINSTEIN CA (1999) New lasers for skin resurfacing: erbium-YAG/CO$_2$ systems. Perspect Plast Surg 13 (1): 57–82

[43] WEINSTEIN C (1998) Computerized scanning erbium:YAG laser for skin resurfacing. Dermatol Surg 24 (1): 83–89

[44] WEISS MA, ET AL (1999) Periorbital skin resurfacing using high-energy erbium: YAG laser: results in 50 patients. Lasers Surg Med 24 (2): 81–86

# III-7.2
# Laser Treatments of Skin Malignancies

C. Fritsch, K. W. Schulte, K. Lang, W. H. G. Neuse, P. Lehmann and T. Ruzicka

**Contents**

## Introduction

Laser treatment is gaining increasing importance in the therapy of skin malignancies. In particular, photodynamic therapy (PDT), photovaporization and in situ photocoagulation are used to cure or to palliatively treat epithelial precancerous lesions and tumours of the skin. The laser systems mostly applied in PDT were tunable dye lasers. However, in recent years, incoherent non-laser light sources have been used more frequently for PDT in dermatology. For this reason, the PDT chapter mainly contains data on the use of incoherent light sources.

In dermatological PDT, δ-aminolevulinic acid (ALA) is topically applied inducing high porphyrin formation in the skin, preferentially in tumour tissues. The porphyrin-enriched tumour tissues are irradiated with red or green non-coherent light. Porphyrins are activated by light and then induce cell death via formation of singlet oxygen and other free radicals. PDT is highly efficient in the treatment of solar keratoses and superficial basal cell carcinomas. Initial squamous cell carcinomas also show good response rates to topical ALA-PDT.

The ALA-induced porphyrins can also be used for a diagnostic procedure which may be termed fluorescence diagnosis with ALA-induced porphyrins (FDAP). Here, the irradiation of porphyrin-enriched tissues with Wood's light leads to the emission of a homogenous intensive red fluorescence in tumours and precancerous lesions of the skin. The histopathological extension of the tumours correlates very well with the borders detected by the tumour-specific fluorescence. The main indications of FDAP are the delineation of clinically ill-defined skin tumours and the control of the efficacy of any other tumour therapy. In preoperative planning, FDAP is a valuable method to determine the peripheral borders of a given tumour.

For photovaporization in particular $CO_2$ lasers are used in curative or palliativel treatment of malignant as well as benign skin tumours. Photovaporization is highly effective in the treatment of precancerous lesions of the skin (e.g. Bowen's disease), mucocutaneous malignancies (e.g. cheilitis actinica or leucoplakia) or other neoplastic tissue affecting the skin (e.g. extramammary Paget's disease).

The photocoagulation technique generally includes the light of a Nd:YAG laser. It has been shown to effectively treat voluminous cutaneous tumours.

## Indication

### Photodynamic Therapy (PDT)

Photodynamic therapy (PDT) refers to the light activation of a photosensitizer with subsequent generation of highly reactive oxygen intermediates. These intermediates irreversibly oxidize essential cellular components, causing tissue injury and necrosis. For the treatment of gastrointestinal, cerebral or bronchopulmonary tumours the compounds are administered orally or intravenously [54]. Skin tumours, endometrial tumours, and bladder carcinomas (intravesical instillation) are treated by topical application of the sensitizer [21, 54, 83].

In 1900, Raab demonstrated for the first time that dyes, e.g. acridine, in combination with light were able to sensitize or to kill micro-organisms, e.g. Paramecia [59]. Some years later the oxygen dependency of this reaction was postulated and the term photodynamic action was described [74]. Successful treatment of skin disorders, e.g. condylomata lata, lupus vulgaris and various skin tumours, with eosin and white light proved the efficacy of PDT [35, 75]. Until 1980, haematoporphyrin dervative (HpD) was the most frequently applied porphyrin compound in PDT. Photofrin (a mixture of dihematoporphyrinester and -ether) (Ipsen Pharma GmbH, Ettlingen, Germany) and Photosan 3 (Seehof Laboratorium GmbH, Wesselburen, Germany) are the only approved drugs for systemic PDT [54]. However, the systemic administration of photosensitizers induces generalized phototoxicity. The tumour-selective PDT with porphyrins was modified in 1990 by applying the porphyrin precursor ALA topically [37]. Exogenous administration of ALA bypasses the rate-limiting enzyme of haem synthesis, ALA synthase, which synthesizes ALA from glycine and succinyl CoA [21]. Thus, ALA treatment induces an increase in tissue porphyrin levels, especially in neoplastic tissue. The efficacy of ALA-PDT in the treatment of skin tumours was proven by several groups [10, 11, 20, 21, 37, 83]. Especially solar keratoses were shown to be highly sensitive to topical ALA-PDT [21, 25].

### Indications for PDT

In dermatology, the following (pre)malignant diseases can effectively be treated by PDT: solar keratosis (SK; single or multiple ones), superficial basal cell carcinoma (BCC), superficial squamous cell carcinoma (SCC) and Bowen's disease (BD). Nodular BCCs respond poorly to PDT. However, recent studies showed that the debulking of the exophytic tumour tissue prior to PDT leads to higher response rates in these tumours. Additionally, in the case of large and destructively growing tumours, PDT can be performed at least to reduce the tumour mass and to facilitate other subsequent treatment procedures (e.g. surgery). T-cell lymphoma such as plaques of mycosis fungoides do also partly respond to PDT. Cutaneous dysplasias (SK, SCC) in immunosuppressed patients (e.g. after kidney or bone marrow transplantation) can be treated by PDT; however, they often need a higher number of total PDT sessions as compared to non-immunosuppressed individuals.

### Fluorescence Diagnosis with δ-Aminolevulinic Acid Induced Porphyrins (FDAP)

In this chapter, we also discuss the significance of a novel presurgical method, fluorescence diagnosis with ALA-induced porphyrins (FDAP). This procedure has been developed in parallel to PDT. In FDAP, the fluorescence of porphyrins is used to detect neoplastic tissues.

Policard used the characteristic brick-red fluorescence of porphyrins, e.g. haematoporphyrin, for tumour detection for the first time in 1924 [58]. The predominant porphyrin fluorescence in tumour tissues was confirmed in humans and animals by several investigators [15, 60]. HpD consisting of porphyrin derivatives was shown to be even more preferentially stored in carcinomas [32, 44]. Thus, in the case of early bonchogenic carcinomas, the fluorescence of HpD or Photofrin has been effectively used to detect the neoplastic tissue [39, 41]. FDAP with ALA-induced porphyrins was shown to be capable to differentiate bladder carcinomas from the adjacent normal tissue [40, 68], to detect malignant gliomas intraoperatively [70], and to distinguish oral and palatial SCC from normal mucosa [43].

We examined the ALA-induced porphyrin fluorescence in various dermatological disorders, particularly in skin tumours. Correlation was made between the clinically detected fluorescence extension and the tumour margins which were examined histopathologically [23].

### Indications for FDAP

FDAP can be effectively used for the following indications: detection of neoplastic tissue, demarcation of neoplastic skin in actinically damaged or multiple pretreated skin, delineation of clinically ill-defined tumours to facilitate preoperative planning, and post-treatment control of any tumour therapy to prove its effectivity.

### Photovaporization with the CO₂ Laser

In general, any lesion which is superficially localized and sharply defined can be treated by carbon dioxide ($CO_2$) laser vaporization. In the case of several premalignant (precancerous) lesions, in particluar enoral, genital and phalangeal dermatological diseases, the use of the $CO_2$ laser is the treatment of choice [13]: Bowen's disease, cheilitis actinica, leucoplakia, lichen sclerosus et atrophicus, erythroplasia Quéyrat. In addition, the vaporization technique can be performed as an alternative treatment in: extramammary Paget's disease [5], BCC, and cutaneous in situ SCC. Of course the $CO_2$ laser showed also high efficacy in the therapy of benign tumours of the skin (e.g. syringoma, trichoepithelioma, hyperplasia of the sebaceous glands, adenoma sebaceum, dermal or epidermal nevus, xanthelasma, neurofibroma), viral-induced skin diseases (e.g. condylomata acuminata, focal endothelial hyperplasia (Heck's disease), bowenoid papulosis, verrucae vulgares), and other partly cosmetic indications such as rhinophyma, tattoos or facial wrinkles [13].

### In Situ Laser Coagulation

In situ coagulation is mostly known from the endoscopically guided treatment of bronchial, hepatic, primary breast cancer or advanced head and neck tumours using bare fibres. Laser coagulation is attractive, since it is minimally invasive and carries a low morbidity. In dermatological oncology, this technique covers only a small part of the therapeutic strategies. However, the Nd:YAG laser may be used for coagulation of BCC, SCC, Kaposi's sarcoma and metastatic skin tumours, in particular for palliative reasons. Voluminous vascular anomalies are still mostly treated with conventional surgery, although haemangiomas, preferentially subcutaneously growing ones, have been effectively treated by Nd:YAG or also by potassium-titanyl-phosphate (KTP) lasers.

## Practical Performance

### Photodynamic Therapy (PDT)

In order to perform PDT in skin tumours, the porphyrin precursor ALA is topically applied under occlusive foil. Irradiation should be performed when the optimal ratio of photosensitizer levels between tumour and tissue is reached (in the case of ALA, 2–6 h after application) [24]. The type of light source (laser or incoherent light) and the required fluence depend on the photosensitizer used as well as on the type and localization of the lesion. In ALA-PDT, ALA concentrations of 10% (SK) and 20% (BCC, SCC, etc) are recommended. In the treatment of large skin lesions, incoherent light devices are superior to laser systems due to larger irradiation fields. In early clinical studies incandescent lamps or even slide projectors were used; 635 nm was revealed as the most effective wavelength in ALA-PDT [72]. Reduction of fluence rate or fractionated irradiation may increase PDT efficacy due to increased singlet oxygen levels in regions of spare capillary density and re-accumulation of porphyrins in pretreated and sensitizer-bleached tissue. Optimal therapeutic results were achieved by using green light (PDT Saalmann, Saalmann GmbH, Herford, Germany; 545 nm; 25 mW cm$^{-2}$, 30 J cm$^{-2}$) for SK, BCC and SCC or alternatively red light (PDT 1200, Waldmann GmbH, – 150 mW cm$^{-2}$, 150–180 J cm$^{-2}$).

Irradiation of porphyrin-sensitized skin areas causes erythema, severe burning and pain, particularly if performed on the face. These symptoms possibly last several hours after irradiation with decreasing severity [21]. The application of creams containing antibiotics (e.g. fusidinic acid or gentamycine) and/or corticosteroids for 1 to 4 days is advisable to treat and/or to prevent an inflammation and/or a bacterial infection. Formation of crusts is common in treated areas and they resolve after a few days. Removal of crusts is facilitated by the application of urea-containing creams. Healing occurs without a scar within 10 to 14 days with cosmetic results equal or superior to other treatment methods such as cryosurgery or surgery. Hyperpigmentation is common but subsides within several months in all patients [10, 20, 21]. The cosmetic and therapeutic outcome can be evaluated 2–3 weeks after PDT. In the case of superficial and nodular tumours, the consecutive performance of two to three PDT sessions at intervals of 1 week is recommended, independent of the clinical outcome after the first PDT session.

Although PDT is a highly effective technique, clinical and, in selected cases, also histopathological controls are necessary twice a year.

## Fluorescence Diagnosis with δ-Aminolevulinic Acid Induced Porphyrins (FDAP)

ALA is applied in an ointment vehicle (10–20%; 50–200 mg ALA cm$^{-2}$) for cutaneous lesions under occlusive foil to enhance tissue penetration and to avoid photobleaching. After 4–6 h, intralesional porphyrin formation is evaluated by the emission of red fluorescence during irradiation with Wood's light (370–405 nm; Fluolight®, Saalmann, Germany) (Figs. 1–8) [19, 20, 23]. The fluorescent area can be marked with a surgical pen to optimize a planned surgical excision of a certain tumour. If the subsequent therapeutic step is not PDT or if the tumour therapy is not immediately performed after FDAP, it is important to cover the ALA-treated skin area with a light protecting tape for 24 hours to avoid any phototoxic reaction of the treated skin. Otherwise a phototoxic reaction may be provoked by even only a short exposure to daylight (even in the winter months). As a consequence, the skin reacts with an erythema which might harm a planned surgery or any other therapeutic strategy.

## Photovaporization with the CO$_2$ Laser

Prior to any laser treatment, the diseased skin is cleansed on isopropanol. Depending on the localization of the lesion, local superficial (e.g. in the case of a BCC on the trunk) or block anaesthesia (e.g. in the case of verrucae vulgares on the fingers) is performed using a 1% Mepivacain-HCl solution.

In general, the vaporization effect can be achieved by focused, prefocused and defocused beams of the CO$_2$ laser. The most widely used modus is the defocused beam application, which allows homogeneous and gentle vaporization of the superficial tissue layers. The focused beam is mainly used to cut tissue and allows a plane vaporization only in combination with scanner systems. The prefocused beam shows its focus in the depth of the tissue and leads to high power and heat in this tissue area. As this technique increases the risk of scarring, its use is limited mainly to the treatment of viral papillomas.

The wave modus of the laser beam used depends on the thickness and the size of the lesions. For large skin lesions (> 5 mm; e.g. BCC, BD or leucoplakia) the continuous-wave modus is recommended. Small, and thin lesions (e.g. xanthelasma, verrucae planae juvenils) show good and sufficient response rates to the superpulse modus with 0,05–0,2 s. impulse duration. A beam diameter of 2 mm and intensities from 5 to 20 W depending on the thickness of the diseased skin, are recommended.

The tissue is vaporized stepwise layer by layer. The carbonization layer with whitish epithelial residues which develop after each vaporizing cycle is removed by gauzes soaked with H$_2$O$_2$ (3% in water solution). The vaporization procedure should be repeated as many times as necessary, until the desired extent of ablation is achieved. Posttreatment, tapes containing polyvidon iodine or fusinic acid ointments are applied for 3–7 days to improve wound healing and to avoid bacterial infections. After several days crusts are formed which can be easily removed with dexpanthenol ointments. In general, it is possible to evaluate the therapeutic outcome 3 to 4 weeks after laser therapy. However, to assess a cosmetic result it takes about 3 months.

**Fig. 1.** (**a**) Superficial BCC on the back. Approximately 3 x 4 cm, well-defined hyperpigmented macule with crusts. (**b**) Six hours after topical application of ALA (20%) under irradiation with Wood's light (370–405 nm) = FDAP: Sharp delineation of the tumour-bound porphyrin fluorescence from the adjacent tissue. Tumour borders histologically corresponded to the extension of clinical fluorescence. There are some dark blue coloured areas (arrows), in particular on the right lower edge of the lesion, which corresponded to lentigo simplex lesions. There is no fluorescence in these benign lesions

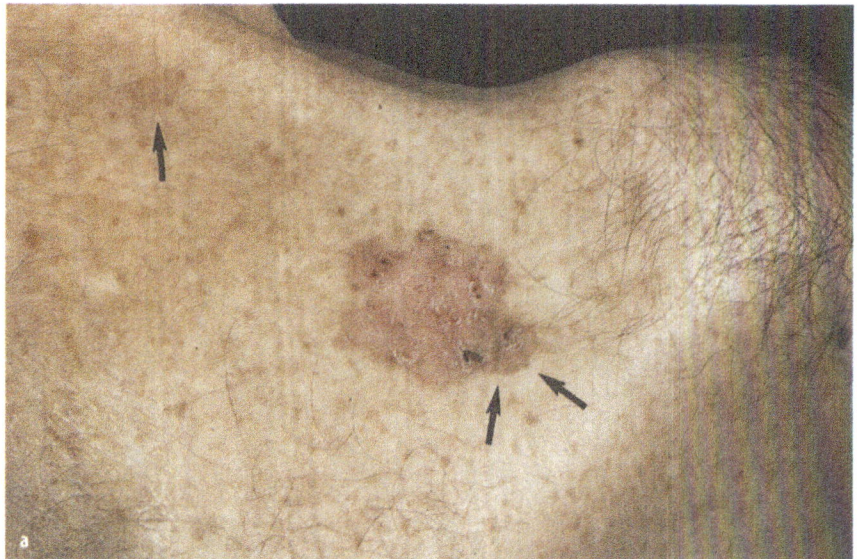

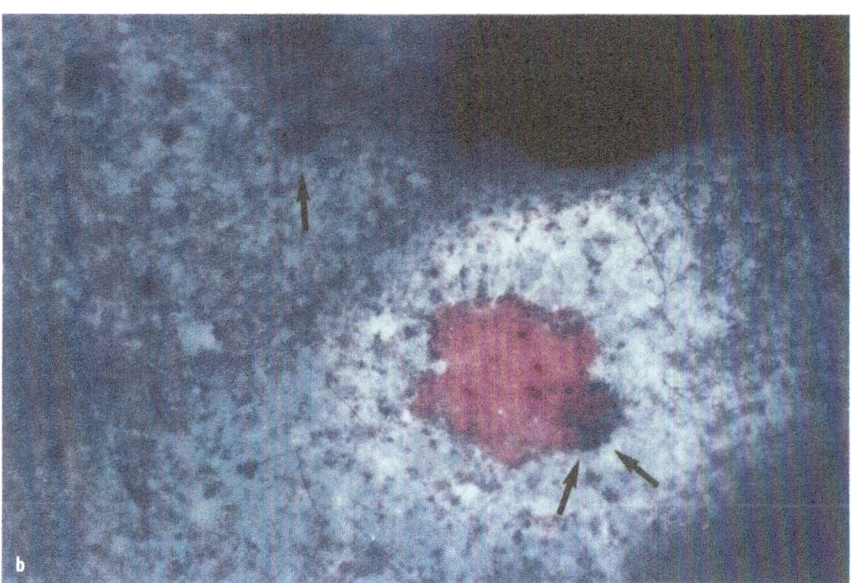

## In Situ Laser Coagulation

Depending on the localization and the size of the lesion, treatments are carried out under local or general anaesthesia. The Nd:YAG laser is the most important surgical laser. It can not only be used for cutting tissues but also for precise focal coagulation. Tumour tissue can be destroyed by non contact irradiation. Laser energy is alternatively delivered by means of interstitial placement of bare fibres. Depending on the response, coagulation therapy must to be repeated two or several times. Control of therapeutic procedures can be done by ultrasound or magnetic resonance imaging (MRI) [6, 79]. There might be some ulceration during the healing phase. Complications may also include wound infection.

## Criteria for Exclusion

### Photodynamic Therapy (PDT)

PDT using topical ALA should not be performed in large nodular BCC or SCC. In particular, tumours which infiltrate anatomically difficult areas of the face should be treated primarily by surgical excision. In addition, malignant melanoma (MM) may not be treated by ALA-PDT, although cells of MMs show sufficient porphyrin accumulation upon incubation with ALA [7].

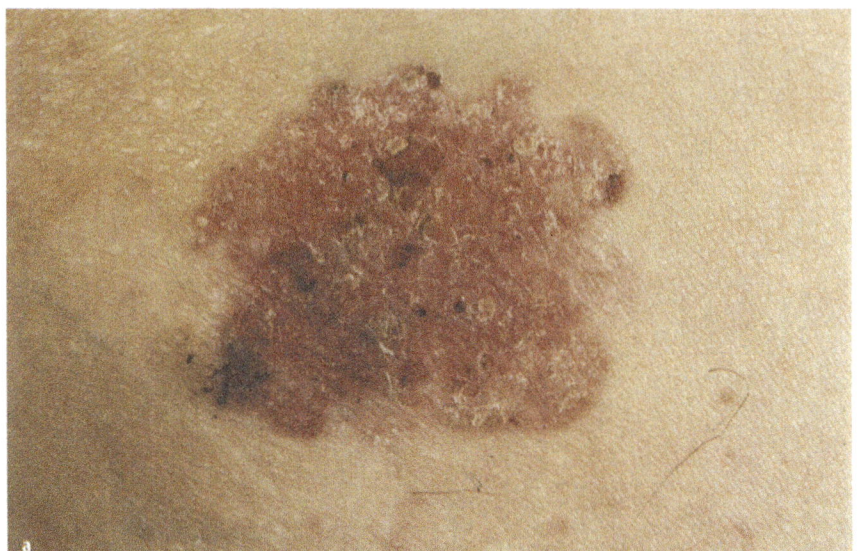

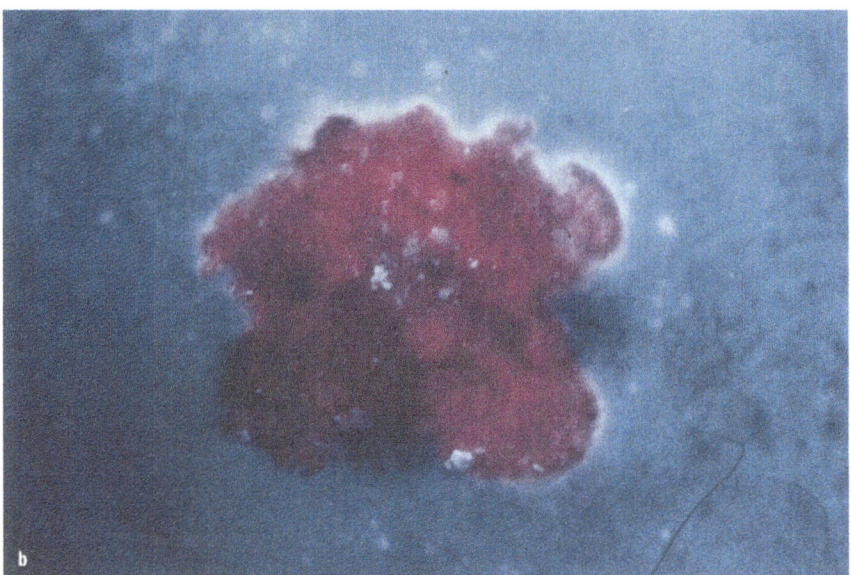

**Fig. 2.** (**a**) Bowen's disease on the right gluteal area: 2.5 x 3 cm, in part sharply bordered erythematous plaque. (**b**) FDAP: the porphyrin fluorescence allows a sharp demarcation of the tumour tissue

## Fluorescence Diagnosis with δ-Aminolevulinic Acid Induced Porphyrins (FDAP)

Since ALA was introduced to the PDT technique, neither our own experience in the 6 years nor the data of the literature shows any report on hypersensitivity to ALA. Thus, there is no contra indication for topical ALA application.

In the case of sclerodermiformic BCCs, the expressiveness of FDAP is partly limited.

## Photovaporization with $CO_2$ Laser / In Situ Laser Coagulation

$CO_2$ laser vaporization or Nd:YAG coagulation should not be performed in individuals who show an increased risk of scarring or keloid formation. In patients suffering from an allergy to local anaesthetics, $CO_2$ laser beams can be administered only in a limited manner. Laser vaporization or coagulation should be avoided if the lesion is clinically ill-defined or has an anatomically difficult localization due the lack of histopathological examination.

**Fig. 3.** (**a**) Multiple BCCs on the back. Several lesions were already pretreated by surgery and cryosurgery (white areas represent scars). The clinically suspect areas impress as red macules. (**b**) FDAP allows the detection and clear demarcation of all BCC lesions - present

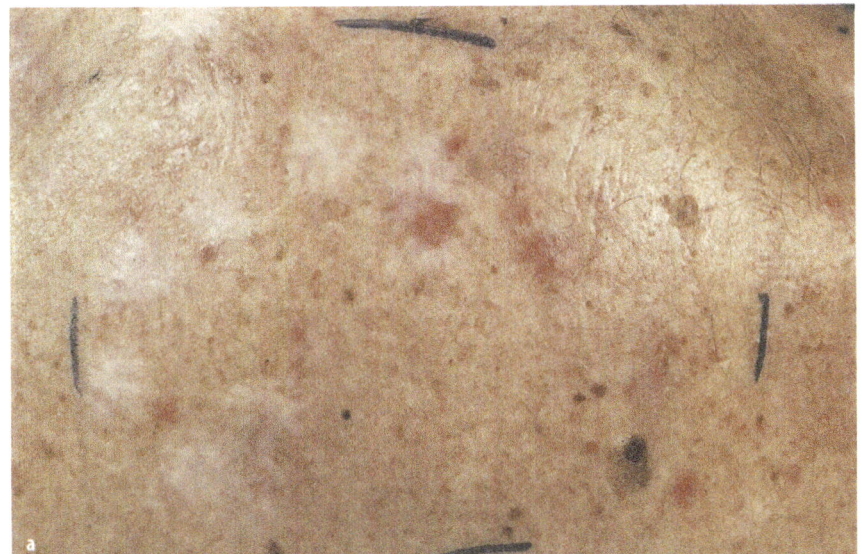

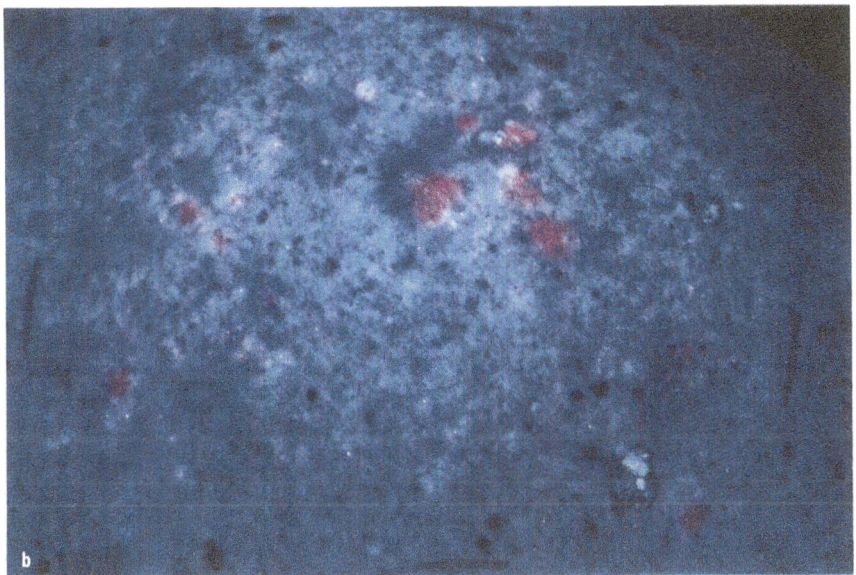

## Comparison with Alternative Methods

### Photodynamic Therapy (PDT)

PDT must be compared with all other treatment strategies for skin tumours: in contrast to surgery, PDT provides no histopathological data proving the total excision of a certain tumour. In comparison to cryosurgery, surgery or $CO_2$ laser therapy – also methods without any histopathological control – PDT does not lead to any scarring with a generally superior cosmetic result. Other tumour treatment techniques such as intralesional chemotherapeutics or topical fluorouracil, induce other side effects, e.g. local inflammation, fever or weakness.

## Fluorescence Diagnosis with δ-Aminolevulinic Acid Induced Porphyrins (FDAP)

FDAP is the only known technique giving data on the horizontal extension of a neoplastic tissue prior to surgical excision. Mohs' micrographic surgery allows the determination of the tumour borders only after its excision. FDAP can replace neither Mohs' surgery nor any other micrographically controlled surgery. However, FDAP gives important data on the delineation of clinically ill-defined tumours. This is essential in the management of skin tumours, in particular if they are localized in anatomically specific areas such as the nose or the ear.

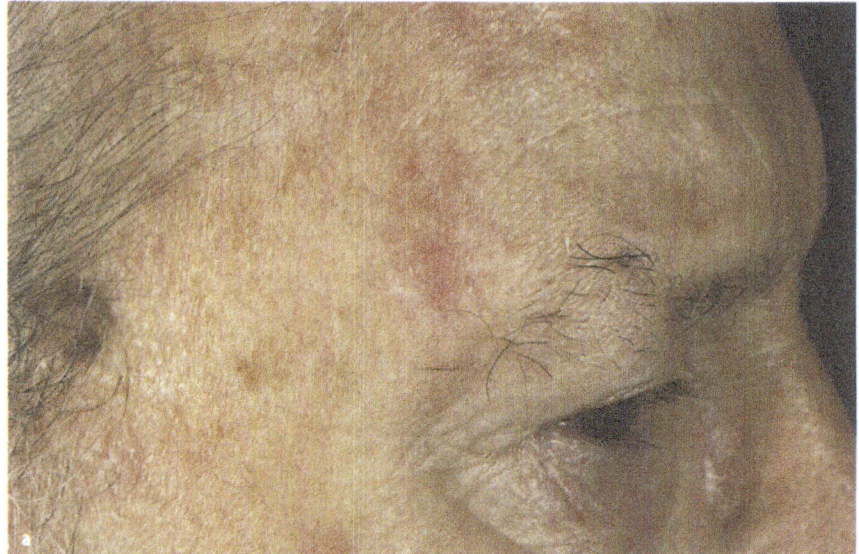

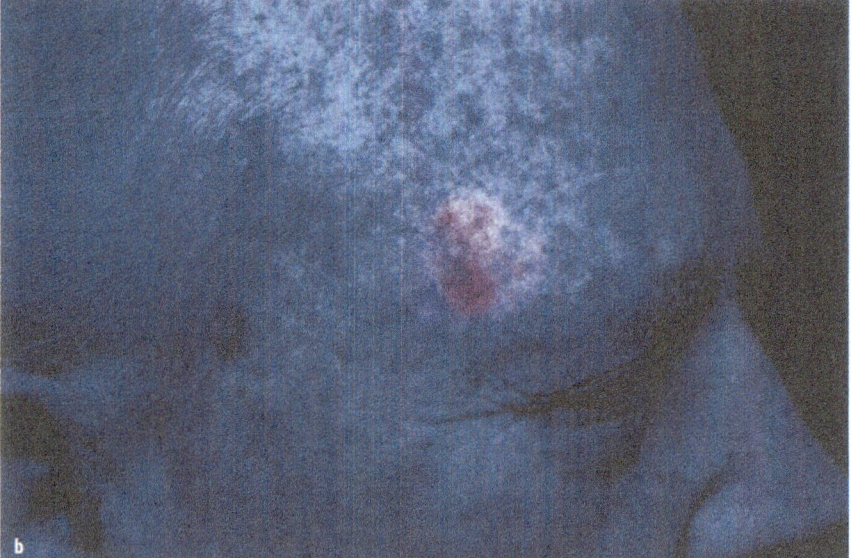

**Fig. 4.** (a) This lesion on the right temporal area was histopathologically verified to be a sclerodermiformic BCC. Clinically it is impossible to detect the borders of the tumour. (b) In FDAP, however, there is a relatively sharp delineation of the deep red fluorescence indicating the borders of the tumour. In addition to the tumour-specific fluorescence, there is also a pink-whitish fluorescence close to the strong intralesional fluorescence. This background fluorescence represents the porphyrin enrichment in the normal skin. Nevertheless, the tumour can be distinguished from the tumour-adjacent normal skin due to the fluorescence intensity. The borders of the fluorescent area were marked by a surgical pen, the lesion was excised according to the marking, and a complete excision of the lesion was histopathologically demonstrated

## Photovaporization with CO₂ Laser

$CO_2$ laser therapy represents the therapy of choice for specific indications such as leucoplakia, lichen sclerosus et atrophicus or erythroplasia Quéyrat with excellent functional and cosmetic results. In the case of cheilitis actinica, therapy results are also very encouraging, so that it might partly replace the more invasive surgical vermilonectomy [33]. Lesions such as BD can be treated effectively by laser vaporization, however, with the consequence of scarring and the lack of histopathology.

## In Situ Laser Coagulation

In situ photocoagulation is generally a palliative treatment strategy for tumours. Intralesional photocoagulation of haemangiomas with KTP or Nd:YAG lasers may be preferred to superficial laser treatment for the decrease of cutaneous skin damage and the more effective reduction of bulky deep lesions [1].

**Fig. 5.** (**a**) Multiple verrucae vulgares on the back of the hands and fingers. (**b**) There is no fluorescence detectable upon 6 h ALA treatment

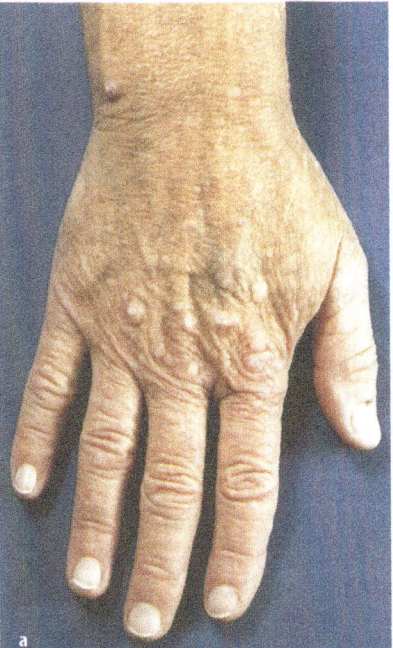
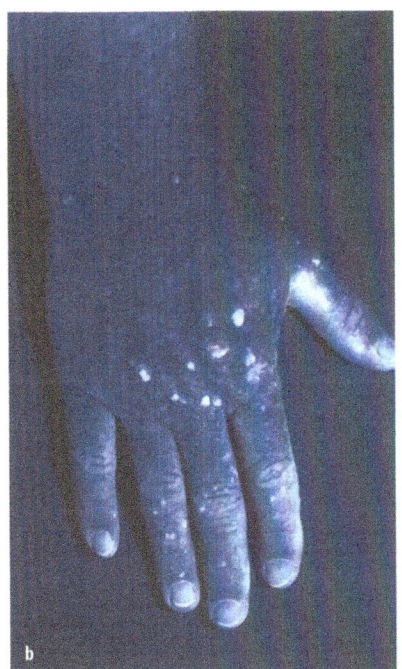

**Fig. 6.** (**a**) Multiple SK on the scalp. They were in part histopathologically proven to be initial SCCs. (**b**) FDAP clearly demarcates the extension of the lesions. The extended fluorescent areas on the lower part of the lesions were histologically proven to be SCCs. The fluorescing areas were efficiently treated twice by PDT (10% ALA and green light, 30 J cm⁻², 543–548 nm; PDT Saalmann, Herford, Germany). Irradiation was performed in four single sessions to treat all lesions in each quadrant of the capillitium.

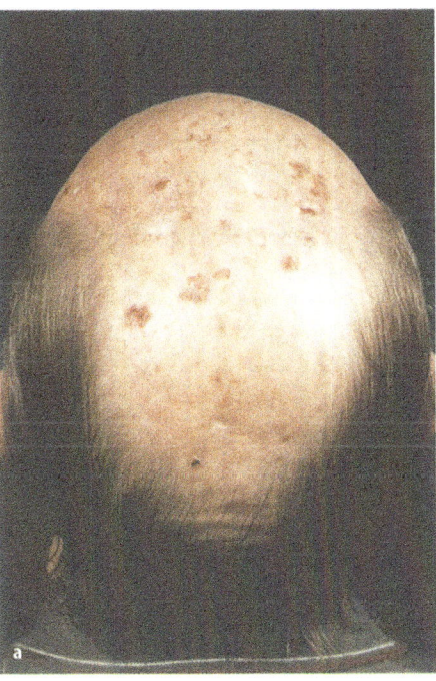
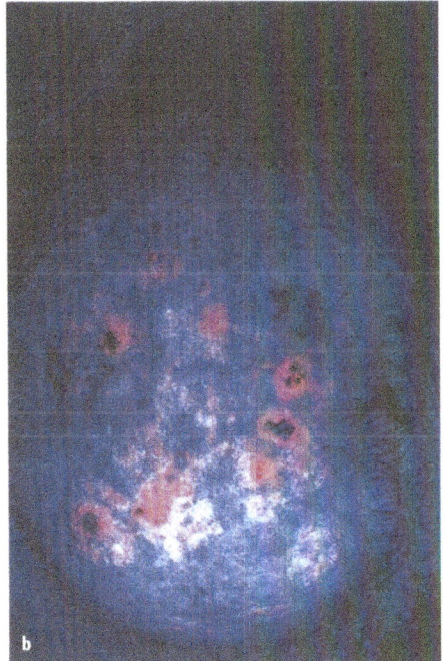

## Results

### Photodynamic Therapy (PDT)

In general, surgical excision is the most effective and preferential treatment for epithelial skin tumours. However, alternative methods are necessary for extensive or multiple disseminated lesions such as superficial BCC and actinic keratoses to improve functional and cosmetic results. In general, the outcomes of clinical studies on PDT treatment of skin tumours are difficult to compare since different parameters of PDT were used [21]. The following discussion will review mainly data on topical ALA-PDT.

### Basal Cell Carcinoma (BCC) (Figs. 1, 3, 4)

For systemically administered porphyrins, complete response rates (CR) from 31–100% were reported [2, 9, 12, 14, 49, 56, 62, 77, 80]. Topical ALA-PDT resulted in CR between 10–100% [11, 14, 16, 20, 21, 37, 47, 50, 65, 71, 77, 78, 80, 83] depending on the size and the thickness of the BCCs, and the CR was proportional to the applied light intensity [21]. In contrast to early enthusiastic clinical reports [76], histological and long-term follow-up studies showed a less favourable outcome especially in large tumours probably due to insufficient and inhomogeneous ALA penetration and inhomogeneous light irradiation [11, 48, 56]. In large tumours, primary surgery remains the treatment of choice; however, pretreatment with PDT may improve the cosmetic and functional outcome in difficult localizations [20].

### Bowen's Disease (BD) (Fig. 2)

In several patients, successful treatment of BD with porfimer sodium and an argon-PDL was described [9, 36, 62]. BD also showed a good initial response to PDT with ALA, but long-term results vary considerably (CR: 30–100%) [10, 16, 36, 71]. Even repeated PDT treatments yielded a CR of only 50–75% [16, 21]. Incomplete response of BD may be due to the thickened epithelial layer with reduced ALA-penetration [9, 10, 11, 16, 21, 36, 49, 51, 62, 67, 71, 76, 78].

### Solar Keratoses (SK) (Figs. 6, 7)

SKs represent one of the best indication for (ALA-) PDT in dermatology at present. In most clinical studies, a CR of 80 to 100% was achieved using 10 to 20% ALA [11, 16, 21, 34, 37, 73, 83]. The burning pain experienced by most patients during irradiation of multiple lesions on the scalp can be significantly reduced by using 10% ALA [21] and irradiating with green instead of red light [25]. SK on the arms and hands responded less well to ALA-PDT than lesions located on the scalp [73].

### Squamous Cell Carcinoma (SCC) (Fig. 6)

With systemic photosensitizers 40–100% remission of SCCs could be obtained [2, 14, 49, 56]. In topical ALA-PDT the CR was 60 to 92% for superficial SCCs [11, 21, 36, 83] and 0–67% for nodular SCCs [11, 21, 47]. Initial stages of SCCs could be treated effectively by topical ALA-PDT, and even in nodular SCCs promising remission rates could be obtained after repeated PDT [11, 21].

### Other Neoplastic Skin Disorders

PDT also seems to be a promising method in the treatment of premalignant epithelial lesions and SCCs of the oral mucosa [31, 52], genital precancerous stages like erythroplasia Queyrat [66], actinic cheilitis [69], Paget's disease (PD) [57], and tumours in xeroderma pigmentosum [82].

### Malignant Melanoma (MM)

So far, there is little information on the efficacy of ALA-PDT in the treatment of primary and metastatic MM, and the results are contradictory [12, 49, 83]. The strong pigmentation of melanoma tissues may be the limiting factor, by inhibiting light penetration.

### Cutaneous and Subcutaneous Metastases

Some clinical PDT studies focused on the treatment of breast cancer and other metastases, however, with minor benefit (systemic PDT: 4–75% CR [2, 28, 38, 45]; topical PDT with ALA or TPPS 0–83% CR [10, 11, 37, 42, 63].

### Mycosis Fungoides

Topical ALA application and subsequent exposure to polychromatic or laser light was effectively used to treat plaque-stage cutaneous T-cell lymphoma [65, 71, 81]. However, the apparent clinical cure was not confirmed histologically [3].

### Kaposi's Sarcoma

Classical and AIDS-related oral Kaposi's sarcoma lesions have been successfully treated showing early and late CR [12, 64].

**Fig. 7.** (**a**) Extended dissemination of SK on the forehead left of a 58-year-old male patient. (**b**) FDAP demarcates very impressively the dimension of skin affection and the exact localization of all lesions. (**c**) The fluorescing area was treated once by PDT (10% ALA and green light, 30 J cm$^{-2}$, 543–548 nm; PDT Saalmann, Herford, Germany). The therapeutic result is excellent, showing a plane skin of the scalp with some remaining postinflammatory hyperpigmentations

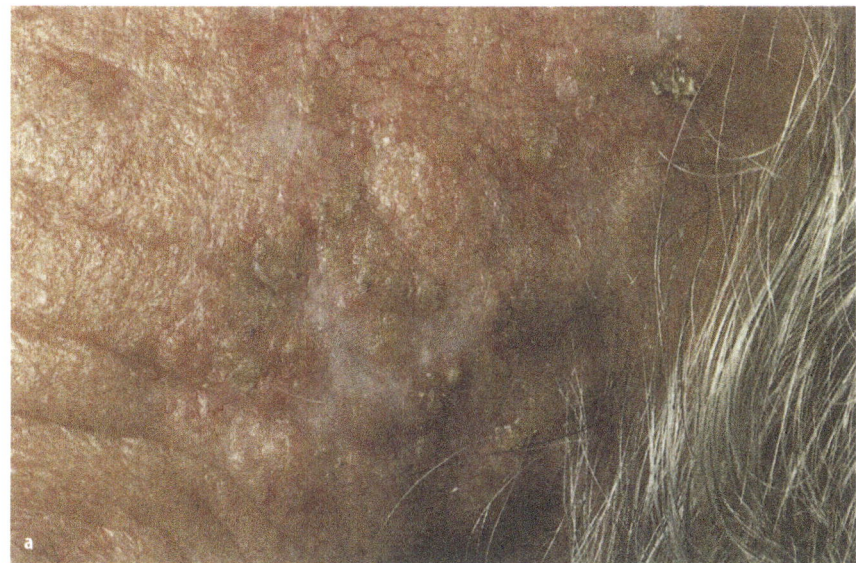

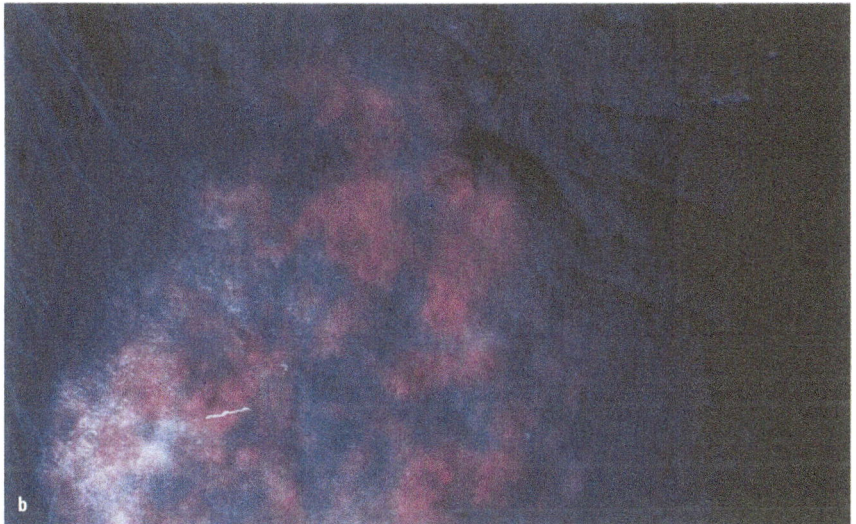

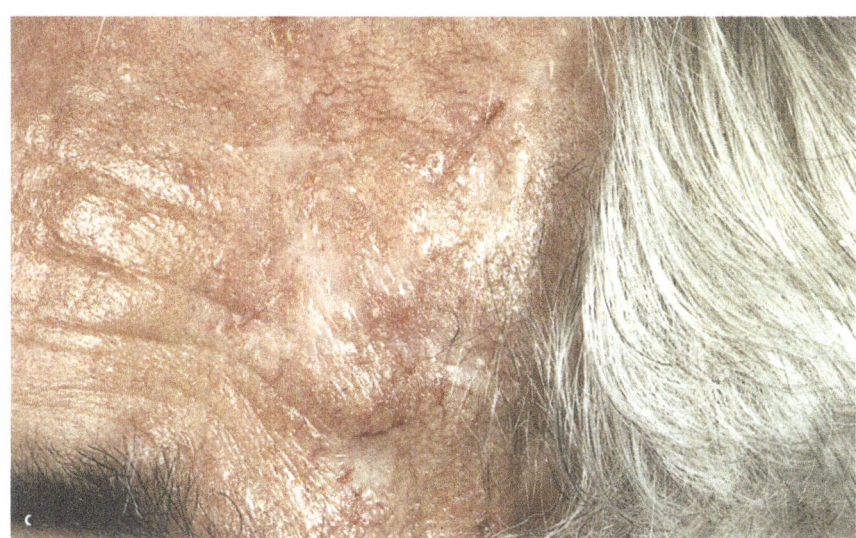

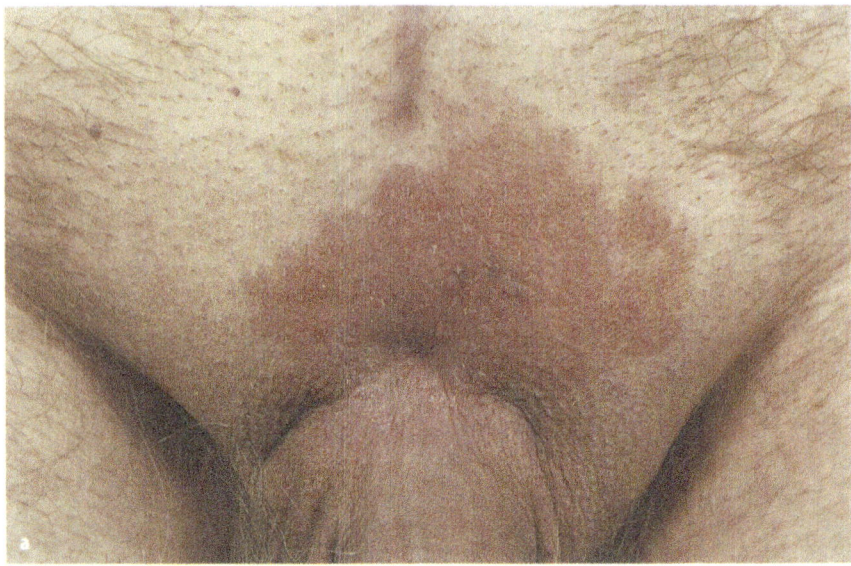

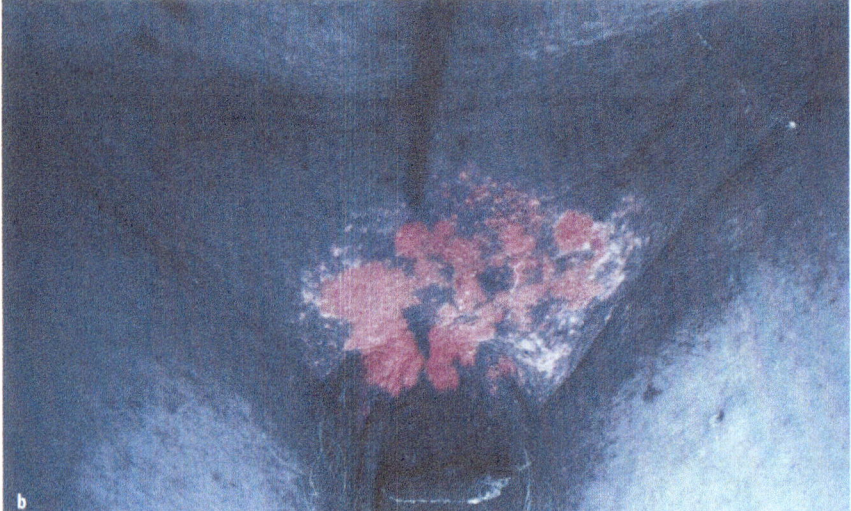

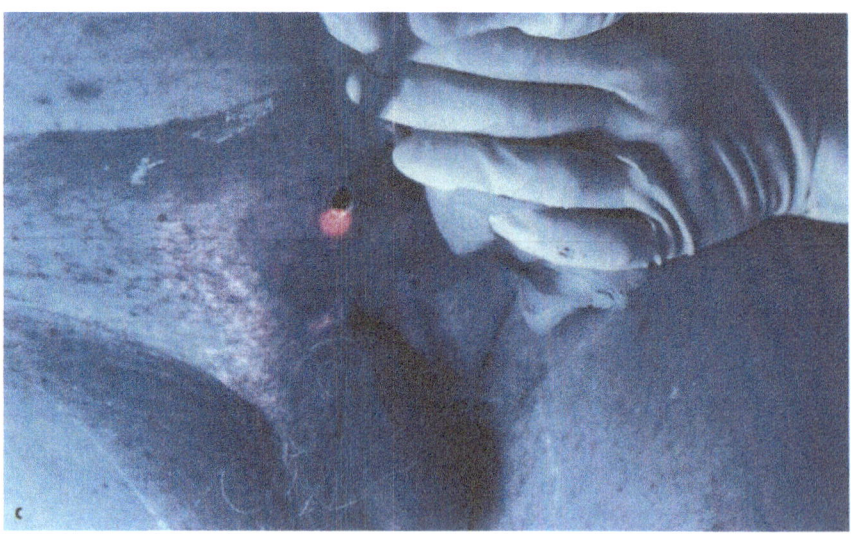

**Fig. 8. (a)** Extramammary Paget's disease in the suprapubic region of a 71-year-old patient. **(b)** FDAP: the fluorescing areas mark the neoplastic regions as proven by biopsies taken from the deeply red fluorescing areas. Under Wood's light irradiation, all fluorescing skin was treated by $CO_2$ laser vaporization (Limmer, Appen, Germany). We used the power of 15–20 W and a defocused beam with a diameter of 2-mm spot size. Vaporization was performed to a depth of approximately 2 mm and horizontally 1 cm beyond the fluorescent area. Postoperatively, a good granulation was evident, and, after 4 weeks, a good cosmetic result was achieved. **(c)** FDAP: 3 months after first laser therapy. This picture shows FDAP-guided $CO_2$ vaporization of a remaining lesion. The partly vaporized tissue impresses as a dark spot (upper part of the red lesion). The still untreated tumour area shows brick red fluorescence. The mild fluorescence shown in the ALA-treated area is located in the normal skin and is probably induced by propionibacteria. Biopsies taken from these areas showed no tumour cells

## Fluorescence Diagnosis with δ-Aminolevulinic Acid Induced Porphyrins (FDAP)

Epidermal neoplasms such as BCC (Figs. 1, 3, 4), SCC (Fig. 6), BD (Fig. 2), SK (Figs. 6, 7) and extramammary Paget's disease (Fig. 8) show an intensive uniform red fluorescence. All other melanotic or amelanotic, benign or malignant tumours, such as malignant melanoma, lentigo senilis, verruca seborrhoica or nevus cell nevus demonstrate no or only minimal fluorescence. Fluorescence is absent in all verrucae vulgares (Fig. 5). In plaques of mycosis fungoides an intermediate fluorescence intensity is demonstrable [23].

Uninvolved skin shows different fluorescence intensities depending on the anatomical area and ALA application time. The higher fluorescence intensity on face, axilla or groin as compared to trunk or extremities is probably due to an increased bacterial flora producing also relatively high porphyrin levels [46]. In these body areas, fluorescence in normal skin may interfere with that in neoplastic skin, but differentiation is still possible due to the higher fluorescence intensity in tumour tissue. The fluorescence intensity in normal skin is increased by a longer ALA exposure time [24], and preliminary data show that shorter application times of ALA, e.g. 3 h, or pretreatment with an erythromycin-containing cream facilitate and improve the differentiation between tumour and normal tissue fluorescence.

### FDAP for Preoperative Planning and Control of Tumour Therapy

In anatomically difficult sites such as the nose or the ear and especially in pretreated skin the detection of (regrowing) tumour areas can be facilitated by using FDAP (Figs. 3, 4). In general, all high intensive fluorescing areas probably represent neoplastic tissue, as proven in various studies by histopathology [19, 20] (Figs. 3, 4, 8). The clinical fluorescence corresponds well with the histological borders of the tumour [19, 20]. FDAP allows one to delineate clinically ill-defined tumours and to detect tumour relapse or new tumours which were clinically not visible [19].

### Photovaporization with the CO₂ Laser

For leucoplakia, lichen sclerosus et atrophicus and erythroplasia Quéyrat excellent functional and cosmetic results were achieved. The lesions in cheilitis actinica were effectively cured by the use of laser vaporization. However, patients were partly suffering from formation of crusts and pain until complete re-epithelialization 3 to 4 weeks after therapy. Rarely,

bacterial and viral (herpes simplex) superinfections occurred as side effects. The final outcome was sometimes influenced by scarring or grey-whitish discoloration of the lips.

In addition, the combination of vaporization technique with FDAP was shown to be very effective as an alternative technique in the treatment of extramammary Paget's disease (Fig. 8) [5]. The lesions of Paget's disease were nicely demarcated by FDAP. Under irradiation with Wood's light, all intensive fluorescing tissue was treated by laser vaporization. We also used this combined technique successfully in the treatment of BCC and SCC. Of course the $CO_2$ laser proved highly efficient in the therapy of benign lesions such as syringoma, trichoepithelioma, hyperplasia of the sebaceous glands, adenoma sebaceum, dermal or epidermal nevus, xanthelasma, neurofibroma, condylomata acuminata, focal endothelial hyperplasia (Heck's disease) and verrucae vulgares. For the modelling of extensive rhinophyma, a $CO_2$ laser with sufficient intensity ($> 20$ W) is necessary to achieve good cosmetic results. The therapeutic outcome of tattoos depended on the infiltration depth of the colour pigments and the amount of vaporized cutaneous tissue.

Hypertrophic scarring or keloid formation can be found in approximately 1% of the patients treated.

### In Situ Laser Coagulation

Interstitial laser photocoagulation induces successful palliation of pain and functional disabilities in patients with advanced and recurrent head and neck cancer [53]. In the treatment of haemangiomas by intralesional coagulation, in general, good to dramatic response rates are reported, although there seems to be a subset of haemangiomas which responds only poorly to laser therapy [1].

## Discussion

### Photodynamic Therapy (PDT)

PDT with ALA is very effective in the treatment of SK, superficial BCCs, and initial SCCs. Results of PDT in epithelial skin tumours should, however, be viewed critically due to the methodological shortcomings of many studies. Histological examinations demonstrated tumour tissue in a large proportion of tumours despite clinical regression after a single PDT cycle, and tumor recurrence was common after long term follow-up [11, 21]. In conclusion, follow-up periods in most published studies were too short. The lack of topical PDT efficacy in more deeply localized and remaining tumour parts may be due to the limited photophysical properties of PDT, the limited permeability for ALA due to overlying normal skin and the presence of encapsulated tumour cell islands resistant to ALA permeability [24, 55]. Thus, retreatment is useful before normal skin can cover possible tumour remnants in deeper tissue layers. Debulking of the superficial (exophytic) tumour parts prior to ALA application was shown to increase the efficacy of ALA-PDT, due to better and more homogeneous sensitization of the deeper tumour layers.

Recently, the topical application of ALA methylester proved to be highly effective in PDT. Its selectivity in lesional porphyrin enrichment is higher as compared to free ALA [22]. ALA methylester is now distributed as Metrix® by Galderma.

## Fluorescence Diagnosis with δ-Aminolevulinic Acid Induced Porphyrins (FDAP)

Topical application of ALA in amounts from 0.05 to 0.2 g cm$^{-2}$ (total: 0.05 to 7.0 g ALA) does not lead to measurable systemic porphyrin levels in humans [26]. Therefore, no systemic side effects are observed in topical ALA application techniques. In contrast, in PDT with systemically administered ALA, transient rises in serum aspartate aminotransferase, nausea, vomiting, headache, circulatory failure and prolonged photosensitivity have been reported [17, 21, 61]. This might partly be due to the prolonged increase in hepatic protoporphyrin levels which was shown to be 50-fold in hamsters treated with 500 mg ALA/kg b.w. i.v. although ALA and porphyrins were cleared rapidly from the blood and the skin [17].

Our experiences with FDAP underline an increased ALA-induced porphyrin biosynthesis in neoplastic tissues. It is postulated that ALA treatment induces a high ALA enrichment and a selective accumulation of porphyrin metabolites in tumours [24, 55]. The mechanism of preferential intratumoral uptake of precursors and photosensitizers is still not fully understood. In the case of ALA, active transport is the most likely explanation, but passive diffusion may be operative as well. Enzymatic differences between normal and neoplastic tissue such as a lower activity of the ferrochelatase, which in erythropoietic protoporphyria (EPP) leads to an accumulation of protoporphyrin [29], seem to be less effective in ALA-induced porphyrin sensitization. Basal porphyrin levels in various human tissues did not reveal any changed values in tumour tissue in comparison to normal tissues [18, 30]. The metabolite mainly formed upon topical ALA application is protoporphyrin [17, 18]. The level of synthesized porphyrins mainly depends on the of ALA amount penetrating through the skin into the neoplastic cells. Thus, we assume a reduced ALA pen-

etration or uptake in lesions such as verruca vulgaris and verruca seborrhoica. In the case of psoriatic lesions the hyperkeratotic areas may limit ALA penetration and additionally, light excitation. Due to the photophysical properties of the skin [8], the efficacy of FDAP is limited to the horizontal dimension of lesions.

The clinical detection of the borders of BCCs and SCCs, particularly in anatomically difficult sites such as the face, is a frequent problem. Multiple surgical procedures can become necessary for complete tumour removal. FDAP has been demonstrated to be very effective in detecting and demarcating also clinically ill-defined tumour tissue (Figs. 3, 4, 6) [23]. Thus, we recommend FDAP as a useful easy technique to visualize and detect the extension of the tumour preoperatively and to control the therapeutic efficacy of any tumour treatment [4, 19, 20].

## Photovaporization with the $CO_2$ Laser

The induction of relatively high thermic damage of the tissue surrounding the irradiated area is one (side)effect of the standard $CO_2$ laser. Nevertheless, this technique allows a gentle vaporization of a benign or a malignant skin tumour. The heat coagulation of the epidermal and dermal layers facilitates lesion removal without bleeding. The lesions to be treated by $CO_2$ laser photovaporization should be selected carefully. In the case of malignant skin tumours such as progressive BCC or SCC, the palliative destruction of tumour tissue is reasonable. However, for the cure of initial BCC or SCC in the face, techniques such as PDT or FDAP-guided surgery should be preferred.

Aesthetical operations for, e.g. periorbital resurfacing or correction of a rhinophyma, are subjects of present interest. Here the use of short- and superpulsed $CO_2$ lasers promises good results with even few undesired effects on the surrounding skin.

## In Situ Laser Coagulation

Intralesional photocoagulation treatment with the KTP and Nd:YAG lasers is effective and safe for the treatment of haemangiomas (e.g. periorbital) in the majority of patients with minimal complications. Additionally, in situ laser coagulation offers an effective method to cure or reduce enlarged cutaneous neoplasms. Combining intratumoral chemotherapy (e.g. cisplatinum) with intralesional laser phototherapy may even increase the efficacy of palliative laser therapy for advanced head and neck cancer [53]. It may allow treatment of deep and difficult-to-reach tumours e.g. in the head and neck and cutaneous areas. For optimum therapeutic control improved non invasive monitoring techniques of laser–tissue interactions must to be developed.

## Final Comment

Laser systems are being increasingly used in the field of dermatological medicine [27].

Future progress will be achieved with the development of more effective light sources, fractionated light irradiation, the use of new, promising compounds such as esterified ALA derivatives [22] or second-generation photosensitizers such as porphycenes and systemic administration of ALA may enhance FDAP and PDT efficacy. In addition, debulking of the superficial tumour parts may optimize FDAP and PDT efficacy, especially in nodular tumours.

## Acknowledgements

We thank Mrs. K. Kleinert for assistance in the preparation of the manuscript.

## References

[1] ACHAUER BM, CHANG CJ, VANDERKAM VM, BOYKO A (1999) Intralesional photocoagulation of periorbital hemangiomas. Plast Reconstr Surg 103: 11–16

[2] ALLEN RP, KESSEL D, THARRATT RS, VOLZ W (1992) Photodynamic therapy of superficial malignancies with Npe6 in man. Elsevier; Amsterdam, PP: Photodynamic therapy and biomedical lasers. 441–445

[3] AMMANN R, HUNZIKER T (1995) Photodynamic therapy for mycosis fungoides after topical photosensitization with 5-aminolevulinic acid. J Am Acad Dermatol 33:541

[4] BECKER-WEGERICH P, FRITSCH C, NEUSE W, SCHULTE KW, RUZICKA T, GOERZ G (1995) Effektive Kryochirurgie oberflächlicher Hauttumoren unter photodynamischer Diagnostik. Zeitschr. Hautkr. H+G 70: 891–895

[5] BECKER-WEGERICH P, FRITSCH C, SCHULTE KW, MEGAHED M, NEUSE W, GOERZ G, STAHL W, RUZICKA T (1998) Carbon dioxide laser treatment of extramammary Paget's disease guided by photodynamic diagnosis. Br J Dermatol 138. 169–172

[6] BLACKWELL KE, CASTRO DJ, SAXTON RE, NYERGES A, CALCATERRA TC, SCHILLER V, GRANT E, SOUDANT J, HIRSCHOWITZ S, HAWKINS R (1993) MRI and ultrasound guided interstitial Nd:YAG laser phototherapy for palliative treatment of advanced head and neck tumours: clinical experience. J Clin Laser Med Surg 11: 7–14

[7] BOLSEN K, LANG K, VERWHOLT B, FRITSCH C, GOERZ G (1996) In vitro incubation of porphyrin biosynthesis in various human cells after incubation with d-aminolevulinic acid. Arch Dermatol Res 288: 320

[8] BRULS WAG, SLAPER H, LEUN VAN DER JC, BERRENS L (1984) Transmission of human epidermis and stratum corneum as a function of thickness in the ultraviolet and visible wavelengths. Photochem Photobiol 40: 485–494

[9] BUCHANAN RB, CARRUTH JAS, MCKENZIE AL, WILLIAMS SR (1989) Photodynamic therapy in the treatment of malignant tumours of the skin and head and neck. Eur J Surg Oncol 15:400–406

[10] CAIRNDUFF F, STRINGER MR, HUDSON EJ, ASH DV, BROWN SB (1994) Superficial photodynamic therapy with topical 5-aminolevulinic acid for superficial primary and secondary skin cancer. Br J Cancer 69: 605–608

[11] CALZAVARA-PINTON PG (1995) Repetitive photodynamic therapy with topical δ-aminolaevulinic acid as an appropriate approach to the routine treatment of superficial non-melanoma skin tumours. J Photochem Photobiol B: Biol 29:53–57

[12] DOUGHERTY TJ (1981) Photoradiation therapy for cutaneous and subcutaneous malignancies. J Invest Dermatol 77:122–124

[13] DOVER JS, ARNDT KA, DINEHART SM, FITZPATRICK RE, GONZALEZ E, GUIDELINES/OUTCOMES COMMITTEE (1999) Guidelines of care for laser surgery. J Am Acad Dermatol 41: 484–495

[14] FEYH J, GUTMANN R, LEUNIG A (1993) Die photodynamische Lasertherapie im Bereich der Hals-, Nasen-, Ohrenheilkunde. Laryngo-Rhino-Otol 72:273–278

[15] FIGGE FHJ, WEILAND GS, MANGANIELLO LOJ (1948) Cancer detection and therapy: affinity of neoplastic, embryonic, and traumatized tissues for porphyrins and metalloporphyrins. Proc Soc Exp Biol Med 68: 640–641

[16] FIJAN S, HÖNIGSMANN H, ORTEL R (1995) Photodynamic therapy of epithelial skin tumours using deltaaminolevulinic acid and desferrioxamine. Br J Dermatol 133: 282–288

[17] FRITSCH C, ABELS C, GOETZ AE, STAHL W, BOLSEN K, RUZICKA T, GOERZ G, SIES H (1997) Porphyrins preferentially accumulate in a melanoma following intravenous injection of 5-aminolevulinic acid. Biol Chem 378: 51–57

[18] FRITSCH C, BATZ J, BOLSEN K, SCHULTE KW, ZUMDICK M, RUZICKA T, GOERZ G (1997) Ex vivo application of δ-aminolevulinic acid induces high and specific porphyrin levels in human skin tumours: possible basis for selective photodynamic therapy. Photochem Photobiol 66:114–118

[19] FRITSCH C, BECKER-WEGERICH PM, MENKE M, RUZICKA T, GOERZ G, OLBRISCH RR (1997) Successful surgery of multiple recurrent basal cell carcinomas guided by photodynamic diagnosis. Aesthetic Plast Surg 21: 437–439

[20] FRITSCH C, BECKER-WEGERICH PM, SCHULTE KW, NEUSE W, LEHMANN P, RUZICKA T, GOERZ G (1996) Photodynamische Therapie und Mamillenplastik eines großflächigen Rumpfhautbasalioms der Mamma. Effektive Kombinationstherapie unter photodynamischer Diagnostik. Hautarzt 47: 438–442

[21] FRITSCH C, GOERZ G, RUZICKA T (1998) Photodynamic therapy in dermatology. A review. Arch Dermatol 134: 207–214

[22] FRITSCH C, HOMEY B, STAHL W, LEHMANN P, RUZICKA T, SIES H (1998) Preferential relative porphyrin enrichment in solar keratoses upon topical application of δ-aminolevulinic acid methylester. Photochem Photobiol 68: 218–221

[23] FRITSCH C, LANG K, NEUSE W, RUZICKA T, LEHMANN P (1998) Photodynamic diagnosis and therapy in dermatology. Skin Pharmacol Appl Skin Physiol 11: 358–373

[24] FRITSCH C, LEHMANN P, STAHL W, SCHULTE KW, BLOHM E, LANG K, SIES H, RUZICKA T (1999) Optimum porphyrin accumulation in epithelial skin tumours and psoriatic lesions after topical application of δ-aminolaevulinic acid. Br J Cancer 79: 1603–1608

[25] FRITSCH C, STEGE S, SAALMANN G, GOERZ G, RUZICKA T, KRUTMANN J (1997) Green light is effective and less painful than red light in photodynamic therapy of facial solar keratoses. Photodermatol Photoimmunol Photomed 13: 181–185

[26] FRITSCH C, VERWOHLT B, BOLSEN K, RUZICKA T, GOERZ G (1996) Influence of topical photodynamic therapy with 5-aminolevulinic acid on the porphyrin metabolism. Arch Dermatol Res 288: 517–521

[27] FUCHS B, BERLIEN HP, PHILIPP C (1999) Lasers in medicine. Z Ärztl Fortbild Qualitätssich 93: 259–266

[28] GILSON D, ASH D, DRIVER I, FEATHER JW, BROWN S (1988) Therapeutic ratio of photodynamic therapy in the treatment of superficial tumours of skin and subcutaneous tissue in man. Br J Cancer 58:665–667

[29] GOERZ G, BUNSELMEYER S, BOLSEN K, SCHÜRER NY (1996) Ferrochelatase activity in patients with erythropoietic protoporphyria and their families. Br J Dermatol 134:880–885

[30] GOERZ G, LINK-MANNHARDT A, BOLSEN K, ZUMDICK M, FRITSCH C, SCHÜRER NY (1995) Porphyrin concentrations in various human tissues. Exp Dermatol 4: 218–220

[31] GRANT EW, HOPPER C, MACROBERT AJ, SPEIGHT PM, BOWN SG (1993) Photodynamic therapy of oral cancer: photosensitisation with systemic aminolaevulinic acid. Lancet 324: 147–148

[32] GREGORIE HG JR, HORGER EO, WARD JL (1968) Hematoporphyrin-derivate fluorescence in malignant neoplasms. Ann Surg 167: 820–828

[33] HOHENLEUTNER S, LANDTHALER M, HOHENLEUTNER U (1999) $CO_2$-Laservaporisation der Cheilitis actinica. Hautarzt 50: 562–565

[34] JEFFES EW, MCCULLOUGH JL, WEINSTEIN GD, FERGIN PE, NELSON S, SHULL TF, SIMPSON KR, BUKATY LM, HOFFMAN WL, FONG NL (1997) Photodynamic therapy of actinic keratosis with topical 5-aminolevulinic acid. A pilot dose-ranging study. Arch Dermatol 133:727–732

[35] JESIONEK A, TAPPEINER VON H (1905) Zur Behandlung von Hautcarcinome mit fluorescierenden Stoffen. Arch Klin Med 82: 72–6

[36] JONES CM, MANG T, COOPER M, WILSON BD, STOLL HL (1992) Photodynamic therapy in the treatment of Bowen's disease. J Am Acad Dermatol 27:979–982

[37] KENNEDY JC, POTTIER RH, PROSS DC (1990) Photodynamic therapy with endogenous protoporphyrin IX: basic principles and present clinical experience. J Photochem Photobiol 6: 143–148

[38] KHAN SA, DOUGHERTY TJ, MANG TS (1993) An evaluation of photodynamic therapy in the management of cutaneous matastases of breast cancer. Eur J Cancer 29A: 1686–1690

[39] KINSEY JH, CORTESE DA, SANDERSON DR (1978) Detection of haematoporphyrin fluorescence during fiberoptic bronchoscopy to localize early bronchogenic carcinoma. Mayo Clin Proc 53: 594–600

[40] KRIEGMAIR M, BAUMGARTNER R, KNÜCHEL R, STEPP H, HOFSTÄDTER F, HOFSTETTER A (1996) Detection of early bladder cancer by 5-aminolevulinic acid induced porphyrin fluorescence. J Urol 155:105–109

[41] LAM S, PALCIC B, MCLEAN D, HUNG J, KORBELIK M, PROFIO E (1990) Detection of early lung cancer using low-dose Photofrin II. Chest 97: 333–337

[42] LAPES M, PETERA J, JIRSA M (1996) Photodynamic therapy of cutaneous metastases of breast cancer after local application of meso-tetra-(para-sulphophenyl)-porphin (TPPS4). J Photochem Photobiol B 36: 205–207

[43] LEUNIG A, RICK K, STEPP H, GUTMANN R, ALWIN G, BAUMGARTNER R, FEYH J (1996) Fluorescence imaging and spectroscopy of 5-aminolevulinic acid-induced protoporphyrin IX for the detecion of neoplastic lesions in the oral cavity. Am J Surg 172: 674–677

[44] LIPSON RL, BALDES EJ, OLSEN AM (1961) The use of a derivate of hematoporphyrin in tumour detection. J Natl Cancer Inst 26: 1–4

[45] LOWDELL CP, ASH DV, DRIVER I, BROWN SB (1993) Interstitial photodynamic therapy. Clinical experience with diffusing fibres in the treatment of cutaneous and subcutaneous tumours. Br J Cancer 67: 1398–1403

[46] LUCCHINA LC, KOLLIAS N, GILLIES R, PHILLIPS SB, MUCCINI JA, STILLER MJ, TRANICK RJ, DRAKE LA (1996) Fluorescence photography in the evaluation of acne. J Am Acad Dermatol 35: 58–63

[47] LUI H, SALASCHE S, KOLLIAS N, WIMBERLY J, FLOTTE T, MCLEAN D, ANDERSON RR (1995) Photodynamic therapy of nonmelanoma skin cancer with topical

aminolevulinic acid: a clinical and histologic study. Arch Dermatol 131: 737–738

[48] MARTIN A, TOPE WD, GREVELINK JM, STARR JC, FEWKES JL, FLOTTE TJ, DEUTSCH TF, ANDERSON RR (1995) Lack of selectivity of protoporphyrin IX fluorescence for basal cell carcinoma after topical application of 5-aminolevulinic acid: implications for photodynamic treatment. Arch Dermatol Res 287: 665–674

[49] MCCAUGHAN JS, GUY JT, HICKS W, LAUFMANN L, NIMS TA, WALKER J (1989) Photodynamic therapy for cutaneous malignant neoplasms. Arch Surg 124: 211–216

[50] MORTON CA, MACKIE RM, WHITEHURST C, MOORE JV, MCCOLL JH (1998) Photodynamic therapy for basal cell carcinoma: effect of tumour thickness and duration of photosensitizer application on response [letter]. Arch Dermatol 134: 248–249

[51] MORTON CA, WHITEHURST C, MOSELEY H, MCCOLL JH, MOORE JV, MACKIE RM (1996) Comparison of photodynamic therapy with cryotherapy in the treatment of Bowen's disease. Br J Dermatol 135: 766–771

[52] NAUTA JM, VAN LEENGOED HLLM, STAR WM, ET AL (1996) Photodynamic therapy of oral cancer – a review of basic mechanisms and clinical applications. Eur J Oral Sci 104: 69-81

[53] PAIVA MB, GRAEBER IP, CASTRO DJ, SUH MJ, PAEK WH, ESHRAGHI AA, SAXTON RE (1998) ((Laser and cisplatinum for treatment of human squamous cell carcinoma)) Laryngoscope 108: 1269–1276

[54] PASS HI. Photodynamic therapy in oncology (1993) Mechanism and clinical use. J Natl Cancer Inst 85: 443–456

[55] PENG Q, WARLOE T, MOAN J, HEYERDAHL H, STEEN HB, NESLAND JM, GIERCKSKY KE (1995) Distribution of 5-aminolevulinic acid-induced porphyrins in noduloulcerative basal cell carcinoma. Photochem Photobiol 62: 906–913

[56] PENNINGTON DG, WANER M, KNOX A (1988) Photodynamic therapy for multiple skin cancers. Plast Reconstr Surg 82: 1067–1071

[57] PETRELLI NJ, CEBOLLERO JA, RODRIGUEZ-BIGAS M, MANG T (1992) Photodynamic therapy in the management of neoplasms of the perianal skin. Arch Surg 127: 1436–1438

[58] POLICARD A (1924) Etude sur les aspects offerts par des tumeurs expérimentales examinées à la lumière de Wood. Cr Soc Biol 91: 1423–1424

[59] RAAB O (1900) Über die Wirkung fluorescierender Stoffe auf Infusoriera. Z Biol 39: 524

[60] RASSMUSEN-TAXDAL DS, WARD GE, FIGGE FHJ (1955) Fluorescence of human lymphatic and cancer tissues following high doses of haematoporphyrin. Cancer 8: 78

[61] REGULA J, MACROBERT AJ, GORCHEIN A, BUONACORSI GA, THORPE SM, SPENCER GM, HATFIELD ARW, BOWN SG (1995) Photosensitisation and photodynamic therapy of oesophageal, duodenal, and colorectal tumours using 5-aminolevulinic acid-induced protoporphyrin IX – a pilot study. Gut 36: 67–75

[62] ROBINSON PJ, CARRUTH JAS, FAIRRIS GM (1988) Photodynamic therapy: a better treatment for widespread Bowen's disease. Br J Dermatol 119: 59–61

[63] SANTORO O, BANDIERAMONTE G, MELLONI E, MARCHESINI R, ZUNINO F, LEPERA P, DE PALO G (1990) Photodynamic therapy by topical meso-tetraphenylporphyrinesulfonate tetrasodium salt administration in superficial basal cell carcinomas. Cancer Res 50: 4501–4503

[64] SCHWEITZER VG, VISSCHER D (1990) Photodynamic therapy for treatment of AIDS related oral Kaposi's sarcoma. Otolaryngol Head Neck Surg 102: 639–649

[65] SHANLER SD, WAN W, WHITAKER JE, MANG TS, JONES C, WILSON BD, STOLL HL, PINCUS S, OSEROFF AR (1993) Topical δ-aminolevulinic acid for photodynamic therapy of cutaneous carcinomas and cutaneous T-cell lymphoma. J Invest Dermatol 101: 406a

[66] STABLES GI, STRINGER MR, ASH DV (1995) The treatment of erythroplasia of Quéyrat by topical aminole-

vulinic acid photodynamic therapy. Br J Dermatol 133(Suppl45): 30

[67] STABLES GI, STRINGER MR, ROBINSON DJ, ASH DV (1997) Large patches of Bowen's disease treated by topical aminolaevulinic acid photodynamic therapy. Br J Dermatol 136: 957–960

[68] STEINBACH P, KRIEGMAIR M, BAUMGARTNER R, HOFSTÄDTER F, KNÜCHEL R (1994) Intravesical instillation of 5-aminolevulinic acid: the fluorescent metabolite is limited to urothelial cells. Urology 44: 676–681

[69] STENDER IM, WULF HC (1996) Photodynamic therapy with 5-aminolevulinic acid in the treatment of actinic cheilitis. Br J Dermatol 135: 454-456

[70] STUMMER W, STOCKER S, WAGNER S, STEPP H, FRITSCH C, GOETZ C, GOETZ AE, KIEFMANN R, REULEN HJ (1998) Intraoperative detection of malignant gliomas by 5-aminolevulinic acid-induced porphyrin fluorescence. Neurosurgery 42: 518–526

[71] SVANBERG K, ANDERSON T, KILLANDER D, WANG I, STENRAM U, ANDERSSON-ENGELS S, BERG R, JOHANSSON J, SVANBERG S (1994) Photodynamic therapy of non-melanoma malignant tumours of the skin using topical δ-amino levulinic acid sensitization and laser irradiation. Br J Dermatol 130: 743–751

[72] SZEIMIES RM, ABELS C, FRITSCH C, KARRER S, STEINBACH P, BÄUMLER W, GOERZ G, GOETZ AE, LANDTHALER M (1995) Wavelength dependency of photodynamic effects after sensitization with 5-aminolevulinic acid in vitro and in vivo. J Invest Dermatol 105: 672–677

[73] SZEIMIES RM, KARRER S, SAUERWALD A, LANDTHALER M (1996) Photodynamic therapy with topical application of 5-aminolevulinic acid in the treatment of actinic keratoses: an initial clinical study. Dermatology 192: 246–251

[74] TAPPEINER VON H, JESIONEK A (1903) Therapeutische Versuche mit fluoreszierenden Stoffen. Münch Med Wochenschr. 50: 2042–2044

[75] TAPPEINER VON H, JODLBAUER A (1904) Ueber die Wirkung der photodynamischen (fluorescierenden) Stoffe auf Protozoen und Enzyme. Arch Klin Med 80: 427–487

[76] WALDOW SM, ROCCO VL, KOHLER IK, WALLK S, FRITTS TF (1987) Photodynamic therapy for treatment of malignant cutaneous lesions. Lasers Surg Medicine 7: 451–456

[77] WARLOE T, PENG Q, MOAN J, QVIST HL, GIERCKSKY KE (1992) Photochemotherapy of multiple basal cell carcinoma with endogenous porphyrins induced by topical application of 5-aminolevulinic acid. In: Spinelli P, Dal Fante M, Marchesini R (eds.). Photodynamic therapy and biochemical lasers. Elsevier, Amsterdam, pp 449-453

[78] WENNBERG AM, LINDHOLM LE, ALPSTEN M, LARKO O (1996) Treatment of superficial basal cell carcinomas using topically applied delta-aminolaevulinic acid and a filtered xenon lamp. Arch Dermatol Res 88: 561-564

[79] WERNER JA, LIPPERT BM, GOTTSCHLICH S, FOLZ BJ, FLEINER B, HOEFT S, RUDERT H (1998) Ultrasoundguided interstitial Nd:YAG laser treatment of voluminous haemangiomas and vascular malformations in 92 patients. Laryngoscope 108: 463–470

[80] WILSON BD, MANG T, STOLL H, JONES C, COOPER M, DOUGHERTY TJ (1992) Photodynamic therapy for the treatment of basal cell carcinoma. Arch Dermatol 128: 1597–1601

[81] WOLF P, FINK-PUCHES R, CERRONI L, KERL H (1994) Photodynamic therapy for mycosis fungoides after topical photosensitization with 5-aminolevulinic acid. J Am Acad Dermatol 31: 678–680

[82] WOLF P, KERL H (1991) Photodynamic therapy on a patient with xeroderma pigmentosum. Lancet 337: 1613–1614

[83] WOLF P, RIEGER E, KERL H (1993) Topical photodynamic therapy with endogenous porphyrins after application of 5-aminolevulinic acid: an alternative treatment modality for solar keratoses, superficial squamous cell carcinomas, and basal cell carcinomas? J Am Acad Dermatol 28: 17–21

# III-7.3
# Pigmented and non-Pigmented Benign Skin Tumours

R. Kaufmann and Ch. Beier

## Contents

## Introduction

Non malignant skin tumours are extremely common and comprise congenital malformations, nevi and a large variety of benign lesions, developing throughout life. They can originate from epithelial, melanocytic, vascular or other connective tissue cells. The treatment of vascular lesions is discussed separately within this book. Therefore, this chapter deals exclusively with pigmented lesions and non vascular benign skin tumours. Usually, their removal is requested for cosmetic reasons. Since these tumours all are easily accessible, a vast array of therapeutic options have evolved for their management. Indeed, most lesions can be alternatively treated by a number of methods, including simple approaches such as curettage, dermashaving or cryotherapy, and somewhat more sophisticated techniques, such as scalpel surgery, electrosurgery, topical chemotherapy or photodynamic treatment (Table 1) [56, 64, 66]. However, in many of the epithelial lesions, as well as in several other benign skin tumours, a more selective destruction will be achieved by laser techniques. Though selected cases of premalignant or even in situ skin malignancies can also be cured by surface-ablative techniques, it has to be kept in mind that with laser destruction a tissue specimen is not available. Grob and Dummer recently reported a case in which a diagnosis of amelanotic melanoma was delayed for 1 year due to previous $CO_2$ laser-treatment [40]. Though pigmented cells in such in situ neoplasms can be effectively targeted by modern Q-switched laser irradiation, leading to remissions in appropriate cases [88], not all of the tumour cells are melanin-containing, and moreover they also tend to invent deeper adnexal structures. Because of the frequency with which benign lesions can mimic cancer and vice versa, diagnosis should usually be established by biopsy prior to any destructive procedure whenever it cannot be clearly made by skin examination [56, 66]. Also, surgical removal should be the treatment of choice in all skin tumours requiring a histological control of complete eradication.

**Table 1.** Alternative treatment options for benign skin tumours

| Techniques | Examples of indications |
| --- | --- |
| 1. Surgical | |
| Excision | Melanocytic nevi |
| shaving, curettage | Seborrhoic keratosis, warts |
| 2. Thermal destruction | |
| Electrodessication | Warts |
| Cryotherapy | Actinic keratosis, haemangioma |
| Laser vaporization | Warts, benign epithelial tumours |
| Laser coagulation | Angioma |
| Laser ablation | Epidermal nevus, syringoma |
| Selective photothermolysis | Vascular lesions, pigmented lesions |
| 3. Others | |
| Photodynamic therapy | Actinic keratosis |
| Topical chemotherapy, chemocaustics | Actinic keratosis |

Due to the increasing versatility of laser treatment options, the range of applications in dermatology has been rapidly expanding. Hence, for epidermal, vascular and pigmented tumours, several laser systems can be used alternatively, with preference depending on the depth, volume and type of tumour growth (Table 2). Skin tumours are destroyed as a whole using either coagulation or, more specifically, selective photothermolysis. Superficially located and circumscribed lesion can also be removed stepwise by vaporization or ablation. The resulting superficial wounds, even on a convex surface such as the forehead or a nose, will usually heal well, since there is virtually no granulation tissue formation, and contraction as observed with full-thickness excisional wounds. However, the cosmetic result is not only dependent on the depth, size and location of the wound, but is also influenced by the skin colour and the age of the patient. Therefore, usually a test treatment should be recommended first, demonstrating the beneficial outcome also in critical areas or cases.

## Laser Treatment of Pigmented Benign Skin Tumours

Pigmented skin tumours usually contain melanin or hemosiderin as targets and can originate from several cell types and in different levels of the skin. The most commonly seen non vascular types are pigmented melanocytic nevi and pigmented seborrhoic keratosis (including senile lentigo as a superficial variant). Additional lesions that usually present in childhood and may benefit from laser treatment comprise epidermal nevi, café-au-lait macules, nevi of Ota, congenital melanocytic nevi, nevus spilus, lentigines, epidermal nevi or Becker's nevus [19, 82]. Melanin absorbs in a broad spectral range varying from 250–1200 nm and is therefore a good target for several lasers. This is why virtually any laser that produces light in the ultraviolet, visible or infrared can remove unwanted cutaneous pigment to some degree [25]. However, laser with emission lines that are both preferentially absorbed by melanin over other cutaneous chromophores and penetrate to the depth of the targeted pigment can be utilized to eliminate cutaneous

**Table 2.** Laser treatment of benign pigmented and non-pigmented skin tumours

| Diagnosis | Ablation | Vaporization | Coagulation | Selective Photothermolysis |
| --- | --- | --- | --- | --- |
| 1. Epidermal and adnexal tumours | | | | |
| Epidermal nevi | ++ | + | + | |
| Syringoma | ++ | | | |
| Sebacoeus gland hyperplasia | | | | |
| Viral warts | (+) | ++ | | |
| Seborrhoic keratosis | ++ | | | |
| Actinic keratosis | ++[a] | ++[b] | | |
| 2. Melanocytic tumours | | | | |
| Congenital melanocytic nevi | (+) | | | |
| Benign dermal nevi | (+) | | | |
| Nevus of Ota | | | | ++ |
| 3. Vascular tumours [b] | | | | |
| Senile angioma | | | ++ | ++ |
| Spider nevus | | | ++ | ++ |
| Haemangioma | | | ++ | |
| 4. Others | | | | |
| Neurofibroma, fibroma | (+) | (+) | | |
| Xanthelasma | ++ | + | + | |
| Osteoma cutis | ++ | | | |

(+) to be considered in selected cases, + possible option, ++ good indications.
a   As alternative option (of Table 1) and after histology-proven diagnosis.
b   Discussed in detail in a separate chapter.

melanin more selectively. Such a spectral window fulfilling both prerequisites lies between 630 and 1100 nm. By using exposure times that are less than the thermal relaxation time of melanin, the pigment targets are selectively destroyed by photothermolysis and therefore pulsed or Q-switched systems have an advantage over continuous-wave devices, used earlier for such purposes. Today, different Q-switched and pulsed laser systems are used to target melanin pigment at different wavelengths (504 and 510 nm dye laser, 511 nm copper vapour laser, 694 nm ruby laser, 755 nm alexandrite laser, 1064 nm Nd:YAG laser [15, 23, 32, 33, 25, 58, 60, 106]. Owing to their wavelength characteristic and pulse duration they can more or less selectively destroy pigment-containing organelles or cells by selective photothermolysis, allowing the melanin chromophore target to dissipate its heat without burning normal adjacent structures. Thus, a variety of pigment-specific lasers can be used to achieve significant clinical improvement, if not complete removal, of the majority of these lesions with greatly reduced risks of textural or pigmentary changes and scar formation [19]. Particularly in more macular lesions such as nevus of Ota or flat seborrhoic keratosis (senile lentigo) they mostly yield excellent results. In other pigmented skin tumours, however, they should be used with great caution and only after an exact diagnosis and confirmation of dignity has been established. As an alternative to selective photothermolysis, some of these pigmented skin tumours (e.g. pigmented epidermal nevus, lentigo, congenital melanocytic nevus) can be likewise or even better removed by a superficial etching using ablative laser systems [59].

## Congenital Melanocytic Nevi

Apart from the typical congenital nevus-cell nevi with their "giant" type and the more frequent smaller variants, these include distinct entities, among which the nevus of Ota is the most common deeper-located dermal melanocytosis and yields a high prevalence among individuals with Asian background.

Within the past decade, several studies have focused on the use of pigmented lesion lasers in the nevus of Ota (oculodermal melanosis). Meanwhile, virtually all available systems have been studied for this particularly challenging indication. Basically, each report more or less confirmed the excellent results in pigment removal.

The Q-switched ruby laser treatment (694 nm) is used with 6–10 J cm$^{-2}$ and can be expected to yield good to excellent results in approximately 2/3 of the patients after up to eight sessions [21, 31, 70, 87, 108]. The Q-switched alexandrite laser treatment (755 nm)

using 4.75–7.0 J cm$^{-2}$ in 8–12 week intervals showed complete lesional clearance after an average of five treatments in the majority of patients and no recurrences up to 1 year. Histology revealed elimination of upper dermal pigmentation without epidermal disruption and only rare melanophages and pigmented spindle cells in the deeper reticular dermis [1]. In a study from Korean patients, postinflammatory hyperpigmentation was seen in 55%, which resolved within 4 months. The therapeutic outcome was not affected by colour but by the depth of the nevus; while nevi of less than 1 mm showed excellent or good results, those with deeper nests were less likely to respond [53]. Also the Q-switched yttrium-aluminium-garnet laser (both 532 and 1064 nm) has been introduced as a therapeutic tool in patients suffering from nevus of Ota and showed improvement also after argon laser treatment failures [5]. We use it preferentially with energy densities of 2–3 J cm$^{-2}$ in repetitive sessions at its wavelength of 532 nm, better absorbed by melanin (Fig. 1). Apart from the Q-switched systems, also flashlamp-pumped dye lasers of appropriate wavelengths have been used successfully [81].

Hence, today, in contrast to earlier argon laser treatment or other types of removal (e.g. dermabrasion) used for this indication, treatment of nevus of Ota by pulsed or Q-switched laser systems is widely accepted as an effective and safe therapy [22]. Usually, with this type of treatment no scarring, textural changes, or pigmentary side effects are observed [1, 76]. Usually, several treatment sessions are required and effective rates tend to be higher in patients who not only underwent more irradiation treatments but also were observed for a longer period of time [102]. However, sometimes only partial benefit can be achieved. Nevertheless, even this can be considered a major help, since any improvement will be suited to relieve much of the psychological trauma usually associated with all disfiguring types of nevi.

In contrast to the nevus of Ota, other types of congenital melanocytic nevi are treated not only for cosmetic-aesthetic reasons but also because of their known increased life-time risk for malignant transformation, especially associated with larger types (giant nevi) [78]. Lesions of appropriate size and location for excisions (mostly serial operations or skin expander) are usually eradicated surgically, while larger one can be removed only partially or superficially. Since these nevi mostly exhibit dense pigmentation and also disfiguring hair growth (naevus pigmentosus et pilosus), potential targets for a laser destruction include melanocytic nevus cells, melanin pigment and hair follicles. However, as yet nothing is known about the long-term risk of laser irradiation on these types of melanocytic disorders, and therefore this therapeutical approach should be regarded

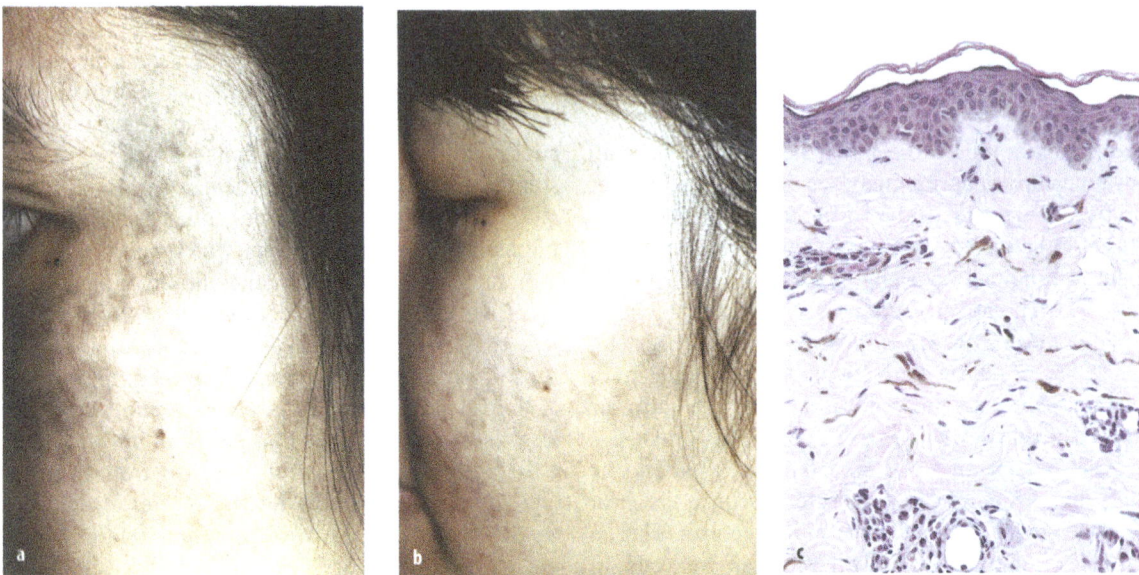

**Fig. 1a–c.** Selective photothermolysis of nevus of Ota before **a** and after **b** treatment (total of six sessions, 532-nm Q-switched Nd:YAG laser). **c** Histology with deeper spindle-shaped melanocytic nevus cells and pigmented melanophages

as still investigational, with the exception of superficial ablation, that can be used as an alternative to dermabrasion [Hohenleutner, Kaufmann 43, 65, 90] (Fig. 2). Since many of the nevus cell nests are still located in more superficial layers of the skin in newborns, it has been generally recommended to remove them early within the first few months of life [90]. This can be achieved by dermabrasion of superficial skin layers, which, as a whole, contain both pigment and nevus cells. However, deeper parts or excessive hairy follicles cannot be eliminated by this approach. Recently, dermabrasion has been more and more replaced by laser vaporization or laser ablation in these patients [7, 43, 67]. For this purpose, we prefer the use of the erbium: YAG laser, which, in our experience, has replaced dermabrasion in almost all of its former indications (Table 2). Removal of these nevi by vaporization or ablation later in life has the disadvantage that even with deeper removal the likelihood of recurrence and the risk of scar formation is increased. Moreover, following incomplete removal as with destructive techniques, $CO_2$ laser vaporization can lead to the formation of pseudomelanoma, resembling clinical and histologic features of malignant melanoma, but behaving benignly [110].

Congenital nevi of different size have also been treated later in life with the intention of eliminating cells and/or pigments by different pigment-specific laser systems. In normal-mode ruby laser treatment using 10–30 J cm$^{-2}$ and four treatment sessions, significantly reduced pigmentation almost to the level of surrounding skin, and no recurrence in a 18–39-

month study period were observed [111]. Also Imayama and coworkers found significant improvement in the majority of patients treated by the same technique, though even in those cases with good cosmetic outcome the presence of residual nevus cells was seen both in short-term and long-term histological observations [49]. They concluded that multiple treatments produced immediate thermal damage to the superficial nevus-cell nests and a subsequent subtle microscopic scar, masking the underlying residual nevus cells. After at least 8 years of follow-up, they found no evidence of malignant changes in the treated areas [49]. For a more selective approach, Q-switched lasers have been used in congenital nevi, including both the ruby laser [35, 114] and the Nd:YAG laser [38]. Though both systems affect the superficial as well as the deep portions of the congenital melanocytic nevi, neither laser is able to destroy all nevomelanocytes, particularly in the deeper, less pigmented portion of the lesions. In ex-vivo experiments, Kopera et al. demonstrated selective vacuolization of pigmented structures (melanin granules, melanophages, pigmented melanocytes, pigmented keratinocytes) after quality-switched ruby laser irradiation of cutaneous melanocytic lesions, while deeper-pigmented melanocytic cells persisted [68]. In comparison, the Q-switched ruby laser appears to be more effective in removing superficial nevomelanocytes [38]. In a medium-sized biopsy-proven congenital melanocytic nevus, even a complete clinical removal with no recurrence, scarring, dyspigmentation or atrophy after 5 years has been reported with

**Fig. 2a–b.** Laser ablation of large congenital nevus before **a** and after **b** treatment in two sessions. Left 1 month post-op: right 6 weeks post-op

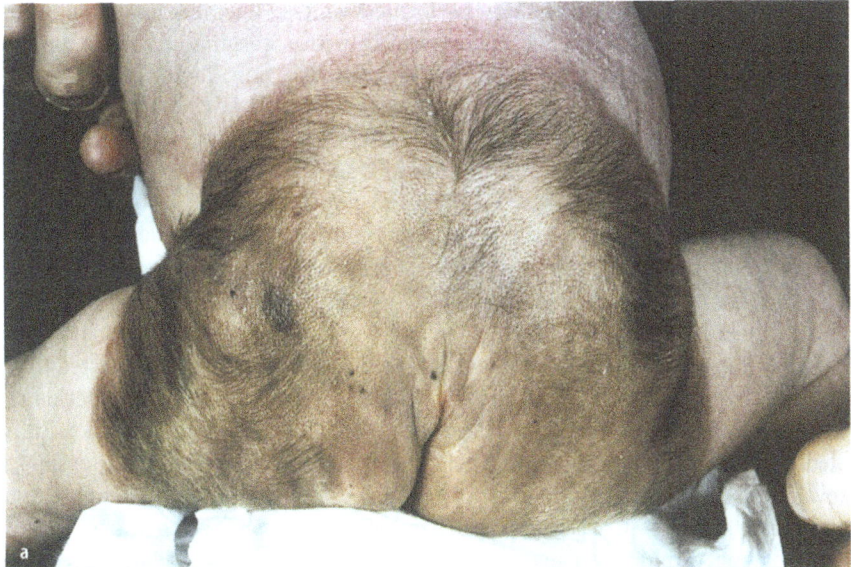

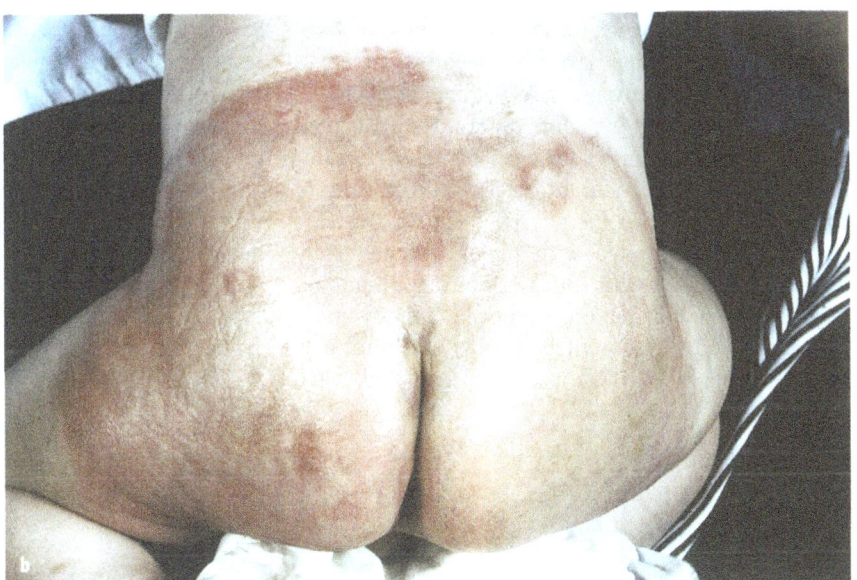

this laser [84]. Clinical experience has also been gained with the PLDL-1 pigmented lesion dye laser (pulsed dye laser at 510 nm, pulse duration 300 ± 50 ns) in the treatment of different pigmented birthmarks in 20 patients aged 2–17 years; 45% showed excellent results, 10% showed some lightening, 30% showed no improvement and 15% some hyperpigmentation at the test-patch sites which had not disappeared at 6 months' follow-up. No scarring or change in the clinical behaviour could be observed within that period [99].

## Acquired Melanocytic Nevi

The vast majority of nevocellular nevi develop throughout childhood and adolescence or even later in live. According to their major distribution pattern of cell aggregates, they are basically grouped into the three large families of junctional, compound or dermal nevi. Among these, atypical or "dysplastic" nevi fulfill criteria of suspicious pigmented lesions which need further dermatological evaluation or histology and thus are no candidates for any type of destructive laser therapy. In normal nevi, however, different laser systems have been used either to target the pigment or to simply remove them by a stepwise vaporization or ablation. In a comparative study performed with normal mode- and Q-switched ruby laser, as with congenital nevi, Duke and coworkers found a clinical improvement after several treatment sessions but not a complete elimination of nevus cells [26]. Also in a series treated by Vibhagool et al., one third of lesions showed only incomplete response [113]. Partial elimination and lightening with histological reduction in epidermal pigmentation and melanocytes has also been observed following therapy with Q-switched alexandrite or Nd:YAG lasers [97]. Also in laser vaporization of normal nevi, the likelihood of recurrence due to an incomplete superficial etching with subsequent formation of a pseudomelanoma cannot be ruled out [27]. On the other hand, with deeper removal, the risk of unpleasant scar formation or even keloid development increases, especially in locations prone to hypertrophic scarring.

Among the special variants of melanocytic nevi, which can be clinically distinguished from the common types, are blue nevi and nevus spilus. In both types laser treatments have also been attempted to achieve clinical improvement. As with nevus of Ota, in blue nevi the nevoid melanocytes are deeply located and the pigment can be targeted by selective photothermolysis. Successful long-term elimination of common blue nevi on the nasal skin of two patients has been demonstrated with the Q-switched ruby laser [79]. Also in nevus spilus exhibiting inhomogenous pigment distribution, significant improvement was achieved using Q-switched ruby laser and, to a lesser degree, by Nd:YAG laser treatment [39, 112].

However, following initial response, with complete elimination of the junctional or compound nevi portion, Taylor et al. found no improvement in the café-au-lait portion and a return of the background pigment by 1 year [107, 109].

## Other Melanocytic Disorders

Further melanocytic nevoid or melanin-containing macular pigmentary disorders, in which a removal is frequently requested for mainly cosmetic reasons, include ephelids, Becker's nevus, café-au-lait macules and lentigo simplex. In principle, laser therapy of all these lesions is possible, but again accurate diagnosing and excellent treatment techniques are vital because of the possibility of severe side effects [78]. In some patients, excellent results have been reported for lentigines, café-au-lait macules, nevus spilus and ephelides using a Q-switched ruby laser [69, 84, 79, 91, 93, 109]. Becker's nevus, however, shows inconsistent results and poses a special challenge, since it characterizes a pigmented hamartoma with larger areas usually located unilaterally after it develops, predominantly in adolescent males over the entire shoulder, and not only shows hyperpigmentation but also relevant hypertrichosis. Here, long-pulse ruby laser has been effectively used in order to remove pigments and significantly reduce hair density over several months after repetitive sessions with fluences in the range of 18–22 J cm$^{-2}$ [83].

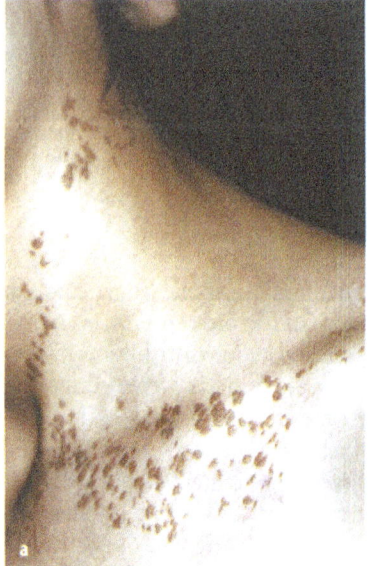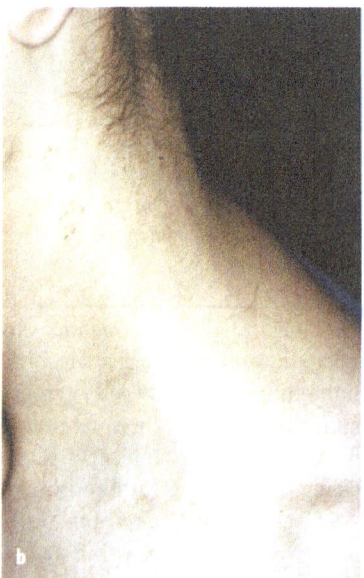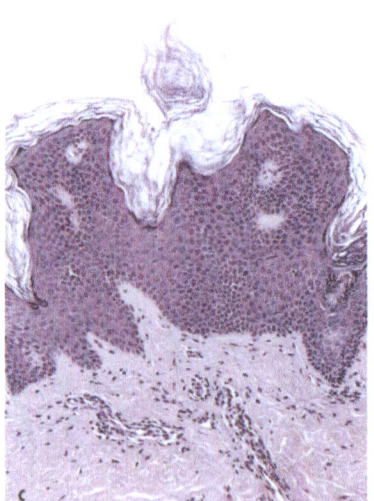

**Fig. 3a–c.** Laser ablation of pigmented epidermal nevus before **a** and after **b** treatment. **c** Histology shows the epithelial character of the lesions without any nevus cell nests, and typical acanthotic epidermis, that can be entirely ablated

## Epidermal Nevi

Different variants of this epithelial disorder show significant amounts of pigmented cells. They have to be distinguished from congenital melanocytic nevi by clinical examination and, in some cases, also by additional histology. In contrast to congenital melanocytic lesions, they present no risk of malignant transformation and, owing to their epithelial character, are well suited as candidates for a superficial removal. Therefore, dermabrasion has been a popular therapy option for these lesions. However, in recent years, laser coagulation, vaporization or skin-resurfacing, procedures have begun to replace the dermabrasion devices also in this indication. After initial attempts with the argon-laser coagulation for the treatment of softer papular lesions [44] or carbon-dioxide laser vaporization for more widespread hyperkeratotic variants [43, 92], a more tissue-sparing removal could be achieved with the introduction of pulsed erbium technology [62]. In contrast to dermabrasion, laser ablation allows a precise and individual removal of smaller papules frequently aggregated to larger plaques in these lesions. Also it can be applied irrespective of the anatomical site on the skin surface. For this reason, laser ablation is especially advised in critical areas, such as the eyelids, neck, or genitoanal skin (Fig. 3). In darkly pigmented epidermal nevi also the melanin deposits can be targeted by the laser treatment. In such cases, ruby lasers have been used successfully [9].

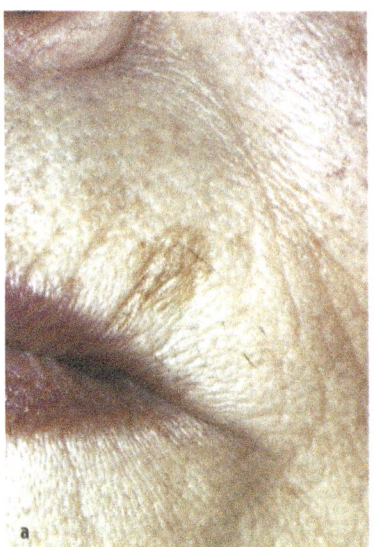
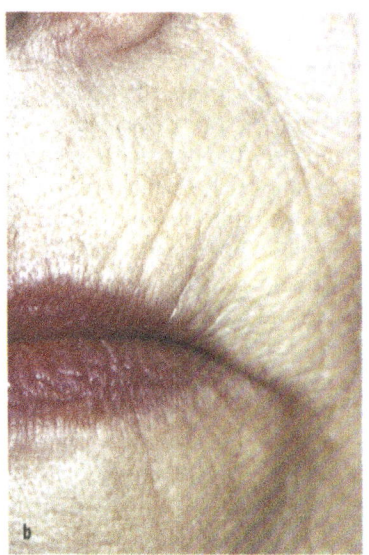
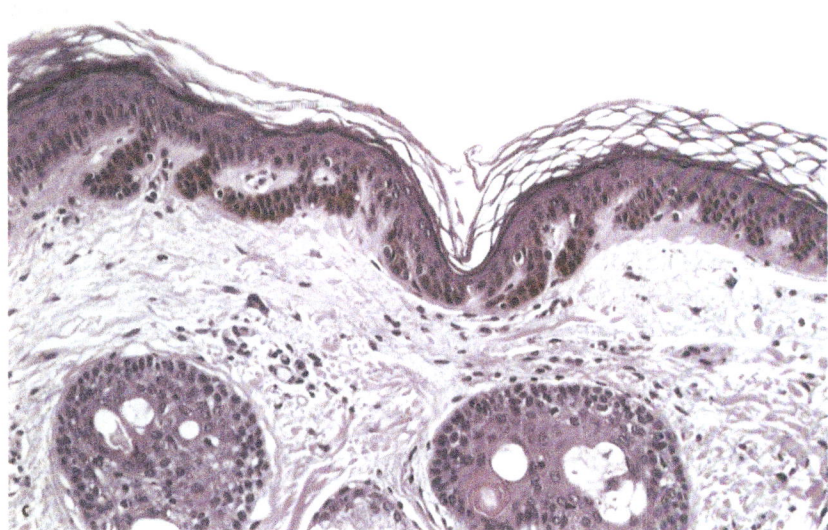

**Fig. 4a–c.** Laser ablation of lentigo senile (flat seborrhoic keratosis). A Clinical appearance before and **b** after treatment. **c** Typical histology of comparable lesion revealing hyperpigmented basal keratinocytes with few clear cells in between representing melanocytes

## Seborrhoic Keratosis and Senile Lentigines

Both lesions are variants of the same epidermal disorder that can lead to pigmented macules or warty lesions of typical appearance on clinical examination and epiluminescence microscopy. In these indications, superficial laser ablation as well as selective photothermolysis of melanin pigments with destruction of chromophore-containing superficial cells can be used to treat the lesions effectively [61]. With ablative techniques, a gentle stepwise and superficial removal using the 2940-nm Er:YAG laser should be preferred in order to avoid unpleasant depigmentation following deeper removal, especially in delicate skin areas or atrophic tissue of elderly patients (Fig. 4).

## Laser Treatment of Non-Pigmented Benign Skin Tumours

Many benign non-pigmented skin tumours have been successfully treated by $CO_2$ laser vaporization at 10 600 nm in the defocused mode [29]. Benign appendageal tumours such as adenoma sebaceum, syringoma, trichoepithelioma and apocrine hydrocystoma are often effectively removed with the $CO_2$ laser, but also hyperplastic proliferations such as rhinophyma or sebaceus gland hyperplasia. Verrucae vulgares and condylomata acuminata resistant to other destructive methods can be vaporized as an alternative or adjuvant to curettage. Actinic cheilitis [17] is a premalignant condition for which the $CO_2$ laser has been recommended as a treatment of choice, leaving good function and cosmesis. Nevertheless, because of the uncertainty of focal malignant transformation and the possibility of histological control, we prefer vermilionectomy as a surgical standard approach. Successful $CO_2$ laser treatment has also been reported with neurofibromas, digital myxoid cysts, xanthelasma and others [29]. However, excess heat damage has to be considered with regard to wound healing and scar formation, especially in deeper lesions or on critical anatomic sites prone to keloid development. While superficial coagulation is still popular and can produce good or excellent results in smaller vascular tumours of the skin surface [10, 57, 61, 71, 72], most of the benign non-pigmented skin tumours are meanwhile being treated with laser systems avoiding heat injury. Therefore, today, thermal vaporization using the cw $CO_2$ laser is more or less restricted to few exceptions (e.g. viral warts) and replaced by either pulsed $CO_2$ laser-technology [30] or Er:YAG ablation [24, 63]. Moreover, latest developments of ablative lasers can also produce additional coagulative pulses, if required for superficial haemostasis, and the majority of the benign superficial non-pigmented skin tu-

mours can be easily ablated without tissue coagulation. Among the most frequent indications treated by laser ablation are common non-pigmented benign skin tumours, such as viral warts, sebaceous gland hyperplasias or xanthelasmata, but also less common tumours, such as syringoma, and also some rare and delicate lesions, such as osteoma cutis.

## Viral Warts

Different types of viral warts can develop depending on age groups, location and viruses, comprising common warts (verrucae vulgares), plane juvenile warts, condylomata acuminata and molluscum contagiosum. Basically, they develop as benign tumours of epithelial cells in response to incorporation of papilloma or molluscum contagiosum viral genome into epidermal cellular DNA. As with electrosurgical treatment, concerns about the risk of virus contamination with papilloma viruses following laser vaporization cannot be completely ruled out and special cautions are mandatory (masks, suction devices) [3, 16, 101]. While PCR studies have detected HPV-DNA in treated warts and in samples from monitoring medical personnel during treatment of genital warts with $CO_2$ laser and electrocoagulation, indicating that there is a risk of contamination of the operator [16, 74], no such virus particles were detected in one study examining the plume of erbium ablation for HPV-DNA by PCR [47]. Sawchuk et al. found greater amounts of viral DNA in laser-derived vapours than in the electrosurgical probes from the same warts and found surgical masks to be capable of removing virtually all laser-derived viruses, strongly suggesting that such masks can protect operators from potential inhalation exposure [101].

Though a spontaneous resolution without any treatment is not unusual and has to be considered when evaluating cure rates, common warts can become a therapeutical problem in special areas, such as around the nails, or on the plantar skin. Laser treatment of common verrucae either aims at a thermal or ablative destruction of the warts or intends to target the vascular supply. To maintain a proliferative growth, neovascularization with prominent capillary vessels is induced in the dermal papillae underlying the acanthotic tumours. Therefore, different lasers have been used for these purposes. Heat injury achieved by argon or KTP laser resulted in clearing rates of up to 50% of patients [18, 36]. Hyperthermia induced by Nd:YAG laser treatment was able to eliminate viral DNA, while cryotherapy was not [28]. Most experience has been, however, collected with $CO_2$ laser vaporization in this indication. Laser vaporization is, in fact, a valuable method, but it cannot be ex-

**Fig. 5a, b.** $CO_2$ laser vaporization of pariungual warts before **a** and after **b** treatment

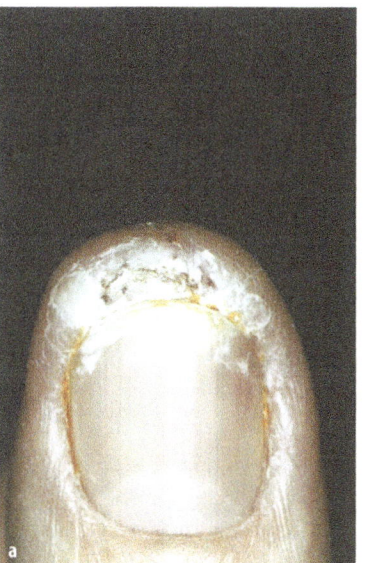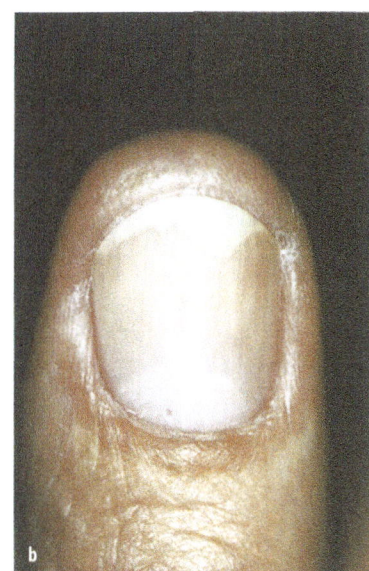

pected to cure all such warts [4, 46]. Usually, several treatment sessions are required or a combined approach is warranted [77, 89]. An overall cure rate of about 60% can be expected with treatment [103]. Even in problematic sub- and periungual warts cure rates of about 50–70% were reported in previously recalcitrant lesions [75, 104] and up to 80% in those treated as first line [75] (Fig. 5). More recently, $CO_2$ lasers have been applied as pulsed systems [32] and also the erbium:YAG laser has been used by us and others to ablate veruccae successfully [73].

Furthermore, pulsed dye laser treatment has been recommended as an option in recalcitrant warts by several investigators in order to destroy the deeper vessels of the tumours using a 585 nm flashlamp-pulsed tunable dye laser with fluences in the range of approximate 8–10 J $cm^{-2}$ applied repetitively in several sessions (Fig. 6). A clearance varying from 45% to 95%, with best responses seen in warts located on the body and palmar skin, has been reported [51, 52, 98], along with marked agglutination of red blood cells and vessel wall necrosis within the treated areas [105]. However, especially in resistant plantar or periungual verrucae, other investigators were unable to show major therapeutical effect with pulsed dye laser treatment. They found either only a moderate response in term of size reduction [48] or a larger proportion of resistant lesions, particularly in more fibrotic and less vascularized types [12]. This less favourable outcome is also in agreement with our experience in these locations and after previous treatments had failed.

Apart from common warts, two other types are preferentially found in children and adolescents.

These include mollusca contagiosa and plane juvenile warts. In both indications a careful superficial ablation can be performed. Older children will generally tolerate the procedure well after topical application of anaesthetic EMLA cream. Condylomata acuminata show a preferential location in the genitoanal skin and mucous surface and can also be alternatively treated by electrosurgery or laser coagulation and vaporization. For vaporization, $CO_2$ lasers are widely used and have also been reported as a safe and effective option in lesions appearing in childhood [52]. Laser vaporization can be combined with electrosurgery for larger plaques or curettage of ablated or coagulated tissue material. As with electrosurgery, these procedures should be performed only by experienced laser surgeons and with great caution, avoiding excessive trauma to the underlying cutaneous structures. In order to prevent recurrence, systemically administered interferon alfa-2a has been suggested [41]. Coagulation with the neodymium: YAG laser has also been reported [116], but due to the relatively unspecific heat damage of larger and deeper tissue volumes, the risk for unwarranted necrosis and scar formation is increased [100, 45].

## Syringomas

Syringomas (syn: hydradenoma) originate from adnexal structures and are composed of small cystic ducts in the upper and mid dermis. They appear as skin-coloured or slightly yellowish, and form papules usually located around the eyelids. Removal of facial syringomas by ablative or vaporizing laser devices

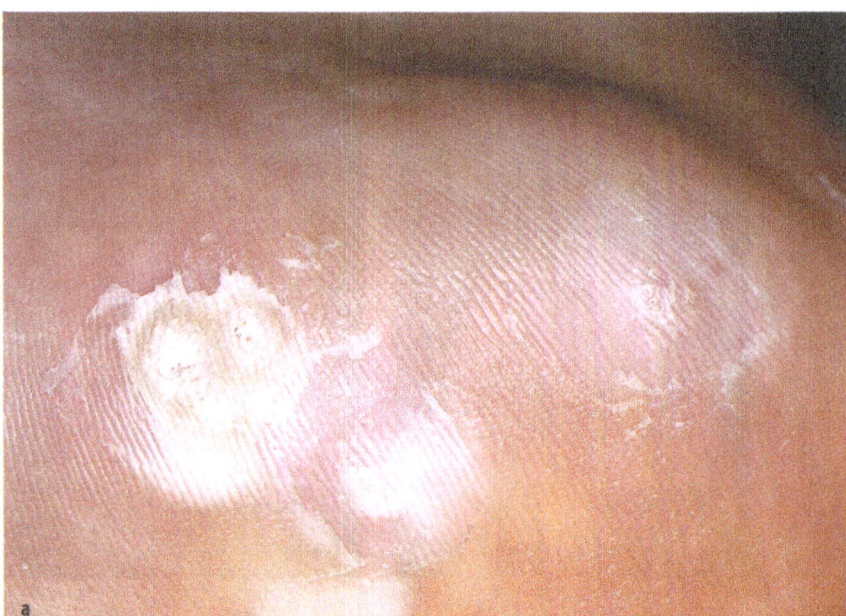

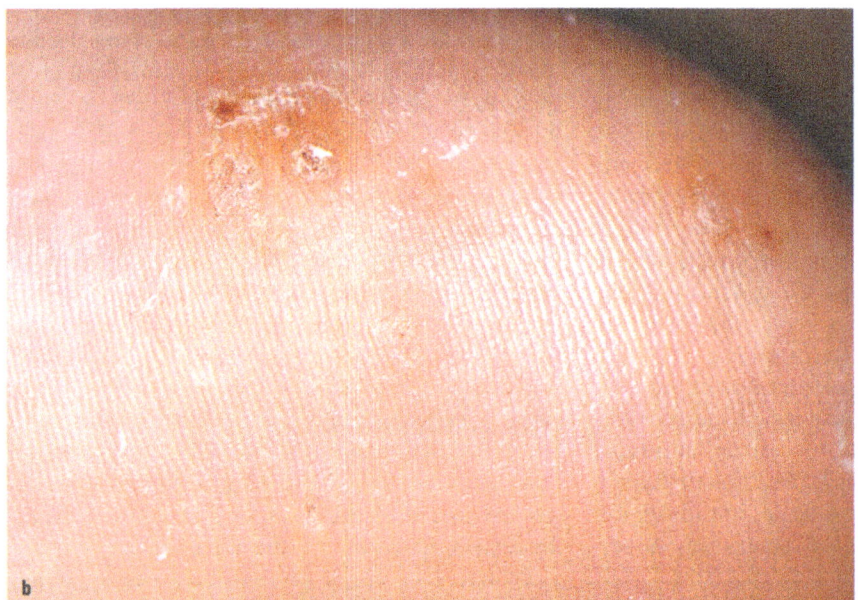

**Fig. 6a, b.** Pulsed dye laser treatment of viral warts **a** with minimal response following only two sessions at 8 J cm$^{-2}$ each **b**

has been suggested by several investigators [20, 63, 95, 115]. Castro et al. used a pulsed $CO_2$ laser connected to an operating microscope [20]. Kaufmann and others reported the use of an erbium:YAG system to achieve a highly controlled ablation in this particularly subtle indications [63, 95] (Fig. 7).

## Xanthelasma

Although palpebral xanthelasma occur regularly in dyslipoproteinemic patients, the majority of cases are normolipemic patients requesting removal of these disturbing yellow plaques simply for cosmetic reasons. In spindle-shaped lesions, they are easily to excise also in combination with blepharoplasty. In larger or multiple plaques, however, we prefer a laser vaporization or stepwise ablation with secondary healing of the resulting defects, that yields good results after both $CO_2$ or Er:YAG laser treatment [2, 6, 63, 94]. In a larger series of 52 periorbital xanthelasmas treated by Raulin et al., they report no visible scarring, and only transient hyper- or hypopigmentation in up to 13% of cases treated with the pulsed CO laser.

**Fig. 7a, b.** Laser ablation of syringoma of the lower eyelid before **a** and after **b** treatment

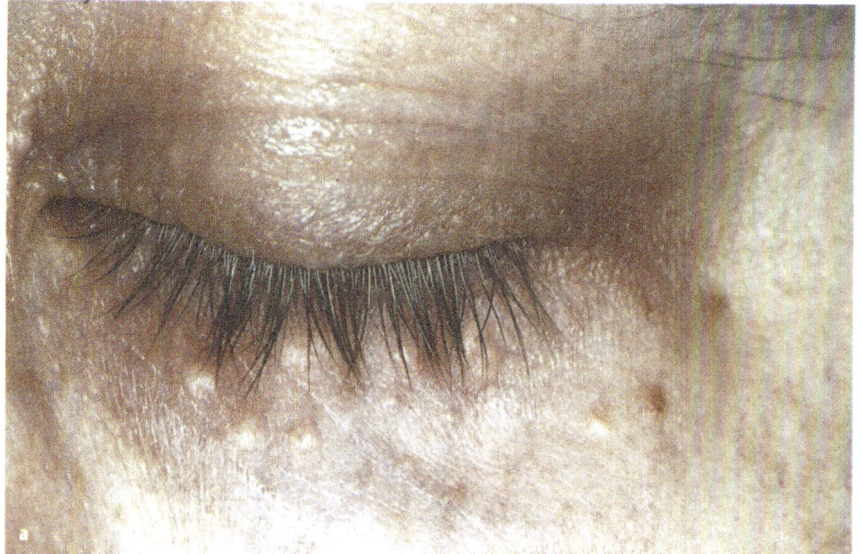

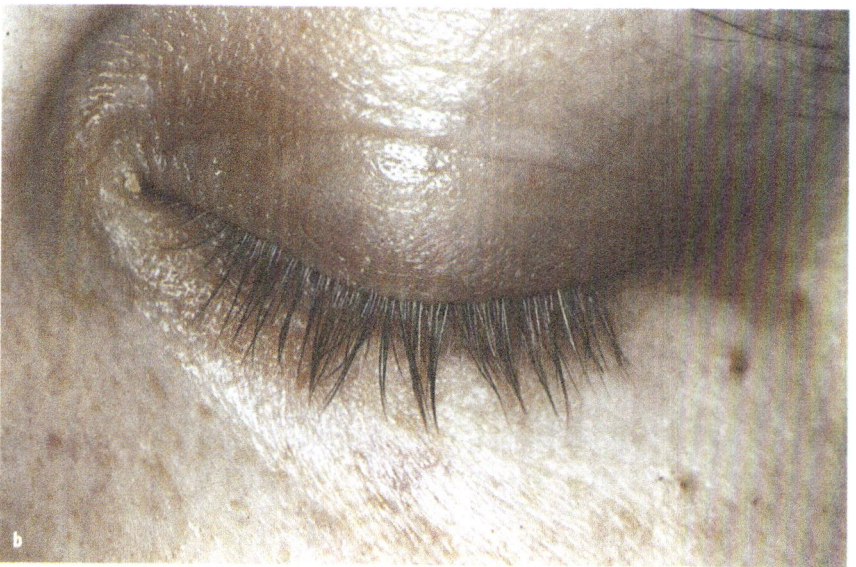

## Sebaceous Gland Hyperplasia and Rhinophyma

In both conditions, vaporization as well as tissue ablation have been used. Sebaceous gland hyperplasia can mimic other adnexal tumours such as basal cell carcinoma, and might require careful examination by an experienced dermatologist prior to treatment. After confirmation of diagnosis it can be easily removed by stepwise ablation. Deeper crater lesions, however, should be avoided in order to minimize the risk of atrophic scar formations. Rhinophyma is usually associated with rosacea, where a combined conservative and surgical approach is generally required. Laser applications include the use of systems appropriate to treat vascular components of the disease (teleangiec-

tasias), while $CO_2$ laser vaporization has been widely employed as an alternative to surgical shaving techniques or electrosurgery in rhinophyma [37]. Stepwise laser vaporization can avoid the bleeding frequently associated with dermashaving or dermabrasion. As an alternative to $CO_2$ laser vaporization, erbium lasers can be used in less angiomatous tumour formations or with systems allowing a combined haemostasis function.

## Neurofibroma

These benign nerve sheath tumours can be seen as a solitary lesion or in the context of neurofibromatosis.

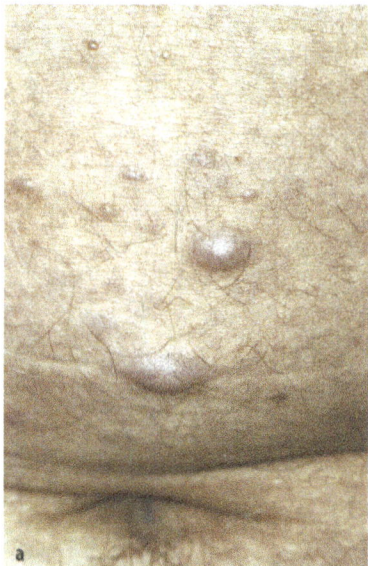 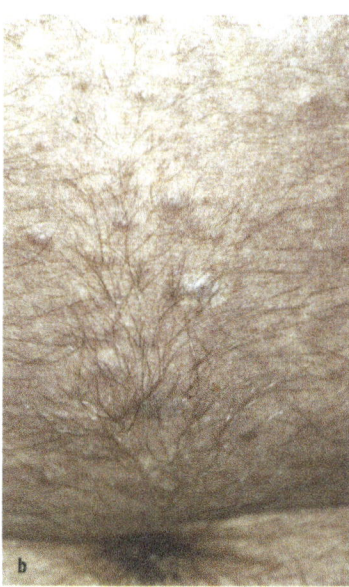

**Fig. 8a, b.** Vaporization of multiple smaller neurofibromas in NF1 disease before **a** and after **b** treatment

Those seen by dermatologists are located in the dermis and subcutaneous tissue presenting as protuberant, flesh-coloured papules and nodules that are typically rather soft on palpation and sometimes pedunculated. More superficial types have been removed by continuous vaporization or partial laser incision and subsequent extirpation of the tumours using $CO_2$ laser vaporization or laser incisions [54]. However, due to the depth of the created wounds, a secondary healing leading to scar formations is unavoidable. Therefore, laser vaporization is better combined with scalpel excisions and should be better limited to patients presenting with disseminated eruptions of smaller and medium-sized superficial papules after test treatment has proved beneficial [14, 95] (Fig. 8). Both techniques are used complementarily, and the choice between them will depend on the size, location and depth of the individual neurofibroma.

## Nevus Sebaceus

Though also this adnexal nevoid malformation has been vaporized by carbon dioxide laser treatment [8], it is more recommended to completely excise these lesions due to the increased association with development of malignant tumours and also because of their depths, requiring penetrating laser wounds with an enhanced likelihood of subsequent scarring or pigmentary changes [13].

## Others

Baginski and Arpey used a $CO_2$ laser in combination with curettage in order to remove facial osteoma cutis, and reported an excellent outcome in their patient [11]. In the same year, the use of erbium ablation, together with mechanical removal in this same delicate condition, was suggested by Ochsendorf and Kaufmann, who had previously treated three cases by this combined approach [86]. Adenoma sebaceum is an angiofibrotic benign tumour that can appear independently or in association with Pringle's disease. Usually, multiple papules are aggregated in the midface. Due to their high content of vessels, they can be targeted by vascular-specific lasers [55]. In more fibrotic lesions also vaporization or ablation is an appropriate approach.

## References

[1]  ALSTER TS, WILLIAMS CM (1995) Treatment of nevus of Ota by the Q-switched alexandrite laser. Dermatol Surg 21 (7): 592–596
[2]  ALSTER TS, WEST TB (1996) Ultrapulse $CO_2$ laser ablation of xanthelasma. J Am Acad Dermatol 34(5 Pt. 1): 848–849
[3]  ANDRE P, ORTH G, EVENOU P, GUILLAUME JC, AVRIL MF (1990) Risk of papillomavirus infection in carbon dioxide laser treatment of genital lesions. J Am Acad Dermatol 22 (1): 131–132
[4]  APFELBERG DB, DRUKER D, MASER MR, WHITE DN, LASH H, SPECTOR P (1989) Benefits of the $CO_2$ laser for verruca resistant to other modalities of treatment. J Dermatol Surg Oncol 15 (4): 371–375
[5]  APFELBERG DB (1995) Argon and Q-switched yttrium-aluminium-garnet laser treatment of nevus of Ota. Ann Plast Surg 35 (2): 150–153
[6]  APFELBERG DB, MASER MR, LASH H, WHITE DN (1987) Treatment of xanthelasma palpebrarum with the carbon dioxide laser. J Dermatol Surg Oncol 13 (2): 149–151

[7] ARONS MS (1998) Successful treatment of a giant congenital melanocytic naevus with high-energy pulsed CO₂ laser. Br J Plast Surg 51: 570–571

[8] ASHINOFF R (1993) Linear nevus sebaceus of Jadassohn treated with the carbon dioxide laser. Pediatr Dermatol 10 (2): 189–191

[9] BABA T, NARUMI H, HANADA K, HASHIMOTO I (1995) Successful treatment of dark-coloured epidermal nevus with ruby laser. J Dermatol 22 (8): 567–570

[10] BAHMER FA (1991) The neodym YAG laser in dermatology. In: Steiner R, Kaufmann R, Landthaler R, Braun-Falco O (eds) Lasers in dermatology. Springer, Berlin Heidelberg New York, pp 73–84

[11] BAGINSKI DJ, ARPEY CJ (1999) Management of multiple miliary osteoma cutis. Dermatol Surg 25: 233–235

[12] BAKUS AD, GARDEN JM (1993) Pulsed dye laser treatment of plantar verrucae. Laser Med Surg (Suppl 5): 59

[13] BEER GM, WIDDER W, CIERPKA K, KOMPATSCHER P, MEYER VE (1999) Malignant tumours associated with nevus sebaceous: therapeutic consequences. Aesthet Plast Surg 23 (3): 224–227

[14] BECKER DW JR (1991) Use of the carbon dioxide laser in treating multiple cutaneous neurofibromas. Ann Plast Surg 26 (6): 582–586

[15] BEKKOR PS (1995) Removal of cutaneous pigmented lesions with Q-switched lasers. Dermatol Surg 21 (11): 990

[16] BERGBRANT IM, SAMUELSSON L. OLOFSSON S, JONASSEN F, RICKSTEN A (1994) Polymerase chain reaction for monitoring human papillomavirus contamination of medical personnel during treatment of genital warts with CO₂ laser and electrocoagulation. Acta Dermato-Venereol 74 (5): 393–395

[17] CALLEN JP, BICKERS DR, MOY RL (1997) Actinic keratosis. J Am Acad Dermatol 36: 650–653

[18] CARLSON BA (1992) Argon laser treatment tackles recurrent plantar verrucae. Clin Laser Mon 10 (4): 61–62

[19] CARPO BG, GREVELINK JM, GREVELINK SV (1999) Laser treatment of pigmented lesions in children. Semin Cutan Med Surg 18: 233–243

[20] CASTRO DJ, TARTELL PB, SOUDANT J, SACTON RE (1993) The surgical management of facial syringomas using the superpulsed CO₂ laser. J Clin Med Surg 11: 33–37

[21] CHANG, CJ, NELSON JS, ACHAUER BM (1996) Q-switched ruby laser treatment of oculodermal melanosis (nevus of Ota) Plast Reconstr Surg 98 (5): 784–790

[22] CHAN HH, KING WW, CHAN ES, MOK CO, HO WS, VAN KREVEL C, LAU WY (1999) In vivo trial comparing patients' tolerance of Q-switched alexandrite (QS Nd:YAG) lasers in the treatment of nevus of Ota. Laser-Surg Med 24: 24–28

[23] DINEHART SM, WANER M, FLOCK S (1993) The copper vapor laser for treatment of cutaneous vascular and pigmented lesions. J Dermatol Surg Oncol 19: 370–375

[24] DMOVSEK-OLUP B, VEDLIN B (1997) Use of Er:YAG laser for benign skin disorders. Lasers Surg Med 21 (1): 13–19

[25] DOVER JS, KANE KS (1997) Lasers for the treatment of cutaneous pigmented disorders. In: Arndt KA, Dover JS, Olbricht S (eds) Lasers in cutaneous and aesthetic surgery. Lippincott Raven, Philadelphia, pp 165–187

[26] DUKE B, BYERS HR, SOBER AJ, ANDERSON RR, GREVELINK JM (1999) Treatment of benign and atypical nevi with the normal mode ruby laser and the Q-switched ruby laser: clinical improvement but failure to completely eliminate nevomelanocytes. Arch Dermatol 135 (3): 290–296

[27] DUMMER R, KEMPF W, BURG G (1998) Pseudomelanoma after laser therapy. Dermatology 197: 71–73

[28] EL-TONSY MH, ANBAR TE, EL-DOMYATI M, BARAKAT M (1999) Density of viral particles in pre and post-Nd:YAG laser hyperthermia therapy and cryotherapy in plantar warts. Int J Dermatol 38 (5): 393–398

[29] FITZPATRICK RE, GOLDMAN MP (1994) CO₂ laser surgery. In: Goldman MP, Fitzpatrick RE (eds) Cutaneous laser surgery. Mosby, St. Louis, pp 198–258

[30] FITZPATRICK RE, GOLDMAN MP, RUIZ-ESPARZA J (1994) Clinical advantage of the CO₂ laser superpulsed mode. Treatment of verruca vulgaris, seborrhoic keratoses, lentigines, and actinic cheilitis. J Dermatol Surg Oncol 20 (7): 449–456

[31] GERONEMUS RG (1992) Q-switched ruby laser therapy of nevus of Ota. Arch Dermatol 128 (12): 1618–1622

[32] GERONEMUS RG, KAUVAR AN, MC DANIEL DH (1998) Treatment of recalcitrant verrucae with both the ultrapulse CO₂ and PLDL pulsed dye lasers. Plast Reconstr Surg 101 (7): 2010

[33] GOLDBERG DJ (1993) Benign pigmented lesions of the skin. Treatment with the Q-switched ruby laser. J Dermatol Surg Oncol 19 (4): 376–379

[34] GOLDBERG DJ (1997) Laser treatment of pigmented lesions. Dermatol Clin 15 (3): 397–407

[35] GOLDBERG DJ, STAMPIEN T (1995) Q-switched ruby laser treatment of congenital nevi. Arch Dermatol 131: 621–623

[36] GOOPTU C, JAMES MP (1999) Recalcitrant viral warts: results of treatment with the KTP laser. Clin Exp Dermatol 24 (2): 60–63

[37] GREENBAUM SS, KRULL EA, WATNICK K (1988) Comparison of CO₂ laser and electrosurgery in the treatment of rhinophyma. J Am Acad Dermatol 18(2 Pt 1): 363–368

[38] GREVELINK JM, VAN LEEWEN RL, ANDERSON RR, BYERS RH (1997) Clinical and histological responses of congenital melanocytic nevi after single treatment with Q-switched lasers. Arch Dermatol 133: 349–353

[39] GREVELINK JM,GONZALEZ S. BONOAN R, VIBHAGOOL C, GONZALEZ E (1997) Treatment of nevus spilus with the Q-switched ruby laser. Dermatol Surg 23 (5): 365–370

[40] GROB M, SENTI G. DUMMER R (1999) Delay of the diagnosis of an amelanotic malignant melanoma due to CO₂ laser treatment – case report and discussion. Schweiz Rundsch Med Prax 88: 1491–1494

[41] GROSS G, ROUSSAKI A, BAUR S, WIEGAND M, MESCHEDER A (1996) Systemically administered interferon alfa-2a prevents recurrence of condylomata acuminata following CO₂ laser ablation. The influence of the cyclic low-dose therapy regimen. Results of a multicentre double-blind placebo-controlled clinical trial. Genitourin Med 72 (1): 71

[42] HOHENLEUTNER U, LANDTHALER M (1999) Laser treatment of nevus cell nevi in the newborn infant. Hautarzt 50: 221

[43] HOHENLEUTNER U, WLOTZKE U. KONZ B, LANDTHALER M (1995) Carbon dioxide laser therapy of a widespread epidmermal nevus. Lasers Surg Med 16 (3): 288–291

[44] HOHENLEUTNER U, LANDTHALER M (1993) Laser therapy of verrucous epidermal naevi. Clin Exp Dermatol 18 (2): 124–127

[45] HREBINKO RL (1996) Severe injury from neodymium: yttrium-aluminium-garnet laser therapy for penile condylomata acuminata. Urology 48 (1): 155–156

[46] HRUZA GJ (1997) Laser treatment of warts and other epidermal and dermal lesions. Dermatol Clin 15 (3): 487–506

[47] HUGHES PS, HUGHES AP (1998) Absence of human papillomavirus DNA in the plume of erbium: YAG laser treated warts. J Am Acad Dermatol 38 (3): 426–428

[48] HUILGOL SC, BARLOW RJ, MARKEY AC (1996) Failure of pulsed dye laser therapy for resistant verrucae. Clin Exp Dermatol 21 (2): 93–95

[49] IMAYAMA S, UEDA S (1999) Long- and short-term histological observations of congenital nevi treated with the normal-mode ruby laser. Arch Dermatol 135: 1211–1218

[50] JACOBSEN E, MC GRAW R, MC CAGH S (1997) Pulsed dye laser efficacy as initial therapy for warts and against recalcitrant verrucae. Cutis 59 (4): 206–208

[51] JAIN A, STORWICK GS (1997) Effectiveness of the 585 nm flashlamp-pulsed tunable dye laser (PTDL) for treatment of plantar verrucae. Lasers Surg Med 21 (5): 500–505

[52] JOHNSON PJ, MIRZAI TH, BENTZ ML (1997) Carbon dioxide laser ablation of anogenital condyloma acuminata in pediatric patients. Ann Plast Surg 39 (6): 578–582

[53] KANG W, LEE E, CHOI GS (1999) Treatment of Ota's nevus by Q-switched alexandrite laser: therapeutic outcome in relation to clinical and histopathological findings. Eur J Dermatol 9 (8): 639–643

[54] KATALANIC D (1992) Laser surgery of neuofibromatosis 1 (NF1) J Clin Laser Med Surg 10: 185–192

[55] KAUFMAN AJ, GREKIN RC,GEISSE JK, FRIEDEN IJ (1995) Treatment of adenoma sebaceum with the copper vapor laser. J Am Acad Dermatol 33: 770–774

[56] KAUFMANN R (1998) Management of epithelial dermatologic neoplasia. Onkologie 21: 36–43

[57] KAUFMANN R, LANDES E (1992) Dermatologische Operationen. 2. Auflage. Thieme, Stuttgart, pp 19–50

[58] KAUFMANN R, HARTMANN R, BOEHNCKE WH (1995): Einsatz gepulster Laser bei Pigmentläsionen der Haut. In: Tilgen W, Petzoldt D (Hrsg) Operative und konservative Dermatoonkologie. Springer, Berlin Heidelberg New York, S. 345–352

[59] KAUFMANN R (1995) Gepulste Laser – Stellenwert zur Hautablation und Therapie pigmentierter Hautveränderungen. In: Tebbe B, Goerdt S, Orfanos CE (Hrsg) Dermatologie – Heutiger Stand, Thieme, Stuttgart, S. 357–358

[60] KAUFMANN R (1997) Stellenwert der Lasertherapie bei Pigmentläsionen. In: Landthaler M, Hohenleutner U (Hrsg) Operative Dermatologie im Kindes- und Jugendalter. Blackwell, Berlin, S. 23–28

[61] KAUFMANN R (1998) Laseranwendungen in der Dermatologie. Fortschr Med 116: 26–32

[62] KAUFMANN R, HIBST R (1990) Er:YAG laser skin ablation: experimental results and first clinical application. Clin Exp Dermatol 15: 389–393

[63] KAUFMANN R, HIBST R (1996) Clinical evaluation of Er:YAG lasers in cutaneous surgery. Laser Surg Med 19: 324–330

[64] KAUFMANN R (1998) Comparison of different procedures for the treatment of benign and malign skin tumours. Min Invas Ther Allied Technol 7: 511–517

[65] KAUFMANN R (1999) Klassische Dermabrasion versus Laserverfahren. In: Plettenberg, Meigel W, Moll I, Dermatologie – Aktueller Stand von Klinik und Forschung. Springer, Berlin Heidelberg New York, S 669–672

[66] KAUFMANN R (2000) Surgery for tumours of the skin. In: Burg G (Hrsg) Cancer of the skin. WB Saunders, Philadelphia

[67] KAY AR, KENEALY J, MERCER NS (1998) Successful treatment of a giant congenital melanocytic naevus with the high-energy pulsed $CO_2$ laser. Br J Plast Surg 51: 22–24

[68] KOPERA D, HOHENLEUTNER U, STOLZ W, LANDTHALER M (1997) Ex vivo quality-switched ruby laser irradiation of cutaneous melanocytic lesions: persistence of S-100-, HMB-45- and Masson-positive cells. Dermatology 194 (4): 344–350

[69] KOPERA D, HOHENLEUTNER U, LANDTHALER M (1997) Quality-switched ruby laser treatment of solar lentigines and Becker's nevus: a histopathological and immunohistochemical study. Dermatology 194 (4): 338–343

[70] KUNACHAK S, LEELAUDOMLIPI P, SIRIKULCHAYANONTA V (1999) Q-switched ruby laser therapy of acquired bilateral nevus of Ota-like macules. Dermatol Surg 25 (12): 938–941

[71] LANDTHALER M, HOHENLEUTNER U (1997) The Nd:YAG laser in cutaneous surgery. In: Arndt KA, Dover JS, Olbricht SM (eds) Lasers in cutaneous and aesthetic surgery. Lippincott Raven, Philadelphia, pp 124–149

[72] LANDTHALER M, HAINA D, BRUNNER R, WAIDELICH W, BRAUN-FALCO O (1986) A 5-year experience with laser therapy in dermatology. Curr Probl Dermatol 15: 272–281

[73] LANGDON RC (1998) Erbium:YAG laser enables complete ablation of periungual verrucae without the need for injected anesthetics. Dermatol Surg 24 (1): 157–158

[74] LI H, ZHU W, XIA M (1997) Detection of human papillomavirus DNA in condylomata acuminata treated with $CO_2$ laser by polymerase chain reaction. Chin Med J (Engl) 110 (1): 78–80

[75] LIM JT, GOH CL (1992) Carbon dioxide laser treatment of periungual and subungual viral warts. Australas J Dermatol 33 (2): 87–91

[76] LOWE NJ, WIEDER JM, SAWCER D, BURROWS P, CHALET M (1993) Nevus of Ota: treatment with high-energy fluences of the Q-switched ruby laser. J Am Acad Dermatol 29 (6): 997–1001

[77] MANUSO JE, ABRAMOW SP, DIMICHINO BR, LANDSMANN MJ (1991) Carbon dioxide laser management of plantar verruca: a 6-year-follow-up survey. J Foot Surg 30 (3): 238–243

[78] MARGHOOB AA, SCHOENBACH SP, KOPF AW, ORLOW SJ, NOSSA R, BART RS (1996) Large congenital nevi and the risk for the development of malignant melanoma. Arch Dermatol 132: 170–175

[79] MICHEL S, HOHENLEUTNER U, BAUMLER W,LANDTHALER M (1997) Q-switched ruby laser in dermatologic therapy. Use and indications. Hautarzt 48 (7): 462–470

[80] MILGRAUM SS, COHEN ME, AULETTA MJ (1995) Treatment of blue nevi with the Q-switched ruby laser. J Am Acad Dermatol 32: 307–310

[81] MIXTER RC, CARSON LV, WALTON BJ, GERSON RM (1996) Treatment of nevus of Ota with the Candela PLDL and PLTL lasers. Plast Reconstr Surg 98 (6): 1112–1113

[82] MORELLI JG (1998) Use of lasers in pediatric dermatology. Dermatol Clin 16 (3): 489–495

[83] NANNI CA, ALSTER TS (1998) Treatment of a Becker's nevus using a 694-nm long-pulsed ruby laser. Dermatol Surg 24: 1032–1034

[84] NELSON JS, APPLEBAUM J (1992) Treatment of superficial cutaneous pigmented lesions by melaninspecific selective photothermolysis using the Q-switched ruby laser. Ann Plast Surg 29 (3): 231–237

[85] NELSON JS, KELLY KM (1999) Q-switched ruby laser treatment of a congenital melanocytic nevus. Dermatol Surg 25 (4): 274–276

[86] OCHSENDORF FR, KAUFMANN R (1998) Erbium:YAG laser-assisted treatment of miliary osteoma cutis. Br J Dermatol 138: 371–372

[87] ONO I, TATESHITA T (1998) Efficacy of the ruby laser in the treatment of Ota's nevus previously treated using other therapeutic modalities. Plast Reconstr Surg 102: 2352–2357

[88] ORTEN SS, WANER M, DINEHART SM, BARDALES RH, FLOCK ST (1999) Q-switched neodymium:yttrium-aluminium-garnet laser treatment of lentigo maligna. Otolaryngol Head Neck Surg 120: 206–302

[89] PETERSEN CS, NURNBERG BM (1994) Carbon dioxide laser vaporization combined with perilesionally injected interferon alfa-2b in the treatment of a hyperkeratotic verruca vulgaris on the upper eyelid. Arch Dermatol 130 (11): 1369–1370

[90] PETRES J, ROMPEL R (1992) Konnatale Nävuszellnävi (1992) In: Burg G, Hartmann AA, Konz B (eds) Onkologische Dermatologie. Springer, Berlin Heidelberg New York, S 220–229

[91] PFEIFFER N (1993) Q-switched ruby laser used to remove pigmented lesions. J Clin Laser Med Surg 11 (3): 147–148

[92] RATZ JL, BAILIN PL, WHEELAND RG (1986) Carbon dioxide laser treatment of epidermal nevi. J Dermatol Surg Oncol 12 (6): 567–570

[93] RAULIN C, SCHONERMARK MP, GREVE B, WERNER S (1998) Q-switched ruby laser treatment of tattoos and benign pigmented skin lesions: a critical review. Ann Plast Surg 41 (5): 555–565

[94] RAULIN C, SCHOENERMARK MP, WERNER S, GREVE B (1999) Xanthelasma palpebrarum: treatment with the ultrapulsed $CO_2$ laser. Lasers Surg Med 24 (2): 122–127

[95] RIEDEL F, WINDBERGER J, STEIN E, HORMANN K (1998) Treatment of peri-ocular skin lesions with the erbium:YAG laser. Ophtalmologe 95 (11): 771–775

[96] ROENIGK RK, RATZ JL (1987) $CO_2$ laser treatment of cutaneous neurofibromas. J Dermatol Surg Oncol 13 (2): 187–190

[97] ROSENBACH A, WILLIAMS CM, ALSTER AT (1997) Comparison of Q-switched alexandrite (755 nm) and Q-switched Nd:YAG (1064-nm) lasers in the treatment of benign melanocytic nevi. Dermatol Surg 23: 329–245

[98] ROSS BS, LEVINE VJ, NEHAL K, TSE Y, ASHINOFF R (1999) Pulsed dye laser treatment of warts: an update. Dermatol Surg 25 (5): 377–380

[99] SCHEPERS JH, QUABA AA (1993) Clinical experience with the PLDL-1 (pigmented lesion dye laser) in the treatment of pigmented birthmarks: a preliminary report. Br J Plast Surg 46 (3): 247–251

[100] SAKKA G, KARAGIANNIS A, KARAYANNIS D, DIMOPOULOS K (1995) Laser treatment in urology: our experience with neodymium:YAG and carbon dioxide lasers. Int Urol Nephrol 27 (4): 405–412

[101] SAWCHUK WS, WEBER PJ, LOWY DR, DZUBOW LM (1989) Infectious papillomavirus in the vapor of warts treated with carbon dioxide laser or electrocoagulation: detection and protection. J Am Acad Dermatol 21 (1): 41–49

[102] SHIMBASHI T, HYAKUSOKO H, OKINAGA M (1997) Treatment of nevus of Ota by Q-switched ruby laser. Aesthet Plast Surg 21 (2): 118–121

[103] SLOAN K, HABERMAN H,LYNDE CW (1998) Carbon dioxide laser-treatment of resistant verruca vulgaris: retrospective analysis. J Cutan Med Surg 2 (3): 142–145

[104] STREET ML, ROENIGK RK (1990) Recalcitrant periungual verrucae: the role of carbon dioxide laser vaporization. J Am Acad Dermatol 23 (1): 115–120

[105] TAN OT, HURWITZ RM, STAFFORD TJ (1993) Pulsed laser treatment of recalcitrant verrucae: a preliminary report. Lasers Surg Med 13: 127–37

[106] TAN OT, MORELLI JG, KURBAN AK (1992) Pulsed dye laser treatment of benign cutaneous pigmented lesions. Lasers Surg Med 12: 538–542

[107] TAYLOR CR, ANDERSON RR (1993) Treatment of benign pigmented lesions by Q-switched ruby laser. Int J Dermatol 32: 908–912

[108] TAYLOR CR, FLOTTE TJ, GANGE RW, ANDERSON RR (1994) Treatment of nevus of Ota by Q-swiched ruby laser. J Am Acad Dermatol 30: 743–751

[109] TAYLOR CR, ANDERSON RR (1993) Treatment of benign pigmented epidermal lesions by Q-switched ruby laser. Int J Dermatol 32 (12): 908–912

[110] TRAU H, ORENSTEIN A, SCHEWACH-MILLER M, TSUR H (1986) Pseudomelanoma following laser therapy for congenital nevus. J Dermatol Surg Oncol 12 (9): 984–986

[111] UEDA S, IMAQAMA S (1997) Normal-mode ruby laser for treating congenital nevi. Arch Dermatol 133 (3): 355–359

[112] VAN LEEUWEN RL, BASTIAENS MT, GREVELINK JM (1997) Management of nevus spilus with laser. Pediatr Dermatol 14 (2): 115–156

[113] VIBHAGOOL C, BYERS HR, GREVELINK JM (1997) Treatment of small nevomelanocytic nevi with a Q-switched ruby laser. J Am Acad Dermatol 36: 738–741

[114] WALDORF HA, KAUVAR NB, GERONEMUS RG (1996) Treatment of small and medium congenital nevi with the Q-switched ruby laser. Arch Dermatol 132: 301–304

[115] WANG JI, ROENIGK HH JR (1999) Treatment of multiple facial syringomas with the carbon dioxode ($CO_2$) laser. Dermatol Surg 25: 136–139

[116] ZEMTSOV A, CHOKHAVATIA S (1991) Nd:YAG laser treatment of extensive recalcitrant anal condyloma acuminata. Genitourin Med 67 (5): 432

# III-7.4
# Laser Treatment of Scars and Keloids

D. Scharschmidt

## Contents

## Summary

Despite increasing knowledge of wound healing and collagen metabolism, so far no universally accepted and completely satisfying method in the treatment of hypertrophic scars and keloids has been established.

Hypertrophic scars and keloids not only represent a cosmetic problem, but often also cause pruritus, dysaesthesia and pain or can form strictures. Both hypertrophic scars and keloids become raised and red, but whereas a hypertrophic scar remains within the confines of the original skin damage, a keloid grows beyond them. And whereas a hypertrophic scar might flatten spontaneously in the course of 1 or more years, keloids remain elevated.

In the past, numerous different treatment methods have been tried with differing results [1, 2]. Especially the response of keloids is often unsatisfactory and they often reoccur after therapy. Established treatments for scars or keloids are excisional surgery and cryotherapy, adjunctive intralesional corticosteroid application, pressure therapy and covering with silicon gel sheets. Radiation therapy following surgical removal should be obsolete, due to the possible long-term carcinogenity.

Besides traditional therapies, laser treatment of hypertrophic scars and keloids has been in use for several years. Different laser systems (Argon, Nd:YAG-, CO$_2$ laser) and techniques have been used with a range of success [3–8]. A comparison of the reported results, including when using the same laser type, is difficult, in the case of different laser parameters and application techniques, combined therapies, different follow-up times and no exact declaration of the type or appearance of scars or keloids.

During the last years the flashlamp-pumped pulsed dye laser (FPDL) was frequently used to treat hypertrophic scars and keloids with very good and promising results by well-controlled studies. Alster et al.[9,10] demonstrated a 57 or 83%, Dierickx[11] an 80% improvement after one to two FPDL treatments.

## Clinical Measurements

Clinical measurements were carried out by blood-flow measurements of microcirculation by imaging with a laser Doppler flowmeter (moorLDI-VR Laser Doppler Imager System, Moor Instrument Ltd). Scar thickness and volume were measured by ultrasonic and digital image processing scanner with optical profilometry (Primos, GFMesstechnick GmbH). Larger vessels were detected by colour-coded duplex sonography (Siemens Elegra, Siemens). Sequential clinical and photographic analyses were performed before and after treatment.

Patients' subjective complaints of symptoms like pain or pruritus, and estimation of satisfaction with treatment results in flattening, lightening and pliability of the scars or keloids were monitored.

The measurements were done to select the most suitable laser system and to verity the succes of laser therapy.

## Clinical Appearance and Selected Laser Type

Clinical appearance and measurements of scars and keloids with their individual colour, shape, size or vascularization are necessary for choosing the best laser type or technique with its specific effects (Tables1, 2). Despite clinical and histological differences, hypertrophic scars and keloids are sometimes difficult to differentiate. Both can be erythematous or more pale and very different in volume or extent.

Besides hypertrophic scars and keloids there is the other group of athrophic scars, those after surgery or traumatic. Acne scarring can also take several clinical forms, including atrophic ice-pick scars, sharp wall scars or more superficial macular scars. Acne scarring can also result in hypertrophy or keloid scar formation.

Because of the very different clinical appearance, there could not be only one laser for all different types of scars or keloids.

It is imperative to properly categorize the type of scar and to determine which laser is best to treat it. According to the different clinical appearance of scars, the correct laser type with its specific and characteristic tissue effects has to be chosen for best treatment and satisfying results [12]. Furthermore, we have to consider the skin type of patients and the localization of the lesion.

The $CO_2$ laser, wavelength of 10 600 nm, with its specific absorption in water, is used for excision or together with a scanner for superficial vaporization of tissue. The 1064-nm Nd:YAG laser, with no selective absorption, is used for tissue coagulation or vaporization. The argon laser at 514 nm has a selective absorption in haemoglobin and leads to coagulation of superficial vessels. The haemoglobin selective 585 nm flashlamp-pumped pulsed-dye laser, well known in the treatment of port wine stain, destroys superficial vessels smaller than 0.5 mm (Table 2).

In many cases, only combined treatment with different laser types can lead to optimal and satisfying results.

**Table 1.** The clinical appearance and measurements indicates the used laser type

| | |
|---|---|
| • Erythema, vascularisation, teleangiectatic appearance | – Pulsed dye laser, (argon laser) |
| • Proliferative, high vascularisation, large vessels, larger size | – Nd:YAG laser transcutaneous or interstitial |
| • Non-erythematous, non-proliferative, slightly elevated hypertrophic scars (and keloids) and atrophic acne scars | – $CO_2$ Laser |

## Flashlamp-Pumped Pulsed Dye Laser (FPDL)

The pulsed dye laser with its mechanism of selective photothermolysis with irreversible destruction of microvessels has been used in the treatment of the port-wine stain for many years.

**Table 2.** Different laser types and their specific effects in treatment of scars and keloids

| Laser | Wavelength (nm) | Absorption | Effects | Target tissue | Results |
|---|---|---|---|---|---|
| $CO_2$ | 10 600 | Specific (water, high) | Vaporization | Epidermis, soft tissue | Flattening |
| Nd:YAG | 1 064 | Unspecific (tissue) | Coagulation, vaporization | Soft tissue, vessels | Lightening, flattening, shrinking |
| Argon | 514 | Selective (haemoglobin, melanin) | Coagulation | Vessels (<1 mm) | Lightening |
| FPDL | 585 | Selective (haemoglobin) | Photothermolysis | Vessels (<0,5 mm) | Lightening, flattening, improved pliability and texture |

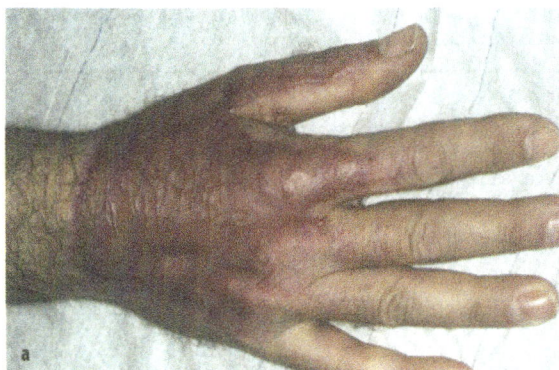

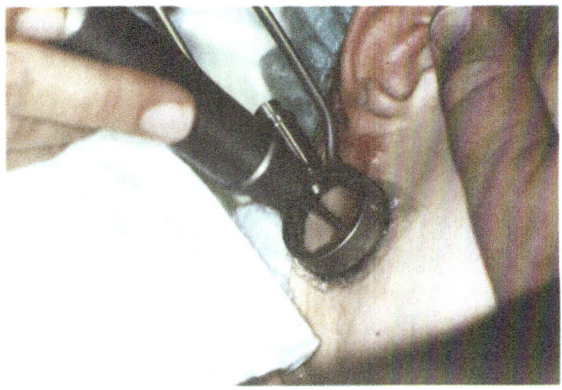

**Fig. 2** Dye laser treatment of a keloid with a cooling chamber for reduction of epidermal side effects and for pain reduction

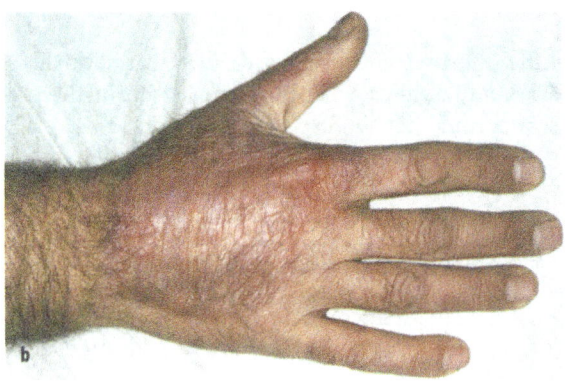

**Table 3.** FPDL in treatment of scars and keloids – indications and parameter

Indications

- Erythematous hypertrophic scars and keloids
- Erythematous burn scars
- Dermal contractures

Parameter

| | |
|---|---|
| Wavelength | : 585 nm |
| Pulse duration | : 300 μsec |
| Spot diameter | : 5 mm |
| Energy density | : 6–8 J $cm^2$ |

**Fig. 1a, b** A 2 year old burn scar of the hand (**a**) and the result after six dye laser treatments with good lightening and improvement of skin texture (**b**)

For many years, the dye laser has also been used in treatment of hypertrophic scars and keloids [9–14]. Indications for the vascular specific FPDL are erythematous hypertrophic scars and keloids and especially erythematous burn scars and keloids.

The pulsed dye laser treatment of hypertrophic scars and keloids reduces scar microcirculation and leads to a reduction of erythema with lightening of scars and keloids. Not expected at the beginning was the significant alteration in scar texture, bulk and pliability (Fig. 1a, b). Also symptoms like pain and pruritus react very well.

We use the flashlamp-pumped pulsed dye laser (Vasognost, Fa. Baasel Lasertec) with a wavelength of 585 nm and with fluency per pulse of 6–8 J/cm², a pulse duration of 300 μs and a spot size of 5 mm with an overlap of 10–20% (Table 3). In the case of an intense erythema we prefer a continuous cooling of the skin surface using a cooling chamber. With the cooling chamber, thermal side effects at the epidermis and the possible snapping pain can be sufficiently reduced (Fig. 2). Multiple treatments are necessary and can be repeated after 6–8 weeks.

Anaesthesia is not necessary in most cases, use of Emla cream or the cooling chamber in treatment of sensitive regions or childrens is sufficient.

A purpuric tissue response develops immediately after pulsed dye laser irradiation, which disappears within 10–15 days (Fig. 3). In a few cases, small bubbles or crusts develop in the treated area, which heal in some days without intervention. Other rare side effects are hyperpigmentation or infection. The recurrence rate is very low and only seen in treatment of proliferate keloids. We did not see a worsening.

Optimal results with 80–90% improvement in scar or keloid appearance require an average of three to five treatments. Cases of more fibrotic or proliferative scars or keloids require up to eight treatment sessions.

In contrast to other investigations, we saw partial recurrences after FPDL treatment in some cases of proliferative keloids.

In general, the better results were seen with erythematous hypertrophic scars (Fig. 4a, b), especially hypertrophic burn scars, less elevated keloids and those with a shorter history. Less satisfying results occurred with prominent and fibrotic keloids, proliferate keloids, dermal contractures within the scars or lesions older than 2 years [13–15]. In these cases, a primary transcutaneous or interstitial Nd:YAG laser treatment promises better results.

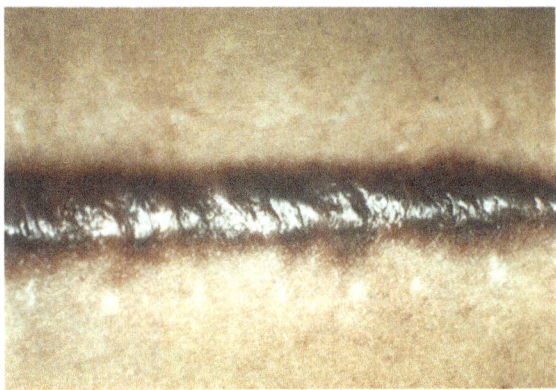

**Fig. 3** The typical appearance after dye laser treatment with purpura and without other side effects

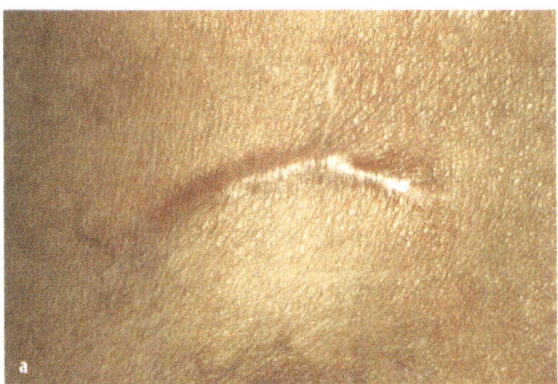

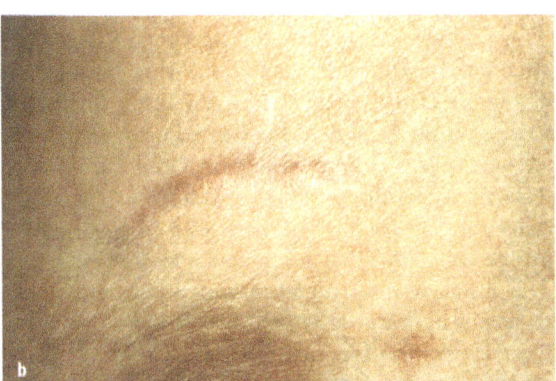

**Fig. 4a, b** Five year-old hypertrophic scar of the breast after surgery (a) and the result after three dye laser treatments with flattening, lightening, size reduction (b)

Therefore patients who are prone to develop hypertrophic scars and keloids or with a history of many recurrences after other therapies should be treated with laser as early as possible after surgical intervention, to prevent new keloid formation [16]. Our results in early treatment for prevention of recurrences are promising. Data with a longer observation time are necessary.

**Table 4.** Argon laser in treatment of scars and keloids – indications and parameter

Indication

• Residual teleangiectatic vessels after dye laser treatment

Parameter

| | |
|---|---|
| Wavelength | : 514 nm |
| Power | : 2 W |
| Focus | : 0,5–1 mm |
| Pulse duration | : 0.1–0.2 s |

## Argon Laser Treatment

After the first encouraging studies in treatment of scars and keloids with the argon laser, further research could not confirm the first results [3, 4].

With the argon laser treatment, a reduction of scar erythema can be seen, but there is no significant improvement in scar bulk and pliability or in skin texture.

We use the argon laser (Aesculap Meditec DL 5000, Fa. Aesculap Meditec) in only few cases of residual telengiectatic vessels after dye laser application or in primary treatment of teleangiectatic appearance in pale and flat scars (Table 4). Side effects of this treatment are very low, with acute redness for some hours and the development of small crusts. No recurrence was seen.

## Nd:YAG Laser Treatment

Investigations of biological effects showed that the Nd:YAG laser damages deep dermal blood vessels without destroying surrounding connective tissue [17]. Furthermore, the study showed a decreased collagen synthesis and unchanged collagenolytic activity, which indicates that the Nd:YAG laser could be successful in treatment of hypertrophic scars and keloids.

Former investigations of Nd:YAG laser treatment in scars and keloids created a tissue infarction with charring and sloughing of the treated tissue with following scar recurrence after second intention-healing [7, 8].

The effect on blood vessels has been successfully used in transcutaneous or interstitial Nd:YAG laser treatment of haemangiomas and vascular malformations for many years [18, 19].

Similar to the treatment of haemangiomas and vascular malformations a transcutaneous or interstitial application in treatment of scars and keloids is possible [12, 13]. It is necessary for the treatment to avoid blanching, coagulation or charring of the tissue or a defect of the epithelium or deeper tissue layers with following neovascularization and secondary intention healing followed by a higher risk of recurrence or deterioation.

## Transcutaneous Application

In cases of large and rigid keloids, in the case of proliferate and growing keloids or in special areas, like the face, and especially if in colour-coded duplex sonography larger vessels within the keloid could be seen, a transcutaneous Nd:YAG laser application is used.

With this treatment a reduction of pathologic vessels is achieved, by thermal damage of the endothelium with a following vascular occlusion of pathologic vessels. The result is a stop of scar growth, a lightening and a reduction in size. Also pain and pru-

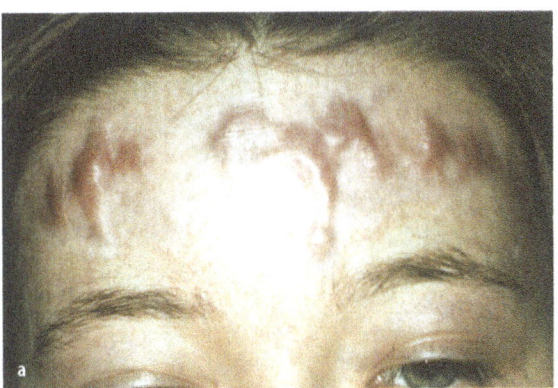

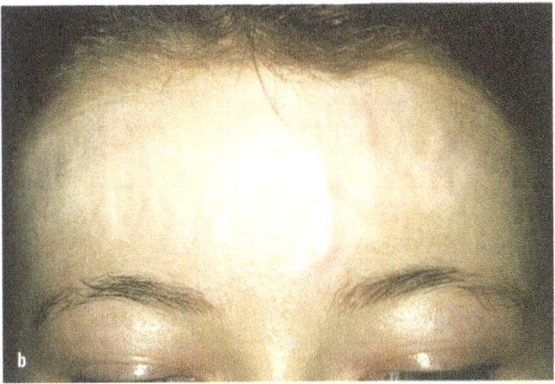

**Fig. 6.** One year-old proliferate keloid at the forehead after burn (**a**) and the result after two transcutaneous Nd:YAG laser treatments and five dye laser treatments with flattening and lightening (**b**)

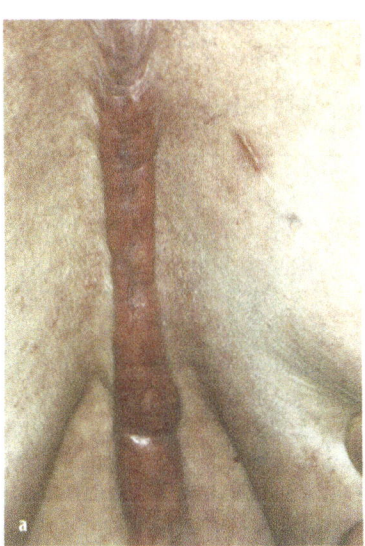

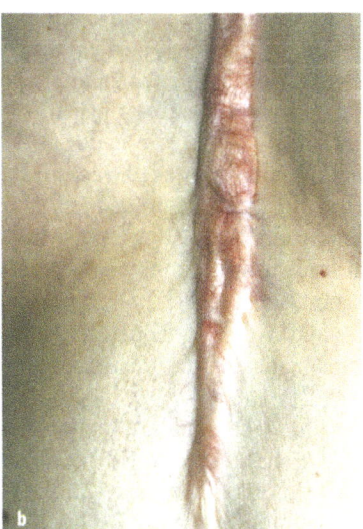

**Fig. 5a, b.** A four year old and very painful and active keloid after sternotomic (**a**) and the result after two transcutaneous Nd:YAG laser treatments with reduction of size and lightening (**b**), but especially a very good non-visible reduction of the symptoms

ritus were successfully reduced. There is no complete reduction in keloid size with this method (Fig. 5 a, b), but often, with a following dye laser treatment, further improvement is possible (Fig. 6a, b).

For transcutaneous Nd:YAG laser application (My 60, Fa. Martin) the power used is 35 W or more continuous wave or chopped mode (Table 5). The power depends on the keloid's colour, the more erythemateous the keloid the lower the power. A continious ice-cube cooling of the surface is necessary under laser treatment to reduce thermal side effects. The ice cube has to be clear and without inclusions of air or dirt to guarantee a good laser light transmission and less absorption of the ice (Fig. 7). Coagulation points with blanching of the treated tissue should be avoided because of a higher risk of development of areas with thermal necrosis. A sign for optimal treatment under laser irradiation is a longer reperfusion time of the keloid. After treatment, the keloid develops a glassy and shining appearance with swelling. The treatment is painful and only possible with a local or general anaesthesia.

Side effects are similar to the interstitial Nd:YAG laser treatment with swelling and induration. The development of small bubbles or thin crusts is possible.

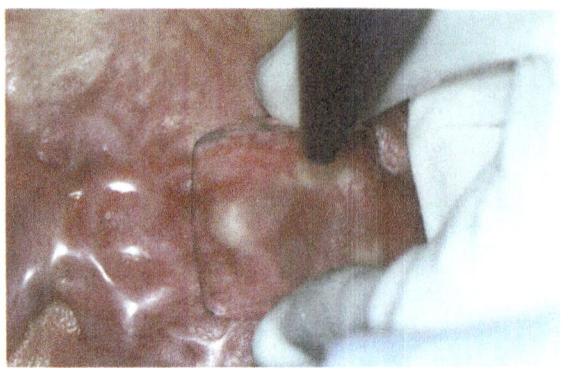

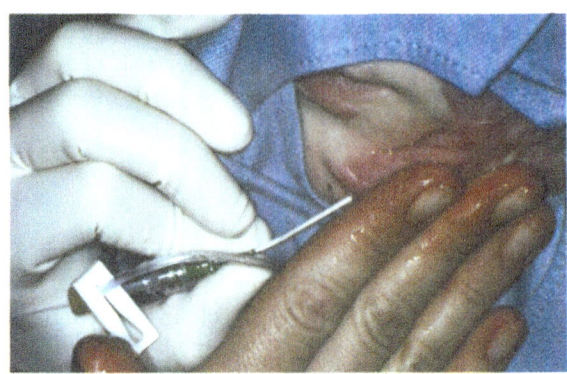

**Fig. 7.** Technique of the transcutaneous Nd:YAG laser treatment with ice-cube cooling of the surface

**Fig. 8.** Technique of the interstitial Nd YAG laser vaporization of keloids: puncture system with an i.v. line device and laser bare fibre, temperature control of the skin with the ungloved hand during vaporization

**Table 5.** Transcutaneous application of Nd:YAG laser with ice cube cooling of the surface in treatment of scars and keloids – indications and parameter

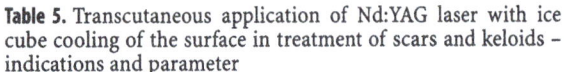

Indication

- Large proliferate keloids
- Large fibrotic keloids
- Deeper, larger vessels

Parameter

| | |
|---|---|
| Wavelength | : 1064 nm |
| Power | : 35–50 W |
| Exposure | : cw |

**Table 6.** Interstitial application of Nd:YAG laser in treatment of scars and keloids – iIndications and parameter

Indication

- Huge, fibrotic and non-proliferate keloids (>1 cm diameter or more)

Parameter

| | |
|---|---|
| Wavelength | : 1064 nm |
| Power | : 30–35 W |
| Exposure | : 0.1–0.3 s/chopped |

The risk of a thermal damage with following necrosis is low if blanching or thermal defects have been avoided. Other side effects, like infection or hyper- or hypopigmentation, are rare.

The result of the irradiation can be judged at the earliest after 6–8 weeks, at which time also a new treatment session is possible. Also in Nd:YAG-Laser-treatment three to five sessions are necessary to reach the best results, and depend on the primary extent or activity of the keloid.

## Interstitial Application

Distinct and rigid, more fibrotic and pale keloids with a diameter of 1 cm or more can be treated by an interstitial Nd:YAG laser vaporization. With this technique, a reduction of the scar extent by vaporization of the inner keloid fibrosis without epidermal injury is possible.

With this technique a 400 or 600 μm laser bare fibre is positioned in the centre of the lesion by means of an i.v.-line device. The position of the bare fibre can be controlled with ultrasound or located at the laser pilot beam. Under continuous digital control of the skin surface temperature with the ungloved hand

(Fig. 8), the inner part of the keloid tissue is then vaporized using 30–35 W in a chopped or continuous wave mode and under slow retraction of the system (Table 6). The speed of the retraction depends on the temperature of the skin. If keloid extension is massive, new punctures in different directions followed by interstitial vaporization can be performed at the same session.

Using interstitial Nd:YAG-laser vaporization a shrinking and flattening of the thick and extremely rigid keloids can be achieved (Fig.9), also an improvement in colour and a reduction of pain or pruritus. A completely regression of the keloid size is not possible. Treatment can be repeated after 8–10 weeks if the keloid volume allows this. After good reduction of keloid size, a transcutaneous Nd:YAG or a pulsed dye laser treatment will lead to more improvement with further keloid reduction (Figs. 10, 11).

Interstitial Nd:YAG laser therapy is painful, and a local anaesthesia with Lidocaine or a general anaesthesia are necessary.

The side effects of this technique are higher than with the dye laser or transcutaneous Nd:YAG laser treatment with ice-cube cooling. Normal reactions after therapy are swelling and induration. Despite control of the skin temperature, the development of small

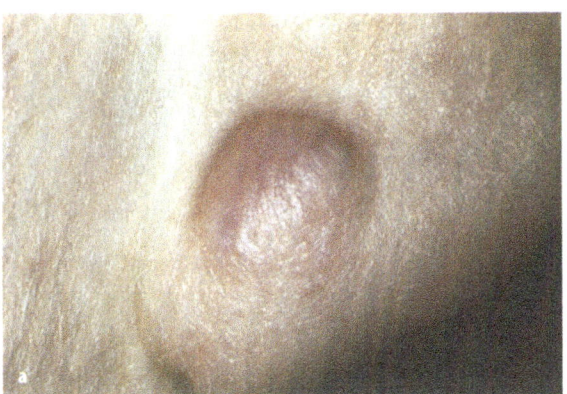

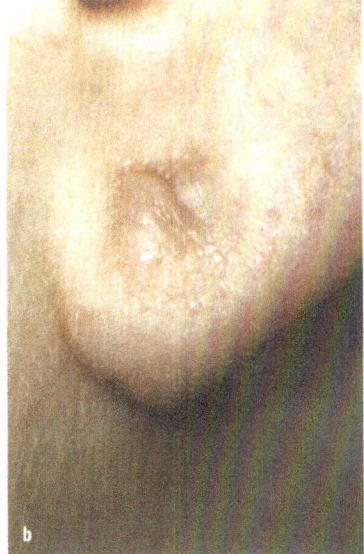

**Fig. 9.** An 8-year-old fibrotic keloid of the ear lobe with a diameter of 12 mm (**a**) and the result after two interstitial Nd YAG laser treatments with good size reduction (**b**)

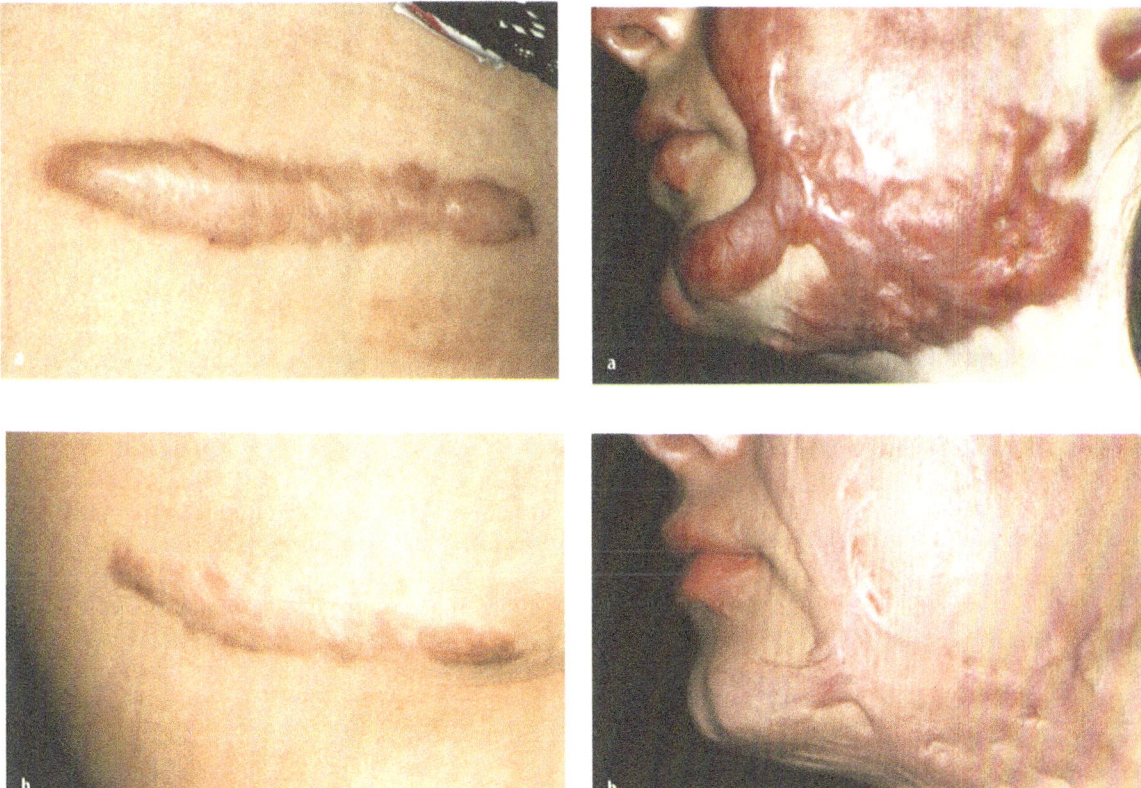

**Fig. 10a, b.** Keloid at the thorax after surgery (**a**) and the result after one interstitial Nd:YAG laser vaporization and following four dye laser treatments with lightening and flattening (**b**)

**Fig. 11a, b.** Growing keloids of the face after burn (**a**) and the result after combined treatment with two interstitial Nd:YAG laser vaporizations of the thicker keloid parts (chin, medial cheek) and later combined with transcutaneous Nd:YAG laser treatment and dye laser treatments of the other parts with good flattening, lightening after ten treatment sessions (**b**)

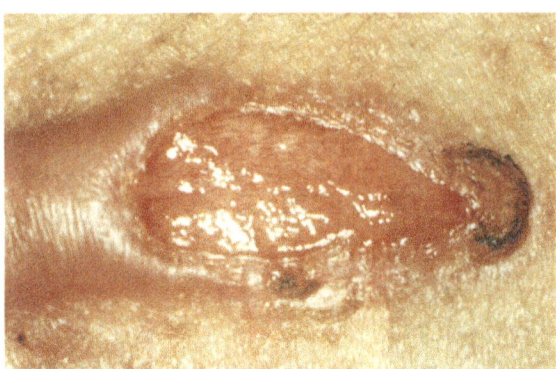

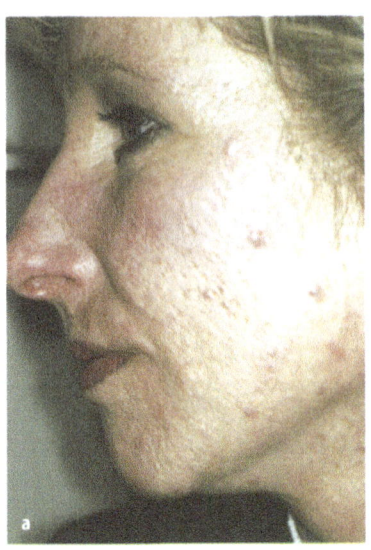

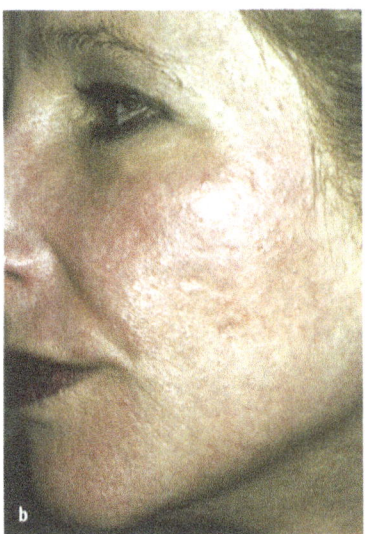

**Fig. 13a, b.** Acne scars of the cheek (**a**) and the result after two CO₂ laser treatments with scanner with smoothening of the surface (**a**)

**Fig. 12.** Necrosis of a part of the keloid after interstitial Nd:YAG laser treatment because of insufficient temperature control

bubbles and crusts are possible. We recommend an immediate ice-cube cooling after laser application. Without sufficient control of the skin temperature with the ungloved hand, thermal damage with the development of necrosis can occur with a high risk of keloid recurrence or deterioration (Fig. 12). An other possible side effect is an infection; hyper- or hypopigmentation are not seen.

The interstitial laser treatment requires much experience in laser application and technique on the part of the physician. The result of treatment with this method can be easily minimized through "over treatment" or insuffient treatment.

## Posttreatment

In posttreatment after all these laser applications, we recommend ice packs and only a greasy ointment, Vaseline for example. The patients should avoid intense sun exposure for 6–8 weeks and mechanical irritation of the treated area. Wound dressing is recommended only for regions where mechanical alterations are possible. Antibiotic or antiviral medications or creams are not necessary. The patients can wash the area carefully with mild soaps or have a shower, bathing is not allowed for 2 weeks.

## CO₂ Laser Treatment

We use the scanner-attached CO₂ laser system regularly in the treatment of atrophic acne scars and in a few cases of light, older and slightly elevated hypertrophic scars. Because of a high risk of recurrence or worsening, we avoid the use of the CO₂ laser in treatment of keloids.

High-energy pulsed or scanned CO₂ laser systems allow controlled superficial vaporization of thin layers of skin while minimizing damage to surrounding structures, resulting in smoothing the skin and a return to normal skin surface markings (Fig. 13). Different investigations show equivalent clinical results for pulsed or scanned CO₂ laser systems [20, 21].

We work with a continuous wave CO₂ laser (Smartoffice, Fa. Deka M.E.L:A.s.r.l.) with an attached microprocessor-controlled scanner (Optomedic-Scannersystem) to ensure that the thermal relaxation time – the time it takes the tissue to cool to 50% of its peak power temperature – is less than 1 ms. We use a power of 10 to 15 W and the radiation time on any given point of focused beam is 500 µs by moving in different geometric pattern and sizes (Table 7). Depending on the lesion, we perform two to three passes on the same region. More passes result in deeper thermal damage with following longer healing time ore higher risk of side effects, and do not lead to more clinical or histological improvement.

**Table 7:** $CO_2$ laser treatment of acne and hypertrophic scars – indications and parameter

Indication

• Atrophic acne scars

• Older non erythematous and non proliferative hypertrophic scars and burn scars

Parameter

| | |
|---|---|
| Wavelength | : 10 600 nm |
| Power | : 10–15 W |
| Exposure | : cw |
| Radiation time | : 500 μs |

In most of the cases, a topical anaesthesia with Emla cream or a local anaesthesia with Lidocaine is sufficient. General anaesthesia is recommended only for larger areas or in full-face treatment. If the treatment of only parts of the face is necessary, the aesthetic units of the face should be borne in mind.

In posttreatment we prefer an open technique with a Vaseline ointment and the application of cold black tea compresses five to eight times a day and cool packs. A semi-occlusive dressing is also possible (Polyurethan, Hydrogels). The development of a hard crusting cover over the superficial lesion is to be avoided. Sun protection or sun avoidance up to 4 months is recommended to minimize the risk of hyperpigmentation. We recommend an antibiotic medication only in treatment of the full face and an antiviral medication only in the case of a history of herpes. If a hyperpigmentation is seen, patients can be treated with azelaic acid cream or hydroquinon-containing products.

Side effects after therapy are an edema and swelling, a serous wound exudation and the development of crusts. Complete re-epithelization takes 7–10 days. The subsequent erythema disappear within 6–8 weeks. Complications at the acute phase are bacterial or viral wound infection, at the later phase a prolonged erythema or a hyper- or hypopigmentation. A hypopigmentation is rare and more often seen after treatment of older burn scars. Rare side effects are scars or deterioation.

Especially in the treatment of atrophic acne scars a nearly complete adjusting with smoothing of the skin is possible. Depending on the primary appearance up to three treatment sessions with an interval of 6 months or more are necessary to reach the best result. In the case of hypertophic scars and hypertrophic burn scars mostly only an improvement is seen and no complete reduction.

## Conclusion

Laser therapy is becoming more and more an alternative procedure with satisfying results, low side effects and easy application in an outpatient setting.

Pain or pruritus nearly always disappear after laser treatment. Most cases see a reduction of erythema. Scar rigidity and height can often be reduced.

Nevertheless, a complete regression is mostly not possible, and it is certainly important to tell the patient that the scar will not become completely invisible.

The pulsed dye laser gives the best results and least side effects, and is best used to treat erythematous hypertrophic scars, especially burn scars, and slightly elevated keloids. This treatment is a good alternative to traditional therapies and for some cases the better treatment method.

The combination with an interstitial or transcutaneous Nd:YAG laser treatment is possible in the case of distinct or growing keloids. It is likely that the increasing combination of different laser treatments and adjuvant therapies leads to better treatment results. In the treatment of problematic cases, more successful results can be obtained through a multi-method treatment. In these cases we combine laser therapy with intralesial application of corticoid steroids or we recommend the use of silicon gel sheets during the treatment intervals [22–24]. The concomitant use of other treatment options seems to show not only adjunctive, but summative results.

Long-term follow-up is necessary to confirm the promising results in laser and combined treatment. An observation time of 2 years is necessary to rule out recurrence. The exact examination and comparison of the results is complicated by the very different and always very individual appearance of this clinically inhomogenous group of lesions.

The pulsed or scanned $CO_2$ laser treatment is now a safe and very sufficient method to treat atrophic acne scars and slightly elevated scars. The very good results in many investigations and studies show, that it can be the better and safer alternative in treatment of these lesions in comparison to traditional methods like dermabrasion or chemical peeling.

The laser is not a wonder tool in treatment of scars and keloids, but the results show that it can compete with traditional treatment methods, and in the case of correct indication and choice of the best laser system, and with the knowledge and experience of the user, it can be in some cases often the better treatment option.

# References

[1] BERMAN B, BIELEY HB (1995) Keloids. J Am Acad Dermatol 33 (1): 117–123

[2] LAWRENCE WT (1991) In search of the optimal treatment of keloids: report of series and a review of the literature. Ann Plast Surg 27:164–78

[3] HENDERSON DL, CROMWELL TA MES LG ET AL (1984) Argon and carbon dioxide laser treatment of hypertrophic and keloid scars. Lasers Surg Med 3:271–277

[4] HENNING JPH, ROSKAM Y, VAN GEMERT MJC (1986) Treatment of keloids and hypertrophic scars with an argon laser. Lasers Surg Med 6: 72–75

[5] KANTOR MR, WHEELAND DG, BAILIN PL, ET AL (1985) Treatment of earlobe keloids with carbon dioxide laser excision: report of 16 cases. J Dermatol Surg Oncol 11: 1063–1067

[6] APFELBERG DB, MASER MR, WHITE DN (1989) Failure of carbon dioxide laser excision of keloids. Lasers Surg Med 9: 382–388

[7] APFELBERG DB, SMITH T, LASH H ET AL (1987) Preliminary report on use of the neodynium-YAG laser in plastic surgery. Lasers Surg Med 7: 189–198

[8] SHERMAN R, ROSENFELD H (1988) Experience with Nd:YAG laser in the treatment of keloid scars. Ann Plast Surg 21: 231–235

[9] ALSTER TS (1994) Improvement of erythematous and hypertrophic scars by the 585-nmn flashlamp-pumped pulsed dye laser. Ann Plast Surg 32: 186–190

[10] ALSTER TS, WILLIAMS CM (1995) Treatment of keloid sternotomy scars with 585 nm flashlamp-pumped pulsed-dye laser. Lancet 345: 1198–1200

[11] DIERICKX C, GOLDMANN MP, FITZPATRICK RE (1995) Laser treatment of erythematous/hypertrophic and pigmented scars in 26 patients. Plast Reconstr Surg 95: 84–89

[12] ALSTER TS, HANDRICK C (2000) Laser treatment of hypertrophic scars, keloids and striae. Semin Cutan Med Surg 19 (4): 287–292

[13] SCHARSCHMIDT D, ALGERMISSEN B, PHILIPP C, BERLIEN HP (1998) Prinzipien der Laserbehandlung von Narben und Keloiden. Journal DGPW 16: 7–9

[14] SCHARSCHMIDT D, ALGERMISSEN B, BERLIEN H P..Die Laserbehandlung von Narben und Keloiden In: BERLIEN HP, MÜLLER G (eds) Angewandte Lasermedizin, Lehr- und Handbuch für Praxis und Klinik. III–3.10.2.1. ecomed, Landsberg

[15] PAQUET P, HERMANNS JF, PIERARD GE (2001) Effect of the 585nm flashlamp-pumped pulsed dye laser for the treatment of keloids. Dermatol Surg 27 (2): 171–174

[16] MCCRAW JB, MCCRAW JA, BETTENCOURT N (1999) Prevention of unfavorable scars using early pulse dye laser treatments: a preliminary report. Ann Plast Surg 42 (1): 7–14 a.L., München

[17] CASTRO DJ, ABERGEL RP, JOHNSTON KJ ET AL (1983) Wound healing: biological effects of Nd: YAG Laser on collagen metabolism in pig skin in comparison to thermal burn. Ann Plast Surg 11 (2) 131–140

[18] POETKE M, PHILIPP C, BERLIEN HP (1997) Ten years of laser treatment of haemangiomes and vascular malformations: techniques and results. In: BERLIEN HP, SCHMITTENBECHER PP (eds) Laser surgery in children. Springer, Berlin eidelberg New York, pp 82–91

[19] POETKE M, BÜLTMANN O, URBAN P, BERLIEN HP (1998) Vaskuläre Malformationen im Kindes- und Erwachsenenalter. Therapie mit dem Nd:YAG-Laser. Vasomed 10: 338–347

[20] ALSTER TS, NANNI CHA, WILLIAMS CM (1999) Comparison of four carbon dioxide resurfacing lasers. A clinical and histolopathologic evaluation. Dermatol Surg 25: 153–159

[21] TRELLES MA, RIGAU J, MELLOR TK, GARCIA L (1998) A clinical and histological comparison of flash scanning versus pulsed technology in carbon dioxide laser facial skin resurfacing. Dermatol Surg 24 (1): 43–49

[22] CONNELL PG, HARLAND CC (2000) Treatment of keloid scars with pulsed dye laser and intralesional steroid. J Cutan Laser Ther 2 (3): 147–150

[23] FULTON JE JR (1995) Silicone gel sheeting for the prevention and management of evolving hypertrophic and keloid scars. Dermatol Surg 21(11): 947–951

[24] GOLD MH (1993) Topical silicone gel sheeting in the treatment of hypertrophic scars and keloids. A dermatologic experimentation. J Dermatol Surg Oncol 19: 988–992

# III-7.5
# Long-Lasting Epilation of Hair with Laser and Light Devices

B. Algermissen

## Contents

## Introduction

In the animal kingdom hair has a sensory and protective function, while in human culture, the reduced "coat of hair" occupies only a psychosocial position.

The wish to model hair freely by dyeing, cutting or by removal at unwanted body sites is understandable. The procedures for hair removal have hardly changed during the last hundreds of years. Hair has been tweezed mechanically, shaved, chemically depilated or, in this century, electrocaustically epilated by different procedures. The demand and the market for epilation, as also for hair growth substances, are immense and sales world wide per annum are estimated at around several billion Euros. Since laser systems have been efficiently used for the treatment of several vascular or other diseases, now the conquest of the market for "permanent hair removal" with easily modified laser and intense pulse light systems is being undertaken. Unfortunately, without sufficient scientific investigation, but with unprecedented advertising expenditure "customers" (physicians) to buy the systems and clients (formerly patients) to use them are being recruited. Some of the originally euphorically praised and applied devices have been considered more critically in the meantime and responsible physicians do not hesitate to explain possible side effects and real changes of success.

Nevertheless, the aim of all systems is to realize permanent hair removal in the desired localization. Some laser and light systems have approval for this indication from the Food and Drug Administration (FDA) of the USA as devices for "permanent" hair removal. The word permanent comes form the Latin verb permanere, translated last, remain. The FDA already assigns approval for a device for "permanent" hair removal, if 30% of the hair has not regrown after treatment with the epilation device in an observation period of 9 weeks. This basic definition of the restriction allows many of systems to be sold with FDA approval. Therefore, the FDA permission entails device demonstrating its efficiency soon after in clinical studies.

The basic principle of epilation with laser and light systems is the destruction of the hair follicle after selective absorption of the light through the chromophore melanin in the areas of the hair sheath and hair shaft.

With the exception of the Q-switched Nd:YAG laser, which directly or after incorporation of charcoal particles, works via mechanical disruption of the hair follicle, all other systems cause thermal damage to the hair follicle. In contrast to the melanocytes in the epidermis, with 1µs thermal relaxation time, the large hair follicle has a clearly higher time of approximately 100–200 µs. The irradiation of the hair follicle with pulse lengths of 3–50 ms is presumed to cause a selective heating up of the follicle structure, while the melanocytes and/or pigment-containing keratinocytes close to the basal membrane zone are able to conduct the heat continuously to the surroundings. These effects are described by the terms selective thermokinetics and selective photothermolysis.

Next to these effects for hair removal, the hair cycle (hair biology) plays a pivotal role. Every hair can be divided coarsely into three regions – the infundibulum, hair sheath and hair bulbus. The hair bulbus consists of a mesenchymal part covered in a half-moon shape with matrix keratinocytes interspered with melanocytes. The components of the mesenchymal include a vascular plexus and neural structures, and are essential for hair formation. The hair sheath ends at the insertion of the arector pili muscle in close vicinity to the bulge region, an area where the epidermal stem cells are located and the cell division rate rises in the early anagen phase. The infundibulum combines this area with the epidermis and plays no essential role during hair formation.

Every hair passes periodically through a complex growth cycle, which is divided into anagen, telogen or catagen phases. The length of a phase depends, among other things, on the localization of the hair on the body. Every stage can be subdivided into several parts with typical morphological and physiological changes of the hair and its environment. In the early anagen phase, epidermal cells of the bulge area and the mesenchymal part come into close contact and the formation and growth of the new hair is initiated.

This phase seems to be the most vulnerable for hair destruction. As could be demonstrated in mice hair, in the early anagen stage complete destruction could be permanent, but hair in other phases showed only growth delay and/or miniaturization of the hair (thinner). While in rodents the growth is synchronized, within the human being all growth phases are close to each other. Furthermore, different cycle lengths can be found within the human being according to body localization. Even with complete treatment of all hairs in the upper lip area, in which a maximum of 10–15% of the hairs are in the early anagen

phase, only this part of the anagen hairs can be damaged permanently. For the axillary region with anagen phases of more than 4 months, the percentages should be still lower. The consequences are that an epilation of one area always requires several sessions with an interval for 6 to 8 weeks, and does not lead to a satisfactory result after one treatment.

Next to the hair biology, further essential factors must be considered. As already stated, melanin is the main chromophore of the hair. Up to now, patients with light blond or grey hair cannot be treated efficiently without loss of selectivity and induction of side effects. Furthermore, depending on the body site, hairs are localated in the deep corium or subcutaneous fat. A penetration depth of light of up to 6–8 mm, for example for axillary hairs, is necessary.

A laser system which emits a defined wavelength must be selected regarding the limitations of penetration and absorption.

Due to the high demand on the cosmetic market, older laser systems were adapted for epilation but also new laser and light systems have been developed for this indication. The entire offer is becoming increasingly unmanageable and many systems which are offered today are only inadequately evaluated. Unsatisfying results of treatment and patients disappointed with the results are the consequences. Also the physician who bought such system with high capital expenditure in order to expand the treatment spectra within his practice or clinic, knows only after the purchase how efficiently his system guarantees the permanent epilation of hair.

However, for some laser and flashlamp systems, considerable efforts in the past years have been undertaken to provide clinical and histological studies. Considering these results, it can be mentioned that nowadays approximately 10–15% of the treated hairs could be epilated permanently or for longer periods and that the growth of the remaining hairs can be delayed after treatment and/or the remaining hairs regrow as thinner hairs. Satisfactory results can be achieved after several sessions at intervals of 6–8 weeks. Up to now no other alternative procedure has achieved a complete treatment of a large area with the same low side-effect profile. Further long-term studies with observation periods of up to 3 years demonstrated that with no other alternative procedure such as for example, electrolysis, long-lasting epilation of up to 30% of the irradiated hairs after three treatment cycles can be achieved. With this, laser epilation with scientifically evaluated systems has been proved a serious procedure and should, like previous procedures, such as, for example, electrolysis, be accepted by the insurance schemes.

## Anatomy and Biology of Hair

### Physiological and psychosocial Functions of the Hair

To be able to carry out efficient destruction of the hair follicle, one needs basic knowledge of hair anatomy and the structures necessary for hair origin within a hair follicle. Evolutionarily, the hair follicle is a further development of the fish and reptile scale. For mammals, the hair coat is of great importance for thermal regulation or protection of the skin, e.g. from mechanical or ultraviolet light. Furthermore, the coat of hair plays a role in social communication within a species. While for the human being the body hair growth has lost most of its functional importance, only sexual hair and, especially, head hair dominate the social communications within a culture. The fashionable hairstyle and hair care (styling, colour) stand at the centre of body care and reflect the culture and/or standard of living of the people, together with the latest fashion. Troubles through loss of hair or unusually localized hair growth can induce avoidance or rejection.

Hirsutism develops in areas with typical vellus hair existence [1, 2]. Strong terminal hairs grow as a result of systemic or local elevated concentrations of androgens, which produce beard, breast hairiness and modifications of the pattern of pubic hair. Another problem can arise from congenital hypertrichosis. As a consequence of these changes, faulty communication with the environment results. The patients report on problems in their professional life and in their private environment, suffer from enormous emotional stress, even to depressions and suicide. Therefore, each patient attempts to change the condition and is continuously searching for therapeutical options.

### Anatomy and Biology of the Hair Follicle

At the end of the 2nd month of gestation, the neogenesis of the hair follicles begins and is completely finished at birth. In later life, no further neogenesis of the hair follicle occurs [3–5].

The first hairs are the lanugo hairs, that already intrauterine and/or shortly after birth fall out and are replaced by vellus and/or terminal hairs during growth, according to localization.

Vellus hairs have a cross-sectional diameter of 30 µm or less and a miniaturized to hypoplastic, mostly unpigmented hair shaft [6–9]. The terminal hairs are pigmented (genetically determined) and have a cross-sectional diameter between 200 and 350 µm with a hair shaft diameter ranging from 40 to 120 µm [10–14]. The type of hair developed out of any follicle can change and is influenced, for instance, by hormones [15]. These influences can be observed during puberty or diseases such as polycystic ovary [6].

For the neogenesis and later formation of the hair follicle, the communication of a specialized dermal (later dermal papilla) and epidermal component (bulge area) is necessary [16]. The developed hair follicle can be distinguished roughly into three segments: infundibulum, isthmus and proximal part with dermal papilla. The infundibulum extends from the follicle orifice up to the entrance of the sebaceous gland and the isthmus region of the sebaceous gland entrance up to the insertion of the arrector pili muscle. The inferior part of the hair follicle, which extends from the insertion of the arrector pili muscle to the dermal papilla, includes the hair bulb, the segment which is responsible for the production and formation of the hair. In the region of the insertion of the arector pili muscle, a thickening of the external root sheath is found, called the bulge region. This region contains the epidermal stem cells of the hair follicle necessary for the reconstruction of a new hair [17–19]. The external root sheath coats the hair shaft coil-shaped and feeds the inner root sheath that is between hair shaft and external root sheath [20, 21]. The proximal bulbus lies onion-like over the dermal papilla. This highly vascularized region consists of specialized mesenchymal cells standing in close contact to the matrix keratinocytes that surround the proximal papilla. Here are also interspersed with the y melanocytes which are responsible for the pigmentation of the hair and which transfer the melanosomes to the hair matrix keratinocytes via dendrites. Follicular melanocytes differ from epidermal melanocytes. Firstly, the ratio of melanocyte:keratinocytes increases in the follicle from 1:30 in the epidermis to 1:5. Secondly, the synthesized melanosomes in the hair shaft are larger and more numerous than in the epidermis and are not so degraded. Thirdly, production of melanosomes depends on hair cycles. Only in the anagen stage II–IV are the melanocytes active and synthesize the melanosomes. Which hair colour will develop is genetically determined and depends on the ratio of the brown-black eumelanin to the reddish pheomelanin [22–24]. Characteristic for the function as a chromophore is that eumelanin absorbs 30 times more light of a wavelength in the range of 694 nm than pheomelanin which, in addition, possesses a very small absorption in the wavelength range more than 700 nm [25–27]. Black or dark brown hair contains large eumelanosomes, red hair pheomelanosomes. This explains the better results with black as opposed to reddish hair in laser epilation. Blond hair possesses fewer melanosomes, grey hair has reduced numbers of melanocytes and poorly melanized melanosomes, and in white senile hair no dopa-positive melanocytes are found [22].

The matrix keratinocytes continuously produce the hair shaft, which can be divided into three parts: the external cuticula, one subordinate cortex and an inner air-filled medulla. Vellus and lanugo hairs do not have a medulla, which plays an important role in the elasticity of the hair. Depending on localization and hair cycle, the entire hair follicle is localized in the corium up to the subcutaneous fat tissue.

As already mentioned, hair growth passes through different cyclical phases, the anagen, catagen and telogen phases [15, 16, 28]. The anagen phase represents the growth phase of the hair, the catagen phase the dying phase and telogen the resting phase until the loss of the hair shaft. The duration of the individual phases depends on body localization [1, 29]. Variable long hairs grow according to body field and duration of the anagen/telogen phase [30]. The longest anagen phase exists in the field of the head hair with anagen phases of 4–6 years [29–32]. The catagen phase is the shortest with up to 1 to 4 weeks. The telogen phase lasts mostly several months (3 to 9 months) [30]. In head hair approximately 80–90% of the hairs are found in the anagen phase, approximately 2% in the catagen phase and 10–15% in the telogen phase [6, 33]. On the other hand, in other body areas, for example the legs, more than 50% of the hairs are in the telogen phase [6, 29, 33, 34].

In the catagen phase as a result of apoptotic processes, the matrix keratinocytes and the inner root sheath die. The hair shaft loses contact to the dermal papilla and moves in the direction of the isthmus until it falls out. The external root sheath folds itself accordeon-like and the telogen phase follows the hair shaft. In the early anagen phase, the dermal papilla is very highly localized in the corium and comes close to the bulge area. Together with the epithelial stem cells from the bulge region, the formation of a new anagen hair is initiated [9]. In the following anagen stages, the new hair germ penetrates again down to the level of subcutaneous fat, developing the structures of an anagen hair. In the early anagen phase the dermal papilla appears more strongly pigmented and the bulge area shows a raised rate of cell division [15].

Both alterations (pigmentation, raised cell division rate) and the localization of the dermal papilla in close vicinity to the bulge area make the hair follicle very vulnerable in this phase, and thermal destruction of the hair follicle by light or laser systems can cause permanent removal of the hair follicle, while in later phases only partial changes such as growth delay or thinner hair regrowth are achievable [26]. It is presumed that the bulge area and the dermal papilla represent the central target regions which must both be destroyed for permanent epilation [34].

The essential condition to achieve optimal epilation is the hair anatomy and physiology, which also

**Table 1.** Important anatomical and physiological parameters for the selection of light and laser system and laser parameters for epilation

| Hair type/-diameter | – Terminal hair<br>– Vellus hair |
|---|---|
| Hair pigmentation | – Ratio of pheomelanin/eumelanin (reddish/fair/brown/black colour) |
| Hair depth | – 2–7 mm (to subcutaneous fat) |
| Hair cycle phase | – Anagen-/catagen-/telogen phase<br>– Distribution within the phase according to body localization |

indicate the boundaries of the maximum epilation that can be achieved. In selecting the light or laser system and the parameters for epilation these must be considered in planning and carrying out treatment (Table 1).

## Physical Basics of Epilation of Hair Follicle with Light and Laser Systems

The principle of epilation with light and/or laser systems is based on the effects of selective photothermolysis and selective thermokinetics. As described in the chapter [laser tissue interactions], the light absorption of a chromophore primarily produces a heating of the chromophore and then secondly, as a result of the heat diffusion, a heating of the surrounding tissue [35]. How much of the heat is yielded to the surrounding tissue depends on many factors.

Related to the light and/or laser system, common parameters are important such as wavelength and/or range of wavelength (intense pulse light systems), pulse length, pulse peak power, power density as well as focus diameters. In addition to general tissue interactions such as reflection, scattering and penetration of the light, the absorption coefficient of the chromophore, the distribution of the chromophore in the tissue, the chromophore density and the volume of the target tissue play a pivotal role in achieving the necessary effects. For epilation, the chromophore melanin that was generated in melanocytes of the epidermis and the hair follicle has the essential role in the destruction of the hair follicle. In accordance with the common models, light penetrates, first of all, into the epidermis and more deeply into the corium, depending on specific optical tissue properties. In the case of laser-induced epilation, light is absorbed by melanin-(melanosome-) containing cells (melanocytes/keratinocytes) and/or structures (hair shaft) which are heated, after which the heat is conducted to the environment. The speed with which a structure can yield heat via conduction to the environment is called thermal relaxation time. The thermal relaxation time (TRZ) for a structure can be estimated with the equation TRZ = d2 gk$^{-1}$, where $d$ stands for diameters, $g$ for a geometrical factor of the structure and $k$ for the thermal diffusion constant of the structure. If one considers the thermal relaxation times of the melanin- and/or melanosome-containing cells and/or structures of the skin, the term selective photothermolysis can be explained and the treatment parameters adapted to the skin type. Melanocytes have a very short relaxation time of approximately 1 µs, which means that heat developing by absorption can be dissipated to the surroundings very fast. The "voluminous" hair follicle has a clearly extended relaxation time of several 100 µs, therefore a light pulse below the relaxation time of the hair follicle would lead only to a warming up of the hair shaft without serious destruction of the surrounding structures. For laser epilation, pulse lengths clearly over the relaxation times (for example 3 ms) of a hair shaft are necessary to induce high temperatures (> 60 °C) into the hair follicle and in the surrounding tissue. The term selective thermokinetics describes different warming-up properties of the hair follicle in comparison to keratinocytes and melanocytes of the epidermis. The heat that primarily rises in the melanocytes and in the hair shaft by absorption is yielded to the surrounding tissue via conduction with different kinetics. The result is that keratinocytes and melanocytes can conduct the heat faster to the surroundings due to the shorter thermal relaxation than the hair follicles which are being continuously heated. Using long pulses, the structures close to the follicle are additionally heated. In the targeted hair follicle a crucial temperature rises, with the effect of thermal cell damage. These different processes of heating characteristics are described as selective thermokinetics. In the light of these conditions, parameters for the treatment can be selected and adapted. Patients with light skin (skin types I–II) can be treated with short pulse lengths and high-power densities without severe side effects, for instance thermal destruction of the epidermis. On the other hand, the pulse length in the case of darker skin types (III–V) should be increase and the power densities fitted to avoid complications such as, for example, blister and crust formation. Another possibility exists in the application of short impulses with higher pulse peak power and pulse durations below the relaxation time. The large structure of the hair follicle is continuously heated, but the melanocyte or keratinocyte in the epidermis can dissipate the heat taken up.

Another method is cooling the epidermis with ice cubes, high-powered spray cooling systems, precooled air or hydrated gels [36–41]. These methods decrease or remove the heat during treatment and can improve the so-called index matching.

Another possibility using the principle of the selective photodestruction is the application of very short pulses, for instance the Q-switched Nd:YAG laser device. In the treatment of tattoos, this procedure is based on photomechanical effects. The chromophore melanin takes up the high-energy pulse of a length of ns. The hair follicle and other targets become athermically destroyed. To improve this effect, and also for treatment of hair shafts low amount of melanin, a suspension of charcoal particles with diameters of approximately 100 µm was applied to the skin. The particles should enrich the hair follicle or could be inserted into deeper structures of the hair shaft using low-power pulses of a Q-switched Nd:YAG laser prior to treatment. After superficial removal of the particles, the destruction of the hair follicle occurs passively with photomechanical destruction of the inserted particles. In the literature results of this technique are discussed very controversial [42, 43].

## Techniques and Methods for the Removal of Unwanted Hair

### Alternative Methods for the Epilation and/or Depilation of Unwanted Hair

For millennia, unwanted hairs have been epilated [44]. The most widespread and simplest procedure is mechanical epilation, that means picking out the hairs, for example with tweezers, application of wax or specific epilation plasters. Mostly, only the hair and parts of the hair shaft are removed, but the dermal papilla or hair bulge remain in the follicle structure. In general, each hair regrows with some delay. Besides pain during tweezing and redness afterwards, frequent complications are irritations of the skin and/or folliculitis with possible scarring or keloid formation [45–50]. Another possibility for hair removal is the use of keratolytic substances which dissolve the hair shaft chemically. The substances destroy the disulphide bridges of cystine, the hair loses its structure and can be easily removed. The effects are, however, only of short duration, since only the hair in the region of the infundibulum is removed [51, 52]. Chemical bleaching is also frequently used by women with dark to black hair. The hairs lighten and become more inconspicuous. Typical side effects are toxic and/or allergic dermatitis and disturbances of skin pigmentation (hypo- and/or hyperpigmentation) [53, 54].

During the last century, electrical epilation with diathermy needles was developed as the treatment of first choice for hair removal [55,56]. Most health insurance frequently refer to this „established" procedure if a patient tries to clain money from the insurance for laser treatment of unwanted hairs. The main

**Table 2.** Spectrum of some laser and light systems which are offered for the epilation of hairs at that time

| System name | Source | Wavelength (nm) | Pulse length (ms) | Power ($J\,cm^{-2}$) | Focus (mm) |
|---|---|---|---|---|---|
| EpiLaser | Ruby | 694 | 3 | 10–75 | 10–12 |
| Epitouch | Ruby | 694 | 8 | 5–10 | 4–6 |
| Apogee | Alexandrite | 755 | 20 | 10–50 | 7–12 |
| Epitouch | Alexandrite | 755 | 2 | 10–40 | 5–7 |
| SightSheer | Diodenarray | 810 | 5–30 | 10–40 | 9 |
| Smartepil | gep. Nd:YAG | 1064 | 3–20 | 10–140 | 6 |
| Softlight | QS Nd:YAG | 1064 | 10–20 ms | 2–3 | 7 |
| EpiLight | Flash | 590–1100 | 2–7 | 10–50 | Var. |
| Ellipse | Flash | 590–1000 | 1–50 | 10–50 | Var. |

argument is that the method laser epilation has not up to now been sufficiently evaluated.

During electrolysis, the hair follicle is punctured with a fine diathermy needle and heated, together with the surrounding tissue, by receiving a direct current and being thermally destroyed. The process is very time-consuming and painful. In one session only small areas can be partially treated. Approximately 15–20% of the hairs treated in a session (50–200) are permanently destroyed [57]. Many women with hirsutism who we see in our outpatient department have been previously treated by a beautician or dermatologist with electrolysis over months to years. Nevertheless, hundreds of hairs could be recognized next to mostly atrophic scars in the region of the chin and upper lip. It must be doubted that this procedure can still be regarded as the "golden standard" for hair removal. Also the cost-benefit relation appears worthy of discussion in the case of unsatisfactory treatment lasting for years in many sessions [58]. Typical side effects are pain und burning during and after the treatment as well as redness, oedema and crusts. In addition, the highest incidence of folliculitis and mostly atrophic scarring was found in comparison with other techniques. A further procedure, that itself has been developed from electrolysis is the blend technique [59]. With this procedure, a solution is applied onto the skin and the hair follicle also punctured with a fine needle. Then, a small direct current is put on and stopped according to the reaction of the hair (thermal electrolysis) after several seconds. This procedure is supposed to be milder to the skin and induce a lower incidence of side effects compared to electrolysis. Approximately 50% of the treated hairs should be permanently epilated [60]. Nevertheless, also slight pain occurs during the treatment and typical side effects such as redness, oedema and, to a much lesser intent, folliculitis and scarring. This procedure allows, however, only partial treatment in many sessions over a long period. To what extent it

is more efficient than laser epilation has to be investigated in future studies. It could possibly be a method for the treatment of light blond or white hairs, which could not be efficiently removed by laser epilation.

## Light and Laser Systems for Hair Removal

In the last decade numerous systems were developed for hair destruction, and introduced onto the market [61, 62].

Almost monthly, new systems are added, recently particularly diode lasers, so that it is impossible to make a complete list of all the systems offered on the market for this indication at the moment. An essential problem, due to the short life time of the systems, is the scientific evaluation of a system. Because an observation interval of 1–2 years should be required and the systems have to be sold very fast, long-term experience is lacking for many systems. Mostly only one study – if any it all – is made for a laser device to obtain permission from the FDA for laser epilation.

Therefore in this chapter the basic differences of the light and laser systems can be only shortly presented without going into too great detail about an individual system. Finally, it is left to the customer to make his own experience about the technical operation and quality of a system and the achievable effects. For individual laser systems and techniques that are employed in our clinic the treatment parameters are given and the experience is shortly described. The parameters for the therapy of the employed laser systems can be found in the chapter Therapeutic Guidelines see page 7.

### Preparation of Patients for Laser Epilation

In the experiments with epilation with a Q-switched Nd:YAG laser, the preparations and subsequent treatment procedures are identical to all other laser systems. It is very important to inform each patient fairly and sufficiently about alternative methods, the suggested and other laser techniques, about the preparations, effects and side effects during and after laser therapy. Further, no excessive expectations should be aroused with too optimistic prognosis, and each patient should be selected as carefully as for other aesthetic treatments. The time spent for the first consultation will be repaid later during the treatment. The highest aim should be to treat a well-informed, motivated and reliable patient.

The general side effects of laser epilation do not essentially differ from other treatments. Typical side effects are: remaining hairs, regrowth of hairs, blister and crust formation, scar formation, wound infection and wound healing troubles and pigment disturbances (hypo- and hyperpigmentation). According to the system employed or the localization, the general side effects must be complemented with particular side effects, for example injury/flare of the lip-red. A permanent epilation should never be promised to the patient. Rather, one should speak of long-lasting hair removal. Per session, and this seems to be independent of the laser system used, a maximum of 10 to 20 % of the hairs can be longlasting epilated. Also six to ten sessions are mostly necessary for each area. The interval between treatments is still under discussion. In our clinic, the patients are treated after 6 week during the first treatment periods and then the interval varies according to the regrowth or growth activity of the hair.

Rarely, the possibility will arise to select between several laser systems in a practice or clinic. Patients with dark hair and dark skin (Fitzpatrick type IV, V) should be treated rather with an alexandrite than with a ruby laser. Up to now, patients with light hair should be treated with an infrared laser; for example pulsed Nd:YAG or diode laser. If it is impossible to select between different lasers, the laser parameters must be adapted to the skin type and the localization. Before each treatment, accessories, for example scanner systems, should be selected because in the field of the upper lip-/chin area scanners are rather hindering, but in the field of the back or the legs very helpful.

After anamnesis, clinical examination and further exclusion of accompanying diseases, for example polycystic ovary or diabetes mellitus, a careful photo documentation of the areas (closeup, details) to be treated should always be made in order to have a control for the progress of epilation and to demonstrate it to the patient.

For every epilation the hairs should protrude a maximum of approximately 1 mm from the infundibulum. Too-long hairs can burn during laser application and damage the epidermis and infundibulum

thermally. In addition, the smoke and odour formation are disturbing. According to hair growth, the patients should shorten the hairs by shaving 1–3 days before the epilation. Tweezing or wax epilation should be avoided at regular laser epilation in the interval because otherwise the response rate of hair removal will clearly be reduced. Most female patients will repeat that they do not want to shave the hairs because otherwise the hairs even grow faster, even become stronger and more abundant. This old myth, that shaving increases hair growth and induces a thickening of the hairs, is complete nonsense. Therefore if a hair is too long prior to epilation, it should be shaved directly before the treatment.

Laser epilation is, on the one hand, painful, on the other, side effects can occur as a result of thermal alterations of the epidermis. It is always advisable to apply superficial cooling. Some devices have cooling systems, mostly built in into the hand piece in the form of a Pietzo element or connections for spray cooling or precooled air systems. Other simple possibilities are the thin application of gel (e.g. ultrasound gel) or the precooling of the area with ice cubes. In addition, prior to treatment, local anaesthesia can be achieved by topical application of EMLA cream®.

After irradiation, the skin should be cared for with hydrophilic ointments and chilled if necessary with cool packs. Each patients should be informed to avoid sunbathing (UV light) and intensive sun exposure, and should use sun protective creams with high sun protecting factor (SPF 20–60) according to skin type and sun intensity in order to avoid long-lasting hyperpigmentation after laser therapy. If hyperpigmentation occurs, the epilation has to be discontinued and interrupted for approximately 6–9 months. The hyperpigmentation lights faster after daily local application of azelainic acid-containing creams with/without a combination with vitamin-A acid. Hydroquinones are under suspicion of inducing cancer and may no longer be used in Germany.

### Epilation of Hairs with Ruby Laser Devices

Ruby lasers were one of the first laser systems used for laser epilation. Numerous publications described the response rates [63–74] and attempted to investigate the basics of hair epilation [63, 75–82], that are valid not only for ruby laser, but also for all other procedures and systems for laser epilation. The ruby laser was mainly used in the Q-switched mode for the photomechanical destruction of melanosoms (lentigo senilis [83–87], N. Ota [88–93] etc.) or particles (tattoo/accidentical ingestions of foreign particles) [94–96]. At first, the free-running mode found little attention until it was recognized that with a pulse length of 0.5 to 2 ms hair follicles can thermally destroyed. A high rate of pigmentation of the hair and a low pigmentation of the epidermis are the condition, however, to cause only minor side effects with maximum effect. For epilation, high-power densities which can lead to a high heating of the epidermis as a side effect, however, are necessary. The results are blistering, crust formation and possible development of scars and wound infections. Therefore, if one employs high-power densities, the skin surface must be effectively chilled. Different techniques are described for this purpose. With power densities around 60 J cm$^{-2}$ and under cooling of the skin surface (gel, ice, dynamic spray cooling), 15 to 20% reduction of hairs is described after one treatment after 6 months. After 2 years approximately 30–80% of the "removed" hair, depending on localization, demonstrated no regrowth [64]. Further studies showed that firstly high-power

densities between 50–60 Jcm$^{-2}$ are necessary, secondly no difference in the result exists whether pulse length of 0.3 or 3.0 ms, as well as single and double pulse per hair were applied [97]. A majority of the hairs showed delayed regrowth as well as a change from telogen to vellus hair [34]. Effective treatment results were achieved only for dark brown to black hairs with a skin type II–III (Fitzpatrick) [66].

In our department, the ruby laser was employed for the first time for epilation in 1997. The employed laser had a maximum focus diameter of 3 mm, 1 ms of pulse duration and maximum power density of 48 J cm$^{-2}$ at 1 Hz. The irradiations were carried out without a scanner device. Mainly, women with hirsutism and hyperandrogenaemia and unwanted hairs in the face were treated (Fig. 1A, B). Depending on the

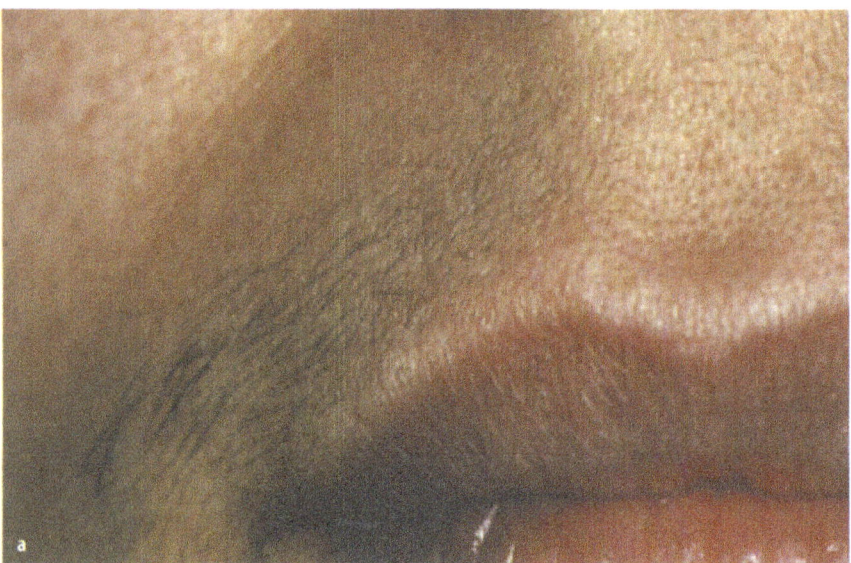

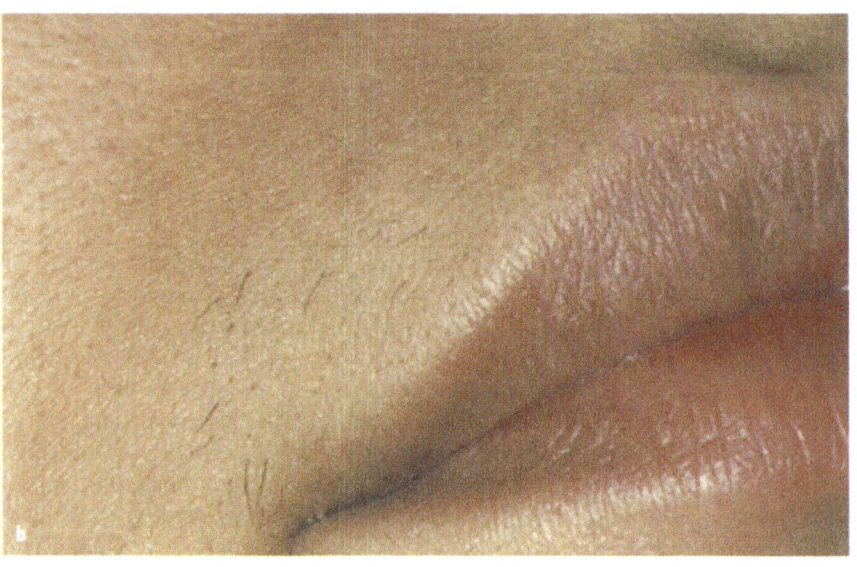

**Fig. 1a, b.** Epilation of unwanted hairs in the region of the upper lip with a ruby laser. Before treatment **a** a lot of dark hairs could be seen which disappeared 3 months after the fourth treatment **b**

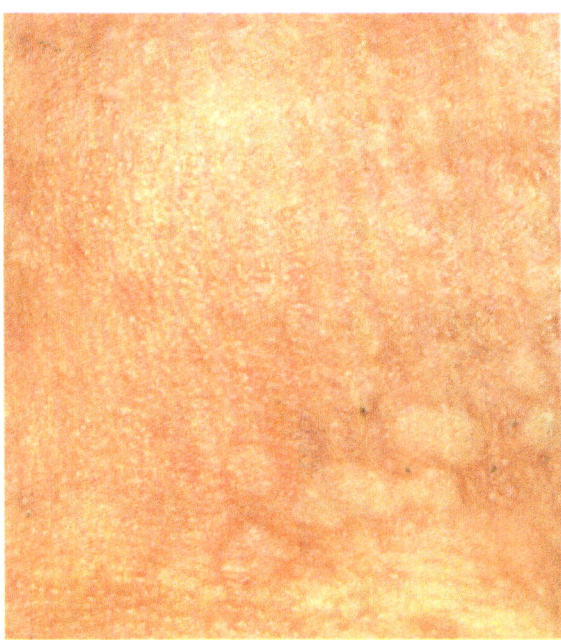

**Fig. 2.** Acute side effects after laser epilation with ruby laser. Shortly after irradiation with a ruby laser, erythema, edema and weals developed. These typical side effects disappeared after several hours

skin type, for the first sessions power densities of 30 to 48 J cm$^{-2}$ were used and treatments repeated after approximately 6 weeks. Typical side effects during irradiation were burning to pricking pain, redness and sometimes urticaria-like edema (Fig. 2). The two latter symptoms could persist for some hours, e.g. until the evening. The skin was chilled during the processing with either ultrasound gel or ice cubes. Cooling of the skin decreased the pain sensations and the side effects. Particularly patients with skin type IV or V reported much longer-lasting swelling and partial crust formation. Blistering or scarring were not observed. Furthermore, hypo- or hyperpigmentations, as described by other authors, were never found [98, 99].

Especially dark hairs respond quite well in the area of the cheeks and lower jaw. The hairs in the region of the upper lip and chin proved to be most stubborn. The regrowing hairs showed, however, a clear delay in growth and became increasingly finer (Fig. 1A, B). Most women found laser epilation more pleasant than wax depilation or electrolysis. Above all, they appreciated that the hairs in the beard region could be completely removed in one session, and that a further treatment of the regrowing hairs was necessary only after 6–8 weeks. Therefore, for a long period other methods of hair removal, such as tweezing, were not necessary.

The small spot diameter and the time-consuming procedure due to the small spot diameter and low

repetition rate of 1 Hz were disadvantageous. Furthermore, brown hairs showed little to no response to epilation with a ruby laser.

### Epilation of Hairs with Alexandrite Laser Devices

In analogy to the ruby laser, alexandrite lasers were firstly employed for the removal of pigmentation or particles in the skin using the Q-switched mode [94, 100–103] and then evaluated for hair removal [38, 104–112]. Unlike the ruby laser, the pulses of the alexandrite laser are not continuous, but each pulse consists of many short pulses (spikes), so that one should talk of a burst rather than a pulse. The advantage of the alexandrite laser consists in the possibility of producing a long pulse duration, up to several 100 ms. Through the burst pulse – at least theoretically – the thermal damage in the region of the epidermis can be reduced while the follicular structures can be damaged more efficiently.

The alexandrite laser emits light in the wavelength of 755 nm and has approximately the same working principle as the ruby laser, but a deeper penetration into the dermis [113]. The pulse lengths vary from 2–500 ms, the power densities of 5–50 J cm$^{-2}$ and the spot diameters are between 8 and 12 mm. Most systems with alexandrite laser employ hexascanners with repetition rates up to several Hz. For this reason a large area (e.g. 4 × 4 cm) can be uniformly irradiated in a very short time. Large areas, e.g. the back or legs, become treatable. While 1999 only some few publications were known, already in the following year the number of publications rose to more than 50 publications, so that next to the ruby laser the alexandrite laser appears to be evaluated as excellent for laser epilation. Due to the variability of the pulse length, epilation of skin types of IV–VI becomes possible, because long pulse durations (for example 30 ms and more) and corresponding cooling of the skin surface prevent or reduce direct damage to the epidermis [114].

The alexandrite laser showed similar response rates with 15–20% reduction of hair/treatment (Fig. 3A, B). According to localization and pigmentation, six to eight treatments are mostly necessary. The rates of side effects are, however, by far lower compared to the ruby laser. Pain sensation and crusts are clearly minimized. Redness and urticarial alterations of the skin regressed already after few hours. Blistering and scars were not observed in our department. The parameters which are used in our clinic for epilation with an alexandrite laser (Episcan, Baasel, Starnberg, Germany) are 15 ms pulse width, 15–20 J cm$^{-2}$ at 3 Hz repetition rate and cooling of the skin with ice cubes, cold air or ultrasound gel. According to statements of the patients, direct precooling of the skin was consid-

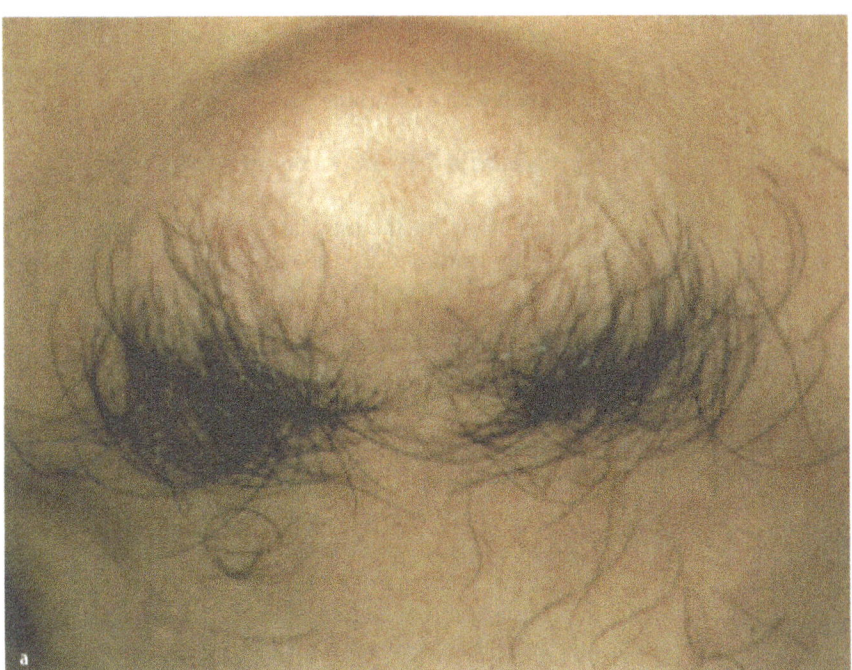

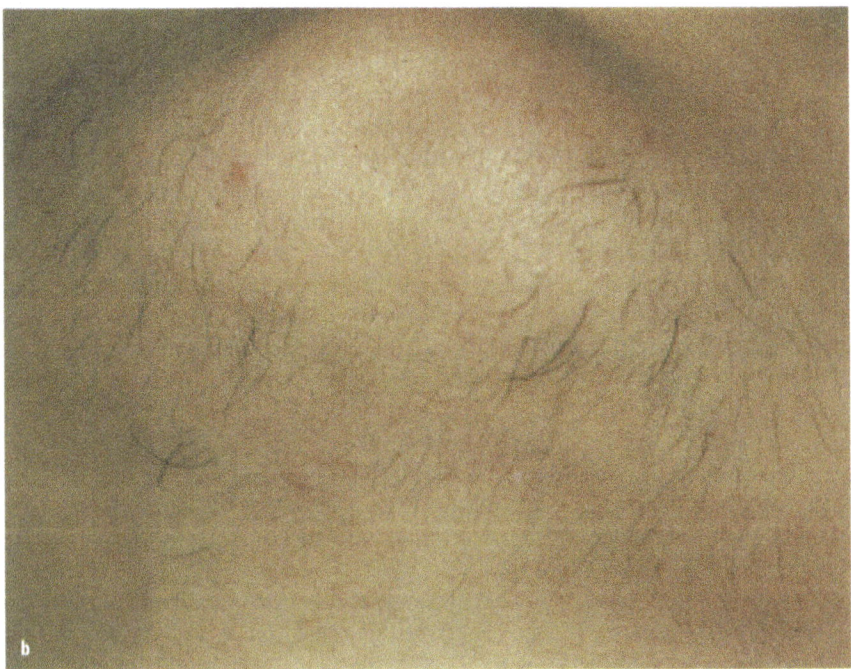

**Fig. 3a, b.** Epilation of unwanted hairs of the chin with an alexandrite laser. Before **a** and 3 months after the second treatment **b**. Some of the not removed hairs are regrowing as thinner hairs

ered the most effective in minimizing pain sensations and side effects before laser treatment.

### Epilation of Hairs with Nd:YAG laser devices

Nd:YAG laser are used in the cw mode for coagulation or cutting of tissue or in the Q-switched mode for photomechanical disruption of particles. Firstly, this latter was used to attempt damage to hair follicles

[115]. Since the amount of "particles" in the hair follicle is too low, very small carbon-based particles were inserted into the follicle by topical application. The irradiation with an Nd:YAG laser in Q-switched mode with ns pulse causes photomechanical disruptions of the hair shaft and the inserted particles [43]. Thus, it was supposed that the acoustic waves induce passive destruction of the target tissues in the follicle and long-lasting destruction of the hair follicle. The re-

sults are controversial and ranged from no effect up to more than 25% hair reduction after one treatment. Additional studies are planned using 100 nm large particles, which should at first be inserted by Q-switched Nd:YAG at low power densities into the follicles and then the destruction of the hair follicle should be induced at high power densities [116]. Whether this technique will raise the response rates seems to be questionable.

Considerable more successful for epilation proved to be the pulsed Nd:YAG systems with a wavelength around 1064 nm and pulse duration of 5–50 ms and energy densities of 10–140 J $cm^{-2}$ [117–120]. These laser systems also allow the treatment of vascular lesions (superficial leg veins etc.) so that, particularly at clinics or in practices with phlebologic indications, these laser systems have advantages over ruby or alexandrite laser devices. Now corresponding studies for the epilation of hairs are being published. The advantages of this system are in the deeper penetration (thermal alterations to 1.1 cm [118]) and the small absorption by melanin. However, to avoid thermal damage in the skin at higher power densities, sufficient cooling of the epidermis should supplied. Also dark skin types can be irradiated without essential damage to the epidermis. At least in theory the absorption of the light occurs independently of melanin. According to the principle of the selective thermokinetics, the stronger heating of the follicle occurs due to its larger structure and not due to the melanin content. Fair hairs can also be damaged which do not respond efficiently to an irradiation with ruby or alexandrite laser [120]. The disadvantages consist of the higher rates of side effects due to thermal damage of the skin. Compared to the alexandrite laser, the treatment is more painful and the appearance of redness and edema and the generation of blister and crust formation is more frequent. Double exposures with high power densities should be avoided because of the risk of inducing atrophic scares.

### Epilation of Hairs with Diode Laser Devices

The emitted wavelengths of the different diode lasers vary according to the systems between 800 and 980 nm, but most of the studies use the 810-nm diode laser [121–125]. Other systems with different wavelengths are under investigation. In analogy of the pulsed Nd:YAG lasers, the diode lasers could also be used for the treatment of vascular lesions. According to the wavelength, different penetration depths can be expected. The essential advantage is the compact construction of the laser compared to the large noisy laser devices. Diode lasers are mostly smaller but do not differ, however, in purchase price. Treatment pa-

rameters, response rate and side effect profiles correspond to those of the pulsed Nd:YAG laser systems and alexandrite laser [126].

### Epilation of Hairs with Intense Pulse Light Systems (IPL)

Intense pulse light systems emit light in the visible field in a broad range of 550–1100 nm. Using the so-called cutoff filters, the melanin-absorptive ranges were filtered out. For the absorption of the infrared component of the emitted light, water is set in for absorption in order to extend the life duration of the flashlamp (cooling) and to avoid unwanted thermal effects (infrared light) at high-power densities [127]. Up to now, some studies have been published [128–134]. In two studies, with an observation period of 3 months after one treatment in the face region, a reductions of hairs between 45 and 60% or 80%, respectively, or of 40 and 50% of the hairs in the field of the tribe and the extremities were described [133, 135]. Regarding knowledge of hair biology and the very short observation period of 3 months, it seems that these results reflect rather a short-term success than a long-lasting epilation of the hairs, and must be scrutinized critically. According to the statements of the manufacturers, further studies for evaluation of the long-term effects and complications such as blistering and/or scar formation are currently running.

### References

[1] MESSENGER AG (1993) The control of hair growth. J Invest Dermatol 101, 4S–9S
[2] ROOK A (1965) Endocrine factors on hair growth. Br Med J 1: 609
[3] HASHIMOTO K (1970) The ultrastructure of the skin of human embryos-V: the hair germ and perifollicular mesenchymal cells hair germ mesenchyme interation. Br J Dermatol 83: 167–176
[4] HASHIMOTO K (2001) The ultrastructure of the skin of human embryos-IV: formation of intradermal hair canal. Dermatologica 141: 49–54
[5] PINKUS H (1958) Embryology of hair. In: MONTAGNA W, ELLIS R (eds) The biology of hair growth. Academic Press, New York, pp 1–32
[6] PINKUS H (1947) The story of the hair root. J Invest Dermatol 9: 91
[7] WITZEL M, BRAUN-FALCO O (1963) Über Haarwurzelstatus am menschlichen Capilitium unter physiologischen Bedingungen. Arch Klin Exp Dermatol 221: 216
[8] VAN SCOTT E, EKEL T (1958) Geometric relationships between the matrix of the hair bulb and is dermal papilla in normal and alopecic scalp. J Invest Dermatol 29: 281–287
[9] HEADLINGTON JT (1984) Transverse microscopic anatomy of human scalp: a basis for a morphometric approach to disorders of the hair follicle. Arch Dermatol 120: 449–456
[10] NARISAWA Y, KOHLDA H, TANAKA T (1997) Three-dimensional demonstration of melanocyte distribution of human hair follicle: special reference to the bulge area. Acta Dermato-Venereol 77: 97–101
[11] BASSUKAS ID, HORNSTEIN OP (1989) Effects of plucking on the anatomy of the anagen hair bulb. A light microscopic study. Arch Dermatol Res 281: 188–192

[12] BIRBECK M, MERCER E (1957) The electron microscopy of the human hair follicle. J Biophys Biochem Cytol 3: 203–213

[13] HAYASHI A (1975) Trichogram. J Invest Dermatol 80: 70–75

[14] STENN KS, MESSENGER AG, BADEN HP (1991) The molecular structural biology of hair. Ann JY Acad Sci 642: 1–519

[15] PAUS R, HANDJISKI B, CZARNETZKI BM, EICHMÜLLER S (1994) Biologie des Haarfollikel. Hautarzt 45: 808–825

[16] PAUS R, CZARNETZKI BM (1992) Neue Perspektiven in der Haarforschung: Auf der Suche nach der "Biologie Uhr" des Haarzyklus. Hautarzt 43: 264–271

[17] COTSARELIS G, SUN TT, LAVKER RM (1990) Label-retaining cells reside in the bulge area of pilosebaceous unit: implications for follicular stem cells, hair cycle and skin carcinogenesis. Cell 61: 1329–1337

[18] LAVKER RM, MILLER SJ, SUN TT (1993) Epithel stem cells, hair follicles, and tumor formation. Recent Results Cancer Res 128: 31–43

[19] SUN TT, COTSARELIS G, LAVKER RM (2001) Hair follicular stem sells: the bulge-activation hypothesis. J Invest Dermatol 96: 77S–78S

[20] HASHIMOTO K, SHIBAZAKI S (1970) Ultrastructural study on differentiation and function of hair. In: KOBORI T, MOTANGA W (eds) Biology and disease of hair. University Tokyo Press, Tokyo, 23–57

[21] ITO M (2001) The morphology and cell biology of the hair apparatus: recent advances. Acta Med Biol 38: 51–67

[22] CESARINI J (1990) Hair melanin and colour. In: ORFANOS C, HAPPLE R (eds) Hair and hair disease. Springer, Berlin Heidelberg New York, pp 91–104

[23] ITO M, JIMBOW K (1983) Quantitative analysis of eumelanin and pheomelanin in hair and melanomas. J Invest Dermatol 80: 268–272

[24] ORTONNE JP, PROTA G (1993) Hair melanins and hair color: ultrastructural and biochemical aspects. J Invest Dermatol 101: 82S–89S

[25] ANDERSON RR, PARRISH JA (1981) The optics of human skin. J Invest Dermatol 77: 13–19

[26] MARGOLIS RJ, DOVER JS, POLLA IL (1989) Visible action spectrum for melanin-specific selective photothermolysis. Lasers Surg Med 9: 389–397

[27] MENON IA, PERSAD S, HABERMANN HF (1983) A comparative study of the physical and chemical properties of melanins isolated from human black and read hair. J Invest Dermatol 80: 202–206

[28] HARDY MH (1993) The secret life of the hair follicle. Trends Genet 8: 55–61

[29] MYERS B, HAMILTON J (1951) Regeneration and rate of growth of hairs in human. Ann NY Acad Sci 53: 562–568

[30] SAITOH M, UZUKA M, SAKAMOTO M (1970) Human hair cycle. J Invest Dermatol 54: 65–81

[31] KLINGMAN AM (1961) Pathological dynamics of human hair loss. Arch Dermatol 83: 175

[32] KLINGMAN AM (1959) The human hair cycles. I Invest Dermatol 33: 307–316

[33] TROTTER M (1924) The life cycle of hairs in selected regions of the body. Am J Phys Anthropol 7: 427

[34] GROSSMAN MC, DIERICKX C, FARINELLE W, FLOTTE T, ANDERSON RR (1996) Damage to hair follicles by normal-mode ruby laser pulses. J Am Acad Dermatol 35: 889–894

[35] ROSS EV, LADIN Z, KREINDEL M, DIERICKX S (1999) Theoretical considerations in laser hair removal. Dermatol Clin 17: 333–355

[36] RAULIN C, GREVE B, HAMMES S (2000) Cold air in laser therapy: first experiences with a new cooling system. Lasers Surg Med 27: 404–410

[37] HAAS AF (2000) Use of a unique cooling gel applied prior to laser hair removal. Dermatol Surg 26: 1045–1046

[38] ONO I, TATESHITA T (2000) Histopathological changes in the hair follicle after irradiation of long-pule alexandrite laser equipped with a cooling device. Eur J Dermatol 10: 373–378

[39] ALTSHULER GB, ZENZIE HH, EROFEEV AV, SMIRNOW MZ, ANDERSON RR, DIERICHX C (1999) Contact cooling of the skin. Phys Med Biol 44: 1003–1023

[40] ANVARI B, TANENBAUM BS, MILNER TE, KIMEL S, SVAASAND LO, NELSON JS (1995) A theoretical study of the thermal response of skin to cryogen spray cooling and pulsed laser irradiation: implications for treatment of port wine stain birthmarks [published erratum appears in Phys Med Biol 1996 July: 41(7): 1245]. Phys Med Biol 40: 1451–1565

[41] ZENZIE HH, ALTSHULER GB, SMIRNOV MZ, ANDERSON RR (2000) Evaluation of cooling methods for laser dermatology. Lasers Surg Med 26: 130–144

[42] NANNI CA, ALSTER TS (1998) A practical review of laser-assisted hair removal using the Q-switched Nd:YAG, long-pulsed rub, and long-pulsed alexandrite lasers. Dermatol Surg 24: 1399–1409; discussion 1405

[43] NANNI CA, ALSTER TS (1997) Optimizing treatment parameters for hair removal using a topical carbon-based solution and 1064-nm Q-switched neodymium:YAG laser energy. Arch Dermatol 133: 1546–1549

[44] SCOTT MJ, SCOTT MJ, SCOTT AM (1990) Epilation. Cutis 46: 216–217

[45] YOUNG HS, COULSON IH (2000) Cranuloma annulare following waxing-induced pseudofolliculitis-resolution with isotretinoin. Clin Exp Dermatol 25: 274–276

[46] MIMOUNI-BLOCH A, METZKER A, IMOUNI M (1997) Severe folliculitis with keloid scars induced by wax epilation in adolescents. Cutis 59: 41–42

[47] DE ARGILA D, ORTIZ-FRUTOS J, IGLESIAS L (1996) Occupational allergic contact dermatitis from colophony in depilatory wax. Contact Dermatitis 34: 369

[48] TOMAS VS, TORNE CJ (1992) Folliculitis caused by *Pseudomonas aeruginosa* after hair removal with wax. Rev Clin Exp 190: 104

[49] GOLDBERG NS, ZALKA AD (1989) Retin-A and wax epilation. Arch Dermatol 125: 1717

[50] VILLA-REAL R (1989) Pustular lesions following wax epilation. Guilty: a foam. Enferm Infecc Microbiol Clin 7: 59

[51] FOUSSERAU J, MELAVILLE J, GROSSHANS E, ARAUJO A, MAILLOT C, BASSET A (1967) Allergic eczema caused by thioglycerin in depilatory creams. Disadvantages of this substance in cosmetology. Bull Soc Fr Dermatol Syphiligr 74: 762–766

[52] STEPHENS FO, CONOLLY WB (1966) The use of depilatory creams in surgery. Med J Aust 2: 886–888

[53] BREITING V, HELLBERG S (1981) Chemical depilation as an alternative to shaving. A comparative study of preoperative skin preparation. Ugeskr Leaeger 143: 1646–1647

[54] EARHART RN (1978) Depilation treatment of hypertrichosis. J Pediatr 93: 721

[55] RICHARDS RN, MEHARG GE (1995) Electrolysis: observations from 13 years and 140 000 hours of experience. J Am Acad Dermatol 33: 662–666

[56] WAGNER RF (1993) Medical and technical issues in office electrolysis and thermolysis. J Dermatol Surg Oncol 19: 575–577

[57] RIDLEY CM (1969) A critical evaluation of the procedures available for the treatment of hirsutism. Br J Dermatol 81: 146–153

[58] GORGU M, ASLAN G, AKOZ T, ERDOGAN B (2002) Comparison of alexandrite laser and electrolysis for hair removal. Dermatol Surg 26: 37–41

[59] URUSHIBATA O, KASE K (1995) A comparative study of axillar hair removal in woman: plucking versus the blend method. J Dermatol 22: 738–742

[60] PEEREBOOM-WYNIA JD, STOLZE E, VAN JOOST T, KLEIMAN H (1985) A comparative study of the effects of electrical epilation of beard hairs in women with hirsutism by diathermy and by the blend method. Arch Dermatol Res 278: 84–86

[61] HAEDERDAL M, MATHEN P, WULF HC (2000) Laser epilation. A systematic review of evidence-based clinical results. Ugeskr Laeger 162: 6809–6815

[62] DIERICKX CC (2000) Hair removal by lasers and intense pulsed light sources. Semin Cutan Med Surg 19: 267–275

[63] MORLEY S, GAULT D (2000) Hair removal using the long-pulsed ruby laser in children. J Clin Laser Med Surg 18: 277–280

[64] GAULT DT, GROBBELAAR AO, GROVER R, LIEW SH, PHILIP B, CLEMENT RM, KIERNAN MN (1999) The removal of unwanted hair using a ruby laser. Br J Plast Surg 52: 173–177

[65] ZACHARIAE H, BJERRING P, LYBECKER H (1999) Laser depilation using a free-running long pulse ruby laser. Exp Dermatol 8: 301–302

[66] LIEW SH, GROBBELAAR A, GAULT D, SANDERS R, GREEN C, LINGE C (1999) Hair removal using the ruby laser: clinical efficacy in Fitzpatrick skin types I–V and histological changes in epidermal melanocytes. Br J Dermatol 140: 1105–1109

[67] WILLIAMS RM, CHRISTIAN MM, MOY RL (1999) Hair removal using the long-pulsed ruby laser. Dermatol Clin 17: 367–372

[68] SOLOMON MP (1998) Hair removal using the long-pulsed ruby laser. Ann Plast Surg 41: 1–6

[69] BJERRING P, ZACHARIAE H, LYBECKER H, CLEMENT M (1998) Evaluation of the free-running ruby laser for hair removal. A retrospective study. Acta Dermato-Venereol 78: 48–51

[70] VANDERKAM VM, ACHAUER BM (1997) Hair removal with the ruby laser (693 nm). Plast Surg Nurs 17: 144–145, 137

[71] TOPPING A, LINGE C, GAULT D, GROBBELAAR A, SANDERS R (2000) A review of the ruby laser with reference to hair depilation. Ann Plast Surg 44: 668–674

[72] WIMMERSHOFF MB, SCHERER K, LORENZ S, LANDTHALER M, HOHENLEUTNER U (2000) Hair removal using a 5-ms long-pulsed ruby laser. Dermatol Surg 26: 205–210

[73] CAMPOS VB, DIERICKX CC, FARINELLI WA, LIN TY, MANUSKIATTI W, ANDERSON RR (2000) Ruby laser hair removal: evaluation of long-term efficacy and side effects. Lasers Surg Med 26: 177–185

[74] LIEW SH, CERIO R, SARATHCHANDRA P, GROBBELAAR AO, GAULT DT, SANDERS R, GREEN C, LINGE C (1999) Ruby laser-assisted hair removal: an ultrastructural evaluation of cutaneous damage. Br J Plast Surg 52: 626–643

[75] LIEW SH, GROBBELAAR AO, GAULT DT, GREEN CJ, LINGE C (1999) The effect of ruby laser light on cellular proliferation of epidermal cells. Ann Plast Surg 43: 519–522

[76] HAYWOOD RM, WARDMAN P, GAULT DT, LINGE C (1999) Ruby laser irradiation (694 nm) of human skin biopsies: assessment by electron spin resonance spectroscopy of free radical production and oxidative stress during laser depilation. Photochem Photobiol 70: 348 352

[77] LIEW SH, LADHANI K, GROBBELAAR AO, GAULT DT, SANDERS R, GREEN CJ, LINGE C (1999) Ruby laser-assisted hair removal success in relation to anatomic factors and melanin content of hair follicles. Plast Recontr Surg 103: 1736–1743

[78] MCCOY S, EVANS A, JAMES C (1999) Histological study of hair follicles treated with a 3-ms pulsed ruby laser. Lasers Surg Med 24: 142–150

[79] LIEW SH, GROBBELAAR AO, GAULT DT, SANDERS R, GREEN CJ, LINGE C (1999) The effect of ruby laser light on ex vivo hair follicle: clinical implications. Ann Plast Surg 42: 249–254

[80] THOMSON KF, SOMMER S, SHEEHAN-DARE RA (2001) Terminal hair growth after full thickness skin graft: treatment with normal mode ruby laser. Lasers Surg Med 28: 156–158

[81] WALTHER T, BAUMLER W, WENIG M, LANDTHALER M, HOHENLEUTNER U (1998) Selective photothermolysis of hair follicles by normal-mode ruby laser treatment. Acta Dermato-Venereol 78: 443–444

[82] TOPPING A, GAULT D, GROBBELAAR A, GREEN C, SANDERS R, LINGE C (2000) The temperatures reached and the damage caused to hair follicles by the normal-mode ruby laser when used for depilation. Ann Plast Surg 44: 581–590

[83] ARNDT KA (1984) Argon laser treatment of lentigo maligna. J Am Acad Dermatol 10: 953–957

[84] ASHINOFF R, GERONEMUS RG (1992) Q-switched ruby laser treatment of labial lentigos. J Am Acad Dermatol 27: 809–811

[85] KOPERA D, HOHENLEUTNER U, LANDTHALER M (1996) Q-switched ruby laser application is safe and effective for the management of actinic lentigo (topical glycolic acid is not). Acta Dermato-Venereol 76: 461–463

[86] RAULIN C, PETZOLDT D, HELLWIG S (1996) Lentigo benigna. Removal with the Q-switched ruby laser. Hautarzt 47: 44–46

[87] SHIMBASHI T, KAMIDE R, HASHIMOTO T (1997) Long-term follow-up in treatment of solar lentigo and café-au-lait macules with Q-switched ruby laser. Aesthet Plast Surg 21: 445–448

[88] WATANABE S, TAKAHASHI H (1994) Treatment of nevus of Ota with the Q-switched ruby laser. N Engl J Med 331: 1745–1750

[89] TAYLOR CR, FLOTTE TJ, GANGE RW, ANDERSON RR (1994) Treatment of nevus of Ota by Q-switched ruby laser. J Am Acad Dermatol 30: 743–751

[90] LOWE JN, WIEDER JM, SAWCER D, BURROWS P, CHALET M (1993) Nevus of Ota: treatment with high-energy fluences of the Q-switched ruby laser. J Am Acad Dermatol 29: 997–1001

[91] GERONEMUS RG (1992) Q-switched ruby laser therapy of nevus of Ota. Arch Dermatol 128: 1618–1622

[92] GOLDBERG DJ, NYCHAY SG (1992) Q-switched ruby laser treatment of nevus of Ota. J Dermatol Surg Oncol 18: 817–821

[93] YANG HY, LEE CW, RO YS, YU HJ, KIM YT, KIM JH (1996) Q-switched ruby laser in the treatment of nevus of Ota. J Korean Med Sci 11: 165–170

[94] MORENO-ARIAS GA, CAMPS-FRESNEDA A (1999) Use of the Q-switched alexandrite laser (755 nm, 100 ns) for eyebrow tattoo removal. Lasers Surg Med 25: 123–125

[95] SCHEIBNER A, KENNY G, WHITE W, WHEELAND RG (1990) A superior method of tattoo removal using the Q-switched ruby laser. J Dermatol Surg Oncol 16: 1091–1098

[96] HELLWIG S, SCHONERMARK M, RAULIN C (1996) Accidental dirt tattooing. Removal with Q-switched ruby laser. HNO 44: 592–594

[97] DIERICKX CC, GROSSMAN MC, FARINELLI WA, ANDERSON RR (1998) Permanent hair removal by normal-mode ruby laser (see comments). Arch Dermatol 134: 837–842

[98] HASAN AT, EAGLSTEIN W, PARDO RJ (1999) Solar-induced postinflammatory hyperpigmentation after laser hair removal. Dermatol Surg 25: 113–115

[99] HAEDERDAL M, EGEKVIST H, EFSEN J, BJERRING P (1999) Skin pigmentation and texture changes after hair removal with the normal-mode ruby laser. Acta Dermato-Venereol 79: 465–468

[100] JANG KA, CHUNG EC, CHOI JH, SUNG KJ, MOON KC, KOH JK (2000) Successful removal of freckles in Asian skin with a Q-switched alexandrite laser. Dermatol Surg 26: 231–234

[101] ALSTER TS, WILLIAMS CM (1995) Treatment of nevus of Ota by the Q-switched alexandrite laser. Dermatol Surg 21: 592–596

[102] HAKOZAKI M, MASUDA T, OIKAWA H, NARA T (1997) Light and electron microscopic investigation of the process of healing of the naevus of Ota by Q-switched alexandrite laser irradiation. Virchows Arch 431: 63–71

[103] ROSENBACH A, WILLIAMS CM, ALSTER TS (1997) Comparison of the Q-switched alexandrite (755 nm) and Q-switched Nd:YAG (1064 nm) lasers in the treatment of benign melanocytic nevi. Dermatol Surg 23: 239–244

[104] RAULIN C, GREVE B (2000) Temporary hair loss using the long-pulsed alexandrite laser at 20 milliseconds. Eur J Dermatol 10: 103–106

[105] LAUGHLIN SA, DUDLEY DK (2000) Long-term hair removal using a 3-millisecond alexandrite laser. J Cutan Med Surg 4: 83–88

[106] ASH K, LORD J, NEWMAN J, MCDANIEL DH (1999) Hair removal using a long-pulsed alexandrite laser. Dermatol Clin 17: 387–399, ix

[107] BOSS WK JR, USAL H, THOMPSON RC, FIORILLO MA (1999) A comparison of the long-pulse and short-pulse alexandrite laser hair-removal systems. Ann Plast Surg 42: 381–384

[108] ROGERS CJ, GLASER DA, SIEGFRIED EC, WALSH PM (1999) Hair removal using topical suspension-assisted Q-switched Nd:YAG and long-pulsed alexandrite lasers: a comparative study. Dermatol Surg 25: 844–844

[109] FINKEL B, ELIEZRI YD, WALDMAN A, SLATKINE M (1997) Pulsed alexandrite laser technology for noninvasive hair removal. J Clin Laser Med Surg 15: 225–229

[110] GOLDBERG DJ, AHKAMI R (1999) Evaluation comparing multiple treatments with a 2-ms and 10-ms alexandrite laser for hair removal. Lasers Surg Med 25: 223–228

[111] MCDANIEL DH, LORD J, ASH K, NEWMAN J, ZUKOWSKI M (1999) Laser hair removal: a review and report on the use of the long-pulsed alexandrite laser for hair reduction of the upper lip, leg, back, and bikini region. Dermatol Surg 25: 425–430

[112] LLOYD JR, MIRKOV M (2000) Long-term evaluation of the long-pulsed alexandrite laser for the removal of bikini hair at shortened treatment intervals. Dermatol Surg 26: 633–637

[113] NANNI CA, ALSTER TS (1999) Laser-assisted hair removal: side effects of Q-switched Nd:YAG, long-pulsed ruby, and alexandrite lasers. J Am Acad Dermatol 41. 165–171

[114] GARCIA C, ALAMOUDI H, NAKIB M, ZIMMO S (2000) Alexandrite laser hair removal is safe for Fitzpatrick skin type IV–VI. Dermatol Surg 26: 130–134

[115] GOLDBERG DJ, LITTLER CM, WHEELAND RG (1997) Topical suspension-assisted Q-switched Nd:YAG laser hair removal. Dermatol Surg 23: 741–745

[116] GOLDBERG DJ, SAMADY JA (2000) Evaluation of a long-pulse Q-switched Nd:YAG laser for hair removal. Dermatol Surg 26: 109–113

[117] ALSTER TS, BRYAN H, WILLIAMS CM (2001) Long-pulsed Nd:YAG laser-assisted hair removal in pigmented skin: a clinical and histological evaluation. Arch Dermatol 137: 885–889

[118] GOLDBERG DJ, SILAPUNT S (2001) Hair removal using a long-pulsed Nd:YAG laser: comparison at fluences of 50, 80, and 100 J cm$^{-1}$. Dermatol Surg 27: 434–436

[119] LITTLER CM (1999) Hair removal using and Nd:YAG laser system. Dermatol Clin 17: 401–430, x

[120] BENCINI PL, LUCI A, GALIMBERTI M, FERRANTI G (1999) Long-term epilation with long-pulsed neodymium:YAG laser. Dermatol Surg 25: 175–178

[121] ADRIAN RM, SHAY KP (2000) 800 nanometer diode laser hair removal in African American patients: a clinical and histological study. J Cutan Laser Ther 2: 183–190

[122] BAUGH WP, TRAFELI JP, BARNETTE DJ JR, ROSS EV (2001) Hair reduction using a scanning 800-nm diode laser. Dermatol Surg 27: 358–364

[123] GREPPI I (2001) Diode laser hair removal of the black patient. Lasers Surg Med 28: 150–155

[124] CAMPOS VB, DIERICKX CC, FARINELLI WA, LIN TY, MANUSKIATTI W, ANDERSON RR (2000) Hair removal with an 800-nm pulsed diode laser. J Am Acad Dermatol 43: 442–447

[125] LOU WW, QUINTANA AT, GERONEMUS RG, GROSSMAN MC (2000) Prospective study of hair reduction by diode laser (800 nm) with long-term follow-up. Dermatol Surg 26: 428–432

[126] HANDRICK C, ALSTER TS (2001) Comparison of long-pulsed diode and long-pulsed alexandrite lasers for hair removal: la long-term clinical and histological study. Dermatol Surg 27: 622–626

[127] BJERRING P, CRAMERS M, EGEKVIST H, CHRISTIANSEN K, TROILIUS A (2000) Hair reduction using a new intense pulsed light irradiator and an normal mode ruby laser. J Cutan Laser Ther 2: 63–71

[128] LASK G, ECKHOUSE S, SLATKINE M, WALDMAN A, KREINDEL M, GOTTFRIED V (1999) The role of laser and intense light sources in photo-epilation: a comparative evaluation. J Cutan Laser Ther 1: 3–13

[129] TROILIUS A, TROILIUS C (1999) Hair removal with a second generation broad spectrum intense pulsed light source – a long-term follow-up. J Cutan Laser Ther 1: 173–178

[130] SADICK NS, WEISS RA, SHEA CR, NAGEL H, NICHOLSON J, PRIETO VG (2000) Long-term photoepilation using a broad-spectrum intense pulsed light source. Arch Dermatol 136: 1336–1340

[131] SCHROETER CA, RAULIN C, THURLIMANN W, REINEKE T, DE POTTER C, NEUMANN HA (1999) Hair removal in 40 hirsute women with an intense laser-like light source. Eur J Dermatol 9: 374–379

[132] TSE Y (1999) Hair removal using a pulsed-intense light source. Dermatol Clin 17: 373–385, ix

[133] WEISS RA, WEISS MA, MARWAHA S, HARRINGTON AC (1999) Hair removal with a non-coherent filtered flashlamp intensive pulsed light source. Lasers Surg Med 24: 128–132

[134] WEIR VM, WOO TY (1999) Photo-assisted epilation – review and personal observations. J Cutan Laser Ther 1: 135–143

[135] GOLD MH, BELL MW, FOSTER TD, STREET S (1997) Long-term epilation using the EpiLight broad band, intense pulsed light hair removal system. Dermatol Surg 23: 909–913

# III-7.6

# Laser Treatment of Virus-Assisted Skin Diseases

U. Müller and H.-P. Berlien

## Contents

## Introduction

Virus-assisted skin diseases take many forms and affect people from different age groups, but one thing which all these diseases have in common is their tendency to recur, which sometimes requires long and tiresome courses of treatment for the patient. The fact that there is a wealth of therapeutic options available may be seen as a sign that an optimum treatment has yet to be found.

The development of medical lasers has ushered in a promising new era. The rates of recurrence are lower, and the burden on the patient caused by postoperative bleeding, pain and prolonged recovery times is kept to a minimum.

In this chapter, laser therapy for major virus-assisted skin diseases is discussed. These are essentially molluscum contagiosum and diseases associated with the human papilloma virus, such as condylomata acuminata, giant condyloma (Buschke-Löwenstein tumour), Bowen's disease, and bowenoid papulosis, as well as common warts, plantar warts and flat or juvenile warts.

Molluscum contagiosum is caused by a strictly epidermotropic DNA poxvirus which infects the epidermis and adjoining structures. The virus is transmitted by skin contact, and the incubation period ranges from 14 days to several months. Sometimes minor epidemics have been observed among children and adults with an atopic predisposition. The clinical signs range from solitary forms through endophytic, exophytic and pedunculated molluscs to giant molluscs which, in the differential diagnosis, may be suggestive of a tumour. The most frequent signs are multiple, exophytic, skin-coloured to waxy yellow papules with a dimpled centre. Spontaneous regression is possible. However, if left untreated, the disease may spread; superinfections are also possible. The diagnosis is established on the basis of clinical or histological evidence. The histological picture shows a cystic structure with numerous small molluscous bodies.

Various therapeutic approaches may be considered. If the number of lesions is manageable, they may be scraped out with a sharp scoop after the surface has been disinfected; sometimes – especially in children – a local anaesthetic is called for. The risk of recurrence with this method is considerable, because it leaves open wounds. For the same reason, the practice of expressing the contents of the papules, which was once common, should also be avoided.

A more effective form of treatment is the use of cryotherapy; this leads to the formation of subepidermal vesicles with haemorrhagic vesicles. Crusts develop, which heal after a few days. Cryotherapy carries a risk of scarring and should therefore be used only by experienced practitioners.

Good results have also been achieved with the topical application of agents such as 5-fluorouracil or inorganic and organic acids, but these may be applied only to a limited extent. If used incorrectly, they can result in caustic burns to the skin.

The application of creams containing vitamin A likewise leads to healing through irritation of the skin with subsequent desquamation; this approach is contra indicated, however, in acute eczema and rosacea, and side effects may also occur in the form of circumscribed hypopigmentation. Since little is known about percutaneous absorption and systemic side effects, the indication must be strictly applied in infants and pregnant women.

Another effective method is to paint the lesions twice a day with liquid polyvinylpyrrolidon-iod or to use the same substance as ointment under an occlusing dressing.

Our first choice is the treatment with the Argon-laser under local anaesthesia with Emla cream. With a wavelength of 514 nm, 5 W and a 0.2 s pulse length, most of the lesions dry up within a few days. In cases of extensive findings in children we sometimes prefer a general anaesthesia.

In unsuccessful cases we treat with the Nd:YAG-laser using 20 W and 0.2 s in a repeat mode. The Nd:YAG-laser penetrates deeper into the tissue, but the risk of scar formation is higher, so it should not be used by untrained doctors. The treatment is painful and often we also need a general anaesthesia.

For some years, $CO_2$ laser therapy has been used to treat the molluscs. The lesions are vaporized under local anaesthetic (or, in the case of extensive papules, under a general anaesthetic) with 8 W of energy delivered in continuous-wave mode. The immediate slight carbonization that occurs ensures that the lesion is sealed, and there is no dissemination of the infectious content. It is essential that an efficient fume hood be used, because considerable smoke may be generated. Fears have often been expressed of possible infection by virus particles flying around in the ambient air, but these are unfounded, since the smoke is aspirated directly over the lesion, and the heat which is generated during laser treatment also leads

to a denaturation of the infected material. In the course of 5 years, we have not observed a single case of an operator becoming contaminated. The method does, however, carry a certain risk of scarring for those unpractised in the method.

In the literature the flash-pumped dye laser is also mentioned as being successful in treating molluscum contagiosum.

## Warts

Warts are benign, epithelial virus-assisted hyperplasias which occur frequently and affect predominantly children and adolescents. They sometimes show an epidemic distribution pattern, and vary widely in severity. In patients with congenital or acquired immune deficiency and immunosuppressed patients (e.g. recipients of organ transplants), for example, several hundred lesions are often found; malignant transformation is possible, although the HPV types are not among the high-risk groups in terms of their oncogenic risk.

Flat warts (verrucae planae juveniles) occur mostly in children and adolescents. They are found on the hands, arms and face and may be mistaken for seborrhoeic warts. They have a high spontaneous remission rate, so that a wait-and-see approach is justified. In refractory cases, lesions may be treated as described in the following.

The common wart (verruca vulgaris) is the most frequently occurring form. HPV types 1, 2, 4, 26 and 29 are detectable in these lesions. They may be found anywhere on the body, but show a predilection for regions with poor blood supply, such as the feet and hands. Autoinoculation through scratching is possible, as are spontaneous remissions.

The treatment options available include conservative medication (topical agents containing vitamin A, immune response modifiers, virostatics and keratolytic agents) as well as invasive approaches such as cryotherapy, electrotherapy, the use of a sharp scoop and laser therapy. The aim of any therapy should be to achieve a functionally and cosmetically satisfactory result.

We prefer a combined approach to treatment. Our patients first apply Collomack – a mixture of salicylic acid and lactic acid – once daily for 4 weeks. This treatment is then followed by a topical application of chloroacetic acid once a week for 4 weeks. This may only be applied to closed lesions, because otherwise it may lead to inflammatory irritations. The local reaction should therefore be controlled.

Pretreatment causes minor lesions to resolve, and the hyperkeratosis regresses, leaving sharply circumscribed areas of wart for subsequent laser therapy and allowing the resulting lesions to be minimized.

Essentially, three laser systems are available for therapy, depending on the localization and extent of the lesions: flash-pumped dye laser (FDL), $CO_2$ laser, and Nd:YAG laser.

We use the FDL to treat flat isolated lesions on the hands and in the paronychial region. Treatment is possible without anaesthesia, and we deliver an energy per unit area of 8 J $cm^2$. The treatment serves to cut the blood supply to the wart, which then regresses. Oedema and haematoma can occur, which are perceived by the patient as unpleasant side effects. An advantage is that the treatment can be administered in an outpatient setting and can be repeated as often as necessary; the cost is equally low for patient and doctor alike. Among our own patients, FDL laser therapy has not proved to be so efficient in immunosuppressed patients.

Using Nd:YAG laser therapy delivering 25 W with an exposure time of 0.2 s in repeated-pulse mode, individual raised lentil-sized warts are coagulated until blanching occurs. Within 14 days, the wart becomes demarcated and drops off. Here, too, the treatment is sometimes accompanied by oedema with vesicle formation; the cosmetic result is good, and the risk of concomitant damage to the matrix of the nail in the case of paronychial lesions is minimal.

Extensive and deep wart areas are vaporized and excised by $CO_2$ laser under local or general anaesthesia. We use 20 W of energy in continuous-wave mode and with an adequate fume hood for smoke extraction. During the laser procedure, the lesion must be kept moist to avoid excessive carbonization, because otherwise it is difficult to assess whether the ablation is complete. Lesions are primarily sealed by carbonization and show little wound secretion, thus minimizing the possibility of autoinoculation. These heal in 1 to 3 weeks, depending on the extent of the lesions.

All laser methods have the advantage that, on account of the low risk of bleeding, they show a minimal recurrence rate and are thus superior to other methods.

We do not use the erbium laser, which is favoured by a number of centres, because it does not ensure adequate ablation in deeper tissue and thus leads to increased bleeding with higher rates of recurrence. For flat lesions we thus use the in situ treatment options described above.

Plantar warts (verrucae plantares) occur as flat lesions, which can develop into very painful, deep-rooted warts. They are recognizable by the small dark spots which appear as a result of thrombosed blood vessels and they have a hard hyperkeratitic wall. After conservative pretreatment, they are vaporized and ex-

cised by $CO_2$ laser delivering 25 W in cw mode; here, too, an efficient fume hood is essential. The healing phase may be prolonged owing to the depth of the lesions and the poor blood supply.

## HPV-Related Skin Diseases

Infections with HPV are among the most frequent of all infections seen in humans. The human papilloma virus belongs to the family of Papovaviridae and has a double-stranded circular DNA with icosahedral symmetry and a genome of almost 8000 base pairs. It is a species-specific virus and induces squamocellular and fibroepithelial tumours. Infection is generally transmitted sexually or by local contact.

The clinical and histological picture and the natural history of HPV are largely dependent on the HPV type. To date, 100 different HPV types have been identified. Typing is performed by molecular biological techniques. Of clinical relevance is the fact that the subtypes are responsible for a variety of diseases and show differences in their oncogenic potential. For this reason, a thorough healing of all lesions and long-term follow-up are essential.

Condylamata acuminata are the most common HPV-induced papillomas in the genital/perianal region. These venereal warts are regarded as the most common sexually transmissible viral infection. According to a study in the USA in 1995, the virus affects 15% of the sexually active population aged between 15 and 49. A quarter of these patients show clinically manifest genital warts. In most cases, tissue samples reveal the presence HPV 6 or 11. Especially in patients showing a change in their immune status, e.g. people with HIV infection or recipients of organ transplants, several subtypes are found concomitantly. Infection occurs during sexual intercourse, when the virus enters the subepithelial membrane via microlesions. The warts develop after a latency period of 6 months or more; many infections run a subclinical course. The main risk factors are early onset of sexual activity and frequent changes of sexual partner. Congenital or acquired immune deficiencies can also predispose to the development of warts. During pregnancy, for example, a latent infection can become clinically manifest.

In rare cases (in our own patient population, three patients in the course of 4 years), children are also affected. The infection in these cases occurs through skin contact or a maternal infection; sexual abuse should be excluded.

Genital warts are in most cases easy to detect. Blanching with acetic acid may be helpful, but sometimes leads to false-negative results.

**Table 1.** Rates of clearance and recurrence with different treatment methods, according to Beutner and Ferency (1997) and Beutner and Wiley (1997)

| Treatment | Clearance rate (%) | Recurrence rate (%) |
|---|---|---|
| Trichloroacetic acid | 64–81 | 36 |
| Podophyllin | 38–79 | 21–65 |
| Podophyllotoxin | 68–88 | 16–34 |
| 5-Fluorouracil (5-FU) | 68–97 | 0–8 |
| Interferon | | |
| Intralesional | 36–53 | 21–25 |
| Systemic | 7–82 | 23 |
| Topical | 6–90[a] | 6[a] |
| Cryotherapy | 70–96 | 25–39 |
| Laser therapy | 72–97 | 6–49 |
| Electrocautery | 94 | 25 |
| Electric loop | 72 | 51 |
| Surgical excision | 89–93 | 19–22 |

All anogenital regions are susceptible, including the perineum and the surrounding skin as far as the thighs. Intra-anal lesions extend rectally as far as the dentate line. An exception in our own patient population is the immunodeficient patient, who shows extensive tumour-like lesions also extending above the dentate line. Perioral and intraoral as well as peri auricular and intranasal lesions have been observed. Isolated lesions also occur, such as flat exophytic reddish brown warts, which are mostly asymptomatic. More rarely, symptoms such as pruritus, burning, bleeding, pain, dyspareunia, ulceration and secondary infection may occur.

There are many treatment strategies available, and rates of recurrence vary in the literature between 8 and 65%, depending on the treatment method, although the extent of the primary lesion and the closeness of follow-up with immediate after treatment certainly also play a part here (Table 1).

Since the end of the a 1990s, a new substance, imiquimod (Aldara), has been available for the topical treatment of genital warts. Imiquimod is a so-called immune response modifier. Its efficacy in small perianal lesions has been demonstrated with careful application in multicentre studies; intra-anal, intravaginal and urethral lesions have yet to be investigated in controlled studies.

## Laser Therapy

External lesions are treated by vaporization with a $CO_2$ laser (wavelength 10 600 nm). We use a setting of 10–20 W, depending on the thickness and extent of the lesions, in continuous-wave mode. The laser permits exact layer-by-layer preparation, so that the epidermis can be selectively removed without causing damage to the underlying tissue and subsequent scarring. An adequate fume hood for the extraction of smoke is absolutely essential.

The lasered lesions are sealed by slight carbonization and show little postoperative weeping. Patients are hampered very little, thanks to the minimal degree of pain.

In the case of intra-anal, intra-vaginal, cervical and intra-urethral lesions, we use the Nd:YAG laser (wavelength 1064 nm), to which a bare fibre is attached as a light guide. The bare fibre is very flexible, has a diameter of 600 µ and can be coupled with an endoscope, allowing both access to parts of the anatomy which are otherwise difficult to reach and adequate illumination of these parts by means of the light sources attached. Treatment is administered with 20 W and an application time of 0.2 s in repeat mode until the lesion shows slight blanching. The fibre must not come into contact with the tissue. Intraluminal application calls for considerable experience on the part of the operator. At all events, deep-seated lesions with the risk of a stenosis formation or even perforation must be avoided. In the treatment of extensive lesions, to prevent postoperative adhesions of the urethra resulting from swelling, it is recommended that an indwelling catheter be inserted for 3 days.

Circular preputial condylomata, which cover the preputial skin as an exophytic growth, show a marked tendency to recur and can also cause cosmetic problems postoperatively, so that in isolated cases a circumcision by $CO_2$ laser should be performed.

Patients are closely monitored for 6 months after treatment. The rates of recurrence after laser therapy of up to 49% as reported in the literature deviate considerably from our own experience. In our patient population, we see a recurrence rate after laser therapy alone of 9% with all cases, including extensive lesions, so that laser therapy represents a superior approach when used adequately.

Giant condyloma (or Buschke-Löwenstein tumour) is a relatively rare form of condylomata acuminata. In addition to HPV 6 and 11, this form of the disorder also shows HPV 16 and 18. It leads to cauliflower-like tumours of the perianal and vulvar region which tend to form abscesses and fistulas. Lesions can then also spread via an epithelialized fistula to deeper layers of tissue.

If the condition is allowed to persist, it may lead to a squamous cell carcinoma. Metastases are rare. The therapeutic objective must be to remove the lesions completely while retaining organ function. To achieve this, several sessions at short intervals are necessary in most cases. The exophytic lesions are ablated using the bare fibre of the Nd:YAG laser in tissue contact at 30 W in continuous-wave mode. Thicker vessels are coagulated step by step using a non-contact procedure, so that the risk of substantial bleeding is minimal. Here, too, a fume hood for smoke extraction is necessary.

The ablated tissue must be sent for histological examination to ensure that squamous cell carcinoma is diagnosed in good time.

Bowen's disease is an intraepithelial spinocellular carcinoma which shows a tendency towards invasive growth. Once the basement membrane has been crossed, a Bowen's carcinoma is present which metastasizes via the lymphatic system and has a poor prognosis. The lesions in this disorder frequently show HPV 16, which is one of the human papilloma viruses with relatively high oncogenic potential. Exposure to arsenic and sunlight are also being discussed as predisposing factors. Bowen's disease almost always occurs after the age of 40 and can affect any part of the body. Men are affected slightly more often than women. A frequent coincidence with secondary carcinomas has been reported several years after the diagnosis was established. Tumours may occur, for example, in the respiratory, gastrointestinal and urogenital tracts. The causal link remains unclear. Bowen's disease manifests itself clinically as a flat, scaly plaque and may be mistaken for psoriasis. Isolated lesions may occur, as may multifocal lesions. Apart from occasional pruritus, it is virtually asymptomatic. Owing to the fact that the disease never regresses spontaneously and can become carcinomatous after a number of years, careful staging is always called for when the diagnosis is established to ensure that the appropriate treatment can be planned and any secondary tumours detected.

The aim of any treatment must be to clear all the foci of disease once the diagnosis has been confirmed by histological examination. Various approaches are available to achieve this, such as surgical excision, curettage with or without electrodesiccation and cryotherapy. For inoperable patients, local therapy with 5-fluorouracil or fractionated soft radiation therapy may be considered. For the latter, single daily doses of 3–5 Gray up to a total dose of 40–60 Gray are administered.

Various methods may be considered for laser therapy. For example, the foci may be precisely detected and biopsied after photodynamic diagnosis and then, after confirmation by histological examination, excised or vaporized by $CO_2$ laser. This is performed with 15 W in continuous wave mode under either local or general anaesthesia and using a fume hood for smoke extraction. The technique has no side effects, can be repeated as often as necessary and causes the patient hardly any postoperative pain.

A disadvantage of the technique is that the removed tissue does not always allow an accurate histological examination as a result of the heat-induced changes. We therefore follow up our patients very closely and always perform control biopsies 3 months later following photodynamic diagnosis. Photodynamic diagnosis, which is discussed in another chapter, has a 90% success rate in the discovery of macroscopically invisible foci and is therefore used routinely in our department.

A further option for laser therapy lies in photodynamic therapy. Once an invasive carcinoma has been excluded, 20% 5-ALA cream is applied to the diseased areas of skin for 4 h on the evening before treatment. The areas must be occluded from light. The photoactive substance accumulates in the dysplastic cells, and laser irradiation the next day leads to apoptosis, which is discernible by the erythema and oedema which occur with subsequent wound secretions and crusting. The laser therapy is carried out using a laser at wavelength 633 nm using either continuous wave (cw) dye laser or flashlamp-pumped dye laser and delivering an energy per unit area of 100 J $cm^{-2}$. The procedure may be repeated as often as necessary and produces good cosmetic results because there is rarely any scar tissue. The elaborate cosmetic procedures that are sometimes required following surgical excision are not necessary because the healing process is accompanied by re-epithelialization.

Bowenoid papulosis occurs as flat, reddish brown papules, sometimes confluent, in the genital region. Predominantly younger people tend to be affected; malignant transformation is relatively rare, and spontaneous remissions have been reported. HPV types 16, 34, 37 and 42 are detectable in the lesions; histologically, it is difficult to distinguish from Bowen's disease.

In addition to cryotherapy, diathermy and the application of podophylline and 5-fluorouracil, treatment with $CO_2$ laser may be considered. Once the lesions have been histologically confirmed to exclude a carcinoma, we vaporize the lesions with 10 W in continuous-wave mode. A fume hood for smoke extraction is likewise essential here. The patients show little wound secretion or dysaesthesia, and here, too, there is virtually no formation of scar tissue.

Using Nd:YAG laser therapy delivering 25 W with an exposure time of 0.2 s in repeated-pulse mode, individual raised lentil-sized warts are coagulated until blanching occurs. Within 14 days, the wart becomes demarcated and drops off. Here, too, the treatment is sometimes accompanied by oedema with vesicle formation; the cosmetic result is good, and the risk of concomitant damage to the matrix of the nail in the case of paronychial lesions is minimal.

Extensive and deep wart areas are vaporized and excised by $CO_2$ laser under local or general anaesthesia. We use 20 W of energy in continuous wave mode and with an adequate fume hood for smoke extraction. During the laser procedure, the lesion must be kept moist to avoid excessive carbonization, because otherwise it is difficult to assess whether the ablation is complete. Lesions are primarily sealed by carbonization and show little wound secretion, thus minimizing the possibility of autoinoculation. These heal in 1 to 3 weeks, depending on the extent of the lesions.

All laser methods have the advantage that, on account of the low risk of bleeding, they show a minimal recurrence rate and are thus superior to other

## References

BEUTNER KR, FERENCY A (1997) Therapeutic approaches to genital warts. Am J Med 102: 28–37
BEUTNER KR, WILEY DJ (1997) Recurrent external genital warts. Papillomavirus Report 8: 69–74

## Further Reading

BRAUN-FALCO O 1984 Dermatologie und Venerologie.
ORFANOS CE, GARBE C 2002 Therapie der Hautkrankheiten. Springer-Verlag
BARRASSO C 1977 Human papilloma virus infection. Ullstein, Mosby
GOLLNICK P (2001) HPV-induzierte Erkrankungen in der Gynäkologie. 53. Kongress der DGGG, München 2001
MEYER T (2001) Bedeutung des HPV-Nachweises. 53. Kongress der DGGG, München 2001
BERLIEN H-P, MÜLLER U (1994) Angewandte Lasermedizin. Ecomed, Landsberg
Müller U, ET AL (2001) Laser therapy of human papilloma virus. Laser Med 16
HAMMES S, Greve B, Raulin G et. al. 2001 Molluscum contagiosum: Treatment with pulsed dye laser. Hautarzt 52: 38–42
HINDSON C, COTTERILL J 1997 Treatment of molluscum contagiosum with the pulsed tuneable dye laser. Clin Exp Dermatol 22: 255

# III-7.7
# Laser Applications in Phlebology

U. Müller and H.-P. Berlien

## Contents

## Introduction

This chapter deals exclusively with such acquired venous diseases as are accessible to laser therapy. This refers to incompetent perforating veins without truncal varicosis, to small reticular varices and spider leg veins.

## Lasers in the Treatment of Perforating Veins

Not so many years ago, isolated incompetent perforating veins on the lower leg were considered to be of minor importance. With the development of the endoscopic dissection of perforating veins, such incompetence has increasingly been ligated on extended ulcerations and dilatations, which resulted in distinctly improved symptoms, so that the subfacial interruption of perforating veins can be deemed to be an efficacious therapeutic principle.

Like other forms of therapy, laser therapy requires a careful diagnosis of the venous system. An extended truncal varicosis, for example should primarily be referred to surgical intervention. In our opinion, there is an indication for lasering in the case of extended non-closing ulcerations on multimorbid and, consequently, inoperable patients as well as in the case of isolated incompetent perforating veins which cause clinical complaints.

## Method

At first, an ultrasonic examination is performed. A 7.5-mHz ultrasonic head locates the incompetent perforating veins and marks the points of passage at the fascia on the skin. After skin disinfection and sterile covering of the environment, the marked skin area is subjected to local anaesthesia, which also extends to the subcutaneous fatty tissue. The facia should remain sensible to pain to prevent the heat from extending to deeper regions during the laser procedure so that any damage to the underlying venous system can be avoided.

An ultrasound-controlled indwelling Teflon cannula is positioned directly above the facia up to a few millimetres of the vessel to be treated. Thus, any vein puncture should be avoided. After withdrawal of the mandril, the 600-μm bare fibre of a 1064-nm Nd:YAG laser is guided right to the vessel through the indwelling cannula. The fibre must project from the indwelling cannula by approximately 3 mm to avoid any lasering inside the cannula. The position of the fibre tip can be controlled either by the transcutaneously translucent NeHe pilot beam or by colour-coded duplex sonography. While the complete catheter system is continuously withdrawn, the laser energy is applied. The laser is used in continuous mode at a power of 5 W. A crepitation by disaggregation of gases released from the tissue can be felt or heard, if any, as a beginning response. The colour-coded ultrasound displays this response as so-called colour bruit. During the entire laser procedure, the dermal temperature must be controlled by palpation by the surgeon to avoid any excessive heating and coagulation with subsequent dermal necrosis. Occlusion of the puncture track and stopping of bleeding in the case of unintended vein puncture is effected by intermittent laser irradiation at an exposure time of 0.5 s, each, while the fibre system is withdrawn. The puncture track must not be lasered directly to the cutis to prevent any uncontrolled coagulation of the skin. Furthermore, the skin should not be excessively compressed during the temperature control by palpation in order not to provoke a necrosis, which may occur by pressing against the laser fibre. The laser procedure is stopped when the patient reports a thermal pain, which is provoked at the non-anaesthesized fascia, or when a distinct coagulation in the form of a solid node can be felt. In ultrasound a tissue compression is then detectable directly at the facial exit of the perforating vein, and within a period of 1 or 2 weeks a vascular occlusion occurs. The intervention is made without ischaemia. It can be performed on several areas in one session.

For larger convolutes, rinsing with 0.9% NaCl solution is preferred so that more laser energy can be applied while reducing the thermal strain on the skin and the fascia. The energy is then increased to 8 W.

For rinsing, a NaCl infusion solution is connected through a three-way tap to the positioned indwelling cannula, the diameter of which permits the fluid to pass even if the bare fibre is in horizontal position.

After the treatment is finished, the patient must wear a compression stocking with positioned truss pads above which bandages are tightly wrapped that are removed by the patient on his own maximally after 6 h, or earlier if there is any pain, sensibility disturbance or any other sign of imperfect leg perfusion. The compression stocking is worn for 3 days for 24 h,

**Table 1.** Procedure the treatment of incompetent perforating veins

| |
|---|
| Local anaesthesia of skin and subcutis, of fatty tissue, but not fascia/muscles |
| Epifascial, paravasal application |
| 600-μm fibre, 16 G Abbocath |
| Close to fascial gap |
| Fibre must freely project from the catheter |
| 5 W power (without rinsing), cw, 8 W power (with rinsing), cw |
| Skin temperature controlled by palpation |
| Pressure bandage for 6 hours above compression stocking |
| Much physical exercise/sports post-operatively |
| Success control and/or repeat treatment after 6 weeks |

each, and for another 4 weeks in the daytime. The immediately effective pressure is decisive for the adhesion of the vein walls within the postoperative perivasculitis and for prevention of recanalization.

Consequently, the principle of laser treatment is an epifascial coagulation of the convolute and not a thrombus formation, as with sclerotherapy. As a result, a perivasculitis occurs with shrinking of the vein convolute and scarring of the fascial gap as an incompetence point. The coagulation developed is slowly resorbed so that the patients feel a node in the treated regions for several weeks. The patients are able to work immediately after treatment and can do physical exercises. The treatment success is controlled by ultrasound examinations. If no occlusion or even recanalization occurs within a period of 6 weeks, the treatment can be repeated.

## Reticular Varicosis

For reticular varicosis, sclerotherapy has proved to be an efficacious and simple-to-perform treatment method so that alternative procedures are restricted to cases which do not respond to sclerotherapy, e.g. cases with a known intolerance against sclerosants or cases of general inoperability.

The procedure is similar to that of lasering incompetent perforating veins. Here, too, indication requires a thorough diagnosis of the venous system. The fibre is applied at only a few millimetres' distance from the truncal vein exit. However, in the treatment of reticular varicosis, contrary to that of perforating veins, the attempt should always be made to puncture the vessel for intravasal lasering, as the occlusion rates are higher due to the sometimes long and extremely wavy veins. Puncture must be very careful to avoid any extended wall injury with subsequent bleeding into the tissue. Such intravasal lasering al-

**Table 2.** Truncal varicosis procedure

Local anaesthesia of skin and subcutis

Intraluminal application, if any, 16 G Abbocath, 600 μm fibre

Fibre must project freely from the catheter

Intravasal: rinsing; 10 W, cw

Extravasal: rinsing; 8 W, cw

Fibre withdrawal at 0.5 to 1.00 mm s$^{-1}$

Skin temperature controlled by palpation

Pressure

ways requires rinsing and is effected at 10 W in continuous-wave mode. Without rinsing, the fibre tip in the haemoglobin would immediately carbonize, and no vascular occlusion would occur.

If the vascular puncture fails, the paravasal method may be applied instead. This method also requires rinsing because the reticular veins often run very close under the skin which may be damaged before the desired laser effect occurs in the tissue. Similarly to the treatment of perforating veins, a power of 8 W in continuous-wave mode is applied.

The convolutes being formed during lasering of reticular varices often run very close under the skin. They may be sensible to pressure and patients sometimes feel disturbed by them. A thrombus formation cannot always be prevented by intravasal application, so that similar to the procedure following sclerotherapy, small stab incisions may be required. As in the case of perforating veins, lasering offers the advantage of imposing little strain and quick recovery.

## Laser Treatment of Spider Leg Veins

Spider veins are purely intradermal dilatations of small veins, which frequently run fan-shaped in small grooves on the external surface of the thigh. All spider veins are connected to a feeding vascular system. The feeding region is hypertensioned with resulting vascular dilatation or neoangiogenesis. There are red and blue spider veins which differ by their oxygen content (76 and 69%, respectively) and by their wall structure. Dark spider veins have almost no muscularis and only a few elastic fibres. 50% of these spider veins have reflux by perforating veins. Both types of spider veins represent a considerable cosmetic problem, and are sometimes just the tip of the iceberg as they may indicate more serious venous diseases. For this reason, it is very important to diagnose the venous system carefully before treatment. The truncal and reticular veins should always be lasered first in order to avoid a recanalization of the dermal vessels through these veins.

Now, as previously, sclerotherapy is the procedure of choice. Laser therapy should be applied to vessels which are no longer accessible to puncture or to cases with a known intolerance against sclerosants. Another indication is telangiectatic matting after sclerotherapy. Fewer cases of matting and pigmentation are observed following laser therapy than following sclerotherapy. The effect is purely local, i.e. exactly where the laser beam hits; no systemic side effects occur.

Various types of laser systems are commercially offered, with a pulsed Nd:YAG laser delivering the best results. The beam diameter is 2.5 mm, and the parameter settings are selected depending on the diameter (0.5 to 1.5 mm) and depth of the vessel, as well as on the skin type. Accordingly, the energy ranges between 100 and 130 J cm$^{-2}$ at 2.5-mm beam diameter, while the pulse length ranges between 3 and 8. As a general rule, it should be considered that the darker the skin the larger the breaks between the individual pulses to avoid any surface coagulation (variants ranging between 10 and 20 ms). Coagulation is further prevented by skin cooling. For this purpose, various systems are available. Cooling is possible by air, by gels, by simple ice cubes which are used for skin precooling and by a cooling cuvette through which the laser beam passes. Each cuvette is provided with a supply and a drainage facility and light source adapter. Through its chamber flows water with glycol. The top surface of the cuvette consists of glass, while the face, which contacts the skin, consists of a flexible diaphragm. This permits it to be optimally adapted to any anatomic requirement. Thanks to its dilatability, both the filling level of the cuvette and the pressure to compress particularly thicker vessels are variable. The cooling system minimizes the scattering inside the skin while only slightly reducing the effect of the laser beam in the centre.

During the laser procedure, the HeNe pilot beam must be switched on so that it is always clearly visible that the laser beam actually hits the vessel. The beam then moves over the entire length of the vessel spot by spot, with the vessel being appropriately cooled. Considering the risk of skin damage, an overlap is to be avoided, but, on the other hand, the laser pulses must be set close together to ensure that the complete vessel is occluded. Lasering is always performed from the periphery to the centre, i.e. from the smaller calibre to the larger. On light spider veins, it can be seen how the vessel becomes bloodless, while on the dark ones this effect is less clearly visible because of the smaller vasoconstriction. Immediately after lasering, the treated vessels develop reddening with wealing that recedes within 24 h. An intracutaneous scab is formed later, which may remain for maximally 10 days. Only thereafter is it possible to determine whether or not the vessel is occluded or a recanalization has occurred. Contrary to sclerotherapy, no pressure bandage is applied after lasering of teleangiectatic legs. Lasering causes the endothelium to distend and subsequently the vessel to occlude; this process takes an extended time and cannot be accelerated by means of a pressure bandage.

Recanalization is mainly due to feeding vessels which feed spider vein nests and make them appear not to respond to treatment. Consequently, these feeders must be detected and treated first. We trace such a feeding vessel using a 13.5-mHz ultrasonic head and approach this vessel by the bare fibre of the Nd:YAG laser under ultrasonic control, as is done in the case of lasering perforating veins. Lasering is performed at 5 W in continuous-wave mode until flux is no longer visible. Here, too, the skin temperature must be continuously controlled by palpation to prevent any co-agulation. Once the occlusion is successful, the cutaneous area is often already becoming pale, the remains are lasered according to the above procedure at intervals of 2 weeks each.

In this way, 94% of all spider leg veins can be successfully occluded. The treatment imposes no strain on the patients and can be repeated as often as required.

## References

[1] Müller U, ET AL (1994) Laser treatment of perforating veins. Lasermedizin 10: 150–154
[2] SOKOLL U, ET AL (1995) Treatment of teleangiectases with a new cooling system. Lasermedizin 11: 204–211
[3] CHESS C, CHESS Q (1993) Cool laser optics. Treatment of large teleaniectasia of the lower extremities. J Dermatol Surg Onc 19: 74–80
[4] WEBER M, MAY R, WEBER J (1990) Funktionelle Phlebologie. Thieme, Stuttgart, pp 74, 319, 328, 332, 346, 380

# III-8
# Laser Treatment in Urology

B. P. Shumaker and S. A. Selman

**Contents**

## Introduction

Laser treatment of urologic disease and conditions has been present almost since the first clinical lasers were used. It was apparent from the beginning that some unique properties of the genitourinary tract presented special technical and anatomic problems that needed to be overcome. Urologists are primarily endoscopists, and require lasers that will work through fiberoptic material. Additionally, this energy must be delivered in a fluid environment. It must be delivered to the target tissue via the smallest possible delivery system because of the physical anatomic limits of urethra, ureters, etc. The energy must also be delivered in a very precise manner due to the close proximity of other vital structures. In spite of these apparent limitations, laser use in urology has been strong, active and ongoing.

To present this material in an organized fashion, urologic applications will be discussed by laser type, starting with the longer-wavelength lasers.

## CO$_2$ Laser

The carbon dioxide laser offers many advantages as a device. It is relatively cheap to manufacture, provides excellent vaporization of target tissue, and can be very precisely aimed and focused down to micron size. Its greatest limitation is that it will only work in an air medium; due to its long wavelength (in the far infrared), it will not transmit energy through a fibre. Thus, its use in urology has been confined to cutaneous lesions. It works extremely well for condyloma accuminata due to papilloma virus and superficial skin cancers. The laser can be used either with a hand piece or through an operating microscope/colposcope. Because the energy can be finely focused, the precision with which tissue can be ablated is unequalled with any other laser [1, 2].

## Ho: YAG Laser

The holmium:YAG laser has been used much more extensively in recent years [3]. Originally touted as the most capable laser for orthopaedic use, it has been used increasingly in urology in two areas: The destruction of urinary tract stones and the ablation of soft tissue.

The best soft tissue model to illustrate the use of this laser is the prostate. Since treatment of prostatic obstruction is a very significant part of a urologic practice, a brief overview of the surgical treatment of benign prostatic hyperplasia (BPH) is warranted.

To understand what the laser can do, a quick explanation of the "gold standard" is in order. Transurethral prostatectomy (TURP) was developed in the 1930s. It was consistently refined until the 1960s, when it had reached a level of universal acceptance by clinicians and patients, and has become the measure by which all other treatments for BPH have been compared.

The average TURP takes less than 1 h. An electric loop inside a cystoscope (called a resectoscope) is passed into the urethra and strips of prostatic tissue are shaved away under direct vision. Blood vessels are electrically coagulated. Although all arterial bleeders have been presumably fulgurated intraoperatively, the smaller blood vessels retract under the influence of the electrocautery. As the contraction stimulus wears off, the vessels open and begin to ooze. A catheter remains in place of 24 h, until the urine is free of clots and almost clear. It can than be removed and the patient discharged. The advantages to the patient are several: (1) there is no open incision; (2) blood loss is usually minimal with transfusions rare; (3) the procedure is essentially painless postoperatively; (4) other complications associated with the procedure are by themselves minimal (fluid absorption, retrograde ejaculation, stricture formation, etc.), but collectively yield a complication rate of about 18%.

It should be apparent that for lasers to become an effective and viable treatment for BPH, they must be cheaper, safer and cost-effective. Hospital stay, patient discomfort and anaesthesia risks must be minimal, and long-term results must be as good as or better than conventional techniques.

The Ho:YAG wavelength (2140 nm) has a short absorption length (0.4 mm), and a very narrow zone of coagulation (2–3 mm). Thus, this laser causes vaporization rather than coagulation. During Ho:YAG prostate surgery, prostate tissue can be resected in a relatively bloodless manner. Simplistically, the technique used is to make cuts at the bladder neck, and then make cuts out to the distal end of the prostate. The tissue between the cuts is then gradually undermined and the tissue is cut free at the bladder neck.

The large fragments of tissue need to be cut into smaller ones to facilitate removal through the scope. A catheter is inserted and removed the next day, prior to patient discharge [4, 7].

The laser combines the advantages of conventional TURP with somewhat better haemostasis. However, the need for availability of the instrumentation, a short but steep learning curve, and the lack of a good tissue morsellator (to grind up and remove the fragments) remain major impediments to its widespread use.

## Other Soft Tissue Uses

Since the laser cuts cleanly and precisely, it is an effective tool for the ablation of urethral strictures and vesical neck contractures. With little spread of the laser energy, there seem to be minimal problems with recurrence of these urologic lesions. Small bladder or ureteral tumours respond well to this laser. Indeed, the Ho:YAG should be considered as a possible treatment option for almost any urologic soft tissue surgery.

## Stones

Urinary calculi respond extremely well to this laser. It can be used wherever a calculus can be approached under direct vision and a laser fibre can reach. Consequently, bladder and ureteral stones are quite amenable to the laser. Unlike other lasers used for stones (to be discussed subsequently), the Ho:YAG can fragment any stone, no matter how hard. Large bladder calculi are often first cut up into smaller pieces. Ureteral and accessible renal stones respond well, but great care must be taken that the laser does not damage the adjacent soft tissue wall. The laser will ablate soft tissue with equal impunity.

## Nd: YAG Laser

The Nd:YAG laser (1064 nm) was one of the earliest lasers used by urologists. Primarily and endoscopic laser, it allowed fibre delivery in a fluid environment. There are two major ways to deliver energy with this laser: contact and non-contact (or free beam).

Non-contact means that the laser energy is delivered from the tip of the optical fibre, through the fluid environment, and then into the target tissue. This technique provides significant coagulation and very little cutting or vaporization. It was used initially for the treatment of bladder tumours. However, the fact that it could be absorbed deeply into darker, underlying tissues (with subsequent adjacent organ damage) has limited its use. Still, it became apparent that, used appropriately in the prostate, it could potentially offer significant advantages over TURP.

This technique became known as VLAP (visual laser ablation of the prostate). Rodi and Arretz [5] demonstrated the feasibility of this technique in an animal model. It suggested that non-contact laser coagulation of the prostate was safe and effective. The procedure is quite straightforward. A special fibre is passed through a standard cystoscope. The fibre, which has been designed to emit energy out the side of the fibre tip, is directed into the prostatic tissue. Although no clinical trials have clearly established, the optimal power/time setting, 50–60 W at 60–90 s has been typical. Several different areas in the prostate are treated about every 2 cm. The tissue may "pop" and blanche during treatment, but it takes several weeks for the full effect of this coagulative necrosis to take effect. A catheter is placed and usually removed on the first or second postoperative day. The destroyed tissue gradually breaks down and is sloughed out over time.

Multicentre prospective studies clearly show long-term efficacy [6]. However, the most significant adverse event in these patients has been postoperative dysuria, frequency and urgency related to the slough of necrotic tissue.

At first view, the published data would seem to support continued utilization of this method. However, the urologic community has abandoned the procedure for several reasons. Concerns such as over cost, symptom morbidity and the appearance of other equally efficacious treatments for BPH have relegated the VLAP procedure to primarily a historic one.

The other laser technique has been the use of contact laser prostatectomy [6]. Contact laser energy delivery is where the laser energy is concentrated into a synthetic sapphire or quartz bulb attached to the end of a laser fibre. High-energy densities are reached, which causes no tissue effect unless placed directly against the tissue. The effect is limited to about 1 mm of penetration (as opposed to non-contact). High-power contact laser causes tissue vaporization; lower powers will coagulate.

In this technique, the laser energy is concentrated into a synthetic sapphire or quartz tip. High-power contact lasers cause tissue vaporization; lower powers will coagulate. The TURP can be performed either by directly vaporizing the obstructive prostatic tissue or by haemostasing a raw prostatic fossa after conventional TURP (laser-assisted prostatectomy).

Patients have a catheter placed and are discharged home and the day following removal of the catheter. The patient has minimal dead tissue remaining, so is able to void immediately with minimal irritative symptoms. Still, several disadvantages exist. The technique of primary ablation is slow, particularly if the prostate is large. The bleeding propensity is the same, especially if the laser-assisted technique is used. Perhaps the biggest problem is that the results reported have either been in small, uncontrolled series of patients or have been anecdotal reports. Regardless, the urologic community rarely uses contact laser prostatectomy. Sporadic use of the contact laser is still indicated in selected applications, such as small bladder or ureteral tumours, urethral strictures and other urologic soft tissue applications where precision energy delivery is required.

## Solid-State Diode Laser

Wavelengths ranging from 800 to 1000 nm have been shown to demonstrate relatively deep penetration into tissue. Although various diode lasers have been evaluated, a common one at present is the indigo 830-mm laser. This device, along with its fibre, provides precise feedback to the laser to keep the temperature at the treatment tip of the fibre within a very narrow range. Again, the prostate is used as the soft tissue example. The acronym for the treatment of BPH with this laser is ILC (interstitial laser coagulation). ILC has a technical advantage over other lasers in that the power is quite low (2–20 W), and it is a small, solid-state laser. During the technique, the target cells are heated to the point where they undergo protein denaturation and subsequent coagulation necrosis.

ILC is invariably performed in an outpatient setting and normally takes only about 15 min. Through a standard cystoscope, the fibre is passed and actually pierces the obstructing prostatic adenoma. The fibre is advanced into the tissue until the treatment tip is buried in the gland and away from the urethral wall. Since each treatment site needs only about 3 min of laser time, the procedure is quite fast. The target zone is about 2 cm$^2$. A catheter is placed, and due to the fact that significant oedema takes place in the urethra, may not be able to be removed for several days. It takes several weeks for the treated prostate tissue to gradually involute, so patients notice little change in voiding symptoms until this happens.

ILC is a new technique. It offers great promise but has several limitations, as pointed out. Its use in patients with very large prostates or those in urinary retention seems to show a marginal response. Overall, it will be some time before the urologic community can make a final decision.

## Tunable Dye Lasers (600–680 nm)

Tunable dye lasers are used almost exclusively in the treatment of malignancies of the urinary tract. The technique is called photodynamic therapy (PDT).

## Photodynamic Therapy in Urology

The use of photoactivated substances for the treatment and detection of disease is the basic for both photodynamic therapy, PDT, and photodynamic diagnosis, PDD. The concept is not new. Whitmore and colleagues suggested that tetracycline could be used in conjunction with UV light to aid in the detection of bladder cancer [9]. In 1976, Kelly and Snell employed haematoporphyrin as an agent to be used in conjunction with visible light for both diagnosis and treatment of bladder cancer [10]. The development of medical lasers and flexible optical delivery systems has further spurred the growth of photodynamic therapy.

## Concept and Mechanism

Photodynamic therapy and photodynamic detection depend on the administration of exogenous photosensitizers localizing to an area of diagnostic or therapeutic interest. Subsequent to photosensitizer sequestration, light of the appropriate wavelength directed at the targeted photosensitized tissues initiates events leading to photodynamic destruction or photodynamic detection. Photodetection depends upon photosensitizer fluorescence, while photodynamic therapy (with tissue destruction) depends on an interplay between the excited photosensitizer and molecular oxygen resulting in the production of toxic oxygen species (singlet oxygen).

Tumour destruction during photodynamic therapy is probably the result of direct cell killing and anoxia caused by vascular shut down. Originally, it was felt that cellular death resulted from activation of intracellular photosensitizer. However, several groups demonstrated that photodynamic therapy results in collapse of the tumour microvasculature [11, 12]. This may occur secondarily to endothelial cell injury or activation of platelets with intravascular thrombosis. Undoubtedly, direct cell kill also occurs, perhaps by the initiation of apoptotic pathways [13].

## Bladder Cancer

Haematoporphyrin derivative (HpD), a mixture of fluorescent porphyrins, was the first modern photosensitizer to find clinical applications. The "active component", a mixture of dihaematoporphyrin ether and esters, is currently marketed under the name Photofrin and is approved in Canada for the treatment of superficial bladder cancers [14]. Light of different wavelengths is capable of exciting this photosensitizer; the shorter wavelengths, around 405 nm, are used for photodetection while longer wavelengths, around 630 nm, are used for treatment, since they are capable of deeper tissue penetration. Haematoporphyrin derivative or Photofrin are administered systemically and 24–48 h later the targeted area is illuminated. HpD and Photofrin are the most widely investigated compounds used in PDT of bladder tumours. There is fairly extensive experience in the treatment of bladder cancer [23–29].

A number of other photosensitizers, cloraluminum sulfonated phthalocyanine (CASPc), meso-tetrahydroxyphenylchlorin, mTHPC, tin etiopurpurin (SnET2) and bacteriochlorin (BCA), so-called second-generation sensitizers, have been developed to obviate the perceived problems with HpD. In particular, prolonged skin photosensitivity, relatively poor absorption in the red region of the visible spectrum and the uncertain nature of the biologically active component within HpD has driven the development of these compounds. None of these photosensitizers has received clinical approval for use in the treatment of bladder cancer.

Kennedy and associates suggested that aminolevulinic acid (ALA) could be used as a pharmacologic agent in a non-classical approach to photodynamic detection and photodynamic therapy [15]. ALA is the chemical precursor or protoporphyrin, an active photosensitizing agent when irradiated at 630 nm. Certain tissues and tumours accumulate protoporphyrin when exposed to excess ALA. Thus, by administering exogenous ALA, selected tissues accumulate protoporphyrin and are targeted for photodynamic detection or therapy. ALA has been investigated as an agent for the photodetection and photodynamic treatment of blatter cancer [16]. Its effect, after local application, avoids the cutaneous photosensitivity associated with the classic photosensitizers. Thus, intravesical installation of ALA can be used for the localization and treatment of superficial intravesical neoplasms.

## Prostate

A number of investigators have been exploring the feasibility of using photodynamic therapy for the treatment of both prostate cancer and benign prostatic hyperplasia [17–19]. Tin etiopurpurin, Photofrin, m-TPMC and ALA have all been investigated in experimental systems. As a treatment for benign prostatic hyperplasia, sensitized prostate is treated with a transurethral light application. Histologic studies have demonstrated glandular atrophy within the light-treated volume. The urethral mucosa is not spared, but regeneration has been documented within 3 weeks.

Since prostate cancer is a multifocal disease, the goal of treatment is total ablation of the glandular epithelium. The feasibility of using multiple interstitially placed light-delivery fibres for treatment of the prostate has been demonstrated in the canine model.

Clinical experiences with PDT of the prostate have begun to appear in the literature. Windahl reported on the first two patients with prostate cancer treated with PDT [20]. In both patients the prostate was resected transurethrally and, 6 weeks later, one course of PDT with transurethral light 2 to 3 days after HPD or DHE sensitization was given. Decreasing PSA values documented the short-term treatment effect, but no long-term data were reported. Stroka et al. [21] treated one patient with a 72a carcinoma (PSA 10.4 ng ml$^{-1}$). After oral administration of 40 mg kg$^{-1}$ of ALA, multiple interstitial diffusers were placed via the perineum and a light dose of 225 J was applied at each irradiation site. During the 7-month follow-up, a PSA decrease to < 2.5 ng ml$^{-1}$ was measured. The largest clinical study to date was reported by Nathan et al. [22] in patients with clinically localized recurrent prostate cancer after radiation therapy. Twelve patients received 0.15 ng ml$^{-1}$ mTHPC intravenously 3 days before light treatment with two to eight interstitial diffusers inserted through the perineum under ultrasound guidance. The total light doses delivered ranged from 300 to 1600 J at a wavelength of 652 nm. Treatment effects were observed as non-enhancing lesions during magnetic resonance imaging that corresponded to necrosis on biopsy. PSA values decreased in seven men two patients achieving nadirs of < 0.5 ng ml$^{-1}$.

## Future

Photodynamic therapy has slowly gained acceptance as a treatment for selected indications. As new photosensitizers and light-delivery systems are developed, the indications for its use will undoubtedly encompass a spectrum of urologic disease. Photodynamic therapy will eventually find its place in the arena of minimally invasive therapies in urology.

## Frequency-Doubled Nd:YAG Laser (KTP)

The KTP laser is an interesting device. It takes advantage of the fact that a standard Nd:YAG laser can be made to emit laser energy at half the wavelength (double the frequency) using crystal technology. A small crystal of KTP (postassium titanyl phosphate) is placed in front of the standard Nd:YAG beam; the output is a green light laser wavelength of 532 nm. The unique characteristics of light at this wavelength make it a device similar to the argon laser in tissue effects. This effect allows for precise cutting and some coagulation; the depth of penetration is about 1–3 mm. It can be used in a contact mode, which allows for soft tissue incisions, or non-contact, which allows some coagulation of blood vessels. Since the green is preferentially absorbed by blood, it is quite effective in coagulating small blood vessels. The other advantage of this device is that a switch will change the laser back to the standard Nd:YAG wavelength.

This laser is quite useful for urethral stricture and vesical neck contractures, as well as other soft tissue applications in the urinary tract. Some of the newest, high-power KTP lasers have been used to vaporize prostatic tissue in an essentially bloodless field. It has also been used as an "engine" in tunable dye systems for PDT.

## Flashlamp Dye Laser

This laser, also known as the Candela laser (for the company that pioneered its development), has one use only: to fragment urinary tract calculi. First developed in the mid-1980s, it presented urologists with the unique ability to safely fragment ureteral and bladder stones. This is done by a tremendous blast of energy, delivered over an extremely short period. The energy creates a delivered over an extremely short period. The energy creates a plasma jet in the stone-water interface, and the expanding explosion starts to fragment the stone. Since the laser light is green, there is minimal tissue absorption and almost negligible tissue damage. Thus, it is safe for stones in the ureters and even kidney. The stones is fragmented under direct vision, and the delivery fibre is about 0.5 mm in diameter, which is small enough to be passed through the smallest endoscope.

Because the energy effects take place only against structures with a crystalline makeup, soft tissue is basically unaffected. Thus, even if the laser fibre fires repeatedly against the uretheral wall, very little tissue damage takes place. The Candela laser, although already a dated technology, is still used in many clinical settings. It is one of the few lasers widely accepted, used and considered an integral part of the urologic armamentarium.

## Other Lasers

It needs to be understood that in a brief overview such as this, only the most cursory presentation can be made. There are other lasers (i.e. alexandrite, free electron, low-level biostimulation) being used for urologic applications that are either common in other countries, experimental or as yet unproven to have true clinical application. It is hoped that when this chapter is updated in several years, exciting new treatment paradigms for urologic disease are evident. These may well be based on technological advances in laser design, wavelength and delivery systems.

## References

[1] BENSON RC (1986) Laser use in open surgery and external lesions. Urol Clin North Am Aug 13 (3): 421–434

[2] SCHAEFFER RJ (1986) Use of the $CO_2$ laser in urology. Urol Clin North Am Aug 13 (3): 393–403

[3] JOHNSON DE, CROMMEENS DM, ET AL (1992) Use of the Ho:YAG laser in urology. Lasers Surg Med 12: 353–356

[4] BHATTA KM (1995) Lasers in urology. Lasers Surg Med 16 (4): 312–330

[5] RODI RA, ARRETZ TH (1991) Transurethral ultrasound guided laser-induced prostatectomy (TULIP procedure): a canine prostate feasibility study. J Urol 146: 1128–1135

[6] SHUMAKER BP (1994) Contact laser ablation of the prostate in the treatment of benign prostatic hyperlasia. Semin Urol XII, 3 (Aug): 170–173

[7] GILLING PF, HENKE R-P, ET AL (1996) Holmium laser resection of the prostate: preliminary results of a new method for the treatment of benign prostatic hyperplasia. Urology 47: 48–51

[8] KABALIN JN, Gill HS, ET AL (1997) Prospective multicenter prolase II clinical trail of Nd:YAG laser prostatectomy. Urology 50: 63–65

[9] WHITMORE WF, BUSH IM (1968) Ultraviolet egstoscopy. JAMA 203 (12): 1057

[10] KELLY JF, SNELL ME (1976) Hematoporphyrin derivative: a possible aid in the diagnosis and therapy of carcinoma of the bladder. J Urol 115: 150

[11] Selman SH, KREIMER-BIRNBAUN M, KLAUNIG JE, GOLDBLATT PJ, KECK RW, BRITTON SL (1984) Blood flow in transplantable blatter tumor treated with hematoporphyrin derivative and light. Cancer Res 44 (5): 1924

[12] HENDERSON BW, WALDON SM, MANG TS, POTTER WR, MALONE PB, DOUGHERTY TJ (1985) Tumor destruction and kinetics of tumor cell death in two experimental mouse tumors following photodynamic therapy. Cancer Res 45: 572

[13] KESSEL D, LUO Y (1998) Mitochondrial photodamage and PDT-induced apoptosis. Photochem Photobiol 42: 89

[14] DIAMOND I, GRANELLI S, MCDONAGH A, NIELSEN S, WILSON C, Jaenicke R (1972) Photodynamic therapy of malignant tumors. Lancet 2: 1175

[15] KENNEDY JC, POTTIER RH (1992) Endogenous protoporphyrin 1X, a clinically useful photosensitizer for photodynamic therapy. Photochem Photobiol B 14: 275

[16] KRIEGMAIR M, BAUMGARTNER R, LUMPER W, WAIDELICH R, HOFSTETTER A (1996) Early clinical experience with 5-aminolevulinic acid for the photodynamic therapy of superficial bladder cancer. Br J Urol 77: 667

[17] SELMAN SH, KECK RW (1994) The effect of transurethral light on the canince prostate after sensitization with the photosensitizer tin II etiopurpurin dichloride: a pilot study. J Urol 152: 2129

[18] SHETTY SD, PEABODY JO, BECH ER, CERNY JC, AMIN MH, Richter A (1999) Transurethral photodynamic therapy in benign prostatic hyperplasia: a canine pilot study using benzoporphyrin derivative. SPIE 3590: 151

[19] CANG S-C, BUONACCORSI GA, MACROBERT AJ, BOWN SG (1997) Interstitial photodynamic therapy in the canine prostate with disulfonated aluminum phthalocyanine and 5-aminolevulinic-acid-induced protoporphyrin 1X. Prostate 32: 89

[20] WINDAHL T, ANDERSON SO, LOFGREN LA (1990) Photodynamic therapy of localized prostate cancer. Lancet 336: 1139

[21] STROKA R, STEEP H, MUSCHTER R, KNUCHEL R, PERMUTTER R, BAUMGARTNER R (1998) 5-ALA-assisted photodynamic therapy on prostates. 7th Biennial Congr of the Int Photodynamic Association (IPA), Nantes, France, July 1998 (Abstr RC 112)

[22] MANYAK MJ (1995) Practical aspects of photodynamic therapy for superficial bladder carcinoma. Tech Urol 1, 2: 84

[23] NSEYO UO, DOUGHERTY TJ, SULLIVAN L (1987) Photodynamic therapy in the management of resistant lower urinary tract carcinoma. Cancer 60: 3113

[24] JOCHAM D, BAUMGARTNER R, Steep H, Unsold E (1990) Clinical experience with the integral photodynamic therapy of bladder carcinoma. Photochem Photobiol 6: 183

[25] NAITO K, HISAZUMI H, UCHIBAYASHI T, AMANO T, HIRATA A, KOMATSU K, ISHIDA T, MIYOSHI N (1991) Integral laser photodynamic treatment of refractory multifocal bladder tumors. J Urol 146: 1541

[26] WINDAHL T, LOFGREN LA (1993) Two years' experience with photodynamic therapy of bladder carcinoma. Br J Urol 71: 187

[27] UCHIBAYASHI T, KOSHIDA K, KUNIMI K, HISAZUMI H (1995) Whole bladder wall photodynamic therapy for refractory carcinoma in situ of the bladder Br J Cancer 71: 625

[28] NSEYO O, DEHAVEN J, DOUGHERTY TJ, POTTER WR, MERRILL DL, Lundahl SL, LAMM DL (Feb. 1998) Photodynamic therapy (PDT) in the treatment of patients with resistant superficial bladder cancer: a long-term experience. J Clin Laser Med Surg 18(1): 61–8

[29] SHUMAKER BP (1990) Photodynamic therapy in the treatment of bladder cancer. The Malnati Symposium of October, 1990, Chicago. Academic Press, New York

# III-9
# Laser Application in Orthopaedic Medicine

W. Siebert

## Contents

## Introduction

Throughout history, humans have suffered from diseases of the musculoskeletal system. Humanoid bones dating back to the Stone Age show signs of skeletal affliction; Hippocrates described the symptoms of spinal degeneration, which is still considered a disease of modern society.

Albert Einstein first described the theory of stimulated emission in 1917 [32]. He hypothesized that accelerated electrons produce electromagnetic beams of a certain wavelength. In 1960, Maiman, a young American physicist, constructed the first functional laser system, a ruby laser which emitted visible light at a wavelength of 694.3 nm [91]. In 1960 the first medical applications in dermatology and ophthalmology were conducted by Goldmann and other clinicians [16, 47, 46]. Since then, medical lasers have become powerful and indispensable surgical tools in almost every medical field, including orthopaedic medicine.

The appeal of surgical laser systems in orthopaedics is based on the universal drive to minimize surgical techniques for better access of pathological tissues with descret and precise treatment and at the same time reduce tissue trauma. The laser's application instruments are so small that the surgeon is able to reach very narrow spaces. The laser is also used to stop the flow of blood, and ensures better visual control. Laser irradiation of tissue samples, in orthopaedics these are usually articular tissues, leads to coagulation and tissue ablation [132, 38].

These effects have been put to use in orthopaedic surgery since the mid-1980s, mostly for treatment of joint disorders of the knee, shoulder, ankle etc. (e.g. laser-assisted arthroscopy) and spinal disorders (e.g. percutaneous laser disc denaturation, endoscopic laser foraminoplasty [132. 38].

## Laser Systems in Orthopaedic Medicine

There are many medical laser systems available today, but they all use the principal of selective photothermolysis, which means getting the right amount of the right wavelength of laser energy to the right tissue to damage or destroy only that tissue, and nothing else.

The most frequently used laser systems in orthopaedic medicine – in alphabetical order – the $CO_2$ laser, diode laser, erbium:YAG laser, excimer laser, holmium:YAG laser, KTP laser and neodym:YAG laser. These lasers function in the ultraviolet (e.g. excimer) or in the infrared (e.g. $CO_2$, holmium:YAG laser, neodym:YAG laser) and have the optimal combination for orthopaedic surgery: tissue vaporization with no or hardly any thermal effect and the ability to treat small hard-to-reach spaces [138, 38].

## $CO_2$ Laser in Orthopaedic Medicine

Often referred to as the surgical laser, the action of the $CO_2$ laser most resembles traditional surgery. The $CO_2$ laser was the first laser widely used by surgeons, and is still the most used of all the medical lasers. Strongly absorbed by water, which constitutes of 80% of soft tissue, the $CO_2$ laser emits continuous wave or pulsed far infrared light at 10 600 nm, which can be focused into a thin beam and used to cut like a scalpel, or defocused to vaporize, ablate or shave soft tissue. The $CO_2$ laser may be operated in pulsed mode or used with scanning devices to precisely control the depth and area of ablation. The $CO_2$ laser derives its energy from the electrical excitation of $CO_2$ gas with resulting emission of photons. The $CO_2$ laser is highly absorbed by water with minimal scattering; therefore, gaseous insufflation is required to prevent absorption of the laser by the liquid distension media.

The $CO_2$ laser is a good cutting instrument with some coagulation effects. The tissue is rapidly heated above boiling and is obliterated by the $CO_2$ laser with a small residue of ash. The thermal effect is less than 200 μm and thermal necrosis less than 50 μm [65, 135].

The $CO_2$ laser primarily creates a vertical incision with minimal lateral necrosis. The smoke must be evacuated from the joint during the procedure and the carbon ash residue should be removed with irrigation [133].

Other histological studies revealed that the pulsed $CO_2$ laser produces thermal tissue injury to a distance of 350 μm from the laser tissue interface and also reduces the surface char [118, 135].

The $CO_2$ laser cannot be transmitted through standard fiberoptic cables and requires an articulating arm with a channel of sensitive reflecting mirrors. Typically, the $CO_2$ laser uses a helium-neon laser beam because the $CO_2$ laser light is invisible.

One of the earliest evaluating the $CO_2$ laser was Terry Whipple et al. (1984) who conducted arthrotomy and subtotal meniscectomy on rabbits [146]. The authors concluded that the carbon ash residue, although not particularly harmful, caused a severe synovial response.

In an experimental in vivo study published in 1995 by Vangsness et al., the authors found no adverse effects after arthroscopic $CO_2$ laser irradiation [144].

Philandrianos and other authors used the $CO_2$ laser as a routine procedure in arthroscopic surgery of the knee joint [101, 135].

The $CO_2$ laser requires gas insufflation of the joint, which can result in subcutaneous emphysema [135]. Smith et al. described a technique for $CO_2$ laser arthroscopy, in which the jet of carbon dioxide gas creates a bubble on the meniscal surface so that the laser beam passes through the gas to the target tissue without being absorbed by the liquid media used to distend the joint, which largely eliminates the problems of postoperative joint effusions and emphysema formations. The authors reported good experience with this technique and pointed out that no specific complications occurred [134, 104].

Lee used a microscopic $CO_2$ laser system for treatment of lumbar disc herniations. In a clinical study of 300 patients treated by $CO_2$ laser-assisted open microdiscectomy, he reports on over 90% good to excellent results based on the MacNab criteria after a follow-up period of 2 years [85].

Lee also used the $CO_2$ laser combined with a surgical microscope to treat cervical ossification of the posterior longitudinal ligament. Although he describes the system as being somewhat cumbersome, he found 80% good to excellent results and 20% fair results; no poor results were observed [84].

Despite modifications in the operative techniques that lead to a significant improvement in $CO_2$ laser-assisted meniscectomy, main problems in orthopaedic practice remain the cumbersome light-delivery device and the poor visualization of the meniscus due to bubble formation. For this reason, the $CO_2$ laser has not gained significant world-wide acceptance in orthopaedic surgery [65, 116, 127].

## Diode Lasers in Orthopaedic Medicine

Diode lasers are solid-state devices similar in construction to LEDs. The familiar "laser pointers" are, in fact, diode lasers. Clinical diode lasers emit near-infrared light in the 800–900 nm range. Currently, their principal application is in ms-range pulsed mode. The use of diode laser in orthopaedic medicine is still in the experimental stage.

Tatay et al. have been assessing the effect of the diode laser in the intervertebral disc. In an experimental in vivo study they found osteophytic proliferation, necrosis and mild cell activation. The position of the optic fibre appears to be important [140, 141].

Hendrich et al. tested the diode laser for photodynamic laser therapy for treatment of rheumatoid arthritis [58, 57].

Glinkowski et al. used a non-contact pulsed diode laser (904 nm, 15 W peak power) to evaluate low-energy laser therapy mechanisms and to explain why some laser therapists prefer to begin low-power lasing by scanning followed by local irradiation [43].

In another study, Glinkowski et al. treated 18 patients with delayed union of long bone fractures with the pulsed diode laser (904 nm, 10–15 W peak power). Treatment consisted of series of laser irradiation every alternate day, 10 min in contact mode above the fracture gap, locally with a frequency of 2.4–4 kHz. Fourteen patients experienced consolidation of fracture just after the therapy. No side effects or osteolysis around implants was observed in the examined group of patients [44].

Gevargez evaluated the diode laser (980 + 30 nm wavelength, Ceralas D, CeramOptec and light fibres with a 240 µm outside diameter) for CT-guided percutaneous laser disc decompression (PLDD) on 13 patients with herniated lumbar discs using a thin coaxially bent memory canal. Laser radiation was performed in single shots lasting 0.1–1 s and 1-s breaks. Output was 4 W, 1800–4000 J; 84% (VAS) of the patients had excellent or good results [41].

Grönemeyer also used the diode laser with a wavelength of 980 + 30 nm to treat disc herniation with nerve root compression, and procured good initial results [49].

In an experimental study, Morimoto applied diode laser radiation in contact mode for fusion of ruptured pig menisci and found promising results [98].

Recently, Hoteya et al. conducted double blind studies using a GaAlAs diode laser therapy device (wavelength 830 nm) for treatment of 596 patients with chronic pain resulting mainly from sports activity. The diagnoses were tennis elbow, jumper's knee, throwing shoulder and achilles tendinitis. Good results were found in cases with superficial disorders, e.g. medial epicondylitis humeri, patellar tendinitis, peripheral inflammation of the achilles tendon. On the other hand, when pain resulted from complex pathologic conditions or the disorder was relatively deep (e.g. shoulder impingement syndrome, tarsal syndrome) the success rate was lower [62].

Al-Awami treated 40 patients with Raynaud's phneomenon with low-level laser treatment by means of a 250-mW, 640-nm continuous-wave diode laser (Holbo Lasers, Gallspach, Austria). Their preliminary data suggested very promising short- to medium-term effects [3].

Because of their relative simplicity and low maintenance requirements, diode lasers and diode-pumped solid state lasers will be used more in the near future as more wavelengths become available. However, more long-term studies are needed to validate the feasibility of diode lasers in orthopaedic medicine.

## Erbium:YAG Lasers in Orthopaedic Medicine

The erbium:YAG laser emits a mid-infrared beam at 2940 nm, which coincides with the absorption peak for water. Its principal use is for precise cutting and coagulation in ophthalmology, dermatology and laryngeal surgery.

In orthopaedics, the erbium:YAG laser has been tested for tissue welding of torn menisci. The Er:YAG laser is well absorbed by water in menisci so that the ablation threshold is very low. Histological analysis of acute Er:YAG laser lesions reveals precise cutting effects with minimal thermal damage.

Dews et al. reported on tissue welding of a torn meniscus using an Er:YAG laser in 1988 [28].

As Siebert and other researchers have demonstrated, the Er:YAG laser can be used for osteotomies with minimal thermal necrosis [131, 65, 122].

Yonezawa et al. tested a new Er:YAG laser device and fibre to sapphire fibre transmission system for percutaneous laser disc decompression (PLDD) and bone ablation. This system would enable the use of contact tissue ablation under water circulation with fewer fibre problems [147].

In an experimental histological and SEM study by Birnbaum et al., Er:YAG laser irradiation was used to extract polymethylmetacrylate cement (PMMA) on the boundary surface between the cement and human femoral bone [15].

Asai tested the erbium:YAG laser for its usefulness in bone resection, e.g. removal of bony spurs and ossification of the posterior longitudinal ligament. His results suggest that the procedure may be clinically applicable [4].

The Er:YAG laser may be quite effective to cut and ablate articular tissue. So far, however, a widespread clinical application is limited by the necessity of an extremely rigid zirconium-chloride light guide that is too fragile for routine use [104, 116, 60].

## Excimer Lasers in Orthopaedic Medicine

The name excimer is given to a series of lasers that operate in the same manner using noble gas halides [argon-fluoride (Arf) 193 nm, krypton-fluoride (Krf) 248 nm, xenon-chloride (XeCl) 308 nm, for xenon-fluoride (XeF) 351 nm]. Excimer lasers emit invisible ultraviolet (UV) light that triggers a photochemical reaction on the target tissue which leads to a bond breaking in tissue proteins converting them to a gaseous plasma. This mode of operation contrasts with any photothermal mechanism of infrared lasers, which depend on the conversion of light to heat through absorption phenomena [65].

Excimer gases are activated by an electrical discharge and react with halogen molecules to form compounds. These compound molecules are known as excimers (excited dimers) and exist only briefly in the high-energy state. A high-energy pulse vaporizes a very thin tissue layer without affecting the surrounding tissue. The result is predominantly a surface effect with no measurable effect beyond 5–10 µm. At energy levels above 200 mJ cm$^{-2}$, the excimer laser acts as a very precise, non-heating contact-free scalpel [65].

The most common medical application is the ArF laser used for vision correction, e.g. LASIK (laser in situ keratomilieusis). Excimer laser radiation also shows great promise for lithoptripsy, but is currently limited by the lack of durable UV-capable fiberoptic delivery devices.

In orthopaedic medicine, numerous studies of the excimer laser in arthroscopic surgery and percutaneous nucleotomy have been conducted [20, 45, 48, 67, 88, 103, 107, 114].

Excimer laser ablation of cartilage is very precise without damage to adjacent tissue [35]. The maxima of absorption in human menisci are at 280- and 340 nm wavelengths, attributed to the presence of tryptophan and hydroxypyridinoline, respectively.

The tissue effects of the XeCl excimer laser on fibrocartilage have been evaluated by Dressel et al., Kroitzsch et al., and Raunest et al. [104, 105, 31, 81].

As the laser beam is delivered fiberoptically by 600 to 1000-µm light guides, the practicability in arthroscopic meniscectomy is excellent. There is substantially no relevant photothermal effect with its use and the remaining surfaces of fibrocartilage following ablation are smooth and well delineated [31].

In a controlled in vivo study by Raunest et al. employing a rabbit model, the authors found a precise cutting profile following excimer laser-assisted partial meniscectomy and a significantly reduced degree of reactive synovitis [103].

The XeCl excimer laser has also been used for meniscectomy in clinical practice. Advantages exist due to the small diameter of quartz fibres delivering the laser beam and creating good access even to narrow joint compartments. Clinical data comparing excimer laser-assisted meniscectomy to conventional meniscus surgery prove a reduced incidence of postoperative swelling and effusion, reduced pain and an earlier return to normal function in the laser group [1].

One concern over excimer lasers is the possibility of a mutagenic effect. The ultraviolet wavelength of 248 nm is the most mutagenic wavelength. However, as studies have proved so far, the 308-nm spectral range shows no adverse consequences. The 248-nm KrF excimer laser has no clinical importance in meniscal surgery [80].

The initial excitement over excimer lasers has dwindled due to its inadequate ablation rate [67, 69]. The relatively low amount of energy that can be delivered through a fused silica fibre restricts the ablation rate and makes excimer laser meniscectomy unsuited for clinical practice. The relatively low energy density transmitted through the fibre leads to a reduced ablation efficiency, which causes a time-consuming prolongation of the ablation process [104].

## Holmium:YAG Lasers in Orthopaedic Medicine

The holmium:YAG laser is an infrared laser that emits light at a wavelength of 2100 nm and is absorbed by tissue water which converts light energy to heat. The emission band of the ho:YAG laser does not correspond directly to the typical water absorption peaks; however, this laser proved to be an excellent tool for cutting and ablation of cartilagenous tissues, provided that the power density exceeds the ablation threshold of cartilagenous tissues. The holmium:YAG laser can be transmitted through conventional optical fibres. It can precisely and rapidly resect, cut, coagulate, vaporize and ablate, while providing superior haemostatic control [65, 120, 137, 132, 38].

This laser can function in a saline medium and may be used in direct contact with the tissue for direct surgeon feedback. However, usually a near-contact mode with a free beam is utilized [74, 83]. Many of the advantages of the laser are related to its small application instruments and its ablative and haemostatic effects [132, 38].

The ho:YAG laser produces smooth tissue defects. The transitorial zones of thermal alterations in the meniscal tissue are well controlled and do not exceed 150 µm at light microscopy sections under condition of optimal output parameters. There is no gross evidence of thermal injury, coagulation necrosis or carbonization. In a study by Trauner et al., the authors found zones of thermal damage extending only 550 µm from ablation sites. A delayed biological response was not observed [143].

With regard to the laser parameters to be employed in meniscus surgery, pulse width seems to be the most decisive parameter for determining ablation and cutting efficiency. Morphometrical studies have shown that a pulse width of 250 ns is considered an optimum value for clinical application [104, 143].

Interestingly, the extent of the lateral damage zone is independent of the pulse width. Experimental studies on the Ho:YAG laser by Imhoff et al. demonstrated that the highest ablative efficiency is procured with a pulse rate of 10 Hz and 1.5 J or 15 Hz and 1.0 J. A higher pulse rate of 20 Hz did not increase the ablation rate [65].

Based on experimental results, further improvement of the laser and clinical applications, several clinical studies were conducted in the 1980s and 1990s, which prove the merit of Ho:YAG laser arthroscopy [17, 30, 130, 128, 132, 38, 39].

In a report by Saunier et al., the authors state that since 1990, the Ho:YAG laser has been used in more than 500 000 arthroscopic procedures [109].

The Ho:YAG laser is utilized for numerous orthopaedic procedures, e.g. meniscal-cartilage lesions, chondromlalacia, rheumatoid synovialitis, chronic patellofemoral instability, osteoarthritis of the knee, laterization of the patella, shoulder instability, ankle joint disorders (synovialitis, exostosis of the distal tibia, cicatrizations, cartilage defects) and spinal disc herniations.

Numerous authors describe good results after Ho:YAG laser-assisted arthroscopy of the knee (Figs. 1, 2; Table 1), ankle and shoulder [11, 9, 10, 33, 100, 117, 128, 1, 2, 12, 19, 30, 40, 52, 66, 54, 94, 89, 108, 148, 87, 90, 132, 38, 71, 114].

**Table 1.** In a prospective study, Lübbers and Siebert compared Ho:YAG laser arthroscopy of the knee to conventional arthroscopy. Following disorders were treated: meniscal lesion, chondromalacia, rheumatoid synovialitis, femoropatellar pain syndrome (lateral release) and combined meniscal-cartilage injury. After 2 years, the modified Lysholm score was significantly increased in comparison to preoperative scores (e.g. meniscal lesion p < 0.0002). The authors conclude that, when used correctly, the holmium:YAG laser is the more moderate and exact instrument for arthroscopic surgery with several advantages over conventional methods [90]

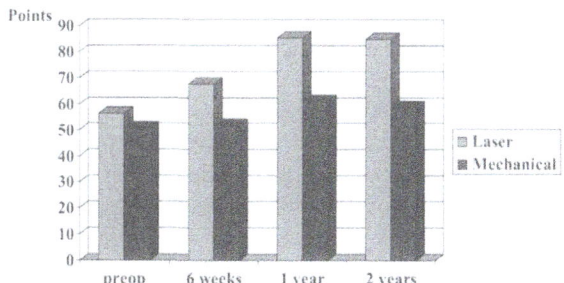

Rheumatoid Synovitis
Modified Lysholm score

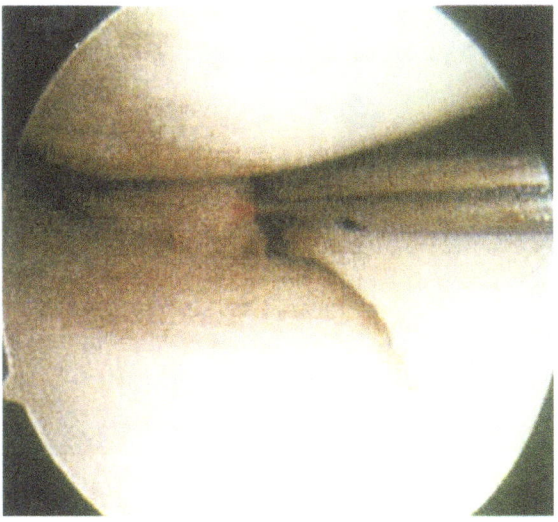

**Fig. 1.** Ho:YAG laser-assisted arthroscopy of the knee. Meniscal cartilage smoothening

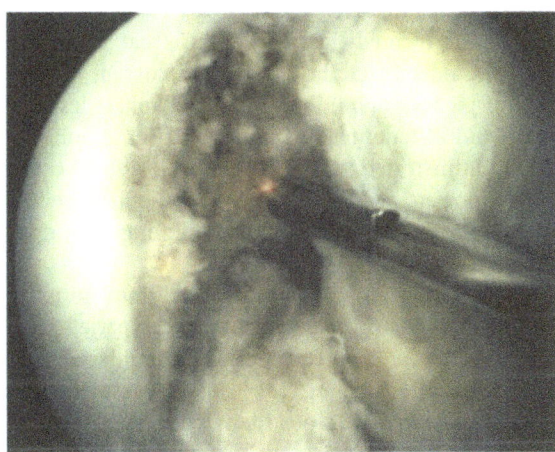

**Fig. 2.** Ho:YAG laser-assisted arthroscopy of the knee: the laser is used for haemostasis during a notch plastic procedure

Most recently Saunier reported on good results after performing 600 2.1 Ho:YAG laser-assisted knee arthroscopies. He additionally conducted MRIs to assess possible bone necrosis and found no substantial proof for Ho:YAG-induced necrosis [109].

Gerber et al. compared laser-assisted arthroscopy to conventional arthroscopy. One hundred received either Ho:YAG laser-assisted meniscectomy (hook, punch and coherent versapulse 60 W) or conventional meniscectomy (hook, punch, shaver). They found significantly better results for the laser group, with less effusion. There was no case of osteonecrosis after surgery in either group [37].

Several research groups [50, 64, 82, 34, 113, 136, 87, 68, 142] have been conducting Ho:YAG laser-assisted capsular shrinkage (LACS) for treatment of shoulder

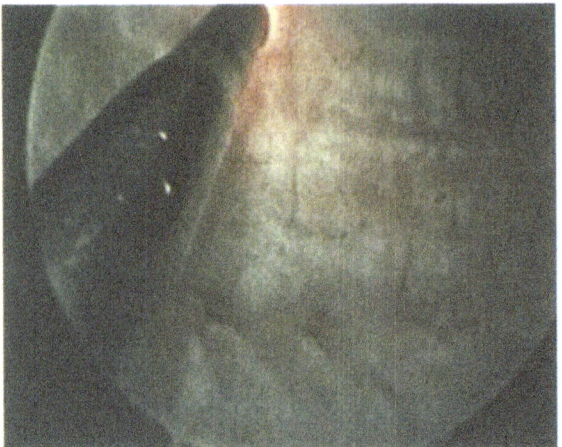

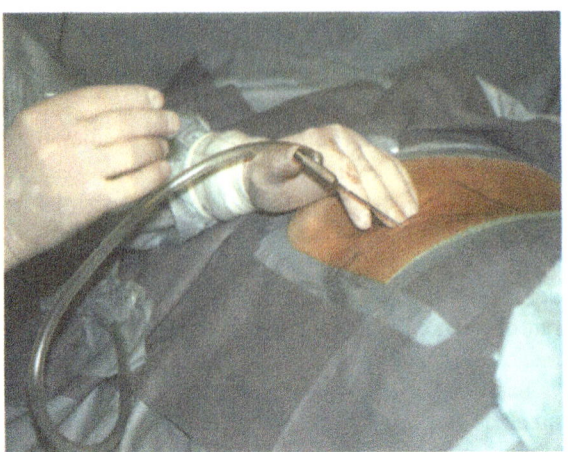

**Fig. 3.** Ho:YAG laser-assisted capsular shrinkage of the shoulder joint (LACS)

**Fig. 4.** Ho:YAG laser-assisted percutaneous laser disc decompression (PLDD): the patient is in a prone position on his stomach. The surgeon has inserted the laser fibre through the guide cannula

instabilities, based on the basic research of Markel and Hayashi [93, 51, 54, 53].

Pisot performed Ho:YAG LACS to treat 57 patients with shoulder instability (48 with recurrent dislocation of glenohumeral joint without Bankart lesion, 9 with multidirectional instability). Laser irradiation was administrated at an energy level of 1.0 J and a repetition rate of 10 Hz (10 W); 51 patients were very satisfied with the results and had full range of motion [102].

Sommerfeld and Siebert procured good to excellent results in approximately 90% of the patients suffering from uni- or multidirectional shoulder dislocations after performing the Ho:YAG laser-assisted LACS procedure on more than 200 patients (Fig. 3) [136].

The effect of the Ho:YAG laser on spinal disc cartilage has also been analyzed and discussed in numerous publications [92, 110, 111, 121, 125, 149, 139, 132, 38].

The 2.1 Ho:YAG laser in combination with numerous techniques, e.g. percutaneous laser disc decompression (PLDD) (Fig. 4) and endoscopic laser foraminoplasty (ELF), is also used to treat spinal disc disorders, especially lumbar disc herniations [34].

Most recently, Baskov et al. reported on the regenerative effect of the Ho:YAG laser on intervertebral disc cartilage in an experimental in vivo study. The authors found that the laser radiation promoted both fibrous and hyline carilagenous growth [13].

In an experimental study, Segura et al. used the Ho:YAG laser at a lower non-ablative energy level for collagen and disc shrinkage, and found promising results [115].

Nishijima et al. procured good results (80% success rate) after performing Ho:YAG laser-assisted endo-

scopic percutaneous lumbar disc decompression on 436 patients suffering from lumbar herniated nucleus pulposus. His inclusion criteria were a contained lumbar herniated nucleus pulposus, and failed conservative treatment for at least 3 months. Exclusion criteria were, for example, non-contained discs, spinal stenosis, intraspinal canal ossification, previous disc surgery, and tumours [99].

Lee utilized the Ho:YAG laser to treat cervical disc herniations. Positive indications: sustained radiculopathy with positive correlation to CT and MRI findings and failed conservative treatment; contra-indication: myelopathy with upper motor neuron signs, ataxic gait and paraparesis. He reported a success rate of almost 90% after treating several hundred patients [86].

Based on the prior research and experience of several authors, Ho:YAG laser foraminoscopic surgery was developed to treat herniated nucleus pulosus in the lumbar spine without the morbidity and complications of open spinal surgery [61, 73, 72, 96, 95, 21, 132, 38]. Contained and non-contained lumbar disc herniations that are endoscopically accessible can be treated. Foraminal laser endoscopic disc ablation (FLEDA) and endoscopic laser foraminotomy (ELF) are two procedures that have procured good results, with success rates ranging from 75 to 89%. However, these techniques necessitate a long learning curve [21, 22, 96. 129].

Stücker reported a series of 85 patients who were treated for non-contained herniated disc by a relatively new endoscopic technique which utilizes Ho:YAG laser vaporization in the epidural space to remove sequestered disc material. Indications were young patient age, pathology at disc level, no previous disc surgery and disc space narrowing of less than

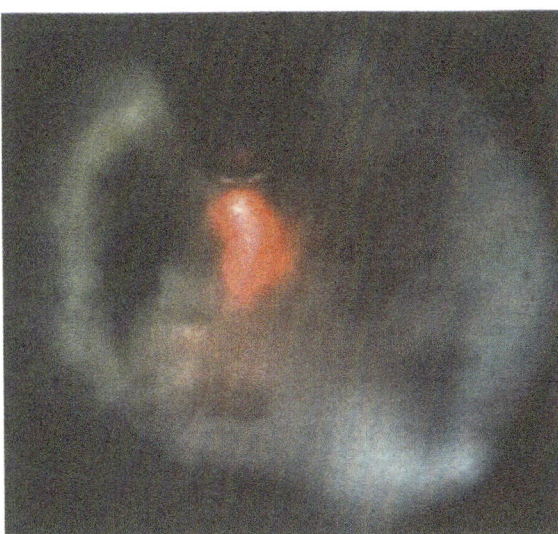

**Fig. 5.** Ho:YAG laser-assisted transforaminal endoscopy of the lumbar disc

50%. Contra-indications were sequestered disc material not located at disc level, lateral recess stenosis and cauda equina syndrome. Treatment of the level L5/S1 was described as problematic if the pelvic crest is very high. The Ho:YAG laser energy was used for tissue ablation or to obtain haemostasis. According to the Mac-Nab criteria, 80% of the results were good to satisfactory. No adverse effects due to the Ho:YAG laser energy were observed [138].

Siebert et al. also performed Ho:YAG laser-assisted transforaminal endoscopy to treat sequestered and non-sequestered lumbar disc prolapses with a decreased risk of postdiscectomy syndrome with a success rate of more than 75%. They used the 2.1-nm holmium:YAG laser (energy level between 0.6 and 1.6 J, frequency 6–20 H) with a side-firing tip for haemostasis and tissue ablation (Fig. 5). Thermal damage is avoided by continuous cool rinsing. The side-firing laser allows tissue treatment under visual control and avoids possible damage to neural structures. By using high-energy levels, laser ablation of bone and/or calcified tissue is possible without damage to neurovascular tissue. The transforaminal endoscopic technique does not endanger neurovascular structures. Due to the lateral access, manipulation to the dura and radix, as occures in open and dorsal entries, is minimized [124, 124, 126, 129].

In contrast to other intradiscal techniques that have a more or less indirect unfocused effect through pressure reduction in the nucleus pulposus (chemonucleolysis, percutaneous discectomy, automated percutaneous lumbar discectomy, percutaneous laser disc decompression), transforaminal endoscopy is a focused procedure with direct control by the surgeon. Extra-intraforaminal, mediolateral and medial lum-

bar prolapses can be treated. Very large lumbar prolapses, filling the spinal canal more than 50%, are considered to be a contra-indication [139].

Controversies exist that concern adverse secondary effects of the Ho:YAG laser in articular tissues, especially in subchondral bone. Several authors have conducted biophysical experiments and MRI examinations to observe the modifications of the musculoskeletal tissues after Ho:YAG laser irradiation. All authors have excluded the possibility of significant thermal damage due to Ho:YAG laser irradiation when the laser is used with parameters appropriate for cartilage laser surgery [109, 8, 59, 7, 6, 119, 63, 132, 38, 39].

In a thoroughly investigated literature review, Saunier found only 11 cases out of several thousand Ho:YAG laser-assisted arthroscopies of substantiated bone necrosis [109].

The holmium:YAG laser does not cause osteonecrosis during arthroscopic surgery if it is used with the correct technique and with the optimal setting.

## KTP Laser in Orthopaedic Medicine

When Nd:YAG laser light at 1064 nm is passed through a potassium-titanyl-phosphate (KTP) crystal, the wavelength is halved to 532 nm and a brilliant green light is emitted. This laser is used in continuous-wave mode to cut tissue, in pulsed mode for treatment of vascular lesions. Delivery is through an insulated fibre, scanner or microscope for cw/pulsed mode, and articulating arm for Q-switched mode. Output power typically reaches 20 W. The KTP laser has significant advantages: it requires no fibre cooling and the energy can pass through a flexible fibre. KTP laser light is absorbed by tissue pigments such as haemoglobin and melanin. In orthopaedic surgery, the KTP laser is utilized to treat spinal disorders [65].

Pathology studies indicate that KTP LDD attenuates intradiscal inflammation and in addition has a thermoplastic annealing effect on the damaged disc wall [76, 77].

Knight conducted a prospective study on 34 patients with discal radial tears and annular leaks, confirmed with discography, who were treated by KTP laser disc decompression and posterior wall reconstruction and reviewed after 2 years. He found that KTP laser disc decompression may resolve persistent annular leaks [79].

Knight also performed KTP laser-assisted percutaneous cervical laser disc decompression in patients with soft cervical disc protrusions. He concluded that this technique procures promising results in patients with neck pain and prachialgia and can ablate inflammatory disc tissue and anneal the annulus [78].

After performing over 550 KTP laser spinal disc decompressions for treatment of disc protrusions and annular tears, Knight found satisfactory results, with a complication rate less than that reported for open discectomy, Initially, indications were confined to small contained disc protrusions but, due to promising results, the application was widened to larger disc protrusions and to leaking discs. The procedure was performed after MRI scanning though the posterolateral approach following provocative spinal probing and discography, using a 2-mm side-firing probe inserted under image-intensifier control and under neuroleptic anaesthesia. The aware state avoided compromise of neural structures. The tissue effect was applied to a 9-mm diameter core over a 25-mm length. Patient feedback identified that the correct structures contributing to individual elements of the symptom complete were being addressed (viviprudence). Knight concluded that KTP laser disc decompression is effective for treating contained lumbar disc prolapses causing sciatica in patients with wide-based disc protrusions and annular tears and leaks. This technique is not suitable for hourglass disc protrusions, disc segments that have received previous surgery or lateral recess stenosis [75].

## Neodym:YAG Laser in Orthopaedic Medicine

A true workhorse, the neodym:YAG emits a near-infrared invisible light at 1064 nm. The beam is produced by means of excitation of the neodymium-impregnated yttrium-aluminium-garnet (YAG) crystal. It may be delivered in continuous-wave mode through a fibre to a sapphire tip to cut tissue or, because of its deeper penetration, used to coagulate tissue directly. The Nd:YAG laser is relatively poorly absorbed by water, but is highly absorbed by protein, e.g. tissue chromophres and pigments, with a significant coagulation zone of 4–5 mm. Fibrocartilage has a low content of chromophres; the continuous-wave Nd:YAG laser induces extended thermal changes in the meniscus with a transitorial zone presenting coagulated and carbonized tissue [65, 104].

Concerning meniscus surgery, the continuous-wave Nd:YAG laser is not suited for tissue ablation and cutting; however, some therapeutic potential may be inherent using this laser for welding procedures at energy densities far below the ablation threshold [65, 104, 14].

To avoid extensive thermal injury contact, Nd:YAG lasers have been modified for meniscal surgery. A sapphire crystal attached to the tip of the light guide converts light energy to heat. The fibre then functions as a heater probe, which can possibly cut and ablate fibrocartilage with a significantly reduced extent of thermal damage [97].

Zweifel and other researchers extensively tested the neodym:YAG laser under laboratory conditions and demonstrated that, when using a neodym:YAG laser with an average energy of 20 to 50 W, the application time must be less than 5 s to avoid irreversible thermal damage within a distance of 5 mm from the laser tip. More than 50 W are not recommended. Breaks should be longer than the time for lasing. In contrast, the Ho:YAG laser with 50 W caused no thermal damage within 4 mm from the laser tip as long as there is a continuous flow of saline solution irrigation [111, 112, 149, 132, 38].

In the late 1970s Glick tested the 1.06-μm neodym:YAG laser in a free beam mode for meniscectomy, the results of which were rather disappointing [42].

Inoue et al. found the Nd:YAG lasers to be ineffective for arthroscopic surgery [70].

Brillhart was the first to use the 1.44-μm Nd:YAG laser in arthroscopic surgery in 1992. The tissue effects of the 1.44-μm Nd:YAG are comparable to that of the Ho:YAG laser and should not to be confused with the 1.06-μm Nd:YAG laser [18].

Due to the more appropriate properties of the Ho:YAG laser, the use of the Nd:YAG laser in arthroscopic joint surgery has diminished. Based on the results of several basic research investigations, it can be concluded that the Nd:YAG laser, which is deeply absorbed in the tissues, is not appropriate for arthroscopic surgery [36, 132, 38].

However, a number of orthopaedic surgeons utilize the Nd:YAG laser, especially for treatment of spinal disorders, and obtain good results. In fact, the Nd:YAG laser was the first laser to be used for intradiscal laser therapy in an operation performed by Ascher and Chy in 1986 [5, 25, 26].

Choy has extensively used the Nd:YAG laser for percutaneous laser disc decompression (PLDD) and successfully treated several hundred patients with lumbar disc herniations with a success rate of 75% and a complication rate of less than 0.5% [23, 24].

Hellinger also performs the Nd:YAG assisted non-endoscopic percutaneous laser disc decompression and nucleotomy to treat spinal disc herniation of the cervical thoracic and lumbar spine. In a prospective study he described the results after more than 10 years' experience. From 1989 to 1999 he treated more than 4900 patients (> 300 cervical, > 350 thoracic cases). Inclusion criteria were discogenic-vertebragenic vertebral pain syndrome (local, pseudoradicular, radicular, medullar and vegetative), positive radiographic correlation and postnucleotomy syndrome. Contra-indications were non-vertebragenic discal pain syndrome and free disc sequester. Subjective results of patients with lumbar syndrome were 80% positive, 86% for the cervical group and 90% for the thoracic group. Paresis syndrome was reduced in

90% of the patients, starting with the first postoperative day. The overall complication rate was less than 1% [56].

In order to assess the reason for the success of Nd:YAG percutaneous laser disc decompression, Hellinger measured pre- and postoperative CT density in 21 patients with lumbar radicular pain syndromes due to protruded or extruded discs. Preoperative density measurements averaged 83.43 HU (Houndsfield units). Postoperative values were significantly reduced to 66.33 HU ($p = 0.001$). Postoperative density was reduced by an average of 20% [55].

Diehl et al. also performed over 100 percutaneous Nd:YAG laser disc decompressions for treatment of monosegmental and multilevel cervical radiculopathy without paresis, and procured good to excellent results in 90% of the first follow-up examinations of 40 patients. In all cases, no infection or any other severe permanent complication occurred [29].

## Conclusion and Future Outlook

Each new medical technological advance or therapeutic measure is most promising when a specific treatment effect or an exclusive technical advantage is achieved, which has so far not been obtained by other means. Almost every efficacious therapy has side effects. The safety of a method depends predominantly on the compliance to the principles of a procedure and the adherence to indications and contra-indications, that should be limited within the stated goal.

In order to apply this principle to orthopaedic laser technology, the specific characteristics of the laser should be exposed: depending on the wavelength, the different absorption effects of laser energy can lead to an ablative or to a non-ablative tissue effect, which is mostly thermic. Induced by laser light, such a non-ablative action can be directed and modulated much more precisely than with ultrasound or other high-pressure procedures. The ablation and cutting function of a laser beam is regulated by the intensity and design of the fibre tip in the hand piece, which can be of such a small dimension that an exact three-dimensional cutting effect can be exploited by the possible alignment into small joint angles otherwise difficult to reach.

According to the above-described investigations and reports, the $CO_2$ laser is not appropriate for comprehensive use in arthroscopic joint surgery. The primary tissue interactions with the diode, excimer, erbium and holmium lasers are much more favourable.

For treatment if spinal disorders, e.g. disc herniation, the Ho:YAG laser, Nd:YAG laser and KTP laser procure satisfactory results when applied correctly within a strict range of indications.

The other lasers and other possible applications, such as laser-assisted tissue welding and laser-assisted photodynamic therapy of rheumatoid arthritis, need to be researched more thoroughly before their establishment in orthopaedic medicine.

As with any other orthopaedic surgical instrument, it is imperative to use laser systems correctly and with care. Sufficient education and training through experienced surgeons and appropriate education centres is absolutely mandatory. A number of well-established societies, such as the International Musculoskeletal Laser Society (IMLAS) offer such training courses (for more information contact imlas@imlas.org). Clinical use of laser systems through incompetent and unqualified surgeons is reprehensible and must be prevented. Only those orthopaedic surgeons with qualified training and experience will be able to exploit the beneficial effects of laser systems.

## References

[1] ABELOW SP (1993) Use of lasers in orthopedic surgery: current concepts. Orthopedics 16: 551–556
[2] ABELOW SP (1997) Laser capsulorrhaphy for multidirectional instability of the shoulder. Op Tech Sports Med 5: 244–248
[3] Al-AWAMI M (2000) Low level laser-treatment of primary and secondary Raynaud's pheomenon. In: Proc 7th Int Congr of the International Musculoskeletal Laser Society (IMLAS). Vienna 2000, Abstr 46
[4] ASAI T (2000) The future of laser therapy in orthopedic surgery: use of the erbium YAG laser for treating bone. In: Proc 7th Int Congr of the International Musculoskeletal Laser Society (IMLAS). Vienna 2000, Abstr 42
[5] ASCHER PW, HOLZER P, SUTTER B, TRITTHART H (1991) Nukleus-pulopsus-Denaturierung bei Bandscheibenprotrusionen. In SIEBERT WE, WIRTH CJ (eds) Laser in der Orthopädie. Thieme, Stuttgart, pp 169–172
[6] ASSHAUER T, RINK K, DELACRÉTAZ G (1994) Accoustic transient generation by holmium-laser-induced cavitation bubbles. J Appl Phys 76: 5007–5013
[7] ASSHAUER T, RINK K, DELACRÉTAZ G, SALATHÉ RP, GERBER B, FRENZ M, PRATISTO H (1994) Acoustic transient generation in pulsed holmium laser ablation underwater. SPIE 2134A: 423–433
[8] ATIK OS (2000) Histological and MRI alterations after partial meniscectomy using holmium:yag laser. In: Proc 7th Int Congr of the International Musculoskeletal Laser Society (IMLAS). Vienna 2000, Abstr 41
[9] ATIK OS, SENER E, BOLUKBASI S, CILA E, ALTUN N, SIMSEK A, BASKAN T (1994) Laser arthroscopy in knee injuries. J Neurol Orthop Med Surg 15: 26–27
[10] ATIK OS, SENER E, BOLUKBASI S, SIMSEK A, UZUMCUGIL O (1996) Arthroscopic chondroplasty using laser in knee, shoulder and ankle joints. J Arthroplast Arthroscop Surg 12: 14–15
[11] ATIK OS, SENER E, BOLUKBAS? S, CILA (1992) Arthroscopic laser surgery – Early results. J Arthropast Arthroscop Surg 5: 1–3
[12] BAKER CL, GRAHAM JM JR (1993) Current concepts in ankle arthroscopy. Orthopedics 16: 1027–1035
[13] BASKOV A, SHECHTER A, BASKOV V, SOBOL E, VOROBIEVA N, OMELCHENKO A, ZAKHARKINO O (2001) Study of regeneration processes in the intervertebral disc cartilage under non destructive laser radiation. In: Proc 8th Int Congr of the International Musculoskeletal Laser Society (IMLAS). Cartagena 2001, Abstr 13

[14] BICKERSTAFF DR, WYMAN A, LAING WR, SMITH TW (1991) Partial meniscectomy using the neodymium:YAG laser. An in vitro study. Arthroscopy 7: 63

[15] BIRNBAUM K, GUTKNECHT N, HELLER K-D (1996) Er:YAG laser irradiation on the boundary surface between polymethylmetacrylat (PMMA) and human femoral bone. A tried and tested method for cement extraction within hip joint replacement? In: Proc 3rd Int Congr of the International Musculoskeletal Laser Society (IMLAS). Kassel 1996, Abstr 10

[16] BLANCARD P, SORATO M, BLANLUET G, IRIS L, LIOTET S (1964) Therapeutic laser experiments. Bull Soc Ophtalmol Fr 64: 1009–1016

[17] BRILLHART AT (1991) Arthroscopic laser surgery. Am J Arthros 1: 5–12

[18] BRILLHART AT (1993) Ablation efficiency determination using the 1.44-micron neodymium:YAG laser. SPIE Proc 1880: 29–30

[19] BRILLHART AT (ed) (1994) Arthroscopic laser surgery. Clinical applications. Springer, Berlin Heidelberg New York

[20] BUCHELT M, PAPIOANNOU T, FISCHBEIN M, PETERS W, BEEDER C, GRUNDFEST W (1991) Excimer laser ablation of fibrocartilage: an in vitro and in vivo study. Lasers Surg Med 11: 271–279

[21] CASPER GD (2001) Laser foraminoscopic surgery. In: GERBER G, KNIGHT M, SIEBERT WE (eds) Lasers in the musculoskeletal system. Springer, Berlin Heidelberg New York, pp 317–319

[22] CASPER GD, HARTMAN VL, MULINS LL (1996) Foraminal laser endoscopic disc ablation for the treatment of lumbar disc disease: preliminary findings. In: Proc 1rd Eur advanced spinal surgery forum. Manchester 1996, Abstr 28

[23] CHOY DSJ (1995) Clinical experiences and results with 389 PLDD procedures with the Nd:YAG laser, 1986 to 1995. J Clin Laser Med Surg 13 (3): 209–213

[24] CHOY DSJ (1996) Clinical experiences with percutaneous laser disc decompression – 101/2year follow-up in 570 procedures in 419 patients. In: Proc 3rd Int Congr of the International Musculoskeletal Laser Society (IMLAS). Kassel 1996, Abstr 10

[25] CHOY DSJ, ASCHER PW, CASE RB, KAPLAN M, ERON L (1986) Percutaneous laser nucleolysis of lumbar disc. Nd:YAG laser in medicine and surgery. Fundamental and clinical aspects. In: OGURO Y, ATSUMI K, JOFFE S (eds) Professional postgraduate services. Tokyo, Japan, pp 363–369

[26] CHOY DSJ, CASE RB, FIELDING W (1987) Percutaneous laser nucleolysis of lumbar disc. N Engl J Med 317: 771–772

[27] DE SIMONI C, LEDERMANN T, IMHOFF AB (1996) Holmium-YAG laser in outlet impingement of the shoulder. Mid-term results. Orthopäde 25: 84–90

[28] DEW D, SUPIK L, DARROW C, PRICE G (1993) Tissue repair using lasers: a review. Orthopedics 16 (5): 581: 587

[29] DIEHL K, RUFFING A, WOLLNY A (1996) Experiences in 120 cases with percutaneous non endoscopic laser disc decompression (PLDD) using the Nd:YAG laser system in the cervical spine – retrospective study of 40 cases. In: Proc 3rd Int Congr of the International Musculoskeletal Laser Society (IMLAS). Kassel 1996, Abstr 37

[30] DILLINGHAM MF, FANTON GS (1990) The use of the holmium:YAG laser in operative knee arthroscopy – a double blind prospective study using a new arthroscopically guided laser system. Arthroscopy 6: 152–153

[31] DRESSEL M, JAHN R, NEU W, JUNGBLUTH KH (1991) Studies in fiber-guided excimer laser surgery for cutting and drilling bone and meniscus. Lasers Surg Med 11: 569

[32] EINSTEIN A (1917) Zur Quantentheorie der Strahlung. Physiol Z 18: 121–128

[33] FANTON G, DILLINGHAM M (1992) The use of the holmium laser in arthroscopic surgery. Semin Orthop 7: 102–116

[34] FREDRICH H (2000) Biomechanical and histomorphological investigation of LACS and ETACS. In: Proc 7th Int Congr of the International Musculoskeletal Laser Society (IMLAS). Vienna 2000, Abstr 40

[35] FREEDLAND Y (1999) Use of the excimer laser in fibrocartilaginoous excision from adjacent bone stroma: a preliminary investigation. J Foot Surg 27 (4): 303

[36] GERBER B (1996) Basic research of arthroscopic laser treatment of the locomotor system: the clinical context. In: Proc 3rd Int Congr of the International Musculoskeletal Laser Society (IMLAS). Kassel 1996, Abstr 1

[37] GERBER B (2000) What works in arthroscopic laser use and why. In: Proc 7th Int Congr of the International Musculoskeletal Laser Society (IMLAS). Vienna 2000, Abstr 27

[38] GERBER B, KNIGHT M, SIEBERT WE (eds) (2001) Lasers in the musculoskeletal system. Springer, Berlin Heidelberg New York

[39] GERBER B, ZIMMER M, ASSHAUER T, PREISS S, NORBERG M, DELACRÉTAZ G, PRATISTO H, FRENZ M (2001) In vitro and experimental animal research on arthroscopic laser treatment of cartilage – setup near to clinical application conditions. In: GERBER B, KNIGHT M, SIEBERT WE (eds) Lasers in the musculoskeletal system. Springer, Berlin Heidelberg New York, pp 42–60

[40] GERBER BE, ASSHAUER T, DELACRÉTAZ G, JANSEN T, OBERTHÜR T (1996) Biophysikalische Grundlagenuntersuchungen zur Wirkung der Holmium-Laserstrahlung am Knorpelgewebe und deren Konsequenzen für die klinische Applikationstechnik. Orthopäde 25: 21–29

[41] GEVARGEZ A (2000) CT-guided, percutaneous radiofrequency thermocoagulation of the cervical and lumbar zygapophyseal joint such as sacroiliac joint for non-radicular cervical pain and low back pain syndrome. In: Proc 7th Int Congr of the International Musculoskeletal Laser Society (IMLAS). Vienna 2000, Abstr 45

[42] GLICK J (1981) YAG laser meniscectomy. Proc Triannual Meet of the International Arthroscopy Association. American Arthroscopy Association of North America. Rio de Janeiro, 1981

[43] GLINKOWSKI W, MAZIK Z (1996) Reflectance changes of skin irradiated by IR LLLT. Multispectral analysis of image analysis study. In: Proc 3rd Int Congr of the International Musculoskeletal Laser Society (IMLAS). Kassel 1996, Abstr 66

[44] GLINKOWSKI W, OHOWIAK R, WASILEWSKI L (1996) Infrared low-level laser irradiation as a treatment of delayed union. In: Proc 3rd Int Congr of the International Musculoskeletal Laser Society (IMLAS). Kassel 1996, Abstr 66

[45] GLOSSOP N, JACKSON R, KOORT H, REED S, RANDLE J (1995) The excimer laser in orthopaedics. Clin Orthop 310: 72–81

[46] GOLDMAN L, ROCKWELL RJ JR (1968) Laser systems and their applications in medicine and biology. Adv Biomed Eng Med Phys 1: 317–382

[47] GOLDMAN L, SILER VE, BLANEY D (1967) Laser therapy of melanomas. Surg Gynecol Obstet 124: 49–56

[48] GRIFKA J (1993) Arthroskopische Therapie der Gonarthrose in Abhängigkeit vom Grad der Chondromalazie. Arthroskopie 6: 201–211

[49] GRÖNEMEYER DWH (2000) CT-guided, percutaneous laser disc decompression with Ceralas D, a diode laser with 980 nm wavelength and 200 µm fiber optics. In: Proc 7th Int Congr of the International Musculoskeletal Laser Society (IMLAS). Vienna 2000, Abstr 27

[50] HARDY P, THABIT G III, FANTON GS, BLIN JL, LORTAT JACOB A, Benoit J (1996) Arthroscopic management of recurrent anterior shoulder dislocation by combining a labrum suture with antero-inferior holmium:YAG laser capsular shrinkage. Orthopäde 25: 91–93

[51] HAYASHI K, NIECKARZ JA, THABIT G III, BOGDANSKE JJ, COOLEY AJ, MARKEL MD (1997) Effect on nonablative laser energy on the joint capsula: an in vivo rabbit study using a holmium:YAG laser. Lasers Surg Med 20: 164–171

[52] HAYASHI K, THABIT G III, VAILAS, AC, BOGDANSKE JJ, COOLEY AJ, MARKEL MD (1996) The effect on nonablative laser energy on joint capsular properties. An in vitro histologic and biochemical study using a rabbit mode. Am J Sports Med 24: 640–646

[53] HAYASHI K, THABIT G III, VAILAS, AC, BOGDANSKE JJ, COOLEY AJ, MARKEL MD (1996) The effect on nonablative laser energy on joint capsular properties. An in vitro histologic and biochemical study using a rabbit mode. Am J Sports Med 24: 640–646 ((DOPPELT!!))

[54] HAYASHI K, THABIT G III, BOGDANSKE JJ, MASCIO LN, MARKEL MD (1996) The effect of nonablative laser enegery on the ultrastructure of joint capsular collagen. Arthroscopy 12. 474–481

[55] HELLINGER J, LINKE R, HELLER H (2001) A biophysical explanation for Nd:YAG percutaneous laser disc decompression success. J Clin Laser Med Surg 19: 235–238

[56] HELLINGER J, STERN S (2000) Nonendoskopische PLDN-Nd-YAG 1064 mm ((muss das nicht nm heißen??)) – Eine 10-Jahres-Bilanz als Megastudie und Metaanalyse. In: Proc 86th German Orthopedic Society Meet (DGOT). Wiesbaden 2000, Abstr V-240

[57] HENDRICH C (2000) Experimental photodynamic laser therapy for rheumatoid arthritis using a second generation photosensitizer. In: Proc 7th Int Congr of the International Musculoskeletal Laser Society (IMLAS). Vienna 2000, Abstr 34

[58] HENDRICH C, HÜTTMANN G, SEARA J, SIEBERT W (1996) Photodynamic laser therapy for treatment of rheumatoid arthritis – an overview. In: Proc 3rd Int Congr of the International Musculoskeletal Laser Society (IMLAS). Kassel 1996, Abstr 23

[59] HENDRICH C, JAKOB PM, BREITLING T, BERDEN A, KRENN V, HAASE A, SIEBERT WE (1996) MRI measurement of the temperature distribution in cartilage following laser therapy. Orthopäde 25: 17–20

[60] HIBST R (1992) Mechanical effects of erbium:YAG laser bone ablation. Lasers Surg Med 12: 125

[61] HIJIKATA S, YAMIAGISHI M, NAKAYAMA T (1975) Percutaneous discectomy,: a new treatment method for lumbar disc herniation. Todan Hosp 5: 5–13

[62] HOTEYA K (2000) The effect of low reactive level laser therapy in the field of orthopaedics. In: Proc 7th Int Congr of the International Musculoskeletal Laser Society (IMLAS). Vienna 2000, Abstr 44

[63] IKUO K, OKAMOTO S, YONEZAWA T, ABE M (1997) Bone changes on a postoperative MRI after Ho:YAG laser-assisted knee arthroscopy. In: Proc 4th Int Congr of the International Musculoskeletal Laser Society (IMLAS). Kyoto 1997, Abstr OP-40

[64] IMHOFF A (1996) Laser-assisted arthroscopic subacromial decompression – clinical results. In: Proc 3rd Int Congr of the International Musculoskeletal Laser Society (IMLAS). Kassel 1996, Abstr 29

[65] IMHOFF A (2001) Basics, laser physics and safety for the clinician's requirements. In GERBER B, KNIGHT M, SIEWERT WE (eds) Lasers in the musculoskeletal system. Springer, Berlin Heidelberg New York, pp 11–17

[66] IMHOFF A, LEDERMANN T (1995) Arthroscopic aubacromial decompression with and without the holmium:YAG-laser. A prospective comparative study. Arthroscopy 11 (5): 549–556

[67] IMHOFF A, LEU H (1991) Arthroscopic operations with the excimer laser. Initial experiences. In: SIEBERT W, WIRTH C (eds) Laser in der Orthopädie. Thieme, Stuttgart, pp 48–53

[68] IMHOFF A, ROSCHER E, KÖNIG U (1997) Arthroskopische Schulterinstabilisierung. Differenzierte Behandlungsstrategie mit Suretac, Fastak, Holmium-YAG Laser und Elektrochirurgie. Orthopäde 8: 518–531

[69] IMHOFF A, SCHREIBER A (1988) Etiology and pathogenesis of synovitis villonodosa pigmentosa. Orthopäde 8: 518–531

[70] INOUE K (1984) Arthroscopic laser surgery. IAA, London, pp 29–30.9

[71] JANIS LR, KRAVITZ RD, WAGNER SS (1994) The pulsed holmium: yttrium-aluminum-garnet laser. Applications to ankle arthroscopy. Clin Podiatr Med Surg 11 (3): 483–498

[72] KAMBIN P (1991) Arthroscopic microdiscectomy, minimal intervention in spinal surgery. Urban & Schwarzenberg, München

[73] KAMBIN P, BRAGER MD (1987) Percutaneous posterolateral discectomy. Anatomy and mechanism. Clin Orthop 223: 145–154

[74] KAUTZKY M, SUSANI M, SCHENK P (1992) The holmium:YAG infrared laser and the UF excimer. Effects of lasers on the oral mucosa. Laryngorhinootologie 71: 347–352

[75] KNIGHT M, GOSWAMI A (2001) Lumbar percutaneous KTP 532-nm laser disc decompression and disc ablation in the management of discogenic pain. In: GERBER B, KNIGHT M, SIEBERT WE (eds) Lasers in the musculoskeletal system. Springer, Berlin Heidelberg, New York, pp 212–316

[76] KNIGHT MTN (2001) Past, present and future of minimalist invasive spine surgery. In: Proc 8th Int Congr of the International Musculoskeletal Laser Society (IMLAS). Cartagena 2001, Abstr 21

[77] KNIGHT MTN, VAJDA A, PANTOJA S, JAKAB G (1996) KTP 532 laser disc decompression – 4 years experience. In: Proc 3rd Int Congr of the International Musculoskeletal Laser Society (IMLAS). Kassel 1996, Abstr 51

[78] KNIGHT MTN (2000) Cervical percutaneous laser disc decompression: results of a prospective outcome study. In: Proc 7th Int Congr of the International Musculoskeletal Laser Society (IMLAS). Vienna 2000, Abstr 50

[79] KNIGHT MT (2000) Treatment of radial tears and annular leaks with posterior wall reconstruction. In: Proc 7th Int Congr of the International Musculoskeletal Laser Society (IMLAS). Vienna 2000, Abstr 8

[80] KOCHEVAR IE (1989) Cytotoxicity and mutagenicity of excimer laser radiation. Lasers Surg Med 9: 440

[81] KROITZSCH U, LAUFER G, EGKHER E, WOLLENEK G, HORVATH R (1989) Experimental photoablation of meniscus cartilage by excimer laser energy. Arch Orthop Trauma Surg 108: 44

[82] KUNZ M, JOHANN K (2001) Arthroscopic treatment of shoulder joint instability using laser assisted capsular shift (LACS). In: GERBER B, KNIGHT M, SIEWERT WE (eds) Lasers in the musculoskeletal system. Springer, Berlin Heidelberg New York, pp 177–180

[83] LANE G, SHERK H, MOOAR, P, LEE S, BLACK J (1992) Holmium:YAG laser versus carbon dioxide laser versus mechanical arthroscopic debridement. Semin Orthop 7 (2): 95–101

[84] LEE SH (2001) Anterior cervical decompression and fusion using carbon dioxide laser attached to surgical microscope in treatment of cervical ossification posterior longitudinal ligaments. In: Proc 8th Int Congr of the International Musculoskeletal Laser Society (IMLAS). Cartagena 2001, Abstr 37

[85] LEE SH (2001) Usefulness of $CO_2$ laser in lumbar discectomy. In: Proc 8th Int Congr of the International Musculoskeletal Laser Society (IMLAS). Cartagena 2001, Abstr 45

[86] LEE SH, GASTAMBIDE D (2001) Perkutane endoskopische Diskotomie der Halswirbelsäule. In PFEIL J, SIEBERT W, JANOUSEK A, JOSTEN C (eds) Minimal-invasive Verfahren in der Orthopädie und Traumatologie. Springer, Berlin Heidelberg New York, pp 41–61

[87] LEVY O, WILSON M, WILLIAMS H, BRUGERA JA, DODENHOFF R, SFORZA G, COPELAND S (2001) Thermal capsular shrinkage for shoulder instability. J Bone Joint Surg [Br]: 640–645

[88] LÖHNERT J, RAUNEST J (1994) Operationstechnik in der arthroskopischen Knorpelablation mit dem 308 nm Xenon-Chlorid-Excimer-Laser. Arthroskopie 7: 170–173

[89] LÜBBERS C, SIEBERT WE (1996) Die arthroskopische Holmium:YAG-Laseranwendung im Vergleich zu konventionellen Verfahren am Knie. Zweijahresergebnisse einer prospektiven Studie. Orthopäde 25: 64–72

[90] LÜBBERS C, SIEBERT WE (2001) Holmium:YAG laser assisted arthroscopy versus conventional methods or treatment of the knee. In: GERBER B, KNIGHT M, SIEBERT WE (eds) Lasers in the musculoskeletal system. Springer, Berlin Heidelberg New York, PP 88–96

[91] MAIMAN TH (1960) Stimulated optical radiation in ruby. Nature 6: 493–494

[92] MANNMEISS DD, GUYER RD, HOCHSCHULER STH (1994) Laser disc decompression. The importance of patient selection. Spine 19: 2054–2059

[93] MARKEL MD, HAYASHI K, THABIT G 3rd (2001) Basic properties of collagen shrinkage and laser-collagen interactions. In: PFEIL J, SIEBERT WE, JANOUSEK A, JOSTEN C (eds) Minimal-invasive Verfahren in der Orthopädie und Traumatologie. Springer, Berlin Heidelberg New York, pp 162–169

[94] MARKEL MD, HAYASHI K, THABIT G 3rd, THIELKE RJ (1996) Veränderungen am Gelenkkapselgewerbe durch Holmium:YAG-Laserexposition in nichtablativen Energiedichten. Eine potentielle Anwendungsmöglichkeit zu Stabilisierungseingriffen. Orthopäde 25: 37–41

[95] MATHEWS HH (1996) Transforaminal endoscopic microdiscectomy. Neurosurg Clin North Am 7: 59: 63

[96] MATHEWS HH, FIROE M, MOLLIGAN H, LONG B (1996) Foraminoscopic approach to lumbar disc sequestrum: a surgical technique. In: Proc 9th Meet of the North American Spine Society, Minneapolis 1996, Abstr 65

[97] MILLER DV, O'BRIEN SJ, ARNOCZKY SS, KELLY A, FEALY SV, WARREN RF (1989) The use of the contact Nd:YAG laser in arthroscopic surgery: effects on articular cartilage and meniscal tissue. Arthroscopy 5: 245

[98] MORIMOTO Y (2000) Laser-assisted meniscus fusion. In: Proc 7th Int Congr of the International Musculoskeletal Laser Society (IMLAS). Vienna 2000, Abstr 38

[99] NISHIJIMA Y, ISHIZIKA H, AIKAWA H, TSUCHISHIMA ((Xinitial fehltX)), Toda N (2001) Endoscopic percutaneous lumbar disc decompression with the Ho:YAG laser (Ho:YAG EPLDD). In: GERBER B, Knight MTN, SIEBERT W (eds) Lasers in the musculoskeletal system. Springer, Berlin Heidelberg New York, PP 330–336

[100] O'BRIEN SJ, GARNIC JG, JACKSON RW, SHERK HH, SMITH CF, WANGNESS CT (1991) Lasers in orthopedic surgery. Contemp Orthop 9: 61–69

[101] PHILANDIRANOS G (1985) Le laser à gaz carbonique en chirurgie arthroscopique du genou. Presse Med 143: 2103

[102] PISOT V, MARCAN R (2000) Five years experience with LACS in shoulder instability with holmium laser. In: Proc 7th Int Congr of the International Musculoskeletal Laser Society (IMLAS). Vienna 2000, Abstr 18

[103] RAUNEST J (1995) Laseranwendung in der Gelenkchirurgie – Experimentelle Ergebnisse zum arthroskopischen Einsatz thermischer und gepulster UV-Laser. Hefte zu Der Unfallchirurg, Heft 247, Springer, Berlin Heidelberg New York

[104] RAUNEST J (2001) Laser surgery on the meniscii. In: GERBER B, KNIGHT M, SIEBERT WE (eds) Lasers in the musculoskeletal system. Springer, Berin Heidelberg New York, pp 140–153

[105] RAUNEST J, DERRA E (1995) Morphologische, biomechanische und experimentelle In-vivo-Untersuchungen zur laserassistierten Meniskusresektion. Langenbecks Arch Chir 380: 12

[106] RAUNEST J, LÖHNERT J (1989) Arthroskopische Synovektomie unter Anwendung des Neodym-YAG-Laser. Chirurg 60: 782–787

[107] REED S, JACKSON R, GLOSSOP N, RANDLE J (1994) An in vivo study of the effect of excimer laser irradiation on degenerate rabbit articular cartilage. Arthroscopy 10: 78–84

[108] SAUNIER J, INDERMÜHLE F, COMPÈRE J (1993) Use of the holmium 2.1 laser in surgical arthroscopy. Rev Med Suisse Romande 113: 129–132

[109] SAUNIER JA, KINDYNIS P, JUILLERAT E, INDERMÜHLE F (2001) Mri modifications after ho 2.1 laser knee arthroscopy. In: GERBER B, KNIGHT M, SIEBERT WE (eds) Lasers in the musculoskeletal system. Springer, Berlin Heidelberg New York, pp 133–139

[110] SCHLANGMANN B (1994) In vitro Untersuchungen zur Wirkung eines Holmium:YAG Lasers an Bandscheibengewebe. Dissertation, Medizinische Hochschule Hannover, Hannover

[111] SCHLANGMANN B, BERENSEN BT, SIEBERT WE (1996) Temperatur-Ablationsmessungen bei der Laserbehandlung von Bandscheibengewebe. Orthopäde 25: 1–3

[112] SCHMOLKE S (1994) Experimentelle Untersuchungen zum Einsatz von Neodym:YAG Lasern bei der perkutanen Laser Diskus Dekompression. Dissertation, Medizinische Hochschule Hannover, Hannover

[113] SCHMOLKE S (2000) Laser-assisted capsular shrinkage (LACS) in addition to multiple suture repair in the arthroscopic therapy of shoulder instability. In: Proc 7th Int Congr of the International Musculoskeletal Laser Society (IMLAS). Vienna 2000, Abstr 23

[114] SCHREINER CH (1994) Laser in der Arthroskopie. Arthroskopie 7: 148–153

[115] SEGURA J, RUGELES JG, HERRERA M, IZQUIERDO M, RAMIREZ JF (2001) Holmium laser thermodiscoplasty effect in intervertebral disc. A protocol presentation. In: Proc 8th Int Congr of the International Musculoskeletal Laser Society (IMLAS). Cartagena 2001, Abstr 14

[116] SHERK H (1993) The use of lasers in orthopedic procedures – current concepts review. J Bone Joint Surg 75 [Am]: 768–776

[117] SHERK HH (1991) Orthopedist using lasers in surgery. Am J Arthrosc 9: 7–8

[118] SHERK HH, BLACK JD, PRODOEHL JA, DIVEN J (1995) The effect of lasers and electrosurgical devices on human meniscla tissue. Clin Orthop 310: 14

[119] SHERK HH, MAES K (1996) Osteonecrosis of the knee following laser-assisted arthroscopic surgery. In: Proc 3rd Int Congr of the International Musculoskeletal Laser Society (IMLAS). Kassel 1996, Abstr 17

[120] SHI W, VARI S, VAN DER VEEN M, FISCHBEIN M, GRUNDFEST W (1993) Effect of varying laser parameters on pulsed Ho:YAG ablation of bevine knee joint tissues. Arthroscopy 9: 96–102

[121] SIEBERT W (1993) Percutaneous laser disc decompression (PLDD): the European Experience. Spine – State Art Rev 7: 103–133

[122] SIEBERT W (1996) Laser-Osteotomie mit experimentellen Lasersystemen. Dr Kovac, Hamburg

[123] SIEBERT W (1999) Endoscopic spine surgery. Min Invas Ther Allied Technol 8: 303–308

[124] SIEBERT W (2000) Diagnostic and therapeutic mini-endoscopy in orthopedics and traumatology. In: Proc MICRO.tec 2000, Applications, Trends, Visions. VDE, Berlin, pp 303–308

[125] SIEBERT W, BERENDSEN BT, TOLLGAARD J (1996) Die perkutane Laser Diskus Dekompression (PLDD). Erfahrungen seit 1989. Orthopäde 25: 42–48

[126] SIEBERT W, KAISER J (1999) Die endoskopische Operation des Bandscheibenvorfalles. Arthroskopie 12: 74–78

[127] SIEBERT WE (1992) Laseranwendung in der Arthroskopie. Orthopäde 21: 273–288

[128] SIEBERT WE (1994) Overview of arthroscopic laser surgery in Europe. In: BRILLHANT AT (ed) Arthroscopic laser surgery. Springer, Berlin Heidelberg New York, pp 175–178

[129] SIEBERT WE (1996) Eondoscopic foraminoscopic surgery with laser supplementation. In: Proc 1st Europ advanced spinal surgery forum. Manchester 1996, Abstr 26

[130] SIEBERT WE, SAUNIER J, GERBER B, LÜBBERS C (1994) Ho:YAG-Laser in der arthroskopischen Chirurgie des Kniegelenkes. Arthroskopie 7: 182–192

[131] SIEBERT WE, STANKE M (2001) Bone healing after erbium:YAG laser osteotomy of sheep tibia. In: GERBER B, KNIGHT M, SIEBERT WE (eds) Lasers in the musculoskeletal system. Springer, Berlin Heidelberg New York, pp 75–80

[132] SIEBERT WE, WIRTH CJ (eds) (1991) Laser in der Orthopädie. Thieme, Stuttgart

[133] SISTO D, BLAZINA M, HIRSH L (1993) The synovial response after $CO_2$ laser arthroscopy of the knee. Arthroscopy 9: 574–575

[134] SMITH CF, JOHANSEN EL, VANGSNESS CT, Marshall GT, SUTTER LV, BONAVOLET T (1992) Gas bubble technique in arthroscopic surgery. Semin Orthop 7: 86

[135] SMITH CF, JOHANSEN EL, VANGSNESS C, SUTTER L, MARSHALL GT (1989) The carbon dioxide laser. A potential tool for orthopedic surgery. Clin Orthop 242: 43–59

[136] SOMMERFELD F, SIEWERT W (2001) Minimal-invasive Eingriffe am Bewegungsapparat. Lasereinsatz in der Schulterchirurgie. In: PFEIL J, SIEBERT W, JANOUSEK A, JOSTEN C (eds) Minimal-invasive Verfahren in der Orthopädie und Traumatologie. Springer, Berlin Heidelberg New York, pp 88–92

[137] STEIN E, SEDLACEK T, FABIAN R, NISHIOKA N (1990) Acute and chronic effects of bone ablation with a pulsed holmium laser. Lasers Surg Med 10: 384–388

[138] STÜCKER R (2001) Percutaneous transforaminal laser-assisted surgery in the epidural space for noncontained hernaited discs. In: GERBER B, KNIGHT M, SIEBERT WE (eds) Lasers in the musculoskeletal system. Springer, Berlin Heidelberg New York, pp 337–339

[139] STÜCKER R, KRUG, CH, REICHELT A (1997) Endoskopische Behandlung Sequestrierter Bandscheibenvorfälle. Der perkutane trasforaminale Zugang zum Epiduralraum. Orthopäde 26: 280–287

[140] TATAY JR (2000) Acute and chronic effects induced by the diode laser in intervertebral discs of rabbits; histological, immunohistochemical and ultrastructural study. In: Proc 7th Int Congr of the International Musculoskeletal Laser Society (IMLAS). Vienna 2000, Abstr 28

[141] TATAY JR, TATAY A, GARCIA PM, TATAY JR (2001) Histological study of the effects induced by the diode laser in intervertebral disc of rabbits. In: Proc 8th Int Congr of the International Musculoskeletal Laser Society (IMLAS). Cartagena 2001, Abstr 17

[142] TIBONE JE, SHRADER TA (1998) Glenohumeral joint translation after arthroscopic, nonablative, thermal capsulolasty with a laser. Am J Sports Med 26:495–498

[143] TRAUNER K, NISHIOKA N, PATEL D (1990) Pulsed holmium:yttrium-aluminium-garnet (Ho:YAG) laser ablation of fibrocartilage and articular cartilage. Am J Sports Med 18: 316–320

[144] VANGSNESS C, SMITH C, MARSHALL G, SWEENY J, JOHANSEN E (1995) The biological effect of carbon dioxide laser surgery on rabbit articular cartilage. Clin Orthop 310: 48–51

[145] VANGSNESS CT, WATSON T, SAADATMANESH V, MORAN K (1995) Pulsed Ho:YAG laser meniscectomy: effect of pulsewidth on tissue penetration rate and lateral thermal damage. Lasers Surg Med 16: 61

[146] WHIPPLE T, CASPARI R, MEYERS J (1984) Synovial response to laser-induced carbon ash residue. Lasers Surg Med 3: 291–295

[147] YONEZAWA T (2000) The possibility of erb:YAG laser device and transmission system for PLDD. In: Proc 7th Int Congr of the International Musculoskeletal Laser Society (IMLAS). Vienna 2000, Abstr 32

[148] ZANGGER P, GERBER BE (1996) Laseranwendung in der Arthroskopie des oberen Sprunggelenks. Indikationen, Technik, erste Ergebnisse. Orthopäde 25: 73–78

[149] ZWEIFEL K, PANOUSSOPOULOS A (1996) Laser und Bandscheibenchirurgie. In: BERLIN HP, MÜLLER G (eds) Angewandte Lasermedizin. Lehr- und Handbuch für Praxis und Klinik. ecomed, Landsberg, pp III-3.11.3

# III-10
# Laser Applications in Pulmology and Thoracic Surgery

T. Okunaka and H. Kato

## Contents

## Introduction

Since the late 1970s, there has been a dramatic expansion in the role of the bronchoscope as a therapeutic instrument. A major factor in developing its therapeutic role has been the introduction of high-power lasers which can be transmitted through optical fib, or articulated optical systems, for the palliative resection of inoperable tumours, recurrence or cases in which poor cardiopulmonary function precludes invasive surgery. The purpose of most procedures is excision or vaporization of lesions occupying or obstructing the airway lumen and, by so doing, re establish the patency of the airway. There are primarily two methods at present to perform bronchoscopic laser surgery: (1) debulking of tissue by high-energy lasers, such as $CO_2$ gas, Nd-YAG (neodymium-yttrium, aluminium, garnet) lasers, (2) causing tumour necrosis by photochemical reaction (photodynamic therapy: PDT; the activation of drugs by light to produce a cytotoxic effect) using low-power lasers, such as argon-dye, excimer dye and diode lasers.

With recent developments in lasers, it is now possible to debulk such tumours, leading to significant improvement in symptoms and promoting the period and quality of survival. Candidates for endoscopic treatment include the following: (1) tuberculous or post-trauma cicatricial stenosis, (2) granulomatous change following tracheoplasty or bronchoplasty, (3) benign and low-malignancy tumours of the trachea and bronchus, (4) all primary, recurrent and metastatic tumours causing stenosis or obstruction of the trachea or bronchus in postoperative or inoperable cases and (5) inoperable central-type early-stage squamous cell carcinoma.

The application of lasers in thoracic surgery has also been explored in the past 10 years. Some thoracic surgeons have developed laser thoracoscopy for the treatment of spontaneous pneumothorax, diffuse bullous emphysema and malignant pleural effusion.

A wide range of minimally invasive therapies using lasers are thus available through the bronchoscope and thoracoscope. In this chapter, a comprehensive guide to minimally invasive laser therapy in pulmology and thoracic surgery will be discussed.

# Bronchoscopic Thermal Resection with High-Power Lasers

Most interventional bronchoscopists have found endobronchial effects of other endoscopic techniques, such as electrocauterization [1] or cryosurgery [2]. In the late 1979s, however, bronchoscopic laser resection occupied an important place in the treatment of patients with thoracic neoplasia [3]. Used alone or in association with other therapeutic methods, laser photoablation of intraluminal tumours in patients with severe main airway obstruction provides symptomatic improvement, may restore ventilatory function and probably improves the quality of life. Within a few years, there were several reports of large series of patients with central tracheobronchial obstruction who were treated successfully by laser photoablation, and palliation of airway disease has been accomplished with little morbidity [4–7].

Because technological advances have made many laser procedures possible via flexible quartz optic fib, many bronchoscopic interventions today are performed with both rigid and flexible bronchoscopes. The relatively recent comeback of therapeutic rigid bronchoscopy during the past decade relates to several factors: improved telescope optics resulting in improved visualization, new video technology, refinements in rigid instrumentation and a growing recognition of the usefulness of therapeutic bronchoscopy in managing patients with malignant or benign obstructive lesions of the tracheobronchial tree.

## Lasers in Therapeutic Bronchoscopy

Three laser systems are used principally for bronchoscopic tissue resection in thoracic neoplasia. The first report of a laser beam employed in combination with bronchoscopy was that of Strong et al. in 1974, who successfully vaporized papilloma of the airway and treated bronchial stenosis by means of a $CO_2$ laser via rigid bronchoscope under general anaesthesia [8]. The $CO_2$ system has a wavelength of 10 600 nm and delivers laser energy in the visible, far infrared region. Light is converted immediatelly to thermal energy, raising tissue temperatures to more than 100 °C, which causes vaporization of the water content of the target tissue. This laser has been used successfully for resection of tracheal papillomatosis and removal of granulation tissue from the larynx and trachea for patients with postintubation subglottic tracheal stenosis. The $CO_2$ laser system has two important drawbacks that impede ore extensive bronchoscopic applications in the lower airways. The first is its biophysical properties. Although ideal for precise mucosal

vaporization, the $CO_2$ laser has limited haemostasis capability. Its depth of penetration is less than 1 mm, making it an excellent scalpel but an extremely ineffective coagulator. The second disadvantage of the $CO_2$ laser is that the laser energy cannot be transmitted through a flexible fiberoptic delivery system. Despite some advances in the search for a flexible fiberoptic delivery and wavelength, the $CO_2$ laser is not suitable in the case of bronchoscopic ablation of potentially haemorrhagic neoplasms obstructing the tracheobroncheal tree [9].

The argon laser delivers a visible, blue-green light with a wavelength of approximately 514 nm. Energy can be delivered through flexible quartz fib, facilitating endoscopic administration. The argon laser has been used in bronchoscopic tissue ablation; however, because the wavelength is absorbed strongly by haemoglobin, and secondarily due to unpredictable effects on soft tissues and delayed tissue healing, this is not an ideal method to be used widely.

The first clinical application of an Nd-YAG laser beam with a fiberoptic endoscope was made by Toty et al., who in 1979 discovered the bronchial tumour [3]. The Nd-YAG laser is still the most popular laser system used for bronchoscopic surgery. This system has a wavelength of 1064 nm and delivers radiation in the invisible, near-infrared lesion. The Nd-YAG laser is excellent for deep coagulation of bulky tissue because scattering of laser radiation exceeds absorption, especially in lightly pigmented tissues. This wavelength is readily absorbed by haemoglobin as well as water, and laser energy penetrates several millimetres into target tissue. Haemostasis is achieved by direct photocoagulation of blood vessels during laser therapy and because vaporization of water contained in tissues surrounding blood vessels ultimately results in vascular constriction. At a high-power density, tissue vaporization is also accomplished. Thus, thermal energy delivered by the Nd-YAG laser system provides superficial and deep haemostasis, as well as tissue necrosis.

## Procedure Technique

The Nd-YAG laser was introduced by either the rigid or the flexible bronchoscope. Bronchoscope selection depends on multiple factors, including the availability of general anaesthesia, location of lesion and, most importantly, technical expertise and experience of the bronchoscopist [10].

In selected instances, laser resection can be performed safely through the flexible fiberoptic bronchoscope [11]. One reason for advocating its use is facile performance under local anaesthesia, thereby sparing patients the cost and potential dangers of general anaesthesia. Many practitioners today are also more adept at flexible fiberoptic bronchoscopy than at rigid bronchoscopy. Another advantage of the flexible bronchoscopy is that it can be oriented into a smaller bronchi, should there be an indication for laser debulking within an upper-lobe or lower-lobe segmental bronchus [12].The many potential problems associated with flexible bronchoscopic laser resection probably outweigh its benefits. Multiple sessions are usually necessary for complete tissue ablation. Smoke evacuation its cumbersome because simultaneous laser resection and suctioning are not possible. Most importantly, complications are more difficult to control than through the rigid tube. For example, palpation of endobronchial lesions is difficult, and without the tactile feedback provided by a rigid suction catheter or by the tip of the rigid bronchoscope, it is difficult or the operator to know when to cease laser firing.

There are many reasons, however, for preferring the rigid bronchoscope for laser resection [13]. Incubation with the rigid tube assures maintenance of a clear airway. Tumour ablation is performed under general anaesthesia while simultaneous suction of smoke is achieved using suction catheters. Large forceps are used to remove tissue debris, and the tip of the rigid tube may be used to remove tissue debris, or to shear off tumour from the airway wall. Bleeding is controlled readily by saline lavage, local administration of epinephrine, tamponade of bleeding area with the rigid bronchoscope and laser photocoagulation. Another advantage of the rigid tube is that concomitant therapeutic procedures such as stent insertion or balloon dilation can be performed at the same time.

Most bronchoscopists, therefore, prefer to use the rigid tube for laser resection. The flexible bronchoscope is almost always used in conjunction with the rigid tube for visualization of the distal and peripheral airways and for bronchial cleansing; thus, proficiency in both techniques is essential. Procedures are usually performed in the operating room under general anaesthesia, provided with jet ventilation-assisted spontaneous ventilation using an open system.

After incubation with the rigid bronchoscope and administration of additional topical anaesthesia to the airway, tracheobronchial inspection is performed. Care is taken to clear both the right and left endobronchial trees of secretions before commencing laser resection. The most suitable power setting for the Nd-YAG laser is 50–70 W, since this will give good haemostasis and vaporization. It is best to set the maximum exposure at 2 s, i.e. the laser will cut out if the foot switch is depressed for longer than this. In practice, it is difficult to fire for a long time period without straying from the target because of respiratory movements. The laser cannot seal vessels greater than 1 mm in diameter. There is always a risk that the tumour may be eroding into a major vessel and that life-threatening haemorrhaging may occur. During coagulation, the tumour shrinks and often branches. Palpitation with the suction catheter allows the bronchoscopist to feel the effect of the laser on the tumour tissue. Care is taken to avoid laser firing perpendicular to the tracheobronchial wall. Tissue destruction is achieved by charring and vaporization of tumour tissue or by resecting necrotic tumour catheters. Careful attention is given to haemostasis throughout resection and vaporization. Figure 1 shows a case of recur-

**Fig. 1.** A case of recurrent squamous cell carcinoma in both main bronchi. The tumour appears to have developed from the right main bronchus to beyond the bronchial wall and was thought to extend to the left main bronchus. Nd-YAG laser ablation was performed with rigid bronchoscopy under general anaesthesia; subsequently, Dumon's Y stent was inserted into the carina

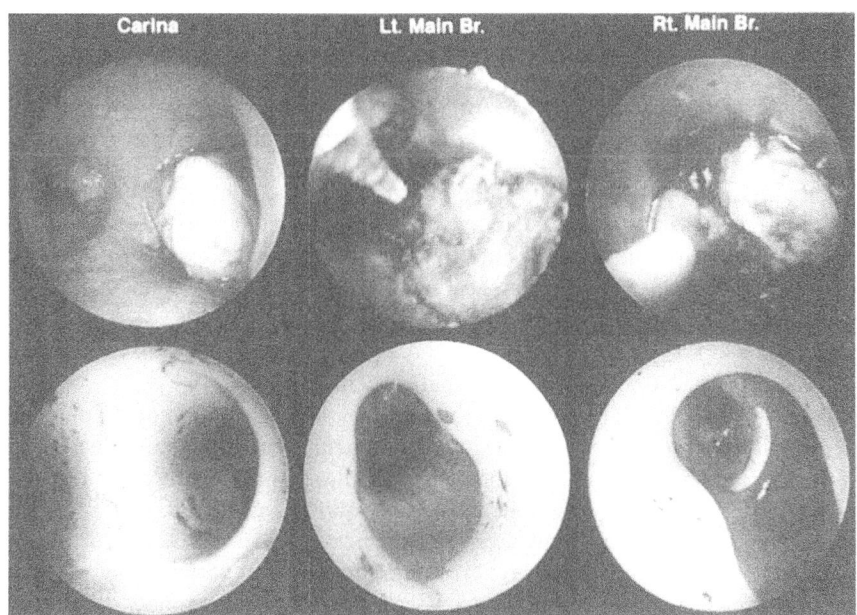

rent squamous cell carcinoma in both main bronchi. The tumour appears to have developed from the right main bronchus to beyond the bronchial wall and was thought to extend to the left main bronchus. The patient suffered from severe dyspnea and fever accompanied by obstructive pneumonia. Nd-YAG laser ablation was performed with rigid bronchoscopy under general anaesthesia; subsequently, Dumon's Y stent was inserted to carina to keep the bronchial lumen opened.

### Indication of Nd-YAG Laser Resection

Endoscopic Nd-YAG laser treatment can be indicated only in those cases in which the respiratory tract distal to the obstruction or stenosis is physiologically viable from the point of view of ventilation and circulation. Indeed, laser resection is a mode of palliative, not curative, therapy and usually is not indicated until other conventional methods have failed [14, 15]. However, immediate debulking or stenting should be considered in symptomatic patients with severe malignant airway obstruction and in selected patients with high-grade airway obstructions before referral for external beam radiation or chemotherapy, if warranted by the patient's clinical status. On the other hand, in cases in which two or more cartilage arches have been destroyed by a tumour, even if this procedure widens the airway, it will collapse, therefore such cases are considered contra indications.

Cases of lung cancer in which bronchoscopic Nd-YAG laser is indicated are naturally primarily limited to lesions in large airways with no peripheral lesions and also in cases having little tendency toward submucosal invasion or intramural lymphatic invasion, in which axial length is less than 4 cm and the bronchial lumen is fully visible [16, 17]. From this point of view, cases of squamous cell carcinoma in the trachea or main stream bronchus, adenoid cystic carcinoma, mucoepidermoid carcinoma or carcinoid originating in the main bronchi can be indications. It is also indicated as a palliative treatment to widen the airway and improve the general condition in cases of lung cancer located from the trachea to the lober bronchi in which the peripheral respiratory tract is functional.

Indications of Nd-YAG laser treatment in benign diseases include benign tumours or cicatricial stenosis located from the trachea to the segmental bronchi. The most favourable indication is the benign tracheal tumour, and this method is now considered quite valuable to replace thoracotomy as a single radically therapeutic procedure. In the case of cicatricial stenosis, only scars with a base in the bronchia wall of less than 1 cm in length are indications for the procedure

[11]. The procedure can also not be indicated in cases which do not show complete cicatrization and have some remaining inflammatory aspects, especially active tuberculous granuloma or cases positive for tuberculosis bacilli, because they will exhibit remarkable granular proliferation following treatment.

Bronchoscopic Nd-YAG laser treatment can be indicated in cases of bleeding. Haemostatic effects can be obtained throughout a fairly wide area using a power output of 40–50 W. Treatment of haemostasis can be performed from the trachea to the segmental bronchi, regardless of whether the case be malignant or benign. However, cases of widespread bleeding, as in necrotizing bronchitis, are not indications.

### Comparison with Other Methods

Endobronchial tumours may be controlled by a variety of different methods, used either singly or as a combination. Besides laser resection, the commonly reported methods are biopsy recanalization, electrocoagulation, cryosurgery, photodynamic therapy, endobronchial prosthesis, external beam radiotherapy and endobronchial brachytherapy for cases with both a high and low dosage rate [18–20].

Brachytherapy, the most commonly used alternative method, which uses high-energy isotopes within the lumen of hollow tumours, was first explored in the 1950s; however this was not practicable because they were too bulky for easy placement in the airway and required long exposures. Modern equipment using iridium192 in high dosage rates after loading machines now permits treatment in a few minutes with the fiberscope and local anaesthesia. Unlike lasers and other resection techniques, brachytherapy can help to control anaesthesia. Unlike lasers and other resection techniques, brachytherapy can help to control tumours which extrinsically compress the airways [20]. Unfortunately, the response to brachytherapy is delayed, but life-threatening airway obstruction can be palliated with the insertion of an stent. A variety of stents are now available and some can be inserted with the use of a fiberoptic bronchoscope alone.

### Outcome

Evaluation of the outcome of Nd-YAG laser treatment is difficult. Endoscopic appearance and symptoms have more commonly been used as criteria, having been more commonly used as criteria for response; however, these are more difficult to quantify or to compare among different studies, because several studies simply describe the early experience of a particular group and often involve relatively poor results as they describe the learning curve of the authors. Furthermore, some reports have combined their results for the treatment of malignant and non-malignant cases. Hetzel arranged the symptomatic response to treatment in the 21 studies [21]. The report represented 4247 cases, of which approximately 3000 appeared to have been treated for malignant tumours. Symptomatic response rates for improvement in dyspnea ranged from 63 to 95%, but the average figure is about 75%. In those studies which subdivide results for partial and complete airway obstruction, results for complete obstruction are generally poorer and range more widely from 32 to 88%, the average result being about 50%. This reflects the greater technical difficulties in finding distal bronchi beyond an totally collapsed airway and the problem that the lung, which has been collapsed for some time, may be irrevocably damaged.

Although laser resection often improves symptoms, ventilatory function and radiographic appearance, the effect on patient survival has not been well documented [22, 23]. In part, this lack of documentation results from the ethical dilemma of conducting prospective randomized studies in terminally ill patients who have compromised airways. The study design is difficult because survival is affected by tumour response to radiation or chemotherapy. Survival may also depend on whether patients are referred for laser resection at an early or late stage in the course of their disease. Successful laser resection and stenting depends on the extent of the disease, the importance of underlying neoplasms and the extent of extrinsic compression, along with the duration of illness, residual lung function and the technical expertise and experience of the bronchoscopist.

## Complications

Many of the causes of complications are also potential causes of death. The most common complication is haemorrhaging. Pneumonia predominantly complicates re-expansion of collapsed lobes. Smoke inhalation occasionally causes acute pulmonary oedema, especially for the patients who had received pneumonectomy. Hetzel reported in a retrospective study of 200 cases under local anaesthesia and 20% under general anaesthesia. The mortality rates were 14.2 and 3,5%, respectively. For individual treatments, the complication rate were 14.2 and 9% and mortality rates 6 and 1.5%. respectively, for local and general anaesthesia. Thus, these results emphasize the increased safety of the rigid bronchoscope [21, 24].

## Nd-YAG Versus PDT Ablation for Advanced Lung Cancer

Nd-YAG versus PDT ablation for partially obstructive lung cancer have been compared retrospectively in our institution [25]. Cases in which the tumour size or degree of bronchial obstruction decreased by over 50% were classified as effective in terms of tumour response rate. Ineffective cases were those in which tumour or degree of obstruction was reduced by less than 50; 258 lesions were treated, 81 cases by PDT and 177 by Nd-YAG laser. All were evaluated 1 month later. PDT achieved effective results in 61 out of 81 lesions (75%). In the Nd-YAG laser group, 143 out of 177 (81%) showed effective results. In this study, tumour location was also examined to evaluate effective tumour response. When the tumour was located in the trachea or main bronchi, effective results were obtained in 73% (19/26) of cases treated by PDT and 92% (64/69) treated by Nd-YAG laser. However, in cases in which the tumour was found in lober or segmental bronchi, the tumour response was effective in 70% (42/55) of PDT patients and 73% (79/108) of Nd-YAG laser treatment patients. No fatal complications occurred in any of the PDT-treated cases. However, in PDT-treated cases 91% (71/81) developed skin photosensitivity; none of the cases required treatment. Severe complications, including massive bleeding in 10 cases (6%) and bronchial perforation in 4 cases (3%), occurred in Nd-YAG-treated cases. As a result of these complications, 3 patients (1.3%) expired. Toilet bronchoscopy following laser therapy was different in the PDT-treated group and the Nd-YAG treated group. Clean-up after PDT involved the simple removal of gelatinous or fibrinous plugs form the bronchus in large pieces. Nd-YAG therapy, however, usually required time-consuming piecemeal forceps removal of charred and coagulated tissue resulting from the treatment. The therapies also differed in that the fib used in PDT could be inserted safely into the tumour for treatment without perforation or harm to adjacent vessels. With the Nd-YAG laser, however, bronchial wall perforation or massive haemorrhaging due to blood vessel injury sometimes occurred. Our study shows that in the bronchial wall, tumour necrosis is not induced as effectively by Nd-YAG laser treatment as it is by PDT. This is the case especially in the smaller bronchi due to the lesser margin for error with the Nd-YAG laser before entering the vessel. Nd-YAG treatment in distal bronchi is extremely difficult and dangerous. Haemorrhage from pulmonary vessels was one of the major complications of using the Nd-YAG laser, but did not occur with PDT. With a mortality rate of 0%, PDTs greatest advantage over Nd-YAG is safety [26].

A prospective, randomized trial of PDT versus Nd-YAG laser ablation for partially obstructive lung cancer has been reported [27]. This included data from 15 cent over Europe (141 patients) and 20 cent in the US and Canada (70 patients). Tumour response was similar for both therapies at 1 week, but at 1 month, 61 and 42% of PDT patients were still responding in the European and US/Canada trials, respectively, whereas for the Nd-YAG, 36 and 19% were responding in the two trials. Improvement in dyspnea and cough were superior for PDT over Nd-YAG in the European group. It was concluded that PDT is superior to Nd-YAG for relief of dyspnea, cough and haemoptysis. Overall, adverse reactions were similar for PDT and Nd-YAG (73% PDT, 64% Nd-YAG) and 20% of patients in the PDT group experienced a photosensitivity reaction due to lack of compliance with precautions.

## Photodynamic Therapy for Endobronchial Malignancy

When the earliest reports on the use of light-absorbing chemicals to cause photoreactions in biological systems was made by Raab et al. in 1900 [28], probably nobody anticipated that the efficacy and application of photodynamic therapy (PDT) technology would so expand in many medical fields. Since the first systematic clinical studies of PDT were initiated by Dougherty and colleagues in 1976 [29], while Hayata

et al. were the first to apply fiberoptic endoscopic laser irradiation to treat early-stage bronchogenic carcinoma with PDT in 1980 [30], increasing attention has been paid to the potentials of PDT in the treatment of malignant tumours. The past few years have witnessed an increasing interest in PDT and 10 000 papers on the subject have been published. This method has been used to treat a wide variety of malignancies in over 5000 patients worldwide [31]. Nine hundred patients have undergone PDT for endobronchial malignancy. PDT is now on the verge of becoming an established lung cancer treatment method (Table 1).

In 1993, Canada received approval from the Board of Health for the use of Photofrin-mediated PDT for treating recurrent superficial bladder cancer. In Japan, PDT with Photofrin and excimer dye laser was granted government approval in October 1994 and finally obtained national insurance reimbursement status in April 1996 for early-stage lung, oesophageal, gastric and cervical cancer for the first time in the word [32]. Further regulatory submissions for a variety of applications have been made in the 18 countries including the US, UK and France. The most promising treatment sites may be those where there is limited thickness in the tumour, such as early-stage carcinomas involving the aerodigestive tract, bronchus or genito-urinary tract. Other potential uses include those in which PDT can be combined with surgery or

**Table 1.** Photodynamic therapy in lung cancer

| Author | Year | Lesions | (results) | Comments | Journal |
|---|---|---|---|---|---|
| Hayata Y, Kato H | 1982 | Early 10 | (CR: 70%) | First report | Chest |
| Vincent RG | 1984 | Advanced 17 | (PR: 76%) | | Chest |
| Li JH | 1984 | Advanced 24 | (PR: 83%) | | Lasers Surg Med |
| Balchum OJ | 1984 | Advanced 22 | (PR: 90%) | | Clin Chest Med |
| Kato H | 1985 | Advanced 15 | (PR: 73%) | Preoperative | JTCS |
| Keller GS | 1985 | Advanced 15 | (PR: 93%) | | Arch Otolalyngol |
| Cortese DA | 1986 | Early 14 | (CR: 93%) | | Chest |
| McCaughan JS | 1986 | Advanced 26 | (PR: 96%) | | Lasers Surg Med |
| Hugh-Jones | 1987 | Advanced 15 | (PR: 80%) | | QT Med |
| Lam S | 1987 | Advanced 5 | (PR: 100%) | | Photo Photo |
| Edell ES | 1989 | Advanced 40 | (PR 93%) | | Mayo Clin Proc |
| Pass HI | 1989 | Advanced 15 | (PR: 87%) | | Ann Thoracic Surg |
| Horai T | 1990 | Early 9 | (PR: 78%) | | Seijinbyou |
| Kubota K | 1991 | Early 39 | (CR: 59%) | | Biotherapy |
| Sutedja T | 1992 | Advanced 11 | (CR: 91%) | | Eur J Cancer |
| | | Advanced 11 | (CR: 73%) | | |
| Karg O | 1992 | Early 15 | (CR: 89%) | | Pheumologie |
| Furuse K | 1993 | Early 59 | (CR: 85%) | Multicentric analysis | J Clin Oncol |
| Kusonoki Y | 1994 | Early 20 | (CR: 65%) | Multiple lung cancer | Respiratory |
| Imanura S | 1994 | Early 39 | (CR: 89,7%) | With radiation | Cancer |
| van den Bergh | 1997 | Early 33 | (CR: 85%) | Using m-THPC | Head Neck Surg |
| Mognissi K | 1997 | Advance 17 | (PR: 100%) | With Nd-YAG | Thorax |
| Kato H | 1997 | Early 116 | (CR: 84%) | 5-year survival, 68% | Lasers Surg Med |

chemotherapy, or treatment techniques that are common in pleural mesothelioma or novel applications of PDT, such as preoperative PDT for advanced bronchogenic carcinoma.

## Mechanism of PDT, Photosensitizers

In order to bring about cytotoxicity during therapy, a sensitizer, light and oxygen must e simultaneously present. Choosing the senitizers for clinical trials is a highly selective process. Great care must be taken in selecting fluorescent materials that are safe and that have an affinity for malignant tissues. Haematoporphyrin derivative (HpD), synthesized from haematoporphyrin by an acetylation procedure that produces a mixture of as many as 20 components, is now the most widely used photosensitizer. The many studies conducted on fluorescent materials have shown HpD to be a highly appropriate sensitizer in terms of safety, stability, low toxicity and tumour selectivity.

Research on new photosensitizers that efficiently absorb longer wavelengths of light and show more effective cytocidal activity is now underway. Among the new photosensitizers being developed are pheophorbide, bacteriopheophorbide, chlorine 6 [33], phthalocyanine [34] and many porphyrin derivatives.

Although Photofrin clears rapidly from most tissue within several hours after intravenous administration, it remains in malignant tissues, the liver, kidneys, spleen and skin for several days [35]. The mech-

anism by which Photofrin remains in tumours is not yet fully understood. However, some connection between Photofrin selectivity and the "micro-environment of tumours" is suspected [36]. Photofrin clearance may be inhibited when molecular aggregates of Photofrin accumulate around tumour neovasculature and lymphatic drainage within neoplasia is poor [37]. The isolated molecular aggregates then disassociate and Photofrin is distributed into cell membranes due to the hydrophobic nature of monomeric Photofrin. Although cellular and mitochondrial membranes are major sites of photodynamic activity [38], recent studies show that tumour vasculature is the primary site of Photofrin susceptibility. Anoxia from vascular damage result in tumour necrosis. Studies have shown that platelet aggregation occurs, followed by complete haemostasis and microvascular haemorrhage, seconds after tumour vasculature is exposed to light [39].

The mechanism of the cytotoxicity induced by PDT is not sufficiently understood. Nevertheless, it is known hat when a photon is absorbed by a photosensitizer molecule and is then transferred to another molecule in the presence of oxygen, a photochemical reaction takes place. Two types of reactions occur when the photosensitizer is excited: type 1, a process of electron transfer and type 2, an energy transfer with an oxygen molecule. The first type of reaction yields free radicals that react with oxygen to produce various oxidized materials which may activate free radical chain reactions [40, 41]. In the second type, a

**Fig. 2.** Chemical structure and fluorescence spectrum of Photofrin

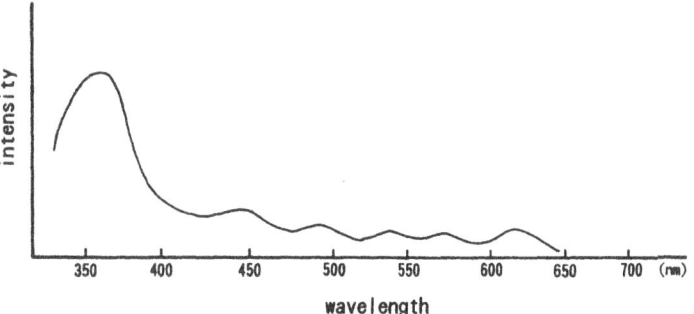

Bis-1-(8-(1-hydroxyethyl)deuteroporphyrin-3-yl)ethyl ether

reaction between photosensitizer molecules in an excited state and ground state oxygen produces singlet oxygen, which is highly reactive and induces cytotoxicity [42]. When the sensitizer is excited by light with a wavelength within its absorption band, photodynamic activity occurs. In aqueous solution, Photofrin reaches maximum absorption in the ultraviolet light range near 365 nm. There are also several lesser peaks of absorbance, including one near 639 nm in the red range. Absorbance of Photofrin is much greater at 363 than 630 nm. However, PDT utilizes red light near 630 nm because it has better tissue penetration and a lower rate of interference from absorbance by haemoglobin and other cellular (Fig. 2) components.

## Laser Systems

PDT can be performed with any light source with an appropriate spectrum. Through the emission of a monochromatic form of intense collimated light energy, lasers offer an especially effective way to deliver light. The argon dye laser is commonly used to provide the red beam required for photoradiation of tumour tissue sensitized by photoactive drugs or dyes [43], In this process, a dye, usually rhodamine B or Kiton red, is pumped into an appropriate optical cavity to produce red laser light tunable to a specific wavelength. This laser has drawbacks, however, primarily its limited ability to penetrate tissue. A new diagnostic and therapeutic endoscopic laser system was developed by the authors and their colleagues in 1982. This system utilizes an excimer dye laser capable of emitting a high-energy pulsed laser beam [44]. This laser appears to have many advantages over other PDT delivery systems. As demonstrated in the murine m-KSA sarcoma model, the excimer dye laser was shown to have greater tissue penetration than the argon dye laser [45]. High-energy photons capable of exciting Photofrin in tumour tissue to significant levels are provided by pulsed dye-laser systems in as little as 10 ns. Another pulsed beam laser, the gold vapour laser, was developed in the hope of achieving greater tissue penetration, but maintaining the laser equipment to ensure optimal performance is difficult.

The characteristic high-energy beam of the excimer laser is generated from a gas mixture of 0.9% Xe, 0.1% HCI and 99% He at 2 atm, pressure. At 308 nm, the optimal performance of the excimer laser is 30 mJ-pulse at one half peak power for 10.9 ns. Generation of the 630-nm high-energy beam used in PDT requires coupling the excimer laser to a system containing 2 molar rhodamine B dye in ethanol [46].

Recently, a new high-power red laser diode and system (Panasonic, Osaka, Japan) was developed for PDT. This system has a wavelength of 664 nm an a power output of 500 mW $cm^{-2}$ (cw) in tissue. It is compact ($49 \times 20 \times 40$ cm), light-weight (20kg), easy to operate and reliable. This system needs only 100 VAC $\cdot$ 3 A power supply, and requires no maintenance.

## The Bronchoscopic PDT Procedure

Approximately 48 h following intravenous injection of 2.0 mg $kg^{-1}$ bodyweight of Photofrin, bronchoscopic PDT is performed under topical anaesthesia. Patients should avoid direct sunlight for at least 2 weeks after Photofrin injection. The laser beam is transmitted through a quartz fiber (400 µm) inserted through the instrumentation channel of a fiberoptic bronchoscope. Held 1–2 cm perpendicularly from the target, the fib tip produces a 4–8 mm circular area of illumination. When using the argon dye laser, power output at the fib tip is adjusted to 80–400 mW. Energy is adjusted to 4 mJ-pulse at 30 Hz frequency when using the excimer dye laser. Illumination time generally ranges from 10–40 min at energy densities of 100–200 J $cm^{-2}$ for surface irradiation in the treatment of early-stage lung cancer (Fig. 3). With the fib tip inserted in the tissue, interstitial radiation was performed in advanced cases of cancer with endobronchial obstruction [25]. We now employ two types of fiberoptic tips for laser light delivery: the microlens-tipped fib, which provides a uniform field of light for front-surface illumination; and the cylindrical diffuser-tipped fib, which permits delivery of light around the wall of the bronchus when a microlens-tipped fib cannot be aimed directly at a lesion, and/or is inserted directly into the tumour tissue for effective, tumour-specific interstitial light delivery, especially in cases of advanced obstructing tumours. After PDT, necrosis of the tumour occurs, and the necrotic debris and associated secretions must be removed through toilet bronchoscopy some days after initial treatment. In lung cancer cases, complications that occurred following PDT included skin photosensitivity (90%), obstructive pneumonia (5%) and massive bleeding (1%) [47]. Complications after PDT can be avoided or minimized of patients avoid direct sunlight for a minimum of 2 weeks and receive regular toilet bronchoscopy. Light dosage should also be restricted in cases of large tumours reaching beyond organ walls.

## Use of PDT in Early-Stage Lung Cancer

An increasing number of early-stage lung cancer cases are being detected as a result of improved survey and diagnostic techniques. In cases of early detection, it is generally possible to perform curative resection. However, there is high surgical risk for many pa-

**Fig. 3.** How to do PDT

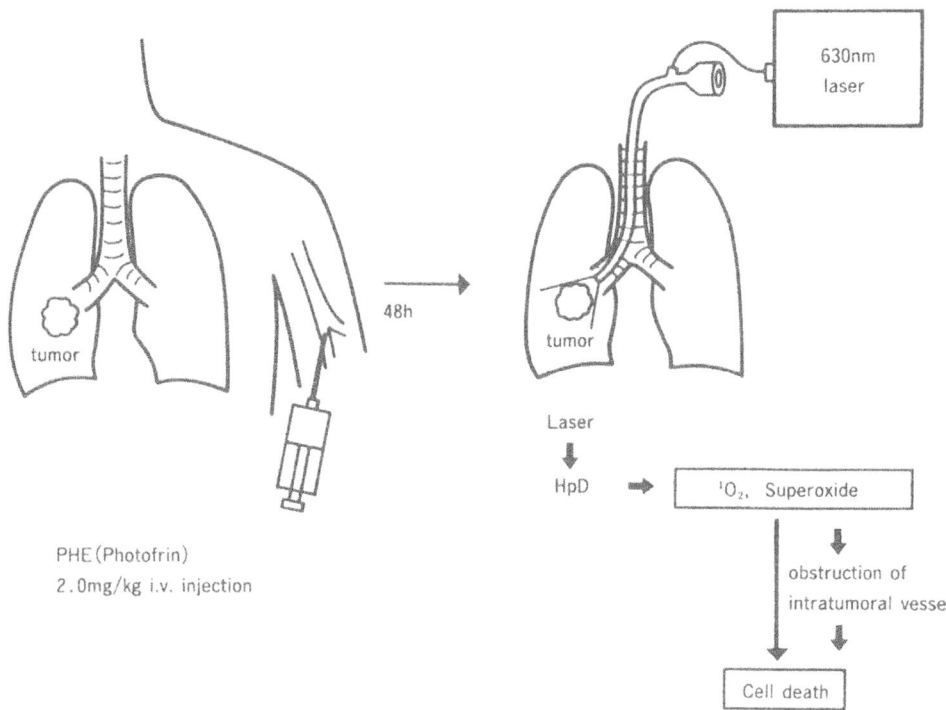

PHE (Photofrin)
2.0mg/kg i.v. injection

tients due to concurrent cardiovascular or chronic obstructive pulmonary disease. Conservative treatment of initial early-stage lung cancer in order to preserve lung tissue is essential for the quality of life of the patient. We used PDT to treat 149 lesions in 113 cases of endoscopically detected early-stage lung cancer (stage 0) between 1978 and 1999. The age of the patients ranged from 36 to 85 years, with a mean age of 65. All but one were male. Except for one adenocarcinoma, all lesions were squamous cell carcinomas. Though the treatment of choice for early-stage lung cancers is usually surgical resection, PDT was performed. Many patients have poor pulmonary function or refuse surgery, and thus do not undergo resection. In the treatment of these patients, we used an argon or excimer laser coupled to a dye laser employing rhodamine-B dye to generate 630 nm light.

Three grades of tumour response were noted: complete remission (CR): no visible presence of a tumour through biopsy and/or brushing cytology for at least 4 weeks; partial remission (PR): over 50% reduction in tumour volume but cancer still detectable on biopsy or brushing for at least 4 weeks after therapy; and no change (NC): tumour size remained the same and cancer was still recognizable on biopsy or brushing. Tumour response to PDT was evaluated endoscopically, roentgenographically and histologically 1 month after treatment. Endoscopic and histological examinations were conducted on the treated areas in surgicallly resected or autopsied cases.

The results of PDT in endoscopically detected cases of early-stage lung cancer are shown in Table 2. Ninety eight cases (128 lesions) out of 113 cases (149 lesions), or 85.9%, achieved complete remission; however, in 15 other cases (21 lesions), the entire extent of the lesion could not be seen endoscopically Therefore, radiotherapy (prevent recurrence and ensure a curative effect. Nineteen cases (14.1%) experienced recurrence and were treated by second PDT, surgery and radiotherapy.

In Japan, a multicent study was conducted on 59 early-stage lung cancers. After initial PDT, 50 (84.7%) were classified as CR. Of these, CR was obtained in 28 carcinomas with a longitudinal extent of 1 cm or less [48].

**Table 2.** Results of PDT for early stage lung cancer

| Cases (Lesions) | CR | PR | Rec. After CR |
|---|---|---|---|
| 113 (149) | 98 (128) (85.9%) | 15 (21) | 19 (19) (14.1%) |
| | Combined therapy | | |
| | Surgery | 15 | 5 |
| | Radiation | 4 | 2 |
| | PDT | 2 | 7 |
| | Chemotherapy | 0 | 1 |
| | No therapy | 0 | 4 |

CR: complete response, PR: partial response, Rec: recurrence

There has been remarkable consistency in results of various investigators, showing CR + PR rates ranging from 70 to 100% [49]. Mucosal tumours or early-stage (stage 0) lung cancers showed the best results. In treatment of superficial lesions, it is possible for PDT to be used as a substitute for surgery, having an overall CR rate 66.7% [50]. A study of patients at the Mayo Clinic with early-stage lung cancer (including those with in situ carcinomas), who underwent PDT, , showed a 71.4% CR rate in 14 tumours treated in 13 patients [51]. During the first 2 years of follow-up, there was recurrence in 23% of the 10 tumours that showed a complete response. Surgical resection was performed on two and a second PDT on the third, and all again achieved CR.

A typical CR case is presented in Fig.4. In this 79-year-old-man, squamous cell carcinoma of the lung was initially diagnosed based on positive sputum cytology. The tumour was polypoid, located in the orifice of the right B6 bronchus and 0.5 x 0.5 cm in size (Fig.4a). Since the patient's pulmonary function was very poor, he was subsequently treated by PDT. Figure 4b shows the same site 3 months after PDT; the lesion was completely disappeared. He is now apparently disease-free 48 months after PDT.

The results of our study have shown that the following conditions are essential for successful PDT in early-stage lung cancer. (1) The entire lesion must be visible endoscopically, (2) the tumour must be situated where sufficient laser beam photoradiation can be delivered, (3) the lesion should be superficial and 1 cm or less in longest dimension, (4) the histological type should be squamous cell carcinoma and (5) here should be no lymph node involvement. Identification of cases with tumours limited to the bronchial wall and without involvement of lymph nodes is the most difficult aspect of this treatment method. We examined resected specimens to investigate the effect of the presence or absence of lymph node involvement in early-stage central-type lung cancer. No involvement of lymph nodes was found in 13 lesions of carcinoma in situ resected at our hospital [52]. Nagatomo reported that in resected specimens from 92 patients with roentgenographically occult lung cancer, no lymph node involvement was present when the tumour diameter was less than 2 cm [53].

We histologically examined the basis of resected specimens of lung cancer cases including an early-stage case treated by PDT that did not show complete remission. Complete remission was not obtained when lesions were (1) at an anatomical site difficult to photoirradiate, (2) located subumucosally and photoradiation could not be performed at a 90° angle to the surface of the lesion, (3) located beyond the cartilage and (4) extensive. To overcome these difficulties,

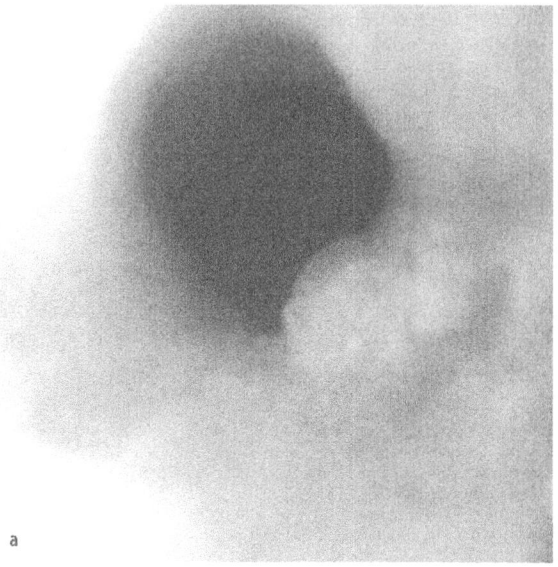

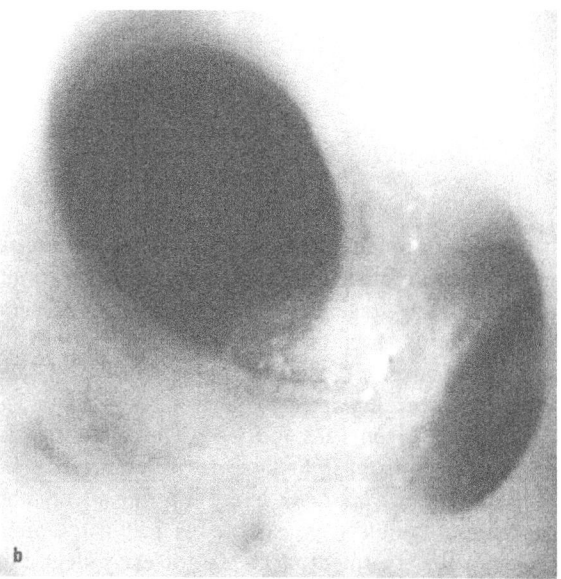

**Fig. 4a, b.** Bronchoscopic findings of 79-year-old male with a polypoid squamous cell carcinoma, 0.5 x 0.5 cm in size in orifice of the right B6 bronchus (**a**). **b.** The same site 3 months after PDT. Complete remission was obtained

increased laser power and PDT using cylindrical quartz fib with 360¡ diffusion should be used [54]. PDT can be an effective alternative to surgical resection as the primary treatment of patients with early-stage central-type lung cancer. The effectiveness of utilizing PDT with Photofrin should be a considered when deciding on cancer treatment. A phase-III study on operable cases of central type early-stage lung cancer with tumour invasion of less than 1 cm is now being conducted by the authors.

## Use of PDT in Advanced Tracheo-Bronchial Malignancies

Little can be done in cases of advanced-stage lung cancer. Only 12% of stage-III patients survive for 2 years regardless of therapy [55–56]. Transbronchial resection by Nd-YAG laser has become well accepted after the initial reports of its efficacy. Transmitted through optical fib, the Nd-YAG laser provides excellent vaporization, low absorption by haemoglobin, and good tissue penetration 5 to 10 mm from the focal point with high-energy output. Good results were reported in cases of symptomatic endobronchial tumour obstruction treated with the Nd-YAG laser [57]. The drawback of this laser, however, is its non-specificity. The Nd-YAG laser produces a beam that not only vaporizes or coagulates tumour tissue, but may also harm normal bronchial tissues and thus lead to severe bleeding or perforation of the bronchial wall. Treatment by Nd-YAG laser may improve ventilation of patients with endobronchial tumours, but only partial removal of the tumour should be attempted using this method.

With PDT, the endobronchial tumour, if small enough, can be completely eliminated with no fear of complications. In patients in the US with lung cancer or endobronchial obstruction in cases where intrinsic lesions of the bronchus cause partial or complete obstruction [58]. Symptomatic improvement was used as a basis to evaluate patients in some studies. Generally reported as positive responses were clearance of airway obstruction or improved operability, and 340 out of 376 sites 90.4%) showed a positive response (CR+PR) [50]. The responses of the series of advances disease patients were more difficult to interpret, however. Among these patients were those with a variety of combinations of T and N factors and other prognostic elements, such as performance score and weight loss. Signs of distant metastasis, sometimes shortly after treatment, were seen in a significant percentage of the test population during follow-up. Frequent and repeated treatment was determined to be feasible and safe in studies by McCaughan et al. [60]. In these studies, 30 treatments were performed on a series of 18 patients in 26 different areas of the lung. Complete or partial response was experienced at 1 month by 96% of the patients, and a total of 61 showed clinical improvement. The mean survival period was 8.3 months. A study by Hugh-Jones on a series of 15 patients with advanced squamous carcinoma showed that adequate palliation was achievable [53]. Initial therapy resulted in a 100% response by patients, with 12 (60%) achieving 50% reduction in tumour size and three (20%) achieving complete tumour elimination. After PDT, one patient was disease-free for 2 years.

In prospective independent clinical trials conducted to test the safety and efficacy of PDT with Photofrin on endobronchial lesions at three US cent between 1983 and 1988, 370 endobronchial sites were treated in 170 patients [49]. For obstructive lesions and mucosal lesions, the overall CR rates were 50 and 75%, respectively. CR + PR (total response) rates were 75 and 78%, respectively. Possible treatment-related side effects included dyspnea (19% of all patients), fever (20%), photosensitivity (8%) an haemoptysis (8%). Eight deaths occurred due to haemoptysis (one during post-PDT debridement and seven an 2–10 weeks after PDT). However, PDT appears to cause no additional toxicity in patients who had previously undergone either chemotherapy or radiation therapy. Thus, PDT can be safely combined with standard treatment-methods.

The advantages and disadvantages of PDT compared with the Nd-YAG laser were summarized by McCaughan et al. The disadvantages of PDT are photosensitizer injection, from which skin photosensitivity may result, (2) the required waiting period between injection and treatment and (3) the need to perform frequent toilet bronchoscopy. The advantages of PDT treatment are: (1) the technical ease allowed by its safety: there is little chance of perforation and little risk of intra-operative haemorrhaging, (2) no endobronchial smoke and (3) freedom to insert the fib blindly into tissue [61]. Although we have emphasized the advantages of PDT for advanced obstructing bronchial malignancy, the first choice for patients with severe respiratory distress is immediate Nd-YAG laser therapy, because PDT requires a 2–3-day waiting period for selective retention of Photofrin and also may cause mucosal edema of tissue. Furthermore, patients treated with PDT may be ambulatory, but because they are instructed to avoid direct sunlight, their activity is severely limited for at least 2 weeks of their short remaining life span. When deciding among alternative therapies, physicians of patients with late-stage lung cancer should give careful consideration to this problem.

### Combined Use of Photodynamic Therapy and Surgery

As we have described, in lung cancer, PDT is now mainly performed for the early-stage lesions, in which the results are satisfactory. However, in spite of recent advances in the diagnostic techniques and establishment of mass screening systems in Japan, most lung cancers are still detected at an advanced stage in our country. As would be expected, better therapeutic results were obtained in all resected cases of lung cancer than in non resected cases [56]. Therefore, it is necessary to increase the number of operable cases of lung cancer to improve the survival rate. In addition, even after curative resection, 15% of patients eventually died of poor postoperative pulmonary function. In the authors' institution, preoperative PDT is employed as one of the options to increase the operability and reduce the extent of resection in lung cancer cases [62].

The present authors reported that 27 lung cancer patients underwent preoperative PDT in order to either reduce the extent of resection or increase operability [63]. Bronchoscopic PDT was performed under topical anaesthesia approximately 48 h after the intravenous injection of 2.0 mg kg$^{-1}$ body weight of Photofrin. Operation was performed 2–9 weeks after PDT.

Among three cases with tracheal invasion, sleeve lobectomy of the right upper lobe was performed in one case, pneumonectomy in one case and tracheoplasty in the remaining one. Among three cases of carinal invasion by tumour, one underwent right upper sleeve lobectomy and one patient underwent left pneumonectomy, but the remaining one patient ended up receiving exploratory thoracotomy because of extensive hilar lymph node involvement. In 21 cases with tumour invasion to the main bronchi, lobectomy was performed in 9 cases, and sleeve lobectomy in 10 cases, in order to preserve pulmonary function. However, pneumonectomy was performed in the remaining two cases because of extensive hilar lymph node involvement. Therefore the purpose of PDT, i.e. either the reduction of extent of resection or the conversion of inoperable disease to operable status, was achieved in 24 out of the 27 patients treated. Of 5 patients initially with inoperable disease, 4 were converted to operable status by PDT. In 22 patients who were originally candidates for pneumonectomy, it became possible to reduce the extent of resection to lobectomy or sleeve lobectomy in 20 cases. The overall survival curve for 10 patients who were classified as p-stage IIIA, and who received preoperative PDT with subsequent lobectomy, is shown in Fig.1. The survival curve, calculated by the Kaplan-Meier method, showed a 5-year survival rate of 59.3%.

This new option of PDT as preoperative treatment may contribute to the management of advanced lung malignancy and expand the indications of PDT for lung cancer patients.

## Laser Applications Thoracic Surgery

This section mainly discusses the range of minimally invasive laser techniques currently possible through the thoracoscope. Thoracoscopy was introduced for the surgical management of tuberculosis in the prechemotherapy era. In later years, although the number of reports has been low, it has been shown to have considerable potential, particularly for palliation of pleural tumours, in the management of pneumothorax and emphysematous bullae in patients who would not easily tolerate thoracotomy. This technology has been further facilitated by the via tharacoscope [64].

The application of lasers to thoracic surgery was not explored seriously until a few years ago. However, since Toth et al. and Boutin et al. reported the use of the lasers in thoracoscopy, for the treatment of spontaneous pneumothorax, diffuse bullous emphysema and malignant pleural effusions has been reported frequently [65–67]. An application of PDT for malignant methothelioma at open thoracotomy will also be discussed in this section.

## Laser Pleurectomy for Pleural Effusion

Recurrent massive pleural effusions, whether benign or malignant, are not uncommon. They usually present shortness of breath on exertion and the pleural effusion is readily demonstrable by chest X-ray. Boutin et al. reported a study in which thoracoscopy was needed in 215 of 1000 patients with pleural effusions after cytology of the pleural fluid and needle biopsy of the pleura failed to establish the diagnosis [65]. Malignancy was diagnosed in 131 of them. For the treatment of pleural malignancy, surgical intervention such as thoracotomy and pleurectomy has been carried out; however, although surgical pleurectomy has a high success rate, patients with metastatic malignant pleural effusions maybe too ill to receive such a major operation [68]. As an alternative, thoracoscopic laser pleurectomy was developed to remove the partial pleura followed by vaporization of the remaining tumour on both the partial and visceral pleura using the free beam Nd-YAG laser [69].

## Laser Ablation for the Treatment of Bullous Emphysema

Bullous Emphysema is commonly seen among elderly chronic smokers. In the formation of these bullae, chronic bronchiolitis and an inflamed mucosa may act as a one-way valve which traps inspired air. This trap increases inside the bulla, thus destroying more alveoli. During expiration, the tension inside the bulla increases and compresses adjacent bronchioli or bronchi, thus increasing their resistance.

The procedure is performed under general anaesthesia and single-lung ventilation. Pleural adhesions are very commonly found in association with bullae and these must be completely freed. Type-1 bullae (classical giant bullae without trabeculae inside) are simply excised using the contact Nd-YAG laser scalpel at 14–17 W in a continuous-wave mode. The bullaous wall is grasped by the instruments and tension is applied. When it is punctured, the trapped air empties instantly, thus collapsing the bulla. The bullous wall is widely opened, bronchial communications are identified, and closed with sutures. In most patients, several bullae are found and all of them are removed in a similar fashion. Type-3 bullae (diffuse bullae with multiple trabeculae) are contracted with either the defocused $CO_2$ laser beam at 2–12 W in a continuous-wave mode or with a contact Nd-YAG laser round probe at 10–16 W. If the wall of the bulla becomes very redundant, it is plicated with sutures.

Wakabayashi reported 500 laser thoracoscopic treatments for bullous emphysema. The majority of patients treated have reported an improvement in their breathing and physical capacity. Objective assessments, based on pulmonary function tests and the treadmill stress test, have confirmed a statistically significant improvement [70, 71].

On the other hand, Barker and associates have reported an early experience with unilateral video-assisted thoracoscopic laser bullectomy in which they observed prolonged anaesthesia time, universal and significant postoperative air leakage, routine requirement for postoperative ventilatory assistance, prolonged intensive care unit stay and a significant mortality rate [72]. Furthermore, data relating to objective assessment of results have been limited, and those that have been available fail to show the same degree of improvement observed with the procedure described [73]. Cooper et al. reported using sternotomy and bilateral resection for this procedure to achieve maximum benefits in one operation, with minimum morbidity [74]. They have previously used this approach for resection of enphysematous bullae with good results. At the onset of this series, consideration was given to the possible role of a video-assisted approach to minimize morbidity. However, they have not adopted this approach because several factors, including the desire to operate on both lungs at one sitting with a minimum of anaesthesia time, the inability to palpate and examine all areas of the lung before selecting areas for resection and the ability with an open procedure to minimize air leak and to more accurately identify and secure any air leaks that do occur during the procedure were expected.

## PDT for Malignant Mesothelioma

Although the yearly incidence of malignant pleural mesothelioma in the United States is 3000~4000 cases [75], the treatment of patients with pleural malignan-cies, specially mesothelioma, remains a problem. The need for innovative treatment is clear because there is no universally accepted standard treatment for malignant mesothelioma, and the efficacy of current therapies yields median survivals of only 6–16 months [76]. Radical resection can seldom be performed. Macroscopically, the resection may appear complete, but microscopically, tumour cells are often evident at the surgical margin. For cases that are considered to be surgical candidates, adjuvant treatment such as radiation and/or chemotherapy have generally been given. Despite some positive results, overall survival did not significantly improve, whereas adverse reactions increased. In these backgrounds, PDT for malignant mesothelioma has been challenged under thoracoscopy and/or thoracotomy.

A phase-I study of intracavitary PDT for peritoneal carcinomatosis first reported by Pass et al. [77] attracted much attention. Takita et al. performed a phase-II study of surgery and intracavitary photodynamic therapy for a total of 31 patients with malignant pleural mesothelioma [78]. Two $mg\,kg^{-1}$ of Photofrin was administrated intravenously, and 48 h later the patient was taken to the operating room and the bulky tumour was excised by a pleuropneumonectomy or pleurectomy. The patient then received $20 . 25\ J\ cm^{-2}$ of 630-nm light energy with an argon dye laser. The overall median survival of patients of all stages was 12 months. Survival of stage-III and stage-IV patients was 8 months; however, the median survival of 9 patients with stage-I and –II disease 21 months.

In the largest study by Pass et al., 42 patients were treated with PDT using HpD in a phase-I study [79]. Thirty-one of the 42 patients (74%) died and no increased survival (mean 12.4%) was observed. Forty eight h after administration of 2 mg-kg of Photofrin, PDT was performed with two argon dye lasers. The actual laser administration time was 68 min to achieve a total light dose of $25\ J\ cm^{-2}$.

In all previous studies with PDT, its potential has been recognized; however, the lack of high-power lasers, effective photosensitizers and the lack of understanding of dosimetry and the inability to administer adequate light doses in a short period of time has prevented its more general use. Ris et al. performed a pilot study in eight patients with 0.3 mg-kg of m-THPC and $10\ J\ cm^{-2}$ of laser irradiation. Seven patients obtained good local control of their thoracic disease but distant metastasis developed after 4–18 months. In one patient, who expired of pulmonary embolism 8 days after resection, postmortem examination showed extensive necrosis in the remaining tumour but showed no damages to normal structures such as the heart and oesophagus [80]. Bass et al. also treated five patients with a pleural malignancy using

PDT, performed with light of 652 nm from a high-power diode laser, and m-THPC as the photosensitizer. The light delivery to the surface of the thoracic cavity was monitored by in situ isotropic light detectors. The position of the light-delivery fib was adjusted to achieve optimal light distribution, taking account of reflected and scattered light in the thoracic cavity. With this system, light delivery to large surfaces for adjuvant PDT is feasible in a relatively short period of time. In situ dosimetry ensures optimal light distribution and allows total doses to be monitored at different positions within the cavity [81]. This combination of light delivery and dosimetry is well suited for adjuvant treatment with PDT in malignant pleural tumours. There is little doubt that further experimentation with this technique will help us to use it more effectively.

## Conclusions

Many of the laser techniques described in this chapter are still relatively early stages of development and have been used in a rather simple manner. With more understanding of their effects on tissues, it appears that much more could be realized in terms of both better palliation and increased chances of cure for patients untreatable by traditional surgery. Better laser systems, with lower expenses and tunable, maintenance-free devices are necessary. The success of PDT for treatment of various cancers in clinical trials offers encouragement for its future use. Research to develop new photosensitizers that reach emission peaks at long wavelengths and have minimal phototoxicity should be conducted. Whether used curatively for early-stage cancer, palliatively for local improvement of lesions in advanced cases or in combination with surgery or ionizing radiation and chemotherapy., laser therapy provides sound benefits and holds great potential in the treatment of pulmonary diseases.

## References

[1] HOOPER RG JACKSON FN (1985) Endobronchial electrocautery . Chest 87: 712–714
[2] WALSH DA, MAIWAND MO, NATH AR ET AL (1990) Bronchoscopic cryotherapy for Advanced bronchial carcinoma. Thorax 45: 509–513
[3] TOTY L, PERSONNE CL, HERTZOG P, ET AL (1979) Utilisation d'un faisceau laser (YAG) à conducteur souple, pour le traitement endoscopique de certaines lesions trachéobronchiques. Rev Fr Mal Respir 7: 475–482
[4] DUMON JF, REBOUND E, GARBE L. ET AL (1982) Treatment of tracheobronchial lesions by laser photoresection. Chest 81: 278–284
[5] ARABIA A, SPAGNOLO SV. (1984) Laser Therapy in patients with primary lung cancer. Chest 86: 519–523
[6] PERSONNE C, COLCHEN A, LEROY M, ET AL. (1986) Indications and technique for endoecopic laser resections in bronchology: a critical analysis based upon 2284 resections. J Thorac Cadiovasc Surg 91: 710–715

[7] CAVALIERE S, FOCCOLI, P, FARINA, PL. (1988) Nd-YAG laser bronchoscopy: a 5-year experience with 1396 applications in 1000 patients. Chest 94: 15–21
[8] STORONG MS, VAUGHAN CW, POLANYI T, ET AL. (1974) Bronchoscopic carbon-dioxide laser surgery. Ann Otol Rhinol Laryngol 83: 769–798
[9] MCELVEIN, RB, ZORN, GL JR (1984) Indications, results, and complications of bronchoscopic carbon dioxide laser therapy. Ann Surg 199: 522–525
[10] GEORGE PJM, GARRETT COP, NICON C, ET AL. (1987) Laser treatment for endobronchial tumours: local or general anaesthesia? Thorax 42: 656–660
[11] OHO, K, OGAWA, R, AMEMIA, T ET AL (1983) Indications for endoscopic Nd-YAG laser surgery in the trachea and bronchus. Endoscopy 15: 302–306
[12] JOYNER, LR, MARAN, AG YAKBOSKI, A. (1984) Neodymium YAG laser treatment of intrabronchial lesions. Chest, 85: 418–427
[13] DUMON, JF. (1990) Technique of sae laser surgery. Lasers Med Sci, 5: 171–180
[14] PARR, GVS, UNGER M, TROUT, RG ET AL (1984) One hundred neodymium-YAG laser ablations of obstructing tracheal neoplasms. Ann Thorac Surg, 38: 374–380
[15] WOLFE, WG, SABISTON, DC. (1986) Management of benign and malignant lesions of the trachea and bronchi with the neodymium-yttrium aluminum garnet laser. J Thorac Cadiovasc Surg 91: 374–380
[16] CORTESE, DA, EDELL, ES. (1993) Role of phototherapy. Laser therapy, brachytherapy, and prosthetic stents in the management of lung cancer. Clin Chest Med, 1: 149–159
[17] DUMON JF, SHAPSHAY S, BOURECEREAU J, ET AL. (1984) Principles for safety in application of neodymium: YAG laser in bronchology. Chest, 86: 163–168
[18] MEYLAN, D, STRUBLER, K, UNAL, A, ET AL (1983) Transbronchial brachytherapy of recurrent bronchogenic carcinoma; new approach using flexible fiberoptic bronchoscope. Radiology, 147: 253–254.
[19] SCHRAY MF, MCDOUGALL JC, MARTINEZ A, ET AL (1988) Management of malignant airway compromise with laser and loe dose rate brachytherapy. Chest, 93: 264–269
[20] BEDWINEK J, PETTY A, BRUTON C, ET AL. (1992) The use of HDR endobronchial brachytherapy to palliate symptomatic endobronchial recurrence of previously irradiated bronchogenic carcinoma. Int J Radiat Oncol Biol Phys, 22: 23–30.
[21] HETZEL, MR (1995) Outcome, morbidity of Nd-YAG laser photoresection. In Hetzel, MR, (ed) Minimally invasive techniques in thoracic medicine & surgery. Chapman & Hall Medical, London, pp 137–146,
[22] ROSS, DJ, MOHSENIFAR Z, KOERNER, SK. (1990) Survival characteristics after neodymium: YAG laser photoresection in advanced stage lung cancer. Chest, 98: 581–585
[23] KVALE, PA, EICHENHORN, MS, RADKE, JR ET AL (1985) YAG laser photoresection of lesions obstructing the central airways. Chest, 86: 283–287.
[24] HETZEL, MR, NIXON, C, EDMONSTONE, W ET AL (1985) Laser therapy in 100 tracheobronchial tumours. Thorax, 40:341–345
[25] FURUKAWA, K, OKUNAKA, T, YAMAMOTO, H ET AL (1999) Effectiveness of photodynamic therapy and Nd-YAG laser treatment for obstructed tracheobronchial malignancies. Dig Ther Endosc 5: 161–166
[26] OKINAKA, T, KONAKA, C, TSUSTSUI, H ET AL (1994) Present status of endoscopic treatment of lung cancer. Jpn J Bronchol 16: 712–722
[27] DOUGHERTY, TJ, GOMER, CJ, HENDERSON, BW. (1998) Photodynamic therapy. J Nat Cancer Inst, 90: 889–905
[28] RAAB, O. (1900) Über die Wirkung von fluoreszierenden Stoffen. Infusoria Z Biol, 39:524
[29] DOUGHERTY, TJ, LAURENCE, G, KAUFMAN JG ET AL (1979) A. Photoradiation in the treatment of recurrent breast carcinoma. J Natl Cancer Inst, 62: 231–237
[30] HAYATA, Y, KATO, H, KONAKA, C ET AL (1982) haematoporphyrin derivative and laser photoradiation in the treatment of lung cancer. Chest, 81: 269–277

[31] MANYAK, MJ, RUSSO, A, SMITH PD ET AL (1988) Photodynamic therapy. J Clin Oncol, 6 (2): 380–391

[32] KATO, H, OKUNAKA, T, SHIMATANI, H. (1996) Photodynamic therapy for early-stage bronchogenic carcinoma. J Clin Laser Med Surg, 14: 235–238

[33] NELSON, JS, ROBERTS, WG, BERNS, JW. (1987) In vivo studies on the utilization of mono-L-aspartyl chlorin (Npe6) for photodynamic therapy. Cancer Res, 47: 4681–4685

[34] TRALAU, CJ, BARR, H, SANDERMAN, R ET AL (1987) Aluminum sulfonated phthalocyanine distribution in rodent tumour of the colon, brain and pancreas. Photochem Photobiol, 46: 777–781

[35] KESSEL, D. (1984) Chemical and biochemical determinants of porphyrin localization. In: Doiron DR, Gomer JC (eds) Porphyrin localization and treatment of tumours. Liss, New York, pp 405–418

[36] BUGELSKI, PJ, CW, DOUGHERTY, TJ, ET AL (1981) Autoradiographic distribution of haematoporphyrin derivative in normal and tumour tissue of the mouse. Cancer Res, 41: 4606–4612

[37] MOAN, J. (1984) The photochemical yield of singlet oxygen from porphyrin in different states of aggregation. Photochem Photobiol, 39: 445–449

[38] GIBSON, SL, HILF, R. (1983) Photosensitization of mitochondrial cytochrome c oxidase by haematoporphyrin derivative. Cancer Res 43: 1994–1999

[39] ZHOU, C. (1989) Mechanisms of tumour necrosis induced by photodynamic therapy. J Photochem Photobiol B: Biology 3: 299–318

[40] HIFF, R, WARNE, NW, SMAIL, DB ET AL (1984) Photodynamic inactivation of selected intracellular enzymes by haematoporphyrin derivative and their relationship to tumour cell viability in vitro. Cancer Lett 24: 165–172

[41] TAKEMURA, T, NAKAJIMA, S. AND SAKATA, I. (1994) Tumour-localizing fluorescent diagnostic agents without phototoxicity. Photochem Photobiol 59, 366–370

[42] WEISHAUPT, KR, GOMER, CJ DOUGHERTY, TJ. (1976) Identification of singlet oxygen as the cytotoxic agent in photoactivation of a murine tumour. Cancer Res, 36: 2326–2329

[43] FULLER, TA. (1983) Fundamentals of lasers in surgery and medicine. In: Dixon JA (ed) Surgical application of lasers. Yearbook, Chicago, pp 11–28

[44] YAMAMOTO, H, KATO, H, OKUNAKA, T ET AL (1991) Photodynamic therapy with the excimer dye laser in the treatment of respiratory tract malignancies. Lasers Life Sci, 4: 125–133

[45] OKUNAKA, T, KATO, H, C ET AL (1992) A comparison between argon-dye and excimer-dye laser for photodynamic effect in transplanted mouse tumour. Jpn J Cancer Res, 83: 226–231

[46] HIRANO, T, ISHIZUKA, M, SUZUKI, K ET AL (1989) Photodynamic cancer diagnosis and treatment system consisting of pulse lasers and an endoscopic spectro-image analyzer. Lasers Life Sci, 3: 99–116

[47] KATO, H, KONAKA, C, KINOSHITA, K ET AL (1990) Laser endoscopy with photodynamic therapy in the respiratory tract. Gann Monogr Cancer Res, 37, 139–151

[48] FURUSE, K, FUKUOKA, M, KATO, H. et al (1993) A prospective phase II study on photodynamic therapy with Photofrin II for centrally located early-stage lung cancer. J Clin Oncol, 11: 1852–1857

[49] MARCUS SL, DUGAN M. (1992) Global status of clinical photodynamic therapy: the registration process for a new therapy. Lasers Surg Med, 12: 318–324

[50] KATO, H, OKUNAKA, T. (1995) Photodynamic therapy in early tumour. In: Hetzel MR (ed) Minimally invasive techniques in thoracic medicine and surgery. Chapman & Hall Medical, London, pp 149–178

[51] EDELL, ES, CORTESE, DA. (1989) Bronchoscopic localization and treatment of occult lung cancer. Chest, 96: 919–924

[52] HAYATA, Y, KATO H, KONAKA, C ET AL (1993) Photodynamic therapy in early-stage lung cancer. Lung Cancer, 9: 287–294

[53] NAGATOMO, N, SAITO, Y, OHATA, S ET AL (1989) Relationship of lymph node metastasis to primary tumour size and microscopic appearance of roentgenographically occult lung cancer. Am J Surg Pathol, 13: 1009–1013

[54] KATO H, KAWATE N, KINOSHITA K. (1990) Photodynamic therapy of early-stage lung cancer. In: Bock G, Harnett S. (eds) Photosensitizing compounds: their chemistry, biology and clinical use. John Wiley, New York, pp 531–535

[55] HARA, N, OHTA, M, TANAKA, K, ET AL (1984) Assessment of the role of surgery for stage-II bronchogenic carcinoma. J Surg Oncol, 25: 153–158

[56] MOUNTAIN, CF. (1985) The biologic operability of stage III non-small cell lung cancer. Ann Thorac Surg, 40: 60–64

[58] MCCAUGHAN, JS, WILLIAMS, TE, BETHEL, BH. (1986) Photodynamic therapy of endobronchial tumours. Lasers Surg Med, 6: 336–345

[58] BALCHUM, OJ, DOIRON, DR. (1985) Photoradiation therapy of endobronchial lung cancer: large obstruction tumours, nonobstructing tumours, and early-stage bronchial cancer lesions. Clin Chest Med, 6: 255–275

[59] MCCAUGHAN, JS, WILLIAMS, TE, BETHEL, BH. (1986) Photodynamic therapy of endobronchial tumours. Lasers Surg Med, 6: 336–345

[60] HUGH-JONES, P, GARDNER, WN. (1987) Laser photodynamic therapy for inoperable bronchogenic squamous cell carcinoma. Q J Med, 64: 565–581

[61] MCCAUGHAN, JS, HAWLEY PC, BETHEL, BH ET AL (1988) Photodynamic therapy of endobronchial malignancies. Cancer, 62: 691–701

[62] KATO, H, KONAKA, C, ONO, J ET AL (1985) Preoperative laser photodynamic therapy in combination with operation in lung cancer. J Thorac Cardiovasc Surg, 90: 420–429

[63] OKUNAKA, T, HIYOSHI, H, FURUKAWA K, ET AL (1999) Lung cancer treated with photodynamic therapy and surgery, Diag and Therap End 5: 155–160

[64] BOUTIN, C. (1989) The laser in thoracoscopy. Pneumologie, 43: 96–97

[65] BOUTIN, C, ASTOUL, P. SEITZ, B. (1990) The role of thoracoscopy in the evaluation and management of pleural effusions. Lung, 168: 1113–1121

[66] TOTH, T, SZOTS, I. AND UGHY, T. (1989) Use of laser knife in thoracic surgery. 47: 887–889

[67] TORRE, M. BELLONI, P. (1989) Nd:YAG laser pleurodesis through thoracoscopy: new curative therapy in spontaneous pneumothorax. Ann Thorac Surg 47: 887–889

[68] MARTINI, N, BAINS, MS, BEATTIE, EL. (1975) Indications for pleurectomy in malignant effusion. Cancer, 35: 734–738

[69] WAKABAYASHI, A. (1991) Expanded application of diagnostic and therapeutic tharacoscopy. J Thorac Cardiovasc Surg, 102: 721–723

[70] WAKABAYASHI, A. (1994) Thoracoscopic partial lung resection in patients with severe chronic obstructive pulmonary disease (COPD): preliminary report. Arch Surg 129: 940–944

[71] WAKABAYASHI, A, BRENNER, M, KAYALEH, R.A, ET AL (1991) Thoracoscopic carbon dioxide laser treatment of bullous emphysema. Lancet, 337: 881–883

[72] BARKER, SJ, CLARKE, C, TRIVEDI, N, ET AL (1993) Anaesthesia for thoracoscopic laser ablation of bullous emphysema. Anaesthesiology, 78: 44–50

[73] BRENNER, M, KAAYALEH, RA, MILNE, EN, ET AL (1994) Thoracoscopic laser ablation to pulmonary bullae: radiographic selection an treatment response. J Thorac Cardiovasc Surg 107: 883–890

[74] COOPER, JD, TRULOCK, EP, TRIANTAFILLOU AN, ET AL (1995) Bilateral pneumectomy (volume reduction) for chronic obstructive pulmonary disease. J Thorac Cardiovasc Surg, 109: 106–119

[75] QUA JC, RAO UNM, TAKITA H. (1993) Malignant pleural mesothelioma: clinicopathological study. J Surg Oncol 54: 47–50

[76] ANTMAN KH, PASS HI, RECHT A. (1989) Benign and malignant mesothlioma. In: Devia V (ed) Cancer; principles and practice of oncology. JB Lippincott Philadelphia, PA, pp 1399–1417

[77] PASS HI, TOCHNER Z, DELANEY TF, ET AL (1990) Intraoperative photodynamic therapy for malignant mesothelioma. Ann Thorac Surg 50: 687–688

[78] TAKITA H, DOUGHERTY TJ. (1995) Intracavitary photodynamic therapy for malignant pleural mesothelioma. Semin Surg Oncol 11: 368–371

[79] PASS HI, DELANEY TF, TOCHNER Z, ET AL (1994) Intrapleural photodynamic therapy: Results of phase I trial. Ann Surg Oncol 1: 28–37

[80] RIS HAS-BEAT, ALTERMATT HJ, NACHBUR B, ET AL (1996) Intraoperative photodynamic therapy with m-THPC for chest malignancies. Laser Surg Med 18: 39–45

[81] BASS P, MURRER L, ZOETMULDER FAN, ET AL (1997) Photodynamic therapy as adjuvant therapy in surgically treated pleural malignancies. Br J Cancer 76: 819–826

# III-11
# Transmyocardial Laser Revascularization

Th. Krabatsch and R. Hetzer

## Contents

## Introduction

Transmyocardial laser revascularization (TMLR) is a surgical procedure which is used as an ultimaratio method in the treatment of severe diffuse coronary disease. Transmyocardial channels are created in the free wall of the left ventricle by use of a laser. After such an operation, up to 80% of the patients report a significant relief of angina symptoms combined with increased physical endurance levels.

Initially, it was assumed that after TMLR small amounts of blood are pressed through the channels into the intramyocardial vascular network during every contraction of the heart, thus increasing the myocardial blood supply independently of the diseased coronary arteries. More recently, when it became clear that the laser channels occlude immediately after their creation, other theories were debated, which included angioneogenesis, destruction of free nerve endings or whole nerves that are responsible for the conduction of angina pain signals.

Transmyocardial laser revascularization has been used in about 5000 patients worldwide over the past 5 years. At the Deutsches Herzzentrum Berlin we have performed TMLR operations on 191 patients who suffered from severe angina pectoris that was unresponsive to medical therapy from 7/1994 to 9/1999. In all patients, conventional bypass surgery or PTCA did not seem to offer the possibility of success. The relief of angina symptoms and the improved quality of life after TMLR seems to justify the use of the method in patients who present with urgent indications for CABG surgery, although their coronary vessel status does not seem to offer success for this conventional procedure.

However, considering its experimental foundation and long-term effects it seems that a conclusive assessment of TMLR is not yet possible.

## Historical Background

For more than 60 years it has been known that in warm-blooded species the myocardium is primarily perfused via coronary arteries, which provide a capillary network, but there are direct connections between the ventricular chamber and the capillaries by Thebesian veins and sinusoids [62].

Until coronary artery bypass grafting became the standard method for treating coronary disease several surgical approaches were common, which provided an additional blood flow to the myocardium via extracoronary pathways. The Vineberg procedure was such a method, where the internal thoracic arteries were directly implanted into the left ventricular myocardium without any vascular anastomoses [57]. To the end of the use of this procedure it remained controversial, even though indications had been found that in some cases there was a measurable improvement of myocardial perfusion. later, the implantation of perforated plastic tubes or segments of the carotid artery with one end reaching into the ventricular chamber was under investigation for carrying blood from the left ventricular chamber into the myocardium [15, 34].

It was the German surgeon Paul Walter [58 – 61], who, based on reports in 1956 from the Indian surgeon the intramyocardial vascular network. With the development of CABG in the late 1960s, this method became forgotten.

For the first time in the early 1980s, Mirhoseini and coworkers created transmyocardial channels with the help of a laser [35–42]. The use of a laser was initially presumed to ensure that the channels did not occlude within an early period after their creation. Okada and coworkers performed the first clinical TMLR procedure in 1970 [45].

In Germany there have been about 2000 TMLR operations over the past 5 years. The first transmyocardial laser revascularization in Germany was performed at the Deutsches Herzzentrum Berlin in the summer of 1994.

## The TMLR Procedure

The operative setting for a laser revascularization depends on whether a sole TMLR of a TMLR combined with a CABG operation has to be performed.

When there are no suitable target vessels for bypass grafting at all, s sole TMLR is the only treatment option. Then it is not necessary to arrest the heart, and extracorporeal circulation is not required. The creation of transmyocardial channels on the beating heart is possible by the use of a laser that can release a high amount of energy within a short time. The ma-

jority of institutions performing TMLR use the Heart-laser, a carbon dioxide laser (wavelength $\lambda = 10\,600$ nm) that delivers invisible radiation with an energy output of 800 W. For the creation of one channel this laser takes less than 40 ms. For this operation the patient is placed in the right lateral position and a left anterolateral thoracotomy in the 5th intercostal space is carried out. This gives the surgeon an optimal approach to the whole left ventricle. The use of this anterolateral surgical approach has the advantage of the absence of complication adhesions. The majority of patents for sole TMLR have CABG operations in their history, which has usually led to large retrosternal adhesions, which do not have to be passed.

A TMLR in addition to bypass grafts is performed during a CABG operation. The operation itself does not differ from a regular CABG procedure (i.e. median sternotomy, use of extracorporeal circulation), only that the laser is used in regions that are not amenable for bypass grafting.

Recently, cardiologists have performed TMLR as a catheter technique. Compared with the surgically operated TMLR patient group, there are no intermediate or long-term results of this approach available [22, 23].

Only a few institutions have experience with thoracoscopic TMLR as an minimally invasive endoscopic technique [56].

## Indication

The patient had to meet the following criteria before we decided to perform a sole TMLR:
- severe diffuse coronary artery disease is present,
- patient suffers from severe angina (CCS class 3 or 4) that has led to marked deterioration in the quality of life,
- maximum antianginal therapy has been carried out,
- the patient is not a candidate for PTCA, CABG or heart transplantation.

When in cases of severe diffuse multivessel disease where only some regions are suitable for bypass surgery, while others have no bypass target vessels, these regions can be treated by adjunctive TMLR.

## Criteria for Exclusion

We consider TMLR as not indicated when the patient's left ventricular contractility is severely diminished (LVEF < 30%) due to one or more myocardial infarctions. It is our experience that these patients do not benefit from the procedure, even when the perioperative mortality of the TMLR operation is acceptably low. These patients often suffer more from dyspnea than from angina, which is a result of the severely diminished left ventricular function. However, LV function seems not to be improved by a TMLR.

## Comparison with Alternative Methods

TMLR is currently the only surgical treatment option for patients with diffuse, inoperable coronary artery disease. It can reduce the patient's complaint significantly for up to 3 years, and can be repeated [27]. For these patients, a number of non-surgical alternative modes of treatment have been recommended, such as long-term intermittent urokinase therapy, neuronal stimulation in the form of transcutaneous electric nerve stimulation (TENS) and spinal cord stimulation (SCS) [5, 8, 49]. However, up to now there no major controlled studies have been published that compare the results of these methods in the treatment of end-stage coronary patients.

Only for the comparison of TMLR versus a continued medical management have the results of controlled, randomized studies been published [1, 12, 50]. The authors mostly conclude that TMLR is superior with regard to angina symptoms, physical endurance and the quality of life. Single groups claim that TMLR even improves myocardial perfusion [12].

## Results

After initial clinical application of TMLR in the USA within an FDA phase-I trial, this procedure was more and more used by cardiosurgical sites worldwide. At the end of 1998 there were 27 European heart surgery units using the $CO_2$ laser, and another group of teams that used other laser systems, such as the excimer laser or a holmium-YAG laser. Table 1 summarizes the experience with the $CO_2$-Heart Laser in European countries.

There are a large number of clinical studies on TMLR published. The results from these studies differ only marginally. Generally, the findings can be summarized as follows [1, 4, 7, 9, 11, 12, 18, 19, 26, 30, 33, 43, 44, 47, 50, 55]:
- there is a significant decline in angina symptoms after TMLR,
- physical endurance is improved after the procedure,
- there seems to be no proof of an increased myocardial perfusion,
- despite occasionally seen improvement of regional myocardial function, global left ventricular ejection fraction seems to stay unimproved by TMLR,

**Table 1.** Number of performed TMLR operations using the $CO_2$ Heart Laser system until 12/1998 in different European countries

| Country | No. of TMLR sites | No. of operations |
|---|---|---|
| France | 2 | 27 |
| Germany | 12 | 1066 |
| Greece | 1 | 7 |
| Italy | 3 | 84 |
| Norway | 1 | 50 |
| Poland | 1 | 101 |
| Spain | 2 | 37 |
| Switzerland | 3 | 315 |
| United Kingdom | 4 | 159 |
| Total | 29 | 1846 |

**Table 2.** Characterization of the 186 patients for TMLR

| | Group A (TMLR) | Group B (TMLR + CABG) |
|---|---|---|
| No. of patients | 163 | 28 |
| Gender (m:f) | 6.1 : 1 | 2.2 : 1 |
| Age (years) | 64.0 ± 7.0 | 61.9 ± 9.0 |
| Patients with three-vessel-disease (%) | 84.0 | 96.4 |
| Previous CABG (%) | 88.3 | 10.7 |
| No. of PTCA per patient | 0.5 | 0.27 ± 0.3 |
| No. of infarcts per patient | 1.5 ± 0.7 | 1.0 ± 0.3 |
| Angina class (CCS) | 3.5 ± 0.9 | 3.1 ± 0.9 |
| Maximum work load (W) | 58.0 ± 29.7 | 83.2 ± 39.2 |
| Left ventricular EF (%) | 44.0 ± 11.6 | 43.1 ± 5.6 |

- TMLR channels seem to occlude within the first few days after the procedure, such that mechanisms other than blood flow via TMLR channels have to be considered as responsible for clinical TMLR effects.

The following section will summarize the experience at the Deutsches Herzzentrum Berlin in the treatment of TMLR patients.

Within the period from 7/1994 to 9/1999, a total of 191 patients underwent TMLR at the Deutsches Herzzentrum Berlin. In 28 patients TMLR was an adjunct to coronary bypass grafting. Within the remaining 163 patients we performed the laser procedure as sole therapy. Anamnaestic and clinical data from the entire 191 operated patients are given in Table 2. The patient group A (isolated TMLR) was investigated following a protocol consisting of echocardiography, thallium perfusion scan under physical exercise, left heart catheterization and a special questionnaire on complaints and quality of life of the patients, clinical assessments were made prior to discharge (echocardiography) and at 3, 6, 12, 18, 24 and 36 months postoperatively.

We consider it impossible to differentiate between CABG effects and TMLR effects in patients who received TMLR as an adjunct to coronary bypass grafting. For this reason, these patients (group B) were excluded from further investigations. The following results are based on the 163 patients after TMLR as the sole therapy.

## Angina Symptoms

Postoperatively, the absolute majority of patients reported a distinct decline in their angina symptoms. One year after the operation 79% and dyspnoea was reduced in 67.2 % of the patients who had reported these complaints preoperatively, the decline of complaints was defined as diminished frequency or intensity of angina or equivalents or their occurrence on a distinctly higher level of physical stress.

Pre- and postoperatively all patients were classified according to the classification of the Canadian Cardiovascular Society (CCS) into one of four angina classes. The results are summarized in Table 3.

In order to characterize the patients who benefited from the TMLR procedure when compared with those who reported no changes, both subgroups were analysed for significant differences with regard to 50 different anamnaestic and clinical factors. Significant differences between both groups were fund only in preoperative left ventricular ejection fraction. While more than 75% of the patients with a preoperative LVEF of 30% or more benefited from TMLR, only 34% of the remaining patients reported significantly diminished angina or dyspnoea (p < 0.05).

## Left Ventricular Contractility

The ejection fraction of the left ventricle as an expression of global left ventricular contractility postoperatively tended to increase, but the difference did not reach a statistical significance. The highest value was measured 3 months after TMLR. The increase of LVEF to an average 109% of the initial value barely missed the level of significance (p = 0.082).

For a detailed analysis of left ventricular wall motion we divided the left ventricle into 20 areas. For each area we determined the extent of wall motion disturbance. Based on this, we calculated a left ventricular wall motion disturbance index by adding the values for the 20 areas. This index changed in accordance with the LVEF. However, the difference again did not reach statistical significance; but in this global index the septal contractility was involved, even though we did not create laser channels in septal areas. A specific wall motion disturbance index, calculated only for treated areas, showed an improvement of myocardial contractility in the regions of interest in 64% of our patients 3 months postoperatively (Table 4).

## Thallium Perfusion Scan with Physical Endurance

The myocardial perfusion scan using 201 thallium revealed ischemic areas in only 60% of the patients. In the remaining patients, we found more or less scars in the left ventricle. Among the patients with preoperative evidence of ischemia, we found in 44% of the postoperative perfusion scans a reduced extent of the myocardial ischemia. In 42% we saw unchanged scans, and the remaining 14% of the patients pre-

**Table 3.** Pre- and postoperative angina class (CCS)

| Preoperatively | 3 months postop. | 6 months postop. | 12 months postop. | 18 months postop. | 24 months postop. | 36 months postop. |
|---|---|---|---|---|---|---|
| $3.4 \pm 0.9$ | $2.4 \pm 0.8$ | $2.0 \pm 0.9$ | $2.4 \pm 1.0$ | $2.7 \pm 1.0$ | $2.9 \pm 1.1$ | $3.0 \pm 1.0$ |

**Table 4.** Regional wall motion in laser-treated areas (% of investigated patients)

| | After | | | | |
|---|---|---|---|---|---|
| | 3 months | 6 months | 12 months | 18 months | 24 months |
| Improved | 64 | 38 | 41 | 67 | 50 |
| Unchanged | 8 | 19 | 18 | 0 | 0 |
| Worsened | 28 | 43 | 41 | 33 | 50 |

**Table 5.** Average ergometrical endurance before and after TMLR

| | Preoperatively | Examination after | | | | |
|---|---|---|---|---|---|---|
| | | 3 months | 6 months | 12 months | 24 months | 36 months |
| Max. work load (W) | 58.0 | 73.3 | 77.1 | 78.2 | 86.0 | 82.4 |

sented with increased ischemic perfusion defects. Among the patients with improved perfusion scans were five patients with less ischemia presenting in the area of septum, despite the fact that the latter was not treated.

During the thallium scan the patients underwent physical exercise testing on a bicycle ergometer. The results from this tests are summarized in Table 5. It became evident that the majority of the patients postoperatively tolerated a significantly higher level of endurance. As a cause of withdrawal from the examination, angina occurred less and less, but dyspnoea became the primary cause for stopping the physical stress test.

### Perioperative Morbidity and Mortality

Twelve of the 191 patients died during the postoperative course, one of them had undergone a combined procedure (TMLR + CABG), and 11 underwent TMLR as sole therapy. This results in a 6.3% mortality. As with the majority of the other TMLR groups, our perioperative mortality has continuously decreased over the years. Indeed, for the last 50 patients there was only one perioperative death (2% mortality). The 1-year mortality rate seems to be unfavourably high in patients with a preoperative left ventricular ejection fraction of less than 30%. This has increased our reluctance to operate on this subgroup.

In 3 of the 163 patients (isolated TMLR group), we observed an increasing mitral incompetence during the postoperative course. It may be assumed that the laser beam injured the chordae during the TMLR operation. Some weeks later, chordae rupture occurred and mitral incompetence was established. However, during routine echocardiography examination prior to discharge there was no evidence of mitral regurgitation in any of the three patients. On the other hand, the mitral incompetence could have been of ischemic origin.

### Histological Findings

The relatives of 8 of the 11 patients who died after sole TMLR gave their consent to a postmortem examination. In cooperation with the Rudolf Virchow Hospital's Institute of Pathology, detailed histological analysis was performed on the hearts of these patients in whom a total of 250 channels had been created. In the process several hundred individual slides were prepared [25].

The specimens were stained with haematoxylin eosin (HE). Furthermore, immunohistological staining methods that used the avidin-biotin method were performed in addition to HE. As primary antibodies we chose CD 68 for reaction with macrophages, CD 31 for reaction with endothelial cells, a polyclonal collagen tape-IV staining agent for collagenic fibres and MIB 1 for reaction with the Ki-67 antigen, a proliferation-associated nuclear antigen.

We found the TMLR channels which had been created by a $CO_2$ laser to be surrounded by a zone of necrosis with an extent of about 500 µm. In the hearts from patients who died in the early postoperative period (1 to 7 days postoperatively) almost all channels were closed by fibrin clots, erythrocytes and macrophages. There were no obvious connections between the channels and the ventricular cavity.

In specimens from patients who died 2 weeks or more after the procedure, a granular tissue with high macrophage and monocyte activity was observed. Within this tissue we observed a developing network of capillaries. Otherwise, the tissue filling the channels did not substantially differ from scar tissue. We did not observe connections between the ventricular cavity and the new capillaries.

### Repeat TMLR

In five of our 191 TMLR patients, after an initial angina-free interval of 1–3 years, we decided to repeat the laser operation because the patient reported a reoccurrence of the angina symptoms [27]. We consider a repeat TMLR after an initial period of angina relief to be indicated only in single cases. The occlusion of a previously patent bypass graft can be an indication for such a redo-TMLR. Today, four out of the five patients are well and experience significantly less angina than prior to the second TMLR operation.

## Discussion

Today, there is little or no doubt that severe angina can be treated effectively by TMLR. The data that have been published seem to show that this is the case, at least for the fist and second postoperative year. The quality of life and physical endurance are also improved. Still to be answered is the question of the underlying mechanism, and especially as to whether improvement of myocardial perfusion by the laser operation occurs, which some study groups claim [7, 11, 12] and others reject [51, 52].

A significant subgroup of patients who underwent TMLR report an immediate relief of angina after the procedure. For this phenomenon there are three possible explanations. One is a placebo effect, the second is myocardial denervation by the laser, and the third is the "classic" perfusion theory. The last postulates that there is an improvement in myocardial perfusion via the created laser channels. However, the data published in the literature on this topic are not in agreement. On the one hand, the majority of the experimental study groups are unable to prove any beneficial effects of TMLR regarding myocardial perfusion [14, 17, 29, 63], but other investigators like Horvath and coworkers report that there is a distinct protection of the myocardial areas treated by TMLR [20, 64]. Recent clinical data published by Frazier and coworkers [12] seem to support this thesis. However, it is our opinion that the question of whether there is blood flow via the TMLR channels is still open.

The majority of the published experimental data originated from canine models of acute ischemia. In contrast, we favour porcine animal models because there are only a limited number of collateral vessels in pig hearts and the dimensions of the pig heart are comparable to that of a human. In our opinion this is a significant limitation of the canine TMLR studies, because the canine heart is known to have excellent collateralization as well as a profound interindividual variability in its myocardial vasculature. Studies on rat hearts [16] we consider unhelpful, because the relationship between the myocardial wall thickness, the channel diameter and the erythrocytes differs significantly from circumstances found in clinical TMLR.

Recently, preliminary results have been published that suggest that an additional application of endothelial growth factors, especially a gene transfer, into the vicinity of the laser channels during the TMLR procedure could enhance the long-term patency of the channels [2]. This places the question about the principle possibility of myocardial perfusion via channels direct from the left ventricular cavity back at the centre of the debate. In the past, here were major doubts as to whether a blood flow via laser channels was possible, because for decades it was well known that in a healthy myocardium the IMP exceeds the intraventricular pressure throughout the heart cycle. The existing pressure gradient should, therefore, on average prevent blood from flowing from the ventricular cavity into the myocardium [46]. However, we were able to demonstrate that this is not true in the case of severe ischemia. In fact, in a severely ischemic myocardium, the intraventricular pressure can exceed the IMP. The IMP rose after TMLR in our experiment, which can be interpreted as indirect evidence of relief of the ischemia, otherwise this effect would counteract the blood flow via the channels. Therefore, we assume that a steady state should be established at a certain level.

Our own clinical results have tentatively shown that all TMLR channels are occluded by thrombotic material after 2 to 3 days [25]. However, the results of other experimental and clinical studies make it probable that there is a blood flow via TMLR channels immediately after the laser procedure. This seems to be of special interest because recent studies with TMLR, in combination with the application of vascular growth factors, found a higher patency rate of the TMLR channels after growth factor injection into the myocardium [2].

## References

[1]  Allen KB, Dowling RD, Fudge TL, Schoettle GP, Selinger SL et al. Comparison of transmyocardial revascularization with medical therapy in patients with refractory angina. N Engl J Med 1999; 341: 1029–36

[2]  Brilla CG, Rybinski L, Rupp H, Moosdorf R, Transmyokardiale Laserrevaskularisation und Applikation von VEGF oder VEGF-Gentransfer beim Schwein. Z Kardiol 1998; 87 (suppl I): 202

[3]  Burkhoff D, Fisher PE, Apfelbaum M, Kohmoto T, De Rosa CM, Smith CR. Histologic appearance of transmyocardial laser channels after $4^{1}/_{2}$ weeks. Ann Thorac Surg 1996; 61: 1532–5

[4]  Burns SM, Sharples LD, Tait S, Caine N, Wallwork J, Schofield PM. The transmyocardial laser revascularization international registry report. Eur Heart J 1990; 20: 31–7

[5]  Chauhan A, Mullins PA, Thuraisingham SI, Taylor G, Petch MC, Schofield PM. Effects of transcutaneous electrical nerve stimulation on coronary blood flow. Circulation 1994; 89: 694–707

[6]  Cooley DA, Frazier OH, Kadipasaoglu KA, et al.: Transmyocardial laser revascularization. Anatomic evidence of long-term channel patency. Tex Heart J 1994; 21 (3): 220–4

[7]  Cooley DA, Frazier OH, Kadipasaoglu KA, Lindenmeir MH, Pehlivanoglu S, Kolff JW, Wilansky S, Moore WH, Transmyocardial laser revascularization: clinical experience with twelve-month follow-up. J Thorac Cardiovasc Surg 1196; 111: 791–9

[8]  De Jongste MJ, Nagelkerke D, Hooyschur CM, et al. Stimulation characteristics, complications and efficacy of spinal cord stimulation systems in patients with refractory angina. A feasibility study. Pace 1994; 17: 1751–60

[9] Donovan CL, Landolfo KP, Lowe JE, Clements F, Coleman RB, Ryan T. Improvement of inducible ischemia during dobutamine stress echocardiography after transmyocardial laser revascularization in patients with refractory angina pectoris. JACC 1997; 30: 607–12

[10] Emond M, Mock MB, Davis K, Fischer LD, Holmes DR, Chaitman BR, Kaiser GC, Aldermann E, Killip III T. Long-term survival of medically treated patients in the coronary artery surgery study (CASS) registry. Circulation 1994; 90: 2645–57

[11] Frazier OH, Cooley DA, Kadipasaoglu KA, Pehlivanoglu S, Lindenmeir M, Barasch E, Conger JL, Wilansky S, Moore WH. Myocardial revascularization with laser. Preliminary findings. Circulation 1995; 92 (suppl II): II-58-II-65

[12] Frazier OH, March RJ, Horvath KA. Transmyocardial revascularization with a carbon dioxide laser in patients with end-stage coronary artery disease. N Engl J Med 1999; 341: 1021–8

[13] Gassler N, Wintzer HO, Stube HM, Wullbrand A, Helmchen U. Transmyocardial laser revascularization. Histological features in human nonresponder myocardium. Circulation 1997; 95: 371–5

[14] Goda T, Wierzbicki Z, Gaston A et al.: Myocardial revascularization by $CO_2$ laser. Eur Surg Res 1987; 19: 113–7

[15] Goldman A, Greenstone SM, Preuss FS, Chang ES. Experimental methods for producing a collateral circulation to the heart directly from the left ventricle. J Thorac Surg 1956; 31: 354–74

[16] Guo JX, Pan L, Ma M, Chen Z, Xu H, Shi AY. Experimental studies of laser myocardial revascularization in rats. Chin Med J 1993; 106: 665–7

[17] Hardy RI, James FW, Millard RW, Kaplan S. Regional myocardial blood flow and cardiac mechanics in dog hearts with $CO_2$ laser-induced intramyocardial revascularization. Basic Res Cardiol 1990; 85: 179–97

[18] Horvath KA, Cohn LH, Cooley DA, Crew JR, Frazier OH, Griffith BP, Kadipasaoglu K, Lansing A, Mannting F, March R, Mirhoseini MR, Smith C. Transmyocardial laser revascularization: results of a multicenter trial with transmyocardial laser revascularization used as sole therapy for end-stage coronary artery disease. J Thorac Cardiovasc Surg 1997; 113: 645–54

[19] Horvath KA, Mannting F, Cummings N, Shernan SK, Cohn LH, Transmyocardial laser revascularization: operative techniques and clinical results at 2 years. J Thorac Cardiovasc Surg 1996; 111: 1047–53

[20] Horvath KA, Smith WJ, Laurence RG, Schoen FJ, Appleyard EF, Cohn LH. Recovery and viability of an acute myocardial infarct after transmyocardial laser. JACC 1995; 25(1): 258–63

[21] Kadipasaoglu KA, Pehlivanoglu S, Conger JL et al. Long- and short-term effects of transmyocardial laser revascularization in acute myocardial ischemia. Lasers Surg Med 1997; 20: 6–14

[22] Kim CB, Kesten R, Javier M, Hayase M, Walton AS, Billingham ME, Kernoff R, Oesterle SN. Percutaneous method of laser transmyocardial revascularization. Cathet Cardiovasc Diagn 1997; 40: 223–8

[23] Kim CB, Oesterle SN. Percutaneous transmyocardial revascularization. J Clin Laser Med Surg 1997; 15: 293–8

[24] Kohmoto T, Fischer PE, Gu A, et al., Physiology, histology, and 2-week morphology of acute transmyocardial channels made with a $CO_2$ laser. Ann Thorac Surg 1997; 63: 1275–83

[25] Krabatsch T, Schäper F, Leder C, Tülsner J, Thalmann U, Hetzer R. Histological findings after transmyocardial laser revascularization. J Card Surg 1996; 11(5): 326–31

[26] Krabatsch T, Tambeur L, Lieback E, Schäper F, Hetzer R. Transmyocardial laser revascularization in the treatment of end-stage coronary artery disease. Ann Thorac Cardiovasc Surg 1998; 4: 64–71

[27] Krabatsch T, Tambeur L, Lieback E, Hetzer R. Secondary transmyocardial laser revascularization in the treatment of end-stage coronary artery disease J Card Surg 1998; 13: 93–7

[28] Kwong KF, Kanellopoulos GK, Nickols JC, Pogwizd SM, Saffitz JE, Schuessler RB, Sundt. T. $3^{rd}$. Transmyocardial laser treatment denervates canine myocardium. J Thorac Cardiovasc Surg 1997; 114(6): 883–9

[29] Landsreneau R, Nawarawong W, Laughlin H et al., Direct $CO_2$ laser "revascularization" of the myocardium. Lasers Surg Med 1991; 11: 35–42

[30] Lutter G, Frey M, Saurbier B, Nitzsche E, Hoegerle S, Brunner M, Martin J, Lutz C, Spillner G, Beyersdorf F, Behandlungsstrategien bei therapierefraktärer Angina pectoris: Transmyokardiale Laserrevaskularisation. Z Kardiol 1998; 87(Suppl II): 199–202

[31] Lutter G, Martin J, Koster W, Grawitz AB, Esenwein P, Geiger A, von Specht B, Beyersdorf F. Analysis of the new indirect revascularization method by determining objective parameters of clinical chemistry, histochemistry and histology. Eur J Cardiothor Surg 1999 May; 15(5): 709–16

[32] Mack CA, Patel SR, Rosengart TK. Myocardial angiogenesis as a possible mechanism for TMLR efficacy. J Clin Laser Med Surg 1997; 15: 275–9

[33] March RJ. Transmyocardial laser revascularization with the $CO_2$ laser: one year results of a randomized, controlled trail. Semin Thorac Cardiovasc Surg 1999; 11: 12–8

[34] Massimo C, Boffi L. Myocardial revascularization by a new method of carrying blood directly from the left ventricular cavity into the coronary circulation. J Thorac Surg 1957; 34: 257–64

[35] Mirhoseini M, Cayton M, Shelgikar S, Fischer JC. Laser myocardial revascularization. Lasers Surg Med 1986; 6(5): 459–61

[36] Mirhoseini M, Cayton M. Revascularization of the heart by laser. J Microsurg 1981; 2: 253–60

[37] Mirhoseini M, Cayton M. Transmyocardial laser revascularization: historical background and future directions J Clin Laser Med Surg 1997; 15(6): 245–53

[38] Mirhoseini M, Fischer J, Cayton M. Myocardial revascularization by laser: a clinical report. Lasers Surg Med 1983; 3(3): 241–5

[39] Mirhoseini M, Muckerheide M, Cayton MM. Transventricular revascularization by laser. Lasers Surg Med 1982; 2(2): 187–98

[40] Mirhoseini M, Shelgikar S, Cayton M. Clinical and histological evaluation of laser myocardial revascularization. J Clin Laser Med Surg 1990; 6: 73–8

[41] Mirhoseini M, Shelikar S, Cayton MM, Transmyocardial laser revascularisation: a review. J Clin Laser Med Surg 1993; 11: 15–9

[42] Mirhoseini M, Shelgokar S, Cayton MM. New concepts in revascularization of the myocardium. Ann Thorac Surg 1988; 45(4): 415–20

[43] Moosdorf R. Transmyokardiale Revaskularisation. Z Kardiol 1997; 86 (Suppl 1): 115–24

[44] Nägele H, Stubbe HM, Nienaber C, Rödiger W. Results of transmyocardial laser revascularization in non-revascularizable coronary artery disease after 3 years follow-up. Eur Heart J 1998; 19: 1525–30

[45] Okada M, Ikuta H, Shimizu K, Horii H, Nakamura K. Alternative method of myocardial revascularization by laser: experimental and clinical study. Kobe J Med Sci 1986; 32: 151–61

[46] Pifarre R. Intramyocardial pressure during systole and diastole. Ann Surg 1968; 168: 871–75

[47] Raffa H, Memon F, Jabbad H, Moinuddin M, Kayali MT, Langer J, Ramadan M, Kumar D, Makarem H. Transmyocardial laser revascularization: Saudi experience. Asian Cardiovasc Thorac Ann 1996; 4(2): 75–9

[48] Reuthebuch O, Bauer EP, Berwing K et al.: Transmyocardial laser revascularization: Evidence of channel-perfusion by means of contrast-echocardiography. Thor Cardiovasc Surg 1997; 45, suppl. 1: 145

[49] Schoebel FC, Leschke M, Jax TW, Stein D, Strauer BE. Chronic-intermittent urokinase therapy in patients with end-stage coronary artery disease and refractory angina pectoris: a pilot study. Clin Cardiol 1996; 19: 115–20

[50] Schofield PM, Sharples LD, Caine N, Burns S, Tait S, Wistow T, Buxton M, Wallwork J. Transmyocardial laser revascularization in patients with refractory angina: a randomized controlled trail. Lancet 1999; 353: 519–24

[51] Sen PK, Daulatram J, Kinare SG, Udwadia TE, Parulkar GB. Further studies in multiple transmyocardial acupuncture as a method of myocardial revascularization. Surgery 1968; 64(5): 861–70

[52] Sen PK. Transmyocardial acupuncture – a new approach to myocardial revascularization. J Thorac Cardiovasc Surg 1965; 50: 181–9

[53] Sigel JE, Abramovich CM, Lytle BW, Ratliff NB. Transmyocardial laser revascularization: three sequential autopsy cases. J Thorac Cardiovasc Surg 1998; 115: 1381–5

[54] Spanier TB, Smith CR, Burkhoff D. Angiogenesis: A possible mechanism underlying the clinical benefit of transmyocardial laser revascularization. J Clin Laser Med Surg 1997; 15: 269–73

[55] Sundt 3rd TM, Rogers JG. Transmyocardial laser revascularization for inoperable coronary artery disease. Curr Opin Cardiol 1997; 12: 441–6

[56] Vincent JG, Bardos P, Kruse J, Maass D. End-stage coronary disease treated with the transmyocardial $CO_2$ laser revascularization: a chance for the "inoperable" patient. Eur J Cardio-thorac Surg 1997; 11: 888–94

[57] Vineberg A. Clinical and experimental studies in the treatment of coronary artery insufficiency by internal mammary artery implant. J Int Coll Surg 1954; 22: 503–18

[58] Walter P, Hempelmann G, Hundeshagen H, Borst HG. Blutflussmessungen im ischämischen Myokard und nach Ausführung einer Revaskularisationsmethode mit Hilfe von Rubidium-86. Mitteilungen der Gesellschaft für Nuklearmedizin 1970; 12: 1–14

[59] Walter P, Hundeshagen H, Borst HG. Treatment of acute myocardial infarction by transmural blood supply from the ventricular cavity. Eur Surg Res 1971; 3: 130–8

[60] Walter P, Lamprecht W, Hundeshagen H, Borst HG. Myocardial blood flow and alterations of LDH isoenzymes in infarcted heart muscle and after transmural punctures. Cardiology 1971; 56: 371–6

[61] Walter P, Zazvorka F, Hundeshagen H, Borst HG. Experimental evaluation of transmural puncture as a treatment of acute myocardial infarction. Bulletin de la Societe Internationale de Chirurgie 1973; 1: 2–11

[62] Wearn JT, Mettier SR, Klumpp TG, Zschiesche LJ. The nature of the vascular communications between the coronary arteries and the chambers of the heart. Am Heart J 1933; 9: 143–64

[63] Whittaker P, Kloner RA, Przyklenk K. Laser-mediated transmural myocardial channels do not salvage acutely ischemic myocardium. JACC 1993; 22: 302–9

[64] Whittaker P, Rakusan K, Kloner RA. Transmural channels can protect ischemic tissue. Assessment of long-term myocardial response to laser- and needle-made channels. Circulation 1996; 93: 143–52

# III-12
# Lasers in General Surgery

# III-12.1
# Laser Treatment of Neurofibromatosis Type 1

Sabine D. Schmitz and B. Algermissen

## Contents

## Extended Summary for Non-Experts

In 1882, von Recklinghausen gave the definitive account of the clinical and pathologic features of this disease [1]. Neurofibromatosis is a polysymptomatic hereditary neuroectodermal systemic disease. Nearly 50% occurred as a spontaneous mutation and the inheritance is autosomal dominant [2]. In the past decades the entity of neurofibromatosis was divided into eight different clinical disorders with somewhat overlapping features (Table 1) [3, 4, 5].

The two most common types are designated neurofibromatosis type 1 (NF1) and neurofibromatosis type 2 (NF2) localized on different chromosomal loci: neurofibromatosis type 1 on the long arm of chromosome 17, neurofibromatosis type 2 on chromosome 22 [6, 7]. The incidence of neurofibromatosis type 1 is 1: 3500–4000 [8, 9], of neurofibromatosis type 2 about 1–2: 100.000 [10].

Neurofibromatosis type 2 is characterized by the presence of bilateral acoustic neurinomas, which were found in more than 90% of patients with neurofibromatosis type 2. Other cranial nerve tumours, meningiomas and only a few skeletal abnormalities are typical. Cutaneous or subcutaneous neurofibromas are rare, and the occurrence of diffuse plexiform neurofibromas unusual.

Table 1. Different clinical neurofibromatosis subtypes according to Riccardi [2-4]

| | |
|---|---|
| 1. Classical form, neurofibromatosis type 1 (>90%) | NF I |
| 2. Central form (fam. acoustic-neurinoma) | NF II |
| 3. Mixed form (1 and 2), multiple cranial nerve tumours | NF III |
| 4. Diffuse café-au-lait spots and large neurofibromas NF IV | |
| 5. Segmental café-au-lait spots and neurofibromas | NF V |
| 6. Diffuse café-au-lait spots, no neurofibromas, mental deficiency | NF VI |
| 7. Late-onset NF (3rd decade, deep neurofibromas) | NFVII |
| 8. Unspecific neurofibromas | NF VIII |

**Table 2.** Report of a National Institute of Health Consensus Conference (1998): two or more of the following findings strongly support the diagnosis of NF1

1. The presence of six or more café-au-lait maculas with a diameter ≥ 5,0 mm in children less than 6 years of age and > 15 mm in older individuals

2. Two or more neurofibromas of any type or one plexiform neurofibroma

3. Freckling in the axilla or inguinal region

4. An optic nerve glioma

5. Two or more Lisch nodules

6. Dysplasia of the sphenoid bone or thinning of the cortex of long bones, with or without pseudarthrosis

7. A first-degree relative exhibiting these changes

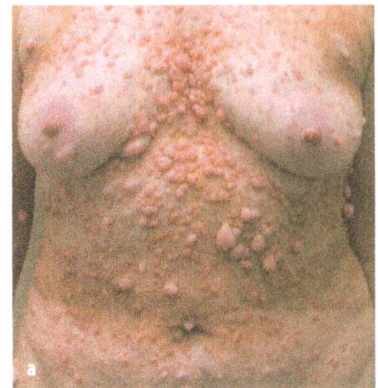

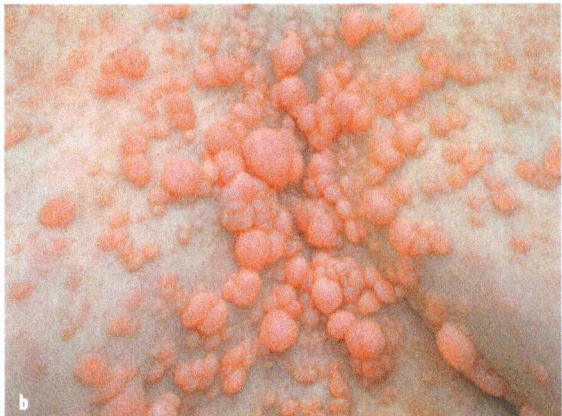

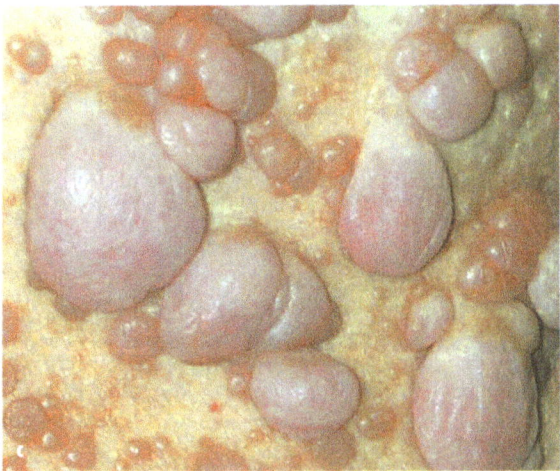

**Fig. 1a–c.** Neurofibromas can appear on all parts of the body, preferentially involving the trunk (**a, b**). The clinical picture of cutaneous neurofibromas shows a high number of variations from sessile, pedunculated or flattened to lobulated form (**c**). Mostly always all sizes and forms of neurofibromas are present at the same time

In contrast, neurofibromatosis type-1 presents a high rate of cutaneous features (café-au-lait spots and freckling) and cutaneous, subcutaneous (100% of all neurofibromatosis type-1 patients) or plexiform (30% of all neurofibromatosis type-1 patients) neurofibromas [11, 12]. Early manifestations are opticus gliomas (more than 15% of neurofibromatosis type-1 patients) and skeletal abnormalities such as dysmorphies, scoliosis or pseudarthrosis. The clinical phenotype can show considerable variations. In the majority of patients, café-au-lait spots and cutaneous or subcutaneous neurofibromas are the first clinical findings to confirm the diagnosis neurofibromatosis type-1 (Table 2) [8].

Per DNA analysis from patient's blood mutations up to 70% could be detected. If there is any suspicion of neurofibromatosis type-1 (minor clinical findings), molecular genetics diagnostic can help to confirm it.

Neurofibromas may be present in late childhood or early adolescence and develop at any time thereafter, generally with a steady increase in number and size with age [13]. Not only do the early-onset lesions tend to become more obvious and serious, new and varied types of lesions characteristically develop at different time periods. Neurofibromas can occur in all parts of the body and anywhere on the skin surface, preferentially involving the trunk, in accordance with the distribution of skin temperature [14, 15]. They may be of discrete occurrence or else may be spread over the entire body (more than 5000), so that social and clinical disfigurement and stigmatization in the private or professional surroundings are one of the outstanding problems of neurofibromatosis type-1 patients (Fig 1a,b,c). Also in regions where neurofibromas are growing very close together, daily hygiene is aggravated and complications like inflammation (mostly with corynebacterium or candida species) and malodour are frequent [16] (Fig. 2a,b).

The pathogenesis of the development of neurofibromas has not yet been enlightened. The most com-

mon theory is that neurofibromas are benign tumours arising from small to large sensible nerves or nerve branches. Each neurofibroma is composed primarily of Schwann cells, fibroblasts, perineural cells,

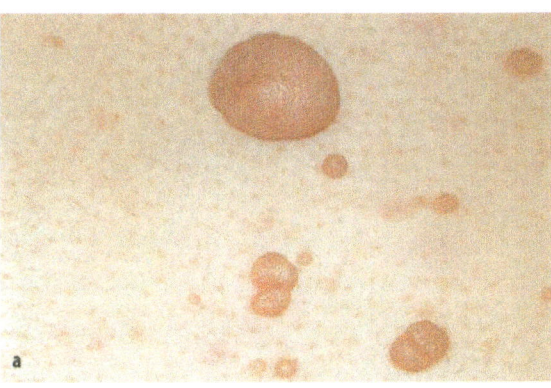

**Fig. 2a, b** . Cutaneous and subcutaneous neurofibromas can be spread separately over the whole body (**a**) or massive (**b**) mm – cm in size, sometimes so close side by side that hygiene is aggravated and complications like inflammation are programmed

endothelial cells, mast cells and a small number of nerves. Neurofibromas are apparently the same in terms of their histology and biological behaviour whether they occur as part of neurofibromatosis or as solitary tumours, distinct from neurofibromatosis.

An often reported symptom of neurofibromatosis patients is occasional itching over the ensuing days or weeks [17] accompanying the development or the growth of neurofibromas after an acute or intense trauma. In 1992 Riccardi described [13] the wound healing theory, that traumas of various types disrupt the normal relationships between Schwann cells and neurons and initiate wound healing of the nerve. Usually, in patients without neurofibromatosis, the result is a re-establishment of the normal state or formation of a scar. In the presence of a local somatic mutation with the loss of the neurofibromatosis type-1 allele, the wound healing persists and a local singular neurofibroma develops. In patients with neurofibromatosis, the abnormality is the absence of the signal that the wound-healing phase is complete. Riccardi affirmed that the origin of this signal is the key to understanding the basic defect in neurofibromatosis [13]. But this theory raises some questions, for example the development of neurofibromas at body sites

with a low number of mechanic trauma, and the findings of neurofibromatosis type-1 -/- Schwann cells in neurofibromas [18].

Up to now, no recommendable systemic therapy of neurofibromatosis type-1 exists. Based on the large amount of mast cells in neurofibromas and the clinical findings of symptoms, such as itching before neurofibromas growth, Riccardi hypothesised that mast cells contribute directly to neurofibroma development and growth [19, 20]. The first clinical studies used ketotifen, which functions as an antihistamine and additionally as a weak inhibitor of mast-cell degranulation. Patients were treated orally with 2–4 mg ketotifen per day for 30–43 months [20]. Neurofibroma growth and symptoms, for example pruritus, were documented before, during and after the therapy. These preliminary studies underline the hypothesis and a follow-up study with an open label protocol and double blind protocol was initiated. The results of the last study were controversial to the preliminary data. It was not possible to demonstrate a significant effect on the growth of neurofibromas, but only on clinical symptoms such as pruritus. The patients reported also an unexpected improvement in overall sense of well being, productivity and general performance during therapy that could be explained rather in the typical antihistaminic functions of the H1-blocker than in the function of mast cell stabilization of ketotifen. Therefore, despite a lack of mast cell degranulation inhibitors which could be used for systemic treatment, the hypothesis is up to now relevant and should be the basis for further studies.

Another target for systemic treatment may be the cellular ras activity [21]. The protein neurofibromin is expressed in several tissues and has the function of a tumour-suppressor regulating the activity of the oncogen ras [22, 23]. Neurofibromin belongs to the GTPase activating proteins (GAP) [24]. Therefore, the active form of ras (ras-GTP) was transformed to the inactive form via binding to the GAP domain of neurofibromin [25]. A mutation of the neurofibromatosis gene caused an increase in active forms of ras, and the imbalance could play a critical role in the development of clinical findings of neurofibromatosis type-1. Currently, phase I and II studies with different ras inhibitors are running, and in the future these strategies could offer new options for the systemic treatment of plexiform neurofibromas or severe forms of dermal neurofibromas development.

At this moment, only surgical treatments of neurofibromas are available.

Surgical excision with the scalpel of a large number of neurofibromas is time consuming and often accompanied by side effects such as bleeding [26]. In contrast, laser therapy allows the complete removal of up to 1000 neurofibromas in one session without major side effects [26–34]. The outstanding complications could be hypertrophy scars or, seldom, keloids with a rate of 0.5% and recurrence of 1–5 %. These side effects were also found after non-laser surgical treatments.

## Clinical Manifestations of Neurofibromatosis Type-1

Dermal benign neurofibromas are the most common skin manifestation of von Recklinghausen neurofibromatosis (NF-1). From the clinical finding, they could be divided into several subtypes of neurofibromas:

1. Cutaneous neurofibromas. These tumours move when the skin is moved and are often soft and fleshy to the touch and rarely painful. They vary in size from some mm to cm or more. Also, they may assume many forms, for example sessile, pedunculated or flattened and stay separately or grouped. The colour of the skin over these neurofibromas is often lilaceous (Fig. 1c).

2. Subcutaneous neurofibromas. These tumours are deeper to the dermis, so that they do not move when the skin above is moved. They have a spherical or ovoid shape, the consistency of a rubber, and may be painful (Fig. 3). Some tumours penetrate from the subcutis through a small opening in the connective tussue like a hernia and can be pushed back with the finger (phenomenon of „button-holing") [7].

3. Plexiform neurofibromas. They combine elements of cutaneous and/or subcutaneous neurofibromas and, if the skin is involved, often hyperpigmentation and/or hypertrichosis are seen [10, 11]. Sometimes, plexiform neurofibromas evolve enormous folds of the skin (dermatochalasis), which appear as a pendulous dewlap (dermatolysis) (Fig. 4). They are localized in the craniofacial region in the skin and are sometimes paraspinal or gastrointestinal, and can reach enormous size. Therefore, a complete surgical removal depends on body location and size and is mostly impossible, so that only a particular reduction could be achieved.

The café-au-lait spots [35] are one of the characteristic types of hyperpigmentation of neurofibromatosis type-1. They are often the first clinical manifestation of neurofibromatosis type-1 and occur until the first birthday. The size varies with age and after childhood they are sometimes better visible. The pigmentation ranges from light to dark brown (café-au-lait) but more or less uniform. Ten % of the normal population have one or more lesions of this type, but if a child has more than six café-au-lait-spots exceeding 5 mm, they nearly always prove to be a neurofibromatosis type-1 [8] (Fig. 5).

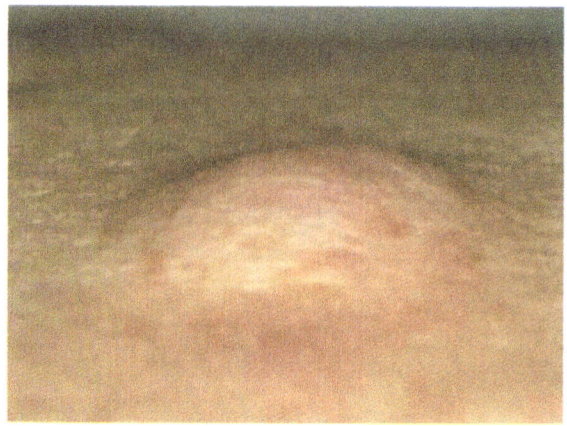

**Fig. 3.** Subcutaneous neurofibromas mostly have a spherical or ovoid form and can reach enormous depth. The skin over these neurofibromas is not influenced in colour

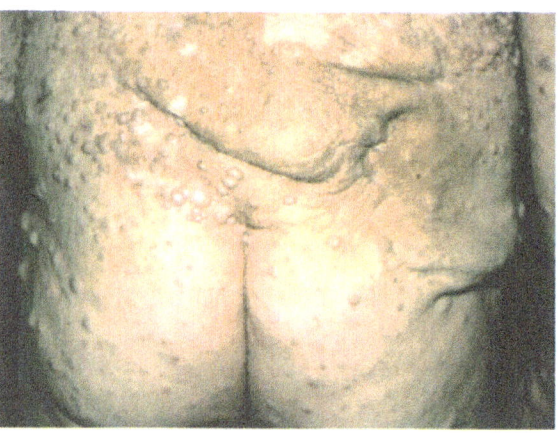

**Fig. 4.** Plexiform neurofibromas often show hyperpigmentation of the overlying skin and sometimes hypertrichosis. Dermatolysis and dermatochalasis make them look like pendulous dewlaps. They can reach enormous sizes like backpacks or more, and infiltrate into the nearby tissue

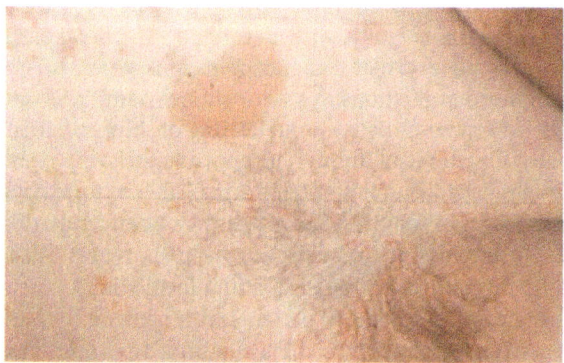

**Fig. 5.** Café-au-lait-spots appear over the whole body without preference and are significant for the diagnosis, even in childhood. Their colour is more or less uniform and looks like milk coffee. Axillar and inguinal are typical localizations for the little brown freckling spots

Another typical clinical finding is freckling. The very small hyperpigmented maculas (1 to 3 mm) are most commonly seen in the axilla and/or inguinal region (Fig. 5).

Iris Lisch nodules [36, 37] are pigmented hamartomas of the iris [38]. They are directly visible by examination as greyish brown nodules and can be found by the mid 60s in nearly 100% (presence and number is a function of age).

Optic nerve gliomas are observed in about 15% of neurofibromatosis type-1 patients, only about a third are associated with loss of vision [39].

Bone skeletal abnormalities like cyphoscoliosis, short statuce, cystic dilatations of the long bones, abnormal fragility, talipes equinus, genu valgum and varum, pseudarthrosis (tibia), thickening and abnormal length and spina bifida could be developed.

Mental deficiency, learning difficulties and precocious puberty appear to be significant in neurofibromatosis type-1.

Malignant transformation of a neurofibromatosis type-1 lesion, mostly plexiform neurofibromas, occurs in approximately 5 % of neurofibromatosis type-1 patients, the most common tumour is a neurofibrosarcoma [40, 41]. Furthermore, the higher risk of developing, for example, Wilm's tumours, rhabdomyosarcomas or leukaemia has also been noted for neurofibromatosis patients.

Besides the severe manifestations of this disease, the main problem daily facing each patient is the stigmatization or disfigurement due to multiple neurofibromas, leading to attendant psychosocial complications. In addition, handicapped hygiene and resulting local infections in areas of high density of large cutaneous neurofibromas aggravate the situation. Another problem is the initiation of uncontrolled growth of neurofibromas after different trauma or surgical intervention. In areas where mechanical trauma could appear due to backpacks, clothes or other causes, the injured neurofibroma could start to grow. Therefore, the removal of the cutaneous neurofibromas is not only an aesthetic indication but it prevents the development of further side effects and complications and helps the patient to greater self-confidence and acceptance in daily life.

## Laser Treatment

### Introduction/History

Only in the past decades have serious attempts been undertaken to enlarge upon the long-standing idea of treating neurofibromatosis surgically by laser and to optimize this therapy option. Early investigations of the treatment of neurofibromas of the skin (the treatment of single tumours) were reported by Aronoff (1977) [42] and by Oshiro (1980) using the $CO_2$ laser. Fifteen years ago, Dr. Katalinic started neurofibromas removal with an argon laser [43]. In the following years, he used additional two lasers and defined the indication for the different laser systems regarding the three groups of clinical styles [44]:

1. The stage at the outbreak of the disease, characterized by multiple minuscule fibromas of the skin. The usually reddish brown colour and the newly formed tissue make them suitable for coagulation with an argon laser. Thereby an enormous erythema of the treated regions developed, which disappeared in the following hours.
2. The stage in which the developed phacomas present an extremely varied picture. Their size ranges from that of a pinhead to 2–3 mm. In this stage, Dr. Katalinic preferred the $CO_2$ laser.
3. This stage includes large and expansive tumours. Intermingled with small or medium-sized neurofibromas, they present a broad field of application for the $CO_2$ laser in combination with conventional surgical techniques.

Later Dr. Katalinic introduced a combination between $CO_2$ laser and Nd:YAG laser, whereby for the treatment the $CO_2$ laser, with precise cutting, and the Nd:YAG laser, with a wider range of coagulation, were combined.

In the immediate postoperative phase, Dr. Katalinic used a scanned helium neon and an infrared laser for low-level laser therapy of the extended wounds twice a day for 20 min in each session to accelerate wound healing [46].

### $CO_2$ Laser

Since it was developed in the 1960s, the carbon dioxide ($CO_2$) laser has taken an important role in surgical therapy. It belongs to a group of gas lasers with a mixture of the medium $CO_2$ and nitrogen. Its principal wavelength emission is 10600 nm and the emitted light of the middle infrared is completely absorbed in glass or water, so that nearly the whole energy is absorbed on the tissue surface and transformed to heat. Vaporization of the tissue (temperature > 300 °C) can be achieved with relatively low output power. The $CO_2$ laser light requires transmission through articulating arms with mirrors, and cannot be transmitted through flexible quartz fibres like Nd:YAG and argon light. Using specific lens systems, the carbon dioxide laser light can be focused so precisely that it is suitable as a „knife" for common and microsurgical operations with an operation microscope system. In the defocused mode, this laser can be used for vaporization of the tissue.

## Methods

### Pretreatment

The removal of neurofibromas with the $CO_2$ laser is a selective intervention, which should be well planned together with the patient. Therefore, before the treatment, information about possible complications, management of wound care and the clinical appearance of the scars after the treatment must be given. With an indication for the surgical removal of neurofibromas, each neurofibromatosis type-1 patient is routinely examined for the following conditions. The patient must be mentally and physically able to be set under general anaesthesia. Therapy with oral anticoagulants with acetylsalicylic acid (ASA) should be interrupted a week before. Patients taking phenprocoumon should be treated only when a pressing indication for the removal exists, like pain or loss of function of anatomical structures. The anticoagulant has to be changed with heparin. Furthermore, each patient should be checked for routine parameters. Allergic reactions to substances given during and after the surgery have to be asked before, for example iodine or antibiotics (cephalosporin). Because of the large area of wounds, a current tetanus vaccination should always be required.

In addition, each neurofibromatosis type-1 patient should be photographed to document the presurgical development of the neurofibromas.

### $CO_2$ Laser Treatment

By using the $CO_2$ laser we can remove from 300 up to 1000 cutaneous and subcutaneous neurofibromas of various size and location under general anaesthesia during one intervention. The duration of the surgery ranges between 2 and 4 h. During the operation, systemic antibiosis with 2 g cefazolin intravenous is initiated, a second dose of 2 g cefazolin was given on the evening after laser intervention.

To limit the postoperative discomfort, only one side of the patient should be treated (back or front) so that it allows him to lie after the operation on the untreated side.

Sterile conditions in the operation room are required. The area to be treated is sterile washed (PVP iodine) and the rest of the body is covered with sterile sheets. Furthermore it is absolutely necessary to cover the articulation arm of the laser and the flexible tube of the smoke evacuator with a sterile coating. Mostly two surgeons working parallel, placed on different sides of the patient to reduce the time of operation. Additionally, at least one theatre nurse has to handle the two high-powered smoke evacuators (X Plume, Sharplan Lasers GmbH, Freising, Germany). The im-

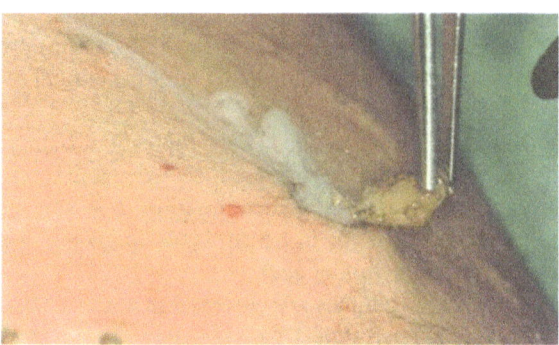

**Fig. 6.** Sufficient smoke suction is absolutely necessary during operation with a $CO_2$ laser. Every single laser shot with vaporization or excision of tissue produce so much smoke that the overview during the operation without suction is not optimal

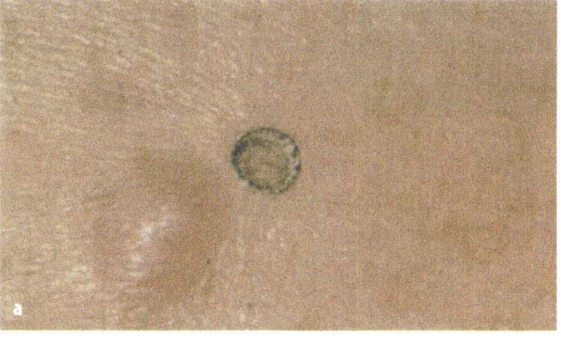
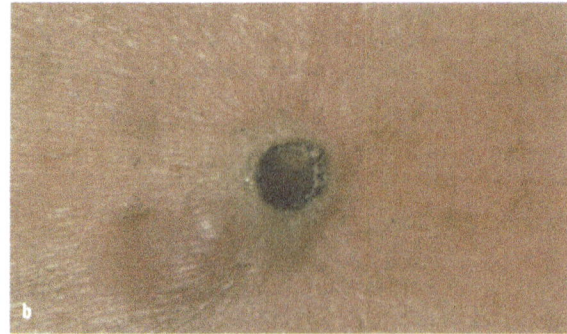

**Fig. 7a, b.** If vaporization time is too short, remains of neurofibroma tissue left in the wound (a) and recidives are frequent. Only if the fatty tissue is visibly fluid and the wound lip is raised is the tumour tissue definitely destroyed (b)

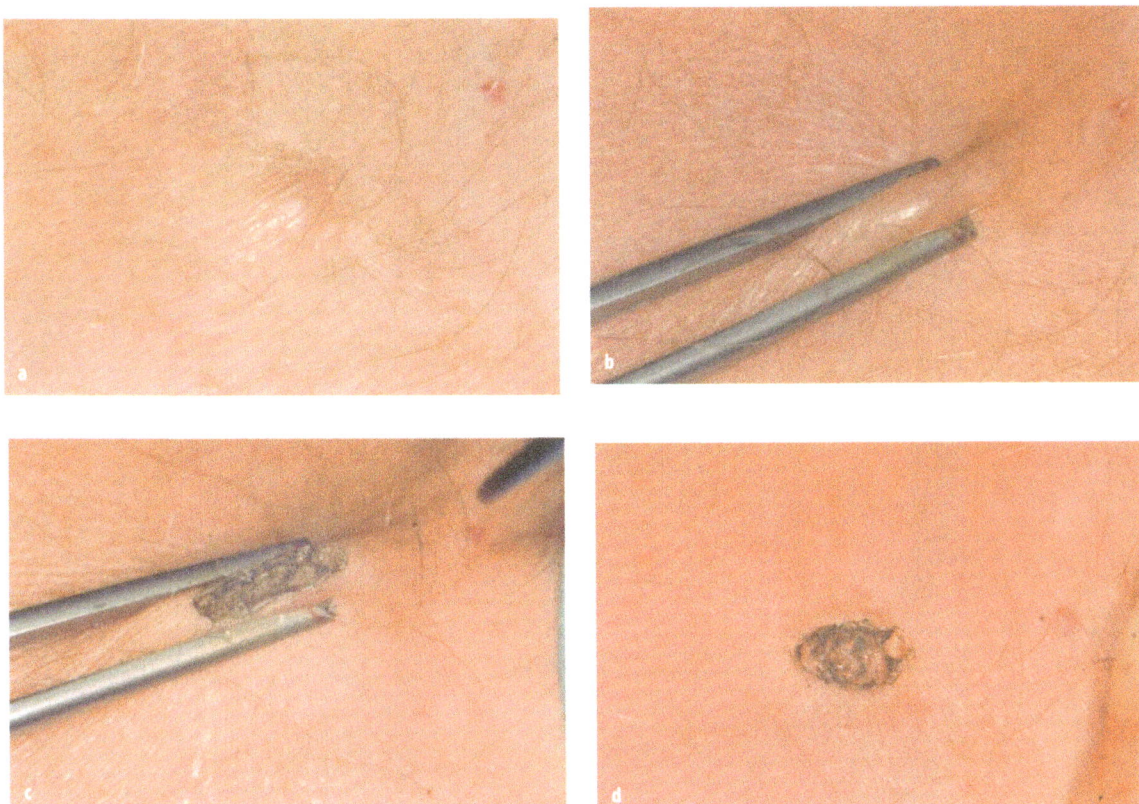

**Fig. 8a–d.** If neurofibromas are localized at difficult places, near large vessels, the tumour (**a**) is grasped between the two branches of a forceps (**b**), pulled up to distinguish it from the background and is vaporized in the given manner (**c, d**)

mense amount of smoke during vaporization and excision with the $CO_2$ laser requires high-powered smoke evacuation, which is absolutely necessary for the view in the treated area and to protect the operation team (Fig 6a, b).

Depending on the clinical features of the neurofibromas, carbon dioxide laser treatment is handled in different modalities.

Cutaneous neurofibromas up to 5 mm of size are easily vaporized in the slightly defocused mode and a power between 25 and 50 W/continuous mode by slow uniform movements of the hand piece in a circular manner to achieve the complete neurofibroma. Also, during treatment, the lips of the wound are raised slightly while the rest of the wound is pressed down. Only if the underlying fatty tissue in the wound becomes visibly fluid is the tumour tissue destroyed definitely (Fig. 7a, b). The laser beam should not be moved too slowly, otherwise damage of the deeper anatomic structures will be caused. Also, the laser hand piece must be hold almost vertical during treatment, so that no undermining of the wound lips or unnecessary removal of normal skin occurs. Nevertheless, it is important to treat the edge of the neurofibroma as well as the neurofibroma up to the fatty tissue to prevent and minimize the rate of recurrence.

If neurofibromas are localized in areas near for example cervical vessels, nerves or on the forearm, where the subcutaneous fat is low, the use of other techniques seems to be safer. The surrounding tissue of the tumour should be grasped with anatomic forceps and elevated, so that the neurofibroma can be treated between the two branches of the forceps while the ground is protected (Fig 8a–d). In this case, the power of the laser is reduced to 10 to 15 W/continuous wave.

Neurofibromas over 5 mm in size or larger subcutaneous tumours must be prepared and excised with the $CO_2$ laser in the focused mode using a power between 25 to 50 W/continuous wave.

1. The pedunculated form of neurofibromas is grasped with surgical forceps at the base and excised with the focused laser beam around the base. Additionally the wound ground is treated in the defocused mode until the fatty tissue becomes visibly fluid.

2. The subcutaneous and sessile form mostly has the bulk of the tumour mass in the fatty tissue. After ovoid cutting of the skin around the neurofibroma with the focused laser beam, one side of the neurofibroma is grasped with surgical forceps. Step by step the neurofibroma is separated from the surrounding fatty tissue with the focused $CO_2$ laser beam (Fig 9a–d). The neurofibroma could be easily

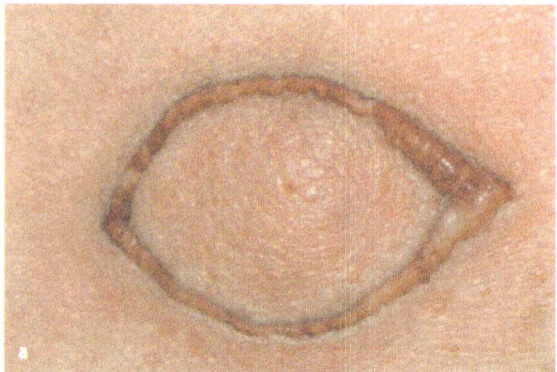
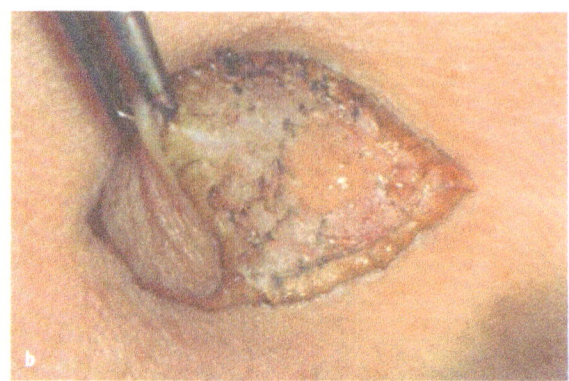
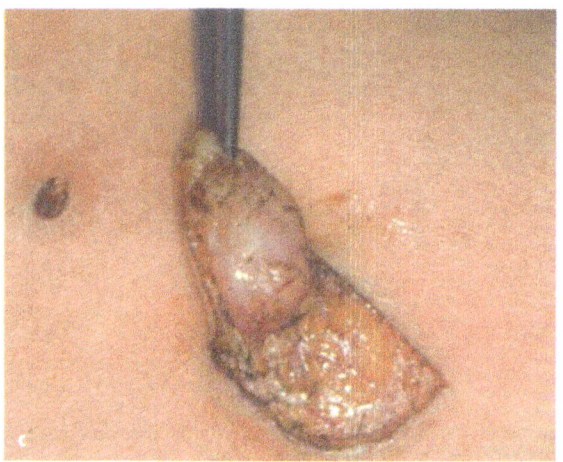
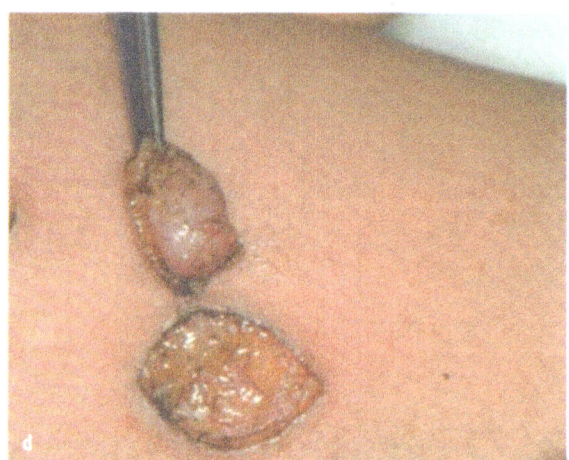

**Fig. 9a–d.** Subcutaneous and sessile neurofibromas must be prepared by an ovoid opening of the overlying skin with the laser beam (**a**) and afterwards separated step by step (**b, c**). The colour and consistency of neurofibromas is also macroscopically different from fatty tissue (**d**), so that exact excision can be controlled afterwards. The wound defect must be closed by suture

differentiated from the surrounding tissue because of its light, ivory shining structure compared to the yellow fatty tissue. It is important to prepare the subcutaneous part very exactly, because neurofibromas can grow, worm-like, deep into the surrounding fatty tissue (Fig. 10).

3. Neurofibromas localized near vessels, nerves or other important structures that have to protected should be removed with an alternative technique. The skin over the tumour will be cut lineally with the focused laser beam and a power between 8 and 10 W/continuous wave. Then the subcutaneous portion of the tumour is squeezed with manual pressure on either side through the opened skin. Now the protruded portion can be easily grasped by forceps and elevated over skin level to be cut at the base. Remaining parts of the tumour should be vaporized.

**Fig. 10.** Exact preparation of subcutaneous tumours is important, because sometimes the deep part of the tumour can reach unexpected sizes

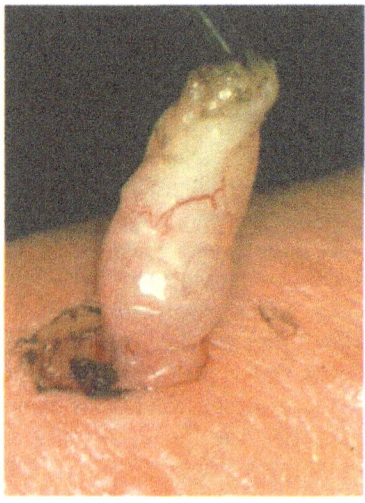

In comparison with other methods, the treatment of neurofibromatosis by laser surgery is a nearly bloodless to largely bloodless interventional procedure [46] because vessels until 0.5 mm were immediately closed, so that ligations of blood vessels are rarely necessary. If a vessel can not be closed immediately, a coagulation with the defocused laser beam after removing the blood from the wound ground with a

**Fig. 11.** Duration of laser treatment is limited only by the wounds a patient can manage. Depending on size and clinical type of neurofibroma, 300 up to 1000 tumours are removed in one session under general anaesthesia

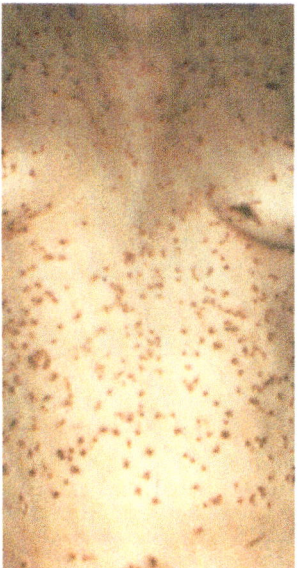

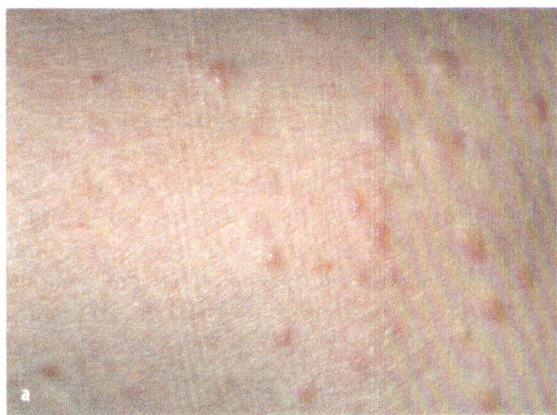

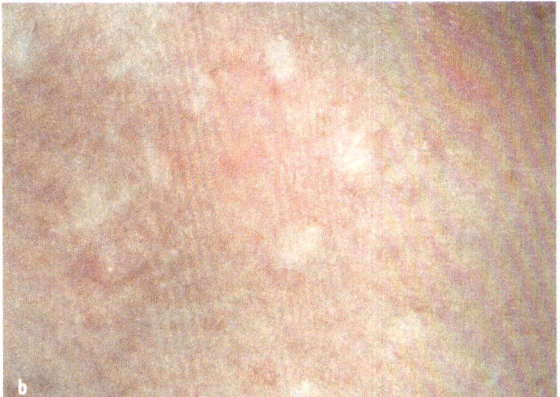

**Fig. 12 a, b.** Good cosmetic results with flat smooth white scars nearly the size of the treated neurofibroma are seen after approximately 6 month. (**a**) Small cutaneous neurofibromas before treatment. (**b**) Scars 6 month after laser treatment

swab can be attempted. Bleeding vessels over 1 mm in size should be ligated.

Wounds should be closed primarily with skin sutures (Resolon), depending on the anatomical location, when the defects are over 1 cm of size. But in case of the carbonized wound lips, the sutures should not be closed too tightly so that only an adaptation of the wound lips would be achieved. Exceptions to the rule are wounds in the region of the face and the mamilla. They should be closed primarily with very thin thread (up to 5/0 Resolon).

The final size of the defect is small, corresponding closely to the actual size of the treated neurofibroma basis. Between the resulting skin lesions normal skin areas/bridges must remain to optimize the wound healing. The second benefit of these skin bridges between the multiple scars resulting after $CO_2$ laser treatment is to maintain the elasticity and flexibility of the skin. Likewise, when neurofibromas are large in number and placed closely side by side, not every NF can be treated during one intervention.

All wounds closed primarily with skin sutures or minor defects for secondary wound healing were covered with PVP-iodine ointment and bandaged with large wadding pads at the end of laser treatment.

The duration of vaporization was determined clinically. The number of wounds the patient can manage is the only limit to the number of tumours treated (Fig. 11).

## Postoperative Care

To cope with the postoperative pain with an extended wound surface, a reasonable pain management for the following days is required. On the first postoperative day, the systemic antibiosis should be changed to an oral therapy with 3 x 1g cephalexin, continued for the following 2 weeks. Daily wound treatment with PVP-iodine ointment for the large wounds and PVP-iodine spray for the smaller wounds topped with absorbent wadding pads was carried out. Patients should be asked to take showers beginning at the 4th postoperative day to clean the wounds of exudates and to remove remaining ointment. The release from indoor treatment was planned in our department for the 5th postoperative day, if no complications like wound infection occurred.

The cutaneous sutures (Resolon) should be removed after 10–14 days. Then, the treated regions should be managed with a combination of 1% erythromycin in triamcinolone-16α, 17α-acetonid 0.025% hydrophile ointment (for example Volonimat N, Schering, Berlin, Germany) to prevent local infection and the development of hypertrophy scars and keloids until the wounds are closed. Later, different ointments can be applied to care for the skin. Extensive sun or UV light should be avoided because of the risk of hyperpigmentation in the area around the scars. Even the use of sun creams with a high sun protection factor is recommended for the following 3 months after treatment, if the treated regions are sun-exposed.

Re-epithelization after laser treatment is usually completed within 3–6 weeks with second intention healing, dependent on size and location of the wound, similar to wound healing after treatment with the electrocute. In the early phase, the scars appear red and change on average after 4–8 months into depigmented, flat, smooth and almost ovoid areas (Fig. 12a, b). The development of keloids is rare if the described postoperative wound care is followed.

Mostly a retreatment of the other neurofibromas is aimed at especially if the clinical picture was expanded or neurofibromas were standing adjacent side by side. A new date for the next laser treatment can be about 4 to 5 months later after the wound healing is definitely finished.

## Criteria for Exclusion

Criteria for exclusion are large infiltrating plexiform neurofibromas. They infiltrate finger like into the normal tissue or important internal anatomic structures, which makes it very difficult to remove the tumour mass completely. A great disadvantage is that the remaining parts could start to grow more aggressively and even faster after surgical intervention. Even when they are located paraspinally or gastrointestinally, laser treatment is not the first choice of treatment.

## Comparison with Other Methods

The removal of neurofibromas in neurofibromatosis type-1 can be achieved with different methods [26]. Each surgical intervention has advantages and disadvantages. Therefore the indication and the aim for treatment helps to find the best treatment method for the individual patient.

Classic surgical intervention with the scalpel is an indisputable indication in the removal or reduction of some neurofibromas or plexiform neurofibromas [26], but the removal of numerous neurofibromas with scalpel or circular punch biopsy is time-consuming, with a higher risk of bleeding, because of the hypervascularization of the tumours (Fig. 13). The wounds must be closed primarily, so only a few neurofibromas can be treated in one session.

Using the $CO_2$ laser, small vessels are coagulated and the necessary manual ligation of blood vessels can be avoided, so the overview during the operation is given at all times (Table 3).

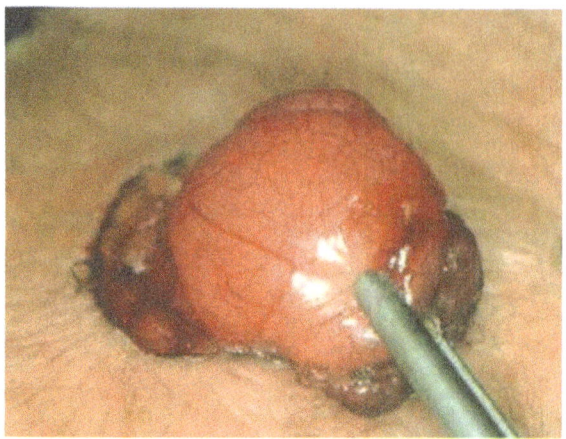

**Fig. 13.** Neurofibromas are tumours of high vascularization, so the immediately closure of small vessels up to 0.5 mm by $CO_2$ laser with bloodless overview of the operation area is a great advantage to classical surgical excision by scalpel

**Table 3.** Comparison of $CO_2$ laser and scalpel treatment – advantages and disadvantages. D. Katalinic (1996) [47]

| Advantages of $CO_2$ laser treatment | Disadvantage of scalpel treatment |
|---|---|
| 1. Less bleeding | 1. Intense bleeding |
| 2. No postoperative swelling | 2. Postoperative swelling |
| 3. Good treatment of small NF | 3. No treatment of mm sized NF |
| 4. Large number of NF treated in one session (> 600) | 4. Only a few NF treated in one session (<10) |
| 5. Short operation time per NF | 5. Prolonged operation time per NF |

Another device to remove neurofibromas is the monopolar diathermy loop by passing the tumour through the loop and clipping it at the base. The wound healing in this case is like the wound healing after $CO_2$ laser treatment by second intention. The removal of neurofibromas with the diathermy loop also allows a fast removal of neurofibroma but it is limited to papillomatous or pedunculated neurofibroma subtypes [2, 26]. In addition, the $CO_2$ laser causes less thermal damage to the surrounding uninvolved dermis than the electrocute.

In comparison with burn wounds, the clinical signs and pathophysiological mechanisms are different to the $CO_2$ laser treatment. The risk of shock symptoms and the loss of body fluids (plasma, blood) with a disorder of electrolytes as observed with large burn wounds has never been found after $CO_2$ laser treatment. Even the risk of septic fever and scar tissue contraction does not appear after laser treatment (Table 4).

**Table 4.** Comparison of $CO_2$ laser wounds and burn wounds. Katalinic; [48]

| $CO_2$ laser treatment | Burn wounds |
|---|---|
| 1. No signs of degree 1, 2 or 3 defect healing (degree 4) | 1. Redness (degree 1), bubbles (degree 2) necrosis (degree 3), defect healing (degree 4) |
| 2. No shock | 2. Danger of shock, hypovol. shock |
| 3. No loss of fluidity | 3. High loss of fluidity |
| 4. Normal electrolytes | 4. Disorder of electrolytes |
| 5. No fever | 5. Septic fever |
| 6. Less pain | 6. Lot of pain |
| 7. Minimal rate of keloids | 7. High rate of keloids |
| 8. No scar tissue contraction | 8. Scar tissue contraction |
| 9. Discharge on 5th post-operative day | 9. Long stationary treatment |

## Conclusion

Neurofibromatosis type-1 is a relatively common (1:3500–4000) autosomal dominant disorder with an extraordinary variability of clinical manifestations. Cutaneous and subcutaneous neurofibromas are the most common skin manifestations of Recklinghausen neurofibromatosis. They firstly develop in late childhood or early adolescence and grow in number and size at any time thereafter. Many patients exhibit hundreds to thousands of these tumours over the whole entire body, so that stigmatisation and social problems in private or professional life are major problems and cause loss of self-confidence. Additionally, neurofibroma formations which grow in groups close together complicate the normal body care and cause local infections, for example with corynebacteria or candida species.

Currently, no systemic treatment exists to prevent or minimise the growth of neurofibromas. Several single neurofibromas can be treated by the classical surgical method with the scalpel, but considering the high number of tumours one patient can generate, the removal is time consuming and discouraging. In addition, the well-vasculized tumours bleed easily and, besides diminished overview in the operation area, ligation of all the vessels is also time-consuming. So only a few neurofibromas can be treated with this method in one session.

With the $CO_2$ laser the neurofibromas were vaporized or excised depending on localization and configuration in the focused or defocused mode. Small vessels to 0.5 mm in size were closed by the laser immediately, vessels of a diameter more than 1 cm must be ligated. Because of the nearly bloodless treatment, a good overview and the rapid $CO_2$ laser removal of the tumours, hundreds of neurofibromas (300–1000 NF per session in 3–4 h) can be achieved in a short time under general anaesthesia. Keloids are rare with 0.5%. The low rate of recurrence and the good aesthetic results of plain white scars at the end can only be achieved when the laser technique and the postoperative treatment care are handled correctly, as described. This helps the patients to become more self-confident and minimizes the stigmatization in private or professional life. Scars are more accepted than neurofibromas. Therefore the removal of so many neurofibromas is not only an aesthetic result. The flattened skin after laser treatments in the place of large and closely located tumours minimizes further complications and the progress of the disease (Fig. 14a, b).

The $CO_2$ laser treatment is the treatment for choice of cutaneous and subcutaneous neurofibromas, because of the high rate of speed, the high number of neurofibroma removal, extremely low rate of recurrences, excellent improved aesthetic results, few side effects and the high number of satisfied patients.

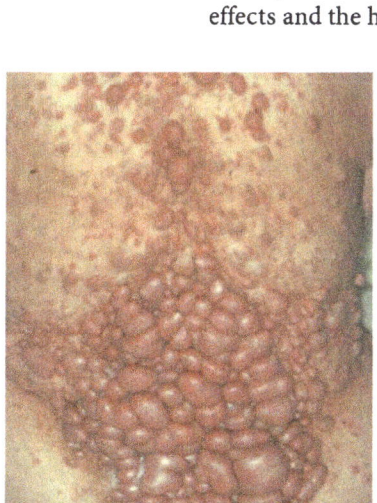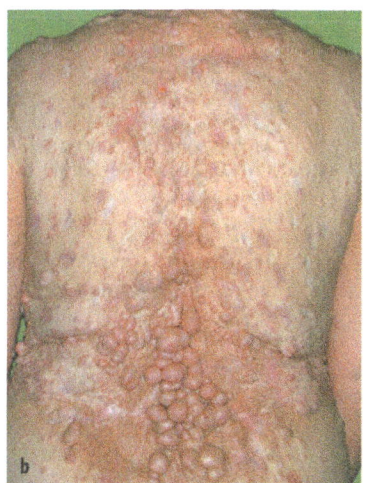

**Fig. 14 a, b.** The replacement of neurofibromas, especially when they are standing extremely close side by side (**a**), by flat, plant scars, optimizes the hygiene in this area and at the same time limits the risk of inflammation. The benefit patients reach after laser treatment is not only a better acceptance and more self-confidence, at least the progress of the disease is suppressed. (**b**) Most of the large tumours were removed after several sessions

## References

[1] CRUMP T (1981) Translation of case reports, Über die multiplen Fibrome der Haut und ihre Beziehung zu den multiplen Neuromen by v. Recklinghausen. Adv Neurol 29: 259–275

[2] GUTMANN DH, COLLINS FS (1993) The neurofibromatosis type-1 gene and its protein product, neurofibromin. Neuron 10: 335–343

[3] RICCARDI VM (1982) The multiple forms of neurofibromatosis. Pediatr Rev 3: 293

[4] RICCARDI VM (1982) Neurofibromatosis: clinical heterogenity. Curr Probl Cancer 7 (2): 1

[5] RICCARDI VM (1982) Early manifestations of NF: diagnosis and management. Comp Thr 8 (10): 35

[6] ROULEAU GA, ET AL (1987) Genetic linkage of bilateral acoustic neurofibromatosis to a DNA marker on chromosome 22. Nature 329:246

[7] SEIZINGER RR, ET AL (1987) Genetic linkage of von Recklinghausen neurofibromatosis to the nerve growth factor receptor gene. Cell 49: 589

[8] CROWE FT, ET AL (1956) A clinical, pathological and genetic study of multiple neurofibromatosis. Charles C Thomas, Springfield, IL

[9] GOLDBERG Y, DIBBERN K, KLEIN J, RICCARDI VM, GRAHAM JM Jr (1996) Neurofibromatosis type-1 – an update and review for the primary pediatrician. Clin Pediatr (Phila) 35: 545–561

[10] CONSTANTINO PD, ET AL (1989) Neurofibromatosis type II of the head and neck. Arch Otolaryngol Head Neck Surg 115: 380

[11] BEAUCHAMP GR (1995) Neurofibromatosis type one in children. Tran Am Ophthalmol Soc 93: 445–472

[12] KORF BR (1999) Plexiform neurofibromas. Am J Med Genet 89: 31–37

[13] RICCARDI VM (1992) Neurofibromatosis phenotype, natural history and pathogenesis 2nd edn. Johns Hopkins University Press, Baltimore

[14] KAUFMANN D, TINSCHERT S, ALGERMISSEN B (2001) Is the pattern of dermal neurofibromas in Neurofibromatosis type-1 (NF1) related to the pattern of the skin surface temperature? Eur J Dermatol (in press)

[15] BROMLY AR, SHERMAN JE, GOULIAN D (1982) Neurofibromatosis – distribution of lesions and surgical treatment. Ann Plast Surg 8, 272–276

[16] RICCARDI VM (1982) Neurofibromatosis: Clinical heterogeneity. Curr Probl Cancer 7: 1–34

[17] RICCARDI VM (1981) Von Recklinghausen neurofibromatosis. N Engl J Med 305: 1617

[18] SERRA E, ROSENBAUM T, WINNER U, ALEDO R, ARS E, ESTIVILL X, LENARD HG, LAZARO C (2000) Schwann cells harbor the somatic NF1 mutation in neurofibromas: evidence of two different Schwann cell subpopulations. Hum Mol Genet 9, 3055–3064 Springfield, IL

[19] RICCARDI VM (1981) Cutaneous manifestations of neurofibromatosis: cellular interaction, pigmentation, and mast cells. Birth Defects Orig Art Ser 17 (2): 129–145

[20] RICCARDI VM (1987) Mast-cell stabilisation to decrease neorifobroma growth Preliminary experience with ketotifen. Arch Dermatol 123 (8): 1011–1016

[21] REED N, GUTMANN DH (2001) Tumor genesis in neurofibromatosis : new insights and potential therapies. Trends Mol Med 7: 157–162

[22] REY I, TAYLOR HARRIS P, VAN ERP H, HALL A (1994) R-ras interacts with rasGAP, neurofibromin and c-raf but does not regulate cell growth or differentiation. Oncogene 9: 685–692

[23] BOLLAG G, MCCORMICK F, CLARK R (1993) Characterisation of full-length neurofibromin: tubulin inhibits Ras GAP activity. EMBO J 12: 1923–1927

[24] XU GF, O´CONNELL P, VISKOCHIL D, CAWTHON R, ROBERTSON M, CULVER M, DUNN D, STEVENS J, GESTELAND R, WHITE R, ET AL (1990) The neurofibromatosis type-1 gene encodes a protein related to GAP. Cell 62: 599–608

[25] SCHEFFZEK K, AHMADIAN MR, WIESMULLER L, KABSCH W, STEGE P, SCHMITZ F, WITTINGHOFER A (1998) Structural analysis of the GAP-related domain from neurofibromin and its implications. EMBO J 17: 4313–4327

[26] BECKER DW, JR (1991) Use of the carbon dioxide laser in treating multiple cutaneous neurofibromas. Ann Plast Surg. 26, 582–586

[27] RICCARDI, VM (1992) Neurofibromatosis, 2nd edn. ((Wo erschienen??)), pp 343–345

[28] MORENO JC, MATHORET C, LANTIERI L, ZELLER J, REVUZ J, WOLKENSTEIN P (2001) Carbon dioxide laser for removal of multiple cutaneous neurofibromas. Br J Dermatol 144: 1096–1098

[29] QUERINGS K, FUCHS D, KUNG EE, HAFNER J (2000) $CO_2$ laser therapy of stigmatising cutaneous lesions in tuberous sclerosis (Bourneville-Pringle) and in neurofibromatosis 1 (von Recklinghausen). Schweiz Med Wochenschr 130: 1738–1743

[30] KATALINIC D (1996) Laser surgical treatment of neurofibromas. Khirurgiia (Mosk) 52–54

[31] ANDRE P, CHAVAUDRA J, DAMIA E, GUILLAUME JC, AVRIL MF (1990) Lasers in dermatology. Ann Dermatol Venerol 117: 377–395

[32] ROENIGK RK, RATZ JL (1987) $CO_2$ laser treatment of cutaneous neurofibromas. J Dermatol Surg Oncol 13:187–190

[33] KATALINIC D (1992) Laser surgery of neurofibromatosis (NF1). J Clin Laser Med Surg 185–192

[34] ALGERMISSEN B, MÜLLER U, KATALINIC D, BERLIEN HP (2001): $CO_2$ laser treatment of neurofibromas of patients with neurofibromatosis type-1: five years experiences. Med Laser Appl (in press)

[35] PERRY HD, FONT RL (1982) Iris nodules in von Recklinghausen neurofibromatosis: Electron microscopic confirmation of the melanocytic orign. Arch Ophthalmol 100: 1635–1640

[36] RICCARDI VM (1980) The pathophysiology of neurofibromatosis: IV. Dermatologic insights into heterogeneity and pathogenesis. J AM Acad Dermatol 3: 157–166

[37] LEWIS RA, RICCARDI VM (1981) Von Recklinghausen neurofibromatosis: prevalence of iris Hamartomata. Ophthalmology 88: 348–354

[38] LISCH K (1937) Beteiligung der Augen; insbesondere das Vorkommen von Irisknötchen bei der Neurofibromatose (Recklinghausen). Z Augenheilkd 93: 137–143

[39] RICCARDI VM, EICHNER JE (1986) Neurofibromatosis: phenotype, natural history, and pathogenesis. Johns Hopkins Press, Baltimore

[40] HOPE DG, MULVIHILL JJ (1981) Malignancy in neurofibromatosis. Adv Neurol 29: 33–55

[41] HERRERA GA, DE MOREAES HP (1984) Neurogenetic sarcomas in patients with neurofibromatosis (von Recklinghausen´s disease). Virchows Arch [A]    403: 361–376

[42] ARONOFF BI (1977) Laser surgery, I-II. 2nd International Laser Symposium Dallas, pp 191–216

[43] KATALINIC D (1988) Morbus of Recklinghausen treated with Argon laser. Laser Jan–April: 9–12

[44] KATALINIC D (1991) Die erfolgreiche Lasertherapie der Neurofibromatose. Dermatologie 5: 10

[45] KATALINIC D (1992) Laser surgery of neurofibromatosis 1, Laser Applications in Medicine and Surgery, Proceedings of the 3rd World Congr Int Soc for low Power Laser Applications in Medicine, Bologna, Italy, Sept 9-12, 1992. Moduzzi Editor, pp 463–467

[46] DISCLAFANI A, WILKIN JK, ROBERTSON JT (1984) Neurofibromatosis. J Tenn Med Assoc 77: 143–148, 150

[47] KATALINIC D (1996) Therapie der Neurofibromatose (NF 1) im Gesicht. Ästhetische Chirurgie. Einhorn, pp 374–377

[48] KATALINIC D (1996) Großflächige Laseranwendung vs. Verbrennungstrauma. Hautnah Dermatol 7: 73–77 Schweiz

# III-12.2
# Interstitial Laser Treatment
# of Liver and Soft Tissue Metastasis

T. J. Vogl, M. G. Mack, R. Straub, K. Eichler, K. Engelmann, S. Zangos,
S. Heß and A. Roggan

## Contents

## Extended Summary for Non-Experts

The liver as a large solid organ is very suitable for interventional methods, due to a short transcostal access and its excellent functional capacity. It is the most common site of metastatic tumour deposits, especially for colorectal cancer, which is the third leading cause of death in Western communities, outnumbered only by lung and breast cancer [80]. At the time of death, approximately two-thirds of patients with colorectal cancer present liver metastases [79]. As a high number of metastases grow in the liver, new treatment protocols are currently under investigation. Laser-induced thermotherapy (LITT) has recently been applied for a minimally invasive technique in local treatment of liver metastases [48, 76]. A percutaneous approach avoiding laparatomy, local anaesthesia and an outpatient therapy management are the main advantages of this new kind of therapy.

Laser energy is transmitted via thin optical fibres resulting in a well-defined area of coagulative necrosis. This means a destruction of tissue by direct heating, while greatly limiting damage to surrounding structures [72]. Magnetic resonance imaging (MRI) has proved an ideal clinical instrument for the exact positioning of the optical fibres in the target area, real-time monitoring of the thermal effects and the subsequent evaluation of the extent of induced coagulative necrosis [3, 19, 31, 32, 34, 45].

Clinical results of MR-guided LITT are evaluated on the experience of a collective of more than 1000 patients and show local tumour control rates of 98.1 and 97.3% after 3 and 6 months. Median survival of 42.6 months in the patients with liver metastases of colorectal cancer were documented, indicating an improvement of survival time compared to data from the literature.

## Indications

The primary therapy goal is defined as local tumour control in patients with restricted hepatic malignant tumour deposition. A majority of patients are suffering from liver metastases of colorectal cancer, but also solitary hepatic manifestations from other primary tumours, like breast cancer, carcionoids and others, are treatable. According to our inclusion criteria, patients who are eligible for this treatment have five or less than five lesions with a maximum diameter of 50 mm, are unfit for surgical resection, suffer from irresectable tumours in both hepatic lobes or refuse surgical resection. Patients after a partial resection of one hepatic lobe developing a new lesion in the remaining liver part are also suitable for this therapy. Basic requirements for this treatment are the complete resection of the primary tumour and absence of extrahepatic metastases.

1. LITT of liver tumours
   - Max. five lesions.
   - Max. diameter of 50 mm.
   - Patients with tumour recurrence after surgery, radiation or chemotherapy.
   - New metastases after liver resection.
   - No response to chemotherapy.
   - Metastases in both liver lobes.
   - Lesions in high-risk locations e.g. near the bile duct.
   - LITT as replacement of the oncological therapy in case of patient refusal.

2. LITT of soft tissue tumours
   - Tumour recurrence in the head and neck region.
   - Tumour recurrence in the pelvis.
   - Lymph node metastases in the abdomen, the head and neck region and the retroperitoneum.

   Criteria for Exclusion
   - Extensive extrahepatic tumour spread.
   - Contra indications for MRI (pace-maker).
   - Ascites, apparent infections, coagulation disorders.
   - Diffuse and multiple pattern of metastasis.

## Comparison with Alternative Methods

### Surgery

Surgical resection of liver metastases is still considered one of the best options for a radical treatment of malignant tumours, but only 20% of the patients are suitable for surgical resection. Clinical conditions like the presence of lesions in both hepatic lobes or the reduced clinical condition of a patient exclude surgical

treatment. Additional liver surgery is associated with a mortality rate of approximately 3–8%, but the main problem is tumour recurrence of up to 70% after surgical resection. Only in some selected cases reresection is possible, so that further surgery is restricted for the majority of patients [1, 4, 6, 7, 13, 14, 21, 22, 25, 27, 28, 44, 52, 61, 63]. MR-guided LITT offers the option of several retreatment courses. For new hepatic lesions the same inclusion criteria are applied as at the time of the initial LITT.

### Percutaneous Alcohol Injection (PAI)

MR-guided ethanol ablation is used to reduce tumour bulk and cancer pain, but mostly in low-risk regions, and has been used especially to treat hepatocellular carcinoma (HCC). Among the limitations of this technique is the lack of an instrument for monitoring the effects induced in both normal and pathological tissue. Comparisons between interstitial laser coagulation (ILP) and percutaneous alcohol injection (PAI) to treat colorectal hepatic metastases have shown that there were no major complications after ILP or PAI, but that pain during PAI was more severe and also the efficacy for tumour control was lower [2, 37, 40, 65, 66]. In summary, alcohol injection is an accepted therapy method for small HCC nodules, but no treatment option for metastases.

### Cryosurgery

At present, cryosurgery is carried out using laparatomy and ultrasound guidance. Single and multiprobe arrays are possible. Tumour tissue will be frozen to less than minus 190 C and defrosted. The double-freezing technique results in a reliable destruction of tumour cells. Inclusion criteria are similar to LITT concerning size and number of lesions. Cryosurgery under MR guidance has been investigated in animal studies and MR imaging is promising here for real-time monitoring of the progress of freezing and thawing during cryosurgery. One drawback is the need for laparatomy with its complications, so that the minimal invasive aspect is missing. Hospital stays of about 7 to 10 days are normal [5, 8, 15, 33, 47, 49, 54, 60, 69, 78, 81, 82].

### RF-Surgery

Radiofrequency (RF) ablation is another method for thermal coagulation of tumour tissue. Heating is caused by a high tissue impedance between two applicator tips. Solbati demonstrated currently the feasibility of tumour-destruction with conventional and cooled-tip monopolar RF electrodes in patients with metastatic gastrointestinal carcinomas [67]. The studies included up to 29 patients treated with mono- and multiprobe arrays and also with liquid cooled-tip techniques. The same inclusion criteria are applied, but there are problems with the monitoring technique. Interferences between the RF applicators and the MRI device prevent sufficient imaging [20, 35, 36, 38, 39, 42, 58].

## Focused Ultrasound

Non invasive surgery using focused, high-intensity ultrasound beams was first proposed as a therapeutic method for destruction of central nervous system tissue. Recently, a novel solution was demonstrated for real-time monitoring of focused ultrasound using only MR imaging for visualizing the temperature increase during sonication and for delineating regions of tissue necrosis. This promising method is at present under in vitro evaluation for soft tissue tumours in the brain and breast. Problems are the rapid increase of temperature and to reach larger volumes with a small focus [9,–11, 24, 50, 77]. So far, focused

ultrasound has no indication for treatment of liver metastases due to breathing artefacts.

## Transarterial Chemo-Embolization (TACE)

Transarterial chemo embolization is defined as a local hepatic deposition of chemotherapy via tumour-providing vessels and a transfemoral entry for the treatment of HCC nodules and hepatic metastases of neuroendocrine tumours. In other liver metastases there is a palliative intention to reduce tumour bulk and to decelerate tumour progression. Indication is justified due to low percentage of side effects, outpatient management and nearly no impairment of the quality of life [41, 59].

## Systemic and Regional Chemotherapy

All the above-mentioned techniques are restricted regionally to the liver, so the adjunct use of systemic chemotherapy is essential in most cases. Protocols, doses and intervals depend on the primary tumour and data from staging.

**Table 1.** Literature overview – interventional treatment methods for liver metastases

| Reference | Year | Technique | Primary | Median survival |
|---|---|---|---|---|
| Rossi [58] | 1996 | RF | Colorectal cancer and others | 11 months |
| Lencioni [35] | 1998 | RF, cooled tip | Colorectal cancer, cancer of the stomach, neuroendocrine tumours | 6.5 months |
| Ravikumar [55] | 1991 | Cryotherapy | Colorectal cancer | 5 months–5 years |
| Vogl [75] | 1997 | Laser induced thermotherapy (LITT) | Colorectal cancer, cancer of the stomach, breast, pancreatic gland, thyroid gland, melanoma | 40.8 months |
| Lopez [41] | 1997 | Chemo embolization | | 10 months |
| Martin [46] | 1995 | Chemo embolization | Neuroendocrine tumours | 33 months |
| Sanz-Altamira [59] | 1997 | Chemo embolization | Colorectal cancer | 10 months |
| Amin [2] | 1993 | PEI, ILP | Colorectal cancer | ILP: 27 months PEI: 7 months |

**Table 2.** Overview: local treatment modalities for liver metastases

| Technique | Curative effect | Max. no. | Max. size (mm) | Effort | Complica- tions (%) | Local tumour control (%) | Survival |
|---|---|---|---|---|---|---|---|
| Laser (LITT) | (+) | 5 | 50 | +++ | 1–3 | 96–99 | 5 J-ÜL 21–34% |
| Radiofrequency (RF) | (+) | 5 | 50 | ++ | 2–5 | 91 | 5 J-ÜL 15–31% |
| Alcohol (PEI) | (+) | 5 | 60 | + | 2–6 | 93–99 | 5 J-ÜL 30–50% (HCC) |
| Chemo embol. (TAE) | – | – | – | ++ | 3–4 | 80–91 | 3 J-ÜL 10% (HCC) |
| Cryotherapy | (+) | 5 | 70 | +++ | 20 | ~80 | 5 J-ÜL 10–30% |

Tables 1 and 2 present an overview of technical features, indications and results of the described techniques.

## Results

### In Vitro Studies

In vitro studies using pig liver demonstrated reproducible loss of signal intensity in MRI corresponding to increasing tissue temperatures. Using a power of 5 W and an application time of 12 min, the maximum diameter of the region with signal loss was 25 mm. This effect was best monitored using the Thermo-TurboFlash sequence at TI values of 300 to 400 ms, providing a nearly linear, inverse correlation between signal intensity and temperature as well as the Thermo-FLASH-2D sequences.

A mean size necrosis of 2 cm$^3$ of ellipsoid morphology may be achieved by an application time of approximately 20 min and a power of 5–6 W using one applicator system. There are two possibilities of enlarging the necrosis: first the pullback technique, which allows are more longitudinal enlargement of necrosis, and second the multi-applicator technique, where the positioning of two or three applicator systems results in a necrosis size up to 17 cm$^3$.

The use of an internally cooled power laser applicator at power settings between 25 and 30 W over 10 to 20 min additionally results in a significant enlargement of the obtained coagulative necrosis.

### Patients

So far, from 1993 to 2002, we treated 1112 patients with a total of 2318 liver metastases of colorectal carcinoma, oesophageal, gastric, pharyngeal, testicular and pancreatic tumours. A total of 8214 laser applications was performed. The laser-induced necrosis was quantified by comparing the pre- and posttherapeutic plain- and contrast-enhanced MR images. Parameters like size, morphology and contrast enhancement were directly compared to pretherapeutic MRI. Successful therapy is defined by a geometrically shaped necrotic area without enhancement in the adequate topographic position and reliable safety margin.

Local tumour control rates for data from 1997 to 1999 are 98.1% after 3 months and 97.3% after 6 months in liver lesions smaller than 50 mm. Mean survival time for all patients is 45 months using the Kaplan-Meier method (95% confidence interval: 40.9 to 49.2 months, median 39.8 months, maximum survival 74.6 months) (Fig. 1).

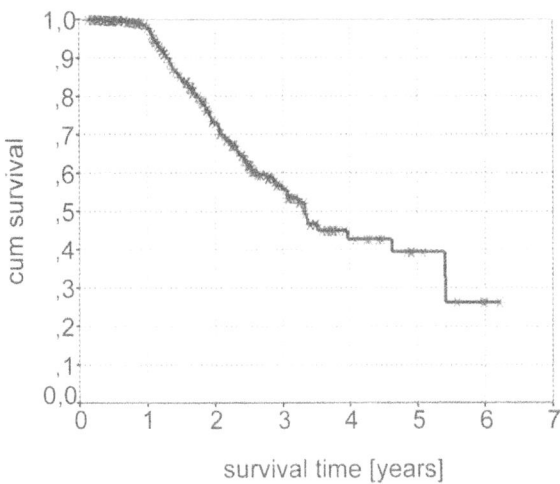

**Fig. 1.** Survival curves for all patients treated from 1993 to July 1999. Mean survival time is 45 months (95% confidence interval: 40.9 to 49.2 months, median 39.8 months, maximum survival 74.6 months

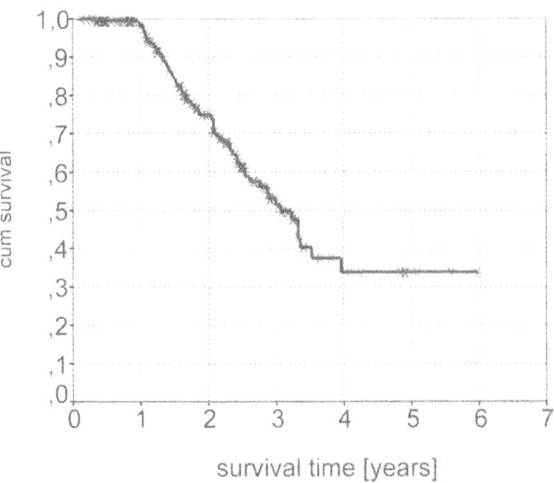

**Fig. 2.** The most homogeneous patient group is patients with liver metastases of colorectal cancer. Mean survival time is 42.6 months (95% confidence interval: 37.7 to 47.5 months, median 36.7 months

The most homogeneous patient group are patients with hepatic metastases of colorectal primaries. Mean survival time for these 274 patients in 42.6 months (95% confidence interval: 37.7 to 47.5 months, median 36.7 months) (Fig. 2).

Survival did not differ significantly (p > 0.05) between male and female patients, or between patients with colorectal metastases and those with metastases of other primary tumours.

With respect to the development of the LITT procedure since 1993, the patients were subdivided for

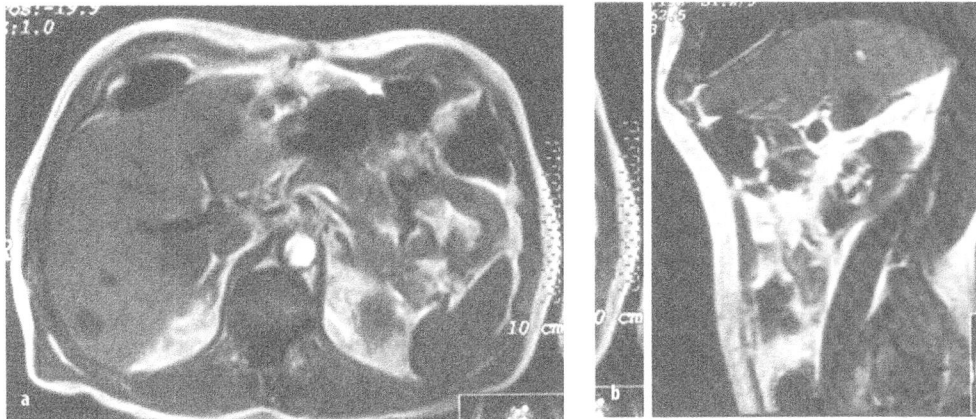

**Fig. 3a.** Liver metastases from colorectal carcinoma: 60-year-old male patient. GRE sequence, axial slice orientation, plain. Hypointense 30-mm lesion in segment 6 near the portal and caval vein. Second lesion in lateral liver parenchym. **b** GRE sequence, saggital slice orientation. Note the relationship to the portal vessel

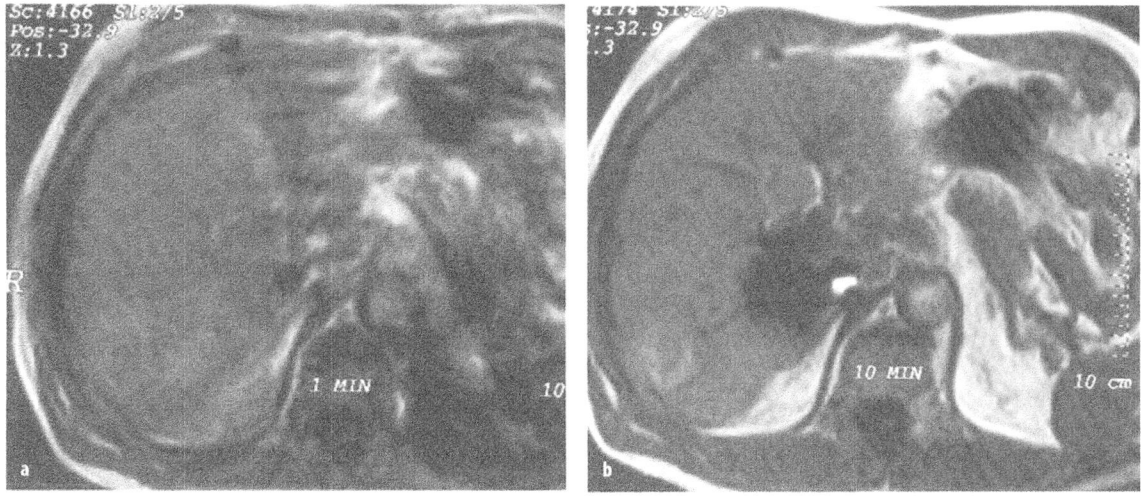

**Fig. 4.** MR-guided LITT: **a** thermosensitive GRE sequence at the beginning of intervention. **b** Thermosensitive GRE sequence 15 min during LITT. Note the signal decrease in the treated area

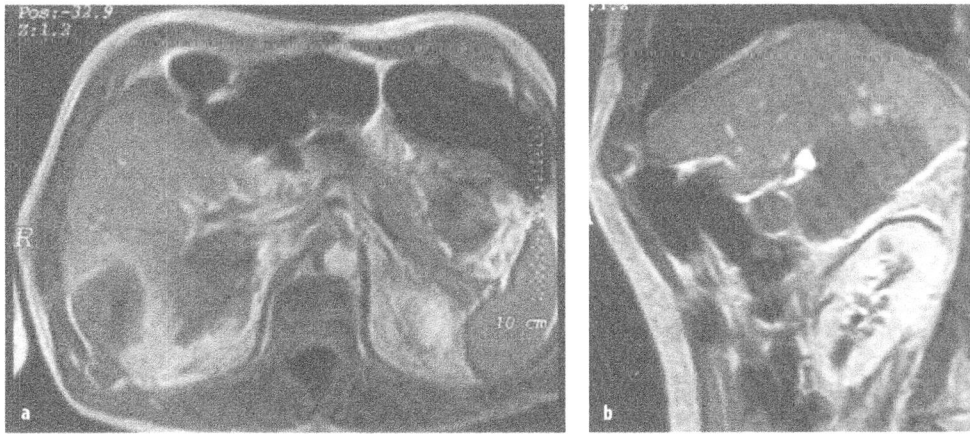

**Fig. 5.** Twenty four h after MR-guided LITT: axial GRE sequence, contrast-enhanced. Coagulative necrosis, 70 x 80 mm, geometrical shape, no enhancement, indicating a reliable safety margin all around the lesion. The second was treated in another session

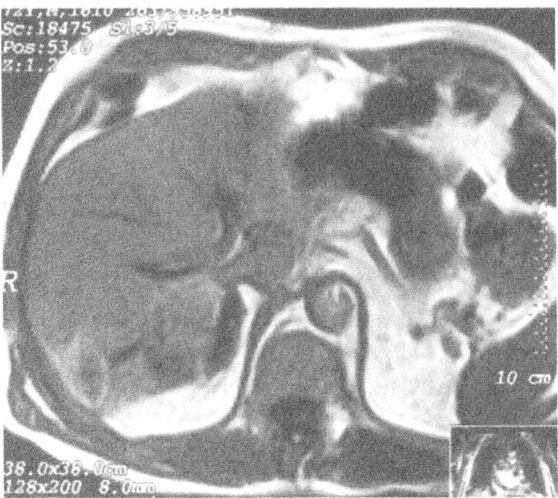

**Fig. 6.** Three months control after MR-guided LITT: GRE sequence contrast-enhanced. No enhancement in the area of the coagulative necrosis

the evaluation of the local tumour control rate in to three groups (group 1: patients 1–100, group 2: patients 101–175, group 3: patients 176–322). The 3 months' MRI control demonstrated a local tumour progression in group 1 in 35.1% versus 20.3% in group 2 versus 0.8% in group 3. The 6 months' control showed in group 1 a local tumour progression in 61.7% versus 32.9% in group 2 versus 2.2% in group 3 (successful local tumour destruction was defined on plain and contrast-enhanced MR imaging (Figs. 3–6).

## Complications

All patients tolerated the procedure under local anaesthesia well. The following side effects could be visualized on clinical examinations or imaging studies: pleural effusion in 7.28%, subcapsular haematoma in 2.46%, intrahepatic abscess in 0.11%, intra abdominal bleeding in 0.11% and local infection on puncturing site in 0.2%. All complications except the following were clinically not relevant and only visible on imaging studies. In 0.36% of the cases, the pleural effusion had to be treated by percutaneous drainage. Two patients died within 4 weeks after LITT. One patient was operated for suspicion of a colonic perforation.

## Head and Neck and Extrahepatic Tumours

Sixteen patients (six woman and ten men, mean age 62 years, range 57–77 years) with recurrent tumours of the head and neck were treated with LITT. In all patients, the primary tumour was located in the head

and neck region. In the majority of patients, there was a recurrent squamous cell carcinoma. Two patients with pleomorphic adenoma in the parapharyngeal space were treated for tumour recurrence after primary surgery. All patients tolerated the procedure well under local anaesthesia. No side effects were observed. The procedure was judged as being successfully performed if postinterventional MRI revealed a sharply demarcated necrosis which was found in each control examination. A relevant reduction of clinically important symptoms, such as pain, was observed in 11 patients, and a reduction of swallowing problems and symptoms of nervous compression in 6 patients.

In an other group of 14 patients, abdominal tumorous lymph-node involvement or recurrent primary pelvic tumours had been successfully treated using MRI-guided LITT. In this minor group, we were able to reduce clinical symptoms in 68% of the patients.

## Discussion

The liver is one of the most frequent deposits of tumour cells, so many therapeutic strategies are accessed or are currently under investigation. At the time of death, approximately two-thirds of patients with colorectal cancer have hepatic metastases [73, 79]. At present, surgical resection of hepatic metastases is the well-accepted method for curative treatment for this kind of cancer, if extrahepatic spread is excluded, but only 20% of patients are candidates for resection according to clinical conditions like presence of lesions in both hepatic lobes or extrahepatic tumour spread or a poor clinical status [16, 18, 27, 29, 43, 51, 53, 56, 62, 63, 68]. Surgery is associated with a mortality rate of approximately 5% and present studies show that two-thirds of the patients develop recurrences in the remaining liver part [21, 23, 26, 28, 70, 71]. So the development of alternative treatment methods is required, such as are described in the preceding section.

Ultimate goal of the MR-guided LITT is defined by 100% of local tumour control. There is no effect to the surrounding tissue, and the MRI online thermometry allows exact guidance of the interventional procedure. MRI provides unparalleled topographic accuracy due to its excellent soft tissue contrast and high spatial resolution. Therefore, the early detection of local complications like bleeding and haemorrhage and treatment effects like coagulative necrosis is possible. Dynamic contrast-enhanced MR images represent the superior parameter for the evaluation of the treated lesions especially for the short- and long-term evaluation [12, 17, 30, 49, 57, 64, 74].

In the literature, Stangl followed up 1099 consecutive patients with colorectal liver metastases [68], 566 of whom (51.1%) received no treatment for their hepatic metastases; 340 (31%) underwent hepatic resection; 123 (11.2%) received regional chemotherapy; and 70 (6.4%) received systemic chemotherapy. Thirty-four patients died within 30 days as a result either of postoperative complications or advanced disease; 48 were excluded, because they developed a second primary cancer. After hepatic resection, 60% of all patients survived 5 years (median survival 30 months). In patients who underwent regional or systemic chemotherapy, the median survival was 12.7 and 11.1 months, respectively. In the untreated group, 31.3% of the patients were alive at 1 year, 7.9% at 2 years, 2.6% at 3 years and 0.9% at 4 years. Considering these data, our results are comparable to those of surgery. Lack of perioperative mortality and low rate of side effects, combined with an outpatient management, are responsible for high patient tolerance.

Patients with irresectable liver tumours need a multimodal therapy concept in palliative intention, while preserving a good quality of life. LITT alone or combined with systemic or locoregional chemotherapy, transarterial chemoembolization and surgical resection improves local tumour control and tumour-free survival in patients with several primary tumours.

# References

[1] ADSON MA, HEERDEN VAN J, ADSON MH, WAGNER JS, ILSTRUP DM (1984) Resection of hepatic metastases from colorectal cancer. Arch Surg 119: 647–651

[2] AMIN Z, BOWN SG, LEES WR (1993) Local treatment of colorectal liver metastases: a comparison of interstitial laser photocoagulation (ILP) and percutaneous alcohol injection (PAI). Clin Radiol 48 (3): 166–171

[3] ANZAI Y, DESALLES AA, BLACK KL, SINHA S, FARAHANI K, BEHNKE EA, CASTRO DJ, LUFKIN RB (1993) Interventional MR imaging. Radiographics 13 (4): 897–904

[4] BOZZETTI F, DOCI R, BIGNAMI P, MORABITO A, GENNARI L (1987) Patterns of failure following surgical resection of colorectal cancer liver metastases. Ann Surg 205 (3): 264–269

[5] BREWER WH, AUSTIN RS, CAPPS GW, NEIFELD JP (1998) Intraoperative monitoring and postoperative imaging of hepatic cryosurgery. Semin Surg Oncol 14: 129–155

[6] BUTLER J, ATTIYEH FF, DALY JM (1986) Hepatic resection for metastases of the colon and rectum. Surg Gynecol Obstet 162 (2): 109–113

[7] CADY B, STONE MD, MCDERMOTT WV, JENKINS RL, BOTHE A, LAVIN PT, LOVETT EJ, STEELE G (1992) Technical and biological factors in disease-free survival after hepatic resection for colorectal cancer metastases. Arch Surg 127: 561–569

[8] CHARNLEY RM, DORAN J, MORRIS DL (1989) Cryotherapy for liver metastases: a new approach (comment in: Br J Surg 1990; 77: 354). Br J Surg 76: 1040

[9] CLINE HE, HYNYNEN K, HARDY CJ, WATKINS RD, SCHENCK JF, JOLESZ FA (1994) MR temperature mapping of focused ultrasound surgery. Magn Reson med 31(6): 628–636

[10] CLINE HE, HYNYNEN K, WATKINS RD, ADAMS WJ, SCHENCK JF, ETTINGER RH, FREUND WR, VETRO JP, JOLESZ FA (1995) Focused US system for MR imaging-guided tumour ablation. Radiology 194 (3): 731–737

[11] CLINE HE, SCHENCK JF, HYNYNEN K, WATKINS RD, SOUZA SP, JOLESZ FA (1992) MR-guided focused ultrasound surgery. J Comput Assist Tomogr 16 (6): 956–965

[12] CLINE HE, SCHENCK JF, WATKINS RD, HYNYNEN K, JOLESZ FA (1993) Magnetic resonance-guided thermal surgery. Magn Reson Med 30 (1): 98–106

[13] DOCI R, GENNARI L, BIGNAMI P, MONTALTO F, BOZZETTI F (1991) One hundred patients with hepatic metastases from colorectal cancer treated by resection: analysis of prognostics determinants. Br J Surg 78: 797–801

[14] EKBERG H, TRANBERG K-G, ANDERSSON R, HÄGERSTRAND I, RANSTAM J, BENGMARK S (1986) Determinants of survival in liver resection for colorectal secondaries. Br J Surg 73: 727–731

[15] FEIFEL G, SCHÜDDER G, PISTORIUS G (1999) Kryochirurgie – Renaissance oder echter Fortschritt? Chirurg 70: 154–159

[16] FORTNER JG, SILVA JS, GOLBEY RG, COX EB, MACLEAN BJ (1984) Multivariate analysis of a personal series of 247 consecutive patients with liver metastases from colorectal cancer. Ann Surg 199: 306–315

[17] FRIED MP, MORRISON PR, HUSHEK SG, KERNAHAN GA, JOLESZ FA (1996) Dynamic TI-weighted magnetic resonance imaging of interstitial laser photocoagulation in the liver: observations on in vivo temperature sensitivity. Lasers Surg Med 18 (4): 410–419

[18] GAYOWSKI TJ, IWATSUKI S, MADARIAGA JR (1994) Experience in hepatic resection for metastatic colorectal cancer: analysis of clinical and pathological risk factors. Surgery 116: 703–710

[19] GEWIESE B, BEUTHAN J, FOBBE F, STILLER D, MULLER G, BOSE LANDGRAF J, WOLF KJ, DEIMLING M (1994) Magnetic resonance imaging-controlled laser-induced interstitial thermotherapy. Invest Radiol 29 (3): 345–351

[20] GOLDBERG SN, GAZELLE GS, SOLBIATI L, LIVRAGHI T, TANABE K, HAHN PF, MUELLER PR (1998) Ablation of liver tumours using percutaneous RF therapy. Am J Roentgenol 170: 1023–1029

[21] GRIFFITH KD, SUGARBAKER PH, CHANG AE (1990) Repeat hepatic resections for colorectal metastases. Surgery 107: 101–104

[22] HARNED RKN, CHEZMAR JL, NELSON RC (1994) Recurrent tumor after resection of hepatic metastases from colorectal carcinoma: location and time of discovery as determined by CT. Am J Roentgenol 163 (1): 93–97

[23] HERFARTH C, HEUSCHEN UA, LAMADE W, LEHNERT T, OTTO G (1995) Rezidiv-Resektionen an der Leber bei primären und sekundären Lebermalignomen. Chirurg 66: 949–958

[24] HILL CR (1994) Optimum acoustic frequency for focused ultrasound surgery. Ultrasound Med Biol 20 (3): 271–277

[25] HOHENBERGER P, SCHLAG P, SCHWARZ V, HERFARTH C (1988) Leberresektion bei Patienten mit Metastasen colorektaler Carcinome. Ergebnisse und prognostische Faktoren. Chirurg 59: 410–417

[26] HOLM A, BRADLEY E, ALREDETE JS (1989) Hepatic resection of metastases from colorectal carcinoma. Morbidity, mortality and pattern of recurrence. Ann Surg 209 (4): 428–434

[27] HUGHES KS, SIMON R, SONGHORABODI S, ADSON MA, ILSTURP DM, FORTNER JG, MACLEAN BJ, FORSTER JH, DALY JM, FITZHERBERT D, SUGARBAKER PH, ET AL (1988) Resection of the liver for colorectal carcinoma metastases: a multi-institutional study of indications for resections. Surgery 103 (3) 278–288

[28] HUGHES KS, SIMON R, SONGHORABODI S, ADSON MA, ILSTRUP DM, FORTNER JG, MACLEAN BJ, FOSTER JH, DALY JM, FITZHERBERT D, SUGARBAKER PH, IWATSUKI S, STARZL T, RAMMING PH, LONGMIRE WP, O'TOOLE K, PETRELLI NJ, HERRERA L, CADY B, MCDER-

MOTT W, NIMS T, ENKER WE, COPPA GF, BLUMGART LH, BRADPIECE H, URIST M, ALDRETE JS, SCHLAG P, HOHENBERG P, STEELE G, HODGSON WJ, HARDY TG, HARBORA D, MCPHERSON A, LIM C, DILLON D, HAPP R, RIPEPI P, VILLELLA E, ROSSI RL, REMINE SG, OSTER M, CONNOLLY DP, ABRAMS J, ALJURF A, HOBBS KEF, LI MKW, HOWARD T, LEE E (1986) Resection of the liver for colorectal carcinoma metastases: a multi-institutional study of pattern of recurrence. Surgery 100 (2): 278–284

[29] IWATSUKI S, SHAW BW, STARZL TE (1983) Experience with 150 liver resections. Ann Surg 197 (3): 247–253

[30] JOLESZ FA, BLEIER AR, JAKAB P, RUENZEL PW, HUTTL K, JAKO GJ (1988) MR imaging of laser–tissue interactions. Radiology 168 (1): 249–253

[31] JOLESZ FA, BLUMENFELD SM (1994) Interventional use of magnetic resonance imaging. Magn Reson Q 10 (2): 85–96

[32] KAHN T, BETTAG M, ULRICH F, SCHWARZMAIER HJ, SCHOBER R, FURST G, MODDER U (1994) MRI-guided laser-induced interstitial thermotherapy of cerebral neoplasms. J Comput Assist-Tomogr 18 (4): 519–532

[33] LEE FT, MAHAVI DM, CHOSY SG, ONIK GM, WONG WS, LITTRUP PJ, SCANLAN KA (1997) Hepatic cryosurgery with intraoperative US guidance. Radiology 202: 624–632

[34] LEE TH, ANZAI Y, LUFKIN RB (1993) Magnetic resonance imaging-guided needle biopsy of head and neck lesions. West J Med 159 (1): 69

[35] LENCIONI R, GOLETTI O, ARMILLOTTA N, PAOLICCHI A, MORETTI M, CIONI D, DONATI F, CICORELLI A, RICCI S, CARRAI M, CONTE PF, CAVINA E, BARTOLOZZI C (1998) Radio-frequency thermal ablation of liver metastases with a cooled-tip electrode needle: results of a pilot clinical trial. Eur Radiol 8: 1205–1211

[36] LEWIN JS, CONNELL CF, DUERK JL, CHUNG Y, CLAMPITT ME, SPISAK J, GAZELLE GS, HAAGA JR (1998) Interactive MRI-guided radiofrequency interstitial thermal ablation of abdominal tumors: clinical trial for evaluation of safety and feasibility. JMRI 8: 40–47

[37] LIVRAGHI T, GIORGIO A, MARIN G, SALMI A, DE SIO I, BOLONDI L, POMPILI M, BRUNELLO F, LAZZARONI S, TORZILLI G, ET AL (1995) Hepatocellular carcinoma and cirrhosis in 746 patients: long-term results of percutaneous ethanol injection. Radiology 197 (1): 101–108

[38] LIVRAGHI T, GOLDBERG N, LAZZARONI S, MELONI B, SOLBIATI L, GAZELLE GS (1999) Small hepatocellular carcinoma: treatment with radio-frequency ablation versus ethanol injection. Radiology 210: 655–661

[39] LIVRAGHI T, GOLDBERG SN, MONTI F, BIZZINI A, LAZZARONI S, MELONI F, PELLICANO S, SOLBIATI L, GAZELLE GS (1997) Saline-enhanced radio-frequency tissue ablation in the treatment of liver metastases. Radiology 202: 205–210

[40] LIVRAGHI T, LAZZARONI S, MELONI F, TORZILLI G, VETTORI C (1995) Intralesional ethanol in the treatment of unresectable liver cancer. World J Surg 19 (6): 801–806

[41] LOPEZ RL, PAN SH, LOIS JF, MCMONICLE ME, HOFFMANN AL, SHER LS, LUGO D, MAKOW L (1997) Transarterial chemoembolization is a safe treatment for unresectable hepatic malignancies. Am Surg 63: 923–926

[42] LORENTZEN T (1996) A cooled needle electrode for radiofrequency tissue ablation: thermodynamic aspects of improved performance compared with conventional needle design. Acad Radiol 3 (7): 556–563

[43] LORENZ M, ROSSION I (1995) Adjuvante und palliative regionale Therapie von Lebermetastasen kolorektaler Tumoren. Dtsch Med Wochenschr 120: 690–697

[44] LORENZ M, STAIB-SEBLER E, ROSSION I, KOCH B, GOG C, ENCKE A (1995) Ergebnisse der Resektion und adjuvanten Therapie von Lebermetastasen kolorektaler Primärtumoren – eine Literaturübersicht. Zentralbl Chir 120: 769–779

[45] LUFKIN RB, ROBINSON JD, CASTRO DJ, JABOUR BA, DUCKWILER G, LAYFIELD LJ, HANAFEE WN (1990) Interventional magnetic resonance imaging in the head and neck. Top Magn Reson Imaging 2 (4): 76–80

[46] MARTIN M, TARARA D, WU YM, UKAH F, FABREGA A, CORWIN C, LANGE E, MITROS F (1996) Intrahepatic arterial chemoembolization for hepatocellular carcinoma and metastatic neuroendocrine tumors in the era of liver transplantation. Am Surg 62: 724–732

[47] MASTES A (1990) Cryotherapy for liver metastases [letter comment]. Br J Surg 77 (3): 354

[48] MASTES A, STEGER AC, LEES WR, WALMSLEY KM, BOWN SG (1992) Interstitial laser hyperthermia: a new approach for treating liver metastases. Br J Cancer 66 (3): 518–522

[49] MATSUMOTO R, OSHIO K, JOLESZ FA (1992) Monitoring of laser and freezing-induced ablation in the liver with TI-weighted MR imaging. J Magn Reson Imaging 2 (5): 555–562

[50] MCDANNOLD N, HYNYNEN K, WOLF D, WOLF G, JOLESZ F (1998) MRI Evaluation of thermal ablation of tumors with focused ultrasound. JMRI 8: 91–100

[51] NORDLINGER B. GUIGUET M, VAILLANT JC, BALLADUR P, BOUDJEMA K, BACHELLIER P, JAECK D (1996) Surgical resection of colorectal carcinoma metastases to the liver. A prognostic scoring system to improve case selection, based on 1568 patients. Assoc Fr Chir Cancer 77 (7): 1254–1262

[52] NORDLINGER B, VAILLANT J-C (1994) Repeat resections for recurrent colorectal liver metastases. Kluwer, Boston Dordrecht, pp 57–61

[53] OHLSSON B, STENRAUM U, TRANBERG KG (1998) Resection of colorectal liver metastases: 25 years experience. World J Surg 22: 268

[54] ONIK GM, ATKINSON D, ZEMEL R, WEAVER ML (1993) Cryosurgery of liver cancer. Semin Surg Oncol 9: 309–317

[55] RAVIKUMAR TS, KANE R, CADY B, JENKINS R, CLOUSE M, STEELE G (1991) A 5-year study of cryosurgery in the treatment of liver tumors. Arch Surg 126: 1520–1524

[56] RINGE B, BECHSTEIN WO, RAAB R, MEYER H-J, PICHLMAYR R (1990) Leberresektion bei 175 Patienten mit kolorektalen Metastasen. Chirurg 61: 272–279

[57] ROBERTS HRS, PALEY M, SAMS VR, WILKINSON ID, LEES WR, HALL-GRAGGS MA, BOWN SG (1997) Magnetic resonance imaging control of laser destruction of hepatic metastases: correlation with post-operative helical CT. Min Invas Ther Allied Technol 6: 53–64

[58] ROSSI S, DI STASI M, BUSCARINI E, QUARETTI P, GARBAGNATI F, SQUASSANTE L, PATIES C, SILVERMAN DE, BUSCARINI L (1996) Percutaneous RF interstitial thermal ablation in the treatment of hepatic cancer. Am J Roentgenol 167: 759–768

[59] SANZ-ALTAMIRA PM, SPENCE LD, HUBERMANN LS, POSNER MR, STEELE GJ, PERRY LJ, STUART KE (1997) Selective chemoembolization in the management of hepatic metastases in refractory colorectal carcinoma: a phase II trial. Dis Colon Rectum 40: 770–775

[60] SARANTOU T, BILCHIK A, RAMMING KP (1998) Complications of hepatic cryosurgery. Semin Surg Oncol 14: 156–162

[61] SCHEELE J, ALTENDORF-HOFMANN A, STANGL R, SCHMIDT K (1996) Surgical resection of colorectal liver metastases: gold standard for solitary and completely resectable lesions. Swiss Surg Suppl 4: 4–17

[62] SCHEELE J, STANGL R, ALTENDORF HOFMANN A, GALL FP (1991) Indicators of prognosis after hepatic resection for colorectal secondaries. Surgery 110 (1): 13–29

[63] SCHLAG P, HOHENBERGER P, HERFARTH C (1990) Resection of liver metastases in colorectal cancer – competitive analysis of treatment results in synchronous versus metachronous metastases. Euro J Surg Oncol 16: 360–365

[64] SCHWARZMAIER HJ, KAHN T (1995) Magnetic resonance imaging of microwave induced tissue heating. Magn Reson Med 33 (5): 729–731

[65] SHIINA S, TAGAWA K, UNAMA T, FUJINO H, UTA Y, NIWA Y, HATA Y, KOMATSU Y, SHIRATON Y, TERANO A, SUGIMOTO T (1990) Percutaneous ethanol injection therapy of hepatocellular carcinoma: analysis of 77 patients. Am J Roentgenol 155: 1221–1226

[66] SIRONI S, DE COBELLI F, LIVRAGHI T, VILLA G, ZANELLO A, TACCAGNI G, DELMASCHIO A (1994) Small hepatocellular carcinoma treated with percutaneous ethanol injection: unenhanced and gadolinium-enhanced MR imaging follow-up. Radiology 192 (2): 407–412

[67] SOLBIATI L, GOLDBERG SN, LERACE T, LIVRAGHI T, MELONI F, DELLANOCE M, SIRONI S, GAZELLE SG (1997) hepatic metastases: percutaneous radio-frequency ablation with cooled-tip electrodes. Radiology 205 (2): 367–373

[68] STANGL R, ALTENDORF HOFMANN A, CHARNLEY RM, SCHEELE J (1994) Factors influencing the natural history of colorectal liver metastases. Lancet 343 (8910): 1405–1410

[69] STEELE G JR (1994) Cryoablation in heptaic surgery. Semin Liver Dis 14(2): 120–125

[70] STEELE G, OSTEEN RT, WILSON RE, BROOKS DC, MAYER RJ, ZAMCHECK N, RAVIKUMAR TS (1984) Patterns of failure after surgical cure of large liver tumors. Am J Surg 147: 554–559

[71] SUGIHARA K, HOJO K, MORIYA Y, YAMASAKI S, KOSUGE T, TAKAYAMA T (1993) Pattern of recurrence after hepatic resection for colorectal metastases. Br J Surg 80 (8): 1032–1035

[72] VAN HILLEGERSBERG R (1997) Fundamentals of laser surgery. Eur J Surg 163: 3–12

[73] VIADANA E, BROSS IDJ, PICKREN JW (1978) The metastatic spread of cancers of the digestive system in man. Oncology 35: 114–126

[74] VOGL TJ, MACK MG, STRAUB R, MÜLLER P, EICHLER K, ENGELMANN K, FELIX R (1998) MR-guided laser-induced thermotherapy of malignant liver lesions: technique and results. Onkologie 21: 412–419

[75] VOGL TJ, MACK MG, STRAUB R, ROGGAN A, FELIX R (1997) Magnetic resonance imaging – guided abdominal interventional radiology: laser-induced thermotherapy of liver metastases. Endoscopy 29: 577–583

[76] VOGL TJ, MULLER PK, HAMMERSTINGL R, WEINHOLD N, MACK MG, PHILIPP C, DEIMLING M, BEUTHAN J, PEGIOS W, RIESS H, ET AL (1995) malignant liver tumors treated with MR imaging-guided laser-induced thermotherapy: technique and prospective results. Radiology 196 (1): 257–265

[77] VYKHODTSEVA NI, HYNYNEN K, DAMIANOU C (1994) Pulse duration and peak intensity during focused ultrasound surgery: theoretical and experimental effects in rabbit brain in vivo. Ultrasound Med Biol 20 (9): 987–1000

[78] WEAVER ML, ASHTON JG, ZEMEL R (1998) Treatment of colorectal liver metastases by cryotherapy. Semin Surg Oncol 14: 163–170

[79] WEISS L, GRUNDMANN E, TORHORST J, HARTVEIT F, MOBERG I, EDER M, FENOGLIO PREISER CM, NAPIER J, HORNE CH, LOPEZ MJ, ET AL (1986) Haematogenous metastatic patterns in colonic carcinoma: an analysis of 1541 necropsies. J Pathol 150 (3): 195–203

[80] WINGO PA, TONG T, BOLDEN S (1995) Cancer statistics. CA Cancer J Clin 45 (1): 8–30

[81] YEH KA, FORTUNATO L, HOFFMANN JP, EISENBERG BL (1997) Cryosurgical ablation of hepatic metastases from colorectal carcinomas. Am Surg 63 (1): 63–67

[82] ZHOU X-D, TANG Z-Y, YU Y-Q, WENIG J-M, MA Z-C, ZHANG B-H, ZHENG Y-X (1993) The role of cryosurgery in the treatment of hepatic cancer: a report of 113 cases. J Cancer Res Clin Oncol 120: 100–102

# III-13
# Open and Endoscopic Laser Treatment in Paediatrics

J. Waldschmidt

**Contents**

## Introduction

In 1981, the first period of laser treatment in children was started in our department. In the beginning the thermal laser-tissue interaction of the laser light was utilized. In paediatric surgery, improvements could be made by focusing the laser beam with special lens systems.

Haemostasis by coagulation, tissue cutting by vaporization and finally regressive processes in haemangiomas through induction of a vasculitis could be achieved by non-contact application.

The second phase followed with the introduction of interstitial laser treatment in the form of LITT (laser-induced interstitial thermic therapy). It started with bare fibre punction (Berlien 1984) for the treatment of haemangiomas and vascular malformations and was extended and perfected in 1991 through the development of cooled probes for power laser-application, which can also be used with the diode laser (800–1000 nm) and Nd:YAG laser [35].

The puncture can be controlled by ultrasound, CT or NMR, and checked afterwards by CCDS and MRI [1, 3, 8, 35, 40, 60]. By introduction of additional laser systems with specific modes and the development of the HPD, the range of indications could finally be completed in the 1990s, so that even diseases for which Nd:YAG cannot be used can now be treated with laser.

In our children's hospital the Nd:YAG laser is used in nearly 50% of all operations, completed by treatments with argon laser, $CO_2$ laser and dye laser. For the transmission of Nd:YAG laser light quartz bare fibres are used. The laser light is applied open, with hand pieces, in contact mode (bare fibre, sapphire tips) or non-contact mode (focused or with bare fibre) and via endoscopy with the 400 or 600 nm probes. The intralesional therapy can be made by percutaneous or endoscopic puncture with bare fibre or power-laser application probes.

## Application Forms and Indications for Use of the Nd:YAG Laser

The Nd:YAG laser can be used in 90% of all indications in paediatrics [2, 5, 6, 21, 63, 66]. It is complemented by the $CO_2$ laser for several dermatological diseases and intervention in pharynx and larynx as well as by the dye laser for capillary vascular malformations. The HPD is used in children with larynx papillomatosis, with endobrachyesophagus and with AIDS-related Kaposi's sarcoma [10, 13, 17, 59, 62].

## The Nd:YAG Laser in Pediatric Surgery

The Nd:YAG laser has simplified and improved a great number of operations in paediatric surgery. Some of them would not even be possible without laser. This applies not only to the treatment of haemangiomas and to vascular malformation, but also to several endoscopic operations on the airways, the digestive tract and urogenital system, as well as laparoscopic and thoracoscopic operations in neonates and infants. It is possible to coagulate and to cut precisely without any damage to the adjoining structures [23]. This precision can not be reached even with bipolar forceps, which unavoidably cause electric loops in the diathermic field. Another advantage is the free choice of application – open in non-contact or contact mode, continuous or interrupted, interstitial and intraluminal with bare fibre – depending on the therapeutic requirements. Bare fibre permits its use in all endoscopic techniques and in minimal invasive surgery (MIC). The Nd:YAG laser can be recommended as universal tool in endoscopic and open laser surgery in children. The advantages of operating with the Nd:YAG laser can be of use especially in operations of highly perfused organs (lung, tongue, liver, spleen, pancreas, kidney), in infected wounds and in paediatric oncology [20]. Moreover the sealing of artificial stomata or fistulas is possible, especially as it can be done under local anaesthesia. The bloodless, dry operating field makes a gentle and tissue preserving procedure possible in endoscopic as well as in open laser surgery. The wound healing is faster, the rate of infections is significantly lower compared to conventional surgery. The postoperative pain is reduced and the time of reconvalescence is shortened. The use of a drainage can frequently be avoided. Laser surgery can be seen as the "gold standard" in numerous operations these days.

## Indications for Treatment with the Nd: YAG Laser in Children

### Open Laser Application

Skin: Haemangiomas, spider naevi, vascular malformations, hyperkeratosis, mollusca, condylomata, cutaneous lymphangiomas, warts.

External genitalia: Haemangiomas, lymphangiomas, cysts, tumours and viral diseases of the vulva, the scrotum and penis.

Proctology: Haemorrhoids, fissures, marisques, polyps, anal fistulas, rectovaginal fistulas, tumours and haemangiomas of the rectum.

Umbilical diseases: Granulomas, fistulas, cysts, embryonal remnants.

Thoracic cavity: Lung resections, "wedge" resections, cysts, haemangiomas, AV-aneurysma, traumatic injuries.

Pleura: Decortication, tumours and metastases, vaporization and resections of the diaphragm, thymus, pericardium and other mediastinal processes.

Abdominal cavity: Posttraumatic injuries, resection of tumours and cysts of the liver, spleen, pancreas, omentum, mesenterium, hydatid cysts, amoebic abscesses, destruction and resection of metastases in all organs and in peritoneal carcinomatosis and sarcomatosis

Stoma occlusion: Occlusion of enterostoma, colostoma, tracheostoma, gastrostoma and cutaneous fistula.

Internal genitalia: Cysts and tumours of the Fallopian tube, ovaries and uterus, endometriosis

Retroperitoneum: Kidney cysts, pole resection, other partial resections, tumour exstirpation, resection of an explanted kidney, horseshoe kidney, renal duplication. Suprarenal glands (tumours, cysts), retroperitoneal lymphangiomas, haemangiomas and other retroperitoneal tumours, urinary bladder

Urinary bladder: Tumours, partial resections.

Extremities: Soft tissue tumours. lymphangiomas and haemangiomas, muscle angiomatosis, different forms of varicosis, Proteus-syndrome, Klippel-Trenaunay syndrome

### Endoscopic Laser Application

1. Pharynx, Choanal atresia, tumours, cysts, angiomas, polyps, granulomas, adenomas, stenoses
2. Tracheobronchial system Haemangiomas, lymphangiomas, polyps, adenomas, tumours, cysts, acquired stenosis, scar tissue, congenital stenosis due to hypoplasia, ring cartilage, web, fibrous tissue, heterotopic tissue, laryngeal cleft, tracheal diverticulum
   tracheoesophageal fistula
3. Gastrointestinal tract see later section
4. Urinary tract Strictures, urethral valves, congenital stenosis, fistula, depilation, nephrotomy, haemangiomas and tumours of the bladder, ureterocele, diverticula
5. Minimally invasive surgery
   Thoracoscopy: see later section
   Laparoscopy: see later section

## Laser in Thoracic Surgery in Children

### Lungs

Technique: Nd:YAG laser 1064 nm, focusing hand piece, non-contact mode, spot 1.5 mm, 90–100 W, cw.

The advantages of the Nd:YAG laser in thoracic surgery can be seen in the bloodless dissection with simultaneous sealing of the cutting surface [66].

The cutting surface is waterproof and airtight, so that incisions and even resections can be done on the ventilated lung. The sealing effect on the lung tissue is faster and better than on the parenchymatous organs of the abdomen [16].

Instead of unilateral intubation, both lungs are ventilated during the whole operation. So even premature babies and other risk children with restriction of the pulmonary function (pulmonary fibrosis, cystic fibrosis, multiple metastases, polycystic lung disease, diffuse bronchiectasis, etc.) can be operated. [63, 66] The cutting surface is dry (Fig. 1). Only the large vessels and bronchi have to be ligated [34]. Also in cases of multiple manifestations the number of parenchyme sutures can be reduced and limited to interrupted sutures. They can be placed in the indurated laser-treated tissue more securely than in the native lung tissue. Due to the coagulated peripheral area, the spreading of tumour cells and bacteria in the infected tissue by blood or lymph is significantly diminished. The development of adhesions can be avoided because there is no bleeding and no fibrin exsudation. A possibly necessary second operation is simplified. The time sparing effect is mainly increased and clarified in cases of multiple metastases, haemangiomas, bronchiectasis, cysts and other round focuses.

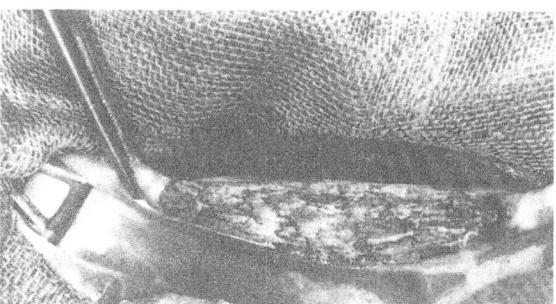

**Fig. 1.** Pulmonary section surface. The small vessels and bronchi of the lung are closed only by laser (FK hand piece, 80 W, noncontact) sutures were not needed

## Mediastinum

In processes of the mediastinum it is particularly advantageous that there is no bleeding, whereby the operating field is dry and the view not impaired. The anatomical structures in the mediastinum like oesophagus, ductus thoracicus, sympathic cord, N. vagus, aorta, V. cava and other large vessels, can be protected reliably. Extensive en-bloc resections including the thoracic wall are feasible in oncology with parenchyme resections of the lung, not following the anatomical borders (Fig. 2). Extended resections of the pericard, diaphragm and Sibson's fascia can be carried out more radically and effectively than in conventional operations. This refers particularly to benign processes like lymphangiomas, haemangiomas and other vascular malformations, which penetrate the foramina intervertebralia or grow diffusely surrounding the mediastinal structures. In the same way, operations of pseudotumours, of the lymph nodes and the complete extermination of actinomycosis, aspergillosis and other chronic suppurations can be done more easily and more radically.

## Pleura

The main indication for pleural processes is the decortication in cases of fibrothorax covering the lungs. The laser effect is not diminished by suppuration from pleuraempyemas. Multiple pleura metastases and pleural manifestations in sarcomatosis can be destroyed palliatively by vaporisation and coagulation. This is above all a great advantage in recurrent haemorrhagic effusions.

Leakage of the pleura and the parenchyme of the lungs however is treated thoracoscopically and is no indication for open surgery in children. This will be discussed in a later section. In individual cases laser treatment proves to be useful also in traumatic haemothorax, especially in larger lesions of the parenchyma and lacerations with diffuse bleedings.

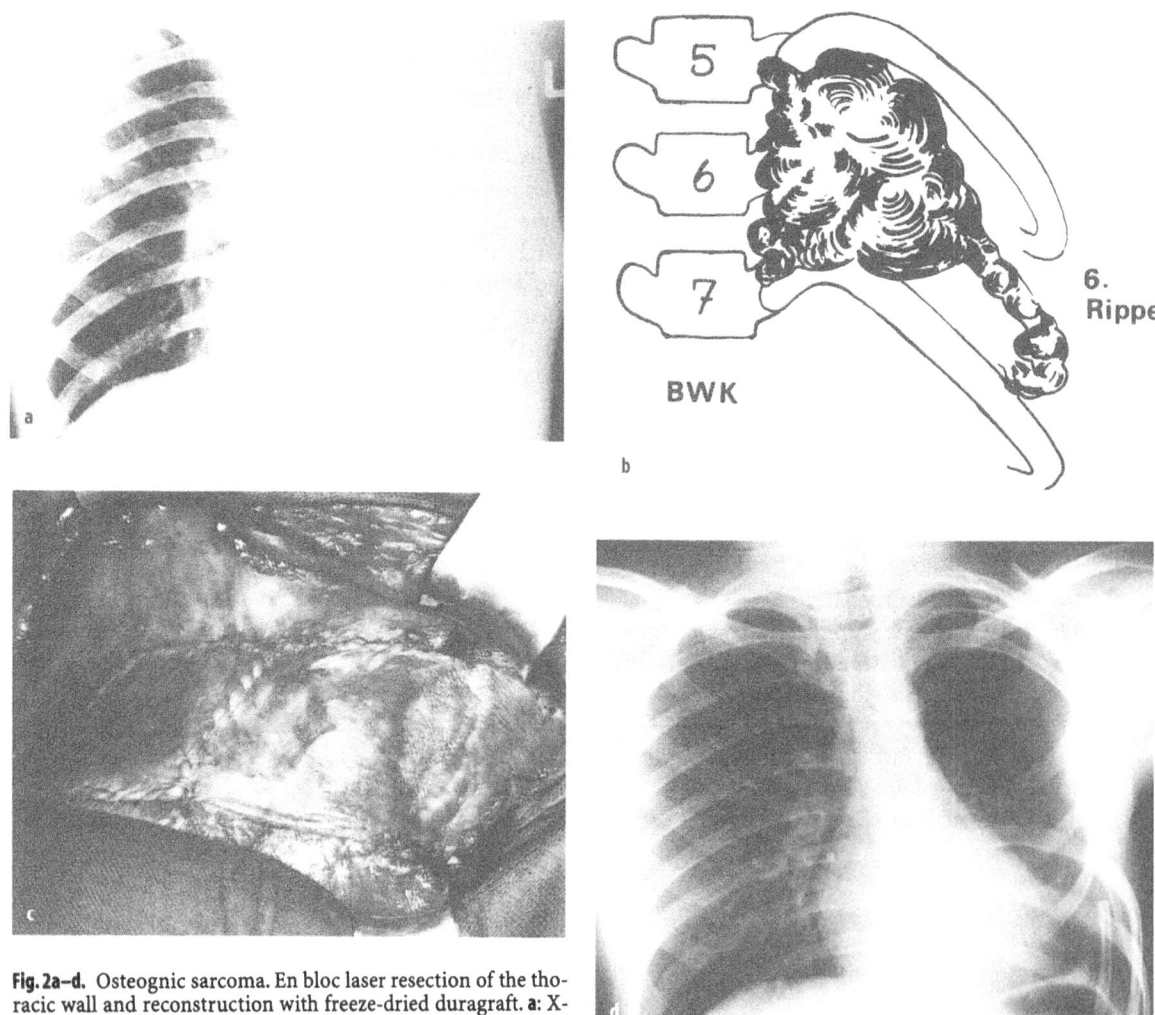

**Fig. 2a–d.** Osteognic sarcoma. En bloc laser resection of the thoracic wall and reconstruction with freeze-dried duragraft. **a**: X-ray before and **d** after tumour extirpation. **c** the implanted duragraft, **b** the extend of tumor

## Chylothorax

Thoracoscopy is the treatment of choice for chylothorax. Only in cases of recurrent and chronic chyleous effusion may open surgery be necessary. In these children the closure of the leakage with the laser in combination with humane fibrin glue is safer and easier than conventional treatment. The advantages of the laser can be seen especially in the treatment of extensive leakages and of Gorham-Stout-disease. In cases of right-sided chylothorax the fistulas are located in the posterior inferior mediastinum, in left-sided fistulas mainly in the cupola of the left thoracic cavity. An oral dose of 50 ml cream given at the beginning of the intervention is of great help to find the leakage.

## Laser in Abdominal Surgery

### Laser Resection in Parenchymatous Organs

Technique: Nd:YAG laser 1064 nm, focusing hand piece, non-contact, spot 0.5–1.5 mm, 90–100 W, cw.

Loss of blood has to be avoided in neonates and infants. Even smaller imbalances may lead to circulatory shock and are followed by permanent damages. Therefore, the use of the Nd:YAG laser is particularly important and of great significance in all operations of the liver, spleen, pancreas, kidneys and suprarenal glands. Alternative techniques like cusa, infrared or diathermy are not suitable. On these organs ultracision cannot replace the laser but it can complement it in some way. This also applies to the HF-argon beamer, which is very helpful for fast haemostasis in extensive bleeding like laceration of the spleen or liver, but the thin coagulation zone does not guarantee a permanent haemostasis, therefore a laser coagulation in non-contact mode and with defocused beam has to follow.

Before extensive resections are done we put a preliminary loop around the supplying vessels, so that a tourniquet can be applied if needed. The digital compression of the dissected tissue can help to show the larger vessels. Veins with a diameter of more than 3 mm and arteries of more than 1.5 mm have to be ligated additionally.

If bleeding occurs in spite of precautions like tourniquet, compression and spreading of the parenchyma, the cutting surface has to be rinsed with physiological sodium chloride solution, so that the blood cannot absorb the laser light and prevent the coagulation of the tissue. Intermittent rinsing under pressure with a small cannula is necessary, because continuous rinsing hinders carbonization.

## Liver

Knowledge of the anatomy of the liver is also necessary while operating with the laser, but there is no need to hold to the anatomic structures exactly. In major hepatic resection the hepatic artery portal vein and bileduct are isolated for an exact ligature.

In biopsies and partial and segmental resections the supporting compression of the cutting surface is sufficient to diminish the circulation (Fig. 3).

Even multiple resections can be done like in this may without any problems. The cutting surface is carbonized and therefore dry. Thus, the outflow of lymph and the development of fistulas can be prevented. Even in multifocal processes, postoperative adhesions can be diminished. Haemihepatectomy, segmental resections and extended segmental resections cannot be done without ligature of the larger vessels. Therefore, they should be preliminarily exposed and a loop should be applied. This is done also with the V. cava inf., supra- and infrahepatic vessels, as well as with the common bile duct. Therefore the diaphragmal ligaments should be dissected completely to mobilize the liver. For extended segmental resections, atypical resections and hemihepatectomy we seal the large cutting surfaces additionally with human fibrin glue and cover them with an omentum patch. This is an additional protection against the development of postoperative biliary leakage, wound haematomas and seromas, and it prevents the development of ascites.

Indications for open surgery and laser resection in children are tumours, cysts, haemangiomas, AV-malformations, solitary or multiple metastases, hydatid cysts, amoebic abscess and posttraumatic injuries [5, 6, 9, 27, 37, 38, 49, 50].

## Pancreas

Operations on the pancreas are rare in childhood. The technique and the parameters should be chosen as in liver resection. In operative procedures on the pancreatic head, the duodenum with the common bile duct, the duodenal papilla Vateri and also the venous confluence of V. lienalis and V. mesenterica superior with the caval vein have to be dissected. This applies especially to the pancreatic segmental resection in pancreas divisum and segmental alterations of the processus uncinatus. Here, one has to take care at the tripus Halleri. In procedures on the pancreatic tail the spleen vessels have to be exposed exactly. Indications for laser resections are: tumours, cysts and malformations of the pancreas, nesidioblastosis, stenosis of the pancreatic duct, stones and segmental pancreatitis

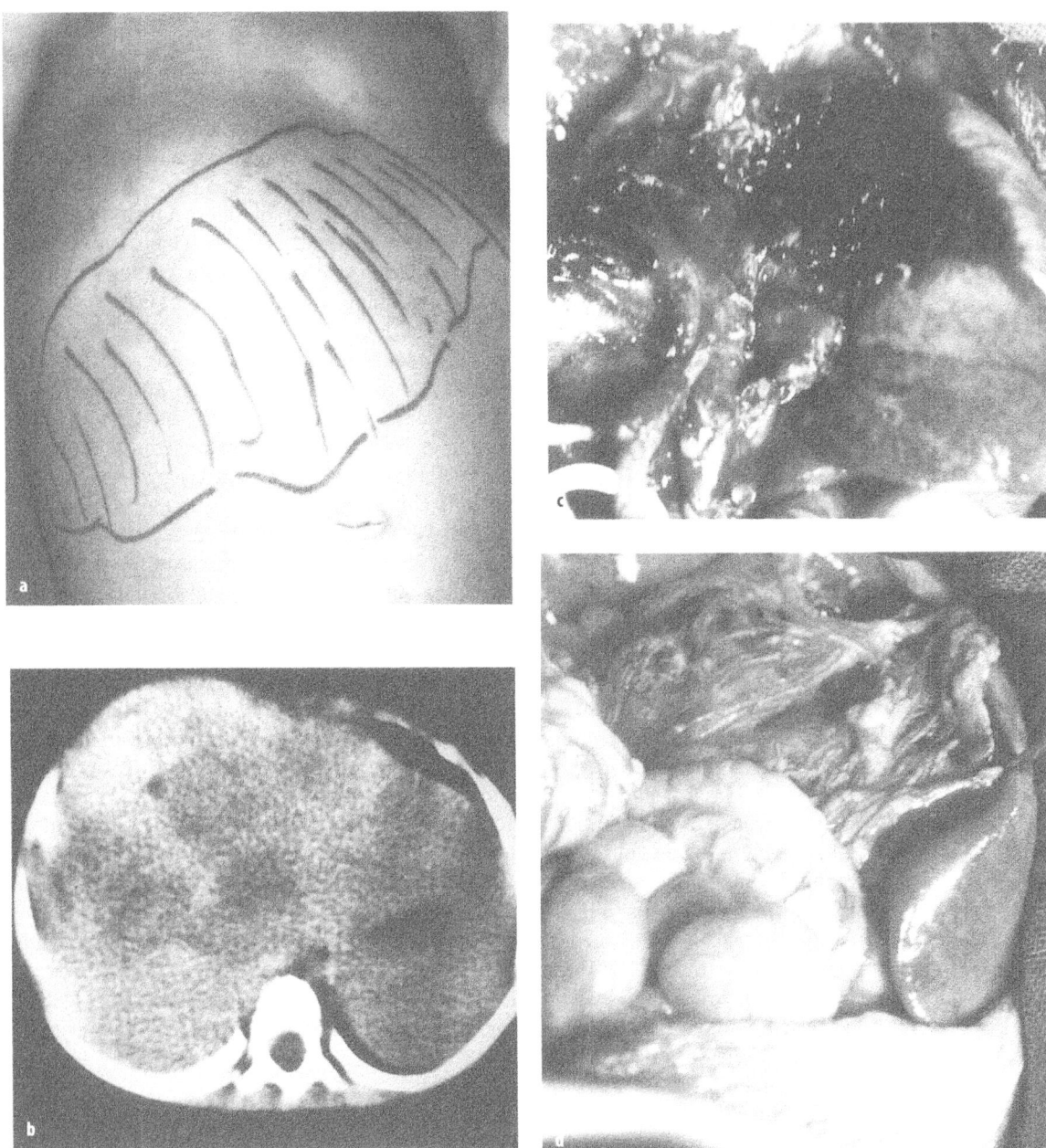

**Fig. 3a–d.** Laser resection of a malignant hepatoblastoma in a 2-year-old girl (**a**). The cutting surface before (**c**) and after (**d**) coverage with an omentum patch. **b** shows the MRI of the tumour.

(Rossi, etc.), insulinomas, teratomas and posttraumatic rupture [4, 32, 52, 70].

## Spleen

The most important aim in spleen surgery is to preserve the organ in order to avoid overwhelming infections [15, 24, 63]. Resections and biopsies should be limited to the essentials. Laser surgery of the spleen includes segmental resection, pole resections and enucleation of cysts and tumours. In all cases, the ligaments of the spleen have to be dissected, to mobilize the spleen. The vessel bundle has to be exposed every time, so that a temporary tourniquet can be placed if necessary. Ligatures, clips and dissection of the arteries and veins have to be limited to the affected segments and should be done preliminarily.

Indications for laser resections are: tumours, cysts, abscesses, hydatid cysts, aspergillosis and posttraumatic injuries. Occasionally, the volume has to be reduced because of splenomegaly [27, 49, 53, 54, 57].

## Laser Surgery of Tracheal and Bronchial Stenosis

Technique: Nd:YAG 1064 nm bare fibre 0.4 mm. Resection (scar, tumour): contact 25 W, 0.2 s.

Coagulation (haemorrhage, angioma): non-contact 25 W, 0.2 s. Ablation (scar, tumour): contact, 20 W, 1 s, single impulse. ITT: intralesional, 4–6 W, cw.

### Cases and Findings

Stenosis and stricture of the larynx, trachea and bronchi in childhood are feared because of possible complications. There is a high risk of closure of the residual lumen by secretion, swelling of the mucosa, granulation, blood clots or kinking. Mechanical ventilation is no longer possible in these cases. Considering the size of anatomical structures in premature and neonate infants, the time for diagnosis and treatment is often very limited.

From 1984–1999, we treated tracheal and bronchial stenoses in a total of 513 children (weight: starting at 680 g, age: 1, day up to 14 years, mean age: 1.4 years).

Three types and four stages were distinguished. Type A is a localized lesion, type B extends down to the bifurcation, and type C includes the larynx and/or the bifurcation and the more distal bronchial tree (Fig. 4). According to a classification by Cotton, stage I refers to a stenosis of less than 70%, stage II to a stenosis between 70 and 90%, stage III to more than 90%, and stage IV to a complete obstruction [14, 67] (Figs. 5, 6, 7).

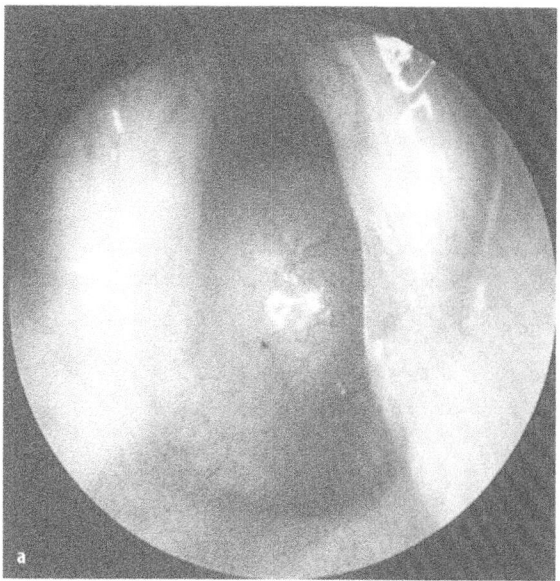

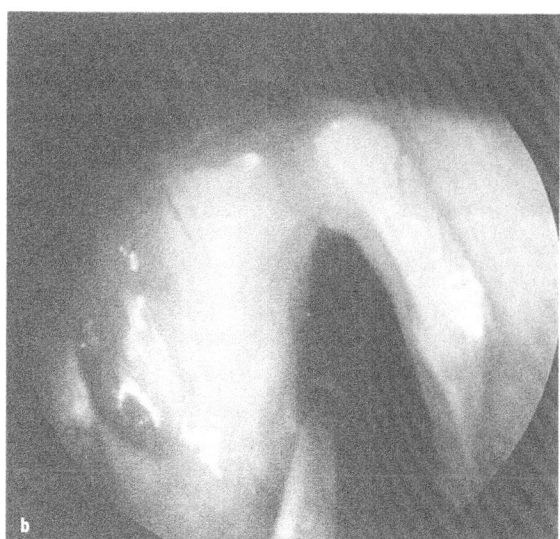

**Fig. 5a, b.** A 6-month-old girl. Larynx cyst, before (**a**) and after (**b**) resection, 25 W, interrupted, 0.2 s

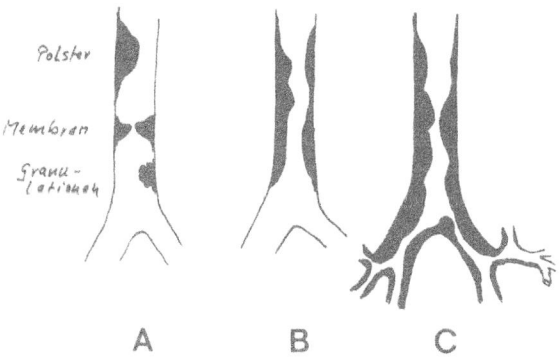

**Fig. 4A–C.** Types of intrinsic stenosis

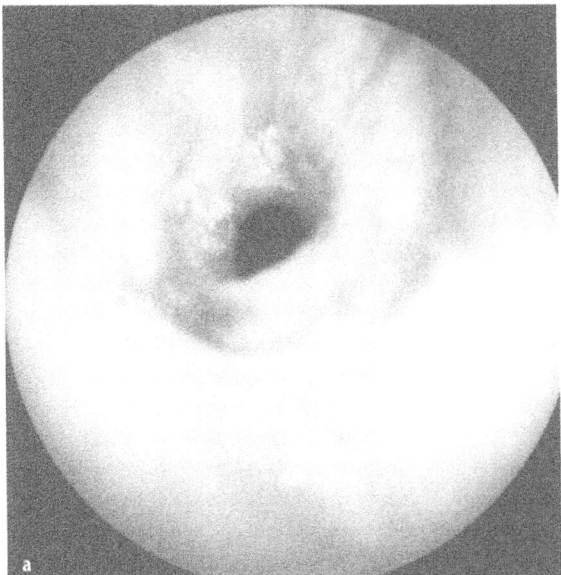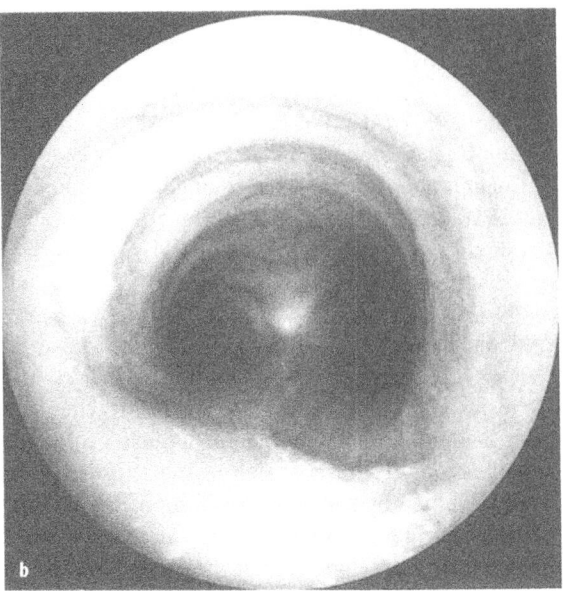

**Fig. 6a, b.** Intrinsic stenosis of the thoracic trachea, before (**a**) and after (**b**) one session laser resection

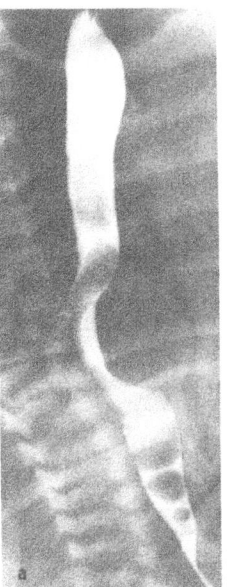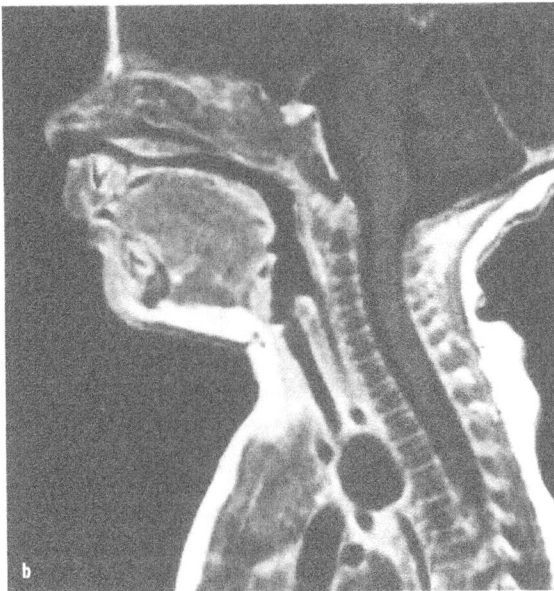

**Fig. 7a, b.** A newborn at the 15th day of age with trachea cyst

## Techniques

Therapy must be chosen according to individual requirements. The laser application will often be combined with various endoscopic disobliteration techniques, including the proven methods of balloon dilatation and scaling with the ring knife. Other methods such as dilatation with a bougie, cryosurgery or electroresection, are now rarely used – mainly when no laser equipment is available. Laser may be used in all intrinsic circumscribed (type A) stenoses as well as in long segmental and multiple-type B and C stenoses and strictures [63].

Aetiology and localization varied; 154 children suffered from postintubational or postinflammatory scarring, 231 from granulations, 51 from angiomas and so on (see Table 1 and 2).

The endoscopic procedure is performed under general anaesthesia with the rigid tracheobronchoscopy equipment by STORZ. The length, localization, diameter and cause of the stenosis are determined.

Almost half of the children required adjuvant techniques with a stent, open surgery or endoluminal dis-

**Table 1.** Causes of airway stenosis in 513 children's (UKBF 1984–1999)

| | |
|---|---|
| Scar formation | 154 |
| Granuloma | 231 |
| Angioma | 51 |
| NTB | 18 |
| Cleft | 13 |
| Cartilage hypoplasia | 7 |
| Smoke inhalation | 8 |
| Tumour | 4 |
| Cyst | 22 |
| Papillomatosis | 5 |

obliteration by balloon dilatation, ring knife disobliteration or balloon extraction [10, 14, 48, 55, 56, 58, 63, 67, 69] (Figs. 8, 9).

Most of the cases required multiple laser applications, most frequently in granulations and laryngeal scarring (see Table 3). A single laser session was sufficient in children with haemangiomas, only. Tracheal scarring required an average of 3.5 sessions, and granulomas 4.1 sessions.

**Table 2.** Location of airway stenoses in 513 childrens (UKBF, 1984–2000)

| | |
|---|---|
| Choana | 12 |
| Epiglottis | 15 |
| Larynx | 139 |
| Vocal cord | 27 |
| Subglottic | 132 |
| Thoracic trachea | 106 |
| bronchi | 82 |

**Table 3.** Number of laser sessions in 513 children with stenoses of the airways (UKBF, 1984–1999)

| | |
|---|---|
| Scar: | 3.5 |
| Granuloma | 4.1 |
| Angioma | 1.4 |
| other | 2.5 |

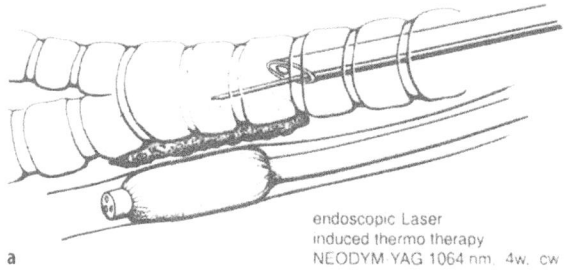

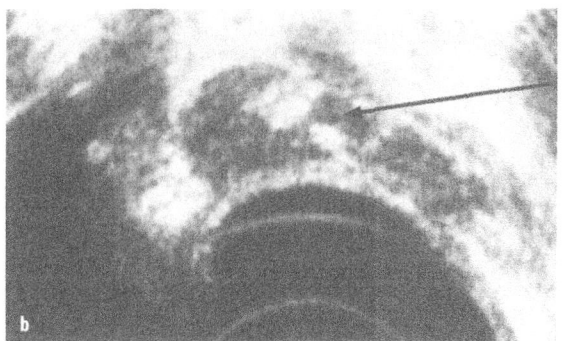

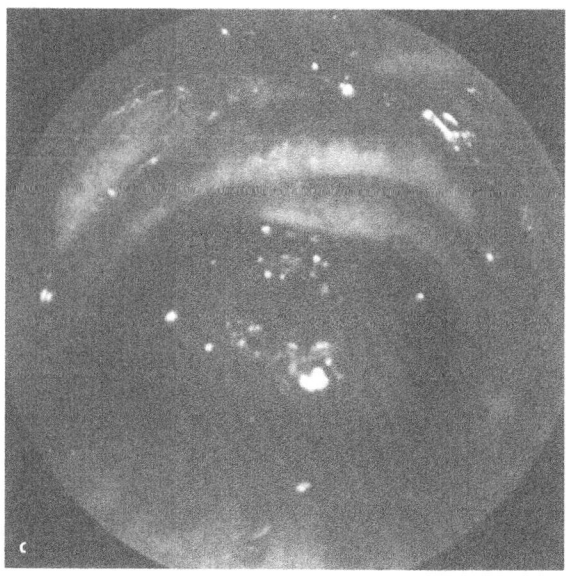

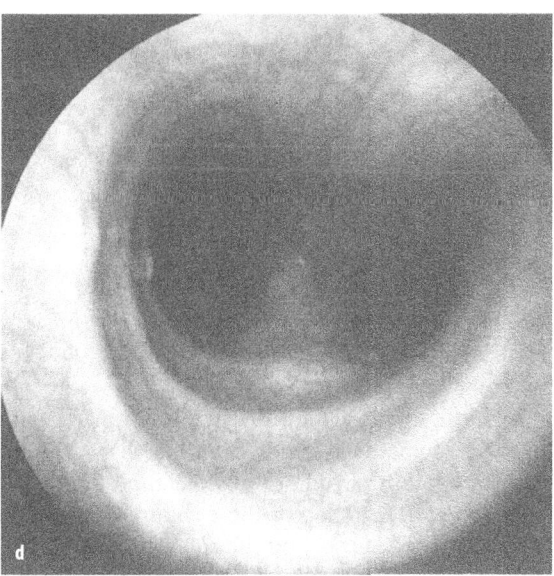

**Fig. 8a–d.** An 11 year old girl with an angiofibroma of the carina of trachea before (**c**) and after (**d**) transesophageal ultrasound (**b**) controlled LITT, 4 W, cw Fig **a** shows the tracheascopic guided bare fiber and the introduced ultra sound balloon

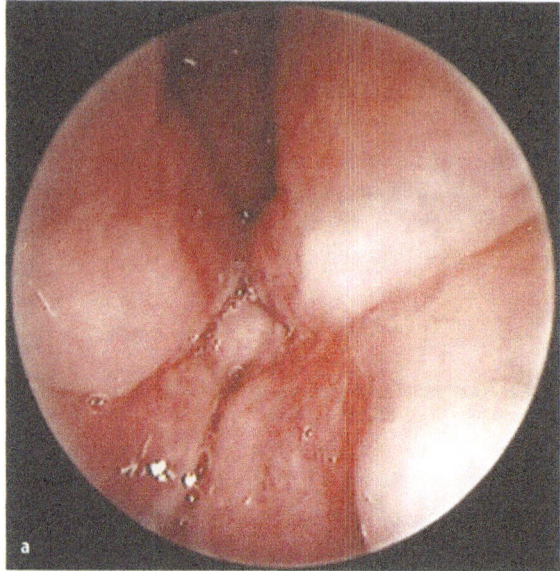

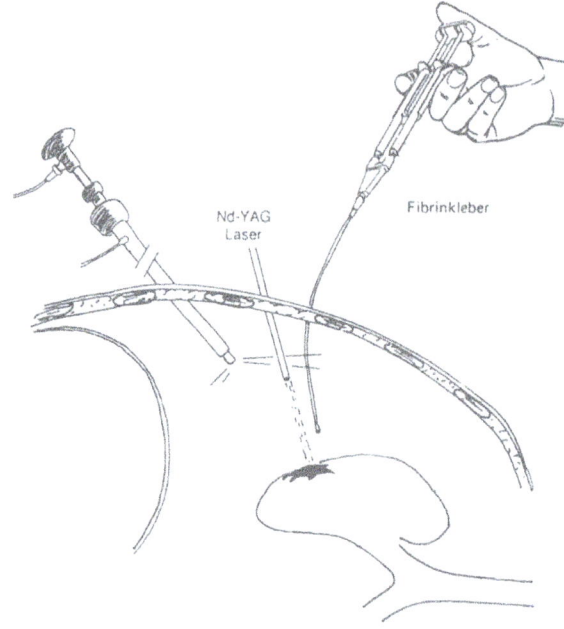

**Fig. 10.** Thoracoscopic laser application

## Thoracospcopic Laser Application

Technique: Nd:YAG, bare fibre 0.6 mm. Cutting: contact mode, fibertom-mode, 30 W, cw. Coagulation: non-contact standard mode, 30 W, 0.3 s interrupted. ITT: intralesional 5 W cw

### Introduction

Thoracoscopy is a very good tissue-preserving technique for diagnostic and operative therapy of various intrathoracic diseases. It can be performed at any age and offers special advantages in the newborn with reduced lung capacity and in older children with advanced alterations in combination with cystic fibrosis and bronchopulmonary dysplasia.

We perform thoracoscopy in relaxed children under general anaesthesia with positioning on the contralateral side. Elevation of the arm and padding of the axilla draws the scapula cranially and widens the intercostal spaces. Cleansing, disinfection and coverage are managed in such a way that the thoracoscopy can be extended to a standard thoracotomy at any time. The artificial pneumothorax is usually initiated at the 5th intercostal space in the midaxillary line with the Veress cannula. In infants and older children the thoracoscopy is performed with the lung spontaneously collapsed. In most neonates, however, the lung does not collapse adequately, particularly in cases of RDS and in the presence of inflammatory in-

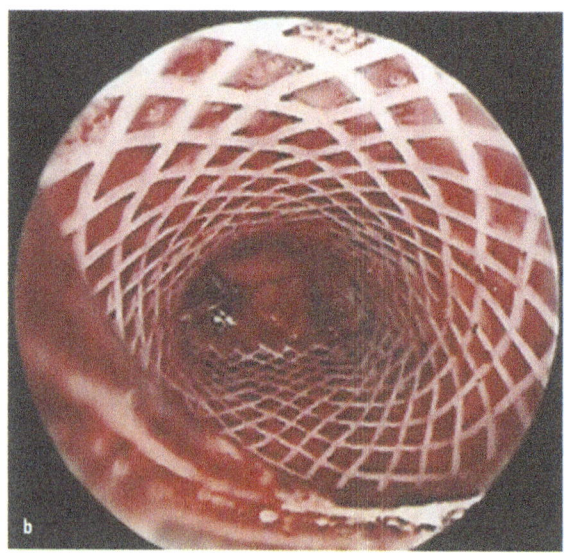

**Fig. 9a, b.** A 19-month-old girl with intrinsic stenosis by granulations (A).After laser resection and implantation of a nitinol stent

### Results

All children with intrinsic stenosis were primarily treated endoscopically with the Nd:YAG laser 1064 nm, some were operated later.

In most of the emergency tracheotomized children, the stoma had already been established in another hospital prior to laser treatment in our department. In more than half of these children, the stoma has been closed in the meantime.

filtrates, emphysema and cysts. The newborn are therefore submitted to a tension pneumothorax at 4 to 6 mmHg with the $CO_2$ insufflator. After removal of the Veress cannula, the initial optical trocar is introduced at the same puncture site; it is 5 mm in size with a 4-mm optical system. The other working trocars are then placed under guidance. Incision and haemostasis are always done with the Nd:YAG laser. It can be used to close resections and fistulas on the parenchyma, bronchus and thoracic duct. For defect closure, we additionally use human fibrin glue. The laser fibre can be inserted separately through a thin puncture needle [14 g], [25, 39, 66] (Fig. 10).

## Cyst Puncture

For smaller cysts, puncture will be sufficient. A sample is drawn for cytology and bacteriology. Following puncture and reduction of cyst size, it is advantageous to destroy the epithelial lining. The cyst is punctured again and evacuated to complete collapse with subsequent scarring.

## Fenestration

For large cysts puncture alone will not be enough: the cyst encompasses the whole lobe, the surrounding parenchyma of the lung is compressed. Cyst resection would thus mean removal of the complete lobe. We fenestrate the cyst by cutting technique with fibertom mode and vaporize the inner surface using the interrupted technique, non-contact. The whole organ is thus preserved (Fig. 11).

## Lung resection

Benign tumours, solitary or multiple metastases and multicystic destroyed segments will be resected by wedge resection. The lung tissue is first incised using the fibertom contact technique. The resection surface is than coagulated using the non-contact procedure, and the fistulas are sealed in endoluminal mode. Finally, we excise a wedge-shaped segment in the middle of the resection surface, which permits good folding and approximation of the resection edges. These are then joined together with human fibrin glue. Pulmonary parenchymal biopsy can be obtained using a forceps and bare fibre in the same manner.

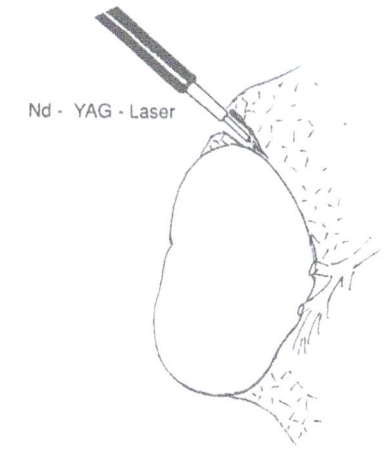

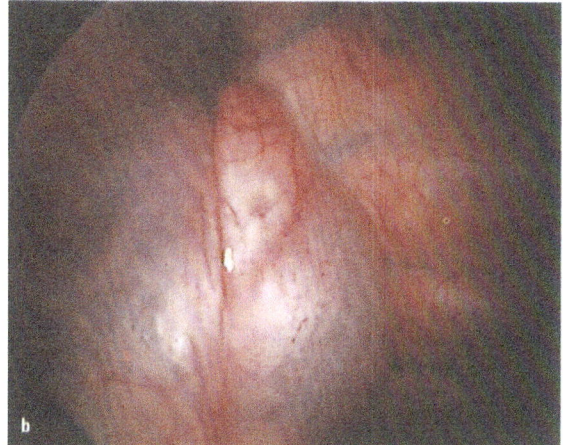

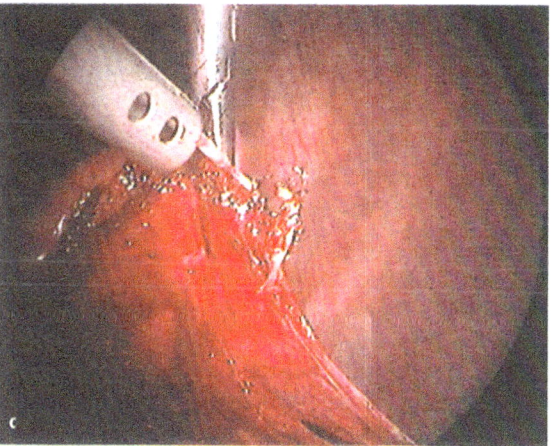

**Fig. 11a–c.** Thoracoscopic laser resection of a lung cyst by fibertom mode, contact, cw

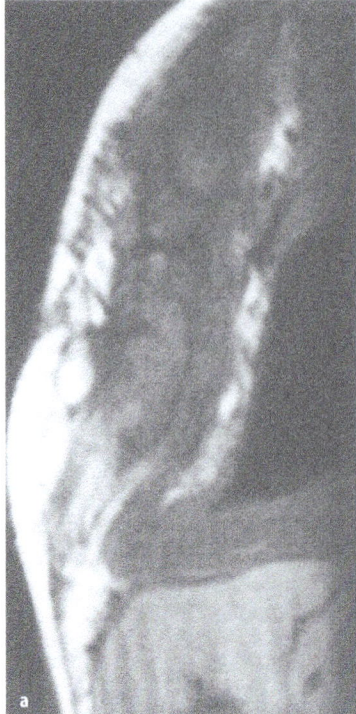
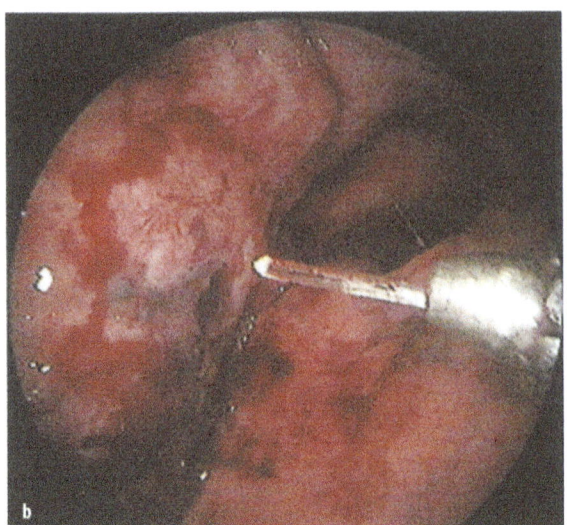
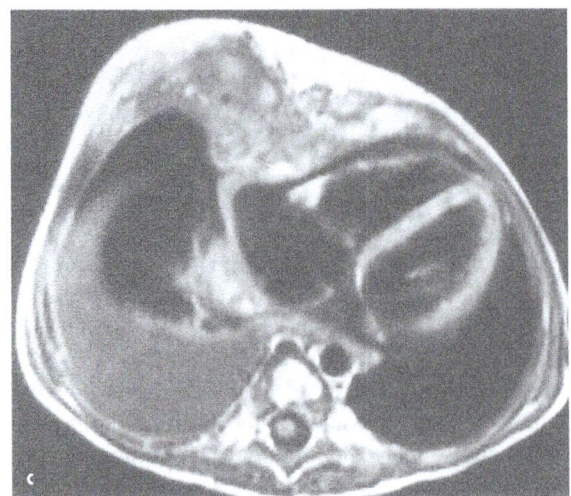

**Fig. 12a–c.** Haemangioendothelioma of the pleura, diaphragm and pericardium. Thoracoscopic laser application with ablation and LITT

## Indication

Thoracoscopic surgery is indicated in the bronchus and trachea for cysts and fistulas and in the lung parenchyma for biopsies, wedge resection in cases of metastases and emphysematous bullae, cyst resection, fistula closure, posttraumatic lung rupture and haemangiomas. Further indications are interventions in the pleura: adhesiolysis, decortication, pleurodesis, encapsulated emphysema and metastases.

It is also indicated in the mediastinum for resection of tumours, cysts, angiomas, haemangiomas and lymphangiomas; in the peripheral nervous system for sympathectomy and vagotomy; in the esophagus for achalasia, rupture and cysts; in the thoracic duct for leakage and in the pericardium for fenestration and cysts. (Table 4)

**Table 4.** Indication for Thoracoscopic Laser Surgery

| | |
|---|---|
| Bronchus, Trachea: | Cysts, fistulas, malformations, tracheomalacia, tracheoventropexy |
| Mediastinum | Tumour, cysts, haemangiomas, lymphangiomas, thymectomy, aortic arch anomalies |
| Lung | Videoscopic controlled biopsy, wedge resection (metastases, emphysematous bulla), cysts resection, closure of a fistula, traumatic rupture, haemangiomas |
| Nervous system | Sympathectomy, vagotomy, tumour |
| Oesophagus | Achalasia, rupture, cysts |
| Pleura | Adhesiolysis, pleurodesis, loculated empyema, metastases |
| Thoracic duct | Leakage |
| Pericardium | Fenestration, cysts |

## Our Own Experience

Since 1984 79 children have undergone thoracoscopic laser procedure at our hospital. Most of the cases were cystic lymphangiomas followed by unknown lung diseases, bronchial fistulas and mediastinal masses. The mean age was 1.7 years. The youngest child was a 1-day-old newborn with brochogenic cyst. The lowest body weight of a premature baby with CCAM was 1280 g. In a 1-year old girl, who had a haemangioendothelioma of the thoracic wall, mediastinum and diaphragm, multiple steps were necessary. Intrathoracal and extracorporal LITT was used.

Another newborn infant weighing 3110 g had an extensive multicystic, partially solid lymphangioma of the mediastinum and left thoracic cavity continuing extrathoracically to the neck, pharynx and thoracic wall. Respiration was already so impaired that the child had to be intubated directly after birth. Bleeding into the cysts in several places had rapidly increased the tumour size. Respiration was impaired by the structures of the neck to such a degree that the pharynx and larynx with the trachea were shifted to the right. The tumour spread caudally into the mediastinum and to the diaphragm. In the anterior mediastinum, it surrounded the thymus and in the left thoracic cavity the apex of the lung, the percardium and the other mediastinal structures. In the third week of life, the endopharyngeal and laryngeal part of the tumour was resected, followed by the thoracoscopic procedure. We used the Nd:YAG laser 1064 nm with a 0.6 mm bare fibre. The boy could already be extubated on the 2nd postoperative day. Six weeks later, a further cyst in the right anterior upper mediastinum was percutaneously treated in the same manner with the ITT technique. The boy is now more than 2 years old and free from complaints. There has been no tumour recurrence.

## Results

Seventyseven of 79 thoracoscopies were performed without any complications. Complications developed in two children:

In one child, an intercostal artery was injured during insertion of the chest tube, which resulted in bleeding that required transfusion. In the second child with bronchogenic fistula the fistula persisted and thoracotomy was necessary. Four recurrences were noted, two involved pneumothoraces with fistulas, the other two developed a new cystic lesion and/or second metastatic tumour in a child with bone sarcoma, and thoracotomy was required. In three children, the lesions were so extensive that a thoracoscopic approach was inadequate, and the thoracoscopy was converted to thoracotomy. The three children with conversions were 3 weeks (bronchial cyst within chest), 3 months (CCAM) and 12 years of age (car accident, polytrauma, multiple lesions with lung ruptures). In all cases the patients subsequently underwent a second thoracoscopy or thoracotomy without adhesions to the parietal pleura and with only moderate adhesions to the lung tissue (Table 5).

## Discussion

We have treated 79 intrathoracic lesion thoracoscopically with the Nd:YAG laser 1064 nm in more than 15 years. The overall recurrence rate in this series was 9%. Two recurrent cysts developed after 2 years and a recurrent pneumothorax a few days after the initial procedure. In one case a second thoracoscopy and in the three other cases a thoracotomy were performed. The complications that developed were mostly avoidable. The Nd:YAG laser 1064 nm is particularly well suited to thoracoscopic treatment. The precise cutting effect in fibertom mode enables endoscopic resections of tumours, cysts, angiomas, metastases and

**Table 5.** Results in thoracoscopic laser operation ($n = 79$)

| Intraoperative finding | | Outcome | | Intraoperative finding | | Outcome |
|---|---|---|---|---|---|---|
| Cystic lesion | 11 | Recurrence | 1 | Pericardial effusion | 1 | Well |
| Fistulas | 7 | Recurrence | 1 | Metastases | 1 | Well |
| Biopsy | 7 | Well | | Esophageal cysts | 1 | Well |
| Tumours | 4 | Well | | Esophageal achalasia | 1 | Well |
| Haematothoraces | 3 | Well | | Mediastinal tumour | 3 | Well |
| CCAM | 5 | Conversion | 3 | Decortication | 5 | Well |
| Septic granulomatosis | | | 1 | Castleman syndrome | 1 | Well |
| Lymphangioma | 12 | Recurrence | 1 | Lung sequestration | 1 | Well |
| Chylothorax | 4 | Recurrence | 1 | Empyema | 3 | Well |
| Haemangiomas | 4 | Well | | Emphysema | 1 | Well |

**Fig. 13a–d.** A newborn, 2nd day of life, with a multicystic lymphangioma of the throat, face, neck, mediastinum and left pleural cavity. LITT by percutaneous and thoracoscopic laser application, fenestration by fibertom mode (Fig. **c**)

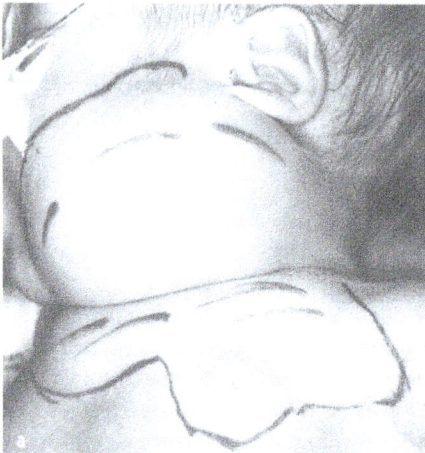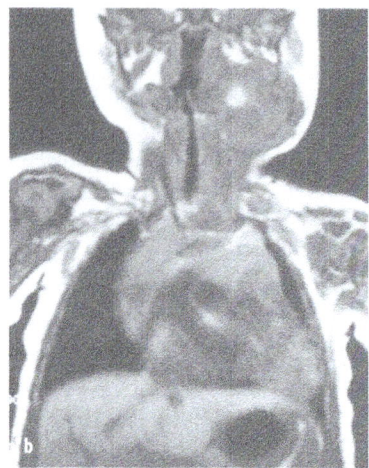

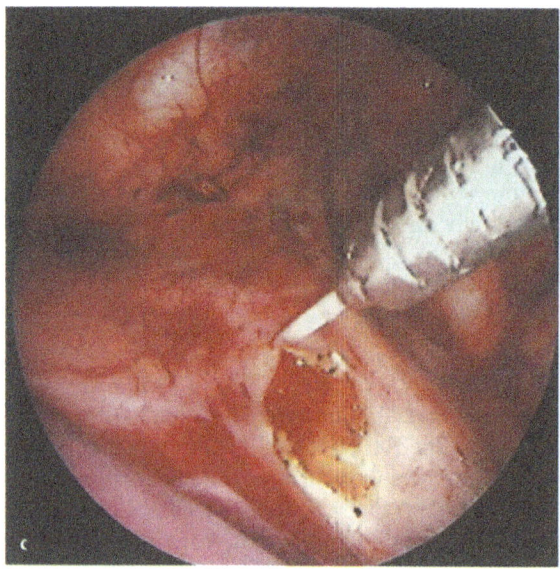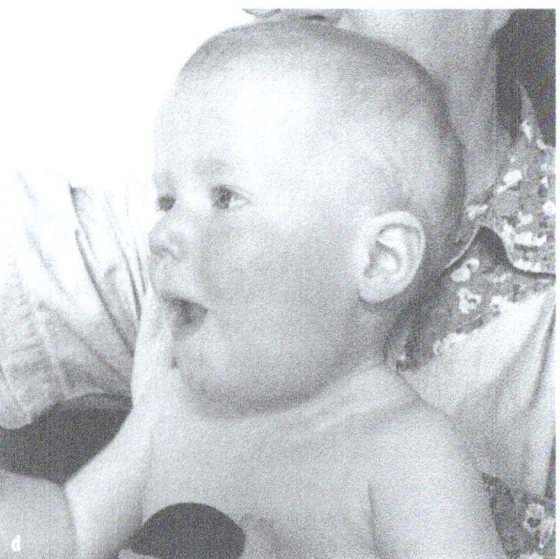

other pulmonary and mediastinal lesions. In addition, the Nd:YAG laser has a high penetration depth, thus permitting coagulation with a good volume effect. ITT of angiomas, lymphangiomas, tumours and metasteses can be performed applying 5 W in the cw technique with long pulse durations.

For thoracoscopic resections, we apply the fibertom mode. Besides the well-known characteristic property of Nd:YAG application, the coagulation effect, the fibertom mode has an exciting new feature, that is cutting with bare fibre tips. In this operation mode, the microprocessor uses the backscattering of radiation for optimum control of the temperature at the tip by contamination, and thus offers the surgeon an excellent cutting tool. Also, the tissue does not adhere to the fibre tip in the contact procedure, which makes cutting unproblematic (Fig. 13).

**Table 6.** Endoscopic laser treatment at the digestive tract (rigid, flexible)

| | |
|---|---|
| Mouth | Haemostasis, excision, desobliteration, resection, hemangiomas, lymphangiomas, cysts, tumours of the tongue, lips and cheek |
| Throat | Choanal atresia, angiomas, tumours, cysts, congenital stenoses, strictures by corrosion and burning |
| Esophagus | Haemostasis in ulcer, haemangiomas and varicosis, recanalization of tumours, strictures and congenital stenosis, closing of congenital and acquired fistulas |
| Stomach | Bleeding ulcer, polyps, haemangiomas, membranous stenosis and atresia |
| Duodenum | Bleeding ulcer, varicosis, membrane stenosis and atresia, duodenal cysts, spherical duplications, intervention at the papilla Vateri |
| Colon | Polyps, tumours, haemangioma, VMF, angiodysplasia, strictures, recanalization |

The thoracoscopic laser application offers great advantages in the endoscopic treatment of mediastinal and other intrathoracic lesions. Since there are clearly less postoperative adhesions, the intervention can, if necessary, be performed repeatedly within a short time. Bleeding is avoided by applying the laser. Thus the general view is not disturbed, which permits non-traumatic and very precise dissection. The laser also ensures that the cutting surfaces of the lymph vessels are completely welded together in order to avoid a postoperative seroma and pleural effusion. Drainage is therefore required for a few days only or not at all. Thoracoscopic laser surgery is a valuable addition to our therapeutic repertoire in neonates and children. We would therefore recommend laser-assisted thoracoscopic treatment as an effective and uncomplicated procedure for managing intrathoracic and mediastinal lesions.

## Endoscopic Laser Application on the Oesophagus and the Gastrointestinal Tract

Technique: Nd:YAG 1064 nm bare fibre 0.4mm. Resection (scar, tumour): contact 25 W, 0.2 s, interrupted. Coagulation (haemorrhage, angioma): non-contact 25 W, 0.2 s, interrupted. Ablation (scar, tumour): contact, 20 W, 1 sec, interrupted. ITT: intralesional, 4–6 W, cw.

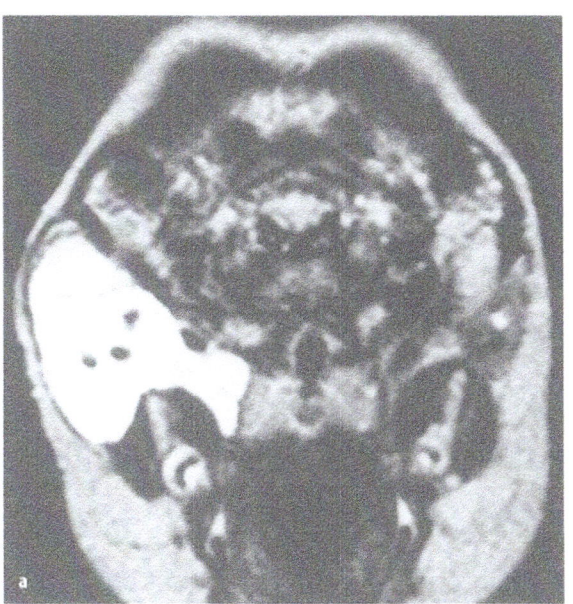

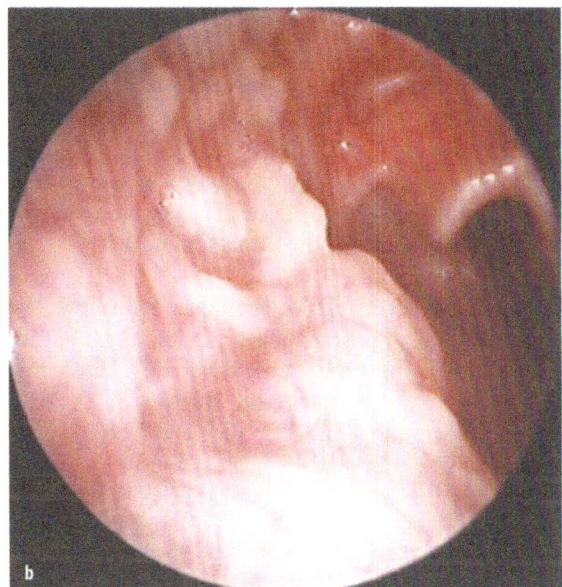

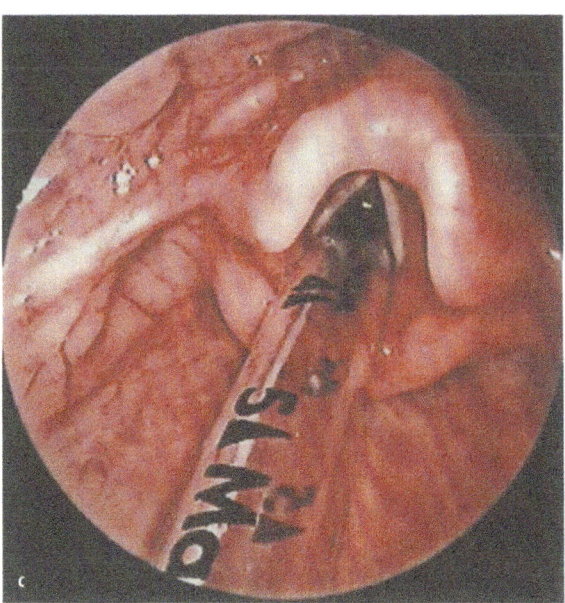

**Fig. 14 a–c** Lymphangioma of the pharynx, before (**a, b**) and after (**c**) laser ablation

Even more than in open surgery, the laser has revolutionized endoscopic surgery in children. There is hardly one other area of laser medicine where the advantages of laser treatment can be used as consisrently. This includes the gastrointestinal tract, as well as the airways and the MIC. (Table 6).

## Mouth

The most common operations with laser in the mouth are haemangiomas, lymphangiomas, cysts and tumours of the lips and tongue. In cases of haemangiomas, we proceed similar to the treatment of haemangiomas of the body surface.

Lymphangiomas appear in the mouth in three different forms:
1. Small eruptive cysts on the tongue's surface.
2. Diffuse lymphangiomatous infiltration of the tongue up to the radix linguae with extension to the vallecula.
3. Lymphangiomas on the floor of the mouth with or without involvement of the pharynx and the adjoining soft parts of the throat and neck.

They are treated with ITT technique. 4–6 W, cw. The eruptive cysts on the tongue surface and the tuberous penetrations of the pharynx and the vallecula can be ablated using the ablation mode. With diffuse infiltration, however, a partial resection of the tongue in fibertom mode is needed. In this, the resection is easier if a tourniquet is laid to reduce the blood flow temporarily. In cases of operations of the mouth or the throat, preventive antibiotics have to be given.

## Throat

Indications for endoscopic laser surgery are choanal atresia [22], haemangiomas and lymphangiomas, adenoids, tonsillar hyperplasia and other benign processes [60,63]. Very effective is the resection or ablation of scar tissue. Especially after corrosion through leaching, a subtotal, in some of cases even a complete obliteration, of the meso- and hypopharynx may appear. The desobliteration is done in several steps, possibly with the use of temporary stents. The nutrition of these children is maintained through gastrostoma, the breathing is guaranteed by involvement of the epiglottis or the larynx through a tracheotomy (Fig. 14).

## Oesophagus

Good indications for laser treatment at the oesophagus in children are congenital membranous stenosis [30, 42] and short distant strictures after corrosions and operations. These have to be combined normally with bougienage [19]. Other indications are: haemostasis at bleeding haemangiomas, ulceration and varicosis of the oesophagus in Barett oesophagus [18, 29, 30], but in these cases the sclerosing treatment and pressure-reducing operation techniques have gained acceptance in paediatrics.

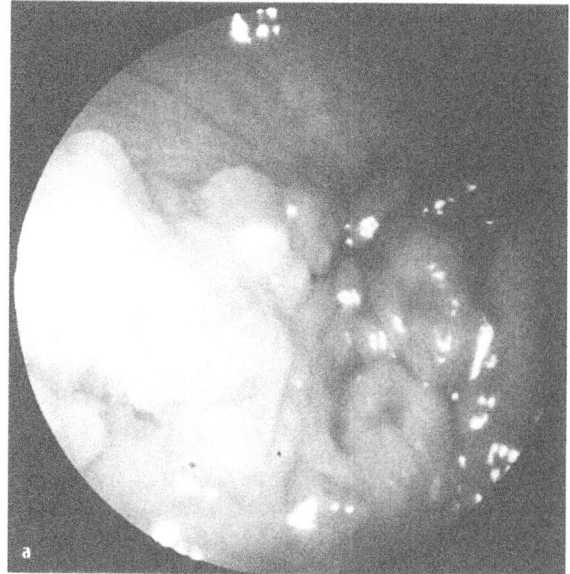

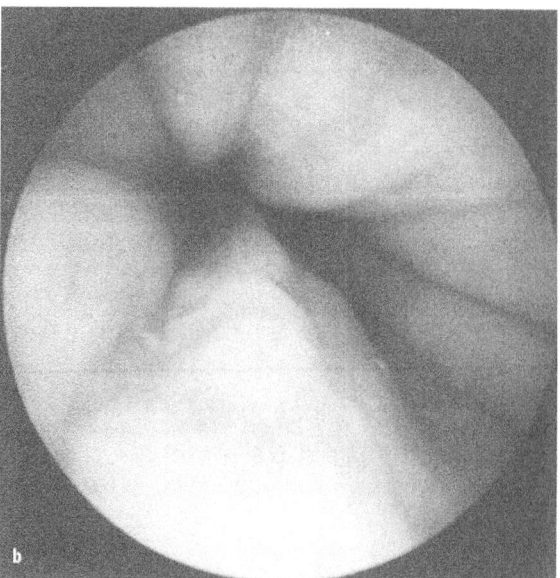

**Fig. 15a, b.** A 13-year-old boy with Goltz-Gorlin syndrome. oesophageal stenosis by papillomata, before (**a**) and after (**b**) laser resection

**Fig. 16a, b.** Oesophago-mediastinal fistula, before (**a**) and after (**b**) laser occlusion and fibrin glue application

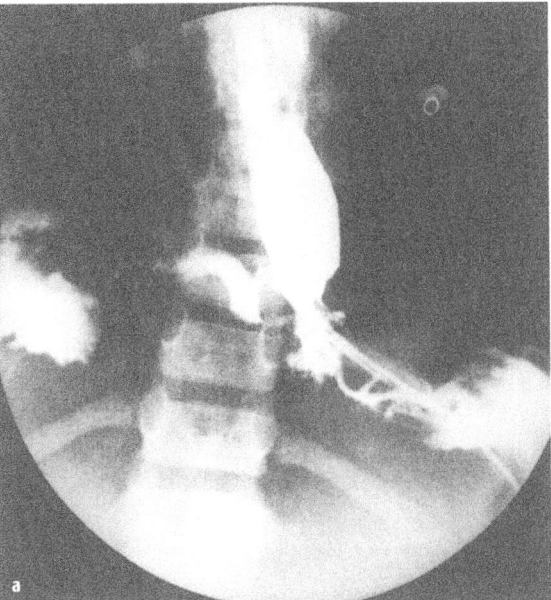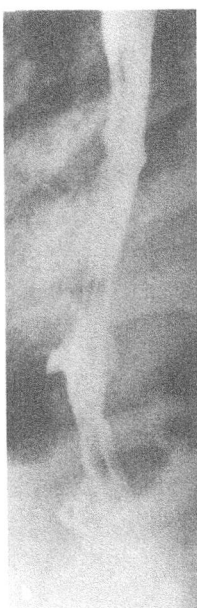

Seldom indications are also the ablation of polyps, papillomas and benign endoesophageal tumours.

This can be done with rigid as well as with flexible fibre endoscopes. Preferred parameters are 25 W interrupted with an exposure time of 0.2 s in non-contact mode. The styptic treatment, however, is carried out with defocused beam and prolongation of the exposition time. All resections and excisions are made in contact mode (Fig. 15).

### Oesophago-Pleural and Oesophago-Mediastinal Fistulas

The closing of oesophageal fistulas is normally combined with the application of fibrin glue. The laser fibre is inserted into the fistula and then pulled back, applying a power of 5 W and with a velocity of 1 mm s$^{-1}$. Very wide fistulas and pseudocysts filled with fluid have first to be cleaned carefully by rinsing. The deepilized fistula duct is then cleaned again after treatment, sucked off and then filled with a lead seal of human fibrin glue.

The same procedure is chosen in cases of traumatic and artificial injuries and ruptures of the oesophagus. In cases of accompanied pleura empyema or a pneumothorax the pleural cavity is relieved additionally by a Buelau drainage (Fig. 16).

### TEF

The proceeding with tracheoesophageal fistulas is similar, but the closing of these fistulas has to be done always through the trachea. It is performed by laryngo-tracheoscopy and is therefore described in the chapter on airway disease [51, 61, 63] (Fig. 17).

### Stomach

Indications at the stomach are even more seldom than on the oesophagus. They are limited to haemostasis and ablations of polyps, the polyps having always to be acquired for histological examinations. Other indications are haemangiomas and congenital stenosis or obturations due to membranes and pathological folding [26, 31, 41].

### Duodenum

Indications are haemostasis in ulcers or haemangiomas and excision of membrane stenosis and atresia (Fig. 18) [28, 29a, 29b, 71]. Other rare indications are the fenestration of spherical duplications and duodenal cysts.

### Congenital Duodenal Membrane

Congenital duodenal membrane causes an ileus even in newborn. For the endoscopic description special mini-duodenoscopes with a sufficient working canal are needed. This is relatively simple if the membrane has a small central opening. Then forceps or a small balloon can be brought forward through this opening into the duodenum distal of the membrane (Fig. 18). By filling the balloon or spreading the forceps, the thickness and anatomical relations of the membrane

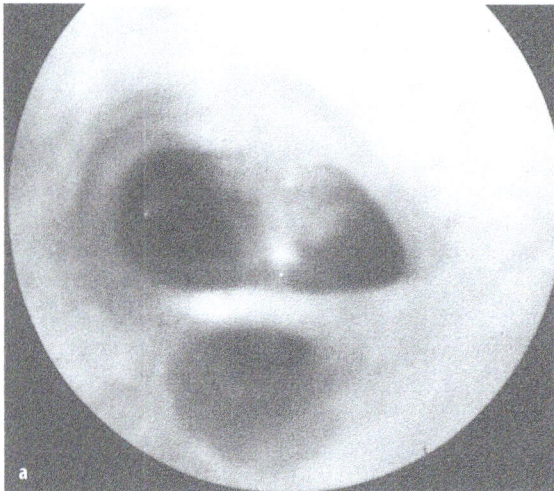

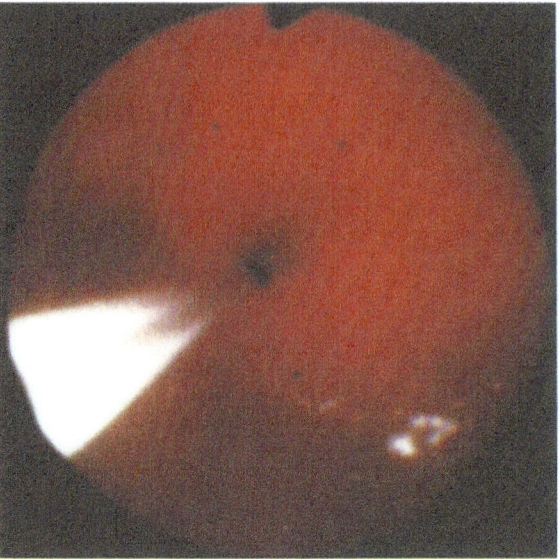

**Fig. 18.** Endoscopic laser resection of a duodenal membrane. By using a biopsy forceps in the opening to pull back the membrane, the risk of damaging the duodenal wall by the laser beam is reduced

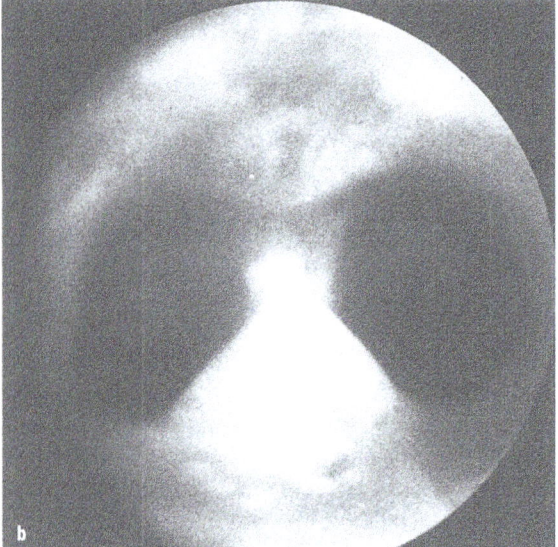

and the duodenal wall can be made visible while drawing back the instruments. Excision or star-shaped incisions of the duodenal wall have to begin at the large curvature, so that there will be no injury of the Ductus Wirsungianus and the Papilla Vateri. Also the mural part of the Ductus choledochus can run through the membrane, sometimes even with building of a spiral, so that the bilary duct may be injured, if the course isn't exactly shown before. If necessary, the excision of the membrane should be refused and has to be operated conventionally.

## Laparoscopic Laser Application

Technique: Nd:YAG 1064 nm, bare fibre 0.6 mm. Resection: contact, standard 25 W, 0.2 s, interrupted. Coagulation: non-contact, 20–30 W, 0.5 s, interrupted. Cutting: contact, 30 W, fibertom, cw. ITT: intralesional, 5 W, cw. Ablation: contact, 20 W, 1 s, single impulse.

Laparoscopy allows inspection and surgical procedures of the organs of the abdomen, pelvis and retroperitoneal space with minimal tissue trauma. Age and body weight do not represent contrainidications to the procedure. In neonates and infants, laparoscopy should be performed using general anaesthesia with muscle relaxation. Stomach, rectum and bladder must be decompressed.

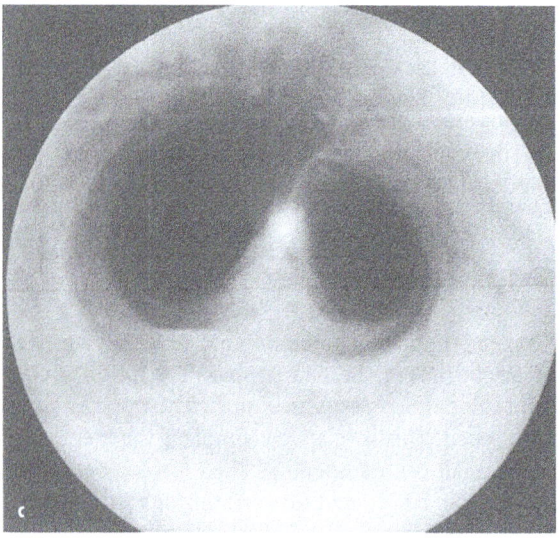

**Fig. 17a–c.** Oesophago-tracheal fistula, before (**a**) and after (**b**) laser and fibrin glue. **c:** After 6 weeks

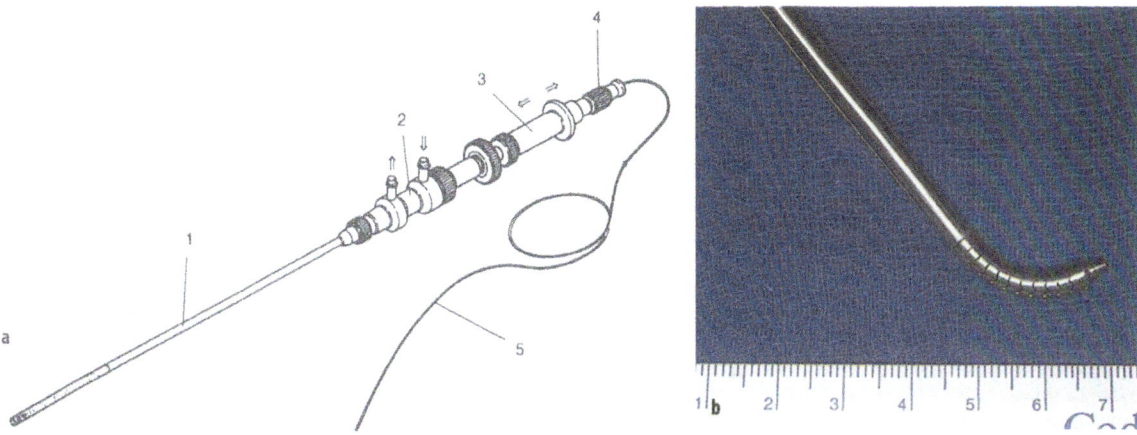

**Fig. 19a, b.** Laparoscopic laser-hand pieces from MBB-Dornier (**a**) and Jakoubek-Dufner (**b**)

## Procedure

We use the Storz pediatric laparoscope with a 2- or 5-mm telescope and instruments. These are supplemented with clip applicators, the Roeder sling, and the laparoscopic laser hand pieces of Jakoubek and MBB-Dornier. Positioning and arrangement of the instruments and the equipment are the same for all procedures. The surgeon and the nurse stand on the left side of the child, the assistant on the right. The Veress cannula and the optical trocar are inserted at the lower edge of the omphalos and at the Munro point in the neonates, respectively (Fig. 19).

In infants and newborn the umbilical region must be avoided, because accompanying umbilical abnormalities can be expected, and injuries could be inflicted due to the high bladder and the fragile umbilical arteries.

## Indications

Specific indications for laparoscopic laser application are listed in the following Table 7.

## Laparoscopic Appendectomy

Appendectomy is the most common laparoscopic operation in children. We use the three point technique. Veress cannula and the telescope are introduced at the umbilicus. The two working trocars are inserted at the right and left lower quadrant of the abdomen under laparoscopic visual control. The incision and haemostasis of the meso-appendix is carried out by laser using the fibertome mode cw. Then the appendix is ligated with two Roeder slings and divided between these two ligatures by laser. The advantages of the laser application are a complete sealing and oblit-

**Table 7.** Indications for laparoscopic laser application

| | |
|---|---|
| Female genitalia | Diagnostic biopsy and gonadectomy in intersex<br>Ovarectomy, salpingectomy, hysterectomy<br>Cysts and neoplasm of the ovary and Fallopian tube, Stein-Leventhal syndrome<br>Cong. ligaments<br>Torsion of the ovary and other complications<br>Ovaropexy in oncology<br>Endometriosis |
| Male genitalia | Cryptorchidism<br>Varicocele<br>Resection for Müllerian duct remnants |
| Adhesiolysis Bowel | Cong. and acquired adhesions |
| Bowel | Appendectomy, coecopexy, Meckel's diverticula, bowel duplicature, endorectal pull-through for Hirschsprung's disease |
| Stomach, duodenum | Perforated ulcer, adenomas, angiomas, cysts, polyps |
| Cholecystectomy | |
| Liver | Cyst, angioma, AV aneurysm<br>Posttraumatic lesions<br>Biopsy, resection<br>Metastasis – ITT coagulation and vaporization<br>Tumour<br>Hydatid cysts |
| Spleen | Biopsy, cysts, angiomas |
| Pancreas | Lavage and drainage in pancretitis, fistel occlusion, cyst fenestration, resection |
| Peritoneal cavity | Cysts and tumours, mesentery, omentum, peritoneum |
| Oncology | |
| Retroperitoneum | Tumours, haemangioma, lymphangioma, diseases of the kidney and adrenal gland |
| Organopexy | Coecum mobile, coloptosis, wandering spleen, nephroptose (combined with fibrin glue) |
| Intra abdominal abscesses | |

eration of the appendix lumen and a small coagulation zone without damage to the ligature.

## Resection of Meckel's Diverticulum

Resection of a Meckel's diverticulum or an omphaloenteretric duct is technically similar to appendectomy. The diverticular artery and vein are transsected using the laser. The tip of the diverticulum is grasped and elevated with a forceps. The diverticulum is stretched and ligated along its base with two Roeder slings and transsected by laser between the two endolops. Then the mucosa at the stump is destroyed with the laser, resulting in a small pinpoint scar. In addition to a Meckel's diverticulum, other associated embryonic remnants may be resected as well and may be followed by appendectomy. A vitelline duct cyst should be evacuated before laser extraction (Fig. 20).

## Organopexy

Combined with human fibrin glue, laparoscopic pexis by laser represents an efficient and safe method for cecopexy of the mobile cecum, colopexy for mesoileocolicum commune and coloptosis type B (broad gastrocolic ligament without gastroptose). Other indications include hepatopexy in Chilaiditi's syndrome and splenopexy for the wandering spleen.

The technique of organopexy includes creation of an extended deserosalized area between the parietal peritoneum and the organ involved, which is best achieved with the Nd:YAG laser (fibertome mode 30 W). In addition, this area should be covered with fibrin glue. In colopexy, the deserolization of the parietal peritoneum alone will not suffice. The transverse fascia must be exposed as well [64, 65]. Otherwise, the peritoneum will be left like a mesentery and the pathologic mobility will recur with torsions, kinking and strangulation.

## Lysis of Adhesions

Congenital and acquired adhesions often cause bowel obstruction and chronic abdominal pain in childhood. Congenital adhesions may be found in the ileocecal region as a pathologic ileoparietal ligament. Acquired adhesions occur in early childhood following appendectomy, peritonitis, NEC, adnexitis and other previous surgery. These adhesions of the bowel mesentery and omentum can be transsected by laser without bleeding and fibrin exsudation. The dissected areas shrink, and rapid reperitonealization occurs and may prevent formation of new adhesions. We prefer in this case the fibertom mode 20 W cw.

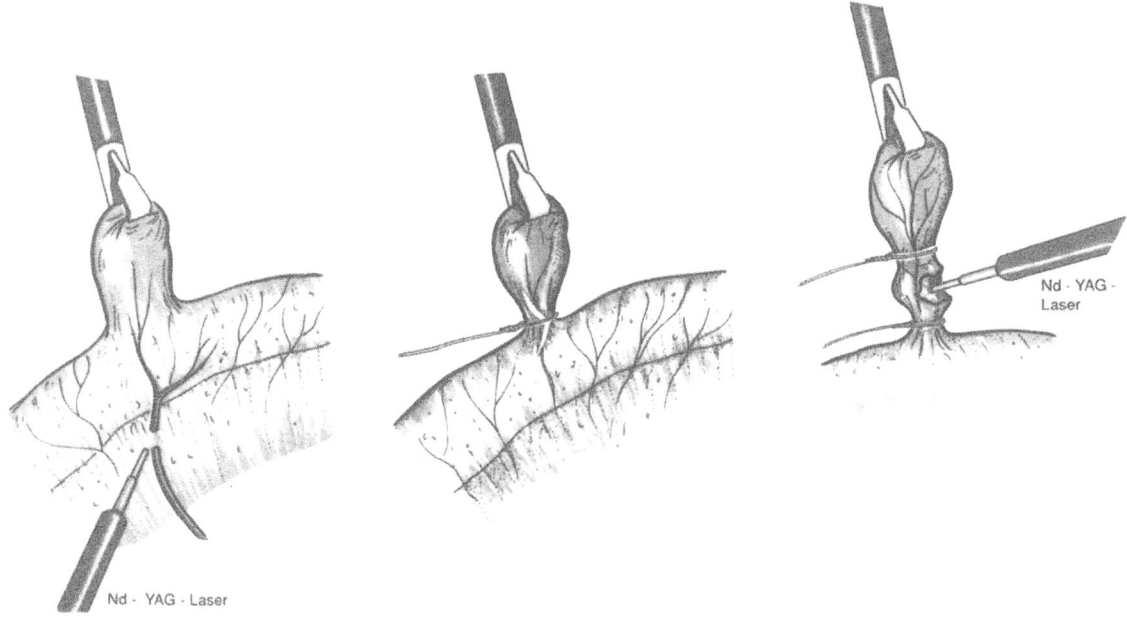

**Fig. 20.** Resection of a Meckel's diverticulum. Before the diverticulum is severed it has to be ligated with two endoloops

## Female Internal Genitalia

All parts of the internal genitalia are easily accessible for laparoscopic laser intervention. Indications are genital malformations, ambiguous genitalia, ovarian cysts and tumours, adnexel torsion, endometriosis and further disturbances of the ovary and Fallopian tube. Most of the cases are fenestrations and resections of ovarian cysts, gonadectomy and biopsy. For gonadectomy, the ovary is elevated from mesosalpinx with a grasping forceps and resected along the mesovarium using the bare fibre 0.6 mm, fibertom mode cw. The Fallopian tube is preserved with this technique.

### Ovarian Cysts

Functional cysts of neonates contain a serous liquid and regress spontaneously in many cases. In cysts with a diameter larger than 3 cm complications often threaten: haemorrhage, rupture, torsion. Secondary aganodism and bowel obstruction may result.

Management of the ovarian cysts depends on the laparoscopic findings. For smaller cysts, puncture and decompression are sufficient. Following reduction of the size, the epithel lining is destroyed with the Nd:YAG laser (7 W cw, 8–10 s). For larger multiloculated cysts this procedure is often insufficient. The cyst may encompass the entire ovary, with the stroma becoming part of the cyst capsule. Cystectomy would mean removal of all the normal ovarian tissue. Therefore, fenestration of the cyst with coagulation of lining epithelium is necessary using the interrupted technique, non-contact 25 W 0.5 s. Thereafter, the cyst will subsequently shrink and form a new gonad. Thus, the whole organ can be preserved along with the ovarian stroma within the cyst wall. Should necrosis of the cyst and ovary occur following torsion and for cysts with solid components that require histological examination, ovariectomy is indicated. An endolope is placed and the torquired stalk divided with the laser.

### Laparoscopic Hysterosalpingectomy

Uterine anomalies are a rare cause of Müllerian duct malformations in intersexuality. Early hysterectomy and salpingo-ovarectomy in the first year of life is necessary to minimize future obstetric and gynaecologic complications. Standard laparoscopy is performed with 1.7-mm instruments and 0.6-mm bare fibre. First, the infunibulo-pelvic ligaments are stretched and divided in fibertom mode. As a second step, the peritoneal sheet of mesosalpinx is incised followed by mobilization of the fallopian tube and streak gonads. Then the uterovesical pouch is opened and the vesicouterine and vesicocervical ligaments divided in contact technique. Finally, the cervix and vagina are mobilized and cut by laser at the urethral junction. The vaginal stump will be ligated, and hysterosalpingectomy and ovarectomy completed (Fig. 21).

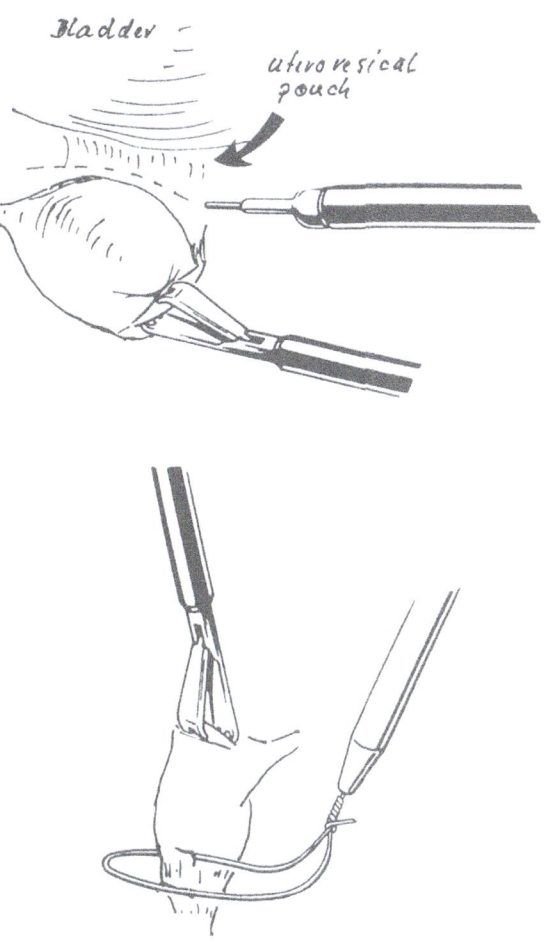

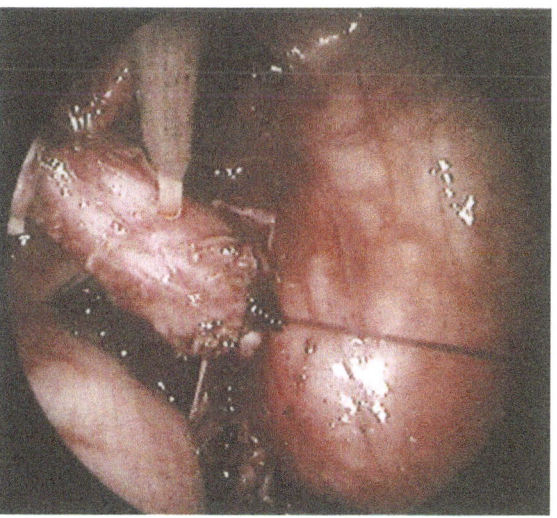

**Fig. 21.** Laparoscopic hysterectomy and salpingoovarectomy in a 6 month old infant with intersex

## Cryptorchidism

Preliminary laparoscopic laser dissection (PLLD) of the internal spermatic vessels is indicated when the internal spermatic artery and vein are so short that they pass to the gonads stretched out in a straight line and without a loop. An elongation of the vessels with a transfer of the testicle into the scrotum is then no longer possible by conventional techniques.

Alternatives are the free graft, the two-stage procedure according to Corkery and the Fowler technique of vessel dissection. We achieved the best results with respect to testicular atrophy and growth by applying the Fowler technique of vessel dissection. Of the boys, 15% could not be submitted to the Fowler operation, however, since perfusion via collateral vessels was not adequate (Fig. 22). This collateral circulation can be improved if tension is relieved on the gubernaculum and its vessels by severing the internal spermatic artery with the laser. After the dissection, the gubernaculum retracts immediately, thus spontaneously drawing the testicle to the internal inguinal ring. In the further course of time, the testicle even enters into the inguinal canal. At the time of the second session, 6–8 weeks later, the testicle is easy to localize in the inguinal canal. It has increased considerably in size and is very well vascularized by the collateral circulation between the deferential artery, the external spermatic artery and the gubernacular vessels. The gubernaculum and collateral vessels are spared in the subsequent dissection. The vas deferens is mobilized in the usual manner. The vas deferens is always long enough in the cranially situated testis to permit tension-free transfer into the scrotum and to obviate epigastric-vessel tunneling. [63, 64, 65].

## Varicocele

Operative management of a varicocele is indicated if embolization therapy is not adequate or not possible (15% of approximately 650 embolizations in our group of patients). We previously performed high ligature of the spermatic vein according to Paloma in these boys. The laparoscopic technique is better and less traumatic. The view is excellent; even small collateral veins are clearly recognizable by the telescope. The spermatic artery can easily be lifted from the venous plexus with a small forceps and should therefore be spared. Closure of the thick main vessel is achieved with clips. The collaterals are obliterated with the laser in non-contact technique.

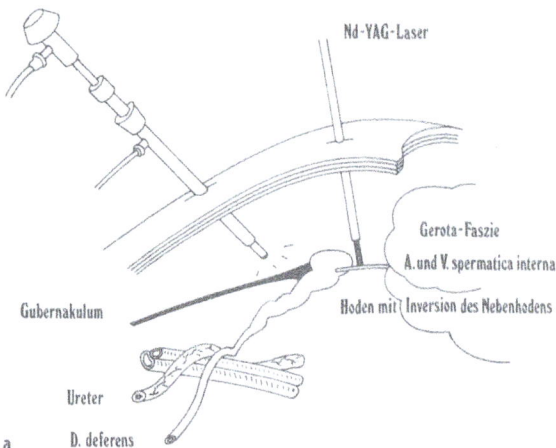

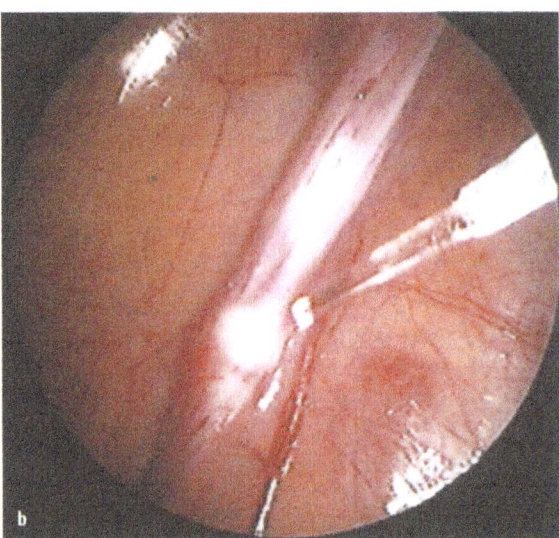

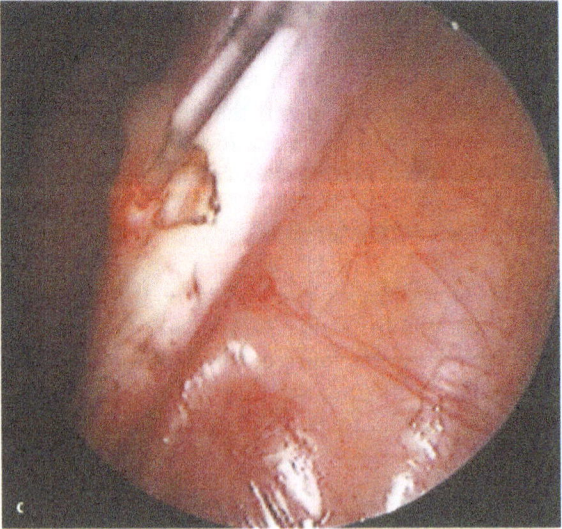

**Fig. 22a–c.** Laparoscopic laser dissection of spermatic artery in cryptorchidism. At first artery and vein have to be occluded in non-contact (30 W, 0.5 s) and then transected in contact using fibertom technique

## Cholecystectomy

The indication for cholecystectomy is rarer in children than in adults. Sonographically detected concrements usually dissolve spontaneously, particularly in neonates, but even calcareously incrusted, i.e. radiologically recognizable, stones commonly cannot dissolve spontaneously. Nevertheless, all complications of stone migration (impaction, obstructive jaundice, chronic dropsy, pancreatitis) and bile-duct infection (cholecystitis, pericholecystitis, empyema, cholangi-

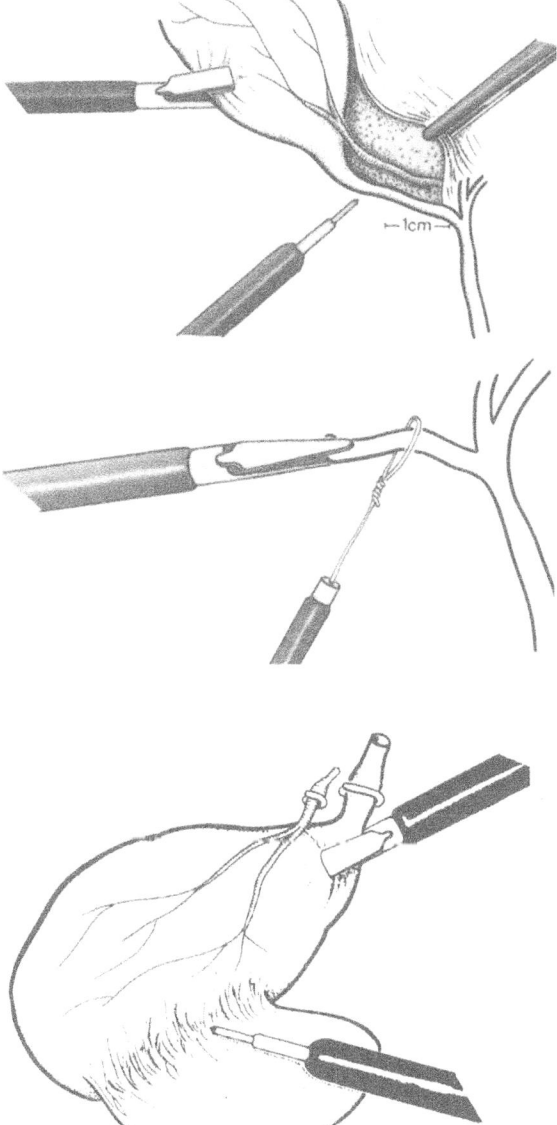

**Fig. 23.** Laparoscopic cholecystectomy. The cystic artery and duct can be temporarily occluded with the laser and will be transected afterwards, followed by ligation with an endoloop and enucleation of the gall bladder

tis, cholangiolitis) may be expected to develop in childhood. Any child with symptomatic cholecystolithiasis should therefore be treated. (Our youngest laparoscopically laser-cholecystectomized child with a symptomatic cholecystolithiasis was 6 months of age.) (Fig. 23).

Indications are thus all complications of gall stone migration, bacterial infection and local irritation. The technique of laparoscopic cholecystectomy has been standardized by surgeons for adults. We apply it in older children but had to make modifications for newborns and infants because of the short working distances and small proportions. Application of the laser plays an important role in this connection. It permits the dissection in Calot's triangle without bleeding. The cystic artery and duct can be temporarily occluded with the laser and then transected. The central stumps of the cystic duct and the cystic artery are then ligated with a Roeder loop. The same applies to the neck of the gall bladder. In this way, clips, which are too large and unwieldy for neonates could be dispensed with. In the UKBF children's hospital the laser is also used for the further dissection and subserous enucleation of the gall bladder from the liver bed. Advantageous is the good haemostasis and sealing of the liver bed to prevent bile leakage through accessory cystic ducts. We can thus dispense with drainage. As in laser-assisted appendectomies, the suction rinsing device is not needed, so that we can manage with 5-mm trocars in newborn and do without the 10-mm trocars. The cystic duct and cystic artery, however, are occluded with clips in older children and transected with the laparoscopic scissors.

## Ileocecal Duplication

Ileocecal duplications are the most frequent duplications of the gastrointestinal tract. They are usually spherical and lie directly in the ileocecal region. In many cases, their resection is already necessary in neonates, since they often become symptomatic and compress the ileocecal valve due to a rapid increase in size. These duplications are frequently detected prenatally in the form of abdominal cysts (Fig. 24). Early surgical treatment is indicated in view of the high risk of complications. The newborn underwent laparoscopy in the first week of life. The umbilical region was moist and sticky. Therefore, we inserted the Veress needle and the 2mm optical trocar at Munro's point in the left lower abdominal quadrant. [63, 64].

First, the widely branching mesenteric vessels were coagulated by non-contact lasering at 25 W. We then punctured the cyst and emptied about one fourth of the vitreous, colourless mucous fluid. It was then easy to stretch the cyst so that the border to the normal

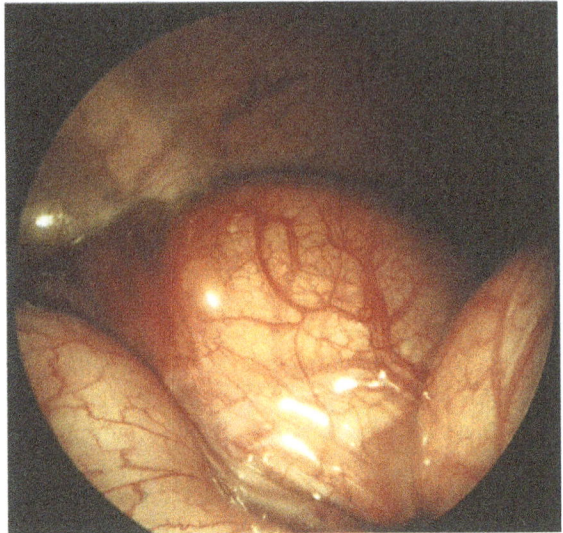

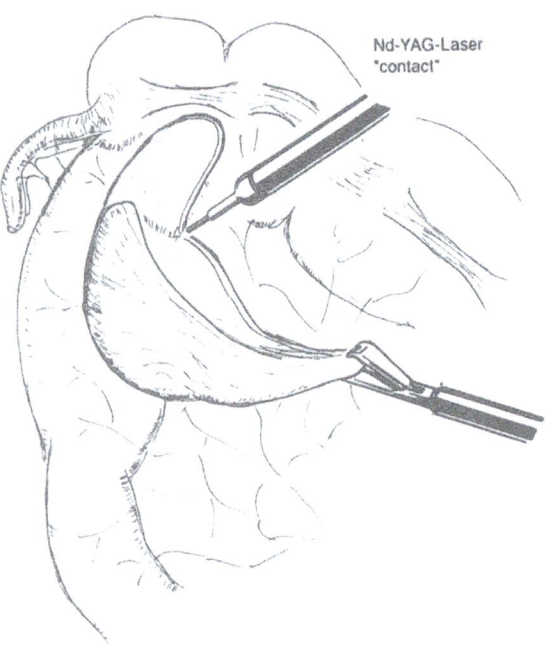

Nd-YAG-Laser
"contact"

**Fig. 24.** Laparoscopic laser resection of an ileocecal duplication in a newborn

ileum became visible through a groove. The serosa of the duplication merged into that of the ileum. After making an incision at this point, the muscles were separated as far as possible by non-contact lasering and laparoscopic forceps. Then the cyst was opened and its wall resected. A small portion of the wall remained on the ileum contact surface. The mucosa of this residual cystic wall fragment was coagulated, again by non-contact lasering. The mesenteric defect was closed with fibrin glue.

Complications such as bleeding, infections, invaginations or volvulus require an emergency resection. Since this generally results in a loss of the ileocecal valve, numerous late complications must be expected. An intestine preserving resection procedure should therefore be applied in treating these duplications. Demucosation of the residual cystic wall fragment bordering the normal intestine is essential to avoid a mucocele, peptic ulceration or malignant degeneration. As ORR has shown, these originate in the mucosa. Resection and complete removal of the mucosa is thus of utmost importance (Fig. 25).

## Rectosigmoidectomy with the Swenson's Pull-Through for Hirschsprung Disease

Laparoscopic pull-through procedure is the standard technique for Hirschsprung disease. It can be performed without colostomy. An optical trocar and three further working trocars are inserted around the umbilicus. The sigmoid colon and proximal rectum are mobilized by diathermy or laser. Only the large arteries and veins are cut between two endoclips. The sigmoid colon is then pulled down using telescope technique in continuity, divided above the transition zone by the Nd:YAG laser (fibertom mode 20 W cw). With this one-stage technique colostomy can be avoided, and bougienage is not necessary in most children. Blood transfusions are unnecessary. Our youngest infant was 6 weeks of age. The advantages of this technique are: no contamination of the peritoneal cavity and no loss of blood, in all cases it is possible to establish a deep anastomosis.

## Proctology

Technique: Nd:YAG 1064 nm, focusing hand pieces. Haemorrhoids II, fissures: Non-contact, defocusing, spot 2–3 mm, 30 W, 0.3 s, interrupted.

Haemorrhoids III: Ice cube, focusing, 45 W, cw.

Tumour recanalization, strictures, stenoses: Bare fibre 0.6 mm, contact, ablation mode 20 W, 1 s

Anorectal diseases are a wide field of laser application in children. Besides the Nd:YAG laser the $CO_2$ laser is used. Indications for the $CO_2$ laser are viral diseases: condylomata, mollusca, papillomata, warts and naevi. If necessary, the lesions can be treated repeatedly in layers, in which the remaining vaporized tissue has to be wiped off with physiological NaCl solution.

The indications for Nd:YAG laser are haemorrhoids, fissures, marisques, fistulas, anorectal abscesses and anorectal diseases like stenosis, haemangioma, polyp, adenomata, and strictures [12, 43, 44,

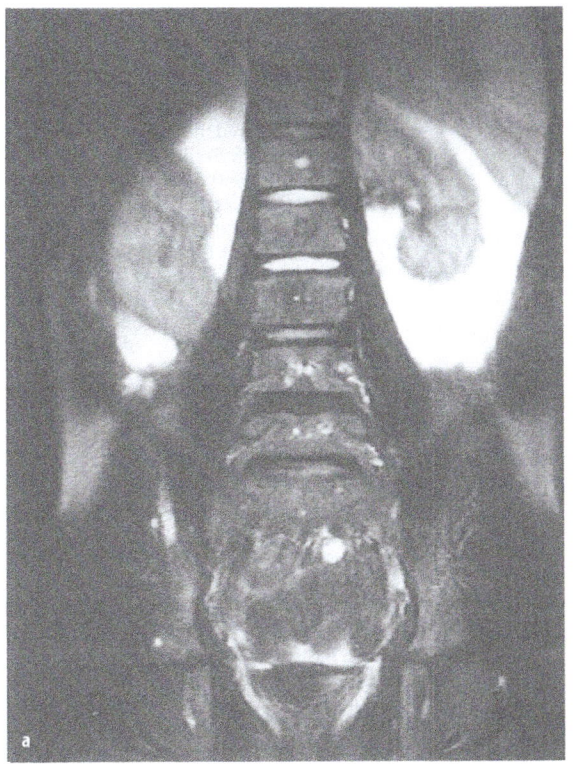
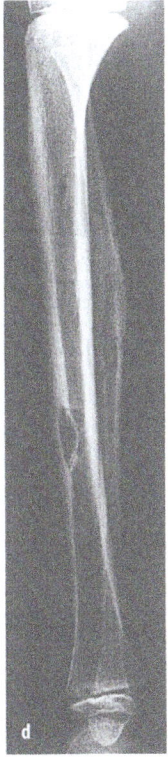
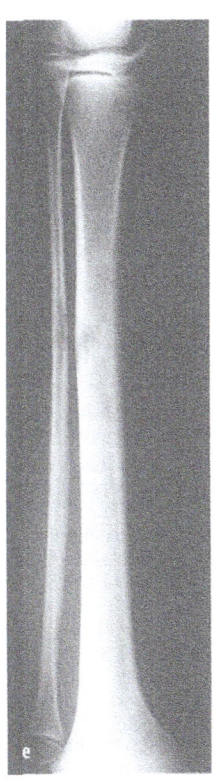
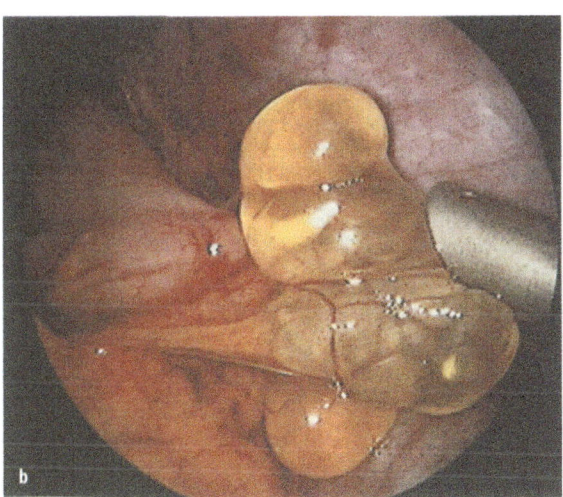
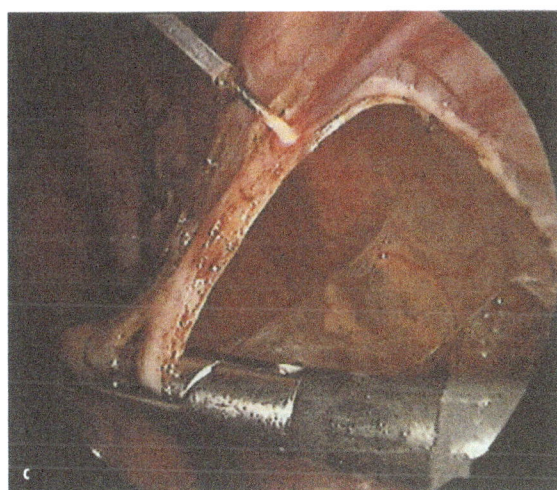

**Fig. 25a–e.** Gorham-Stout disease. Lymphangioma in the retroperitoneum on both sides and in different bones. Laparoscopic laser fenestration and resection, and X-ray-guided, percutaneous LITT of the intraosseous bone cysts

45). The treatment of haemangiomas is carried out in the same technique and parameters as other vessel tumours. In haemorrhoids and fissures we use the Nd:YAG focusing hand piece, non-contact, interrupted, defocused with a spot size of 2–3 mm, cw. In haemorrhoides III¡ ice-cube technique is necessary,

45 W, cw, non-contact for occluding the inf. haemorrhoidal artery. Endorectal tumour recanalizaton and scar excision of strictures are done in bare fibre ablation mode. The prolapsed haemorrhoids and the secondary nodes are coagulated with the Nd:YAG laser in non-contact mode.

## References

[1] ALBRECHT D, GERMER CT, ISBERT C, RITZ JP, ROGGAN A, MÜLLER G, BUHR HJ (1998) Interstitial laser coagulation: evalution of the effect of normal liver blood perfusion and the application mode of lesion size. Laser Surg Med 23: 40–47

[2] AZIZKHAN RG (1992) Laser in pediatric surgery. Surg Clin North Am 72: 1315–1333

[3] BEUTHAN J, MÜLLER G, SCHALDACH B ZUR CH (1991) Fiber design for interstitial laser treatment. SPIE 1420: 234–341

[4] BERLATZKY Y, MUGGIA-SULLAM M, MUNDA R, JOFFE SN (1985) Use of Nd:YAG laser in pancreatic resections with duodenal preservation in the dog. Laser Surg Med 5 507–514

[5] BERLIEN HP (1987) Resection of parenchymatous organs and bloody tumors with the laser. In: KONSTADINOS S (ed) Laser in medicine. Kolimpari Proc, pp 243–248

[6] BERLIEN HP, MÜLLER G, WALDSCHMIDT J (1990) Laser in pediatric surgery. Progress Pediatr Surg 25: 5–22

[7] CHATLANI PT, KRASNER N (1991) Laser treatment of colorectal adenomas. In: KRASNER N (ed) Laser in gastroenterology. Chapman, London, pp 191–204

[8] CHOLEWA D, WACKER F, ROGGAN A, WALDSCHMIDT J (1998) Magnetic resonance imaging: controlled interstitial laser therapy in children with vascular malformation. Laser Surg Med 23: 250–257

[9] DALY CJ (1989) Laser cholecystectomy. In: Joffe SN (ed) Lasers in general surgery. Williams & Wilkins, Baltimore pp 40–46

[10] DES LOOVERFE C, ILBERG V C (1998) Laser treatment of recurrent respiratiory papillomatosis. In: Berlien HP, Schmittenbecher PP (eds) Springer, Berlin Heidelberg New York, pp 123–132

[11] DIERKESMANN R, HUZLY A (1983) Die Anwendung des Nd:YAG Lasers in der Bronchoskopie. Prax Klin Pneumol 37: 989–990

[12] EDDY HJ, EDDY E (1989) Surgical techniques in perianal and other disease. In: Joffe SN (ed) Laser in general surgery. Williams & Wilkins, Baltimore pp 150

[13] ELL C, BAUMGARTNER R, GOSSNER L, HÄUSSINGER K, IRO H, JOCHAM D, SZEIMIES RM (2000) Photodynamische Therapie Dtsches Ärztebl 97: B 2804–2810

[14] ENGERT J (1998) Diagnosis and therapy of pediatric airway obstructions. In: Berlien HP, Schmittenbecher PP (eds) Laser surgery in children. Springer, Berlin Heidelberg New York pp 92

[15] ERAKLIS A, FILLER RN (1972) Splenectomy in childhood: a review of 1413 cases. J Pediatr Surg 7 383–388

[16] FANTA J, REHAK F, HORAK L, KABAT J, ADAMEK S, MAREK J (1989) Open lung surgery with Nd:YAG laser. Laser Med Sci 4 pp 13–15

[17] FEYH I, SCHMITTENBECHER PP (1998) The treatment of laryngeal papillomas with the aid of photodynamic laser therapy. In: Berlien HP, Schmittenbecher PP (eds) Laser surgery in children. Springer, Berlin Heidelberg New York, pp 133–137

[18] FLEISCHER D (1985) Endoscopic Nd:YAG laser therapy for active esophageal variceal bleeding. A randomized controlled study. Gastrointest Endosc 31: 4

[19] FRIEDL P, STERN J (1987) Mechanische Dilatation und Nedodym-YAG-Laser-Koagulation zur kombinierten Therapie peptischer Ösophagusstenosen. Verhandlungsbericht der DGLM EBM, München Zürich 3, pp 149–156

[20] GANS SL (1998) Tumor remoral by laser. In: Berlien HP, Schmittenbecher PP (eds) Laser surgery in children. Springer, Berlin Heidelberg New York, pp 181–186

[21] GANS SL, AUSTIN E (1988) The use of lasers in pediatric surgery. J Pediatr Surg 23: 695–704

[22] HEALY GB, SIMPSON GT (1982) Laser correction of choanal atresia. Ear Nose Throat I 61: 30

[23] HEPPNER F (1982) Der medizinische Laser und seine Vorläufer. Verhandlungsbericht der DGLM 1, pp 2–6

[24] HIRA N, STEGER AC, MOORE KC (1987) Use of Nd:YAG laser in an emergency partial splenectomy. Laser Med Sci 2: 127–129

[25] HOFFMANN K, CHOLEWA D, WALDSCHMIDT J (1996) Anwendung des Lasers bei der Thorakoskopie im Säuglings- und Kindesalter. In: Waidelich W, Staehler G, Waidelich R (eds) Laser in der Medizin. Springer, Heidelberg New York Berlin, pp 360–365

[26] JOFFE SN (1989) Applications in gastrointestinal bleeding. In: Joffe SN (ed) Laser in general surgery. Williams & Wilkens, Baltimore, pp 173–183

[27] JOFFE SN (1989) Liver resection. In: Joffe SN (ed) Laser in general surgery. Williams & Wilkens, Baltimore, pp 82–95

[28] KAY GA, LOE TE, CUSTE MD (1992) Endoscopic laser ablation of obstructing congenital duodenal webs in the newborn: a case report of limited saccess. J Pediatr Surg 27: 279–281

[29a] KIEFHABER P, KIEFHABER K, HUBER F, NATH G (1986) Endoscopic neodymium:YAG laser coagulation in gastrointestinal haemorrhage. Endoscopy 18 Suppl 2: 46

[29b] KIEFHABER P, NATH G, MORITZ K (1977) Endoscopical control and massive gastrointestinal hemorrhage by interaction with a high power Nd:YAG laser. Prog Surg 15: 140–155

[30] KREVSKY B, PUSATERI JP JV (1989) Laser lysis of an esophageal web. Gastrointest Endosc 35: 451–453

[31] LAIZON LA (1991) Endoscopic laser treatment for peptic ulcer and variceal haemorrhage In: Krasner N (ed) Laser in gastroenterology. Chapman & Hall, London, pp 75–108

[32] MEIER H, DIETL KH, STÖHR G, WILLITAL GH (1986) Experience of Nd:YAG laser in pediatric surgery. Laser 2: 68–74

[33] MERGUERIAN PA, SEREMETIS G (1994) Laser-assisted partial nephrectomy in children. J Pediatr Surg 29: 934–936

[34] MOGHISSI K, DENCH M, NEVILLE E (1989) Effect of the non-contact mode of Nd:YAG laser on pulmonary tissues and its comparision with electrodiathermy: an anatomopathological study. Lasers Med Sci 4: 17–23

[35] MÜLLER G, ROGGAN A (1995) Laser induced interstitial thermotherapy. Bellingham SPIE Press, pp 1–549

[36] NOTH G, GORISH W, KIEFHABER P (1973) First laser endoscopy with a fiberoptic transmission system. Endoscopy 5 203–218

[37] ORTH K, RUSS D, DUERR R, MATTFELDT T, STEINER R, BEGER HG (1997) Laser coagulation zones induced with the Nd:YAG laser in the liver. Lasers Med Sci 12: 137–143

[38] PHILIPP C, POETKE M, BERLIEN HP (1998) Basis of laser resection in parenchymatous organs. In Berlien HP, Schmittenbecher PP (eds) Laser surgery in children. Springer, Berlin Heidelberg New York, pp 170–180

[39] RODGERS DA, PHILIPPE PG, LOBE TE, KAY GA, GILCHRIST B, SCHROPP KP, RAO BN (1992) Thoracoscopy in children: an initial experience with an evolving technique. J Laparoscopic Surg 2: 7–14

[40] ROGGAN A, ALBRECHT D, BERLIEN HP, BEUTHAN J, FUCHS B, GERMER C, MESECKE V. RHEINBABEN I, RYGIEL R, SCHRÄNDER S, MÜLLER G (1995) Application equipment for intraoperative and percutaneous laser-induced interstitial thermotherapy. In: Müller G, Roggan A (eds) Laser induced interstitial thermotherapy. SPIE Press, Bellingham, pp 224–248

[41] ROHDE H, THON K, FISCHER M, ET AL (1980) Results of a defined therapeutic concept of endoscopic Nd:YAG laser therapy in patients with upper gastrointestinal bleeding. Br J Surg 67: 360

[42] ROY GT, COHEN RC, WILLIAMS SJ (1996) Endoscopic Laser division of an esophageal web in a child. J Pediatr Surg 31: 439–440

[43] RUTGEERTS P, VANTRAPPEN G (1991) Non-acute bleeding: Angiodysplasia and other vascular anomalies. In: Krasner N (ed) Laser in gastroenterology. Chapman & Hall, London, pp 109–123

[44] SANKAR MY (1989) Contact Nd:YAG laser hemorrhoidectomy. In: Joffe SN (ed) Laser in general surgery Williams & Wilkens, Baltimore, pp 137–149

[45] SOHN N (1989) Anorectal disorders. In: Joffe SN (ed) Laser in general surgery. Williams & Wilkens, Baltimore, pp 125

[46] SPINELLI P, DAL FANTE M (1987) Laser in medicine: endoscopic applications and proposals. Proc 1st Eur Workshop Kolimpari, pp 24–35

[47] SPINELLI P, DAL FANTE M, MANCINI A, CASELLA G (1998) Endoscopic laser therapy in pediatric gastrointestinal disorders. In: Berlien HP, Schmittenbecher PP (eds) Laser surgery in children. Springer, Berlin Heidelberg New York, pp 147–154

[48] SHAPSHY SM (1987) Laser applications in the trachea and bronchi: a comparative study of the soft tissue effects using contact and noncontact delivery systems. Laryngoscope 97 Suppl 41: 1–26

[49] SULTAN RA, FALLOUH H, LEFEBVRE-VILARDEBO M, ET AL (1986) Separate and combined use of Nd:YAG and $CO_2$ lasers in liver resection: a preliminary report Lasers Med Sci 1: 101–105

[50] SCHMITTENBECHER PP (1990) The neodymium YAG laser in surgery of parenchymatous organs in childhood. Prog Pediatr Surg 25: 23–31

[51] SCHMITTENBECHER PP, MANTEL K (1998) Laser treatment of TEF. In: Berlien HP, Schmittenbecher PP (eds) Laser surgery in children. Springer, Berlin Heidelberg New York, pp 138–146

[52] SCHRÖDER T, RÄMÖ OJ (1989) Laser in pancratic surgery. In: Joffe SN (ed) Lasers in general surgery. Williams & Wilkins, Baltimore, pp 96–99

[53] SCHULTZ LS, HICKÖK DF, GRAVER JN (1989) Advanced intraabdominal tumours: In: Joffe SN (ed) Laser in general surgery. Williams & Wilkins, Baltimore, pp 55–64

[54] STEGER AC, LEES WR, BOWN SG (1989) Interstitial laser hyperthermia: a new approach to local destruction of tumours. BMJ 299: 362–365

[55] STEINER W (1986) Einsatzmöglichkeiten von Lasern im Bereich des oberen Aero-Digestivtraktes. Laser Med Surg 2: 77–87

[56] STEINER W, AMBROSCH P (1996) Laser im Kopf- und Halsbereich. In: Müller GJ, Berlien HP (eds) Fortschritte in der Lasermedizin.ecomed, Landsberg, 13 pp 127–136

[57] TRANBERG KG, MÖLLLER PH, LINDBERG L, HENRIKSSON PH, PERSSON BRR (1994) Energy delivery and monitoring in interstitial laser thermotherapy. Minim Invas Med 5: 36–41

[58] TOTY L, PERSONNE C, COLCHEN A, VOURC'H G (1983) Bronchoscopic management of tracheal lesions using the Nd:YAG laser. Thorax 36: 175–178

[59] UNSÖLD E, JOCHAM D (1988) Grundlagen photodynamischer Laser-Therapieverfahren. Chirurg 59: 76–80

[60] VOGL T, MACK MG, MÜLLER PK, PHILIPP C, BÖTTCHER H, ROGGAN A, JUERGENS M, DEIMLING M, KNÖBBER D, WUST P, FELIX R 1995) Recurrent nasopharyngeal tumours: preliminary clinic results with interventional MR imaging controlled laser induced thermotherapy. Radiology 196: 725–733

[61] WAAG KL, JOPPICH J, MANEGOLD BC, DEL SOLAR E (1985) Endoskopischer Verschluß ösophago-trachealer Fisteln. Z Kinderchir 27 Suppl 93–96

[62] WAIDELICH R, KRIEGMAIR M, BAUMGARTNER R, STEPP H, HOFSTETTER A (1998) Photodynamische Therapie mit 5-Aminolävulinsäure bei Urethralcarcinomen des oberen Harntrakts. In: Waidelich W (ed) Laser in der Medizin. Springer, Berlin Heidelberg New York, pp 170–173

[63] WALDSCHMIDT J, SCHIER F, HAUCK G (1991) Laseranwendung in der Kinderchirurgie. Lasermedizin 7: 115–125

[64] WALDSCHMIDT J, SCHIER F (1993) Laparoscopic procedures in neonates and infants.In: Holcomb GW (ed) Pediatric endoscopic surgery. Appleton 4. Lange, Norwalk, pp 67–76

[65] WALDSCHMIDT J, CHOLEWA D, HOFFMANN K (1995) Laser in der Minimal-Invasiven Chirurgie (MIC). Langenbecks Arch Chir [Suppl II]: 888–891

[66] WALDSCHMIDT J, EL DESSONKY M, CHOLEWA D (1996) Laseranwendung in der offenen und endoskopischen Chirurgie der Lunge beim Kind . In: Müller GJ, Berlien HP (eds). Fortschritte in der Lasermedizin 13 ecomed, Landsberg, pp 169–175

[67] WALDSCHMIDT J, CHOLEWA D, STROEDTER L, DOEDE TH (1997) Laser surgery of tracheal and bronchial stenoses. In: Antypas G (ed) Lasers at the down of the third millemeum. Monduzzi, Bologna, pp 321–325

[68] WALDSCHMIDT J, HOFFMANN K, WAHEEB S (1997) Technique and results of thoracoscopic laser application in newborn and children. Lasermedizin 13: 3–9

[69] WALDSCHMIDT J, STROEDTER L, DOEDE TH, KISCHKEL A (2000) Treatment of vascular malformation of the gastrointestinal tract. Prog Biomed Optics Imag 1: 1–9

[70] ZIEGLER K, SCHIER F, WALDSCHMIDT J (1992) Endoscopic laser resection of a duodenal membrane. J Pediatr Surg 27: 1582–1583

# III-14
# Laser Applications in Gynaecology

M. S. Ismail and C. M. Philipp

**Contents**

**Fig. 1.** Laser procedures as a function of power density

| | |
|---|---|
| Power density | Removal/cutting<br>    Photo fragmentation<br>    Photo disruption<br>    Photo ablation<br>    Photo vaporization |

Removal/cutting
    Photo fragmentation
    Photo disruption
    Photo ablation        Necrosis
    Photo vaporization                        Therapy

Thermotherapy
    LIC
    TDR                    Apoptosis          Absorption
    LIHT

Photochemical
    PDT
    LIF
        xeno
        auto

Optic tomography
    Diaphanoscopy                         Diagnostic    Without absorption
    Transmission tomography
    Remission tomography

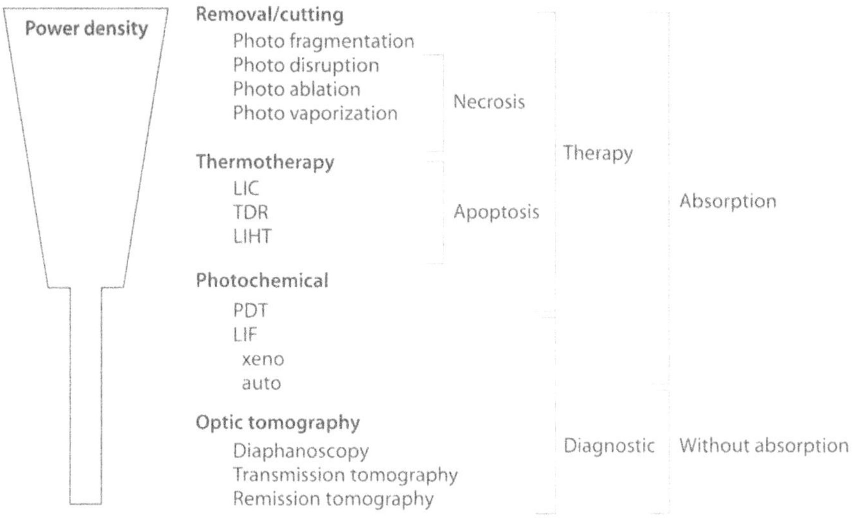

**Fig. 2.** Laser application modes and controlling methods

| Superficial<br><br>Intracorporal<br>– endoscopic | Direct control<br>(visualization/digitally) |
|---|---|
| – non-endoscopic<br>– intraluminal application (ILA)<br>– interstitial (ITT) | Indirect control |

## Introduction

The first source of medicinal light was the sun. The ancient Egyptians of 4000 years ago were sun worshippers. In the fourteenth century BC, the Pharaohs began to identify themselves with the sund (Sun god Amen-Ra) and attached RA to their names. In Japan, as in ancient Egypt, the mystery of the sun is elevated to a theistic plane.

Laser is an acronym which stands for Light Amplification by Stimulated Emission of Radiation. So, laser ist not more than light but it is not an ordinary light. It represents an intense beam of light with particular properties. These particular properties of laser light alow selective tissue effects with a precision not achieved using traditional surgical instruments.

## Laser Procedures as a Function of Power Density

At higher power densities a laser will cut or vaporize. At lower power density, only tissue coagulation can be effectively achieved for thermal coagulation of tumours or vessels. At lower power density a photochemical reaction can be achieved. A laser beam of lower power density can be used to excite fluorescent dyes for tumour localization or destruction as defined by the concept of LIF (laser-induced fluorescence) and PDT (photodynamic therapy). Boxes 6 (Fig. 1)

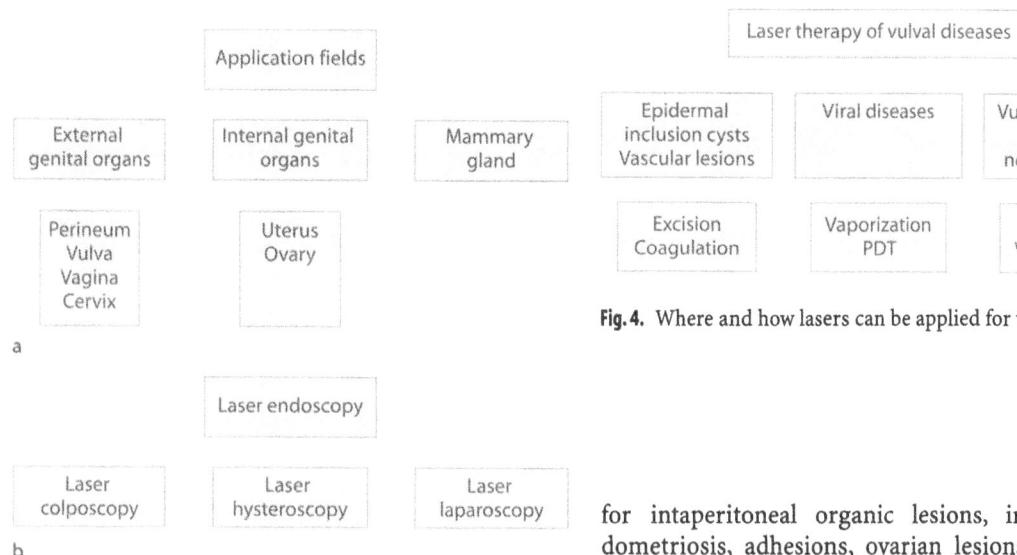

Fig. 4. Where and how lasers can be applied for vulval lesions

**Fig. 3a-b.** **a** Laser application in gynaecological lesions, and **b** Gynaecological endoscopies

## Methods of Laser Application in Gynaecology

In gynaecology the laser is applied in three main different ways, superficially or directly, endoscopically and interstitally (Fig. 2). Also it may also be applied in contact and non-contact modes.

### Superficial Application

For superficial lesions lasers are mainly applied for lower genital tract including vulval, vaginal and cervical lesions. Other superficial laser applications in gynaecology include treatment of endometriosis in the peritoneal cavity and treatment of breast and skin lesions.

### Endoscopic Application

Laser application in endoscopic surgery is by means of colposcopy for vulval, vaginal and cevical lesions, hysteroscopy for intrauterine lesions and laparoscopy

for intaperitoneal organic lesions, including endometriosis, adhesions, ovarian lesions, uterine lesion, Fallopian tube surgery and liver metastasis.

### Interstitial Application

Interstitial applications include the use of laser-induced thermotherapy and photodynamic therapy for treatment of momas, vulval, vaginal, peritoneal lesions and liver metastasis.

## Gynaecological Lesions

Gynaecological lesions are an important field for laser application, owing to the frequent occurrence and wide variety of lesions which are easily accessible to light sources, possibly coupled to fiberoptic endoscopy. Laser applications include treatment lesions of external genital organs, e.g. perineum and vulva, vaginal and cervical lesions. Laser applications for treatment of lesions in internal organs include uterine and ovarian lesions, breast lesions (Fig. 3A). Most applications of lasers in gynaecology are by endoscopy, including colposcopy for perineum, vulval, vaginal and cervical lesions, hysteroscopy for uterine lesions and laparoscopy for uterine, ovarian and peritoneal lesions (Fig. 3B).

**Table 1.** Types of lasers applied for different vulval lesions

| Lesion | CO$_2$ laser | Nd-YAG/Diode lasers | Argon | Flashlamp dye laser | Dye laser for PDT |
|---|---|---|---|---|---|
| Condyloma | +++ | ++ | + | | + |
| Haemangioma | | +++ | ++ | ++ | |
| Vulval dystrophy | ++ | | | +++ | |
| Vulval skinectomy | +++ | + | | | |
| Vulval tumour | | +++ | | | ++ |

## Vulval and Perineal Lesions

The following lists 1A, 1B shows where and how lasers can be applied for vulval and perineal lesions (see Fig. 3). Table 1 shows the different types of laser for these applications.

## Vulval and Perineal Viral Lesions

Sexually transmitted viral diseases like human papilloma virus, herpes simplex and moluscum contagiosum are accepted as having a major role in the etiology of lower genital tract neoplasia. The treatment of most viral-associated lesions in the anogenital region is primarily for the control of these lesions rather than as a prophylaxis against gynaecological malignancy. Lesions may be unicentric or multicentric, and their severity ranges from minute to massive. Soaking with 5% acetic acid will usually produce prominent aceto-whitening of skin that had appeared normal to naked eye examination.

With optimal equipment in the hands of a skilled surgeon, extended laser ablation offers a highly effective method for destroying any volume of viral-associated vulvar or perineal disease under precise depth control. Morbidity is low, and the risk of scarring is negligible. The use of the $CO_2$, diode or Nd:YAG laser offers a means of removing large condylomatous aggregations, simultaneously debulking the entire viral load and destroying any foci of dysplastic epithelium to an appropriate depth.

The first surgical plane: Destruction of the first plane removes only the surface epithelium, to the level of the basement membrane. This plane is reached by limiting the penetration of the laser crater to pickel cell layer, and is accomplished by rapid oscillation of the micromanipulator. When done correctly, the helium-neon spot will describe a roughly parallel series of lines, each pass of the beam revealing bubbles of silver opalescence beneath the squamous charred surface. This manoeuvre should be accompanied by a distinctive cracking sound, due to the explosion of the individual epithelial cells. Inadvertent penetration of the basement membrane is signalled by the loss of these two characteristic signs.

Lasing to the prickel cell layer shears the basal cells from the basement membrane, thereby producing a plane of cleavage. These detached basal cells are easily removed by wiping with moistened gauze, thus exposing the smooth, intact surface of the papillary dermis. Such wounds heal completely within 6–7 days with very good cosmetic results.

The second surgical plane: Destruction to the second surgical plane will also remove the papillar dermis (the loose network of fine collagen and elastin fibres lying between the dermal papillae). When done correctly, the scorched surface should show a finely roughened contour and a yellowish colour, somewhat reminiscent of a chamois cloth. This clinical appearance indicates that a zone of coagulation necrosis will lie within the papillary dermis. The second plane is the preferred level of ablation for extensive condylomas. Such wounds also heal completely within 6–7 days with very good cosmetic results.

The third surgical plane: Laser ablation to the mid-reticular level is ideal for vulval intraepithelial neoplasia, as it often extends into the pilosebaceous ducts and involvement is generally limited to the superficial portions of the ducts.

Destruction of the third surgical plane removes the epidermis, the upper portions of the pilosebaceous ducts and a part of the reticular dermis. Ablation to the mid-reticular layer uncovers coarse collagen bundles that can be seen through the operating microscope as white fibres resembling water-logged cotton threads. Wiping away the surface char will then reveal a pattern of starkly white collagen plates, interspersed by a horizontally orientated network of deep dermal vessels. Hair follicles and sweat glands are readily visible through the operating microscope, being seen as tiny refractile granules that resemble grains of sand. The third surgical plane represents the deepest level from which optimal healing will occur.

Extended laser ablation: The aggressiveness of the treatment must be counter-balanced against the degree of disease expression in the individual patient. Indications for extended laser ablation are very extensive condyloma formation (coalescent lesions occupying 30% of the vulvular surface), refractory benign condyloma and vulvar intraepithelial neoplasia (VIN).

### Preoperative Procedure

We recommend the use of laser-induced fluorescence with application of 20% ALA to evaluate the condition and determine the biopsy sites. Topical application of ALA (photosensitizer is 2 – 4 h. Figure 5 shows the D-light telescope for LIF used for the stimulation of the photosensitizers and detection of fluorescence signals.

### Instruments

- D-light telescope for fluorescence diagnosis.
- Laser equipment: $CO_2$, Nd-YAG, diode laser and dye lasers with laser light emission at wavelength 630 nm for photodynamic therapy.
- Operating microscope and micromanipulator for $CO_2$.

**Fig. 5.**
Vulval
dystrophy

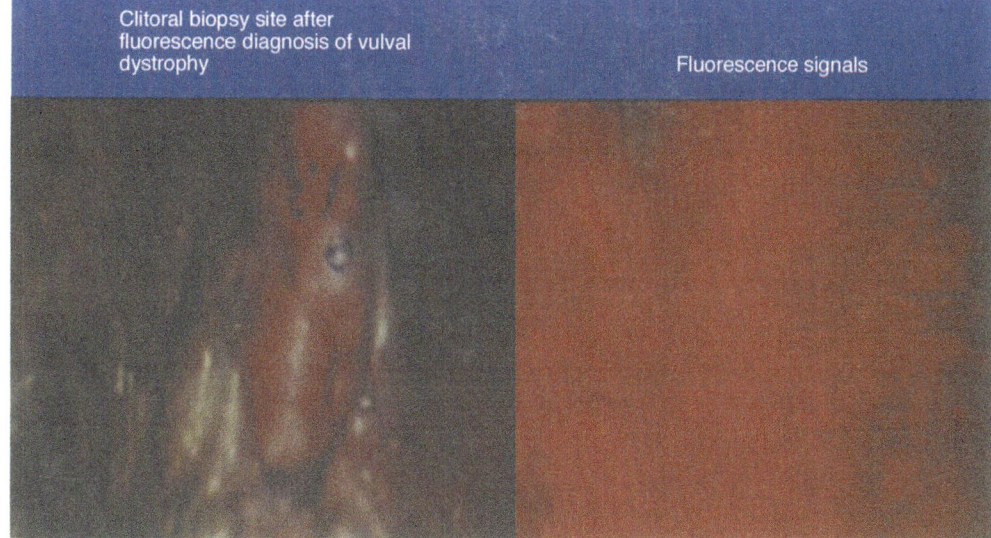

**Fig. 6.**
Exophytic
acuminate
warts matted
together
replacing
the perianal
area. This
patient was
treated with
an Nd-YAG
laser

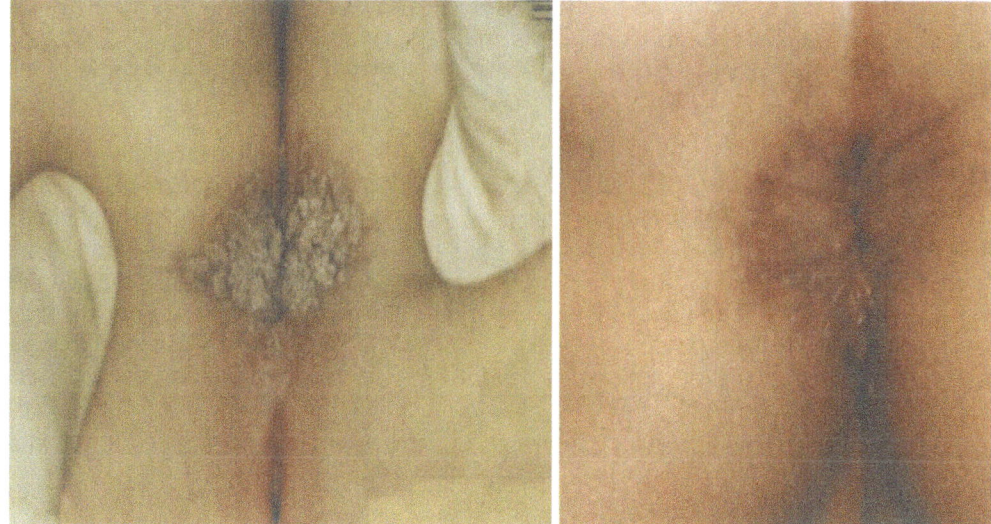

- Colposcopy.
- Laser fibres (400 – 600 µm) for Nd-YAG and dye lasers in PDT.
- Aluminium foil for PDT.
- Gas evacuation or suction tube.
- Applicator with moistened cotton.
- Saline 0.9%, ascetic acid 3%, toluidine blue 1%.

### Anaesthesia

- No anaesthesia (PDT), general, regional or local.

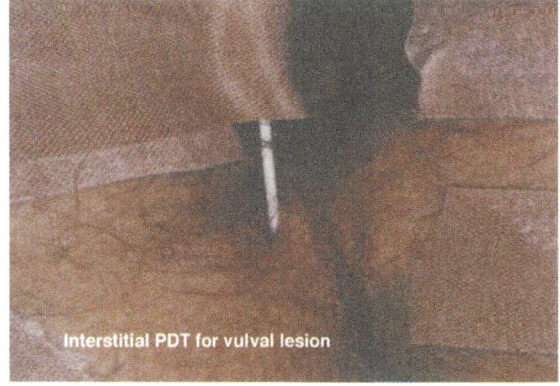

**Fig. 7.** Interstitial PDT for deeply situated vulval lesion

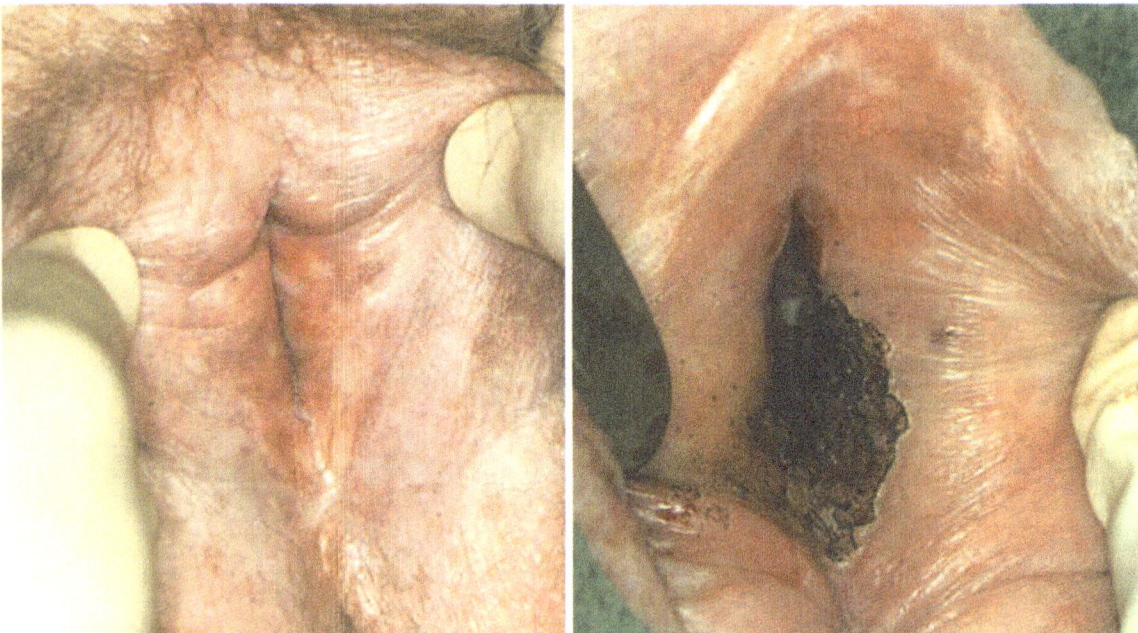

**Fig. 8.** Laser vaporization using the $CO_2$ laser for superficial vulval lesion using the bruising technique. The laser vaporization must be down to the level of the papillary dermis. The small amount of tissue carbonization can be washed away with normal saline.

## Technique

- Disinfection of vulva, vagina and anal region.
- Biopsies for histopathological examination (guided by fluorescence signals).
- Precooling of the treated region by pressure with an ice bag (not obligatory).
- For $CO_2$ vaporization (skin penetration depth 5–10 nm 2-mm bordering area in healthy tissue. Micromanipulator, focus diameter 1–2 mm, power 10–30 W). Ablation by rapid hatching movements of the laser beam with micromanipulator, moving from dorsal to ventral. Haemostasis through compression with moistened gauze or punctual coagulation by defocusing the laser beam (Fig. 6).
- For microsurgical lesion excision (Nd-YAG, diode, 20–40 W, cw, bare fibre 400 μm), and coagulation (Nd-YAG, diode 10–30 W, cw bare fibre 400 μm) at penetration depth of 2–4 mm as well as a 2-mm bordering area in healthy tissue. Tangenital movements of the laser beam to the area in order to avoid deep-seated tissue necrosis.
- LITT for tumour coagulation (Nd-YAG laser, 5–10 W). Controlling the procedure with visual and thermal palpation, CCDS.
- PDT (superficial application, handpiece/interstitial PDT (Fig. 7), (cylindrical laser fibre applicator). Topical application of ALA 10% 2–4 before treatment or systemic photosensitizer application HPD 48–72 h before application of laser dye light 630 nm to the treated area. Healthy area to be covered with aluminium foil. Avoid exposure of the patient to direct light or sunlight in the case of systemic application of the photosensitizer.
- For partial resection of the vulval skin (vulval dystrophy, pruritus vulva) a $CO_2$ laser focus 1.5 mm 15 W, cw, is used. Microsurgical vaporizing of the epithelium by quick circulatory movements with the laser beam. Ablation to a depth is reached at which very small blood points appears which due to the circulating laser beam, fade immediately (first surgical plane). Bed rest 5–7 days postoperatively is needed (Fig. 8). PDT can be used with better results.
- For skinning vulvectomy (vulval intraepithelial neoplasia, Bowen's disease), a $CO_2$ laser 1.5 diameter, 15 W, cw, is used. Microsurgical vaporizing of the epithelium by quick circulatory movements with the laser beam. Ablation to a depth of 3 mm in hairy areas and 1–2 mm in non-hairy areas. Additional removal of a healthy cutaneous zone of 3 mm. Ablation of the whole vulval skin is indicated in cases with dense multifocal or generalized extension in two sessions, one for each side. PDT can be used with better results.
- For vulvectomy (invasive vulval cancer). A $CO_2$ laser with power 40 W, focus diameter 1.5–2.0 mm. External and internal oval incisions of the vulva with the laser beam. Infiltration of the tissue to be transected with vasopressin solution. Excision of

**Fig. 9.** Therapeutic timing for haemangioma and vascular malformation

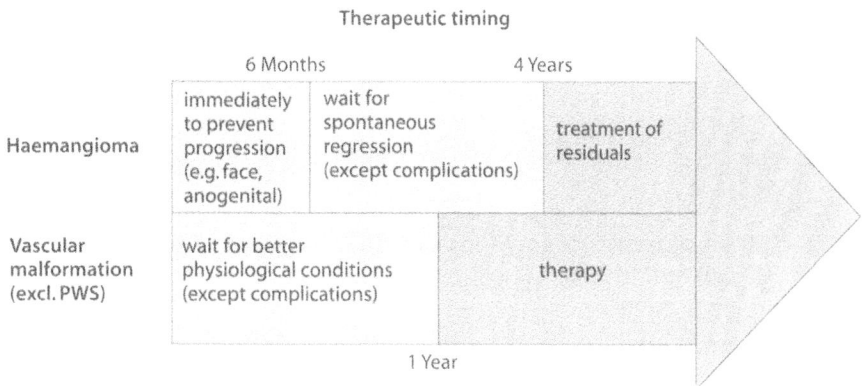

Therapeutic timing

|  | 6 Months | | 4 Years |
| --- | --- | --- | --- |
| Haemangioma | immediately to prevent progression (e.g. face, anogenital) | wait for spontaneous regression (except complications) | treatment of residuals |
| Vascular malformation (excl. PWS) | wait for better physiological conditions (except complications) | therapy | |

1 Year

**Table 2.** Characteristic features and findings of vascular birthmarks

| Parameter | Haemangioma | VM |
| --- | --- | --- |
| Definition | Vascular tumour | Structural anomalies (inborn error of vascular morphogenesis) |
| Sex incidence | F:M 3:1 | F:M 1:1 |
| Clinical features | Usually nothing seen at birth, 30% present as red macule, visible development starts mostly in the first weeks of life (rapid postnatal) | Present at birth, sometimes not full size; growth rate may expand as a result of trauma, sepsis, hormonal modulation. Subsequent hypertrophy of body region extremities often shown |
| Course | Spontaneous involution and acute progression are possible | No spontaneous involution, but often enlargement due to pathological flow. Size increase corresponds to natural growth, the volume can change |
| Hematological | platelet trapping: thrombocytopenia (Kasbach-Merritt syndrome) | stasis (venous): localized consumption coagulopathy |
| Angiography | Well-circumscribed intense lobular-parenchymal staining with equatorial vessels | Diffuse, no parenchyma, enlarged, tortuous arteries with arterio-venous shunting |

the vulva in ventrodorsal and lateromedial direction. Selective ligature of the clitoral and pudendal vessels and careful haemostasis. Suturing of the external front edge of the wound Y-shaped, of the external with the internal edge O-shaped by interrupted sutures with Vicryl 2–0.

### Postoperative Care

- Infiltration of the treated areas with 0.25% lignocain in 1:100 000 adrenalin.
- Patients who require large areas of ablation within the vestibule may be offered a urinary catheter.
- Analgaesics and antiinflammatory drugs may be advised.
- Application of silver sulphadiazine or nitrofurazone cream as prophylactic against labial conglutination.
- Annual Papanicolaou smear is mandatory.
- Adjuvant regimens: topical 5-fluoruracil (5-FU), Interferone therapy are under discussion.

### Vulval and Perineal (Anogenital) Vascular Anomalies (Haemangiomas and Vascular Malformations)

Two major types of vascular birthmarks have been described: the haemangiomas, demonstrating endothelial hyperplasia, and malformations, with normal endothelial turnover. Haemangioma, the most common tumour in infancy, is a hypercellular tumour of vascular origin. It is not a collection of dilated or malformed vessels, and can proliferate to establish a large cell mass. It is subdivided into capillary and tuberous haemangiomas. The second major category of vascular birthmarks is malformations (VM), lesions that exhibit a normal rate of endothelial cell turnover throughout their natural history. These lesions are true structural anomalies, inborn errors of vascular morphogenesis.

The characters that distinguish haemangiomas from vascular malformations during infancy and childhood are summarized in Table 2 and Fig. 9.

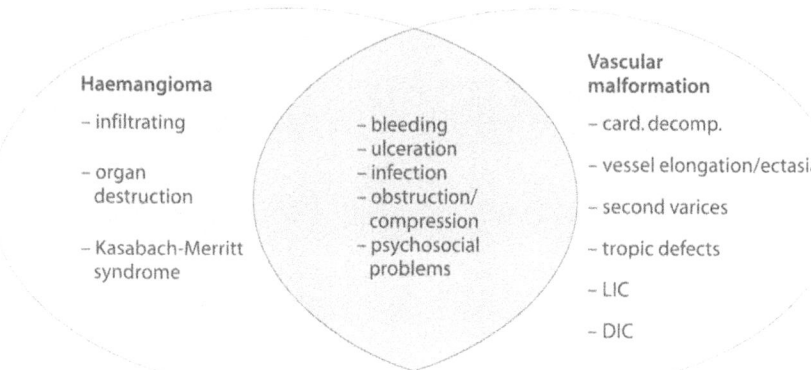

**Fig. 10.** Complications of haemangioma and vascular malformation

## Indications for Treatment

Somatic complications appear at least in 8%. Beginning complications are a clear indication for treatment and should lead immediately to a suitable therapy. The Kasbach-Merritt syndrome is a generalized bleeding disorder that results from a profound thrombocytopenia and is associated with large or extensive haemangioma, classically occurring during the early postnatal period of rapid growth of the haemangioma. The psychological aspect has to be taken seriously into account as an indication for beginning the treatment. Even if capillary haemangiomas can regress spontaneously over several years, they lead to a long period of stress. Often the regression is incomplete, with skin changes which correspond to the largest size of the haemangioma. In the cavernous types of haemangioma, a spontaneous regression is not seen in all cases and can come late and with cosmetically unsatisfactory residuals which have to be corrected in later operations (Fig. 9, 10)

## Technique

Argon, flashlamp pumped dye lasers and Nd:YAG lasers are the currently lasers used for the treatment of haemangiomas. The Nd:YAG laser is characterized by emission of light with a wavelength 1064 nm which will be absorbed significantly in blood, lees in the surrounding tissue. An interaction depth can be reached up to 10 mm according to the tissue and irradiation parameters. According to form and degree of development of the vascular disorder, different irradiation parameters and application procedures were used with the intention of either definite coagulation or the induction of stasis in the vessels with minimum lateral damage and subsequent regression. The selection of application technique depends mainly on the classification and size of the lesion. Since 1984, H-P. Berlien and colleagues have been using two tech-

niques of Nd:YAG laser treatment for voluminous and deep-located lesions: transcutaneous irradiation with permanent ice-cube cooling of the skin and percutaneous interstitial or intraluminal irradiation of deeper lesions. Permanent ice-cube cooling provides safe protection for the skin and is most suitable for haemangiomas with combined subcutaneous and cutaneous portions. Compression is used to enhance the depth of the laser effect. The interstitial is suitable for large or only subcutaneous haemangiomas. We mostly use 600 µm fibres, introduced through an Abbocath G16 and 5–10 W of Nd:YAG laser power for maximum 180 s per single application. During further applications in the same region, the time is reduced to maximum 120 s. In cases of very superficial (intracutaneous) lesions, argon or flashlamp dye lasers can be used for the treatment. The different laser types for the treatment of haemangioma are:

- Prodromalphases, capillary, intracutaneous
  - Flash lamp-pumped dye laser (FDL)
  - Argon-laser
  - Nd:YAG-laser
- Thickness 1–5 mm, intra- and subcutaneous portion
  - Nd:YAG laser with continuous surface cooling
- Subcutaneous, deep seated
  - Nd:YAG-laser with ITT technique

For the treatment of vascular anomalies the following laser types are used:
Transcutaneous (depth 0–1 mm)
Nd:YAG laser (1064 nm), 20 W, 0:1 s, 1 mm spot diameter
Transcutaneous with ice cube cooling
(depth 7–8 mm)
Nd:YAG laser (1064 nm), 35 W, cw
1–3 mm spot diameter, focussing handpiece irradiation through ice cubes, good contact to skin obligatory
Percutaneous (depth >2 mm)
interstitially, Nd:YAG laser (1064 nm), max. 5 W, cw
bare fibre, fresh broken

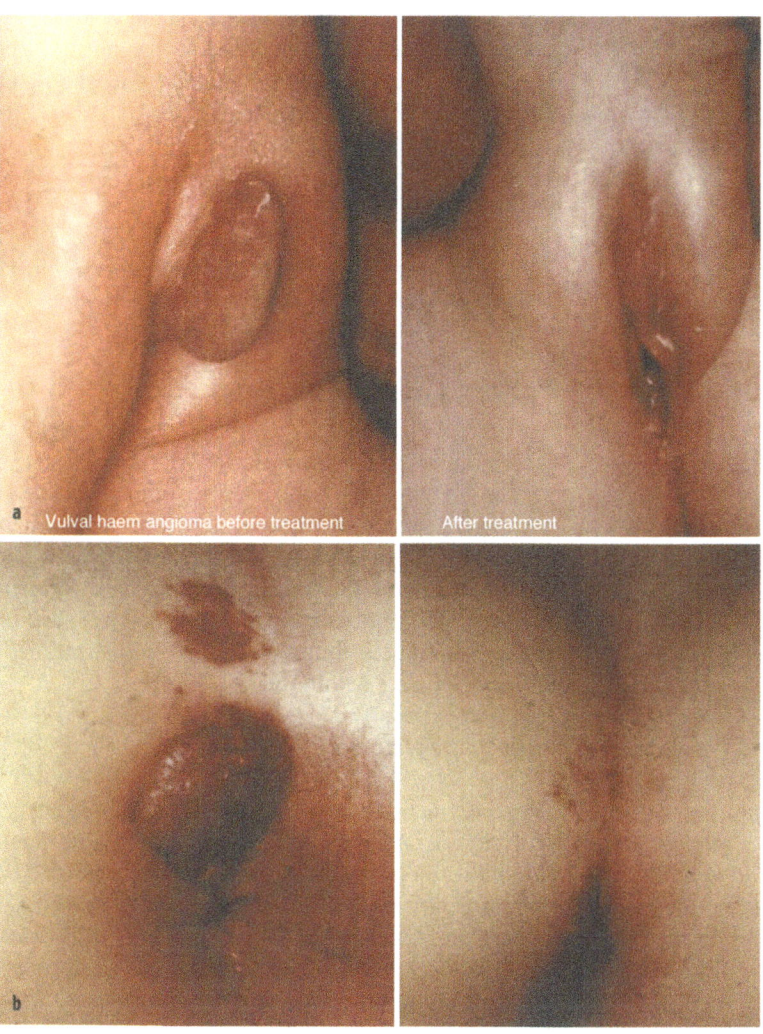

The characteristics of laser coagulation under the ice-cooling system are as follows:

- Used with Nd:YAG laser
- Fixed contact temperature (0 °C)
- Melting energy 335 J/g$^{-1}$
- Cooling and compression
- Max. coagulation depth 7 mm (enhanced by compression)
- VIS and IR transparent

### Postoperative Care

Postoperatively, the irradiated area will swell considerably as a sign of the induced inflammation reaction. This is no complication but an intended tissue reaction. After treatment, a therapy-free interval of between 6 and 8 weeks is necessary to enable the complete healing of the induced inflammation reaction and a restitution of the skin. After transcutaneous ir-

**Table 3.** Side effects in laser application for treatment of vulval and perianal vascular anomalies

| Side effect | Management |
| --- | --- |
| Superficial treatment | |
| – skin burn II degree: | instant cooling for reduction of thermal effects |
| – hyperpigmentation: | mostly temporary (1–2 years), active reduction of pigmentation possible with argon or Q-switched Nd:YAG laser (532 nm) |
| – hypopigmentation: | only seen in argon laser use with wide spot |
| Interstitial treatment | |
| – swelling | intraoperative corticoid iv. in patients with orotracheal lesions |
| – paraesthesia | control examination |

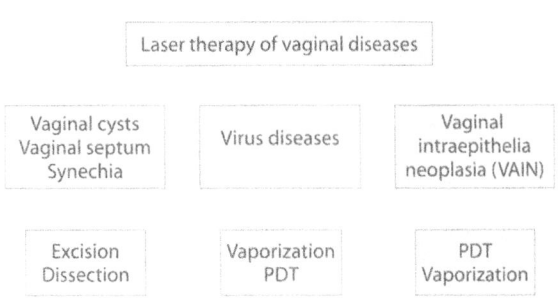

- Vaginal cysts
- Uretheral polypectomy
- Adhesiolysis of vaginal scars
- Excision of vaginal septum and synechia
- Papilloma, endometriosis
- Condyloma acuminata
- Vaginal intraepithelia neoplasia (VAIN)
- Tumour resection; tumour coagulation

**Fig. 12.**  Where and how t apply lasers for vaginal lesions

radiation the treated body area should be protected from mechanical irritation and by strain excessive sunlight. Eventually, blisters appear and should not be opened because the restitution should not be endangered. A small crust formation is possible after transcutaneous irradiation. Washing hair, taking a shower and washing is possible, but should be brief to avoid soaking of the skin. Therefore, no swimming is allowed in a period of 2 to 3 weeks. The treated areas should be swabbed or air-dried. In the case of interstitial or intraluminal irradiation, this advice is not valid. Sport is possible, except for combative or heavy weight events. Especially after interstitial laser application, postoperatively the irradiated area can swell considerably, but regresses within the next few days. If the swelling does not decrease within 3 days and additional inflammation is seen, treatment with antibiotics should start immediately.

### Results

With these laser techniques adequate treatment of haemangiomas located on the vulva and perianal regions has been established as well as for some types of vascular malformations. We have treated 36 patients with one or more anogenital haemangiomas. The treatment was performed on either in- or outpatient basis according to age, localization and type of hae-

mangioma. Lesion regression was usually achieved after two to three treatments, depending on the type (capillary or tuberous), with effective and good cosmetic results (Table 3; Fig. 11A, B).

### Vaginal Lesions

Figure 12 illustrates where and how lasers could be applied for vaginal lesions.

### Vaginal Intraepithelial Neoplasia

Vaginal intaepithelial neoplasia (VAIN) demonstrates similar histopathological features similar to those seen in cervical intraepithelial neoplasia. Colposcopy allows the distribution of VAIN to be delineated in most cases and, following appropriate diagnostic biopsies, lasers provide the ideal way to treat such lesions. Scarring is minimal and re-epithelization is usually complete within weeks.

#### *Preoperative procedure*

We recommend the use of laser-induced fluorescence with 20% ALA to evaluate the condition and determine the biopsy sites. Topical application of ALA (photosensitizer) is 2–4 h. A D-light telescope for LIF is used to stimulate the photosensitizers and detect the fluorescence signals. Laser treatment can be applied to the required location and depth with great precision.

#### *Instruments*

- D-light telescope for fluorescence diagnosis.
- Laser equipment: $CO_2$, Nd-YAG, diode laser, dye lasers with light emission at wavelength 630 nm for photodynamic therapy.
- Operating microscope and micromanipulator for $CO_2$.
- Colposcopy.
- Laser fibres (400–600 μm) for Nd-YAG and dye lasers in PDT.
- Gas evacuation or suction tube.
- Applicator with moistened cotton.
- Saline 0.9%, ascetic acid 3%, toluidine blue 1%.

*Technique*

- Local or general anaesthesia (GA) (the latter is preferred during the treatment of large lesions).
- Disinfection of vulva, vagina and anal region.
- Biopsies for histopathological examination (guided by fluorescence signals)
- Precooling of the treated region by pression with an ice bag (not obligatory).
- For $CO_2$ vaporization (mucosal penetration depth 2–3 mm, 10-mm bordering area in healthy tissue. Micromanipulator, focus diameter 1–2 mm, power 10–30 W). Ablation by rapid rotary movements of the laser beam with micromanipulator, moving from dorsal to ventral. Haemostasis through compression with moistened gauze or punctual coagulation by defocusing the laser beam. The full extent of epithelial destruction is signalled by the occurrence of tissue fluid weeping from the stroma.
- For microsurgical lesion excision (Nd-YAG, diode 20–40 W, cw, bare fibre 400 μm), and coagulation (Nd-YAG, diode 10–30 W, cw, bare fibre 400 μm) at penetration depth of 2–4 mm as well as a 10-mm

bordering area in healthy tissue (Fig. 13). Tangental movements of the laser beam to the area avoid deep-seated tissue necrosis.
- LITT for tumour coagulation (Nd-YAG laser, 5–10 W, cw). Controlling the procedure visually and with thermal palpation, CCDS.
- PDT (superficial application, handpiece/interstitial PDT, cylindrical laser fibre applicator). Topical application of ALA 10% 2–4 h before treatment or systemic photosensitizer application HPD 48–72 h before the application of laser dye light 630 nm to the treated area. Healthy area to be covered with aluminium foil, avoid exposure of the patient to direct light or sunlight in the case of systemic application of photosensitizer (Figs. 14, 15)

## Cervical Lesions

Figure 16 illustrates where and how lasers can be applied for cervical lesions.

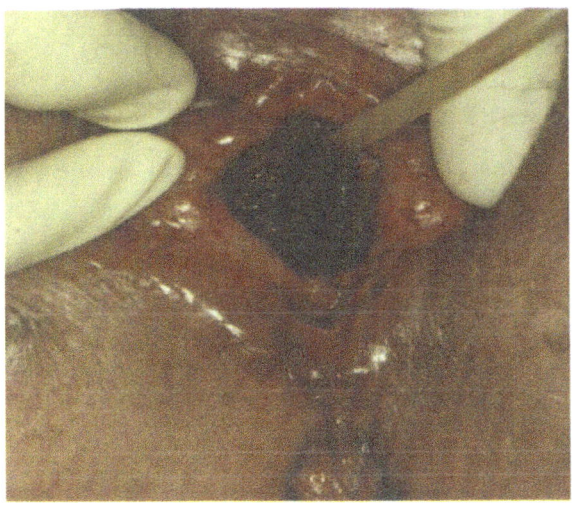

**Fig. 13.** Laser excision of vaginal mass with good haemostasis

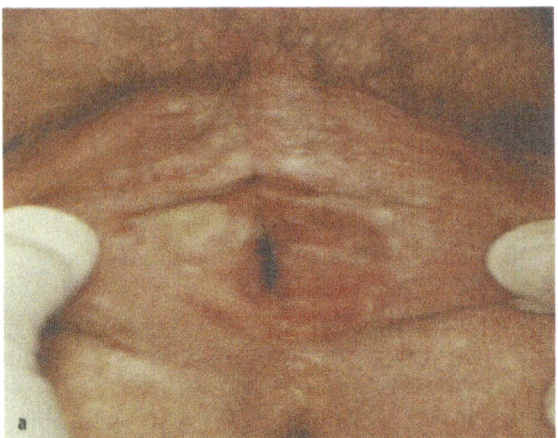

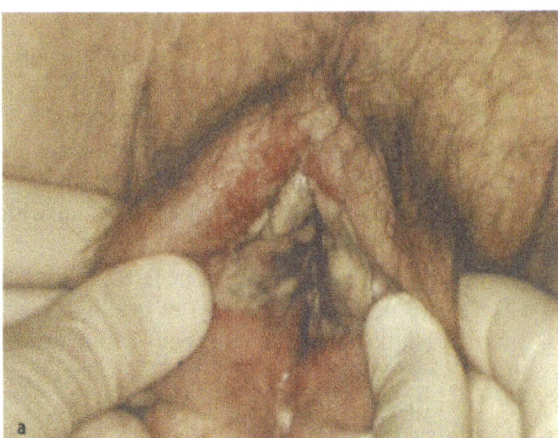

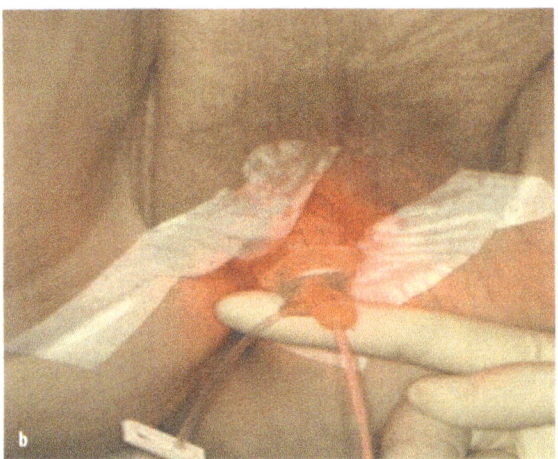

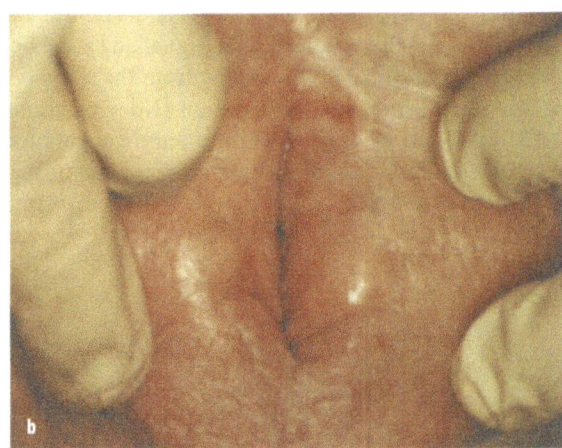

**Fig. 15a, b.** 72 h post-PDT. **a** 2 weeks post-PDT. Biopsy revealed complete remission of VAIN

**Fig. 14a-c.** PDT VAIN **a** Lesion before PDT. **b** During laser exposure. **c** Applicator with bottle filled with intalipid as light scattering agent to be sure of a homogenous light delivery pattern to the tissues

## Vaginal Intraepithelial Neoplasia

After colposcopy and directed biopsy have provided a diagnosis of intraepithelial neoplasia, local ablation of the abnormal transformation zone may be undertaken. The use of local ablative therapy is based on the principle that, following the destruction of abnormal epithelium, healthy tissue will regenerate to cover the cervix. The exact treatment modality of the cervical lesions requires destruction in depth, and laser vaporization has been applied with good effect.

Cervical intraepithelial neoplasia (CIN) may be treated by a range of techniques varying from local ablative methods such as laser vaporization or cryotherapy, or may be treated with excisional procedures such as laser or cold knife conization.

### *CO₂ Laser Vaporization*

The $CO_2$ laser is the laser most commonly used for cervical lesions. The $CO_2$ laser energy may be focused and applied to the cervix in a precise way by means of the micromanipulator and guided by colposcopic magnification. The power density of the laser beam will determine the rate at which vaporization occurs. Thermal damage to adjacent tissues is related to the duration of exposure and can be minimized by selecting high-power density resulting in rapid vaporization. Removal of the entire transformation zone to a depth of 5–7 mm including any disease in the cervical crypts is indicated. Laser vaporization or conization may be performed on an out-patient basis and side effects are minimal. Patient discomfort may be almost abolished by local anaesthetic infiltration. Secondary haemorrhage may occur; cervical stenosis is very unusual and pelvic inflammatory disease very rare. The healed cervix shows minimal distortion. The cure rate of CIN after laser excisional cone is 97%.

### *Technique for Vaporization*

- Local anaesthesia injection of a mixture of lignocaine and vasopressin with a fine-gauge needle. Submucosal injection at several sites around the transformation zone may be more effective than paracervical block. General anaesthesia may be used.
- Effective smoke evacuation equipment is essential to maintain a satisfactory view of the cervix during treatment, and for health and safety reasons.

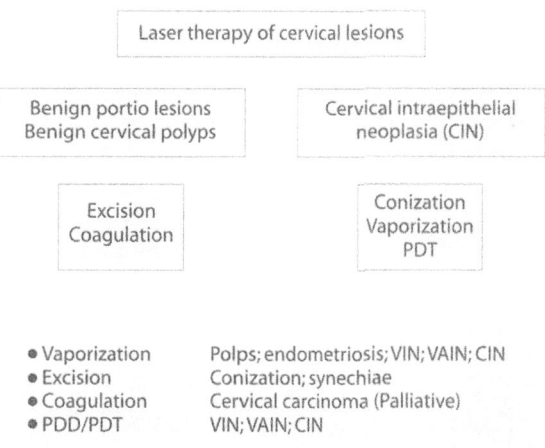

**Fig. 16.** Where and how to apply the lasers for cervical lesions

- Colposcopic examination to confirm the original findings and to demonstrate the entire transformation zone.
- Staining of the cervix with Schiller iodine to highlight the transformation zone (TZ)
- $CO_2$ with spot size 1–2 mm and power output 20 W.
- The cervix is lasered 2–3 mm peripherally to the TZ to outline the area.
- The depth of tissue destruction in this peripheral area is deepened to around 5–6 mm. This will reduce the blood supply to the central cervical area.
- Vaporize the transformation zone by quadrants, starting on the posterior aspect of the cervix until the desired depth of tissue destruction has been achieved. The appearance of clear bubbles of mucous in the vaporization area means that epithelium removal has not yet been completed.
- To control any haemorrhage, defocus the laser beam and reduce the laser power output. This will generate more heat during vaporization and may produce the desired coagulative effect. Simple pressure, vasoconstrictor injection or cervical suture may be helpful in some cases.

### *Techniques for Excisional Conization*

The same steps from 1–5 as in vaporization procedures.

The cervix is fixed with skin hooks and the cervical canal length is measured with the uterine sound or even Hegar's dilator.

Ligation of the descending branch of the uterine artery paracervically at 03 and 07.00 – 2. Position.

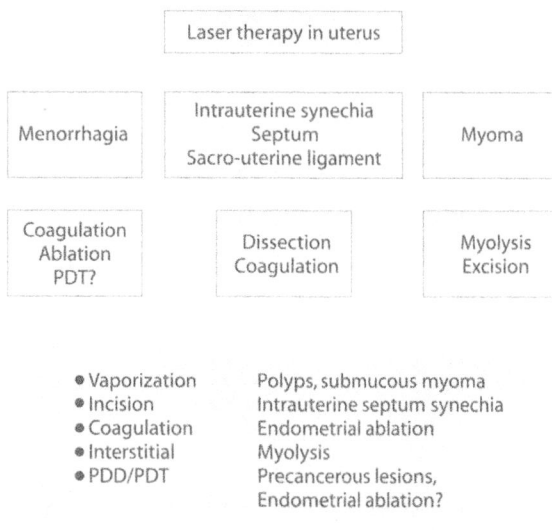

Fig. 17. Where and how to apply the lasers for cervical lesions

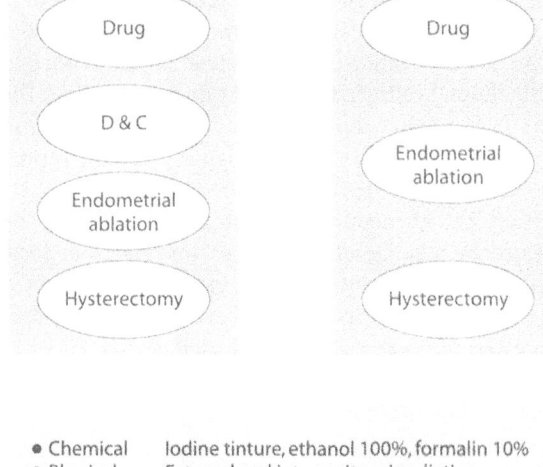

Fig. 18. Changing treatment strategy for menorrhagia

The outline incision is deepened to 1.0–1.5 cm, at which point the laser is directed towards the endocervical canal and a cone or cylinder of tissue 1.5–2.0 cm in depth is removed. The $CO_2$ beam (30 W power) is directed in the appropriate manner by a combination of manipulation of the cervix.

### Followup

The wound takes 3–4 weeks to heal. Cytological and colposcopical examinations follow 8–12 weeks after the intervention and thereafter every 6 months.

### Palliative Treatment of Invasive Cervical Carcinoma

Palliative treatment of invasive cervical carcinoma includes laser ablation of the exophytic lesions and coagulation of the endophytic lesion.

### Instruments

- Laser equipment: $CO_2$, Nd-YAG, diode lasers.
- Operating microscope and micromanipulator
- Colposcopy
- Laser fibres (400–600 µm)
- Gas evacuation or suction tube
- Applicator with moistened cotton
- Saline 0.9%, ascetic acid 3%, toluidine blue 1%

### Technique

- Excision of the exophytic tumour through contact laser with the bare fibre
- Coagulation of the base of the lesion and the endophytic tumour with laser of 40–50 W in a non-contact technique.
- The procedures can be repeated.

### Uterine Lesions

Figure 17 shows where and how lasers could be applied for uterine lesions.

### Endometrial Ablation for Dysfunctional Excessive Uterine Bleeding (Menorrhagia)

Menorrhagia means excessively heavy menstruation and has been defined as a measured menstrual blood loss of more than 80 m$^{-1}$ month. In the past, when hormonal therapy failed, this was treated primary by hysterectomy. Dysfunctional, uterine bleeding (DUB) is a term used to describe all abnormal uterine bleeding in women with no underlying organic disease and no hormonal abnormalities.

Over the past decade, hysteroscopic techniques of endometrial destruction have been developed that allow some of these women with dysfunctional menorrhagia to avoid a hysterectomy or the need for hormonal therapy while controlling their abnormal bleeding.

The aim of endometrial ablation in the treatment of menorrhagia is the deep destruction of the basal endometrial layer, leading to the subsequent loss of its regeneration capacity.

Indications for endometrial ablation in the treatment of DUB are menorrhagia, polymorrhea and metrorrhagia. The advantages are minimal invasive procedure and as an alternative to hysterectomy.

### Development of lasers for Endometrial Ablation

With the development of optical fibres capable of transmitting laser energy, lasers have steadily gained acceptance over the past decade for use in performing endoscopic and hysteroscopic surgical procedures. One such laser, the Nd:YAG is particularly well suited for photocoagulation of the endometrium as a result of its high power and transmission through fibreoptics. Its deep-tissue penetration and extensive heating distinguish it from the $CO_2$ and argon lasers.

The advent of hysteroscopic laser surgery has provided an alternative to hysterectomy in women with menorrhagia refractory to other forms of treatment.

Laser endometrial ablation has been used with considerable success to reproduce the signs and symptoms of Asherman's syndrome. Experience in patients treated hysteroscopically with an Nd-YAG laser has demonstrated that excellent results were achieved. Endometrial ablation will change the strategy of management of menorrhagia (Fig. 18).

### Methods of Endometrial destruction

For many years gynaecologists have had the idea of endometrial destruction. Chemical, physical, surgical and, recently, photochemical methods have been used as given in the following list:
● Chemical
  Iodine tincture, ethanol 100%, formalin 10%
● Physical
  External and intracavitary irradiation, cryocoagulation, Heated water (thermal ballon), electrocoagulation, conventional laser (Nd:YAG)
● Surgical
  Resection
● Photochemical
  PDT

### Preoperative Patient Preparation

Efficient treatment of menorrhagia through endometrial ablation depends on the ability to destroy the entire endometrium; for this preoperative endometrial suppression is critical. This can be accomplished with either hormonal therapy including the use of gonadotropin-releasing hormone (GnRh) analogues, Danocrine (Danazol) or medroxyprogesterone acetate. All these drugs should be administered for a minimum of 1 month preoperatively, and probably postoperatively as well in younger patients. Mechanical endometrial preparation can be achieved through endometrial curettage 1 week before ablation.

### Anatomical Uterine Considerations for Laser Endometrial Ablation

The human uterus is a pear-shaped hollow organ. It is divided into a triangular body (or corpus) above and a fusiform cervix below, joining at the isthmus. The cavity of the body has a smooth lining and is triangular in shape. Both the anterior and posterior walls are in apposition. The human uterus is characterized by a relatively thick myometrium (usually >1.5 cm) and a thin endometrium (1–3 mm), thus providing an ideal environment for laser surgery, as the thick myometrium which surrounds the endometrial layer serves as a natural barrier which protects pelvic organs during illumination. The fact that only 5% of the uterine thickness must be ablated in order to destroy the endometrium provides a substantial safety factor. The uterus is tilted forwards on the vagina (anteversion) and is bent forwards at the isthmus (anteflexion). However the intrauterine light distribution and the accessibility of the entire endometrium is not a straightforward and easy task. According to the anatomical geometry of the uterus, the endometrial ablation with the terminal end laser fibre or even with the roll ball or resectoscope either will be not complete (with subsequent high recurrence rate) or will need more and vigorous instrumental manipulation (with subsequent risk of perforation and haemorrhage), or will take more operative time (with subsequent fluid overload and anaesthetic risks). So the development of an applicator or a delivery fibre which can be easily applied in accordance with the uterine anatomical structure will significantly increase the efficiency of the technique and decrease complications.

### Patient Selection

Patient counselling should include a discussion of the fact that the procedure may not eliminate the bleeding, but will reduce the bleeding and improve patient acceptance. The patient must understand that the outcome may be in the form of amenorrhaea, normomenorrhaea, hypomenorrhaea or, in some cases, the condition may not improved. The patient should be warned about the possibility of pregnancy. Patients with dysfunctional (no intrauterine pathology) abnormal uterine bleeding disorders (menorrhagia) are good candidates for endometrial ablation after failure of medical treatment. Careful history, general and gynaecological examinations should be done with a normal-sized uterus and no gross pathological pelvic lesions. Pelvic ultrasonography should confirm the condition. Dilation and curettage should be done with histopathological examination and diagnostic hysterescopy 1 week before the endometrial ablation procedures.

### Instruments

Good light source, a video camera, operative hysteroscopy, cervical grasper, cervical dilators, equipment (Nd-YAG or diode with output power of up to 60 W).

### Operative Procedure

- Anaesthesia. General, spinal or paracervical blockage.
- Operative hystroscopy. A continuous-flow operative three-channel hysteroscope with 30° optics. This allows constant infusion of the distending media and constant drainage of fluid. A third channel is used to introduce the 600-μm laser fibre.
- Distension media. Ringer lactate or NaCI 0.9% (saline) can be used as a distension media, (3-l container of NaCI 0.9% with continuous flow). The fluid flows in by gravity and is removed by low suction to provide a constant flow of fluid at low pressure. At the end of the procedure the fluid volume that has been recovered from the inflow port and catch basin should be measured.
- Laser. Nd:YAG or diode laser.
- Laser fibres. Bare fibre (400–600 μm), side-fire fibre (400–600 μm).
- Bladder evacuation.

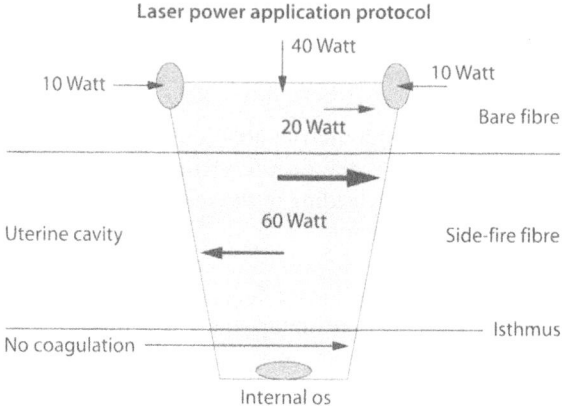

**Fig. 19.** Laser power application control

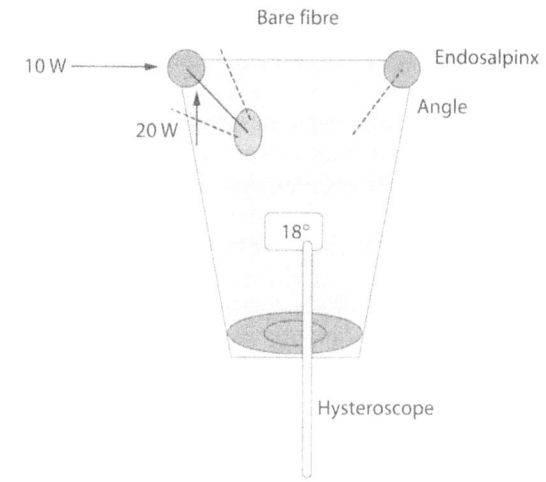

**Fig. 20.** Use of the bare-fibre method

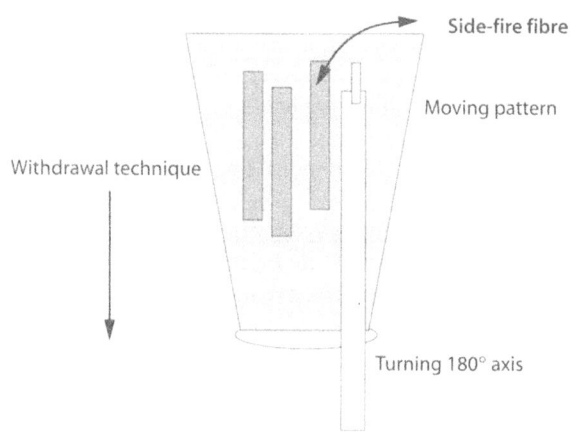

**Fig. 21.** Movement pattern and withdrawal of the side fibre during coagulationof the endometrium

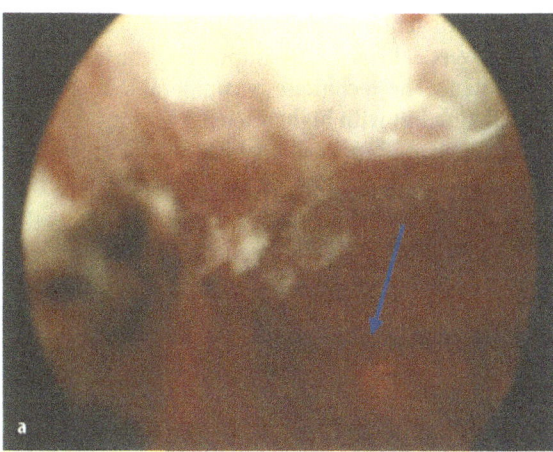

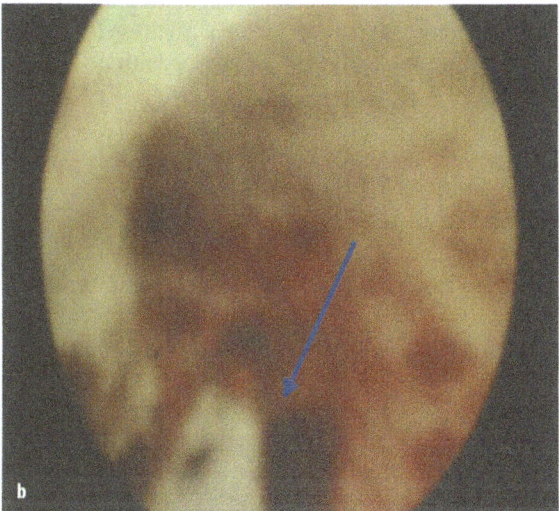

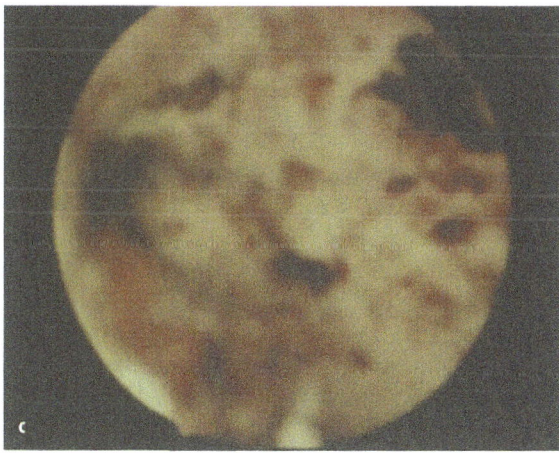

**Fig. 22a-c. a** The bare fibre (arrow) with complete coagulation of the endosalpinx and cornea angle. **b** Coagulation of the anterior uterine wall with the side-fire fibre (arrow). **c** Hysteroscopic picture of the entire endometrial cavity after its complete coagulation.

- Non-contact-endometrial Nd-YAG laser vaporization and coagulation to be performed under direct visualization (5 mm distance between the tips of the bare fibre and the coagulated area).
- Use a power output of 10 W for endosalpinx coagulation, 20 W for corneal coagulation, 40 W for fundus coagulation and 60 W for the rest of the endometrium (Fig. 19).
- The entire endometrial lining is coagulated. Using the bare fibre (bare fibre 9M–6065, quarz, 600 μm, 1050 μm outer diameter) (Martin Medizin-Technik, Tuttlingen. Germany), the endosalpinx is coagulated at the start and with a withdrawing movement of the bare fibre we coagulate the endometrium of the corneal angle at the right side and after that at the left side (Fig. 20).
- Turning the instrument on an axis of 180° facilitates reaching the coagulated area without changing the position of the hysteroscope. Care must be taken to avoid perforation near the tubal ostia, the thinnest portion of the uterus.
- Laser coagulation is extended across the fundus with the same bare fibre. Change the laser delivery fibre with the side-fire fibre (quarz 600 μm, 1800 μm outer diameter, to complete the ablation from the fundus down the lateral walls; (Fig. 21) and both the anterior and posterior walls, terminating 4 cm from the external cervical os.
- Also, turning the instrument on an axis of 180° facilitates reaching the coagulated walls without changing the position of the hysteroscope. An average distance of 5 mm between the fibre and the uterine walls should always be preserved. A homogenous total endometrial coagulation should be achieved (Fig. 22).

The production of gas bubbles using this parameter is minimal. Thus, there is a continuous clear view to observe the coagulation effect.

The procedure takes an average time of 25 min.

- Postoperative observation for uterine bleeding with blood electrolyte profile should be carried out.
- Patients can be discharged $1^{-2}$ day after the ablation. Pelvic ultrasonography to be done 2 and 4 weeks later.

### Complications in Endometrial Ablation and the Advantages of Using the New Laser technique

The major complications associated with this procedure have included:

- Uterine perforation
- Haemorrhage
- Gas embolism

- Fluid overload
- Infection
- Electrical and laser injury
- Recurrency of bleeding

Most of these complications occur mainly with resectoscopic and electrocautery endometrial ablation, and can be avoided with laser endometrial ablation.

Currently, laser endometrial ablation is performed using the bare fibre. We have started to use the side-fire laser delivery fibre in combination with the terminal bare fibre to optimize laser energy delivery to the uterine walls during endometrial ablation in trial to provide an ideal intrauterine laser energy delivery according to the geometry of the uterine cavity and the light dosimetry distribution.

For HF electrosurgery the uterine cavity must be distended by a non-conducting fluid such as 32% dextran 70. Hysteroscopic laser endosurgery can be performed with conducting fluid like Nacl 0.9%, which is isotonic to plasma and therefore may be used for every endoscopic procedure, except when electrocautery or electroresection are needed. Because of its isotonicity, normal saline solution will not cause hyponatremia following resorption into the circulation. With electroendometrial ablation, large underlying vessels may be transsected during tissue resection, which causes bleeding and necessitates increased intrauterine pressure. This leads to increased intravasation of the distension medium with the subsequent hazard of fluid overload. This hazard is avoidable using a laser technique. We started to use the side-fire laser fibre in combination with the bare fibre for ablation of the side walls and both anterior and posterior walls in all candidates for endometrial ablations. This laser fibre offered an appropriate applicator for direct laser apposition with the endometrium, easy accessibility of all endometrial angles especially the corneal angles, and a very short procedure. The advantages of using the side-fire fibre are:

Appropriate applicator for direct laser apposition
   with the endometrium
Easy accessibility for uterine walls
Short time (45 vs 25)
Efficient light penetration
Safe procedure
Non-contact coagulation
Complication-free, high efficiency, no haemorrhage,
   no fluid overload, no perforation
In all patients, the procedure was not accompanied by any complications and the followup period of 4 months revealed no recurrence of symptoms. All patients were discharged on the second day postoperative.

Thus efficient endometrial penetration and vaporization were achieved by manipulating the laser delivery system. The technique is comparable to roll-ball coagulation but with high efficiency and the certainly of complete endometrial destruction due to the deep homogenous penetration depth of the Nd-YAG irradiation and the great advantage of performing the procedure in non-contact and under visual control. We advise the use of our new technique for laser endometrial ablation for the following reasons: our operating time was averaged at 25 min, which leads to a reduction in the patient's anaesthesia time. There is no risk of endometrial perforation. Coagulation of the endosalpinx leads to a reduced possibility of ectopic gestation. The use of the side-fire fibre provides a much greater endometrial area for coagulation and gives a homogenous coagulation pattern. All these factors account for a rapid, efficient procedure with no complications.

## Laser-Induced Myolysis (LIM): New Trend in the Therapy of Myomata

Uterine leiomyomas are the most common of all pelvic tumours. They arise from the myometrium, are frequently multiple and may grow to a huge size. Although their aetiology is unknown, they appear to be oestrogen-dependent.

In 27% of women with infertility, myomas are said to be the major causative factor. Other symptoms of which the patient may complain are menorrhagia, pelvic discomfort, pressure symptoms and pregnancy loss.

Management of the symptomatic leiomyoma is traditionally open surgery. Unfortunately, the myoma's pseudocapsule has a rich blood supply, so myomectomy tends to be a haemorrhagic affair and dense adhesion formation is a common sequale. For these reasons, hysterectomy is preferred for those patients who are not concerned with fertility. In the 600 000 hysterctomies performed annually in Germany and the USA, leiomyomata is an indication in about 25%. During the last 15 years, some endoscopic gynaecologists have been satisfactorily removing small or medium-sized leiomyomas by operative laparoscopy or hysteroscopy. Nevertheless, haemorrhage can still be a problem when large subserous leiomyomas are excised by endoscopic knife or laser, and subsequent adhesion formation is common despite the meticulous closure of the overlying periteneum, particularly if diathermy is used to control haemorrhage. In addition, interstitial myomas of almost any size are very difficult, and often impossible, to treat by normal operative laparoscopic measures. In general there remain problems, both with dissection techniques for large masses, and also in harvesting such tumours without the need to perform colpocoeliotomy or minilaparotomy, which would lead to losing advantages of the laparoscopic approach. These two problems still limit laparhoscopic removal of fibroids, which remains a time-consuming procedure. The procedure involves using the energy delivered by a bipolar coagulation needle or, recently, by a laser bare fibre to coagulate symptomatic interstitial leiomyomas.

## Patients Selection

- Perimenopausal women with symptomatic sessile subserosal and intramural leiomyomas.
- High-risk patients for hyserectomy or myomectomy.
- Corneal leiomyomas or those obstructing the tubes.

### Laser Myolysis

- KTP, Nd:YAG and diode lasers, with wavelengths of 532, 1064 and 805 nm are the currently used light sources.
- The laser light is delivered in a continuous wave via a quartz laser fibre generally attached to a specially designed long, thin cannula called a needle microstat. Laser power of 5–8 W and exposure of 10–30 s were used.

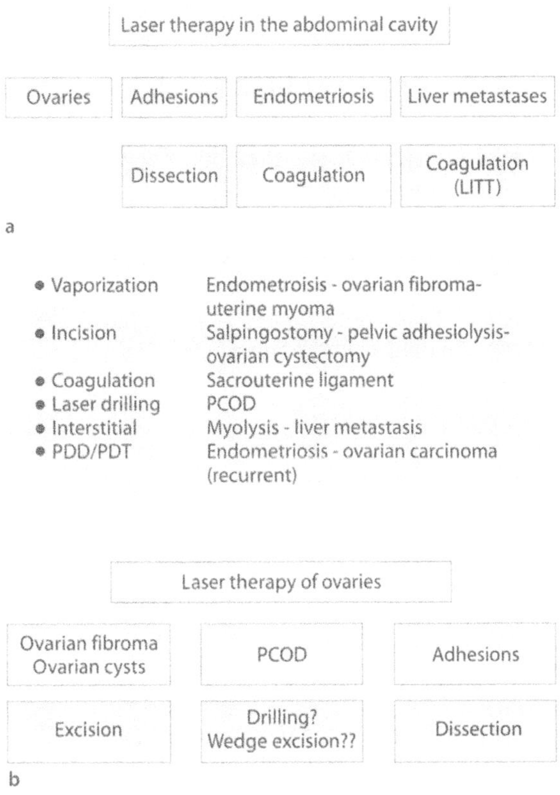

### Disadvantages of Laser Myolysis

There is no immediate visual control of the coagulation effects.
There is no control of the depth of destruction.
Interstitial laser coagulation of the tissue resembles unipolar electricoagulation, producing uncertain and unpredictable tissue destruction.

### High-Frequency Electrocoagulation

Experience with laparoscopic tubal coagulation pointed the way to this new therapy. The postoperative course after destruction of the Fallopian tube is usually completely unremarkable. The tube musculature completely regresses in the denatured area, and adhesions are seldom found subsequently.

To overcome the problem of unipolar electrocoagulation, a bipolar electrocoagulation was used through a bipolar coagulation probe.

Various sizes of electroprobes can be selected, depending on the diameter of the myoma. Using such a

a

- Vaporization — Endometroisis - ovarian fibroma - uterine myoma
- Incision — Salpingostomy - pelvic adhesiolysis - ovarian cystectomy
- Coagulation — Sacrouterine ligament
- Laser drilling — PCOD
- Interstitial — Myolysis - liver metastasis
- PDD/PDT — Endometriosis - ovarian carcinoma (recurrent)

b

**Fig. 23 a, b.** Where and how to apply the lasers for a intra abdominal lesions, b for ovarian lesions

probe, puncture of the base of the myomata can be performed easily. Electrocoagulation is carried out with computer-controlled output to a maximum of 120 W. Using this high wattage, haemostasis is immediately achieved without any bleeding and without the electro needle becoming encrusted with coagulated blood.

### Results

- Myomata up to 7 cm in diameter can be treated. In this range different sizes of probes (electrocoagulation) and of laser fibres could be used.
- Postoperatively, regression of coagulated tissue after denaturation was completely unremarkable and there were no signs of myometrial necrosis-induced pains.
- Reduction in size of the denaturized myoma could be followed by ultrasound over the following weeks.

- After coagulation of larger myomata, ultrasound follow up showed an average reduction of myomata volume of 70–70% (these figures are similar to those following GnRH treatment, but after myoma coagulation the effect obtained is permanent, with no regrowth, and every fibroid appears to respond to coagulation).
- For greater volume reduction, a combination of hormonal GnRH treatment and coagulation is possible. In cases of menorrhagia associated with myomata, the combination of myolysis and endometrial ablation has been reported.

### Disadvantages of Myolysis

- Lack of histological material for examination.
- Myolysis, especially in larger intramural fibroids, may lead to cystic formation, which is usually without symptoms.
- Stability of the uterine wall in pregnancy is uncertain.

## Laser Applications in the Peritoneal Cavity

Where and how lasers could be applied for intrauterine abdominal cavity, are as follows (see also Fig. 23):

- Vaporization    Endometriosis – ovarian fibroma – uterine myoma
- Incision    Salpingostomy – pelvic adhesiolysis – ovarian cystectomy
- Coagulation    Sacrouterine ligament
- Laser drilling    PCOD
- Interstitial    Myolysis – liver metastasis
- PDD/PDT    Endometriosis – ovarian carcinoma (recurrent)

### Laparoscopic Pelvic Adhesiolysis

Pelvic adhesiolysis is commonly practised and is indicated for the treatment of both pelvic pain and infertility. The ineffectiveness of adhysiolysis by laparotomy means there is little place for this type of surgery in the treatment of chronic abdominal pain. Laparoscopic adhesiolysis is a different technique and seems to be worthwhile. Gross anatomical distortion caused by extensive pelvic adhesions causes infertility. The most common risk factors associated with this problem are previous surgery and salpingitis. Ovulation is less likely to take place in patients with tubo-ovarian adhesions. In addition, tubo-ovarian adhesions may restrict movements of the tubes and ovary, and thus most likely further impair function.

Laparoscopic adhesiolysis offers a useful alternative to laparotomy techniques. This type of surgery

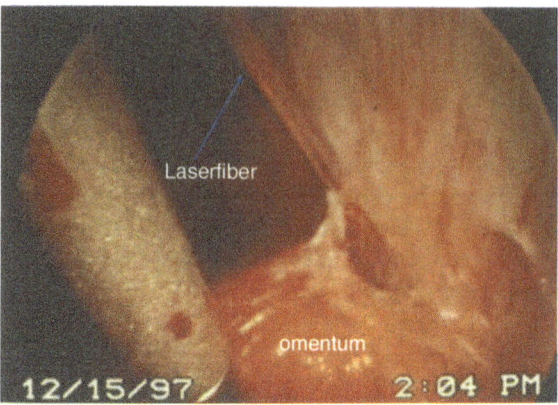

**Fig. 24.** Lutting and coagulation during adhesiolysis with a Nd:YAG-Laser fibre

can be performed at the time of diagnostic laparoscopy; it is a simpler technique and seems to be as effective as microsurgery. From the point of the patients inconvenience and postoperative morbidity, laparoscopic surgery appears more acceptable than techniques requiring laparotomy. Various techniques can be employed to perform laparoscopic adhesiolysis.

### Mechanical Laparoscopic Adhesiolysis

Including the use of scissors is the simplest technique, but this method is not without problems. Scissors have no haemostatic function and bleeding can occur, making visualization and continuation of the procedure difficult. It diathermy is used to stop bleeding, particularly if monopolar diathermy is used, tissue necrosis will occur and this will inevitably increase the risk of further adhesion formation. Bipolar laparoscopic diathermy forceps are available and will induce haemostasis with much less tissue necrosis.

### Laser Laparoscopic Adhesiolysis

An alternative and easier technique for performing adhesiolysis is to use lasers. Although the equipment is expensive, adhesiolysis can be performed rapidly and safely with minimal tissue necrosis. Bleeding is rarely troublesome. The $CO_2$ laser, KTP, Nd:YAG or diode lasers can all be used for adhesiolysis.

$CO_2$ laser was the first laser used in laparoscopic surgery. One disadvantage of the $CO_2$ laser its that it cannot for practical purposes be transmitted through a flexible fibre. The beam can only be transmitted in a straight line and reflection is needed to alter its direction. Access to many locations is difficult. Wave guides have been developed to direct the laser beam, but they are straight instruments and therefore do not improve access. In order to avoid causing damage to tissue distal to the adhesion, a backstop is inserted on the end of the adhesiolysis probe. In certain situations this can be difficult to place, and an alternative is to position a flexible blackened probe through a third or fourth portal trocar or to stretch the adhesion over a pool of fluid which will absorb laser energy. During salpingolysis the operator should use relatively high-power densities, preferably on a superpulse mode, to achieve precise cutting. Care must be taken between the tube and ovary where the mesosalpinx is extremely vascular and troublesome bleeding is encountered if the thin-walled veins are accidentally punctured. In this situation the bleeding is difficult to stop with electrodiathermy without causing damage to the tube. The blunt-tipped endocoagulator probe is more effective here, but has to be cooled by running irrigant fluid to prevent if from sticking to the tissues and forming a coagulum, which starts to bleed again when the probe is pulled off.

Nd:YAG, diode and KTP-532 lasers can be transmitted through a flexible fibre. These lasers will also divide vascular adhesions in a bloodless manner. They have both tissue-cutting and coagulating effects, unlike the $CO_2$ laser, which is principally a vaporizing or cutting instrument. Nd:YAG and diode lasers have replaced the use of the $CO_2$ laser in the treatment of adhesiolysis. Compared to microsurgical diathermy, which produces significant tissue damage and adhesions, Nd:YAG and diode lasers produced less tissue injury and no adhesions.

### Ovarian Drilling for Polycystic Ovarian Disease (PCOD)

Polycystic ovarian disease (PCOD) is a disorder of chronic anovulation and infertility. The ovaries are symmetrically enlarged with a smooth, thickened capsule and multiple ovarian follicular cysts 4–8 mm in diameter. Wedge resection of the ovaries has been used to treat such cases, but its main drawback was the extensive postoperative adhesions with subsequent effection fertility. A number of alternatives to wedge resection by laparotomy have been proposed to decrease the morbidity and the associated postoperative adhesion formation. Laser ovarian drilling is a procedure to treat such PCOD in a minimally invasive way without drawbacks. The mechanism for this drilling effect is unclear, but may involve reduction in ovarian androgen production through a decrease in stromal mass or disruption of parenchymal blood flow. Ovarian drilling (cauterization) should be limited to patients who have failed to ovulate under the standard regimes for controlled stimulation of ovulation.

$CO_2$ laser was used initially for treatment of PCOD. However, the volume of tissue vaporized produced a tremendous amount of smoke, in addition to the difficulty of haemostasis if any bleeding occurs, and the rigidity of the applicators of $CO_2$ systems limited their use. With the availability of the lasers transmitted through fibres the $CO_2$ laser was replaced by these, including the Nd-YAG, diode, KTP and Argon lasers.

### Instruments

- Laser equipment: Nd:YAG, diode, KTP, argon and $CO_2$ lasers.
- Micromanipulator and operating microscope when using the $CO_2$ laser in the laparatomic approach.
- Operating laparoscopy.
- Laser fibres (400–600 µm).

### Technique

- General anaesthesia (GA).
- Disinfection of skin of the abdomen.
- Application of laparoscopy in a three-puncture sites (subumblical, bilateral iliac fossa trochars. Laser procedures can be done under water seal (artificial ascites using saline or Ringer's solution) (not obligatory).
- For $CO_2$ drilling: (Penetration depth 2–3 mm, 10–15 drilling points. Micromanipulator, focus diameter 1–2 mm, power 10–30 W), haemostasis through compression with moistened gauze or punctual coagulation by defocusing the laser beam.
- For visible light laser drilling (Nd:YAG, diode, KTP, argon, 20–40 W, cw, bare fibre 400 µm), perforating hole diameter 2 mm in 10 to 15 points of the tubal averted site of the ovary.
- Prevention of adhesions by means of perfect intraperitonal irrigation, leaving 200 ml of Ringer's solution after the procedure or insertion of an irrigation suction catheter for continuous irrigation and suction for 24–48 h.

### Management of Ectopic Pregnancy

Many approaches in the management of ectopic pregnancy can be effected through laparoscopy instead of laparotomy approach. Laparoscopy has long been widely accepted as an effective tool for diagnosis and treatment of ectopic pregnancy. The introduction of laser laparoscopic surgery allows more precise incisions with controlled depth of penetration ($CO_2$ laser) and more coagulative and haemostatic properties (Nd:YAG, diode, KTP, argon lasers). Linear salpingetomy, partial salpingectomy, total salpingectomy and aspiration are further techniques performed through laparoscopy.

### Instruments

- Lasers equipment: Nd:YAG, diode, KTP, argon and $CO_2$ lasers.
- Micromanipulator and operating microscope when using $CO_2$ laser in the laparatomy approach.
- Operating laparoscopy.
- Laser fibres (400–600 µm).

### Technique

- General anaesthesia (GA).
- Disinfection of skin of the abdomen.
- Application of laparoscopy at three puncture sites (subumbilical, bilateral iliac fossa trochars. Laser procedures can be carried out under water seal (artificial ascites using saline or fingersolution) (not obligatory).
- Aspiration of any blood in the peritoneal cavity.
- Irrigation and suction to make the field clear.
- Haemostasis of any active marked bleeding points.
- Injection of vasopressin into the mesosalpinx and in the tube at the site of incision.
- Manipulation of the affected tube with traumatic grasping forceps; linear incision is made on the dorsal side (proximal portion) of the tube using one of the lasers. Initial rinsing into the tubal ostium to mobilize the gestational sac and blood clots. Removal of any adherent trophoblastic tissue. Tubal incision left open after good haemostasis to heal by secondary intention.
- For $CO_2$ laser: Focus diameter 0.5 mm, power 10–15 mm, cw), haemostasis through compression with moistened gauze or punctual coagulation by defocusing the laser beam.
- For VIS and NIR lasers: Nd:YAG, diode, KTP, argon, 20–40 W, cw, bare fibre 400 µm.
- Prevention of adhesions by means of perfect intraperitoneal irrigation, leaving 200 ml of Ringer's solution after the procedure or insertion of an irrigation suction catheter for continuous irrigation and suction for 24–48 h.

## Treatment of Endometriosis

Pelvic endometriosis is an ideal condition to be treated with laser. The laparoscopy has enabled the gynaecologist to make the diagnosis, stage of the disease, and by introducing the laser fibre, the surgeon can destroy the endometriotic foci and perform the adhysolysis. Endometrioma can be opened, drained and vaporized with correction of all the associated

pathology at the time of the procedure. It is important that the peritoneal surface of the pelvis be examined carefully and all the lesions be treated. Laser laproscopic treatment of endometriosis is an alternative and easier technique. Although the equipment is expensive, the procedure can be performed rapidly and safely with minimal tissue necrosis. Bleeding is rarely troublesome. The $CO_2$, KTP, Nd-YAG or diode lasers can all be used for endometriosis.

### Instruments

- Laser machines: Nd-YAG, diode, KTP, argon $CO_2$ and dye lasers for PDT.
- Micromanipulator and operating microscope when using the $CO_2$ laser in the laparatomic approach.
- Operating laparoscopy.
- Laser fibres (400–600) µm.
- D-light telescope for fluorescence diagnosis.

We recommend the use of laser-induced fluorescence with systemic injection of HPD or methylene blue to evaluate the condition and determine the biopsy sites. A D-light telescope for LIF is used to stimulate the photosensitizers and detect the fluorescence signals. Laser treatment can be applied to the required location and depth with great precision.

### Technique

- General anaesthesia (GA).
- Disinfection of skin of the abdomen.
- Application of laparoscopy at three puncture sites (subumblical, bilateral iliac fossa trochars. Laser procedures can be done under water seal (artificial ascites using saline or Ringer's-solution) (not obligatory).
- D-light telescope for fluorescence diagnosis.
- For $CO_2$ vaporization: Micromanipulator, focus diameter 1–2 mm, power 10–30 W). Haemostasis through compression with moistened gauze or punctual coagulation by defocusing the laser beam. Vaporiziation or excision down to the level or healthy tissue is more useful for larger lesions, although deep lesions are more accurately excised than vaporized. Excision is started by cutting through the peritoneum into the loose connective tissue below with the scissor $CO_2$ laser beam or the laser fibre. The lesion is outlined and the loose connective tissue and fat noted. A blunt probe, forceful irrigation (aquadissection), or spreading of the scissor blades is then used to dissect these layers. Injection of normal saline or Ringer's lactate subperitoneally, to push the peritoneum away from the vessels, bowel, bladder or the ureter. This technique will allow the laser light to be absorbed

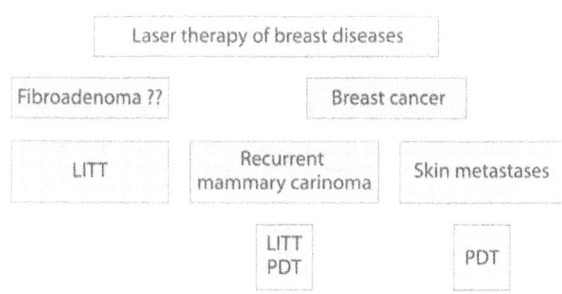

**Fig. 25.** Where and how to apply lasers for breast diseases.

by the water and prevents damage to the underlying organs. Rinsing of the ablation sites with heparinized Ringer's lactate (5.000 IU l$^{-1}$). Suction and irrigation of all the tissue debris blood clots are essential. Continuous irrigation and suction for 24–48 h are advised.

- For VIS and NIR lasers Nd-YAG, diode, KTP, argon, 20–40 W, cw, bare fibre 400 μm.
- For laser resection of the ovarian endometriotic cyst. Injection of 1–2 ml of vasopressor solution into the surrounding normal ovarian tissue. The cyst is opened using the laser. Suction of the cystic fluid. Excision of an eliptical portion of the cyst wall and superficial vaporization of the complete endothelial lining of the cyst by laser beam of 20 W.
- PDT. Superficial application, handpiece, cylindrical laser fibre applicator. Topical application of ALA 10% 2–4 h before treatment or systemic photosensitizer application HPD 48–72 h before the application of laser dye light 630 nm to the treated area. Avoid exposure of the patient to direct light or sunlight in the case of systemic application of photosensitizer.
- Prevention of adhesions by means of perfect intraperitoneal irrigation, leaving 200 ml Ringer's solution after the procedure or insertion of irrigation suction catheter for continuous irrigation and suction for 24–48 h.

## Treatment of Dysmenorrhaea and Chronic Pelvic Pain

Lasers can be used for treatment of severe dysmenorrhaea and chronic pelvic pain not responding to conventional methods such as medical treatment. In these cases laser ablation of uterosacral ligaments can relieve the condition. The procedure involves transecting the sensory nerve fibres as they leave the uterus on the way to the pelvic plexuses. Because of the divergence of these nerve fibres in the uterosacral ligaments, it has been recommended that this procedure be performed as close to the uterus as possible.

The procedure is like that done for adhesiolysis in regard to the required instruments and technique. Laparoscopic ablation of a segment (1–2 cm long and 1 cm deep) of the uterosacral ligaments in their cervical insert region with the laser beam (focus diameter 0.5–0.1 mm, power 15 W, cw for $CO_2$ and with bare fibre and laser power of 20–40 W for the VIS and NIR light lasers).

## Treatment of Breast Disease

Figure 25 illustrates where and how lasers could be applied for breast diseasis.

The superficial location of the breast and its associated lesions, plus the easy localization, palpation and visualization of lesions within the breast or the chest wall area make it an ideal site for laser therapy. Local recurrence of metastasis may be slightly more common after breast conservation surgery rather than mastectomy.

For palliative reasons, extensively pretreated subcutaneous chest wall recurrences in eight patients suffering from breast cancer have been treated by the LITT method under the control of colour-coded duplex sonography (CCDS).

### Instruments

- Nd-YAG, diode, dye lasers for PDT.
- Laser fibres (bare fibre, cylindrical fibre).
- Ultrasonic machine.
- Teflon canula.

### Imaging

- Ultrasonography is used to measure the dimensions of the metastases and the largest diameter is to be noted.
- Colour-coded duplex sonography, with a transducer frequency of 7.5 MHz, is used for both preoperative visualization of the tumour anatomy and for control of the therapeutic coagulation effects.
- The power and amplification of the signal are set individually according to the required application.

### Laser Parameter

- A continuous-wave Nd:YAG (wavelength 1064 nm) or diode laser.
- A sterilized, single freshly cleaved, 600 μm fibre with a bare tip to deliver laser energy to the tumour.

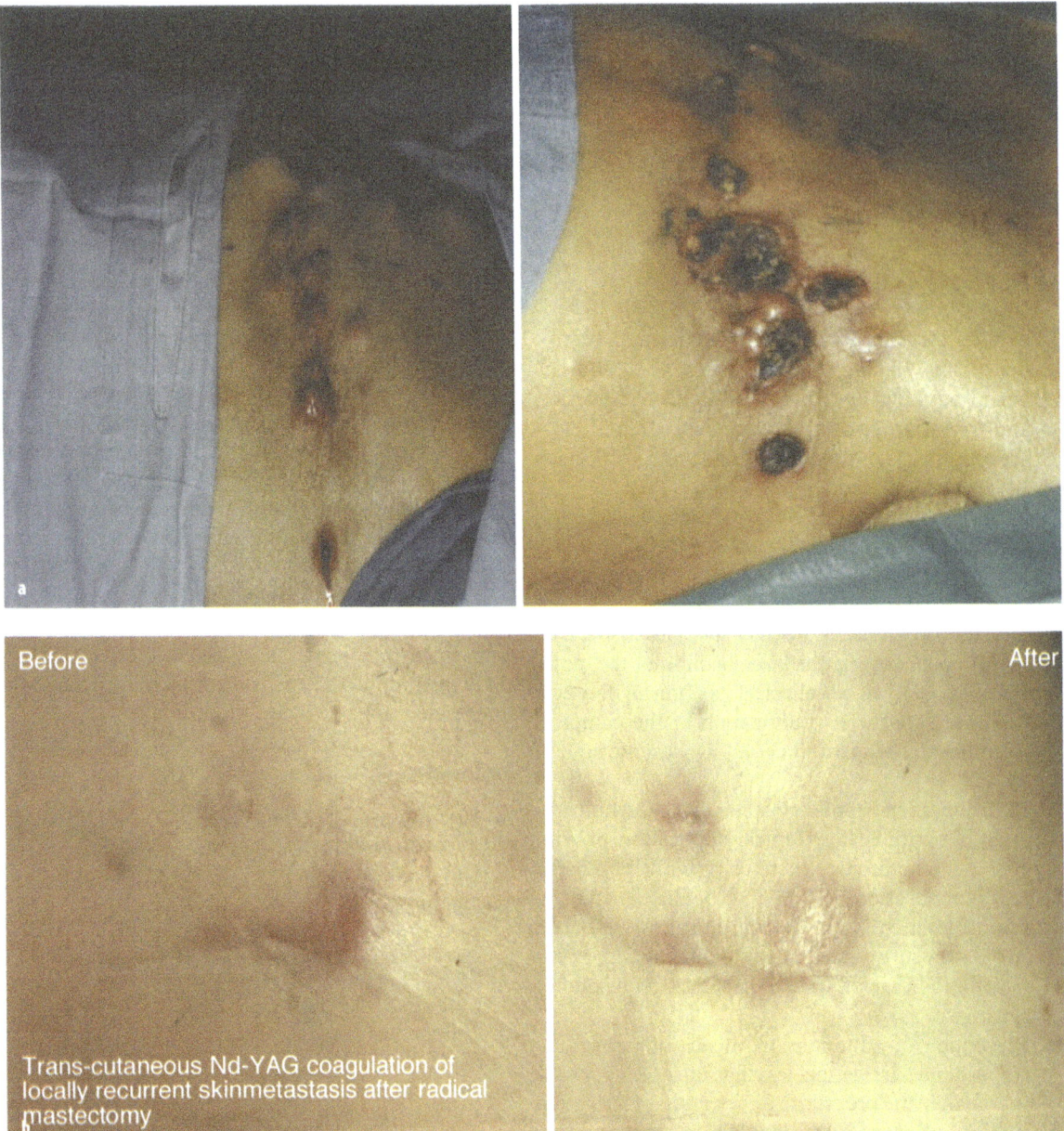

**Fig. 26a,b. a** Chest wall with local recurrences of breast cancer. Status after bilateral mastectomy and axillary lymph node dissection and irradiation about the actual heat development. Puncture of a metastasis with a Teflon catheter (Abbocath G 16). The puncture site is always in the healthy skin area. There were more than 15 chest wall metastasis near each other. Skin ulceration and delayed healing resulted. **b** Small metastases on the line before and after laser treatment.

## Procedure (Fig. 26)

- Anaesthesia. The interstitial laser treatment is performed under local anaesthesia for small metastases (diameter <10 mm) or under general anaesthesia for multiple or large-sized metastases (>10 mm), where multiple punctures are needed at a distance of 10 mm.

- Under sterile conditions and after visualization of the diseased area with CCDS imaging (a sterile sheet over the ultrasound probe), the area to be treated is punctured with a Teflon catheter, to be placed within the tissue under B-scan control.

- The Teflon catheter is always introduced through a healthy skin area about 1 cm from the margin of the metastases to avoid direct skin puncture over the lesion.
- In each case, a freshly cleaved bare fibre is inserted through the Teflon catheter, which is pulled back for 10 mm over the fibre, and care is taken to avoid thermal interaction between laser and catheter material.
- The Nd:YAG laser is applied with a power setting of 5 W for a range time of 120–600 s (600–3000 J).
- In this way it is possible to achieve a coagulation zone of approximately 15 mm in diameter, which extends cylindrically and retrograde along the fibre.
- Heat expansion is intermittently controlled digitally and monitored by ultrasound and CCDS, respectively.
- Attention is paid that, in cases with metastasis directly near the skin, the cutis in the above-laying layers of the metastases are always saved from laser thermal effects to using continuous ice-cooling.
- The positioning of the catheter and bare fibre inside the metastatic lesion is easy to visualize under the B-scan. In cases with subcutaneous lesions without skin infiltration, precise coagulation of the tumour without destruction of the skin or ulceration is achieved.
- After a few seconds of laser irradiation, a spherical hyperechogenic zone is monitored under the B-scan around the laser delivery fibre tip. This is followed after 30 s by a signal in CCDS imaging in the form of random mixtures of red and blue coloration. The hyperechoic area is expanded to reach a peak within 100–300 s (according to laser power) and to remain approximately the same size and shape until the end of therapy. Under the CCDS and as photocoagulation progresses, the colour signal is enlarged but is not obligatory equal in size and placement to the hyperechogenic zone. The onset of colour signals is dependent on the power setting.
- After a varying time, depending on the irradiation parameters, these signals are decreased and no further changes are observed in CCDS imaging. During further irradiation, the hyperechogenic zone under the B-scan is shifted along the puncture channel towards the periphery. At this time gas bubbles pass along the Teflon catheter and are observed on its superficial end. The extent of the coagulation zone is visible, using the B-scan several minutes after laser exposure. The cutis in the above-lying layers of the metastasis are always saved without any sign of necrosis. The control CCDS 2 weeks after primary treatment showed all the signs of fibrosis and reduction of vascular perfusion in these areas.

## Experience with LITT Versus PDT for Palliative Treatment of Locally Recurrent Skin Metastasis After Radical Mastectomy

Until now, only a small number of studies have been reported on LITT for breast cancer. In two recent studies, interstitial laser photocoagulation was used for the evaluation of tumour coagulation efficiency prior to surgical resection. The assessment of response to therapy was carried out with contrast-enhanced magnetic resonance (MR) imaging, which showed that LITT can achieve complete tumour ablation in primary breast cancer and the postcontrast MR images can define the extent of both laser-induced necrosis and residual tumour following LITT of breast cancer. The same group have used LITT for treating breast fibroadenoma in 14 patients: 40% reduction in tumour volume in all treated fibroadenomas after 8 weeks postlaser application and >90% tumour had disappeared. LITT was applied in 13 patients with mammography-detected, well-defined tumours (6–18 mm in diameter). Total tumour ablation was achieved in nine cases.

For several years in our centre, LITT and PDT procedures in the treatment of subcutaneous lesions of different pathological origin have been carried out. Eight patients suffering from subcutaneous pretreated chest wall recurrences of breast cancer were treated by the LITT for palliative reasons. Material and methods were as follows:

- Laser:
  Nd:YAG laser, bare fibre, 4–6 W, 120–600 s
- Controlling method:
  B-scan, CCDS
- Heat expansion control:
  Digitally (warming, crepitation); CCDS (intermediate signals of red and blue coloration)
- Anaesthesia: Local/General

On the whole, LITT was performed 11 times in these patients. In this physical condition, it was no longer possible to treat these lesions surgically by a TRAM-flap or a latissimus dorsi muscle flap. In all cases, the areas of concern had been pretreated by chest wall irradiation after mastectomy and/or systemic chemotherapy. The tissue changes in the areas treated could be excellently visualized with CCDS. Thus, it became possible to change the positioning of the bare fibre, if necessary, and to coagulate the metastasis totally.

This is the first clinical report on the application of interstitial laser therapy as a mode of treatment for locally advanced recurrences of breast cancer. It is also the first study using CCDS as a controlling imaging method to achieve the complete disappearance of local recurrences. In our centre, two cases with skin infiltration and multiple recurrence lesions near each other were complicated by skin ulceration and delayed healing. In the first 6 months' followup two patients died of distant metastasis. Three patients with solitary metastasis and metastases with a large surface of healthy skin between each lesion showed disappearance of the coagulated metastasis with no skin ulceration. The life quality of these patients was improved as a consequence the disappearance of chest wall metastasis and also from the psychological aspect. In two patients, a residual tumour was noted after treatment. Patients with local anaesthesia experienced minimal pain and discomfort during therapy

and in the early postlaser therapy period, which was readily controlled by oral analgaesia. In the digital control of the heat expansion, elevation of the temperature through the treated tumour and its skin covering was detected. We also detected a crepitation caused by vapour and degasification of $CO_2$ in the tissues as consequence of tissue temperature elevation. This reaction was also monitored in CCDS in the form of random noise of red and blue coloration. Our results were not promising in achieving the complete disappearance of local breast recurrences in patients with multiple, intracutaneous metastasis and with metastases near each other, as in these cases a necrotic skin ulcer with prolonged and bad healing followed laser coagulation. This meant the conversion of a closed lesion into an ulcerated lesion. It is also noted that in these severly advanced cases, the quality of life and reduction of suffering were not improved. LITT is ideal for cases with solitary subcutaneous lesions and

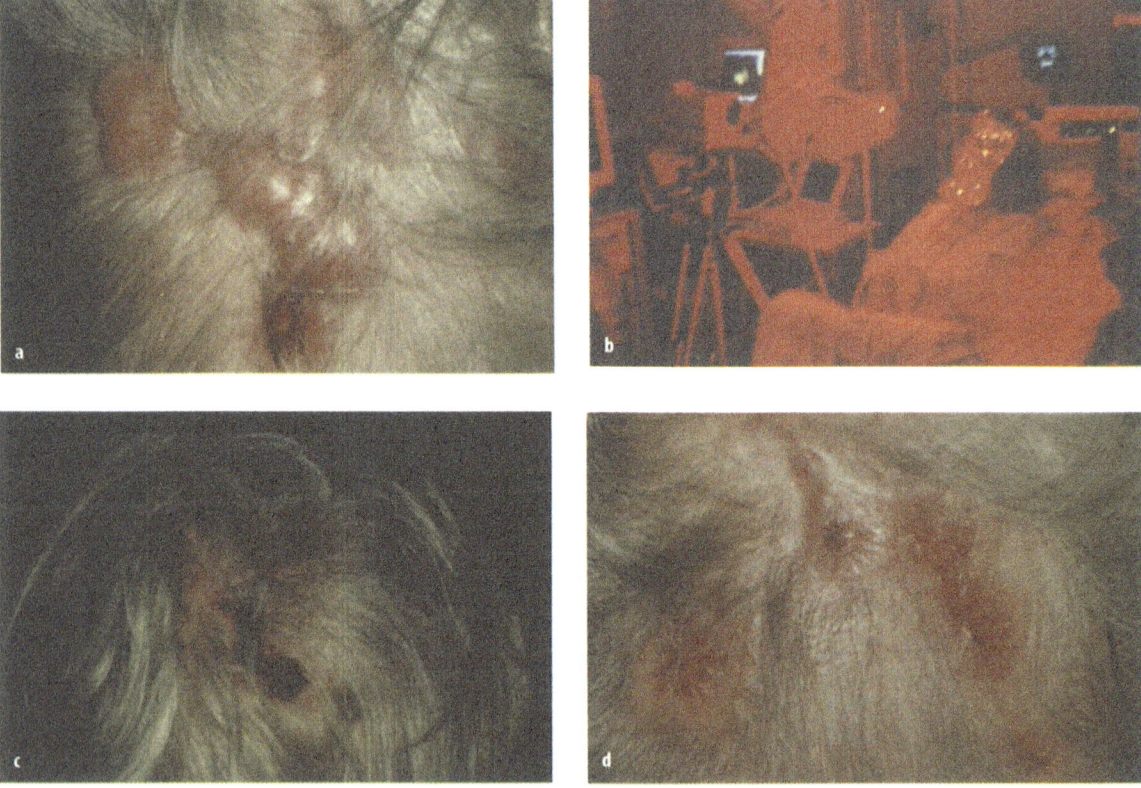

**Fig. 27a-d.** A case of multiple scalp metastasis secondary to primary breast cancer. **a** Scalp metastasis before PDT. **b** PDT during the process. The healthy tissues should be covered with alu-minium foil. A Thermocamera is used to control the temperature of the treated area. **c** picture 1 month after first session of PDT. **d** After second session of treatment

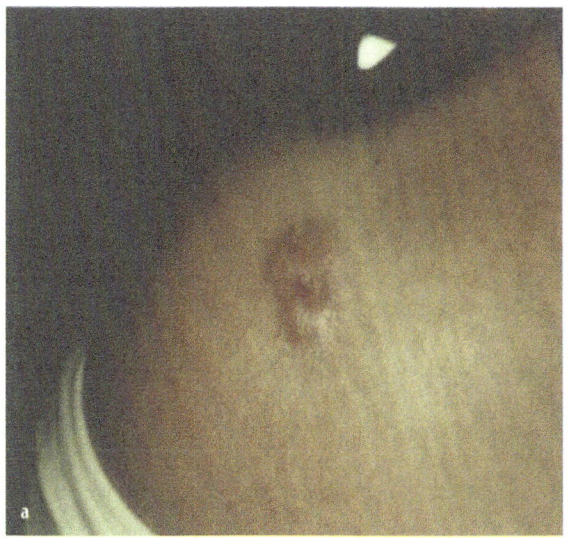

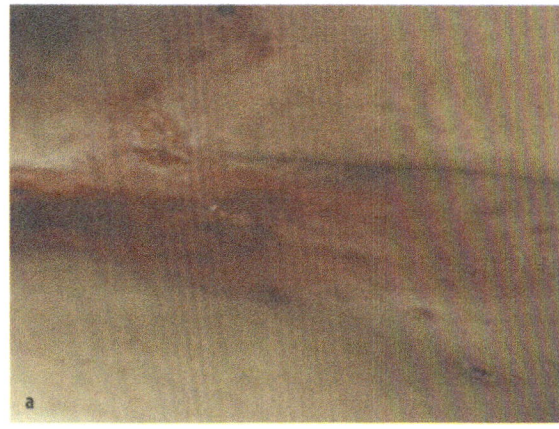

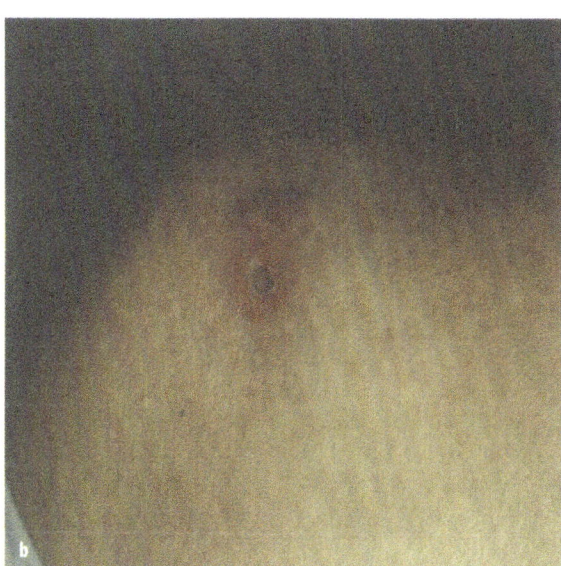

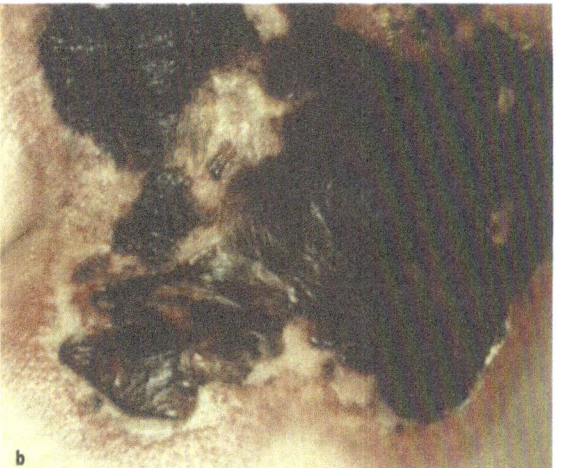

**Fig. 28a-b.** A case of skin metastasis secondary to primary breast cancer. **a** Skin metastasis before PDT. **b** 1 week after first session of PDT

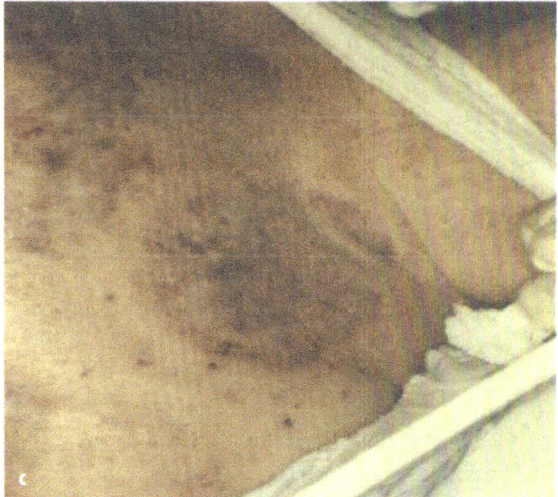

**Fig. 29a-c.** A case of skin metastasis secondary to primary breast cancer with lymphatic carcinomatosis. **a** Before PDT. **b** 4 days post-PDT. Extensive skin reaction due to over-accumulation of the photosensitizer in the cells secondary to lymphatic affection. **c** 1 month after first PDT daily dressing of the treated area.

when the metastases are far from each other, enabling a good chance of healing without exulceration of the metastasis. As in two patients skin ulceration developed, this ulceration was a consequence of two factors, the first was the tumour skin infiltration and in this case the heating effects of laser in the skin layer could not be avoided, as avoidance of skin coagulation means incomplete destruction of all the malignant cells. The photodynamic therapy could perhaps offer less incidence of skin ulcerations in cases with intracutaneous affection.

For the successful application of LITT in clinical practice, an imaging modality is required in the treatment to control the application procedure and treatment parameters and follow up the treatment results. Several methods are currently available for imaging control in laser applications including visual and digital control, ultrasound, MRI and the steriotaxic system CCDS. Each of these methods has its advantages and disadvantages and cannot satisfy all aspects of control and evaluation of the LITT procedure. A more easily applicable technique, which generates real-time images of the LITT procedure and its effects on tissue, is required to make interstitial laser application a common clinical procedure. Initial interest in ultrasound in imaging the effects of LITT was based on the simplicity of its use its portability and the fact that imaging is performed at the same time and place as the treatment. In view of the severity of the disease and the escape phenomena by other treatment modalities, laser-induced thermotherapy (LITT) can play a role in treatment of repeated local recurrences of breast cancer especially by helping to prevent the exulceration of these multiple tumours. These high-risk patients may thus derive an advantage from this method, above all as far as their quality of life is concerned. Further clinical trials are warranted. CCDS is a reliable and easy technique for pre- and postoperative on-line control of the LITT procedure.

Materials and methods for PDT for treatment of skin metastasis are as follows:

- Photosensitizer: HPD (Photosan-3, Photofrin)
- Laser:KTP dye laser, 630 mm, 2–7 W, microlens, surface illumination
- Light dose: 40–100 J/cm$^{-2}$, 100 mW-cm$^{-2}$
- Time interval: 48, 72, 96, 120 h post-PS
- Heat expansion controlling: Thermocamera
- Anaesthesia: none

Patients and methods

- 14 patients suffering from multiple subcutaenous chest wall recurrences and skin secondaries of breast cancer were treated with LITT (8) and PDT (6)
- Pretreatment modality:
- Mastectomy, radiotherapy, chemotherapy, Hormonaltherapy

- palliative reasons:
- No alternative treatment

Results

- Precise destruction of the metastatic lesions
- Extent of the coagulation zone was visible under B-scan US during LITT
- patient experienced minimal pain and discomfort during and in postlaser period (LITT > PDT)
- The control CCDS 2 weeks after the laser treatment showed all signs of fibrosis and lag of perfusion of treated lesions

Postoperative observations

LITT

- Pain (++)
- Eschar (++)
- Skin ulceration (++)
  (if skin is infiltrated)

PDT

- Pain (+)
- Eschar (+)
- Skin ulceration (+)
  (can be avoided through fractional irradiation)

Follow up

PDT

- 1 patient died of metastatic disease at 4 months post-treatment
- 1 patient had CR (CR = Complete Remission) of treated site within 4 months (one lesion) post-treatment
- 4 patients had PR (PR = Portial Remission > 50% in volume) of the treated sites and new metastatic development

LITT

- 3 patients died of metastatic disease within 6 months post-treatment
- 1 patient had CR of treated site at 2 months (one lesion) post-treatment
- 3 patients had CR at 6 months (multiple lesions)
- 2 patients had PR of the treated sites and new metastatic development

Advantages and Disadvantages

LITT

- Thermal effect
- On-line monitoring
- Anaesthesia
- Short duration of procedure
- Appropriate for large and subcutaneous metastases
- Hospitalization: short, possible on outpatient base
- No risk of photosensitivity
- Cost: inexpensive

PDT

• Photochemical reaction
• No on-line monitoring of photoreaction process
• No needs for anaesthesia
• Long duration of procedure
• Hospitalization: approximately 1 week
• Risk of photosensitivity
• Good cosmetic results
• Cost: expensive

Conclusions

• LITT is effective in treatment of local recurrences and skin secondaries of breast cancer mainly the subcutaneous (deep lesions) and large sized lesions (> 1 cm)
• PDT can play the same role mainly for cutaneous and small sized lesions (< 1 cm)
• Both modalities can be combined
• This group of high risk patients derives great advantages mainly concerning their quality of life

**Further Reading**

[1] ABSTEN GT (1992) The physics of light and lasers. In: Sutton C. Lasers in gynaecology. Chapman & Hall Medical, London
[2] ALFANO R, TANG G, PRADHAM A, LAM W, CHOY D, OPHER E (1987) Fluorescence spectra from cancerous and normal human breast and lung tissues. IEEE j Wuant Electron 10: 1806–1811
[3] ANDERSSON-ENGELS S, BERG R, JOHANSSON J, SVANBERG S ET AL. (1992) Laser spectroscopy in medical diagnosis. In: Henderson B, Dougherty T (eds) Photodynamic therapy: basic principles and clinical application. New York, pp 388–424
[4] ANZAI Y, LIFKIN R, SACTON R, ET AL. (1991) Nd-YAG interstitial laser phototherapy guided by magnetic resonance imaging in an ex vivo model: dosimetery of laser MR tissue interation. Laryngoscope 101: 775–600
[5] AMIN Z, DONALD J, MASTER A, KANT R, STEGER A, BOWEN S, ET AL. (1993) Hepatic metastasis: interstitial photocoagulation with real-time US monitoring and dynamic CT evaluation of treatment. Radiology 168: 249–253
[6] AMIN Z, HARRIES S, LEES W, BOWN S (1993) Interstitial tumour photocoagulation. Endosc Surg 1: 224–229
[7] ASCHER PW (1990) Interstitial thermotherapy of brain tumours with Nd-YAG laser under real-time MRI control. SPIE Proc 1200: 242–245
[8] ASCHER PW, JUSTICH E, SCHROETTNER O (1991) A new surgical but less invasive treatment of central brain tumours with Nd-YAG laser under real-time monitoring by MRI. J Clin Laser Med Surg 1: 79–83
[9] BAGGISH MS (1982) Treating viral venerial infections with $CO_2$ laser. J Reprod Med 27: 737–742
[10] BAGGISH MS, DORSEY JH, ANDERSON M (1989) A ten-year experience treating cervical intraepithelial neoplasia with $CO_2$ laser. Am J Obstet Gynaecol 161: 60–80
[11] BAGGISH MS, CHONG SP (1983) Intra-abdominal surgery with the $CO_2$ laser. J Reprod Med 28: 269
[12] BERLIEN HP, MÜLLER G, WALDSCHMIDT J (1986) Correct selection of different types of laser treatment of surface and deep located vessel anomalies. 3rd Congress of the European Laser Association (ELA), Amsterdam, November
[13] BERLIEN HP, WALDSCHMIDT J, MÜLLER G (1988) Laser treatment of cutaneous and deep vessel anomalies. In: Waidelich W. (ed) Laser optoelectronics in medicine. Springer, Berlin Heidelberg New York, pp 526–528
[14] BERLIEN HP, MÜLLER G, WALDSCHMIDT J (1990) Lasers in pediatric surgery. In: Angerpointer X (ed) Progress in pediatric surgery, vol 25. Springer, Berlin Heidelberg New York pp 5–22
[15] BERLIEN HP, PHILIPP C, FUCHS B, ENGEL-MURKE F (1992) Therapeutische Leitlinien zur Laserbehandlung. In: Berlien HP, Müller G (eds) Angewandte Lasermedizin, Lehr- und Handbuch für Praxis und Klinik. Ecomed, Landsberg
[16] BERLIEN HP, PHILIPP C, ENGEL-MURKE F, FUCHS B (1993) Laser in der Gefäßchirurgie. Zb Chir 118: 383–389
[17] BERLIEN HP, PHILIPP C, ROHDE E, FUCHS B, ISMAIL MS (1995) Laser in der Medizin. 112. Kongress der Deutschen Gesellschaft für Chirurgie. Demeter Verlag
[18] BOSMAN S, PHOASK, VAN GEMERT M (1991) Effect of percutaneous interstitial thermal laser ablation on normal liver of pig: sonographic and histopathological correlation. Br J Surg 78: 575
[19] BOTTIROL G, DAL FANTE M, MARCHESINI R, CROC A, ET AL. (1992) Naturally occurring fluorescence of adenocarcinoma, adenoma and non-neoplastic mucosa of human colon. Endoscopic and spectrofluorometry. In: Photdynamic therapy of biomedical lasers. Expectra Medica
[20] BOWN SG (1983) Phototherapy of tumours. World J Surg 7: 700–709
[21] BUYALOS RP (1992) Principles of endoscopic laser surgery. In: Azziz R, Murphy AA (eds) Practical manual of operative laparoscopy and hysteroscopy. Springer, Berlin Heidelberg New York
[22] CREMER H (1992) Gefäßveränderungen im Kindesalter. Kinderarzt 23: 24–26
[23] PHILIPP C, POETKE M, BERLIEN HP (1992) Klinik und Technik der Laserbehandlung angeborener Gefäßerkrankungen. In: Berlien HP, Müller G (eds) Angewandte Lasermedizin, Lehr- und Handbuch für Praxis und Klinik. Ecomed, Landsberg
[24] DACHMANN AH, MCGEHEE J, BEAM T, BURRIS J, POWELL D (1990) US guided percutaneous laser ablation of liver tissue in a chronic pig model. Radiology 176: 129–133
[25] DANIELL JF (1988) Advanced operative laproscopic laser techniques. Laser Med Surg News 16
[26] DANIELL JF, HERBERT CM (1984) Laproscopic salpingostomy utilizing the $CO_2$ laser. Fertil Steril 41: 558–563
[27] DANIELL JF, MILLER W (1989) Polycystive ovaries treated by laparoscopic laser vaporization. Fertil Steril 51: 232
[28] DE SWIET M, CHAMBERLIN G (1992) Anatomy of the female genital tract. In: Basic science in obstetrics and gynaecology. Churchill Livingstone, Edingburgh pp 53–102
[29] DEQUESNE J (1987) Hysteroscopic treatment of uterine bleeding with the Nd-YAG laser. Laser Med Sci 2: 73
[30] DEQUESNE J (1987) Laser hysteroscopic surgery and focal ablation of the endometrium. In: Sutton C (ed) Lasers in gynaecology. Chapman & Hall Medical, London
[31] DIAMOND MP (1992) Assessment of results of laparoscopic laser surgery. In: Sutton C (ed) Lasers in gynaecology. Chapman & Hall Medical, London
[32] DONNEZ J (1992) $CO_2$ laser laparoscopy in endometriosis. In: Sutton C (ed) Lasers in gynaecology. Chapman & Hall Medical, London
[33] DRESSLER C, ISMAIL MS, NOWAK C, HERTER R, SENZ R, HEGEMANN R, RODER B, KOEPPE P, BERLIEN HP (1995) On the pharmcokinetics of the far red-absorbing octa-butyloxy zinc phthalocyanine in Lewis lung carcinoma-bearing mice. In: Denis A, Cortese A (eds) 5th International Photodynamic Association Biennial Meeting. Proc SPIE, vol. 2325: 220–227
[34] DRESSLER C, ISMAIL MS, NOWAK C, HERTER R, SENZ R, HEGEMANN R, RODER B, KOEPPE P, BERLIEN HP (1995) Pharmakokinetische Untersuchungen von Oktbutyloxy-Zinkphthalocyanin am Lewis lung Tumormodell. 5. Wissenschaftswoche Universitätsklinikum Benjamin Franklin der Freien Universität Berlin, pp 56–57
[35] DRESSLER C, STROBELE S, DASKALAKI A, ISMAIL MS, PHILIPP C, BERLIEN HP, LIEBSCH M, SPIELMANN H (1995) Studies on the optimized fluorescence diagnosis of

tumours by comparing 5-ALA-induced xenofluorescence intensities and autofluorescence intensities of a murine tumour-non-tumour tissue system cultivated on the CAM. Joint Meeting of the European Laser Association and the Biomedical Optics Society, 12–16 Sept. Barcelona, Spain. SPIE, vol 128

[36] DRESSLER C, ISMAIL MS, STROBELE S, NOWAK C, HERTER R, SENZ R, HEGEMANN R, KOEPPE P, RODER B, BERLIEN HP (1995) Absorption spectroscopic analysis of the pharmacokinetics of octa-butyloxy zinc phthalocyanine in Lewis lung carcinoma-bearing mice. In: Denis A, Cortese A (eds) 5th. International Photodynamic Association Biennial Meeting. Proc SPIE, vol. 2371: 544–553

[37] FESTE J (1992) Gyaecological microsurgery using the CO2 laser. In: Sutton C (ed) Lasers in gynaecology. Chapman & Hall Medical, London

[38] GALLINAT A, LUEKEN RP, MOLLER CO (1989) Die Neodym-YAG Laser Applikation in der gynaekologischen Endoskopie. LaserBrief 14

[39] GOLDRATH MH (1992) Neodymium-YAG laser hysterscopy: total endometrial ablation. In: Sutton C (ed)

[40] GOLDRATH MH, FULLER TA, SEGAL S (1981) Laser photovaporization of endometrium for the treatment of menorrhagia. Am J Obset Gynaecol 104: 14

[41] GROSS GE (1987) Virusinfektionen der Vulva. Arch Gynecol Obset 241

[42] HERRMANN U (1989) Atlas on laser surgical operations in gynaecology. In: Müller G, Berlien (eds) Advances in laser medicine. Ecomed, Berlin

[43] HOFSTETTER A (1991) Interstitielle Thermokoagulation (ITK) von Prostatatumoren. Lasermedizin 7: 179–183

[44] HUNDT G, BÖHRER H, WALLWIENER D (1992) Anaethesiological consideration concerning the choice of fluid for irrigation during hysteroscopy with Nd:YAG laser. In: Bastert G, Wallwiener D (eds) Laser in gynacology, possibilities and limitations. Springer, Berlin Heidelberg New York, pp 243–244

[45] ISMAIL MS (1996) Untersuchungen zur Pharmakokinetik verschiedener Photosensibilisatoren für die PDT. Promotionsarbeit (MD), Fachgebiet Lasermedizin, Freie Universität Berlin

[46] ISMAIL MS, WEITZEL H, BERLIEN HP (1996) New photosensitizers classification in the field of photodynamic therapy (Abstract). Laser Surg Med 159: 27

[47] ISMAIL MS (1997) Endometriosis: an infertility problem. Arab Ärztebl Eur 3: 58–59

[48] ISMAIL MS, SEROUR G (1997) Principles of constructive surgery for infertile patients. Arab Ärztebl Eur 3: 41–43

[49] ISMAIL MS, DRESSLER C, KOEPPE P, PHILIPP C, RODER B, WEITZEL H, BERLIEN HP (1997) $13^2$-Hydroxy-bacteriopheophorbide, a methyl ester-specific uptake ratio (SUR), in Lewis lung carcinoma. Lasermedizin 13: 41–44

[50] ISMAIL MS, DRESSLER C, KOEPPE P, SENZ R, PHILIPP C, RODER B, WEITZEL H, BERLIEN HP (1997) Pharmacokinetics analysis of octa-butyloxy zinc phthalocyanine in mice bearing Lewis lung carcinoma. J Clin Laser Med Surg 15:4: 157–161

[51] ISMAIL MS, DRESSLER C, KOEPPE P, SENZ R, PHILIPP C, RODER B, WEITZEL H, BERLIEN HP (1997) Specific uptake ratio (SUR) of octa-butyloxy zinc phthalocyanine in Lewis lung carcinoma. Lasermedizin 13: 4

[52] ISMAIL MS, DRESSLER C, WEITZEL H, BERLIEN HP (1997) Pharmacokinetic comparison of $13^2$-hydroxy-bacteriopheophorbide a menthyl ester and octa-butyloxy zinc phthalocyanine in mice bearing Lewis lung carcinoma. Laser 97, München 1997

[53] ISMAIL MS, DRESSLER C, STROBELE S, DASKALAKI A, PHILIPP C, WEITZEL H, BERLIEN HP, LIEBSCH M, SPIELMANN H (1997) Modulation of 5-ALA-induced xenofluorescence intensities of a murine tumour-non tumour tissue system cultivated on the CAM. Laser Med Sci 12: 218–225

[54] ISMAIL MS, DRESSLER C, RODER B, WEITZEL H, BERLIEN HP (1997) A pharmacokinetic evaluation of 132-hydroxy-bacteriopheophorbide, a methyl ester and octa-butyloxy zinc phthalocyanine for future application in photodynamic therapy. BIOS, 97, San Remo, Italy

[55] ISMAIL MS, MULLER U, URBAN P, PHILIPP C, SERGIUS G, WEITZEL H, BERLIEN HP (1997) Improvement of the intrauterine laser delivery for endomerial ablation using the side-fire laser fiber. Lasermedizin 13: 55–59

[56] ISMAIL MS, MULLER U, BERLIEN HP (1997) New laser applicator and new laser power utilization for endometrial destruction. J Coel Chir 24: 39–42

[57] ISMAIL MS, PHILIPP C, WEITZEL H, BERLIEN HP (1997) Minimal laser invasive procedure for treatment of haemangioma in the gynaecological field. Laser 97, München

[58] ISMAIL MS, PHILIPP C, BERLIEN HP (1997) Laser therapy of haemangioma in pediatric and adolescent gynaecology. Indications, techniques and results. Wien

[59] ISMAIL MS, TORSTEN U, EL-SHEHRRY A, PHILIPP C, WEITZEL H, BERLIEN HP (1997) Laser-induced thermotherapy (LITT) for treatment of local recurrences in patients with breast cancer. BIOS, 97, San Remo, Italy

[60] ISMAIL MS, TORSTEN U, AFIFY YM, SEOUR G, WEITZEL H, BERLIEN HP (1997) Mechanical versus thermal endometrial ablation: results. Min Invas Med 8: 1–2

[61] ISMAIL MS, TORSTEN U, PHILIPP C, WEITZEL H, BERLIEN HP (1997) Laser-induced thermotherapy (LITT) for treatment of local recurrences in patients with breast cancer. Laser 97, München

[62] ISMAIL MS, WEITZEL H, BERLIEN HP (1997) Applied photosensitizers classification for clinical photodynamic therapy. Laser 97, München

[63] ISMAIL MS, WEITZEL H, BERLIEN HP (1997) Improvement of the intrauterine laser delivery for endometrial ablation using the side-fire laser fiber. Laser 97, München

[64] ISMAIL MS (1998) Development and characterization of new photosensitizers with the application of the photodynamic process in the gynaecological field. Habilitation Thesis, Frauenheilkunde und Geburtshilfe, Fachgebiet Humanmedizin, Freie Universität

[65] ISMAIL MS, DRESSLER C, KOEPPE P, SENZ R, PHILIPP C, RODER B, WEITZEL H, BERLIEN HP (1998) Specific uptake ratio (SUR) of octa-butyloxy zinc phthalocyanine $13^2$-hydroxy-bacteriopheophorbide a methyl ester in Lewis lung carcinoma. Lasermedizin 13: 1

[66] ISMAIL MS, DRESSLER C, KOEPPE P, RODER B, NOWAK C, LEMM M, BERLIEN HP (1998) $13^2$-hydroxy-bacteriopheophorbide, a methyl ester pharmacokinetics in Lewis lung carcinoma. Laser Med Sci 13: 78–81

[67] ISMAIL MS, PHILIPP C, TORSTEN U, WEITZEL H, BERLIEN HP (1998) Laser-induced thermotherapy (LITT) for retreatment of locally advanced recurrences of brest cancer. Laser Med Sci 14: 136–142

[68] ISMAIL MS, TORSTEN U, PHILIPP C, WEITZEL H, BERLIEN HP (1998) Color-coded duplex sonography (CCDS, simple imaging procedure for monitoring of laser-induced thermotherapy (LITT) in treatment of locally recurrences breast cancer. J Gynaecol Surg 14: 2: 65–73

[69] ISMAIL MS, TORSTEN U, SEROUR G, WEITZEL H, BERLIEN HP (1998) Is endometrial ablation a safe contraceptive method? Pregnancy following endometrial ablation. Europ J Contracept Reproduct health Care 3: 99–102

[70] ISMAIL MS, TORSTEN U, DRESSLER C, DIEDERICHS JE, HUSKE S, WEITZEL H, BERLIEN HP (1999) Photodynamic therapy of malignant ovarian tumours cultivated on CAM. Laser Med Sci 14: 91–96

[71] JOBSON V, HOMESLEY H (1983) Treatment of vaginal intraepithelial neoplasia with the $CO_2$ laser. Obstet Gynaecol 62: 90–93

[72] JOPPICH I, SCHIELE U (1973) Die Indikation zur operativen Behandlung von Hämangiomen im Säuglingsalter. Kinderarzt 19: 619–625

[73] KISTENER RW (1975) Endometriosis in infertility. In: Kistener RW, Behrmann SJ (eds) Progress in infertility. Brown & Co, Boston, pp. 345–364

[74] MACK M, VOGEL T, MÜLLER P, SCHOLZ W, WEINHOLD N, PHILIPP C ET AL (1996) Laserinduzierte Thermotherapie (LITT) von Kopf-Halstumoren. Fortschr Röntgenstr 164–172

[75] MARLOW JL (1987) Lasers in intra-abdominal gynaecologic surgery. In: Fuller TA (ed) Surgical lasers, clinic guide 104–108

[76] MCINDOE G, ROBSON MS, TIDY J ET AL (1989) Laser excision rather than vaporization: the treatment of choice for cervical intraepithelial neoplasia. Obstet Gynaecol 64: 451–458

[77] MOELLER P (1989) Externe Anwendung des Nd:YAG Laser: Warzen, Marsupiliasation, Malformation, benigne Portioveränderungen. Fortbildungsveranstaltung mit Workshop, Lasers in der Gynaekologie, Hamburg, Germany

[78] MULLIKEN J (1988) Classification of vascular birthmarks. In: Mulliken J, Young A (eds) Vascular birthmarks, hemangioma and malformations. W.B. Saunders

[79] MUMTAZ H,HALL-CRAGGS M, WOTHETSPOON A, PALEY M ET AL. (1996) Laser therapy for breast cancer: MR imaging and histopathologic correlation. Radiology, 200: 651–658

[80] NEZHAT C, CROWGEY S, GARRISON CP (1986) Surgical treatment of endometriosis via laser laproscopy. Fertil Steril 45: 778

[81] PHILIPP C, ROHDE E, BERLIEN HP (1995) Treatment of congenital vascular disorders (CVD) with laser-induced thermotherapy (LITT). In: Müller G, Roggan A (eds) Laser-induced interstitial thermotherapy. SPIE proc 267–278

[82] RIED R (1985) Superficial laser vulvectomy I. The efficacy of extended superficial ablation for refractory and very extensive condylomas. Am J Obstet Gynaecol 151: 1047–1052

[83] RIED R (1985) Superficial laser vulvectomy III. A new surgical technique for appendage-conserving ablation of refractory condylomas and vulvar intraepithelial neoplasia. Am J Obstet Gynaecol 151: 504–509

[84] RIED R (1985) Superficial laser vulvectomy II. The anatomic and biophysical principles permitting accurate control over the depth of dermal destruction with the carbondioxide laser. Am J Obstet Gynaecol 152: 261–271

[85] RIED R (1989) Laser therapy of human papillomavirus infections. In: Keye W (ed) Laser surgery in gynaecology and obstetrics. Yearbook Publisher, Chicago, pp 46–99

[86] RIED R (1992) Laser surgery on the vulva. In: Sutton Lasers in gynaecology. Chapman & Hall Medical, London

[87] RIED R, DORSEY J (1990) Laser therapy for vulval and vaginal intraepithelial neoplasia. In: Coppleson M (ed) Gynaecologic oncology. Fundamental principles and clinical practice. Churchill Livingstone, Edingburgh

[88] ROHDE E, PHILIPP C, BERLIEN HP (1995) Monitoring of interstitial laser induced thermotherapy (LITT) with colour-coded duplexsonography (CCDS). In: Müller g, roggan a (eds) laser-induced interstitial thermotherapy. SPIE Proc 267–278

[89] SAUNDERS N, SHARP F, JORDAN J (1992) Laser treatment of cervical and vaginal intraepithelial neoplasia. In: Sutton C (ed) Lasers in Gynaecology. Chapman & Hall Medical, London

[90] SHUH M, NSEYO U, POTTER W, DAO T, DOUGHERTY T (1987) Photodynamic therapy for palliation of locally recurrent breast carcinoma. J Clin Oncol 5: 1766–1760

[91] STEPP H, BÄUMGÄRTNER R, BEYER W ET AL. (1995) Fluorescence imaging and spectroscopy of ALA-induced protoporphyrin IX preferentially accumulated in tumour tissue. SPIE, Proc Optical Biophysics 2627, 23–32

[92] STROBELE S, ISMAIL MS, DRESSLER C, KOEPPE P, BERLIEN HP (1995) Regressionanalyse zur Pharmacokinetik des 8-a-ZnPc am lewis lung Carcinom. Lasermedizin 11: 92

[93] SUTTON CJ (1989) Laser laproscopic uterine nerve ablation. In: Donnez J (ed) Operative laser laparoscopy and hystroscopy. Nauwelaerts Publishers, Belgium, pp 43–53

[94] TADIER Y, TROMBERG B, WYSS P, STEINER R, MADSEN S,SVAASAND L, VILLALON V, BERNS M (1995) Photomedicine of the female genital tract. In: Asch R, Studd J (eds) Progress in reproductive medicine. The Parthenon Publishing Group, London, pp 139–148

[95] ZOLTÁN J (1988) Hämangiome-Lymphangiome. Handbuch der Plastischen Chirurgie, Vol II, Spezielle Plastische Chirurgie. De Gruyter, Berlin

[96] TADIER Y, TROMBERG B, WYSS P, STEINER R, MADSEN S,SVAASAND L, VILLALON V, BERNS M (1995) Photomedicine of the female genital tract. In: Asch R, Studd J (eds) Progress in reproductive meParthenon Publishing Group, London, pp 139–148

[97] ZOLTÁN J (1988) Hämangiome-Lymphangiome. Handbuch der Plastischen Chirurgie, Vol II, Spezielle Plastische Chirurgie. De Gruyter, Berlin

# III-15
# Laser Applications in Diagnostics

P. Urban

**Contents**

## Introduction

Lasers have been a part of medicine and surgery since the late 1960s. In the past 10 years, however, there has been growing interest in using lasers as diagnostic devices, an area of research that has been termed optical diagnostics. Optical diagnostic techniques seek to provide diagnostic information about tissue by using light in a probing, yet non-destructive fashion (Nishioka).

## Infrared Diaphanoscopy (IRD)

Diaphanoscopy has long been a recognized procedure for the screening of thin tissue layers by means of visible light for the presentation of lower-level structures. Already in 1929 and 1931, Cutler [1, 2] presented the method in diagnosis of pathological mamma alterations. Due to the great improvements in X-ray technique and sonography, however, the process found broader application only in sections of medical diagnosis, for example examination of the scrotum. In the 1980s models were developed calculating the light propagation in tissues [3–7], which led to a deeper understanding of the tissue optics, which, with the simultaneously improved sensor technique, allowed further development in the field of optical tomography. The basis of all considerations is the fact that the spreading of light in a tissue is determined by absorption and scattering, and that pathological modifications of the tissue biology here lead to a particularly strong change in the scattering qualities [3].

While at the beginning, the efforts mainly consisted of implementing morphological illustrations under avoidance of a X-ray radiation exposure, newer findings now provide information about the functional state of the tissue, as inflammations modify the density in the tissue, with correlating changes of the scattering qualities [8].

At the outset of the method, strong sources of light in the visible range were applied for transillumination, so that the assessment had to occur in a darkened room. With the evolution of infrared diaphanoscopy (IRD), which uses light within a so-called optical window at a wave band of 650 to 1100 nm whereby strength of the light emergent from the body is detected by means of a NIR-sensitive CCD (charge-coupled device) camera and made visible after amplification on a monitor, these investigations have now become feasible with room light.

Besides applications in ophthalmology, gynecology and urology, IRD find its broadest field of application in diseases of the ear, nose and throat [9], where the method is used mainly for diagnosis of paranasal sinuses. In the case to external IRD, the applicator is placed in the pharynx below the hard palate. The light penetrates the nose sinuses and is shown after inflammation-dependent scattering and absorption as an infra-orbital sickle-shaped bright area on the monitor. Inflammation processes induce increased scattering to decreased transmission of the emergent light, recognizable as a decreased intensity of the optical image. Thus, this procedure is especially suitable for the observation of the course of paranasal sinus inflammation. A further indication field in ENT medicine is endoscopic IRD to facilitate demonstration of natural entrance ways to the paranasal sinuses in order to avoid insufficient planning during operative proceedings [10]. Here, the applicator is placed from outside on the frontal sinus or maxillary sinus; the infrared light can be seen after scattering in the paranasal sinus by an endoscope in the nasal cavity in the area of the natural connection between nasal carity and paranasal sinus. For this, as additional technical equipment, an endoscopic camera with beam splitter for the changeover between visible light for endoscopic orientation and detection of NIR light is necessary.

A further interesting area of application for IRD is diagnosis of early-stage inflammatory rheumatic degenerative arthrosis [3]. Using light at the wavelengths 675 and 900 nm, a clear rise in tissue absorption can already be verified at early inflammatory changes. The articular capsule and synovia, marked as a reduction of local flare distribution in the field of a diseased interphalangeal joint [11, 12].

Besides the above described methods of time-integral cw-IRD, also a time-dependent and a strength-modulated procedure is possible [3]. Here, after very short light pulses of a few ps, the different runs of differently scattered photons passing a tissue are measured or, after irradiation of a continuously amplitude-modulated light, the amplitude and phase displacement of the infiltrating light can be determined. However, these measurements require considerable technical and methodical expenditure and thus have not, up to now found application in clinical routine.

## Optical Coherence Tomography (OCT)

This method was developed under the keyword optical biopsy with the aim to receive diagnostically relevant pictures of the micropattern of a tissue without removing this by means of excision and further processing. The procedure is analogous to the ultrasonic or radar technique [13], and allows the presentation of the tissue structure up to cellular structures in a sectional view by measuring the backscattered and reverberated light. Unlike ultrasound, OCT is, however, feasible in the non-contact mode and shows considerably better results. The illustration results from a focused ray of light, that is emitted from a laser diode onto the tissue to be viewed, measuring the reverberation delay period arising with reflection at the tissue structures until the light hits a detector. The sectional view results from a scanner-like transverse shift of the ray of light above the tissue. The spatial solution and representation of the intensity distribution of the light becomes possible, since the light emerging from the laser is first divided up in a beam splitter in equal parts, one half being guided onto a reference mirror. According to the reflection also of this light component onto the detector, by comparison of the different running times and strengths of the two light components, a spatial picture is constructed by a computer programme. Due to this optical procedure, OCT characterizes measurements at the eye especially well, but in not translucent tissues reasonable applications also result since this is relatively permeable for light in the near-ultraviolet range, so that at optimized wavelengths at 1300 nm, presentation levels of 1–2 mm are possible for most tissues. The axial solution amounts typically to 10–20 µm and can be increased by use of ultrashort pulsated laser to 2–4 µm [14]. The obtained image appears either in grey scales or as a false-colour picture similar to a histological section. Here, the solution is so good that fasciculus of peripheral nerves and the elastic membrane of microvessels could be shown [15].

## Laser Doppler Perfusion Imaging (LDPI)

For more than 15 years, this method has been described to investigate the local superficial blood flow in the dermis and the upper subdermal region [16, 17]. The principle in this measurement uses the frequency shift of scattered light at the surface of moving particles such as red blood cells following the Doppler effect, also known from the colour-coded duplex sonography. Such a device emits photons of monochromatic light, which undergo frequency shifts as a result of interaction with flowing erythrocytes in the detected tissue volume. Most of the backscattered

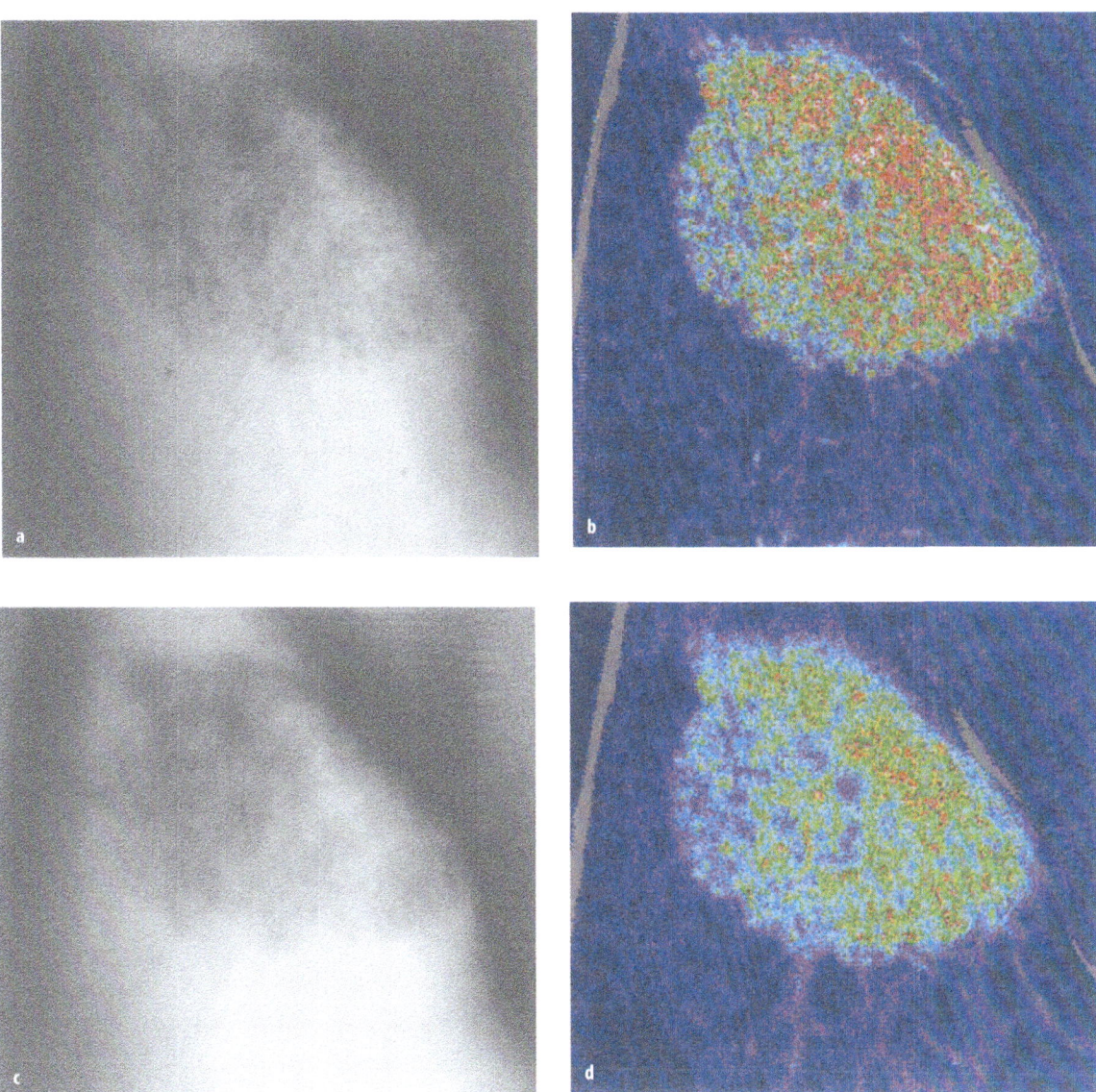

**Fig. 1a–d.** LDPI of a vascular malformation of the right hip. Superficial pathological vessels; above 633 nm; below 780 nm

light, which is measured in photodiodes, has been scattered by static tissue, which does not impart a Doppler shift. A small proportion of the detected light has been scattered by moving blood and is Doppler-shifted, where it is converted to a voltage output proportional to the velocity of the moving red cells and their concentration [18]. The term commonly used to describe blood flow measurement by the laser Doppler technique is flux, a quantity proportional to the product of the average speed of the blood cells and their concentration, expressed in arbitrary perfusion units.

In former laser Doppler perfusion measurements, only point measurement in a restricted area of approximately 1 mm was possible, although skin blood flow is known to vary greatly between adjacent sites [19]. In addition, direct contact of the glass fibre with the skin was necessary, with the risk of blood flow being altered by the applicated probe. To solve these problems, a continuously scanning beam technique was developed and first described by Essex and Byrne in 1991 [20]. In this device, a low-intensity laser light beam of less than 2 mW is scanned over the surface by means of mirrors that are moved by microprocessor-controlled step-motors and that also serve to return the backscattered light to the detector. The great advantage is that not only point measurements are feasible, but also the sampling of blood flow over a wide area without direct contact with the tissue. With special computer programmes, data from the

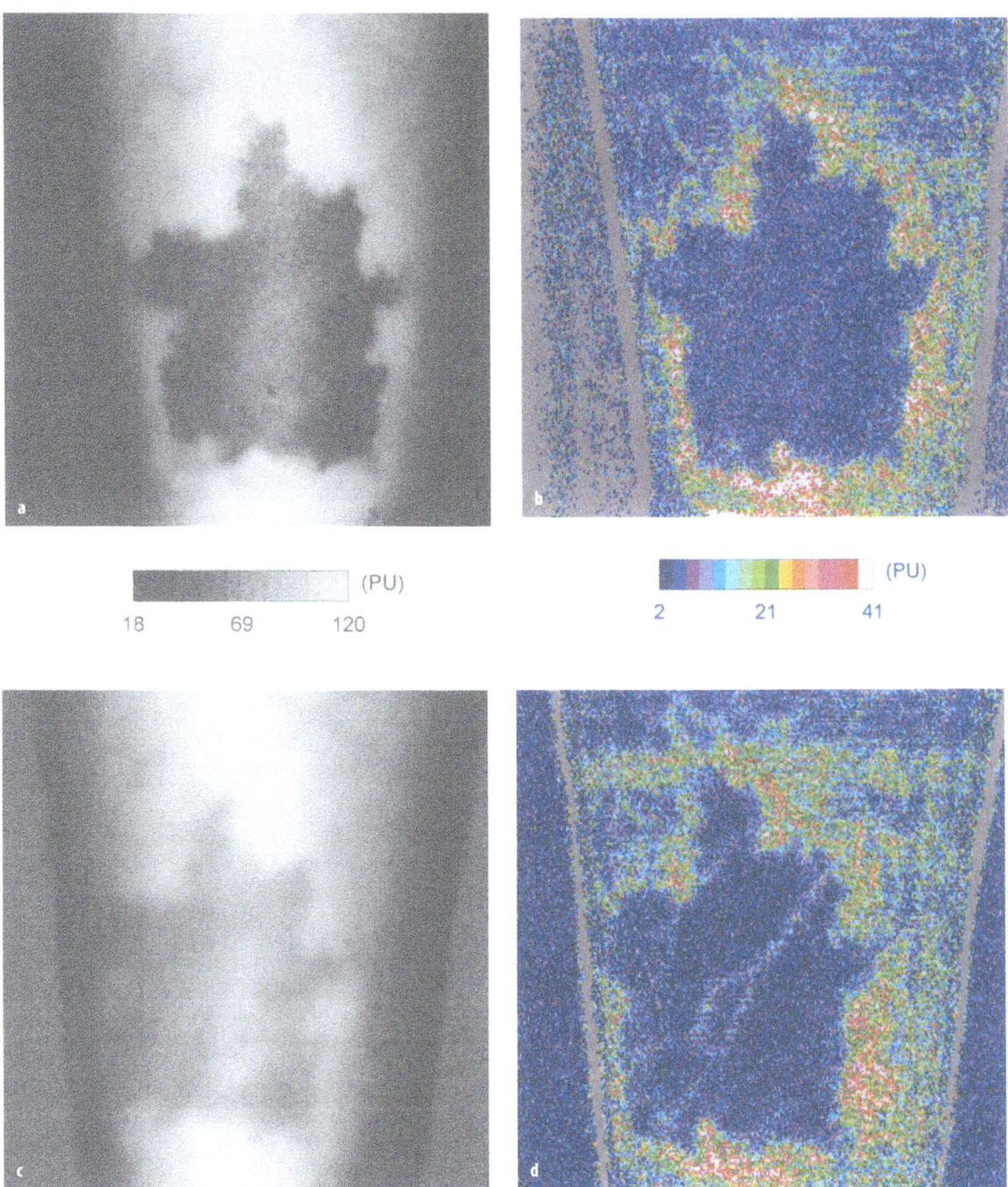

**Fig. 2a–d.** LDPI in a patient with a.v. malformation of the thigh. Demarcation of a superficial necrosis, status 3 days after embolization; above 633 nm; below 780 nm

Doppler-shifted light are corrected for noise and beam angle and displayed as a colour-coded LDI flux image with additional LDI photorealistic black and white images, using the intensity of backscattered light. A further option is the non-contact monitoring of flux and concentration of the moving particles, sampled at one spot in a curve of temporal resolution.

The problem of determining the depth of a detected flux signal is still unsolved [21].

Most systems use a red HeNe laser with a wavelength of 633 nm for the more superficial layers. This laser penetrates into the skin from 600 to 1500 μm, depending on the characteristics of the tissue bed; this means that it penetrates deeply into the dermis and even reaches underlying fatty subcutaneous tissue. For further penetration, diode laser of 780 or 830 nm in the near-infrared are available. We found in our own investigations, correlated with findings from colour-coded duplex sonography, that even vessels lying 3 mm beneath the skin surface could be detected [22] (Fig. 1).

The method of LPDI has found applicability in all fields of medical routine where information on microcirculation is needed (Fig. 2): burn depth assessment [23, 24], investigation of autologous tissue transplants [25] and island flaps in reconstructive surgery [18, 26], follow-up in laser treatment of vascular anomalies, e.g. portwine stains [27], monitoring of local effects under photodynamic laser therapy with topical or systemic sensibilization [29], investigation of inflammatory diseases in dermatology [29, 30], quantification of symptoms in various kinds of vasculitis, vascular damage in arterial hypertension [31] and Raynaud phenomenon, evaluation of microcirculatory effects of numerous drugs [32], examination in patients with arterial occlusive disease [33], and experimental investigations in neurosurgery [34, 35].

## Fluorescence Diagnostics

In addition to the above measurements of transmission, reflection and scattering, the qualitative and quantitative measurement of fluorescence is a further field for laser-supported diagnostics. Next to a quantitative recording of specific metabolism products, distinguishing their function is also possible, since, for example, the coenzyme nicotinamide-adenine-dinucleotide (NAD) is stimulated in the near-ultraviolet range only in its reduced state as NADH. Via a change in the redox condition with different fluorescence signals, defects in the mitochondrial respiratory chain are recognizable [36].

In the autofluorescence measurement of endogenous fluorophores, besides definition of the status of the cell metabolism, the presentation of the local metabolism distribution is possible. Main representatives of this group are proteins such as tryptophane and tyrosine in an emission spectrum of 300–390 nm, the coenzymes NADH and NADPH at 440–470 nm, extracellular components like collagen and elastin at 380–470 nm and, in particular, porphyrins at 590–700 nm [37, 38]. The clinically most important field of application of xenofluorescence contains, in particular, porphyrins, phthalocyanines and pheophorbides, which reveal not only an exact definition of localization and spread by photodynamic diagnosis (PDD) in the case of malignant tumours and their preliminary stages, but in the same application step also allows

treatment by means of photodynamic therapy (PDT) [39].

The basis for diagnostic and also therapeutic application is a time- and concentration-dependent enrichment of protoporphyrin IX (PPIX) in the pathologically changed tissue, since there is an impairment of the normal metabolism of haemsynthesis. PPIX can be provided intravenously, be given orally in the form of the precursor delta-aminolaevulic-acid (ALA) or be applied locally. Both experimental [40] and clinical [41] studies could show that an unambiguous differentiation between healthy and tumorous infiltrated tissue is possible by a decrease of the normal tissue autofluorescence in the range around 490–500 nm and an increase in the induced fluorescence at 635 nm.

We ourselves utilize in our team the laser-induced endoscopic autofluorescence diagnostics system (LENA) developed by the Laser- und Medizin-Technologie gGmbH, Berlin (LMTB) [42], which allows point measurements of both PPIX and NADH fluorescence for diagnostic tissue differentiation. The system consists of two lasers, a short-pulsed nitrogen laser of 337 nm and an HeNe laser of 632 nm, and allows simultaneous measurements up to depths of 200 μm, which corresponds to approximately ten layers of cell tissue. The emission of the stimulating laser light as well as the measurement of the resultant fluorescence occurs in contact mode via a multifibre applicator, so that in principle also an endoscopic use is possible. On the one hand, by correlating the measured values for NADH as well as the PPIX channel, a delimitation of different tissue states is possible (healthy tissue: tumour edge : tumour centre), on the other hand, recording the PPIX fluorescence after topical ALA application allows the monitoring of precancerous stages and skin tumours during a photodynamic therapy (PDT).

We could show that after correct ALA application before PDT, in comparison with the zero values, the PPIX concentration in the centre of the lesions rose four to ten fold, whereas we found a drop down to the level before ALA application after sufficient PDT [43] (Fig. 3). Other teams described further possible applications of the method. Thus, after instilling ALA solution into the urinary bladder, under stimulation by a krypton laser at 413 nm, the fluorescence diagnostic of tumours is possible [44].

The first commercial system for endoscopic autofluorescence diagnosis in the tracheobronchial tree (LIFE) was introduced already in 1993, and uses a helium cadmium laser as light source [45]. In addition to these laser-supported procedures of fluorescence detection, for reasons of cost-saving and practicability in surface diagnostic, numerous procedures have

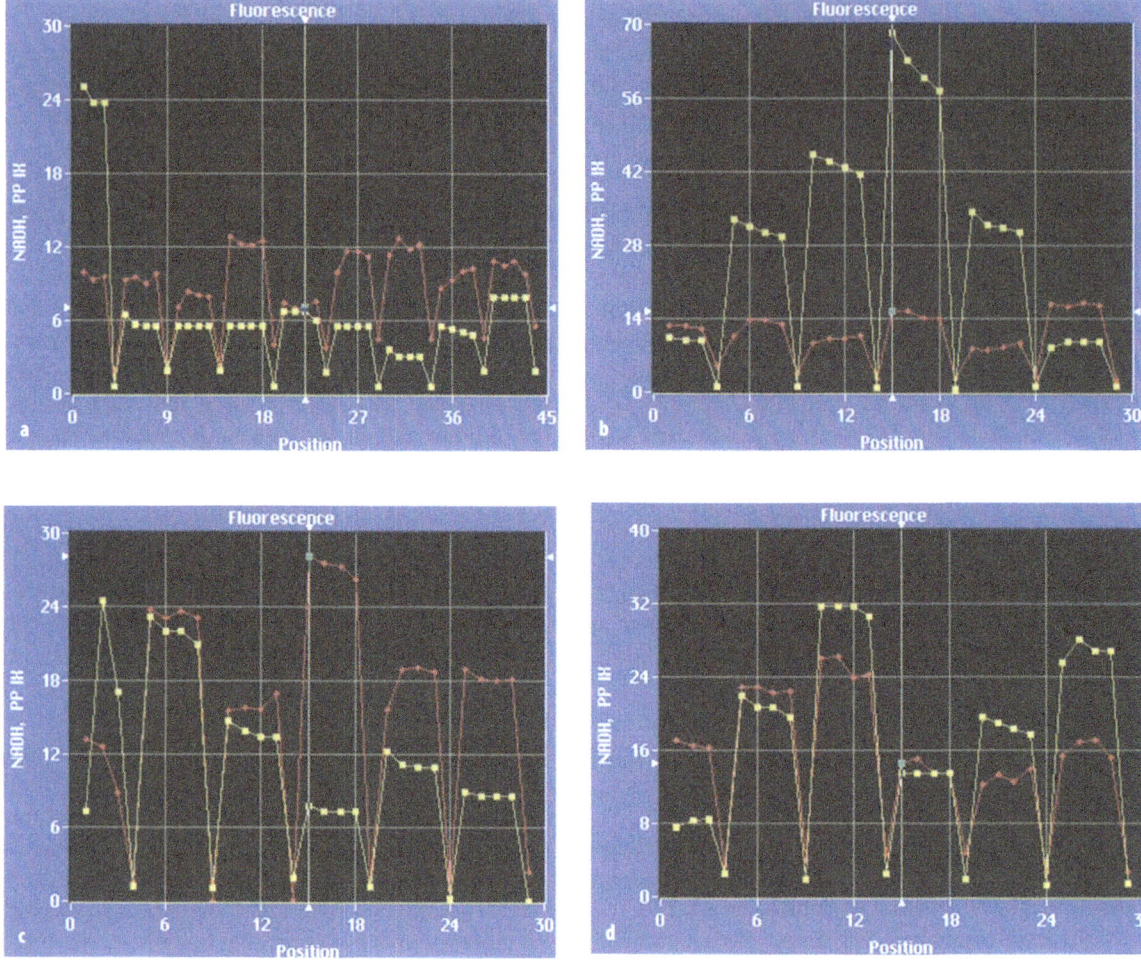

**Fig. 3a–d.** LENA – PDT monitoring. PPIX point-measurement (yellow squares) in a patient with BCC. **a** Before ALA application. **b** 4 h after ALA application, before PDT. **c** Directly after PDT. **d** 1.5 h after PDT

been established that use incoherent sources of light in the form of specific lamp systems for the stimulation of fluorescence. Under application of a Wood's rays lamp at 370–405 nm, neoplastic skin modifications show typical carmine porphyrin fluorescence after corresponding ALA application, so that calculated biopsies or connected therapy forms (PDT, laser therapy, surgical excision) are possible in confirmed affected localizations [46].

More extensive procedures use the D-light system (Karl Storz GmbH, Tuttlingen) where by means of a xenon high-pressure lamp and a bandpass filter, light at 380–430 nm is generated and transfered to a liquid light guide with a lens system, for the reason of parallelism.

Besides direct examination, the emitted fluorescence can be separated from the stimulation light by an edge filter and recorded by a CCD camera; after digital transformation presentation is then possible on a PC monitor. In addition, the corresponding clini-

cal picture can be documented by an RGB camera connected in parallel [47].

In the field of clinical and experimental optical biopsy we use in our group the spectroscopic imaging system (SIMAS) from the Laser- und Medizin-Technologie gGmbH Berlin (LMTB). Here, the light of a mercury vapour lamp with changeable emission bands of about 365 and 420 nm is inserted into a conventional operation microscope (Zeiss). Recording and measurement of the fluorescence occurs with a CCD camera; due to a preconnected filter changer, different measuring bands for different questions are recorded automatically within the scope of one measurement: as an expression of the cellular metabolism in addition to the presentation of bound NADH at 442 nm, the detection of free NADH at 458 nm is recorded, the PPIX fluorescence can be measured separately at both 620 nm and 690 nm, structural information is received through the collagen fluorescence band at 405 nm, with indications of areas of hypervas-

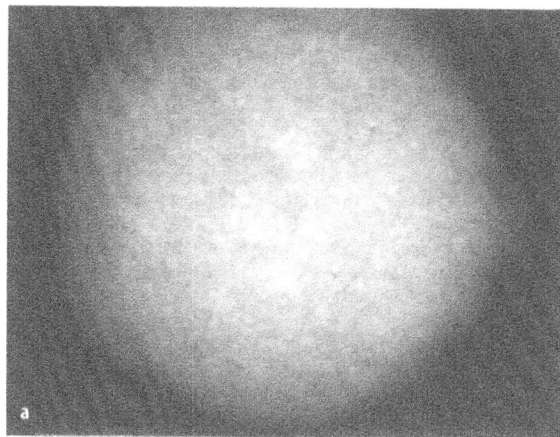

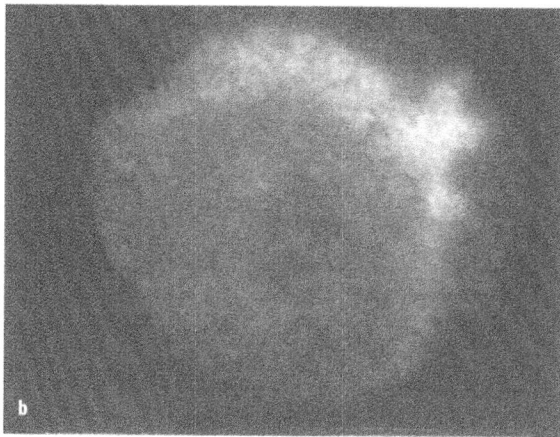

**Fig. 4.** SIMAS – PDT monitoring. PPIX fluorescence in a patient with BCC. **a** Before PDT. **b** After PDT; crescent-shaped remaining fluorescence in the upper margin outside the field of laser irradiation

cularization through haemoglobin absorption at 550 nm. On this occasion, the different fluorescences can either be represented separately, or be overlaid as false-colour picture to a colour image recorded by an RGB camera in parallel. In particular in tumour borders a more accurate depiction of possible infiltrations and tumour islands otherwise concealed is possible [48]. Besides the representation of lesions and tissue differentiation, this system is used also in monitoring during PDT application [43] (Fig. 4).

A further range of application of coherent light is fluorescence spectroscopy, by which under irradiation with defined wavelengths, specific spectra can be recorded. On the one hand, this method characterizes in vivo the differentiation between normal and premalignant or malignant changed tissue, for example in cervical diagnostics [49]. On the other hand, functional investigations are possible on a cellular level, as, for example, mitochondrial NADH on stimulation with light of 340 nm shows a fluorescence band with a

maximum at 470 nm; NAD, however, has no absorption band beyond 300 nm and thus has no fluorescence in the above-mentioned wavelength range [50].

In addition to investigation of the effects of cytotoxins on the respiratory chain, the progress of mitochondrial products of metabolism can be followed generally [51] and further findings can be won about porphyrins and phthalocyanins used in the application of PDT [52, 53]. Further refinements of the method by time-resolved procedures are being tested.

## References

[1] CUTLER M (1929) Transillumination as an aid in the diagnosis of breast lesions, with special reference to its values in cases of bleeding nipple. Surg Gynecol Obstet 48: 721–729
[2] CUTLER M (1931) Transillumination of the breast. Ann Surg 93: 223–224
[3] PRAPAVAT V (1997) Anwendung lichtoptischer Verfahren: Diagnose früher entzündlich-rheumatischer Gelenkerkrankungen. In: BERLIEN HP, MÜLLER G (eds) Angewandte Lasermedizin. Ecomed, Landsberg, 14 Erg.Lfg. 12/97, III-2.3: 5–16
[4] ISHIMARU A (1978) Wave propagation and scattering in random media. Vol 1: Single scattering and transport theory. Academic Press, New York
[5] ISHIMARU A (1978) Wave propagation and scattering in random media. Vol 2: Multiple scattering, turbulence, rough surfaces, and remote sensing. Academic Press, New York
[6] STAR W, MARIJNISSEN JPA, VAN GEMERT MJC (1988) Light dosimetry in optical phantoms and in tissue: I. Multiple flux and transport theory. Phys Med Biol 33 (4): 437–454
[7] WILSON BC, SEVICK EM, PATTERSON MS, CHANCA B (1992) Time-dependent optical spectoscopy and imaging for biological applications. Proc IEEE 80: 918–930
[8] MASTERS BR (1993) Functional imaging of cells and tissues: NAD(P)H and flavoprotein redox imaging. In: MÜLLER G, CHANCE B (eds) Medical optical tomography – functional imaging and monitoring. SPIE Inst Ser IS11, SPIE-Press, Washington, pp 555–576
[9] BEUTHAN J, MÜLLER G (1992) Infrarot-Diaphanoskopie – Renaissance einer vergessenen Methode. Med Tech 3 (1): 13–17
[10] MESECKE- VON RHEINBABEN I, PRAPARVAT V, LINNARZ M (1995) In vitro studies of the specifications for methological and technical equipment requirements of endoscopic infrared diaphanoscopy. Biomed Tech 40: 255–262
[11] BEUTHAN J, PRAPAVAT V, NABER RD (1996) Diagnosis of inflammatory rheumatic diseases with photon density waves. SPIE 2676: 43–53
[12] PRAVAVAT V, ET AL (1998) Evaluation of early rheumatic disorders in PIP joints using a cw-transillumination method: first clinical results. SPIE 3196: 71–78
[13] FUJIMOTO JG, TEARNEY G, BOUMA B (1995) Biomedical imaging with optical coherence tomography. Proc DGLITT, München, pp 573–579
[14] HEE MR, JZATT JA, AQNAON EA (1995) Optical coherence tomography of the human retina. Arch Ophthalmol 113: 325–332
[15] BREZINSKI ME, TEARNEY GJ, WEISSMANN NJ (1997) Optical biopsy with optical coherence tomography: feasibility for surgical diagnostics. J Surg Res 71 (1): 32–40
[16] MARKS NJ, TRACHY RE, CUMMINGS CW (1984) Dynamic variations in blood flow as measured by laser-Doppler velocimetry: a study in rat skin flaps. Plast Reconstr Surg 73: 804–810

[17] FISCHER JC, PARKER PM, SHAW WW (1985) Laser-Doppler flowmeter measurements of skin perfusion changes associated with arterial and venous compromise in the cutaneous island flap. Microsurgery 6: 238–243

[18] ARNOLD F, HE CF, JIA CY, CHERRY GW (1995) Perfusion imaging of skin island flap blood flow by a scanning laser-Doppler technique. Br J Plast Surg 48: 280–287

[19] TENLAND T, SALERUD G, NILSSON GE (1983) Spatial and temporal variations in human blood flow. Int J Microcirc Clin Exp 2: 81–90

[20] ESSEX TJH, BYRNE PO (1991) A Laser-Doppler scanner for imaging blood flow in skin. J Biomed Eng 13: 189–194

[21] JAKOBSSON A (1992) Sampling depth in laser-Doppler flowmetry. Linkoping studies in science and technology. Thesis 307. Linkoping University

[22] URBAN P, PHILIPP C, WEINBERG l, BERLIEN HP (1997) Bildgebende Verfahren in der Beurteilung Vaskulärer Läsionen. Proc 13th Int Congr Lasermed, München, p 5

[23] NIAZI ZBM, ESSEX TJH, PAPINI R (1993) New laser doppler scanner, a valuable adjunct in burn depth assessment. Burns 19 (6): 485–489

[24] ATILES L (1995) Laser Doppler flowmetry in burn wounds. J Burn Care Rehabil 16 (4): 388–393

[25] CLINTON MS (1991) Establishment of normal ranges of laser-Doppler blood flow in autologous tissue transplantats. Plast Reconstr Surg 73: 804–810

[26] TUOMINEN HP, ASKO-SELJAVAARA S, SVARTLING NE, HÄRMÄ MA (1992) Cutaneous blood flow in the TRAM flap. Br J Plast Surg 45: 261–269

[27] URBAN P, GROSSEWINKELMANN A, PHILIPP C, BERLIEN HP (1998) Vascular lesion imaging with non-radiological, computer-aided methods. Proc 13th Int Workshop ISSVA, Berlin

[28] URBAN P, GROSSEWINKELMANN A, ALGERMISSON B (1999) Monitoring von ALA-PDT Effekten durch Infrarot Thermographie, Laser-Doppler Flussmessung und zwei Fluoreszenz-Detektions-Systeme. Laser Med 14 (3): 78

[29] QUINN AG, MCLELLARD J, ESSEX T, FARR PM (1991) Measurement of cutaneous inflammatory reactions using a scanning laser-Doppler velocimeter. Br Dermatol 125: 30–37

[30] SPEIGHT EL, ESSEX T, FARR PM (1993) The study of plaques of psoriasis using a scanning laser-Doppler velocimeter. Br Dermatol 128: 519–524

[31] NOVO S, FAILLA G, LIQUORI M (1991) Vascular damage in arterial hypertension: its noninvasive assessment. Cardiologia 36: 323–337

[32] BELCARO G, RULO A, CANDIANI C (1989) Evaluation of the microcirculatory effects of Venoruton in patients with chronic venous hypertension by laserdoppler flowmetry, transkutaneous $PO_2$ and $PCO_2$ measurements, leg volumery and ambulatory venous pressure measurements. Vasa 18 (2): 146–151

[33] RANFT J (1988) The status of laser Doppler examination in patients with arterial occlusive disease. Herz 13 (6): 382–391

[34] EBEL H (1996) Vasomotion, regional cerebral blood flow and intracranial pressure after induced subarachnoid haemorrhage in rats. Zentralbl Neurochir 57 (3): 150–155

[35] ARBIT E, DIRESTA GR (1996) Application of laser Doppler flowmetry in neurosurgery. Neurosurg Clin North Am 7 (4): 741–748

[36] SCHNECKENBURGER H, GSCHWEND MH, KÖNIG K (1996) Laser in der Diagnostik am Beispiel der Fluoreszenz-Diagnostik und Laser Mikroskopie. In: REIDENBACH HD, Lasertechnologien und Lasermedizin, MÜLLER G, BERLIEN HP (eds), Fortschritte in der Lasermedizin 13, ecomed, Landsberg, pp 93–96

[37] ANDERSSON-ENGELS S, JOHANSSON J, SRANBERG K (1991) Fluorescence imaging and point measurement of tissue: applications to the dermarcation of malignant tumours and atherosclerotic lesions from normal tissue. Photochem Photobiol 53: 807–814

[38] KÖNIG K, SCHNECKENBURGER H (1994) Laser-induced autofluorescence for medical diagnosis. J Fluoresc 4: 17–40

[39] URBAN P, ALGERMISSEN B, PHILIPP C, WEINBERG L, BERLIEN HP (1998) Monitoring of ALA-PDT effects using infrared thermography, laser Doppler perfusion imaging and a fluorescence detection system (SIMAS). Proc 7th Biennial Congr Photodynamic Association, Nantes, France

[40] JOHANSSON J, BERG R, SVANBERG K, SVANBERG S (1997) Laser-induced fluorescence studies of normal and malignant tumour tissue of rat following intravenous injection of delta-amino levulinic acid. Lasers Surg Med 20(3): 272–279

[41] WANG I, CLEMENTE LP, PRATAS RM (1998) Evaluation of early rheumatic disorders in PIP joints using a cw-transillumination method: first clinical results. SPIE 3196: 71–78

[42] BOCHER T, BENTHAN J, SCHELLER M (1995) Combined quantitative and qualitative two-channel optical biopsy technique for discrimination of tumour borders. SPIE Proc 2627: 118–124

[43] URBAN P, ALGERMISSEN B, PHILIPP C, WEINBERG L, BERLIEN HP (2000) Monitoring of ALA-PDT effects using infrared thermography, laser-Doppler perfusion imaging and two fluorescence detection systems. Lasers Surg Med Suppl 12: 42

[44] HEIL P, STOCKER S, SROKA R, BAUMGARTNER R (1997) In vivo fluorescence kinetics of porphyrins following intravesical instillation of 5-aminolefulinic acid in normal and tumour-bearing rat bladders. J Photochem Photobiol 38: 158–163

[45] LAM S, MAC AULAY C, HUNG J (1993) Detectiono dyplasia and carcinoma in situ in lung imaging fluorescecendpvice. J Thorac Cardiovasc Surg 10: 10540

[46] LANG K, FRITSCH C, SCHULTE KW (2000) Möglichkeiten und Grenzen der Fluoreszenzdetektion mit Delta-Aminolävulinsäure-induzierten Porphyrinen (FDAP) bei Hauttumoren. In: LIPPERT BM, SCHMITT S, WERNER JA (eds) Fluoreszenzdiagnostik und Photodynamische Therapie. Shaker, Aachen, pp 15–21

[47] SZEIMIES RM, ACKERMANN G, LANDTHALER M, ABELS C (2000) Fluoreszenzdiagnostik epithelialer Tumoren. In: LIPPERT BM, SCHMIDT S, WERNER JA (eds) Fluoreszenzdiagnostik und Photodynamische Therapie. Shaker, Aachen, pp 7–13

[48] BOCHER T, LUHMANN T, BAIER S (1997) Multispectral fluorescence imaging device for malignancy detection. SPIE Proc 3197: 60–67

[49] AGRAWAL A, UTZINGER U, BROOKNER C (1999) Fluorescence spectroscopy of the cervix:influence of acetic acid cervical musus, and vaginal medications. Lasers Surg Med 25: 237–249

[50] MORITZEN V, SCHRAMM W, HÖHNE W, KRONFELDT HD (1995) Untersuchung der Wirkung von Zellgiften auf die Atmungskette mittels NADH-Fluoreszenz. Proc DGLITT, München, pp 618–621

[51] GESCHWEND MH, LOBANOV OV, STRAUß WSL, STEINER R, SCHNECKENBURGER H (1995) Zeitaufgelöste Energietransferspektroskopie mitochondrialer Stoffwechselprodukte. Proc DGLITT, München, pp 606–611

[52] SCHNECKENBURGER H, TREGUB I, SAILER R, RÜCK A, STRAUß WSL (1995) Time-resolved fluorescence spectroscopy and imaging of porphyrins. Proc DGLITT, München, pp 627–630

[53] RÜCK A, BECK G, BACHOR R (1996) Dynamic fluorescence changes during photodynamic therapy in vivo and in vitro of hydrophilic Al(III) phthalocyanine tetrasulphonate and lipophilic Zn (II) phthalocyanine administered in liposomes. J Photochem Photobiol 36: 127–133

# Part IV
# Laser Safety in Medicine

# IV
# Laser Safety in Medicine

D. H. Sliney

**Contents**

## Introduction

Safety has always been important for any laser application in surgery and medicine, and several issues dominate. These issues include: wearing eye protectors, dealing with the plume of vaporized tissue, and controlling potential fire hazards. No one denies that laser can pose a serious hazard to the eye, but the decision to wear eye protectors in all procedures has been frequently questioned. The degree of effort needed to minimize the very serious risk from chronic breathing of vaporized tissue also requires judgement. Aside from a few eye injuries from laser beam exposure, most serious accidental injuries (and even a few deaths of patients) reported to date from the laser beam itself can be traced to the ignition of surgical drapes and airway tubes.

With the increasing variety of lasers and the number of wavelengths now available, safe laser use has become a more complex issue [1]. The extent of potentially hazardous reflections, the type of eye protection and the ancillary hazards can vary considerably with the type of laser and the procedure. However, in all cases, the laser surgeon and hospital staff must be concerned with the protection of both the patient and the operating room staff (including the surgeon). Patient safety is assured by limiting needless exposure to adjacent tissues (by choice of wavelength and to a large extent by surgical technique), using non combustible materials adjacent to the beam and by protecting the patient's eyes. Safety of the operator and assistants requires concern for both system safety design and means to limit potentially hazardous reflections. Environmental hazards from the smoke produced by vaporizing tissue must be minimized by local exhaust ventilation or fume extractors. The pathogenicity and chemical toxicity of vaporized tissue have been the subject of a number of investigations, as discussed later. Safety standards for medical laser applications have been issued that consider all of these potential hazards and their control measures [1–8]. The current consensus standard in the United States, The American National Standard Z136.3–1998, Safe Use of Lasers in Health Care Facilities [4], and similar user guidelines have been crafted to provide a realistic and balanced approach to medical laser safety, without needless control measures. This can be achieved only if the entire community of laser surgeons, nurses, biomedical engineers and hospital administrators participate in the development of consensus standards and codes of practice.

As with other electrical or electronic medical equipment, surgical lasers in the clinical environment may pose safety problems. Potential hazards of electric shock exist, requiring appropriate grounding, and other electrical safety laser use, and biomedical engineers and medical electronics technicians familiar with safe installation of electrical and electronic equipment in hospital and health-care environments should have no difficulty in providing guidance for the safe electrical installation and use of laser equipment [9].

As with electrosurgical techniques, a laser can produce potentially hazardous airborne contaminants from the photovaporization of tissues. Unfortunately, the vaporized tissue (smoke) from laser surgery has often been referred to as laser smoke or the laser plume, suggesting that it is unique to laser surgery. This emphasis on the laser origin has frequently led to the result that vaporized tissue fragments from bone saws and pyrolysis products of tissue from electrosurgery have been overlooked as having the same degree of hazard. Vaporized tissue in sufficient quantities must receive special attention, and local exhaust ventilation will almost always be required [9, 10]. The pyrolysis products are similar to those resulting from the barbecuing of meats. They contain toxic by-products and known carcinogens such as nitrosamines. A number of studies have measured the concentration of potentially hazardous airborne contaminants in the laser operating room, with the result that concentrations are shown to be kept below permissible concentrations with appropriate exhaust ventilation [1].

The one hazard that is truly unique to the laser and that requires special attention results from the laser beam itself – the optical radiation hazard. Unlike other light sources, the laser beam may be collimated and directed over some distance; hence, the area of potential hazard may not be limited to the immediate surgical site. Unwarranted fears often accompany the introduction of lasers into the surgical theatre or the clinical environment. Therefore, proper appreciation of the real laser beam hazard is necessary for each member of the medical staff so that realistic safety precautions are followed [10–18]. In most countries, occupational safety and health regulations emphasize the critical importance of informing and educating the worker on workplace risks, and this clearly important with regard to laser use [19].

Laser hazards depend upon the laser in use, the environment and the personnel involved with the laser operation (the operator, ancillary personnel and patient). The laser hazard is roughly defined by the hazard classification [1–4], whereas the other factors must be analyzed in each situation. A basic understanding of laser biological effects and hazards is necessary to assess intelligently each laser hazard in the operating room. Once the hazards are understood, the safety measures are obvious.

## Biological Hazards of Laser Beams

### Eye Hazards

Because of its special optical properties, the human eye is considered to be the most vulnerable to laser light. Aside from the oral mucosa, the only living tissue exposed to the environment is the cornea and conjunctiva. Without the comparative protective features of the stratum corneum of the skin, the eye is exposed to the harsh environment of the sun, wind, dust, ultraviolet radiation and intense light. The eye has a natural protective mechanism, in its lid reflex, which limits the retinal exposure to very intense visible light or to intense exposure from infrared rays that raise the temperature of the cornea. However, some laser beam intensities are so great that injury can occur faster than the protective action of the lid reflex, which occurs between 0.2 and 0.25 s [9].

Laser hazards to the eye depend most predominantly upon wavelength, as shown in Fig. 1. Obviously, laser energy cannot damage tissue unless the light energy is able to penetrate to and be absorbed in that structure. For this reason, rays in the visible and near-infrared (visible and IR-A-band), which can be transmitted through clear ocular media and be absorbed in the retina, can, in sufficient intensity, damage the retina. The high size of such a point at the retina is of the order of 10–20 μm (smaller than the diameter of a human hair). For this reason, lasers operating between 400 and 1400 nm are particularly dangerous to the retina. This spectral region is often referred to as the retinal hazard region, since the increased concentration of light after entering the eye and falling on the retinal is of the order of 100 000. Hence, a collimated beam of 1 W cm$^{-2}$ at the cornea will focus to a small spot with an irradiance of 100 kW cm$^{-2}$. Although damage to such a small region of the retina may seem insignificant at first, it is important to realize that certain parts of such a small region e.g. the central retina, the macula and its fovea (centre of the macula), are extremely small areas responsible for critically important high-acuity vision. If these areas are damaged by laser radiation, substantial loss of vision can result. The argon, krypton, KTP, copper-vapour, helium-neon (He-Ne), diode and neodymium:YAG, all emit in the retinal-hazard spectral region (~400 to ~1400 nm).

The image area alone may not be the only site of damage, but as a result of heat flow and mechanical (acoustic) transients, the tissue surrounding the image site may also be damaged, leading to more severe consequences upon visual function. For example, it has not been uncommon for an individual to lose almost total function in an eye exposed to a very small amount of energy (several hundred microjoules) when a Q-switched laser has been accidentally imaged on the fovea. Instead of a normal visual acuity of 6/6 (20/20 in the USA), the visual acuity in such accidental situations has often been recorded as 6/60 (20/200) following an accident. Fortunately, in most accidents, only one eye is exposed to a collimated beam. However, after some recovery, any visual loss remaining after 60 days is generally permanent, since the neural tissue of the retinal has very little ability for repair.

At wavelengths outside the retinal hazard region – in both the ultraviolet and far-infrared regions of the spectrum–injury to the anterior segment of the eye is possible. Certain spectral bands may injure the lens (notable at wavelengths between 295 and 320 nm and wavelengths between 1 and 2 µm infrared wavelengths may actually pose a greater risk for permanent injury than the 10.6 µm $CO_2$ laser wavelength [9].

Excimer lasers operating in the ultraviolet spectral region are no longer unusual in the surgical arena. Certain excimer lasers pose a particular hazard to the cornea, and the 308-nm XeCl excimer laser can be considered additionally dangerous as it can produce an immediate cataract of the lens. By contrast, the 193-nm ArF excimer laser wavelength cannot ever penetrate deep into a single cell; hence, the biological consequences of scattered radiation are not at all serious, even if the conservative MPE were to exceeded by 100-fold at this extremely short wavelength. The surface cells (wing cells) of the cornea have an average lifetime of only 48 h and are quickly sloughed off after being damaged, thus leading to no sequellae. The holmium:YAG, hydrogen-fluoride, carbon dioxide and carbon monoxide lasers are all potentially hazardous to the cornea because wavelengths that cause corneal damage are not reconcentrated by the eyes as are wavelengths in the retinal hazard region. The thresholds for injury of the cornea are generally much higher than those that may injure the retina. Table 1 lists permissible occupational exposure limits for most of the commonly used surgical lasers [2–8].

**Table 1.** Selected occupational exposure limits (MPEs) for some lasers[a]

| Type of laser | Principal wavelength(s) | MPE (exe) |
|---|---|---|
| Argon-flouride laser | 193 nm[b] | 3.0 mJcm$^{-2}$ over 8 h |
| Xenon-chloride laser | 308 nm | 40 mJcm$^{-2}$ over 8 h |
| Argon ion laser | 488, 514.5 nm | 3.2 m Wcm$^{-2}$ for 0.1 s |
| Copper vapour laser | 510, 578 nm | 2.5 mWcm$^{-2}$ for 0.25 s |
| Helium-neon laser | 632.8 nm | 1.8 mWcm$^{-2}$ for 1.0 s |
| Gold vapour laser | 628 nm | 1.0 mWcm$^{-2}$ for 10 s |
| Krypton ion laser | 568, 647 nm | |
| Neodymium:YAG laser (primary ) | 1064 nm | 5.0 µJcm$^{-2}$ for 1 ns to 50 µs<br>Nbo MPE for t<1 ns 5 mWcm$^{-2}$ for 10s |
| Neodymium:YAG laser (secondary .....) | 1334 nm | 40 µJcm$^{-2}$ for 1 ns to 50 µs<br>40 mWcm$^{-2}$ for 10 s |
| Pulsed Nd:YAG (1.44 µm), | 1.44 µm | 0.1 Jcm$^{-2}$ for 1 ns to 1 ms |
| Pulsed holmium laser | 2.1 µm | |
| CW holmium laser | 2.1 µm | 100 mWcm$^{-2}$ for 10 s to 8 h, limited area |
| CW carbon monoxide laser | ... 5 µm | 10 mWcm$^{-2}$ for >10 s for most of body skin |
| Carbon dioxide laser | 10.6 µ | |

[a] All standards/guidelines have MPEs at other wavelengths and exposure durations.
[b] Sources: ANSI Standard 136–1–1003; AGGIH TLVs (1993) and IRPA.
Note: to convert MPEs in mWcm-2 to mjcm$^{-2}$, multiply by exposure time t in seconds, e.g. the He-Ne or argon MPE at 0.1 s is 0.32 mJcm$^{-2}$.

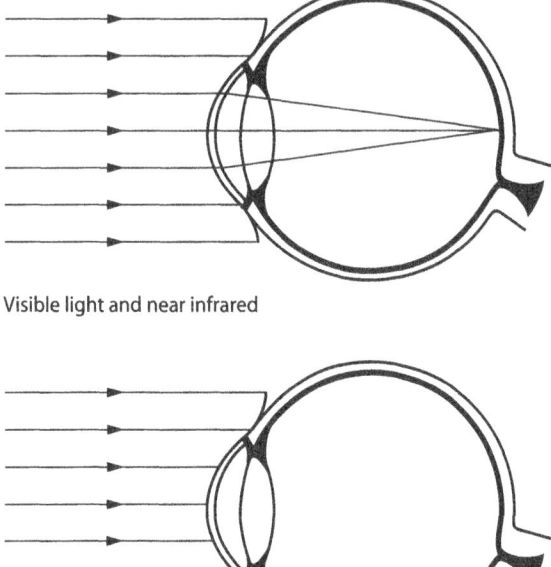

Visible light and near infrared

Far UV and infrared (UV-B, C, IR-B & C)

**Fig. 1.** Retinal hazards (left) and corneal hazards (right). The biological effects of optical radiation depend upon the spectral absorption properties, which vary greatly with spectral region. The lens is the principal absorption site in the near-ultraviolet spectral region (UV-A)

## Skin Hazards

The skin is less vulnerable to injury than the eye. However, it should be remembered that the probability exposure to some part of the skin from a reflected laser beam is far greater than to the small area occupied by the eye. Injury to the skin can occur from either photochemical damage mechanisms (predominant spectrum). For example, erythema (sunburn)results from injury to the epidermis – and to some extent, to the dermis as well – and originates from a photochemically initiated event. Skin protection can be much easier for UV excimer lasers if the hazard is understood [20]. First, second and third-degree skin burns can be induced by visible and infrared laser beam exposure, producing thermal injury, and the same irradiances that can produce a severe thermal burn can also ignite fabrics and burn plastics [1].

The severity of the injury depends upon the length of exposure and the penetration depth of the laser radiation. Generally, if the exposure lasts for a second or more, a pain response elicits a jerking movement to move the exposed tissue away from the laser beam, thereby limiting the exposure duration to a second or less. High-power laser beam exposure will not result

in a deep tissue burn at $CO_2$ wavelengths if the exposure time is extremely short, since the penetration depth of the $CO_2$ laser beam is very shallow (of the order 20 µm) and, in fact, does not penetrate the normal thickness of the stratum corneum. Injury to the epidermis from the $CO_2$ laser is by heat conduction from the stratum corneum to deeper layers. However, short-pulsed exposure to 1064-nm Nd:YAG laser radiation, which penetrates several millimetres into tissue, can cause a deep, severe burn at a radiant exposure just above burn threshold, albeit at a much higher threshold than for a $CO_2$ laser burn. A holmium laser (2.1 µm) and KTP (doubled Nd:YAG, 532 nm) or argon (488 and 514.5 nm) laser would produce a burn depth intermediate between $CO_2$ and Nd:YAG [9].

Clearly, the focal spot of a focused surgical laser or the concentrated beam irradiance at the tip of an optical fibre is designed to ablate or vaporize tissue and will be hazardous to skin if located near the focal spot. Skin injury also can occur as a result of ignition of clothing by a reflected laser beam with tragic consequences. Significant skin injuries from accidental exposure to industrial or medical lasers rarely occur; at least, they are rarely reported. Actual thresholds of injury to the skin are normally of the order of $Jcm^{-2}$, and this level of exposure does not occur outside the focal zone of a surgical laser.

## Reflections and Probability of Exposure

An examination of laser accident records indicates that the source of accidental ocular exposure is most frequently a reflected beam. Figure 2 illustrates the types of mirror like (specular) laser beam reflections that can occur from the flat or curved surfaces, which are characteristic of metallic instruments used in other surgical procedures. Skin injury of the hand holding an instrument is also possible. Normally, the collimated beam is considered the most hazardous type of reflection, but a very close range, a diverging beam may pose a greater likelihood of striking the eye [1, 20, 21].

A number of steps can be taken to minimize the potential hazards to both the patient and surgical staff. Preventive measures will depend upon the type of laser. One of the most commonly employed lasers in surgical applications today may still be the $CO_2$ laser. Since the $CO_2$ laser wavelength of 10.6 µm is in the far-infrared spectral region – and invisible – the presence of hazardous secondary beams could go unnoticed. This added hazard, resulting from an infrared laser beam's lack of visibility, is common to other infrared lasers, such as the 2.1 µm holmium or the 1064 nm Nd:YAG laser. Because there have been a

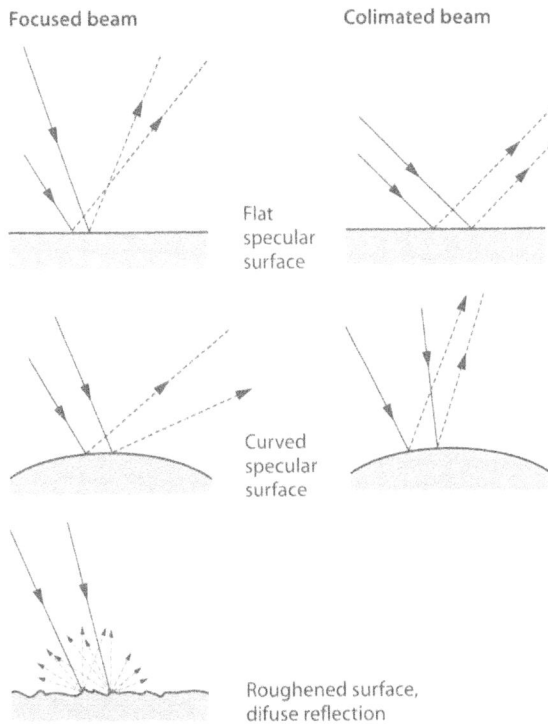

Focused beam Colimated beam

Flat
specular
surface

Curved
specular
surface

Roughened surface,
difuse reflection

**Fig. 2.** The hazard of reflections greatly depends upon surface characteristics. The hazards from specular mirror-like reflections are generally more enhanced because beam characteristics are maintained

number of serious retinal injuries caused by improper attention to safety with Nd:YAG lasers [10], the use of the Nd:YAG laser must be approached with even greater caution than the $CO_2$ laser. By contrast, the argon laser and the second-harmonic Nd:YAG (sometimes referred to as the KTP) laser emit highly visible, blue-green (488, 514.5 and 532 nm) beams and in some ways pose a lesser potential hazard.

Most current surgical lasers, such as the $CO_2$ Nd:YAG, holmium, or argon are continuous-wave (cw), or nearly so. Even the so-called super-pulse laser is quasi cw compared to single-pulse ophthalmic laser photodisruptors, dermatological pulsed dye and ruby lasers or some excimer ablative lasers. The biological effects and potential hazards from high-peak power-pulsed lasers are quite different from those of cw lasers. This is particularly true of lasers operating in the retinal hazard region of the visible (400–760 nm) and near-infrared spectrum (IR-A; 760–780 to 1400 nm), as shown in Fig. 1. The severity of retinal lesions from a visible or near-infrared (IR-A) cw laser is normally considered to be far less than from a Q-switched laser. Another major factor that influences the potential hazard is the degree of beam collimation. Almost all surgical lasers are focused, thereby

limiting the hazardous area (referred to as the nominal hazard zone in IEC 60825-1 and ANSI Z136.1–2000 [2]]. An exception is the highly collimated beam from many lasers having articulated arms, which may remain hazardous at some distance from the instrument [20].

Reflections are most serious from flat mirror like (specular) surfaces–characteristic of many metallic surgical instruments. Many surgical instruments now have black anodized or sand blasted, roughened surfaces to reduce (but not eliminate) potentially hazardous reflections. The strong curvature and surface roughening spread the reflected energy and greatly reduce the reflection hazard. The surface roughening is generally more effective than the black (ebonized) surface, since the beam is diffused. However, in some cases, combining a special black surface with roughening provides increased protection, and adding a black polymer finish has been shown by experiment measurements to offer the greatest protection at the $CO_2$ wavelength–despite initial scepticism by investigators [21]. However, other groups argue against blackening the surface, since the instrument will become hotter than without for visible wavelengths. Therefore, the use of the special blackened surfaces must be approached with caution for each application.

It should be noted that both the surface finish and reflectance seen in the visible spectrum do not indicate those qualities in the invisible, far-infrared spectrum. In fact, a roughened surface that appears to be quite dull and diffuse at a shorter, visible or IR-A wavelength will always be more specular at far-infrared wavelengths (e.g. at 10.6 μm). This results from the fact that the relative size of the microscopic structure of the surface relative to the incident wavelength determines whether the beam is reflected as a specular or diffuse reflection [1, 9, 21].

A specularly reflected beam with only 1% of the initial beam's power can still be quite hazardous. Hence, the rougher the surface of an instrument likely to intercept the beam, the safer the reflection. For example, even a 1% reflection of a 40-W laser beam is 400 mW! It is somewhat surprising that there have been few cases reported of eye injuries to residents and other persons observing Nd:YAG laser surgery without eye protectors. Hazardous specular reflections from a laser beam emerging from an endoscopic optical fibre are limited in extent because the beam rapidly diverges, as shown in Fig. 3.

Most invisible beam surgical lasers have a visible alignment beam. Infrared lasers most often make use of a low-power coaxial He-Ne (632.8 nm) or diode (e.g. 635 nm) red laser. It is desirable, where feasible, for this alignment beam to be 1 mW or less, since the maximum cw, visible laser beam power that can safely

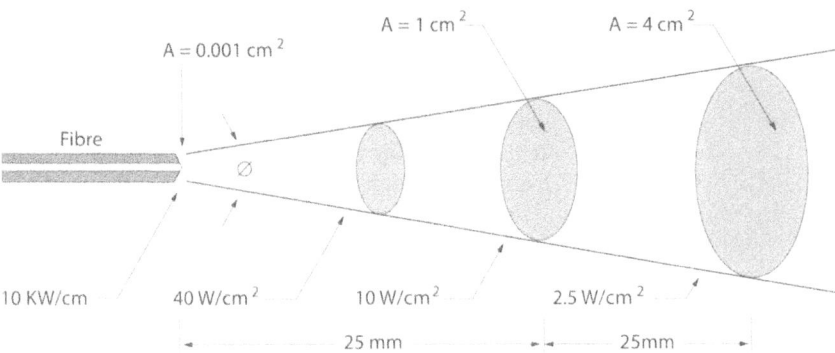

**Fig. 3.** The beam spreads rapidly from a non-lensed fibre tip or broken fibre. The beam irradiance rapidly decreases with distance from the distal tip of the fibre

enter the eye within the aversion response (i.e. within the blink reflex, etc. of 0.25 s), is 1 mW. This type of laser is then classified as class 2, and poses a very low risk to the user.

## Patient Safety

Most laser safety regulations do not apply to the exposure of the patient at the target site for surgery. However, accidental exposure of the patient by misdirection of the laser beam should be of concern and can result in injury to eye and skin [1]. Ignition of drapes can be particularly hazardous to the patient who is under anaesthesia and unable to warn the operating room (OR) staff of the sensation of heat. Details of accidents are often not published because of litigation, but anecdotal reports indicate that misfiring of a laser when not in use, or undetected breakage of optical fibres, has led to fires from the ignition of surgical drapes, with serious injury to patients. Procedural methods, such as the use of the standby switch or proper placement of the laser foot switch, can reduce the number of such accidents, but never completely eliminate them. Preparations for extinguishing fires or the moistening of drapes must always be part of the OR safety standing operating procedure (SOP).

Accidental injury of the eye is of particular concern when lasers are used in or near the eyes and where exposure of the eye itself is not intended. Special eye shields are available for patient protection (Fig. 4).

Patient safety has been of particular concern to head and neck surgeons using the laser near the trachea. As a result of a number of fatal injuries to patients from ignition of airway tubes, the means to reduce ignition have been addressed by anaesthesiologists and manufacturers of endotracheal tubes [22–27]. The reduction of ignition depends upon avoiding the use of combustible anaesthetic gasses or by using intravenous anaesthesia, and by avoiding the

**Fig. 4.** Eye shields that reflect incident energy accidentally directed toward the eye are essential for laser surgery near the eye

use of combustible materials such as polyvinylchloride (PVC), e.g. by the use of metal tubes. Unfortunately, at least some part of a metal airway (e.g. a PVC conduit inside the metal tube) frequently remains as a plastic or rubber element. Training of the surgical staff in prevention of endotracheal tube fires becomes essential in the field of head and neck surgery. Compound tubes developed in Germany provide particularly good protection by layering composite materials with moistened foam to reduce radial heat transfer to contiguous tissues.

## Safety of the Surgeon

Normally, the surgeon views the target issue through the optics of an endoscope, an operating microscope, colposcope, slit-lamp biomicroscope etc., and the reflections are safely attenuated by protective filters within the optics. Under such indirect viewing conditions, the surgeon or laser operator is normally not highly susceptible to injury due to proper design of the laser instrument. However, if the laser is accidentally actuated when the surgeon is not looking through the viewing optics, he or she will be at risk, as any other person in the room. Additionally, with hand-held laser delivery systems, one should remember that the surgeon's hand is the closest to the laser target and therefore closest to potentially hazardous reflections from adjacent surgical instruments (e.g. metal retractors).

### Safety of the Surgical Staff

Nurses, other surgical assistants and operating room staff are potentially exposed to misdirected laser beams. Lasers have been accidentally initiated when the beam delivery system was directed other than at the patient, a foot switch was accidentally pressed, or similar errors have occurred, and the beam directed at a person. Accidental firing of a laser has also occurred because of confusion created by multiple foot switches positioned below an operating room table. The laser foot switch should be covered and clearly identified. Assistants are potentially exposed to secondary reflections from surgical devices, whereas the surgeon's eyes are protected by filtration in the viewing optics. Reflections from the cornea or the contact lens used in ophthalmic surgery may be hazardous to assistants or bystanders in line of view of the contact lens to a distance of 1–2 m. The operating microscope used in laser microsurgery by a number of specialities would protect the eyes of the surgeon if properly designed, whereas assistants and bystanders may be exposed to potentially hazardous reflections from surgical instruments inserted into the beam path.

### Safety of Other Bystanders

Bystanders in the surgical facility or outpatient laser facility who are present to observe or to calm the patient (e.g. a patient's relative) may be susceptible to exposure from reflected laser beams in the same manner as a surgical assistant or nurse. In addition, because of lack of training or knowledge about the laser surgical procedure, bystanders may be a greater risk by inadvertently placing themselves in a dangerous position. These individuals should always be provided with laser eye protectors.

### Service Personnel

Service personnel are particularly susceptible to laser injury, since they often gain access to collimated laser beams from the laser cavity itself or by opening up the beam delivery optics and gaining access to collimated laser beams prior to the beam focusing optics or fiberoptic beam delivery system. Once the laser beam leaves the delivery system and comes rapidly into a focus, it then diverges again, or if emerging

from a fibre, it also rapidly diverges. The zone where the beam is concentrated to a level sufficient to pose severe hazard to the eyes or skin (the nominal hazard zone or NHZ) is normally a limited zone of 1–2 m near the beam focal point. However, a collimated laser beam, as the raw beam for most laser cavities, or a specular reflection from a running mirror or Brewster window in the laser console may be emitted from the laser cabinet (protective housing) when the service person gains access. Several serious eye injuries have occurred to service personnel exposed to secondary, collimated, invisible 1064-nm Nd:YAG laser beams when the service personnel gained access to the laser cavity.

### Occupational Exposure Limits

Relevant ELs for lasers of interest are given in Table 1 and are calculated or measured at the cornea. If the laser beam is <7 mm in diameter, it is assumed that the entire beam could enter the dark-adapted pupil and one can express the maximal safe power or energy in the beam (in the 0.4–1.4 µm retinal hazard region); it is the EL multiplied by the area of a 7-mm pupil, i.e. 0.4 cm$^2$. For example, for the visible cw lasers, an exposure limited by the natural aversion response of 0.25 is 2.5 Wcm$^{-2}$, and this EL multiplied by 0.4 cm$^2$ results in the limiting power of 1.0 mW. This 1-mW value has a special significance in laser safety standards, since it is the accessible emission limit (AEL) of class 2, i.e. the dividing line between two laser safety hazard classifications: class 2 and class 3 [2–8]. A 3.5-mm aperture is applied with cw infrared lasers operating at wavelengths greater than 1400 mm and a 1-mm aperture is applied in the UV spectral region for brief ocular exposures.

### Laser Hazard Classification

As noted, any cw visible laser (400–700 nm) that has an output power <1 mW is termed a class-2 (low risk) laser and could be considered more or less equivalent in risk with staring at the sun, at a tungsten-halogen spotlight or at other bright lights that can cause a photic maculopathy (central retinal injury). Only if one purposely overcomes the natural aversion response to bright light, can a class-2 laser pose a real ocular hazard. An aiming beam or alignment laser operating at a total power above 1.0 mW would fall into hazard class 3, and could be hazardous even if viewed momentarily within the aversion response time. A subcategory of class 3, termed class 3R (formerly 3A in the US), consists of lasers from 1–5 mW in power, and these lasers pose a moderate ocular hazard under viewing conditions where most of the beam enters the eye. Class 3B is when the subcategory that comprises, among certain pulsed lasers, cw-visible lasers that emit 5–500 mW output power. Even momentary viewing of class-3B lasers is potentially hazardous to the eye.

Only lasers that are totally enclosed or that emit extremely low output powers fall into class 1 and are safe to view. Any cw laser with an output power above 0.5 W (500 mW) falls into class 4. Class-4 lasers are considered to pose skin or fire hazards as well as severe eye hazards if not properly used. The purpose of assigning hazard classes to laser products is to simplify the determination of adequate safety measures, i.e. class-3A measures are more stringent than class-2 measures and class-4 measures are more stringent than class-3B measures. Virtually all surgical lasers fall into class 4, although the ophthalmic Nd:YAG photodisruptor is one example of a class-3B surgical laser.

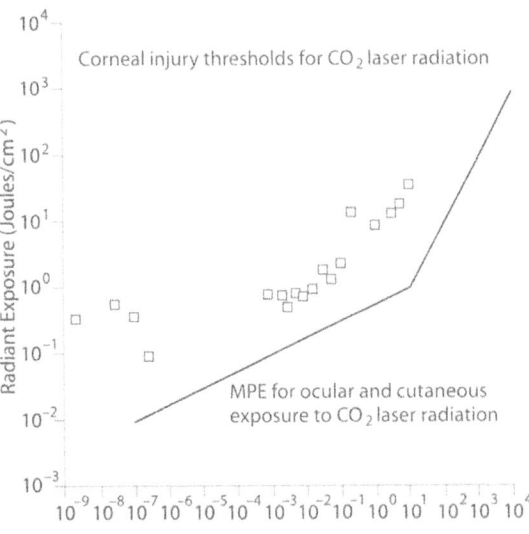

**Fig. 5.** The thresholds for corneal and skin damage are of the order of 10 $Wcm^{-2}$ for a 1-s laser exposure. The thresholds for laser eye-protector surface damage are higher

## Laser Eye Protectors

Laser eye protectors provide the principal means to assure against ocular injury from the direct or reflected laser beams in the operating room [28–42]. Although the eyes of the laser operator may be inherently protected by viewing optics, this should always be ascertained from the manufacturer of the viewing optics. In this regard, ordinary optical glass in compound lens systems protects substantially against all wavelengths shorter than about 300 nm and greater than approximately 2700 nm [9], although for certainty at wavelengths greater than 4000 nm. Laser-protective filters may be obtained for endoscopes and other viewing optics for the spectral region between these two spectral bands. Eye protectors are available as spectacles, wrap-around lenses, goggles, and related forms of eyewear. It is important that the eyewear be marked with the wavelengths and optical densities provided at those wavelengths. The markings must be clearly understood by all of the operating room staff. The proper use of eyewear and the meaning of the eye-protector markings are key subjects for laser safety training of the staff.

Clear plastic goggles or spectacles with side shields, which are known to be made of polycarbonate, are normally suitable for use with the $CO_2$ laser, but should be marked by the laser safety officer with an indication of the optical density, e.g. OD-4 at $CO_2$ wavelength of 10.6 µm. Some LSOs may be uneasy about marking eye protectors not sold as laser eye protection because of perceived (or very real) legal concerns. Studies of plastic eye protectors show clearly that polycarbonate is far superior in burn-

through resistance than other plastics, and such a marking has been argued to be quite justifiable for use with $CO_2$ lasers having power output up to about 100 W [28]. In some countries, the marking of eye protection by anyone other than the manufacturer is not legally recognized, and the LSO has no alternative but to obtain similar polycarbonate eye protectors that have been certified and labeled by the manufacturer. Some manufacturers of laser eye protection and some laser safety specialists have made a major issue of the importance of damage resistance of eye protectors, and this concern is evident in eye-protection standards in Europe [EN 207, EN 208]. These standards require burn resistance for a static beam for 10 s. However, burn-through times of plastic eye protectors appear to be of little concern in some quarters, as in the US [36]. Those who are sceptical about these concerns of burn-through argue that, with the powers generally used in laser surgery of 100 W or less, burn-through are unrealistic, since the wearer would certainly move his head within a second after detecting a flame shooting from the goggle. Indeed, the skin will incur a serious burn, as would the unprotected, exposed cornea (Fig. 5), and clothing would ignite at levels below plastic burn-through irradiances [28, 29, 41]. Goggles manufactured of special glass can frequently be designed to withstand irradiances higher than the ~100W $cm^{-2}$ order of magnitude typically required to burn through polycarbonate lenses in 10 s [28]. In any case, the eye protector requirements vary from country, and the user is under an obligation to be informed of the legal requirements in his or her locality.

A more serious problem associated with laser eye protection occurs when more than one laser eye protector must be worn for some procedures, as in tattoo removal. Several different visible wavelengths may be required to remove the different tattoo inks, and choosing the wrong eye protector during a procedure nearing the end of a tiring day has reportedly injured some dermatologists. Unclear labelling has been one contributing factor. Again, the application of a customized label related to the specific laser has been recommended in some countries (as in the US). However, in other countries, placing labels not provided by the manufacturer on eye protectors may not be approved. To compound the problem, the issue of laser eye protectors having only a coded indication of the protection (e.g. L4-1064) is strictly forbidden in the US, because of the great importance placed upon informing the user in an understandable statement! Hopefully, future safety standards will be harmonized on this subject.

## Laser Skin Protection

Skin protection is seldom a serious concern other than for the hands located near the focal zone of an open-beam laser. Ignition of clothing can be of concern if $CO_2$ lasers are used carelessly and the beam is accidentally misdirected toward clothing. Again, the same attention related to the potential burn-through of plastic goggles should apply to clothing, it this is a realistic concern. If Xe-Cl, 308-nm lasers are employed, scattered UV radiation would be of concern, and the skin should be covered [43–48]. Most tightly woven fabrics will have an attenuation factor exceeding $10^4$, i.e. an optical density exceeding 4.0 [43]. The potential UV skin hazards from scattered 193-nm, ArF laser radiation – as used in corneal corneum, and does not even penetrate a single cell layer of the corneal surface; however, the MPEs remain more conservative than the MPE for the 308-nm wavelength. While MPEs will probably be altered for these short wavelengths in the future, the wearing of surgical gloves is advisable when the surgeon's hands are near the patient's eye and close to the beam path.

## Respiratory Protection

Probably no issue has caused more concern in surgical laser safety than the potential hazards from breathing airborne contaminants produced during the vaporization of tissues. Although photocoagulation does not produce a smoke plume, any laser (or electrosurgical) cutting of tissues will produce gasses and airborne particulates that must be evaluated as a potential respirable hazard. A number of careful studies of these airborne contaminants and how they are produced show that potential hazards exist if these are in the breathing zone of either the patient or surgical staff. The studies of both the chemical toxicity of pyrolysis products and the potential viability of infectious particulates (e.g. viral fragments) have shown real cause for concern unless very good exhaust ventilation and respiratory protection are employed [45–53].

Several medical equipment suppliers offer splash-proof, full-face protection to protect against splash of patients with viral infection [54]. Although normally not designed as laser protection, the shields (particularly if polycarbonate) will offer some protection against exposure from reflected $CO_2$ laser radiation. However, thin, lightweight transparent plastic filters may be more susceptible to burn-through. The ultimate respiratory protection would, of course, be afforded by wearing a self-contained respirator. As protection against airborne contaminants, several manufacturers provide variously configured self-contained respiratory devices, but these are very cumbersome and have only rarely been seriously suggested (or worn) in a surgical setting. The best protection can never be foolproof, and with appropriate training, most surgical staffs recognize the importance of using good fume extraction. If additional concern exists because of pathogenicity, the wearing of a high-performance face mask provides a still greater level of protection. Hazards from laser gasses are normally controlled by system enclosure, and in some instances with exhaust ventilation or chemical scrubbers, as with excimer laser systems [55].

## Other Potential Ancillary Hazards to the Patient

Besides breathing some airborne contaminants, a patient may experience increased blood levels of methaemoglobin even during endoscopy. It has been suggested by Ott [56] that this increase results through peritoneal absorption as well as through the respiratory tract. Clearly, a balanced review of benefits vs. risks will provide the best choice in surgery, and the benefits of laser surgery have generally far outweighed the new risks. However a review is always required with the introduction of new techniques in any type of surgery.

## Safety Administration and Training

The practical implementation of a laser safety program, which includes a laser safety training program, cannot be treated in detail here. Other reviews of the subject treat these aspects in detail [1, 4, 10–18]. Clearly the design of a safety program depends largely upon the size of the institution and the variety and number of lasers in use. An office practice might have only a safety SOP and a designated LSO; a large institution frequently benefits from a laser safety committee. In the end, the importance of a well-trained staff cannot be overemphasized. Accidents can only be prevented by a well-trained staff and an administrative policy that encourages a sustained effort toward safe laser use.

## Conclusions

The potential exposure levels to the eye and skin from scattered laser radiation from most surgical laser applications are substantially below a threshold for injury, and only the direct beam or specular reflections are of concern. Only with UV lasers should one be seriously concerned with chronic exposure and delayed effects. The surgical laser user can be assured that today a consensus exists almost worldwide controls requiring the use of appropriate eye protection when needed and control of vaporized tissue byproducts require both a well-trained surgeon and OR staff. As with many other applications of lasers in industry and research, laser safety training is of crucial importance.

## References

[1] SLINEY DH, TROKEL SL (1992) Medical lasers and their safe use. Springer, Berlin Heidelberg New York
[2] ACGIH TLV's (1993) Threshold limit values and biological exposure indices for 1993–1994. American Conference of Governmental Industrial Hygienists, Cincinnati, OH
[3] ANSI (2000) Safe use of lasers. Standard Z136.1-2000. American National Standards Institutes, Laser Institute of America, Orlando, FL
[4] ANSI (1996) Safe use of lasers in health-care facilities. Standard Z136.3-1996. American National Standards Institute, Laser Institute of America, Orlando, FL
[5] BRITISH STANDARDS ORGANISATION (1984) Radiation safety of laser products and systems. Standard BS4803, London, BSI
[6] WORLD HEALTH ORGANIZATION (WHO) (1982) Environmental health criteria no 23, lasers and optical radiation. Joint publication of the United Nations Environmental Program, the International Radiation Protection Association and the World Health Organization, Geneva
[7] ICNIRP (2000) International commission on non-ionizing radiation protection. Guidelines for limits of human exposure to laser radiation. Health Physics
[8] INTERNATIONAL ELECTROTECHNICAL COMMISSION (2001) Safety of laser products. Part 1: Equipment classification, requirements and user's guide, IEC 60825-1: 1993 + A2: 2001

[9] SLINEY DH, WOLBARSHT ML (1980) Safety with lasers and other optical sources. Plenum Press, New York
[10] BALL KA (1991) Medical lasers: the perioperative challenge. Mosby, St. Louis 1991
[11] LUNDERGAN D, SMITH S (1983) Nurses administrative responsibilities for lasers. AORN 217–222
[12] MAKETY CJ (1984) The laser committee: Its role in quality assurance. Health Care Strategic Management
[13] PFISTER J (1983) A guide to lasers in the OR: a manual for OR personnel. Education Design
[14] STEFFERS V (1984) Safety checklist documents responsibilities of the laser nurse. Clin Laser Month, 2: 55
[15] FISHER JC (1985) Principles of safety in laser surgery and therapy. In: BAGGISH MS (ed) Basic and advanced laser surgery in gynecology
[16] HOLMES JA (1986) Summary of safety considerations for the medical and surgical practitioner. In: APFELBERG DB (ed) Surgical lasers. Springer, Berlin Heidelberg New York, pp 69–95
[17] CAYTON MM (1983) Nursing responsibilities in laser surgery. Med Instrum 17: 419
[18] CARRUTH JAS, MCKENZIE AL, WAINWRIGHT AC (1983) Clinical laser safety. In: ATSUMI K (ed) New frontiers in lasers in medicine and surgery. Elsevier, ORT??
[19] US Department of Labor, title 29, Codes of Federal Regulations, Occupational Health and Safety
[20] SLINEY DH, MAINSTER MA (1987) Potentially hazardous reflections to the clinician during photocoagulation. Am J Ophthalmol 103 (6): 758–760
[21] WOOD RL, SLINEY DH, BASYE RA (1992) Laser reflections from surgical instruments. Lasers Surg Med 12: 675–678
[22] FRIED MP, MALLAMPATI SR, LIU FC, KAPLAN S, CAMINEAR DS, SAMONTE BR (1991) Laser-resistant stainless steal tracheal tube: experimental and clinical evaluation. Lasers Surg Med 11: 301–306
[23] SCHRAMM VL JR, MATTOX DE, STOOL SE (1991) Acute management of laser-ignited intratracheal explosion. Laryngoscope 91: 1417–1426
[24] BIRCH AA (1974) Anaesthetic considerations during laser surgery. Anaesth Analg 52: 53–58
[25] NOLT ML, DEVOS V (1978) New endotracheal tube for laser surgery of larynx. Ann Otol Rhinol Laryngol 87: 554–557
[26] WOO P, VAUGHN C (1983) Small metal cuffless venturi ventilation system for use of laser surgery. Otolaryngol Head Neck Surg 91: 497
[27] WONG KC, OYKMAN PF (1983) Anesthetic considerations in laser surgery. In: Dixon J (ed) Surgical applications of lasers. Yearbook Medical, Chicago
[28] FOTH H-J (1998) Laser resistance of endotracheal tubes. I: Experimental results of a compound tube in comparison to a metallic tube. Lasers Med Sci 13: 242–252
[29] FOTH H-J (1999) Laser resistance of endotracheal tubes II: observed temperature rise and theoretical explanation. Lasers Med Sci 14: 24–31
[28] SLINEY DH, SPARKS SD, WOOD RL (1992) The protective characteristics of polycarbonate lenses against $CO_2$ laser radiation. J Laser Appl 5 (1): 49–52
[29] ENVALL KR, COAKLEY JM, PETERSON RW, LANDRY RJ (1975) Preliminary evaluation of commercially available laser protective eyewear. US Dept of Health, Education, and Welfare, Bureau of Radiological Health DHEW Publ (FDA) 75-8026: 32
[30] ERIKSON P, GALOFF PK (1989) Measurements of laser eye protective filters. Health Phys 56 (3): 741–742
[31] GALOFF PK, SLINEY DH (1988) Evaluation of laser eye protectors in the ultraviolet and infrared. In: COURT L, DUCHENE A, COURANT D (eds) Proc Int Symp on Laser Biological Effects and Exposure Limits (Lasers et Normes de Protection), Paris, Nov 24–26, Commissariat a l'Energie Atomique, Paris, Fontenay-aux-Roses 356–366
[32] HOLST GC (1973) Proper selection and testing of laser protective materials. Am J Opt 50: 477–483

[33] LAYON TL, MARSHALL WJ (1986) Nonlinear properties of optical filters – implications for laser safety. Health Phys 51: 95–96

[34] ROBINSON AA, MARSHALL WJ,DUDEVOIR SG (1990) Study of saturation in commercial laser goggles. SPIE Proc 1207: 202–213

[35] SCHERR AE, TUCKER RJ, GREENWOOD RA (1996) New plastics absorb at laser wavelengths. Laser Focus 5: 26–48

[36] SLINEY DH, LE BODO H (1990) Laser eye protectors. J Laser Appl 2: 9–13

[37] SPENCER DJ, BOXLER HA (1972) IR laser radiation eye protector. Rev Sci Instrum 43: 1545–1546

[38] SWEARENGEN PM, VANCE WF, COUNTS DS (1988) A study of burn-through times for laser protective eyewear. Am Ind Hyg Assoc J 49: 608–612

[39] SWOPE CH (1970) Design considerations for laser eye protection. Arch Environ Health 20: 184–187

[40] WILLIAMS DR (1970) Some comments on the properties of absorptive lenses. J Am Opt Assoc 41: 82–91

[41] YEO R (1989) Laser eye protection I. Optics Laser Tech 21 (4): 257

[42] ZWICK H, BELKIN M, BEATRICE ES (1988) Effects of broadbanded eye protection on dark adaptation. In: COURT L, DUCHENE A, COURANT D (eds) proc Int Symp on Laser Biological Effects and Exposure Limits (Lasers et Normes de Protection), Paris, Nov 24–26, 1986, Commissariat a l'Energie Atomique, Paris, Fontenay-aux-Roses

[43] SLINEY DH, ET AL (1987) Transmission of potentially hazardous actinic ultraviolet radiation through fabrics. Appl Ind Hy 2 (1): 36–44

[44] PARISH JA, PATHAK MA, FITZPATRICK TB (1981) Protection of skin from germicidal ultraviolet radiation in the operating room by topical chemicals. N Engl J Med 284: 1257–1258

[45] BAGGISH MS, POIESZ BJ, JORET D, WILLIAMSON P, REFAI A (1991) Presence of human immunodeficiency virus DNA in laser smoke. Lasers Surg Med 11: 197–203

[46] BAGGISH MS, ELBAKRY J (1987) The effect of laser smoke on the lungs of rats. Am J Obstet Gynecol 156: 1260–1265

[47] BELLINA JH, STJERNHOLM RL, KURPEL JE (1982) Analysis of emissions after irradiation with carbondioxide laser. Reprod Med 27: 268

[48] DICKES J (1989) Face masks as protection from laser plume. AORN J 50: 520–522

[49] GARDEN JM, O'BANNION KM, SCHEINITZ LS, PINSKI KS, BAKUS AD, REICHMANN ME, SUNDBERG JP (1988) Papillomavirus in the vapor of carbon dioxide treated verrucae. JAMA 259: 1199–1202

[50] SAWCHUK WS, WEBER PH, LOWRY DE, DZUBOWQ LM (1989) Infectious papilomavirus in the vapor of warts treated with carbon dioxide laser or electrocoagulation: detection and protection. J Am Acad Dermatol 21: 41–49

[51] VOORHIES RM, LAVYNE MH, STRAIT TA, ET AL (1984) Does the $CO_2$ laser spread viable brain tumor cells outside the surgical field? Neurosurgery 60: 892

[52] MILLER GW, GERACI J, SCHUMRICH DA (1983) Smoke evacuation for laser surgery. Otolaryngol Head Neck Surg 92: 582

[53] KOKASA JM, EUGENE J (1989) Chemical composition of laser tissue interaction smoke plume. J Laser Appl 1: 59–63

[54] AMERICAN NATIONAL STANDARDS INSTITUTE (1998) Safety Code for Head, Eye, and Respiratory Protection. ANSI Z-2.1, American National Standards Institute, New York

[55] SLINEY DH (1991) Safety of ophthalmic excimer lasers with an emphasis on compressed gases. Refract Corneal Surg 7: 308–314

[56] OTT D (1993) Smoke production and smoke reduction in endoscopic surgery: Preliminary report. Endosc Surg 1: 230–232

# IV-1
# Guidelines for Safe Clinical Laser Applications

B. Fuchs, C. Philipp, H.-P. Berlien

## Contents

## Introduction

Given the continuous development of new lasers and application techniques, laser safety does not represent a static problem, but has to be adapted to these changes. National and international organizations like the ISO (International Organization for Standardization) and the IEC (International Electrotechnical Commission) deal with questions of laser safety and have therefore issued a number of recommendations and guidelines, which are continuously being brought up to date. A great number of these decisions were first taken regarding the industrial use of lasers and were transferred to medical use later, whereby specific medical requirements have only occasionally been considered. Furthermore, safety guidelines mostly try to protect medical staff – the patient's safety is not always taken into consideration. It is the aim of this chapter to point out typical hazards in using medical lasers in daily routine for both staff and patient and to give practical advice as to how to avoid risks and accidents.

Medical lasers belonging to the classes 3A, 3B and 4 (see below are allowed to be used only within laser protection areas, which are defined as locations where the intensity of laser light can exceed the limit for the maximally tolerable irradiation, i. e. which does not lead to any damage of skin or eye (this limit varies between different lasers and depends on the wavelength, power, pulse rate and other settings). Therefore the laser protection area must not necessarily include the whole operating unit, but only the operating theatre.

A laser protection area can be separated from the operating theatre by a protective curtain (see Fig. 1). Here, protective glasses must be available for the staff before entering that part of the theatre where laser irradiation is applied. If any laser is in use, this should be indicated by a visual warning system outside the operating theatre.

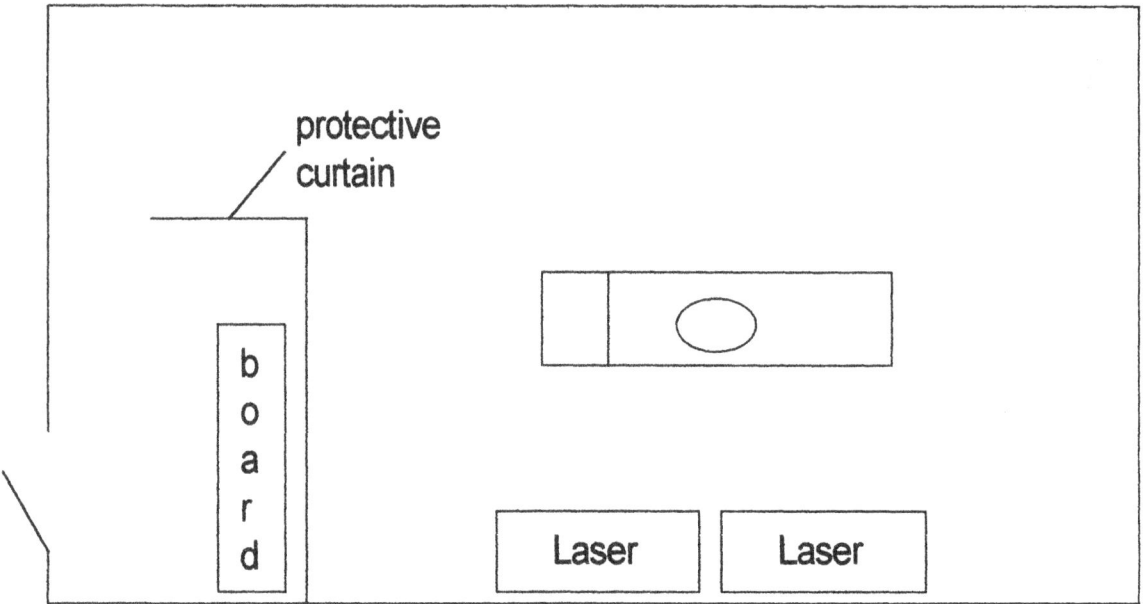

**Fig 1.** Arrangement of an operating theatre with an integrated laser safety area. Laser protective glasses can be kept on the <u>board</u>

In contrast to open laser surgery, where protective glasses must always be worn, the laser area in endoscopic or interstitial procedures can be defined as inside the patient, if

- the fibre is inside the patient`s body,
- the fibre is protected against bending and
- *video*endoscopy is performed in endoscopic procedures.

Under these conditions, the use of protective glasses is not required. Furthermore, to prevent risks, it is advisable to have the laser in Stand by position whenever possible. Questions of laser safety concern not only the direct but also the *reflected* laser beam from a material's surface, which can also be dangerous. The amount of reflected irradiation differs between infrared and visible light: while black surfaces hardly reflect visible light, they can behave like a mirror with infrared light. Furthermore, shiny metals and smooth, unpolished metal surfaces are good reflectors for infrared rays, but non-metal surfaces (e.g. ceramic) are bad ones, even if they appear shiny.

The practical significance of this phenomenon is, in general, limited to the operating area and mostly concerns surgical instruments (e.g.,tweezers) in close proximity to the laser beam. If such close contact between the laser beam and surgical instruments is expected, laser-resistant tools should be chosen. As, on the other hand, the power of a diffusely scattered beam is reduced fourfold if the distance is doubled, the reflection of a metal infusion stand, for instance, is hardly significant.

With respect to the laser apparatus, the actual power emitted should be continuously controlled by an internal checkup system. If a significant deviation is measured, an alarm should be set off followed by an automatic inactivation of the faulty system components. Furthermore, an acoustic signal is helpful if laser irradiation is being applied. According to the accident prevention guidelines, medical lasers are used only under the control of an authorized laser safety representative, who must supervise all aspects of laser safety and should therefore have a broad practical and theoretical knowledge of laser medicine.

## Specific Risks Using Medical Lasers

Specific risks using medical lasers and their importance are summarized in Table 1.

### Risks to the Surgeon

Laser therapy is applied to effect controlled damage to tissue. To realize this, the laser, its parameters, the application technique and specific tissue characteristics must be adjusted.

The most important risks when using medical lasers are related to the surgeon, and include insufficient personal experience, wrong indication of the operative procedure and wrong decisions regarding the laser and its parameters. Trainee programs and literature reviews can minimize these risks.

**Table 1.** Potential risks when using medical lasers

No personal experience
Wrong indication
Wrong laser
Wrong parameters

Eye hazards
Ignition of combustible materials

Earthquake destroys operating room

Electrical hazard

Intoxication by smoke
Intoxication by laser agents
Skin hazards to staff

The laser beam can lead to the ignition of combustible materials such as surgical drapes and/or other protective covers, tubes, cuffs, catheters, disinfectants (for instance ethanol) and anaesthetic gases ($O_2$, $N_2O$ etc.).

With surgical drapes, protective covers and tubes, the risk of ignition is relatively high if the surgeon has to operate inclose proximity to these materials. Under these circumstances laser-resistant equipment, as offered by several companies, should be used. In all other cases, laser-resistant materials are not necessary for every operation, as the risk of ignition is, in general, quite low (if the surgeon is aware of this problem and experienced), so that the high costs of this specific equipment are not warranted.

Changes in the flammable characteristics of combustible materials (e.g., compress,swab) can be effected by rinsing with water.

## Smoke Intoxication

Toxic smoke can develop from heating or vaporization of tissue, plastics or other materials. In the case of a virus infection of the tissue, the smoke can even contain living particles, if the tissue is only removed by pulsed lasers but not vaporized. Therefore all hazardous fumes and vapours must be adequately expelled from the operating area until no smell can be defected.

## Eye Hazards

Direct, reflected or scattered laser light can lead to damage to the eyes during open laser surgery. The range of damage can vary between different lasers and applications modes. Where as ultraviolet or far-infrared irradiation (10 600 nm) is mostly absorbed

in the cornea or the lens, infrared or visible light can pass the lens and is focused on the retina.

## Other Hazards

Some lasers use toxic materials as lasing medium (e.g., poisonous dyes, halogens in the case of the excimer laser). In the case of an accident, these substances can leak out of the laser and cause health injuries. Moreover, such substances must be disposed of according the producer`s guidelines.

The application systems used must be controlled again after finishing the procedure. Parts of the bare fibre, for instance, are not toxic, but can lead to a foreign body reaction. Further components of the LIT-Tapplicator (e.g., the glass dome of the fibre) may seriously damage the affected tissue.

Many lasers require a high-voltage power supply. Charged condensers can be dangerous even if the laser is disconnected from the electrical supply. Pulsed lasers may, additionally interfere electromagnetically with other electronic medical equipment. Electric hazards often go unnoticed.

Injuries to the skin during laser therapy mostly affect the staff and can lead to burns.Uncontrolled movements of the laser by the surgeon, reflected laser light or inattentiveness of the staff may be the reasons. As the irradiation of non-anaesthized skin is, in general, quickly noticed because of pain, the affected area is quickly removed from the laser beam, so that serious injuries are prevented.

Nevertheless, unintended skin injuries to the patient can be very serious, as anaesthesia interrupts this pain-feedback system. A typical complication is necrosis of the skin of the cheek during Nd: YAGlaser therapy of the enoral mucous membrane with ice-cube cooling at the same site; here, the mucous membrane is protected by the ice cube, but the skin on the outside can be heated and damaged. Control of the skin temperature is therefore essential during this procedure. Heat conduction must also be taken into consideration if the surgeon is using metal tools to protect or clamp a special area of tissue.

## Radiation Hazards

Besides the risk of damaging eyes and skin after exceeding a critical limit of irradiation (so-called maximally tolerable irradiation), there are two other important hazards of irradiation. Depending on the wavelength, mutation of DNA can be induced, whereby UV light is more dangerous, because the absorption maximum of DNA is about 260 nm. Further, the intended therapy result will not be achieved if the surgeon chooses the wrong parameters or laser. An overdose can lead to increased side effects and injury to surrounding healthy tissue; but even an underdose may also result in a higher percentage of side effects, as the irradiation time has to be prolonged to reach the desired therapy effect, resulting in potential injury to surrounding tissue.

## Classes of Lasers

According to their potential danger, lasers are divided into four classes:

Class 1    Lasers of this class are definitely or constructionally safe (power <0.39 mW).

Class 2    These lasers emit visible rays (400–700 nm) of low power (<1 mW). Eye protection is ensured by the blink reflex.

Class 3A   The emitted power of these lasers is less than 5 mW and includes all wavelengths. Looking directly into the beam with optical aids (e.g., a binocular microscope) can be dangerous.

Class 3B   With these lasers, looking direct by into the laser beam is dangerous close to the outlet of the applicator/handpiece.

Class 4    Lasers of this class emitting irradiation of more than 0,5 W represent a danger for the eyes due to diffuse reflections and/or direct by looking into the beam.

All medical lasers belong to the classes 3 and 4.

## Laser Protective Glasses

The eyes of both staff and patient must be protected when working with high-energy laser systems. For the patient, it has proven to be of advantage to oc-clude his/her eyes completely with compresses and adhesive stripes (e.g., Leukosilk) instead of glasses. Especially in operations in the face, glasses may hinder the surgeon or can slip out of place. The staff and the surgeon however, should wear goggles or glasses, which must be chosen with respect to the relevant wavelength as well as to the type and intensity of the irradiation.

According to the DIN 58215, protective glasses are characterized by the following four numbers and letters:

| 1 | 2 | 3 | 4 |
|----|----|----|----|
| WL | LP | IC | St |

1. The first number describes the wavelength WL, against which the protective glasses provide protection (e.g., 1060 nm for the Nd:YAGlaser).
2. The second complex gives information about the level of protection LP. Thus, L3A, for instance, means that the intensity of the first-mentioned wavelength is reduced by $10^3$. The capital letter A allows all types of lasers to be used with these glasses, D is for continuous and I for pulsed lasers only.
3. The third part is reserved for the index code IC of the manufacturer.
4. Finally, the last complex describes the relevant standard St, (e.g., DIN) for these glasses.

## Practical Advice on How to Avoid Laser Hazards

To minimize the risk of laser hazards as described in the preceding sections, a clinical laser checkup should be performed before putting the apparatus into operation. The following five-step procedure is recommended:

1. The fibre (260/400/600 μm must be compatible with the adapter/laser. If resterilizing a fibre, its size and resterilization cycle should be recorded.
2. The fibre must be checked for visible defects or contamination before use.
3. The adapter/fibre must be checked for optical permeability. A simple test to control the optical permeability of the fibre is to put one end of the fibre against a light source and to look at the opposite end: if one can see a bright light, the fibre can be used. To clean the adapter, only compressed air, acetone or high by concentrated alcohol (97%) are allowed, used with a dust-free tissue.
4. The complete laser system must then be checked for its optical permeability and the quality of the pilot beam.
   First, the fibre should be carefully connected without touching the surface of the fibre or adapter.

Then the pilot beam is activated; this should appear circular and regular at different positions of the fibre handpiece. The entire length of the fibre must also be controlled for any lateral emergence of the pilot laser,which would indicate a partial or complete break of the fibre. In this case,shortening the fibre is necessary or a new one must be used.

5. Finally, the laser system can be put into operation. Calibration of a focusing handpiece may be performed, if necessary. If the surgeon notices that the efficiency of the laser system is too low, the fibre applicator should first be checked once again, then the power can be gradually increased.

For the practical aspects of laser safety in daily routine, seven golden rules are helpful in preventing hazards to staff and patients.

## Laser Safety in Daily Clinical Routine

Protective measures for the patient must always be adapted to the current treatment situation, and can have the character of a recommendation only. Additionally, they must be established only by those responsible for patient treatment.

To combine laser protective measures for the staff with optimal treatment, the interests of the patient and the prevention of accidents, the following seven golden rule have proved invaluable in years of experience.

1. Keep the laser in the standby mode.
2. Keep distance.
3. Wear glasses in the laser area.
4. The laser is not a pointer.
5. Do not lase instruments.
6. Do not lase combustible materials.
7. Check your system.

### 1. Keep the Laser in the Standby Mode

Accidents can occur at the moment that the fibre or handpiece is put down "only shortly" and the surgeon or another person steps on the foot switch accidentally. Severe injuries can result because during such a break in treatment the OP staff no longer recognize the danger of a laser irradiation, and remove their protective goggles. In addition, fire can be caused, as the applicators are frequently put down on medical textiles. At the end of a laser operation the laser should be switched from ready to standby mode.

### 2. Keep Distance

In contrast to industrial lasers, during the use of fibres and handpieces in most medical lasers a beam divergence exists (exception: focusing handpieces); the power density of the laser irradiation decreases with the square of the distance. The setting of a sufficient safety distance is, as with X-rays, the most important precaution to prevent injuries of any kind through laser irradiation. Moreover, no person should be present in the surgical operating room who is not directly participating. This is also important for reasons of hygiene. Laser surgial applications are not to be confused with a laser show!

### 3. Wear Glasses in the Laser Area

With all open laser procedures in medicine, the wearing of protective glasses is prescribed in the laser area. The laser area in endoscopic or interstitial procedures can be defined as inside the patient if:
• the laser is in standby mode (whenever possible)
• and in the ready mode only if:
  – the fibre is inside the body,
  – the fibre is protected against bending.

Videoendoscopy is obligatory for endoscopic procedures.

Protective glasses are effective against laser radiation only when the emitted wavelength

lies within the specific wavelength spectrum for which they are intended. This can be

determined by reading the label, usually on the side of the glasses.No glasses give protection against all laser wavelengths – this would correspond to a black eye shield, which is needed to look directly into the laser beam with the eye unprotected. Before beginning a laser operation, the surgeon must ensure that all employees and the patient have the appropriate eye protection. For the patient the best solution is to cover the eyes with a medical compress fastened with Leukosilk.

To make it possible to choose the correct protective goggles before entering the laser area, it is helpful to note to laser wavelength used in the operating room. In contrast to open laser applications, wearing of laser protection glasses is not necessary in endoscopic or interstitial laser treatment, if the laser area is shifted into the body cavity of the patient. Then it is a laser of the class I like a CD-Player. This is the case,
• If the laser fibre of a ready-mode laser is in the body of the patient.
• Further the fibre must be protected against kinking; this can result due to a fibre break caused by

escaping radiation. A beam visible outside an endoscope indicates a damaged fibre, which should be replaced.

- Finally, videoendoscopy is obligatory in endoscopic operations. Over the optical fibre bundles of the endoscope no laser irradiation can be transmitted back and can lead to eyehazards. The real risk exists, however, if an awkward movement at the instrumentation channel, near the ocular, causes a fibre break. If videoendoscopy is not possible, laser protective glasses must be worn. The use of a so-called ocular protection filter is senseless.If the unprotected eye of the surgeon is directed to the fibre break, it can lead to an eye hazard in spite of the high divergence, because of the small distance only a few millimetres. The risk in fact arises  not from reflection of laser irradiation through the endoscope but from the fibre break near the ocular of the endoscope. The ocular protection filter protects against a risk that does not exist, but does not protect, however, against the actual risk of a fibre break near the eye.
- Videoendoscopy has two important advantages:
  1. If the head of the endoscope can be held far away from the eye, the distance is so great that in the case of a fibre break at the place of insertion, no direct irradiation of the eye occurs; the break can be recognized immediately and the laser irradiation stopped.
  2. With relaxed working, the risk of a fibre break clear is reduced. If videoendoscopy is not possible, only the surgeon needs to wear protective glasses.

## 4. The Laser Is Not a Pointer – the Staff Is Not Your Target

Even in darkened rooms the visible pilot beam of a laser has a certain optical fascination, but it should not be confused with the medical laser, which has a laser pointer and can only indicate the operation area. Laser irradiation must be completed before the handpiece is removed. At the end of the operation or in the case of interruption, the applicator must be deposited in a safe place. In addition, neither the laser beam nor the pilot beam should be focused directly on the eyes of an employee, even if protective goggles are worn. Employees are not a target! Note that these regulations for clinical safety are valid only because a medical laser is a checked controlled laser unit. Check it!

## 5. Do not lase instruments

Not only the direct laser beam can lead to eye hazard or tissue damage, but also the reflected one. Polished instruments, metallic surfaces or even ceramic tiles can serve as mirrors. Thus black surfaces reflect no visible light, but infrared light, however, very well. So it becomes understandable that only blackening instruments offers no protection against reflection because the reflection of infrared wavelengths is not prevented and the reflected pilot beam is not visible, so that nobody can recognize where the reflected beam strikes the tissue. Additionally, blackened instruments absorb visible light and can become very hot.

Because of the beam divergence, and in contrast to industrial use, where lower energy densities are involved all reflections outside the operating area are of no importance. In the direct environment of laser applications the reflected beam shows the same power density as the initial laser beam and represents a source of danger. This risk can be reduced by using instruments with a rough and metallic surface because then a diffuse reflection of the light occurs. Such instruments are very expensive and only necessary if the beam path repeatedly hits the surface of the instruments.

## 6. Do not Lase Combustible Materials

All combustible materials can be ignited by different causes (laser/HF surgery, light cord etc.). The industry offers a whole palette of products and materials with higher laser safety and less fire risk after exposure to laser light. However their routine use is not necessarily economically justifiable in most cases. Normally, it is sufficient if medical textiles or compresses in the direct environment of the operation area are moistened with saline solution. To counter the risk of a tube fire, the most effective protection  is to keep the tube away from the direct beam path and from the operating area. This should not, however mean hiding the tube under a mountain of medical compresses where it cannot be located. In these cases, it is better to keep the tube in sight and under control to avoid direct irradiation. Additionally, the tube can be protected by moist cotton wool or Merozell Laserguard. In general, a nasal tube should be used where possible in pharyneal and laryngeal operations, otherwise the laser safety tube offers the best, but still however, not 100% protection, against fire. Placing is also recommended a catheter filled with 50 ml saline solution in the operating area to be able to extinguish a fire immediately. If there is a smell of burned synthetic materials, a tube fire must be assumed until the contrary is proved beyond doubt! Smoke and the smell of burned plastic are often the first warning signals of a tube fire.

## 7. Check Your System!

It is a frequently recurring mistake during the application of medical laser systems to increase uncritically the power or energy of the laser in the case of no or only an ineffective tissue reaction.

Before every use, lasers, and in particular the optical applicator, should be checked first. These can represent a weak point in the whole system and are especially sensitive to damage.

Before every application an inspection of the applicators should be carried out. Because this can most easily be carried by the user, he/she is responsible for this. When using flexible light guides, it is necessary:

To check the fibre for external, visible damage and dirtying (ethylene oxide dissolves plasticizers, formaldehyde oxidizes metal).

To test fibres for optical continuity when one end is held against a source of light the other end must appear bright and straight).

The pilot beam must not escape anywhere on the side of the fibre.

To check the pilot beam on a dark surface – it must be circular and under control.

During the use of rigid or inflexible application systems a test exposure should be carried   out – for example on a wooden spatula to determine the correspondence of pilot and reactive beam. If the laser effect does not occur directly at the pilot beam, this means readjusting the mirrors in the articulated arm.

If the desired effect is not achieved at the tissue, first the whole system must be checked, after which the power or energy can be increased. If calibration of the applicator is carried out and a great discrepancy noted between the adjusted and the measured power, the possibility of steaming up of the focusing lense or a defect in the instrument measuring internal power should be considered. Inspection with an external power – measuring instrument can be helpful here. To depend an only one calibration for information on the correct function of a laser system would be negligent.

Medical lasers represent complex systems whose effective service depends on different qualities such as wavelength, pulse duration, etc. To be able to use this service purposefully and safely, one needs not only theoretical knowledge of optical physics, but also an appropriate training. Also from the viewpoint of insurance, participation in specific courses is absolutely recommended. In addition, practical experience should be widened, for example, by hospitations with colleagues/institutes with well-founded laser medical experience. Often there are small tips and tricks which considerably facilitate later contact with the laser and point out its potential dangers. The information from the manufacturer alone cannot replace this education and information!

The rules must be adapted to the respective situation. This adaption presupposes a thorough knowledge of both laser physics. The ensuing laser protection rules and the **typical processing situation**. On this basis it is recommended that in every functional area in which laser operations occur regulary, a responsible medical doctor be designated as local laser safety officer to determine the guidelines under local conditions and with knowledge of  the specific requirements for this field. In addition, however, in particular at larger hospitals a technician should be appointed (or for example a technical department), to be responsible for compliance with structural measures, organization of regular training sessions, announcements, etc.

## Further Reading

Altomare DF, Memeo V (1993) Colonic explosion during diathermy colotomy. Dis Colon Rectum, 3: 291-292

Axelrod EH, Kusnetz AB, Rosenberg MK (1993) Operating room fires initiated by hot wire cautery. Anesthesiology, (5): 1123-1126

Bailey MK, Bromley H, Allison JG, Conroy JM, Krzyzaniak W (1990) Electrocautery – induced airway fire during traceostomy. Anesth Analg, 71: 702-704

Capizzi PJ, Clay RP, Battey MJ (1998) Microbiologic activity in laser resurfacing plume and debris Lasers Surg Med, 23: 172-174

Philipp C, Albrecht H, Hug B, Berlien H-P, Müller G (1992) Significance of laser safety. Lasers in gynecology. Springer Verlag Berlin Heidelberg New York. 435-446

Rockwell RJ (1994) Laser accidents: reviewing 30 years of incidence: what are the concerns - old and new? J Laser Applic, 6: 203-211

# Subject Index

The manufacturer's authorised representative in the EU is Springer
Nature Customer Service Centre GmbH, Europaplatz 3, 69115 Heidelberg,
Germany. If you have any concerns regarding our products, please
contact ProductSafety@springernature.com

Printed and bound by CPI Group (UK) Ltd, Croydon, CR0 4YY
28/04/2026
02098462-0008